EDGE INDEX

HOW TO USE THE EDGE INDEX

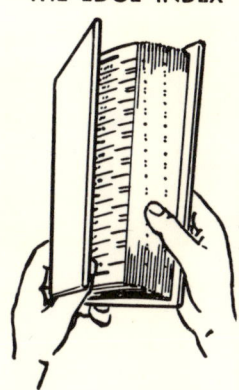

Bend the pages nearly double and hold it in your right hand as shown.

Locate the listing you want in the Edge Index.

Match up the 1- or 2-line symbol next to the listing you have selected with the corresponding 1- or 2-dot symbol on the page edge.

OPEN THERE

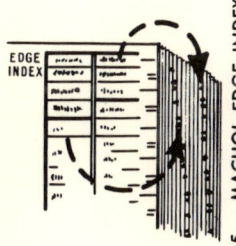

The listings in the left-hand column of the Edge Index will be identified by the dot symbols in the first or left-hand row. The listings in the right-hand column by the dot symbols in the right-hand row.

Contributors		
Foreword and Preface		
Methodology of System Engineering		
SYSTEM ENVIRONMENTS	The Ocean	**SYSTEM THEORY**
	Land Masses	Decision Theory
	Urban Areas	Simplex Method
	Lower Atmosphere	Making Models in Linear Programming
	Upper Atmosphere	Dynamic Programming - Theory and Application
	Space and Astronomy	Queues and Markov Processes
		Feedback Theory
SYSTEM COMPONENTS	Digital Computer-system Characteristics	Adaptive and Learning Control Systems
	Logical Circuits	**SYSTEM TECHNIQUES**
	Analog Elements	Human Information-processing Concepts
	Electronic Analog/Hybrid Computers	Simulation
	Communications Equipment	Reliability
	Communication Systems	System Testing
	Radar	Economics
	Infrared	Management
	Optical Sensors	Radio-telemetry Systems
	Satellite Systems	**USEFUL MATHEMATICS**
	Aerodynamic Systems	Probability
	Guidance	Laplace and Related Transforms
	Propulsion	Numerical Analysis
	Selected Energy-conversion Systems	Propositional Calculus
		SUBJECT INDEX

TEN FOREIGN PATENTS ISSUED
U.S. PATENT NO. 2680630
MACHOL EDGE INDEX. LICENSE NO. 345

system engineering handbook

Edited by
ROBERT E. MACHOL

Head, Department of Systems Engineering
University of Illinois in Chicago
Formerly Vice President, Systems
Conductron Corporation, Ann Arbor, Mich.
in collaboration with
WILSON P. TANNER, JR.
Professor of Psychology and Director of
the Sensory Intelligence Laboratory
The University of Michigan
and **SAMUEL N. ALEXANDER**
Chief of the Information Technology Division
The National Bureau of Standards

This unique handbook answers a long-felt need by providing a single source in which the system designer and practicing engineer can locate the many and diffuse subjects classified under "system engineering."

This book codifies the data and techniques most useful for the design of large, complex systems. It provides practical aid for the man responsible for solving technological problems and establishing engineering projects for which little or no precedent exists. The planning and execution of such a project or "system" may require knowledge of a dozen different disciplines, as well as a number of ideas not yet established as standard practice.

The systems discussed in this book have the following characteristics: (1) they have a single purpose; (2) they are man-made from equipment or "hardware"; (3) they are semiautomatic, involving man-computer interaction; (4) their inputs are stochastic or relatively unpredictable; (5) they are often competitive.

To cope with the staggering problems posed by these characteristics, this book analyzes, in turn, system environments, system components, system theory, and system techniques. A final section offers useful mathematics associated with system engineering.

Here is comprehensive, authoritative, and unique coverage of the many techniques and subsystems most commonly encountered in system engineering. This book will prove invaluable to those engaged in electrical and electronic engineering, aerospace, communication systems, telemetry, engineering management, operations research, military science, reliability, instrumentation and measurement—and to all others seeking orientation in system engineering.

SYSTEM ENGINEERING HANDBOOK

OTHER McGRAW-HILL HANDBOOKS OF INTEREST

AMERICAN INSTITUTE OF PHYSICS · American Institute of Physics Handbook
AMERICAN SOCIETY OF MECHANICAL ENGINEERS · ASME Handbooks:
    Engineering Tables   Metals Engineering—Processes
    Metals Engineering—Design   Metals Properties
ARCHITECTURAL RECORD · Time-Saver Standards
BERRY, BOLLAY, AND BEERS · Handbook of Meteorology
BLATZ · Radiation Hygiene Handbook
BRADY · Materials Handbook
CHOW · Handbook of Applied Hydrology
CONDON AND ODISHAW · Handbook of Physics
CONSIDINE · Process Instruments and Controls Handbook
CONSIDINE AND ROSS · Handbook of Applied Instrumentation
ETHERINGTON · Nuclear Engineering Handbook
FLÜGGE · Handbook of Engineering Mechanics
HARRIS AND CREDE · Shock and Vibration Handbook
HENNEY · Radio Engineering Handbook
HUNTER · Handbook of Semiconductor Electronics
HUSKEY AND KORN · Computer Handbook
JURAN · Quality Control Handbook
KALLEN · Handbook of Instrumentation and Controls
KING AND BRATER · Handbook of Hydraulics
KNOWLTON · Standard Handbook for Electrical Engineers
KOELLE · Handbook of Astronautical Engineering
KORN AND KORN · Mathematical Handbook for Scientists and Engineers
LANDEE, DAVIS, AND ALBRECHT · Electronic Designer's Handbook
LANGE · Handbook of Chemistry
MANTELL · Engineering Materials Handbook
MARKS AND BAUMEISTER · Mechanical Engineers' Handbook
MEITES · Handbook of Analytical Chemistry
ROTHBART · Mechanical Design and Systems Handbook
STREETER · Handbook of Fluid Dynamics
TERMAN · Radio Engineers' Handbook
TRUXAL · Control Engineers' Handbook
URQUHART · Civil Engineering Handbook

SYSTEM ENGINEERING HANDBOOK

Edited by

ROBERT E. MACHOL

*Head, Department of Systems Engineering
University of Illinois in Chicago
Formerly Vice President, Systems
Conductron Corporation, Ann Arbor, Mich.*

IN COLLABORATION WITH

WILSON P. TANNER, JR.

*Professor of Psychology and Director of
the Sensory Intelligence Laboratory
The University of Michigan*

SAMUEL N. ALEXANDER

*Chief of the Information Technology Division
The National Bureau of Standards*

McGRAW-HILL BOOK COMPANY

New York   San Francisco   Toronto   London   Sydney

SYSTEM ENGINEERING HANDBOOK

Copyright © 1965 by McGraw-Hill, Inc. All Rights Reserved. Printed in the United States of America. This book, or parts thereof, may not be reproduced in any form without permission of the publishers. *Library of Congress Catalog Card Number:* 64-19214

39371

CONTRIBUTORS

ISAAC L. AUERBACH, *President, Auerbach Corporation.* (*Digital Computer-system Characteristics*)

GEORGE A. BEKEY, *Electrical Engineering Department, University of Southern California.* (*Simulation*)

RICHARD BELLMAN, *The RAND Corporation.* (*Dynamic Programming—Theory and Application*)

EUGENE W. BIERLY, *U.S. Atomic Energy Commission.* (*The Lower Atmosphere*)

ROBERT H. BODEN, *Rocketdyne Division of North American Aviation, Inc.* (*Propulsion*)

KENNETH E. BOULDING, *The University of Michigan.* (*Economics*)

LUDWIG BRAUN, *Associate Professor of Electrical Engineering, Polytechnic Institute of Brooklyn.* (*Feedback Theory*)

ARTHUR W. BURKS, *Professor of Philosophy, The University of Michigan.* (*The Propositional Calculus*)

NORMAN L. CANFIELD, *U.S. Weather Bureau.* (*The Lower Atmosphere*)

A. CHARNES, *Northwestern University.* (*Elements of a Strategy for Making Models in Linear Programming*)

J. F. CLAYTON, *Bendix Systems Division, Bendix Corporation.* (*Management*)

GEORGE R. COOPER, *School of Electrical Engineering, Purdue University, Lafayette, Indiana.* (*Decision Theory*)

W. W. COOPER, *Carnegie Institute of Technology.* (*Elements of a Strategy for Making Models in Linear Programming*)

L. J. CUTRONA, *Vice President for Applied Research, Conductron Corporation; Professor of Electrical Engineering, The University of Michigan.* (*Communications Equipments*)

GEORGE B. DANTZIG, *Chairman, Operations Research Center, University of California, Berkeley, California.* (*The Simplex Method*)

A. NELSON DINGLE, *Department of Meteorology and Oceanography, The University of Michigan.* (*The Lower Atmosphere*)

CONTRIBUTORS

CHARLES L. DOLPH, *Professor of Mathematics, The University of Michigan. (Properties of the Laplace and Related Transforms)*

W. G. DOW, *Professor and Chairman, Department of Electrical Engineering, The University of Michigan. (Selected Energy-conversion Subsystems)*

RAYMOND L. DUSSAULT, *Itek Corporation. (Optical Sensors)*

LESLIE C. EDIE, *Port of New York Authority. (Urban Areas)*

JEROME FREEDMAN, *Lincoln Laboratory, Massachusetts Institute of Technology. (Radar)*

D. L. GERLOUGH, *Planning Research Corporation. (Simulation)*

JOHN E. GIBSON, *Director, Control and Information Systems Laboratory, School of Electrical Engineering, Purdue University. (Adaptive and Learning Control Systems)*

JOHN A. GOSDEN, *Program Manager, Auerbach Corporation. (Digital Computer-system Characteristics)*

CARL D. GRAVES, *TRW Space Technology Laboratories. (Space and Astronomy)*

ARTHUR D. HALL, *Head, Television Engineering Department, Bell Telephone Laboratories. (Communications Systems)*

WALTER HELLY, *Port of New York Authority. (Urban Areas)*

E. WENDELL HEWSON, *Department of Meteorology and Oceanography, The University of Michigan. (The Lower Atmosphere)*

CONRAD H. HOEPPNER, *President, Industrial Electronetics Corporation. (Radio-telemetry Systems)*

RONALD A. HOWARD, *Associate Professor of Electrical Engineering and Associate Professor of Industrial Engineering, Massachusetts Institute of Technology. (Probability)*

R. M. HOWE, *Department of Aeronautical Engineering, The University of Michigan. (Guidance)*

LESLIE M. JONES, *High Altitude Engineering Laboratory, Department of Aeronautical and Astronautical Engineering, The University of Michigan. (The Upper Atmosphere)*

WALTER J. KARPLUS, *Department of Engineering, University of California, Los Angeles. (Analog Elements)*

GRANINO A. KORN, *Professor of Electrical Engineering, University of Arizona. (Electronic Analog/Hybrid Computers and Their Use in System Engineering)*

RICHARD A. LAING, *Logic of Computers Group, The University of Michigan. (The Propositional Calculus)*

CONTRIBUTORS

VERNON D. LANDON, *Radio Corporation of America, Defense Electronic Products, Astro-Electronics Division, Princeton, New Jersey. (Satellite Systems)*

GERARD A. La ROCCA, *Head of Scientific Space Projects, Space and Information Systems Division, North American Aviation. (System Testing)*

GEORGE K. LEWIS, *Department of Geography, Boston University. (Land Masses)*

WILLIAM A. LYNCH, *Head, Mechanical Engineering Department, Polytechnic Institute of Brooklyn. (Feedback Theory)*

DUNCAN E. MACDONALD, *Itek Corporation. (Optical Sensors)*

ROBERT E. MACHOL, *Head, Department of Systems Engineering, University of Illinois in Chicago; formerly Vice President, Systems, Conductron Corporation, Ann Arbor, Michigan. (Methodology of System Engineering)*

MAURICE A. MEYER, *Vice President, Adcole Corporation. (Logical Circuits)*

PHILIP M. MORSE, *Director, Operations Research Center, Massachusetts Institute of Technology. (Queues and Markov Processes—the Response of Operating Systems to Fluctuating Demand and Supply)*

LLOYD G. MUNDIE, *Lockheed-California Company, Burbank, California. (Infrared)*

JACK N. NIELSEN, *Director of Research and Development, Vidya Division, Itek Corporation. (Aerodynamic Systems)*

R. D. O'NEAL, *Vice President, Aerospace Systems, The Bendix Corporation, Detroit, Michigan. (Management)*

CLAY L. PERRY, *Director, Computer Center, University of California, San Diego, La Jolla, California. (Numerical Analysis)*

RICHARD W. PEW, *Human Performance Center, Department of Psychology, The University of Michigan. (Human Information-processing Concepts for System Engineers)*

DONALD J. PORTMAN, *Department of Meteorology and Oceanography, The University of Michigan. (The Lower Atmosphere)*

AUBREY W. PRYCE, *Acoustics Programs, Office of Naval Research, Washington, D.C. (The Ocean)*

FAZLOLLAH M. REZA, *Syracuse University, Syracuse, New York. (Information Theory)*

HAROLD D. ROSS, JR., *Director of Research Relations, IBM Corporation, Data Systems Division, Poughkeepsie, New York. (Reliability)*

MANUEL ROTENBERG, *University of California, San Diego, La Jolla, California. (Numerical Analysis)*

CONTRIBUTORS

EDWARD RYZNAR, *Department of Meteorology and Oceanography, The University of Michigan. (The Lower Atmosphere)*

MORRIS SCHULKIN, *Westinghouse Electric Corporation, Baltimore, Maryland. (The Ocean)*

SIDNEY STERNBERG, *Vice President and General Manager, Electro-Optical Systems Inc., Pasadena, California. (Satellite Systems)*

ROBERT M. THRALL, *Professor of Mathematics and Operations Analysis, The University of Michigan. (Game Theory)*

JOHN G. TRUXAL, *Vice President for Educational Development, Polytechnic Institute of Brooklyn. (Feedback Theory)*

FOREWORD

In late 1960 Professor Harry H. Goode of The University of Michigan agreed to prepare a System Engineering Handbook for the McGraw-Hill Book Company. He apparently prepared a brief outline and indicated a few of the individuals from whom he intended to request articles. After his untimely death on October 30 of that year, Tanner, who was his literary executor, discovered these papers in his files and conceived the idea of completing this work as a memorial to Professor Goode. At his suggestion, Machol and Alexander agreed to join him in this effort. Completion of the original outline was done by the three of us in 1961.

The large amount of clerical work involved in this undertaking required that our efforts be concentrated in a single city and a single office. The efforts involved, during 1961 and 1962, in convincing outstanding scholars to devote their time to the preparation of articles for this handbook were performed primarily by Machol and Tanner, who were both in Ann Arbor. The editorial efforts in 1962 and 1963, the proofreading and indexing in 1964, and the heavy load of administrative and clerical responsibilities for putting this handbook together in 1962–1964 were Machol's.

In addition to being a memorial to Professor Goode, for whom we still grieve, this handbook has been an eleemosynary effort to benefit the Goode Educational Fund. Most of the authors, as well as the three editors, have waived their honoraria and other financial rights in this work, and the editors wish to express their very deep appreciation to the authors who have contributed so freely of their time to this worthy purpose.

Robert E. Machol
Wilson P. Tanner, Jr.
Samuel N. Alexander

PREFACE

Gratitude having been expressed in the Foreword and the organization of this book briefly outlined at the beginning of Chapter 1, it remains to explain to the reader what I have attempted to do in this handbook and how I have attempted to do it.

The words "system" and "system engineering" mean many things to many people; what I mean by them is explained in Sec. 1-1, and this definition has influenced the choice of subjects, of authors, and of material included. Thus, for the topic of urban environment I chose operations research experts who wrote a chapter (4) of very different flavor than might have been prepared by a sociologist or city planner; and the chapter (31) on human engineering is written from an information-theoretic viewpoint which I believe to be peculiarly pertinent to system engineering.

In the level of mathematical sophistication of the presentation, one runs always between the Scylla of laborious, inefficient, dull presentation on too elementary a level and the Charybdis of passing completely over the reader's head by use of advanced mathematics. It is recognized not only that the mathematical preparation of readers varies greatly, but also that there may be quite a difference between the mathematics to which the reader has been formally exposed and that with which he is facile. In most chapters, familiarity with elementary calculus (such as may be learned in the first year or two of most undergraduate engineering curricula) is assumed; in addition, familiarity with elementary matrix operations (at least to the extent of notation) is assumed in a number of chapters. More advanced techniques such as variational methods and function theory are rarely used. However, this rule is not followed slavishly, and in certain chapters such as Dynamic Programming (27) and Laplace Transform (39), a considerable degree of sophistication is expected of the reader, though still not at the level of the professional mathematician (except perhaps in the appendix to Chapter 39).

Probability theory is a special case. About a dozen chapters require an understanding of this theory; for those who are not familiar with it, Chapter 38 presents it in a very elementary and, in my opinion, extraordinarily lucid fashion.

Length and coverage of this handbook are necessarily compromises, and have been subjects of much consideration. Inevitably some readers will find here things they consider superfluous and fail to find others which they seek. Certainly Part II (Chapters 2 through 7) on environments is unusual in a handbook of this type, but I think it will be exceptionally interesting and useful to many system engineers.

I have endeavored to make the handbook efficient by minimizing overlap and repetition. I have carefully scrutinized and coordinated the outlines, and subsequently the texts, of each chapter, and exchanged many of these among authors (as will be obvious, for example, to anyone who reads Chapters 20 and 21, or 25 and 26, which had considerable possibility of overlap). I have inserted numerous cross references from one chapter to another for this reason. However, this approach was also not followed slavishly; the reader does not wish to turn continually from one chapter to another, and each chapter was therefore made self-contained, within reasonable limits of duplication.

So much time is required to produce a work of this magnitude that, inevitably, some material appeared to lose its freshness. A special effort was therefore made to review the sections and bring them up to date through 1964 by changes either in manuscript or in proof.

Finally, in a book of this magnitude, in spite of editing and proofreading efforts, there will inevitably be errors. I would greatly appreciate having them called to my attention.

Robert E. Machol

CONTENTS

Contributors v
Foreword ix
Preface xi

PART I. INTRODUCTION

Chapter 1. Methodology of System Engineering 1-3

PART II. SYSTEM ENVIRONMENTS

Chapter 2. The Ocean 2-3
Chapter 3. Land Masses 3-1
Chapter 4. Urban Areas 4-1
Chapter 5. The Lower Atmosphere 5-1
Chapter 6. The Upper Atmosphere 6-1
Chapter 7. Space and Astronomy 7-1

PART III. SYSTEM COMPONENTS

Chapter 8. Digital Computer-system Characteristics 8-3
Chapter 9. Logical Circuits 9-1
Chapter 10. Analog Elements 10-1
Chapter 11. Electronic Analog/Hybrid Computers and Their Use in System Engineering 11-1
Chapter 12. Communications Equipment 12-1
Chapter 13. Communication Systems 13-1
Chapter 14. Radar 14-1
Chapter 15. Infrared 15-1
Chapter 16. Optical Sensors 16-1
Chapter 17. Satellite Systems 17-1
Chapter 18. Aerodynamic Systems 18-1
Chapter 19. Guidance 19-1
Chapter 20. Propulsion 20-1
Chapter 21. Selected Energy-conversion Subsystems . . . 21-1

PART IV. SYSTEM THEORY

Chapter 22.	Information Theory	22-3
Chapter 23.	Game Theory	23-1
Chapter 24.	Decision Theory	24-1
Chapter 25.	The Simplex Method	25-1
Chapter 26.	Elements of a Strategy for Making Models in Linear Programming	26-1
Chapter 27.	Dynamic Programming—Theory and Application	27-1
Chapter 28.	Queues and Markov Processes—the Response of Operating Systems to Fluctuating Demand and Supply	28-1
Chapter 29.	Feedback Theory	29-1
Chapter 30.	Adaptive and Learning Control Systems	30-1

PART V. SYSTEM TECHNIQUES

Chapter 31.	Human Information-processing Concepts for System Engineers	31-3
Chapter 32.	Simulation	32-1
Chapter 33.	Reliability	33-1
Chapter 34.	System Testing	34-1
Chapter 35.	Economics	35-1
Chapter 36.	Management	36-1
Chapter 37.	Radio-telemetry Systems	37-1

PART VI. USEFUL MATHEMATICS ASSOCIATED WITH SYSTEM ENGINEERING

Chapter 38.	Probability	38-3
Chapter 39.	Properties of the Laplace and Related Transforms	39-1
Chapter 40.	Numerical Analysis	40-1
Chapter 41.	The Propositional Calculus	41-1

Index follows Chapter 41.

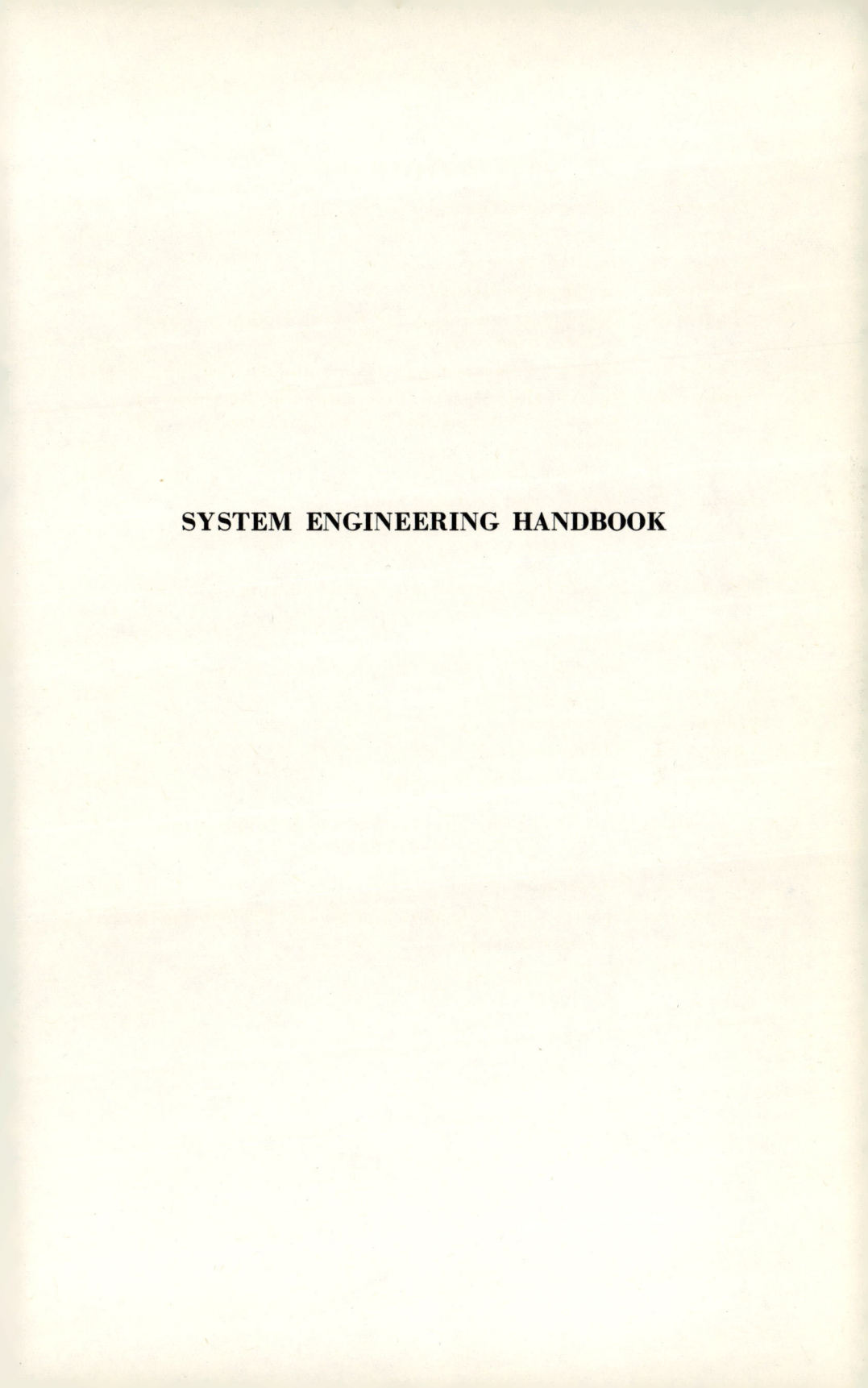

SYSTEM ENGINEERING HANDBOOK

Part I

INTRODUCTION

Chapter 1

METHODOLOGY OF SYSTEM ENGINEERING

ROBERT E. MACHOL, *Head, Department of Systems Engineering, University of Illinois in Chicago; formerly Vice President, Systems, Conductron Corporation, Ann Arbor, Michigan.*

CONTENTS

1-1. Definition	1-4
1-2. Anatomy of System Engineering	1-4
Phases	1-5
Steps	1-6
Tools	1-7
Parts	1-7
Subsystems	1-7
1-3. Principles of System Design	1-7
1-4. The Systems Viewpoint	1-8
1-5. Operations Research	1-10
1-6. The Compleat System Engineer	1-11

The subject of this handbook is system engineering, and it behooves us to define this term so that the reader may expect to know what he will find herein. System engineering, and hence this book, is concerned with the design of a specific class of systems, which are characterized below. The reader is presumably concerned with the design or analysis of such a system; he will find in these pages theory and practice which will aid him in this process. This is organized in the following fashion: First we describe system environments; then we present the most important components used in modern systems; this is followed by certain theoretical tools and techniques commonly used in this field; and we conclude with a few branches of mathematics which are particularly pertinent to the system-design endeavor.

Much of the material in this handbook would have been omitted from a work on "engineering" twenty years ago because it would have been classified as "science," but today it is not only pertinent but a vital part of the armamentarium of the engineer. This results from the change which has taken place in engineering, which in turn is due to the increasing difficulty of the problems being undertaken and the compressing of time scale between discovery of new knowledge and the application of that knowledge in engineering. Nowhere is this more true than in system engineering, where the most advanced scientific discoveries are consciously brought to bear on the most complicated engineering problems facing mankind.

1-1. Definition

Several books on system engineering have appeared in recent years,[2-6] and in each the authors have wrestled with the problem of defining such terms as system, system analysis, and system engineering. On the other hand, the best article on system analysis which I have seen[7] offers no such definitions, and no apology for their omission. A sampling of such definitions is given in the Appendix to this chapter. It does not appear that there is much utility in such definitions, or even that any universally acceptable definition is likely to be available in the near future. About all one can hope for is that the author should delineate the area about which he is discoursing, so that the reader can have an indication of the intersection of their areas of interest. Thus, when I speak of system engineering, I refer to the design of systems, the word "design" being significant. It implies that the output of the system engineer is a set of specifications suitable for use in constructing a real system out of hardware. By system analysis I mean the study of a system which does not yet exist (at least in the modification under study) in an attempt to elucidate its effectiveness or performance, its cost in dollars or other factors, and the effect of parameter variations on these quantities.

It remains to define "system." The following list of seven characteristics will sufficiently restrict the class of systems to constitute a definition of those systems about which I am speaking, and to which the discussions in this and the following chapters are primarily applicable. It should come as no surprise if some of these discussions should be applicable to other systems, but this would not constitute a valid reason for expanding the definition.

1. The system is man-made, from equipment, or "hardware." This eliminates anthills, universes, totemic relationships, and many other interesting "systems."

2. The system has integrity—all components are contributing to a common purpose, the production of a set of optimum outputs from the given inputs. What this purpose is, how we define optimum, and even the nature of the inputs will often be unknown at the start of the system-design process, and their elucidation will be an important part of the task. It follows that a city (Chapter 4) is not, by this definition, a system, although it is important as the environment of some systems.

3. The system is large—in number of different parts, in replication of identical parts, perhaps in functions performed, and certainly in cost; such things as the ignition system in an automobile are thus excluded.

4. The system is complex, which is here taken to mean that a change in one variable will affect many other variables in the system, rarely in linear fashion; in other words, the mathematical model of the system will be complicated. This eliminates systems which are merely large, such as a bridge or highway (apart from considerations of traffic flow thereon).

5. The system is semiautomatic, which means that computers always perform some of the functions of the system and human beings always perform some of the functions of the system. The large, completely manual system is eliminated because it is too inefficient, and the large, completely automatic system because it probably does not exist.

6. The system inputs are stochastic, which leads to an inability to predict the exact load or performance at any instant. In some cases, the rate of input is predictable (e.g., in an automatic factory), but even here the difficulties in design are due to unpredictable variability in such things as environment and raw materials.

7. Most systems, and especially the most difficult systems, are competitive. In military systems, a rational agent (the enemy) is trying to destroy or reduce the effectiveness of our system; in business systems ordinary competition, or in public service systems cheating or mere noncooperation, have similar effects.

1-2. Anatomy of System Engineering

Now consider the design of such a system. Someone hands you $100,000,000 and says "make a system to control air traffic" or to shoot down intercontinental ballistic missiles, or to connect 70,000,000 telephones by direct dialing, or to perform some other function. What do you do next? Part of this problem is discussed under the heading of Management in Chapter 36. Here I wish to stress that the problem of designing a large-scale system is overwhelming if it is attacked all at once; yet if the attack is piecemeal, it is unlikely to be successful. It is necessary to subdivide the problem in a number of ways, both conceptually and organizationally. In particular, there are:

1. The (chronological) phases of system design
2. The (logical) steps of system design
3. The (mathematical and scientific) tools of system design
4. The (functional) parts of a system
5. The (administratively designated) subsystems of a system

These methods of subdivision, while coexistent, are not equivalent or spread along a continuum; they are orthogonal vectors in system space, and this multidimensionality must be borne in mind if the system-design process is to be understood.

Phases. Because of the many different types of systems, and the different environments under which they are designed, the history of one may bear little resemblance to the history of another. Reference 1 states (in Chapter 3) that "the process can best be described by assuming that the design passes through well defined phases in chronological order, realizing, however, that the phase is often unrecognizable until it has passed," and then describes the phases of initiation, organization, preliminary design, principal design, prototype construction, test, training, and evaluation. Reference 2 gives (in Chapter 2) an excellent history of a particular system-design effort. Chapter 36 of this handbook describes four phases of "the system program." Names for certain phases—e.g., "phase one study" and "preliminary design"—have become part of the jargon of system engineering with reasonably well-defined meanings.

In addition to the wide variety of systems, important modifying influences are sponsorship, salesmanship, and urgency. Many large-scale systems are designed by one organization for another. The sponsor may do merely enough study to permit writing an acceptable request for proposal, and the preliminary design is then done by the organization or organizations which seek the system-design contract; or the system-design concept may originate with a contractor who submits an unsolicited proposal. Contractors have spent upward of a million dollars on a single proposal—a sum of money which clearly implies that fairly extensive preliminary design work has been performed. Much of this work is necessarily motivated by the desire to obtain the contract as much as by the desire to obtain an optimum system.

The urgency of the need for a large-scale system often stimulates telescoping of the time schedule for design and development; for example, prototype construction is undertaken before the design effort is completed, or production is started before the prototype has been tested. Within the design effort itself, telescoping may consist of proceeding on the assumption that specified performance will be obtained from a component before that performance has been demonstrated. The obvious disadvantages of such procedures—because they may lead to wasted time and effort and redoing of certain phases—must be balanced against the equally obvious disadvantages of proceeding in an orderly fashion—because the system, when finally complete, may be too late to be useful.

On the one hand, the tempo of technological change has become ever quicker, speeding obsolescence; on the other hand, our systems have become ever more complex, stretching necessary development times. The result is that most systems are in some sense obsolescent, if not obsolete, by the time they are operational. The military suffer most severely from this problem; they have found that complex defensive systems, by the time they are deployed in the field, are effective only against obsolete threats; but of course they also have the converse problem of attempting to deploy offensive systems before a corresponding defense can be implemented. Competitive systems are always faced with this special problem: "For example, an air defense team will sometimes be given the responsibility of designing a system which, *with High Confidence*, will guarantee that the enemy will get less than 20% of his bombers through. In the same building, you will have an air offense team which is given the responsibility of designing a system that guarantees that at least 80% of our bombers get through his defenses, *also with High Confidence.* The astute reader will notice that even though the numbers 80 and 20 add up to 100, they are in this case completely incompatible."[7] The only hope for evading this type of situation is to speed up the system development.

For all these reasons, the chronological development of a complex system is difficult to describe in general terms, and we refer the reader to the references for more specific discussions. Here we raise two important points concerning these phases: their repetitive nature, and the absence of the final phase.

The phases, as they succeed one another, repeat one another. In the first go-around, the general outlines of the system and one-significant-digit estimates of its performance can be drawn up in a few hours by one skilled in the state of the art, using rules of thumb for such things as the noise figure available at X band or the number of pounds of rocket engine required to produce a pound of thrust. In successive phases, the design is refined and taken up in greater and greater detail, with justifications of performance and design of interconnections being carried on with ever greater specificity.

The final phase of system engineering, after the system has been built and tested, is nominally evaluation. The interesting point is that this evaluation phase is essentially null; in good system engineering, evaluation goes on throughout the design process and has largely been completed before the prototype is constructed. The system must, of course, be tested. It is a foregone conclusion that it will not work perfectly the first time it is turned on; but with reasonably competent engineering behind it, it is also a foregone conclusion that the "bugs" can be sufficiently eliminated so that the system can be made to work in some fashion. In fact, unless drastic errors have been made, the performance eventually demonstrated in test will be very similar to that specified. Evaluation is now supposed to determine whether that performance is adequate for the system to fulfill its mission. And the point here is that nothing pertinent to such a decision is known after the test which was not known before. It therefore follows that evaluation should be largely completed before the really expensive phases of prototype construction and test are undertaken. This is one of the outstanding characteristics of modern system engineering (Ref. 1, Sec. 31-2).

Steps. The steps of system design are logical steps, but they are not performed in order. Logically, one must formulate the problem before one solves it; in fact, one does both simultaneously throughout the system-design process. Because the problem cannot be adequately formulated until it is well understood, and because it cannot be well understood until it has been more or less solved, the two are inseparable.

An important group of steps which do permit a useful subdivision of the problem for analysis is based on the realization that what happens to a single input as it passes through the system can to some extent be studied separately from what happens when two or more inputs require service simultaneously. Similarly, the effects of noncooperation can to some extent be studied separately from the operation of the system without interference. This leads to the "steps" of single-thread design, high-traffic design, and competitive design (Ref. 1, Chapter 3).

Another important group of steps is concerned with the model of the system and its manipulation as a substitute for experimentation with the actual system. While certain physical models (mock-ups, wind-tunnel experiments, etc.) are often essential, most important here are the various mathematical and symbolic representations, and their manipulation, analytically or synthetically, with pencil and paper or with a computer. One distinguishing characteristic of system engineering is that the system is usually far too complicated for complete understanding to come through test of the actual system under all the possible environments in which it may have to operate—even if such tests were possible on the basis of time and expense. Instead, one simulates the system, and performs the tests on this simulated model. The simulation may vary from a few simple equations to a sophisticated and detailed description which requires a large computer; and this in turn may be combined with real or simulated portions of actual equipments and/or human operators. The degree to which the design is guided and checked out by analysis and simulation is one of the distinguishing characteristics of system engineering. Mathematical models are further discussed in Sec. 1-5; simulation is the subject of Chapter 32.

Tools. Almost any scientific or mathematical discipline may be a tool of the system engineer. Most of the following chapters are devoted to these tools, and except to underline the fundamental nature of probability theory (Chapter 38) in system engineering, no more needs to be said about them here.

Parts. The parts of the system, such as communications (Chapters 12 and 13), material handling, and control (Chapters 29 and 30) are also discussed in separate chapters in this handbook. What needs to be stressed here is only that material handling, which has been most important in classical engineering, is least important in system engineering. Following Wiener, we may assert as a principle that if the information flow is adequately handled, the material flow will almost take care of itself.

Subsystems. The division into subsystems is somewhat arbitrary, and is largely for convenience in administering the subsystem work. Generally, a subsystem should be in a single geographic location, and should include or exclude (rather than being bounded by) those interfaces or interconnections which cause most difficulty. If the division is well done, a group can be given responsibility for the design of a subsystem and work independently on it, with only occasional (rather than continuous) feedback to other parts of the system-design effort.

Since every system is a subsystem of some larger system, and conversely every subsystem may be considered a system with its own subdivisions, the definition of any particular level as a subsystem cannot be absolute. Hierarchies of terms have been suggested, such as system, subsystem, equipment, component, subassembly, and part, with the word item applying to any of these. Definitions of some of these terms have also been suggested; for example, a component is the lowest functional item, and a part is an item which is destroyed by disassembly. While such terms can be useful in controlling the paperwork of a particular system effort, they do not at present encompass universally accepted or well-defined meanings.

1-3. Principles of System Design

In Ref. 1, Sec. 21-6, certain principles of system design are stated; these are briefly summarized here.

The fundamental principle of system design is simply to maximize the expected value. Obviously this requires considerable interpretation in any particular case, but at least expected value has a succinct and well-understood definition (see Expectation, Sec. 38-4). Where one has the choice of supplying too much of something (resulting in excessive cost) or too little (with the possibility of a penalty if it proves inadequate), this rule gives a guideline, and this kind of thing is done continually in the design process. Thus we have "trade-off analyses," in which, for example, we compare the cost and weight of supplying extra radar performance (which permits us to detect the enemy earlier) with that of extra propulsion (which permits us to get to the enemy more quickly after detecting him). In such a two-parameter analysis, it is conceptually simple to find a maximum; in a realistic situation, we would have to trade also with dozens of other parameters, with pairwise comparisons being totally inadequate (see Suboptimization, Sec. 1-4). This leads to "cost/effectiveness studies," in which we attempt to maximize the effectiveness of the system (or its expected value) for fixed cost, or minimize the cost for fixed effectiveness. Because it is generally impossible to find a single number which realistically represents the effectiveness of a complex system, there is, as indicated below, a good deal of art, as well as science, in system engineering.

The principle of events of low probability seems so obvious that it would not be worth mentioning if there were not so often pressure from the sponsor (especially in military systems) to violate it. The principle states that the fundamental mission of the system should not be jeopardized, nor its fundamental objectives significantly compromised, in order to accommodate events of extremely low probability. Yet one frequently hears: "the most trivial detail may be the key to the entire intelligence picture; therefore, the system must be able to store and process every conceivable intelligence input" in spite of the fact that the resulting system is too complex to be workable; or "the soldier in the foxhole may have urgent requirement for the airborne reconnaissance information; therefore, the entire data-processing system must be airborne, and provision supplied for air drop of finished data to the front lines" in spite of the fact that the resulting allocation of weight to airborne data-processing equipment will seriously compromise the reconnaissance performance. The system engineer can sympathize with the soldier in the foxhole and the commander who is sensitive to his needs, but he should insist on reasonable compliance with the fundamental principle of maximizing the expected value of the entire system.

In many systems, a compromise is possible: the system can be designed to handle most events automatically, and to sound an alarm which calls for manual intervention when an uncommon event occurs which is beyond its capabilities. For example, an automatic mail-sorting system would throw out, for manual sorting, those letters which were not of standard size, shape, or location of address; such a system might handle 95 per cent of the mail automatically, at a cost much less than that of 100 per cent manual handling, and enormously less than 100 per cent automatic handling. Similarly, when you reach a wrong number through (automatic) direct-distance dialing, you simply call an operator for (manual) rectification of the error.

The principle of centralization refers to centralization of authority and decision making, that is, to centralization of information as distinguished from material. Most organizations are built on the principle that routine inputs are handled at a low echelon, with higher echelons being informed so that they may veto specific decisions or change policy on general decisions if it is appropriate (this is sometimes called "management by exception"). Nonroutine decisions are passed to higher echelons for decision, this decision hopefully establishing a policy so that similar decisions in the future will become routine. The difficulty arises when a situation is both nonroutine and urgent; this difficulty arises only when the speed required for making the decision exceeds the speed with which the information may be communicated to the higher echelon and the decision made there and transmitted back. Thus, as speeds of communication and decision making increase, the disadvantages of decentralization decrease. With the improvements in computers (Chapters 8 to 11), communications (Chapters 12 to 13), displays, and theories of decision (Chapter 24), the optimum in the continuum between centralization and decentralization moves more in the direction of centralization in our complex systems.

The principle of suboptimization states that optimization of each subsystem independently will not in general lead to a system optimum and, more strongly, that improvement of a particular subsystem may actually worsen the overall system. Since every system is merely a subsystem of some larger system, this presents a difficult if not insoluble problem, which should at least be faced by the system designer, as discussed below.

1-4. The Systems Viewpoint

Described above have been seven characteristics of systems (Sec. 1-1), a number of traits of system engineering (Sec. 1-2), and certain principles of system design (Sec. 1-3). But system engineering is more than a knowledge of all these, and in what follows an attempt is made to give the reader a feeling for the point of view which makes system engineering different from classical engineering.

The heart of the matter lies in the complexity of the system, and the danger of being unable to see the forest for the trees. The designer must somehow deal with the various subsystems and component parts in such a way as to optimize the cost/effectiveness of the overall system—which means avoiding the dangers of suboptimization. The word "suboptimize" was coined in 1952 by C. J. Hitch,[8] and the following example is taken in part from his article.

An excellent study, one of the classics of operations research (Sec. 1-5), was performed during World War II on the optimum size of a merchant-ship convoy.[9] The problem was the sinking of United States and Allied merchant ships by "packs" of German submarines in the North Atlantic Ocean. There is, of course, never enough data for such problems, because of the statistical variability in such things as sightings and sinkings and the numerous questions of luck and skill involved, but the researchers were able to show (with what most system engineers would agree was reasonable confidence) that the number of ships sunk when a convoy was attacked by a given pack of submarines was independent of the number of ships in the convoy; furthermore, the number of submarines sunk in such an encounter was proportional to the number of escort vessels (destroyers and the like) in the convoy. It follows that the payoff, chosen as the ratio of submarines sunk to merchant ships sunk, varies as the square of the size of the convoy (assuming that the same ratio of escorts to merchant ships is retained). The recommendations from this study were put into effect, and the number of merchant-ship sinkings decreased drastically, contributing importantly to the winning of the Battle of the Atlantic, and consequently to the winning of the war. In fact, the decrease was even more dramatic than predicted; the submarines were so ineffective in the North Atlantic that they were transferred to more profitable missions elsewhere.

This celebrated problem has been the subject of a number of postwar studies, and it now appears that the change in tactics (increasing convoy sizes) was probably right, but for many of the wrong reasons. In fact, the study as described above is remarkable for the number of errors which have been made from the system viewpoint.

In the first place, if one really believed the above conclusions, then he would recommend taking every bottom available to the Allies and putting them into a single giant convoy. This is clearly ridiculous (in system engineering, as in mathematics, extreme cases are often illustrative); the optimum convoy size must consider the disadvantages of increasing size as well as the advantages. And the obvious disadvantages are that the convoy can move no faster than its slowest ship, and that the arrival of a convoy swamps port facilities, greatly increasing turn-around time.

In fact, the study is guilty of suboptimization; what has been optimized is the skirmish between a convoy and a submarine pack, for which the measure of effectiveness is the ratio of submarines sunk to merchant ships sunk; what should have been optimized is the Battle of the Atlantic, for which the proper measure of effectiveness is the goods delivered to the eastern shore of that ocean. Of course, one can quibble about modifications of this measure depending on the length of the war (do we want to maximize goods delivered during the next month or during the next year?), and the desirability of saving human lives (at least on our side), but the principle is clear: if the convoy is too large, it will take so long to assemble, load, sail, unload, and return that the amount of goods delivered may be considerably less even though we lose less shipping.

But even this viewpoint is a suboptimization, because the real objective is not so much to win the Battle of the Atlantic as to win the war. And when the German submarines went elsewhere, they indicated that we had gone too far. It is a principle of competitive situations (and, to the extent that they can be considered games against nature, all stochastic systems) that when we have achieved our optimum strategy, we are indifferent as to what the enemy (or nature) may do; this concept is made formal in game theory (Chapter 23). It follows that if the enemy (or nature) has a clear-cut preference available, we are not at the optimum. In this case, if the German submarines could clearly do better by leaving the North Atlantic, we must have made our convoys too large.

But even this is a suboptimization, because the objective of winning the war should be subordinated to the objective of optimizing the postwar world. In this example, such considerations would be stretching things, but it was, for example, a serious question in the decision to drop the bomb on Nagasaki. And of course, there are even higher objectives. And what is the system designer to do if he is several echelons farther down (for example, designing an antisubmarine guided-missile system aboard one of the escort vessels)?

In answer, Hitch[8] suggests "the relevancy of economics" which "involves the analysis of relations between suboptimizations at lower and higher levels." I would add that, while absorption in the problems of higher levels can lead to paralysis or, worse, severe political repercussions, the designer should always be cognizant of their problems and the effects of his actions on them. Most important, he should know the level of his own sponsor, and select, with him, appropriate criteria.

But the sponsor is not omniscient, and the problem as originally stated is usually wrong; in fact, as asserted above, when the problem is well understood, the solution is frequently obvious. As a tool to aid in understanding the problem as well as posing it at the proper level, in addition to being most useful in solving the problem and in testing the solution, construction of the mathematical model is preeminent. Which brings us to operations research.

1-5. Operations Research

Operations research has had as many definitions as system engineering. Whether or not it is a well-defined entity, it is certainly a going concern. The Operations Research Society of America, one of 17 national operations research societies affiliated with the International Federation of Operations Research Societies, has at this writing over 4,000 members, and there is scarcely a large organization in the country without an officially designated operations research group.

Without attempting to define operations research (there are a dozen recent books and innumerable journal articles which do so), we may state that the viewpoint of operations research is very much like that of system engineering as discussed in the preceding section—in fact, the two citations in that section were to a book and a journal specifically devoted to operations research. The two disciplines differ perhaps primarily in the types of problems which they nominally prefer to attack, the operations researcher being prone to optimize operations, presumably with existing equipment, while the system engineer is prone to design new equipment. However, both would agree that whether new equipment or new tactics or both are appropriate should and does depend on the problem rather than the man called in to solve it.

In addition to the names "operations research" and "system (or systems) engineering," numerous other terms have been introduced in recent years for similar disciplines and lines of endeavor. One of the best known of these is "cybernetics," coined by Norbert Wiener in 1947; whatever he may have meant by the term (see Ref. 10 or, for a brief description, Ref. 1, Sec. 25-3), it has been widely used, especially in the U.S.S.R, to subsume those fields which are here called system engineering and operations research. Various neologisms, such as "synnoetics,"[11] and combinations of familiar words, such as "value engineering," "management science," and "computer-related disciplines," have also been used to describe various educational curricula, professional societies, and administrative departments which have interests more or less akin to the subjects here discussed. No attempt is here made to define and differentiate these entities, except to point out that they have technological content overlapping to a greater or lesser degree the subject of this handbook.

Whatever operations research is, every work on the subject stresses the importance of mathematical modeling, and the professional operations analyst (a synonym for operations researcher) is highly qualified at constructing and applying mathematical models; in many cases he is also adept at solving them, but some operations analysts are not good applied mathematicians, and most applied mathematicians are not good operations analysts.

Operations research cannot take credit for the concept of a mathematical model, since physicists, economists, and others had been using it long before operations research was born during the Battle of Britain in 1940. However, operations research did show that it could be extended to a wide variety of fields where it had not been previously applied, and most especially to operational situations (such as the submarine vs. convoy struggles described above) where human beings are deeply involved and where, for this and other reasons, there are numerous random aspects on which the outcome depends. The concept of defining a quantitative measure of effectiveness such as the ratio of submarines sunk to merchant ships sunk, and then writing equations showing how the expected value of this ratio depends on such parameters as the number of each present in the battle and on various factors such as the weather or the tactical choices of either side, is a breakthrough in the history of human thought, perhaps comparable in importance with some of the major discoveries in physics. This type of mathematical model, its construction and use, lie at the bottom of much modern system analysis.

Construction of mathematical models is discussed in books on operations research, and in a chapter (26) of this handbook concerned primarily with linear-programming models. Solution of models occupies many of the chapters of this handbook, and much of the journal literature on operations research. Frequently the model, once constructed, is too complicated to solve, and so one renders the model tractable by linearizations, by replacing distributions with expected values, and the like; or one discovers new methods of solution, such as fill the journal articles and lead to the mistaken impression that operations analysts are applied mathematicians. These techniques are of great value, and the mathematician who can solve the equations, or the experienced analyst who recognizes how the present problem can be transformed into one previously solved, are invaluable members of the system-design team. But at an earlier level, there is an even bigger problem: to determine which aspects of physical reality must be accurately simulated in the model and those which are comparatively irrelevant for the problem at hand, and then to construct a model which satisfies these constraints. Often the model can be greatly simplified by discarding some aspect which is not pertinent to the particular question being asked; or conversely, the model may be enormously complex, and still useless because it has ignored some vital factor. Thus the system engineer, even if he is not an expert mathematician, must assure himself that the right equations have been written down, so that conclusions derived from them will be valid for the system problem for which he is responsible.

When all the above has been done, it frequently happens that the model cannot be handled by simple paper-and-pencil manipulation of the equations. This may be because of its complexity, or because adequate transfer functions cannot be written down for certain components (especially human beings). In all these cases, simulation (Chapter 32) is resorted to. The ability to simulate a system as yet unbuilt, and thereby to predict its performance realistically, and the regularity with which this is done, are characteristic of modern system engineering.

1-6. The Compleat System Engineer

The system engineer must be a generalist as distinguished from a specialist, but he must not be a dilettante. The ideal system engineer is a "T-shaped man," broad, but deep in one field; the depth is provided by scholarly experience—a Ph.D. or equivalent—and the breadth by extended interests and abilities. He will frequently find that he must become a "6-month expert" in a new field, which may be meteorology or television or electroencephalography, but he will find that because of his background in mathematics and engineering, he can learn enough in a short time to allow him to work with real experts in the field.

The system engineer must know most of the material in this handbook, and much related material, or at least know where to find it and be prepared to assimilate it quickly. In addition to technical material, he must know a good deal about administrative and marketing matters. In particular, he must be a good salesman, because regardless of the merit of his ideas, he must convince some sponsor that his project is more worthy of support than the numerous other proposals which are invariably competing for the limited financing, equipment, or time available. If he follows the experience of many system engineers, he will spend a large portion of his time writing proposals. He must know about the project system of management and about PERT, which are discussed in Chapter 36; he must know about costs and accounting procedures, which are not discussed in this handbook; and he must know about organizational and administrative politics, which probably cannot be learned from any book.

Of course this superman does not exist, but there are thousands of system engineers who come remarkably close. Much of what the practicing or aspiring system engineer needs to approach this ideal, and particularly the technical material, has been made the subject of this handbook. The real definition and methodology of system engineering lie in all the following pages.

APPENDIX

Some Definitions of System

"A system is an assembly of interconnected but separable and independent parts. The system specification is a statement of the social function the system is to perform. System design is a statement of what elements the system will contain and the manner in which they are to be interconnected. This is the task of an individual who is now known as the system engineer."[4]

"A system is something which accomplishes an operational process; that is, something is operated on in some way to produce something. That which is operated upon is usually called input; that which is produced is called output, and the operating entity is called the system. A system is a device, procedure, or scheme which behaves according to some description, its function being to operate on information and/or energy and/or matter in a time reference to yield information and/or energy and/or matter."[3]

"Systems engineering probably is not amenable to a clear, sharp, one-sentence definition. Systems engineering has many facets; therefore, a complete definition of it must have many facets too. To get a reasonably comprehensive and useful view of all sides, we might define the systems engineering function in terms of: (a) Its evolution (b) The systems engineering process (as generalized from case histories) (c) Its objectives (d) The kinds of work done and called systems engineering (e) Organizational arrangements for carrying out the function (f) The tools and techniques which it uses (g) The kind of people who do it (h) Its relation to other fields such as administration, applications engineering, operations research, etc."[2]

"A system may be defined as an array of components designed to achieve an objective according to plan."[5]

"The term systems engineering is used to describe an integrated approach to the synthesis of entire systems designed to perform various tasks in, what is expected to be, the most efficient possible manner. Thus the term "systems engineering" is used to describe an approach which views an entire system of components as an entity rather than simply as an assembly of individual parts; i.e., a system in which each component is designed to fit properly with the other components rather than to function by itself."[6]

Definition. A system $\mathfrak{S}$ is an object

$$\{T, \tau, \Gamma, \Sigma, \Omega, \{\Gamma_R\}, \{\Sigma_R\}, \{\omega_{\gamma\sigma}\}, \{\bar{\omega}_{\gamma\sigma}\}\}$$

subject to postulates 1 to 5 below:

Postulate 1. Γ, Σ, and Ω are sets.
Postulate 2. T is a directed set, $(T, \leq)$, and τ is a set of directed subsets of T.
Postulate 3. For each $R \in \tau$,

$$\Gamma_R \subset \Gamma^R \quad \text{and} \quad \Sigma_R \subset \Sigma^R$$

Postulate 4. If $R \in \tau$, and $\gamma \in \Gamma_R$, and $\sigma \in \Sigma_R$, there is a mapping

$$\bar{\omega}_{\gamma\sigma}: \Gamma \bigotimes \Sigma \to \Omega$$

defined and called the $\gamma\sigma$ correlatant.

Postulate 5. If $R \in \tau$, $S \in \tau$, $f: R \to S$ is an order isomorphism, $\gamma \in \Gamma_R$, $\sigma \in \Sigma_R$, $\hat{\gamma} \in \Gamma_s$, $\hat{\sigma} \in \Sigma_s$, $\hat{\gamma} = f^{-1}\gamma$ and $\hat{\sigma} = f^{-1}\sigma$, then

$$\omega_{\hat{\gamma}\hat{\sigma}} = f^{-1}\omega_{\gamma\sigma}$$

(Ref. 3).

REFERENCES

1. Goode, H. H., and R. E. Machol: *System Engineering*, McGraw-Hill Book Company, New York, 1957.
2. Hall, A. D.: *A Methodology for Systems Engineering*, D. Van Nostrand Company, Inc., Princeton, N.J., 1962.

3. Ellis, D. O., and F. J. Ludwig: *Systems Philosophy*, Prentice-Hall, Inc., Englewood Cliffs, N.J., 1962.
4. Gosling, W.: *The Design of Engineering Systems*, John Wiley & Sons, Inc., New York, 1962.
5. Johnson, R. A., F. E. Kast, and J. E. Rosenzweig: *The Theory and Management of Systems*, McGraw-Hill Book Company, New York, 1963.
6. Dommasch, D. O., and C. W. Laudeman: *Principles Underlying Systems Engineering*, Pitman Publishing Corporation, New York, 1962.
7. Kahn, H., and I. Mann: *Techniques of Systems Analysis*, RM-1829-1, The Rand Corporation, 1957.
8. Hitch, C.: Sub-optimization in Operations Problems, *J. Op. Res. Soc. Am.*, vol. 1, pp. 87–99, 1953.
9. Morse, P. M., and G. E. Kimball: *Methods of Operations Research*, John Wiley & Sons, Inc., New York, 1951.
10. Wiener, Norbert: *Cybernetics*, John Wiley & Sons, Inc., New York, 1948 (2d ed., 1961).
11. Fein, Louis: *Am. Scientist*, vol. 49, pp. 149–168, 1961.

Part II

SYSTEM ENVIRONMENTS

Chapter 2

THE OCEAN

MORRIS SCHULKIN, *Westinghouse Electric Corporation, Baltimore, Maryland.*
AUBREY W. PRYCE, *Office of Naval Research, Washington, D.C.*

CONTENTS

2-1. Man and the Ocean	2-4
Exploitation and Problems	2-4
Instrumentation and Techniques	2-5
Structures	2-6
2-2. Chemical and Physical Properties of Sea Water	2-7
Composition	2-7
Equation of State	2-8
Thermal Properties	2-8
Acoustic and Mechanical Properties	2-8
Optical and Radio Properties	2-9
2-3. Thermodynamics	2-9
Heat Transport	2-9
Temperature in the Ocean	2-10
Temperature-Salinity Relationship	2-10
Water Budget	2-10
2-4. Dynamics of the Ocean	2-10
Ocean Currents	2-10
Sea Waves and Swell	2-11
Tides	2-11
2-5. Geology of the Ocean Basins	2-12
Classification of the Marine Environment	2-12
Sediments	2-12
The Ocean Bottom	2-13
2-6. Biology in the Ocean	2-13
The Sea as a Biological Environment	2-13
Organisms and the Composition of Sea Water	2-14
Marine Life	2-14

The oceans cover almost 71 per cent of the earth's surface. With a mean depth of 3,800 meters they provide a natural habitat, for life, of very much greater volume than the land masses. Man's knowledge of this great water mass is largely of recent origin and confined on the one hand to broad general characteristics and on the other to specific details of limited value to the system engineer. Clearly the study of the oceans cuts across many scientific disciplines and in this brief chapter no more than the salient features of the subject can be given. It is convenient to commence coverage of our knowledge of the oceans with an outline of the physical and chemical properties of sea water which are of importance to all its users and fundamental to its other characteristics. The thermodynamic characteristics of the ocean largely control the earth's heat budget and its weather pattern. Dynamics covers the general water circulation and the behavior of the sea surface. The geology of the ocean basins covers the other boundary of the water mass from the shoreline to the deepest trench. Significantly all these areas are interrelated and all are important to the marine life which is essentially part of the ocean and which contributes to its physical and chemical characteristics.

Since it may be claimed from the practical viewpoint that the ultimate outcome of knowledge is its best use by man, we take up first the subject of man and the ocean before discussing the scientific factors.

2-1. Man and the Ocean

Exploitation and Problems. Man on the shore is exposed to the roughest part of the ocean. Throughout the ages man in his approach to the ocean has been faced with the largely unpredictable violence of the coastal and surface waters exhibited by surface waves, currents, and tidal changes, which is in direct contrast to the general stability of the deep oceans, of which his knowledge has been negligible. Understandably, he has looked upon the ocean with fear, only overcome by the need to fish the inshore waters for food and by curiosity as to the land across the sea, generated more perhaps by the impetus of commercial gain than by scientific interest. From these beginnings coupled with the military use of the sea for offense and defense, large commercial interests in fishing and transportation have developed.

The yearly catch of fish from the oceans is between 30 and 40 million tons and is increasing at a rate of about 5 per cent per annum. Overfishing has long been a problem in certain areas and for some of the more valuable species, particularly the marine mammals which as a class provide such valuable products as fur, oil, and meat, but which reproduce slowly. Whale hunting has had to be rigidly controlled by international agreement, and extension of such controls will undoubtedly become necessary if fishing is expanded into a true farming activity and widespread industrial use is made of the oceans. Individual fishing fleets are now becoming larger and more highly organized. They are provided with modern navigation equipment, frequently with aircraft for spotting and with echo-ranging sonar capable of detecting shoals of small fish or individual fish of the larger species. To date, however, little use appears to have been made of listening or passive sonar to determine the species, although it is well known that certain species emit characteristic sounds. Surprisingly the catching methods—nets, trawls, and lines—have changed very little with the passage of time.

Ship transportation provides the only method of handling the great bulk of international commerce necessary to maintain the exchange of basic raw materials, food supplies, and manufactured goods. The general low cost of this form of transport for bulk shipment is best illustrated by the high level of coastal shipping in many areas of the world where it is in direct competition with overland modes of transportation. The needs of this traffic have resulted in the design of a wide range of specialized vessels, such as tankers, refrigeration ships, and colliers. As yet, all such vessels are surface craft, but recently the cargo submarine has been advanced as a possibility which would have the advantage of reduced drag, shorter routes under the ice, and independence of weather.

Exploitation follows knowledge, and both in terms of the ocean require operations at sea. Most operations at sea have to be conducted from surface ships with all the attendant problems of a moving and unsteady platform from which the handling of even the lightest overboard equipment presents difficulty and requires special techniques and experience. For many purposes accurate knowledge of position and the ability to return to it are of major importance. Once out of visual or radar sight of land, celestial navigation, limited by weather, or radio location methods must be used, and these are frequently inadequate to meet the navigational accuracy required. For work below the surface, currents and their variation with depth and the limited accuracy of depth gauges often preclude any but the roughest determination of the position of equipment relative to the ship. Some positioning problems can be partially solved by anchoring to the bottom, but motion of the vessel still results from changing winds and surface currents. Deep-water anchoring and mooring are recent possibilities and have had to await the availability of suitable cable materials, since most materials will not support their own weight to the depths of the ocean.

Many of the characteristics of sea water which are favorable to development of marine life are major problem areas in terms of operations at sea. The high density of sea water requires pressure equalization of equipment to the depth of submergence or encasement in a pressure capsule. The inertia of the water mass tends to control motion of all but the highest-density equipment. The dilute alkaline solution presents difficult corrosion problems. Marine life itself presents the added problems of marine boring and fouling in the near-surface waters, the possibility of direct physical damage particularly to cables, and interference due to its presence with the operation of equipment and instrumentation. Perhaps the most overlooked problem is that, for all but shallow depths in the clearest water in calm conditions, one loses sight of equipment once it is below the surface, and one must therefore attempt to provide adequate sensors. Underwater television cameras are of some help in this regard, but these and sensors in general require cable to the surface for data transmission and increase the already difficult handling problems.

As knowledge of the ocean-water mass and the bottom has grown, spurred on in part by military needs in subsurface warfare and in particular in pro- and antisubmarine warfare, the potential for exploitation of the oceans has grown enormously. Valuable reserves of petroleum are already being taken from the shallower waters of the continental shelves. In the future lie the possibilities of harnessing the tides for power; mining the sea for particular mineral deposits; extending the fishing activities into a true farming activity—aquaculture—to meet the world's growing food requirements; the making of fresh water from the sea itself, already undertaken in a few special localities, or from ice; and the using of the sea and the ice-bound areas for storage or for the disposal of harmful by-products such as radioactive waste. Probably the greatest advantages of an increasing knowledge of the oceans, their behavior, and their interaction with weather conditions lie in the general area of forecasting. Tide forecasting, of great importance to docking schedules, has long been a reality, whilst ice forecasting has for some years provided an increased measure of safety on the North Atlantic shipping routes. Ocean forecasting in terms of winds, surface waves, and swell offer, through adroit selection of course, the possibility of substantial reductions in transit times and in ship maintenance and repair, all of which could substantially reduce operating costs.

Instrumentation and Techniques. Most of the problems discussed in relation to exploitation apply equally to instrumentation. Except in those rare cases where the eye may be used for direct observation, other sensors must be used. Absorption of energy by the sea limits not only the use of optical techniques but also all electromagnetic methods and to a certain extent acoustical ones. The absorption of low-frequency acoustic energy is small, and under certain conditions sound can be heard for thousands of miles; for example, in the Sofar channel. At frequencies in the lower kilocycle range absorption is still low enough for their use in many applications such as echo sounders and fish-location sonars. The range of underwater acoustic applications will undoubtedly grow as knowledge of the ocean environment increases, but it is difficult to see how acoustic techniques could meet the needs for long-range high-data-rate communications. Virtually all ship-to-ship and ship-to-shore communications are provided by radio. For shore-to-shore purposes, however, cable systems, first installed about 100 years ago, have proved remarkably reliable.

Remote sensors must in general be connected to the observation platform, for the supply of power, to support the instrument package, and for signal return. Difficult cable problems result particularly when one is working in the deep parts of the ocean. Instrumented-buoy systems of the self-recording type or transmitting data by radio link above the surface, programmed self-propelled instrumentation vehicles, and the use of acoustic telemetering are relatively new.

The methods utilized to obtain information on the ocean bottom point up the instrumentation and technique problems experienced at sea. The general contour of the bottom can be obtained with conventional echo sounders, but the wide beam of these devices smooths much of the detail, while corrections have to be made for the varying speed of sound through the water mass. Precision depth recorders have been developed, but their accuracy is limited by our knowledge of the speed of sound. This can be measured directly as a function of depth by reference to a standard, but although the measurements are adequate for determining gradients, the overall accuracy of perhaps 1 part in 5,000 is not sufficient for all purposes. Low-frequency echo sounders are finding increasing use in determining the characteristics of the bottom sediments, particularly the thickness of the layers, but are limited by power, pulse length, and beam-width considerations. Explosions which are rich in low-frequency sound content are also widely used for this purpose. Bottom samplers, in the form of grabs or corers, are frequently used to recover samples of the bottom sediments, but a given sample may change during recovery and storage and may not be typical. Techniques for bottom photography have been developed but can cover only a very limited area. The greatest contributions to our knowledge of the bottom structure have come from the adoption of seismic shot-firing methods.

The time required for deep-sea measurements is frequently long, for it may take many hours to lower and raise equipment, as for a single oceanographic station involving sampling of the water mass at a number of levels for subsequent examination and temperature determination. Special instrumentation such as the bathythermograph has aided considerably in obtaining routine temperature profiles for the upper few hundred feet of the ocean, but even here the time response of the measuring device must be taken into account and the instrument cannot be lowered too rapidly. This instrument is inherently too slow in response and of insufficient precision to provide information on the thermal microstructure, detailed observation of which requires the use of chains of thermistor elements. These, with suitable concern for the supporting technique, can be towed through the water to provide thermal sections in space, or left stationary to provide the same information in time.

Structures. The limitations of the surface ship as a platform for operations at sea have been outlined above. They remain, however, and undoubtedly will remain, the primary vehicle for operations at sea, be this for exploitation in transportation, fishing, etc., or in the building of knowledge of the oceans. Alongside the surface ship, a growing family of other platforms is being exploited to meet special needs. It includes submarines, buoys, towers, seaplanes, hydrofoils, piers, moles, and ice islands. The submarine, developed for military use, and carrying man down into the ocean to depths of a few hundred feet, provides a more stable platform than the surface ship for obtaining information. Man's desire to know about the deep ocean has led to the development of special craft such as the bathyscaph, capable of carrying him into the deepest parts of the ocean for direct observation. Alongside these have been the development of a number of special-purpose small submarines capable of shallow-depth operation. To these developments must be added the self-propelled programmed-instrument packages capable of deep operation and the remotely controlled vehicle capable of performing limited mechanical functions under water.

Stability is perhaps the overriding requirement of a platform for measurement purposes at sea, and the most difficult to meet. One approach to this problem is to design a vehicle with great draft but with small freeboard and cross section which will be only slightly disturbed by the wind and surface waves. The Floating Instrument Platform (FLIP) with a draft of about 300 ft, is perhaps the first structure of this type. In recent years a new type of permanent or semipermanent platform has been developed for use in the shallower waters of the continental shelf in which the operational platform is carried above the height of the waves on a bottom-mounted supporting structure. These platforms, typified by the so-called Texas towers, are already widely used for defense purposes and in the exploitation of offshore oil deposits. They are finding additional uses as stable platforms for research purposes, as, for example, the so-called Argus Island on Plantagenet Bank 20-odd miles from Bermuda, and as replacements for light ships and navigational aids.

2-2. Chemical and Physical Properties of Sea Water

Composition. Sea water is a dilute solution of salts. Its composition in terms of ion concentrations is given in Table 2-1, four ions (Na^+, Mg^{++}, Cl^-, and SO_4^{--}) constituting 97 per cent of the total. The proportions of the different constituents present are nearly constant in the open sea, although the total salinity varies between about 33 and 38‰ (parts per thousand by weight of sea water), depending upon precipitation, evaporation, and ice melting. Considerably lower salinities may be observed in coastal regions and in almost-closed seas with low evaporation and a considerable inflow of fresh water, e.g., the Baltic where the salinity may be as low as 5 to 8‰. Conversely, with little inflow and high evaporation as in the Red Sea and Persian Gulf, salinities reach as high as 40‰. Nevertheless the assumption of a mean salinity of 35‰ for sea water (90 per cent of the ocean differs by less than 3 per cent from this value) is adequate for many engineering applications where high precision is not required and to define many of its physical properties.

Table 2-1. The Principal Constituents of Sea Water of 35‰ (3.5%) Salinity

Cations	g/Kg	Anions	g/Kg
Sodium	10.65	Chloride	19.29
Potassium	0.39	Bromide	0.066
Magnesium	1.30	Sulfate	2.70
Calcium	0.42	Bicarbonate	0.14
Strontium	0.013	Borate	0.027

Trace quantities of at least fifty of the elements have been identified in sea water. Surprisingly, the sea is very poor in iron and, significantly, copper replaces it in the blood pigment of many marine animals. Considerable attention has been given in recent years to the radioactivity of sea water, the radium content averaging about 0.07×10^{-12}‰, which is about 10^{-4} of that in the rocks of the earth's crust.

Gases may enter solution either by purely physical absorption or the gas may combine chemically with the liquid. For the ocean-atmosphere system, physical absorption occurs in the case of oxygen, nitrogen, and the rare gases. Carbon dioxide, on the other hand, reacts with the water and the dissolved salts. In comparison with the oxygen-nitrogen ratio in the atmosphere of 1:4 that of air dissolved in water is more nearly 1:2. The saturation quantities are a function of water temperature and decrease by 50 per cent in going from the polar to equatorial regions. There being ample air available at the sea surface, gas contents in the upper layers are near saturation; and indeed as nitrogen is relatively inert, only minor variations with depth are observed, since all water has been in contact with the atmosphere at some time. The oxygen content is markedly affected by biological factors such as generation by photosynthesis in the upper layers, which occasionally results in supersaturation. Oxygen consumption by animal respiration and bacteriological oxidation reduces the oxygen content and results in a pronounced minimum at depths of 100 to 1,500 m in the middle and lower latitudes. The deeper waters of polar origin are relatively rich in oxygen. In stagnant water, where there is no photosynthesis and oxygen depletion occurs, as at the bottom of the Black Sea, hydrogen sulfide may be formed.

Carbon dioxide is present in sea water as free carbon dioxide (CO_2) and as carbonic acid (H_2CO_3), but a much greater quantity is present as carbonates and bicarbonates. The partial pressure of CO_2 in the sea in contact with the atmosphere tends toward equilibrium with that in the air. However, areas exist where carbon dioxide–rich water is brought to the surface by rising currents, while in other regions the CO_2 partial pressure is frequently low in the spring and summer when plant plankton is assimilating.

Sea water is normally alkaline, as the sum of the cations is greater than that of the anions of strong acids, the alkaline reaction resulting from the carbonates, bicarbonates, and boric acid compounds. The pH values encountered in the sea range between 7.5 and 8.4, generally being high near the surface. Exceptionally, where hydrogen sulfide is formed, pH values may fall below 7 (in the acid range).

Equation of State. The equation of state represents the density of sea water ρ as a function of salinity, temperature, and pressure. This equation is of particular importance in oceanography in that minor changes in density affect the stability, water mass transfer, and turbulent mixing, current flow normally being computed from a knowledge of the density gradients. For these purposes, the oceanographer must know the density to 1 part in 10^5 and must resort to indirect measurements of chlorinity, electrical conductivity, refractive index, or the like for its determination with corrections for depth and temperature. In practice it is customary to use the notation $\sigma = (\rho - 1) \times 10^3$ where ρ is in g/cm^3 and to tabulate σ as a function of the salinity, temperature, and pressure.

The density of sea water at the surface of the open sea ranges from about 1.021 to 1.027 g/cu cm. Generally about 1.025 g/cu cm, it increases with increasing salinity and continually decreases with increasing temperature provided the salinity is greater than 24.7‰. For lower salinities there is a temperature of maximum density as for pure water. There is a small increase of density with pressure, which becomes important because of the high pressures in the deep ocean.

Thermal Properties. Sea water exhibits the unique properties of solutions, namely, osmotic pressure, depression of the freezing point, elevation of the boiling point, and reduction of the vapor pressure. The observed freezing point is $-1.910°C$ for a salinity of 35‰, the freezing-point depression being less than that computed on a basis of complete ionic dissociation by a factor of 0.89. For all practical purposes, the latent heat of evaporation may be taken as that of pure water.

The specific heat at constant pressure c_p is close to 1 cal/g/°C and increases with increasing pressure. The ratio of the specific heat at constant pressure to that at constant volume increases with temperature and pressure. Sea water is compressible, and the effects of adiabatic processes, although small, must be taken into account when studying the vertical distribution of temperature in the great depths of the oceans. Adiabatic processes are those which take place without the flow of heat into or out of the volume of the medium under consideration. When a sample of water is brought in an insulated sampling bottle from an initial depth (pressure) to the surface, a certain amount of thermodynamic cooling occurs because of the change in pressure, irrespective of the temperature of the local environment. The temperature which the sample would have in being brought from the initial depth to the surface without heat flow is called the "potential temperature."

Acoustic and Mechanical Properties. Of the many acoustic parameters associated with the ocean and its boundaries the most important are the speed of sound and the coefficient of absorption. The speed of sound c is given by $c^2 = 1/(\rho \kappa_a)$ where κ_a is the adiabatic compressibility and ρ the density. The speed of sound is 1,471.46 m/sec for sea water of salinity 35‰ at atmospheric pressure at 5°C. It increases by about 1.0 m/sec per 1‰ increase in salinity, by 3.5 m/sec per °C increase in temperature, and by 0.16 m/sec per atmosphere increase in pressure. These variations result in marked refraction in the ocean. The absorption of sound increases rapidly with the acoustic frequency and is dependent on salinity, temperature, and pressure. It may be taken as approximately 0.01 db/Km at 10^3 cps and 1 db/Km at 10^4 cps.

In the ocean the important mechanical mixing property is the eddy viscosity, which is the effective viscosity occurring on a macroscopic scale in large bodies of fluid. This concept is to be differentiated from the molecular viscosity, which is measured on a small sample of fluid in the laboratory. While the dynamic viscosity is defined as the ratio of the shearing stress to the resulting velocity gradient and expressed in grams per centimeter per second, it is common in fluid mechanics to use the kinematic viscosity, which is the dynamic viscosity divided by the density and expressed in square centimeters per second. The molecular viscosity for water is about 0.019 sq cm/sec at 0°C and approximately half this at 20°C; the eddy viscosity is substantially larger and may range from 2 to 7,500 sq cm/sec depending upon the conditions. Surface tension is also of interest in that it controls the capillary waves from which the larger surface waves receive energy from the wind. Its value is about 76 dynes/cm at 0°C and 73 dynes/cm at 20°C.

Optical and Radio Properties. The reduction of the intensity of optical radiation in the sea is of major biological interest. In general, light is confined to the upper few hundred feet of the ocean. Its extinction is due to both absorption and scattering, the former being greater at the red end of the spectrum and the latter at the blue. The extinction is largely a function of the amount of suspended material and is considerably greater in coastal waters. The refractive index of sea water depends strongly on salinity and is used as its measure (with appropriate temperature corrections). The electrical conductivity rises rapidly with increase of salinity, and as a result electrical-conductivity measurements are used also for the determination of salinity (again there is a strong temperature dependence and a correction must be made). The conductivity, which is approximately 4 mhos/m, limits the transmission of radio waves. As sea water has a relative permittivity of 80 for radio waves, it is a conductor below 10 Mc and a dielectric above 10^5 Mc. The skin depth δ at which radio waves decay to $1/e$ of their surface field intensity value is given by $\delta = 250/\sqrt{f}$ meters where f, the frequency, is in cps. At 1 Mc, $\delta = 0.25$ m. Ten kilowatts of power in a plane wave at 16 Kc is attenuated to 1 microwatt at a depth of 23 m. The speed of radio waves is reduced considerably by the conductivity, being equal to $2\pi f \delta$. Thus while at 1 Mc the speed equals 1.5×10^6 m/sec, it is 1.5×10^3 m/sec at 1 cps, or about the same as the speed of sound.

2-3. Thermodynamics

Heat Transport. Since the temperature of the earth as a whole is not changing appreciably, the total amount of heat received from the sun must equal the amount lost by radiation back into space. However, at any one time and place, the earth is either warming or cooling. The intensity of solar radiation is largely a function of latitude and season. Much greater differences in temperature between the tropics and the poles would be experienced if it were not for the relatively free circulation of the atmosphere and the oceans, which compose the working substance of a gigantic thermodynamic engine continually smoothing out temperature differences between various parts of the earth (see Sec. 5-2). The most important factors in the heat budget of the ocean are the incoming and outgoing radiation, heat exchange with the atmosphere, and the evaporation or condensation of water vapor. In addition, heat is conducted through the sea bottom from the earth's interior and is also produced by conversion from kinetic energy, by radioactive disintegration, and by chemical and biological processes.

The large-scale heat transport is due to circulation. In the tropics, the water is heated. As it decreases in density, it expands and spreads northward and southward along the surface, gradually cooling as it approaches high latitudes. As the warm surface water is drained off from the tropics it is replaced by cold bottom water that flows in from the polar regions. Thus the deep-water layer of all the oceans is derived from high latitudes and has about the same characteristics of salinity and temperature as the surface waters of those regions.

Convection is also important in the heat exchange. The sea surface may be cooled by evaporation, precipitation, radiation, or conduction, when the cool surface water will sink. On the other hand, it may be warmed by solar radiation. These processes, with the wind, result in vertical mixing and the widespread formation of an isothermal surface layer.

Temperature in the Ocean. Temperatures in the ocean range from -1.5 to $30°C$. The basic thermal structure is essentially a three-layered system: (1) a relatively warm thin surface layer mixed by the wind and convection processes, (2) a very deep mass of much colder water in which the temperature decreases very slightly and uniformly with depth, and (3) a transition layer, known as the main thermocline. This structure is an inherently stable one since the density, which is largely controlled by the temperature, increases with depth. The warm surface layer is thickest in mid-latitudes, as is the thermocline. In latitudes higher than about $50°N$ and S there is no surface layer and the entire water column is the same as the cold deep ocean water. Superimposed upon this stable situation are marked diurnal and seasonal temperature variations which affect the upper layers.

Temperature-Salinity Relationship. In the surface waters the relationship between temperature and salinity is highly variable because of surface heating and cooling, evaporation, and precipitation. In the main thermocline and the deep-water layer, however, there is a direct correlation between temperature and salinity, and knowing the temperature one can predict the salinity with almost as much accuracy as it can be measured by a routine process ($\pm 0.02\‰$). When water masses from widely different areas are compared, slight differences in their temperature-salinity correlations can be distinguished and can be used to identify them.

In general, the temperature at the surface decreases with increasing latitude, whereas surface salinities reach their maximum in belts centering at about latitude $23°N$ and S. These occur along the outer edges of the trade-wind belt where the air is descending and relatively dry and where evaporation is greater than either nearer the equator or in higher latitudes. Near the surface, salinity gradients arise from (1) the dilution of the surface by rainfall, melting ice, and water runoff from the land, (2) evaporation of water from the sea's surface, and (3) the flow of waters of different salinity over one another as the result of ocean currents. In temperate regions above latitudes $40°N$ and S there is usually an excess of rainfall over evaporation, and consequently salinity tends to increase with depth. Along the coasts of such regions the outflow from rivers greatly augments this effect, and substantial density gradients result from the dilution of the upper layers.

Water Budget. The prime source of atmospheric water vapor is evaporation from the ocean surface. Water vapor formed over the ocean either remains there and is returned directly to the ocean, or is blown over the land. Oceanic water vapor and clouds carried inland are precipitated and, if not evaporated again and returned directly to the atmosphere, return to the sea via the streams, rivers, and other forms of runoff, closing the major water cycle. The total water budget of the oceans is estimated to be a loss by evaporation of about 360×10^3 cu Km/year, some 90 per cent of which is returned directly as precipitation, and the remainder by inflow. Since the volume of the oceans is $1,370 \times 10^6$ cu Km, about $1/4,000$ of the water participates in the cycle per year.

A secondary cycle, ice and water, involves both the water and heat budgets. About 7 per cent of the ocean is ice-covered. The ocean coverage includes sea ice formed from sea water and fresh-water ice from rivers and glaciers. The ice entry into the North Atlantic alone from the Arctic Basin has been estimated at 20×10^3 cu Km/year.

2-4. Dynamics of the Ocean

Ocean Currents. Water movements are caused by wind, tidal forces, and changes in barometric pressure and in density. Other forces, such as bottom friction, internal viscosity, and Coriolis and centrifugal forces, modify the water movement. Surprisingly, weak density gradients extending over great horizontal distances are important in transporting warm tropical and subtropical water to the polar regions.

In the sea, pressure increases downward because of the weight of the overlying layers, and at a particular depth varies with the density of the overlying medium.

From a chart of the pressure distribution it is possible to develop the water movement. Clearly the pressure gradient is greatest in the direction perpendicular to the isobaric contours, and the water tends to flow in this direction toward lower pressure. As soon as the water is set in motion, however, it is deflected from its course by the effect of the earth's rotation (Coriolis force) until it is running more or less parallel to the isobaric contours. This technique of determining water flow from the vertical density structure is based on the so-called Bjerknes circulation theorem.

Near the surface the effects of pressure gradients are modified by the wind systems of the earth. Around each of the central basins of the ocean the winds form a great anticyclonic eddy, consisting primarily of the trades and the prevailing westerlies. The force of these winds produces similar eddies in the ocean, and some of the most powerful ocean currents arise when the prevailing winds and density distributions work together.

Sea Waves and Swell. The outstanding feature of the sea's appearance is the presence of periodic or near-periodic surface waves. The shape of the waves resembles a trochoid and presents a peaked appearance, and under the influence of the wind they become still sharper until they break, forming white caps. The waves caused by the action of the wind may outlast the wind system responsible for their growth. Waves still driven by their generating winds are commonly referred to as "sea," and waves that are no longer driven by the wind as "swell." These wave systems are not characterized by a particular wavelength or amplitude but by a whole range of values, and one can speak only in terms of spectra or probability distributions. As an amplitude measure, wave height is taken as the average of the highest third of the amplitudes in the spectrum. Surface waves move at a speed in miles per hour roughly equal to 3.5 times their period in seconds, their wavelength in feet being 5.12 times the square of the period.

The buildup of surface waves depends upon the fetch, i.e., the length of the ocean path over which the wind blows. At any given fetch, the wave amplitude increases with the length of time the wind blows until a steady state is reached at which the wave height remains constant. Swell waves arise from the high-amplitude low-frequency sea waves. Besides having less dissipation than sea waves, the longer-period swell waves travel faster because wave speed increases with wave period for gravity waves. Swell waves have a narrow spectrum. For practical purposes a number of scales of sea state based on visual estimates have been described and are widely used. One such scale has been adopted by the U.S. Navy Hydrographic Office (now Oceanographic Office) and is shown in Table 2-2.

Table 2-2. Sea States and Wave Heights

Sea state	0	1	2	3	4	5	6	7	8	9
Wave height, ft	0	<1	1–3	3–5	5–8	8–12	12–20	20–40	≥40	Confused

However, experienced seamen frequently differ in their estimates by one or more numbers on these scales. Unfortunately, no convenient objective method of specifying sea state is available. Ideally, one should know the distributions of the various wave characteristics, such as wave height and wavelength.

In stratified water or in water in which the density varies with depth, waves may occur inside the water mass. These are called internal waves. Their amplitude is usually considerably larger than that of ordinary waves at the free surface. The largest vertical displacements of the water particles are found at the boundary surfaces between different strata.

Tides. Tides are driven by the gravitational influence of the sun and the moon. As these bodies change their relative positions, the ocean waters are attracted into a bulge that tends to remain facing the moon as the earth turns under it. A similar bulge travels around the earth on the opposite side, resulting in a basic tidal period of 12 hr and 25 min. When the sun and the moon are aligned with the earth, the tides are large (spring tides). When the two bodies are at right angles with respect to the earth, the tides are small (neap tides). As a given configuration does not recur for 18 years, it is necessary to have a series of observations extending over a 19-year period to compute the tides exactly for a given locality. However, a 1-year period of observation is sufficient to determine the principal components. Harmonic analysis is used to describe the tides at any one locality from a series of observations. By using astronomical data it is possible to predict the periods of tides with considerable accuracy. However, the height and time (phase) of the tide at any place are a function of the shape and size of the basin involved, which may result in more or fewer than two tides per day; for example, there are four tides per day at Southampton. A free oscillation of water in a basin is called a seiche. Amplitudes of tides range from a few inches off the coast of Italy to about 50 ft in the Bay of Fundy.

2-5. Geology of the Ocean Basins

Classification of the Marine Environment. The true oceanic basins are situated beyond the continental shelves, starting from about the 200-m line. These are usually taken to be the Atlantic, Pacific, and Indian. Other large areas are sometimes considered as separate units, such as the North Polar Basin, the Antarctic Ocean, and the Norwegian Sea. The shallow continental shelves vary greatly in width, from a few miles off the California coast to some 800 miles in the Siberian Arctic. Below the continental shelf there is a small area of slope from 200 to 3,000 m leading to the oceanic abyssal plain with depths of 3,500 to 6,000 m.

Submarine geology concerns the topography of the sea floor, the composition and physical character of the sedimentary and igneous materials that are found on the ocean bottom, and the processes involved in the development of the topographic relief, which are crustal deformation and erosion. Crustal deformation results in elevations and depressions of considerable dimensions both on land and on the ocean bottom. Large-scale elevations of the ocean bottom are termed ridges, rises, or swells; isolated mountainlike structures, sea mounts; and a submerged elevation separating two basins, a sill. Large-scale depressions are termed troughs, trenches, or basins. The greatest depths are found in the Pacific Ocean and exceed 10,000 m, while the deepest point in the Atlantic, the Puerto Rico Trough, is 8,750 m deep. Erosion, deposition, and biological activity result in shelves, slopes, banks, shoals, reefs, and submarine canyons. The surface waves in the sea produce, by erosion, flat-topped features restricted to shallow depths. Slides and turbidity flows erode the bottom and the slopes. Deposition further modifies the topography of the sea bottom, sedimentary debris accumulating in depressions. Submarine canyons are steep-walled fissures penetrating the slope and cutting across the shelf. The shoreline itself is called "recent" when its configuration has not been appreciably affected by marine agencies, or "mature" when it has.

Sediments. Ocean sediments are broadly classified as terrigenous and pelagic. The former are coarse-grained materials, washed down from the continents, lying on the continental shelves and slopes and at the margins of the deep-sea basins. They are extremely variable, ranging from very compact sand to soft gray clay, and are of wet densities, as measured at the surface, of about 1.3 g/cu cm on the slopes and 2 to 2.3 g/cu cm on the shelf. The pelagic sediments, fine-grained materials that are usually light reddish-brown, cover the deep-sea floor. They generally contain the skeletons of tiny planktonic plants and animals. If such organic remains constitute more than 30 per cent of a sediment it is called an ooze and is named according to the type of plankton remains that predominates. Globigerina ooze, a type of sediment consisting mainly of skeletons of protozoa, covers about 50 per cent of the ocean floor. It contains about 95 per cent calcium carbonate. The particle size ranges from 100 to 200 microns and the surface density from about 1.3 to 1.7 g/cu cm. When the animal and plant materials make up less than 30 per cent, the sediment is called red clay. This covers about 40 per cent of the ocean floor and has a surface density of about 1.3 g/cu cm. Its composition is about 50 per cent silica, 20 per cent aluminum oxide, 13 per cent iron oxide, 7 per cent calcium carbonate, 3 per cent magnesium carbonate, and 6 per cent water, with minor amounts of manganese, nickel, cobalt, copper, and vanadium. Although copper makes up only 0.2 per cent of the clay, this is ten times as much as in igneous rock on land. Like all deep-sea sediments, the clay is fine-grained and in an unconsolidated state. At a conservative estimate, the sediment is 100 m thick. Diatomaceous earth, a variation of red clay, is a relatively pure silica material found in some places on land, and consists primarily of the skeletons of the planktonic plants called diatoms. Wide belts of this material, known as siliceous ooze, stretch across the Pacific. Planktonic animals called radiolaris also give rise to a siliceous ooze. Certain elements such as calcium, carbon, silicon, iron, and manganese, carried to the sea in solution, are precipitated by organic or inorganic processes and are thereby concentrated on the sea floor. For example, manganese nodules, having diameters of 10 cm or more, have been found to cover extensive areas of the Pacific and Indian Ocean floors. They are composed chiefly of MnO_2 (29 per cent), Fe_2O_3 (21½ per cent), and clay. They have a laminated structure with a nucleus of pumice, volcanic glass, rock, or organic material such as a shark's tooth or whale's earbone. They are thought to have originated through electrolytic deposition in the ocean. Phosphate nodules have been found in many coastal regions, and dredging off the southern California coast has shown phosphorite to be the most abundant type of rock collected. The phosphorite nodules range in size from small oolites to masses weighing about 50 Kg. Most of the nodules show a banded cross section indicating that the accumulation has been discontinuous. Calcium phosphate forms about 67 per cent of the nodule.

The Ocean Bottom. A major feature of the earth's surface is the submarine mountain chain that runs for 60,000 Km, mainly across the bottom of the oceans, and covers an area equal to that of all the continents. Down most of its length the ridge is split by a deep canyon, or rift, in which many earthquakes originate, the ridge being the locus of a crack in the earth's crust that runs nearly twice around it.

Early seismograph networks in Europe and America showed that earthquakes occurred quite frequently in the mid-Atlantic and that they were associated with the mid-Atlantic Ridge. In fact, earthquakes occur within a 150-Km-wide belt down the center of the whole 1,500- to 2,000-Km-wide ridge which traverses the Atlantic, Indian, and Pacific Oceans and the Arctic Basin. The earthquakes of the mid-ocean rift occur at the relatively shallow depth of about 30 Km below the ocean bottom. In addition there are deep earthquakes with epicenters as much as 700 Km below the surface which are associated with the submarine trenches and island chains that surround the Pacific Ocean. Earthquakes modify the ocean bottom by the violent shifting of rock and sediments. They may also cause slump slides or set in motion turbidity currents, which are downhill-rushing flows of sediment-laden water that occur when sediment at the top of a steep slope is jarred loose. Such currents are also often triggered by hurricanes and floods. Their very high speed enables them to spread across the abyssal plains. Their destructive effects on ocean cables are well documented.

2-6. Biology in the Ocean

The Sea as a Biological Environment. The sea has several distinct advantages over the land for the development of life. In the first place water, necessary to all life, is immediately available. Beyond this, sea water contains in solution the required gases and other materials for growth, and is very similar in composition and solute concentration to the body fluids of all animals. The resulting near equality of the osmotic pressure of sea water and the body fluids permits the direct intake of nutrients and rejection of waste in some organisms. The buoyancy provided by the high density of sea water reduces the need for skeletal structure, while the high specific heat of sea water reduces the possibility of rapid temperature change. The range of pressures experienced in the ocean as a whole is very great, and many organisms are consequently restricted to specific layers, but all life is not excluded from the deep abyssal regions, recent bathyscaph dives in particular having confirmed the existence of marine life in the deepest ocean.

Overall, a considerable variety of conditions exists in terms of salinity, temperature, illumination, and pressure in the ocean. Variability with time is largely confined to the coastal waters and shallower layers. Here the marine forms must be adaptable to the changes or capable of adjusting themselves to the optimum condition. Light in the surface layer is particularly important for photosynthesis by the floating microscopic plants. Notably there is an adjustment to the lighting conditions by some marine life, resulting in a diurnal migration to and from the sea surface, a phenomenon which has been observed as a scattering layer to underwater sound. As distinct from the regions where marked variability exists, large volumes of the deep water have remarkably constant conditions, and here specific forms appropriate to the particular environment can develop. The circulation processes in the sea are important to life in many ways. They serve to maintain the oxygen content, to replenish the supply of nutrients, and to aid in the dispersal of waste material and the young life in the form of spores, eggs, and young adults. Clearly, however, unexpected and unseasonable disturbances of the circulation pattern can, by moving life into an unacceptable environment, result in considerable destruction.

Organisms and the Composition of Sea Water. Sea water is more than a solution of salts in a certain proportion, for the marine life is an inherent part of the sea and the marine organisms play an important role in determining the chemical and physical properties at any particular location and time. Their effect on the oxygen and carbon dioxide gaseous contents, for example, has already been noted. Beyond this, certain elements occurring in small concentrations in sea water are removed in relatively large amounts by the marine life for skeletal, structural, and other uses. The inorganic compounds of nitrogen, phosphorus, iron, and silicon, the plant nutrients, fall in this class, and their distribution is affected appreciably by biological activity.

The life cycles of the various organisms in the sea are closely interwoven. Plants of all sizes furnish the basic food for all life and grow in the upper illuminated layer. These plants absorb carbon dioxide from solution and set free oxygen. They absorb phosphates and nitrogen salts and convert them into organic matter. However, the inorganic substances which they use to grow must be made available to them near the surface, even though there is a downward movement of particulate matter in the sea, under the influence of gravity. The availability of these materials at the surface results from circulation of the ocean waters, convective overturning, and upwelling from the bottom. The inflow of minerals from the land is a small factor in supplying nutrients for the sea plants. Wherever plant nutrients are abundant, near the surface, large marine populations occur as compared with other parts of the ocean. Such areas can be identified by their color, for water rich in life is greenish whereas "desert" water is blue.

Marine Life. The sea's inhabitants range from a multitude of microscopic one-celled organisms to the 150-ton Antarctic blue whale. The smaller forms, especially the plants, are extremely prolific under proper conditions and furnish the food supply for successively larger forms of animals. In describing the population of the sea, three major groups are distinguished depending upon habitat. These are nekton, benthos, and plankton. Nekton is the name given to swimming animals that can move freely over large distances. The benthos are found at the bottom of the sea and include sessile animals such as sponges, creeping forms such as crabs, and burrowing forms such as worms. Plankton include all floating plants, those animals that live permanently in a floating state, and also the larvae and eggs of the animal benthos and nekton. The plant plankton (diatoms, dinoflagellates, coccolithophores, and flagellates) and animal plankton (globigerina and radiolaria) are so numerous that extensive areas of the ocean floor are covered with oozes (Sec. 2-5) containing their skeletons. There is some question regarding the grouping of dinoflagellates as plants even though they carry on photosynthesis.

In the sea as on land the plants are the organisms capable of developing complex organic substances from the simple compounds dissolved in water. Without them the animal life would be restricted to a small region near the shore where organic material of terrestrial origin exists. Nearly all the marine plants are Thallophyta, primitive plants with little or no true root, stem, or leaf. They include the marine algae and fungi, while the bacteria are frequently included with them. The brown algae vary greatly in size, and plants of this class form the "kelp beds" commonly harvested for commercial purposes. The yellow-green algae are mainly floating forms and include the diatoms, which are unicellular in structure although individuals may form chains, and the dinoflagellata, again typically unicellular but possessing two flagella for locomotion, and second only to the diatoms as producers in the marine plankton. Some of the latter have the capacity to luminesce when disturbed. The mosses (Bryophyta) and ferns (Pteridophyta), two major divisions of the plant kingdom, are not represented in the sea. The highest plants, the Spermatophyta, are represented, however, by some 30 species of flowering plants. They have invaded the sea by way of fresh water.

By contrast to the plant life, all major groups (phyla) of animal life are represented in the sea. Each phylum is subdivided successively into class, order, family, genus, and species. When thinking of the animal life in the sea, one tends to emphasize the higher orders of marine animals such as the whale and the larger fishes, but these are an extremely small part of the total animal life. Of upward of 50 animal classes, only four are exclusively nonmarine.

For the future, in man's exploitation of marine life, the problem is to understand the natural processes sufficiently well so that in utilizing and encouraging the growth of either plant or animal species, he will not destroy nature's balance, which is a delicate one involving all the characteristics of the ocean.

BIBLIOGRAPHY

Conference on Physical and Chemical Properties of Sea Water, Easton, Md., Sept. 4–5, 1958, National Academy of Sciences, National Research Council Publ. 600, 1959.
Darwin, G. H.: *The Tides and Kindred Phenomena in the Solar System*, W. H. Freeman and Company, San Francisco and London, 1962 (original printing 1898).
Defant, Albert: *Physical Oceanography*, vols. I and II, Pergamon Press, New York, 1961.
Harvey, H. W.: *Recent Advances in the Chemistry and Biology of Sea Water*, Cambridge University Press, New York, 1945.
Horton, J. W.: *Fundamentals of Sonar*, 2d ed., U.S. Naval Institute, Annapolis, 1959.
Kuenen, P. H.: *Marine Geology*, John Wiley & Sons, Inc., New York, 1950.
Marine Fouling and Its Prevention, Woods Hole Oceanographic Institution, U.S. Naval Institute, Annapolis, 1952.
Officer, C. F.: *Introduction to the Theory of Sound Transmission*, McGraw-Hill Book Company, New York, 1958.
Perspectives in Marine Biology, University of California, Berkeley, 1958.
Pierson, W. J.: *Practical Methods for Observing and Forecasting Ocean Waves* . . . , U.S. Hydrographic Office Publication 603, 1955.
Proceedings of the Symposium on Aspects of Deep-sea Research, Feb. 29–Mar. 1, 1956, National Academy of Sciences, National Research Council Publ. 473, 1957.
Stratton, J. A.: *Electromagnetic Theory*, McGraw-Hill Book Company, New York, 1941.
Sverdrup, H. W., M. W. Johnson, and R. H. Fleming: *The Oceans*, Prentice-Hall, Inc., Englewood Cliffs, N.J., 1942.
Undersea Weapon Systems, *Space Aeron.*, vol. 33, no. 1, pp. 21–109, July, 1960.
Von Arx, W.: *Introduction to Physical Oceanography*, Addison-Wesley Publishing Company, Inc., Reading, Mass., 1962.

Chapter 3

LAND MASSES

GEORGE K. LEWIS, *Department of Geography, Boston University.*

CONTENTS

3-1. Introduction.. 3-1
3-2. The Earth as a Planet.. 3-2
    Size and Shape of the Earth................................. 3-2
    Land Area... 3-2
    Elevation.. 3-3
    Age and Composition of the Land Masses................. 3-3
3-3. Water on the Land... 3-5
3-4. Geomagnetism... 3-5
3-5. Terrestrial Gravitation..................................... 3-6
3-6. Summary.. 3-6

3-1. Introduction

Man has been accustomed for millennia to live upon that small portion of the earth's surface that lies exposed above sea level. The several major land masses of the world occupy only a scant 29 per cent of the total surface of the geoid. Under unusual conditions, man has been able to survive for relatively short periods of time on or under the surface of the oceans, always returning eventually to a shore base.*
Man has come to know, over the centuries, much about the land that he occupies and about the earth as a whole. A great amount of the more specific and accurate data, however, has been gathered since World War II. The International Geophysical Year of 1957–1958 was an all-out scientific assault upon some of the gaps in knowledge about the earth and its atmospheric circulation. A review of some of the postwar developments, particularly as they apply to problems in system engineering, is set forth below.
Engineers concerned with the design and construction of a wide variety of projects have long been aware of the need for data related to earth activities: accurate streamflow data for bridge or dam design; climatic data for the design and layout of structures, particularly in the tropics and high latitudes; seismographic data for an underground nuclear-test-detection system; atmospheric information for analysis of the

* Recent increased interest in the oceans and the nature of the ocean bottoms is described in Chapter 2.

reentry problem; gravitational data for satellite-orbit projections; earth size and shape measurements for the compilation of ballistic-missile trajectories; and subsurface bedrock characteristics for the design of new communication systems. System engineers are increasingly aware of the effect of certain characteristics of the land masses upon human survival, upon transportation and communication techniques, and upon the spread of man's activities in general into new and often inhospitable environments.

3-2. The Earth as a Planet

Size and Shape of the Earth. Measurements obtained during the International Geophysical Year and from the behavior of earth satellites have yielded the most precise data on the difference between the geoid and a true sphere. The distance from the epicenter of the geoid to either pole is about 13.5 miles shorter than the distance from epicenter to equator. The radius in this latter plane is 6,378,099 meters, or 3,963.0 statute miles, and the radius from epicenter to pole is 6,356,631 meters, or 3,949.5 statute miles. The equatorial circumference of the earth is 24,899 miles, or 40,075 kilometers.

Land Area. The major land masses of the world occupy a total area, including the permanent ice fields of Greenland and Antarctica, of over 57,000,000 sq miles (see Table 3-1).

Table 3-1. Size of Land Masses, Square Miles

Eurasia	21,123,885
Europe	1,902,894
U.S.S.R.	8,602,700
Asia	10,618,291
Africa	11,732,616
North America	9,354,825
South America	6,892,575
Oceania	3,304,854
Subtotal	52,408,455
Antarctica (estimated)	5,000,000– 6,000,000
Total	57,000,000–58,000,000

Human activities can be found over an ever-increasingly large proportion of this area, and there is almost none of it that has not been explored, at least by reconnaissance. Man occupies permanently, however, a relatively small part of the total land-mass surface, restricting his activity to those areas in which a favorable combination of landforms, climate, and soil occur or in which other special inducements to settlement prevail. Here again, we find that man's increasing technological control of his environment is enabling him to occupy or use effectively areas which only a few decades ago were considered totally inhospitable. There is, however, a great imbalance in the distribution of land and ocean over the earth's surface. A high proportion of land lies in the Northern Hemisphere.* Table 3-2 shows the relative proportions of land and ocean at selected parallels of latitude. A maximum extent of land is actually reached near the Arctic Circle, where Alaska stretches westward, and land occupies 80 per cent of the parallel at that point. A minimum of land appears between latitudes 56 and 63°S, ocean completely covering this latitudinal strip.

If one considers the earth devoid of its grid system for a moment, one can observe a "land" hemisphere, i.e., that half of the earth's surface containing the greatest proportion of land. Such a hemisphere would center on a point near Nantes in western France and would contain about 80 per cent of all the world land surface. A similar

---

* This proportion would be even greater if all ice were to melt from Antarctica, where recent measurements have shown even less of Antarctica to be dry land than was heretofore thought.

LAND MASSES

Table 3-2. Proportions of Land and Ocean at Selected Parallels of Latitude*

Latitude	Land, %	Ocean, %
80°N	25	75
70°N	57	43
60°N	62	38
50°N	57	43
40°N	44	56
30°N	43	57
20°N	31	69
10°N	23	77
0°	21	79
10°S	21	79
20°S	23	77
30°S	20	80
40°S	4	96
50°S	3	97
60°S	0	100
70°S	20	80

* After B. F. Howell, Jr., *Introduction to Geophysics*, McGraw-Hill Book Company, New York, 1959.

"ocean" hemisphere exists (the other half of the world) centering on a point in the Pacific Ocean some 1,000 miles southeast of Wellington, New Zealand.

Elevation. Differences in elevation from place to place are obvious. If these are viewed on an areal basis, we find that, for the entire earth surface, including oceanic topographic differences, the mean elevation, or more accurately, depth, of the earth is about 2,440 m below sea level.* If only land areas are considered, the mean elevation is 840 m. If only suboceanic surfaces are considered, the mean depth of the oceans approximates 3,800 m.

Table 3-3 shows the percentage of earth surface at intervals of 1,000 m both above and below sea level. The areal extent of land surface occurring at different elevations decreases with regularity from sea level to 29,028 ft (Mt. Everest). Areas beneath sea level, however, show no such regular behavior from sea level to the great oceanic trenches. The maximum extent of ocean bottom actually occurs between 4,000 and 5,000 m.

Age and Composition of the Land Masses. There is no substantial contrast in the age or type of geologic structure from one land mass to another. All the major land masses display at the surface sizable exposures of Archean crystalline rocks as well as representative exposures from nearly every period from early Paleozoic to the present. Similarly, each of the land masses exhibits large areas of relatively undisturbed sedimentary strata, of igneous rocks (both extrusive and intrusive), and of highly disturbed and altered metamorphic rocks. When individual land masses are studied, wide variations in the geologic pattern are noted, but no land mass is strikingly different from another in gross aspect.

Certain gross structural similarities have been observed from one continent to another. All the major masses have in their interior either broad lowlands underlain with sedimentary rocks or uplifted plateaus of metamorphic rocks. These basins are frequently cut off from the sea by mountain systems of varying width, height, and geological structure. The Andes and the Sierra Nevada–Rocky Mountain systems (matched by the Brazilian upland and the Appalachians, respectively, on the east) are perhaps the best example of this phenomenon, although similar relationships can be found in Australia, Africa, and eastern Asia.

* B. F. Howell, Jr., *Introduction to Geophysics*, McGraw-Hill Book Company, New York, 1959.

Certain basic contrasts do exist, however, between the underlying structure of the land masses and the oceans. The land masses appear to be composed at the surface of rocks rich in silica and in sodium and potassium aluminum silicates and have therefore been termed *sialic*. The sialic crust is quite shallow and "floats" on a base of denser rocks. This denser layer is composed of highly basic rocks, rich in calcium and magnesium. The term *sima* has been applied to these rocks. The principal difference involved is one of density. The granites, for example, have a density of 2.64, and similarly low densities occur in the other sialic rocks—granodiorites, diorites, and syenites. When the thin veneer of sedimentary rocks is also considered, the average density becomes even lower. Sandstone has a density of 2.2, limestone 2.5, and slate 2.7. The sima layer is composed of such rock types as dunite (one of the peridotite minerals) with a density of 3.3. Basalt and andesite are two of the

Table 3-3. Proportions of Total Earth Area at Different Elevations or Depths

Elevation or depth, m	% total earth area	% land-mass area
3,000–4,000	1	4
2,000–3,000	2	7
1,000–2,000	5	17
0–1,000	21	72
Sea level		
0–1,000	9	
1,000–2,000	3	
2,000–3,000	5	
3,000–4,000	13	
4,000–5,000	23	
5,000–6,000	16	
6,000–7,000	1	
Over 4,000 or under 7,000 m	1	

more common rock types found in the sima layer, which is sometimes referred to as the *peridotite mantle*. The zone of contact between the continental crust and the peridotite mantle is called the *Mohorovicic discontinuity* (or, in the recent vernacular, "Moho").

The peridotite mantle not only underlies the less dense continents but also stretches under the oceans as well. Here, however, there is only a thin covering of marine deposits on top of the sima. It has been estimated that the "Moho" lies on the average only about 3 miles beneath the sea floor, whereas it may lie from 15 to 20 miles beneath the surface of the land masses.

This thin upper layer of the earth is often referred to as the "crust." Beneath this lies the so-called "mantle," about 1,800 miles in thickness, separated from the innermost "core" of the earth by the *Gutenberg-Wiechert discontinuity*. The core is approximately 2,170 miles in radius. Little is known of the composition of these two deeper layers, the core being presumably composed of nickel and iron and the mantle of materials transitional between these and the basic silicate rocks of the surface crust.

Many characteristics of the earth change with increased depth in addition to the nature of the rocks. One is the speed at which seismic waves pass through the rocks. Much of our knowledge of the earth comes from the interpretation of marked differences in such travel times from place to place and at different depths. A second factor is temperature, which is known to increase with depth. In general, the temperature of the earth increases 1 degree centigrade with each 90 meters of depth, at least for the first 100 kilometers toward the center of the earth. Beyond this point the rate of temperature change decreases. Deep-shaft gold mines operate with great difficulty in South Africa because of the extreme temperatures encountered at relatively shallow depths.

3-3. Water on the Land

The land masses receive varying amounts of precipitation in the form of rain or snow (or even, in some cases, dew). This precipitation eventually finds its way back to the atmosphere through evaporation from the soil or from water bodies or through transpiration from plants. A large proportion of the precipitation runoff makes its way to the ocean via streams and rivers or sinks into the ground (to become ground water) and thence eventually to make its way to the oceans. The amount of evaporation from the earth or surface water bodies, the amount of runoff, and the amount entering the ground water supply all vary widely from place to place. About one-third of the precipitation on the land masses eventually enters surface water bodies.

Precipitation itself varies widely from region to region in average annual amount. The actual annual amount received also varies from year to year, particularly in regions having less than 20 in. annually. The average annual precipitation for the earth is about 33 in. and ranges from points averaging over 400 in. of rainfall annually to those in which measurable amounts of moisture will not fall for several years (see Sec. 5-4 and Fig. 5-2, p. 5-16).

Surface water collects in streams which move toward the sea, collecting tributaries to form main rivers. A relatively large proportion of surface runoff occurs in a few of the major drainage systems—over 15 per cent in the combined basins of the Amazon, the Congo, the Missouri-Mississippi, the Nile, and the Parana-Paraguay.

Approximately 10 per cent of the land-mass surface (exclusive of Antarctica) has an internal drainage system with no surface streams reaching the sea. These great depressions, which are particularly characteristic of the Sahara (for example, the Lake Chad drainage area), of central Asia (Lop Nor), and of western North America (Great Salt Lake), contain intermittent streams and temporary lakes, often of very high salinity.

Ground water permeates the unconsolidated surface material of humid areas and a deep zone of the bedrock mantle. Ground water may seep to the surface in springs or, rarely, artesian wells. Many semiarid areas contain abundant subsurface water which has percolated many miles through underground channels from rainier areas or distant snow fields.

3-4. Geomagnetism

The earth is situated within a vast magnetic field which operates not only in the earth's atmosphere (about which the space sciences are primarily concerned, see Chapter 7) but also on the surface of the earth and within the earth's crust. In simplest terms, the earth acts as an enormous dipole magnet, with one pole located about 1,190 miles from the actual geodetic North Pole, and the other about 1,540 miles from the geodetic South Pole. These two poles are not fixed, and their location on the earth's surface migrates from year to year. From 1922 to 1945, the North Magnetic Pole migrated several hundred miles toward the northwest.

Man became involved with this magnetic phenomenon initially in his concern for accurate navigation from place to place, and through this interest, he gained considerable knowledge of magnetic phenomena. There are three principal ways in which geomagnetic properties are measured and mapped. The first of these is the direction of the field, usually referred to as magnetic *declination*. In only a few places on the earth is the magnetic field truly parallel to the longitudinal meridians. Accurate maps exist of this pattern of declination around the world. Lines connecting points of equal declination are called *isogonic lines*. In New York City, declination is about 15°W; in San Francisco, it is about 20°E.

The second characteristic of the field is its deflection from a horizontal surface. This is termed *dip* (sometimes confusingly called *inclination*). The line connecting points with 0° dip (on the 0° *isocline*) roughly parallels the equator, and the dip approaches 90° in the vicinity of the two magnetic poles. The third characteristic is magnetic intensity, which is measured in *oersteds*, an oersted being the strength of a field that exerts a force of 1 dyne on a unit magnetic pole. Lines of equal magnetic intensity are called *isodynamic lines*. The area of least intensity lies off eastern Brazil (less than 0.25 oersted), while areas of strong intensity are found in the general area of the two magnetic poles. This is particularly true of the South Magnetic Pole with a force over 0.70 oersted.

Over the surface of the land masses themselves, the intensity of the magnetic field varies from place to place because of differences in the underlying bedrock. The strong magnetic force of the "lodestone" has long been recognized. Many such "lodestones" were undoubtedly the mineral magnetite which, today, is the most vigorous mineral in terms of local magnetization. At the other end of the spectrum, such common rocks as sandstone and limestone exhibit almost no magnetization. These differences in local magnetization are studied over large areas by the use of an airborne magnetometer. This is particularly useful in reconnaissance explorations for potential deposits of iron ore.

The earth's magnetic field is not a purely constant phenomenon. There are evidences of a long-range shift in the position of the field with respect to the earth's surface. The two poles as we now know them shift in location and the magnetic field with them. Interesting theories have arisen as to the position of the field in earlier geological eras, and the possible influence on such events as Pleistocene glaciation. Such major shifts in the earth's field are carefully measured and analyzed for publication for navigators. There are also daily fluctuations that appear to be related to daily (solar radiation), monthly (lunar changes), and annual factors. Magnetic storms are particularly drastic, noncyclic changes and often accompany periods of intense sunspot activity. Such storms wreak havoc with radio transmission, and their effect is usually felt over large areas of the earth's surface.

3-5. Terrestrial Gravitation

The strength of the earth's gravitational force is measured in *gals* (1 gal or unit of acceleration being equal to 1 centimeter per second squared). At sea level on the equator, this measures 978.0373 gals. Variations from this amount occur over the rest of the earth, depending upon the latitude, elevation, time of the day and month, etc. Gravity decreases toward the poles as it also does, in general, with elevation. The force of gravity also changes over time, with a periodic rise and fall analogous to the tides of the ocean. For many years the pendulum was used to measure the force of gravity from place to place, but recently a variety of *gravimeters* have been developed to measure the force more rapidly and equally accurately. Only in the last few decades have sufficient data been available from different points on the earth's surface to enable geophysicists to plot gravimetric anomalies on an areal basis.

3-6. Summary

The several sciences which have come to focus upon the land masses of the earth each have their rich and growing literature. Even the relatively new discipline of geophysics now finds itself subdivided into more specialized internal fields. The system engineer will only rarely find himself concerned with more than one or two of these fields on a single project. A problem in underground-shelter construction may, for example, call for a detailed knowledge of surface materials and subsurface bedrock conditions as well as groundwater occurrence and behavior. In all such cases, it will be necessary to penetrate the methodology and content of the fields involved to a greater depth than present herein. Systems involving transportation elements are particularly restricted by land conditions. Systems involving communications must consider not only the size and shape of the earth but also the effect of the earth's magnetic field. The erection of all types of structures involves obvious land relationships. The survival of persons, equipment, and supplies at any point on the earth's surface is, in part, dependent upon environmental conditions at that point.

BIBLIOGRAPHY

A general reference that can be used to enter specific fields in greater depth:

Howell, B. F., Jr.: *Introduction to Geophysics*, McGraw-Hill Book Company, New York, 1959.

References related to size, shape, and composition of the earth's land masses:

Jeffreys, Sir Harold: *The Earth: Its Origin, History and Physical Constitution*, 4th ed., Cambridge University Press, New York, 1959.
King, Lester C.: *The Morphology of the Earth*, Hafner Publishing Company, Inc., New York, 1962.

Chapter 4

URBAN AREAS

LESLIE C. EDIE, *Port of New York Authority.*

WALTER HELLY, *Port of New York Authority.*

CONTENTS

4-1. Introduction	4-1
4-2. Description—Facts and Methodology	4-2
Definition, Exhibits of Characteristic Properties	4-2
Techniques for Collecting Data	4-4
Descriptive Methodology	4-5
4-3. Operational Optimization of Urban Subsystems	4-7
Design for Optimum Equilibrium	4-7
Subsystem Optimization under Fluctuations	4-10
4-4. The Urban Complex as a Single Unified System	4-12
R. J. Smeed's Land-use Model	4-12
Area-economy Model—The Input-Output Concept	4-14
Area-transportation Model	4-16
4-5. Conclusion	4-18

4-1. Introduction

The large metropolitan area is a most complex accomplishment of human industry and organization. It is therefore not surprising that systematic study of its operational dynamics is a recent endeavor, still tentative and somewhat disorganized. Often the information required for control or planning is not available. At other times, there is so much of it that great ingenuity is required in the design of an economic data-collection and -classification program. The elements of the urban organism interact strongly, and it is usually impossible to optimize performance of any one subsystem without serious penalty to others. Yet quantitative measures of such penalties are often impossible because a well-developed value system does not exist regarding the objectives for an ideal human environment. Even when something is understood and a coherent control plan is formulated, political obstacles often frustrate action. Despite all these difficulties, perhaps even because of them, the urban mechanism is one to which the system engineer can make exceptionally useful contributions.

The first modern application of system engineering to urban problems was the design and administration of large telephone systems, beginning early in the twentieth century. The equipment required to connect caller and called party was too expensive for only occasional use; it had to be shared by many telephone subscribers. Once such sharing was proposed, it became necessary to predict how much shared, or "common," equipment was required to ensure an adequate standard of service. Then, as now, the engineer had to make two basic assumptions: first, that a large group of persons, individually variable, behave in a statistically predictable manner for fairly long periods; second, that the abstraction of "an adequate standard of service" would be translated into explicit, mathematically precise criteria. Because both assumptions proved to be true, telephone system engineering has grown into a large, fruitful applied science. Its present state is reviewed thoroughly by Syski.[34]

Despite 50 successful years in telephony, today the engineer still must overcome excessive skepticism on the validity of these assumptions every time he seeks to contribute to a new area in which human activity is controlled or catered to. It may be that the education of the layman is more difficult than the solution of his ostensible problems.

4-2. Description—Facts and Methodology

Definition, Exhibits of Characteristic Properties. It is difficult to frame a definition for urban areas which differentiates them from the remainder of the inhabited world without any ambiguity. Until quite recently, it was sufficient to characterize the urban area as one with a high population density. Specifically, it was argued that a population density greater than about 2,500 persons per square mile is urban because it usually requires regular streets and fairly closely spaced residential development. Since such population densities do occur occasionally in rural subsistence economies, it was also required that in the urban area the land be used mostly for purposes other than farms, pasture, or woodland.

Modern transportation has made the above definition somewhat obsolete. Urban centers sprawl over increasingly large areas while in many "rural" regions the mass of the population is engaged in other than agricultural pursuits. In the future it will become more and more necessary to view the urban society as a social rather than a physical entity. Before long we shall view a region as urban whenever the employed population is characterized by extreme specialization and whenever the society requires extensive organization to provide services it considers essential.

Despite these considerations, the most pressing urban problems still are those engendered by congestion. While we have given the classical lower bound to urban population density, we are unable to state the upper limit beyond which existence becomes impossible. Ever since the time of imperial Rome, localized population densities in excess of 250,000 per square mile have been found occasionally, while residential densities of 100,000 per square mile are quite common. The intensities of these concentrations can be illustrated by noting that, at 100,000 per square mile, the existing world population of 3 billion would fit into South Carolina or Portugal.

We shall offer some quantitative information to illustrate features unique to urban areas. However, it would be misleading to present averages based on all cities as universally true. The differences in employment, land use, and residence patterns between a Washington and a Detroit, between a Boston and a Los Angeles are much too great for simple generalizations. Consequently we exhibit in Table 4-1 some facts about a few selected areas, noting that similar information is usually available for others.

The selected group consists of a nested set of five regions. The smallest is the Borough of Manhattan in New York City, tremendously congested, with a great residential population and a still greater working population. Next in size is New York City, including Manhattan. This forms one political unit and is one of the world's largest cities. Nevertheless, the city of New York contains only about half the people who inhabit the "New York–Northern New Jersey Standard Consolidated Area" (a Census Bureau term). This larger region, our third example, is a self-

URBAN AREAS

Table 4-1. Urban Areas—Comparative Statistics*

	Year	Manhattan	New York City	New York–North Jersey Standard Consolidated Area	"Megalopolis"	United States
Physical characteristics:						
Area, sq miles...................	1960	22.3	316	3,939	53,575	3,615,210
Proportion used for streets and roads, %.....................	1960	35.5	30.1	......	......	1
Proportion used for farms and pasture, %.....................	1954	0	0.7	10.5	35	61
Population:						
Total resident, thousands.........	1960	1,698	7,782	14,651	37,000	179,323
Change, 1950–1960, %............		−13.9	−1.4	+13.5	+15.3	+18.5
Density, persons/sq mile..........	1960	76,000	24,600	3,700	690	50
Age distribution:						
0–14 years old, %...............	1950	16.7	21.0	24.0	23.7	26.6
15–64 years old, %..............	1950	74.6	71.4	68.2	68.0	65.3
Over 64 years old, %............	1950	8.7	7.6	7.8	8.3	8.1
Education:						
Median number of school years completed by persons aged 25 years or more..................	1950	9.6	9.1	9.5	......	9.3
Employment:						
Total number employed in area, thousands.....................	1956	2,717	4,051	6,150	......	64,980
Proportion who are nonresident, %.	1956	84	25	9	n	n
Proportion in:						
Agriculture and mining, %......	1956	n	n	0.8	......	11.7
Contract construction, %.......	1956	2.2	2.8	3.6	......	4.0
Manufacturing, %..............	1956	19.5	23.5	28.2	......	26.1
Wholesale trade, %.............	1956	9.9	8.5	6.8	......	4.9
Financial community, %........	1956	8.1	6.2	4.8	......	2.8
Consumer trades and services, %.	1956	15.2	18.0	18.6	......	23.0
Others in office buildings, %.....	1956	22.4	16.9	13.4	......	}27.5
Others not in office buildings, %.	1956	22.7	24.1	23.8	......	
Management:						
Number of the 500 largest industrial corporations with headquarters in given area, from *Fortune Magazine* survey...........	1962	133	134	155	208	499‡
Communications and transportation						
Number of telephones per 1,000 residents.....................	1962	1,120	550	460	......	375
Number of automobiles per 1,000 residents.....................	1962	......	210	......	......	410
Median home-to-work commuting time for those who work in area, min...........................	1956	59	......	42		
Proportion of workers in area who commute primarily by:						
Car and taxi, %................	1951–1960	11†	......	41	......	68
Bus, %.......................	1951–1960	5†	......	14	......	}15
Railroad and rapid transit, %...	1951–1960	81†	......	33	......	
Foot and other, %..............	1951–1960	3†	......	12	......	17

n = negligible.
* From Refs. 11, 17, 22, 26, 30, and 31.
† Persons entering central business district on a typical weekday, 7–10 A.M., 1956. Does not include those who live within central business district.
‡ One company is in Puerto Rico.

contained urban area according to the classic viewpoint. However, the present tendency of adjacent cities to interact strongly and grow toward each other is particularly evident for the area of the eastern United States seaboard which extends from Washington to Boston and includes New York. This, our fourth region, was christened "Megalopolis" by Jean Gottmann,[11] whose delineation of its boundaries is used here. Whether "Megalopolis" today is already one urban organism or not is still debatable. It certainly is becoming one rapidly. Finally, we chose the entire United States as our fifth and final region.

A number of the facts shown in Table 4-1 deserve particular comment:

1. A very large proportion of land area is devoted to transportation when the population density is great.

2. The population of "Megalopolis" has not risen so rapidly as that of the country as a whole. This is not true of urban areas in general. For example, the Los Angeles–Long Beach Standard Metropolitan Area increased its population by more than 50 per cent in the decade from 1950 to 1960. However, the uniform progression of change from Manhattan (−13.9 per cent in the decade) to "Megalopolis" (+15.3 per cent) is characteristic of the now almost universal trend of metropolitan regions to spread to more uniform population densities. It is also rather common to find relatively fewer children coupled to a greater proportion of the population in the work force at city centers than elsewhere. Again, the shown progression is significant.

3. It is interesting that the median educational attainment in Manhattan is only slightly above the national average. The figures may be misleading because of heavy recent migration into New York, both from the southeast United States and from Puerto Rico.

4. Many more people are employed in Manhattan than live in it. The daily mass migration for work is characteristic of many, but not all, urban centers. New York is an office and financial center, as shown clearly by the employment figures. In contrast, Detroit would show an entirely different employment pattern.

5. The proportion of corporation headquarters located in Manhattan is unique.

6. A business center requires good communications, and Manhattan clearly offers these, as indicated by both the small area of the region and the extraordinary number of telephones. However, a price is paid for the convenience of congestion. Note the contrast between Manhattan and Greater New York commuting times. Further, the city center is limited in the number of private automobiles it can accommodate. Hence far more people use public transport and fewer own cars than in the country as a whole. This is not the case in all cities. For example, in Los Angeles there are almost 500 automobiles per 1,000 inhabitants. Nevertheless, public transport remains relatively important in all congested urban regions.

Techniques for Collecting Data. The most reliable information is acquired by the direct observation of the activity of interest in all its manifestations. When this is impractical, one studies instead a sampling of size only sufficient to give some acceptable level of statistical significance. In urban studies, it often is necessary to rely on still more degraded information obtained by quizzing people about their activities.

Land usage and vehicular transportation may be taken as representative of activities susceptible to direct observation. A single photograph, made at a height of 10,000 ft, will exhibit a region of 3 sq miles with sufficient detail to identify all land uses and, more remarkable, to identify individual automobiles. Traffic studies have been made at The Port of New York Authority[20] which were based entirely on such photographs taken frequently enough to exhibit the dynamics involved. Similar information can be developed at ground level where greater observational effort may be compensated for by reduced work in the analysis of results. One can detect, count, and measure the speeds of vehicles, pedestrians, or goods with photoelectric cells, with mechanical triggers, or manually. Automobiles also can be detected by induction coils buried in the roadways, by radar, or by ultrasonic devices. Travel patterns may be deduced by photographing all subjects passing a number of check points and then correlating the results. Ticket sales record the elements of commuter traffic on public facilities.

Tracing the production or consumption of physical goods by direct observation is practicable only in certain fields, exemplified by the public utilities, where adequate instrumentation for regional recording is available or already installed.

The national census, together with its supplementary specialized surveys, is the most important source of information about urban areas in the United States. In addition to similar efforts by other countries, the statistical publications of the United Nations[29] are most useful. If an independent survey must be made, the design of questionnaire and sample requires some forethought on the respondents' biases. It is not easy to compensate for their tendencies to underestimate travel times and costs or to overestimate consumption of and time devoted to meritorious activities.

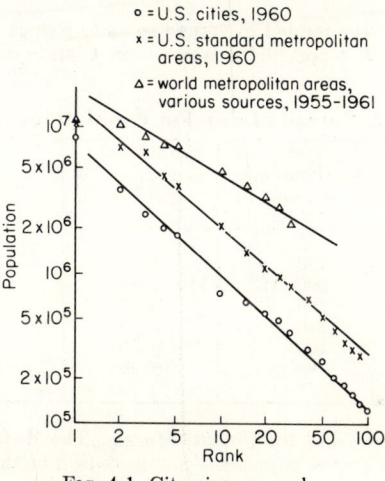

FIG. 4-1. City size vs. rank.

Descriptive Methodology. Systematic study is eased by the codification of raw data. We now discuss some of the techniques for doing this which have had success in the description of urban areas.

The Distribution of City Sizes. Zipf[33] and others have noted that if all cities in a large region are sorted in order of decreasing population, then the population P_j of the jth city may be predicted by

$$P_j = \frac{P_1}{j^a}$$

where the exponent parameter a is dependent on the characteristics of the region selected. Figure 4-1 shows population vs. rank for the largest cities in the United States, for the largest standard metropolitan areas in the United States, and for the largest metropolitan areas in the world. The distribution formula is seen to be moderately successful, and we find that

a = 0.90 for United States cities in 1960
   = 0.83 for United States metropolitan areas in 1960
   = 0.57 for world metropolitan areas in 1955–1960

A relationship such as Zipf's has only limited significance, because the plot of any monotonically decreasing function on logarithmic paper usually will look much like Fig. 4-1.

Population Density as a Function of Distance from the City Center. It is observed that, apart from a relatively unpopulated central business district, the population density of a city tends to decline with distance from the city center. On the basis of a study of 20 cities, ranging from Los Angeles to Budapest, at various stages of growth between 1801 and 1940, Clark[5] expresses the density ρ as

$$\rho = \rho_0 e^{-br}$$

where ρ_0 and b are parameters and r is the distance from the population center. The formula appears to be true except very near the center, where it overestimates the resident population of the central business district, and at the fringes, where it does not account for satellite cities and the rural population.

Weiss[32] suggests that ρ_0 and b are proportional to each other, with time-independent constant of proportionality λ, so that

$$\rho_0(t) = \lambda b(t)$$

With time, most cities have tended to spread so as to reduce population densities at their centers. Table 4-2, reported by Weiss from Clark's data, shows this process for London.

Table 4-2. Spread of London Population, 1801–1939

Year	ρ_0, thousands/ sq mile	b	$\rho_0/b = \lambda$
1801	290	1.35	215
1841	800 (?)	1.40	562
1871	290	0.65	445
1901	210	0.45	467
1921	180	0.35	514
1939	80	0.20	400

Communities of Interest and the Gravity Model. The distribution of population within a city, discussed above, is just one manifestation of the well-known fact that people and their activities tend to cluster together, that human choice usually favors the near over the distant alternative when other considerations are unimportant. Proximity engenders a community of interest, and we are more likely to telephone our neighbors, to patronize a nearby supermarket, and to appeal for police protection in our residential locality than to engage in these activities at great distances from where we reside. The formalization of these tendencies in a manner suitable for prediction is of great value in the prediction of service and consumption demands.

There is a temptation to draw a parallel with the theory of gravitation. Two masses m_1 and m_2 attract each other with a force F_{12},

$$F_{12} = \frac{gm_1 m_2}{r_{12}^2}$$

where r_{12} is the distance between the mass centers, g is a constant, and the direction of the force is between the mass centers. The fact that the gravitational force is inversely proportional to the *square* of the distance between masses makes it possible to introduce a simple gravitational potential $V(r)$ to give a measure of the gravitational attraction at all distances r from a mass such that, for m_1,*

$$V_1(r) = \frac{-gm_1}{r} \quad \text{and} \quad F_{12} = \frac{-dV_1(r)}{dr} m_2$$

The total gravitational potential at any point is the algebraic sum of such potentials due to all masses of the pertinent universe. Unfortunately, the inverse-square attraction, which yields this scalar, additive potential function, is found only rarely in human behavior.

* This is a simplification. The force is actually a vector **F** such that **F** = −**grad** V.

Suppose that there are two locations i and j with populations P_i and P_j, respectively. We wish to predict the frequency F_{ij} of some interaction between individuals in locations i and j. In the absence of any distance factor, we expect that F_{ij} is proportional to the product of the two populations. In the light of the gravitational theory, the likelihood of a distance-dependent community of interest between populations i and j encourages us to try the following function:

$$F_{ij} = \frac{aP_iP_j}{r_{ij}{}^b} \qquad (4\text{-}1)$$

where r_{ij} is the distance between populations i and j and a, b are constants. Here b is not necessarily equal to 2.

Formulas of the form of Eq. (4-1) have often been successful in predicting community-of-interest relationships. In one investigation, by J. D. Carroll and H. W. Bevis,[3] the relative probabilities of trips of various lengths in the Detroit area are found to obey the equation, although the constant b varies according to the purpose

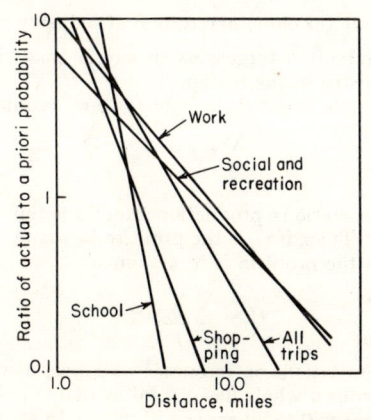

Fig. 4-2. Ratio of actual to expected total person trips as a function of distance—selected trip purposes.

of the trip. Suppose that the a priori probability of a trip is assumed to be aP_iP_j, where P_i is the population of potential travelers at the origin i and P_j is the number of potential destinations at location j. Figure 4-2 shows the ratios of observed to a priori probabilities as functions of distance for work, school, shopping, and recreational trips, where the observed data have been successfully fitted by curves of the form of Eq. (4-1).

Similar relationships have been developed with varying success to predict telephone-call frequencies, migrations, and new residential construction, as well as to plan the delineation of administrative areas and the siting of commercial establishments. An extended discussion is given by W. Isard.[18] D. B. Fairthorpe[42] exhibits some major shortcomings which may arise from the indiscriminate use of such models.

4-3. Operational Optimization of Urban Subsystems

In the urban context as elsewhere, system-engineering practice is focused on two basic problem types. One, embodying initial design, often involves the maximization of some desirable quantity, under specified constraints, in a system assumed to be at equilibrium in its most probable state. The second is concerned with the real-time control of fluctuations in dynamic systems.

Design for Optimum Equilibrium. Problems under this heading include the delineation of competing metropolitan milk sheds, the planning of municipal investment in schools or fire departments, the allocation of resources, and numerous others.

J. C. Tanner[41] provides a particularly thought-provoking example on pricing the use of roads to ensure efficient utilization. We illustrate with two examples: the locating of industrial sites to make efficient use of limited resources, and the planning of routes for transportation or distribution networks.

Locating Industrial Sites in an Environment with Limited Land and Resources. Industrial sites should be located in a systematic, rational manner so as to make efficient use of existing land, transport, labor, water, and power. As the industrial park replaces the individual factory site, planning is done more and more by real estate firms or by government and less by the individual industrial concern. The resultant demand for interindustry planning has evoked interesting techniques for the optimal allocation of available resources. To show the approach generally taken we consider a linear programming formulation for locating industry along the shores of Osaka Bay in Japan, as used by E. Kometani and K. Yoshikawa.[21]

The industrial complex is divided into a number of land areas, with index i, and a number of industries, with index j. Subject to limited supplies of land, water, and power, it is desired to find values of

x_{ij} = output (in monetary terms) of industry j in area i

which meet industry production targets as closely as possible and which maximize the total profit from industry in the region.

To achieve the production target P_j in industry j we require

$$\sum_i x_{ij} = P_j \tag{4-2}$$

It is desired to meet the various production targets as efficiently as possible, i.e., with a maximum profit. Thus, if c_{ij} is the profit to be made from one output unit by industry j in area i, then the problem is to maximize

$$C = \sum_i \sum_j c_{ij} x_{ij} \tag{4-3}$$

subject to (4-2) and to limitations on resources which we now formulate.

Assume that the region as a whole has a total available industrial power supply E which can be distributed at will to all areas within it. In area i let A_i = total industrial land area and W_i = total industrial water supply. Let a_j = land area required, w_j = water required, and e_j = power required per unit of output in industry j. The limitations are

On land: $\quad\sum_j x_{ij} a_j \leq A_i$

On water: $\quad\sum_j x_{ij} w_j \leq W_i \tag{4-4}$

On power: $\quad\sum_i \sum_j x_{ij} e_j \leq E$

Similar limitations might be formulated for manpower and transport, but these were not considered critical to the present problem.

The difficult determination of the coefficients a_j, w_j, e_j, and especially c_{ij} is reported on at length in the original paper. We do not review it here, partly because our space is limited and partly because the information has reduced validity for land-allocation problems in other regions.

Assuming that these coefficients have somehow been determined, the problem can be solved by established linear-programming techniques. We maximize (4-3) subject to (4-2) and (4-4). If no solution exists, the production targets are too ambitious and we must relax (4-2), to wit

$$\sum_i x_{ij} \leq P_j \tag{4-2a}$$

On Planning Efficient Routes for a Transportation or Distribution Network. The calculus of variations is particularly suited to finding the most economical routes for new roads, water mains, or other distribution systems. We show a formulation by P. Friedrich[8] which that author has applied to a number of practical problems.

Suppose that it has been decided to build a new road connecting two points (x_0, y_0) and (x_1, y_1) and that we seek the most economical route $y = S(x)$ between them. The annual costs may be divided into those independent of traffic A_u and those directly dependent on traffic using the road A_w.

If all costs are expressed in terms of the route $S(x)$ then $A_u(S)$ is the product of the annual traffic-independent cost per unit road length a_u and the road length $L(S)$. Similarly, $A_w(S)$ is the product of the annual cost per traffic unit a_w (e.g., in dollars per vehicle mile) by the annual number of traffic units $W(S)$. Here, we have assumed that the cost indices a_u and a_w are fairly uniform over the region studied and that we may use their average values as constants. Thus

$$A_u(S) = a_u L(S) \qquad A_w(S) = a_w W(S) \tag{4-5}$$

$W(S)$ can be expressed in terms of m traffic sources, each producing Q_i vehicles per year $(i = 1, 2, \ldots, m)$, with each vehicle in source i going the distance $l(S)_i$ from origin to destination:

$$W(S) = \sum_{i=1}^{m} l(S)_i Q_i \tag{4-6}$$

The continuum equivalent to (4-6) is

$$W(S) = \iint l(S, x, y) q(x, y) dx\, dy \tag{4-6a}$$

$W(S)$ is a measure both of traffic-dependent annual maintenance costs and of the time expended by the customers in using the facility.

The total annual cost is $J = A_u(S) + A_w(S)$. We formally express this cost in integral form with fixed limits of integration

$$J = A_u(S) + A_w(S) = \int_{x_0}^{x_1} f[x, S(x), dS(x)/dx]\, dx \tag{4-7}$$

because then the minimum-cost route is that $S(x)$ for which $f[x, S(x), dS(x)/dx]$ satisfies the Euler-Lagrange equation

$$\frac{\partial f}{\partial S} - \frac{d}{dx} \frac{\partial f}{\partial (dS/dx)} = 0 \tag{4-8}$$

This is the variational calculus formulation of the optimum-route problem.

Friedrich applies the approach to the siting of a new route through an existing rectangular street grid, the construction of a relief road which forms a new town boundary, and the design of a central arterial road network. Although these applications are extremely interesting, we must refer the reader to the original paper because the mathematical analysis cannot be exhibited in a sufficiently brief manner. Instead we offer a quite artificial example in the hope of clarifying the methodology.

Suppose that a road is to be built between a point A and an extant road $\overline{BC}$, as shown in Fig. 4-3. The problem is to find the road-line equation $S(x)$ which minimizes costs, given end points $y = S(0)$ at $x = 0$ and $y = 0$ at $x = A$,

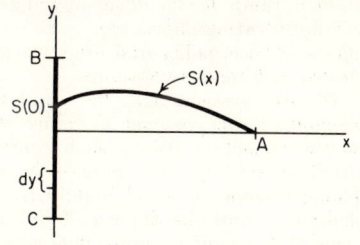

FIG. 4-3. The siting of a new road between an existing road and a fixed point.

and given that the traffic intensity between A and the element dy of $\overline{BC}$ at y is $q(y)dy$. Because we know in advance that $S(x)$ should be a straight line, the present exercise is purely illustrative.

If land values do not vary with location, the traffic-independent cost of $S(x)$ is proportional to the road length,

$$A_u(S) = a_u \int_0^A [1 + (dS/dx)^2]^{1/2} \, dx \tag{4-9}$$

The traffic-dependent annual cost for traffic between A and element dy on $\overline{BC}$ is proportional to the product of $q(y)dy$ and the road distance between A and dy, namely,

$$a_w q(y) dy \left\{ |y - S(0)| + \int_0^{S(0)} [1 + (dS/dx)^2]^{1/2} \, dx \right\}$$

Integrating over $\overline{BC}$, we obtain

$$A_w(S) = a_w \int_C^{S(0)} [S(0) - y] q(y) dy + a_w \int_{S(0)}^B [y - S(0)] q(y) dy$$
$$+ a_w \int_0^B q(y) dy \int_0^A [1 + (dS/dx)^2]^{1/2} \, dx \tag{4-10}$$

Letting $J = A_u(S) + A_w(S)$, we find that the $f(S, dS/dx, x)$ of (4-7) is

$$f = [1 + (dS/dx)^2]^{1/2} \tag{4-11}$$

Application of the Euler-Lagrange equation (4-8) yields $(d^2S/dx^2) = 0$, or

$$S(x) = S(0)\left(1 - \frac{x}{A}\right) \tag{4-12}$$

Thus the best route $S(x)$ is shown to be a straight line.
Hence

$$\int_0^A [1 + (dS/dx)^2]^{1/2} \, dx = [S^2(0) + A^2]^{1/2} \tag{4-13}$$

and the best value of $S(0)$ can now be found by substituting the right side of (4-13) into $J = A_u(S) + A_w(S)$ and solving $dJ/dS(0) = 0$.

Subsystem Optimization under Fluctuations. The capacity of an urban-area subsystem can be increased by physical expansion on the one hand or by better control of an existing plant on the other. The need for expansion is generally forced by peak demands which exceed the mean demand by a factor of 2 or more. When peak demands exceed capacity, or in some cases when they merely approach capacity, there is the likely development of self-aggravating conditions which foster a reduction in system capacity. The reduced capacity then worsens the overload, resulting in a sudden jump in the level of congestion and delay. Urban subsystems in which self-aggravating effects can be observed to arise from load fluctuations include telephone service, police and fire services, power and water supplies, sanitary and storm drains, and roadway systems.

Control systems may be employed to make available plant serve fluctuating demands better, or even to reduce the magnitude of the fluctuations in relationship to the capacity. While such control systems are not yet feasible in all kinds of urban subsystems, they have been effectively used in some. One is the urban telephone system, in which highly complex control and switching systems have been devised to optimize the use of physical plant. A notable practice in telephone systems that would be applicable to other urban subsystems as well is the optimum sharing of certain common facilities. This makes a larger amount of plant available to meet a temporary fluctuation than would be economical without sharing.

Another urban subsystem that has been subjected to considerable study by operations researchers is street traffic. These studies have looked in considerable detail at traffic flow in tunnels, and this subsystem is interesting in the variety of features it displays relative to the problem of optimization under fluctuations. We therefore discuss these tunnel-traffic features in some depth in lieu of a sketchy discussion of a number of subsystems.

An underwater vehicular tunnel is a cast-iron pipe of large cross section—about 30 ft in diameter. It is a pipe carrying two streams of an unusual kind of fluid, which has for particles vehicles weighing in tons. The particles are inelastic and are quite remarkable in avoiding the frequent collisions common to fluids with elastic particles. While compressible, like a gas, it also acts much like a liquid; in fact, during heavy congestion its consistency has much in common with molasses or glue. Another feature our traffic fluid has is waves, which become quite severe at times and produce the accordionlike stop-and-go driving condition which has become familiar to us. To increase the flow of other fluids one can increase the external pressure, but how can one increase the flow of this exceedingly complex kind of fluid?

The first step for a system engineer in dealing with such a problem is to construct a mathematical model describing traffic flow—in quantitative terms if possible, but at least in qualitative terms. A notable qualitative model was developed from fluid mechanics.[23] This model describes the behavior of vehicular traffic at a bottleneck in a roadway, i.e., a curve or merging area, and also the effect of the bottleneck on traffic behavior upstream and downstream from it. A significant feature of the model is its prediction of shock waves upstream from a bottleneck. These waves proceed backward down the roadway from an overloaded bottleneck section with a speed that may be computed from continuity requirements as

$$v = \frac{\Delta q}{\Delta k} \qquad (4\text{-}14)$$

where v is the speed, Δq the change in the flow rate, and Δk the change in concentration occurring with passage of the wave. If q is measured in vehicles/hour and k in vehicles/mile the wave speed is in miles/hour.

Shock waves have been shown to be a self-aggravating feature of roadway systems,[7] and they may be precipitated by minor fluctuations when flow approaches capacity of a roadway section. Once started, the shock wave may be amplified and cause rates of flow to drop far below capacity, thus aggravating the buildup of congestion. Figure 4-4 gives an example of a severe shock wave in the Holland Tunnel.

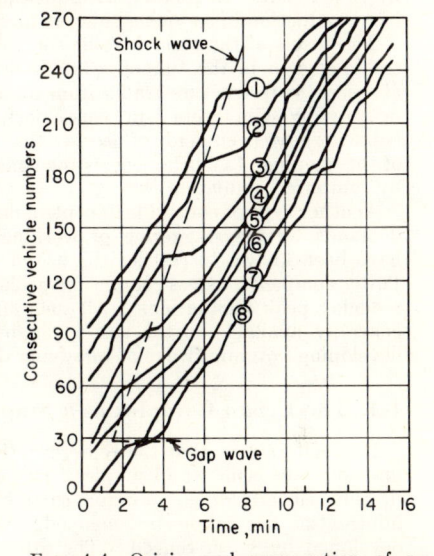

FIG. 4-4. Origin and propagation of a traffic shock wave. Curves show times when successive vehicles pass each of eight observers, located from entrance (1) to exit (8) in an 8,557-ft-long tunnel. Shock wave (low-flow, high-concentration) moves to entrance while gap wave (low-flow, low-concentration) moves to exit. Both originate with vehicle 28 between observers 6 and 7 at 1.5 min.

The behavior of tunnel traffic has been studied theoretically and experimentally. One theoretical construct which has contributed signally to an understanding of traffic flow is that of car-following theories.[6,9,14,15] They describe the interactions between vehicles by studying the accelerations used by drivers, after a time lag, as functions of relative speed, absolute speed, and separation distance. Car-following theories are important because they may be tested experimentally, may be integrated to relate mass flow to speed and concentration, and may be studied for an understanding of traffic-stream stability.[15] It is the increasing instability with increasing flow that limits the capacity of roadways, and the principal parameter for stability in this system, as with servomechanisms (Chapter 29) in general, is the lag time. In traffic flow, lag time is made up of driver reaction time and vehicle response time. Instability is approached when the average lag time between vehicles approaches one-half the average headway time. Data from various experiments show drivers tend to approximate this condition[13] at maximum flow.

The control of tunnel traffic can be exercised to eliminate shock waves and thereby increase flow, reduce breakdowns, and reduce exhaust fumes. This control can be exercised at the entrance to a tunnel to prevent the development of instability by keeping concentration safely below critical values. Although such control may withhold service from a particular motorist for a few seconds, he actually receives better service because the increase in flow clears vehicles ahead of him sooner. The improvement in service of a control system may be many times the improvement in capacity. For example, a 5 per cent increase in capacity when the overload is 5 per cent or less would eliminate all delay. When demands reach a level of 15 per cent above capacity, a 5 per cent increase in capacity could reduce aggregate delays by 70 per cent. In general, with queuing types of situations, an increase in capacity, i.e., a reduction in servicing time, reduces delays more than proportionately.

The result of the tunnel-traffic-control system is a change in the structure of the traffic stream in the tunnel. Traffic systems for street-intersection control, which also seek to restructure traffic from its usual random form, are used in urban areas, notably by Von Stein[14] in Dusseldorf, Germany. Studies of flow concentration behavior have been made of pedestrians[12] for purposes of designing subways. Studies of the queuing of vehicles suggesting control systems for special cases have been made by computer simulation.[13,19]

Another type of control is to control demands. Some urban areas seek to influence demands by the staggering of working hours. Overloaded water-supply systems have been known to restrict the use of water for certain purposes on certain days. Power companies assess charges for peak demands as well as for energy as a means of reducing peak fluctuations. All such efforts meet with limited success and are not generally available to the system engineer. His responsibility continues to lie in developing optimum ways of satisfying demands.

4-4. The Urban Complex as a Single Unified System

It is self-evident that the representation of all activities of the urban area by one unified model is not at all simple. Two approaches may be taken. A drastic simplification, cleverly done, may give some overall structural insight even when detailed information has not been developed. We report as characteristic a land-use model developed by R. J. Smeed. The other approach is to attempt a fairly complete classification of urban activities and to establish realistic rules for each sector's growth or decline and for interactions. To illustrate we discuss an idealized conceptual framework provided by the input-output model as used in a comprehensive study of the New York metropolitan area. We conclude this section with some remarks on the incorporation of the input-output model into a comprehensive, transportation-oriented metropolitan-area model.

R. J. Smeed's Land-use Model. In *The Traffic Problem in Towns*,[27] R. J. Smeed offers an outstanding demonstration of how an extremely simple model is sufficient to show clearly the inevitable effects of various transportation modes on city size and structure. He deduces the minimum ground area required by a person to make a rush-hour trip as a function of the mode of conveyance and length of journey. This, together with appropriate space allotments for disposal of conveyances, for work, and for residence, is sufficient to calculate the size and land-use distribution of the resultant city. We consider Smeed's model and some of his results only for the central business district. The original paper applies the same principles also to residential areas.

To deduce the minimum ground area required for a rush-hour trip by a given mode of transport, we require knowledge of the flow capacity c, in persons per unit time, on a roadway of width w. Then, if the transport system is used at peak capacity for a daily time T, the number of persons who can be carried during the peak period on a unit width of roadway is cT/w. The width required by one person is the inverse of

URBAN AREAS

this, namely, $w/cT \equiv \lambda$. The area of roadway required for one peak-period journey of length L is

$$A = L\frac{w}{cT} = L\lambda \qquad (4\text{-}15)$$

Table 4-3 gives typical values of λ for various modes of transport on the basis of British experience and on the assumption that the busy period T is of 2 hr duration. Also shown are the corresponding values of area required for 1-mile journeys.

Table 4-3. Carrying Capacities of Various Types of Roadway and Associated Areas Required per Person during Peak Period for a 1-mile Journey*

Daily peak busy period = T = 2 hr

Mode of transport	Flow, number of persons per ft width per hr	Speed, mph	λ, ft	Area required per person, sq ft
Automobiles on urban street 44 ft wide with 1.5 passengers per vehicle	67	15	0.0074	39
	95	10	0.00526	28
Automobiles on urban expressway with 1.5 passengers per vehicle................	187	40	0.0027	14
Pedestrian way.........................	800	2.5	0.0006	3
Urban railway line.....................	2,900	18	0.00017	0.9
	2,200	30	0.00023	1.2

* From Ref. 27.

Suppose that the smallest central business district sufficient to accommodate N workers is represented by a circle of radius r. Suppose further that workplaces are uniformly distributed throughout the central district, and that places of residence are all outside this district and are distributed uniformly in all directions. Assume that, quoting Smeed, "in his journey to work, the worker travels along the straight line from his residence to the town center until he reaches radius r. He then travels directly towards the place at which he works, the position of which is not correlated with the place at which he lives or with the point at which he reaches the outer part of the town center."

Smeed deduces that the average of distances from one point on the circumference at radius r to all points within the circle is about $1.36r$, if reasonable provision is made for the restrictions imposed on travel by a rectangular-grid street system. Hence the distance traveled in the central district by N workers is $1.36\ Nr$, and the area required for roadway is $1.36\ Nr\lambda$.

Each worker requires a ground space G for work and an area P for parking. P is assumed to be zero unless the worker travels by automobile. The sum of the areas required for roadway, for work, and for parking is equal to the minimum area of the central business district. Hence

$$1.36Nr\lambda + N(G + P) = \pi r^2$$

Substituting for π and solving for r, we get

$$r = 0.216N\lambda \left\{ 1 + \left(1 + \frac{G + P}{0.147N\lambda^2}\right)^{1/2} \right\} \qquad (4\text{-}16)$$

To examine the general behavior of this equation, Smeed set $G = 100$ sq ft as the ground space required per person for work, $P = 133$ sq ft/person for ground-level parking, and $P = 13.3$ sq ft/person for multilevel parking. The parking figures are based on 200 sq ft of storage space and 1.5 passengers for each automobile.

Table 4-4 lists some of the results obtained by substituting these figures and the appropriate values for λ into (4-16). It is seen that, first, the space required for roadway increases much more rapidly than does the number of workers in the central district and that, second, if all travel is by car, large town centers require more space for roadway than for parking even if all parking is at ground level. The second conclusion is most remarkable. In judging these figures, one should recall that the Manhattan rectangular street grid devotes about 35 per cent of all space to streets and sidewalks.

Table 4-4. Radius r, Miles, of Central Business District and Per Cent of Ground Area Devoted to (a) Roadway, (b) Parking, and (c) Working Space*

Mode of travel to center	Nature of parking	Area P for parking per person, sq ft	Working population in town center 50,000	500,000	5,000,000
Urban railway, $\lambda = 0.00023$	......	0	0.24† 0.3,0,99.7	0.76† 1,0,99	2.44† 4,0,96
Car on expressway, $\lambda = 0.0027$	Multi-level	13.3	0.26† 4,11,85	0.86† 13,10,77	3.16† 34,8,58
Car on expressway, $\lambda = 0.0027$	Ground level	133	0.37† 3,55,42	1.21† 9,52,39	4.24† 25,42,23
Car on 44-ft street, $\lambda = 0.0074$	Multi-level	13.1	0.27† 11,10,79	0.97† 31,8,61	4.48† 67,4,29
Car on 44-ft street, $\lambda = 0.0074$	Ground level	133	0.38† 8,53,39	1.32† 23,44,33	5.47† 55,25,20

* From Ref. 27.

† Key to boxes: | Radius r (in miles) |
                | a, b, c (in %) |

A model of the type described can be adapted to the planning of new or enlarged towns. There it may be extended to cover such possibilities as a circumferential road network just outside the central district or the provision of a large open space in the center. These variations have effects on land use and on mean travel distances which are most straightforward and which can readily be incorporated into a modification of Smeed's model.

Area-economy Model—The Input-Output Concept. To predict the long-term effects of a planned stimulus on the economy of a region, it is necessary to formulate a model that exhibits explicitly the interactions of all major activities that will be affected in any way. The one virtually inescapable formalism for doing this is known as the input-output model.[28] Consisting of statements of continuity which equate the output (or product) of a subsystem to the inputs (or consumption) of all the other subsystems, it can be explained most readily by means of an example. We choose a model developed for the New York Metropolitan Area Study and reported by Barbara R. Berman.[2] The outline below is in the spirit of the reference but has some changes in notation.

The problem is to predict production and employment in each industry, personal incomes, and government expenditures for one metropolitan area. The local area economy is divided into N industries ($N = 43$ in the New York study). For each it is necessary to distinguish between output absorbed by the local market and that contributed to the national market.

Let $x_i(t)$ = output per unit time of ith local industry at time t
$z_{ij}(t)$ = purchases per unit time by jth local industry from ith local industry at time t
$z_{ic}(t)$ = purchases per unit time by consumers from ith local industry at time t
$z_{ig}(t)$ = purchases per unit time by government from ith local industry at time t
$z_{in}(t)$ = purchases per unit time by national market from ith local industry at time t

There are N equations of continuity which together express the required balance between the output rates of the N local industries and the consumption of their products by local and national markets,

$$x_i(t) = \sum_{j=1}^{N} z_{ij}(t) + z_{ic}(t) + z_{ig}(t) + z_{in}(t) \qquad (i = 1, 2, \ldots, N) \qquad (4\text{-}17)$$

In the absence of expansion or contraction of productive capacity, the jth industry's purchases from the ith industry are assumed to be proportional to the output of the jth industry. The constant of proportionality a_{ij} giving amount of raw material i required per unit of product j can be viewed as constant only as long as there is no significant technological change. If industry j expands, purchases from industry i are assumed to be proportional to the rate of change of output in industry j. Again the constant of proportionality c_{ij} remains only in a technologically stagnant society. Thus

$$z_{ij}(t) = a_{ij}x_j(t) + c_{ij}\frac{dx_j(t)}{dt} \qquad (4\text{-}18)$$

It may be assumed that a contracting industry is unable to sell its capital assets at significant prices and hence that $c_{ij} = 0$ when $dx_j(t)/dt < 0$.

A replacement of the derivative in (4-18) by a difference is rather inevitable because the system of equations being developed is too complicated to admit anything but a numerical solution.

$$\frac{dx_j(t)}{dt} \cong \frac{x_j(t) - x_j(t - \Delta)}{\Delta} \qquad (4\text{-}19)$$

Let $P(t)$ = local area population at time t and $y(t)$ = rate of personal income per unit time in the local area at time t. Consumer purchases are assumed to be proportional to $P(t)$ and $y(t)$ with time-independent constants of proportionality m_i and f_i,

$$z_{ic}(t) = m_i P(t) + f_i y(t) \qquad (4\text{-}20)$$

Government purchases from industry i depend on population, and it is assumed that

$$z_{ig}(t) = a_{ig}(t) P(t) + b_{ig}(t) \qquad (4\text{-}21)$$

The parameters a_{ig} and b_{ig} are made time-dependent in the New York study in the explicit recognition that a study of this type will be used to determine the effects of varying government policies on the regional economy.

We substitute (4-18), (4-19), (4-20), and (4-21) into (4-17).

$$x_i(t) = \sum_{j=1}^{N} \left[\left(a_{ij} + \frac{c_{ij}}{\Delta}\right) x_j(t) - \frac{c_{ij}}{\Delta} x_j(t - \Delta) \right] + [m_i + a_{ig}(t)]P(t)$$
$$+ f_i y(t) + b_{ig}(t) + z_{in}(t) \qquad (i = 1, 2, \ldots, N) \qquad (4\text{-}22)$$

The purchases by the national market from local industry $z_{in}(t)$, $i = 1, 2, \ldots, N$ must be estimated by some means outside the present regional model. A similar model of the national economy, as a self-contained unit, might be used for this purpose.

The coefficients a_{ij}, c_{ij}, m_i, f_i, a_{ig}, and b_{ig} are viewed as known quantities and must be obtained by observation of past behavior or by postulate. Thus there are at this stage N equations (4-22) with $N + 2$ unknowns. We need a further formulation to relate population and income to employment, which, in turn, is a function of production.

Let $E(t)$ = total employment in local area at time t
$e_j(t)$, $e_g(t)$, $e_h(t)$ = regional employment in jth local-area industry, in local government, and in local private households at time t

Assume that
$$e_j(t) = h_j(t)x_j(t) \qquad e_g(t) = q(t)P(t) + s(t)$$

and that $e_h(t)$ is determined exogenously. The parameter $h_j(t)$ is the number of employees per unit of product in industry j. $q(t)$ and $s(t)$ are subject to deliberate public policy. Then

$$E(t) = \sum_{j=1}^{N} h_j(t)x_j(t) + q(t)P(t) + s(t) + e_h(t) \qquad (4\text{-}23)$$

Let
$$P(t) = g(t)E(t) \qquad \text{and} \qquad y(t) = k(t)E(t) \qquad (4\text{-}24)$$

with $g(t)$ and $k(t)$ again exogenously determined coefficients.

We now have a set of $(N + 1)$ equations (4-22) and (4-23) and $(N + 1)$ unknowns $x_i(t)$ where $i = 1, 2, \ldots, N$, and $E(t)$. After all the parameters have been determined, either by observation or by planning fiat, these equations may be solved successively for moments (or years) separated by the time interval Δ. The algebraic matrix is likely to be very large and to require use of an electronic computer.

It is obvious that this model is a gross simplification of reality. Economic progress is assumed to be orderly and not affected by panics, wars, or other irregularities. The results cannot be better than the assumptions made; yet, regrettably, the present state of the art of economic analysis is not encouraging to much greater elaboration.

Area-transportation Model. Any transportation model suitable for extrapolation into the future must exhibit the interaction between the economy, represented by land use, and the transportation system, represented by expressways, primary highways, railroads, and other major channels. It is obvious that a change in land use will result in some corresponding change in usage of the transportation system. The reverse is also true, and a change in transport facilities will provoke a response by the economy. For example, a new expressway will encourage residential construction and increase land values along the undeveloped portions of its route and in the area at its outer end.

Comprehensive urban transportation models showing the interactions have been or are being developed for Chicago,[4] Detroit, Los Angeles, Philadelphia, Pittsburgh,[25] and other metropolitan areas. Because these models are rather complicated in their details, we offer here only the basic approach in broad outline.

The first step consists of censuses of households, business enterprises, land usage, and transportation facilities. From these, one develops three basic inventories, of the local economy and population, of the land-use distribution, and of transportation. The primary objective of the area model is to predict these inventories at a future time $t + \Delta t$, given their initial status at time t. To simplify discussion we divide the area model into three subsystems:

1. *Transport-usage Prediction Model.* Given the three basic inventories, this model predicts usage in terms of flow volumes on the transportation network. The model may be experimental in outlook, based on the extrapolation of past origin and destination studies to new situations. A more elegant approach is to postulate that the traffic volume can be predicted with the aid of a gravity model, while its distribution over the available network can be allocated on the basis of a variational approach to minimize total effort or time. This introduction of theory permits meaningful traffic prediction even at a future time when the facilities may be very different from those we can observe today.

2. *Economy and Population Prediction Model.* Given the inventories at t, the corresponding inventories at time $t + \Delta t$ are predicted by use of an input-output model, as discussed above under Area-economy Model.

3. *Land-market Prediction Model.* Given the inventories at time t, and the transport-usage pattern, this model predicts what changes in land usage will occur between t and $t + \Delta t$. The basic process of land development in the United States is on an open-market basis, hence the term "land market." However, public policy, in the form of zoning, has an important place and must be included as a restraint on the otherwise open market.

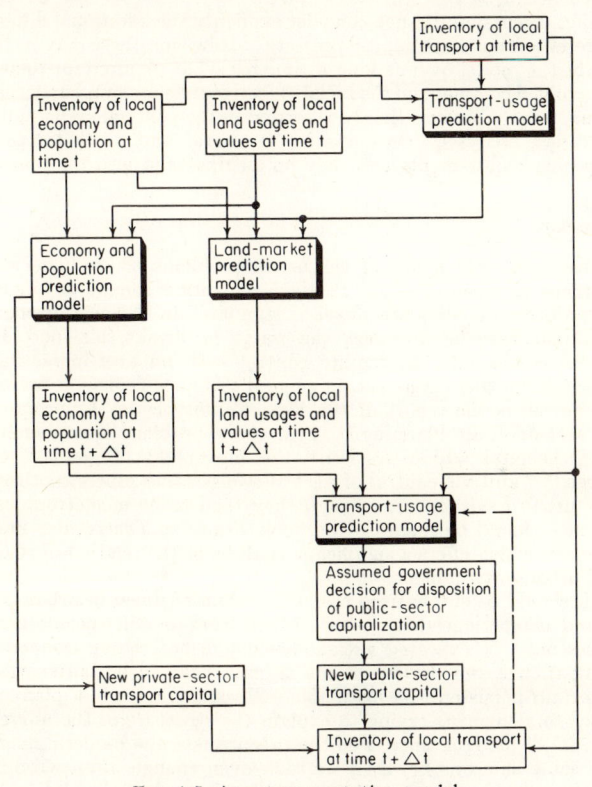

FIG. 4-5. Area-transportation model.

Figure 4-5 shows how these building blocks may be combined to form an area-transportation model. (The diagram would not change very much if our primary interest were in the development of some different aspect of the urban economy.) First, the transport-usage prediction model is applied to the transport inventory at time t. Naturally, if t were in the past or present, this usage could be measured rather than predicted. Second, the economy and population prediction model and the land-market prediction model are used to determine new inventories for the time $t + \Delta t$.

Since the purpose of an area-transportation study is to determine the most effective use of public capital in developing transport, this new capital for the interval Δt is not allocated until the above predictions are made. Instead, the transport utilization at time $t + \Delta t$ is evaluated on the basis of the initial transport inventory of time t, subject to the pressures engendered by the changed economy and land inventory at time $t + \Delta t$. The imbalances and pressures which thus come to light can then be alleviated by appropriate new government capitalization. A tentative allocation of this capital is made and added, along with any private-sector additions, to the old transport inventory to yield a first-trial transport inventory for time $t + \Delta t$.

The process now becomes recursive, as shown by the dotted line in Fig. 4-5. The trial transport inventory of time $t + \Delta t$ replaces the actual inventory of time t in a new assessment of transport usage. The public investment is corrected if necessary and a revised trial transport inventory is calculated. The iteration is repeated until we judge that the public-sector transport capital is budgeted in an optimal manner, When this is done, the entire process of Fig. 4-5 can be repeated to predict the urban inventories at $t + 2\Delta t$, at $t + 3\Delta t$, and so forth.

The foregoing discussion did not consider explicitly the additional difficulties introduced by extremely long-range capital projects. However, these may be planned by a recursive use of the model over as long a time period as required for their conclusion. We have also avoided comment on the social objectives which should underlie long-range urban planning, largely because these are settled in the political arena and are not yet suitable for direct analysis. Only insofar as moral and aesthetic principles are reduced to specific requirements can they be incorporated into the analyst's model.

4-5. Conclusion

We have discussed only a few of the urban problems which have attracted the system researcher in recent years. Particular mention should be made of crucial problems in maintaining adequate physical standards in the city dweller's environment. The atmosphere he breathes, the water he drinks, his food, his survival through natural catastrophe all require control with an ever-increasing efficiency and precision. While there is as yet no general technical review, a useful reference, rather nontechnical, is the report of the First Conference on Environmental Engineering and Metropolitan Planning.[24] Still other problem areas, such as airport management,[35] hospital administration,[36] the movement of goods,[37] queuing for various services,[38,39] and the control of vast administrative superstructures,[40] are not unique to the urban environment, though they tend to be most troublesome there. Finally, reference should be made to the book *Traffic in Towns*, also known as the Buchanan Report, which offers a significant analysis of the relation of road transport to the overall urban environment.

In closing, it should be emphasized that the system engineer in urban planning is a very recent and inexperienced intruder. Much work is still speculative or experimental, failures must be expected with somewhat disheartening frequency, and the consequent impatience and disdain of the layman have to be borne with patience. However, a certain persistence is justifiable. Recently, an urban planner said that "we are similar to alchemists trying to explain the structure of the universe." But the alchemist, in all his ignorance, was the precursor to the modern chemist. So if we are in the same shoes as he, then we too—even though often wrong or pretentious—should press on in the expectation that in time our skills and understanding will be sufficient for the proposed tasks.

REFERENCES

1. Bartholemew, H.: *Land Uses in American Cities*, Harvard University Press, Cambridge, Mass., 1960.
2. Berman, B. R., B. Chinitz, and E. M. Hoover: *Technical Supplement to the New York Metropolitan Regional Study*, Harvard University Press, Cambridge, Mass., 1961.
3. Carroll, J. D., and H. W. Bevis: Predicting Local Travel in Urban Regions, *Papers Proc. Regional Sci. Assoc.*, vol. 3, pp. 194 et seq., 1957.
4. Chicago Area Transportation Study: *Final Report*, Chicago, vol. 1, 1959; vol. 2, 1960; vol. 3, 1962.
5. Clark, C.: Urban Population Densities, *J. Roy. Statist. Soc.*, vol. 114, pt. 4, pp. 490–496, 1951.
6. Edie, L. C.: Car Following and Steady-state Theory for Noncongested Traffic, *Operations Res.*, vol. 9, no. 1, pp. 66–76, 1961.

7. Edie, L. C., and R. S. Foote: Effect of Shock Waves on Tunnel Traffic, *Proc. Highway Res. Bd.*, vol. 39, pp. 492–505, 1960.
8. Friedrich, P.: Die Variationsrechnung als Planungsverfahren der Stadt und Landesplanung, *Schr. Akad. Raumforschung Landesplanung*, vol. 32, Walter Dorn, Bremen-Horn, 1956. Also (in English) B. C. Kahan: Precis of "The Calculus of Variations as a Method for Town and Country Planning," *Research Note* RN/4087/BCK, Road Research Laboratory, Harmondsworth, Middlesex, England, 1961.
9. Gazis, D. C., R. Herman, and R. W. Rothery: Nonlinear Follow-the-leader Models of Traffic Flow, *Operations Res.*, vol. 9, no. 4, pp. 545–567, 1961.
10. Gibbs, J. P. (ed.): *Urban Research Methods*, D. Van Nostrand Company, Inc., Princeton, N.J., 1961.
11. Gottmann, J.: *Megalopolis*, The Twentieth Century Fund, New York, 1961.
12. Hankin, B. D., and R. A. Wright: Passenger Flow in Subways, *Operations Res. Quart.*, vol. 9, no. 2, pp. 81–88, 1958.
13. Helly, W.: Dynamics of Single Lane Vehicular Traffic Flow, *Res. Rept.* 2, Center for Operations Research, M.I.T., Cambridge, 1959.
14. Herman, R. (ed.): *Theory of Traffic Flow*, Elsevier Publishing Company, Amsterdam, 1961.
15. Herman, R., E. W. Montroll, and R. B. Potts: Traffic Dynamics: Analysis of Stability in Car Following, *Operations Res.*, vol. 7, no. 1, pp. 86–106, 1959.
16. Hoch, I.: *Economic Activity Forecast*, Chicago Area Transportation Study, Chicago, 1959.
17. Hoover, E. M., and R. Vernon: *Anatomy of a Metropolis*, Harvard University Press, Cambridge, Mass., 1959.
18. Isard, W.: *Methods of Regional Analysis*, John Wiley & Sons, Inc., New York, 1960.
19. Jennings, N. H., and J. H. Dickins: Computer Simulation of Peak Hour Operations in a Bus Terminal, in H. Gaetzkow (ed.), *Simulation in the Social Sciences*, Prentice-Hall, Inc., Englewood Cliffs, N.J., 1962.
20. Jordan, T. D., J. H. Dickins, and F. N. Caggiano: *Project Sky Count*, Operations Services Department, The Port of New York Authority, New York, 1962.
21. Kometani, E., and K. Yoshikawa: Locating Industry along the Shores of Osaka Bay, *Mem. Dept. Eng. Kyoto Univ.*, vol. 24, pt. 2, 1962.
22. Lichtenberg, R. M.: *One Tenth of a Nation*, Harvard University Press, Cambridge, Mass., 1960.
23. Lighthill, M. J., and G. B. Whitham: On Kinematic Waves, II; A Theory of Traffic Flow on Long Crowded Roads, *Proc. Roy. Soc. (London)*, Ser. A, vol. 229, pp. 317–345, 1955.
24. Logan, J. A., P. Oppermann, and N. E. Tucker (eds.): *Environmental Engineering and Metropolitan Planning*, Northwestern University Press, Evanston, Ill., 1962.
25. Pittsburgh Area Transportation Study: *Final Report*, vol. I, Pittsburgh, 1961.
26. Regional Plan Association: Hub-bound Travel in the Bi-state Metropolitan Area, *Bull.* 91, New York, 1959.
27. Smeed, R. J.: *The Traffic Problem in Towns*, Manchester Statistical Society, 1961.
28. Tinbergen, J., and H. C. Bos: *Mathematical Models of Economic Growth*, McGraw-Hill Book Company, New York, 1962.
29. *United Nations Demographic Yearbook*, United Nations, New York, 1952 and subsequent years.
30. U.S. Bureau of the Census: *County and City Data Book*, 1956, 5th ed., Washington, D.C., 1956.
31. U.S. Bureau of the Census: Statistical Abstract of the United States: 1961 82d ed., Washington, D.C., 1961.
32. Weiss, H. K.: Distribution of Urban Population and an Application to a Servicing Problem, *Operations Res.*, vol. 9, no. 6, pp. 860–874, 1961.
33. Zipf, G. K.: *National Unity and Disunity*, Bloomington, Ind., 1941.
34. Syski, R.: *Introduction to Congestion Theory in Telephone Systems*, Oliver & Boyd, Ltd., London, 1960.
35. Blumstein, A.: The Operations Capacity of a Runway Used for Landings and Take-offs, *Proceedings of the Second International Conference on Operations Research*, English Universities Press, London, 1961.
36. Revans, R. W.: The Hospital as an Organism, *Proceedings of the Second International Conference on Operations Research*, English Universities Press, London, 1961.
37. Hanssmann, F.: Optimal Inventory Location and Control in Production and Distribution Networks, *Operations Res.*, vol. 7, no. 4, pp. 483–498, 1959.
38. Saaty, T. L.: *Elements of Queueing Theory*, McGraw-Hill Book Company, New York, 1961.

39. Morse, P. M.: *Queues, Inventories and Maintenance*, John Wiley & Sons, Inc., New York, 1958.
40. McCloskey, J. F., and J. M. Coppinger: *Operations Research for Management*, vol. II pt. III, The Johns Hopkins Press, Baltimore, 1956.
41. Tanner, J. C.: *Pricing the Use of the Roads*, Laboratory Note No. LN/319/JCT., Road Research Laboratory, Harmondsworth, Middlesex, England, 1963.
42. Fairthorne, D. B.: Description and Shortcomings of Some Urban Road Traffic Models, *Operations Res. Quart.*, vol. 15, no. 1, pp. 17–28, 1964.
43. *Traffic in Towns*, Her Majesty's Stationery Office, London, 1963.

Chapter 5

THE LOWER ATMOSPHERE

DONALD J. PORTMAN, *University of Michigan, Department of Meteorology and Oceanography.*

EUGENE W. BIERLY, *U.S. Atomic Energy Commission.*

NORMAN L. CANFIELD, *U.S. Weather Bureau.*

A. NELSON DINGLE, *University of Michigan, Department of Meteorology and Oceanography.*

E. WENDELL HEWSON, *University of Michigan, Department of Meteorology and Oceanography.*

EDWARD RYZNAR, *University of Michigan, Department of Meteorology and Oceanography.*

CONTENTS

5-1. Planetary Atmospheres	5-2
Physical Characteristics of the Planets	5-2
Thermodynamic and Dynamic Properties of Planetary Atmospheres	5-2
5-2. Atmospheric Thermodynamics	5-3
Radiation Processes	5-3
Thermodynamic Systems in the Atmosphere	5-5
5-3. Large-scale Circulation Features	5-5
Undulating Horizontal Flow	5-5
Variations in the Vertical	5-7
Air Masses and Frontal Systems	5-7
5-4. Clouds and Precipitation	5-7
Clouds	5-8
Rain	5-8
Snow	5-9
Hail	5-10
Weather Modification	5-10
5-5. The Atmosphere Near the Earth's Surface	5-11
Surface-layer Characteristics	5-11
Temperature Profiles	5-12
Flow Properties	5-12
Shear Stress in Neutral Conditions	5-12
Turbulent Transfer in Nonneutral Conditions	5-13

SYSTEM ENVIRONMENTS

    5-6. Synoptic Meteorology and Weather Forecasting................ 5-14
        Data Acquisition, Processing, and Transmission.................. 5-14
        Analysis Techniques.. 5-15
        Forecasting Procedures....................................... 5-15
    5-7. Statistical Properties: Data Sources and Services.............. 5-17
        Statistical Properties.. 5-17
        Data Sources and Services.................................... 5-18

5-1. Planetary Atmospheres*

Comparison of Earth's atmosphere with those of other planets of the solar system affords an opportunity for developing a scientific synthesis and integrated physical picture of planetary atmospheres in general.

Stated broadly, the facts are as follows: The most noteworthy characteristic of Earth's atmosphere in comparison with those of other planets is the presence of oxygen. This gas has not been detected in the atmospheres of other planets. It was suggested by Koene in 1856 that the presence of oxygen in Earth's atmosphere was due to photosynthesis of plants.

Another noteworthy feature of Earth's atmosphere is the relatively large amount of water vapor in it. Present indications are that Mars is the only other planet, with the possible exception of Jupiter, having appreciable amounts of water vapor, with an estimated 2×10^9 tons. Earth's atmosphere has five thousand times as much, amounting to 10^{13} tons.

Physical Characteristics of the Planets. The planets of the solar system may conveniently be divided into two groups. The "warm dwarfs" are the four small planets nearest the sun: Mercury, Venus, Earth, and Mars. The "cold giants," Jupiter, Saturn, Uranus, and Neptune, are much farther from the sun. These and other physical characteristics of the planets of the solar system and of their satellites are set forth in Tables 5-1, 5-2, 7-10, and 7-11.

It is probable that planetary temperatures have not changed greatly from their present values, as specified in Table 5-1, for several billion years. If this is true, we may determine the probable composition of present-day planetary atmospheres. From kinetic theory, the root-mean-square molecular velocity is given by the expression

$$V_{\text{rms}} = \left(\frac{cT}{m}\right)^{1/2} \qquad (5\text{-}1)$$

Thus, the greater the temperature T and the smaller the molecular weight m the greater will be the molecular motion V_{rms}.

It follows that the "warm dwarfs" have lost most of their lighter gases such as hydrogen and helium during geological time, whereas the "cold giants" with their low temperatures and high gravity still retain much hydrogen and helium in their atmospheres. Remote-sensing analytical methods, including molecular spectroscopy and other analyses both direct and indirect, reveal that in fact the atmospheres of Jupiter and Saturn do consist mainly of the two light gases hydrogen and helium. The atmospheres of Uranus and Neptune, on the other hand, are composed mainly of hydrogen and nitrogen, with small amounts of helium and methane. The atmosphere of Venus is four-fifths nitrogen and one-fifth carbon dioxide, whereas that of Mars is believed to be mostly nitrogen, perhaps as much as 95 per cent, with the remainder being carbon dioxide and other gases.

Thermodynamic and Dynamic Properties of Planetary Atmospheres. The behavior of planetary atmospheres is related to the stability of these atmospheres. If the *lapse rate of temperature*, i.e., the rate at which the existing temperature lapses

---
* By E. Wendell Hewson.

Table 5-1. Summary of Planetary Constants[†]

Planet	Distance from sun	Mass	Radius	Density	Albedo	g, cm/sec^2	Avg temp, °K
Mercury	0.3871	0.0543	0.390	5.05	0.063	350	443
Venus	0.7233	0.8136	0.973	4.88	0.59	842	700
Earth	1.0000	1.0000	1.000	5.52	0.36	980	250
Mars	1.5237	0.1080	0.520	4.24	0.15	391	218
Jupiter	5.2028	318.35	10.97	1.33	0.47	2,592	105
Saturn	9.5388	95.30	9.03	0.714	0.46	1,145	78
Uranus	19.19	14.54	3.72	1.56	0.45	1,070	55
Neptune	30.07	17.26	3.50	2.22	0.49	1,381	43
Pluto	39.52	0.0330	0.450	2.0	0.16	160	42
Moon	1.0000	0.012304	0.2727	3.35	0.073	162	275
Io	5.2028	0.0122	0.275	3.24	0.50	158	103
Europa	5.2028	0.0079	0.241	3.12	0.56	133	100
Ganymede	5.2028	0.0260	0.435	1.74	0.35	135	110
Callisto	5.2028	0.0162	0.392	1.48	0.15	103	118
Titan	9.5388	0.0236	0.384	2.30	0.30	157	83
Triton	30.07	0.0220	0.350	2.8	.....	176	

Distance of the earth from the sun = 1.4968×10^{13} cm
Mass of the earth = 5.975×10^{27} g
Mean radius of the earth = 6,370.97 Km
Solar constant = 1.395×10^6 ergs/(sq cm)(sec)

[†] From H. C. Urey, The Atmospheres of the Planets, in S. Flugge, (ed.) *Handbuch der Physik*, vol. 52, pp. 363–418, Springer-Verlag OHG, Berlin, 1959.

or decreases with height, is greater than a critical value, then the atmosphere is unstable and pronounced churning and turbulent mixing occur. On the other hand, if the lapse rate of temperature is less than this critical value, then turbulent mixing is suppressed. This critical lapse rate is known as the *dry adiabatic lapse rate* and represents the rate of adiabatic heating or cooling of a mass of atmosphere as its pressure changes. The dry adiabatic lapse rate is given by the expression

$$\Gamma = \frac{g}{c_p J} \tag{5-2}$$

where g is the acceleration of gravity, c_p is the specific heat of the atmosphere at constant pressure, and J is the mechanical equivalent of heat. Approximate values of the dry adiabatic lapse rates for planetary atmospheres are given in Table 5-2.

Newton's laws of motion must be modified slightly if they are to be used in accelerating frames of reference such as rotating planets. A customary modification is to introduce the inertial Coriolis force into the equations of motion to make allowance for the rotation. The horizontal component of the Coriolis force is given by

$$\text{Coriolis force} = 2\omega V \sin \phi \tag{5-3}$$

where ω is the angular velocity of the planet and ϕ is the latitude at which horizontal motion of velocity V is occurring. The angular velocities of the planets are given in the last column of Table 5-2.

5-2. Atmospheric Thermodynamics[*]

Radiation Processes. Atmospheric processes are influenced by two wavelength bands, solar or short-wave radiation extending from about 0.2 to 4 microns, and terrestrial or long-wave radiation from 4 to 50 microns.

[*] By E. Wendell Hewson.

SYSTEM ENVIRONMENTS

Table 5-2. Thermodynamic and Dynamic Constants

Planet	Specific heat c_p at constant pressure,† cal/(g)(°K)	Dry adiabatic lapse rate Γ, °K/Km	Angular velocity, radians/sec
Mercury	....	...	8.3×10^{-7}
Venus	0.24	8.6	3.2×10^{-7}
Earth	0.24	9.8	7.3×10^{-5}
Mars	0.25	3.6	7.2×10^{-5}
Jupiter	2.82	2.2	1.8×10^{-4}
Saturn	2.82	1.0	1.7×10^{-4}
Uranus	1.72	1.4	1.6×10^{-4}
Neptune	1.72	1.7	1.1×10^{-4}

† Values of c_p and Γ given in this table are reliable only if pressures and temperatures are of the order encountered in Earth's atmosphere. For pressures and temperatures greatly different from these, other values of c_p and Γ must be used.

The sun radiates approximately as a blackbody at a temperature of 6000°K. The solar radiation reaching the earth is given by the *solar constant*, which is the radiant energy falling perpendicularly on unit area at the outer limits of the atmosphere in unit time when the distance between earth and sun is at its mean value. Its value is 2.0 cal/(sq cm)(min). This radiation is believed to be effectively constant, although fluctuations are known to occur in the ultraviolet where the energy is very small. Maximum energy in the solar beam occurs at about 0.48 micron.

Solar radiation entering the earth's atmosphere may be attenuated by three distinct processes. The first of these is Rayleigh scattering by small particles, such as the molecules of air and dust particles. The second attenuation process is absorption: some 6 to 13 per cent of the incident solar energy represented by the solar constant may on the average be absorbed by water vapor in the atmosphere; a smaller amount is absorbed by ozone in the higher atmosphere. The third process is reflection and absorption of solar radiation by clouds. The fraction of incident energy reflected, a quantity known as the *albedo* of the reflector, varies from a few per cent for a very tenuous cloud up to 70 or 75 per cent for a very thick and dense cloud. Up to 20 per cent of the solar radiation incident on a cloud may be absorbed by the water particles and water vapor in it.

Of the solar radiation reaching the earth's surface, some 15 to 20 per cent is reflected by natural objects such as soil and vegetation. Reflection from the oceans depends largely on the elevation of the sun. With the sun near the horizon, the albedo of a smooth sea surface is about 0.4, whereas it is about 0.03 when the sun is near the zenith. The presence of waves causes little change with zenith sun but produces an increase in albedo when the sun is low. About 70 per cent of the radiation entering the water is, on the average, absorbed on traversing the first meter of the ocean. The process of mixing, however, distributes this heat throughout a considerable layer of water, so that the heating of the surface waters is very small. The present state of our knowledge of solar radiation is summarized in Ref. 1.

Terrestrial or long-wave radiation has a peak at about 10 microns, which corresponds to a blackbody temperature of about 273°K. The earth itself radiates effectively as a blackbody (see Chapter 15 and Sec. 16-2). Terrestrial radiation is absorbed by atmospheric water vapor from 5 to 8 microns and at 20 microns and greater wavelengths. There is a strong carbon dioxide absorption band from 13 to 17 microns, and ozone in the higher atmosphere has an absorption band with a peak at 4.6 microns.

Early attempts to analyze atmospheric radiative processes in the terrestrial wavelengths assumed that the atmosphere absorbed and radiated as a gray body, i.e., that it absorbed and radiated a fixed fraction of blackbody radiation at all wavelengths.

This approach proved to be inadequate and has been replaced by graphical or numerical means of integrating the radiative-transfer equations.

Atmospheric long-wave radiation was reviewed in detail in 1959 by C. D. Walshaw;[4] G. N. Plass in 1962[2] presented an excellent discussion of the spectral lines in a band, describing the Elsasser model, the statistical model, and the random Elsasser model. Spectral-band absorptance for atmospheric slant paths was analyzed in detail by Plass in 1963.[3] Goody[5] presents a definitive exposition of the theory of atmospheric radiation.

Thermodynamic Systems in the Atmosphere. The atmosphere as a whole acts as a heat engine, with heat being received at short wavelengths from the sun and emitted by long-wave radiation to outer space. In the process, mechanical work is done by the atmosphere and oceans. There is a net annual gain of radiational energy at low latitudes and a net loss at high latitudes. Our wind and storm systems are an integral part of the mechanism by which most of this energy is transferred horizontally from equator to poles.

On a smaller scale, cloud systems involve complicated thermodynamic processes in which considerable energy conversion occurs, much of it through phase changes of water. The cumulonimbus thundercloud is one such complex thermodynamic system.

Atmospheric thermodynamics is also important on the microscale, as in the processes by which cloud particles develop on condensation and freezing nuclei and later grow to snowflakes or raindrops.

These and other thermodynamic systems such as hurricanes and tornadoes have become highly specialized fields for research.

REFERENCES

1. Allen, C. W.: Solar Radiation, *Quart. J. Roy. Meteorol. Soc.*, vol. 84, no. 362, pp. 307–318, October, 1958.
2. Plass, G. N.: Methods for the Representation of Laboratory Infrared Absorption Measurements, in *Advances in Molecular Spectroscopy*, pp. 1326–1335, Pergamon Press, New York, 1962.
3. Plass, G. N.: Spectral Band Absorptance for Atmospheric Slant Paths, *Appl. Opt.*, vol. 2, no. 5, pp. 515–526, May, 1963.
4. Walshaw, C. D.: Infrared Radiation in the Atmosphere, *Sci. Prog.*, vol. 47, pp. 67–79, 1959.
5. Goody, R. M.: *Atmospheric Radiation—I. Theoretical Aspects*, Oxford University Press, London, 1964.

5-3. Large-scale Circulation Features*

Atmospheric motions on a global scale may be described conveniently in terms of a thin, spherical shell of a variable-density gas in constant rotation. The driving force responsible for its motion relative to the earth is the unequal heating between the equatorial and polar regions. The large-scale motions being so derived effect a longitudinal transport of energy poleward, thereby reducing significantly the north-south temperature gradient that would exist without an atmosphere. The oceans also contribute to the longitudinal heat exchange, but their contribution is probably less than 10 per cent of the total. The great mobility of the atmosphere enables it to transport both sensible and latent heat (of change of phase of water) over thousands of miles in a few days. The energy transfer is, however, uniform in neither time nor space, so that detailed study of the large-scale motion system is required before the observed distributions of temperature, humidity, and cloudiness can be adequately described.

Undulating Horizontal Flow. The major feature of the large-scale atmospheric motion is that, by and large, it is rotating in the same direction as the earth but at a greater speed; that is, west winds (air moving from the west) dominate. Important exceptions exist in the tropics where the circulation is generally from east to west at all levels, and near the poles where it is east to west at or near the surface. It is

* By Donald J. Portman.

apparent, however, that purely *zonal* flow, i.e., flow along the latitude circles, cannot effect significant heat transfer between the equator and the poles, and in fact, such flow is not observed. Instead it is found that the basic west-to-east flow, at any one time, has superimposed upon it a series of undulations which provide the north-south components necessary for longitudinal heat transfer. The basic characteristic is most clearly manifest if one views the global circulation at a height of about 18,000 to 20,000 ft as depicted in Fig. 5-1. Here are shown various idealized types of circulation that may be dominant at any one time. There is a tendency to oscillate between the *high index* (A) and *low index* (D) situations within a period of a few weeks, the intermediate situations persisting during a relatively small portion of the time. It is obvious that for high-index situations the *longitudinal* temperature gradient is large, while for low-index periods large *latitudinal* gradients exist. The undulations, known as long waves (in the westerlies) or Rossby[1] waves, generally number from 3 to 7. They vary in amplitude and general configuration from day to day and ordinarily move eastward at a speed less than that of the wind, although westward motion is not

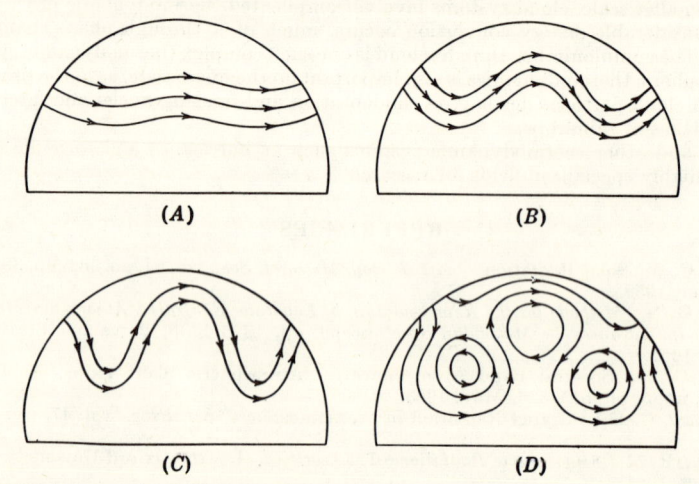

FIG. 5-1. The index cycle. (*After Namias in Dynamical and Physical Meteorology by Haltiner and Martin, McGraw-Hill Book Company, New York, 1957.*)

unknown. Important influences on the characteristics of the undulating, and sometimes cellular, westerly flow are (1) seasonal changes in the longitudinal variation in net radiation, (2) the configuration of continents and oceans, and (3) barriers to the flow due to large mountain ranges.

The simplest analysis of large-scale motion in the free atmosphere is in terms of the *geostrophic* approximation in which the motion is assumed steady, friction forces are absent, and the pressure-gradient force is balanced by the Coriolis force. In many problems, however, the path of the motion has sufficient curvature so that the centripetal acceleration cannot be neglected. The equilibrium motion resulting from a balance of pressure-gradient force, Coriolis force, and the force due to centripetal acceleration is termed the *gradient* wind. Tangential acceleration is assumed to be zero. Thus the pressure pattern (at a given height sufficiently removed from the surface of the earth so that friction forces are insignificant) represents the instantaneous field of *horizontal* motion. The wind direction is parallel to the isobars (lines of constant pressure) and the speed inversely proportional to the isobar spacing—at a given latitude. In the Northern Hemisphere, high pressure is to the right and low pressure to the left of one facing downwind; the reverse relationship holds in the Southern Hemisphere.

Referring again to Fig. 5-1, (D), and assuming that the pattern may represent gradient flow, one may deduce that the two closed circulation cells exhibiting counterclockwise motion are areas of relatively low pressure while the single cell with clockwise circulation is an area of relatively high pressure. The former are termed *cyclones* or simply *lows* and the latter an *anticyclone* or a *high*. They are similar and directly related to the familiar lows and highs of the surface weather map. (In the Southern Hemisphere the direction of flow around a pressure center is opposite to that in the Northern Hemisphere.)

Variations in the Vertical. Vertical motions are usually neglected in the global scale of analysis, and the equation for vertical motion reduces to the hydrostatic expression

$$dp = -\rho g\, dz \qquad (5\text{-}4)$$

where ρ is density, p pressure, and g the acceleration of gravity. Equation (5-4) forms the basis for a transformation from length to pressure for the vertical dimension and results in delineation of flow patterns on constant-pressure surfaces. It is assumed that the atmosphere behaves as an incompressible fluid and, for the simplest model, that it is *autobarotropic*, i.e., that surfaces of constant pressure always coincide with surfaces of constant density and, therefore, that the geostrophic wind is invariant with height. The less restrictive condition that the atmosphere be *baroclinic*—i.e., that the pressure and density surfaces do not coincide and that the geostrophic wind varies with height—is much more realistic but offers serious difficulties for analytical or numerical treatment.

Air Masses and Frontal Systems. Other than for limited regions and times, the atmosphere is, in reality, baroclinic, and the flow patterns represented in Fig. 5-1 are those for a specific level. It follows that lows and highs at the surface do not in general correspond in position and intensity with similar patterns aloft. It is often found, for example, that more than one low-pressure center at the surface is associated with a low center aloft; and in association with the surface lows, or *troughs* of low pressure, are sloping surfaces of density discontinuity. The intersections of these surfaces with the earth are called *fronts;* cold fronts if the colder air is moving and actively displacing the warmer air, and warm fronts if the opposite occurs. They are termed stationary if no motion takes place relative to the earth. Frontal surfaces separate *air masses* having different histories and, importantly, different temperatures and moisture contents. They occur mainly in the mid-latitudes and generally move from west to east in conjunction with the migratory low-pressure centers or troughs.

Wavelike undulations often form on frontal surfaces and may develop into new low centers. The frontal section west of such a center moves as a cold front and that to the east as a warm front. The equatorward section between the two is known as the *warm* sector of the wave. If such a system develops in typical fashion, the cold front moves eastward faster than the warm front, and gradually overtakes the latter (at the center of the low), forming a complex frontal structure called an *occlusion*. Frontal motion invariably causes warm air to be forced upward over cold air, and cooled, thus ordinarily causing cloudiness and precipitation. Thermal stability and moisture content of the air masses involved obviously play an important role in the amount of precipitation and cloudiness. The development, decay, and generally eastward movement of frontal systems, the associated low-pressure centers, and the intervening highs are the major concern of meteorologists for forecasting day-to-day weather in the mid-latitudes.

REFERENCE

1. Rossby, C. G., and collaborators: Relation between Variations in the Intensity of the Zonal Circulation of the Atmosphere and the Displacements of the Semi-permanent Centers of Action, *J. Marine Res.*, vol. 2, June 21, 1939.

5-4. Clouds and Precipitation*

Because the ultimate source of water for all the land areas of earth is precipitation from the atmosphere, a knowledge of this area of meteorology is especially pertinent

* By A. Nelson Dingle.

to water-supply problems in all phases of engineering. In addition, the effects associated with clouds and precipitation (e.g., variations of solar heating and lighting due to cloud cover; and traffic problems, drainage, and wind and lightning damage associated with precipitation in its various forms) upon other activities of mankind are common to many operational systems, and need to be considered in their design.

Clouds. Whereas fog may, especially in certain areas of the globe, be formed by the horizontal transport (advection) of warm moist air over a cold surface, most clouds are formed in the atmosphere by nearly adiabatic cooling of vertically displaced moist air. The vertical displacement of cloud-forming air may be systematic in low-pressure (frontal) systems and in situations where the air is forced to move over a rising land surface (*orographic lifting*). In these cases the nature of the cloud formation is relatively homogeneous over a considerable area. Convective displacement, on the contrary, produces nonuniform cloud cover more or less randomly distributed. Its development generally depends upon the stratification of the atmosphere and the intensity of solar heating of the ground on the one hand, or the rate of cooling aloft on the other, either thermal change being capable of producing convective instability.

The nature of the cloud cover in the temperate zones has a distinct seasonal pattern which is superimposed upon regional patterns determined by the proximity of moisture sources and the character of the land surface. One frequently finds a tendency over land toward dominance by convective cloud in summer and fall and by frontal cloud in winter and spring. The variability of the conditions which produce cloud is so great, however, that climatic data must be consulted for each specific location, and meteorological advice is important to smooth operation at any given time.

The physical properties of clouds vary widely, depending on the temperature and on the principal phase of water substance involved. The vapor pressure of water is an approximately exponential function of temperature:

$$e_s \approx 6.11 \times 10^{\frac{at}{t+b}} \tag{5-5}$$

where e_s is the pressure of water vapor in millibars for $-40° \leq t \leq 60°$ at equilibrium with the solid or liquid phase, at temperature $t°$C, and a and b are constants given by

	Over Ice	Water
$a =$	9.5	7.5
$b =$	265.5	237.3

It is to be expected that the largest amounts of water substance occur in clouds formed from relatively warm air. One of the basic cloud parameters is the liquid-water content. Amounts as high as 10 g/cu m have been measured in heavy cumulonimbus clouds, while very thin and cold water-droplet clouds may have less than 1 mg/cu m. The ice content of clouds becomes most important in connection with the precipitation process, in which ice particles originating in the upper, colder parts of the clouds may become intermingled with supercooled water droplets in the lower portions. Droplets of water normally remain liquid, even at temperatures as low as $-35°$C, unless crystallization is initiated by the presence of a solid particle. The commingling of such supercooled water droplets with ice crystals tends to trigger a freezing process and to cause a rapid growth of the ice particles to sizes such that they fall at appreciable velocities. When they exceed 100 microns in diameter, they are arbitrarily classed as precipitation particles and are no longer considered to be cloud particles, principally because of the fall speeds they attain.

Rain. It is clear that the spectrum of drop sizes in a cloud is a very important parameter influencing the precipitation process. Cloud particles are not infrequently as small as 1 micron in diameter, and extend up to an arbitrary maximum diameter of 100 microns. The broadest range of droplet sizes in a single cloud is found in the heavy cumulus congestus clouds that precede thunderstorm formation. Probably the narrowest is found in stratocumulus clouds associated with advective cooling and low-level mixing processes.

An important consideration in problems of aircraft operation is the size of the supercooled droplets at flight levels in clouds. Unless fairly large supercooled drops are available, ice accretion on aircraft cannot proceed at a very great rate because of the low collection efficiency of the aircraft surfaces for small particles. Observations of cloud-droplet-size spectra and liquid-water content are difficult to make, and probably subject to fairly large error. The compilations presented by Mason[8] and Fletcher[4] constitute the best sources of more detailed information in this area.

Although cloud droplets form a continuum with raindrops in terms of size, the formation of clouds is not necessarily always followed by precipitation. Evaporation in the unsaturated air below cloud bases prevents small drops from reaching the ground; since only those drops which fall to earth are called "rain," the size spectra of rain are considered distinct and separate from those of clouds. For illustrative purposes, it is frequently said that 10^6 average-sized cloud droplets are required to make one average-sized raindrop. The processes whereby this transition is made provide the key to the rainmaking function.

The principal mechanisms whereby the requisite water is collected are (1) *accretion* of cloud particles, and *coalescence* of raindrops with smaller raindrops, both of which result from the sweeping action of the relatively larger drops; and (2) *vapor diffusion* from supercooled droplets to ice crystals (see above). It follows that the largest precipitation particles are associated with clouds of great vertical development whereas smaller ones are characteristic of relatively laminar clouds.

Differences in ability to form precipitation-sized particles also are observed between land and sea. Because continental air tends to be laden with a large number of small nuclei of condensation, whereas maritime air tends to have fewer and larger condensation nuclei, the cloud-droplet spectra characteristic of continental air are narrower than those in maritime air. As a result, relatively shallow maritime clouds are capable of producing rain.

In consideration of the scale of the physical processes just discussed, it is clear that rain systems, within both frontal and convective weather systems, are of a scale small compared with that in which weather data are normally presented and climatological data are compiled. An important consequence of this fact is that climatic precipitation data are extremely irregular. For this reason, it is, for example, infeasible to demonstrate statistically significant changes of rainfall in cloud-seeding experiments.[1,9]

The most intense rainfalls occur generally in narrow shafts of rain and have associated with them the largest of the raindrop sizes. The anticipated exception to this general rule may be found in heavy orographic rainfall.

The drop sizes in rain have been observed only for relatively few storms. Despite a theoretical limit of about 5.8 mm for the maximum stable raindrop diameter, the maximum raindrop observed in nature may be as large as about 7 mm in diameter. Generally speaking, large raindrops are broken up by the action of atmospheric turbulence, the aperiodic forces of which are superimposed upon an oscillatory system which results from the interaction of aerodynamic drag, surface tension, and internal circulation in freely falling drops. Strangely enough, even in the heaviest of rains it is rare that a number density of raindrops greater than 1 per liter is encountered. Details of this subject are presented by Dingle.[3]

Snow. There is no satisfactory way of making a representative measure of the depth of a snowfall. Drifting and movement of snow by the wind produce an uneven layer. Where the depth is known accurately enough, there remains the problem of determining its water equivalent. An old rule-of-thumb water-equivalence figure is 10 per cent, i.e., 1 in. of snow equivalent to 0.1 in. of water. In dry snow, this figure is too high; in wet snow, it is too low. A study[13] of snow at Burlington, Vt., reports a range from 2.5 to 22 per cent water equivalence, with a mode at about

9 per cent. This relationship is quite variable from place to place, and from one storm to another.

The heaviest snowfalls are associated with strong winter cyclonic storms or with location in the lee of an open body of relatively warm water. Many impressive case studies of cyclones (e.g., the Middle Western cyclone of Nov. 10–11, 1940; the Ohio Valley storm of Nov. 25, 1950; the inauguration-day storm of 1961) are available. They have in common a pattern of circulation that involves a strong, sustained flux of warm moist air, and unusually slow translation of the cyclonic system across the affected region. The "snowbursts" of the Great Lakes region are typical of the other class of heavy snowstorms. Their characteristics and climatology have been studied by Wiggin.[12]

A major hazard in connection with precipitation from snow is the visibility problem. Because of the unspecified geometry of the aggregations of snowflakes that fall in a heavy storm, the form of the light-scattering function in such a snowfall has not been adequately described theoretically. The effect of backscattered light from the falling snow is frequently observed to reduce the visual range to vanishingly small values, whereas transmitted light helps to increase it.

An important side effect of snow cover on the surface of the earth during winter is its refrigerating effect on the overriding air masses by virtue of its radiative properties. Snow reflects nearly 100 per cent of incident solar radiation on one hand, and radiates nearly as a blackbody in the infrared on the other. Its thermal-insulation value where it accumulates to some depth is very great, and thus it is not accidental that the major fruit-growing areas in the Great Lakes district, for example, are more or less identical with the areas dominated by heavy snows in the wintertime.

Hail. Precipitation in the form of hail is of economic and operational significance in many cases. Because hail depends both for its formation and for its growth to large size upon the presence of strong vertical upcurrents, it is always associated with strong convective storms. Recent studies of hailstorms indicate that the mean updraft over the most intense 4 mile-diameter portion of a 16 mile-diameter storm may reach as high as 20 m/sec.[5] In addition, it is concluded that the hail cloud differs from ordinary thunderstorms especially with regard to the steadiness and persistence of the updraft and the limited areas of downdraft in the hail cloud.[2,11] Damaging hail therefore occurs primarily in spring and early summer, and is limited to rather narrow shafts. It is frequently associated with violent squall-type winds, lightning, and heavy showers of rain.

Weather Modification. Although the possibility of modifying weather by virtue of introducing the ice phase into supercooled clouds has been clearly demonstrated in the laboratory, and even in the free atmosphere, it has never yet been shown conclusively to change in any systematic way the amount of precipitation received at the earth's surface. Elaborate experiments based upon sophisticated statistical concepts have been conducted at great expense for many years, but the methods used for the measurement of the atmospheric variables and the precipitation produced are not capable of showing a positive statistically significant effect. Thus, although it is fairly clear upon considering the physical factors involved that mankind might modify the precipitation regime within rather narrow limits, it has not been proved conclusively that he has actually done this.

This does not mean that weather cannot be modified on a limited scale and for a specific purpose. There are numerous cases of micrometeorological modification, examples of which are found in the frost-prevention program of fruit growers in the citrus areas of the United States. Here, under critical conditions, frost is regularly prevented by means of mechanical and thermal devices of various kinds, especially synthetic opaque "smudge" clouds which affect the radiation balance. Another important form of weather control which can be exerted very effectively is done by engineering the gravitational drainage of cold air away from sensitive areas.[6]

In summary, it has been shown that controls can be exerted effectively over limited areas near the ground by direct intervention, and over somewhat more extensive areas by major engineering projects (e.g., dams impounding large reservoirs); controls over the vast energy reservoirs of the atmosphere by the use of manageable triggering mechanisms to release the energy stored in metastable states (as was specifically envisioned in cloud seeding, see Langmuir[7]) have been shown to be effective, but measurement of the effects has to date been disappointing.[1,9]

REFERENCES

1. Battan, L., and A. R. Kassander, Jr.: Evaluation of Effects of Airborne Silver-iodide Seeding of Convective Clouds, *Sci. Rept.* 18, N.S.F. Grants G-4175, G-5607, and G-8216, University of Arizona, Tucson, 1962.
2. Browning, K. A., and F. H. Ludlam: Radar Analysis of a Hailstorm, *Imperial College, Dept. Meteorol., Tech. Note* 5 (AFCRC-TN-60-426), 1962.
3. Dingle, A. N.: The Microstructure of Rain in a Summer Shower, *Proc. 8th Weather Radar Conf.*, San Francisco, 1960, pp. 99–106.
4. Fletcher, N. H.: *The Physics of Rainclouds*, Cambridge University Press, New York, 1962.
5. Fujita, T., and H. R. Byers: Model of a Hail Cloud as Revealed by Photogrammetric Analysis, *Nubila*, vol. 5, no. 1, pp. 83–105, 1962.
6. Geiger, R.: *The Climate Near the Ground*, 2d ptg. rev., translated by M. N. Stewart and others, Harvard University Press, Cambridge, Mass., 1957.
7. Langmuir, I.: The Production of Rain by a Chain Reaction in Cumulus Clouds at Temperatures above Freezing, *J. Meteorol.*, vol. 5, no. 5, pp. 175–192, 1948.
8. Mason, B. J.: *The Physics of Clouds*, Oxford University Press, New York, 1957.
9. Neyman, J., E. L. Scott, and M. Vasilevskis: Randomized Cloud Seeding in Santa Barbara, *Science*, vol. 131, no. 3407, pp. 1073–1077, 1960.
10. U.S. Air Force, ARDC, Geophysical Research Directorate: *Handbook of Geophysics*, rev. ed., The Macmillan Company, New York, 1960.
11. Weickman, H. (ed.): *Physics of Precipitation*, American Geophysical Union Monograph 5, 1959.
12. Wiggin, B. L.: Great Snows of the Greak Lakes, *Weatherwise*, vol. 3, no. 6, pp. 123–126, 1950.
13. Wilson, W. T.: Density of New Fallen Snow, *U.S. Weather Bureau, Weekly Weather and Crop Bull.*, vol. 42, no. 51, pp. 7–8, Dec. 19, 1955.

5-5. The Atmosphere Near the Earth's Surface*

It is convenient to divide the lower region of the atmosphere into horizontal layers for analysis and description. The lowest region, roughly 50 m in height, is called the atmospheric *surface layer*. Since it is in contact with the earth's surface, flow in this region is dominated by friction, and Coriolis effects are insignificant. For a steady gradient wind of several meters per second it is possible to treat the surface layer as one in which the vertical flux of momentum (i.e., horizontal shear stress per unit area) is constant with height. For a given gradient wind above a constant-flux layer, the wind structure within the layer is determined primarily by the roughness of the surface and the mean vertical gradient of potential temperature.† Henceforth, the word "profile" will be used to mean the average vertical gradient of a variable such as wind or temperature considered over a given time interval.

Between the surface layer and the free atmosphere is a transitional region, varying from 500 to 1,000 m in thickness, in which the frictional effects diminish and the significance of Coriolis effects increases with increase in height. The result is a change in wind direction with height; the wind direction at the top of the transition layer is usually 15 to 45° clockwise from the surface wind direction in the Northern Hemisphere (and counterclockwise in the Southern Hemisphere). It is possible to estimate the momentum flux through this layer from a knowledge of the gradient wind speed and the height at which it becomes gradient, based on the solution first proposed by Ekman[2] for the analogous problem in ocean currents. Because the transitional layer and the surface layer are significantly influenced by friction, the combined region is termed the *planetary boundary layer*, in keeping with standard hydrodynamic terminology.

Surface-layer Characteristics. Knowledge of the surface layer is much more complete than that of the transitional layer, because it is more accessible to measure-

---
* By Donald J. Portman and Edward Ryznar.

† Potential temperature is the temperature which an air parcel assumes when compressed or expanded adiabatically to a pressure of 1,000 mb.

ment. The physical processes of the surface layer have been described in a textbook by Sutton,[8] and a large body of descriptive material for a wide variety of surface and vegetation conditions has been assembled by Geiger.[3] A brief summary of the major characteristics of this layer and the classical approach used in its analysis follow.

Temperature Profiles. The vertical variations of wind, temperature, and humidity are intimately related. Each is basically controlled, however, by the action of the earth's surface as a source or sink of momentum, heat, and water vapor. Because a clear atmosphere is essentially transparent to solar radiation, during the day the surface of the earth warms by collecting heat from the sun and thereby acts as a heat source. At night the reverse situation prevails; the surface loses heat by long-wave radiation at a rate greater than that of the adjacent atmosphere so that the lowest temperature within the surface layer is the ground itself. Clouds exert a pronounced influence on this regime and, combined with wind, tend to reduce day-night differences. However, for other than windy, overcast conditions, the mean temperature structure near the ground may be described as a *lapse* (decrease of temperature with height) in the daytime and an *inversion* (increase of temperature with height) at night. The absolute magnitude of the height derivative decreases with height, the extreme temperatures occur at the ground surface, and the derivative changes sign twice a day, near the times of sunrise and sunset. The diurnal temperature cycle, however, is asymmetric. The highest surface temperature typically occurs near local noon, but the lowest temperature is commonly observed just before sunrise. For windy and cloudy conditions, particularly over rough terrain, the temperature profile tends to the adiabatic lapse rate (0.01°C decrease per meter) and shows little change through a 24-hr period, even though there may be a sizable diurnal temperature change (at all levels).

It is evident that radiation, roughness, and thermal characteristics of the underlying surface have an influence on the temperature structure in the air. In many instances, however, a more important influence is the presence or absence of water at the surface. This is particularly true in the daytime when evaporation exerts the greatest demand on the heat supply.

Flow Properties. Detailed characteristics of temperature in the surface layer depend on turbulence and wind. The flow near the ground may be broadly classified as either laminar or turbulent, and either aerodynamically smooth or rough. On the other hand, as far as heat transfer is concerned, it is convenient to think of free convection, forced convection, or some combination of the two. The most common condition and the one, perhaps, most successfully explored is fully turbulent, forced convection over rough surfaces. Extreme deviations from this condition occur: laminar flow over a smooth snow surface at night (under the stabilizing influence of a strong inversion) is common, while the opposite condition of free convection in the absence of organized horizontal mean flow frequently occurs at midday in the presence of strong lapse conditions. It is likely, furthermore, that fully turbulent flow at moderate speeds over smooth snow, ice, and water may often be classified as "smooth." The transition between fully forced and free convection is a significant one in terms of analysis and is, in effect, usually treated in terms of the Richardson number (see Ref. 8, p. 153) or a similar parameter indicative of the ratio of the buoyancy forces to the shear forces.

Shear Stress in Neutral Conditions. When buoyancy forces are insignificant and the flow is fully turbulent it is found that the mean wind flow in the surface layer, as proposed by Prandtl,[5] obeys the well-known logarithmic law. Thus

$$\bar{u}(z) = u_* k^{-1} \ln \frac{z}{z_0} \qquad z > z_0 \tag{5-6}$$

in which u_* = friction velocity, cm/sec
$\qquad\quad = (\tau/\rho)^{1/2}$, where τ = horizontal shear stress per unit area, g/(cm)(sec²), and ρ = density, g/cu cm
$\qquad k$ = the von Karman number $\cong 0.4$
$\qquad z$ = height, cm
$\qquad z_0$ = a reference height characterizing surface roughness, cm

Prandtl derived this relationship from the equations of motion with the important assumption that the shear stress was invariant with height. The development includes, also, the concept of the *mixing length* and the assumption that it is proportional to the height.

For practical application of the logarithmic relationship it is convenient to define a drag coefficient

$$C_D = \left(\frac{u_*}{\bar{u}}\right)^2$$

C_D is obviously related to the roughness but can be specifically identified for a given surface only in terms of wind speed measured at a given height. Deacon[1] gives roughness and drag-coefficient data for fully turbulent flow when buoyancy forces are negligible (Table 5-3). His drag-coefficient data are given in terms of the wind speed at 2 m, $\bar{u}(2)$.

Table 5-3. Roughness Parameters of Various Surfaces

Type of surface	z_0, cm	C_D $\left(\dfrac{u_*}{\bar{u}(2)}\right)^2$
Smooth mud flats	0.001	0.0010
Smooth snow on short grass	0.005	0.0012
Desert (Pakistan)	0.03	0.0020
Snow surface, natural prairie	0.10	0.0028
Mown grass:		
1.5 cm	0.2	0.0034
3.0 cm	0.7	0.005
4.5 cm		
$\bar{u}(2) = 2$ m/sec	2.4	0.0085
$\bar{u}(2) = 6$–8 m/sec	1.7	0.007
Long grass, 60–70 cm:		
$\bar{u}(2) = 1.5$ m/sec	9.0	0.016
$\bar{u}(2) = 3.5$ m/sec	6.1	0.013
$\bar{u}(2) = 6.2$ m/sec	3.7	0.010

Turbulent Transfer in Nonneutral Conditions. With very careful and detailed observations, Thornthwaite and Kaser[9] first showed that the logarithmic wind profile observed with neutral thermal stability constituted only a transitional form between thermally unstable (lapse) and stable (inversion) stratification. In a $\bar{u} - \ln z$ diagram it is found that the wind profile is convex toward increasing $\bar{u}$ for lapse conditions and toward decreasing $\bar{u}$ for inversion conditions. Monin and Obukhov[4] have presented, to date, the most satisfactory analysis to account for the interdependency of the wind and temperature fields. (See Sheppard[7] for a summary and discussion of this work.) Under the assumptions of similarity in wind and temperature profiles, and identity of turbulent transfer coefficients for heat and momentum, they derived the relationships

$$\bar{u}(z) = \frac{u_*}{k}\left(\ln \frac{z}{z_0} + \alpha \frac{z}{L}\right) \tag{5-7}$$

in which L is a length related to the stability in terms of the heat flux and shear stress, and α is presumably a universal constant. The relationship can be valid only for $z \ll L$. The equation can be used to describe the vertical profile of any conservative property whose presence does not influence the turbulent motion and whose transport process is dynamically similar to that of momentum. For relatively weak density stratification it may be assumed that buoyancy forces are insignificant in comparison with shear forces, and a relationship similar to (5-7) for potential temperature may be applied. Thus with measurements of wind and temperature profiles one is able to derive the quantity u_* after elimination of L from the two equations. Turbulent heat flux may then be computed following the definition of a turbulent transfer coefficient $K = ku_*z$. Only limited success has been achieved with these relationships.

For an informative summary of free convection processes in the lower atmosphere and of the associated temperature field, the reader is referred to Priestley.[6]

REFERENCES

1. Deacon, E. L.: Vertical Profiles of Mean Wind in the Surface Layers of the Atmosphere, *Geophys. Mem.* 91, London, Air Ministry, Meteorology Office, 1953.
2. Ekman, V. W.: On the Influence of the Earth's Rotation on Ocean Currents, *Arkiv. Mat. Astr. Fysik.*, vol. 2, nos. 1–2, 1905.
3. Geiger, R.: *The Climate Near the Ground*, Harvard University Press, Cambridge, Mass., 1950.
4. Monin, A. S., and A. M. Obukhov: Basic Laws of Turbulent Exchange in the Surface Layer of the Atmosphere, *U.S.S.R. Acad. Sci. Geophys. Inst.*, no. 24, 1954.
5. Prandtl, L.: Meteorologische Anivendung der Stromungslehre, *Beitr. physik. fr. Atmos.*, vol. 19, pp. 188–202, 1932.
6. Priestley, C. H. B.: *Turbulent Transfer in the Lower Atmosphere*, University of Chicago Press, Chicago, 1959.
7. Sheppard, P. A.: Transfer across the Earth's Surface and through the Air Above, *Quart. J. Roy. Meteorol. Soc.*, vol. 84, no. 361, pp. 205–224, 1958.
8. Sutton, O. G.: *Micrometeorology*, McGraw-Hill Book Company, New York, 1953.
9. Thornthwaite, C. W., and P. Kaser: Wind-gradient Observations, *Trans. Am. Geophys. Union*, vol. 24, pp. 168–182, 1943.

5-6. Synoptic Meteorology and Weather Forecasting*

Data Acquisition, Processing, and Transmission. Synoptic meteorology can be defined as the study and analysis of meteorological data obtained simultaneously over a wide area in order to present a comprehensive and nearly instantaneous picture of the state of the atmosphere. Observations to describe the atmosphere are taken and recorded at numerous local stations, coded and prepared for transmission, and sent by land line or radio to other stations and to weather-analysis centers; at the latter, they are plotted on charts, analyzed, and used as the basis for making predictions of the weather. This process is performed on an international scale in such a manner that all countries can understand the observations, analyses, and forecasts.

Many types of meteorological observations are taken, but in general they can be separated into surface and upper-air observations. Surface observations are made at or near ground or sea level using instruments based at the same level. These observations are taken at 0000, 0600, 1200, and 1800, Greenwich Mean Time (GMT), at land stations and aboard certain ships. Details concerning the variables observed and how they are to be observed can be found in Ref. 1. In addition to the scheduled observations, some stations take 3-hourly observations spaced midway between the regular synoptic reports. At many stations throughout the world, airways weather observations are taken every hour, or more often if significant changes occur.

The upper-air observations are made by means of instruments sent into the atmosphere or by observers aboard aircraft. These observations are begun at 0000 and 1200 GMT, although there are selected stations which also take observations regularly at 0600 and 1800 GMT. Details of the radiosonde instrumentation used in obtaining the upper-air data are found in Ref. 2.

Data observed at synoptic times are recorded at the observing station on appropriate forms and on punch cards for future use. Copies of the recorded data are sent to a central point in each country for microfilming and retention as well as for statistical and climatological analyses. The United States center is called the National Weather Records Center and is located at Asheville, N.C. The services and availability of data from this center are described in Sec. 5-7.

The observed data are also coded according to a format promulgated by the World

---
* By Eugene W. Bierly.

Meteorological Organization (WMO).[3] These data are then transmitted according to prescribed schedules,[4,5] so that maximum benefit can be gained from the observations in a minimum amount of time.

Analysis Techniques. The coded observational data are received by many individual stations as well as by analysis centers. At individual stations, the data are decoded and plotted manually on charts, again in a manner set forth by WMO. The plotted charts are then analyzed by meteorologists so that an accurate and detailed picture of the state of the atmosphere can be ascertained. Some of the charts (for example, the surface chart) are analyzed for the pressure field, frontal intersections with the surface, and areas of precipitation and hazards to air navigations, such as fog and thundershowers. Others (for example, upper-level charts) are analyzed for the height of a constant-pressure surface above sea level, temperature distribution, and indirectly the wind field. Some charts show amounts and kinds of precipitation, observed winds at constant heights above sea level, stability indices, pressure change, radar echoes, present weather and cloud amounts, height of the freezing level above the surface, maximum and minimum temperatures, and so on.

At analysis centers, such as the National Meteorological Center (NMC) at Suitland, Md., the coded observations become direct inputs to a digital computer. The computer makes hydrostatic and continuity checks on the data so that only accurate data are included in the analysis. Pressure, temperature, moisture, and wind fields of the initial data are plotted and analyzed automatically, using interpolated data at grid points. The NMC grid in use today contains several thousand grid points and covers all the Northern Hemisphere from the north pole to about 9°N latitude.

Forecasting Procedures. At stations without computer facilities, forecasts are usually made by an experienced meteorologist. The resulting forecast typically is based upon the predicted upper-level height fields and the forecaster's predicted depiction of the surface chart which corresponds to the upper-air movements. The specific forecasts of temperature, precipitation, winds, and clouds are made for individual points or general areas for periods up to 24 hr. The longer the forecast period, the less detail can be given. Also, the larger the area the forecast covers, the more general the forecast has to be.

When a computer is utilized, the initial fields from the observations are advanced by short intervals of time according to one of several atmospheric models until forecast charts for 12, 24, 36, 48, or 72 hr are produced. The current atmospheric model used by the NMC is a modified three-level baroclinic model, that is, a model in which the surfaces of constant density intersect surfaces of constant pressure. Several other models are being tested currently, with the trend pointing toward the more basic or primitive equations of motion which describe the atmosphere and its movement.

At present only upper-level charts are produced by the NMC computer, although the U.S. Navy Numerical Prediction Unit at Monterey, Calif., does produce forecast sea-level charts as well as sea and swell forecasts on a routine basis. Computer prognostic charts of vorticity and vertical motion are also prepared; they are very useful for quantitative precipitation forecasting.

The official position of the American Meteorological Society is quoted as follows to give the reader an idea of what may be expected from weather forecasting.[6]

The American Meteorological Society feels that the preparation of acceptable forecasts requires professionally trained personnel. Forecasts prepared by people so qualified can be expected to achieve the following levels of skill and usefulness:

For periods up to 24 hours, skillful weather forecasts of considerable usefulness are possible. Within this interval detailed weather and weather changes can be predicted. Hour-to-hour variations can be predicted during the early part of the period.

For periods extending to about 72 hours, weather forecasts of moderate skill and usefulness are possible. Within this interval, useful predictions of general trends and weather changes can be made.

Average weather conditions for periods of about a week can be predicted with some skill. Day-to-day or week-to-week forecasts within this time period have not demonstrated skill.

Forecasts for periods of more than one month in advance must be considered as experimental. Although promising research is in progress, skill has not yet been demonstrated.

Until such weather outlooks are of proven value, useful climatological information can be provided in their stead.

Active research is in progress to improve forecasting skill for all time scales and for all altitudes in the atmosphere.

REFERENCES

1. Manual of Surface Observations (WBAN), Circular N, 7th ed. including changes 1–12, U.S. Government Printing Office, January, 1964.
2. Manual of Radiosonde Observations (WBAN), Circular P, 7th ed. including change A-4, U.S. Government Printing Office, May 1, 1964.
3. World Meteorological Organization: Codes, *Weather Reports*, WMO-9TP4, vol. B, Secretariat of the WMO, Geneva, Switzerland, 1954.
4. World Meteorological Organization: Transmissions, *Weather Reports*, WMO-9TP4, vol. C, Secretariat of the WMO, Geneva, Switzerland, 1954.
5. Federal Aviation Agency, Weather Schedules, Air Traffic Service Manual, Oct. 29, 1963.
6. Policy Statements of the AMS, quoted from statement on weather forecasting, *Bull. Am. Meteorol. Soc.*, vol. 44, no. 7, p. 446, July, 1963.

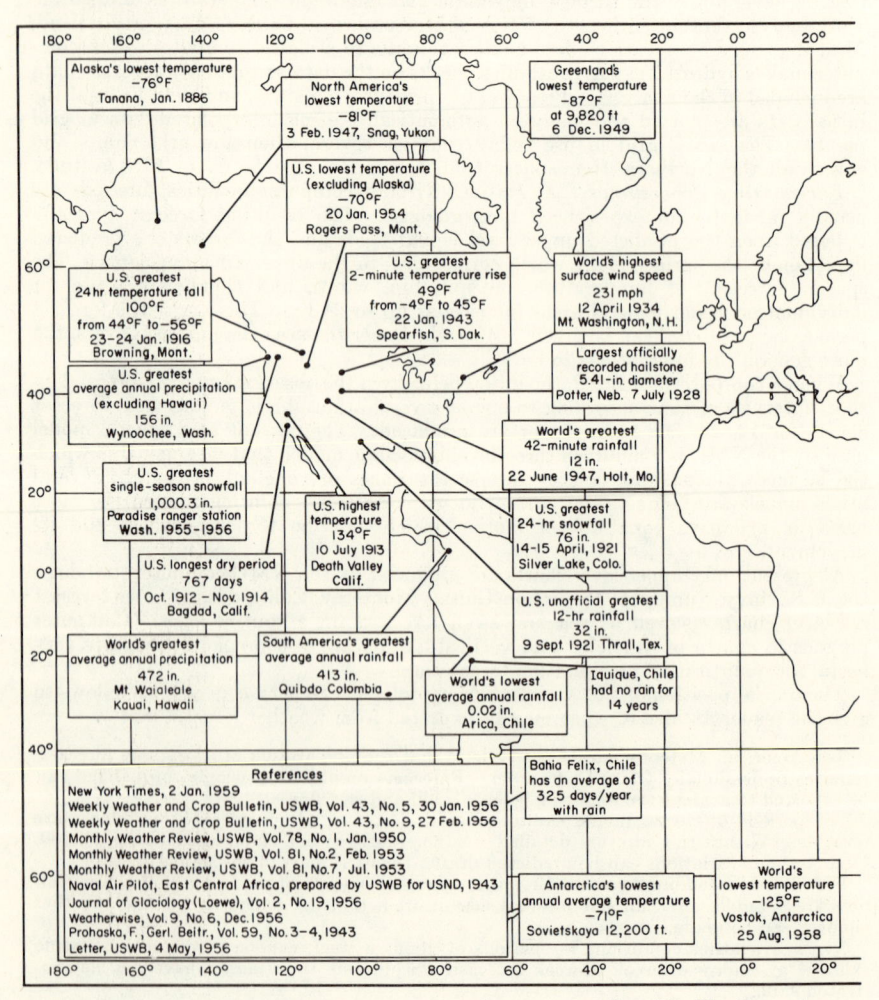

FIG. 5-2. Weather extremes

5-7. Statistical Properties: Data Sources and Services*

Statistical Properties. The statistical properties of the lower atmosphere are being portrayed graphically and cartographically in increasing amount as more measurements become available. Numerous representations applicable to a variety of engineering problems are given by the U.S. Air Force.[4] An earlier general reference work is that of Berry, Bollay, and Beers.[3] Another recent example of a specialized application is the monograph by Bean, Horn, and Ozanich.[2]

To summarize earlier discussions, characteristics of the lower atmosphere, particularly the lower troposphere, are largely determined by the elevation of the sun and the complex interaction of solar radiation with the earth's surface. Significant diurnal and seasonal variations of many meteorological elements result. Both diurnal and seasonal variations of surface temperature are greater over land than

* By Norman L. Canfield.

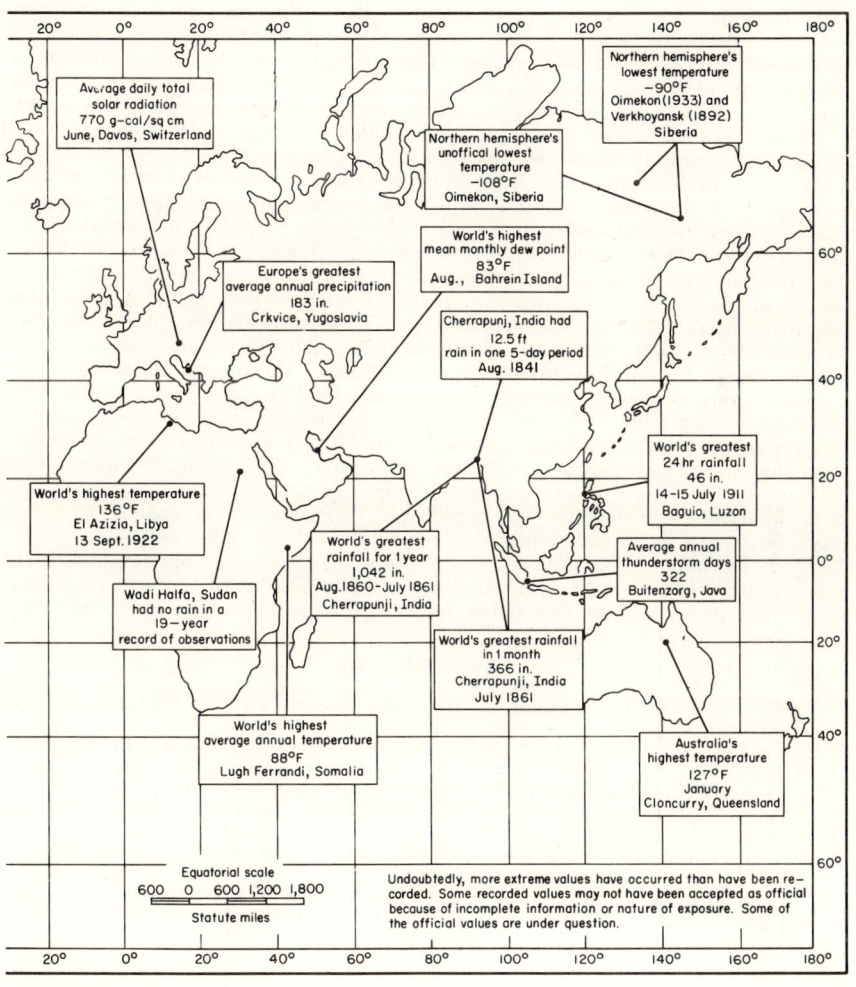

around the world.

over water. Cloudiness and precipitation are greatly affected by topography and proximity to the moisture sources of oceans and large lakes. These are but examples, qualitatively expressed, of a lower atmosphere whose time and space variations are so complex as to require statistical evaluation. Some extremes of familiar atmospheric variables, as observed at the earth's surface, are shown in Fig. 5-2.

Data Sources and Services. Modern engineering problems often are sufficiently varied to limit the utility of climatic summaries or charts previously compiled. Thus new compilations may be needed to meet new problems. In the United States, atmospheric-analysis assistance for the specialized purposes of private enterprise are available from a number of consulting meteorologists and universities recognized by the American Meteorological Society. Analysis services for public or governmental projects can likewise be obtained from these sources and are also available from the U.S. Air Weather Service, U.S. Naval Weather Service, and U.S. Weather Bureau.

A primary source of basic data for the lower atmosphere is the National Weather Records Center, Asheville, N.C. Here is located the official repository for meteorological measurements and visual observations made by the Federal weather services. The center's basic data and previously compiled summaries are available in one or more of several forms, notably typescript, manuscript, microfilm, microcard, 80-column punched card, and magnetic tape. The data are available upon request and at cost of reproduction and/or processing. Machine programming, computer, and analysis services supplement the information and records services of the center.

Some details of the center's activities are given by Barger.[1] This booklet's final pages outline procedures and pitfalls in requesting services.

A considerable amount of hourly, daily, weekly, and monthly data are routinely published by the center. Most is in tabular format but some daily, weekly, and monthly data are analyzed in map form. Most cover only the United States but a few cover the Northern Hemisphere or the world. Each is illustrated in Ref. 5, and brief descriptions of major data periodicals are printed each month on the inside covers of the U.S. Weather Bureau's *Monthly Weather Review*. Of other periodicals reporting results of atmospheric research, one emphasizes papers relating the lower atmosphere to engineering and other specialties. This is the *Journal of Applied Meteorology*, a bimonthly publication of the American Meteorological Society.

Routinely published meteorological data are helpful in many problems where the atmospheric environment is a factor. But the fact remains that new problems often require new computations using data which are already available. Data processing and analysis are relatively straightforward where the problems require only measurements from over the United States or Canada. For other areas of the world, data availability and completeness vary considerably. Processing and analysis costs may be several times those for comparable work with North American data.

Statistical properties of the natural atmospheric environment are relevant to many current problems of engineering and other fields. They will become even more important to evaluations that follow further experiments in weather modification.

REFERENCES

1. Barger, G. L. (ed.): *Climatology at Work*, Government Printing Office, Washington, D.C., 1960.
2. Bean, B. R., J. D. Horn, and A. M. Ozanich, Jr.: *Climatic Charts and Data of the Radio Refractive Index for the United States and the World*, National Bureau of Standards Monograph 22, Government Printing Office, Washington, D.C., 1960.
3. Berry, F. A., E. Bollay, and N. R. Beers: *Handbook of Meteorology*, McGraw-Hill Book Company, New York, 1945.
4. U.S. Air Force, Geophysics Research Directorate: *Handbook of Geophysics*, rev. ed., The Macmillan Company, New York, 1960.
5. U.S. Weather Bureau: *Selective Guide to Published Climatic Data Sources*, Key to Meteorological Records Documentation no. 4.11, Government Printing Office, Washington, D.C., 1963.

Chapter 6

THE UPPER ATMOSPHERE

LESLIE M. JONES, *Department of Aeronautical and Astronautical Engineering, University of Michigan.*

CONTENTS

6-1. Atmospheric Sciences	6-2
6-2. Atmospheric Regions	6-2
6-3. Characteristics of Structural Regions	6-2
Troposphere	6-3
Stratosphere	6-3
Mesosphere	6-3
Thermosphere	6-3
Exosphere	6-3
6-4. Characteristics of Composition Regions	6-4
Homosphere	6-4
Heterosphere	6-5
6-5. The Structure of the Atmosphere	6-5
Scale Height	6-6
Geopotential Altitude	6-6
Molecular Scale Temperature	6-7
Mean Free Path	6-8
Number Density (Concentration)	6-8
Particle Velocity (Speed)	6-8
Collision Frequency	6-8
Sound Velocity (Speed)	6-8
Viscosity	6-8
6-6. Standard Atmospheres	6-9
6-7. Large-scale Variations	6-10
6-8. Aerodynamic Regimes	6-12
Continuum Flow	6-13
Slip Flow	6-13
Transition Flow	6-13
Free Molecule Flow	6-13
6-9. The Ionosphere	6-13
D Region	6-14
E Region	6-15
F Region	6-15

6-1. Atmospheric Sciences

The advent of rockets and satellites has brought the entire atmosphere within reach of measuring instruments. As a first consequence it is now possible to describe the *structure* or gross physical nature of the atmosphere from the surface to hundreds of kilometers. Meteorology is concerned primarily with physical phenomena in the troposphere and lower stratosphere up to 32 Km and is the subject of the preceding chapter. Data in statistical amounts from surface and balloon instruments are available in this region. It is the altitude limit of balloons which has in the past set the upper bound of meteorology. Recently, however, the meteorological significance of events in the upper stratosphere and in the mesosphere has been studied with rockets. In the United States a network of stations providing meteorological data into the mesosphere on a limited synoptic basis has been established. Small standardized rockets are used in this work.

Meteorological satellites such as TIROS (see Chapter 17 and Fig. 17-12), are now giving cloud-cover photographs and have revitalized nephanalysis as a meteorological tool. Further stimulation of methods of inferring meteorological parameters from remote sensing of radiation from the earth and atmosphere may be expected from satellite methods now under development.

Above the mesosphere the photochemical reactions of solar corpuscles and radiation with the atmosphere become significant. The study of the resulting ionization, dissociation, new-compound formation, and the effect of these on atmospheric structure is the province of the science of *aeronomy*. Aeronomical measurements *in situ* have been made with rockets and satellites and, in the case of the ionosphere, with ground instruments in sufficient numbers to describe that region in some detail and also to permit formulation of a tentative standard structural atmosphere to 700 Km.

Measurements out to several earth radii to investigate the reactions among energetic particles (cosmic rays), the earth's magnetic field, and the protonosphere have introduced a new discipline in high-altitude plasma physics following upon the discovery of the Van Allen belts in 1958. These measurements have not, however, been paralleled with comparable structural measurements; so that the magnitudes of the physical parameters of the atmosphere above 700 Km remain a research problem rather than constituting a body of handbook data; these matters are discussed in the following chapter.

6-2. Atmospheric Regions

The most common division of the atmosphere is into several regions based on the temperature vs. altitude profile. Temperature is one of the principal *structural* parameters of the atmosphere, the others being density, pressure, mean molecular weight, and motion (winds and turbulence). Names are also given to certain regions in accordance with their composition. Because of the interdependence of temperature and composition, the boundaries of these regions are also conveniently related to the temperature profile.

The thin layer at the top of a region which marks a cessation in the characteristics of that region carries the suffix *pause*, e.g., *tropopause*.

The ionosphere is that portion of the atmosphere which contains sufficient concentrations of free electrons (and ions) to affect radio-wave propagation. The electron density vs. altitude profile of the ionosphere is a continuous envelope which encompasses several structural regions and which exhibits peaks or inflections due to a tendency to form layers. These features constitute *regions* of the ionosphere which are identified by letters. The ions in the ionosphere vary in their proportion to the corresponding neutral species but do not exceed 1 per cent up to 700 Km.

Figures 6-1 to 6-3 show the principal structural and composition regions of the atmosphere. The altitude bounds and nomenclature of the structural regions follow Nicolet.[1] For a description of the temperature, density, and composition curves, see Sec. 6-6. For a discussion of variations, see Sec. 6-7. The ionosphere is discussed in Sec. 6-9. All curves represent conditions at 45° north latitude at the mean of the solar cycle, except for variations shown in Figs. 6-3 and 6-4.

6-3. Characteristics of Structural Regions

Troposphere. The region from the surface to the tropopause at which height the negative temperature gradient of about 6.5°K/Km ceases. The height of the tropopause varies from 6 Km in polar regions to 18 in the tropics. The troposphere is in convective temperature equilibrium with the surface of the earth and is the region of clouds, precipitation, winds, and other phenomena which constitute weather. For most vehicles, the troposphere is an aerodynamic region of continuum flow (see Sec. 6-8).

Stratosphere. The region from the tropopause to the stratopause at about 50 Km where there is a temperature maximum. The base of the stratosphere (except at low latitudes) is isothermal but the predominant temperature characteristic is a positive gradient. The stratosphere contains significant amounts of ozone, which reaches a peak concentration of 1 part in 10^5 in the 30-Km region. Even here, however, the ozone is not sufficient to change the mean molecular weight, which is the same throughout the stratosphere as at the surface. Winds in the stratosphere seem to follow in general the large-scale patterns of those in the troposphere. The winds are generally zonal in both winter and summer. In the Northern Hemisphere the winds are west in winter but reverse in a complex and often abrupt transition to east in summer. For vehicles, the stratosphere is an aerodynamic region of continuum flow, with slip flow occurring near the stratopause.

Mesosphere. The region of negative temperature gradient between the stratosphere and mesopause. The mesosphere is characterized by turbulence and by winds which are generally of higher velocity than in the stratosphere but which follow similar patterns. Photochemical reactions which affect the temperature and which give rise to airglow and ionospheric phenomena take place in the mesosphere. The gross composition and mean molecular weight, however, are the same as at the surface. Vehicles in the mesosphere for the most part encounter slip flow but will enter the transition regime in the upper mesosphere and through the mesopause.

Thermosphere. The region of increasing temperature above the mesopause which is the last temperature minimum in the atmosphere. The thermal gradient is large from 90 to 190 Km, and then tends to essentially isothermal near 300 Km, the thermopause. The thermosphere is relatively quiescent, permitting diffusive separation which, together with the dissociation of oxygen, causes the mean molecular weight to decrease. The thermosphere encompasses several regions of change in composition noted below. The thermosphere includes two aerodynamic regions for vehicles: transition flow to perhaps 140 Km and free-molecule flow above.

Exosphere. The region of the atmosphere in which the upward mean free path of neutral particles permits them to have ballistic trajectories and those with sufficient thermal energy to escape. Because the escape levels are deep and are different for different gases and for ions and electrons, the location of the exosphere can only be

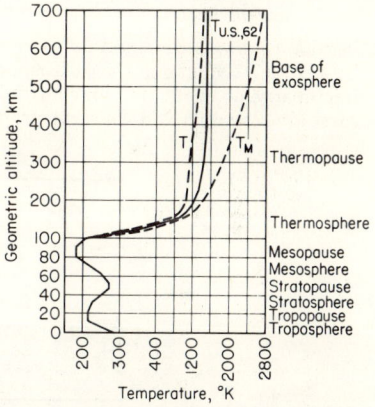

FIG. 6-1. The temperature profile of the atmosphere and structural regions related thereto. $T_{U.S.,62}$ is the kinetic temperature of the *U.S. Standard Atmosphere, 1962*, and T_M the molecular scale temperature of the standard. T is the kinetic temperature of an atmosphere having the same density as the standard but with a mean molecular weight based on recent rocket results.

broadly defined. The base of the exosphere may be taken as 500 Km. The exosphere may be considered to merge with the corona of the sun at an altitude which remains speculative. Analytic treatment of the exosphere is difficult because neither the equations for a continuous medium nor the collisionless Boltzmann equation apply throughout. Some success[2] has been achieved in using the Boltzmann equation with values given to the collision term by asymptotically extending the conditions of the continuous lower atmosphere.

6-4. Characteristics of Composition Regions

Homosphere. The region of the atmosphere up to about 90 Km, wherein the composition and the mean molecular weight are the same as at sea level; this composition is shown in Tables 6-1 and 6-2. At the homopause the composition changes

Table 6-1. Principal Gases in Air

Gas	Mass	% by volume
N_2	28.02	78.08
O_2	32.01	20.94
A	39.96	0.934
Air	28.97	100.00

because of dissociation and diffusive separation. The latter phenomenon is overcome by mixing throughout the homosphere. The cessation of mixing is marked by the mictopause at about the same altitude as the homopause. The noble gases, although rare, are also mixed in constant proportion throughout the homosphere.

Table 6-2. Inert Gases

Gas	Mass No.	ppm
He	4	5.24
Ne	20	18.1
A	40	934
Kr	84	1.14
Xe	132	0.087

Carbon dioxide (44) is relatively constant in the homosphere at about 300 ppm, but near-the-surface decreases of 10 per cent or less are noted over sea water and increases of 10 per cent or less are found in industrial areas.

Water vapor (18) is extremely variable in the atmosphere, ranging from 10 to 10,000 ppm and depending on weather factors, primarily temperature. As a consequence of the latter, water vapor decreases in the stratosphere. Water vapor dissociates in the mesosphere, but not completely as it has been detected higher up.

Ozone in small variable amounts is found near the surface, where it probably appears by the photochemical breakdown of pollutants. In the stratosphere absorbed solar ultraviolet dissociates a fraction of the oxygen, permitting the formation of ozone, which in turn absorbs in the ultraviolet. An ozone peak of 10 ppm in the 20- to 30-Km region exists, but above this altitude the ozone partial pressure decreases more rapidly than the total.

Some chemically active minor constituents are mixed and constant in the troposphere but probably dissociate in the mesosphere and stratosphere. These are CH_4 (1.5 ppm), N_2O (0.5 ppm). Traces of CO, NO_2, and NO have also been detected in the atmosphere.

THE UPPER ATMOSPHERE

Heterosphere. The region where the mean molecular weight decreases because of the combined effect of diffusive separation and photochemical dissociation. The dissociation and diffusion of oxygen start in the 90-Km region. Atomic oxygen exceeds molecular oxygen at about 120 Km, and exceeds molecular nitrogen at perhaps 190 Km, thus becoming the major constituent. Helium, diffusing upward from the surface, and hydrogen, formed in the 100-Km region by the photodissociation of water vapor and methane and also diffusing upward, increase in relative proportion to the heavier gases. Nicolet[3] has discussed the importance of helium as an atmospheric constituent, and Hanson[4] finds that at 1,200 Km helium exceeds oxygen to become the principal constituent, but is in turn surpassed by hydrogen at 5,000 Km. The hydrogen may be as much as 50 per cent ionized (protons) at this altitude and more so above, but the region is nevertheless distinguished from the ionosphere and is called the protonosphere or protosphere. It constitutes a geocorona to the earth and resembles the solar corona with which it eventually merges (Chapter 7).

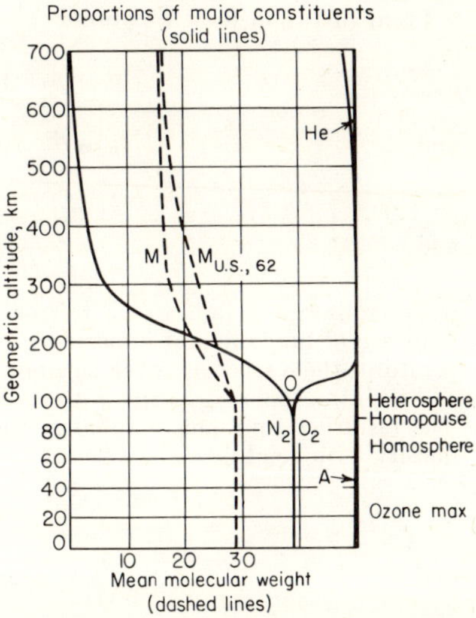

FIG. 6-2. Proportional composition diagram of the atmosphere based on recent rocket measurements. The relative concentrations of N_2, O_2, O, A, and He at any altitude are equal to the relative lengths of segments of a horizontal line at the desired altitude intersecting the solid lines. Thus at 200 Km the atmosphere is 0.47 N_2 and 0.53 O. The dashed lines are the mean molecular weights as read on the lower abscissa as a function of altitude. M is based on recent rocket results and corresponds to the proportional composition diagram given by the solid lines. $M_{U.S., 62}$ is for the *U.S. Standard Atmosphere, 1962*, for which a proportional composition diagram has not been published.

6-5. The Structure of the Atmosphere

The structure of the atmosphere is described by the four principal parameters: pressure p, kinetic temperature T, mass density ρ, and mean molecular weight M,* the latter being one aspect of the larger problem of composition. Motion of the atmosphere (winds and turbulence) is also a structural phenomenon in that motion often may be derived from, or may interfere with, experimental methods for p, T, ρ, and M. Secondary structural parameters are scale height, concentration, mean free path, particle velocity, sound velocity, viscosity, etc.

Two equations relate the four principal parameters. These are the hydrostatic equation

$$dp = -\rho g\, dz \tag{6-1}$$

and the equation of state for an ideal gas

$$pV = RT \tag{6-2}$$

or

$$p = \frac{\rho RT}{M} \tag{6-3}$$

where g is the acceleration of gravity, z the geometric altitude, V the mole volume, and R the universal gas constant.

---

* In the equations of this chapter, M is the gram-molecular weight or the weight of 1 mole of gas in grams. M for air = 28.966.

In general, two of the four parameters must be known to calculate the other two. However, in the region from the surface to 90 Km it has been established that the mean molecular weight is the same as at sea level; so that a knowledge of one is sufficient.

From (6-1) and (6-3) we obtain

$$p = p_0 \exp\left(\frac{-Mg}{R} \int_{z_0}^{z} \frac{dz}{T}\right) \tag{6-4}$$

and

$$\rho = \frac{\rho_0 T_0}{T} \exp\left(\frac{-Mg}{R} \int_{z_0}^{z} \frac{dz}{T}\right) \tag{6-5}$$

and

$$T = \frac{\int_{z_0}^{z} -\rho g \, dz}{\rho R/M} + \frac{\rho_0}{\rho} T_0 \tag{6-6}$$

where subscript 0 indicates a reference level, often sea level. Temperature variations are of fundamental importance in aeronomy, but in the absolute scale of temperature, the variations over substantial altitude intervals are not extremely large and the atmosphere or parts of it may be taken as isothermal for gross considerations. The resulting equations emphasize the basic exponential variation of pressure and density with height.

$$p = p_0 \exp \frac{-(z - z_0)Mg}{RT} \tag{6-7}$$

and

$$\rho = \rho_0 \exp \frac{-(z - z_0)Mg}{RT} \tag{6-8}$$

Certain derived parameters are useful in aeronomy, and values for them are often given in the model, reference, and standard atmospheres.

Scale Height. Further simplification is obtained by substituting the scale height $H = RT/Mg$ in (6-7) and (6-8). Scale height is a convenient concept for visualizing the effect of temperature. In an isothermal region it is the altitude interval in which the pressure and density vary by a factor of e, this interval being larger for a warm atmosphere (or region) than for a cool one. Scale height is also the thickness of a uniform isothermal atmosphere having the same pressure and density as the base of the exponential atmosphere. A definition of scale height is the negative ratio of the parameter to its gradient. In a nonisothermal atmosphere the pressure and density scale heights are different, the former being equal to the atmospheric scale height. Thus

$$H_p = \frac{-p}{dp/dz} = H = RT/Mg \tag{6-9}$$

$$H_\rho = \frac{-\rho}{d\rho/dz} \neq H \tag{6-10}$$

The two scale heights are related by

$$H_\rho = \frac{H_p}{1 + dH_p/dz - 2H_p/(a + z)} \tag{6-11}$$

where a is the radius of the earth.

Geopotential Altitude. In the foregoing equations the acceleration of gravity has been treated as constant with height. To a close degree of approximation, however,

$$g = g_0 \left(\frac{a}{a + z}\right)^2 \tag{6-12}$$

A modified measure of height called geopotential altitude h may be derived in which a change in height of 1 geopotential meter at any altitude results in a change of potential energy of 9.807×10^{-3} joules/g. The relationship between geometric height and geopotential height is given by

$$h = \frac{az}{a + z} \qquad (6\text{-}13)$$

The magnitudes of the differences between z and h at various altitudes are seen in Table 6-3. Height units of potential have advantages over units of distance in some calculations. Logarithmic plots of Eqs. (6-7) and (6-8), for example, will be straight lines in units of geopotential altitude. Errors in the foregoing equations due to neglecting the gravitational attraction of the mass of the atmosphere and "fictitious" forces of the rotating earth and atmosphere are negligible in most cases.

Molecular Scale Temperature. Composition vs. altitude diagrams, such as that of Fig. 6-2, while based to some extent on measurements, have been quite speculative above 90 Km. The consequent uncertainty in the mean molecular weight M has not only prevented the accurate calculation of temperature from experimental pressure and density values (as from satellite drag) but also has caused difficulty in comparing results of calculations using different M's. As an expedient in such situations the concept of molecular scale temperature T_M is used.

$$T_M = T \frac{M_0}{M} \qquad (6\text{-}14)$$

M_0 is the molecular weight at sea level (28.97) and M the uncertain molecular weight at altitude.

The effect is to give temperatures throughout the atmosphere normalized to sea-level molecular weight. The values may then be corrected by (6-14) to true kinetic temperatures for any molecular-weight profile which becomes available. Recent measurements of composition have reduced the uncertainty in M and tend to eliminate the use of T_M.

Mean Free Path. Mean free path is defined as the average distance which a molecule or atom of gas travels between consecutive collisions with other molecules or atoms. For a Maxwellian distribution of velocities,

$$l = \frac{1}{n\sqrt{2}\,\pi\sigma^2} \qquad (6\text{-}15)$$

where n = number density
σ = average collision diameter = 3.65×10^{-8} cm for air
At sea level $l = 6.63 \times 10^{-6}$ cm.

Number Density (Concentration). Number density is the number of neutral atmospheric particles per unit volume. *Electron density* and *ion density* are used separately for these nearly equal concentrations of charged particles, neither of which exceeds 1 per cent of the number density at any point up to the exosphere. The number density $n = Np/RT$ where N is Avogadro's number.

Particle Velocity (Speed). Particle velocity is usually the arithmetic mean velocity of neutral particles in a volume at constant pressure and temperature. It is given by $\bar{u} = (8RT/\pi M)^{1/2}$. For this relation to hold, the velocity distribution must be Maxwellian, in which case the mean velocity $\bar{u}$, the root-mean-square velocity u_{rms}, and the most probable velocity u_w are in the ratio

$$u_{\text{rms}} : \bar{u} : u_w = \sqrt{3/2} : \sqrt{4/\pi} : 1 = 1.22 : 1.13 : 1$$

Collision Frequency. The collision frequency ν for neutral particles is obtained by dividing the mean particle velocity by the mean free path.

$$\nu = 4n\sigma^2 \sqrt{\frac{\pi RT}{M}}$$

Table 6-3. Defining Temperature and Molecular Weights of the U.S. Standard Atmosphere, 1962, and Computed Pressures and Densities (Condensed Version)*

z, Km	h, Km	T_M, °K	L, °K/Km	M	T, °K	p, mb	ρ, g/cu m	M	T, °K
								\multicolumn{2}{c	}{Molecular weights and kinetic temperatures based on recent rocket results[12]}
0.000	0.000	288.15	−6.5	28.966	288.15	10.1325 + 2†	1.2250 + 3†		
11.019	11.000	216.65	0.0	28.966	216.65	2.2632 + 2	3.6392 + 2		
20.063	20.000	216.65	+1.0	28.966	216.65	5.4747 + 1	8.8033 + 1		
32.162	32.000	228.65	+2.8	28.966	228.65	8.6798 0	1.3225 + 1		
47.350	47.000	270.65	0.0	28.966	270.65	1.1090 0	1.4275 0		
52.429	52.000	270.65	−2.0	28.966	270.65	5.8997 − 1	7.5939 − 1		
61.591	61.000	252.65	−4.0	28.966	252.65	1.8209 − 1	2.5108 − 1		
79.994	79.000	180.65	0.0	28.966	180.65	1.0376 − 2	2.0009 − 2		
90.000	88.743	180.65	+3.0	28.966	180.65	1.6437 − 3	3.1698 − 3		
100.000	98.451	210.65	+5.0	28.88	210.02	3.0070 − 4	4.9731 − 4		
110.000	108.129	260.65	+10.0	28.56	257.00	7.3527 − 5	9.8277 − 5		
120.000	117.776	360.65	+20.0	28.07	349.49	2.5209 − 5	2.4352 − 5	26.7	332
150.000	146.541	960.65	+15.0	26.92	892.79	5.0599 − 6	1.8350 − 6	24.8	822
160.000	156.071	1110.65	+10.0	26.66	1022.20	3.6929 − 6	1.1584 − 6		
170.000	165.571	1210.65	+7.0	26.40	1103.40	2.7915 − 6	8.0330 − 7		
190.000	184.485	1350.65	+5.0	25.85	1205.40	1.6845 − 6	4.3450 − 7	22.3	1040
230.000	221.967	1550.65	+4.0	24.70	1322.30	6.9572 − 7	1.5631 − 7		
300.000	286.477	1830.65	+3.3	22.66	1432.10	1.8828 − 7	3.5831 − 8	17.4	1100
400.000	376.312	2160.65	+2.6	19.94	1487.40	4.0278 − 8	6.4945 − 9	16.6	1240
500.000	463.526	2420.65	+1.7	17.94	1499.20	1.0949 − 8	1.5758 − 9	16.3	1358
600.000	548.230	2590.65	+1.1	16.84	1506.10	3.4475 − 9	4.6362 − 10	15.9	1420
700.000	630.530	2700.65	+1.1	16.17	1507.60	1.1908 − 9	1.5361 − 10	15.5	1448

*From Ref. 5.
† Power of 10 by which preceding number must be multiplied.
   h = geopotential altitude
   M = mean molecular weight
   z = geometric altitude
   T = kinetic temperature
   L = gradient of molecular scale temperature = dT_M/dh (below 79 geopotential Km)
                                                          = dT_M/dz (above 79 geopotential Km)
   T_M = molecular scale temperature = TM_0/M
   M_0 = sea-level value of M

Sound Velocity (Speed). The phase velocity of sound waves of small amplitude is given by

$$c = \sqrt{\frac{\gamma RT}{M}}$$

where γ is the ratio of specific heat at constant pressure to specific heat at constant volume. γ for air = 1.40 (sea-level conditions).

Viscosity. Viscosity is the property by which a fluid resists shear stress. It is often called a coefficient but improperly so as it has dimensions given by the expression

$$\mu = 0.673\rho cl$$

The proportionality constant was calculated from kinetic theory by Chapman. Kinematic viscosity is the ratio of the viscosity to the density of the gas: $\eta = \mu/\rho$.

6-6. Standard Atmospheres

The *U.S. Standard Atmosphere, 1962.*[5,6] Model, reference, and standard atmospheres have been used for many years for the purpose of standardizing instrument and vehicle performance and to provide a normalized basis for comparison of theoretical and experimental results. All such "atmospheres" consist of tabulations and/or graphs as a function of altitude of many of the primary and secondary parameters discussed above. Roughly speaking, "model," "reference," and "standard" are used in increasing order of acceptance, the latter usually having legal status. However, the three terms are often used interchangeably in the generic sense. Minzner and Ripley[7] have reviewed the history of various atmospheres from 1919 to 1956. In 1952, the ICAO (International Civil Aviation Organization) atmosphere[8] giving values to 20 Km was adopted in both Europe and the United States and became also the U.S. Standard Atmosphere. Following the advent of rockets, and later satellites, in upper-air research, high-altitude models based on measurements have appeared, of which the most useful have been the Rocket Panel (1952),[9] ARDC 1956,[7] and ARDC 1959.[10]

Table 6-4. Derived Quantities, *U.S. Standard Atmosphere, 1962*[6]
(Condensed Version)

z, Km	h, Km	g m/sec^2	l, m	n, m^{-3}	$\bar{u}$ m/sec	ν sec^{-1}	c, m/sec	μ, Kg/(m)(sec)
0.000	0.000	9.8066	6.6328 − 8*	2.5471 + 25*	458.94	6.9193 + 9*	340.294	1.7894 − 5*
11.019	11.000	9.7727	2.2327 − 7	7.5669 + 24	397.95	1.7824 + 9	295.069	1.4216 − 5
20.063	20.000	9.7450	9.2295 − 7	1.8305 + 24	397.95	4.3117 + 8	295.069	1.4216 − 5
32.162	32.000	9.7082	6.1438 − 6	2.7499 + 23	408.82	6.6542 + 7	303.131	1.4668 − 5
47.350	47.000	9.6621	5.6918 − 5	2.9683 + 22	444.79	7.8146 + 6	329.799	1.7037 − 5
52.429	52.000	9.6468	1.0699 − 4	1.5791 + 22	444.79	4.1573 + 6	329.799	1.7037 − 5
61.591	61.000	9.6193	3.2360 − 4	5.2209 + 21	429.74	1.3280 + 6	269.44	1.216 − 5
79.994	79.000	9.564	4.060 − 3	4.161 + 20	363.4	8.950 + 4		
90.000	88.743	9.535	2.563 − 2	6.591 + 19	363.4	1.418 + 4	269.44	1.216 − 5
100.000	98.451	9.505	1.629 − 1	1.037 + 19	392.4	2.409 + 3		
110.000	108.129	9.476	8.150 − 1	2.073 + 18	436.5	5.356 + 2		
120.000	117.776	9.447	3.233 0	5.226 + 17	513.4	1.588 + 2		
150.000	146.541	9.360	4.114 + 1	4.107 + 16	838.0	2.037 + 1		
160.000	156.071	9.331	6.454 + 1	2.618 + 16	901.0	1.396 + 1		
170.000	165.571	9.302	9.233 + 1	1.830 + 16	940.7	1.019 + 1		
190.000	184.485	9.246	1.668 + 2	1.013 + 16	993.6	5.955 0		
230.000	221.967	9.135	4.429 + 2	3.815 + 15	1064.7	2.404 0		
300.000	286.477	8.942	1.773 + 3	9.528 + 14	1156.8	6.524 − 1		
400.000	376.312	8.679	8.607 + 3	1.963 + 14	1256.7	1.460 − 1		
500.000	463.526	8.428	3.191 + 4	5.294 + 13	1330.2	4.168 − 2		
600.000	548.230	8.187	1.018 + 5	1.659 + 13	1376.1	1.352 − 2		
700.000	630.530	7.956	2.950 + 5	5.726 + 12	1405.0	4.762 − 3		

NOTE: These quantities, being dependent on kinetic temperature T, are subject to correction for new values of T above 100 Km.

z = geometric altitude
h = geopotential altitude
g = acceleration of gravity
l = mean free path
n = number density (concentration)
$\bar{u}$ = mean particle velocity (speed)
ν = collision frequency
c = sound velocity (speed)
μ = viscosity
* Power of 10 by which preceding number must be multiplied.

Recently, new atmospheres based on up-to-date experimental and analytical results and extending to great altitudes have appeared. A high-altitude model extending from 150 to 2,000 Km has been proposed by Nicolet.[3] Another model in the range 120 to 2,050 Km showing the time-dependent variations due to heating has been prepared by Harris and Priester.[11] Two atmospheres extending from the surface to 800 and 700 Km, respectively, have been prepared. One is CIRA 1961[12] (COSPAR International Reference Atmosphere, 1961) proposed by the Committee on Space Research of the International Council of Scientific Unions, but subsequently withdrawn in favor of a revised version now in preparation. The second is a proposed extension to the U.S. Standard Atmosphere by COESA (Committee on Extension to the U.S. Standard Atmosphere). The CIRA and COESA atmospheres, while not identical, are deliberately similar in essential features. The latter will be adopted in the United States as the *U.S. Standard Atmosphere, 1962*.[6] The lower 20 Km are nearly identical with the ICAO standard, and the values to 32 Km will probably be adopted by that organization as a new international standard.

Tables 6-3 and 6-4 give a condensed version of the *U.S. Standard Atmosphere, 1962*. Altitudes are given in both geometric and geopotential kilometers. The values of pressure p and density ρ are a composite of rocket and satellite measurements obtained by several techniques over a period of years. The molecular scale temperature T_M is derived from p and ρ on the basis of a constant molecular weight of 28.966 at all altitudes. In order to calculate the kinetic temperature T for the standard, a profile of molecular weight M was predicted on the basis of theories of dissociation and diffusion. In the region 100 to 150 Km, rocket measurements[13] obtained subsequent to the formulation of the *U.S. Standard Atmosphere, 1962* and CIRA 1961 yield lower values for M and hence for T. The decrease in M is due to significant dissociation of oxygen. The dissociation of nitrogen has also been detected, but measurements[13] of N show it to be less than 5 per cent of N_2 up to 200 Km, and it is thought to be largely consumed in the formation of oxides of nitrogen. Other measurements[14,4] verify the prediction of helium at hundreds of kilometers. Certain derived quantities (in Table 6-4), being dependent on kinetic temperature T, are subject to correction for new values of T above 100 Km.

In Fig. 6-1 the molecular scale temperature T_M and the kinetic temperature $T_{U.S.,62}$ of the *U.S. Standard Atmosphere, 1962* are plotted together with a kinetic temperature T based on the aforesaid rocket measurements of composition. In Fig. 6-2 the proportional composition of a model atmosphere, based also on the recent measurements, is shown together with the resulting mean molecular weight M. A proportional composition diagram has not been published for the *U.S. Standard Atmosphere, 1962*, but numerical values of $M_{U.S.,62}$ are given in Tables 6-3 and 6-4 and are plotted in Fig. 6-2. Figure 6-3 shows a plot of mass density vs. altitude for the standard. Mass density is the quantity most frequently measured with rocket and satellite techniques and may be understood with relatively little difficulty caused by the uncertainty in composition.

6-7. Large-scale Variations

The thermal energy which supports the atmosphere derives from the sun in the form of radiation and corpuscular flow, although not all particles are necessarily from the sun. Some of the heat-producing mechanisms, such as the absorption of ultraviolet by oxygen in various forms, are well established, but others, such as the Arctic warming by the recombination of atomic oxygen proposed by Kellogg[15] remain to be verified. Despite the absence of complete causative detail it is nevertheless reasonable to expect that the atmosphere will rise and fall in both density and temperature with solar output. Considerable success in establishing such correlations has been achieved by Jacchia[16] and others for altitudes above 200 Km using the acceleration in the orbital periods of satellites. In this work the amount of solar radiation at 10.7 cm and the geomagnetic index K_p are taken as indicators (not as causal agents) of the intensity of solar activity. Jacchia has established a diurnal bulge starting at 200 Km and with maximum temperatures 35 per cent higher during the day than at

THE UPPER ATMOSPHERE 6–11

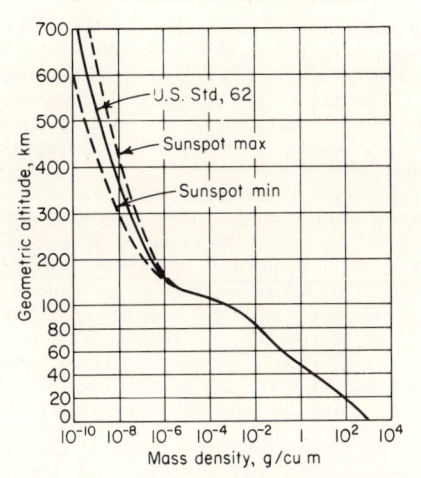

Fig. 6-3. The mass-density profile of the *U.S. Standard Atmosphere, 1962*, and variations *after Johnson*) corresponding to the solar cycle.

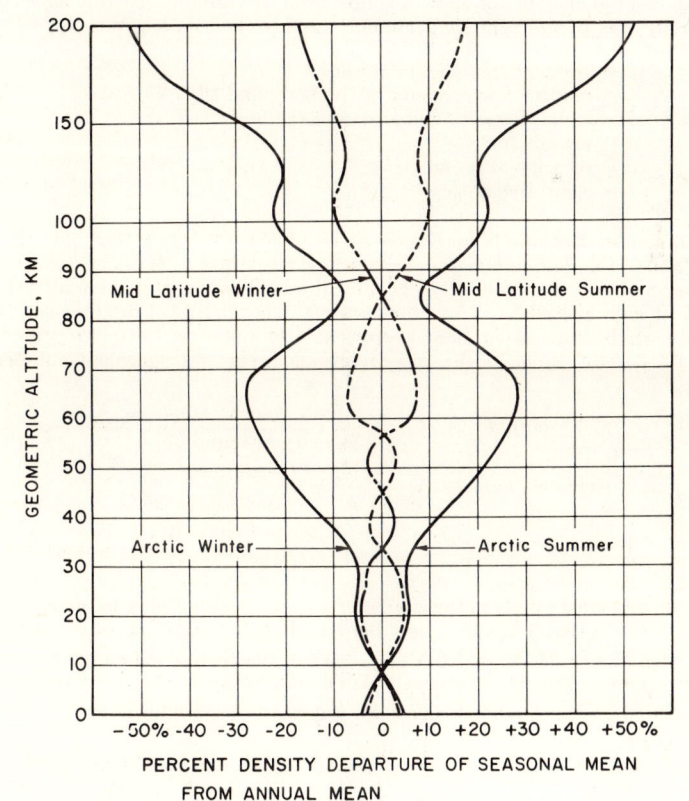

Fig. 6-4. Variations in the mean density of the atmosphere as a function of season and latitude. (*After Quiroz.*)

night. Other correlations with the 27-day period of rotation of the sun, and particularly with secular changes in radiation, have been observed. Variations related to solar flares with the appropriate day or two delay required for the transit of corpuscles have also been shown. The largest slow change is that associated with the 11-year sunspot cycle, a variation of such magnitude as to control the extremes in the structural parameters of the earth's atmosphere. Johnson[17] has calculated limits of variation in several of the structural parameters corresponding to sunspot maximum and sunspot minimum. Figure 6-3 shows the density variation according to Johnson.

At altitudes below 200 Km the statistical population of measurements which satellites give for higher altitudes is not available, except in the realm of meteorology below 32 Km. Vertical rocket firings yield altitude scans but are carried out relatively infrequently so that time correlations are difficult. In one instance a change in density, temperature, and winds at rocket altitudes was definitely related to an explosive warming event[18,19] observed by meteorological balloons, but such tie-ins are rare. By collating rocket results over a period of years it has nevertheless been possible to establish some summer-winter and Arctic–temperate zone variations. Quiroz,[20] for example, has prepared a tentative summary chart, given in Fig. 6-4, of such variations in density.

6-8. Aerodynamic Regimes

With the advent of vehicles and probes penetrating the upper atmosphere, real aerodynamic problems in all flow regimes from continuum to free molecule are encountered. Six parameters are commonly used to describe a given flow situation:

ρ = ambient free-stream density
V = relative flow velocity of the body and the free stream
d = a characteristic dimension of the body
μ = viscous force
c = velocity of sound
l = mean free path

Three dimensionless numbers, Knudsen (K), Mach (M), and Reynolds (Re), are used to "scale" the flow parameters in two flow situations. Actually, two numbers are independent and therefore sufficient, but M and Re are useful at low altitudes and M and K at high altitudes. Thus, for example, a coefficient of drag measured on a sphere in a wind-tunnel experiment is applicable to a spherical satellite provided K, M, and Re are the same in the two situations. The dimensionless numbers are defined as follows:

$$\text{Mach number } M = \frac{V}{c} = \frac{\text{flow velocity}}{\text{velocity of sound}}$$

$$\text{Reynolds number } Re = \frac{\rho V d}{\mu} = \frac{\text{inertia force}}{\text{viscous force}}$$

$$\text{Knudsen number } K = \frac{l}{d} = \frac{\text{mean free path}}{\text{characteristic dimension}}$$

For small Reynolds numbers the characteristic dimension d may be the length L of the body. In this case a Knudsen number based on the body length is appropriate. Thus $K_L = 1.26 \sqrt{\gamma} \, M/Re \approx 1.5 \, M/Re$. When the Reynolds number is large, the viscous effects are confined to a relatively thin boundary layer. In this case a Knudsen number related to the boundary-layer thickness δ is used and, since $\delta \propto d/\sqrt{Re}$, $K_\delta \propto M/\sqrt{Re}$.

Values of Knudsen numbers are used to define the boundaries of four major regimes of aerodynamic flow. These regimes have been discussed by Tsien,[21] Schaaf,[22] and others. More detailed breakdowns for special situations have been put forward, but the following are sufficient for many purposes:

Continuum flow prevails when intermolecular collisions, rather than collisions between molecules and the body, predominate. A common criterion is that $K_\delta < 0.01$. Classical methods of analysis consisting of certain limited solutions to the Navier-Stokes equations are available for this region (see chapter 18).

Slip flow occurs when the mean free path is small but not negligible in comparison with the boundary-layer thickness. The result is that the layer of gas next to the body "slips" or exhibits a tangential component of velocity. Here, K_δ has been given the limits $0.01 < K < 0.1$. Slip-flow analysis is similar to that of continuum flow but with slip corrections applied.

Transition flow is typically taken as occurring in the region where $0.1 < K_\delta$ and $K_L < 10$, that is, where molecular collisions with the body are about as frequent as with other molecules. Few analytical methods for the transition region are available.

Free molecule flow occurs when the rarefaction is very great. Because the molecules reemitted from the surface do not suffer intermolecular collisions except at great distances, the controlling phenomenon is the interaction of the impinging molecules with the surface. Free molecule flow may be said to exist for $K_L > 10$. Analysis consists of solutions to the collisionless Boltzmann equation for certain simple situations.

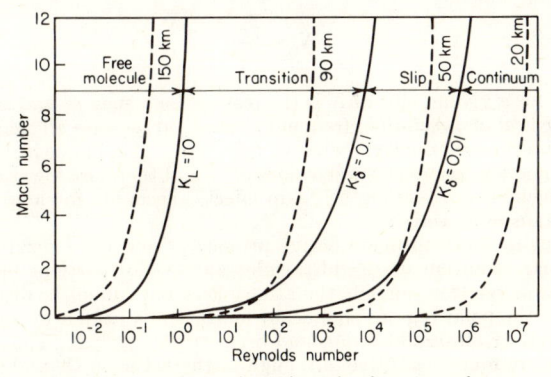

FIG. 6-5. Aerodynamic regimes in the atmosphere.

The solid lines of Fig. 6-5 show the various aerodynamic regimes using the limiting values and definitions of K discussed above. Assuming a characteristic dimension $d = 1.0$ m and values from the *U.S. Standard Atmosphere, 1962*, the conditions existing at important levels of the atmosphere have been calculated and are represented by the dashed lines.

6-9. The Ionosphere

The ionosphere extends from about 60 to 700 or 800 Km in the atmosphere and is characterized by the presence of free electrons in sufficient concentration to affect the propagation of radio waves. The presence of conducting layers in the upper atmosphere was first hypothesized in the 1870s to account for variations in the magnetic field at the surface. Marconi's success in transmitting signals across the Atlantic in 1901 led to further speculation as to conducting layers, and in the following year Kennelly and Heaviside independently postulated the layer which at first carried their names but which is now called the E region. The return to earth of transmitted radio waves is a consequence of the increasing index of refraction due to increasing electron density with height. The reflection also allows for detailed study of ionospheric phenomena with ground-based transmitters and receivers, although rocket and satellite experiments have recently resolved certain questions unanswered by ground techniques. For example, the electron-density profile of the upper part of the ionosphere, which is obscured by the electron-density maximum at $300 +$ Km, has been revealed by the topside sounder satellite Alouette.[23] Ground-based measurements yield electron and ion densities as a function of height, but identification of ion species must be made with *in situ* methods such as rocket-borne ion spectrometers. Inferences of ion species can be made from radar back-scatter observations.

The presence of free electrons increases the phase velocity of radio-frequency waves passing through the ionized medium as compared with their velocity in a vacuum. The index of refraction μ is therefore less than unity:

$$\mu = \sqrt{1 - \frac{Ne^2}{\pi m f^2}}$$

where N = electron concentration
e = charge on the electron
m = mass of the electron
f = frequency of the wave

The waves are refracted in accordance with Snell's law, and for a given value of N, those below a certain frequency will be "reflected" toward the earth. The condition that a wave, having an angle of incidence of 90° upon the ionized layer, be reflected is that $\mu = 0$, from which

$$N = \frac{\pi m f^2}{e^2}$$

The value of f for a given value of N is the plasma frequency f_N and is approximately equal to the *critical* or *penetration* frequency, so-called because all frequencies greater than f_N will pass through the layer without reflection. At oblique incidence, reflection for a given frequency occurs at lower values of N and therefore lower altitude. Multiple oblique-incidence reflection, or "skip effect," accounts for long-distance transmission of certain radio waves.

The electrons in the ionosphere derive primarily from the ionization of the gases present by solar radiation in the ultraviolet and X-ray portions of the spectrum. Some electrons arrive from outside the atmosphere but cannot be distinguished from local ones. In the polar regions significant ionization is caused by solar and cosmic-ray particles and by auroral electrons and bremsstrahlung.[24,25] The ions throughout the ionosphere are mainly positive, although in the D region O^- and O_2^- are present. The ion concentration is about the same as the electron concentration, (except in the lower D region) and the electrons, ions, and neutral particles are in thermal equilibrium to 150 Km, above which measured electron temperatures have exceeded gas temperatures by as much as a factor of 2, although this effect is diminished above 400 Km.[26]

The tendency to form layers is due to the combined effect of the exponential decrease of air density and the nearly exponential increase of ionizing solar radiation with altitude. The peak of electron concentration in a layer occurs at approximately the altitude at which these curves cross. Chapman[27] has examined a simplified model of such a layer. In an exponential, plane-stratified atmosphere with absorption coefficient proportional to density and with ionizing radiation of a single frequency, the electron distribution function is

$$N_2 = N_0 \exp \frac{1 - z - e^z \sec \phi}{2}$$

where z = altitude
N_0 = electron density at reference altitude $z = 0$
ϕ = zenith angle of the sun

The *regions* of the ionosphere are distinguishable from one another by their variability, altitude, and ionizing processes. A *layer* is a peak or maximum of ionization within a region.

D Region. The weakly ionized part of the atmosphere between 60 and 90 Km, with a peak electron density of somewhat over 1,000 per cu cm at 80 Km. The normal D region exists in the daytime only and is thought to be caused primarily by the ionization of a minor constituent (NO) of the atmosphere by solar ultraviolet

(Lyman α) above 70 Km. Below 70 Km, cosmic rays appear to be the ionizing agent. Because of the relatively high air density and consequent electron-air molecule collisions, the D region strongly absorbs r-f radiation and disappears rapidly at night. The typical profile of the D region is subject to disturbances caused by anomalous solar activity. Enhanced D-region ionization following some sun flares is caused by hard X-rays.

E Region. The most clearly defined region of the ionosphere extending from about 85 to 130 Km. The electron density varies from 10^5 to 2×10^5 per cu cm depending on solar cycle, zenith angle of the sun, and season. The E region continues through the night, but the electron density is decreased. The principal ions are O_2^+ and NO^+ created by soft X rays and ultraviolet from the sun. *Sporadic E* is a variable increase in the electron density in a thin layer just above 100 Km. The electron density may double at night in polar regions and in the daytime at low magnetic latitudes. Sporadic E may be caused by several factors including wind shears and meteors.

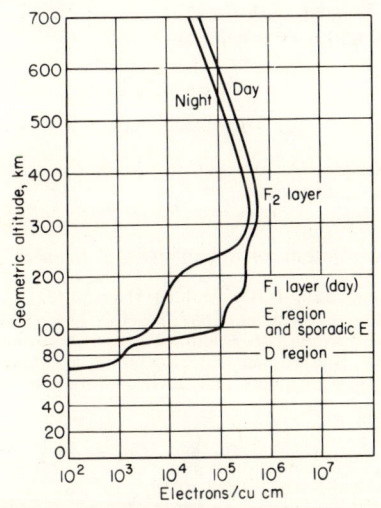

FIG. 6-6. The electron-density profiles of the day and night ionosphere for the mean of the solar cycle compiled from rocket measurements, together with designations of ionospheric regions.

F Region. The most highly ionized part of the ionosphere, apparently caused by ultraviolet lines below 1000A in the solar spectrum. The F region has two layers, F_1 and F_2. The F_1 layer is a slight inflection resulting from the fact that the recombination rate and ionization rates change differently with altitude; it extends roughly from 150 to 200 Km. The electron density is variable between 2.5×10^5 and 4×10^5 per cu cm. The ions are He^+, NO^+, O_2^+, and O^+. The latter predominates at the top of the region. Sometimes the F_1 layer is masked by the F_2, and it disappears at night. The F_2 layer extends from 200 Km to the protonosphere above 1,000 Km where the O^+ ions are outnumbered by protons. The electron density may go as high as 10^6 per cu cm, with the altitude of the maximum varying between 250 and 400 Km. The F region is one of many striking phenomena which cannot be adequately described in a short summary. One important phenomenon is the stretched-out duration of the return r-f pulse from a normal incident one. This is known as *spread F* echo, which is usually attributed to scattering over a large vertical interval. The Alouette measurements[23] have shown that at high latitudes the top of the ionosphere is rough, with columns of intense ionization along magnetic-field lines. The resulting ducted propagation causes a form of spread F.

6-16　SYSTEM ENVIRONMENTS

The regions of the ionosphere are shown in Fig. 6-6 together with electron densities compiled from many sources (mostly measurements) for the 45° north latitude ionosphere in mid solar cycle.

The ionosphere is subject to disturbances and storms caused by solar flares. An S. I. D., or *sudden ionospheric disturbance*, is the result of increased ultraviolet and X radiation from a flare which produces increased ionization, primarily in the D region. The duration of S. I. D.'s is typically 1 or 2 hr, and they are confined to the sunlit side of the earth. The increased electron density together with the high collision frequency of the lower ionosphere can cause complete absorption of r-f waves which normally pass through the D layer to reflect from the upper layers. Longer wavelengths may, during an S. I. D., be reflected from the enhanced D layer.

Solar flares also result in streams of particles (protons and electrons in addition to those in the steady-state solar wind) bombarding the atmosphere after a transit time of 2 or 3 days. The magnetic field and ionosphere, particularly the F region, are then disturbed by a complicated process. One result may be a *blackout* of r-f transmission, especially in polar regions where solar streams are focused by the earth's magnetic field. Such storms can occur on both the dark and sunlit sides of the earth, are often accompanied by auroras, and may last for several days.

REFERENCES

1. Nicolet, M.: *Physics of the Upper Atmosphere*, chap. 2, Academic Press Inc., New York, 1960.
2. Hays, P. B., and V. C. Liu: Department of Aero & Astro Engineering, University of Michigan, private communication.
3. Nicolet, M.: Density of the Heterosphere Related to Temperature, *Smithsonian Inst. Astr. Obs., Spec. Rept.* 75, 1961.
4. Hanson, W. B.: Upper-atmosphere Helium Ions, *J. Geophys. Res.*, vol. 67, no. 1, pp. 183–188, 1962.
5. Committee on Extension to the U.S. Standard Atmosphere: *U.S. Standard Atmosphere*, 1962 (condensed version), *Nature*, no. 4837, pp. 133–134, 1962.
6. Committee on Extension to the U.S. Standard Atmosphere: *U.S. Standard Atmosphere*, 1962 (complete version), Government Printing Office, Washington, D.C., 1962.
7. Minzner, R. A., and W. S. Ripley: The ARDC Model Atmosphere, 1956, *Air Force Surveys in Geophysics*, no. 86, 1956.
8. International Civil Aviation Organization: Standard Atmosphere–Tables and Data for Altitudes to 65,800 ft., *NACA Rept.* 1235, Langley Field, Va., 1955.
9. The Rocket Panel: Pressures, Densities and Temperatures in the Upper Atmosphere, *Phys. Rev.*, vol. 88, no. 5, pp. 1027–1032, 1952.
10. Minzner, R. A., K. S. W. Champion, and H. L. Pond: The ARDC Model Atmosphere, 1959, *Air Force Surveys in Geophysics*, no. 115, 1959.
11. Harris, I., and W. Priester: Time-dependent Structure of the Upper Atmosphere, Goddard Space Flight Center, NASA, Greenbelt, Md., 1962; *J. Atmospheric Sci.*, vol. 19, p. 286, 1962.
12. Committee on Space Research: CIRA 1961–COSPAR International Reference Atmosphere, 1961, *COSPAR Rept.*, North Holland Publishing Company, Amsterdam, 1961.
13. Schaefer, E. J., and M. H. Nichols: Neutral Composition Obtained from a Rocketborne Mass Spectrometer, *Space Research* IV, COSPAR, North Holland Publishing Company, Amsterdam, 1964.
14. Hale, L. C.: Ionospheric Measurements with a Multigrid Potential Analyzer, Abstract, *J. Geophys. Res.*, vol. 66, no. 5, p. 1554, 1961.
15. Kellogg, W. W.: Chemical Heating above the Polar Mesopause in Winter, *J. Meterol.*, vol. 18, pp. 373–381, 1961.
16. Jacchia, L. G.: Variations in the Earth's Upper Atmosphere as Revealed by Satellite Drag, *Rev. Mod. Phys.*, vol. 35, no. 4, pp. 973–991, 1963.
17. Johnson, F. S.: *Satellite Environment Handbook*, chap. 1, Stanford University Press, Stanford, Calif., 1961.
18. Jones, L. M., J. W. Peterson, E. J. Schaefer, and H. F. Schulte: Upper Air Density and Temperature: Some Variations and an Abrupt Warming in the Mesosphere, *J. Geophys. Res.*, vol. 64, no. 12, pp. 2331–2339, 1959.
19. Stroud, W. G., et al.: Rocket-grenade Measurements of Temperatures and Winds in the Mesosphere over Churchill, Canada, *J. Geophys. Res.*, vol. 65, no. 8, pp. 2307–2323, 1960.

20. Quiroz, R. S.: Seasonal and Latitudinal Variations of Air Density in the Mesosphere (30 to 80 Kilometers), *J. Geophys. Res.*, vol. 66, no. 7, 2129, 1961.
21. Tsien, H. S.: Superaerodynamics, Mechanics of Rarefied Gases, *J. Aeron. Sci.*, December, 1946, pp. 653–664.
22. Schaaf, S. A., and P. L. Chambre: *Flow of Rarefied Gases*, Princeton University Press, Princeton, N.J., 1961.
23. Chapman, J. H., and E. S. Warren: The Topside Sounder Alouette, NATO Advanced Study Institute, Skeikampen, Norway, 1963.
24. Popoff, I. G., and R. C. Whitten: Ionization by Auroral Electrons, *Proc. Conf. on Direct Aeronomic Measurements in the Lower Ionosphere*, University of Illinois, 1963.
25. Aikin, A. C., and E. J. Maier: *The Effect of Auroral Bremsstrahlung on the Lower Ionosphere*, NASA Tech. Note D-2096, 1963.
26. Brace, L. H., and N. W. Spencer: Direct Measurements of the Positive Ion Density and Electron Temperature in the F-region by Ejected Electrostatic Probes, NATO Advanced Study Institute, Skeikampen, Norway, 1963.
27. Chapman, S.: The Absorption and Dissociative or Ionizing Effect of Monochromatic Radiation in an Atmosphere on a Rotating Earth, *Proc. Phys. Soc.*, vol. 43, no. 26, p. 484, 1931.

BIBLIOGRAPHY

Bates, D. R. (ed.): *The Earth and Its Atmosphere*, Basic Books, Inc., Publishers, New York, 1957.
Berkner, L. V., and H. Odishaw (eds.): *Science in Space*, McGraw-Hill Book Company, 1961.
Bijl, H. K. K. (ed.): *Space Research*, North Holland Publishing Company, Amsterdam, 1960.
Boyd, R. L. F., and M. J. Seaton (ed.): *Rocket Exploration of the Upper Atmosphere*, Pergamon Press, New York, 1954.
Geophysics Research Directorate, USAF: *Handbook of Geophysics*, The Macmillan Company, New York, 1960.
Goody, R. M.: *The Physics of the Stratosphere*, Cambridge University Press, New York, 1954.
International Council of Scientific Unions: *Annals of the IGY*, Pergamon Press, New York, vol. 1, 1959, and succeeding vols.
Johnson, F. S.: *Satellite Environment Handbook*, Stanford University Press, Stanford, Calif., 1961.
Johnson, J. C.: *Physical Meteorology*, John Wiley & Sons, Inc., New York, 1960.
Kuiper, G. P. (ed.): *The Atmospheres of the Earth and Planets*, rev. ed., University of Chicago Press, Chicago, 1952.
Kuiper, G. P. (ed.): *The Earth as a Planet*, University of Chicago Press, Chicago, 1954.
Landsberg, H. E. (ed.): *Advances in Geophysics*, Academic Press Inc., New York, vol. 1, 1958, and succeeding vols.
Malone, T. F. (ed.): *Compendium of Meteorology*, American Meteorological Society, Boston, 1951.
Mitra, S. K.: *The Upper Atmosphere*, 2d ed., The Asiatic Society, Calcutta, 1952.
Muller, P. (ed.): *Space Research IV*, North Holland Publishing Company, Amsterdam, 1964.
Nawrocki, P. J., and R. Papa: *Atmospheric Processes*, Geophysical Corporation of America, Bedford, Mass., 1961.
Newell, H. E.: *High Altitude Rocket Research*, Academic Press Inc., New York, 1953.
Odishaw, H. (ed.): *Research in Geophysics*, 1; *Sun, Upper Atmosphere, and Space*, M.I.T. Press, Cambridge, 1964.
Priester, W. (ed): *Space Research III*, North Holland Publishing Company, Amsterdam, 1963.
Ratcliffe, J. A. (ed.): *Physics of the Upper Atmosphere*, Academic Press, Inc., New York, 1960.
Van Allen, J. A. (ed.): *Scientific Uses of Earth Satellites*, The University of Michigan Press, Ann Arbor, Mich., 1956.
Van de Hulst, H. C., C. De Jager, and A. F. Moore (eds.): *Space Research* II, North Holland Publishing Company, Amsterdam, 1961. Also succeeding vols.

Chapter 7

SPACE AND ASTRONOMY

CARL D. GRAVES, *TRW Space Technology Laboratories.*

CONTENTS

7-1. Geophysical Space	7-2
Geomagnetism	7-2
Main Dipole Field	7-2
Variations and Perturbations	7-5
The Exosphere and Protonosphere	7-6
The Radiation Belts	7-8
The Inner Zone	7-10
The Outer Zone	7-13
7-2. Interplanetary Space	7-13
The Interplanetary Gas and Magnetism	7-13
Solar Electromagnetic Radiation	7-15
Quiet Sun	7-15
Active Sun	7-16
Solar-flare Particles	7-16
Typical Solar-flare Events	7-16
Relativistic Solar-flare Events	7-18
Directional Characteristics of Protons	7-18
Galactic Cosmic Rays	7-19
Composition	7-19
Geomagnetic Exclusion	7-19
Flux and Energy Spectrum	7-19
Directional Characteristics and Temporal Variations	7-20
Meteoritic Material	7-21
7-3. Interstellar Space	7-22
7-4. Astronomical Data	7-23
Astronomical Terms	7-23
Definitions	7-23
The Measure of Distance	7-24
The Measure of Time	7-24
Coordinate Systems	7-25
The Horizon System	7-26
The Equatorial System	7-26
The Galactic Coordinate System	7-26
The Solar System	7-27
Orbital Data	7-27
Physical Data	7-28

The Sun	7-28
Mercury	7-31
Venus	7-31
Mars	7-31
Jupiter	7-31
Saturn, Uranus, Neptune, and Pluto	7-32
Asteroids, Comets, and Meteors	7-32
Stellar and Galactic System	7-32
Stars	7-32
Galaxies	7-34

Since the orbiting of the first scientific satellite, a significant increase in our knowledge and understanding of the space environment has taken place. A brief summary of the current information on geophysical space, interplanetary space, and astronomical data is presented in this chapter. However, the rapid advances being made in this area mean that much of the material presented will be supplemented by new data in the very near future and many of the outstanding problems resolved.

7-1. Geophysical Space

Geophysical space is that region of space in the immediate vicinity of the earth bounded by the base of the exosphere and the termination of the geomagnetic field. The principal constituents of the environment in this region are the extension of the upper atmosphere, the trapped high-energy protons and electrons, and the geomagnetic field. The interplanetary environment which extends somewhat into this region is treated in Sec. 7-2.

Geomagnetism. *Main Dipole Field.* The source of the main magnetic field of the earth is believed to be a current system resulting from a dynamo action of the earth's molten metallic core.[1] The resulting magnetic field can be closely approximated by the field due to a dipole with a magnetic moment of 8×10^{25} cgs units placed near the center of the earth. The dipole approximation is very good at great heights above the earth but does not represent the irregularities or anomalies which characterize the actual geomagnetic field at the earth's surface. The large (thousands of square miles) anomalies are probably due to "eddies" in the current system, while the smaller local anomalies are due to deposits of ferromagnetic material in the earth's crust. Detailed magnetic charts of the geomagnetic field at the earth's surface as derived from measurements made at many observatories are prepared by the U.S. Navy Hydrographic Office. Current geomagnetic maps and data can also be found in the Air Force's *Handbook of Geophysics*.[2] Figure 7-1 shows the notation commonly used for the components of the earth's magnetic field. Figure 7-2 shows a map of the total intensity F of the magnetic field on the earth's surface. (See also Sec. 3-4.)

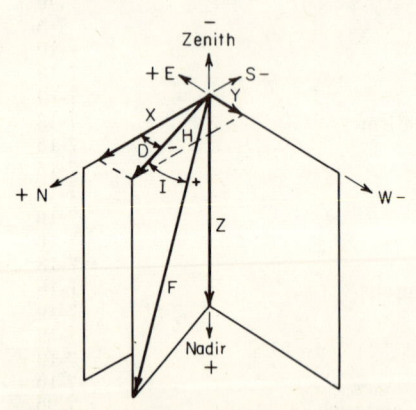

Fig. 7-1. Surface components of earth's magnetic field. D = declination, H = horizontal intensity, Z = vertical intensity, X = north-south component, Y = east-west component, F = total intensity, I = inclination.

The magnetic poles (with respect to the geographic coordinate system) are at about 79°N, 290°E and 79°S, 110°E. The centered dipole is used to define the

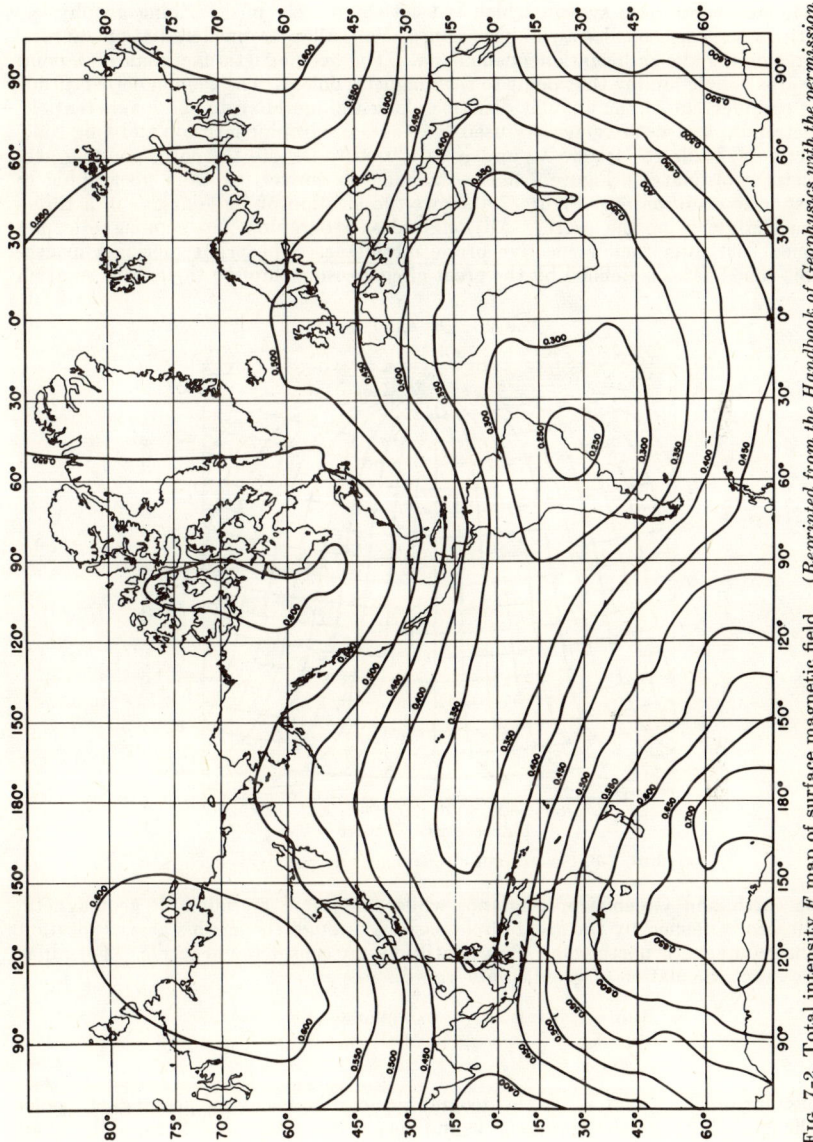

Fig. 7-2. Total intensity F map of surface magnetic field. (Reprinted from the *Handbook of Geophysics* with the permission of the U.S. Air Force Cambridge Research Laboratories, 1961.)

geomagnetic coordinate system which is used extensively in describing geophysical quantities such as cosmic-ray impact zones, Van Allen trapped radiation, auroral displays, and radio-propagation phenomena. The geomagnetic meridian of a point is the great circle joining that point to the magnetic poles. The geomagnetic latitude is the complement of the angular distance along the meridian to the magnetic pole. The letters λ_g and ϕ_g are generally used to denote geographic latitude and longitude, respectively, while the letters λ_m and ϕ_m are used to denote the corresponding geomagnetic coordinates. Figure 7-3 is a nomograph converting from geographic to geomagnetic coordinates. Figure 7-4 shows the location of these angles in a three-dimensional view of the earth. All longitudes, geographic and geomagnetic, are reckoned east from their respective prime meridians. The prime meridian in geographic coordinates is defined by the great circle passing through the axis of rotation

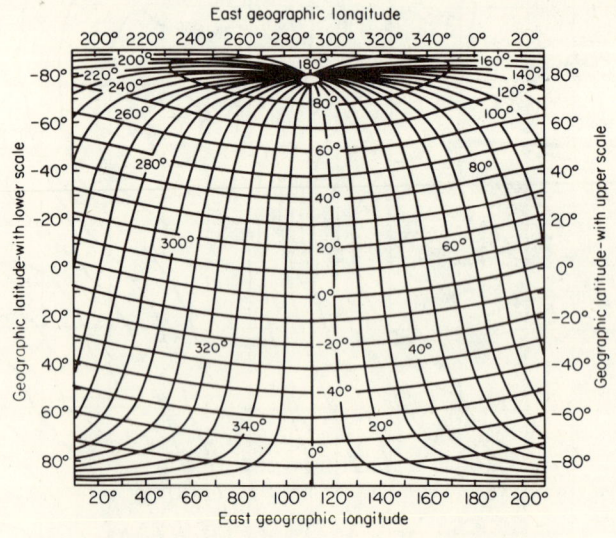

FIG. 7-3. Geographic-to-geomagnetic nomograph.

of the earth and Greenwich, England, while the prime meridian in geomagnetic coordinates is defined by the great circle passing through the earth's axis of rotation and the geomagnetic north pole. South latitudes are considered negative. Formulas for the exact calculation of geomagnetic coordinates are

$$\tan \phi_m = \tan (\phi_g - \phi_0) \sin x \sec (x - \lambda_0)$$
$$\tan \lambda_m = -\cos \phi_m \tan (x - \lambda_0)$$
$$\sin \chi = \cos \lambda_0 \sec \lambda_m \sin \phi_g$$

In these formulas, ϕ_0 and λ_0 are the geographic coordinates of the geomagnetic pole, χ is the angle between the geomagnetic and geographic prime meridians, and x is an auxiliary angle given by

$$\tan x = \cos (\phi_g - \phi_0) \cot \lambda_g$$

Since almost all the geomagnetic field is due to sources inside the earth (there is possibly a few per cent contribution from extraterrestrial current sources), the main magnetic field in the space surrounding the earth can be calculated from the earth-centered source dipole to a good approximation. Furthermore, the effect of the higher-order harmonics necessary to describe the surface-field anomalies[3] becomes negligible at far distances. However, at altitudes of many earth radii, the dipole field is changed by the "solar wind" (see below under The Radiation Belts and The

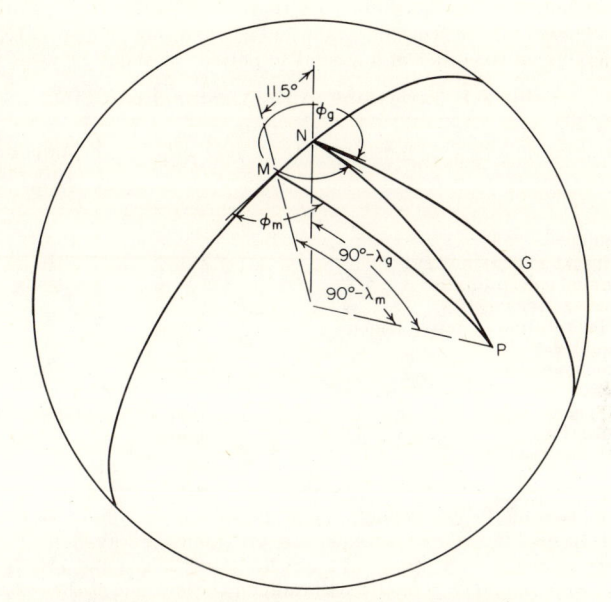

FIG. 7-4. Geomagnetic-coordinate-system angles.

	Geographic	Geomagnetic
North pole	N	M
Prime meridian	NG	NM
Longitude	ϕ_g	ϕ_m
Colatitude	$90° - \lambda_g$	$90° - \lambda_m$

Interplanetary Gas and Magnetism). In the geomagnetic-coordinate system, the magnetic-field components H_r and H_θ in gauss at any point are

$$H_\theta = \frac{a^3}{r^3} k \sin \theta \qquad (7\text{-}1)$$

$$H_r = \frac{a^3}{r^3} 2 k \cos \theta \qquad (7\text{-}2)$$

where r = radial distance to field point
θ = colatitude (polar angle)
a = radius of the earth
k = dipole constant

In the dipole approximation, the magnetic field is symmetrical for rotations about the poles; i.e., there is no longitudinal dependence. In calculating the field, higher accuracy can be obtained by including the higher-order harmonics.[4] Figure 7-10 shows the form of the dipole field where the lines in the figure show the direction of the field.

Variations and Perturbations. There are two primary types of variation in the geomagnetic field. One type is a gradual change in the main magnetic field due to changes in the source of the magnetic field; i.e., changes in the current system inside the earth's core. Thousands of years are required for significant changes of this type.

On the other hand, there are continuous fluctuations in the field due primarily to solar influences. These range in size from giant magnetic storms to micropulsations and in time from periods on the order of a second to periods of about 11 years (Table 7-1).

Table 7-1. Variations in the Geomagnetic Field

Sources	Period	Amplitude, 10^{-5} gauss
Periodic:		
Diurnal solar radiation	1 day	10–100
Annual solar radiation	1 year	50
Solar-activity cycle	11 years	
Solar rotation of active regions	27 days	
Lunar day	25 hr	3
Lunar month	29 days	10
Sporadic:		
Magnetic storms	3 days	100–500
Pulsations	10 min	1–10
Micropulsations	Seconds	0.01–1

The termination of the geomagnetic field due to the solar wind pressure has been studied analytically, but very few experimental results are available.[5] Mariner II interplanetary-solar-wind results[6] would indicate that the geomagnetic field would be terminated at approximately 8 to 10 earth radii on the sun side of the earth and would have a long "tail" on the dark side. Measurements by Explorers XII and XIV of Van Allen belt electrons confirm this (see below under The Outer Zone and Fig. 7-16).

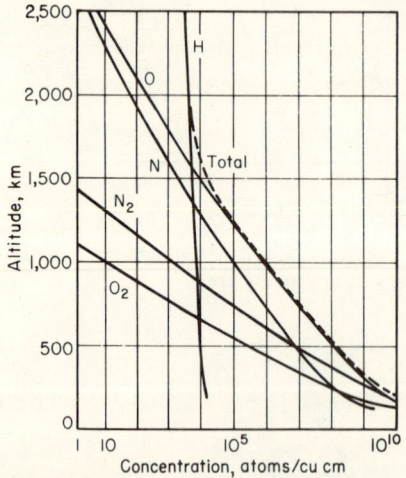

FIG. 7-5. Exosphere neutral-particle concentration—sunspot maximum. (*Reprinted from Satellite Environment Handbook edited by Francis S. Johnson with the permission of the publishers, Stanford University Press. Copyright 1961 by Lockheed Aircraft Corporation.*)

The Exosphere and Protonosphere. As discussed in Chapter 6, the neutral portion of the upper atmosphere consists primarily of nitrogen and oxygen (predominantly molecular at altitudes below a few hundred kilometers and predominantly atomic above) with hydrogen becoming the dominant constituent above about 1,000 Km (sunspot minimum). Embedded in this neutral gas are ionized components consisting of electrons, and equal concentrations of positively charged atoms and molecules. The ionized component becomes important at about 60 Km, the beginning of the ionosphere, with the concentration increasing to a maximum in the region 200 to 400 Km and then gradually falling off in the protonosphere to concentrations at many earth radii equal to the interplanetary density of less than 10 electrons/cu cm.

The exosphere is that region of the upper atmosphere (above about 500 Km) where the gas density is so low that collisions are negligible and particle trajectories are meaningful. It is essentially isothermal (1000 to 1500°K) out to several thousand kilometers and is in diffusive equilibrium. Definitive experimental data are not available, and the concentrations of the various constituents are derived primarily from extrapolations and analysis and are, consequently, uncertain.

SPACE AND ASTRONOMY 7-7

Figures 7-5 and 7-6 show the average neutral-component concentration of the exosphere. It is subject to variations due to the sunspot cycle as great as an order of magnitude. Above about 1,000 Km atomic hydrogen is believed to predominate, and Fig. 7-7 shows its concentration with altitude. Similarly, ionized hydrogen (protons

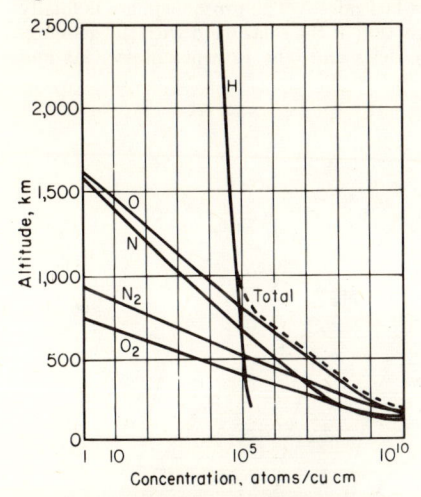

FIG. 7-6. Exosphere neutral-particle concentration—sunspot minimum. (*Reprinted from Satellite Environment Handbook, edited by Francis S. Johnson with the permission of the publishers, Stanford University Press. Copyright 1961 by Lockheed Aircraft Corporation.*)

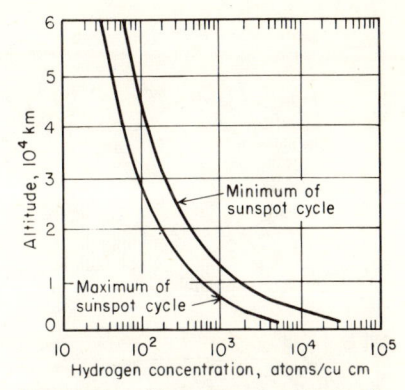

FIG. 7-7. Upper-exosphere atomic-hydrogen concentration. (*Reprinted from Satellite Environment Handbook, edited by Francis S. Johnson with the permission of the publishers, Stanford University Press. Copyright 1961 by Lockheed Aircraft Corporation.*)

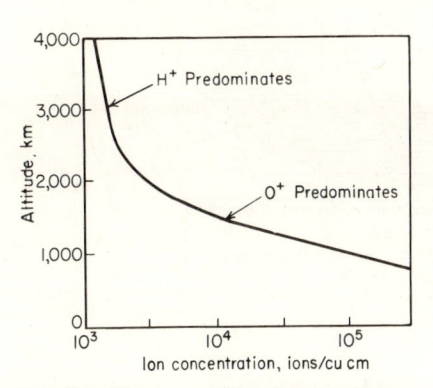

FIG. 7-8. Upper-exosphere ion concentration. (*Reprinted with permission from Progress in the Astronautical Sciences edited by S. F. Singer. Copyright 1962, Interscience Publishers, Inc.*)

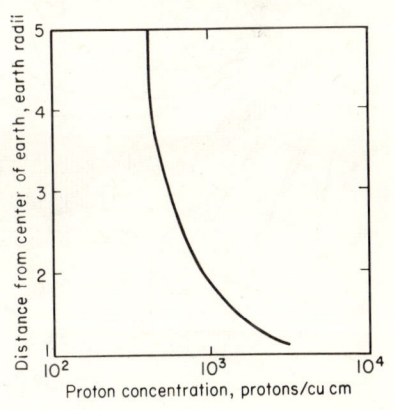

FIG. 7-9. Protonosphere concentration. (*Reprinted with permission from Progress in the Astronautical Sciences, edited by S. F. Singer. Copyright 1962, Interscience Publishers, Inc.*)

and electrons) is believed to be the major ionic constituent above about 1,000 Km (Fig. 7-8). This region has been called the protonosphere to distinguish it from the normal ionosphere with its heavy ion concentration. The hydrogen ions and electrons are produced at lower levels (below 800 Km) by charge exchange between atomic oxygen ions and neutral atomic hydrogen and diffuse upward. Figure 7-9 shows the proton distribution based upon a modified hydrostatic law[7] which includes the effect of a centrifugal-force term due to the fact that the geomagnetic field requires the ions and electrons to rotate with the earth. Thus Fig. 7-9 applies to concentrations along magnetic-field lines in the lower and medium latitudes. The protonosphere is finally terminated at many earth radii by the interaction of the solar wind with the geomagnetic field (see below under The Radiation Belts and The Interplanetary Gas and Magnetism).

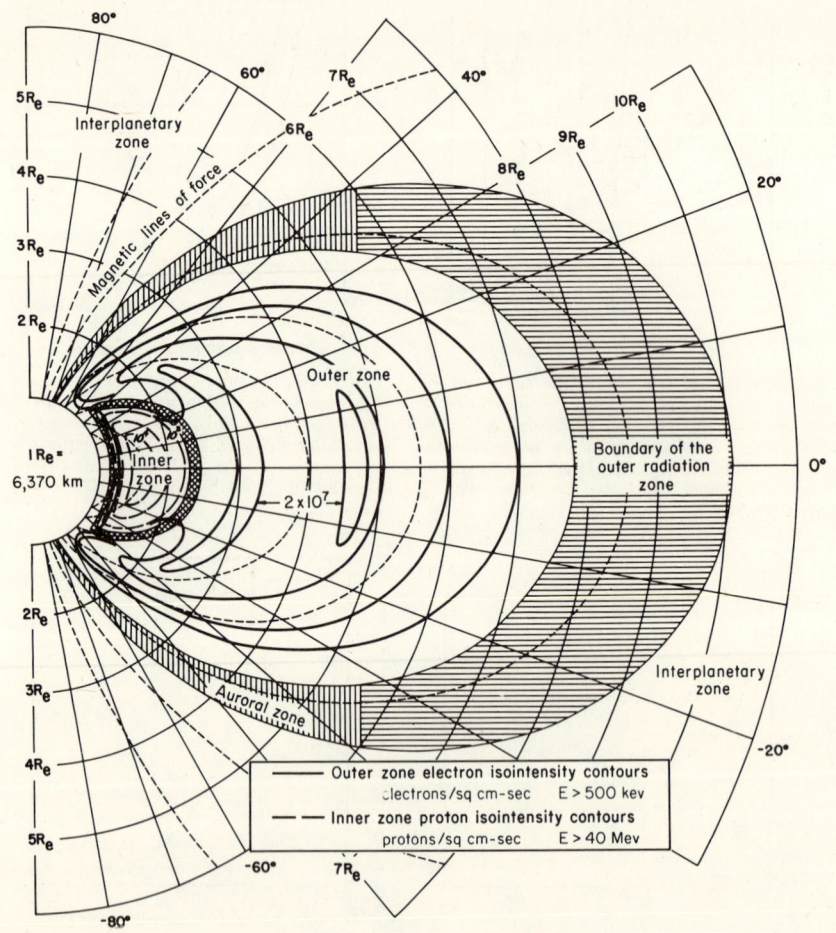

Fig. 7-10. Radiation-belt boundaries.

The plasma concentration at many earth radii has been estimated using the results of radio-whistler propagation studies. The propagation of electromagnetic waves through the plasma of the protonosphere and ionosphere is quite complex and results in diverse phenomena such as fading, Faraday rotation, refraction, and dispersion. These topics, along with the detailed properties and variations which characterize this medium, are discussed in detail in Refs. 8, 9, and 10.

The Radiation Belts. The discovery of trapped radiation in the region close to the earth is one of the most startling results so far of satellite space research. The

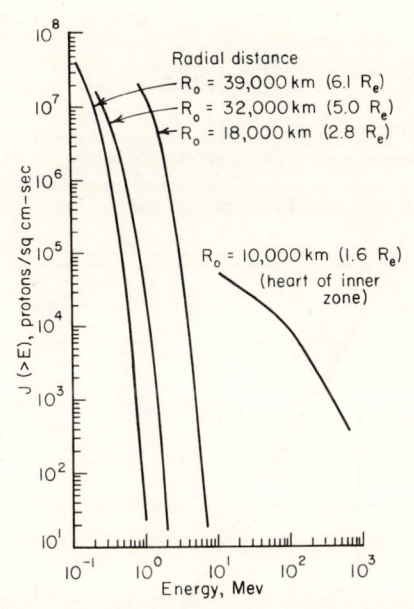

FIG. 7-11. High- and low-energy proton spectrum.

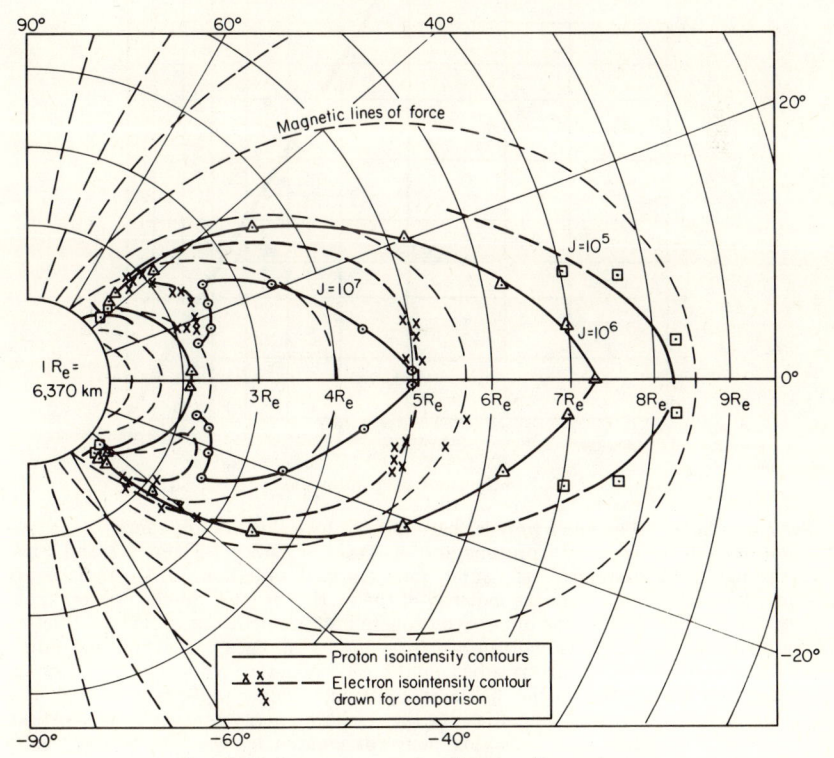

FIG. 7-12. Inner-zone proton isointensity contours.

trapped radiation consists almost exclusively of protons and electrons. Trapping is made possible by the existence of the geomagnetic field which constrains the particle trajectories. Two zones of trapped radiation exist, characterized, respectively, by a penetrating flux (inner zone) and nonpenetrating flux (outer zone). Figure 7-10 defines the boundaries of the two zones. The division into two zones is somewhat artificial since they are indistinguishable at the lower electron and proton energies. A summary of the basic characteristics of the trapped radiation is presented here. Detailed reviews of the inner-belt data[11] and the outer-belt data[12] give analyses of both experimental and theoretical results and contain extensive bibliographies.[32]

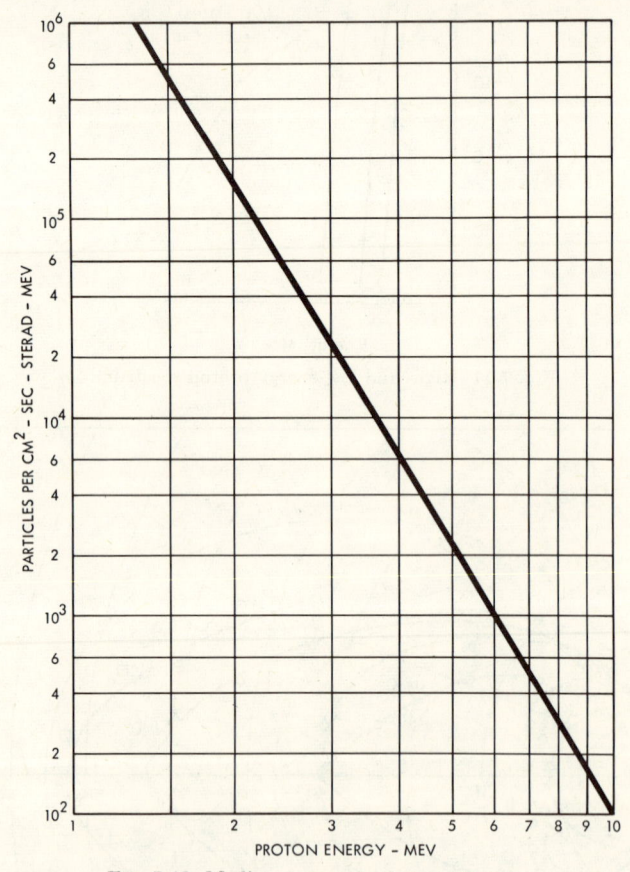

Fig. 7-13. Medium-energy proton spectrum.

The Inner Zone. The inner zone is characterized by a flux of high-energy penetrating protons and lower-energy protons and electrons. Tables 7-2 and 7-3 and Figs. 7-11 through 7-14 summarize the inner-zone trapped radiation. The high-energy protons almost certainly originate in decay of the earth's neutron albedo. Very-high-energy cosmic rays striking the upper atmosphere produce neutrons. A small fraction of the neutrons escape, traveling upward from the earth, decaying eventually into a proton and electron which are trapped in the geomagnetic field. The flux, energy spectrum, and spatial distribution of the trapped high-energy protons seem to fit this type of source mechanism. The origin of the medium-energy protons is less certain; they probably originate from decaying neutrons created by the lower-energy solar

Table 7-2. Inner-zone Protons

	Peak intensity, cm^{-2}sec^{-1}	Energy spectrum	Spatial distribution	Temporal variation
High energy, $E > 40$ Mev	2×10^4	$E^{-2.72}$ (Fig. 7-11 shows both inner- and outer-zone spectra)	The region of 10^3 to 10^4 Km above the earth at the equator, with strong latitude dependence (Fig. 7-12). At geomagnetic latitudes greater than 37.5° high-energy protons are not observed	3-fold increase in intensity generally correlated with geomagnetic storm and/or solar activity. Stable to a few per cent during quiet periods
Medium energy, 1 Mev $< E < 40$ Mev	1×10^7	$\sim E^{-4}$ (Fig. 7-13)	Region beginning at 4×10^3 Km and extending into the outer zone	Unknown
Low energy, 0.5 Kev $< E < 1$ Mev	3×10^8	Unknown	Distribution probably similar to inner-zone high-energy protons (Fig. 7-12) and extending through the outer zone	Unknown

Table 7-3. Inner-zone Electrons

	Peak intensity, cm^{-2}sec^{-1}	Energy spectrum	Spatial distribution	Temporal variation
Natural-belt electrons	10^7–10^8	Not known, but extending from the low Kev region to greater than 1 Mev and possibly similar to outer-zone electron spectrum for high energies (Fig. 7-15). Flat spectrum for low energies	The region of 10^3–10^4 Km above earth at the equator for energies greater than a few hundred Kev and extending through outer zone for low energies (Fig. 7-14 shows high-energy isointensity contours for both inner and outer zone). (Note Fig. 7-16 shows solar-wind pressure modification of outer zone as indicated by recent data)	Probably same as inner-zone protons
Artificially injected electrons (high-altitude thermonuclear explosions)	2×10^9	Fission spectrum (exponential in energy)		Probably exponential decay with a lifetime of a few years

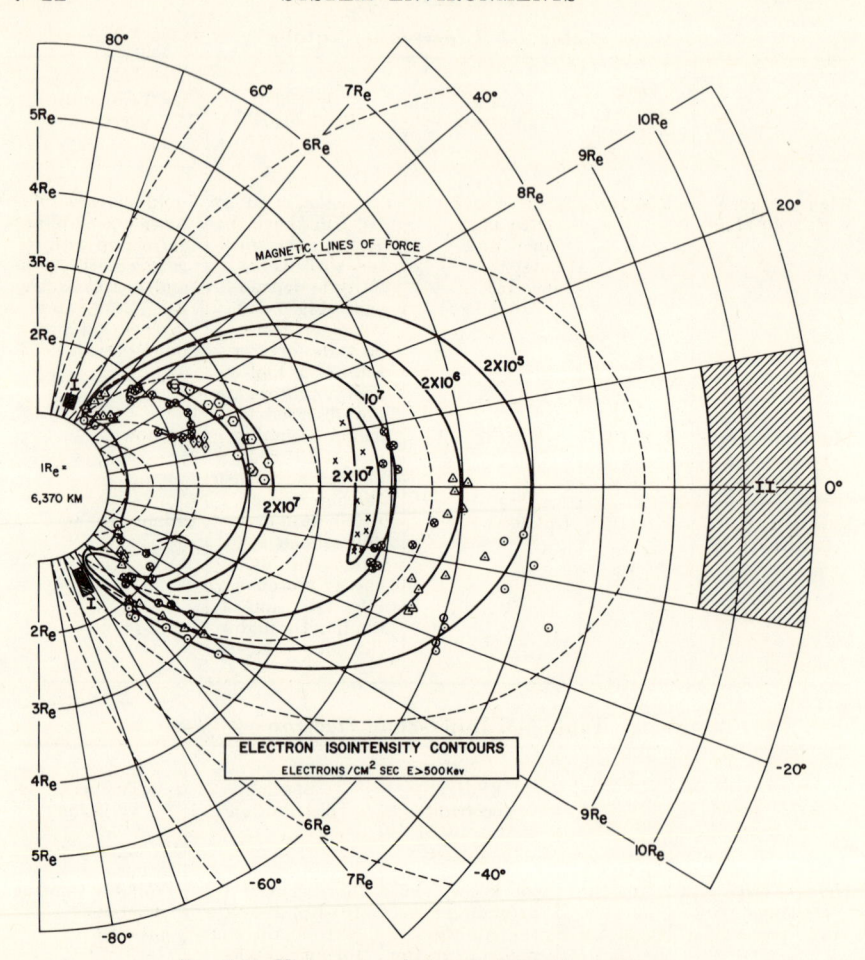

FIG. 7-14. High-energy electron isointensity contours.

Table 7-4. Outer-zone Protons

	Peak intensity, $cm^{-2} sec^{-1}$	Energy spectrum	Spatial distribution	Temporal variations
High energy, $E > 40$ Mev	None			
Medium energy, 1 Mev $< E <$ 40 Mev	10^7	Unknown	Extending from inner zone out to about 3×10^5 Km above the earth	Unknown (possibly large)
Low energy, 140 Kev $< E <$ 5 Mev	7×10^8	See Fig. 7-11	Extending through the outer zone	Unknown (possibly large)

Table 7-5. Outer-zone Electrons

	Peak intensity, $cm^{-2} sec^{-1}$	Energy spectrum	Spatial distribution	Temporal variations
Natural-belt electrons	10^8 ($E > 100$ Kev) $> 10^8$ ($E < 100$ Kev)	See Fig. 7-15	See Fig. 7-16 (trapping region terminated by cut-off in geomagnetic field due to solar wind)	Greater than order-of-magnitude charges correlated with geomagnetic storms. Similar uncorrelated changes have also been observed
Artificially injected electrons	None			

protons striking the polar atmospheric regions. The source of the low-energy protons is unknown. The great amount of energy contained in the large flux of low-energy protons could not be supplied directly from neutron albedo decay. Similarly, the trapped electrons of energy greater than 780 Kev, which is the cut-off energy for neutron decay electrons, could not be supplied directly by the neutron albedo. It is possible that some accelerating processes taking place in the upper atmosphere could be the source, but as yet no clear-cut mechanism is known.

The Outer Zone. The outer zone is characterized by a large flux of high-energy electrons. Tables 7-4 and 7-5 and Figs. 7-15 and 7-16 summarize the outer-zone trapped radiation. Its origin is still uncertain. It is probable that a combination of solar-particle injection and neutron albedo, along with some kind of acceleration mechanism, is responsible. However, the relationship is still not clearly understood.

7-2. Interplanetary Space

The influence of the sun on interplanetary space is overwhelming out to distances of the earth's orbit, 1 astronomical unit (1 A.U.), and perhaps considerably farther. A stream of plasma is continuously ejected from the sun and carries along with it a large-scale magnetic field. A large amount of micrometeoritic material (dust) is trapped in orbit about the sun. The sun is a source of cosmic-ray–like high-energy particles and the principal source of electromagnetic radiation in the solar system. These are the principal constituents of interplanetary space.

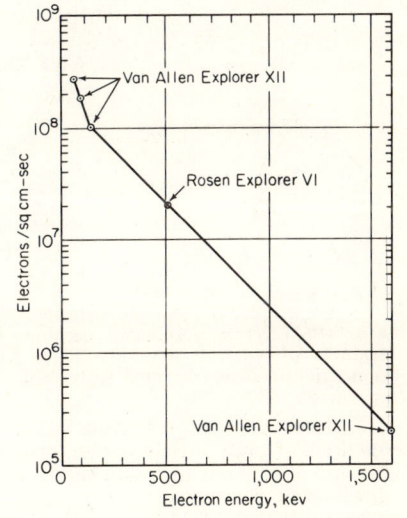

FIG. 7-15. Outer-zone electron energy spectrum.

The Interplanetary Gas and Magnetism. For many years, it was thought that the interplanetary gas (mostly ionized) was in hydrostatic equilibrium about the sun. During the last decade, a significant amount of data has been collected which shows that this is not the case. Biermann observed that comet tails were pushed radially away from the sun as they approached it. He postulated that this was due to corpuscular radiation streaming away from the sun and interacting with the gases making up the comet tail. Parker[13] theorized that the interplanetary gas was not in hydrostatic equilibrium but rather was made up of corpuscular radiation (protons and electrons) in hydromagnetic expansion from the sun. This theory has been dramatically confirmed by the recent flights of Explorer X and Mariner II. Figures 7-17 and 7-18 show the results of the Mariner measurements. The expanding plasma called the solar wind forms the outer corona of the sun. It is composed of a proton and electron flux (with a small percentage of alpha particles) traveling at a speed of about 5×10^7 cm/sec with an intensity of between 10^8 and 10^9 protons/(sq cm)(sec) and thus yielding a density of about 1 to 10 particles/cu cm. Figure 7-19 shows the density of electrons (and protons) as a function of distance from the sun. Other experimental results based on earth-based observations are shown for comparison.[32]

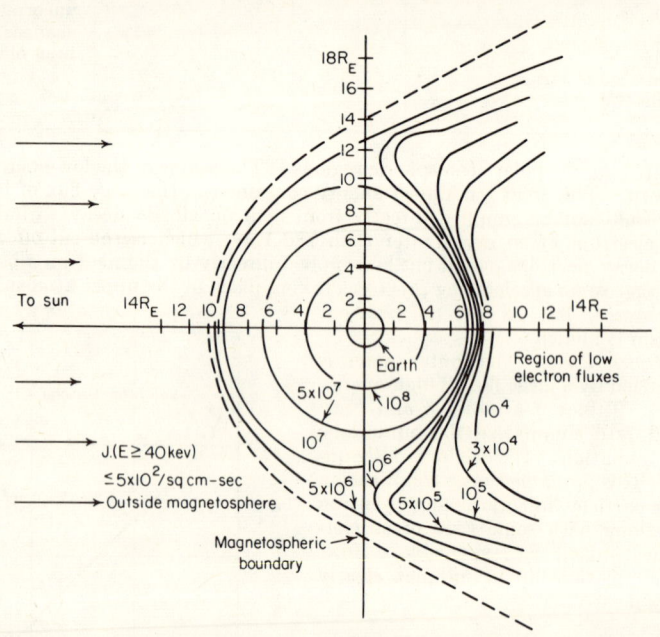

FIG. 7-16. Electron isointensity contours as modified by the solar wind. Quasistationary contours of constant omnidirectional flux of electrons ($E > 40$ kev) in the magnetic equatorial plane as measured with Explorers XII and XIV.

The source of the interplanetary magnetic field is the sun. Magnetic fields of hundreds of gauss have been observed on the sun in localized areas. The main field of the sun is believed to be dipole in character. Our knowledge of the strength and direction of the interplanetary magnetic field is fragmentary. Available evidence indicates that the field is controlled by the plasma (solar wind) ejected from the sun. Because of this, the magnetic field is believed to be primarily radial in the outer regions of the corona. Lines of force are essentially dragged out from the sun by the solar wind. Magnetic fields on the order of at least a few gamma (1 gamma = 10^{-5} gauss) were observed almost continuously throughout the Mariner II flight. During quiet times, the field would typically be about 5 gamma, but it did reach values on the order of 20 gamma during disturbed times with values as low as 2 gamma during very quiet times. The field is quite irregular, with the irregularities having periods on the order of many seconds to many hours.

Available evidence indicates that the interplanetary medium is made up of the streaming plasma of the solar wind. There appear to be few neutral particles and there is no direct evidence for interplanetary stationary gas. However, observations of scattered Lyman-alpha radiation in the night sky made from sounding rockets and from the absorption of direct Lyman-alpha radiation from the sun indicate the existence of an absorbing layer of hydrogen.[14] It is now believed that this absorbing layer of hydrogen is not in interplanetary space, but rather surrounds the earth. Calculations based upon the Lyman-alpha scatter and absorption require approximately 10^{12} to 10^{13} hydrogen atoms per square centimeter column between the earth and sun to explain the observations.

Solar Electromagnetic Radiation. The sun is a source of continuous electromagnetic radiation of great intensity. Although the intensity is wavelength-dependent, the average energy output is remarkably stable (within 1 per cent).

Quiet Sun. The sun is a sphere of hot gas heated by thermonuclear reaction taking place deep in its interior. The outermost regions of the sun can be divided into three regions, the photosphere, the chromosphere, and the corona, each with its own characteristics.

The photosphere is the source of the visible radiation coming from the sun. Below the photosphere the solar gas is opaque. The radiant energy emitted from the photosphere is very close to that of a radiating blackbody at a temperature of 6500°K. Figure 7-20 gives the visible energy radiated as a function of wavelength; the energy at any distance varies as the inverse square of the distance.

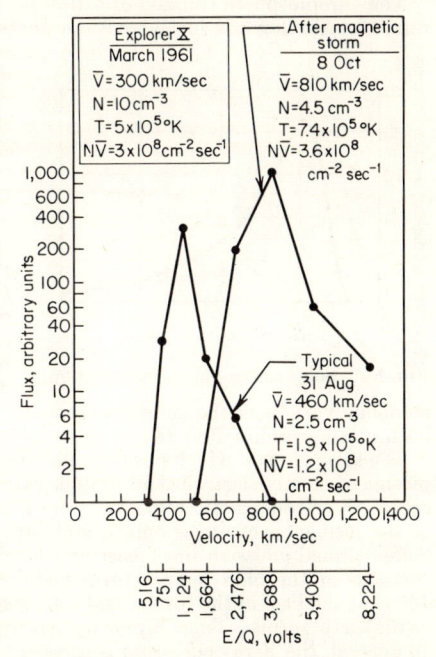

FIG. 7-17. Mariner II solar-wind spectrum.

The rate of electromagnetic radiation from the sun is about 3.8×10^{23} Kw. This corresponds to an integrated energy falling on a 1 sq cm surface normal to the radiation at a distance of 1 A.U. of 2 cal/(sq cm)(min). Although the spectrum of the photosphere, as shown in Fig. 7-20, is essentially continuous in the visible region, it

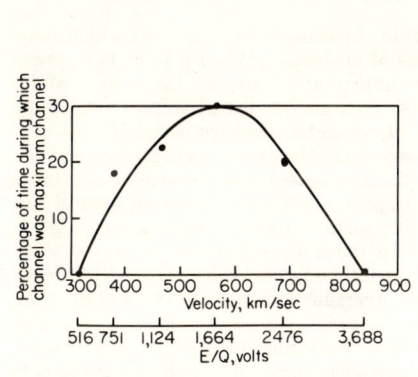

FIG. 7-18. Mariner II solar-wind peak distribution.

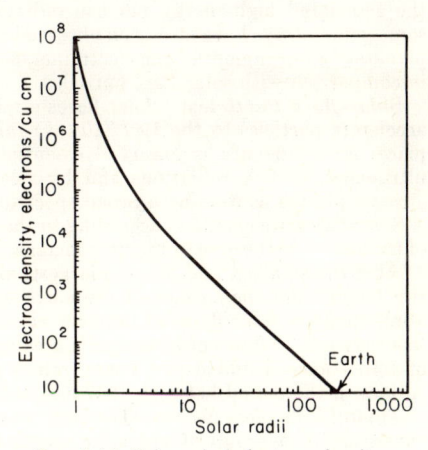

FIG. 7-19. Solar-wind electron density.

does contain a large number of narrow Fraunhofer absorption lines. This absorption takes place in the cooler upper layers of the photosphere.

The chromosphere consists of a thin layer (approximately 10,000 miles) of highly rarefied radiating gas just above the photosphere. Its temperature is considerably less than 6000°K, but its emission spectrum is quite similar to that of the photosphere. At wavelengths shorter than the visible, the radiation from the photosphere falls off sharply until about 1,700 A where it fades completely and the radiation is entirely from the chromosphere. However, only a minute fraction of the energy radiated by the sun is outside the spectrum shown in Fig. 7-20. The detailed characteristics of the energy radiated in the far ultraviolet region are still uncertain. A great deal of experimental data has been taken by rockets and satellites in the last few years. These data indicate that there is a tremendous variety in the emission spectrum in this region, and the observations have been summarized in Ref. 15.

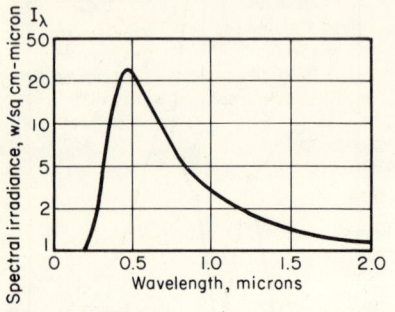

FIG. 7-20. Solar electromagnetic radiation.

The solar corona, which has been discussed above, consists of an extremely rarefied plasma of mostly electrons and protons extending out to distances as far as the earth's orbit. Electromagnetic radiation from it is due primarily to scattered sunlight and in the visible spectrum is only a millionth as strong as that from the photosphere. Some coronal emission lines exist and have been identified as due to highly ionized (because of the high temperature) metallic ions. The temperature is in excess of 10^6°K at the base of the corona and is believed to be greater than 10^5°K even out as far as the earth's orbit. Solar X rays with energies as high as 100 Kev have been observed. In general, the X-ray spectrum is consistent with these high temperatures.

Active Sun. There is a large variety of solar activity both periodic and nonperiodic which affects to some degree the electromagnetic emission from the sun. The active solar phenomena include sunspots, faculae, plages, filaments, prominences, and flares. The largest of these disturbances, solar flares, is discussed below in connection with the associated high-energy particle radiation. Hard X rays (up to 100 Kev) and some gamma rays have also been observed.[16] However, the contribution to the interplanetary environment of the electromagnetic radiation due to solar activity is minute in comparison with solar-flare particles.

Solar-flare Particles. Solar flares are large chromospheric eruptions which may accelerate particles to the Bev (10^9 ev) range of energies. When a solar flare takes place, its occurrence is observed essentially immediately on the earth by visible, ultraviolet, and X radiations and by related ionospheric disturbances. The flare ejects a plasma containing a broad spectrum of energetic particles (see Ref. 32).

Solar flares are classified according to the area of the flare on the visible hemisphere of the sun. As the area of the flare increases, the flare is said to increase in importance, 1 being the smallest and 3+ the largest; an importance 2 flare covers an area of the sun larger than the surface of the earth. The mean number of flares per 24 hr is approximately 8 for flares of importance 1, 2 for flares of importance greater than 1, less than 0.2 for flares of importance greater than 2, and approximately 0.05 for flares of importance 3 and 3+. These values are averages taken over the entire solar surface in the period between 1947 and 1953.

Typical Solar-flare Events. During a solar-flare event, the sun ejects a "solar beam" consisting of a group of energetic charged particles, most of which are protons.[17] However, electrons and gamma rays are also present. The intensity and character of solar beams vary greatly. The more common type of solar beam occurs at a rate of about 10 per year during the solar maximum. The flux of protons with kinetic energies greater than 75 Mev may vary from about 20 times the normal cosmic-ray intensity to about 350 times. The solar protons have an energy range of 1 Mev to about 20 Bev in some cases.

The sequence of events following a normal solar flare is important in understanding the processes occurring. This sequence is generally as follows:

1. If the plasma contains a large number of high-energy particles ($E > 40$ Mev) they may reach the vicinity of the earth within a few hours (or even in some cases as little as 25 min) after the beginning of the flare. These are generally called the "prompt" radiation. In addition to being detected themselves at high latitude and altitude, they will give rise to several secondary effects, such as increased counts in sea-level neutron monitors, increased absorption of cosmic radio noise, increased ionization of the upper atmosphere causing interference in communications, and magnetic disturbances.

2. Lower-energy (1 to 15 Mev) protons ejected from the flare seem to be contained in the main body of the plasma cloud. Those beginning to be detectable at high latitude and altitude some 20 to 40 hr after the flare represent a leakage from the containment of the main body of the plasma and are a forerunner of its arrival.

3. The sudden commencement of a geomagnetic storm indicates the arrival of the main body of the plasma to the immediate vicinity of the earth. The intensity of low-energy protons increases greatly, but there are only minor changes in the geomagnetic cut-off rigidity values (see below under Geomagnetic Exclusion).

4. At the time of the beginning of the main phase of the storm, the flux of low-energy protons has decreased sharply from its initial phase high point, indicating a rapid dissipation of particles in this energy range. Now, for the first time, considerable changes in the geomagnetic cut-offs are clearly evident; low-energy protons are detectable at latitudes down to 53 and 55°. The cut-off depression is apparently directly related to the magnitude of the main phase of the storm, and in addition depends on the energy of the protons.

5. In the latter part of the main phase of the magnetic storm and during the recovery period, the cut-offs gradually return to their prestorm values, while the flux of low-energy protons dies off rapidly.

The integral energy flux may be represented by $N(>E) = KE^{-\gamma}$ protons/(sq cm)(sec). K varies as kt^{-x} with time t; i.e., the multiplicative constant K is quite variable from flare to flare. In the proton energy region from 85 to about 400 Mev, a power-law representation with the exponent γ between 3.5 and 5 is quite satisfactory (Fig. 7-21). At lower energies the spectrum is less steep, and a lower value of γ is indicated. For a typical large event, which may be expected to happen about once each year, the expression for the integral flux is $N(>E) \simeq 1.5 \times 10^{11} t^{-3} E^{-4}$ particles/(sq cm) (sec) for 85 Mev $\leq E \leq 300$ Mev and $t > 1.2$ days (E in Mev, t in days). The proton peak intensity for the large flare described above was approximately

$$N(>30 \text{ Mev}) = 1.5 \times 10^5 \text{ protons}/(\text{sq cm})(\text{sec}) \text{ at } t = 1.2 \text{ days}$$

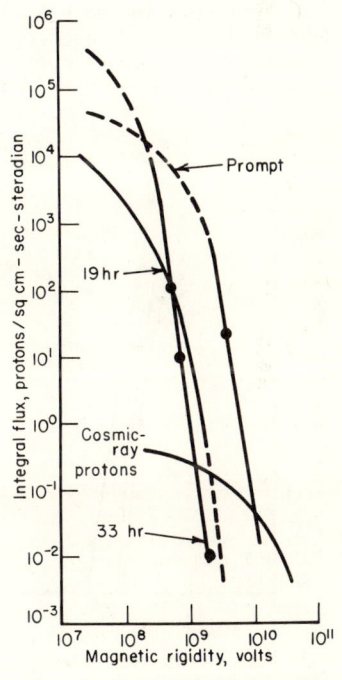

FIG. 7-21. Typical solar-flare spectrum.

Over a 1-day interval, the flux of protons above 30 Mev was approximately 5×10^9 protons/sq cm. The average yearly production rate is on the order of 10^8 protons/sq cm above 100 Mev or 10^{10} protons/(sq cm)(year) above 30 Mev. The corresponding production rates of the sun (over 4π steradians) are 10^{35} protons/year above 100 Mev and 10^{37} protons/year above 30 Mev.

7-18　　　　　　　　　　SYSTEM ENVIRONMENTS

The dose corresponding to 5×10^9 protons/sq cm at an energy of 30 Mev is on the order of 3,000 roentgens in 1 g/cm, which is their range.

Relativistic Solar-flare Events. Only 10 great solar-flare events occurred in the first 25 years after equipment capable of observing these events went into operation (in 1937). The largest event was that of Feb. 23, 1956.

Relativistic solar cosmic-ray events are observed by a number of means. At first only ground-level neutron monitors observed the enhancement of neutrons produced in nuclear reactions by the protons on nitrogen and oxygen in the upper atmosphere. Later the protons themselves were observed with instrumentation flown on high-altitude balloons and rockets.

With only one exception (at 02°S), all the solar flares producing relativistic protons occurred in the northern half of the solar disk, and with only two exceptions all were in the northwestern quarter. One flare was quite exceptional in that solar protons were observed, although it was 23° behind the west limb of the sun. However, its most elevated portions were definitely seen above the limb. This cannot be construed as evidence for protons from an "invisible" flare or from a flare on the back side of the sun since some portions were observable. It appears likely that the cosmic rays extend out into space over a 2π solid angle above the solar surface. The typical events following the onset of a relativistic solar flare are as follows:

1. The proton intensity increases rapidly, maximum intensity occurring within about 15 to 100 min of the observation of the electromagnetic radiation from the flare. For the greatest flare recorded, a peak flux of 10^5 protons/(sq cm)(sec) was observed with values of about 10^3 protons/(sq cm)(sec) being more typical. Once the maximum is reached, the intensity commences a smooth decline to the pre-event level. The time required to attain maximum intensity is considerably less than that required for the intensity to return to the pre-event level.

2. The integral rigidity spectrum of the flare-produced cosmic radiation is very steep, varying approximately as (rigidity)$^{-7}$ (Fig. 7-22).

3. The event occurs at roughly the same time at all points on the earth. The small differences in onset time ($\sim$10 min) that do occur can be explained as follows: The higher-rigidity particles arrive at the earth first, and appear to come from within a relatively small solid angle in the direction of the sun. Thus, at first, only stations within several limited regions (impact zones) on the earth's surface observe a cosmic-ray enhancement. With the passage of time, lower-rigidity particles arrive,

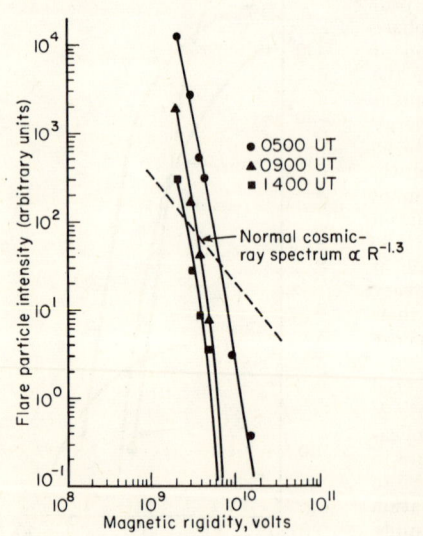

FIG. 7-22. Relativistic solar-flare spectrum.

the radiation becomes isotropic, and all stations for which the rigidity cut-off is low enough observe an intensity enhancement.

Directional Characteristics of Protons. Solar-flare protons are almost always observed by balloon or rocket after impinging on the atmosphere. All the experimental evidence to date points to an angular distribution which is isotropic over the upper hemisphere except perhaps in the first minutes of arrival of the prompt protons from the flare. In space, away from the absorbing atmosphere and the deflecting geomagnetic field, the protons contained in the plasma cloud are probably isotropic over the entire 4π solid angle. Therefore, in space, directional shielding of a spacecraft would be of no advantage.

The prompt protons which arrive at the earth within minutes or at most hours after the onset of the flare may indeed have an angular distribution peaked in the direction of the sun. This conclusion, however, is by no means certain and no data are available.

Galactic Cosmic Rays. High-energy radiation consisting mostly of completely ionized atomic nuclei, the electrons having been stripped away, is present at all times in the interplanetary space.[32] Two distinct sources of this radiation have been identified, the galaxy and the sun. Although the exact mechanism for the generation of galactic cosmic rays is not clear, it is now believed that the primary source, except possibly for particles of very high energy, is supernovae.[18] For cosmic-ray particles with energies as high as 10^{17} ev, it is probably more realistic to assume an extragalactic origin. Solar-flare events which have a much softer particle-energy spectrum originate in active regions on the sun.

Composition. The composition of cosmic radiation is roughly similar to that of matter in the universe as a whole; that is, the fractional abundances of different elements are about the same in cosmic radiation as in the entire universe. The principal constituent is hydrogen nuclei, or protons, and the next most abundant component is helium nuclei, or alpha-particles. The heavy elements are present in somewhat greater abundance in cosmic radiation than in the universe, as deduced from other observations. The elements lithium, beryllium, and boron, which are present in anomalously low concentrations in the universe as a whole, are present in the cosmic radiation observed at balloon altitudes with nearly the abundance which could be expected on the basis of the concentrations of elements which lie next to them in the periodic table. Table 7-6 shows the realtive concentrations as given by Peters.[19] Z is the atomic number.

Table 7-6. Cosmic-ray Relative Abundance

Element	Relative abundance in cosmic radiation at balloon altitudes	Relative abundance in the universe
Hydrogen	100,000	100,000
Helium	7,000	7,500
Lithium, beryllium, boron	35	3×10^{-3}
Carbon	190	10
Nitrogen	120	15
Oxygen	190	50
$10 \leq Z \leq 30$	140	30
$Z > 30$	<0.1	8×10^{-4}

Geomagnetic Exclusion. Since the cosmic-ray particles are electrically charged, their paths are bent by magnetic fields. The deflections of cosmic-ray particles by the geomagnetic field prevents the lower-energy particles from reaching the earth's surface, especially at low latitudes. The controlling factor is the ratio of the particle momentum to the particle charge, because it is a measure of the particle's radius of curvature in a given magnetic field. Extensive calculations on the effect of the geomagnetic field in excluding protons and multiply-charged cosmic-ray particles have been made.[20] Near the geomagnetic equator, protons cannot reach the earth unless they have momenta greater than about 15 Bev/c (1 Bev/c = 5.3×10^{-14} g-cm/sec). The value depends on the angle at which the particles reach the earth, and the value given here is for particles arriving vertically. At higher latitudes, particles with lower momenta are able to make their way through the magnetic field, and alpha-particles (and heavier particles) do not require so much momentum to get through the magnetic field as do protons.

Flux and Energy Spectrum. The free-space flux of galactic cosmic rays in the solar system during the time of sunspot maximum is about 2.5 particles/(sq cm)(sec), and the ionization rate (inside a volume shielded by approximately 1 g/sq cm of material of low atomic number) is about 0.6 mr/hr.

The integral-energy spectrum of the galactic cosmic rays is represented by an inverse-power law in energy. The number of particles with energy greater than E is proportional to E^{-q}; i.e., $N(>E) \alpha E^{-q}$ particles/(sq cm)(sec)(steradian), where q varies from about 1 to 2.5 with the higher values of q corresponding to the higher energies (Fig. 7-23). Measurements of the energy spectrum of various atomic species show that the exponent is nearly equal to that for protons but falls off more rapidly at higher energies. Thus the power law also represents an energy spectrum per nucleon for the heavier components (Fig. 7-24). The remarkable thing about the galactic cosmic rays is that the spectrum extends up to extreme energies of at least 10^{18} ev. The low-energy limit of the galactic spectrum is less than 10^8 ev. The average energy is about 4×10^9 ev/nucleon, and the energy flux falling on the surface of the earth near the pole—that is, from a solid angle of 2π steradians and unaffected by the magnetic cut-off—is about 7×10^{-3} erg/(sq cm)(sec). The energy density of cosmic radiation near the earth is about 0.6 ev/cu cm, which is about equal to that of starlight.

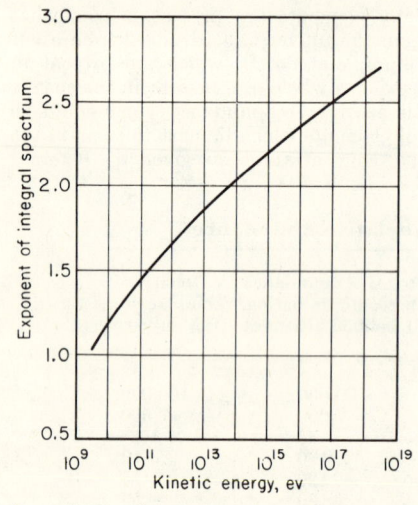

Fig. 7-23. Cosmic-ray integral-spectrum exponent.

Fig. 7-24. Cosmic-ray integral-energy spectrum.

Directional Characteristics and Temporal Variations. Galactic cosmic radiation is spatially isotropic near the earth. Although there are no data pertaining to interstellar space, it seems quite probable that the distribution is isotropic there also.

The flux values given above are typical of galactic cosmic rays during the time of sunspot maximum. The galactic cosmic radiation is modulated during the solar cycle in such a way that the intensity decreases as solar activity increases. The solar-cycle modulation is such as to attenuate the low-energy particles more severely than those of high energy. The overall ionization near the top of the atmosphere at high latitudes changes by a factor of about 2 between the period of sunspot minimum and the period of sunspot maximum. In addition to this 11-year cycle change, the galactic cosmic rays are subject to sudden decreases of intensity called "Forbush decreases." They are caused by large solar eruptions. It is presumed that the modulation occurs because the plasma thrown out during the flare carries a magnetic field with it from the sunspot group, thus producing a decreased intensity within the cloud. Recent measurements with Pioneer V support this conjecture, as relatively huge fields, of about 50-gamma intensity, were observed in space during such a decrease. It is known that these Forbush decreases are not associated with the environment of the earth but exist in the interplanetary medium. The Forbush decreases may be as large as 25 or 30 per cent. They usually occur when solar activity is high—that is, at periods of sunspot maximum—when the galactic cosmic radiation is already attenuated by the solar-cycle effect.

Meteoritic Material. The interplanetary medium contains a significant amount of material in orbit about the sun. Both direct and indirect observations have been made and the terms meteors, meteorites, and micrometeorites (or dust) are used to describe various aspects of these observations. Meteorites are the relatively large pieces of material that fall to the ground from interplanetary space. Their origin is believed to be asteroid fragments. Meteors are the flashes of light observed in the sky when meteoric material interacts with the upper atmosphere. Micrometeorites are tiny dustlike particles in interplanetary space and are possibly fragments of the particles causing the visible meteors and thus probably material from the same origin, probably comets. As comets approach the sun, they are heated and the gases liberated carry off micrometeorite debris (see below under The Solar System). Most of this debris remains in the cometary orbit and accounts for the large increase in meteor activity as the earth passes through cometary orbits.

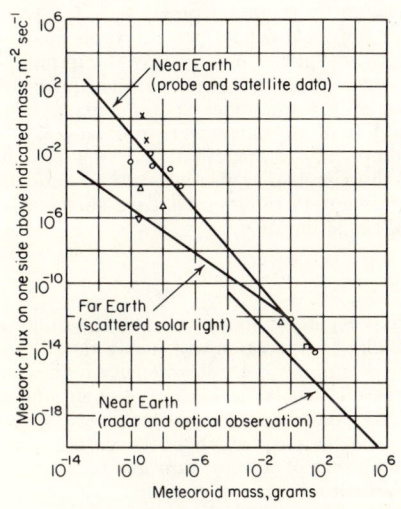

FIG. 7-25. Meteoric flux.

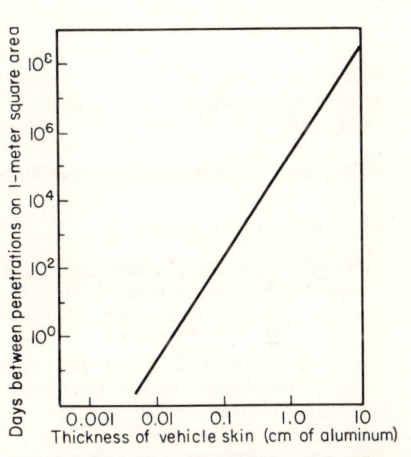

FIG. 7-26. Meteoric-penetration probability.

The material composition of interplanetary space has been studied over a long period of time using earth-based observations of meteor trails and light scattering off the dust particles in the solar corona. Observations from earth-based stations are made by measuring both the visible light from the meteor trail and the radar reflection off the meteor track of enhanced ionization. Light scattering off the dust particles in interplanetary space (corona and zodiacal light) has been studied extensively from high-altitude mountain observatories and airplane flights. Direct measurements of the interplanetary dust or micrometeorites have been made, using rocket probes, earth satellites, and interplanetary spacecraft flights, by detecting the impingement of the micrometeorite particles on a microphone. Although calibration is difficult, it is possible to obtain an estimate of the momentum of the particle by this means as well as to detect its presence. In addition, rocket probes have been flown which collect micrometeorite dust at high altitudes and return it to earth for analysis.[21] An analysis of the micrometeorite particles recovered from rocket probes indicates a varied atomic composition. The presence of aluminum, silicon, iron, titanium, nickel, calcium, and possibly magnesium and copper have been detected.

The experimental evidence, based upon the observations noted above, is somewhat conflicting. Figure 7-25 summarizes the data presently available. These data indicate that there is a concentration of micrometeorite particles near the earth. However, there is no conclusive evidence that the material is in long-term geocentric orbit. Rather, it appears that the concentration may be due to material recently captured into geocentric orbit. A summary of the experimental results and discussion of the data are presented in Refs. 22 and 32. Based upon these data, the meteoric-penetration probability for an earth satellite is shown in Fig. 7-26.

The velocity distribution and size distribution of the micrometeorite particles is also uncertain. Sizes range from a fraction of a micron up to many microns in diameter. The velocity of the micrometeorite particles in a trapped heliocentric orbit near the earth could be as high as 42 Km/sec. For micrometeorites penetrating the earth's atmosphere in retrograde orbit, the earth's orbital velocity of 29 Km/sec would be added to this. However, it is believed that most micrometeorite particles travel in direct orbits about the sun.

Several factors influence the orbital parameters of micrometeorite particles. In addition to the gravitational attraction, there is a repulsive force on these particles due to the impingement of solar electromagnetic radiation. For perfectly absorbing spheres, this would mean all particles of radius greater than $0.6/\rho$ (ρ is the density in g/cu cm) would be swept out of the solar system by radiation pressure. However, this consideration is modified by the Poynting-Robertson effect, which causes dust to spiral into the sun. The micrometeorite orbits are further modified by the impingement of the solar wind, which adds a corpuscular pressure to the radiation pressure and, in addition, sputters away the surface of the micrometeorite. Finally, the gravitational attraction of the planets would tend to concentrate micrometeorite particles in the plane of the ecliptic. The complexity of the situation is such that it is not possible to determine the fraction of micrometeorite particles that would be captured in the geocentric orbit, and the experimental evidence is not yet sufficiently detailed to yield a complete understanding of the situation.

7-3. Interstellar Space

The term interstellar space is applied to that region of space between the billions of stars which make up a galaxy, and in particular to the region of space in the Milky Way galaxy. It is characterized by a relatively uniform, very-low-density hydrogen gas, clouds of ionized and neutral gas of increased density, a mixture of chaotic and uniform magnetic fields, and a distribution of very fine dust particles. Galactic space is that region of space between the galaxies, and little is known about its composition. Theoretical arguments indicate that it is composed of a very rare gas of average density $\rho \approx 10^{-31}$ g/cu cm or 10^{-7} hydrogen atoms/cu cm.

The composition of interstellar space is deduced from optical and radio-frequency observations.[23] The results are sometimes conflicting, and the data presented are still uncertain to an order of magnitude in many cases. Table 7-7 summarizes the

Table 7-7. Composition of Interstellar Space

Average interstellar hydrogen density (in galactic plane)	1 atom/cu cm
Hydrogen density in neutral gas clouds	10 atoms/cu cm
Protons (and electrons) in ionized gas clouds	10 to 10^4/cu cm
Density of neutral gas clouds	10^5 to 10^6/cu kiloparsec
Size of neutral gas clouds	1–10 parsecs
Percentage of space occupied by neutral gas clouds	10
Percentage of space occupied by ionized gas clouds	<1
Average magnetic field	10^{-5} gauss

results of current observations of our galaxy. The clouds of gas are in circular rotation about the galactic center in roughly the same manner as the stars. The magnetic field associated with the clouds is assumed to be chaotic. However, there is some evidence for additional large-scale uniform magnetic fields paralleling the spiral arms of the galaxy. The principal means for obtaining information on the interstellar medium has been through radio-astronomy studies.[24] In particular, the interstellar hydrogen has been intensively studied by this means. The wavelengths corresponding to ground-state transitions of interstellar hydrogen cannot be observed on the earth because of its complete absorption by the atmosphere. This is not the case, however, for the 21-cm line of hydrogen. A ground-state hyperfine structure is formed by the interaction of the intrinsic magnetic moments of the hydrogen proton and electron which split it into two very close levels. By studying the absorption of starlight at this wavelength, it has been possible to determine hydrogen density, temperature, and cloud size and movement in interstellar and to some extent galactic space. In addition, atomic and molecular interstellar absorption lines have been observed by optical means for Na I, K I, Ca I, Ca II, Ti II, Fe I, and CH and CH$^+$.

7-4. Astronomical Data

The astronomical data presented in this section tie in closely with the earth's local space environment. The interplanetary space merges smoothly into interstellar space, and some of the phenomena previously discussed have their source outside the solar system. Indeed the present solar system and its space environment are just one step in a long evolutionary sequence involving billions of stars, some similar and many older or younger.

Astronomical Terms. *Definitions.* The definitions of commonly used astronomical terms are given below. A more complete discussion and definitions of other terms may be found in Ref. 25.

Absolute Magnitude (M). The apparent magnitude a star would have if it were located at a standard distance of 10 parsecs or 32.6 light-years from the earth. It is related to apparent magnitude m by

$$M = +5 - 5 \log r - A$$

where r is distance (in parsecs) of a star and A is the total interstellar absorption (in magnitudes) over the path length r.

Aphelion (Q). Point at which a body passes farthest from the sun.

Apogee. Point at which a body passes farthest from earth.

Apparent Magnitude (m). The relative brightness of a star as seen from the earth. The scale of apparent magnitudes ranges from -26.7 for the sun to $+24$ for the farthest star observable with the 200-in. telescope. The apparent brightness ratio b_2/b_1 of any two stars of apparent magnitude m_2 and m_1 is

$$\log \frac{b_2}{b_1} = 0.4(m_1 - m_2)$$

Apparent Motion of Stars. While the earth revolves around the sun, the near stars seem to describe orbits relative to more distant stars. This apparent orbit varies from a circle for a star at the ecliptic pole to a straight line for a star on the ecliptic.

Argument of Perihelion. Angle along the orbit in the direction of the body's motion between the ascending node and the point of perihelion passage.

Ascending Node. Point at which body passes ecliptic plane going from south to north.

Astronomical Unit (A.U.). The average (of the least and the greatest) distance of the earth's orbit from the sun (1.496×10^{13} cm or 92,900,000 miles).

Celestial Equator. Extension to infinity of earth's equator.

Celestial Poles. Extension of the earth's axis of rotation.

Celestial Sphere. Imaginary extension to infinity of the sphere of the earth.

Constellations. Regions of the celestial sphere set off by arbitrary boundary lines. The visible stars within each region form patterns which are given proper names. It is useful to identify the position of a star as being located in a particular constellation or region of the celestial sphere.

Ecliptic Plane. Plane of the earth's orbit about the sun.

Ephemeris. Tabular position of stars and other objects in space. Reference is made to the annual edition of the *American Ephemeris and Nautical Almanac* and the *Explanatory Supplement* for tables of particular dates and times. The positions of the sun, moon, and planets are reckoned in ephemeris time.

Galactic Coordinate System. See below and Fig. 7-29.

Galaxy. Group of stars, star clusters, gas, and dust which move through the universe as a unit.

H-R Diagram. Hertzprung-Russell plot of the absolute luminosities or magnitudes of stars vs. spectral type or color.

7-24 SYSTEM ENVIRONMENTS

Hubble's Constant (H). Proportionality constant $[24 \times 10^{-19}$ (cm/sec)cm] in the relation which gives a galaxy's velocity v of recession from our galaxy as a function of distance r from our galaxy; $v = Hr$.

Inclination (i). The angle between the plane of an orbit and the ecliptic plane.

Line of Nodes. Line of intersection between plane of the orbit and plane of the ecliptic.

Longitude of Ascending Node (Ω). The angle (celestial longitude) between the ascending node and the direction of the vernal equinox.

Luminosity (L) of a Star. The total power emitted. The luminosity of a star is customarily expressed in terms of the sun's luminosity

$$\log L = 0.4(4.8 - M)$$

where 4.8 is the sun's absolute magnitude and M is the absolute magnitude of the star.

Main Sequence. Smooth curve on which the majority of stars fall on the H-R plot.

Nadir. The point on the celestial sphere directly below the observer.

Parallax (p). The angle subtended by the mean radius of the earth's orbit as seen from the star. The unit of measure of parallax is usually the parsec, which is the reciprocal of the angle of parallax in seconds of arc.

Perigee. Point at which a body passes closest to earth.

Perihelion. Point at which a body passes closest to sun.

Precession of the Equinoxes. A slow circling of the earth's axis of rotation about the poles of the ecliptic (celestial latitude $\pm 90°$) due to action of the sun and the moon. The earth's precessional motion is very slow, requiring approximately 26,000 years for the earth's axis to trace out one complete circle, or 50.2 seconds/year.

Proper Motion (μ). The rate of change in a star's apparent place in the celestial sphere in terms of right ascension and declination.

Star Clusters. Groups of stars which appear to move together for long periods of times.

Stellar Radial Velocity (V). Star's rate of motion directly toward or away from the sun. This is obtained from displacement of the lines of a star's spectrum from their normal position.

Stellar Space Velocity (v). Star's velocity with respect to the sun; $v = \sqrt{V^2 + T^2}$.

Stellar Tangential Velocity (T). Star's velocity perpendicular to the line from the sun.

Vernal Equinox (Υ). One of the two imaginary points in space where the celestial equator intersects the ecliptic (Fig. 7-28).

Zenith. The point on the celestial sphere directly above the observer.

Zodiacal Light. Faint glow across a band in the sky due primarily to sunlight reflected off micrometeorite dust particles.

The Measure of Distance. The various units for measuring distance in space are related as shown in Table 7-8.

Table 7-8. Units of Measure of Distance in Space

	Km	Miles	A.U.	Light-years	Parsecs
Kilometers, 1 Km........	1	0.62137	0.699×10^{-8}	1.057×10^{-13}	0.324×10^{-13}
Miles (statute), 1 mile........	1.6093	1	1.076×10^{-8}	1.701×10^{-13}	0.522×10^{-13}
Astronomical unit, 1 A.U...	1.495×10^8	9.29×10^7	1	1.58×10^{-5}	0.485×10^{-5}
Light-years, 1 l.y.........	9.460×10^{12}	5.88×10^{12}	63,280	1	0.3068
Parsecs, 1 pc....	3.084×10^{13}	1.916×10^{13}	206,265	3.260	1

The Measure of Time

Apparent Solar Time. The local hour angle of the sun's center plus 12 hr. It begins at apparent midnight and runs through approximately 24 hr (Table 7-9, Fig. 7-28).

Ephemeris Time (E.T.). Continuous running arbitrary unit of time starting at the beginning of the year A.D. 1900. The arbitrary unit of measure is the length of that tropical year divided by the number of seconds in that year.

Julian Date Time. Continuous count of days (365.25 days/year) starting ephemeris noon on Jan. 1, 4713 B.C. The fundamental epoch 1900 January E.T. is Julian Ephemeris Date (J.E.D.) 2415020.0.

Table 7-9. The Length of the Apparent Solar Day throughout the Year

	hr	min	sec
Jan. 1	24	0	29
Apr. 1	23	59	42
July 1	24	0	12
Oct. 1	23	59	41

Mean Solar Time. The annual average of the apparent solar time. This is our daily time.

Sidereal Period. The true period of the planet's revolution around the sun.

Sidereal Time. The local hour angle of the vernal equinox (Fig. 7-28). It begins with the upper transit of the vernal equinox and runs about 24 hr until the next upper transit; 1 sidereal day = 0.997 mean solar day.

Sidereal Year. The true period of the earth's revolution about the sun. Its length is 365.25636 days of mean solar time and is increasing at the rate of 0.01 sec/century.

Synodic Period. The interval between two successive conjunctions of a planet with the sun, as seen from the earth.

Tropical Year. The period of the sun's return to the vernal equinox. Its length is 365.2422 days of mean solar time and is decreasing at the rate of 0.53 sec/century.

Universal Time (U.T.). Local mean solar time at the Greenwich meridian.

Coordinate Systems. Various coordinate systems are used for astronomical observations. The three most common are the horizon system, the equatorial system, and the galactic system.[26] See also Chapter 17.

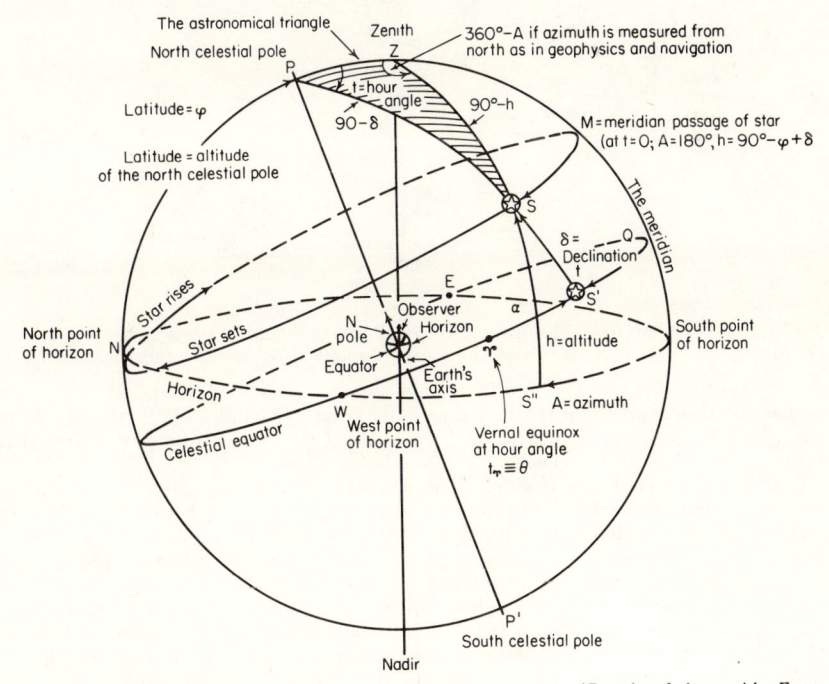

Fig. 7-27. The horizon and equatorial coordinate systems. (*Reprinted from Air Force Surveys in Geophysics with the permission of the U.S. Air Force Cambridge Research Laboratories. Copyright* 1962.)

7-26 SYSTEM ENVIRONMENTS

The Horizon System. This system of coordinates is illustrated by Fig. 7-27. The observer is at latitude ϕ with zenith Z the overhead vertical. The zenith and the celestial pole P define the observer's meridian. The horizon is normal to the vertical, and north and south points on the horizon are defined as azimuth $A = 0°$ and $A = 180°$, respectively; east and west points are at $A = 90°$ and $A = 270°$, respectively. Altitude h is the angular distance $S''S$ of a coordinate point S.

The Equatorial System. The equatorial coordinate system uses the declination δ measured from the celestial equator as shown in Fig. 7-27 and either the hour angle t or the right ascension α for the other coordinate. The hour angle t is the angle QS' with the positive angle measured west from the meridian. The right ascension α

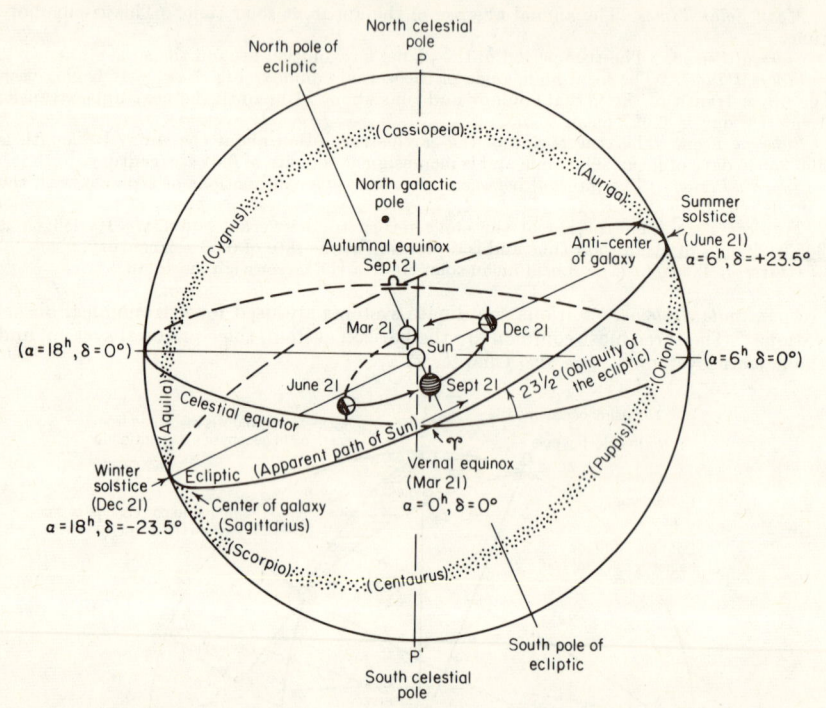

FIG. 7-28. Coordinate relationships. (*Reprinted from Air Force Surveys in Geophysics with the permission of the U.S. Air Force Cambridge Research Laboratories. Copyright* 1962.)

is measured eastward from the vernal equinox (Figs. 7-27 and 7-28). Also shown in Fig. 7-27, the sidereal time t_Υ is the hour angle of the vernal equinox θ. Note that for any sidereal time the relation $t + \alpha = \theta$ holds.

The Galactic Coordinate System. The galactic equator is a great circle on the celestial sphere inclined 62° to the celestial equator. The north galactic pole is at right ascension 12^h 40^m and declination 28°. The galactic longitude l is measured 0 to 360° counterclockwise as seen from the north along the galactic equator from its intersection with the celestial equator. Galactic latitude b is measured from 0 to 90° positive toward the north galactic pole. This puts the galactic center near longitude $l = 370°$. A revision of the galactic coordinate system to place the center of the galaxy at 0° is being undertaken, and current observations are published in the new galactic coordinate system. Conversion tables are being published in the *Annals of the London Observatory.* Figure 7-29 illustrates the new galactic coordinate system.

SPACE AND ASTRONOMY 7-27

The Solar System. The sun is the central body of the solar system, 98 per cent of the total mass. Revolving around it in elliptic orbits are 9 planets with 31 known satellites. In addition, there are tens of thousands of asteroids and numerous comets. These objects comprise the solar system.

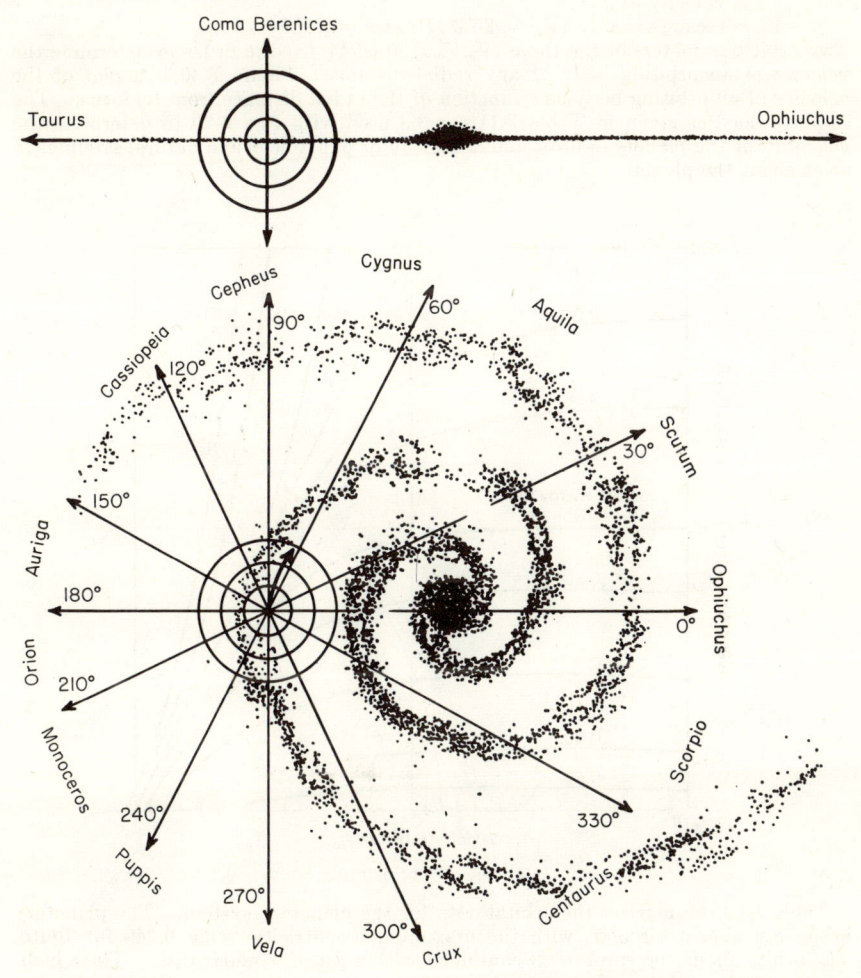

Fig. 7-29. Galactic coordinate system. (*Reprinted from Air Force Surveys in Geophysics with the permission of the U.S. Air Force Cambridge Research Laboratories. Copyright* 1962.)

Orbital Data. Bodies which remain within the solar system maintain elliptic orbits. The orbits of planets, asteroids, and comets are ellipses with the sun at one focus. The orbits of satellites of planets are ellipses with the planets at one focus. The motion is governed by the well-known gravitational law of Newton. The relation between the velocity V of an orbiting body and its distance from the focus is

$$V^2 = GM \left(\frac{2}{r} - \frac{1}{a}\right) = V_0^2 + V_e^2 \left(\frac{R}{r} - 1\right)$$

where G = universal gravitational constant
  M = mass of the central body
  r = radial distance from the focus to the orbiting body
  a = semimajor axis
  R = arbitrary radial distance at which V_e and V_0 are fixed
  V_0 = velocity at $r = R$
  V_e = escape velocity ($V_e^2 = 2GM/R$) at $r = R$

Two orbit parameters of the three (V_e, V_0, a) must be fixed in order to determine the velocity of an orbiting body at any radial distance. Figure 7-30 is a plot of the velocity of an orbiting body as a function of the radial distance from its focus. The escape velocities given in Table 7-11 can be used with Fig. 7-30 to determine the velocities of the planets in orbit about the sun or of artificial and natural satellites in orbit about the planets.

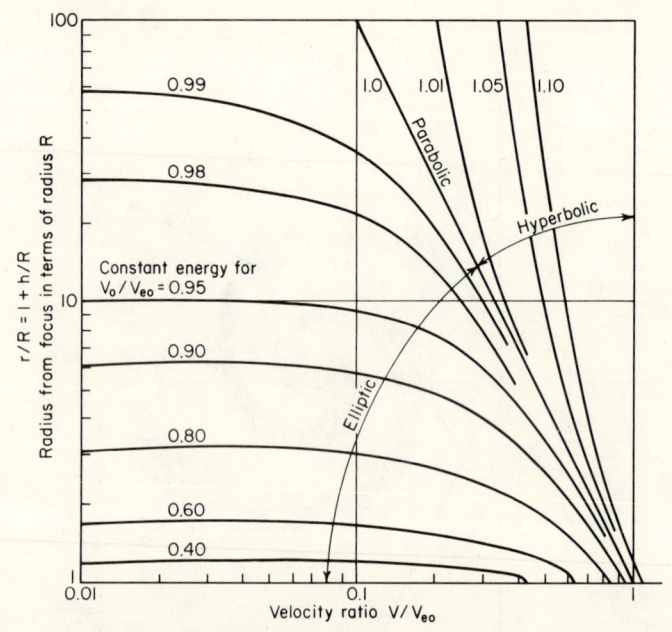

FIG. 7-30. Velocity of orbiting body.

Table 7-10 summarizes the orbital data for the planetary system. The planetary orbits are almost circular, with the maximum eccentricity being 0.249 for Pluto. The orbits about the sun are all contained within a thin circular disk. Their inclinations to the ecliptic plane are only a few degrees, with the exception of Mercury which has an orbital inclination of 7° and of Pluto which has an inclination of slightly over 17°.

Physical Data. Table 7-11 summarizes physical data on the planetary system. Many values in the table are uncertain. The general references included should be consulted for further information. Reference 27 is a recent review of the planetary atmospheres, which are described in Sec. 5-1 of this handbook.

The Sun. The sun is our nearest star. Its mass is so much greater than that of any other body in the solar system that it is usually treated as the center of mass of the solar system. Because of its fluid state the sun rotates about its axis from west to east at a rate which varies from about 25 days at its equator to about 33 days at 75° latitude. As seen from the earth, the sun's apparent rotation takes about 2

SPACE AND ASTRONOMY 7–29

Table 7-10. Solar-system Orbital Data

Body	Symbol	Mean heliocentric distance, A.U. (10⁶ miles)	Sidereal period, days (years)	Eccentricity, e	Inclination, i	Rotation period	Equatorial inclination to orbit plane	Closest approach to earth (10⁶ miles)	No. of satellites
Sun	☉	0 (0)				24.65 days	7°10′		
Moon	☾	1 (92.90)				27.32 days (about earth)	6°41′ (earth)	0.239	
Mercury	☿	0.3871 (35.96)	87.969	0.206	7°0′	88 days	7°	50	0
Venus	♀	0.7233 (67.20)	244.701	0.007	3°24′	10 days (?) to ∞	23°	26	0
Earth	⊕	1.0000 (92.90)	365.256	0.017	0°0′	23 hr 56 min	23°27′		1
Mars	♂	1.5237 (141.7)	686.980	0.093	1°51′	24 hr 37 min	24°	34.6	2
Jupiter	♃	5.2028 (483.3)	(11.862)	0.048	1°18′	9 hr 50 min	3°7′	367	12
Saturn	♄	9.5388 (886.2)	(29.458)	0.056	2°29′	10 hr 02 min	26°45′	745	9
Uranus	⊙̂	19.1820 (1,783)	(84.015)	0.047	0°46′	10 hr 45 min	98°	1,606	5
Neptune	♆	30.0577 (2,794)	(164.788)	0.009	1°46′	15 hr 48 min	29°	2,700	2
Pluto	♇	39.5177 (3,670)	(247.697)	0.249	17°9′				

days longer than the actual rotation, because of the earth's revolution about the sun. The solar sphere consists mainly of hydrogen at pressures from 10^9 atm at the center to 10^{-2} atm at the photosphere. The temperature of the sphere is estimated to range from 20,000,000°K at the center to 6000°K at the photosphere. The bright visible surface of the sun is marked by scattered sunspots which appear as dark areas of lower temperature. Sunspots appear to develop rapidly and decline slowly. They are most abundant between heliocentric latitudes 5 and 30°. Sunspots have associated magnetic fields ranging from 100 gauss in small spots to as high as 3,700 gauss in the great sunspot of 1946. The abundance of sunspots varies from year to year in a roughly periodic manner, reaching its maximum at about 11-year intervals (1956, 1967, etc.). Solar flares appear as sudden increases in brightness near sunspot groups, lasting from minutes to several hours.[28] They are associated with intense ultraviolet radiations, high-energy particles, and radio-noise bursts. Solar-flare characteristics are discussed further in Sec. 7-2.

Just above the photosphere is a cool layer of neutral and ionized gases called the chromosphere (Sec. 7-2). It is considered probable that all known natural elements are present in the sun. The filtering effect of the earth atmosphere prevents study

Table 7-11. Solar and Planetary Physical Data

Body	Mass, earth = 1 (Kg)	Diam, earth = 1 (miles)	Density, earth = 1 (water = 1)	Gravity, earth = 1 (cm/sec²)	Surface temp, °C	Escape velocity, earth = .1 (Km/sec)	Albedo	Max stellar magnitude
Sun	331,950 (1.98 × 10³⁰)	109 (864,000)	0.25 (1.41)	28 2.74 × 10⁴	6000	55 (620)		−26.8
Mercury	0.05 (3.4 × 10²³)	0.38 (2,900)	0.99 (5.46)	0.3 (294)	400 (sun side)	0.37 (4.2)	0.073	−12.6
Moon	0.0123 (7 × 10²²)	0.273 (2,160)	0.60 (3.33)	0.165 (162)	130 to −153	0.21 (2.4)	0.07	−1.9
Venus	0.815 (5.6 × 10²⁴)	0.96 (7,600)	0.90 (4.12)	0.88 (874)	425	0.91 (10.3)	0.76	−4.4
Earth	1 (6.84 × 10²⁴)	1 (7,913)	1 (5.52)	1 (980)	15	1.00 (11.3)	0.39	
Mars	0.1064 (7.2 × 10²³)	0.52 (4,200)	0.75 (4.12)	0.398 (392)	−20 (−100 to 40)	0.45 (5.1)	0.12	−2.8
Jupiter	318.4 (2.2 × 10³¹)	10.97 (86,800)	0.24 (1.35)	2.64 (2,580)	−140	5.38 (61)	0.51	−2.5
Saturn	95.3 (6.5 × 10²⁶)	9.03 (71,500)	0.13 (0.71)	1.17 (1,144)	−110 (max)	3.26 (36.7)	0.50	−0.4
Uranus	14.5 (9.9 × 10²⁵)	3.72 (29,400)	0.28 (1.56)	0.91 890	−130 (max)	1.97 (22.4)	0.66	+5.7
Neptune	17.2 (1.3 × 10²⁶)	3.38 (28,000)	0.45 (2.47)	1.4 (1,100)	−160 (max)	2.24 (25.6)	0.62	+7.6
Pluto	<0.1 (<6.8 × 10²³)	0.45 (3,600)	0.86 (<5.5)	<0.5	−190 (max)	<0.5 (<5.3)	0.16	

of the spectrum in the chromosphere except for parts of the visible and near ultraviolet and infrared. Some of the observed constituents of the solar atmosphere are oxygen, hydrogen, sodium, iron, and calcium. The abundance of hydrogen causes the chromosphere to appear red in color. The chromosphere extends about 10,000 miles above the photosphere.

The solar corona is the outermost layer of the sun and composed of a diffuse gas of ionized atoms of hydrogen, iron, nickel, calcium, and argon at temperatures in the order of 1,000,000°K. It is normally visible during solar eclipse and extends to distances of many solar diameters and possibly out as far as the earth's orbit (Sec. 7-2).

Mercury. The mean density of the planet indicates the probability of a heavy liquid core such as the earth's,[29] probably predominantly iron. It has an extremely diffuse atmosphere, if any. Because its rate of rotation is the same as its period of revolution, roughly the same face of the planet is always directed toward the sun, causing the surface temperature on that side to rise to about 400°C, which is hotter than the melting point of tin and lead and too high to sustain any form of life as we know it. The topography of Mercury appears to be mountainous. No radiation has been detected from the permanently dark side where the temperature is probably extremely low.

Venus. The planet Venus approaches nearer to the earth (26 × 10⁶ miles) than any other planet, and its atmosphere is believed to be mostly N_2. Observation from the earth of the solid surface has been difficult because the planet is surrounded by a thick (about 17 miles) cloud cover which starts about 45 miles above the surface and blocks out all but a small amount of sunlight, making even the bright side of Venus much darker than the earth. Recent studies of Venus by means of the deep space probe Mariner II seem to indicate that the clouds consist largely of condensed hydrocarbons. The dense cloud cover reflects heat back to the surface, and it slowly circulates around the planet, which results in little temperature change between the sunlit and darkened sides. The surface is probably rough (sand and dust) and dry, with negligible amounts of water. The average surface temperature (425°C) is too hot to sustain life as we know it. Venusian pressure is at least 20 atm (294 psi). Magnetometer measurements made by Mariner II show a nonexistent or very weak magnetic field.[6] Recent radar measurements indicate that Venus may not be rotating.[30]

Mars. As viewed from the earth, the planet Mars appears to have white polar caps which vary with the season and are believed to be a thin layer of frost or snow. Two major types of semipermanent surface formations are observed, the bright orange-red continents and the dark grayish maria. The coloration of the maria also has regular seasonal variations, indicating the possibility of a vegetation of primitive plants. The Martian surface is relatively flat, without sharp relief features. However, there do appear to be some low-level plateaus and valleys, with the upper limit on mountain heights believed to be 2 or 3 Km. Three types of haze are observed. A blue haze which is predominantly present in the atmosphere obstructs all detail in the blue, violet, and ultraviolet regions of the spectrum. Yellow dust clouds have also been observed which temporarily and locally obstruct portions of the surface. Frequently on the morning limb a white haze, probably consisting of ice crystals, is observed. The Martian atmosphere is much thinner than the terrestrial, with an estimated surface pressure of 0.01 atm. The most abundant constituent of the Martian atmosphere is probably N_2 (95 per cent) with some A, CO_2, O_2, and H_2O probably present. The mean temperature of the surface is estimated to be about $-40°C$, and large temperature fluctuations indicate the Martian soil is a poor heat conductor, probably composed of dust.[26]

Jupiter. The planet Jupiter is more massive than all the other planets combined. Its major features are easily visible with a small telescope. It is marked by changing bands of color, due possibly to its rapid rate of rotation (10-hr period) which exceeds 25,000 mph at the equator. Irregular clouds have been observed, as has an elliptical-shaped red spot. The atmosphere is quite turbulent, with the most predominant constituent very likely being helium (95 per cent) and the remainder being mostly hydrogen.[27] Methane and ammonia are also present. Electromagnetic radiation in the centimetric wavelength region has been observed, indicating a possibility of cyclotron radiation from high-energy trapped electrons, similar to the earth's Van Allen belts.

Saturn, Uranus, Neptune, and Pluto. The great distance from the earth of the outer planets and their dense cloudy atmospheres has limited the amount of information available about them. Their atmosphere, with the exception of Pluto, contains some nitrogen, carbon, and hydrogen, and absorption bands indicate some ammonia and methane. Pluto, because of its small mass, would have only a very rare atmosphere. None has been detected yet.

Asteroids, Comets, and Meteors. The asteroids are small bodies which orbit about the sun, mostly between the orbits of Mars and Jupiter. They number in the thousands and range in size from hundreds of miles in diameter to less than 1 mile. They are frequently irregular in shape and rotate with periods of several hours. Although some asteroids have been given names, they are generally referred to by their catalog number, which is assigned upon determination of their orbit. The numbered asteroids now exceed 1,600. The direction of revolution of all asteroids is from west to east with periods mainly between 3.6 and 6 years.

Comets are composite bodies postulated to consist of a rather small nucleus composed of solid matter, a gaseous envelope called the coma, and when near perihelion, a less dense gaseous region called the tail. The nucleus is believed to be only several kilometers in diameter, the coma perhaps 10^4 to 10^5 Km in diameter, and the tail region about 10^6 Km long and 10^4 Km wide. Cometary mass is on the order of 10^{17} to 10^{20} g. Observations suggest that the nucleus is made of an "icy conglomerate" of solid dustlike particles and frozen gases that sublime out under the influence of the solar radiation as the comet approaches the sun. Interspersed with the sublimed gases are dust or micrometeorite fragments. The emission from the coma consists of the molecular spectra of a wide variety of neutral free radicals CN, C_2, C_3, NH, NH_2, Na, and the ionized stable molecules CO^+, N_2^+, and CO_2^+. The densities in the coma are believed to range from 10^{10}/cu cm near the nucleus to perhaps 10^3 at the periphery. The surface temperature of the icy nucleus is possibly 10 to 100°K, with the sublimed molecules having a temperature of 100 to 500°K. Two types of comet tails have been observed; only one type is present in any one comet. One type consists mainly of solid particles. The other type consists of ionized stable molecules and is observed to be pushed away from the sun. This is probably due to the interaction of the solar wind with the cometary tail (Sec. 7-2). The density of the comet tail is believed to be about 10^3/cu cm. The orbits of comets are ellipses of generally high eccentricity. Cometary orbits may be grouped into short-period comets (periods less than 200 years) observed on more than one apparition, short-period comets of one apparition only, and long-period comets (periods greater than 200 years). Table 7-12 summarizes some cometary-orbit data. The orbital elements of observed comets are recorded in the *Catalogue of Cometary Orbits* (1960) published by the British Astronomical Association.

Meteors are dense pieces of matter thought to originate from the nucleus of a comet. They are released as a product of disintegration of the comet, which occurs to some degree each time the comet passes perihelion. Meteors usually occur in swarms or streams which travel through the solar system in the same orbit as the comet from which they originated. Most meteor streams travel in or near the plane of the ecliptic, as do most short-period comets (Sec. 7-2).

Stellar and Galactic System. *Stars.* Fifty or more of the brighter individual stars have been given proper names. In some cases a star is designated with a Greek letter followed by the constellation in which it appears. Another method of designation is a number followed by the constellation name. Most stars are cataloged in the *Henry Draper Catalogue* and are identified by catalog letter and number.

The classification of stars according to their spectra has proved to be the most useful method for their systematic evaluation. Table 7-13 summarizes the classification of stellar spectra. Associated with the spectrum classification are the so-called H-R diagrams, which plot the absolute magnitudes or luminosities of stars vs. the spectral class or color. Figure 7-31 shows a typical form of the H-R diagram.

Table 7-12. Cometary Data
Recent Bright Nonperiodic Comets

Comet	Year	Perihelion passage	Perihelion distance, A.U.	Inclination of orbit i
Skjellerup	1927 k	1927 Dec.	0.18	85°
Kyves	1931 c	1931 Aug.	0.06	167
Peltier	1936 a	1936 July	1.10	79
Finsler	1937 f	1937 Aug.	0.86	146
Cunningham	1940 d	1941 Jan.	0.37	52
Paraskevopoulos	1941 c	1941 Jan.	0.79	168
Whipple	1942 f	1943 Feb.	1.36	20
Bester	1947 k	1948 Feb.	0.75	140
	1947 n	1947 Dec.	0.11	138
Honda-Bernosconi	1948 g	1948 May	0.21	23
	1948 l	1948 Oct.	0.14	23
Wilson-Harrington	1951 i	1952 Jan.	0.79	153
Mrkos	1955 e	1955 June	0.54	87
Arend-Roland	1956 h	1957 Apr.	0.32	120
Mrkos	1957 d	1957 Aug.	0.35	94

Some Periodic Comets Recently Observed

Comet	First seen	Last seen	Period, years	Perihelion distance, A.U.
Encke	1786	1957	3.30	0.34
Pons-Brooks	1812	1953	70.88	0.77
Crommelin	1818	1956	27.87	0.74
Pons-Winnecke	1819	1951	6.26	1.23
Faye	1843	1954	7.41	1.65
d'Arrest	1851	1950	6.69	1.38
Temple 2	1873	1956	5.31	1.14
Grigg-Skjellerup	1902	1956	4.90	0.86
Daniel	1909	1950	6.66	1.46
Schaumasse	1911	1951	8.17	1.20
Neujmin	1913	1948	17.93	1.54
Schwassmann-Wachmann	1927	....	16.15	5.52
Oterma	1943	....	7.95	3.41

Most stars, including the sun, fall on the line labeled the main sequence. The relationship between the luminosity L and temperature T of main-sequence stars is

$$L \propto T^{5.5}$$

Main-sequence stars are also related by an empirical mass-luminosity law

$$\log L = 3.3 \log m_s$$

where m_s is the mass in solar units. A number of stars fall off the main sequence, including the white dwarfs and red supergiants. For a given spectral class the supergiants are much more luminous than the dwarfs. The white dwarfs are very dense with extremely small radius (on the order of that of Jupiter). The red supergiants and white dwarfs are steps in the stellar-evolution sequence. Stars first formed undergo a thermonuclear reaction which converts hydrogen to helium. For 10^6 to 10^{10} years, while this conversion proceeds, the star remains on the main sequence

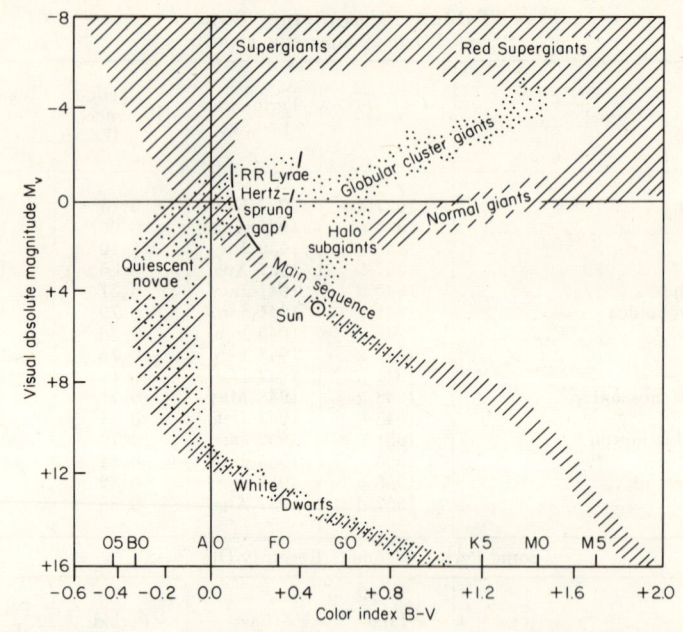

FIG. 7-31. Typical H-R diagram. (*Reprinted from Air Force Surveys in Geophysics with the permission of the U.S. Air Force Cambridge Research Laboratories. Copyright 1962.*)

Table 7-13. Stellar-spectra Classification

Spectrum class	Main sequence Temp °K	Main sequence Color index B–V	Giants Temp., °K	Giants Color index B–V
O5...........	50000			
B0...........	25000	−0.32		
B5...........	15600	−0.16		
A0...........	11000	0.00		
A5...........	8700	+0.15		
F0...........	7600	+0.30		
F5...........	6600	+0.44	6470	
G0...........	6000	+0.60	5300	
G5...........	5500	+0.68	4650	
K0...........	5100	+0.82	4200	+1.01
K5...........	4400	+1.18	3550	+1.52
M0...........	3600	+1.45	3340	+1.56
M5...........	3000	+1.69	2710	

with nearly constant luminosity. When approximately 10 per cent of the hydrogen has been used, the star moves into the red-giant area and from there back across the main sequence into the white-dwarf area with its interior devoid of hydrogen. During this latter evolutionary cycle, the star loses mass by ejection from its surface and often becomes for a short time a very bright nova or supernova.

Galaxies. The universe contains some billions of star groups called galaxies. The galaxies contain, in addition to stars and star clusters, vast amounts of cosmic dust and luminous gas clouds. They are tens to hundreds of kiloparsecs in diameter and rotate with periods of hundreds of millions of years. Galaxies are believed to evolve in an evolutionary sequence beginning with a thin irregularly shaped cluster of stars and progressing into an elliptical shape with huge spiral arms and finally evolving into a spherical shape. The stars composing the galaxies can be divided roughly into two groups, Population I and Population II. The Population I group is composed of relatively new stars and the Population II is composed of older stars which are found primarily in the nucleus of the galaxy and in the halo surrounding the spiral arms.[25]

From Doppler spectroscopic red shifts, it is evident that the galaxies are receding from us at a radial velocity v_r which is directly proportional to their distance r from the solar system. The proportionality constant H is called Hubble's constant:

$$v_r = Hr \tag{7-3}$$

The value of H is uncertain, but approximately $H = 24 \times 10^{-19}/\text{sec}$.[31] The fact that the universe appears to be expanding is of crucial importance in all cosmological theories and means, for example, that the age of the universe is about 13 billion years (i.e., H^{-1} years).

The solar system is part of the Milky Way galaxy and is located about 10 kiloparsecs from the galactic center, or about three-fourths of the way out. The galactic center is a spheroidal concentration of stars about 5,000 light-years in diameter and a strong emitter at radio wavelengths. The star concentration at the center is about two to three times as dense as in the vicinity of the solar system. There are two arms out from the center of the galaxy which spiral around in the same sense and in the same plane. The solar system is located on one of these spiral arms and is revolving about the galactic center at a speed of about 250 Km/sec and with a period of about 200 million years, or 1 galactic year.

REFERENCES

1. Elsasser, W. M.: The Earth's Interior and Geomagnetism, *Rev. Mod. Phys.*, vol. 22, pp. 1–35, 1950.
2. U.S. Air Force Geophysics Research Directorate: *Handbook of Geophysics*, pp. 10–1 to 10–68, The Macmillan Company, New York, 1961.
3. Fanselau, G., and H. Kautzleben: Die analytische Darstellung des geomagnetischen Feldes, *Geofis. Pura Appl.*, vol. 41, pp. 33–72, 1958.
4. Valley, S. L. (ed.): *Space and Planetary Environments*, chap. 2, U.S. Air Force Geophysics Directorate, January, 1962.
5. Frank, L. A., J. A. Van Allen, and E. Macagno: Charged-particle Observations in the Earth's Outer Magnetosphere, *J. Geophys. Res.*, vol. 68, pp. 3543–3554, June 15, 1963.
6. Neugebauer, M., and C. W. Snyder: The Mission of Mariner II: Preliminary Observations, *Science*, vol. 138, pp. 1095–1096, December, 1962.
7. Johnson, F. S.: The Ion Distribution above the F_2 Maximum, *J. Geophys. Res.*, vol. 65, pp. 577–584, February, 1960.
8. Ratcliffe, J. A. (ed.): *Physics of the Upper Atmosphere*, Academic Press Inc., New York, 1960.
9. Singer, S. F. (ed.): *Progress in the Astronautical Sciences*, vol. 1, Interscience Publishers, Inc., New York, 1962.
10. Mitra, S. K.: *The Upper Atmosphere*, Asiatic Society Monograph Series, vol. 5, Calcutta, India.
11. Hess, W. N.: Energetic Particles in the James Van Allen Belt, *Space Sci. Rev.* vol. 1, 1962.
12. Farley, T. A.: The Growth of Our Knowledge of the Earth's Outer Belt, *Rev. Geophys.*, vol. 1, pp. 3–34, February, 1963.
13. Clauser, F. C. (ed.): *Symposium of Plasma Dynamics*, chap. 8, Addison-Wesley Publishing Company, Inc., Reading, Mass., 1960.
14. Kallman, H. (ed.): *Space Research*, pp. 736–745, Interscience Publishers, Inc., New York, 1960.
15. Valley, S. L. (ed.): *Space and Planetary Environments*, chap. 5, U.S. Air Force Geophysics Directorate, January, 1962.

16. Kraushaar, W. L., and G. W. Clark: Search for Primary Cosmic Gamma Rays with the Satellite Explorer XI, *Phys. Rev. Letters*, vol. 8, pp. 106–109, Feb. 1, 1962.
17. LeGalley, D. P. (ed.): *Space Science*, chap. 11, John Wiley & Sons, Inc., New York, 1963.
18. Wilson, J. G., and S. A. Southuysen (ed.): *Progress in Elementary Particle and Cosmic Ray Physics*, p. 339, Interscience Publishers, Inc., New York, 1958.
19. Peters, B.: Progress in Cosmic Ray Research since 1947, *J. Geophys. Res.*, vol. 64, pp. 155–173, February, 1959.
20. Wilson, J. G., and S. A. Southuysen (eds.): *Progress in Elementary Particle and Cosmic Ray Physics*, p. 205, Interscience Publishers, Inc., New York, 1958.
21. Bjork, R. L.: Meteoroids versus Space Vehicles, *Am. Rocket Soc. J.*, vol. 31, pp. 803–807, 1961.
22. Ordway, F. I. (ed.): *Advances in Space Science and Technology*, pp. 273–294, Academic Press, Inc., New York, 1961.
23. Valley, S. L. (ed.): *Space and Planetary Environments*, p. 210, U.S. Air Force Geophysics Directorate, January, 1962.
24. Shklovsky, I. S.: *Cosmic Radio Waves*, Harvard University Press, Cambridge, Mass., 1960.
25. Baker, R. H.: *Astronomy*, 7th ed., D. Van Nostrand Company, Inc., Princeton, N.J., 1959.
26. Valley, S. L. (ed.): *Space and Planetary Environments*, pp. 169–176, U.S. Air Force Geophysics Directorate, January, 1962.
27. Singer, S. F. (ed.): *Progress in the Astronautical Sciences*, pp. 267–339, Interscience Publishers, Inc., New York, 1962.
28. Landsberg, H. E., and J. Van Mieghem (eds.): *Advances in Geophysics*, pp. 1–76, Academic Press Inc., New York, 1961.
29. Kuiper, G. P., and Barbara M. Middlehurst (ed.): *Planets and Satellites*, pp. 159–209, The University of Chicago Press, Chicago, 1961.
30. Goldstein, R. M., and R. L. Carpenter: Rotation of Venus: Period Estimated from Radar Measurements, *Science*, vol. 139, pp. 910–911, Mar. 8, 1963.
31. Sandage, A.: *Astrophys. J.*, 1958, p. 513.
32. LeGalley, D. P., and A. Rosen (eds.): *Space Physics*, John Wiley & Sons, Inc., New York, 1964.

Part III

SYSTEM COMPONENTS

Chapter 8

DIGITAL COMPUTER-SYSTEM CHARACTERISTICS

ISAAC L. AUERBACH, *President, Auerbach Corporation.*

JOHN A. GOSDEN, *Program Manager, Auerbach Corporation.*

CONTENTS

8-1. Introduction	8-5
8-2. Structure	8-6
Basic Units	8-6
The Store	8-6
Central Processor	8-6
Routines	8-6
Basic Connections	8-7
The Basic Computer	8-7
Input-Output Channels	8-7
Off-line Operation	8-7
Matrix-switch Structure	8-8
Multicomputer Structures	8-8
Satellites	8-10
Multiprocessors	8-10
8-3. Internal Storage	8-10
Basic Physical Properties	8-11
Capacity	8-11
Storage Levels	8-11
Alterability	8-12
Access Structure	8-12
Uniform Access Structure	8-12
Cyclic Access Structure	8-12
Serial Access Structure	8-12
Complex Access Structure	8-12
Balanced Access	8-13
Associative Techniques	8-13
Performance	8-13
Smallest Units of Transfer	8-14
Variable-sized Loads	8-14
Transfer Rates	8-14
8-4. External Storage	8-14
Magnetic Tapes	8-16

8-4 SYSTEM COMPONENTS

- Punched Cards and Punched Tape 8-16
- Magnetic Cards 8-16
- Magnetic Disks 8-16
- 8-5. Central Processor 8-17
  - Primary Functions 8-17
  - Control of Sequence 8-17
  - Operations and Operands 8-18
    - Arithmetic 8-18
    - Comparisons 8-18
    - Formats 8-19
    - Instructions 8-19
    - Logic 8-19
    - Tables 8-19
    - Lists 8-20
  - Input-Output Transfers 8-20
    - Synchronizing Aids 8-20
    - Multiple-block Loads 8-20
    - Table-controlled Transfers 8-20
    - Block-size Control 8-21
    - Conversion 8-21
  - Self-analysis 8-21
  - Organization 8-22
    - Central-processor Overlapping 8-22
    - Look-ahead 8-22
    - Multirunning 8-23
    - Multiprocessing 8-24
    - Multisequencing 8-24
    - Interprogram Protection 8-24
    - Intraprogram Protection 8-24
  - Levels of Control of Operations 8-25
- 8-6. Input-Output 8-25
  - General Connections 8-25
    - Circuit Adapter 8-26
    - Code Converter 8-26
  - Special Controller Facilities 8-26
    - Controller Functions 8-26
    - Controller Allocation 8-27
    - Buffer Protection 8-28
    - Multiplexing 8-28
  - Variety of External Storage Media 8-28
    - Printing 8-28
    - Mark Sensing 8-28
    - Character Recognition 8-28
  - Performance Characteristics 8-29
    - Speed 8-29
    - Loading Overheads 8-29
    - Capacity 8-29
    - Common Media 8-29
- 8-7. Simultaneous Operations 8-29
  - Special Units 8-30
  - Different Types of Simultaneous Operations 8-30
  - Independent Operations 8-30
  - Multiple-data Paths and Multiplexing 8-30
  - Partial Overlapping 8-30
  - Restricted Operations 8-30
  - Programmed Time Sharing 8-30
- 8-8. Performance of Computer Configuration 8-31
  - Particular Configuration 8-31

DIGITAL COMPUTER-SYSTEM CHARACTERISTICS   8-5

   Particular Problems.. 8-31
   Implementation.. 8-31
   Operating Environment..................................... 8-31
   Reliability... 8-31
8-9. Comparison Charts...................................... 8-31
8-10. Standard Problems..................................... 8-46
   Generalized File-processing Problems....................... 8-46
      Standard Problem A..................................... 8-47
   Generalized Mathematical Processing........................ 8-48
      Standard Mathematical Problem A........................ 8-48
      Input and Output Records............................... 8-51

8-1. Introduction

In this chapter, the word "computer" is an abbreviation for general-purpose stored-program digital computer. Because of its very nature—namely, that it may be made to manipulate data in many different ways or in fact, to simulate any system at all—a general-purpose computer represents an intriguing study for a system engineer.

There are two fundamentally different viewpoints from which a system engineer may consider a computer system. First, he may consider a computer system from the viewpoint of the designer. Second, he may consider a computer system as a component or tool to perform a specific role as a system or subsystem in order to accomplish some objective of the system designer.

These two viewpoints correspond to those of the producer and the user. The producer's primary objective should be to keep the system general-purpose. The basic advantage of a computer is that it can be useful over a whole family of systems, or even groups of families of systems.[1] The user's objective is to consider how well a particular computer system can be adapted to his specific needs. Naturally, all the facilities that have been incorporated will not be required by all users.

There are two different environments in which a computer system can be considered by a user. First, there is the environment in which a computer system can be assembled from available components; that is, by deciding on a specific configuration of individual units, the designer can arrange to perform the entire range of processes for which the configuration is needed. Second, there is the environment in which a specific configuration of equipment already exists, and it is the objective of the system designer to adapt this particular set of equipment without the prerogative of extending or modifying the system. The first environment involves a choice by the system engineer of the computer to be used; the second involves a choice of technique in using a computer.

The facilities which may be provided by a computer system can be divided into two classes: first, those facilities which are provided directly by the system, or can be constructed somehow from the elements that are provided; and second, those facilities which cannot be provided without extending the system in some way. Data-processing facilities are members of the first class. Normally, computers are general-purpose only in the sense that from a basic set of data-manipulation operations, other data-manipulation operations can be constructed. The most important members of the second class are input-output and control operations. For example, if a paper-tape unit has no provision for backward motion or test for end of tape, these operations cannot be constructed from other operations.

The following sections discuss computer systems as systems. An attempt is made to separate the facilities and functions into logical parts rather than physically separate parts, and to relate the kinds of facilities that are provided with the different types of organization of these facilities. Concentration is on the organization of the units that provide these facilities rather than on their speeds, capacities, and mechanical implementations. These characteristics are summarized in the charts in Sec. 8-9. The system engineer is referred to the references[2,3] for further reading on the mechanical implementation—in particular, the performance of specific computers which affects details of their use, but not the range of facilities provided, the organization, or the conceptual structure.[4]

8-2. Structure

A digital computer is composed of a number of logical units which are usually physically separate. The properties of the whole computer depend not only upon the properties of the individual units, but also upon the way in which they are connected. A computer can be described in terms of both the properties of the basic parts and their basic connecting structure.

Basic Units. The basic units of a computer system have long been defined as the store, the input, the output, the arithmetic unit, and the control. Currently, the arithmetic unit and the control are so integrated that it is sometimes difficult to separate their functions. Input and output units are made so similar that distinguishing them from each other is difficult unless the specific direction of transfer is known.

The Store. The purpose of storage is to provide a place in which to retain data.[3,5,6] There are two basic types of storage, nonreusable and reusable. Nonreusable storage is that in which the data are permanent or can be changed only infrequently by replacing the storage medium itself. Punched holes in cards or tape are examples of such data. Reusable storage is that in which the data can be erased and replaced by other data as required by the system. Increasingly, the more common type of storage is reusable many times (e.g., magnetic tape, magnetic cards, or magnetic disks), but special cases of fixed internal stores are used in the design of several new processors.[7,8]

There are two types of storage which should be distinguished in any system, internal and external. Internal storage is that storage within the system which can be accessed at will by the controlling circuits within the system. External storage is that storage which can be brought to the system or removed from it by external agencies, usually human operators. These two types of store are difficult to distinguish because of recent developments in magnetic tape, disks, and cards. In the case of magnetic tapes, for example, it may be only in the use of the equipment rather than in its design or structure that the distinction between external and internal storage can be made. On the other hand, many internal stores are built into a computer system so that they cannot be removed.

Central Processor. The central processor is a unit that combines the facilities of an arithmetic unit and a control unit.

The arithmetic unit is concerned with the manipulation of data within the computer for the purpose of transforming the data to produce the required results, and using the data to control the decisions that must be made as a program run progresses.

The control unit, on the other hand, has the responsibility for selecting the appropriate instructions in the sequence in which they are to be obeyed, and decoding and initiating them. Usually, this unit also controls the individual operations which comprise each instruction that is executed in the arithmetic unit.

The hardware components of the two parts are not often clearly separated. Because the internal stores of a computer are designed to hold both the working data and the instructions, and to operate on either type of data at will, the individual operations cannot always be definitely assigned either to arithmetic functions or to control functions.

Routines. The routines of instructions inside a computer executed by the control unit form an important logical part of a computer system. In almost all computers, these routines may be changed easily. The variety and extent of the procedures that can reasonably be carried out are functions of the operation repertoire of the computer. In computers, however, many of the infrequent operations which might be undertaken by hardware are optionally executed by routines. For example, less expensive computers may not provide floating-point hardware even though they use such arithmetic extensively in programs. Although standard routines of this type are not physically a part of the hardware, they are often maintained as such and are part of the overall system design.

Basic Connections. The various units of a computer can be connected in many different ways by data channels. None of the types that have been used appear to have been discarded or entirely replaced by other types. Each has its own particular use, for reasons of either speed, economy, or simplicity, though most of the developments have occurred in the design of the more powerful and complex computers. There are relatively few basic types of connection, although the different types have many variations.[9,10]

The Basic Computer. The simplest form of a computer is shown in Fig. 8-1. In this computer, the central processor controls store and the input-output transfers. The center of the system is the arithmetic unit, which sends or dispatches words to different parts of the system or else manipulates the data.

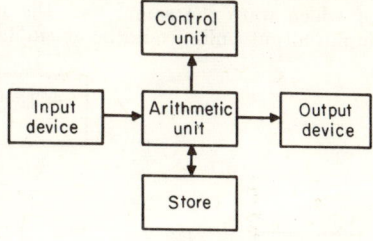

FIG. 8-1. Basic computer.

Normally, such a simple system performs all its transfers in basic data units of one word, controlling them one at a time by individual instructions.

Input-Output Channels. The major disadvantage of the basic computer system is the fact that each of the operations is performed serially; the computer may be devoting the largest part of its time to the organization of input and output, separate transfers of single words, and control and timing of the input-output devices. In order to improve the performance of the computer by a factor of 2 or 3, the system may be organized so that the store becomes the center of the network (see Fig. 8-2). In such cases, the control unit sends instructions to individual controllers on each input-output channel. These controllers are miniature special-purpose computers, specially designed to deal with the particular requirements of their related input-output devices and to organize the transfer of single words or entire blocks of data into or out of the store. Although the central processor may still be delayed for the duration of time that the controllers require access to the main store, this is less than the amount of time the processor would require in gaining access to and waiting to execute the instructions concerned with individual parts of input-output transfers.[11] The particular design of controllers has altered greatly as different types of storage and different types of input-output units have been introduced. These are discussed in more detail in Secs. 8-4 and 8-6, respectively. Figure 8-2a illustrates a simple form of buffered input-output. Figure 8-2b illustrates the later development of time-shared access to a working store, in which auxiliary storage is connected in a way logically similar to input-output. In complex cases, there may be several working stores.

Off-line Operation. As central-processor development increased the speed at which manipulations took place, it was soon found that a disproportionate amount of time was required by input and output operations, particularly where these involved such slow input-output media as punched cards and printing. In many cases, a severe imbalance in the use of different units of a computer resulted. The central processor could be idle for a large part of many runs, even though the input-output channels were working at full speed. To reduce this imbalance, the concept of off-line conversion was introduced. Figure 8-3 illustrates the organization of a computer for on-line and off-line use. In this case, all the punched-card input to the computer can be first transcribed to magnetic tape off-line. Later, the computer output can be transcribed off-line to a printed medium or punched cards. Even if there is a severe imbalance in the computer system, it is not always necessary to have several off-line operations running concurrently for one computer because the flow of work for a given day might be uneven or the off-line equipment can be run during second and third shifts. Some controllers were designed so that they could be connected either to the computer or in pairs to each other to perform off-line conversions.[12] In recent years, it has become popular to use a small general-purpose computer as the heart of an off-line processor.[10] Such a computer is able to perform not only simple transcriptions, but also data editing and other useful functions.

Matrix-switch Structure. As the size of computers increased, designers struggled to transform the apparent complexity into generality. They introduced the concept of modularity, in which a computer was made up of a large number of unit parts, many of which were the same; e.g., the input-output controllers for different types of input-output units would be as similar to each other as possible, and a store would

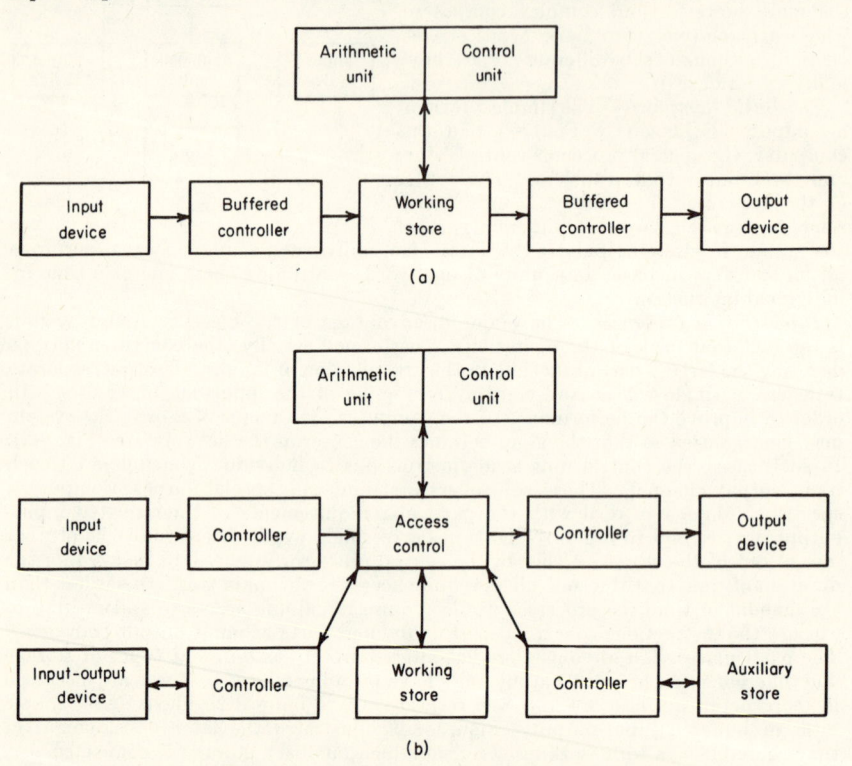

FIG. 8-2. Computers with simultaneous input-output.

be divided into several identical parts. As the number of units to be connected increased, the concept of a store with a number of channels connected to it (see Fig. 8-2b) became awkward to implement. An alternative approach was introduced, in which numbers of units are connected by a large central switching unit (illustrated in Fig. 8-4). In such a system, there could be several modules of stores, several modules of central processors, and a variety of input-output controllers.[13,14] By suitable setting of switches in the switching unit, a store could be connected either to a central processor or to an input-output unit. Each of these connected pairs of units could be operating in parallel with other pairs, in a way similar to simultaneous operations in a multichannel structure.

Multicomputer Structures. As the need for greater reliability in computer systems evolved, it became necessary to design computers with at least two of each part, so that when one failed, the other would be able to continue working. The most obvious examples of systems of this type were duplex computers, pairs of computers

DIGITAL COMPUTER-SYSTEM CHARACTERISTICS

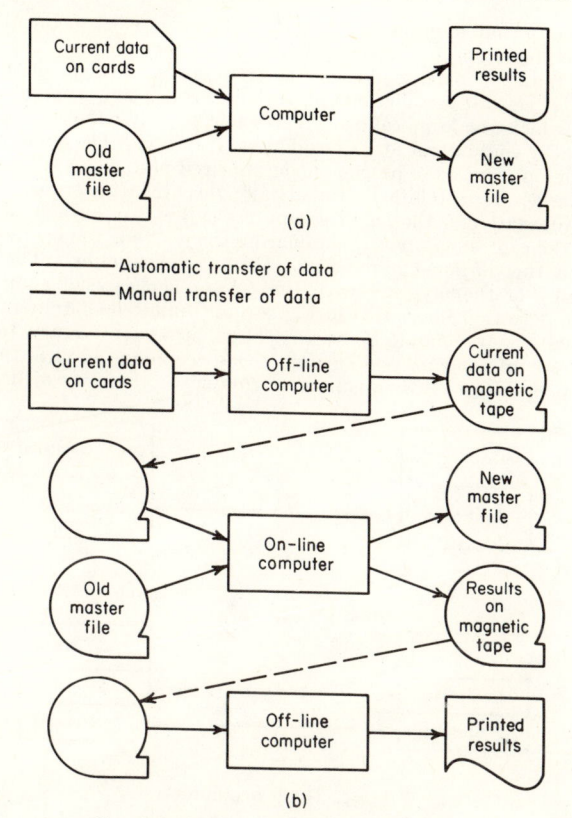

FIG. 8-3. On-line and off-line operation.

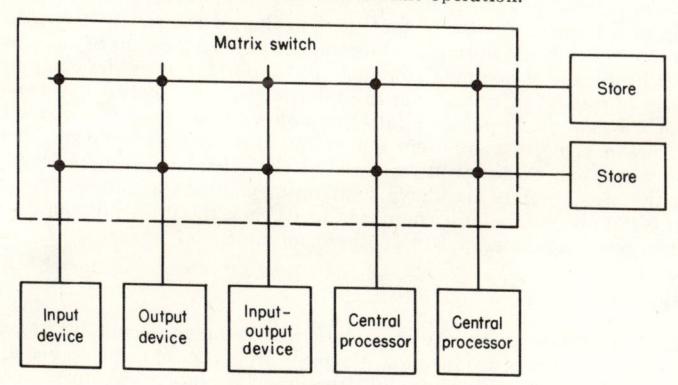

FIG. 8-4. Matrix-switch structure.

connected in such a way that both could work in parallel or, if one failed, the other could carry on.[14,15] Alternatively, there are systems using the matrix-switch concept, in which there could be one extra unit of each type of module connected to a suitable switching-control arrangement so that when any one module fails, a spare can be switched in to take its place (see p. 8-10).[16] Figure 8-5a illustrates the type of duplex unit in which the connections between the two systems are made by exchanging data from time to time in order to relate the progress of each to the other.[17] Figure 8-5b illustrates the major alternative, the use of common stores.

Satellites. With the increase in size of computers and the acquisition by individual organizations of several installations of various sizes, a new type of operation was conceived, in which one large central processor is directly connected to a number of satellites (smaller computers) at a considerable distance. The small computers can be used in different centers to process moderate-sized problems. When the problems become too large, the satellites automatically dispatch the problems to the large computer. Alternatively, the small computers can act as off-line conversion units connected to the central computer, eliminating the need for the operators to transfer the data from the off-line equipment to the main computer. The satellite may transmit a copy of the program with the data. The type of connection used is usually the same as that illustrated in Fig. 8-5 for duplexed computers.[16]

Multiprocessors. As computers became large, it was necessary to increase the computing power of each processing unit. One technique used is the inclusion of several central processors in one computer complex. One way of doing this is by

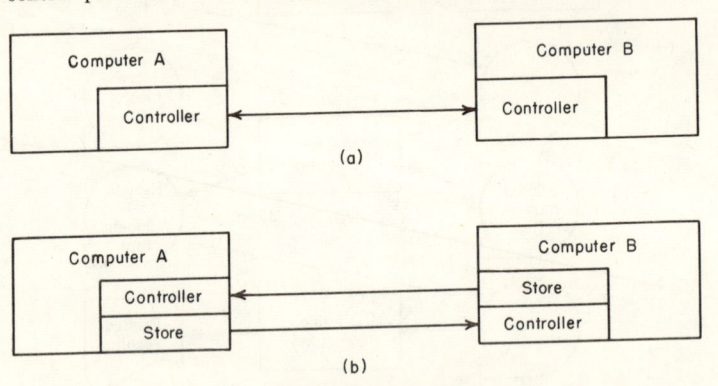

FIG. 8-5. Duplex computers.

means of the polymorphic structure shown in Fig. 8-4.[18] Another is to connect two computers in a manner similar to that used in duplex computers. A pair of computers so connected would normally communicate either by means of an input-output channel or by means of common storage. In the former case, the computers must synchronize mutually at certain times and transfer data between them. In the latter case, the computers put data into common stores and work in a loosely linked way. Common cases of computers linked by stores are those having a common disk store which can be switched from one to the other. In a more loosely linked system, pairs of computers may have common magnetic-tape units. One computer first records data on a tape unit connected to it; then the tape unit can be switched by program, or even manually, to the other computer.

8-3. Internal Storage

As related to system design, the characteristics of storage in computer systems can be divided into three basic groups: first, the basic properties of capacity and structure, including the physical homogeneity and structure of the individual stores; second, the access structure, which is principally concerned with the means of input and output to the stores; and third, the overall performance of the storage system. The performance of a store depends partly on its physical structure, partly on its access structure, and partly on the tasks which it is required to perform.

There are two types of internal storage, working and auxiliary. Working storage can be accessed directly—for operands used in the arithmetic and logical operations in the central processor, or for instructions. Auxiliary storage is all internal storage other than working storage, plus special-purpose stores such as buffers in input-output units and special registers in the central processor.

The internal storage of a computer is usually composed of more than one store arranged in a hierarchy of increasing size and access time. A working store whose access time is comparable with the operation time of the central processor is often called an immediate-access store. An auxiliary store whose capacity is comparable with that of magnetic tapes is often called a mass store. An auxiliary store whose access structure is such that any location can be accessed rapidly (within a time of the order of a second) is often called a random-access store. The access time of such a store is contrasted with the potentially long access times of a serial store such as magnetic tape.

Basic Physical Properties. *Capacity.* The principal characteristic of a store is its capacity. Capacity of storage is measured in many different units, usually associated with the structure of particular stores. The most common data units are words and characters, although the only common unit of measure of storage is a bit. The bit is not a commonly used data-transfer unit but is used to specify the sizes of other basic units.

In general, word sizes range from about 12 to 60 bits. In the larger computers, 32- to 48-bit words are most common. Characters are most commonly 6 bits in size—that is, a sextet of bits; but octets, septets, quintets, quartets, and trios are not unusual. The last two sizes are restricted to use for numeric digits and may be referred to as BCD (binary-coded decimal) and octal, respectively. BCD provides 4 bits to specify a decimal digit; octal, on the other hand, is a synthetic scheme to make binary manipulations compact and easier for human beings.

The important characteristics of a store are the maximum and minimum sizes; the intermediate sizes are usually multiples of a minimum size. The size of a store is frequently measured in the particular word size of its individual computer system. Stores are usually available in modules, each a given number of words in size (usually in the range of 2,048 to 16,384 words for working storage, or 100,000 to 1,000,000 for auxiliary storage). The actual values are either powers of 2 in binary computers, or simple multiples of powers of 10 in decimal or character-oriented computers.

Word sizes vary widely from computer to computer, and it is usual to measure the capacity in equivalent decimal digits ($\log_2 10 \approx 3.3$ bits per decimal digit) or alphabetic characters (6 bits) for purposes of comparison.

Individual sizes of stores vary from small special-purpose high-speed stores in the arithmetic unit (e.g., sets of 16 registers composed of flip-flops or thin film),[8,19] through core stores and drums ranging from a few hundred to hundreds of thousands of words, to drum and disk systems ranging up to tens of millions of words.

Storage Levels. Within any one computer system, there may be several different stores with different characteristics, usually with increasing capacities and increasing access-time lengths. Even in one physical store, there may be different levels of storage. For example, there may be a drum on which for some tracks there is one head per track, so that the access time may be as much as one complete revolution, and other tracks with more than one head each, in which the maximum access time may be reduced to a fraction of a revolution. Usually there are two heads spanning a short arc of a track. That arc is used as a short cyclic store continuously writing at one head and reading at another. In the larger systems, it is usual to have a small amount of extremely fast storage in the central processor, and a moderate to large working store for basic data and instructions being operated upon. A large auxiliary or backing store, which holds large volumes of data that are moved into and out of the working store in pieces (often called segments), is also necessary. From a user's viewpoint, the simplest storage is a one-level homogeneous store which may be divided physically into many pieces.

The problems in using multilevel storage systems are those of access technique and efficient allocation of data to various levels. As the access times differ significantly between levels, the choices of access technique and allocation can affect performance considerably.[20,21] The value of alternative techniques cannot be determined without consideration of the job to be performed.

Alterability. In most computer systems, the contents of all the stores can be changed, but some computers have fixed stores which may be used for retaining standard routines. They have the advantage that their contents cannot be destroyed by a program fault, and represent an extension of the hardware of a computer system. Fixed stores have advantages of economy and speed over erasable stores. Such stores are also used in computer systems to provide microcoding or macrocoding facilities (see Levels of Control of Operations in Sec. 8-5).[7,22] In the sense that a switch stores 1 bit, all the logic wired into a computer may be considered as fixed switches and, by extension, considered as nonerasable storage.

Access Structure. The access structure of a store or series of stores affects not only the access times of the store but also the complexity of the organization that may be necessary to overlap the use of the store efficiently with other parts of the system. There are four basic types of access structure: uniform, cyclic, serial, and complex.

Uniform Access Structure. Core storage is a good example of the simplest type of access structure used in computers. In this type of storage, there is one access device for a large number of words, each of which is equally available at any one time; that is, the latency or the waiting time to obtain access to any particular location is independent of the location being called. Such stores are usually characterized by two important parameters. The first is the cycle time of the store, or the frequency with which the store can deliver individual words successively in response to individual access calls. The second is the waiting time, which is the interval from the beginning of one such cycle when a request is made until the time when the data are provided. Such stores are usually asynchronous. Delays occur only when accesses are requested at intervals shorter than the cycle time.

Cyclic Access Structure. A large variety of stores (in particular, delay lines, disks, and drums) have a cyclic access structure. In such devices, individual parts of a store have a number of access points. Each such point has access to a number of locations which appear cyclically at the time of access. Therefore, the waiting time to obtain access to any particular location depends upon the position of the location in the cycle at the time of access. The principal timing characteristic of such stores is the cycle time—that is, the period between successive instants of availability of any particular location at its access point. In computers in which the cycle time is an order of magnitude greater than the instruction execution time, devices such as "minimum latency coding" have been used. In such coding, operands and instructions are carefully placed to reduce the total waiting time for access to such stores (see also Control of Sequence in Sec. 8-5).

Serial Access Structure. A few stores often have been composed of strips of magnetic tape. A conventional magnetic-tape unit and a tape mounted on it frequently are treated as internal storage. This is a serial store. The access time depends upon the distance of the access mechanism from the particular location to be accessed. Usually there is no continuous movement of the storage medium, as in a cyclic store. Thus the actual time of request, relative to some periodic cycle, is not important. The access time is determined by the distance between one required location on the storage medium and the next.

To provide for special types of use, some serial stores automatically return the access device to a home position with either one end or the middle of a serial store located at the access device.

Complex Access Structure. A drum with one fixed head for each track is a combination of a uniform access to all tracks and a cyclic access to the individual locations in a track. Small delays in switching between heads are not usually significant. If each track contained one location—a full track of data—the access structure would be essentially uniform.

A set of disks with a single moving arm[23] is a combination of a serial access to a disk, a serial access between tracks on a disk, and a cyclic access to locations within a track. In this case, the serial access times are not linear functions of the "distance" to the locations.[24]

A set of magnetic cards (see Sec. 8-4), arranged so that one selected card at a time can be wrapped on a drum for reading or writing, is a combination of uniform access to a card and uniform access to the tracks on a card. Note that in the uniform access to a track on a card, a relatively long time is involved, but this time is independent of the instant of asking; on the other hand, cyclic access time within a track, once a card is available, is relatively short and is essentially independent of the track chosen. This depends, however, upon the time of asking relative to the current position of the card on the drum.

Another common form of complex access structure is the interleaving of various parts of uniform access stores. For example, banks of core stores[26] are often arranged so that, if accesses to separate banks are made successively, it is possible to increase the rate at which words can be delivered by the group of banks, provided accesses are distributed among banks. Thus it is possible to overlap the read half cycle of one store with the write half cycle of another. Where two banks are used in such stores, odd addresses are arranged in one bank and even addresses in another in an attempt to alternate the access to each bank automatically.

Balanced Access. Most stores have balanced access; that is, the access times for reading and the access times for writing are not significantly different from each other. However, some types of store in which reading is much more frequent than writing have been biased so that a read access is faster than a write access. The most important example of unbalanced stores involves auxiliary storage. In many applications, users program two accesses when writing, one to write data and a second to read back the data to check the success of recording. Some stores provide special check-read operations to reduce the programming overheads of checking.

Associative Techniques. Because of the increasing importance and extensive use of tables in data processing (see Operations and Operands in Sec. 8-5), special types of store have been developed to improve the performance of computers on such operations.[27,28,29,30] Such stores are called associative memories. The basic principle of organization is that each location of the store is divided into two parts, an associated address and data; access is organized so that when an associative address is specified the store responds by producing the contents of any location in which the address matches the accessing address. This occurs directly without serial searching of locations within the same order of magnitude of time as a simple access to a store constructed of similar components. There need be no location with any specific address, but duplication of addresses is not permitted.

Associative addressing is very attractive in some applications, and the proportion of time that can be saved increases directly as the sizes of arrays that can be handled. However, high costs have been a deterrent to commercial acceptance.[31]

Performance. The measure of performance of a store in the execution of the functions which it provides is divided into two important parts: first, the waiting time that is necessary after a request has been generated and before transfer of data is initiated; and second, the transfer time that elapses while the data are transmitted between two stores. When writing into a store, the waiting time is normally the more important factor, because if the transfer time is significant, buffers can be interposed to allow other parts of the system to operate independently. Then the only effect of the transfer time is to establish a period during which the store cannot begin a subsequent operation. When reading, both times are important, because the necessary waiting time before the data can be operated upon is usually the sum of the waiting time and the transfer time.

When the volume of data transmitted at a given time is small, the waiting times dominate performance; but when the volume transmitted at one time is large, transfer times dominate. Some stores are designed to deliver small units of data with little delay and are relatively slow in delivering large units. Other stores may have a long waiting time before delivering small units of data but are relatively fast when dealing with very large blocks because they have high transfer rates. Important considerations in measuring the performance of a storage unit are the waiting times and transfer times for both the smallest units of data that can be transferred and the largest units that may be needed. When units larger than those provided for in the system are required, they must be obtained as a set of lesser units. Therefore, the third important consideration in storage performance is the different sizes of loads that can be transferred in one operation.

Smallest Units of Transfer. Most of the fast-working stores now in use in computers, particularly core stores, are designed to transfer one unit of data at a time (one word or one character, or sometimes a pair of characters). In the larger stores, particularly those of a cyclic nature, it is usually convenient to be able to transfer large groups of words, particularly one complete cycle of data (a quantity limited by one delay line or one track of a drum). For simple devices which have relatively long access times, a complete track or complete cycle of data is often the smallest volume that can be transferred in one load.

Variable-sized Loads. One of the facilities often provided by sophisticated auxiliary storage is the ability to read or write an arbitrary number of locations in one operation. In addition, although an auxiliary store is usually divided into locations which are dealt with as complete units, it is often possible to read or write incomplete units. This is called a cut-off. When reading, the rest of the location is ignored, but it is not erased and may be read later. Few stores provide for variable-length locations in the store. Different sizes can be chosen for each job to make efficient use of total storage capacity. Many stores permit only variable lengths within some natural sectional unit. Only in the more sophisticated systems can variable length extend across and between sections. In many large auxiliary stores, the facility of obtaining a block or load, which is composed of many sections, is obtained by programming the access to one section after another.

Transfer Rates. When dealing with large auxiliary stores in which it is often necessary to move considerable volumes of data into and out of the working store, the important times are not the waiting times, but a combination of the waiting time and transfer time. A large unit may require many separate transfers which have to be controlled by a set of instructions. Such data would normally be arranged in ascending sequence of addressed locations of store, and it is necessary to take into account the housekeeping time between successive transfers. Some computers are especially designed to handle these well.[32] The store access may be arranged so that there is a period of inactivity between successive sections of a cyclic store to provide time for a program to do the necessary housekeeping operations between one section and another, and thus make a transfer in every cycle without missing cycles. An alternative scheme used in a few cyclic stores enables a transfer of a complete cycle to begin anywhere in a cycle, thus almost eliminating waiting time.

The comparison charts in Sec. 8-9 show that different auxiliary stores with the same ranges of capacity and similar access times nevertheless have widely different transfer rates for large volumes of data.

8-4. External Storage

Many of the media used for input and output to computers are widely used to extend the external storage capacity of a computer. (External storage is defined as storage that can be removed from the control of a computer.) In approximate chronological sequence of first use with computers, the five important media are:

Punched cards
Punched tape
Magnetic tape
Magnetic cards
Magnetic disks

A potentially important medium is photographic plates or film.

External storage makes it possible to extend the capacity of storage while tolerating some increased access time. Thus increased storage is made available without the need for more access devices.

Table 8-1 summarizes the important characteristics of different types of external storage.

Table 8-1. Properties of Different Types of External Storage

	Equipment cost ratio	Material cost ratio per million characters	Reusable material	Input-output in one unit	Input rate, characters/sec	Output rate, characters/sec	Non-serial access	Back-space	Compatibility with non-computer systems	Storage capability for 1 million characters	Dominant access structure	Human legibility
Punched cards:												
Slow	0.4	2.0	N	N*	100	100	N	N	Y	100,000 cards	Serial	Y
Fast	1.0	2.0	N	N*	2,500	400	N	N	Y	100,000 cards	Serial	Y
Punched tape:												
Slow	0.3	0.1	N	N	150	10	N	N	Y	40,000 ft	Serial	Y
Fast	0.7	0.1	N	N	1,000	300	N	N	Y	40,000 ft	Serial	Y
Magnetic tape:												
Slow	1†	1†	Y	Y	10,000	10,000	N	Y	N	2,400 ft	Serial	N
Fast	3	0.3	Y	Y	200,000	200,000	N	Y	N	480 ft	Serial	N
Magnetic cards	1.2	5.0	Y	Y	100,000	100,000	N	Y	N	50 cards	Uniform	N
Magnetic disks	8.0	48.0	Y	Y	100,000	100,000	Y	Y	N	0.1 disk	Cyclic	N
Photographic media	....	....	N	N	......	......	Y	..	N	....	Uniform	N

* Such units do exist but are not normally used as such in computer configurations. They are actually combined input and output units.
† The slow magnetic tape has been adopted as the standard for comparison of cost ratios.

8–15

Magnetic Tapes. Magnetic tape is a convenient form of external storage and is widely used on all the larger computer configurations.[33] The greatest advantage of magnetic tape is its comparatively high speeds for serial input and output. A second advantage is the fact that it is reusable many times. Another advantage is that the data can usually be freely structured in a variable format.[34] Still another is that tape can sometimes be read both forward and backward (or at least on many machines can be rewound to the beginning and started again). Finally, after the magnetic tapes have been loaded, all these operations can be controlled without human intervention. These advantages are very useful in off-line operation (see Sec. 8-2) in which freely structured data are later transcribed to other devices at slower speeds.

Once a magnetic tape is mounted onto a tape deck, it can be considered as part of the internal storage of the computer system. It is also easy to build up a large external storage library of magnetic tapes which can be loaded relatively quickly. One operator loading a high-density magnetic tape can load 50,000,000 characters of data (which is equivalent to nearly 1,000,000 punched cards) in one operation.

Magnetic tapes are used as stores in many ways. Most are used simply as serial stores for data files; others, such as library tapes and master tapes, contain the routines used in the general organization of a computer operating system and are accessed in a relatively random way for items required intermittently. An access delay of up to tens of seconds is not of extreme importance in such applications. Tapes containing routines are normally shorter than data-storage tapes, or only partly filled.

Punched Cards and Punched Tape. Although punched cards and punched tape have the advantages of cheapness of equipment and of material, they have several compensating disadvantages as external stores. First, they are permanent and not reusable, and must therefore be discarded when the recorded data are no longer required. Second, punched cards and punched tape have a relatively low rate of input and a very low rate of output. As a result, they are used extensively only on less powerful computer systems. Third, they cannot easily be read in a backward direction and do not usually incorporate facilities analogous to the rewind on magnetic tape. They cannot backspace and rewrite or backspace and reread; these disadvantages inhibit automatic handling of errors. Using high-speed equipment for repeated rereading of punched cards and punched tape limits their useful life. Punched cards and punched tape have two unique advantages: first, they are legible to human beings; and second, they can be used with other important noncomputer systems facilities based on punched card or tape storage.

Magnetic Cards. The facilities provided by magnetic cards combine the convenience and facilities of magnetic tapes with an access structure similar to that of disks. In the best-known system, individual cards can be called to drop onto and wrap around a drum, after which the several different tracks of data on a card can be read or recorded as required. The dropping of the card is analogous to the moving of a head mechanism. Once the card is wrapped around the drum, its rotation on the drum surface is similar to the rotating of disks or drums.

Its major advantages are that it can be treated as an external store and cartridges of many cards can be inserted and removed in a time comparable with that required for manual mounting of magnetic tapes. The performance of magnetic cards when reading in a serial mode is similar to that of magnetic tapes.

An additional advantage of magnetic cards is that they have better random access than tape. In particular, rewind time is effectively zero, as is skipping backward or forward to some other special position on the tape. By use of its uniform access to cards, a single unit can be made to behave like several tape units, at a relatively moderate decrease in performance. This is accomplished by considering that the cards in one cartridge are divided into groups, one group for each tape, and accessing the next card from each group as needed.[25,35]

Magnetic Disks. One variety of magnetic disks is available that has groups of removable disks that can be considered as an external storage. Their use is similar to that of magnetic cards.[32]

8-5. Central Processor

The central processor is the one part of the computer where diversities of approach, multiplicities of design, and variations in sophistication have been most marked. In general, the arrangements of input-output devices, of the input-output transfers and controllers, and of the storage devices have been simplified and contain a great deal of relatively straightforward, common structure compared with the structure and design of the central processor. This is largely because the specification of the work that the other units of the computer are required to do is relatively straightforward, and the ingenuity in design has been concentrated on simplification, reducing costs, and improving the performance and reliability of such units. On the other hand, the requirements of the central processor are so varied that different types of design have branched out in many directions. The differences lie, first, in the degree to which various functions are built into the hardware rather than left for a programmer to generate; second, in the orientation of the hardware toward commercial, scientific, process-control, or real-time problems, or other uses; and third, in the structure of the control sequences to match the different types of storage being used.[36]

There has also been a general development of levels of control within a computer system itself—that is, in the facilities by which a master or supervisor program in a computer can be aware of the progress and condition of other programs inside the computer, and in the facilities by which the computer is able to look at its own status and condition to be able to monitor and adapt its own behavior in response to different situations.[18,37]

Primary Functions. The central processor of a computer is a combination of the control and arithmetic unit functions. Because of its control function, the central processor cannot be considered in isolation from other units. Its power to control is directly related to the facilities in the other units that are under its control (see also Self-analysis, Sec. 8-5).

More precisely, the term arithmetic operations has come to cover all the data-manipulation operations in a computer except some operations explicitly concerned with input-output, code, and sometimes, format conversion. In even the most elementary computers, any data-processing operation is possible by combinations of other operations.

The range of facilities necessary in both the control and the arithmetic unit is partly determined by the degree of autonomy vested in the controllers of the input-output units and auxiliary stores.

Control of Sequence. One of the most basic operations in the central processor is obtaining instructions from the store and executing them in the required sequence. Some of these instructions will themselves, directly or indirectly, control the sequence of the instructions to be obtained later.

In the most simple form, there is a sequence counter which is advanced by one unit after each instruction is executed. The counter is used to specify the address of the location of the next instruction. Thus the instructions are obtained serially, one at a time, from series of locations ascending in sequence, whose addresses are in the store.[9]

In the earliest forms of delay line and drum computers, in which access times were significantly longer than execution times, it was desirable to modify this type of approach. The "$N + 1$ address instruction" was used, in which the address of the location of the next instruction was specified directly in each instruction. Thus it was possible to place an instruction in a position where it would have a very short access time after the known completion time of the previous instruction. Occasionally, other methods have been used to reduce the waiting time for instructions. Some systems arrange the locations of a sequence of instructions at fixed intervals in a cyclic-structured store at intervals in which a reasonably short latency is achieved most of the time with average-time-length instructions.

Although both these forms of sequencing were developed very early in the computer era, there have been no substantial deviations from them.[38] There have been only the minor variations necessary to deal with variable-length instructions and variable-format instructions.

Operations and Operands. In addition to the function of sequencing, central processors have the responsibility of providing the repertoire of data-manipulation operations. In this respect, there are two major areas, the different operations provided in the repertoire of the computer, and the different types of operands for which they may be provided. The operands may be fixed or variable in size, they may be in pure binary or in a binary-coded character format,[39] or they may be organized by characters or words. This is the area of processor specifications wherein the greatest variation of arrangements is available. The key to the great power and general utility of a general-purpose digital computer is the fact that data operations can be performed on operands that may be directly or indirectly used by control to execute a program.

Arithmetic. Only addition and subtraction appear to be a common subset of arithmetic operations in computers. Many computers have no multiplication or division facilities, although these facilities are often provided as optional extras. In small or minimum computer systems, multiplication and division are often performed by subroutine, particularly in the areas where computers have replaced punched-card installations. In such cases, multiplication and division facilities are not always essential and the equipment being replaced often did not provide multiplication. In more sophisticated computers, several different types of multiplication and division are provided. There may be separate operations to produce doublelength and single-length products, rounded and unrounded. In division, there is always the problem of deciding on the size of the quotient, whether or not it should be rounded, and the sign of the remainder. More sophisticated machines, particularly those in business and commercial use, provide a variety of division features with or without corrected remainders.

The earliest scientific machines did not provide a floating-point arithmetic; this was provided by separate routines. However, floating point is a standard option on most scientific machines. Some such machines even have restricted fixed-point arithmetic. Scientific calculations often involve computation algorithms, where cumulative round-off and/or truncation errors may be appreciable. The degree of accuracy is often improved by the use of multiple-length arithmetic, in which two or more words are used to hold a number requiring greater precision than the normal word length of the computer affords. The operations are usually provided by subroutine, but the operation repertoire in a machine is sometimes especially arranged so that these operations are easily provided. In some machines, the operations required to perform the algorithm are incorporated directly into the hardware.

Comparisons. Although one of the greatest powers of the computer is its ability to make comparisons based upon results of certain operations or to choose the sequences of instruction that are employed on given groups of data, the facilities for comparison which are built into computers are not normally very extensive. In the simplest types of computers, decisions must depend upon testing results in an accumulator, after performing a subtraction. Usually the tests available are limited to the sign of the accumulator, positive or negative, or its zero or nonzero state. Several nontrivial differences exist between types of machines, particularly those which have different representations for plus zero and minus zero. There is no consistency as to whether the case of zero is included in the test for positive values. When dealing with alphabetical data, the simple use of subtraction does not always provide useful precedence rules for the representation of the letters of the alphabet and the other special symbols. Some computers do have special alphabetic comparison instructions which give correct high, low, and equal comparisons of alphabetic data. Some computers have a special comparison, that is, a subtraction in which the result is not generated, but indicators are set to show whether the result of the comparison is high, low, or equal. Subsequent instructions test these indicators. Some computers set such indicators in every arithmetic operation. In cases where comparisons need to be programmed, either because they are complex in themselves or because

DIGITAL COMPUTER-SYSTEM CHARACTERISTICS   8-19

they are complex relative to the facilities provided in the computer, the so-called logical operations are often useful.

Formats. One of the most important properties of the data used in operations is the implied format. The whole repertoire of operations provided depends to a large extent on the format of the data. The hardware may be designed to recognize many forms: straightforward binary representations of numbers; binary-coded character representations; representations in which each bit of a word is treated separately; variable format selection in which masks or other data may be employed to indicate operation on only a part of a word; variable-length operations in which counters or delimiters are used to define the extent of data; operations on blocks of words; and special conventions such as half-length, double-length, and triple-length wording.[40] In floating-point operations, the hardware has to distinguish the exponent separately from the fixed-point part.

As computers have developed, there has been a tendency to standardize formats. Some computers have homogeneous sizes of parts of words, such as multiples of 6 bits or multiples of 4 bits.[41] The former are more useful for alphabetic input-output, the latter more useful in decimal arithmetic. Sometimes alphabetic data are redundantly represented as 8 bits (two sets of 4 bits) in order to maintain a homogeneity of 4-bit representation. The parts of such words are often called "bytes."[39]

The data inside a computer usually have a fairly rigorously specified format which is chosen for convenience and efficiency of processing. On the other hand, it is often important to have a fairly flexible format for input and output. It is usually desirable that the format of input and output documents should be designed for ease of preparation and use of the document rather than for the internal convenience of the computer. Special instructions are often necessary if sophisticated editing and format control are to be provided. In mathematical applications, simple rearrangements in a few formats may be sufficient, but there should be a facility for zero suppression and control of plus or minus signs. In commercial applications, it is frequently desirable to have special insertion characters in editing both alphabetic and numeric data. Many diverse formats may be required and special facilities such as floating dollar signs and check-protection characters may be provided.

Instructions. The instructions in a computer are often closely related in size to the word length; there are usually either one or two instructions per instruction word. In computers with variable word lengths, usually based on characters, either a fixed instruction length of, say, five characters may be assumed, or the instructions themselves may be of variable length. The diversity in instruction formats is greater than that in the data formats. An instruction of 28 bits may be broken up into parts of 3, 5, 13, and 7 bits. The number of addresses in an instruction may be partially dictated by the type of sequence control being used, one or more addresses for operands and either one or no address to designate the next instruction location for normal-sequence operation. The instructions themselves are subject to much manipulation by program, particularly in elementary modern computers. Hardware modification of instructions may be extensive when recursive indexing and indirect addressing are provided. It is usually the address only that is subject to modification before the instruction is executed.

Logic. Logical operations are conventionally those operations in a computer which are based upon operations on pairs of bits. The normal Boolean operations of AND and OR are usually arranged so that many bits in one word are associated in pairs with the corresponding bits in another word and the multiple results are produced in one composite operation. This is usually conveniently done by modifying the carry mechanism in the arithmetic circuits. The range of logical operations provided does not have to be large. From two logical operations, the others can easily be formed by combinations.

Tables. It was soon recognized that, in much data processing, considerable time was consumed in sequentially searching tables in storage for certain types of entry, particularly in the case of nonnumerical operations.[42] Special operations on tables are now included in the larger or more sophisticated computers. Table look-up instructions provide operations which typically scan a series of locations, looking either for keys to match a prescribed pattern or for keys whose values are higher or lower than some prescribed value. The more sophisticated operations of looking for the greatest or for the least value in an unsorted set are occasionally available.

Table searching can often be implemented by simple use of repeat-operation instructions, which specify the repeated execution of a given instruction or pair of instructions a fixed number of times. These instructions can be very powerful if they automatically advance index registers each time they are used and thus increment their effective addresses. Direct access to data by means of table mechanisms incorporated in associative memories eliminates the time needed for a serial search (see Access Structure in Sec. 8-3).

Special kinds of tables are also being used for operations in which one block of data consists of many pieces held in many different places in a store. A table specifies the addresses of the different sets of locations where the data are. Such tables are most frequently used in input-output, and the transfer operations that use them have been given the names "scatter-read" and "gather-write."[43] These operations are not really input-output functions so much as data-rearrangement functions.

Lists. Some of the more interesting uses of computers in applications related to language logic and heuristic techniques have resulted in the development of special types of operation known as "list processing." In these cases, the data are arranged in the form of a tree or graph. This type of arrangement is represented in a computer by maintaining lists in which each item of data contains a link denoting the location of the next piece or pieces of data in that particular list. There may be more than one link if the list represents a linkage from the trunk up toward the branches or twigs of a tree. Alternatively, if the trees are considered from the twigs to the trunk, several links may lead to one address. Although certain special facilities do contribute to the easy manipulation of the lists, presently no computers are specifically designed for operations on list structures.

Input-Output Transfers. The facilities for input-output are very important and are covered more extensively in Sec. 8-6. Sometimes the operations are covered completely by the central processor by the use of series of instructions, but sometimes the central processor only initiates an operation by sending signals to a controller which then carries out the detailed execution of the individual stages of data transfer. In the most elementary form, the central processor has the ability to start or stop an input-output device and then must transmit, at the proper time, a single character or a small unit of data to be recorded or read from the external medium.

Synchronizing Aids. When the central processor has to provide timing and synchronization, there are several possible levels of sophistication. At a basic level, the processor must initiate each instruction and wait for a lock-out until the input-output unit is ready. At an intermediate level, it may be able to test whether or not a unit is ready; by making frequent tests, it is able to synchronize the transfer. At a higher level, there may be interruption signals from a unit when it is ready, in which case the central processor can do other work without having to make frequent tests.

These same levels of sophistication are also possible when the central processor is able to transmit a complete block of data, input or output, with individual instructions. In these cases, the controller deals with the individual units (words or characters) of data. The execution and timing of successive blocks are left to the central processor. In order to maintain high speeds, the central processor may again either have to wait for a lock-out, or keep testing for busy, or rely on interruption.

Multiple-block Loads. A few computers provide a facility on input-output devices to write many blocks in one load. This occurs mostly on very high-speed computers. when units such as card readers and paper-tape devices are used on-line. It is also of use to reduce stop-start overloads where fixed length blocks are employed on magnetic tape.

Table-controlled Transfers. Input-output is usually a straightforward transfer of a block of data between an internal store and an external store. Special cases such as scatter-read and gather-write are especially useful in eliminating the time spent in moving data within internal storage in order to organize buffer input and output areas by programming. In high-speed computers, the proportion of time taken up by the accesses to a store by the input-output units are high and prevent the central processor from running at full speed. These overheads, called central processor penalty times in the comparison charts (see Sec. 8-9), can be much reduced by the use of tables for input-output control.

Block-size Control. The size of blocks can be controlled in several different ways. There may be:

A fixed block size
A fixed block size with a cut-off
A variable block size determined by a program-controlled counter specifying the number of words to be transferred
A variable block size controlled by a delimiter in the data itself or a delimiter in the store
A combination of methods 3 and 4 (in the most flexible input-output schemes) in which the first condition to be satisfied prevails

When input-output control must be programmed, any of the methods may be employed except for restrictions imposed by fixed block sizes that are built into the system.[34,36]

Conversion. Although in a well-designed computer system, all input-output interfaces with the internal storage are similar, there are significant differences in the data representations and coding among magnetic tape, punched cards, paper tape, and the printed output, although the codes on paper tape and magnetic tape may be closely related.[44,45,46,47,48] In the simplest systems, an input-output transfer produces in an internal store a simple image of the binary patterns in the external store. In all but the most elementary systems, the controllers should perform an automatic code translation between the basic external codes and the internal codes, at least to the extent that there is a common internal code after input and before output, even if this is not the code used for the arithmetic unit. Some units such as paper-tape readers provide for different codes to be wired into the controllers, and on many magnetic-tape systems there are two different codes for binary-coded characters or binary data wired into the controllers.

A central processor itself often includes conversion facilities because they are not present in a controller, either for the cases of special codes which cannot be economically included in a controller, or for all the codes in a cheap machine where no conversion has been put into the controller at all.

There are two types of conversion to be considered: first, radix conversion, such as between binary and decimal; and second, code conversions, such as between 5, 6, and 7-bit codes. In some cases, the two may be combined. Whereas code conversion is often provided in controllers and in the central processor of some powerful systems, radix conversion, if provided, is invariably a central-processor function, often combined with editing.

Self-analysis. In the most straightforward computers, interlocks are provided so that no operation can begin until a previous conflicting operation has been completed. In the case of input-output operations, the computer can spend a large part of its time waiting for previous operations to be completed. In order to increase the efficiency of computers, facilities have been incorporated which permit a program to determine which parts of the computer are busy so that it can organize its work efficiently.[4] In the simplest cases, tests are incorporated in the instruction repertoire for the idle or busy conditions of input-output units or controllers. Another condition which may hinder the efficient operation of simple computers is a stop that may occur when errors are discovered. In such cases, input-output units stop. The central processor stops when overflow occurs. In more sophisticated systems, error indicators are set and the program tests these error indicators and attempts to recover from the error either by rereading (on input-output units such as magnetic tape), or by executing parts of the program again, or by taking other actions specified by the programmer.[49]

When many error indicators can be set, much time can be spent testing the indicators; and it is necessary to insert the instructions in each program. A more sophisticated level of operation occurs when interruption facilities are provided. In these cases, the tests are made continuously by the central processor in parallel with the execution of programs. When one of the conditions exists and an interrupt is executed, the interrupt causes the execution of the current series of instructions to be stopped. Then control is transferred to a supervisor routine, which investigates the cause of interruption, takes the necessary corrective or remedial action, and returns control to the program that was interrupted.[37] Interruptions themselves can take a large part of the time of the computer. In many cases, there are special mask registers which can be set by the program to decide which of the different types of interruptions are being allowed to occur. Special arrangements must be made so that, when an interrupt causes entry into a supervisor routine, interruption of that routine itself is either prevented completely or restricted to special superinterruption conditions. It is usual to make internal central-processing interruptions superior to input-output interruptions. Central-processing interruptions usually indicate faults, whereas input-output interruptions most frequently indicate that equipment is available to be used for further operations. A few computers have been built so that a queue of input-output operations is maintained automatically by the computer and executed when individual units become available,[50] but the normal design of such systems relies on special supervisor programs being maintained in the store of the computer. It is expected that these will be put into fixed stores, where special access and nonerasable facilities are associated with them.

Organization. In the earliest computers, central processors had an essentially simple organization in which one operation could not be carried out until after another was completed. As the specific operations which a central processor could perform were increased, simple organizations of this type became inefficient because most of the circuits were idle most of the time. One of the most challenging objectives of the designers of recent computers has been to increase the amount of parallel operation inside the computer so that greater efficiencies are possible. Two basic techniques for increasing efficiency have been overlapping operations and central-processor look-ahead, the second being a refined version of the first.

The problem of parallel operation occurs not only in the central processor but among all the units of a computer system. Therefore, principles of organization to perform multiple running and multiple processing have also been developed, the details of which are closely related to central-processor organization.[51,52,53]

Central-processor Overlapping. The simplest but most dramatic improvement in the form of overlapping of operations in the central processor was the change from serial arithmetic (operating on one bit or one character after another) to parallel operation (operating on an entire word of many bits or characters at one time). This change introduced problems of carry propagation times. Special circuits have been devised to speed up this type of operation. This technique has also introduced complications when logically serial operations concerned with data editing and variable-length wording are attempted, but in general has been extremely useful.

A second, but rather rare, technique has been to supply special separate circuits for multiplication and division. These operations normally are much longer than other operations, such as add or subtract. In this case, a multiplication operation would be started in a separate part of the arithmetic unit and the remainder of the arithmetic unit could continue with other simpler arithmetic not involving multiplication and division. The result of a multiplication or division would be available several instructions later.

Look-ahead. As soon as attempts were made to overlap many operations in the central processor, it became obvious that the burden on the programming of attempting the optimizing by overlapping (as in the case of overlapping multiplication with addition operations) became very difficult. In practice, relatively small savings were obtained. Then the concept of look-ahead was introduced, in which the central processor was arranged essentially as several subprocessors in a sequence. In a typical case, the first subprocessor would select instructions from the store, the second would examine them and perform indexing operations, the third would obtain operands from the store, the fourth would do the arithmetic, and the fifth would transfer the results to the store as required. In this example, the first, second, and third units are performing look-ahead operations for the fourth and fifth units. Consider a simple case of one multiplication followed by several additions. The first, second, and third units could be building up a queue of addition and subtraction instructions to be executed, while the fourth unit is performing the multiplication. The second unit requires addition and subtraction facilities in order to perform the indexing on addresses, and thus extra circuits can be automatically overlapped in operation. The advantages of look-ahead are most obvious in the fast core-storage computers in which the access time of the stores has been reduced by having independent banks and overlapping accesses to independent banks. Unless the addresses are reasonably distributed, not necessarily randomly, the overlapping will not be effective. Look-ahead can even out the overlapping access times and provide a type of asynchronous access with a loose form of linkage.[36]

Many computers have a trivial type of look-ahead in which they obtain access to pairs of instructions in a single word from the store, but their design is not of the same order of complexity as true look-ahead, in which operands are obtained in advance.

Effective look-ahead has been restricted to a limit of about four instructions plus their operands, for two reasons. The first difficulty arises with branch instructions. One computer attempts to evaluate these branches in advance if they depend upon counters or index registers; if no data are available, it takes the straight-through sequence, and must then backtrack if it has made the wrong decision. The second difficulty arises when frequent interruption is also provided in the computer. Interruption and look-ahead are incompatible in their design, because each interruption requires complex unwinding of the look-ahead.

Multirunning. Each program run upon a computer has its own particular characteristics for the amount of time that it requires of each of the individual units. Many programs do not make balanced use of the computer; some units are not used at all; others are used only a small fraction of the time. The concept of multirunning was introduced to overcome this type of imbalance.[51,54,55,56,57,58]

Multirunning means having several independent programs running in one computer at one time. The central processor takes one program at a time and executes a series of instructions in it, then a series of instructions in another, and so on. The rules for deciding when to change programs vary from implementor to implementor. As a simple example, consider two programs—a "central-processor-limited" file operation, and a "printer-limited" magnetic tape to printer operation. In this case, the central processor would attempt to execute the second program, predominantly magnetic-tape-input and printer-output operations, until the printer inhibits progress. It would then return to the first, the "central-processor-limited" program, until the printer was again free. Then it would revert to the second program, using only a few instructions to keep that one going, and return to the first program. In this way, extra units (a magnetic-tape input and a printer output) can be kept fully busy in return for a very small penalty in central-processor time on the central-processor-limited program. In some computers it is not unreasonable for the central processor to test continually to see which units are not busy and change programs accordingly, but a more convenient method is the use of interruption (see above under Self-analysis).

Interruption is a facility included in the circuits of a control unit that may be enabled to force the natural sequence of instructions in a routine to be interrupted and control to be transferred to a specified location. Most systems provide instructions to determine which events from a given set may or may not interrupt. A master supervisor program is entered whenever interruption occurs, discovers which unit is idle, issues instructions to enter the program which controls that unit, and arranges the change-overs between the programs in a way that tends to optimize the use of the equipment.

There are two methods of implementing multirunning. In one, a set of separate sequence counters, one for each program, maintained in the central processor, selects instructions from each program in turn. A three-address instruction is required in this method so that no central-processor registers are involved, or a set of registers is allocated exclusively to each program. A special facility can be used to suppress some sequence counters when appropriate.[59] The other, and more frequently implemented, method has a special supervisor program which decides upon a priority and transfers control to the different programs as needed. Interruption signals are used to return control to the master program when conditions change. The simplest general rule is to arrange the programs in a priority sequence in which the programs which use the central processor least are placed at the top of a queue. This is a complex subject, and the rules can be more clearly seen in Ref. 51.

Multiprocessing. In some computers, the problems of imbalance are not those of keeping the peripheral units running by having a sufficient number of programs being multirun, but those involving limitations of the speed of the central processor. In these cases, more than one central processor may be incorporated in the computer system (Basic Connections in Sec. 8-2).[18,21,57,60]

Multisequencing. When using multiprocessors, it is possible to use a special technique to improve the performance on a program that is central-processor-limited. In this case, the program may be executed by several processors, each processor executing a separate sequence of instructions from one common program. This is a multisequencing technique[60,61] in which parallel sequences of instructions are executed and control positions are required when the sequences have to be completed together. Little experience has been gained with this type of technique, and new language facilities will be required to take advantage of it.

Interprogram Protection. When more than one program is running in a computer, the obvious danger exists that faults in one program or its data can cause it to corrupt the data or instructions of another program in the computer. In general, there are three cases where protection is required: protection of the registers being used by a program, protection of the store being used by a program, and protection of the input-output units being used by a program. In general, when changing from one program to another, the contents of the arithmetic registers are emptied and stored temporarily in the storage area used by that program, so that the first case becomes the second.[55] The protection of input-output devices is not yet very far developed. In some cases, however, the access to an input-output device is made by indirect addressing through a word in store, so that the third case also reduces to the second.

There are two types of technique which have been devised for protecting the store of each program.[57,62,63] In the first, two limit counters specify the highest and lowest address of the storage area allocated to that program.[64,65] Every access to the store during execution of that program is compared with both registers, and any area that lies outside the range causes an interrupt or an alarm to occur. In the second, the areas of store are marked to correspond to the program to which they are relevant. One machine uses four extra bits with every word (a value of 0 to 15 to indicate which program may have access to it). In such a case, it is possible to specify an area, say code 0, which is common to all programs, and an area, say code 15, which is accessible only to a master program.[51] In one variation of this type, one bit for each block of store is set to "one" or "zero" whenever there is a change of program, to indicate whether the current program may or may not have access to that block of store.[51]

Intraprogram Protection. Because of the general purpose design of computers, there is limited scope for protection of a program against itself. Nearly every possible operation can be required in some conceivable situation. Most intraprogram protection is unique to one program or a class of programs and is provided within the programs themselves. Such a protection is usually provided by consistency checks on the data.

A few devices are sometimes incorporated in the hardware, however. The most common is an indicator or alarm to indicate overflow in a register. This condition is so often not a fault condition that a program can usually suppress or at least ignore the facility.

A second group of facilities is associated with the fact that not all patterns of

DIGITAL COMPUTER-SYSTEM CHARACTERISTICS   8-25

instructions or data are legal. Alarms may be provided to show when nonexistent operation codes are formed in instructions or when illegal characters are detected in data operations (e.g., a character A in an addition operation, even though "arithmetic" on characters is often required).

A third group of facilities is provided by any interprogram protection in a computer. In such a computer there may also be facilities to detect infinite loops (loops of instructions which are being executed repeatedly in error, preventing the program from progressing or terminating as it should). Such loops can be detected by a technique using an interval timer to cause interrupts. At various points in a program the timer is set at selected values. If the timer steps down toward zero, an interrupt occurs. Such an interrupt indicates an unusually slow program because the values are selected so that the timer should be reset well before it reaches zero.

Levels of Control of Operations. Several different types of structure have been designed for the execution of individual operations. Special circuits and wiring are arranged to control the execution of each operation in the instruction repertoire of a computer, the only level higher than this being the arrangement of the programs of instructions by the programmer.[66] As programs became sophisticated, and subroutines were invented, hierarchies of subroutines and subsubroutines were used by the programmers to simplify their own organization, introducing a series of levels of control within their programs.

The design of hardware of the central processor has also begun to follow this pattern. The introduction of microprogramming of individual instructions from the operation repertoire has become popular.[22] The design of one computer includes the design of macroinstructions, which are instructions of a higher level than ordinary instructions. They are special routines coded in the ordinary instruction repertoire but held in special areas of fixed (nonerasable) storage with fast access.[67] In addition to using ordinary locations, a store may have pseudo-central-processor registers. Some computers have used the store to hold tables of arithmetic constants in order to simplify and reduce the expense of arithmetic circuits. Special types of arithmetic circuits and special types of control have been introduced in order to obtain the advantages of variable data lengths in instructions so that the storage space required for instructions can be significantly reduced. More recently, the style of organization of arithmetic operations contained within individual instructions with an operand has been altered to an arrangement more in common with Polish notation.[68]

8-6. Input-Output

Input and output are the mechanisms of communication between a computer and its external storage. From the point of view of external storage as a medium, the properties of some of the different types of storage to which both input and output functions are appropriate have been discussed in Sec. 8-4. Some other forms of external storage, such as printing (see below), are presently used only as media for communication with human beings. This section is concerned mainly with the connections of external storage to a computer system.

The organization of input and output can be divided into four different areas of interest: the general system of connections between external storage and the computer, the special types of facilities provided for the input-output transfers, the different types of external storage media, and the characteristics which affect the performance of different types of input-output.

General Connections. Connection of external storage to the computer system by means of input-output units has been discussed in general in Sec. 8-2. A series of elements is usually provided in the controllers between the external storage and the internal storage of a computer (see Fig. 8-6). First, there are the physical heads which read from and/or record onto the external storage medium, and the transporting mechanisms that may be necessary. Second, there are the control circuits which adapt the particular signals used by the read and write heads to the signals acceptable by the general computer system, and circuits which initiate the stopping and starting and other control aspects of the input-output mechanism. Third, there are the conversion circuits which may be inserted to convert data between the codes used in the external storages and the codes required for use in internal storage and any rearrangement between different word sizes. Fourth, there is the mechanism for controlling at the appropriate times the transfer of data with the internal store. Fifth, in some machines, there may be a part of the computer which is concerned with multiplexing, or sharing time, or interlocking the access of the different input-output channels to the central store.

Circuit Adapter. In order to maintain a uniform system, most units are designed to serve a generalized function. In input-output, the only parts which are specially designed for individual equipments are the circuit adapters whose purpose it is to provide a matching of the general computer interface with a specific device.

Code Converter. Although the logical design of a code converter is different for different input-output devices, a common type of unit can be used throughout a computer. Some units provide alternative code transformations, either transformations controlled by a switch (program or manual) between one or two codes, or a large variety of transformations from which one transformation can be established by a plugboard. Such a unit may be combined with a large buffer to deal with row

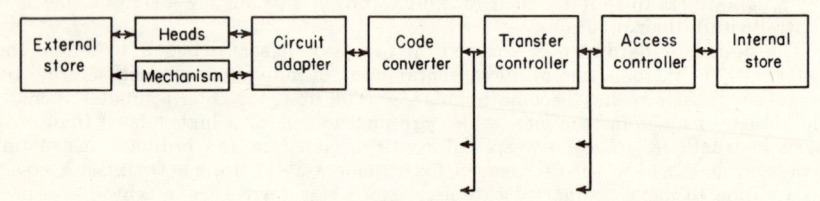

FIG. 8-6. Controller elements.

or column reading or punching of cards. One of the transformations may be a simple bit-for-bit mapping, called a binary image pattern. This enables a program to perform any special or peculiar code transformations that may be desired (Sec. 8-5).

A code converter and a circuit adapter are often combined to form a physical unit called an adapter, which may include switching facilities to enable it to deal with several units at one time.

The types of transfers that the input-output circuits are required to perform are also standardized for many units where possible, in spite of the fact that many of the units are naturally of a fixed size (such as a line of printing or the 80 columns of a card). The general principle of variable-length input-output is almost universal. The total amount of data that may be transferred under one operation specified by one instruction is called a load. A variable load can range from a single character to many units, such as many lines of print, or even many parts of lines of print. In the most sophisticated case, an input or output instruction would specify a number of blocks comprising one load, each block being a separately specified size. Thus the number of blocks and the size of each block may be varied. In less sophisticated computers, there may be fixed block sizes specially determined by the input-output units.

Special Controller Facilities. In addition to the organization of the transmission of loads of data between internal and external storage, the controllers often incorporate other special facilities, usually dependent upon the different types of units which are connected.

Controller Functions. The most common controller functions are rewinding magnetic tape, switching off input-output units, testing input-output units and their controllers for busy or idle status, and testing indicators for error conditions. A few controllers incorporate automatic error-correcting facilities, ranging from redundant codes which are used for automatic error correction to automatic sequences for rereading or rewriting blocks of magnetic tape.[69] A controller is often expected to do all the checking of input and output by reading recorded data at a second station, comparing these data with a copy of the original output or, more often, testing it by parity checks and other convenient schemes. Some magnetic-tape units recognize special control signals for end-of-file conditions and are able to set appropriate indicators. Other magnetic-tape units have search facilities incorporated so that a tape can be rapidly scanned until a certain specified key is discovered in a block. Backward reading is normally characteristic of magnetic tapes only, and is a rare facility even in these.

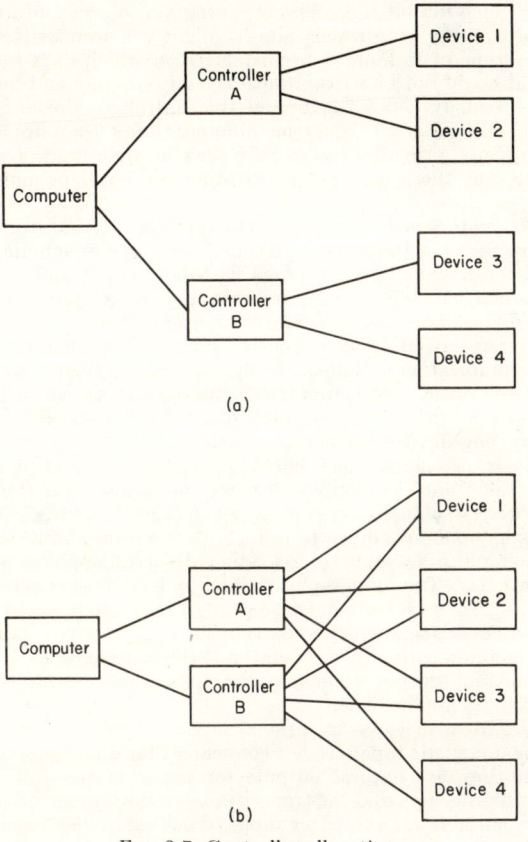

Fig. 8-7. Controller allocation.

Controller Allocation. The number of controllers in a computer system usually determines the number of simultaneous operations possible. In the simplest case (Fig. 8-7a), the allocation of controllers to devices is restricted, and full use of controllers A and B is possible only if one of devices 1 and 2 and one of devices 3 and 4 are working. Figure 8-7b illustrates a more useful scheme of free allocation. In this case, both controllers can be used if any two devices want to work. It is a necessary condition that the controllers are interchangeable. In the weakest case of all, the devices are of the same type; in the strongest case, many different devices use individual adapters so that all may use any controller.

In a few systems, special circuits have been included to allocate any free controller automatically.[41] In most cases, the program must control the allocation, but this may reduce the efficiency due to poor choices and a need to maintain a list of, or search for, free controllers.

Buffer Protection. Scatter-read and gather-write is one sophisticated type of buffer operation (Sec. 8-5). The controllers in earlier computers contained buffer stores which were filled from the input medium or emptied into the external medium, and then filled or emptied by a rapid transfer with the internal store. In most computers, these expensive buffers have been discarded and the controller selects one word at a time from the central store as it is required. In such cases, a program may not know which parts of the internal store buffer have been filled or emptied. In some computers, sophisticated lock-outs have been provided in which bits or registers are set which inhibit the access of a program to areas into which data have not been read or to areas from which data have not yet been written. In one computer, a "weak" type of lock-out was introduced in which data in an output area could be read but could not be written until the output copy had been recorded.[64,65]

Multiplexing. Usually it is a function of the controller to arrange the individual accesses to the internal store from the different units that are complete. This function is commonly provided by some means of time-multiplexing accesses to the central store, but there may be several different levels of multiplexing among simpler controllers.

Variety of External Storage Media. The types of external storage media which were discussed in Sec. 8-4 (punched cards, punched tape, magnetic tape, and magnetic cards) are those which are concerned with both input and output operations. Other types, discussed in this section, are naturally input only or output only, and are not generally considered as storage for computer systems.

Printing. Printed output from computer systems is normally a form of output intended for communication to human beings, and not intended as data to be subsequently reentered into a computer (with exceptions noted below). There are two principal classes of printer output, high speed and low speed.[70]

High-speed printing devices are characterized by line-at-a-time mechanisms using drums, type wheels, or chains which enable these devices to print at speeds of the order of 100 to 1,000 lines a minute. A newer nonmechanical type, typified by a xerographic printer, is able to operate at much higher speeds. The xerographic printer also has the inherent ability to print both the form and its data at the same time, thus avoiding the necessity of reloading different supplies of stationery for different applications. The xerographic printer is less economical when producing multiple copies because each copy must be printed separately.

Low-speed printers are characterized by typewriterlike devices and are used mostly for low-volume communication to an operator at the console, or as a cheap means for providing intermittent output to many remote stations, as in reservation, data retrieval, or command applications.[71]

Mark Sensing. Most of the conventional input devices of a computer (punched cards, paper tape, magnetic tape) are not convenient for direct production by human beings, although they are natural outputs for many mechanical devices such as cash registers. In order to avoid the transcription of documents produced by human beings to a form suitable for computer input, much effort has been devoted to the development of devices to read marks easily or naturally or cheaply produced by human beings. There are three principal types of readers: optical readers which sense the light from a document, magnetic readers that recognize magnetic marks made by special inking or magnetic-pen devices, and mark-sensing devices which use the carbon in a pencil mark as an electrical conductor. Because a human being does not produce well-formed characters which are easy to recognize, most simple sensing devices recognize only the presence or absence of marks in specially designated places on a form. Such marks signify, as do ticks on a checklist, whether or not particular types of data are indicated. Such input devices are not used much on-line. Off-line preparation of such data is more convenient, particularly when the problems of paper handling and data errors are considered.

Character Recognition. Some input devices have been designed to read characters used by human beings. These are for documents which have been printed to be read by both human beings and computers. Such devices may be either optical or magnetic readers. One reader has been designed to read the output of a particular line-at-a-time printer. Another example is the magnetic-ink characters imprinted on checks which bear a strong resemblance to the characters used by human beings but are specially formed to make machine identification easier. Little progress has been made in practical recognition of characters formed by human beings.

Performance Characteristics. For input-output devices, the main characteristics of performance are the speeds at which they can operate and the overheads which are incurred in each type of operation.

Speed. The most usual measure of speed is a transfer rate, usually measured in characters per second. This is not always convenient, because it is sometimes possible to obtain higher rates if only decimal digits are used. More important, it is sometimes difficult to use when the rating is restricted to unit terms, such as cards per minute; the number of characters per card varies depending upon the density with which cards are punched. If the number of unused columns on the card is known, then an effective speed in characters per second can be computed. Effective speeds on units such as magnetic tape can be substantially affected by the interblock gaps and stop and start times, which can considerably reduce the peak or nominal speed of a device. With very high magnetic tape speeds, it is essential to have very large blocks (about 1,000 characters) if the overheads are not to degrade the speed by more than 50 per cent. (See the charts in Sec. 8-9, where block size varies from 100 to 1,000 characters.) The effective speed of a printer depends not only upon the number of characters per line but also upon the distance between lines, inasmuch as time is used to feed the paper from one line to another. (See the charts in Sec. 8-9, i.e., entries referring to line spacing.)

Loading Overheads. There are two types of overhead to be considered in loading a device. First, there is the simple replenishment. A reel of tape has a limited capacity, as do the hoppers and stackers of card-reading devices. At the end of a reel or the end of a stack of cards, it is necessary to unload the device, often rewinding the magnetic tape, and to replace it with a new supply. There are two times involved: the maximum period which the unit can run without being replenished, and the time required for replenishment. The second time is involved when it is necessary to make adjustments to the input-output unit for different types of input-output which are made necessary by change-over from one job to another (for example, a different supply of stationery on the printer). In most modern computers, the use of plugboards has been eliminated and such setup times are now mostly confined to printer units. The newest types of punched-card equipment can now be loaded and unloaded without stopping.

Capacity. An important characteristic of an input-output medium is the quantity of data which can be stored in a single physical unit. A reel of magnetic tape may, at different densities, hold a maximum of between 5 and 20 million characters. A tray of 7,000 cards may hold approximately 500,000 characters. A reel of paper tape is often limited to 1,000 ft, or 120,000 characters when fully punched.

Common Media. Compatibility between different input-output units or different input-output systems requires not only that the recording media be physically compatible (as, for example, ½-in. tape), but also that the registration (of rows and tracks), formats, and coding be compatible. The interblock gaps, and even the recording scheme (e.g., non-return-to-zero), must also be compatible. Sometimes there are concealed differences in the particular block sizes that are used in different variable-length systems, e.g., a block having an integral number of words and the different word sizes 36 and 48 bits. Even different thicknesses of punched cards may cause problems. Nevertheless, there is widespread intensive use of punched tape, punched cards, and magnetic tapes among computers of all makes.

8-7. Simultaneous Operations

The performance of a given configuration of units in a computer system is significantly dependent upon the number of simultaneous operations that can take place. For any task to be performed by means of a computer, there is a certain amount of work assigned to each of the separate units. The time required by each unit to complete its work can be established. If there are no simultaneous operations, the time for the system to do the complete job would be the sum of the times for the individual units to do theirs. Most computer systems provide facilities which permit many of the units to run autonomously or simultaneously, thus allowing many of these times to be overlapped, reducing the total performance time.

The number of simultaneous operations possible in a system is determined by the way computer units are connected and by the way their control is arranged. The number of simultaneous operations that actually occur in a job depends upon the problem, and the skill of the programmer or analyst in arranging the work to permit simultaneous operations. It is obvious that the greatest advantage in simultaneous operations occurs when several units are equally loaded, that is, when there is a reasonable balance of work among the various units. It is one of the rules of efficient system design to attempt to balance work so that the idle time in units of a computer system is minimized.

Special Units. Simultaneous operations often depend upon the inclusion of special units: optional parts of a central processor, integral parts of peripheral controllers, or special multiplexing or time-sharing units; see Sec. 8-5. The most important consideration is usually the number of controllers that permit autonomous data transfer for input and output.

Different Types of Simultaneous Operations. Five major techniques may be used to provide simultaneous operations:

Independent operations
Multiple-data paths and multiplexing
Partial overlapping
Restricted operations
Programmed time sharing

Independent Operations. In general, independent operations are those using units which are not concerned with any other operations at the same time. A typical independent operation is rewinding a magnetic tape. Other typical independent operations are a magnetic-tape search by a controller for a block with a specific key, and data input or output via a buffer which is a separate store or a buffer which is part of a store being used for other operations, but which has its own separate access device, e.g., a track of a drum.

Multiple-data Paths and Multiplexing. Simultaneous data transfers, usually for input-output, are possible when each uses a separate data path. A special case exists when several paths share one cable but are multiplexed, or time-shared, in such a way that one cable is logically equivalent to several cables (see Sec. 8-2). Another common form of multiplexing is the time sharing, or intermingling, of accesses to a store from many units. This is equivalent to multiplexing several data paths into a store.

Partial Overlapping. In many simple systems, input-output transfers cannot generally be overlapped with other operations. In some operations, such as card reading, card punching, and on-line printing, only part of an input-output cycle is needed for data transfers. There are usually periods at the start and end of each card cycle, or during paper movement, during which no access to storage is required. In these cases, some systems prevent (interlock) the operation of other units, usually the central processor, only during the transfer period, and permit computing during the other periods in input-output cycles.

Restricted Operations. In a few rare cases, an input-output operation may only partially restrict the operation of a central processor. One example is the use for input-output transfers of a special register that is also used for multiplication; input-output then inhibits only multiplication and related operations. Another special case is one requiring observation of some timing restrictions and permitting execution only of instructions whose execution time is less than a certain limit.

Programmed Time Sharing. Programmed time sharing, in which a routine is used to arrange data transfers for input-output in small units, is possible in some systems.[52] These transfers must be carefully timed. The penalty paid in central-processor time devoted to this type of activity depends upon the speed and type of facilities available. In preferred order, the facilities are automatic interruption, a "test busy" facility, or nothing (see Sec. 8.5). In the last case, instructions must

DIGITAL COMPUTER-SYSTEM CHARACTERISTICS   8-31

be carefully timed by the programmer. Programmed time sharing can simulate multiplexed access to storage.

8-8. Performance of Computer Configuration

Although the performance of individual units of a computer system can be established fairly easily, the performance of an entire system is a complex consideration. In order to be specific and have some reasonable basis for measurement, five factors must be taken into consideration: the particular configuration of equipment, a particular problem, the implementation techniques, the operating environment, and the reliability of the system.[72,73,74,75]

Particular Configuration. Computers can be assembled in a large variety of configurations. The most important considerations are usually the number of simultaneous operations, the ability to move data easily within the computer, the editing of output data, and particularly for scientific computations, the speeds of the basic arithmetic operations. Where a computer may have a large variety of input-output devices with different speeds, the performance can be affected considerably by the choice of units, particularly the choice of controllers and buffers, as well as the application of other methods of providing a proper balance of work within a system.

Particular Problems. The performance of a computer, compared with that of its competitors, may be significantly affected by the types of problems it is required to solve, whether they be scientifically oriented, business oriented, or consist of much special data manipulation. It is not always so obvious that variations within given problem classes also affect the performance of a computer significantly, usually because there is an imbalance within a particular job; magnetic-tape speeds or central-processor speeds may cause a limitation. In the charts in Sec. 8-9, performance figures are given for some typical comparative configurations of equipment on some standard problems. Standard problems are described in Sec. 8-10.

Implementation. The running time of a sophisticated application or the efficient use of a sophisticated computer is largely dependent upon the ability of the programming staff and/or the compiling program. Good quality programming is often necessary to take advantage of the simultaneous operations in a computer; e.g., maintaining queues if necessary, and suitably packing and blocking the input-output data to take advantage of the system. Computers vary considerably in the advantages that can be gained by moving from adequate to good coding, and this variation is not necessarily related to the size of the computer. Some of the small drum computers are the most difficult to exploit advantageously.

Operating Environment. In addition to the time required for a computer and its peripheral units to run a program, there are many intervals of ineffectiveness. These include unavoidable delays such as reloading tape units where alternates are not being used, the replenishment of printers with stationery, and the time to load and unload a particular job on an installation. These times vary widely, depending upon the data load and operating systems available.[76,77] In situations in which the total execution times are approximately equal, many small jobs obviously carry greater overheads than a few very large jobs.

Reliability. The reliability of a computer system is dependent in a complex way on the faults occurring in the system. Faults in a computer system may cause not only down time but also considerable wasted time. They also require rerun time to recover. The amount of rerun time necessary depends upon the operating system, and the amount of down time may well depend upon the number of spare units made available. Delays may occur because of faults in the data, in the programs themselves, in the machine hardware, or in the actions of the operator.

8-9. Comparison Charts*

Charts 1 through 12 show in tabular form the general characteristics of a selection of computer systems generally available and in wide use today. These tables indicate

Text continued on p. 8-45.

* Adapted from Auerbach *Standard EDP Rept.* (see Ref. 72).

SYSTEM COMPONENTS

CHART 1
COMPARISON OF SMALL TO MEDIUM GENERAL PURPOSE SYSTEMS
HARDWARE: CENTRAL PROCESSOR AND WORKING STORAGE

	System Identity		GE 225	IBM 1410	NCR 315
DATA STRUCTURE	Word Length	Binary Bits	19 + sign + parity	6 + parity + word mark	12 + parity
		Decimal Digits	5.7 + sign	1	3
		Characters	3	1	2
	Floating Point Representation	Radix	Binary	Decimal	Decimal
		Fixed Point Size	30 bits + sign	8 digits (s)	11 digits + sign
		Exponent Size	8 bits + sign	2 digits (s)	3 digits
CENTRAL PROCESSOR	Model Number		225	1411	315-1 to 5
	Arithmetic Radix		Binary; decimal*	Decimal	Decimal
	Operand Length, Words		1 or 2	1 to N char	1 to 8
	Instruction Length, Words		1	1 to 12 char	2 or 4
	Addresses per Instruction		1	2	1
	Likely Fixed Point Execution Times, μ sec (5 Digits Min. Precision)	c = a + b	108	226	168
		c = ab	414	1,206	466
		c = a/b	567	2,440	1,312
	Likely Floating Point Execution Times, μ sec	c = a + b	580*	3,999 (s)	1,066(s)
		c = ab	874*	5,430 (s)	3,066(s)
		c = a/b	1,178*	8,790 (s)	3,266(s)
	Checking	Data Transfers	Parity	Parity, char validity	Parity
		Arithmetic	None	None	None
	Number of Index Registers		3; 96*	15	30
	Indirect Addressing		None	None	Jumps only
	Special Editing Capabilities	Mathematical	Good	Excellent	Good
		Commercial	Good	Excellent	Good
	Boolean Operations		OR, AND	None	None
	Table Look-up		None	Good	Good
	Console Typewriter	Input	Optional	Yes	Yes
		Output	Yes	Yes	Yes
	Features and Comments		Program-compatible with GE 215 and 235	Speeds are about 23% higher with Accelerator feature.	
WORKING STORAGE	Model Number		225	1411	316
	Type of Storage		Core	Core	Core
	Number of Words	Minimum	4,096	10,000	2,000
		Maximum	16,384	80,000	40,000
	Maximum Total Storage	Decimal Digits	93,388	80,000	120,000
		Characters	49,152	80,000	80,000
	Access Time, μ sec	Minimum	9	2.5	4.0
		Average (Random)	9	2.5	4.0
		Maximum	9	2.5	4.0
	Cycle Time, μ sec		18	4.5; 4.0*	6.0
	Effective Transfer Rate, char/sec.		24,000 83,400*	111,000	85,000
	Checking		Parity	Parity	Parity
	Features and Comments				

* With optional equipment.
(s) Using subroutine.

© 1963 Auerbach Corporation and Info, Inc.

DIGITAL COMPUTER-SYSTEM CHARACTERISTICS 8-33

CHART 2
COMPARISON OF SMALL TO MEDIUM GENERAL PURPOSE SYSTEMS
HARDWARE: AUXILIARY STORAGE AND MAGNETIC TAPE

	System Identity		GE 225		IBM 1410		NCR 315	
	Model Number		M640A		1301	1311	353 (CRAM)	
	Type of Storage		Magnetic discs		Magnetic discs	Magnetic discs	Magnetic cards	
AUXILIARY STORAGE	Maximum Number	Units On-Line	32		5	10	16	
		Read/Write Operations	1		2	2	1	
		Seek Operations	32		10	2 or 10*	16	
	Number of Words per Unit	Minimum	6,290,000		28,000,000	2,000,000	2,777,600	
		Maximum	6,290,000		56,000,000	2,980,000*	44,441,600	
	Maximum Total Storage	Decimal Digits	1,107 × 10⁶		280,000,000	29,800,000	133,324,800	
		Characters	604 × 10⁶		280,000,000	29,800,000	88,883,200	
	Waiting Time, msec.	Minimum	0		0	0	3	
		Average (Random)	225		177	170	57	
		Maximum	357		214	288	269	
	Effective Transfer Rate, char/sec.		60,000		42,000 or 82,000*	38,200	91,000	
	Transfer Load Size, char.		192 to 3,072 by 192		1 to 112,000	100 to 20,000 by 100	2 to 3,100 by 2	
	Checking		Parity		Parity, write check	Parity, write check	Parity, read after write	
	Features and Comments		Programmed error correction is possible			Changeable storage medium	Changeable storage medium	
MAGNETIC TAPE	Model Number		MTH680	MTH690	7330	729 series	7340 Mod. 2	332-202	332-203, 332-204
	Maximum Number of Units	On-Line	64	64	20	20	8	16	16
		Reading/Writing	8	2	2	2	2	2	2
		Searching	0	0	0	0	0	0	0
		Rewinding	64	64	20	20	8	16	16
	Demands on Processor, %	Reading/Writing	9	25	100 or 11*	100 to 22*	?	100 or 8.0	100 or 13.3
		Starting/Stopping	0	0	54 or 0*	83 to 0*	?	100 or 0	100 or 0
	Transfer Rate, Kilo-char/sec.	Peak	15.0	41.6	20.0	41.6 to 90.0	34.0	40.0	66.0
		1,000-char blocks	12.7	29.0	14.2	27.3 to 50.2	?	29.2	40.5
		100-char blocks	5.3	6.9	4.0	6.7 to 10.1	?	8.5	9.5
	Data Tracks		6	6	6	6	6 or 8	6	6
	Data Rows per Block		3 to N	3 to N	1 to N	1 to N	1 to N	2 to 15,998	2 to 15,998
	Data Rows per Inch		200	200, 556	200, 556	200, 556, 800	1,511	200, 333	200, 333, 556
	IBM Compatible		Yes		Yes		No	Yes	
	Checking	Reading	Track & row parity		Track & row parity		Dual-row parity	Track & row parity	
		Writing	Read after write		Read after write			Read after write	
	Features and Comments		Can read backward		7340 (Hypertape) is cartridge-loaded and can read backward				

* With optional equipment.

© 1963 Auerbach Corporation and info, inc.

SYSTEM COMPONENTS

CHART 3
COMPARISON OF SMALL TO MEDIUM GENERAL PURPOSE SYSTEMS
HARDWARE: PUNCHED CARD AND TAPE INPUT-OUTPUT

	System Identity	GE 225		IBM 1410	NCR 315	
PUNCHED CARD INPUT	Model Number	D225B	D225C	1402	383	380
	Maximum Number On-Line	1	1	2	1	1
	Peak Speed, cards/min.	400	1,000	800	400	2,000
	Demands on Processor, %	1.0	3.8	1	7	45
	Code Translation	Automatic	Automatic	Automatic	Programmed	Automatic
	Checking	None	Read check	Hole count, validity	None	Validity
	Features and Comments	Decimal or 10- or 12-row binary formats		Three stackers	Errors cause program jumps	
PUNCHED CARD OUTPUT	Model Number	E225K	E225M	1402	IBM 523	IBM 544	IBM 7550
	Maximum Number On-Line	1	1	2	4	4	4
	Peak Speed, cards/min.	100	300	250	100	250	250
	Demands on Processor, %	2.9	8.4	0.4	1	1	1
	Code Translation	Automatic		Automatic	Automatic		
	Checking	Double punch, blank col.	Hole count	Hole count	Echo		
	Features and Comments	Decimal or 10- or 12-row binary formats		Three stackers	Usable for off-line gang punching; 544 and 7550 are no longer offered		
PUNCHED TAPE INPUT	Model Number	Paper Tape System		1011	362	
	Maximum Number On-Line	1		2	1	
	Peak Speed, char/sec.	250 or 1,000		500	1,000	
	Demands on Processor, %	3 or 11		0.5	100	
	Code Translation	Programmed		Plugboard wiring	Matched or programmed	
	Checking	Parity		Parity	Parity	
	Features and Comments					
PUNCHED TAPE OUTPUT	Model Number	Paper Tape System			371	
	Maximum Number On-Line	1			1	
	Peak Speed, char/sec.	110			110	
	Demands on Processor, %	1.2			47	
	Code Translation	Programmed			Matched or programmed	
	Checking	None			None	
	Features and Comments					

* With optional equipment.

© 1963 Auerbach Corporation and Info, Inc.

DIGITAL COMPUTER-SYSTEM CHARACTERISTICS 8-35

CHART 4

COMPARISON OF SMALL TO MEDIUM GENERAL PURPOSE SYSTEMS
HARDWARE: OTHER INPUT-OUTPUT DEVICES

	System Identity	GE 225	IBM 1410		NCR 315
PRINTED OUTPUT	Model Number	P225A	1403 Mod. 1, 2	1403 Mod. 3	340-3
	Maximum Number On-Line	8	2	2	4
	Speed, lines/min. — Single Spacing	900	600 (1,285 numeric*)	1,100	680 (940 numeric)
	Speed, lines/min. — 1-inch Spacing	601	480 (838 numeric*)	750	400 (470 numeric)
	Demands on Processor, %	2	1 to 3	2.6	1.6 max.
	Number of Print Positions	120	100 or 132	132	120
	Checking	Receipt of data, timing	Echo, validity		Validity
	Features and Comments	On/off-line models available			
MICR READER	Model Number	S12B, S12C	1412	1419	402-3
	Maximum Number On-Line	8	2	2	4
	Peak Speed, documents/min.	1,200 (600 on demand)	950	1,600	750
	Demands on Processor, %	1	?	74.7 or 14.1*	1
	Code Translation	Automatic	Automatic		Automatic
	Checking	Validity	Validity, timing		Validity, timing
	Features and Comments	12 stackers; usable for off-line sorting	13 stackers; usable for off-line sorting.		12 stackers; usable for off-line sorting.
OTHER INPUT-OUTPUT DEVICES	Model Number	Datanet-15	1009		353 (CRAM)
	Name	Data Transmission Controller	Data Transmission Unit		Card Random Access
	Peak Speed	2,400 bits/sec.	300 char/sec.		100,000 char/sec.
	Model Number		1014		
	Name		Remote Inquiry Unit		
	Peak Speed		15.5 char/sec.		
	Model Number		7750		
	Name		Programmed Transmission Control		
	Peak Speed		1,200 bits/sec.		

* With optional equipment.

© 1963 Auerbach Corporation and Info, Inc.

CHART 5
COMPARISON OF SMALL SCIENTIFIC SYSTEMS
HARDWARE: CENTRAL PROCESSOR AND WORKING STORAGE

	System Identity		RECOMP III	PB 250	LGP-30
DATA STRUCTURE	Word Length	Binary Bits	40	22 + parity + guard	31 + spacer
		Decimal Digits	12	6,3	9
		Characters	5 8-bit	3 6-bit	5 6-bit
	Floating Point Representation	Radix	Binary	Binary (s)	Binary (s)
		Fixed Point Size	31 bits	22 bits & others (s)	25 bits & others (s)
		Exponent Size	8 bits	22 bits & others (s)	6 bits & others (s)
CENTRAL PROCESSOR	Model Number		D4-F	PB 250	301
	Arithmetic Radix		Binary	Binary	Binary
	Operand Length, Words		1	1 or 2	1
	Instruction Length, Words		0.5	1	1
	Addresses per Instruction		1	1	1
	Likely Fixed Point Execution Times, μ sec (5 Digits Min. Precision)	c = a + b	36,540 & 6,580 avg.	9,252 & 4,744 avg.	52,000 avg.
		c = ab	43,070 & 17,350 avg.	12,336 & 4,766 avg.	69,000 avg.
		c = a/b	45,320 & 17,510 avg.	12,336 & 4,766 avg.	69,000 avg.
	Likely Floating Point Execution Times, μ sec	c = a + b	37,160 & 7,360 avg.*	19,488 & 7,188 (s)	866,000 (s)
		c = ab	43,880 & 18,160 avg.*	19,176 & 6,876 (s)	716,000 (s)
		c = a/b	46,040 & 20,320 avg.*	19,284 & 6,980 (s)	749,000 (s)
	Checking	Data Transfers	None	Parity	None
		Arithmetic	None	None	None
	Number of Index Registers		1	1	None
	Indirect Addressing		None	None	None
	Special Editing Capabilities	Mathematical	None	None	None
		Commercial	None	None	None
	Boolean Operations		AND	AND, Incl. OR	AND
	Table Look-up		None	None	None
	Console Typewriter	Input	Yes	Yes	Yes
		Output	Yes	Yes	Yes
	Features and Comments		Floating Point optional	Times are based on 2 different sequence control methods.	
WORKING STORAGE	Model Number		Main Memory	Fast Memory	Delay Line Memory	Part of 301
	Type of Storage		Disc	Disc	Nickel delay lines	Drum
	Number of Words	Minimum	4,080	16	2,320	4,096
		Maximum	4,080	16	15,888	4,096
	Maximum Total Storage	Decimal Digits	48,960	192	100,100	36,864
		Characters	20,400	80	47,664	20,480
	Access Time, μ sec	Minimum	810	810	12	260
		Average (Random)	8,270	870	1,548	8,333
		Maximum	17,497	2,430	3,084	16,667
	Cycle Time, μ sec		17,000	2,160	3,072	16,667
	Effective Transfer Rate, char/sec		910	---	247,956	60
	Checking		None	None	Parity	None
	Features and Comments			Regenerated loops.	Access times based on standard 256-word loops.	

* With optional equipment.
(s) Using subroutines.

© 1963 Auerbach Corporation and Info, Inc.

DIGITAL COMPUTER-SYSTEM CHARACTERISTICS 8-37

CHART 6

COMPARISON OF SMALL SCIENTIFIC SYSTEMS
HARDWARE: AUXILIARY STORAGE AND MAGNETIC TAPE

	System Identity		RECOMP III	PB 250		LGP-30
AUXILIARY STORAGE	Model Number					
	Type of Storage					
	Maximum Number	Units On-Line				
		Read/Write Operations				
		Seek Operations				
	Number of Words per Unit	Minimum				
		Maximum				
	Maximum Total Storage	Decimal Digits				
		Characters				
	Waiting Time, m.sec.	Minimum				
		Average (Random)				
		Maximum				
	Effective Transfer Rate, char/sec.					
	Transfer Load Size, char.					
	Checking					
	Features and Comments					
MAGNETIC TAPE	Model Number			MTU 1	MTU 2	
	Maximum Number of Units	On-Line		6		
		Reading/Writing		1		
		Searching		0		
		Rewinding		6		
	Demands on Processor, %	Reading/Writing		100		
		Starting/Stopping		100		
	Transfer Rate, Kilo-char/sec.	Peak		2.0	15.0	
		1,000-char blocks		Depends upon routine		
		100-char blocks		Depends upon routine		
	Data Tracks			7		
	Data Rows per Block			1 to N		
	Data Rows per Inch			200		
	IBM Compatible			IBM Binary		
	Checking	Reading		As programmed		
		Writing		None		
	Features and Comments			Tape timing requires constant program control.		

© 1963 Auerbach Corporation and Info, Inc.

CHART 7
COMPARISON OF SMALL SCIENTIFIC SYSTEMS
HARDWARE: PUNCHED CARD AND TAPE INPUT-OUTPUT

System Identity		RECOMP III		PB 250		LGP-30	
PUNCHED CARD INPUT	Model Number	IBM 024 or 026		CR 2		IBM 024 or 026	
	Maximum number On-Line	1		--		1	
	Peak Speed, cards/min.	(20 col/sec.)		--		(20 col/sec.)	
	Demands on Processor, %	100		--		100	
	Code Translation	Programmed		--		Programmed	
	Checking	None		--		None	
	Features and Comments			No firm specifications available.		Model 321 Control Unit required.	
PUNCHED CARD OUTPUT	Model Number	IBM 024 or 026		Card Punch Coupler			
	Maximum Number On-Line	1		--			
	Peak Speed, cards/min.	(20 col/sec.)		--			
	Demands on Processor, %	100		--		NONE	
	Code Translation	Programmed		--			
	Checking	None		--			
	Features and Comments			No firm specifications available.			
PUNCHED TAPE INPUT	Model Number	Flexowriter Reader	Facitape AETR-500	Flexowriter Reader	HSR-1	360	341, 342
	Maximum Number On-Line	1	1	1	2	1	1
	Peak Speed, char/sec.	10	600 (285 ‡)	10	300	10	200 (50 ‡)
	Demands on Processor, %	100	100	6 to 100	100	100	100
	Code Translation	Programmed	Programmed	Matched or programmed		Matched or programmed	
	Checking	None	None	None		None	
	Features and Comments						
PUNCHED TAPE OUTPUT	Model Number	Flexowriter Punch	Facitape AETP-150	Flexowriter Punch	HSP-1	360	342
	Maximum Number On-Line	1	1	1	?	1	1
	Peak Speed, char/sec.	10	150 (90 ‡)	15	110	10	20
	Demands on Processor, %	100	100	22 to 37	0.05 to 100	2.4 to 100	5 to 100
	Code Translation	Programmed	Programmed	Matched or programmed		Matched	Matched
	Checking	None	None	None		None	None
	Features and Comments						

‡ Maximum effective data transfer rate.

© 1963 Auerbach Corporation and Info, Inc.

DIGITAL COMPUTER-SYSTEM CHARACTERISTICS　　8-39

CHART 8

COMPARISON OF SMALL SCIENTIFIC SYSTEMS HARDWARE: OTHER INPUT-OUTPUT DEVICES

	System Identity		RECOMP III	PB 250	LGP-30
PRINTED OUTPUT	Model Number		Flexowriter	Flexowriter	360 (Flexowriter)
	Maximum Number On-Line		1	1	1
	Speed, lines/min.	Single Spacing	(10 char/sec.)	(10 char/sec.)	(10 char/sec.)
		1-inch Spacing	--	--	--
	Demands on Processor, %		100	30 to 100	2.4 to 100
	Number of Print Positions		110 max.	110 max.	180 max.*
	Checking		None	None	None
	Features and Comments		Includes tape reader, punch, and keyboard	Includes tape reader, punch, and keyboard	Includes tape reader, punch, and keyboard
MICR INPUT	Model Number				
	Maximum Number On-Line				
	Peak Speed, documents/min.				
	Demands on Processor, %				
	Code Translation				
	Checking				
	Features and Comments				
OTHER INPUT-OUTPUT DEVICES	Model Number		--	--	
	Name		RECOMP X-Y Plotter	Digital Graph Recorder	
	Peak Speed		200 increments/sec.	200 increments/sec.	
	Model Number				
	Name				
	Peak Speed				
	Model Number				
	Name				
	Peak Speed				

* With optional equipment

© 1963 Auerbach Corporation and Info, Inc.

CHART 9

COMPARISON OF LARGE SCIENTIFIC SYSTEMS
HARDWARE: CENTRAL PROCESSOR AND WORKING STORAGE

	System Identity		CDC 1604	CDC 1604-A	Philco 2000-212	IBM 7094	
DATA STRUCTURE	Word Length	Binary Bits	48	48	48	36	
		Decimal Digits	14	14	14	10.5	
		Characters	8	8	8	6	
	Floating Point Representation	Radix	Binary	Binary	Binary	Binary	
		Fixed Point Size	36 bits + sign	36 bits + sign	35 bits + sign	27 or 54 bits + sign	
		Exponent Size	11 bits	11 bits	12 bits	8 bits	
CENTRAL PROCESSOR	Model Number		1604	1604-A	212	Model I	Model II
	Arithmetic Radix		Binary	Binary	Binary	Binary	
	Operand Length, Words		1 or 2	1 or 2	1	1 or 2	
	Instruction Length, Words		½	½	½; 1 for I/O	1	
	Addresses per Instruction		1	1	1	1	
	Likely Fixed Point Execution Times, μsec (5 Digits Min. Precision)	$c = a + b$	21.6	21.6	4.65	10	7.0
		$c = ab$	39.6 to 77.2	39.6 to 77.2	7.60	16	9.8
		$c = a/b$	82.6	82.6	12.90	20	14.0
	Likely Floating Point Execution Times, μsec	$c = a + b$	33.2	33.2	4.65	12	8.4
		$c = ab$	50.4	50.4	7.60	16	11.2
		$c = a/b$	70.4	70.4	15.40	22	14.0
	Checking	Data Transfers	**	**	Parity	None	
		Arithmetic	None	None	None	None	
	Number of Index Registers		6	6	8	7	
	Indirect Addressing		Yes; recursive	Yes; recursive	Yes; recursive	One level	
	Special Editing Capabilities	Mathematical	None	None	None	None	
		Commercial	None	None	None	None	
	Boolean Operations		AND, INCL OR	AND, INCL OR	AND, INCL OR, EXCL OR	AND, INCL OR, EXCL OR, NOT	
	Table Look-up		Good	Good	None	None	
	Console Typewriter	Input	Yes	Yes	Yes	No	
		Output	Yes	Yes	Yes	No	
	Features and Comments				Look-ahead feature; can repeat up to 4 indexed instructions	36-bit indicator register; 4 sense lights, 6 sense switches	
WORKING STORAGE	Model Number				2100 Series	2000 Series	Model I	Model II
	Type of Storage		Core	Core	Core	Core	Core	
	Number of Words	Minimum	8,192	8,192	8,192	16,384	32,768	
		Maximum	32,768	32,768	32,768	65,536	32,768	
	Maximum Total Storage	Decimal Digits	458,752	458,752	458,752	917,504	344,064	
		Characters	262,144	262,144	262,144	524,288	196,608	
	Access Time, μsec	Minimum	2.2	2.2	0.7	0.5	0.55	?
		Average (Random)	4.8	4.8	0.7	0.5	0.55	?
		Maximum	6.4	6.4	0.7	0.5	0.55	?
	Cycle Time, μsec		6.4	6.4	1.5	1.0	2.00	1.40
	Effective Transfer Rate, char/sec.		371,200 to 555,552	371,200 to 555,552	5,333,333 max.		3,000,000 max.	4,300,000 max.
	Checking		Pulse-size check	Pulse-size check	Parity		None	None
	Features and Comments		Core storage is divided into two independent halves to improve access time		Overlapped access to each bank		Single bank	Dual banks with overlapped access

** Tolerance on pulse size.   © 1963 Auerbach Corporation and Info, Inc.

DIGITAL COMPUTER-SYSTEM CHARACTERISTICS 8-41

CHART 10
COMPARISON OF LARGE SCIENTIFIC SYSTEMS
HARDWARE: AUXILIARY STORAGE AND MAGNETIC TAPE

	System Identity		CDC 1604-A	Philco 2000 – 212		IBM 7090 & 7094		
	Model Number			2300 series		1301		7320
	Type of Storage			Disc		Disc		Drum
AUXILIARY STORAGE	Maximum Number	Units On-Line		4		5		10
		Read/Write Operations		1		2		2
		Seek Operations		1		10		0
	Number of Words per Unit	Minimum		5,242,880		4,650,000		186,400
		Maximum		5,242,880		9,300,000		1,864,400
	Maximum Total Storage	Decimal Digits		250,000,000		418,500,000		16,776,000
		Characters		187,772,160		279,000,000		11,184,000
	Waiting Time, msec.	Minimum		0		0		0
		Average (Random)		?		177		8.6
		Maximum		170		214		17.5
	Effective Transfer Rate, char/sec.			960,000		83,700		203,000
	Transfer Load Size, char.			0 to 262,144		0 to 111,600		Variable
	Checking			Parity		Parity, write check		?
	Features and Comments							

	Model Number		CDC 606	234	334	729 II	729IV	729V	729VI	7340
MAGNETIC TAPE	Maximum Number of Units	On-Line	48	16	64	80	80	80	80	20
		Reading/Writing	6	4	8	8	8	8	8	2
		Searching	0	4	8	0	0	0	0	0
		Rewinding	48	16	64	80	80	80	80	20
	Demands on Processor, %	Reading/Writing	1.2/3.3	1.12	2.3	1.5 max.	2.3 max.	2.2 max.	3.3 max.	6.2 max.
		Starting/Stopping	0	0	0	0	0	0	0	0
	Transfer Rate, Kilo-char/sec.	Peak	30.0/83.4	90	240	15/41.7	22.5/62.5	15/41.7/60	22.5/62.5/90	170
		1,000-char blocks	24.2/49.0	54.6	120	13/26	19.6/44	13/26/33	19.6/44/56	100
		100-char blocks	8.8/10.4	Fixed size blocks	23	6/6.2	9/12	6/6.2/8	9/12/13	22
	Data Tracks		6	12	12	6	6	6	6	8
	Data Rows per Block		Variable	512	Variable	Variable	Variable	Variable	Variable	Variable
	Data Rows per Inch		200/556	375	1,000?	200/556	200/566	200/556/800	200/556/800	1,511
	IBM Compatible		Yes	No	No	Yes	Yes	Yes	Yes	Yes
	Checking	Reading	Row parity	Track & row parity	Track & row parity	Track & row parity				Dual-row parity
		Writing	Read after write	Read after write	Read after write	Read after write				Read after write
	Features and Comments			Tape has pre-recorded block and sprocket marks		7340 (Hypertape) can read backward and is cartridge-loaded				

* With optional equipment.   © 1963 Auerbach Corporation and Info, Inc.

SYSTEM COMPONENTS

CHART 11

COMPARISON OF LARGE SCIENTIFIC SYSTEMS
HARDWARE: PUNCHED CARD AND TAPE INPUT-OUTPUT

	System Identity	CDC 1604 & 1604-A		Philco 2000 - 210/211/212		IBM 7090 & 7094
PUNCHED CARD INPUT	Model Number	1610 (IBM 088)	1617	258		711
	Maximum Number On-Line	3	3	28		1
	Peak Speed, cards/min.	1,300	250	2,000		250
	Demands on Processor, %	--	0.4 max.	0.33 max.		<1
	Code Translation	None	Automatic or matched	Automatic		None
	Checking	At half speed	None	Dual read		Possible
	Features and Comments	A number of other IBM units can be used		May be any of 7 units on a UBC		Reads binary image of 72 out of 80 columns
PUNCHED CARD OUTPUT	Model Number	1609 (IBM 521)		265		271
	Maximum Number On-Line	3		28		1
	Peak Speed, cards/min.	100		100		100
	Demands on Processor, %	0.06 max.		0.02 max.		<1
	Code Translation	Matched or instruction		Automatic		None
	Checking	None		Read after punch		None
	Features and Comments	A number of other IBM units can be used		May be any of 7 units on a UBC		Can be used to gang punch
PUNCHED TAPE INPUT	Model Number	CDC 350		Part of 240	Part of 241	
	Maximum Number On-Line	1		1	28	
	Peak Speed, char/sec.	350		1,000 & 500		
	Demands on Processor, %	0.06 max.		0.1 max.		
	Code Translation	Matched		Matched		
	Checking	None		Parity		
	Features and Comments			Separate direct connection	May be any of 7 units on a UBC	
PUNCHED TAPE OUTPUT	Model Number	BRPE		Part of 240	Part of 241	
	Maximum Number On-Line	1		1	28	
	Peak Speed, char/sec.	110		60		
	Demands on Processor, %	0.16 max.		0.1 max.		
	Code Translation	Matched		Matched		
	Checking	None		None		
	Features and Comments			Separate direct connection	May be any of 7 units on a UBC	

© 1963 Auerbach Corporation and Info, Inc.

DIGITAL COMPUTER-SYSTEM CHARACTERISTICS 8-43

CHART 12

COMPARISON OF LARGE SCIENTIFIC SYSTEMS
HARDWARE: OTHER INPUT-OUTPUT DEVICES

	System Identity		CDC 1604 & 1604-A	Philco 2000-210/211/212	IBM 7090 & 7094
PRINTED OUTPUT	Model Number		1612	2256	716
	Maximum Number On-Line		24	28	1
	Speed, lines/min.	Single Spacing	667	900	75 to 150
		1-inch Spacing	500	600	75 to 150
	Demands on Processor, %		2.8 max.	0.22 max.	<1
	Number of Print Positions		120	120	120
	Checking		None	None	Programmed echo
	Features and Comments		Increased speed is possible with restricted character set	May be any of 7 units on a UBC	Maximum of 72 characters per print cycle
MICR READER	Model Number				
	Maximum Number On-Line				
	Peak Speed, documents/min.				
	Demands on Processor, %				
	Code Translation				
	Checking				
	Features and Comments				
OTHER INPUT-OUTPUT DEVICES	Model Number		1605	2280 series	1414-VI
	Name		Control Unit for various IBM units working in BCD	Digital Incremental Recorder	Input-Output Synchronizer
	Peak Speed		Depends upon units connected	300 point plots/sec.	Communication equipment
	Model Number		1610	209	7281 II
	Name		Control Unit for various IBM units working in column binary	Console Typewriter Buffer	Data Communications Channel
	Peak Speed		Depends upon units connected	--	Non-IBM Real-Time Equipment
	Model Number			293	
	Name			Accounting Clock System	
	Peak Speed			--	

© 1963 Auerbach Corporation and Info, Inc.

8-44

SYSTEM COMPONENTS

CHART 13
SYSTEM PERFORMANCE COMPARISONS
GENERALIZED FILE PROCESSING PROBLEM A

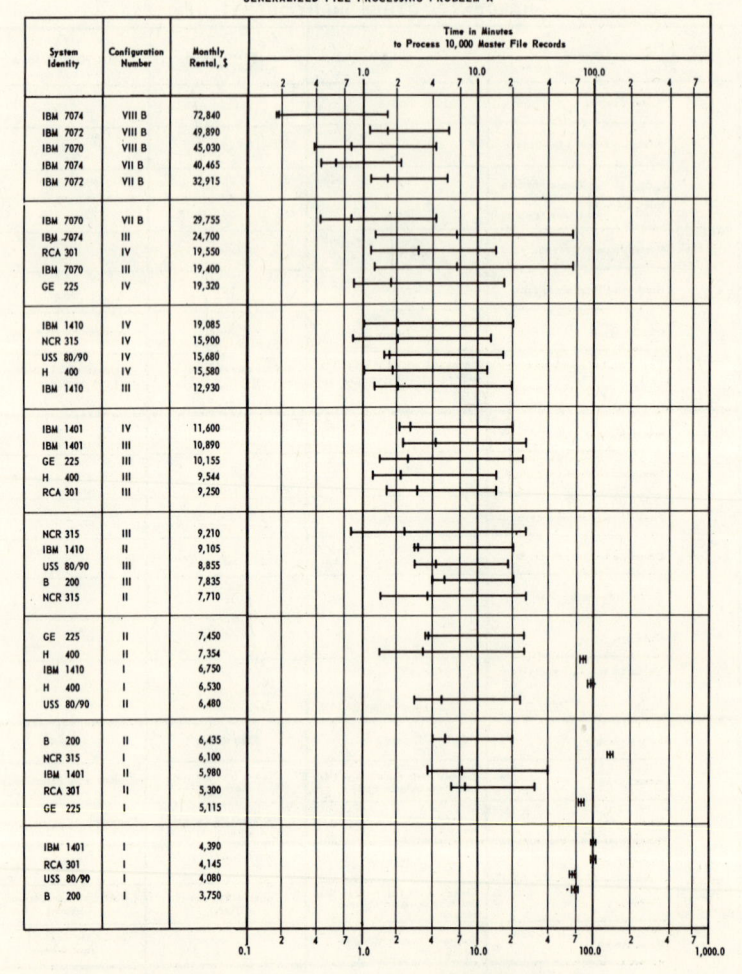

System Identity	Configuration Number	Monthly Rental, $
IBM 7074	VIII B	72,840
IBM 7072	VIII B	49,890
IBM 7070	VIII B	45,030
IBM 7074	VII B	40,465
IBM 7072	VII B	32,915
IBM 7070	VII B	29,755
IBM 7074	III	24,700
RCA 301	IV	19,550
IBM 7070	III	19,400
GE 225	IV	19,320
IBM 1410	IV	19,085
NCR 315	IV	15,900
USS 80/90	IV	15,680
H 400	IV	15,580
IBM 1410	III	12,930
IBM 1401	IV	11,600
IBM 1401	III	10,890
GE 225	III	10,155
H 400	III	9,544
RCA 301	III	9,250
NCR 315	III	9,210
IBM 1410	II	9,105
USS 80/90	III	8,855
B 200	III	7,835
NCR 315	II	7,710
GE 225	II	7,450
H 400	II	7,354
IBM 1410	I	6,750
H 400	I	6,530
USS 80/90	II	6,480
B 200	II	6,435
NCR 315	I	6,100
IBM 1401	II	5,980
RCA 301	II	5,300
GE 225	I	5,115
IBM 1401	I	4,390
RCA 301	I	4,145
USS 80/90	I	4,080
B 200	I	3,750

© 1963 Auerbach Corporation and Info, Inc.

Time in Minutes to Process 10,000 Master File Records.
Activity Factors are designated as follows:
ACTIVITY: 0   0.1   1.0

DIGITAL COMPUTER-SYSTEM CHARACTERISTICS 8–45

CHART 14

FILE PROBLEM A, TIME IN MINUTES

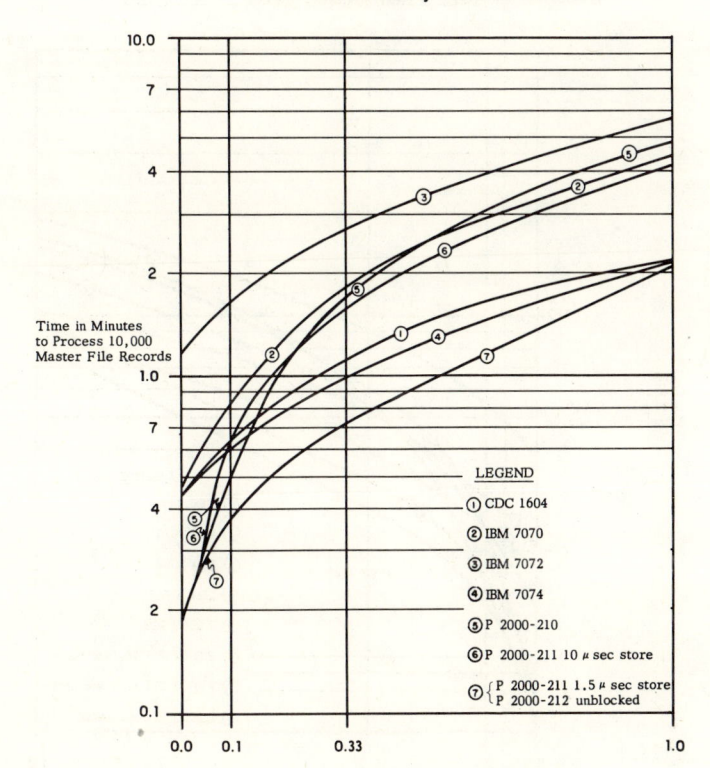

Time in Minutes to Process 10,000 Master File Records

LEGEND
① CDC 1604
② IBM 7070
③ IBM 7072
④ IBM 7074
⑤ P 2000-210
⑥ P 2000-211 10 μ sec store
⑦ { P 2000-211 1.5 μ sec store
    P 2000-212 unblocked }

Activity Factor
Average Number of Detail Records per Master Record

© 1963 Auerbach Corporation and Info, Inc.

the performance details of the various units in the system and indicate the scope of their facilities, usually by highlighting important points. Particular attention in these charts has been given to showing not only peak-performance data but also the overheads involved and the checks provided within the system. Such charts cannot cover the general organization of the system or indicate its overall performance, except in particular cases.

Chart 13 shows in abbreviated form the comparative ratings of a wide range of computers on Generalized File Processing Problem A (see Sec. 8-10). Note particularly the uneven distribution of times between various configurations of one basic model of a computer.

Comparisons of performance for the standard problems described in Sec. 8-10 are shown in Charts 14 through 19.

Charts 14 and 15 show the differences in comparative ratings when time is compared (Chart 14) and when cost is compared (Chart 15). Cost is determined by a specific rental for a determined configuration. The standard used is the Generalized File Processing Problem A.

CHART 15

FILE PROBLEM A, COST IN DOLLARS

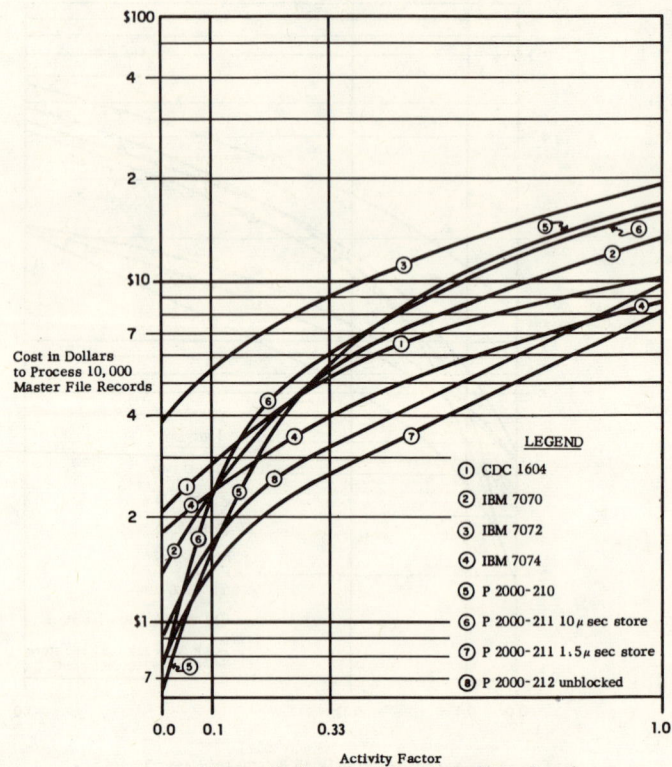

Cost in Dollars to Process 10,000 Master File Records

LEGEND
① CDC 1604
② IBM 7070
③ IBM 7072
④ IBM 7074
⑤ P 2000-210
⑥ P 2000-211 10 μ sec store
⑦ P 2000-211 1.5 μ sec store
⑧ P 2000-212 unblocked

Activity Factor
Average Number of Detail Records per Master Record

© 1963 Auerbach Corporation and Info, Inc.

Charts 16 and 17 show the differences in comparative times when on-line and off-line techniques are compared. Chart 16 represents a single computer technique with on-line card reading and printing. Chart 17 represents off-line operation using auxiliary computers such as the CDC 160, IBM 1401, and Philco Universal Buffer Controller.

Charts 18 and 19 show the differences in scale between the ratings for large-scale scientific computers and those for desk-size scientific computers on the Generalized Mathematical Problem.

8-10. Standard Problems

The generalized file-processing problem and standard mathematical problem described in this section have been adapted from the Users' Guide Section of Auerbach *Standard EDP Reports*.[72]

Generalized File-processing Problems. These are a series of typical commercial data-processing applications. One of the most common jobs in commercial data processing is the processing of a detail file against a master file. The detail file contains data used to update the master file by inserting new records, deleting old

DIGITAL COMPUTER-SYSTEM CHARACTERISTICS 8–47

CHART 16

FILE PROBLEM C, (10-TAPE SYSTEM)
TIME IN MINUTES

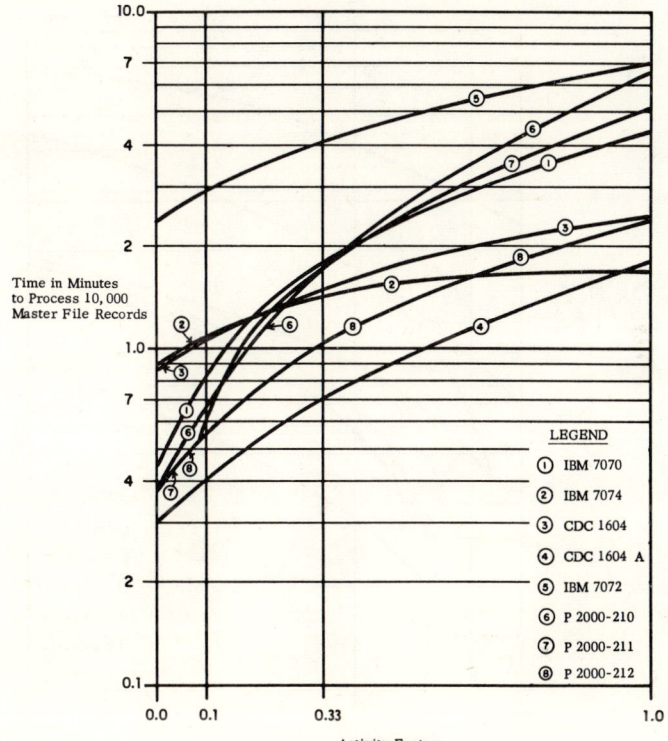

LEGEND

① IBM 7070
② IBM 7074
③ CDC 1604
④ CDC 1604 A
⑤ IBM 7072
⑥ P 2000-210
⑦ P 2000-211
⑧ P 2000-212

Time in Minutes to Process 10,000 Master File Records

Activity Factor
Average Number of Detail Records per Master Record

© 1963 Auerbach Corporation and Info, Inc.

records, and recording changes to records in the file. Usually there is a printed record of the activity. This type of activity occurs, for example, in a payroll routine in which the master file is the payroll file, the detail file contains the details from the time sheets, and the output is largely pay slips.

The application parameters which have the greatest effect on run times in generalized file processing are record sizes, the amount of computation, the ratio of the number of transactions to master records (called the activity factor), and the distribution of transactions. All but the last parameter are considered in a series of standard problems.

Standard Problems A, B, and C vary the record sizes for the master file. Standard Problem D increases the amount of computation. Each problem is estimated for activity factors of zero to unity. In all cases a uniform distribution of activity is assumed.

Low activity occurs in inventory-control applications. Moderate activity occurs in cycle-billing applications. High activity occurs in payroll applications.

Standard Problem A. In Problem A, we use a typical inventory application as a means of making a detailed estimate of the times necessary. Problem A is a typical

CHART 17

FILE PROBLEM C, (20-TAPE SYSTEM)
TIME IN MINUTES

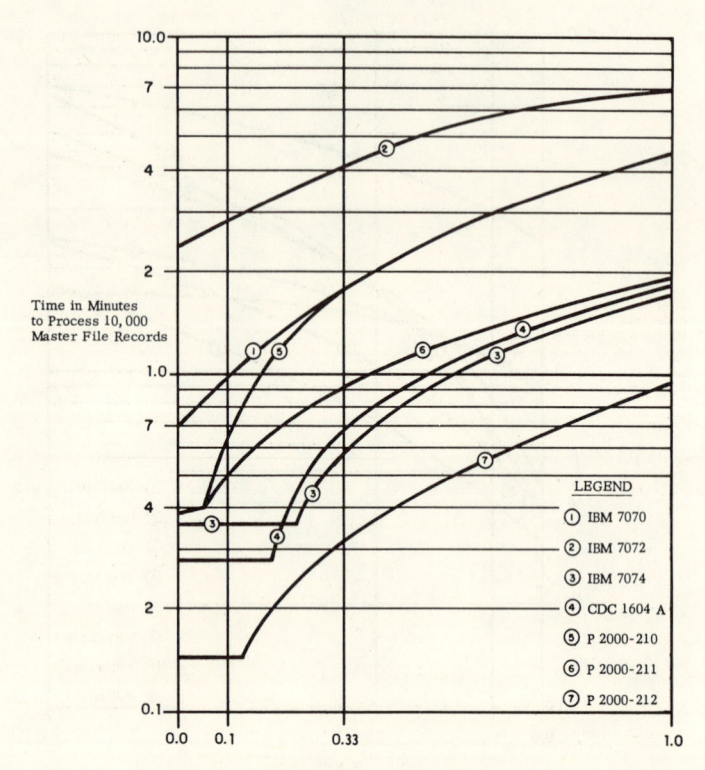

Time in Minutes to Process 10,000 Master File Records

LEGEND
① IBM 7070
② IBM 7072
③ IBM 7074
④ CDC 1604 A
⑤ P 2000-210
⑥ P 2000-211
⑦ P 2000-212

Activity Factor
Average Number of Detail Records per Master Record

© 1963 Auerbach Corporation and Info, Inc.

complete commercial data-processing application and is specified in sufficient detail[72] to enable reliable estimates to be made.

The basic form of the program run is shown in Fig. 8-8.

File 1 is the old master stock-record file. File 2 is the updated master stock-record file. File 3 is the detail file read from cards. File 4 is the report file recording the activity. When considering off-line transcription, it is assumed that Files 3 and 4 contain only one record per block.

The estimated times, quoted in minutes and plotted on a graph against the activity factor, are for a master file of 10,000 records. The activity factor is the ratio of the total number of records in the detail file to the total number of records in the master file. Separate plots are made for each computer configuration.

Generalized Mathematical Processing. *Standard Mathematical Problem A.* This is a straightforward application in which there is one stream of input data, a fixed computation to be performed, and a stream of output results.

The input is a series of records. Each record contains 10 numbers:

$$X_1 \quad X_2 \quad X_3 \quad X_4 \quad X_5$$
$$Y_1 \quad Y_2 \quad Y_3 \quad Y_4 \quad Y_5$$

DIGITAL COMPUTER-SYSTEM CHARACTERISTICS 8-49

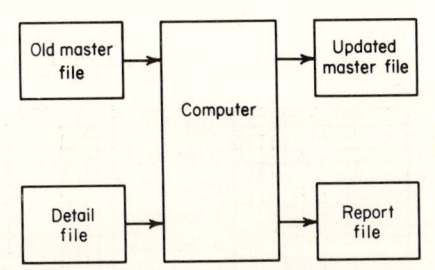

Fig. 8-8. Program for Standard Problem A.

CHART 18

**GENERALIZED MATHEMATICAL PROCESSING
DESK SIZE SCIENTIFIC SYSTEM**

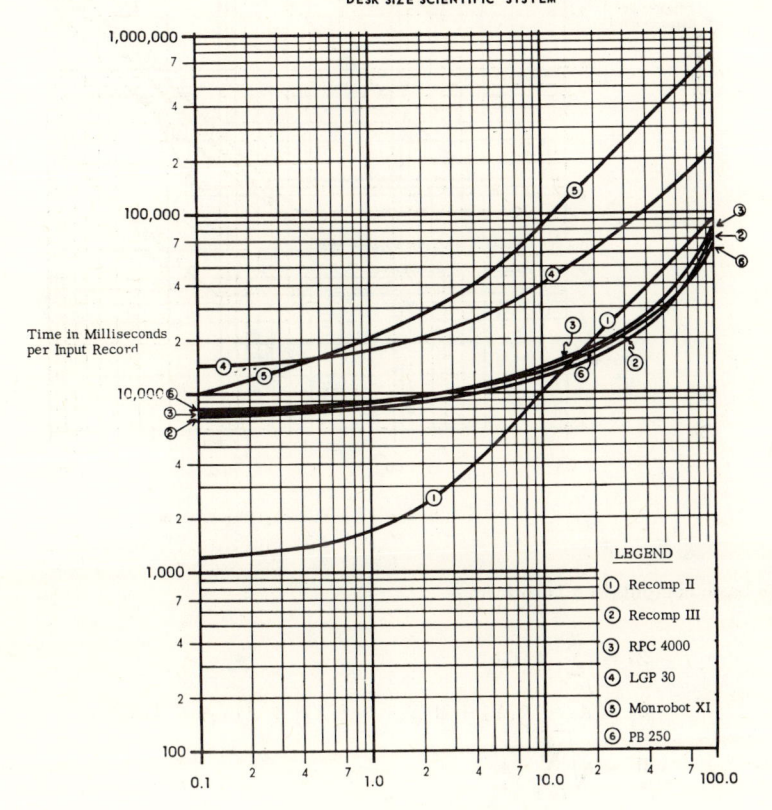

Time in Milliseconds per Input Record

Number of Computations per Input Record

LEGEND
① Recomp II
② Recomp III
③ RPC 4000
④ LGP 30
⑤ Monrobot XI
⑥ PB 250

© 1963 Auerbach Corporation and Info, Inc.

SYSTEM COMPONENTS

CHART 19

GENERALIZED MATHEMATICAL PROCESSING 20-TAPE SYSTEM

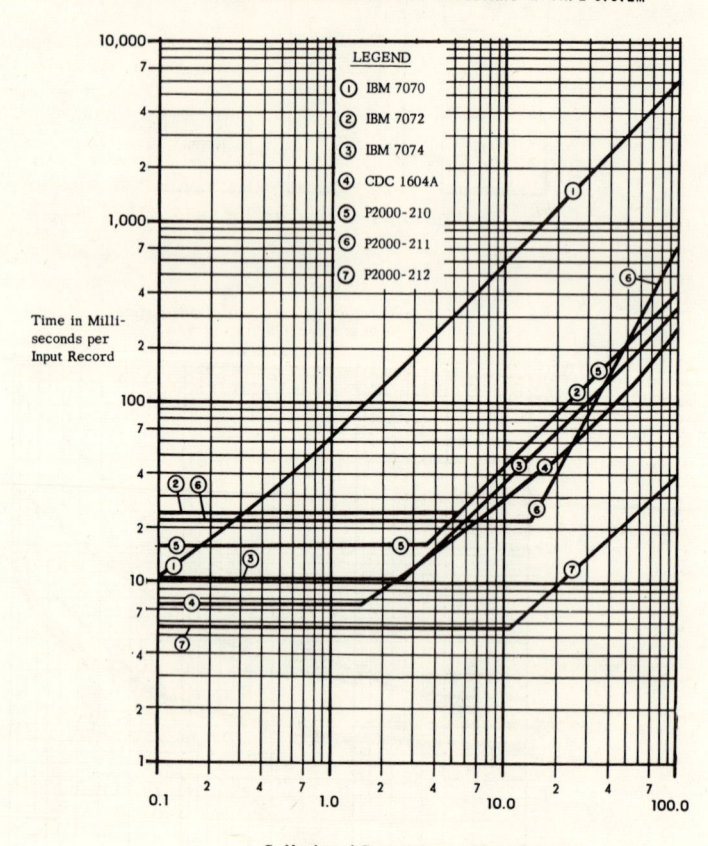

LEGEND
① IBM 7070
② IBM 7072
③ IBM 7074
④ CDC 1604A
⑤ P2000-210
⑥ P2000-211
⑦ P2000-212

Time in Milliseconds per Input Record

C, Number of Computations per Input Record

© 1963 Auerbach Corporation and Info, Inc.

The basic computation is to form

$$Z_i = \sum_{j=0}^{5} A_j X_i^j$$

e.g., $Z_2 = A_0 + A_1 X_2 + A_2 X_2^2 + A_3 X_2^3 + A_4 X_2^4 + A_5 X_2^5$

and $W = \sqrt{\sum_{i=1}^{5} \frac{Z_i}{Y_i}}$

The output is a series of records. Each record contains 10 numbers:

X_1 X_5
Y_1 Y_5
Z_1 Z_2 Z_3 Z_4 Z_5
W

The time quoted is for one input record.

DIGITAL COMPUTER-SYSTEM CHARACTERISTICS 8-51

Two variables are introduced to demonstrate how the time for a job varies with different proportions of input, computation, and output. First, the computation per input record is varied from $0.1T$ to $100T$ where T is the time required to compute W. Second, there are three separate curves on the graph. They correspond to the cases of one output record for each, every tenth, and every hundredth input record.

The times are normally quoted for single-length floating-point operations. Where floating point is provided only by slow subroutines, fixed-point times are also given. Where single-length precision is less than eight decimal digits, double-length times may be given.

Input and Output Records. Each record must be separate from the others, either on separate cards, or on separate lines of print, or delimited on punched tape. Each number may be up to eight digits in size, but the average size is only five digits. Editing style is not critical; the provision of minus signs is essential, but nonsignificant zeros do not need to be suppressed.

REFERENCES

1. Luebert, W. F.: Programming Compatibility in a Family of Closely Related Digital Computers, *Commun. ACM*, vol. 3, no. 7, 420–429, July, 1960.
2. Huskey, H. D., and G. A. Korn (eds.): *Computer Handbook*, McGraw-Hill Book Company, New York, 1961.
3. Grabbe, E. M., S. Ramo, and D. E. Wooldridge (ed.): *Handbook of Automation, Computation and Control*, vol. 2, John Wiley & Sons, Inc., New York, 1958.
4. Bemer, R. W.: A Checklist of Intelligence for Programming Systems, *Commun. ACM*, vol. 2, no. 3, pp. 8–13, March, 1959.
5. Auerbach, Isaac L.: Fast-acting Digital Memory Systems, *Elec. Mfg.*, vol. 52, no. 4, pp. 100–107, October, 1953; vol. 52, no. 5, pp. 136–143, November, 1953.
6. Rajchman, J. A.: A Survey of Computer Memories, *Datamation*, vol. 8, no. 12, pp. 26–30, December, 1962.
7. Devonald, C. H., and J. A. Fotheringham: The ATLAS Computer, *Datamation*, vol. 7, no. 5, pp. 23–57, May, 1961.
8. The UNIVAC 1107 Thin-film Memory Computer, Remington Rand UNIVAC, Division of Sperry Rand, New York, *Rept.* UT-2440, 1961.
9. Burks, A. W., H. H. Goldstine, and J. Von Newmann: Preliminary Discussion of the Logical Design of an Electronic Computing Instrument, *Datamation*, vol. 8, no. 9, pp. 24–31, September, 1962.
10. Bauer, W. F.: Why Multi-computers? *Datamation*, vol. 8, no. 9, pp. 51–55, September, 1962.
11. Penny, J. P., and T. Pearcey: Use of Multiprogramming in the Design of a Low Cost Digital Computer, *Commun. ACM*, vol. 5, no. 9, pp. 473–476, September, 1962.
12. Philco 2000. Input-output Programming System (IOPS), Philco Corporation, Willow Grove, Pa., *Rept.* TM-18, April, 1962.
13. West, G. P., and R. J. Koerner: Communications within a Polymorphic Intellectronic System, *Proc. Western Joint Computer Conf.*, San Francisco (NJCC no. 17); pp. 225–230, May, 1960.
14. Hughes, A.: Multi-computer Data Processing System for a Navy Command and Control System, presented at the 1963 Conference on Military Electronics, Washington, D.C., September, 1963.
15. Dow, J.: Programming a Duplex Computer System, *Commun. ACM*, vol. 4, no. 11, pp. 507–513, November, 1961.
16. The Satellite Computer System, Control Data Corporation, Minneapolis, Minn., *Publ.* 0626.
17. Clark, N., and A. C. Gannet: "Computer-to-computer Communication at 2.5 megabits/sec, IFIP Congress 62, Aug. 27–Sept. 1, 1962, Munich, Germany, pp. 347–353.
18. Anderson, J. P., S. A. Hoffman, J. Shifman, and R. J. Williams: D825—a Multiple Computer System for Command and Control, Fall Joint Computer Conf., Philadelphia (NJCC no. 22), pp. 86–96, December, 1962.
19. Haley, A. C. D.: The KDF9 Computer System, Fall Joint Computer Conf., Philadelphia (NJCC no. 22), pp. 108–120, December, 1962.
20. Holt, A. W.: Program Organization and Record Keeping for Dynamic Storage Allocation, *Commun. ACM* vol. 4, no. 10, pp. 422–431, October, 1961.
21. Fotheringham, J.: Dynamic Storage Allocation in the ATLAS Computer, Including

an Automatic Use of a Backing Store, *Commun. ACM*, vol. 4, no. 10, pp. 435–436, October, 1961.
22. Wilkes, M. V.: Microprogramming, *Proc. Eastern Joint Computer Conf.*, Philadelphia (NJCC no. 14), pp. 18–20, December, 1958.
23. Royse, D.: The IBM 650 RAMAC System-Disk Storage Operation, *Proc. Western Joint Computer Conf.*, Los Angeles (NJCC no. 11), pp. 43–49, February, 1957.
24. Heising, W. P.: Methods of File Organization for Efficient Use of IBM RAMAC Files, *Proc. Western Joint Computer Conf.*, Los Angeles (NJCC no. 13), pp. 194–196, May, 1958.
25. Bloom, L., I. Pardo, W. Keating, and E. Mayne: Card Random Access Memory (CRAM): Functions and Uses, *Proc. Eastern Joint Computer Conf.*, Washington, D.C. (NJCC no. 20), pp. 147–157, December, 1961.
26. Casale, C. T.: Planning the 3600, Fall Joint Computer Conference, Philadelphia (NJCC no. 22), pp. 73–85, December, 1962.
27. Newhouse, V. L., and R. E. Fruin: A Cryogenic Data Addressed Memory, *Proc. Spring Joint Computer Conf.* (NJCC no. 21), pp. 89–99, May, 1962.
28. Lewin, M. H.: Retrieval of Ordered Lists from a Content Addressed Memory, *RCA Rev.*, vol. 23, no. 2, 215–229, June, 1962.
29. Seeber, R. R.: Cryogenic Associative Memory, presented at the National Conference of the ACM, August, 1960.
30. Seeber, R. R., and A. B. Lindquist: Associative Memory with Ordered Retrieval, *IBM J. Res. Develop.*, vol. 6, no. 1, pp. 126–136, January, 1962.
31. Lee, E. S.: Associative Techniques with Complementing Flip-flops, *Proc. Spring Joint Computer Conf.* (SJCC no. 23), pp. 381–394, Denver, Colo., May, 1963.
32. Michlin, G.: The New IBM 1301 Disk Storage Unit, *ICC Bull.* (Centre Internationale de Calcul) Rome, pp. 33–37 (no date).
33. Winsor, P. W., III: Review of U.S. Magnetic Tape Units, *ICC Bull.*, vol. 2, no. 2, pp. 88–98, April, 1963.
34. Graham, J. W., and D. A Sprott: Processing Magnetic Tape Files with Variable Blocks, *Commun. ACM*, vol. 4, no. 12, pp. 555–557, December, 1961.
35. Hayes, R. M., and J. Wiener: Magnacard—A New Concept in Data Handling, 1957 *IRE WESCON Convention Record*, pt. 4, pp. 205–210.
36. Buchholz, W. (ed.): *Planning a Computer System*, McGraw-Hill Book Company, New York, 1962.
37. MCP Master Control Program Characteristics, Burroughs B5000 Information Processing System, Burroughs Corporation, Detroit, Mich., *Rept.* 5000–21003-P, May, 1962.
38. Brooks, F. P.: The Execute Operations—a Fourth Mode of Instruction Sequencing, *Commun. ACM*, vol. 3, no. 3, pp. 168–170, March, 1960.
39. Bucholz, W.: Fingers or Fists, *Commun. ACM*, vol. 2, no. 12, pp. 3–11, December, 1959.
40. Pope, D. A., and M. L. Stein: Multiple Precision Arithmetic, *Commun. ACM*, vol. 3, no. 12, pp. 652–654, December, 1960.
41. New Products—The RCA 601, *Commun. ACM*, vol. 4, no. 4, pp. 197–199, April, 1961.
42. Sherman, P. M.: Table Look-at Techniques, *Commun. ACM*, vol. 4, no. 4, pp. 172, 175, April, 1961.
43. Avery, R. W., S. H. Blackford, and J. A. McDonnell: The IBM 7070 Data Processing System, *Proc. Eastern Joint Computer Conf.*, Philadelphia (NJCC no. 14), pp. 165–168, December, 1958.
44. Smith, H. J., and F. A. Williams: Survey of Punched Card Codes, *Commun. ACM*, vol. 3, no. 12, p. 638, December, 1960.
45. Plotkin, M.: Binary Codes with Specified Minimum Distance, *IRE Trans. Inform. Theory*, IT, vol. 6, no. 4, pp. 445–450, September, 1960.
46. Bemer, R. W.: Survey of Coded Character Representation, *Commun. ACM*, vol. 3, no. 12, pp. 639–643, December, 1960.
47. Ross, H. McG.: Standards—Further Survey of Punched Card Codes, *Commun. ACM*, vol. 4, no. 4, pp. 182–183, April, 1961.
48. Bemer, R. W., H. J. Smith, Jr., and F. A. Williams, Jr.: Design of an Improved Transmission/Data Processing Code, *Commun. ACM*, vol. 4, no. 5, pp. 212–217, May, 1961.
49. Sisson, R. L.: An Improved Decimal Redundancy Check, *Commun. ACM*, vol. 1, no. 5, pp. 10–12, May, 1958.
50. Ling, A. T., and K. Kozarsky: The RCA 601 System Design, *Proc. Eastern Joint Computer Conf.*, New York (NJCC no. 18), pp. 173–177, December, 1960.

51. Kilner, D. E.: The Characteristics of Computers of the Second Decade, *Computer Bull.*, vol. 4, no. 3, pp. 88–112, December, 1960.
52. Miller, A. E., A. B. Shafritz, K. Rose: Multi-level Programming for a Real-time System, *Proc. Eastern Joint Computer Conf.*, Washington, D.C. (NJCC no. 20), pp. 1–16, December, 1961.
53. Gill, S.: Parallel Programming, *Computer J.*, vol. 1, no. 1, pp. 2–10, April, 1958.
54. Codd, E. F.: Multiprogram Scheduling—Parts I and II: Introduction and Theory, *Commun. ACM*, vol. 3, no. 6, pp. 347–350, June, 1960.
55. ATLAS: A New Concept in Large Computer Design, *Commun. ACM*, vol. 3, no. 6, pp. 367–368, June, 1960.
56. Codd, E. F.: Multiprogram Scheduling—Parts III and IV: Scheduling Algorithm and External Constraints, *Commun. ACM*, vol. 3, no. 7, pp. 413–418, July, 1960.
57. Ryle, B. L.: Multiple Programming Data Processing, *Commun. ACM*, vol. 4, no. 2, pp. 99–101, February, 1961.
58. New Products—ORION, *Commun. ACM*, vol. 4, no. 2, pp. 110–113, February, 1961.
59. Lourie, N., H. Schrimpf, R. Reach, and W. Kahn: Arithmetic and Control Techniques in a Multiprogram Computer, *Proc. Eastern Joint Computer Conf.*, Boston (NJCC no. 16), pp. 75–81, December, 1959.
60. Maher, R. J.: Problems of Storage Allocation in a Multiprocessor Multiprogrammed System, *Commun. ACM*, vol. 4, no. 10, pp. 421–422, October, 1961.
61. Dreyfus, P.: Programming Design Features of the GAMMA 60 Computer, *Proc. Eastern Joint Computer Conf.*, Philadelphia (NJCC no. 14), pp. 174–181, December, 1958.
62. Nekora, M. R.: Comment on a Paper on Parallel Processing, *Commun. ACM*, vol. 4, no. 2, p. 103, February, 1961.
63. Strachey, C.: Time Sharing in Large Fast Computers, *Proc. Intern. Conf. Information Processing*, UNESCO Paper B.2.19, June, 1959.
64. Ferranti ORION Computer-programming Exercises (Basic and Symbolic Input), Ferranti Limited, London W1, England, List CS 288A, May, 1961.
65. Ferranti ORION computer system, Ferranti Limited, London W1, England, List CS264, April, 1960.
66. McIlroy, M. D.: Macro Instruction Extensions of Compiler Languages, *Commun. ACM*, vol. 3, no. 4, pp. 214–220, April, 1960.
67. Summer, F. H., G. Haley, and E. C. Y. Chen: The Central Control Unit of the "Atlas" Computer, IFIP Congress 1962, Aug. 27–Sept. 1, 1962, Munich, Germany, pp. 657–663.
68. Allmark, R. H., and J. R. Lucking: Design of an Arithmetic Unit Incorporating a Nesting Store, IFIP Congress 1962, Aug. 27–Sept. 1, 1962, Munich, Germany, pp. 694–698.
69. Lawrance, R. B.: An Advanced Magnetic Tape System of Data Processing, *Proc. Eastern Joint Computer Conf.*, Boston (NJCC no. 16), pp. 181–189, December, 1959.
70. Statland, N.: A Survey of High-speed Printers in the United States, *ICC Bull.*, October, 1963.
71. Vanderburgh, A.: The Lincoln Keyboard—a Typewriter Keyboard Designed for Computer Input Flexibility, *Commun. ACM*, vol. 1, no. 2, p. 4, July, 1958.
72. User's Guide, Auerbach *Standard EDP Rept.*, An Information Service for the Electronic Data Processing Field, vol. 1, Sec. 4, pp. 4:001.001 to 4:210.100, Philadelphia, 1962.
73. Gosden, J. A., and R. L. Sisson: Standardized Comparisons of Computer Performance, IFIP Congress 1962, Aug. 27–Sept. 1, 1962, Munich, Germany, pp. 57–61.
74. Hillegass, J., J. A. Gosden, and R. Sisson: Generalized Measures of Computer System Performance, 1962 *ACM Natl. Conf. Dig. of Tech. Papers*, vol. 1, pp. 120–121, 1962.
75. Gosden, J. A.: Estimating Computer Performance, 1962 *British Computer Society Conference*, Cardiff, August, 1962.
76. Logan, W. A.: The Basic Side of Tape Labeling, *Commun. ACM*, vol. 3, no. 2, pp. 85–86, February, 1960.
77. Ryckman, G. F.: Operational Compatibility of Systems—CONVENTIONS, *Commun. ACM*, vol. 4, no. 6, pp. 266–267, June, 1961.

Chapter 9

LOGICAL CIRCUITS

MAURICE A. MEYER, *Vice President, Adcole Corporation.*

CONTENTS

9-1. Logical-circuit Components	9-2
Semiconductors	9-3
Diodes	9-3
Transistors	9-4
Tunnel Diode	9-6
Magnetic Elements	9-7
Summary	9-8
9-2. Diode Logic	9-8
Single-stage Logic	9-8
Two-stage Logic	9-9
Resistor Diode Logic	9-9
RC Diode Gate	9-9
Matrices	9-9
Reciprocity	9-10
9-3. Transistor Circuits	9-11
Inverters	9-12
Gates	9-13
Direct Coupling	9-13
Resistance Coupling	9-13
Diode Coupling	9-13
Transistor Coupling	9-15
Current Mode Switching	9-16
9-4. Tunnel-diode Logic	9-16
9-5. Magnetic Logic	9-18
PMA Logic	9-19
SMA Logic	9-20
The MAD (Multiaperture Device)	9-22
9-6. Flip-flops	9-22
Static Flip-flops	9-22
Dynamic Flip-flops	9-23
9-7. Shift Registers	9-24
9-8. Counters	9-24
9-9. Logic Packages	9-27
Modules	9-27
Micrologic	9-27

9-1. Logical-circuit Components

The design of logical circuits in recent years has depended largely on semiconductor elements and magnetic elements for their electronic realization. The semiconductors, in particular, have almost completely supplanted the vacuum tube. Initially the advantages of solid-state devices over vacuum tubes were smaller power consumption,

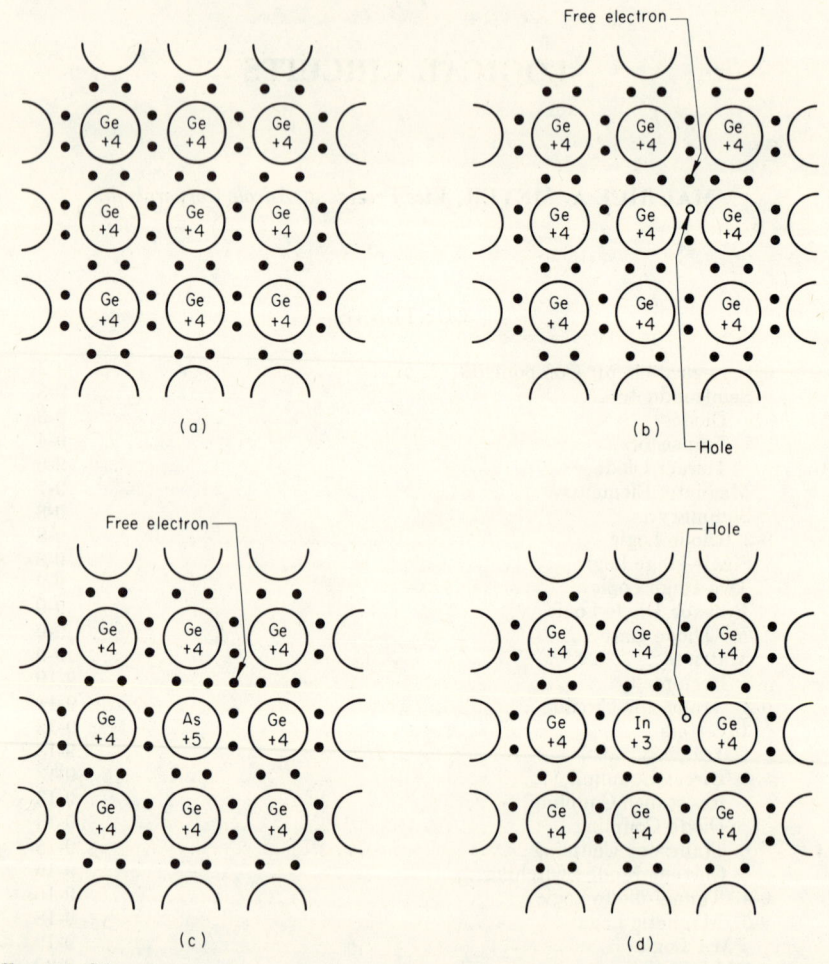

Fig. 9-1. Diagrammatic representation of a germanium lattice section. The pairs of dots represent shared electrons. (a) No free electrons; no holes. (b) Thermal energy frees an electron, leaving a hole. (c) Pentavalent impurity presents a free electron. (d) Trivalent impurity leaves a hole.

size, and weight, and less generation of heat. In recent years it has been possible to obtain semiconductor elements with superior electrical performance (e.g., gain-bandwidth products), with greater reliability than the vacuum tube, and at prices which compete favorably. Consequently, the circuit considerations here will be concerned only with these elements, which are now in wide use.

Semiconductors. The semiconductors are materials which lie between the metals and the insulators in their electrical conductivity. The two most widely used are germanium and silicon. Both these materials have atoms with four valence electrons, and these atoms form crystals by sharing their valence electrons with four neighbors to form four covalent bonds. All the valence electrons are used in the bonds, so that there are no free electrons to move in the crystal, and the crystal is a poor conductor. Thermal energy occasionally causes one of the valence electrons to break away from the atom and move through the crystal; as the temperature increases the conductivity increases. This is discussed quantitatively in Sec. 21-3. An atom which has lost an electron is more positive and can acquire an electron from a neighboring atom. This neighboring atom can in turn acquire an electron from another neighbor. This electron deficiency moves throughout the crystal as easily as the electron; it is conveniently considered as a positively charged particle and is called a *hole*. Whenever a covalent bond is broken by thermal agitation a hole-electron pair is generated; if a hole and electron collide the electron will fill the deficiency the hole represents and both will cease to exist as free-charge carriers. This latter process is known as recombination.

The conductivity of a semiconductor can be increased by "doping" or adding impurities to the crystal in its formation. The impurities contain atoms which

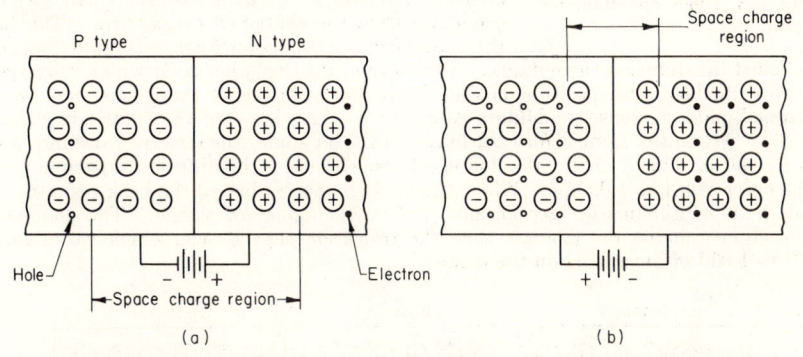

Fig. 9-2. *PN* junction. (*a*) Reverse voltage applied. (*b*) Forward voltage applied.

may have more or fewer valence electrons than Ge or Si. Arsenic, for example, has five valence electrons and takes the place of a semiconductor atom in the crystal, where four of its electrons are used in the covalent bond; the fifth is easily removed to move through the crystal.

When the crystal is doped with pentavalent material, a donor, it has an excess of electrons and is called an *N* (negative)-type semiconductor. When a trivalent, acceptor material is added to the semiconductor it generates free holes by creating an electron deficiency; since conduction occurs by virtue of these free holes the material is called a *P* (positive)-type semiconductor. These arrangements are illustrated in Figs. 9-1 and 21-2.

Diodes. A rectifier or diode is obtained by joining a *P*-type region adjacent to an *N*-type region in the same crystal structure (Fig. 9-2); the boundary between the two regions is called a *PN* junction. When the positive terminal of a voltage source is connected to the *N* region (cathode) and the negative terminal to the *P* region (anode), the electrons in the *N* region (majority carriers) and the holes in the *P* region are attracted away from the junction and a space-charge region builds up in the vicinity of the junction (called the depletion layer) which further retards the flow of the majority carriers. The hole-electron pairs which are thermally generated produce electrons formed in the *P* region, which can flow to the *N* region because of the space-charge potential (ionized donor and acceptor atoms), and holes in the *N* region which can flow to the *P* region; this is called minority-carrier flow. This current is the leakage current under reverse bias of the diode, and it is generally a small value unless the reverse bias is raised enough to impart enough energy to these particles to cause an avalanche multiplication.

When the polarity of the applied voltage is reversed (forward bias) the depletion layer narrows since the majority carriers are pushed toward the junction; thus more majority carriers can diffuse across the junction, where recombination takes place. The majority-carrier current increases markedly in this case so that a large conduction takes place. The diode is characterized by high back resistance in the reverse direction and low forward resistance. For high reverse voltage (the avalanche breakdown voltage), the back resistance becomes very low; a typical curve is shown in Fig. 9-3.

The majority carriers which cross the junction with forward bias become injected minority carriers on the other side of the junction, and these diffuse a certain average distance (diffusion length) before recombination. In this diffusion length, then, there is a cloud of injected minority carriers which must be dissipated when the applied voltage is removed or reversed. These carriers are drawn back to their originating regions, constituting a momentary reverse current which originates from the charge on the carriers. This has the same effect as a capacitance across the junction, and is called the diffusion capacitance. In addition, the depletion layer which is charged with embedded ions, positive on the N side and negative on the P side, acts like a parallel-plate condenser, adding a capacitance which is called the transition capacitance. When the diode is forward-biased in the on mode, the forward resistance is of the order of tens of ohms; the dominant capacitance is the diffusion capacitance, of the order of 0.1 to 1,000 pf. When the diode is reverse-biased the back resistance is hundreds of kilohms for germanium, and much higher for silicon. The dominant capacitance across the diode is then the transition capacitance, which varies from 0.1 to 1,000 pf depending on the diode.

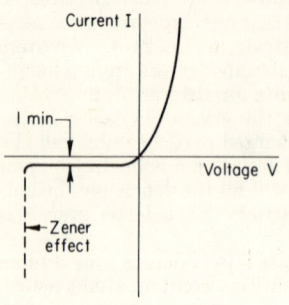

FIG. 9-3. Diode characteristic curve.

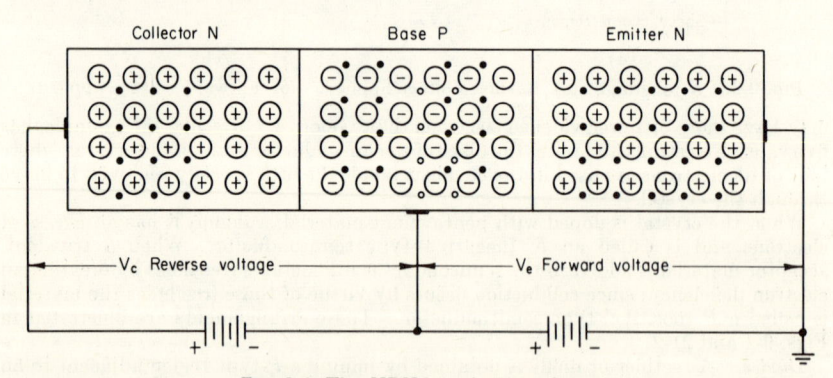

FIG. 9-4. The NPN junction transistor.

Transistors. When a thin P region (the base) is sandwiched between two N regions, an NPN transistor is formed. In the normal mode of operation, the transistor is connected as shown in Fig. 9-4; the three terminals, base, emitter, and collector, are biased so that the base-emitter region is forward-biased, and the base-collector region is reverse-biased. The majority-carrier current of the emitter injects electrons into the base region, the number being controlled by the number of holes in the base region because these determine the space-charge potential. The electrons emitted into the base diffuse across the base, and since they are minority carriers of the base, they will be swept across the collector junction by the collector potential. The

LOGICAL CIRCUITS 9-5

transistor is so constructed, by the doping of the base material and the width of the base material, that it requires only a small base current to produce a larger collector current, the ratio of change of collector current to base current being defined as

$$\beta = \left|\frac{\Delta I_c}{\Delta I_b}\right|_{V_c} = h_{fe}$$

similarly
$$\alpha = \left|\frac{\Delta I_c}{\Delta I_e}\right|_{V_c} = h_{fb}$$

where I_b = base current, I_c = collector current, I_e = emitter current, and V_c is the collector voltage. h_{fe} is known as the forward-current transfer ratio for the common

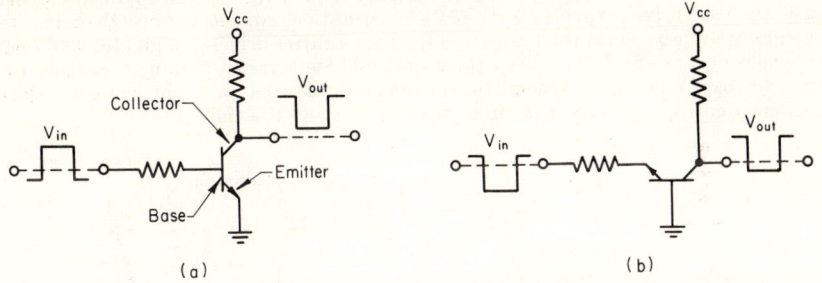

FIG. 9-5. Transistor-circuit arrangements. (a) Common emitter. (b) Common base.

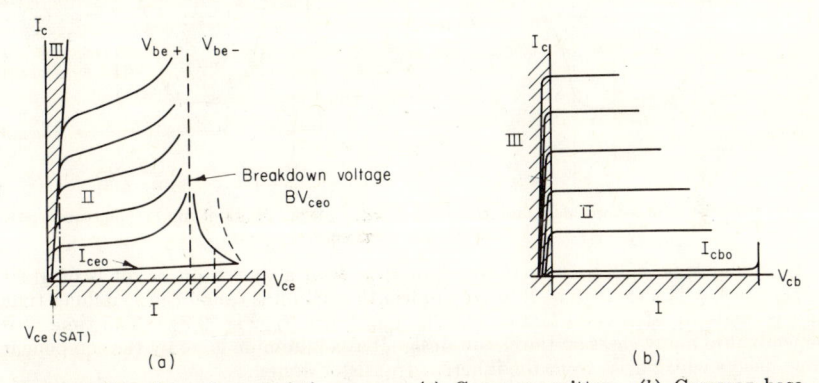

FIG. 9-6. Transistor characteristic curves. (a) Common emitter. (b) Common base.

emitter; h_{fb} is the ratio for the common base configuration. Since $I_e = I_c + I_b$, $\beta = \alpha/(1 - \alpha)$.

For switching-circuit design as used in logical circuits, the large-signal behavior of the transistors is of most interest. By applying appropriate voltages to one or two of the terminals of a transistor, the third terminal can be made to exhibit either a high or a low value of impedance. Two of the most common arrangements are shown in Fig. 9-5, and typical graphs are shown in Fig. 9-6.

The graphs show the three regions of interest, (I) the cut-off region, (II) the active region, and (III) the saturation region. In cut-off, all currents are very small (both junctions reverse-biased); the active region is where the emitter-base junction is forward-biased and the collector-base junction reverse-biased; and in saturation all currents are large (both junctions forward-biased). The quantity I_{co} is the collector-base current leakage current when the emitter is open-circuited. This quantity is of the most concern when the transistor is in the cut-off region. If the emitter is grounded, and the base is left open-circuited, the collector current is generally much larger than I_{co}, being given by $I_{ceo} = I_{co}/(1 - \alpha)$, where α is generally close to unity. If the base-to-emitter junction is reverse-biased, the collector current and the reverse base current are of the order of I_{co}. Improvements in transistor technology are constantly reducing the value of I_{co} for transistor types.

In switching service the active region is of less concern than the cut-off or saturation region, since the device is usually driven completely from cut-off to saturation or back, spending minimum time in the active region, where the small-signal parameters dominate and can be used to predict transient response.

Saturation occurs when the collector-base voltage becomes forward-biased. This happens when the input to the transistor is driven harder than the output can maintain. Such a condition occurs when the collector current $I_c \approx V_{cc}/R_L$ where V_{cc} is the collector supply voltage and R_L is the load resistance. In the common-emitter case, the base drive is then $I_b \geq V_{cc}/\beta R_L$. Saturation happens because the minority carriers are injected into the base, from the base-emitter drive, at a greater rate than the collector-base field can sweep them out. These carriers "pile up," causing the collector-base junction to become forward-biased. The base current increases, since carriers are supplied to the base from the collector and the emitter.

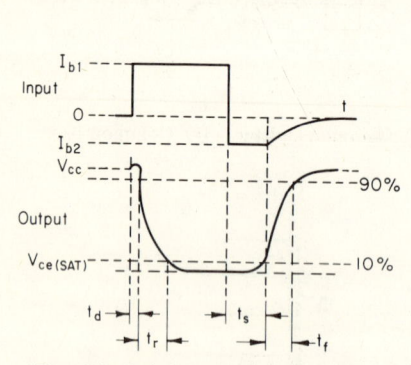

FIG. 9-7. Switching-time parameters.

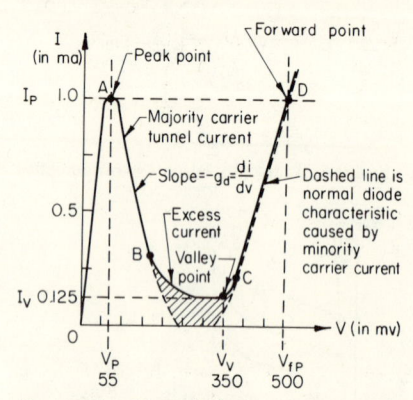

FIG. 9-8. Tunnel-diode characteristic curve.

In digital logic circuits, the time required to turn a switch on or off is of great importance. The total time T to accomplish this action is called the switching time and is made up of several factors $T = t_d + t_r + t_s + t_f$ (Fig. 9-7). The time T is dependent in many cases on the circuit design, but a minimum is set by the component time delays which arise from the inherent transistor design.

The time t_d is called the delay time and is related to the time taken to start collector current flowing in a transistor that is being turned on from the cut-off condition. Such factors as emitter capacitance and the time taken to establish the "on" distribution of minority carriers in the base contribute to the delay time. The delay can be reduced by increasing the base drive current in the common-emitter configuration.

The rise time t_r refers to the time taken for the collector current to reach its final value from the start of current flow. It is approximately given by $t_r = (I_c/I_{b_1})(1/2f_\alpha)$ where I_c is the desired collector current, I_{b_1} the base drive current, and f_α the cut-off frequency of the transistor. f_α is a property of the transistor and is related to the base width, as it is concerned with the time it takes the carriers to diffuse across the base. Similarly the fall time t_f is approximately given by $t_f = (I_c/I_{b_2})(1/2f_\alpha)$ where I_{b_2} is the reverse base current caused by the application of reverse bias.

Tunnel Diode. The tunnel diode is a two-terminal device exhibiting a negative resistance in its characteristic curve (Fig. 9-8). The device is capable of operating at extremely high speeds.

The tunnel diode is formed by utilizing a heavily doped PN junction. As the density of free carriers increases because of the doping, the reverse-breakdown voltage decreases. With heavy doping, the diode can still be in the reverse-breakdown region at a slightly forward bias, giving a relatively large reverse current; when a larger forward voltage is applied, the diode goes out of the reverse-breakdown condition and the current falls to a small value.

The process is explained by a quantum-mechanical effect called tunneling. Even though the particles do not have enough energy to surmount the space-charge barrier, they can almost instantaneously tunnel through it provided there are empty energy states on the other side of the barrier with the same energy as the particles originally have. This process enables the device to be switched at high speed. The static effect of this process is to produce an IV curve for a tunnel diode as shown in Fig. 9-8. Typical current and voltage values are indicated; although tunnel diodes with much larger currents are made, the speed generally decreases as the current increases.

Magnetic Elements. Before the large-scale advent of semiconductor devices such as the transistor, the magnetic element was gaining favor for digital logic devices because of its small size, low power, long life, and economy. Although transistor logical circuits have largely dominated in most circuit designs in recent years, magnetic logical elements at present offer advantages in certain applications which require extremely low power or where especially high resistance to nuclear radiation is required. Although the magnetic device has always been considered highly reliable,

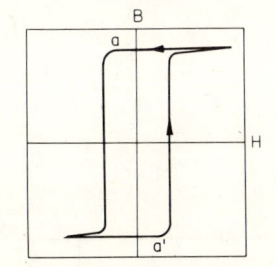

Fig. 9-9. Square-loop BH curve.

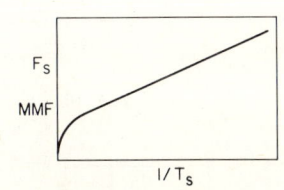

Fig. 9-10. Relation between mmf and switching time.

the fact that it is generally hand-fabricated with many turns of fine wire makes it difficult for the device to meet the reliability of the transistor, which is mechanically much simpler and generally automatically fabricated.

The magnetic devices used are usually toroids of magnetic material such as ferrites or ultrathin metallic tapes wound on ceramic or steel bobbins, upon which are wound coils, as in constructing a transformer. The magnetic material used has retentivity; that is, an mmf applied by passing current in the wire or coil causes the core to become magnetized and stay magnetized, with a flux density B called the remanence, after the mmf is reduced to zero. If the mmf is reversed in direction and applied strongly enough, the direction of the flux will change, and the value after the mmf is removed will be the same but in the opposite direction. This action is characterized by the hysteresis curve of the material, and for digital logic circuits a square hysteresis loop is desired as in Fig. 9-9. The two remanent states are shown as a, a' on the BH curve. The curve is traversed in the direction shown. The application of positive mmf causes the core to switch from the state characterized by a' to a if the value of H exceeds the coercive force, the value of H for which $B = 0$. If H is further increased, B does not increase if the material has a square loop. The change of state can be considered as a flipping or switching action; if a current pulse is applied to a coil on the core, the switching takes place in a time characteristic of the material and the applied current. The switching time decreases for increased currents, according to the curve in Fig. 9-10.

During the switching time, the core acts like a variable resistance which drops to zero once the core reaches saturation. After the core is saturated, the resistance to another pulse applied in the same direction is also zero, and no voltage will be induced in a secondary coil.

Summary. The semiconductors and magnetic elements described in the preceding paragraphs, as well as conventional circuit components such as resistors and condensers, are used in logical circuits. These circuits are the physical realization of some of the abstract considerations of the propositional calculus (Chapter 41).

9-2. Diode Logic

Diodes are used to implement the functions of OR, AND, and combinations of these functions, acting as simple switches. Diodes (other than those which have a negative

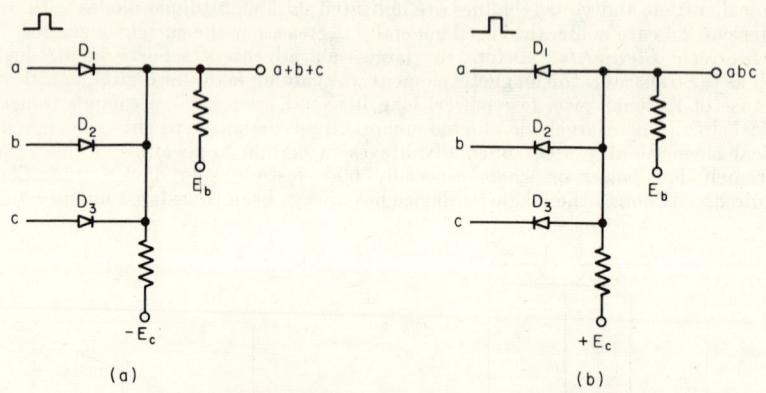

FIG. 9-11. Single-stage diode logic. (*a*) OR. (*b*) AND.

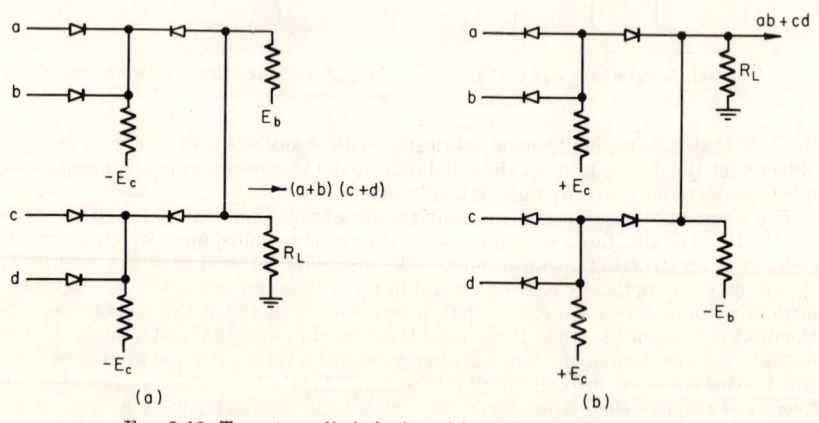

FIG. 9-12. Two-stage diode logic. (*a*) OR/AND. (*b*) AND/OR.

resistance, such as the tunnel diode) have no gain capability. Certain variable-capacitance diodes can function as parametric amplifiers, but these will not be discussed here.

Single-stage Logic. A simple representative diode single-stage logic scheme is shown in Fig. 9-11. Figure 9-11*a* is an OR circuit for positive signals representing a ONE; i.e., for any diode input sufficiently positive, the output will be clamped to that input. If a ZERO is represented by a negative level, the output will be decreased to the ZERO level for all inputs ZERO. Figure 9-11*b* is an AND circuit for positive signals, since a ZERO signal being negative will clamp the output to that level on any input; the output will rise to the ONE level only if all inputs are ONE. If positive signals

LOGICAL CIRCUITS 9-9

represent ZERO, the circuit of Fig. 9-11a becomes an AND and of 9-11b an OR. The d-c design of the circuits is simple; some care must be taken in the design of pulse circuits because of the diode capacity.

Two-stage Logic. Single-stage diode logic circuits can be combined to obtain two-stage logic, resulting in AND/OR or OR/AND combinations. Figure 9-12a shows an OR/AND combination for positive signals and Fig. 9-12b is an AND/OR combination. The logic rules are reversed for negative signals. By proper choice of supply voltages, the output is clamped to either the ONE level or the ZERO level for all values of the input variables. These realizations of two-stage logic can be generalized to more variables than shown. Two-stage logic circuits represent a practical design in that all signal paths are uniform. The clamping criteria are simple and are generally more desirable than multistage logic even though the number of components may be greater. For example, $abe + cde$ is preferable to $(ab + cd)e$, which though logically equivalent, requires a three-stage realization.

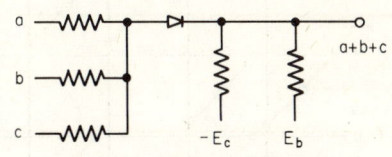

FIG. 9-13. Resistor diode logic.

Resistor Diode Logic. An economical class of circuits is resistor diode logic, in which resistive mixing of the inputs is employed, as in Fig. 9-13. Whenever a sufficient number of inputs is up, the diode conducts and raises the output level. In this scheme a ONE is generally associated with diode conduction; e.g., Fig. 9-13 represents a positive OR. Circuits such as these (threshold circuits) can be used in majority logic.

RC Diode Gate. For some applications of pulse gates, the RC diode gate (Fig. 9-14) is attractive because of its simplicity. The diode conducts whenever a pulse is present on any of the input lines; in the OR circuit, it is normally biased off. Similarly in the AND a sufficient number of pulse inputs causes the diode to conduct.

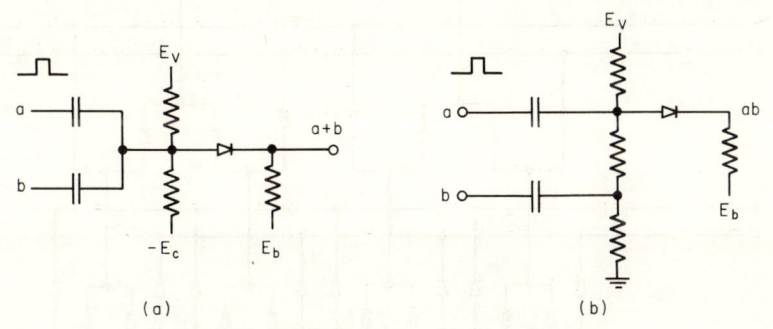

FIG. 9-14. RC diode gates. (a) OR. (b) AND.

Matrices. In many instances, where several binary variables are involved, sets of functions of these variables are required. Rather than implement the reduced logical expressions, it is sometimes more economical, and in some cases necessary, to generate all the minterms or maxterms of the set of binary variables from which any desired function can be generated. The set of minterms of M binary variables is composed of all the possible sums of the binary variables, each sum with M terms containing either the variable or its complement, thus 2^M terms in the set. Similarly the set of maxterms is composed of the possible products of the variables; thus OR gates are used in the case of the minterms, AND gates in the case of maxterms. The generation of these canonical terms is accomplished in diode matrices. The most general one-stage form of such matrices is a rectangular matrix (Fig. 9-15), which can be constructed for any number of variables.

A pyramidal matrix is obtained as a multistage form of the minterm-maxterm realization. For example, the minterms for a, b are generated in rectangular fashion for a single stage; and by using the c variable with the outputs of this rectangular matrix, a pyramid is formed by AND gates (Fig. 9-16). Additional variables create a larger pyramid.

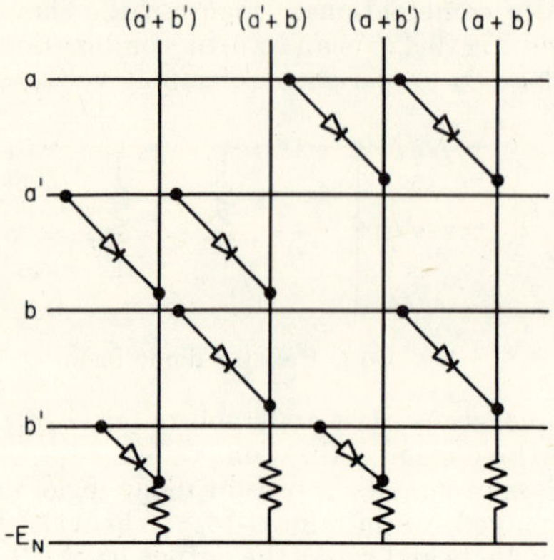

A less costly approach for several variables is the sectional matrix which is formed by using AND gates on the outputs of several rectangular matrices. A variables-to-minterms sectional matrix is shown in Fig. 9-17. Of course, all these realizations can be made in maxterm form by interchanging the AND and OR diode functions. Figure 9-18 is a table giving the numbers of diodes required for various matrices for 2 to 14 variables.

Reciprocity. In some cases it is desired to abstract the set of variables from the minterms (or maxterms). For example, a decoding matrix for BCD (binary-coded decimal) can be generated from a rectangular matrix of the four binary input variables, as various minterms of these variables. On the other hand, the coding matrix obtains the binary variables as functions of the decimal digits. All the rectangular, pyramidal, and sectional matrices are reciprocal, in that by changing the position

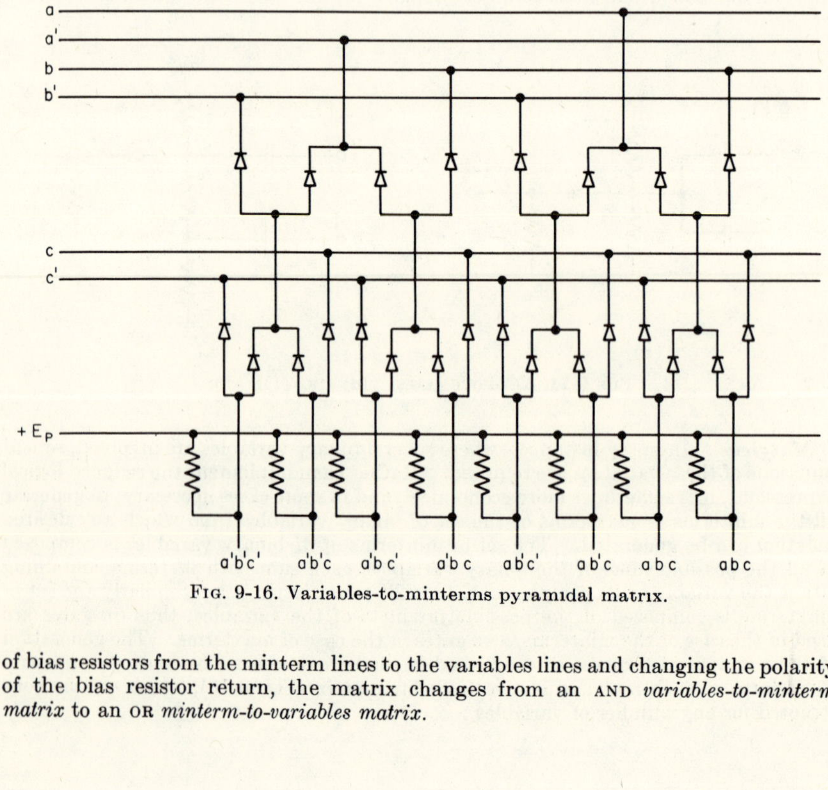

FIG. 9-15. Rectangular matrix.

FIG. 9-16. Variables-to-minterms pyramidal matrix.

of bias resistors from the minterm lines to the variables lines and changing the polarity of the bias resistor return, the matrix changes from an AND *variables-to-minterm matrix* to an OR *minterm-to-variables matrix*.

LOGICAL CIRCUITS 9–11

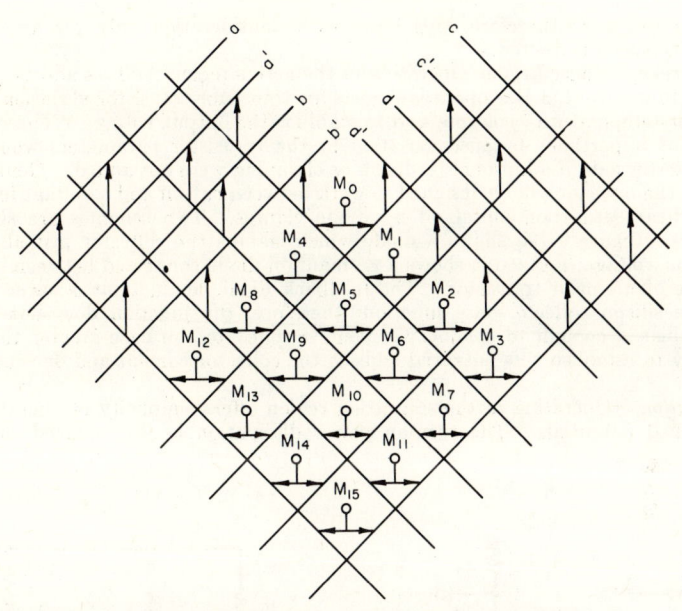

FIG. 9-17. Variables-to-minterms sectional matrix.

No. variables	No. minterms or maxterms	No. diodes.		
		Rectangular	Pyramidal	Sectional
2	4	8	8	8
3	8	24	24	24
4	16	64	56	48
5	32	160	120	96
6	64	384	248	176
7	128	896	504	328
8	256	2,048	1,016	608
9	512	4,608	2,040	1,168
10	1,024	10,240	4,088	2,240
11	2,048	22,528	8,184	4,368
12	4,096	49,152	16,376	8,544
13	8,192	106,496	32,760	16,888
14	16,384	229,376	65,528	33,424

FIG. 9-18. Diode requirements for full matrices.

9-3. Transistor Circuits

Transistors are used as inverters (NOT circuits) and as gates. Inverters can be designed using only the cut-off and saturation regions of the transistor, or only the active region, or both the cut-off and active regions. Operating between cut-off and saturation has the advantage of well-defined voltage levels and simplicity in design. As previously discussed, some sacrifice in operating speed must be made when operating in the saturation region. The speed of transistors has been constantly improving; therefore, this becomes a consideration only for applications where ultraspeed is desired.

Inverters. *Active Region.* Operating in the active region requires additional components to ensure that this operation exists for transistor-transistor variation and for transistor-temperature variation, and to stabilize the output voltage. These factors become of importance because variation in the transistor parameters would vary basic inverter gain if additional feedback or clamping were not added. Generally in this case the inverter will be designed to work between cut-off and a voltage level that is very near saturation, obtained by diode clamps. Two schemes are shown in Fig. 9-19. Figure 9-19a shows a diode which clamps the collector just above the saturation voltage; Fig. 9-19b shows a germanium diode connected between collector and base of a silicon transistor. The feedback diode conducts at a lower voltage than the silicon collector-base junction; therefore, the junction does not become forward-biased enough to conduct; current supplied to the base greater than that necessary to establish this potential adds to the collector current and does not enter the base.

Saturation. Operating in the saturation region offers simplicity of operation and well-defined potentials. The turn-on power dissipation is also reduced, since the

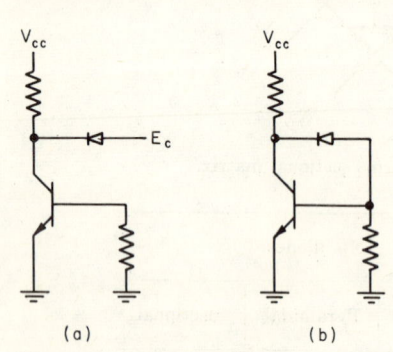

Fig. 9-19. Circuit techniques to avoid saturation. (a) Antisaturation collector clamp. (b) Feedback diode.

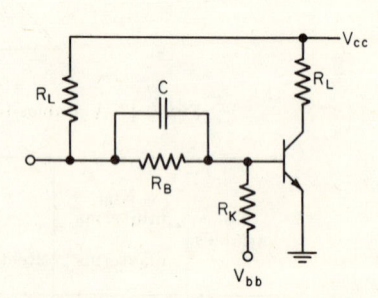

Fig. 9-20. Neutralization of stored charge.

V_{ce} is low. The requirement to keep the transistor in saturation is to supply enough base current for the range of β for a given transistor. Since β can vary by a factor of 4:1 over an extended temperature range, and an additional factor of 3:1 from transistor to transistor, a circuit that will just saturate for the low-low condition will be overdriven for higher β values. An approximate relation for the base drive necessary is

$$I_b \approx \frac{V_{cc}}{R_L \beta_{\min}}$$

By using the given value of storage time t_s the maximum stored charge can be calculated from

$$Q_s = t_s I_b$$

This charge can be neutralized by a speed-up condenser, as in Fig. 9-20, where the value of C is calculated from $C = Q_s/V_{\text{in}}$ and usually the V_{in} is limited by the reverse base-emitter voltage. This can be effectively increased by placing a high-voltage-breakdown diode in series with the emitter lead as in Fig. 9-21a. Switching speed can also be improved by using a clamped diode in the collector as shown in Fig. 9-21b. Here a large V_{cc} can be used so that the collector capacitance is effectively charged with a constant source.

LOGICAL CIRCUITS 9–13

Gates. Transistor gates are generally constructed of combinations of emitter followers or transistor inverters. The inverter configuration shown in Fig. 9-22 can be made to respond to positive or negative inputs, depending on whether it is initally biased to cut-off or saturation. With appropriate groups of transistors the logical operations of AND, OR, NOR, or NAND can be performed, according to the circuit arrangement. AND and OR are generally emitter output devices while the NOR and

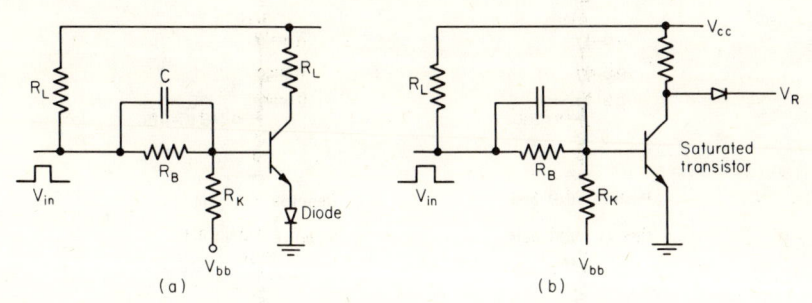

FIG. 9-21. Techniques to increase speed. (a) V_{in} increased by addition of diode. (b) Speed increased with clamped collector.

NAND are the inverter operation. Figure 9-23 illustrates how these are generally accomplished.

The positive OR gate is an emitter-coupled arrangement with all transistors biased off. A positive input on any transistor produces a positive output. Similarly with all transistors biased on, the output is reduced to zero only when all inputs are negative, producing a negative AND gate.

The positive AND gate requires all inputs positive, which turns on all transistors in series to produce an output. If all transistors are normally on, a negative signal on any input will turn it off, producing a negative OR gate. Similar considerations apply for the NOR and NAND circuits.

Direct Coupling. Transistors can be connected directly in the configurations shown in the preceding diagrams, obtaining direct-coupled transistor logic (DCTL) which minimizes the number of component types and reduces power-supply voltages and power. Transistors should be evenly matched to ensure proper operation, and they are always operated in the cut-off or saturation region. Examples of such devices are shown in Fig. 9-24.

Resistance Coupling. As in diode logic, resistors can be used for gating in conjunction with transistors (RTL) as shown in Fig. 9-25, where the resistor-transistor OR inverter is shown. While simple, this configuration is limited in the number of inputs it can accommodate so that the design can take care of transistor parameter variations. The speed is also reduced since the transistor can be more easily overdriven with several on inputs.

FIG. 9-22. Basic inverter. (a) Inverter with bias V_{bb}. (b) Bias voltage cuts off transistor; positive voltage turns it on. (c) Bias adjusted so transistor normally on; negative voltage cuts it off.

Diode Coupling. An approach which utilizes diodes instead of resistors is called diode-transistor logic (DTL); it has the advantage of having a higher "fan-in" capability and can be used for low-level logic; that is, as in resistor-diode, or diode-diode logic, a diode used in place of a resistor does not shunt the signal from another

9-14 SYSTEM COMPONENTS

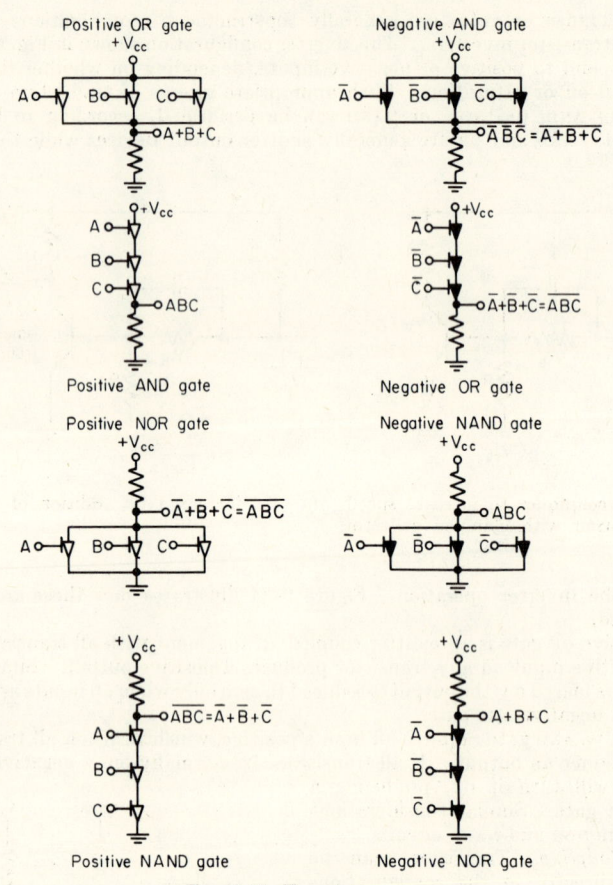

FIG. 9-23. Transistor gates.

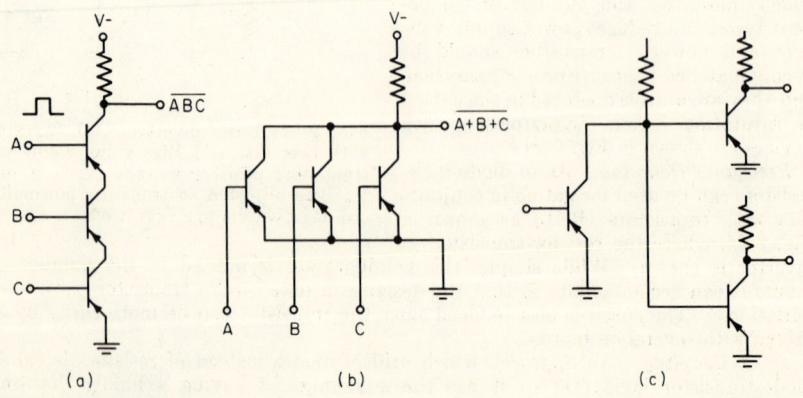

FIG. 9-24. Direct-coupled circuits. (*a*) AND inverter. (*b*) OR inverter. (*c*) Two-stage direct-coupled drive.

LOGICAL CIRCUITS 9-15

Fig. 9-25. Resistor-transistor logic.

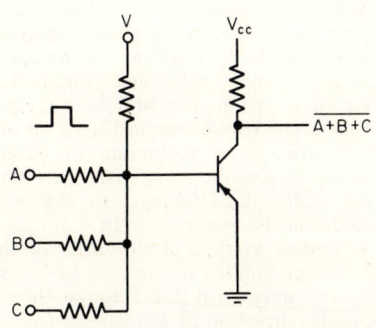

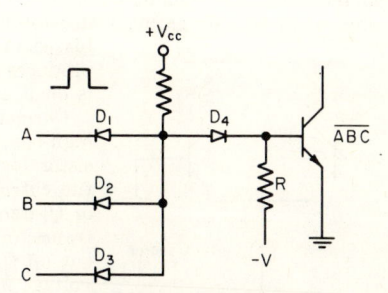

Fig. 9-26. DTL NAND.

input to the same point, except for leakage current; this allows more inputs to be used at a given level, with available voltage savings. A well-known form of DTL is shown in Fig. 9-26, generating the positive NAND function. In this case if at least one of the input lines is connected to the collector of an on transistor, the output transistor is off. The node current flowing through the diode D_4 causes a voltage drop which compensates for the diode drops D_1 to D_3, allowing direct coupling.

Transistor Coupling. An analogous form of this NAND circuit can be constructed using transistors as the gating elements. This approach has the advantage of low power, high speed, and simplicity of design. The contiguous collector (CC) TTL-NAND gate is shown in Fig. 9-27. The logic is the same as that of the diode circuit; if at least one line is connected to the collector of an on transistor, node current will flow to the left. The inverter transistor on the right then has both the saturation voltage of the on transistor and the offset voltage of the switch applied to its base, turning it off. If all input lines are connected to the off collectors, current flows to the base of the inverter transistor.

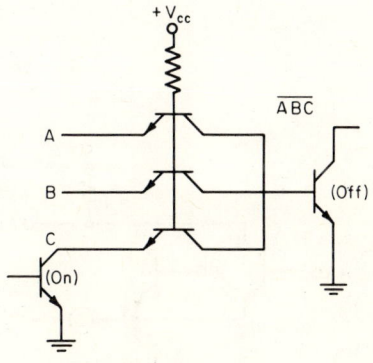

Fig. 9-27. TTL NAND.

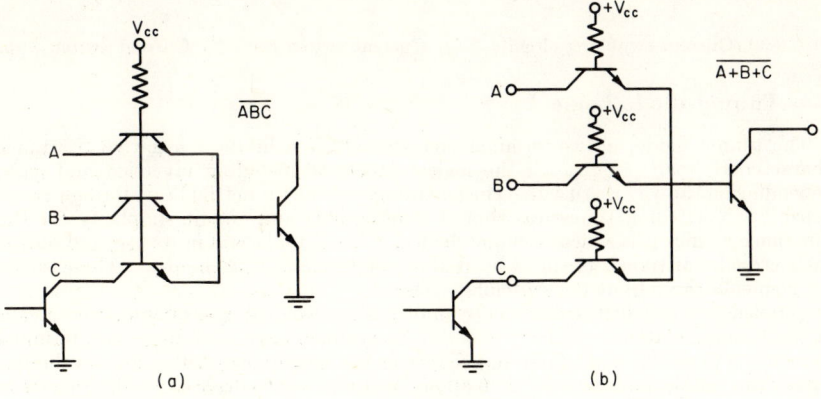

Fig. 9-28. TTL circuits. (a) Positive TTL NAND. (b) TTL NOR.

9–16 SYSTEM COMPONENTS

Other forms of this logic are shown in Fig. 9-28. The positive NAND (Fig. 9-28a) can be obtained by joining the emitters and feeding the collectors, since when an input collector is on, the gating transistor is saturated, and both junctions are forward-biased in either form. The second circuit (Fig. 9-28b) produces the positive NOR; the transistor to the right is on if at least one of the transistors on the left is off.

Current Mode Switching. A technique to obtain high-speed switching by avoiding saturation in transistor logic employs current switching. In Fig. 9-29 the current source feeds the common-emitter junction of Q_1 and Q_2. A positive voltage of the order of the transistor cut-off voltage applied to the base of Q_1 will cut off Q_2 and all the current will flow through Q_1. A voltage in the opposite direction at the base of Q_1 cuts off Q_1 and the current flows through Q_2. The value of the current is set so neither Q_1 nor Q_2 saturates. Logical functions are performed as shown in Fig. 9-30. The positive OR (Fig. 9-30a) is accomplished when the current is switched by raising the potential of either A or B. In the circuit in Fig. 9-30b both transistors must be turned on to switch the current.

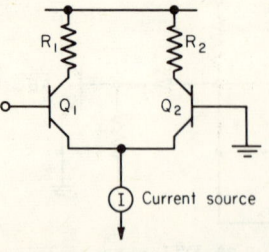

FIG. 9-29. Current switching.

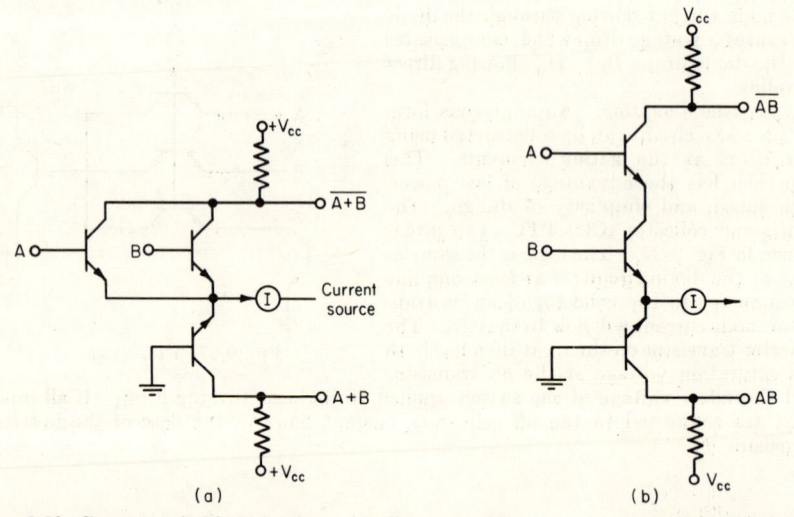

FIG. 9-30. Current-switching circuits. (a) Current switch OR. (b) Current switch AND.

9-4. Tunnel-diode Logic

The tunnel diode, a two-terminal element which exhibits a negative resistance characteristic, can perform all the logical functions including inversion and gain, depending on the way it is used. The advantages of the tunnel diode are its high-speed capability (switching times less than 1 nanosecond) and circuit simplicity for the functions. Since it is a two-terminal device, it is not unilateral in its use, and means such as diodes or transistors must be used to isolate connecting circuits. These isolating elements could limit the switching speed.

Coupled circuits which are driven by multiphase power supplies offer another means of producing isolation. In fact, one way to produce gain with these two-terminal elements is to set the state (high impedance or low impedance) with a low-level input signal; this "impedance switch" then allows pulsed power to drive succeeding stages in a manner similar to the operation of some magnetic logic circuits. Since any assembly

LOGICAL CIRCUITS 9–17

of logic circuits requires gain, a pulsed power supply or reset signal must be used for "all-tunnel-diode" computer logic; it is possible to obtain gain with reactive elements, but this reduces speed. Where tunnel-diode logic circuits are used with other amplifying devices such as transistors, certain circuit simplifications may result.

A simple AND circuit is shown in Fig. 9-31. The inputs are voltage pulses, 0 for ZERO, and V for ONE. When all inputs are present, enough current is supplied to switch the diode to the high-voltage state, producing a ONE pulse out. If the tunnel diode fires for one input, an OR circuit results. These Kirchhoff circuits are also called analog threshold circuits, since the analog sum of the input currents must be large

FIG. 9-31. Tunnel-diode threshold circuit.

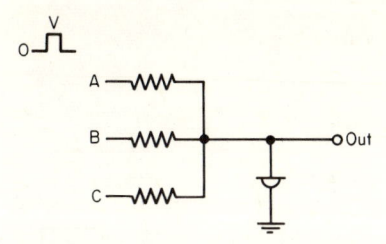

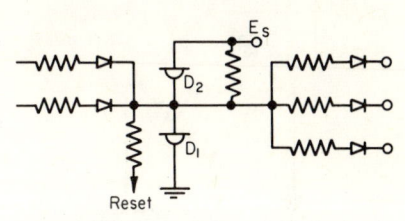

FIG. 9-32. Improved tunnel-diode threshold circuit.

enough to cause the tunnel diode to switch. Diodes are sometimes used in series with the input resistors to reduce the loading effects of the other inputs. The threshold circuit is then sensitive to the variations in input levels, as well as the loading.

A circuit variation which reduces the sensitivity to the number of outputs adds another tunnel diode and by adding reset operations gives logical gain as well (Fig. 9-32). E_S is chosen so that only one tunnel diode can be in the high-voltage state. When D_1 becomes high voltage D_2 becomes low; thus the power supply E_S is coupled to the load through the low tunnel-diode impedance.

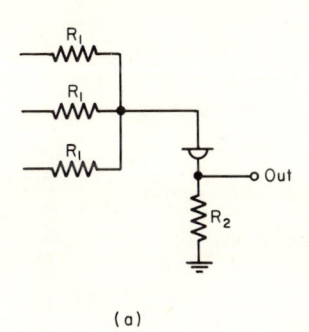

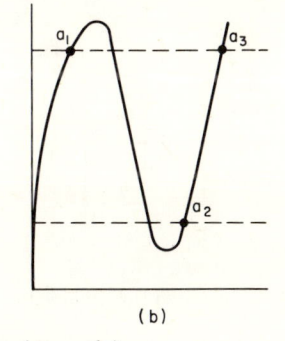

FIG. 9-33. Binary adder (sum only).

A single tunnel diode can also be used to produce the sum function of a full binary adder (Fig. 9-33). The sum output of a binary adder is ONE when only one of the three inputs (X, Y, carry) is ONE, ZERO when any two inputs are a ONE, and ONE when all three inputs are ONE. This is accomplished by using the three operating points shown in Fig. 9-33b. By proper choice of all the resistor values the current will switch between the values shown, and the voltage across resistor R_2 will have the sum output. Thus, when any one input is ONE, the current is high (a_1), and the output is a ONE; when two of the inputs are ONE, the tunnel diode has switched and the current is small (a_2), producing a ZERO output. For three inputs ONE the current is again high (a_3). By using a bridge arrangement (Fig. 9-34) the carry output can also be generated as appearing between the terminals X, Y. The bridge is balanced for only one input, and no output appears. For two or three inputs, the tunnel-diode static impedance at these operating points causes a bridge unbalance, and an output appears.

In "all tunnel-diode" logic circuits, gain and unilateralization can be obtained by using a multiphase power supply. A three-phase power supply with time overlap between phases (Fig. 9-35) allows energy to propagate between groups of circuits whose two power phases overlap while other circuits powered by the third phase are deenergized. In multiphase logic the single-ended threshold gate is similar to the one

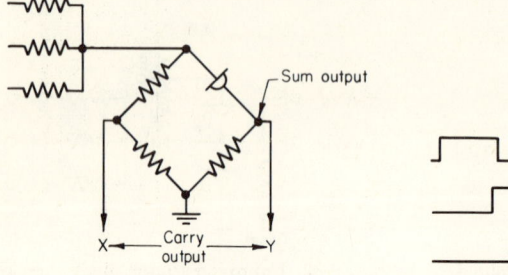

FIG. 9-34. Full adder.

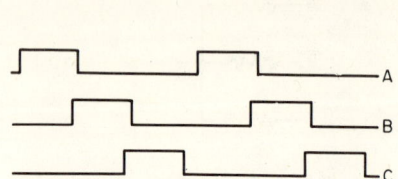

FIG. 9-35. Multiphase power supply.

shown previously (Fig. 9-31) except that a pulse of current corresponding to one phase is also supplied to the tunnel diode with a value less than I_p of the tunnel diode. After the diode has switched, some of this current becomes available to drive succeeding stages; hence the circuit has effective gain.

A balanced threshold gate (Fig. 9-36) is also similar to the gate with reset. The diode D_3 is used to produce a low-impedance voltage source from the current pulse I_S. When a sufficient number of inputs is ONE, current into the junction J is positive; this excess current causes the diode D_1 to break down first when the power phase I_S comes on, for D_1 will reach the I_p before D_2. The output voltage then goes to the ONE state.

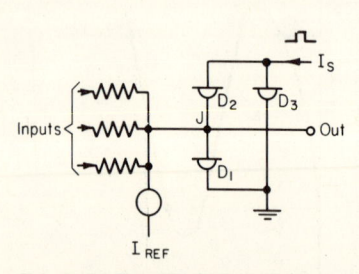

FIG. 9-36. Balanced threshold gate.

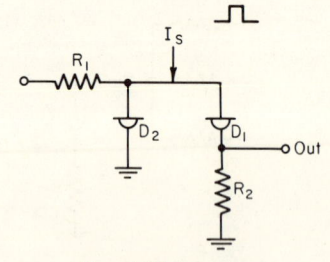

FIG. 9-37. Inverter.

A ZERO output is produced when the current into the junction is negative, causing D_2 to fire first.

A stable inverter circuit is formed as shown in Fig. 9-37. If the input is a ONE, the diode D_1 is driven by the current added to I_S into the high-voltage state. The output is then low, a ZERO, since D_2 and I_S act like a low-impedance voltage source. When the input is ZERO, the diode D_1 remains in the low-voltage state; the output is then high, or a ONE.

9-5. Magnetic Logic

Magnetic elements have been used in a variety of ways to perform digital logic. In many applications they have been used to perform logic as well as gain; in some other applications they have been used principally as amplifiers, and the logic is performed primarily by diodes; they have also been used in conjunction with transistors to perform logic with gain.

Two widely known types of operation are the parallel magnetic-pulse amplifiers (PMA's) and the serial magnetic-pulse amplifers (SMA's). The SMA appears to be more suited for high-speed application than the PMA. The major difference between the two types of operation is that the core transfers power to a load while it is switching in the case of the PMA, while the transfer of power takes place when the core is saturated in the case of the SMA.

Since the core is not unilateral, unilateralization is accomplished by means of multi-phase power supplies and diodes. For example, the PMA's operate from a multiphase clock; a core that is delivering power during its read phase will deliver energy to other PMA's that are in the write phase.

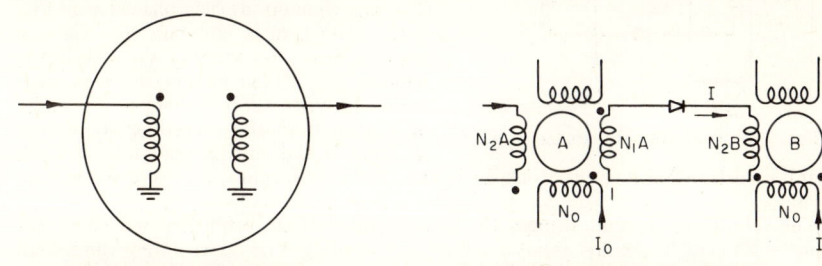

Fig. 9-38. Dot convention.

Fig. 9-39. Single-diode loop.

Before proceeding with the discussion, the dot convention for cores will be described (Fig. 9-38):

1. When a voltage makes the dot end positive or a current enters the dot end, the core is switched to the ZERO state.
2. When a core switches to the ZERO state, all core windings produce positive voltages at the dot ends, tending to drive current from the dot end.
3. When a voltage makes the nondot end positive or a current enters the nondot end the core is switched to the ONE state.
4. When a core switches to the ONE state all core windings produce positive voltages at the nondot ends, tending to drive current from the nondot ends.

PMA Logic. The PMA's are cores with windings which are essentially in parallel with the current-source power supply. A simple transfer loop called a single-diode (SD) loop is shown in Fig. 9-39. The drive currents for successive cores are fed out of phase (1,2). Assume core A is initially in the ONE state and core B is in the ZERO state (clear), as it will always be prior to writing.

Note that the action of reading a current into the dot windings N_0 of any core clears it to the ZERO state in accordance with the dot convention; thus the read action always clears the core. When core A switches to the ZERO state, current I circulates in the loop entering the nondot end, thus switching B to the ONE state, transferring the ONE from A to B. If A were in the ZERO state, it would not switch; so no current would flow and B would remain in the ZERO state. The diode is used to block the flow of current when the data are written into core A at core write time. The current entering the nondot terminal of N_2A produces a current opposite to the easy direction of the diode in the AB loop. The back-loop current (current in N_2A) when A is being read is held small by proper design. This is accomplished by making the turns N_2 smaller than N_1 since the equation $N_1\phi_1 = N_2\phi_2 = \int_0^t v\, dt$ is satisfied. Thus there is a flux amplication from A to B, and a flux reduction from A to its back loop when A is switched.

A second important loop is the conditional-transfer loop shown in Fig. 9-40, which allows a transfer of information only when a particular winding N_0, is excited. In this arrangement the drive current (read) for core A is fed through the balanced diode arrangement, when a transfer is desired. The output voltage caused by the switching of A turns off the upper diode and turns on the lower, allowing current into the nondot winding of B, causing it to switch to the ONE state. If core A is switched by any other winding than N_0, the current induced in N_1 cannot flow in N_2 because of the back-to-back diodes.

The basic primitive set most useful for PMA logic is the OR–NOT–INHIBIT set, i.e., $a + b$, a', ab'. The OR function is easily implemented by parallel single-diode transfer loops feeding the output core, or by using separate windings as in Fig. 9-41. The NOT element is accomplished as in Fig. 9-42. The core is first put into the ONE state at time t_1 by winding N_1. At time t_2 winding N_2, which is reversed from normal, will write ZERO if there is a transfer of a ONE from the preceding stage, and at time t_1 the core is read into B by a conditional-transfer loop; the output is ZERO for a ONE input, and ONE for a ZERO input.

FIG. 9-40. Conditional-transfer loop.

The inhibiting element utilizes the same configuration, requiring nonconcurrent input. Thus at t_1, the first input a is read into winding N_1, and at time t_2 the second input b is read into winding N_2. Thus a ONE is read only if the first input is a ONE and core A is not switched to ZERO by the inhibiting input b being a ONE, giving the function ab'. There are other methods using more complex core arrangements for accomplishing the INHIBIT in two clock phases, and circuits can also be constructed using parallel conditional-transfer loops.

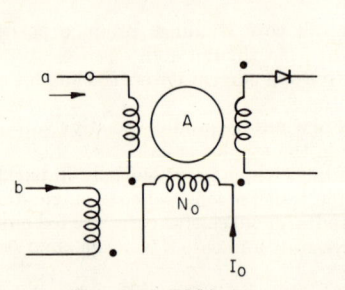

FIG. 9-41. PMA OR.

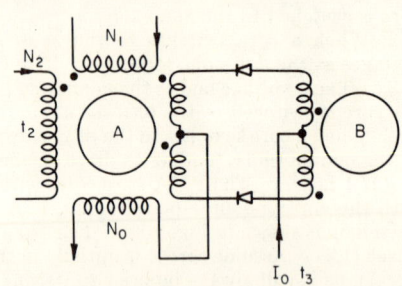

FIG. 9-42. Inverter INHIBIT.

SMA Logic. SMA's are characterized by fast response and high gain. In the amplifier, the load and the output winding are in series with a pulsed-voltage power source. These circuits are generally used as amplifiers, and diodes perform the logic. There are two basic amplifier types, the complementing amplifier (Fig. 9-43) and the asserting amplifier.

A two-phase clock is generally used, with positive and negative amplitude range. When the power pulse is positive, CR_3 conducts and an output can occur; when it is negative, CR_3 is blocked and an input can occur, and no energy can be transferred to the output by transformer action. The power pulse amplitude is chosen just large enough to switch the core from the ZERO state to the ONE state; thus the core will be always cleared to the ONE state. If ZERO input (no power) is received, the core is then in the ONE state at the start of an output period. The core is then saturated for positive pulses, presenting a low impedance to the power pulse, and a large output current flows. When a ONE is received, the core is switched to the ZERO state by the current into N_2. The next output period then switches the core back to the ZERO state, presenting a high impedance to the source; so very little current flows in the output circuit. The circuit then performs as an amplifying complementer with a delay of one clock pulse, or half a cycle. Bias current is supplied to overcome diode losses.

The basic asserting amplifier is similar to the complementing amplifier but with the windings N_2 and NB reversed with respect to the output winding, the small squares replacing the dots in the previous figure. The bias current in this case switches the core from the ONE state to the ZERO state during the input period. With a ZERO

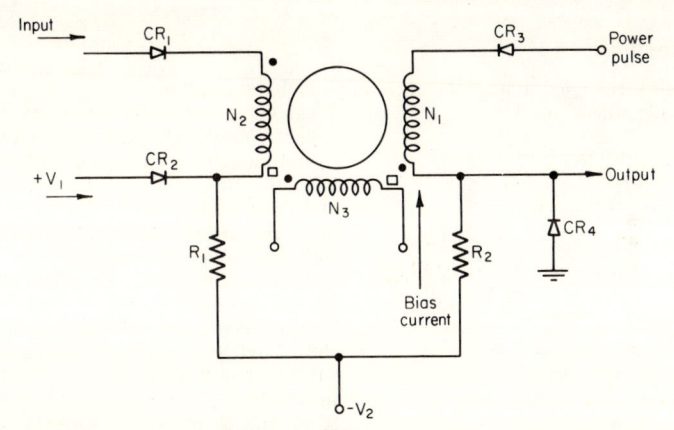

FIG. 9-43. SMA amplifier.

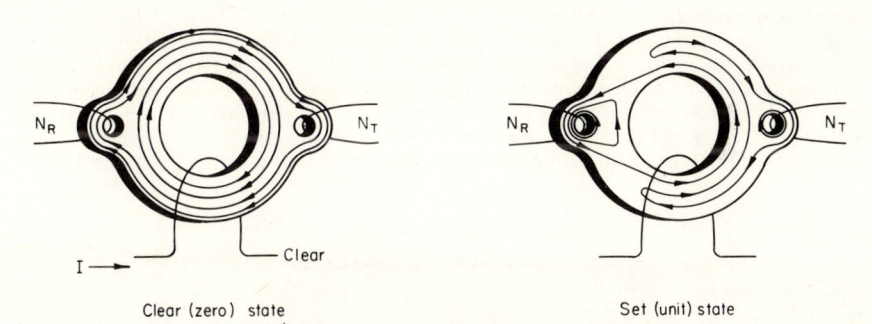

FIG. 9-44. Multiaperture device.

input, the core is in the ZERO state and there is no output. Diode CR_1 is left blocked by the voltage V_1 as the core switches to the ZERO state. A ONE input switches the core to the ONE state, overcoming the bias and holding it in the ONE state until the start of the output cycle. There is then an output during this phase.

In many SMA applications most of the logic is performed externally to the core by diode gating; such functions as AND and OR are easily realized in this way. Special logic functions such as INHIBIT are generated by variations of the basic asserting amplifier. Magnetic gating can be accomplished with SMA's, as with PMA's, but in this case the windings are placed in series between the power pulse and the load; an output will then be obtained if all coils are in the low-impedance state. Magnetic-gating circuits become complex; however, since diodes are used in the basic amplifiers in any event, diode gating seems preferable since diode costs are decreasing and reliability and performance increasing rapidly with recent technological improvement.

The MAD (Multiaperture Device). Magnetic-core circuits with single-aperture cores involve diodes for performing amplification or logic. Disks of square-loop ferrite material with holes of various shapes have accomplished the development of magnetic gating with flux redistribution. A three-hole device with geometry and windings as shown in Fig. 9-44 has the following property: When a large current I is fed into the clear winding, the flux distributes itself so that the local flux paths around the small holes are saturated for small currents and switching does not take place there. If a large *set* current is fed into the N_R winding, the flux is redistributed so that the flux around the small holes can now be switched and a switching resistance exists; also the state of the MAD will not change.

9-6. Flip-flops

A circuit often used in digital computers is called a flip-flop. This is a device which temporarily stores one bit. Such a device can be a circuit with two stable states, called a static flip-flop, or a pulse which recirculates in a loop, called a dynamic flip-flop. Flip-flops can be made with any of the component elements discussed, and form the essential parts of devices such as shift registers and counters.

Static Flip-flops. Direct-coupled. One of the simplest forms of static flip-flop is the direct-coupled transistor flip-flop shown with triggering transistors (Fig. 9-45). The flip-flop has two stable states: in one Q_1 is on and Q_2 is off; in the other Q_2 is on and Q_1 is off. The on transistor, Q_1 say, conducts so that it is saturated; the collector of Q_1 is approximately 0.3 volt, which is low enough to cut off Q_2. The base of Q_1 draws current through R_2 so that its potential, and that of the collector of Q_2, is approximately 0.8 volt; the collector-base saturation voltage of Q_1 is approximately 0.5 volt. A positive-voltage trigger applied to the base of Q_4 draws current through R_2, dropping the potential of the base of Q_1 so that it becomes cut off, which in turn keeps Q_2 turned on. The flip-flop will stay in this state until a trigger is applied to the set terminal Q_3. The collector voltages of Q_1 and Q_2 are complementary static outputs, in that a voltmeter on the collec-

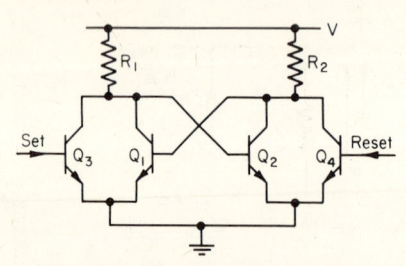

Fig. 9-45. Direct-coupled flip-flop.

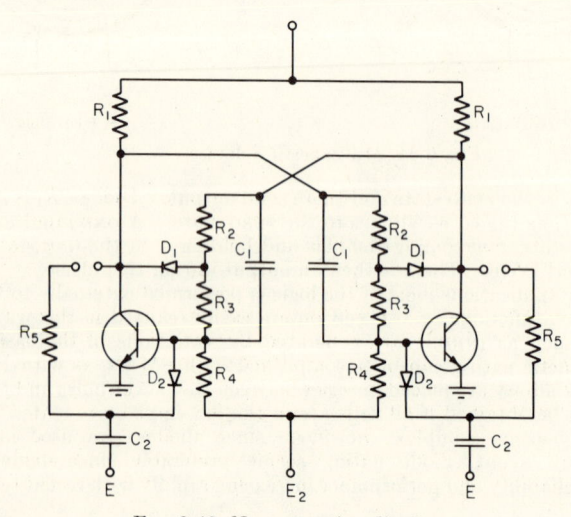

Fig. 9-46. Nonsaturating flip-flop.

LOGICAL CIRCUITS 9-23

tors can indicate the condition of the flip-flop without changing the state. This is in contrast to a magnetic core, where the flux in the core must be changed in order to read its state.

Nonsaturating Flip-flop. Figure 9-46 illustrates the connections for a nonsaturating-type flip-flop which is accomplished by the use of the diode D_1 and resistor R_3. In this case the collector swing is a large portion of the voltage, since the drop in the collector voltage of the off transistor must come only from the resistor divider (R_1, R_2, R_3, R_4) current. The divider values are chosen so that only one transistor is on, and the other is cut off. Tolerance on transistor parameters makes the design difficult to attain for all conditions, however, and there are a considerable number of parts. The triggers are fed to the bases through RC diode gates D_2, R_5, C_2.

Tunnel-diode Flip-flop. A tunnel diode is a bistable static element, but since it is a two terminal device, isolation must be provided. Figure 9-47 illustrates a workable circuit. A hold-on current is fed into the diode lower than the peak current. When the diode is in the low-voltage state the transistor is off. A positive (set) trigger will

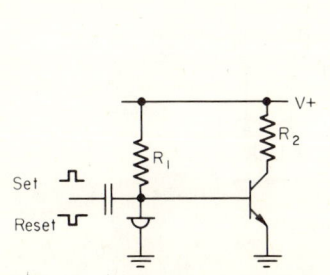

FIG. 9-47. Tunnel-diode flip-flop.

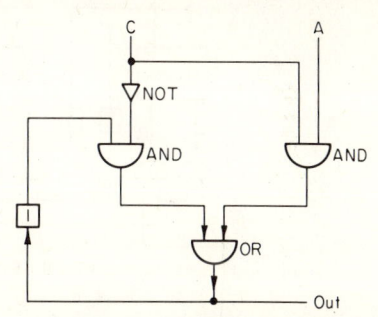

FIG. 9-48. Dynamic flip-flop logic.

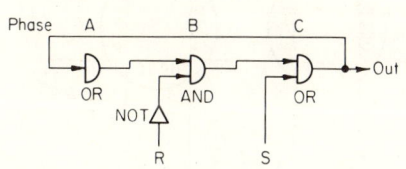

FIG. 9-49. Tunnel-diode dynamic flip-flop.

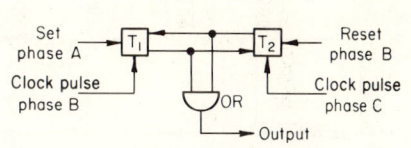

FIG. 9-50. Magnetic flip-flop.

switch the diode to the high-voltage state and it will stay there. A negative trigger reduces the current, and the diode is then reset to the low-voltage state.

Dynamic Flip-flops. A dynamic flip-flop is obtained by recirculating a ONE or ZERO in a delay line of length equal to one clock cycle. The delay line can be logically replaced by utilizing temporary storage during another clock phase, as in the magnetic flip-flop. Figure 9-48 shows the logic of a dynamic flip-flop.

In the figure, a ONE occurring at the signal point A at the same time as a clock pulse at C places a ONE in the recirculating loop. The output will continue to be a ONE with no clock or signal inputs unless the state is changed to ZERO at the time another clock pulse appears. Whenever a clock pulse appears at point C, the flip-flop takes on the signal condition at that time. Amplification is usually provided to overcome the attenuation in the delay loop.

A tunnel-device flip-flop using essentially the same logic but with different clock phases to achieve the delay and isolation is shown in Fig. 9-49. The elements used are tunnel-diode logic elements providing gain, with three clock phases A, B, C. The one-input OR used in phase A takes the place of the unit delay. S and R are the set and reset inputs. The same logical diagram could be used with magnetic devices such as SMA's as the elements.

A magnetic dynamic flip-flop, or ping-pong, using two cores is diagrammatically shown in Fig. 9-50. A ONE set into T_1 at clock phase A is transferred to T_2 at clock phase B by the clock pulse and is transferred back to T_1 at clock phase C. An overriding reset signal injected at phase B can clear the flip-flop. The recirculating ONE can give a level output ORing the two core outputs and rectifying the sum.

9-7. Shift Registers

Shift registers are cascades of flip-flops whose contents can be shifted from one flip-flop to its neighbor. When a shift pulse appears, each flip-flop takes the state that the previous flip-flop had, subject to modification by set and reset inputs.

The basic static flip-flop with set and reset inputs plus the addition of AND gates and small delays, usually introduced by RC elements, forms the static shift-register element, as in Fig. 9-51. ONE input into S makes the right-hand output A, B, C a ONE; and a ONE into R makes A, B, C a ZERO. Therefore, if a ONE is in any flip-flop, a

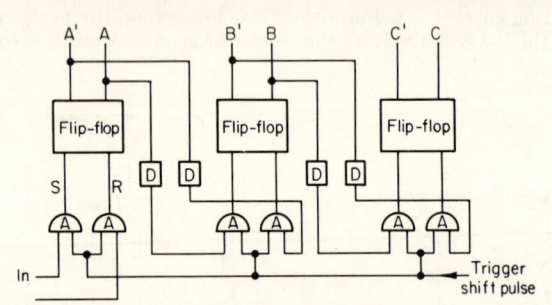

FIG. 9-51. Shift register.

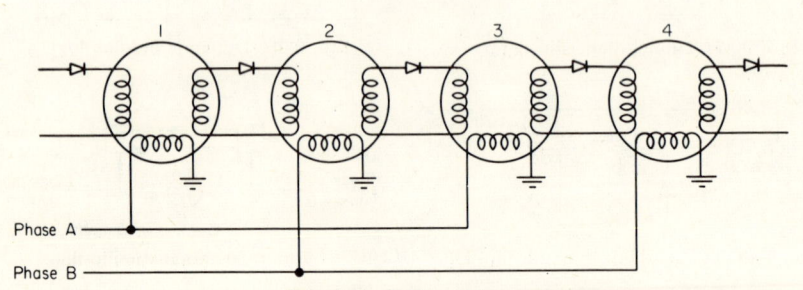

FIG. 9-52. Shift register.

shift pulse enters the S input through the AND gate of the next flip-flop to the right and makes that flip-flop a ONE; if there is a ZERO in the flip-flop it is transferred through the R input. The delay allows the flip-flop to change state without altering the effect of the ANDing of the previous state. The contents of the register then shift to the right.

Dynamic flip-flops can be used with similar logic to produce shift registers. Magnetic elements have long been used in shift registers. A simple PMA using two cores per bit is shown in Fig. 9-52. The phase A shift pulse clears alternate cores and reads the contents of 1 and 3 into 2 and 4. The phase B shift pulse clears 2 and 4 and reads the contents into 3 and 5; thus the contents are shifted to the right. Magnetic elements with delay components have been employed to obtain shift registers with one core per bit.

9-8. Counters

Counters are used for numerous applications such as combining, totaling, or producing an output after a number of events have occurred. There are many ways to obtain the counting function; the choice depends on components and function.

LOGICAL CIRCUITS 9–25

For certain uses a shift register with output connected to its input makes a suitable counter; this is one form of ring counter. A ONE is inserted into the input and is circulated by the shift pulse or counting input. For a magnetic shift-register counter, such a technique is a simple extension of the ping-pong. A ONE circulating in such a manner will produce a multiphase output, in that each state of the counter becomes ONE in a cyclic sequential fashion. A ring counter using magnetic elements in a ring of three is shown in Fig. 9-53.

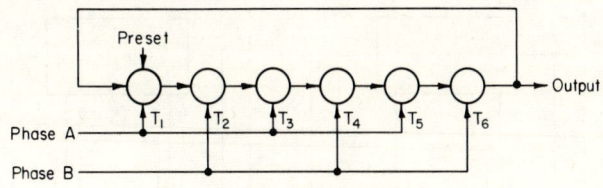

FIG. 9-53. Ring counter.

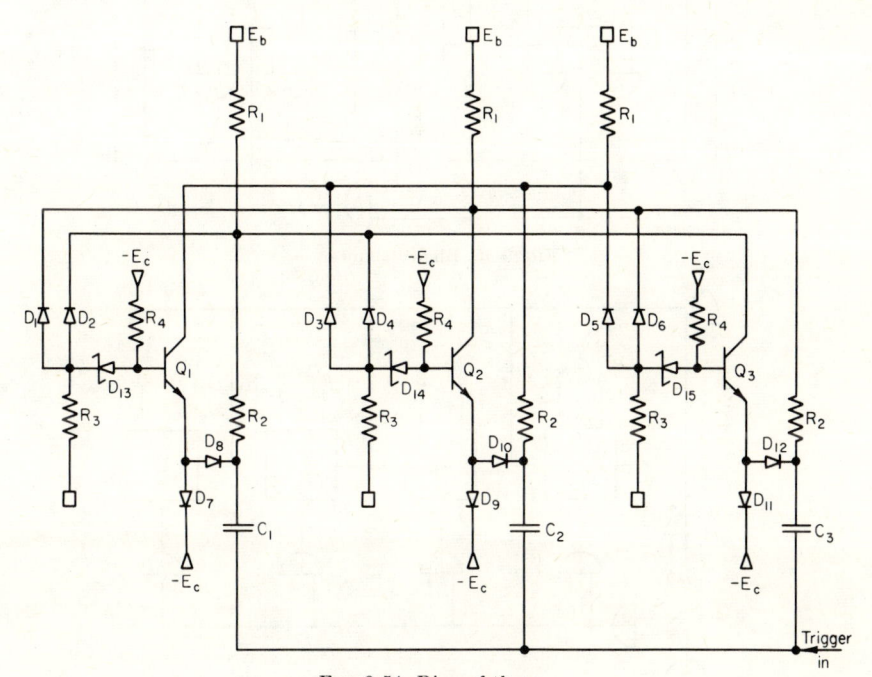

FIG. 9-54. Ring of three.

Ring counters can also be made with transistor shift registers; a very useful transistor counter employs one transistor per bit in static multistate logic. Figure 9-54 shows a ring of three; the transistors are interconnected so that only one transistor in the ring is allowed on at a time; when that transistor is on all others are cut off. A trigger pulse turns off the on transistor, which is arranged to then trigger on the next transistor in succession.

A flip-flop can be connected as a binary counter. When it is connected as shown in Fig. 9-55, the state of flip-flop A will change whenever a trigger pulse occurs. Whenever the state of A is a ONE, a flip-flop B can be triggered.

If all flip-flops of an N-stage binary counter were initially set to ZERO, the last stage would become ONE after 2^{N-1} input pulses, and all stages would be ONE after $2^N - 1$

input pulses. A counter connected in this fashion is counting forward, in that any number initially set in the counter will increase with input pulses entering. If the gating between stages is taken from the complementary output of the previous stage, the counter will decrease its total with increasing pulses. A forward-backward counter can then be made by utilizing a version of this concept. The cascading of

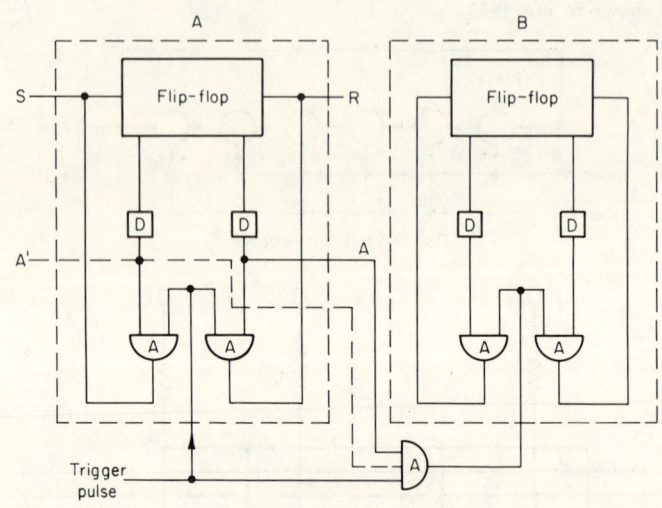

Fig. 9-55. Binary counter.

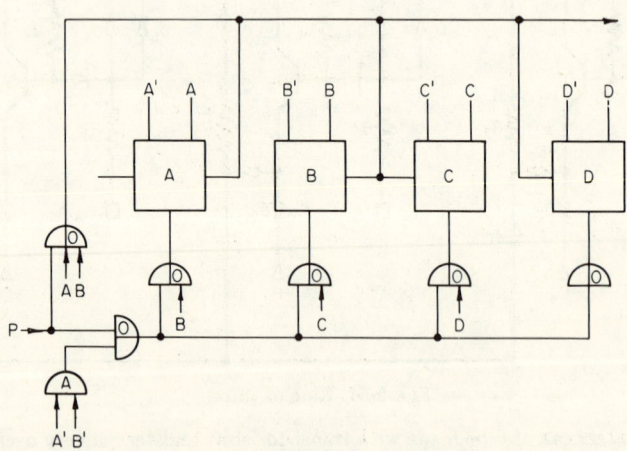

Fig. 9-56. Decimal counter.

stages of such a binary counter can also be accomplished by using the differentiated output of one stage to trigger the next stage, thus eliminating the additional interstage gating.

If it is desired to produce an output after $M = 2^N$ events, gating is performed to detect the desired number, generate the signal, and reset the counter. Figure 9-56 shows a decimal counter which counts to ten. The logic is made very simple by choosing the initial starting count of 0011 (three). The counter counts to binary twelve (1100) which happens after nine input pulses and is then reset back to three on the tenth pulse. The tenth pulse is also inhibited from furthering the stored count. Flip-flops with internal gating (block A, Fig. 9-55) are used; they have a trigger input, a set input, and a reset input, and delayed complementary outputs. Only after a count of nine is $AB = 1$ or $A' + B' = 0$. The count is then inhibited using the OR and AND gates shown. The ANDing of P, A, B also allows the tenth pulse to produce an output and reset the register to 0011.

9-9. Logic Packages

Modules. In computer system or subsystem design it is of advantage to utilize logic packages; these are generally standardized for a given computer. The system engineer today has conveniently available a wide choice of logic modules with which he can assemble any computer function for laboratory evaluation, for special test equipment, etc. These are usually in the form of transistor and diode-transistor logic.

The basic logic elements generally include inverter packages which can be used for gating or gain, diode transistor gates such as the NOR circuit, and a general-purpose flip-flop. Various delay devices (such as one-shot pulse delays and delay lines), shift registers, pulse generators, clocks, and level standardizers can also be obtained. A standard package of a given function will be designed for a certain maximum speed capability. The presently available capabilities vary from 100 Kc to 50 Mc.

A module such as an inverter function will contain a number of inverters (five or more), with all terminals accessible on the module. Thus the inverters can be arranged in numerous ways merely by interconnecting the accessible terminals.

Micrologic. Many companies now offer complete logic packages which are very small in size. In fact, such functional elements as flip-flops, half shift registers, half adders, gates, and buffers are available in a package no larger than a transistor case. These elements are integrated circuits in which transistors and resistors are formed via a basic process on a single semiconductor wafer.

BIBLIOGRAPHY

Meyerhoff, Albert: *Digital Applications of Magnetic Devices*, John Wiley & Sons, Inc., New York, 1960.
Hurley, Richard B.: *Transistor Logic Circuits*, John Wiley & Sons, Inc., New York, 1961.
Ledley, Robert S.: *Digital Computer and Control Engineering*, McGraw-Hill Book Company, New York, 1960.
Renton, C. A., and B. Rabinovici: Tunnel-diode Full Binary Adder, *IRE Trans. Elec. Computers*, vol. EC-11, pp. 213–218, April, 1962.
Lewin, Morton H.: Negative Resistance Elements as Digital Computer Components, *Proc. Eastern Joint Computer Conf.*, December, 1959, pp. 15–27.
Ruegg, H. W., and R. H. Beeson: New Forms of "All Transistor" Logic presented at International Solid State Circuits Conference, Philadelphia, Pa., Feb. 14, 1962.
G.E. *Transistor Manual.*
G.E. *Tunnel Diode Manual.*

Chapter 10

ANALOG ELEMENTS

WALTER J. KARPLUS, *Department of Engineering, University of California, Los Angeles.*

CONTENTS

10-1. Introduction	10-1
Distinction between Analog and Digital Systems	10-1
Types of Analogs	10-2
10-2. Electronic Computing Elements	10-4
The Operational Amplifier	10-4
Multiplication by a Constant	10-7
Addition-Subtraction	10-8
Integration-Differentiation	10-8
Complex Transfer Functions	10-8
Multiplication-Division	10-9
Generation of Functions of the Dependent Variable	10-11
Generation of Functions of Time	10-14
10-3. Mechanical Computing Elements	10-16
Addition and Subtraction	10-16
Integration	10-16
Multiplication by a Constant	10-18
Multiplication of Two Variables	10-19
Function Generation	10-19

10-1. Introduction

Distinction between Analog and Digital Systems. The field of automatic computation and simulation includes a wide variety of devices and techniques which are designed to assist the engineer and the scientist in the numerical solution of mathematical problems and in the analysis, design, and control of physical systems. A basic and useful method of classifying these devices involves the recognition of the manner in which the computer or simulator handles the data or other information being processed. In all computing and simulating devices, quantitative information, the given information of the problem, is applied to the machine. The computer then performs various operations on these data. In some machines, this information is expressed everywhere within the computer in terms of discrete numbers having a specified number of significant figures. Such machines, described in Chapters 8 and 9, are termed *digital* devices. In the other major type of computing system all information is handled in a continuous fashion so that the determination of a variable within the system involves physical measurement rather than counting. These devices are known as *analog* computers or simulators. Common examples of digital computers include the abacus, the desk calculator, as well as such modern machines as the IBM 7094 and the UNIVAC; slide rules, planimeters, and mercury thermometers are analog devices.

From this basic distinction follow a number of important differences between analog- and digital-computing systems. The precision, or number of meaningful significant figures in a solution furnished by analog computers, is determined by the precision with which the continuous variable constituting the data can be sensed and displayed. This, in turn, is related to the quality of the components used in constructing the computer, for example, the tolerance of electrical resistors or mechanical shafts, and the quality of the output equipment. On the other hand, the precision of a digital computer is governed entirely by the number of significant figures carried in the computations. This, in turn, is primarily determined by the size of the digital registers. The precision of digital computers is essentially unlimited, for in order to increase its precision, it is only necessary to enlarge the installation, to provide more toothed wheels or larger electronic memory units. Many digital machines carry over 30 binary digits. By contrast, it is difficult to obtain a precision in excess of 0.05 to 0.1 per cent on an analog device, since this requires electric or mechanical components having extremely exacting tolerances.

For many engineering applications solutions having accuracies of the order of 0.5 per cent are adequate. Under these conditions an analog computer is frequently preferred for a number of reasons, including particularly cost, speed, and convenience of operation. As a direct consequence of the discrete form of data within a digital computer, memory registers play an important part in every digital computation. Such memory devices tend to be expensive; so that most digital computers are designed around one large central memory unit. This unit is time-shared throughout the problem. The computations necessary to obtain a desired solution are arranged in a preprogrammed sequence, and the central memory is utilized in turn for each of these arithmetic operations. This is termed *serial* operation and implies that the more complex a problem, the more calculation steps that are required, the longer the solution time. By contrast an analog computer consists of an assemblage of computing units. A separate and distinct device is used for each mathematical operation (addition, integration, multiplication, etc.), and these units all operate simultaneously to generate the solution. This is termed *parallel* operation. As problem complexity increases, equipment requirements increase, but the solution time remains unchanged. Analog computers are therefore better adapted for automatic-control applications in which real-time operation is essential.

The fact that in an analog computer the solution of the problem becomes available almost immediately permits the design engineer to adjust or to vary any of the design parameters and to observe at once the effect of these variations upon the dynamic response of the system and the study. Furthermore on an analog computer each subunit or component has a direct significance in terms of the system under analysis. This makes the analog computer more conducive to "experimental engineering." By programming the computer and manipulating various controls, for example, by varying electrical potentiometer settings, the engineer frequently is able to gain an insight into the basic operation of the system under study. By contrast, in the case of digital machines the mathematical operations are performed as a sequence of arithmetic operations which generally bear no direct relationship to the system under study.

Types of Analogs. The various devices comprising the general area of analog computers and simulators are best classified according to their basic principles of operation. The systems falling into the resulting categories are subdivided, in turn, according to the type of physical variables which constitute the continuous data within the computer.

One major class of analog devices depends for its operation upon the existence of a direct physical analogy between the analog and the prototype system under study. Such an analogy is recognized by comparing the characteristic equations describing the dynamic behavior of the two systems. An analogy is said to exist if the characteristic equations are similar in form. Such a similarity is possible only if there is a one-to-one correspondence between elements in the analog and in the prototype system. For every element in the original system there must be present in the analog system an element having similar properties, that is, an element having a similar excitation-response relationship; members of this category of analog devices are termed *direct* analogs and are also frequently referred to as *simulators*. Direct analogs may be of either the discrete or the continuous variety. Discrete analogs employ lumped physical elements such as electrical resistors and capacitors interconnected in a one-, two-, or three-dimensional array. Other simulators make use of distributed elements such as sheets of conductive liquids (electrolytic tanks) or electrically conductive papers. The characteristics and applications of direct analogs are considered in some detail in Chapters 11 and 32.

The other major class of analog systems includes mathematical rather than physical analogs. The behavior of the system under study, or the problem to be solved, is first expressed as a set of algebraic or differential equations. An assemblage of computing units or elements, each capable of performing some specific mathematical operation, such as addition, multiplication, or integration, is provided, and these units are interconnected so as to generate the solution to the problem. The resulting computing systems are termed *indirect analogs* or more commonly analog computers.

Indirect analogs are employed principally to solve problems involving algebraic and ordinary differential equations. Computers capable of solving differential equations are frequently referred to as *differential analyzers*. Before World War II mechanical differential analyzers, in which the dependent problem variable is represented by shaft rotation, were in wide use. Since World War II these have been replaced almost entirely by electronic differential analyzers in which the dependent variable is an electric voltage, though mechanical computing devices continue to play an important role in certain special-purpose applications.

Electrical computing elements and electronic computing systems owe their wide acceptance to the convenience that attends their construction and use. Electrical components, particularly resistors and capacitors, having an extremely wide range of magnitudes are mass-produced and are readily available with tolerances as small as 0.01 per cent. Electric circuits can be wired and assembled by relatively untrained personnel, are light in weight and compact, and do not require expensive maintenance. The two dependent variables in electric circuits are voltages across the circuit elements or from node points to ground and the currents through each element. Voltages are used almost exclusively as the computing variable in electric analogs because these can be measured and recorded at any point in the electric circuit without circuit modification. Both a-c and d-c systems can be used in electronic analog computing. A-c analog computers are generally less expensive to build and require less complex auxiliary devices. On the other hand, the errors attending phase shift tend to make this type of computer less accurate than the corresponding d-c computer. For this reason, most large-scale general-purpose installations use d-c voltages as the problem variables.

The d-c electronic differential analyzers now available can in turn be classified into two categories: high-speed and low-speed. High-speed computers have solution times of the order of 5 msec and repeat this solution periodically to facilitate the display of the output transients by means of cathode-ray oscilloscopes. These devices are also termed *repetitive* analog computers. Slow electronic analog computers have solution times ranging from 10 sec to 3 min and employ oscillographs or other graphical output equipment. Such systems are referred to as *one-shot* computers. Repetitive devices lend themselves more readily to certain optimization problems and studies involving random noise and statistical variations. On the other hand, the components employed in constructing a repetitive computer must have wider bandwidths and therefore tend to be more expensive for a given precision than one-shot devices. In recent years most manufacturers of slow analog computers have made provision for repetitive

SYSTEM COMPONENTS

operation. Another recent development is the trend to hybrid systems, employing interconnected analog and digital elements (see Secs. 11-9 through 11-16).

The general classification of analog methods is illustrated diagrammatically in Fig. 10-1. The material presented in the following paragraphs constitutes a survey of the electrical and electronic computing elements and units which form the building blocks of an indirect electronic computer. Mechanical computing elements are

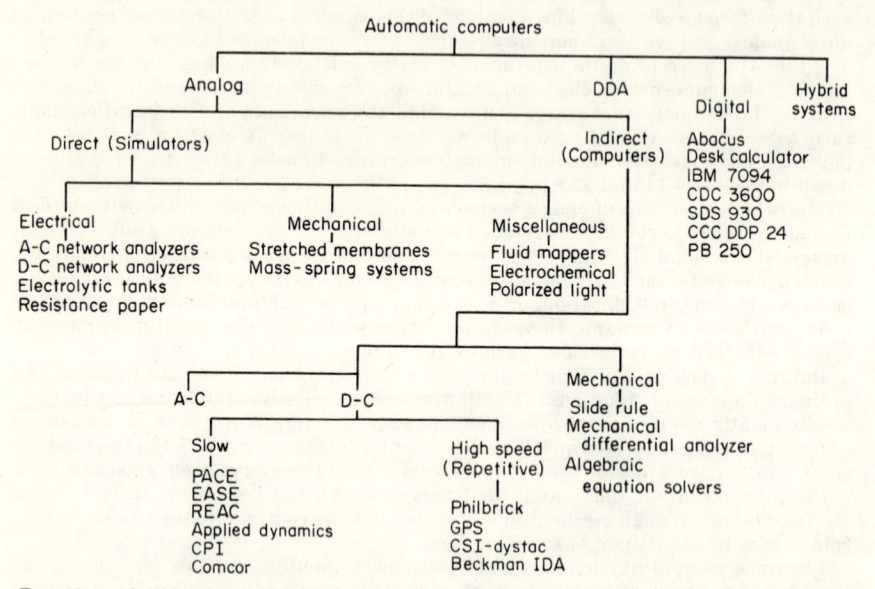

```
                        Automatic computers
         ┌──────────────────┼──────────────┬──────────┐
       Analog              DDA           Digital    Hybrid
         │                                          systems
   ┌─────┴─────┐                         Abacus
 Direct      Indirect                    Desk calculator
(Simulators) (Computers)                 IBM 7094
   │           │                         CDC 3600
                                         SDS 930
                                         CCC DDP 24
                                         PB 250
```

Direct (Simulators):
- Electrical: A-C network analyzers, D-C network analyzers, Electrolytic tanks, Resistance paper
- Mechanical: Stretched membranes, Mass-spring systems
- Miscellaneous: Fluid mappers, Electrochemical, Polarized light

Indirect (Computers):
- A-C
- D-C
  - Slow: PACE, EASE, REAC, Applied dynamics, CPI, Comcor
  - High speed (Repetitive): Philbrick, GPS, CSI-dystac, Beckman IDA
- Mechanical: Slide rule, Mechanical differential analyzer, Algebraic equation solvers

FIG. 10-1. Classification of analog computers. (*Most of the illustrations in this chapter are adapted from W. J. Karplus and W. W. Soroka, Analog Methods: Computation and Simulation, 2d ed., McGraw-Hill Book Company, New York, 1959.*)

reviewed in Sec. 10-3. Various applications of these devices to the solution of engineering problems are considered in Chapter 11. More detailed treatments of these topics will be found in the references.[1-4]

10-2. Electronic Computing Elements

The Operational Amplifier. The heart of the modern electronic differential analyzer is the d-c operational amplifier. Although many mathematical operations can be performed by combinations of passive circuit elements (for example, the resistance-capacitance integrator), such devices are capable of providing accurate results only in the absence of a load. As soon as it becomes necessary to draw a current from a passive analog-computing circuit, the required operation is no longer performed accurately. Since the operation of an analog computer demands the interconnecting and cascading of individual operational units, no-load operation is impossible. It therefore becomes necessary to introduce active elements in the form of operational amplifiers into the circuit.

In order to permit the performance of the desired mathematical operations with sufficient accuracy and precision, the electronic amplifiers used in the computing process must have the following general properties:

1. High gain; generally in excess of 100,000,000
2. Linearity over a wide region of operation; generally from -100 to $+100$ volts at the output

3. Flat frequency response; generally from direct current to several hundred cycles per second but sometimes up to several kilocycles
4. Zero output voltage for zero input voltage
5. Very high input impedance
6. Reversal in polarity between the input and output of the amplifier
7. Very low noise level

The severity of the above requirements, particularly regarding linearity and frequency response, delayed the realization of practical general-purpose electronic analog computers until after World War II. Since then a large number of modifications of the basic d-c amplifier circuit have been introduced. In order to permit operation at direct current it is necessary either to employ direct coupling between successive amplifier stages or to employ elaborate switching circuitry to apply appropriate initial charges to the coupling capacitors. If direct coupling is employed it is necessary to supply a positive as well as a negative high-voltage power supply.

The main source of error attending the use of d-c operational amplifiers is their tendency to *drift*. Prior to use as a computing element the amplifier must be carefully "balanced" by connecting the input terminal to ground and adjusting a bias within the amplifier to give zero output voltage. If this balance setting is left unchanged and the input terminal is maintained at zero, the output terminal will be found to deviate or drift gradually away from zero. The causes for this phenomenon

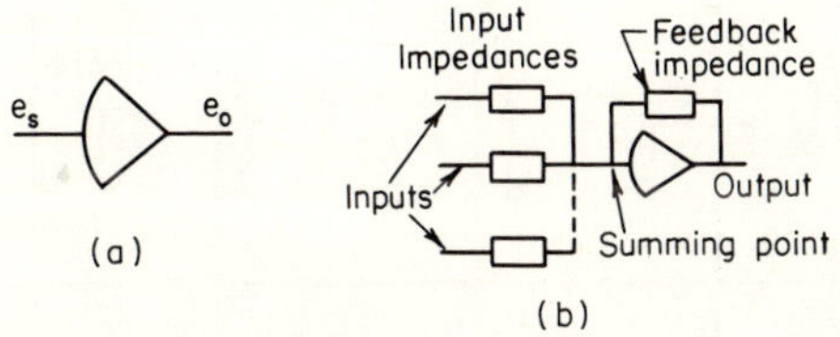

FIG. 10-2. (a) Symbol for d-c operational amplifier. (b) Usual manner of connection.

include variations in the power supply and filament voltages as well as transient variations in the properties of the circuit components. Much effort has gone into the design of operational amplifiers with a minimum of drift errors. The most successful and widely used devices of this type employ so-called *chopper stabilization*. This method is based on the recognition that an a-c amplifier is not subject to drift in the same manner as a d-c amplifier. Accordingly a separate a-c amplifier and an electromechanical vibrator are provided. The voltage appearing at the input of the d-c amplifier is transformed to an alternating signal by means of the vibrator and is applied to the a-c amplifier. The output of the a-c amplifier is therefore a signal whose amplitude is proportional to the voltage at the d-c amplifier input. This output is rectified and filtered to transform it back into d-c voltage and is then applied as a correcting signal to the d-c amplifier chain. The frequency response of the chopper amplifier is such that it reacts to the slow-drift voltage variations but not to the variations of the computing-signal voltages. Another source of error in electronic operations is the grid current drawn by the input stages of the operational amplifier. This can be minimized by the use of special circuits in the first amplifier stage.

The conventional symbol used to denote an operational amplifier is shown in Fig. 10-2a. As in most analog circuit diagrams the ground bus is omitted, it being understood that all voltages in the circuit diagram are referred to the computer ground terminal. The output voltage is equal to the input voltage times a large negative constant. The manner in which operational amplifiers are used in analog-computer circuits makes it unnecessary to specify the magnitude of this constant provided only that it is very large. To fashion analog-computing circuits the operational amplifier is generally connected as shown in Fig. 10-2b. A circuit element, usually a resistor or a capacitor, is connected from the output terminal to the input terminal, often referred to as the *summing point*. A number of other circuit elements, generally fixed resistors,

SYSTEM COMPONENTS

Table 10-1. Basic Electronic Analog-computing Elements

Computer operation	Input e_i	Output e_o	Detailed schematic	Compact schematic
1. Constant multiplication	$f(t)$	$kf(t), k \leq 1$ $-100 < k \leq 0$		
2. Addition	$f_i(t)$	$-\sum_{i=1}^{n} k_i f_i(t)$		
3. Integration	$f_i(t)$	$-\int_0^t \left(\sum_{i=1}^{n} k_i f_i(t) \right) dt + IC$		
4. Complex transfer functions	$f(s)$	$-\dfrac{Z_f(s)}{Z_i(s)} f(s)$		
5. Multiplication	$f(t), g(t)$	$kf(t)g(t)$		
6. Function generation	$g(t)$	$f[g(t)]$		
7. High-gain amplification	$f(t)$	$-gf(t)$		

ANALOG ELEMENTS

are connected to the summing point, and the variable voltages consituting the input to the mathematical operation to be performed are connected at the other terminals of these input impedances. In employing a circuit of the type shown in Fig. 10-2b care must be taken that the specified voltage range of the amplifier not be exceeded. Most frequently, operational amplifiers have a range from -100 to $+100$ volts. Should an output voltage outside this range be demanded in the course of a calculation, serious errors will be committed by the computer unit. To help the computer operator to recognize this eventuality, most analog-computer amplifiers are equipped with overload indicators—neon lights which indicate such an overload condition. One of the major problems in analog-computer programming is the *scaling* of the dependent variables to be handled on the computer to obviate any amplifier overloading. To this end constant multipliers termed *scale factors* are assigned to all dependent variables of the problem.

The most important and widely used electronic analog-computing elements are presented in Table 10-1. Two sets of schematic symbols are shown. The first set used for detailed presentation includes the indication of all input and feedback elements; the second set is useful for the compact representation of complex computer systems.

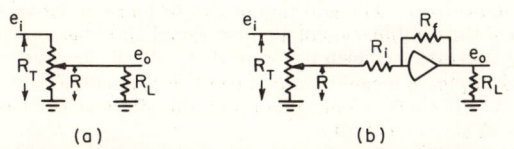

Fig. 10-3. (a) Basic potentiometer circuit. (b) Addition of amplifier to reduce loading error.

Multiplication by a Constant. The basic component for the multiplication of variable electric voltages by a constant is the potentiometer. A potentiometer is a resistance element having terminal connections at each end and a sliding contact which is capable of traversing this element over its entire length, thereby permitting variation in the resistance between the sliding contact and one of the fixed terminals. The principal characteristics which determine the quality of such an element are the linearity and the resolution. Potentiometer linearity deviation is the maximum deviation from the best straight line which can be drawn through the points of a graph of resistance vs. angular position of the sliding contact. This deviation is generally expressed as a percentage of total resistance. The resolution of a potentiometer refers to the accuracy with which any shaft setting can be obtained. Since most potentiometers are of the wirewound type, a graph of resistance vs. rotation has the shape of a staircase rather than a continuous curve. The number of steps, and hence the resolution of the element, is determined by the total number of turns of wire used in winding the potentiometer. Precision potentiometers with linearity tolerances as close as 0.01 per cent, multiturn construction for high resolution, and an extremely low noise level are mass-produced for analog applications.

The elementary multiplication circuit is shown in Fig. 10-3a. The output voltage e_o is to be the product of the input voltage e_i and the angular position of the potentiometer R/R_T. As the result of the presence of the loading resistor R_L, however, this operation is subject to a *loading error* which is a function of the relative magnitudes of R_L and R_T as well as the angular position of the potentiometer. The deviation Δ of e_o from the geometrical setting $e_i(R/R_T)$ is

$$\Delta = \frac{R}{R_T} e_i \left[1 - \frac{1}{1 + (1 - R/R_T)(R/R_L)} \right] \qquad (10\text{-}1)$$

Evidently for accurate calculations it is desirable that R_L be as large as possible. Since it is often necessary to employ potentiometers to drive other analog devices having a relatively low input impedance, it is expedient to isolate the potentiometer from the load by means of an operational amplifier as shown in Fig. 10-3b. Provided that the operational amplifier has the characteristics enumerated in the preceding paragraph and provided that R_i is sufficiently large, it can be shown that the output voltage of this circuit is

$$e_o = - \frac{R}{R_T} \frac{R_f}{R_i} e_i \qquad (10\text{-}2)$$

If the feedback resistor of the amplifier R_f and the input resistor R_i are equal in magnitude (generally of the order of 1 megohm) the output voltage will be determined entirely by the setting of the potentiometer. Note, however, the sign change introduced by the operational amplifier. By making R_f larger than R_i multiplication by negative constants greater than unity can readily be attained.

In order to obviate any error resulting from the loading of the potentiometer by the input resistor R_i, the more elaborate of the modern electronic analog computers are generally equipped with special potentiometer-setting circuits which automatically compensate the loading of the potentiometer. In other applications potentiometers are occasionally supplied with additional taps to permit the connecting of compensating resistors.

Addition-Subtraction. The addition of two or more variables is readily accomplished by means of the combination of an operational amplifier, one feedback resistor, and several input resistors as shown in Table 10-1. If all the resistors have the same magnitude, for example, 1 megohm, the output is the simple sum of all the input voltages times -1. If there is only one input the effect of the circuit is merely to produce a change in sign or polarity.

To subtract one variable voltage from another, the sign of one of the voltages is first reversed, using a sign changer, and is then added to the other voltage by means of a simple adding circuit.

Integration-Differentiation. In order to apply the d-c operational amplifier to the mathematical operation of integration with respect to time, the addition circuit is modified by replacing the feedback resistor by a capacitor, in one-shot computers usually of the order of magnitude of 1 μf. As indicated in Table 10-1, the output voltage is then equal to the negative of the integral with respect to time of the sum of the input voltages. The constant of integration, the output voltage at time $t = 0$, must be applied to the integrator as a separate voltage excitation. Effectively the feedback capacitor is charged to a voltage corresponding to the initial condition, and at time $t = 0$ this excitation is abruptly removed from the circuit by means of a relay.

The operation of differentiation can be accomplished by employing an input capacitor and a feedback resistor. Differentiation circuits suffer from noise difficulties. While any noise, for example, electrical static or power-line interference, at the input terminals of an integrator will tend to be smoothed out and removed in the process of integration, the opposite is true in differentiators. The derivative of a variable function is the rate of change or slope of the function. Any small abrupt variations in the input voltage are therefore greatly magnified in the process of differentiation and may actually tend to mask completely the desired derivative. For this reason every effort is made in employing analog computers to avoid the use of differentiators and to use integrators wherever possible. In the solution of most differential equations, this is possible by a suitable rearrangement of the original equations. This technique is discussed in some detail in Chapter 11.

Complex Transfer Functions. The feedback and input elements of computing circuits utilizing operational amplifiers need not necessarily be limited to single resistors or capacitors. In general, both the input and feedback impedances may be four-terminal networks. As a simple example consider the operational circuit shown in Fig. 10-4. Application of Kirchhoff's law to the summing-point node leads to an expression for the ratio of the output to the input voltage as

$$\frac{e_o(s)}{e_i(s)} = - \frac{z_f(s)}{z_i(s)} = - \frac{R_i}{sC} \frac{(1 + 2sCR)(1 + R_i C_i s)}{sCR} \qquad (10\text{-}3)$$

ANALOG ELEMENTS

where s is the Laplace-transform parameter. Tables listing the transfer ratios of a wide variety of circuits of the type shown in Fig. 10-4 may be found in the references.[1-3]

By using complex input and feedback networks in combination with one operational amplifier, it is possible to obtain functional relationships between input and output voltages which would otherwise require the use of a number of amplifiers. In addition to the economy in operational amplifiers, difficulties attending complex feedback loops can also be avoided by the use of complex transfer impedances. On the other hand, the utilization of complex networks may demand capacitors of odd magnitudes not readily available. An additional problem is the application of appropriate initial conditions to the capacitors comprising the networks.

Multiplication-Division. Frequently it is necessary in the treatment of engineering problems to multiply two quantities, both of which vary with time. Since the basic potentiometric multipliers described above require the manual setting of the sliding arm of a potentiometer, this method can be used only to multiply a variable voltage by a constant. The multiplication of two variables requires additional equipment and technique. Although well over 25 different analog multiplication methods have been proposed and tried in past years, only a small number of these have achieved any considerable importance in general-purpose analog computation. In this paragraph only the three most important of these will be briefly reviewed. More detailed description of these and other methods can be found in the references.[1,2]

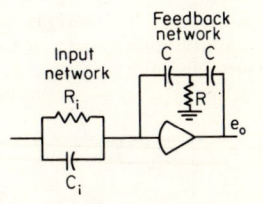

Fig. 10-4. Circuit for generating complex transfer function.

Automatic potentiometric multipliers constitute one of the oldest and most widely used methods of electrical function multiplication. A combination of the following instruments is employed: an operational amplifier, a servoamplifier, a two-phase induction motor, and several multiturn potentiometers the shafts of which are ganged and coupled mechanically to the rotor shaft of the motor. Figure 10-5 is a schematic diagram of such a multiplier. The dotted line indicates mechanical but not electrical connection. One of the variables to be multiplied, e_1, is applied to one input terminal

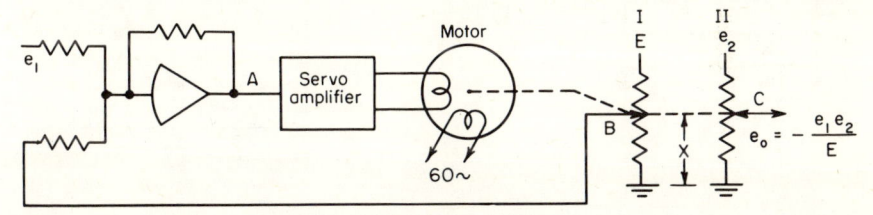

Fig. 10-5. Servomultiplier.

of an electronic adder, while the voltage appearing at point B in the diagram is applied to another input terminal of the same operational circuit. Point B is at the moving arm of a multiturn potentiometer I, one end of which is grounded while the other end is connected to a fixed d-c voltage source E, the polarity of which is opposite to that of e_1. The voltage appearing at B is therefore equal to XE where X is the position of the potentiometer arm as a fraction of full scale. The signal appearing at the output of the operational amplifier at point A is therefore equal to $(-e_1 - XE)$. The servo-amplifier effectively transforms the d-c potential at point A into a 60-cps signal, the magnitude of which is proportional to the voltage at A and the phase of which is determined by the polarity of the voltage at A. This a-c signal is applied to one winding of the motor, while a fixed a-c reference signal is applied to the other winding. The motor characteristics are such that rotation takes place as long as both windings are excited, the direction of rotation being determined by the phase relationship of the voltages in the two windings. As a result of the mechanical coupling of the motor shaft and the potentiometer arm, the rotation of the motor effects an alteration in the potential at point B and, consequently, in the voltage at A. The motor finally comes to rest when the potential at $A = 0$, that is, when $XE = -e_1$. The angular positions X of both ganged potentiometers as a fraction of full scale are then equal to $-e_1/E$. If a variable voltage e_2 is applied to potentiometer II, the potential appearing at its moving arm is equal to $-e_1e_2/E$ and the desired multiplication is accomplished.

The circuit described above is readily modified to permit four-quadrant multiplication. This involves the addition of an operational amplifier and the use of potentiometers with center taps. Servomultipliers are extremely accurate devices provided the variation of the input voltage is sufficiently slow. As a result of the

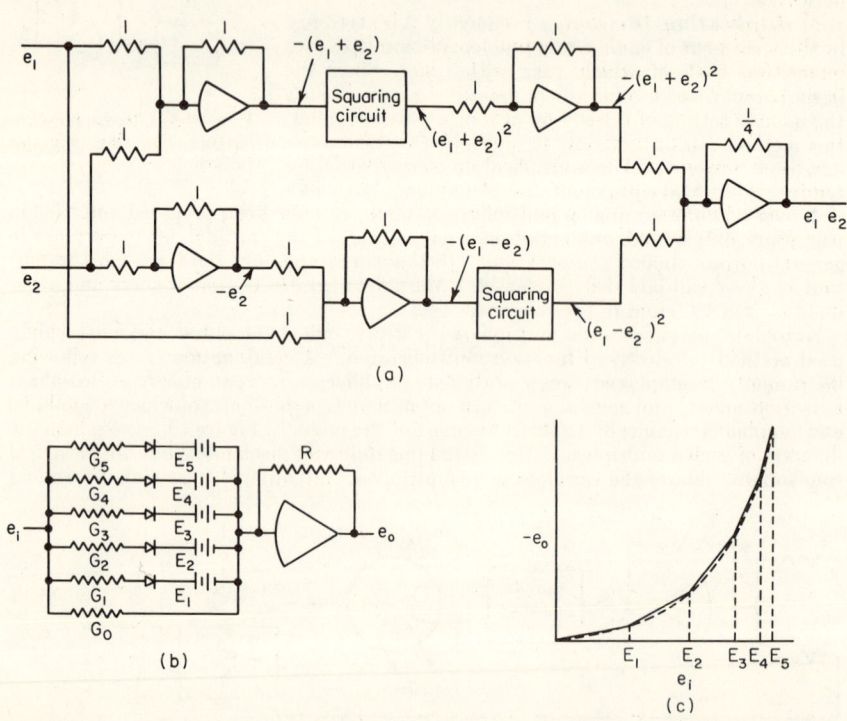

FIG. 10-6. (a) Quarter-square multiplier. (b) Squaring circuit. (c) Output of squaring circuit.

inertia in the motor and potentiometer, the frequency response of servomultipliers is relatively limited.

The principle of operation of another important class of electrical multipliers is based upon the equation

$$xy = \tfrac{1}{4}[(x+y)^2 - (x-y)^2] \tag{10-4}$$

Provided the squares of the sum and the difference of two variables can be generated, the product can be obtained by the simple combination of two such squaring circuits and standard adding circuits, as shown in Fig. 10-6a. A number of devices are available for the squaring operation. The most important of these involve the use of vacuum-tube or semiconductor diodes. An arrangement of diodes, bias supplies, and series conductances necessary to produce an output voltage approximately equal to the square of the input voltage is shown in Fig. 10-6b. The bias voltages

E_1 to E_5 are adjusted in such a way that as the input voltage is increased, successive diodes which are normally cut off are rendered conductive, thereby increasing the slope of the e_o/e_i curve as shown in Fig. 10-6c. A square-law curve may be approximated to within ± 2 per cent with 3 diode sections, to within ± 1 per cent with 5 sections, and to within less than 0.4 per cent with 15 diodes. To permit four-quadrant multiplication it is necessary to employ two sets of diodes and bias supplies—one for positive and the other for negative inputs. It is possible to improve the operation of diode circuits by superimposing a small a-c voltage upon the desired signal. This has the effect of rounding off the corners at the intersection of adjacent line segments, thereby obviating the effect of the sharp changes in the slope of the e_o/e_i curve. While quarter-square multipliers of the type described tend to have a larger static error than servomultipliers, their frequency-response characteristics are much more favorable, because of the absence of moving parts.

The third major category of function multipliers, known as *electronic multipliers*, employs modulation techniques. The two input voltages are made to affect the characteristics of a pulse train generated within the unit. For example, in the so-called time-division multipliers a train of rectangular pulses is employed. The repetition rate of the pulses is fairly high, approximately 10,000 cps. The height of each pulse is made proportional to one of the variables, while the width of the pulse is a function of the other variable. The area under each pulse is then a function of the product of the two inputs and can be found by suitable integration techniques. Other types of electronic multipliers employ triangular waves, frequency modulation, and phase modulation.

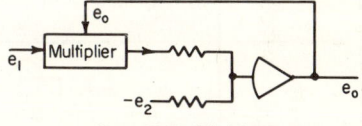

Fig. 10-7. Divider.

Commercially available electronic multipliers suffered for many years from drift problems and other inaccuracies. In recent years, however, electronic multipliers have become available which offer a very favorable combination of accuracy and frequency response.

Separate units to perform the operation of division are generally not supplied in analog-computer systems. Rather a technique termed implicit function generation is employed. Using a circuit of the type shown in Fig. 10-7, an equation of the type

$$e_2 + ke_1e_o = 0 \tag{10-5}$$

is instrumented. The high-gain amplifier is connected in a feedback loop in such a manner that it generates an output voltage which forces the output of the multiplier to be equal and opposite in sign to the input voltage e_2. The output voltage e_o is then proportional to the quotient e_2/e_1.

Generation of Functions of the Dependent Variable. For a wide variety of analog computations, it is necessary to provide operational units which furnish or generate special electrical functions. Such function generators are generally classified as generators of functions of dependent variables if the output voltage bears some specified relationship to the input voltage, and generators of functions of independent variables if the output voltage is some special function of time. Dependent-variable function generators are important particularly in the analysis of nonlinear systems. In general, such nonlinearities arise because the magnitudes of the system parameters or characteristics are governed by the system variables. For the most part the function generators employed for this purpose utilize as basic building blocks the operational adding and integrating circuits described above, as well as biased diodes and multipliers. Summaries of the application of these devices can be found in Refs. 1 to 4. Table 10-2 represents a summary of the most widely used of these circuits.

One large family of function generators is employed to simulate discontinuous functions characterized by abrupt breaks in the input-output voltage curve. Biased diodes are utilized to generate these discontinuities as shown in circuits 1 to 5 in Table 10-2.

Table 10-2. Generators of Functions of the Dependent Variable

Function	Circuit	Input-output relationship
1. Limiter		Slope $\frac{R_2}{R_1}$; $\frac{R_1}{R_2}e_1$, $\frac{R_1}{R_2}e_2$; $R_3=\infty$, $R_3=$ finite, $R_3=0$; $R_4=0$, $R_4=$ finite, $R_4=\infty$
2. Dead zone		Slope $=\dfrac{R_4}{R_i+R_3}$
3. Coulomb friction		$\|e_1\|$, $-\|e_2\|$
4. Absolute value		
5. Hysteresis		$\dfrac{R_2}{R_1}=K$; $e_{i,\min}$, $e_{i,\max}$
6. Square root	Multiplier; $-e_i$, $e_o=\sqrt{e_i}$	$e_o=\sqrt{e_i}$

ANALOG ELEMENTS

Table 10-2. Generators of Functions of the Dependent Variable (*Continued*)

Function	Circuit	Input-output relationship
7. Cube root		$e_o = \sqrt[3]{e_i}$
8. Inverse sine		$e_o = \sin^{-1} e_i$
9. Tangent		$e_o = \tan e_i$
10. Sine-cosine		$e_{o_1} = \cos e_i$ $e_{o_2} = \sin e_i$
11. Tangent		$e_o = \tan e_i$
12. Exponential		$e_o = \epsilon^{-ke_i}$
13. Logarithm		$e_o = \log_\epsilon e_i$
14. Error function		$e_o = \text{erf}(x)$

Implicit-function-generating circuits, of the type shown in Fig. 10-7, are employed to generate another important class of functions including square roots, cube roots, and inverse trigonometric functions, as shown in circuits 6 to 8 in Table 10-2.

A third approach to the generation of functions of the dependent variable involves the utilization of the generalized integration formula

$$\int y\,dx \equiv \int y\frac{dx}{dt}\,dt \tag{10-6}$$

Ordinary analog integrators are capable of integrating only with respect to time. By means of Eq. (10-6), integrations with respect to a dependent variable can be performed. Characteristically circuits making use of this technique include one or more multipliers as well as an integrator with respect to time. Circuits 10 to 14 in Table 10-2 fall into this category.

Since in many analog applications it is necessary to obtain sines and cosines of input voltages, a number of computer manufacturers market special *resolvers*. The general function of these resolvers includes:

1. Rotation from one rectangular system to another
2. Conversion from polar to rectangular coordinates
3. Conversion from rectangular to polar coordinates

Devices for the accomplishment of these functions include a-c induction motors, specially wound potentiometer cards, and mechanical scotch yokes.

Another special category of function generators includes those designed to generate functions of two or more dependent variables. Devices developed for this purpose utilize conductive sheets, such as Teledeltos paper, and complex arrays of diode function generators.

Generation of Functions of Time. Strictly speaking, all analog-computer units generate output voltages which are functions of time. In simulating engineering systems and their excitations, certain time functions are needed so frequently, however, that it is useful to have available a catalog of circuits suitable for the generation of these special functions. Some of the more important of these are presented in Table 10-3. Circuit 1 takes advantage of the fact that the desired exponential function is the solution of a first-order differential equation. A simple one-integrator loop is therefore formed in accordance with the approach described in more detail in Chapter 11. Similarly, sines and cosines are generated by means of the circuit ordinarily used to solve second-order differential equations. Powers of time are generated as shown in circuit 3 using a cascade of integrators. Circuit 4 serves for the generation of square and triangular waves. It consists of a loop including an integrator and adder and the comparator (circuit 3 of Table 10-2); once the switch paralleling the integrator capacitor is opened, the circuit behaves as a free-running oscillator.

Occasionally in analyzing an engineering system it becomes necessary to simulate time delays. For example, in studying transmission processes through hydraulic or pneumatic lines or in studying the motion of radioactive particles in nuclear reactors, it is necessary to produce an output voltage which is identical to an input voltage but displaced in time by a specific interval. A direct method for generating such time delays is to employ magnetic-tape recorders in which the space between the recording and the reading heads is adjustable. A widely used electronic method for the generation of time-delay functions is to employ the so-called Padé approximation. This technique is based on the Laplace-transform expression

$$\mathcal{L}f(t - \tau) = F(s)\epsilon^{-s\tau} \tag{10-7}$$

where τ is the time delay. This exponential is approximated as the ratio of two polynomials in s according to

$$e^{-s\tau} = \frac{P_n(s)}{Q_m(s)} \tag{10-8}$$

ANALOG ELEMENTS

Table 10-3. Generators of Special Functions of Time

Function	Circuit	Output
1. Exponential	(circuit with integrator, input ke, $IC = A$, output $e_o = A\epsilon^{-kt}$)	$e_o = \epsilon^{-kt}$
2. Sine-cosine	(circuit with Pot 1 $\left(\frac{1}{\omega_0}\right)^2$ megohm, integrators producing $\frac{de_o}{dt}$, $-\frac{de_o}{dt}$, e; feedback $2\alpha\frac{de_o}{dt}$ through Pot 2 Positive α / Negative α)	$e_o = \epsilon^{-\alpha t}\sin\omega_d t$
3. Powers	(cascade of integrators, $E = -1$ volt, k_1; outputs $e = k_1 t$, $e = -\frac{k_1}{2}t^2$, $e = \frac{k_1}{6}t^3$, ..., $e = (-1)^n \frac{k_1 t^n}{n!}$)	$e_o = t^n$
4. Square wave, Triangular wave	(circuit with e_B, amplifier B with Switch, amplifier A, E_1, E_2, output e_o)	(square wave between E_1 and E_2; triangular wave e_B between $-E_1$ and $-E_2$)
5. Time delay	(Padé delay network: resistors 5.952, 1.190, 0.5556, 0.5, 1.0; capacitors 0.1τ; feedback resistors 1.0)	$e_o = e_i(t - \tau)$

where the subscripts n and m refer to the highest power of s in the numerator and denominator, respectively. For example, for $n = m = 2$

$$e^{-s\tau} \simeq \frac{a_0 + a_1 s\tau + a_2(s\tau)^2}{b_0 + b_1 s\tau + b_2(s\tau)^2} \tag{10-9}$$

The six coefficients in (10-9) are then chosen to fit the first six terms of the series expansion

$$e^{-s\tau} = 1 - s\tau + \frac{(s\tau)^2}{2} - \cdots$$

This yields $a_0 = b_0 = 1$, $a_1 = -\frac{1}{2}$, $b_1 = \frac{1}{2}$, $a_2 = b_2 = \frac{1}{12}$.

The higher the powers of the numerator and denominator, the more terms of the series can be matched and the smaller the error in the time-delay function. A circuit for generating time delays of from 0.2 to over 30 sec is shown as circuit 5 of Table 10-3. The time delay τ in seconds is determined by the magnitudes of the input resistors to the four integrators. Such a circuit works well only in a limited frequency range (direct current to approximately 50 cps). It is therefore not feasible to apply a step function to this circuit. Furthermore Padé approximation circuits are not applicable to problems in which the time delay τ varies either as a function of time or as a function of dependent variables. A truly satisfactory time-delay unit has yet to be developed.

To analyze the behavior of nonlinear dynamic systems in the presence of noise, it is necessary to employ a special electronic unit having an output voltage whose magnitude is a random function of time. As primary sources of noise, most analog computers utilize either radioactive samples in combination with Geiger-Müller tubes or gas-discharge tubes such as thyratrons. The output voltage from these sources is then amplified and filtered to obtain the desired probability distributions.

10-3. Mechanical Computing Elements

The indirect computing elements described in the preceding paragraphs generate the desired mathematical functions as electrical voltages. In contrast, a second major family of computing elements represents the dependent variable by translational or rotational mechanical displacements. Such units are fashioned from shafts, gears, disks, etc. They have the advantage over electrical elements of ruggedness, simplicity, and dependability. They have, however, the disadvantages of greater weight and bulk, lack of flexibility, and difficulty in assembly. These disadvantages outweigh the advantages to such an extent that mechanical computing elements are no longer used in general-purpose analog computations. There are, however, many special-purpose applications, particularly where reliability and the ability to operate under unfavorable environmental conditions are paramount, in which mechanical elements continue to play an important part. Several of the more important mechanical computing elements are briefly reviewed below. More complete discussions can be found in Ref. 2.

Addition and Subtraction. The variables to be added and their sum usually appear as translations of specified points in a mechanical adder, as angular displacement of its length, or as turns of its shafts. Two common forms of adder are the differential and the tape devices. Variations of the differential adder are shown in Fig. 10-8. Two forms of linkage differential appear in Figs. 10-8a and 10-8b. An adder comprising a pair of sliding racks in contact with a common pinion appears in Fig. 10-8c. The zero positions in each case are shown by dotted lines, the displaced positions by full lines. The bevel-gear differential is shown in Fig. 10-8d, and the spur-gear differential in Fig. 10-8e.

Integration. The most common mechanical integrator is the so-called disk and wheel integrator shown in Fig. 10-9. The turns of the disk (frequently called "turntable") represent the differential variable x to a suitable scale factor. The distance of the wheel center plane from the axis of the turntable represents the integrand y,

ANALOG ELEMENTS

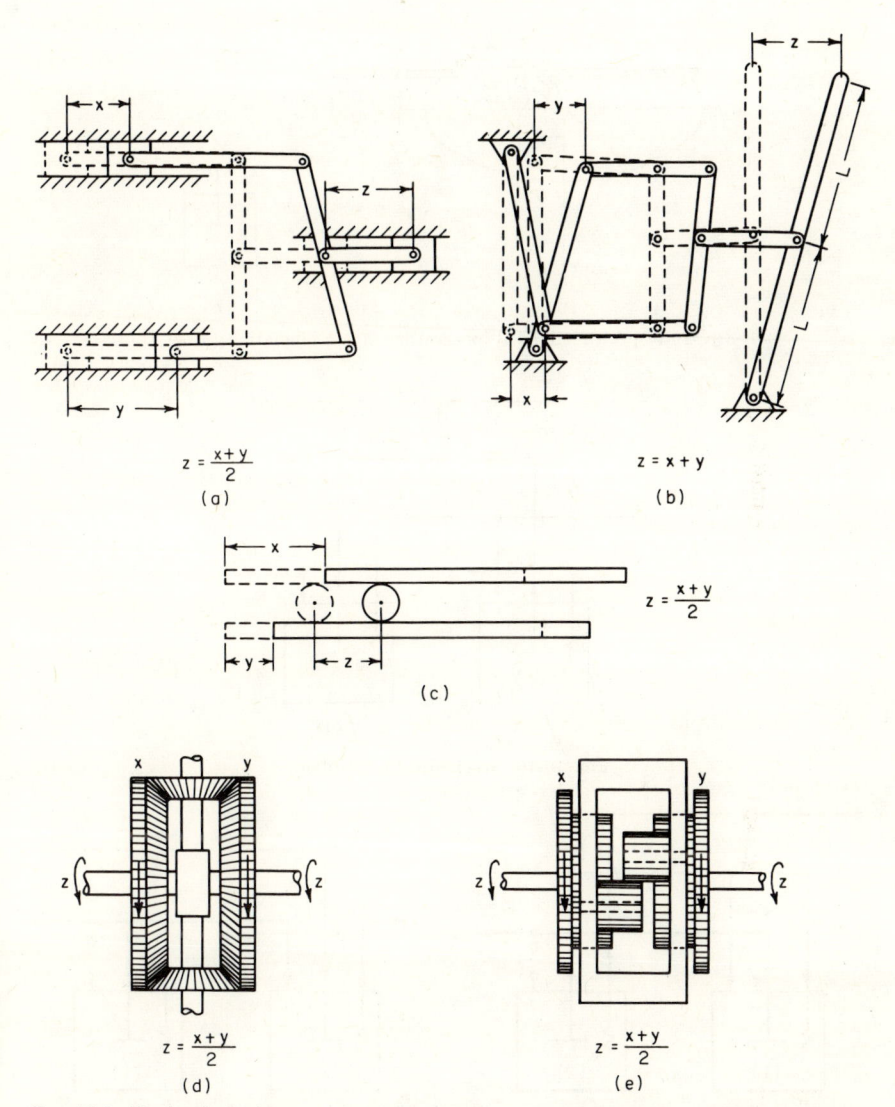

$z = \dfrac{x+y}{2}$
(a)

$z = x + y$
(b)

$z = \dfrac{x+y}{2}$
(c)

$z = \dfrac{x+y}{2}$
(d)

$z = \dfrac{x+y}{2}$
(e)

Fig. 10-8. Mechanical adders. (a) Linkage adder. (b) Alternate form of linkage adder. (c) Rack-and-pinion adder. (d) Differential using bevel gear. (e) Differential using spur gear.

again to some suitable scale factor. These are the two inputs to the integrator. The turns of the integrating wheel represent the value z of the integral to a scale factor predetermined by the two input scale factors and the actual structural details of the unit. This is the output of the integrator. A rotation of the disk through an infinitesimal fraction of a turn dx causes the wheel to turn through a correspondingly small part of a turn dz. For a wheel of radius a

$$dz = \frac{1}{a} y \, dx \qquad (10\text{-}10)$$

10–18   SYSTEM COMPONENTS

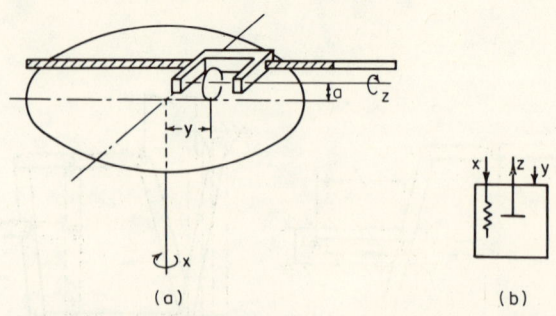

Fig. 10-9. (a) Mechanical integrator. (b) Schematic symbol.

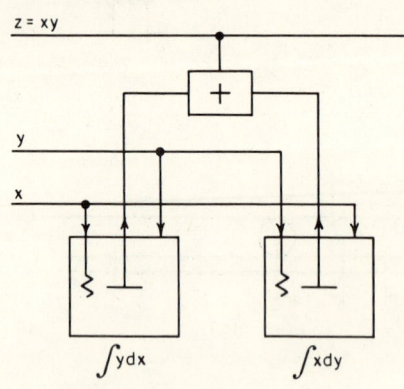

Fig. 10-10. Mechanical multiplier.

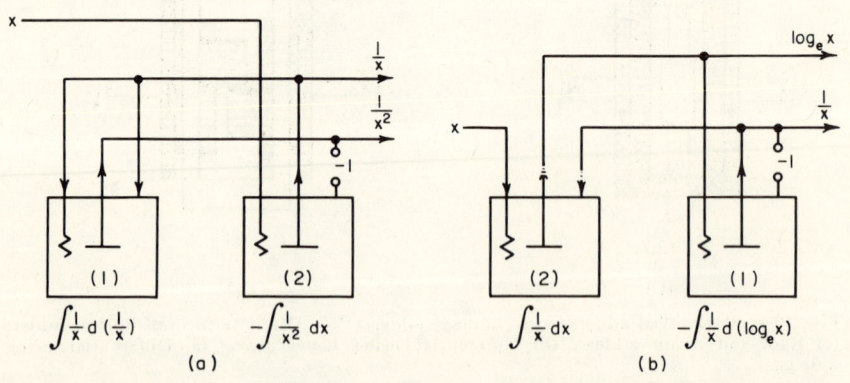

Fig. 10-11. (a) Generator of reciprocals. (b) Generator of logarithm.

During a finite time interval, the x turntable will turn through positive and negative values as called for by the problem. The total number of turns registered by the integrating wheel will then correspond to the integral of y with respect to x.

Multiplication by a Constant. The multiplication of the variable by a constant is accomplished with simple mechanical elements. In the linkage computers the mechanical advantage of the lever arm serves to ratio the displacement up or down. Shaft rotations are stepped up or down with simple gear trains.

Multiplication of Two Variables. Unlike electronic integrators, mechanical integrators are capable of integrating with respect to dependent variables. Multiplication can therefore be accomplished using the familiar formula for integration by parts

$$xy = \int x\,dy + \int y\,dx \qquad (10\text{-}11)$$

The mechanization of these equations then requires merely two integrators and an adder interconnected as shown schematically in Fig. 10-10. Initial conditions on x and y are taken into account as initial settings of the integrating wheel.

Function Generation. Functions of the dependent and independent variables of the problem are readily generated by interconnecting integrators, adders, and constant multipliers in a suitable manner. The generation of two such nonlinear functions of the dependent variable is illustrated in Fig. 10-11.

REFERENCES

1. Fifer, S.: *Analogue Computation* (4 vols.), McGraw-Hill Book Company, New York, 1961.
2. Karplus, W. J., and W. W. Soroka: *Analog Methods: Computation and Simulation*, 2d ed., McGraw-Hill Book Company, New York, 1959.
3. Korn, G. A., and T. M. Korn: *Electronic Analog and Hybrid Computers*, McGraw-Hill Book Company, New York, 1964.
4. Bekey, G. A.: Nonlinear Electronic Computing Elements, in *Handbook of Automation, Computation and Control*, vol. 2, chap. 23, John Wiley & Sons, Inc., New York, 1959.

Chapter 11

ELECTRONIC ANALOG/HYBRID COMPUTERS AND THEIR USE IN SYSTEM ENGINEERING*

GRANINO A. KORN, *Professor of Electrical Engineering, University of Arizona.*

11-1. Complete Computers	11-2
11-2. Operation of Electronic Differential-equation Solvers	11-4
11-3. Programming Differential Equations	11-4
11-4. Amplitude Scaling	11-5
11-5. Time Scaling	11-5
11-6. Representation of Dynamical Systems. Control Systems	11-6
11-7. Special Simulation Techniques	11-7
11-8. The Basic Conversion Elements: Comparators and Switches	11-10
11-9. Precision Limiters and Related Hybrid Circuits	11-10
11-10. Analog Multivibrator Circuits	11-11
11-11. Simulation of Communications and Detection Systems	11-12
11-12. Simulation of Neurons and Synapses	11-14
11-13. Linear Programming and Related Problems	11-14
11-14. Combined Analog/Digital Simulation	11-19
11-15. Faster Analog Computation and Iterative Operation	11-19
11-16. System Design: The ASTRAC System and Monte-Carlo-type Random-process Studies	11-23

In the most general sense, an *analog computer* is any physical system which establishes definite prescribed relations—necessarily stated as mathematical relations—between continuously variable physical quantities; see Chapter 10. It follows that every analog computer, whether used for problem solving, instrumentation, or control, implements a requirement for at least approximate physical realization of a mathematical model, which can be understood and manipulated with relative ease.

---

* Portions of this section are based on material from G. A. Korn and T. M. Korn, *Electronic Analog and Hybrid Computers*, McGraw-Hill Book Company, New York, 1964, G. A. Korn, *Random-process Simulation and Measurements*, McGraw-Hill Book Company, New York, 1965, and H. D. Huskey and G. A. Korn, *Computer Handbook*, McGraw-Hill Book Company, New York, 1962, as well as from the writer's article in the *American Institute of Physics Handbook*, 2d ed., McGraw-Hill Book Company, New York, 1963. The editor and author are grateful to the McGraw-Hill Book Company for their permission to use this material.

11-2　SYSTEM COMPONENTS

Analog computers used to solve problems represent each problem variable by a corresponding physical quantity (voltage, shaft displacement) on a convenient scale (Secs. 11-4 and 11-5). These *machine variables* are made to obey mathematical relations corresponding to those of the given problem. In particular, a *differential analyzer* starts a set of machine variables from specified initial values and employs *analog integrators* (Sec. 10-2) together with other computing elements to enforce a system of prescribed ordinary differential equations.

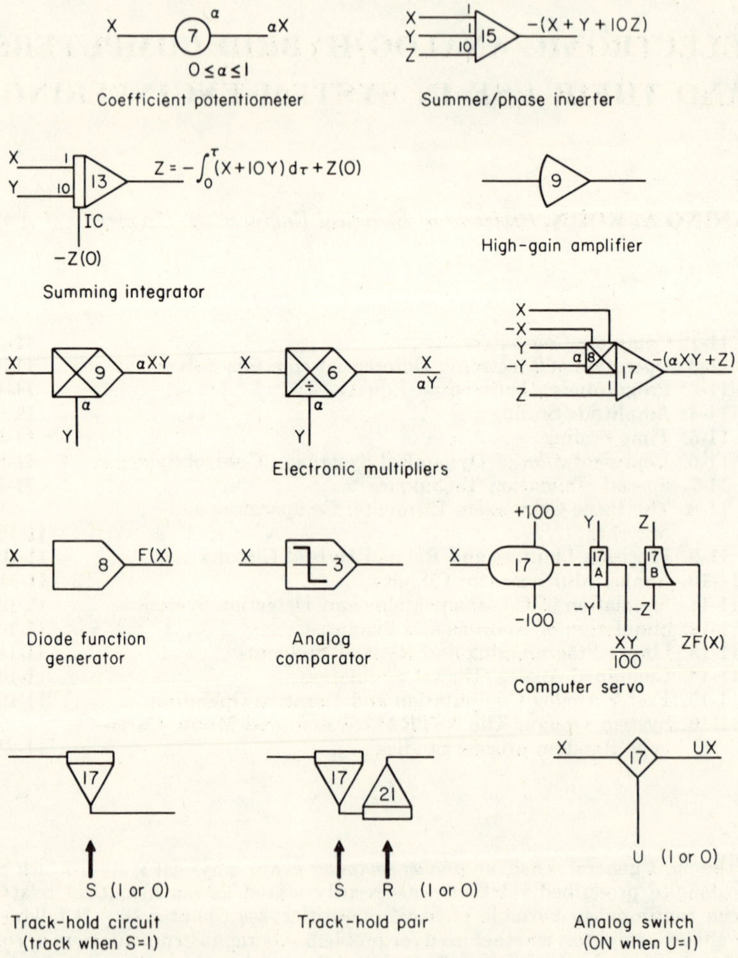

Fig. 11-1. Analog-computer symbols.

11-1. Complete Computers

Besides a set of operational amplifiers and other computing elements (Fig. 11-1) with the necessary power supplies, a complete electronic analog-computer installation will comprise

1. A *patchbay* with computing-element terminations and *patchcords* to set up (program) problems. Modern computers have removable (plug-in) patchboards

which permit problems to be set up away from the computer and also allow for problem storage.

2. *Control switches or relays* to place initial conditions on all integrators (RESET *condition*) and to make the computing elements operative (COMPUTE or OPERATE *condition*), to set potentiometers, to select circuits for metering, etc.

3. *Read-out devices,* such as oscilloscopes, strip-chart recorders, xy recorders, digital voltmeters, and printers.

4. *Reference power supplies* to supply accurate *reference voltages* (usually ±100 or ±10 volts) to be used as computing and initial-condition voltages.

Unlike digital computers, most analog computers employ distinct computing elements for each mathematical operation required to solve a given problem (*parallel operation*).[1] Parallel operation is an essential reason for the *very high computing speeds* possible with d-c analog computers: most problems, regardless of complication, are solved within 2 min by "slow" d-c analog computers and within fractions of a second by repetitive machines (Sec. 11-15). On the other hand, parallel operation imposes *practical limits on the complexity of problems* which can be solved on analog computers. These limits clearly depend on the problem; generally speaking, a 20-amplifier desk-top machine will readily solve the four nonlinear second-order differential equations describing a reasonable four-degree-of-freedom dynamical system, and a medium-sized analog computer (80 amplifiers) can deal with a 10-degree-of-freedom system without special difficulties. Much larger problems, requiring up to 1,000 amplifiers, have been treated successfully in large industrial installations.

The *component accuracies* of individual analog-computing elements for addition, multiplication, integration, etc., vary between 3 *and* 0.005 *per cent of full scale* (note that "full scale" has come to mean 200 volts for a ±100-volt machine). Costs increase rapidly if component accuracies better than 0.2 per cent are desired. Accuracies obtainable in solving a complete problem depend critically on the problem and on the computer set-up as well as on the component accuracies, since errors may cancel or be compounded. In electronic analog computers, the overall accuracy is essentially equal to the component accuracy in problems involving linear differential equations with constant coefficients if the solutions comprise only well-damped modes.

The really important contributions of analog computers to modern system-development techniques go beyond numerical computation as such. *In many applications, the analog approach functions as a direct aid to a research worker's thinking process; an analog-computer set-up serves as a "live mathematical model" which bridges the gap between mathematical symbolism and physical reality.* Simple patchcord connections and potentiometer settings enable the system designer to create a scale model which permits convenient investigation of the performance and interaction of processes and systems. New ideas or hypotheses can be tested at once; blocks of computing elements are readily replaced or assembled into different model designs. Parameters may be changed instantly by corresponding resistance changes, and high computing speeds permit optimization of system performance by successive parameter adjustments.[2]

The *time scale* of analog computations may be slowed down or speeded up within wide limits at the convenience of the experimenter. In particular, analog-computing speeds are sufficiently high to permit representation of most dynamical and electromechanical processes on a 1:1 time scale (*simulation* in the strict sense; see Sec. 32-2). With the computer running on a 1:1 time scale, actual physical components of a system under investigation can take the place of a corresponding block of computing elements (*partial system testing*).

Where computers are used to solve problems (as contrasted with applications of computers as control-system components), it is fair to say that *there is relatively little competition between analog and digital machines;* their areas of application do not overlap so much as is popularly believed. Where the system engineer is free to use both types of machines, his digital-computer work is likely to be an important chore which he would like to delegate to computing-center personnel if at all possible. This fact is not really changed by modern desk-sized digital computers and compiler routines. The designer might, on the other hand, operate the analog computer, if applicable, himself, much as he would use a sketch or block diagram.

11-2. Operation of Electronic Differential-equation Solvers

An *electronic analog computer* represents problem variables by d-c voltages which vary as functions of the time τ (*machine time*, used as the independent variable). The *solution of a set of ordinary differential equations* proceeds as follows:

1. With the machine connected up to solve the given problem, the machine variables (voltages) are set to the correct initial conditions prescribed by the problem (RESET *condition*).

2. The computing elements are made operative and force the voltages in the machine to vary in a manner prescribed by the given differential equations. The voltage variations with time are recorded and constitute solutions of the given problem.

3. The machine is reset to its initial condition and is ready for the next run with changed coefficients, initial conditions, etc.

"Slow" d-c analog differential-equation solvers admit signal frequencies up to about 50 cps; this is sufficient for simulation of most dynamical systems on a 1:1 time scale.

In a *repetitive* differential-equation solver, the steps of the solution outlined above are repeated automatically at a rapid rate (10 to 100 cps) by means of an electromechanical or electronic switching system. During each cycle, each machine variable varies in the prescribed manner and is then reset to its initial value. The results can now be presented on an ordinary cathode-ray oscillograph whose sweep frequency equals the computer repetition rate. *The rapid operation of such machines permits immediate observation of the effects of parameter changes on the solution.* Current trends in electronic analog computation combine repetitive operation with analog memory and hybrid analog/digital automatic programming to permit automatic decisions and iterative subroutines (Sec. 11-15).

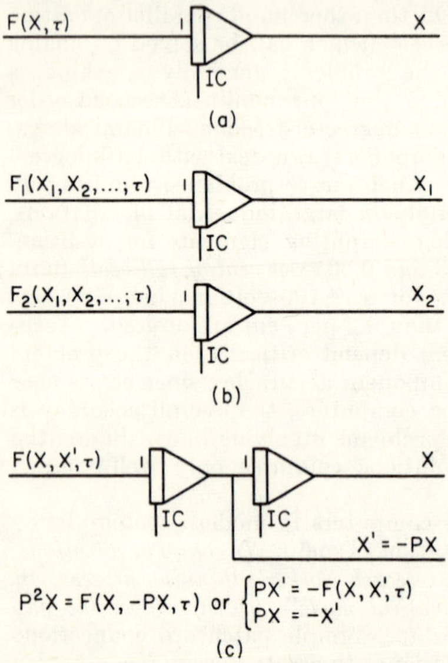

FIG. 11-2. Solution of a first-order differential equation (*a*), of a system of first-order equations (*b*), and of a typical second-order equation of motion (*c*).

11-3. Programming Differential Equations

Figure 11-2*a* illustrates the solution of a first-order ordinary differential equation

$$\frac{dX}{d\tau} \equiv PX = -F(X,\tau) \tag{11-1}$$

by an analog integrator set up to integrate a voltage $F(X,\tau)$ obtained by suitable combinations of coefficient potentiometers, summing amplifiers, multipliers, and function generators; here X is a voltage (machine variable), and the independent variable τ is real time (computer time). Note that feedback of the voltage X around the integrator is involved (implicit computation, classical differential-analyzer technique).

In Fig. 11-2b, n integrators similarly solve n simultaneous first-order differential equations

$$\frac{dX_i}{d\tau} \equiv PX_i = -F_i(X_1, X_2, \ldots, X_n; \tau) \qquad (i = 1, 2, \ldots, n) \qquad (11\text{-}2)$$

Higher-order differential equations are transformed into sets of first-order equations by introducing derivatives PX, P^2X, ... as new variables (Fig. 11-2c).

Input functions of the computer time τ *(forcing functions)* can be generated by solution of suitable differential equations, or special signal generators (e.g., sine and square-wave generators) may be used.

11-4. Amplitude Scaling

Analog computers represent problem variables $x, y, \ldots$ by machine variables (usually voltages, currents, or shaft rotations)

$$X = a_x x \qquad Y = a_y y \qquad \ldots \qquad (11\text{-}3)$$

The dimensional *scale factors* a_x, a_y, ... are chosen so that each machine variable $X, Y, \ldots$ is as large as possible without exceeding its maximum permissible excursion $X_{\max}$, $Y_{\max}$, ... (e.g., 10, 50, or 100 volts) in absolute value.

To scale a given problem for an electronic analog computer capable of supplying ±100-volt machine variables, one may measure $X, Y, \ldots$ in 100-volt *machine units*, but *scaling in volts* or *per cent machine units* is almost universally accepted. Then

$$a_x \leq \frac{100}{|x|_{\max}} \frac{\text{volts}}{\text{problem unit}} \qquad (11\text{-}4)$$

where $|x|_{\max}$ is the largest expected excursion of the problem variable x. For convenience, the numerical value of each scale factor is usually a round number of the form 10^n, 2×10^n, 2.5×10^n, 4×10^n, or 5×10^n.

To obtain the correct *machine equations* relating the voltages $(a_x x)$, $(a_y y)$, ..., write

$$x \text{ as } \frac{1}{a_x}(a_x x) \qquad y \text{ as } \frac{1}{a_y}(a_y y) \qquad \ldots \qquad (11\text{-}5)$$

in each given problem equation. Then set up the block diagram, leaving terms like $(a_x x)$, $(a_y y)$ intact; these terms will appear as voltages whose absolute values cannot exceed 100 volts.

Second-order differential equations (e.g., equations of motion)

$$\frac{d^2x}{dt^2} = f\left(x, \frac{dx}{dt}, t\right) \qquad (11\text{-}6)$$

are best transformed to pairs of first-order equations

$$\frac{d\dot{x}}{dt} = f(x, \dot{x}, t) \qquad \frac{dx}{dt} = \dot{x} \qquad (11\text{-}7)$$

with separate scale factors for x and $\dot{x}$. $|x|_{\max}$ and $|\dot{x}|_{\max}$ are estimated from physical considerations or by exploratory computer runs with small scale factors.

11-5. Time Scaling

If the transfer functions of all computing elements are given in terms of real-time units, d-c integrators (Sec. 10-2) integrate the machine variables *with respect to real time.* Therefore, *real time, henceforth denoted also as machine time τ, is the independent machine variable in d-c analog computers.* The *time scale* is established by writing a transformation equation relating the machine time τ to the mathematical independent variable t, say,

$$\tau = \alpha_t t \tag{11-8}$$

so that the correct machine equations are obtained through the substitution

$$p = \frac{d}{dt} = \alpha_t \frac{d}{d\tau} = \alpha_t P \tag{11-9}$$

where the operator $P = d/d\tau$ denotes differentiation with respect to the real or machine time τ. The dimensional coefficient α_t is called the *time-scale factor* and is often referred to simply as the time scale. The time-scale factor is numerically equal to the number of seconds representing the unit of the independent variable under consideration.

The nature of a d-c analog computer makes it easy to change the time scale of a computer run within certain limits. When the time scale is changed, the voltage changes in the computer remain proportional to the corresponding changes of the mathematical variables, but the rates at which these voltage changes take place may be speeded up or slowed down to improve the accuracy or increase the convenience of the computation. A simple way to increase (slow) the time scale is to increase all capacitors in the computer set-up proportionately.

If the time t occurs *explicitly* in a given differential equation, t will be represented by a machine variable (voltage) $T = a_t t$ in addition to the machine time τ; one must not confuse a_t with the time-scale factor α_t.

11-6. Representation of Dynamical Systems. Control Systems

The example of Fig. 11-3 illustrates our differential-equation-solving procedure and exhibits features essentially common to all analog-computer representations of dynamical (mechanical) systems:

1. Specific blocks of computing elements implement machine equations corresponding to the *equations of motion* for each "generalized coordinate" of the dynamical system.* The output voltages of these blocks represent *generalized displacements* and/or *velocities*. As a rule, the initial values of these quantities are given. For convenient scaling, it is usually best to rewrite second-order equations of motion as pairs of first-order equations (Sec. 11-4).

2. The input voltages "driving" each block will represent *generalized forces* or *accelerations*. In general, each such voltage is a function of the computer time and/or of the machine variables representing generalized displacements and velocities.

3. Coupling forces coupling two degrees of freedom appear as logical interactions between two of the equations of motion and result in corresponding electrical interconnections between two blocks of computing elements.

4. It is usually possible to vary system parameters within reasonable limits and to study the resulting changes in system performance without radical changes in the computer set-up.

The functional analogy between components of an actual dynamical system under consideration and blocks of computing elements not only will furnish numerical solutions of the equations of motion but will also enhance the investigator's intuitive understanding of the system. In a more general sense, dynamical systems can involve electrical and thermodynamic as well as mechanical variables; the corresponding computer set-ups still relate state changes to forcing functions for each system degree of freedom.

The dynamical relationship between the input (stimulus) $x(t)$ and the output (response) $y(t)$ of many control-system components can be represented by a linear

---

* The general formulation of such equations of motion (Lagrange's equations) is discussed, for instance, in R. B. Lindsay and H. Margenau, *Foundations of Physics*, John Wiley & Sons, Inc., New York, 1936.

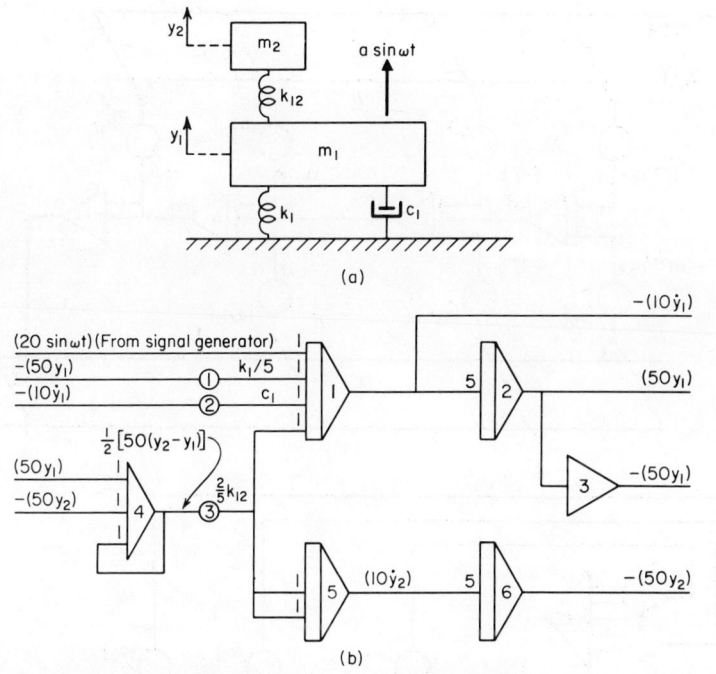

FIG. 11-3. Dynamic vibration absorber (a) and scaled analog-computer representation (b) of the equations of motion

$$m_1 \frac{d^2 y_1}{dt^2} + c_1 \frac{dy_1}{dt} + k_1 y_1 + k_{12}(y_1 - y_2) = a \sin \omega t$$

$$m_2 \frac{d^2 y_2}{dt^2} + k_{12}(y_2 - y_1) = 0$$

with $m_1 = 2m_2 = 1$ slug, $2k = k_{12} = 2$ lb/ft, $c_1 = 0.5$ lb-sec/ft, $a = 2$ lb, and $|y_1|$, $|y_2|$ below 2 ft. A one-to-one time scale is used.

differential equation

$$\frac{d^n y}{dt^n} + a_{n-1} \frac{d^{n-1} y}{dt^{n-1}} + \cdots + a_0 y = b_m \frac{d^m x}{dt^m} + b_{m-1} \frac{d^{m-1} x}{dt^{m-1}} + \cdots$$
$$+ b_o x \qquad (m \leq n) \quad (11\text{-}10a)$$

with constant coefficients a_i, b_k, or in operator form

$$y(t) = \frac{b_m p^m + b_{m-1} p^{m-1} + \cdots + b_o}{p^n + a_{n-1} p^{n-1} + \cdots + a_o} x(t) \qquad \left(m \leq n; p \equiv \frac{d}{dt}\right) \quad (11\text{-}10b)$$

Figure 11-4 shows two analog-computer setups for Eq. (11-10). Frequently, the operator polynomials in Eq. (11-10b) can be factored into first- and second-order terms; the corresponding computer set-ups have been tabulated.[1–4] Blocks of computing elements representing such transfer operators are combined into simulated control systems or "live block diagrams" (Fig. 11-5).

11-7. Special Simulation Techniques

The system designer does not realize the full power of his electronic analog computer if he uses it merely as a differential-equation solver. Special diode circuits and

11-8 SYSTEM COMPONENTS

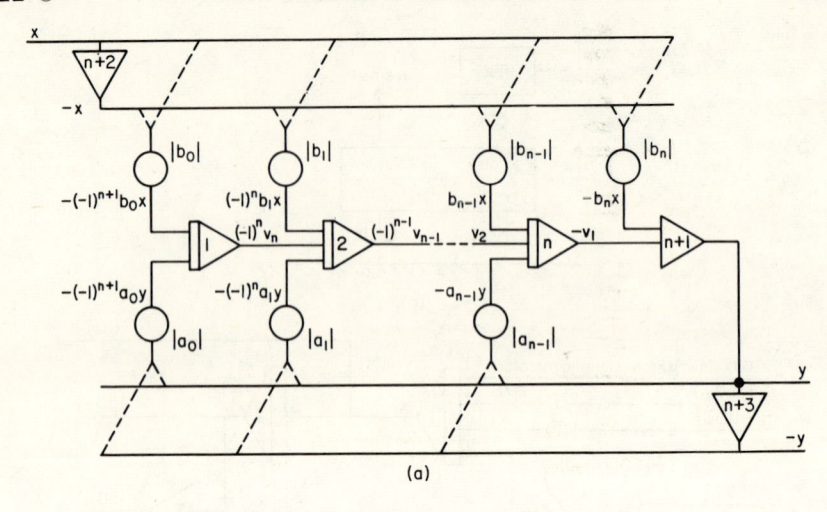

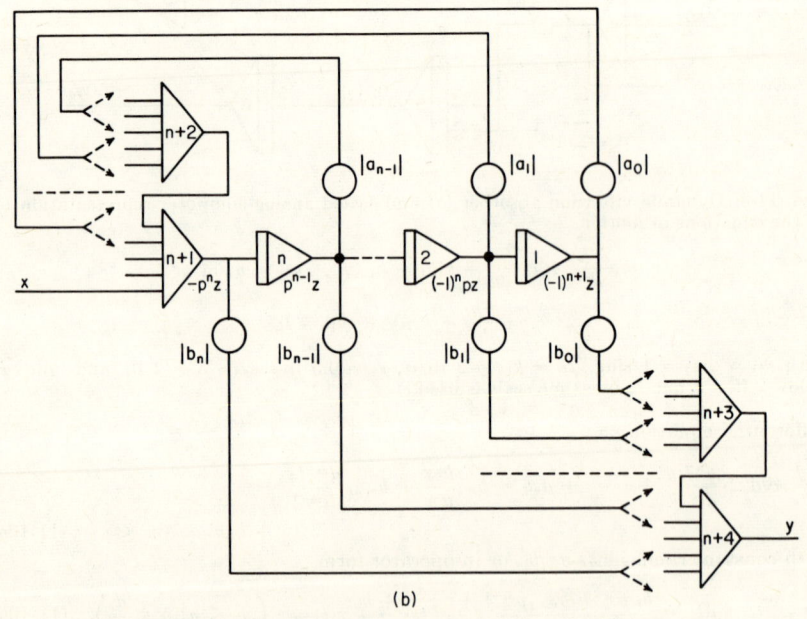

FIG. 11-4. Analog-computer set-ups representing the relation (11-10), or

$$y = \frac{b_m p^m + b_{m-1} p^{m-1} + \cdots + b_0}{p^n + a_{n-1} p^{n-1} + \cdots + a_0} x \qquad \left(m \leq n; \, p \equiv \frac{d}{dt}\right)$$

without explicit differentiation.

modern switching techniques can add some of the memory capacity and decision-making ability of the digital computer to the fast all-parallel operation of the electronic analog computer and permit advanced semiautomatic programming techniques, useful especially in parameter-optimization and random-process studies. Such techniques are most fully exploited in modern high-speed analog computers equipped with special memory circuits and patchable digital-logic modules (Chap-

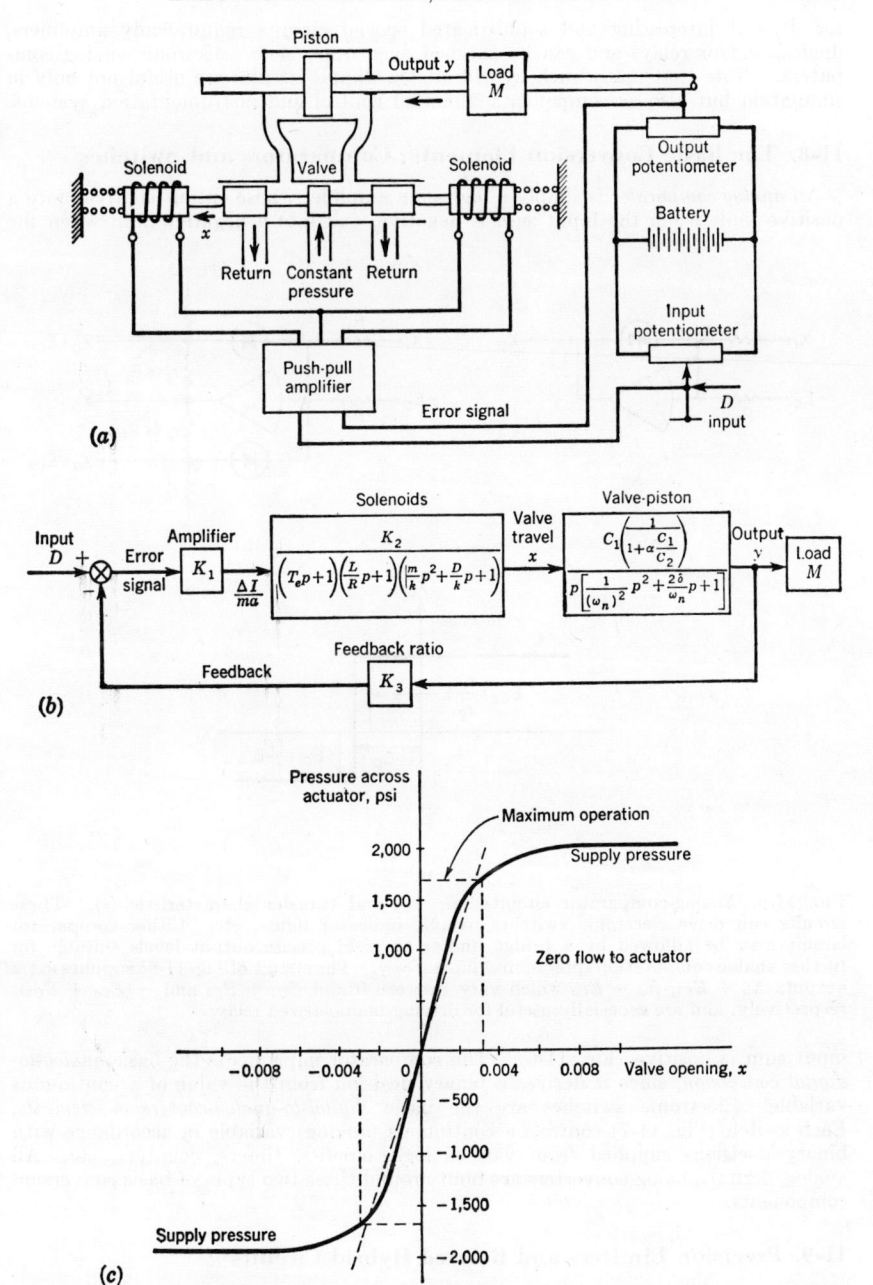

FIG. 11-5. Hydraulic servomechanism (a), and linearized representation (b). A more accurate (nonlinear) simulation would include the displacement-pressure characteristic (c) of the valve by a function generator. (*From J. C. Truxal, Control Engineers' Handbook, McGraw-Hill Book Company, New York, 1958.*)

11-10 SYSTEM COMPONENTS

ter 9); but interesting and sophisticated special circuits require only amplifiers, diodes, and/or relays and can be patched even into "slow" electronic analog computers. Note that many such hybrid analog/digital circuits are useful not only in simulation but also as components of actual control and instrumentation systems.

11-8. The Basic Conversion Elements: Comparators and Switches

An *analog comparator* is simply a high-gain amplifier whose output is driven into a positive limit when the input sum is negative, and into a negative limit when the

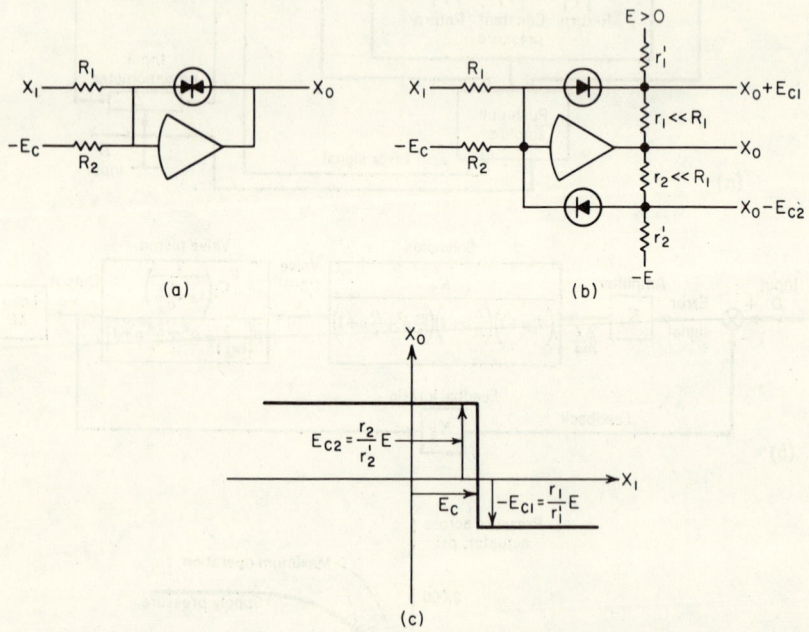

FIG. 11-6. Analog-comparator circuits (a), (b), and transfer characteristic (c). These circuits can drive electronic switches, relays, indicator lights, etc. Either comparator circuit may be followed by a bridge limiter to yield precise output levels suitable for further analog computation (precision comparator)[7]. The circuit of Fig. 11-6b supplies extra outputs $X_o + E_{C1}$, $X_o - E_{C2}$ which vary between 0 and $E_{C1} + E_{C2}$ and $-(E_{C1} + E_{C2})$, respectively, and are especially useful for driving nonpolarized relays.

input sum is positive (Fig. 11-6). The comparator implements the basic *analog-to-digital conversion*, since it derives a binary decision from the value of a continuous variable. Electronic switches are the basic *digital-to-analog-conversion elements*. Each switch (Fig. 11-7) controls a continuous (analog) variable in accordance with binary decisions supplied from various logic circuits, timers, counters, etc. All analog/digital/analog converters are built around these two types of basic conversion components.

11-9. Precision Limiters and Related Hybrid Circuits[6,7]

Figure 11-8 shows a number of *precision-limiter circuits* which may be regarded as basic *partial analog-to-digital converters*, since they supply a binary decision together with a precision-rectified analog output. These circuits and the related *precision-selector circuits* of Fig. 11-9 permit accuracies within a few hundredths of a volt at low frequencies (below 2 cps) and within 1 volt at frequencies up to 20 Kc. In Fig.

ELECTRONIC ANALOG/HYBRID COMPUTERS   11-11

11-10a, a precision limiter is combined with an integrator circuit to produce a precisely timed waveform suitable for generating delays and for analog-to-digital conversion by means of a gated oscillator and counter.[7,8] The *maximum-remembering circuit* of Fig. 11-10b, originally designed for a spacecraft instrumentation application, is useful in many simulation problems. The *dual-slicer circuit* of Fig. 11-10c accu-

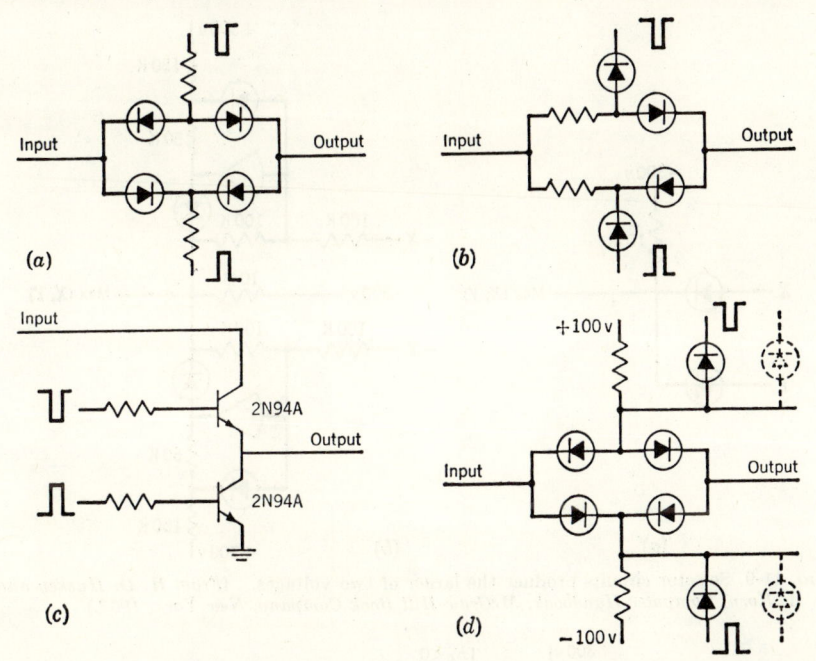

Fig. 11-7. Analog switches. The pulses shown turn the switches *off*. (*From H. D. Huskey and G. A. Korn, Computer Handbook, McGraw-Hill Book Company, New York, 1962.*)

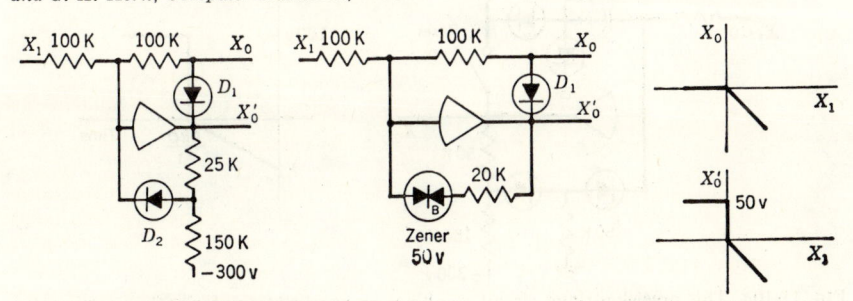

Fig. 11-8. Precision-limiter circuit. A decisive comparator step marks input sign changes, and an accurate limiter output is produced by feedback action. (*From H. D. Huskey and G. A. Korn, Computer Handbook, McGraw-Hill Book Company, New York, 1962.*)

rately marks a class interval $X - \Delta X/2 < x < X + \Delta X/2$ and is useful for amplitude-distribution analysis of random processes.[9,10]

11-10. Analog Multivibrator Circuits

Figure 11-11a shows a simple analog *bistable circuit*.[6,7] The comparator biases itself through a phase-inverting amplifier, so that a voltage $+E_C$ is needed to change

11-12   SYSTEM COMPONENTS

the output state. The input voltage then loses control until it reaches the negative value $-E_C$.

The addition of an integrator to the basic multivibrator circuit of Fig. 11-11a yields a *free-running multivibrator* (Fig. 11-11b) which has many uses as an extraordinarily accurate and inexpensive signal generator. Figure 11-12 shows how

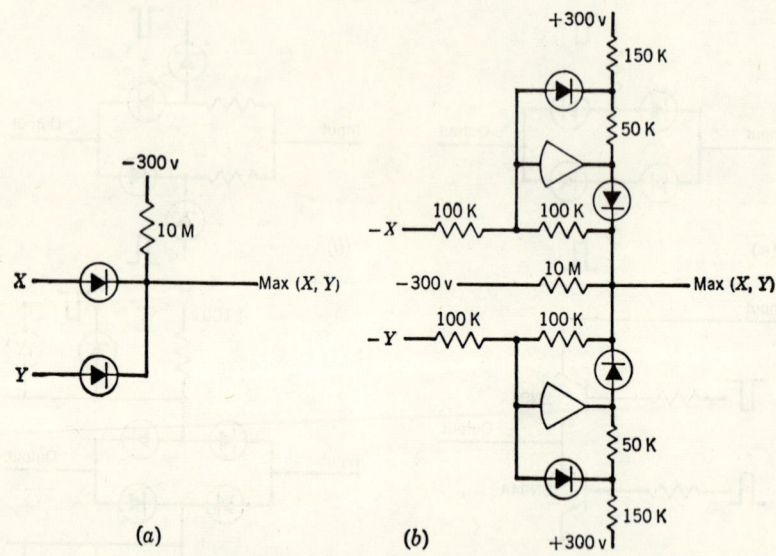

Fig. 11-9. Selector circuits produce the larger of two voltages. (*From H. D. Huskey and G. A. Korn, Computer Handbook, McGraw-Hill Book Company, New York, 1962.*)

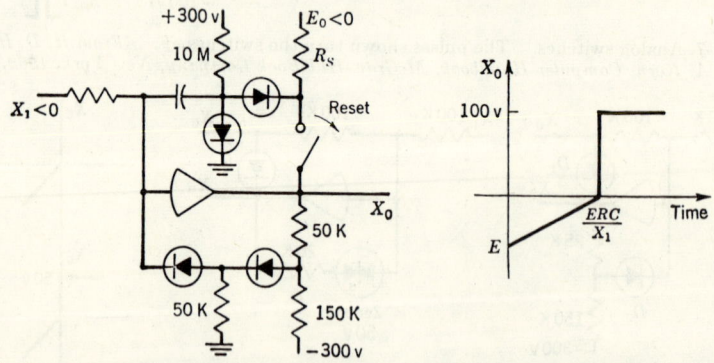

Fig. 11-10a. This precision-timer circuit combines an analog integrator with a comparator to produce a time interval accurately proportional to the input voltage X. (*From H. D. Huskey and G. A. Korn, Computer Handbook, McGraw-Hill Book Company, New York, 1962.*)

easily various types of modulated signals are generated with such a free-running multivibrator.[2]

11-11. Simulation of Communications and Detection Systems

The convenient signal-generator circuits of Fig. 11-12 used in conjunction with suitable random-noise generators permit one to simulate a wide variety of communi-

ELECTRONIC ANALOG/HYBRID COMPUTERS   11-13

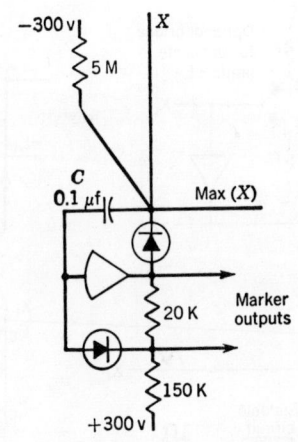

FIG. 11-10b. A maximum-remembering circuit which stores the largest value of X since the start of computation and marks the times of successive maxima with comparator steps. (*From H. D. Huskey and G. A. Korn, Computer Handbook, McGraw-Hill Book Company, New York, 1962.*)

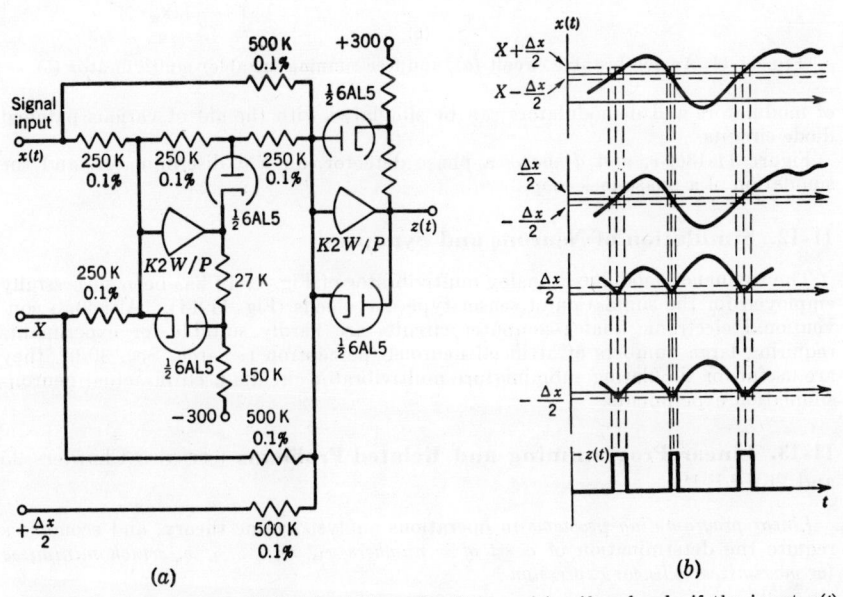

FIG. 11-10c. The output of this dual-slicer circuit is positive if and only if the input $x(t)$ is within $\Delta X/2$ of a set reference level X. This accurate decision circuit is useful, for example, for computing probability estimates. (*From H. D. Huskey and G. A. Korn, Computer Handbook, McGraw-Hill Book Company, New York, 1962.*)

cations and detection systems in their natural environment of noise, channel delays, clutter, etc.[11] One employs a reduced time scale, usually 1 cps/Mc for radar simulation and 1 cps/Kc for communications simulation. Figure 11-13a shows the circuit for a second-order filter with adjustable Q. Such filters may be staggered to simulate all kinds of radio-frequency and intermediate-frequency amplifiers. Many types

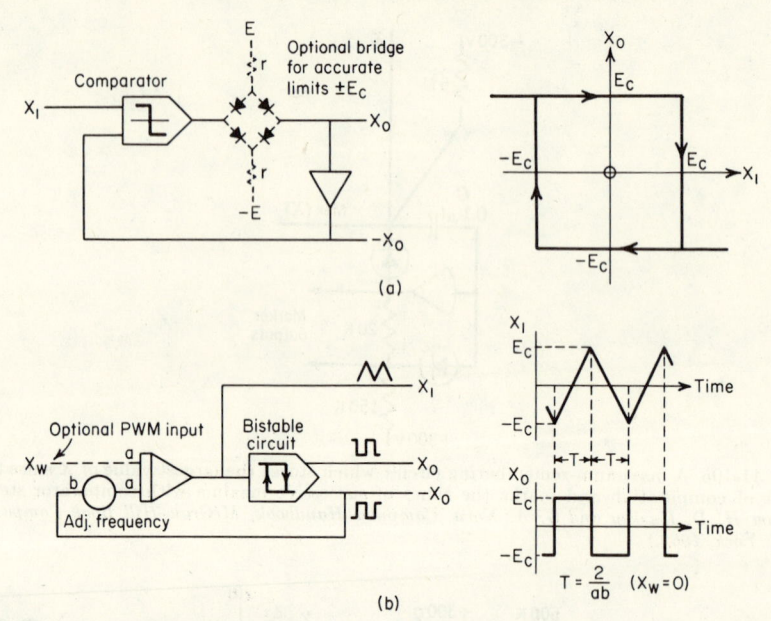

FIG. 11-11. Analog bistable circuit (*a*), and free-running (astable) multivibrator (*b*).

of modulators and demodulators can be simulated with the aid of various patched diode circuits.

Figure 11-13*b*, *c*, and *d* shows a phase detector, an FM discriminator, and the simulation of a phase-lock loop.

11-12. Simulation of Neurons and Synapses[12]

The frequency-modulated analog multivibrator of Fig. 11-12 has been successfully employed for the simulation of sensor-type nerve cells (Fig. 11-14). Although conventional electronic analog-computer circuits are hardly suitable for experiments requiring large numbers of artificial neurons (perceptron research, Sec. 30-9), they are useful for simulating subminiature multivibrator circuits used in actual neuron-simulation experiments.

11-13. Linear Programming and Related Problems (see also Chapters 25 and 26)[1,2,13,14]

Linear-programming problems in operations analysis, game theory, and economics require the determination of *a set of n numbers* $x_1, x_2, \ldots, x_n$ *which minimizes (or maximizes) a linear expression*

$$v(x_1, x_2, \ldots, x_n) = c_1 x_1 + c_2 x_2 + \cdots + c_n x_n \tag{11-11}$$

while satisfying $m \geq n$ *linear inequalities*

$$y_i \equiv a_{i1} x_1 + a_{i2} x_2 + \cdots + a_{in} x_n - y_{i0} \geq 0 \qquad (i = 1, 2, \ldots, m) \tag{11-12}$$

In a typical application, the problem is to buy necessarily positive quantities x_k of n types of raw materials ("inputs") so as to minimize the cost (11-11) while keeping the respective quantities y_i of $m - n$ output products at or above $m - n$ corresponding specified levels a_{i0} ($i = n + 1, n + 2, \ldots, m$).

ELECTRONIC ANALOG/HYBRID COMPUTERS   11-15

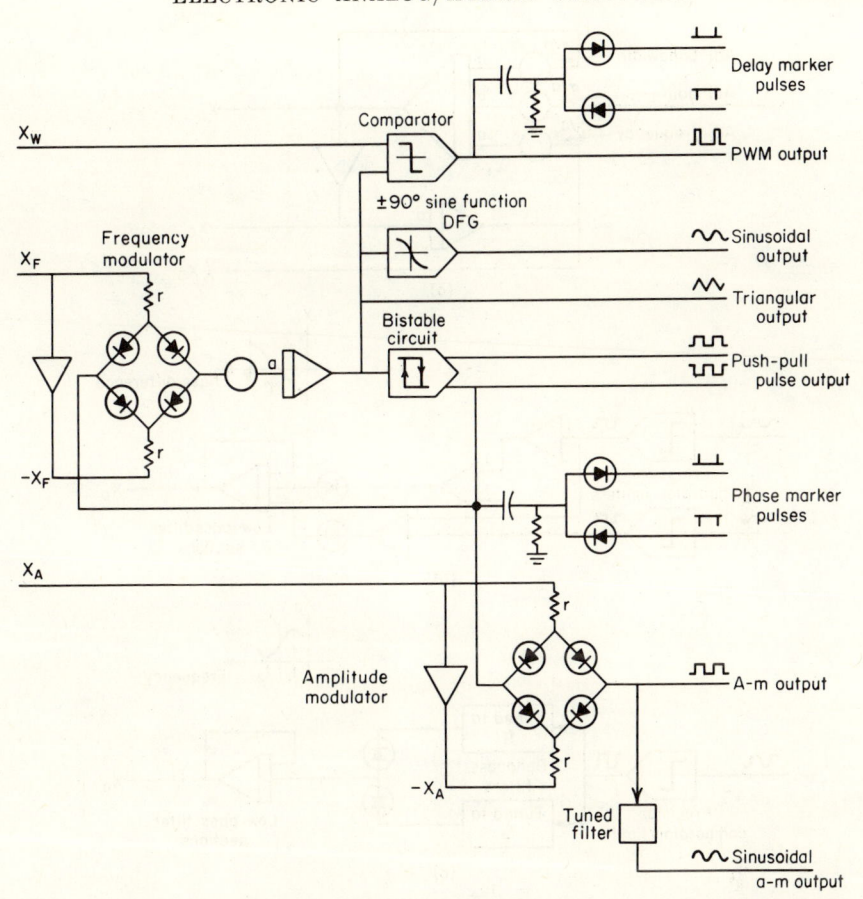

FIG. 11-12. This figure illustrates the generation of various types of modulated signals with the aid of the simple free-running multivibrator circuit of Fig. 11-11b. All outputs permit electrical frequency modulation completely independent from their amplitude or pulse-width modulation. Depending on the components used, the frequency range between 0.0001 and 10 cps can be covered with timing and amplitude accuracies within 0.04 per cent of full scale; accuracies within 0.5 per cent of full scale can be obtained up to 50 Kc. Clearly, only the portion of the circuit specifically needed for a given application need be built or patched.

The linear-programming problem defined by Eqs. (11-11) and (11-12) is equivalent to the minimization (or maximization) of the function

$$F(x_1, x_2, \ldots, x_n) \equiv v + \lambda_1 |y_1| + \lambda_2 |y_2| + \cdots + \lambda_m |y_m|$$

with
$$\lambda_i = \begin{cases} 0 & (y_i \geq 0) \\ B & (y_i < 0) \end{cases} \quad (i = 1, 2, \ldots, m) \tag{11-13}$$

where B is a large positive constant (in practice, B is between 10 and 100). Assuming the existence of a solution, the analog computer minimizes $F(x_1, x_2, \ldots, x_n)$ by implementing the n differential equations

$$\frac{dx_k}{d\tau} = -\alpha \frac{\partial F}{\partial x_k} = -\alpha(c_k + a_{1k}\lambda_1 + a_{2k}\lambda_2 + \cdots + a_{mk}\lambda_m)$$

$$(k = 1, 2, \ldots, n) \tag{11-14}$$

11-16 SYSTEM COMPONENTS

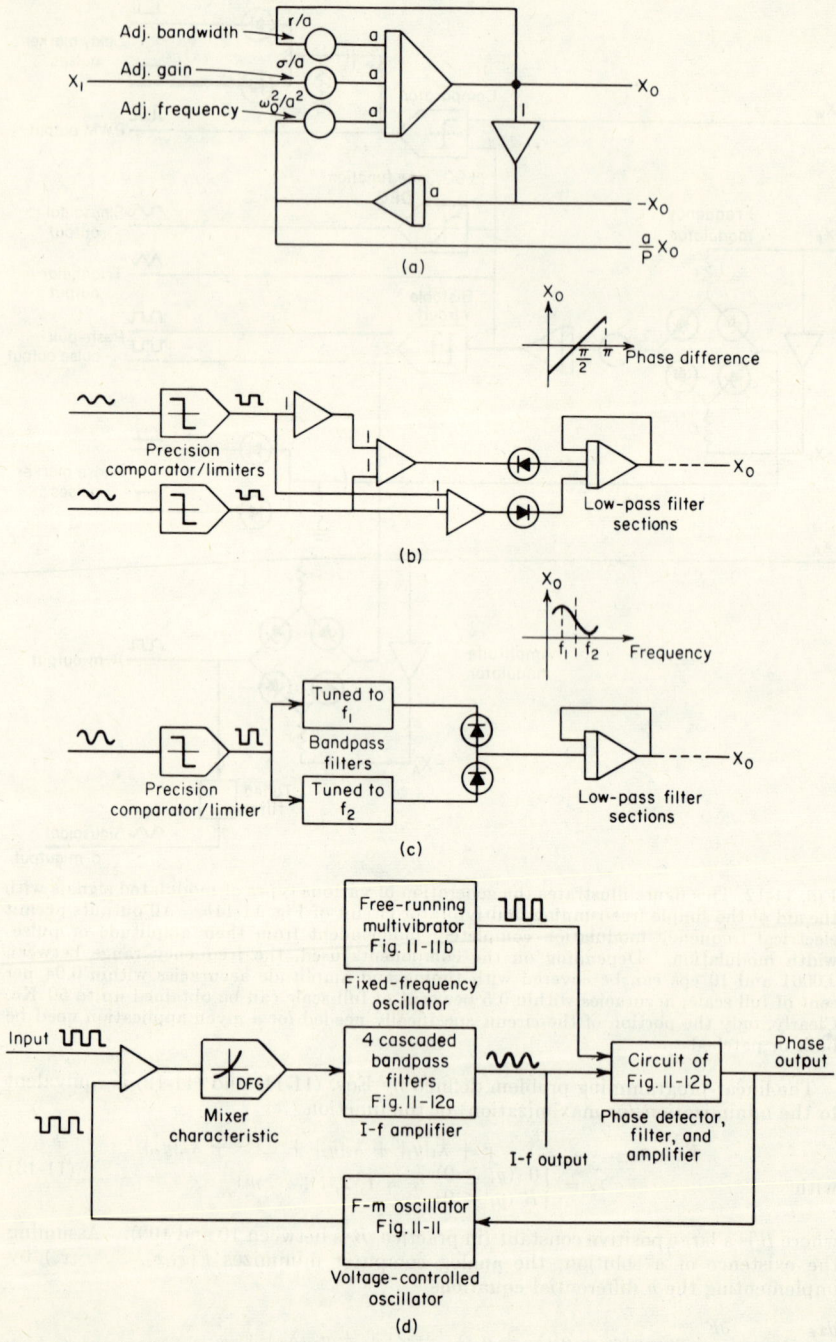

FIG. 11-13. Analog-computer simulation of communications-system components:[2,11] bandpass filter (tuned to $\omega_o = \omega$, 3 db bandwidth $\approx r = \omega_o/Q$) (a), phase detector (b), FM discriminator (c), and simple phase-lock loop (d).

ELECTRONIC ANALOG/HYBRID COMPUTERS   11-17

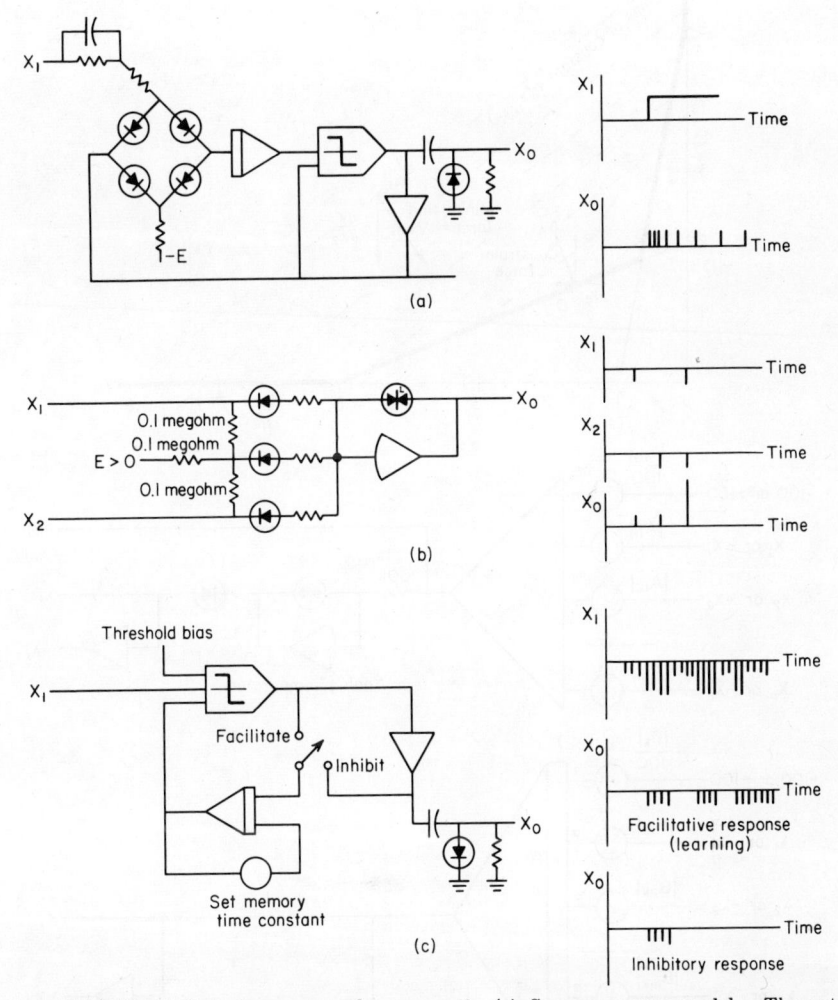

FIG. 11-14. Simulation of neurons and synapses.[12] (a) Sensory-neuron model. The output-pulse frequency increases with the stimulus voltage X up to a certain saturation level. The input filter has the transfer function $\dfrac{as+1}{bs+1}$, so that a step input generates a burst of output pulses which decays gradually to a steady-state pulse rate to simulate adaptation.
(b) Nonlinear summation of two simulated nerve-pulse trains.
(c) Simulated synapse with facilitative or inhibitory adaptation.

where α is a positive constant and τ is the computer time (real time). This *steepest-descent technique* forces each x_k to vary so as to decrease $F(x_1, x_2, \ldots, x_n)$ and hence to minimize v subject to the given constraints (11-12).

In the example of Fig. 11-15, x_1 and x_2 may be regarded as the rectangular cartesian coordinates of a particle moving in a viscous medium. The constraints are satisfied for all points (x_1, x_2) above and to the right of the convex polyhedron of smooth straight-line barriers indicated in heavy lines. The broken lines are lines of equal potential energy $v(x_1, x_2)$. The particle will move *in the direction of the negative gradient vector* $-\nabla v$ as shown until it strikes a barrier, say by violating the inequality

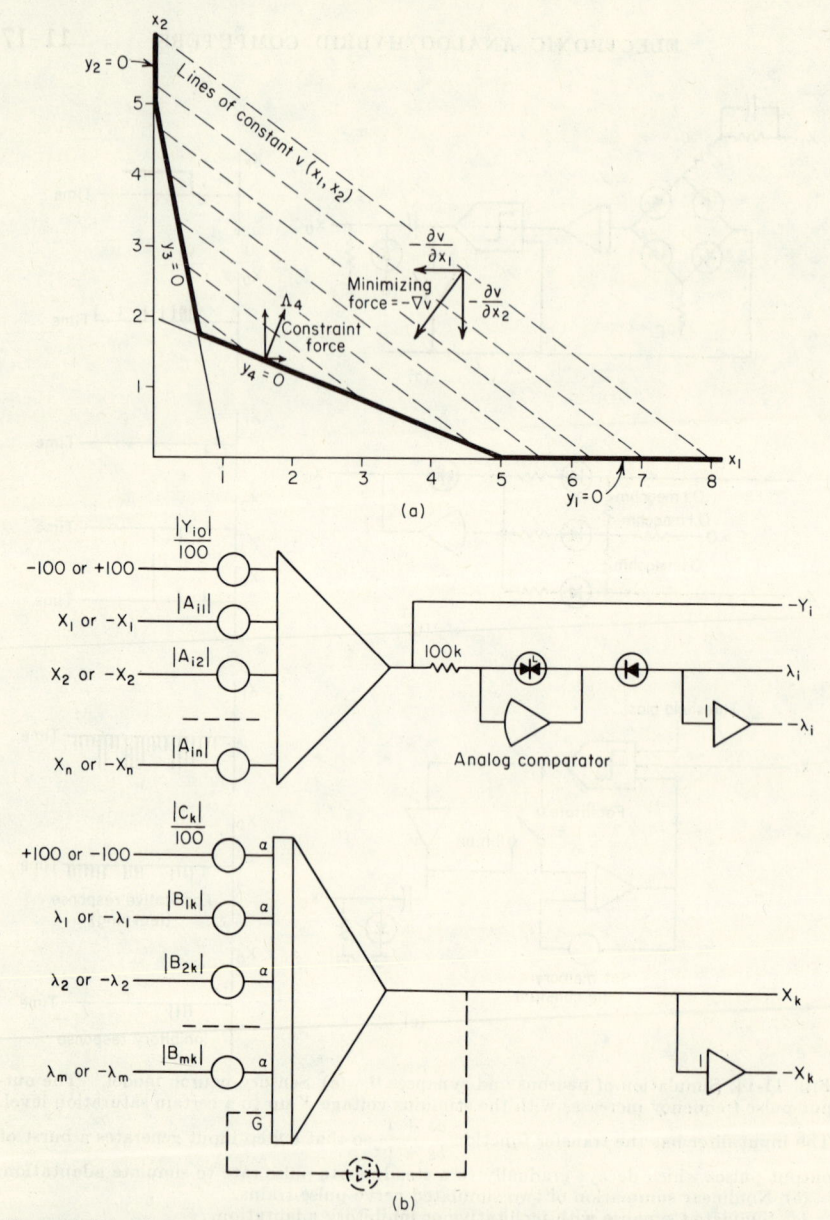

FIG. 11-15. (a) Dynamical-analogy interpretation of a linear-programming problem: steepest-descent minimization of $v(x_1,x_2) \equiv \frac{1}{4}x_1 + \frac{1}{3}x_2$, subject to the constraints

$$y_1 \equiv x_1 \geq 0 \qquad y_2 \equiv x_2 \geq 0$$
$$y_3 \equiv 10x_1 + 2x_2 - 10 \geq 0 \qquad y_4 \equiv 2x_1 + 5x_2 - 10 \geq 0$$

(b) Analog-computer set-up for the solution of linear-programming problems. The inverted input voltages $-X_k$, $-\Lambda_i$, or -1 are used with negative coefficients A_{ik} or C_k. If, as in many practical problems, an inequality $Y_i \geq 0$ takes the simple form $X_j \geq 0$, then the entire block of computing elements generating the corresponding voltage Λ_i is simply replaced by a feedback diode connected across the jth integrator so as to limit its output to nonnegative values (dash lines).

11–18

$y_4 \geq 0$. The particle then experiences an additional force at right angles to the barrier. As a result, the particle will move along the heavy lines to that barrier vertex which corresponds to the lowest potential energy $V(x_1,x_2)$. The problem is degenerate if ∇v is perpendicular to a barrier; in this case, the solution may not be unique.

Maximization problems would be solved similarly, with the minus sign in Eq. (11-14) replaced by a plus sign.

Figure 11-15b shows the analog-computer set-up for the steepest descent equations (11-14). The problem variables x_k, y_i, and v are represented by corresponding voltages $X_k = a_k x_k$, $Y_i = b_i y_i$, so that Eqs. (11-12) and (11-14) take the form of machine equations

$$Y_i = A_{i1}X_1 + A_{i2}X_2 + \cdots + A_{in}X_n - Y_{i0}$$
$$PX_k = -\alpha(C_k + B_{1k}\lambda_1 + B_{2k}\lambda_2 + \cdots + B_{mk}\lambda_m)$$

with
$$(i = 1, 2, \ldots, m;\ k = 1, 2, \ldots, n) \quad (11\text{-}15)$$

$$C_k = a_k c_k \quad A_{ik} = \frac{b_i}{a_k} a_{ik} \quad B_{ik} = a_k a_{ik} \quad Y_{i0} = b_i y_{i0}$$

m voltages $-Y_i$ are produced by summing amplifiers and applied to diode-connected analog comparators to generate the constraint-force voltages λ_i. Each X_k is obtained at the output of a summing integrator. Phase inverters also produce the voltages $-\lambda_i$ and $-X_k$ as needed for multiplication by negative coefficients. Coefficients exceeding unity in absolute value are obtained with the aid of amplifier input gains in the usual manner. Solutions converge within fractions of a second, depending on the integrator gain α used. A voltage proportional to v, if required, could be readily generated with an additional summing amplifier.

The analog-computer method is practical for linear-programming problems with m, $n \leq 20$; the practically instantaneous solution makes it particularly convenient to study effects of coefficient changes on the values of v and of the "slack variables" y_i. The n variables y_i closest to zero correspond to those of the m inequalities (11-12) which actually determine the solution vertex (see also Fig. 11-15a). One can, then, obtain solutions of essentially unlimited accuracy by solving the corresponding n simultaneous *equations* $y_i = 0$ by a suitable iteration method, which could utilize the analog computer in conjunction with a keyboard calculator.

11-14. Combined Analog/Digital Simulation[1,2,15,19]

Real-time simulation problems requiring high computing accuracy (e.g., trajectory computations with three-dimensional coordinate transformations) and simulation of digital system components have led naturally to combination of existing analog and digital computers in many of the larger simulation laboratories.

In the example of Fig. 11-16, a multipurpose digital computer is linked to a real-time analog computer through a commercially available conversion unit. The digital computer performs that part of the real-time computation for which it is best suited, namely, accurate coordinate conversions, trajectory computations, and the simulation of digital control equipment. The analog computer simulates the control-system dynamics, which require greater bandwidth and less accuracy.

The combined simulation technique was pioneered in the aerospace industry; experience clearly indicates that the provision of small multipurpose digital computers is greatly preferable to part-time utilization of large digital computers.

11-15. Faster Analog Computation and Iterative Operation[1,2,15,18,19]

Wherever the inclusion of actual system hardware does not necessitate real-time operation, there is a universal trend toward *faster analog computation*. All-electronic multipliers, function generators, and resolvers have become more accurate as well as faster than computer servomechanisms, whose sole advantage is their relatively low cost (one servo can perform several multiplications). Order-of-magnitude improvements in the speed and offset characteristics of electronic integrator switches have not only improved repetitive computation at 20 to 500 solutions per second but also establish the electronically switched integrator as a novel analog-computing element which can serve as an *analog memory cell* as well as an integrator or summer. Figure 11-17a shows an electronically switched integrator and illustrates its modes of operation.[2,24] Two such integrators are combined into a *comparator-controlled memory pair* (Fig. 11-17b) which can sample fast computer solutions at sampling times determined either by a timing system or by any voltage in the computer. The stepwise output of the memory pair can be introduced iteratively into successive computer runs (iterative differential-analyzer technique).

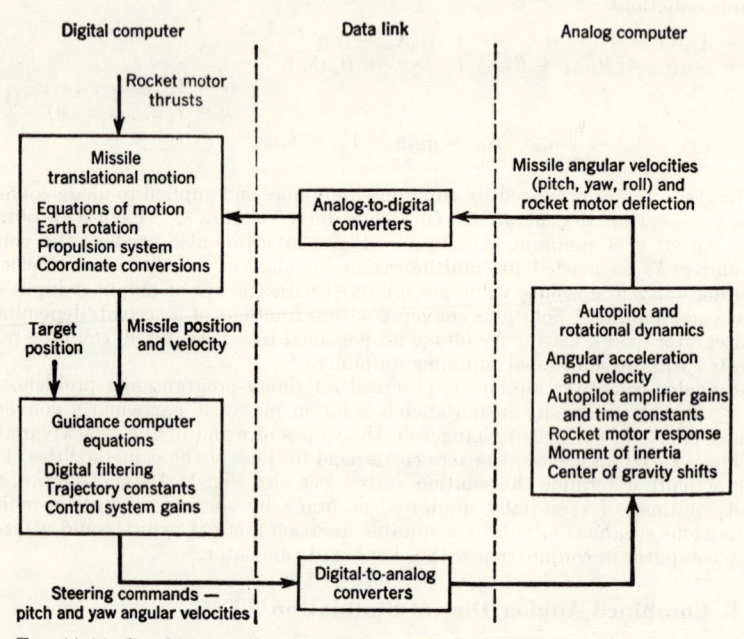

FIG. 11-16. Combined analog/digital simulation of a ballistic-missile system.

The repetitive/iterative differential analyzer with memory permits the introduction of "subroutines" obtained by fast repetitive analog computation into "slow" analog-computer set-ups. Examples of such applications include repetitive computation of the inner integral in a double integral (Fig. 11-18),[16] approximate solution of partial differential equations, and sequential solution of heat- and mass-balance equations at different points of a chemical reactor or distillation column. Another significant application is automatic parameter optimization by various digitally controlled steepest-descent techniques.[19] Automatic optimization of, say, 40 parameters of a simulated control system within a reasonable time makes high-speed repetitive operation very desirable; the need for high computing speed is especially acute when a sample of random-input computer runs must be taken for each combination of parameter values.

As a result, the well-known electronic differential analyzers have evolved into *iterative differential analyzers* which combine digital-computer control and memory features with ultra-high-speed analog solution of differential equations. In addition to conventional analog-computing elements capable of fast repetitive operation, an iterative differential analyzer comprises:

Electronically switched *integrator/memory units* (Fig. 11-17) and relay-switched *capacitor-memory banks*

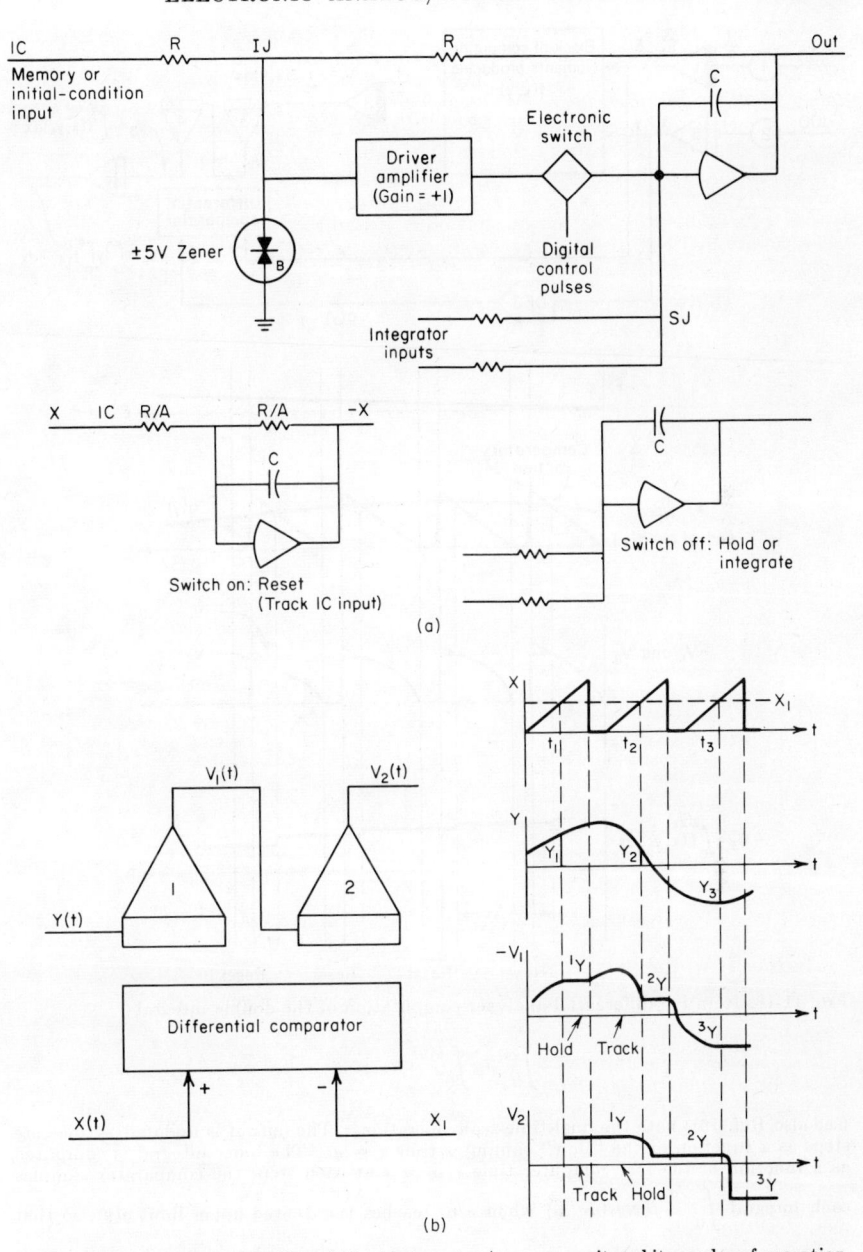

FIG. 11-17. (a) An electronically switched integrator/memory unit and its modes of operation.
(b) A comparator-controlled memory pair. Two electronically switched integrators alternate between TRACK and HOLD to permit solution of a wide variety of problems by finite-difference techniques. The ASTRAC system (see text) also permits digital control of individual integrators or memory pairs.

11-22 SYSTEM COMPONENTS

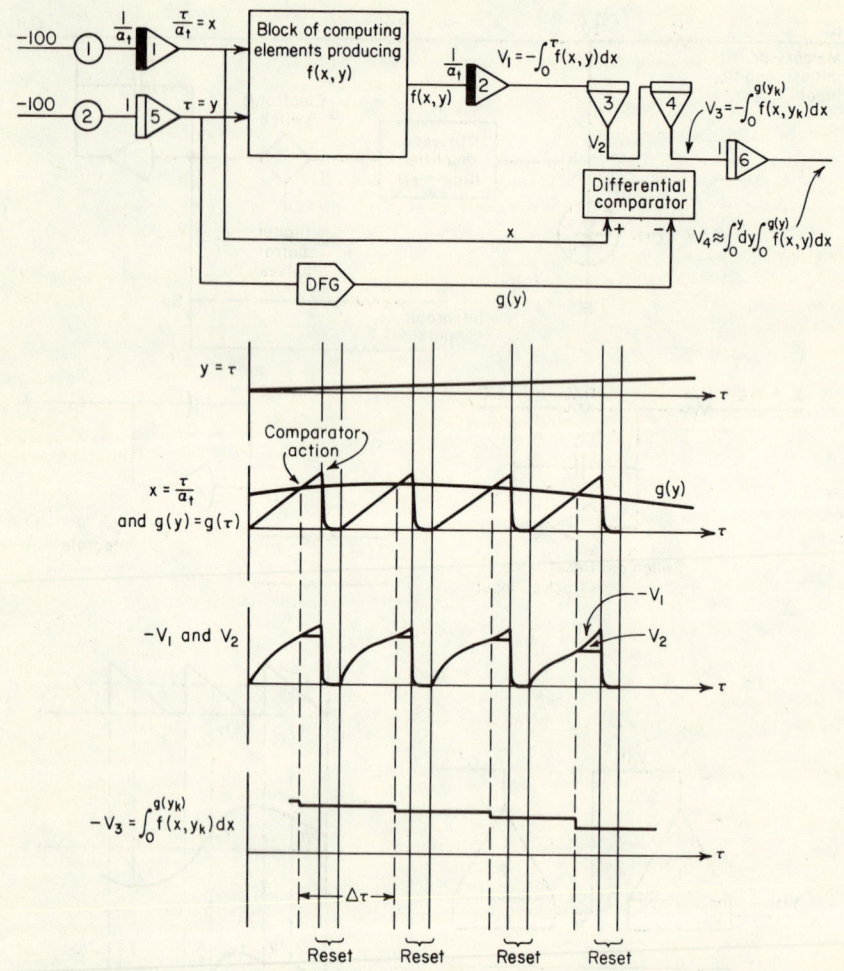

Fig. 11-18. Iterative-differential-analyzer computation of the double integral

$$\int_0^y dy \int_0^{g(y)} f(x,y)\, dx$$

(see also Ref. 16); note the dual-time-scale operation. The output is updated by discrete steps as a function of the "slow" computer time $\tau = y$. The inner integral is computed as a function of the fast computer time $\tau/\alpha_t = x$ at each step; the comparator samples each integral $\int_0^{\tau/\alpha_t} f(x,y)\,(d\tau/\alpha_t)$ when τ/α_t reaches the desired upper limit $g(y)$, so that

$$V_4 = \sum_{k=1}^n \Delta\tau \int_0^{g(y_k)} f(x,y_k)\, dx \approx \int_0^y dy \int_0^{g(y)} f(x,y)\, dx$$

Blacked-in integrators are periodically reset every $\Delta\tau = 0.01$ sec by the computer rep-op control unit.

Analog *comparators* and *switches* to implement solution-controlled program and parameter changes

Digital clocks to time iterative computation cycles and read-out operations

Digital expansion modules (flip-flops, shift registers, and logic gates) which accept comparator or clock pulses to implement complex decisions and/or timing cycles. Special computers of this type may evolve into true hybrid analog/digital computers comprising analog/digital/analog conversion links and special-purpose digital arithmetic units.[22,23] Practically all large analog computers permit set-up of coefficient potentiometers and function generators by digital keyboard control, punched-tape storage of coefficients and functions, and digital print-out of all initial conditions for checking purposes. Such machines are, then, only one step removed from complete automatic set-up by a digital-control computer which would program and change component interconnections as well as coefficient and function settings. As a rule, however, only a part of the necessary large number of component interconnections are made automatically; the rest are still programmed by patchboards.

11-16. System Design. The ASTRAC System and Monte-Carlo-type Random-process Studies[2,20,21]

While most iterative-differential-analyzer systems were developed as accessories for existing real-time computers, the Arizona Statistical Repetitive Analog Computer (ASTRAC) was designed from the start for high-speed operation, although real-time computation is also possible. ASTRAC combines a memory-equipped iterative differential analyzer with digital logic and control. The resulting synthesis of high-speed analog computation with digital automatic programming is of particular interest in connection with Monte-Carlo-type studies of random processes, which serve to illustrate ASTRAC system operation in Fig. 11-19. Here, an analog-computer simulated control system, communications system, queuing problem, etc., is supplied random inputs, initial conditions, or parameters from noise generators with Gaussian or random-impulse output. Reset pulses from a simple digital-control unit cause repetitive simulation of the process under study between 10 and 100 times per second. Accurate sample-hold (analog memory) units read selected process variables at pushbutton preset times t_1, t_2 after the start of each repetitive computer run. A hybrid analog/digital statistics computer accepts these samples to accumulate statistical averages over 100 to 10,000 computer runs, as determined by a preset run counter in the control unit. In this manner, one can estimate ensemble averages such as mean-square delay error, correlation functions, and probabilities for very complicated nonstationary as well as stationary random processes.

Aside from the control unit, various display and recording devices, and ± 300-volt and ± 100-volt ($\pm$ computer reference) power supplies, the ASTRAC system combines *groups of modules*, viz.,

1. *Summer/integrator/comparator group*
   2 summer/integrator/memory amplifiers
   1 summing amplifier
   1 high-gain amplifier/phase inverter
   1 comparator
   5 potentiometers
2. *Free-amplifier group*
   4 free operational amplifiers and bias connections for special diode circuits
3. *Digital-circuit group*
   2 Schmitt trigger/cathode followers
   2 flip-flops
   2 pulse inverter/differentiators
   2 analog switches
   patched diode gates and switches
4. *Multiplier/function-generator group*
   2 quarter-square multipliers
   2 diode function generators

SYSTEM COMPONENTS

ASTRAC I uses vacuum-tube amplifiers having a 0-db bandwidth of 1.5 Mc (model 3) or 6 Mc (model 4), and omits removable patchboards to permit short wiring. A follow-on project, ASTRAC II, employs miniaturized all-solid-state component modules plugged directly into the rear of a shielded patchbay with removable patchboards; this type of construction does away with all wiring between amplifiers and patchbay.

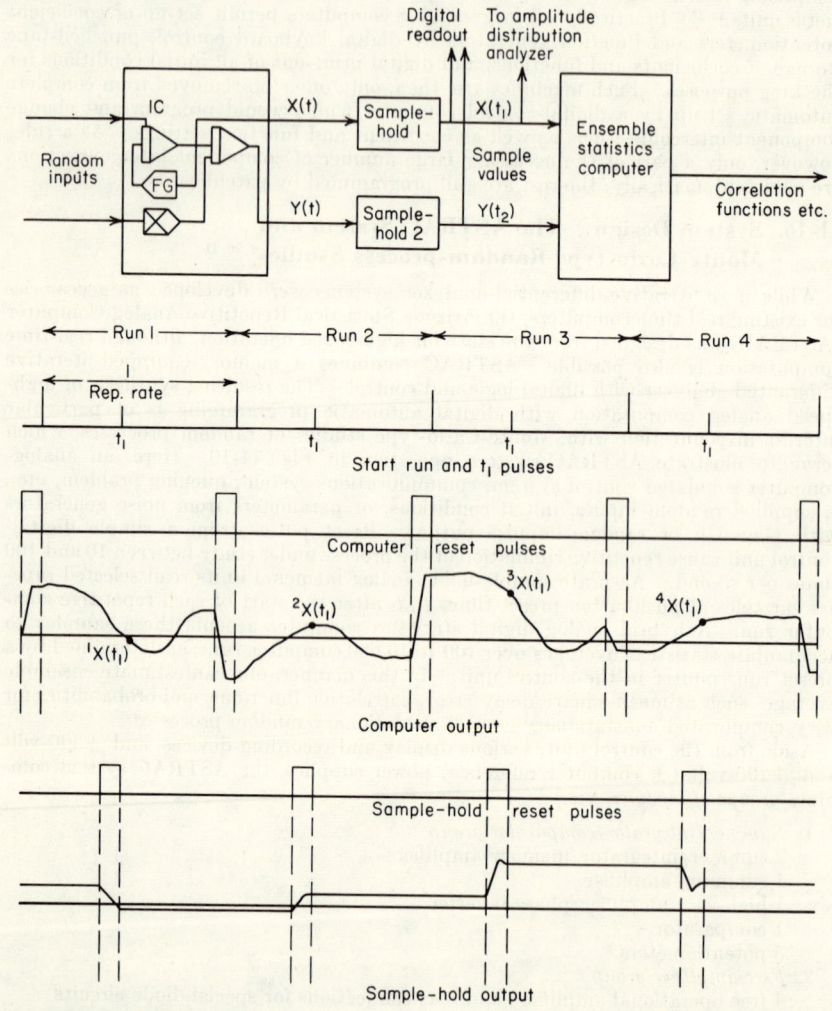

FIG. 11-19. ASTRAC system operation for Monte-Carlo-type studies of random processes.

ASTRAC integrator/memory units (Fig. 11-17) can be comparator-controlled, or they can be switched by suitably patched configurations of commercially available digital-logic modules; note that the circuit of Fig. 11-17 permits even ±100-volt vacuum-tube amplifiers to be controlled with 10-volt switching waveforms from solid-state control multivibrators.

In the ASTRAC digital-control unit,[2,24,25] the analog-computer repetition rate (10 cps to 1 Kc) is counted down from a 10-Kc or 4-Mc crystal oscillator. In practice, the latter is intentionally slightly detuned to avoid statistical sampling at a frequency commensurable with the 60-cps line frequency. Reset pulses marking the end of each computer run are counted in a *run counter* preset to start and stop statistical computations after a sample of 100 to 10,000 runs.

The end of each integrator control pulse also resets additional decimal counters preset to furnish *sampling pulses at pushbutton selected times t_1 and t_2, or $t_1 + \tau$ after the start of each computer run.* These sampling pulses permit very accurate timing of sample-hold read-out circuits (Fig. 11-19).

An additional circuit feature is a SCAN mode which causes the sample-hold read-out time t_2 to advance by a step Δt for each successive run. *This feature is used to read repetitive computer solutions $x(t_2)$, $y(t_2)$ at relatively slow pushbutton-selected rates into accurate recorders, printers, and associated analog or digital computers.*

In addition to its internal automatic operation modes, the control unit provides for control by external reset pulses and external t_1 and t_2 read-out pulses from associated control or instrumentation systems or from other analog or digital computers.

The ASTRAC analog-digital *sample-averaging unit* converts each voltage sample into a pulse of proportional length (Fig. 11-10a) and uses these pulses to gate clock pulses into a type of read-out counter which indicates the resulting sum or sample average on a convenient scale. A stop pulse from the run counter terminates the count. This inexpensive circuit can be switched to average X, $|X|$, or $X^2/100$ to yield estimates of ensemble averages $E\{X\}$, $E\{|X|\}$, or $E\{X^2\}$ for nonstationary as well as stationary random processes. Averages of products XY (in particular, correlation-function estimates) are obtained with the aid of the relation

$$XY = \left(\frac{X+Y}{2}\right)^2 - \left(\frac{X-Y}{2}\right)^2$$

Averages of the function

$$y(x) = \begin{cases} 1 \text{ if } a < x(t_1) \leq b \\ 0 \text{ otherwise} \end{cases}$$

yield estimates of the probability Prob $[a < x(t_1) \leq b]$, and for small values of $b - a = \Delta x$, estimates of the probability density of $x(t_1)$ (*amplitude-distribution analyzer*).[10]

ASTRAC patchable digital modules serve as flexible building blocks for additional control features, as well as for the implementation of complex logically interrelated decisions, especially in the simulation of random processes. An additional important application is the synthesis of special analog- or hybrid-computing elements either for use in a specific computation, or to save breadboarding of such components. Analog and digital components communicate by way of the comparators and analog switches.

REFERENCES

1. Huskey, H. D., and G. A. Korn: *Computer Handbook*, McGraw-Hill Book Company, New York, 1962.
2. Korn, G. A., and T. M. Korn: *Electronic Analog and Hybrid Computers*, McGraw-Hill Book Company, New York, 1964.
3. Jackson, A. S.: *Analog Computation*, McGraw-Hill Book Company, New York, 1960.
4. Smith, G. W., and R. C. Wood: *Principles of Analog Computation*, McGraw-Hill Book Company, New York, 1959.
5. Fifer, S.: *Analogue Computation*, McGraw-Hill Book Company, New York, 1961.
6. Morrill, C. D., and R. V. Baum: Diode Limiters Simulate Mechanical Phenomena, *Electronics*, November, 1952.
7. Koerner, H., and G. A. Korn: Function Generation with Operational Amplifiers, *Electronics*, Nov. 6, 1959.
8. Barker, B., and M. MacMahan: Digital Voltmeter Employs Voltage-to-time Converter, *Electronics*, May 5, 1961.
9. Caldwell, W. F., G. A. Korn, V. R. Latorre, and G. R. Peterson: A Precision Amplitude-distribution Analyzer, *Trans. IRE/PGEC*, June, 1960.
10. Brubaker, T., and G. A. Korn: Accurate Amplitude-distribution Analyzer Combining Analog and Digital Logic, *Rev. Sci. Instr.*, vol. 32, March, 1961.

11. Berger, E. L., and R. M. Taylor: Optimization of Radar in Its Environment by GEESE Techniques, *Proc. Western Joint Computer Conf.*, 1961.
12. Diamantides, N. D.: Artificial Neurons through Simulation, *Proc. 3d AICA Conf.*, Opatija, Yugoslavia, Sept. 3–9, 1961; Presses Académiques Européennes, Brussels.
13. Pyne, I. B.: Linear Programming on an Analog Computer, *Commun. Electronics*, vol. 24, p. 139, May, 1956.
14. Rogers, A. E., and T. W. Connolly: *Analog Computation in Engineering Design*, McGraw-Hill Book Company, New York, 1960.
15. Korn, G. A.: The Impact of Hybrid Analog-digital Techniques on the Analog-computer Art, *Proc. IRE*, vol. 50, pp. 1077–1086, 1962.
16. DYSTAC Applications (Eastern Simulation Council Presentations, Dec. 12, 1960), published by Computer Systems, Inc., Monmouth Junction, N.J., 1960.
17. Gilliland, M. C.: Iterative Differential Analyzer Function and Control, *Instruments and Control Systems*, April, 1961.
18. *Proc. Combined Analog-digital Computer Systems Symposium*, Philadelphia, Pa., Dec. 16–17, 1960; published by Simulation Councils, Inc., 8484 La Jolla Shores Dr., La Jolla, Calif.
19. *Hybrid Computation Course Notes*, Electronic Associates, Inc., Long Branch, N.J., March, 1962.
20. Korn, G. A.: New High-speed Analog and Analog-digital Computing Techniques: the ASTRAC System, *Proc. 3d Intern. Conf. Analog Computations*, Opatija, Yugoslavia, Sept. 3–9, 1961, Presses Académiques Européennes, Brussels.
21. Vander Velde, W. E.: Make Statistical Studies on Analog Simulations, *Control Eng.*, June, 1960.
22. Skramstad, H. K.: A Combined Analog-digital Differential Analyzer, *Proc. Eastern Joint Computer Conf.*, 1959.
23. Schmid, H.: Combined Analog-digital Computing Elements, *Proc. Western Joint Computer Conf.*, 1961.
24. Eckes, H. R., and G. A. Korn: Digital Program Control for Iterative Differential Analyzers, *Simulation*, February, 1964.
25. Korn, G. A.: Fast Analog-hybrid Computation with Digital Control: the ASTRAC II System; *Proc. 4th Intern. Conf. Analog Computations*, Brighton, England, 1964; Presses Académiques Européennes, Brussels.

Chapter 12

COMMUNICATIONS EQUIPMENT

L. J. CUTRONA, *Vice President for Applied Research, Conductron Corporation; Professor of Electrical Engineering, University of Michigan.*

CONTENTS

12-1. Examples of Communication Systems	12-1
12-2. The Components of a Communications Channel	12-2
Source	12-2
Encoder	12-2
Channel	12-3
Decoder	12-4
Destination	12-4
12-3. Multiplexing	12-4
Frequency-division Multiplexing	12-4
Time-division Multiplexing	12-5
Sampling Theorem	12-5
Pulse Time Modulation	12-6
Pulse Code Modulation	12-7
Summary Remarks	12-9
12-4. Communication Channels	12-10
12-5. The Coding-Decoding Problem	12-10

12-1. Examples of Communication Systems

There are many methods for the communication of information from one point to another. Included among these are telegraphy, telephony, radio, television, and data transmission. In each of these cases, a source of information is located at one point in space, whereas the intended destination for the information is at another location. The communication systems used depend in part upon the nature of the information to be transmitted and in part upon the geography of the situation. The messages to be sent may be such that a low-bandwidth channel is sufficient; for instance, telegraphy may be sent over channels whose bandwidth is approximately 800 cps. Similarly, telephony using voice communications can be transmitted over channels having bandwidths of about 3,000 cps. Television signals, on the other hand, require much wider bandwidths, and channels suitable for television must have approximately 4.5 Mc of bandwidth.

12–2   SYSTEM COMPONENTS

The distance between source and destination determines in part whether the link shall make use of wire or radio circuits. Depending upon traffic density, range, need for privacy, etc., choices must be made between radio and wire links. It is the objective of this chapter to discuss the essential components of a communication system and to give a description of the more important modulation and demodulation schemes.

12-2. The Components of a Communications Channel

A communications channel can be represented as shown in Fig. 12-1. As this figure shows, information is generated in a source. The output of the source is fed to an encoder whose function it is to perform operations on the messages from the source to adapt them to the signal channel. There are a number of functions of the encoder, such as modulation, coding for signal-to-noise purposes, and coding for maximizing information per unit power. The output of the encoder is fed to the channel. The channel, which may be wire, coaxial cable, waveguide, a beam of light, or radio, extends from the output of the encoder associated with the transmitter, to the input of a decoder associated with the receiver at the remote location. It is the function of the decoder to perform the inverse of the operations performed by the encoder. The output of the decoder contains the receiver's best estimate of the

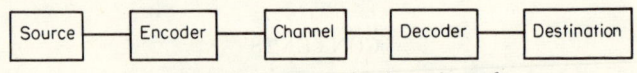

Source ── Encoder ── Channel ── Decoder ── Destination

Fig. 12-1. The communications channel.

message which was sent, or some function of that message. The destination receives the message. The destination may be a viewer watching a television program, a typewriter printing a message, a computer controlling a machine, etc.

Some characteristics and considerations regarding each of these items will be discussed below. Specifically, the problem of multiplexing will be discussed in Sec. 12-3; in this discussion, the roles of the encoder, the channel, and the decoder are important. Section 12-5 will touch on the coding-decoding problem.

Source. The source originates a message or sequence of messages. A tremendous variety of sources exists. A source may be oral, as in telephony; visual, as in television; or a physical variable, as in telemetry. Alternatively, the source may consist of data to be transmitted on punched paper tape, magnetic tape, etc., for purposes of telegraphy. The source is characterized by its statistics, by the rate at which information is generated, and by the nature of the quantity which is varying.

Encoder. It is the function of the encoder to accept the input information from the source and to use the information to modify a carrier of some type, usually electrical—in part to adapt the message to the characteristics of the transmission medium, and in part to accomplish some additional purpose, such as signal-to-noise improvement, error correction, or energy minimization.

In adapting the information to the channel or medium, the information is used to modulate one or more carriers using some method such as amplitude modulation (AM), frequency modulation (FM), or pulse modulation. In most cases, the information is put onto a carrier frequency suitable for radiation (radio), or suitable for transmission over wire lines, coaxial cables, waveguides, etc., for conductive connected circuits.

In some cases, the function of multiplexing is performed by the encoder. This means that the appropriate operations are performed on a multiplicity of messages so that they can be transmitted simultaneously over the same channel. Two important multiplexing philosophies need to be considered, namely, frequency-division multiplexing, and time-division multiplexing. The characteristics of these multiplexing methods are discussed in Sec. 12-3.

The encoder functions thus far described are those which adapt the information to the channel. In addition, encoding is often resorted to for information-theoretic reasons.

COMMUNICATIONS EQUIPMENT 12-3

One of the most important purposes of further encoding a message is signal-to-noise improvement. The improvement in signal-to-noise ratio occurs in the decoder, although this improvement must be anticipated in the design of the encoder. The signal-to-noise ratio of the signal entering the decoder depends upon such factors as radiated power, attenuation through the channel, bandwidth, some characteristics of the medium, and some characteristics of the receiver. If appropriate encoding has been used before transmission of the message, the decoder can give a message at its output whose signal-to-noise ratio will be substantially better than the signal-to-noise ratio which was present at the input terminals of the receiver.

In some cases, encoding is also resorted to for purposes of error correction. Messages can be sent together with sufficient additional information to make possible the location and correction of errors in the transmitted messages. This, too, requires appropriate encoding and decoding.

In some cases it is also a function of the encoder to operate upon the message in such a manner as to achieve privacy or security. For that purpose, the message is modified in such a manner that its decoding is difficult without knowledge of the method of encoding.

Still other purposes for encoding are motivated by the desire to achieve maximum channel capacity. It has been shown by Shannon[1] that, under appropriate conditions, the maximum channel capacity in bits per second is given by

$$C = B \log_2 \left(1 + \frac{S}{N}\right)$$

where B is the bandwidth in cycles per second. Shannon has also indicated the general characteristics of coding schemes to achieve or approximate this rate. In general, long delays both in transmitting the message and in receiving it are necessary to approach this rate of transmission of information. Much work has been done attempting to find efficient codes for this purpose. See Chapter 22.

Channel. The channel is the medium over which the messages in encoded form are conveyed from the source to the destination. Channels may be divided into two classes, depending upon whether conductive paths or radiative paths are provided (compare Table 13-4).

In conduction, wires, transmission lines, coaxial cables, etc., are used to conduct the encoded message from one point to another. Such means have associated with them an attenuation per unit distance. They are also characterized by being directly connected from one specific point to another specific point so that privacy in the messages can be assured more easily.

In radiation, the signals after encoding and impressment upon an appropriate carrier are fed to a transmitting antenna. These signals propagate as electromagnetic waves. A receiving antenna intercepts a portion of this radiated energy. Except for attenuation and distortion in the propagation path and for noise which may have been added by the transmission medium, the received signal duplicates that which was transmitted. Such means have associated with them two types of attenuation: one proportional to an inverse power of the distance (usually inverse square), and one proportional to an exponential function of the distance (as in the conducting case); the latter is negligible in some cases, especially in propagation of electromagnetic radiation through essentially free space.

In the case of radio waves, it is much more difficult to assure privacy, although a limited amount of such privacy can be achieved by building directive antennas and by encoding the messages in an appropriate manner prior to transmission. In the use of radiated signals, frequency assignment becomes important, and because of the limited spectrum available, interference between messages sent over the same frequency bands is often a problem.

In the category of signaling by the use of radiated energy, a new technique is now on the horizon, namely, the modulation of a coherent light beam as a means of communication. In this technique, the coherent light output from a device such as a laser is modulated, and the information is transmitted over the beam of light. Appropriate detection devices extract the message from the light beam.

Decoder. The function of the decoder in a communication system is usually to perform the operation which is inverse or complementary to the operation of the encoder. Generally, the output of the decoder is the equipment's best estimate of the message which was sent. Another function of the decoder is to convert the signal back to a form appropriate for use by the destination. In the case of telephony, for example, the information is converted back to an audio message. In the case of television, the signal is converted to a visual output appearing on the face of a cathode-ray tube. In the case of transmission of data, the destination will generally be some type of recording, such as typed numbers on a page, punched paper tape, or a line graph.

Most frequently it is the function of the decoder to recover the message which was fed into the encoder. However, at times some function of the message (e.g., its derivative or integral) rather than the message itself is required.

An important function of the decoder, in some cases, is improvement of the signal-to-noise ratio. Many systems of encoding exist for which the signal-to-noise ratio at the output of the decoder exceeds that at the inputs to the device. Examples are FM reception, pulse time modulation reception, pulse code modulation reception, and error-correcting codes. The nature and magnitude of this improvement are discussed for special cases in Section 12-3 under Time-division Multiplexing.

Destination. The destination of a message is the receptor of the desired information. Little needs to be said about the destination other than that the signal is required to be in a form acceptable to the destination. For example, if the receptor is the ear, it is necessary to convert the signal to audio form. Similarly, if the destination is to be a printed page of data, the signals must be converted to such a form as to operate the appropriate keys on a printer or typewriter.

12-3. Multiplexing

In communications systems it is often desirable to transmit a multiplicity of messages over the same channel. In this case, it is necessary to perform functions upon the message to combine them in such a manner that their decomposition into the composite messages is possible at the receiver. Two important methods are frequency-division multiplexing and time-division multiplexing.

Frequency-division Multiplexing. In frequency-division multiplexing, a set of messages is combined by translating the individual messages into nonoverlapping frequency intervals. Usually a frequency interval somewhat wider than the bandwidth of the message is used to provide guard bands around the signal.

The frequency translation is generally accomplished by mixing the message with the output of a stable oscillator. The frequency of this oscillator is chosen to be equal to the frequency translation desired. The oscillator signal and the message are mixed in a nonlinear device. In this device a multiplicity of frequencies is generated, given by multiples of the signal frequency (or frequencies). Frequency filtering is used to select the signal components in the desired frequency interval. By proceeding in this manner, it is possible to place the frequency-displaced messages into nonoverlapping frequency bands. A summing of these signals gives the frequency multiplexed signal. This signal may be transmitted as it stands or it may be considered as a composite signal to modulate a carrier, if such is desired.

To recover the messages in frequency-division multiplexing, frequency filters are used to select each of the messages; it is necessary to translate these signals back to the original frequency interval. This is accomplished in a manner similar to the translation used in multiplexing signals, namely, by mixing the frequency-translated signal with an oscillator followed by a filter which selects the pass band of the original message.

Frequency-division multiplexing is the earlier of the multiplexing methods used in communications systems. Systems have been built in which 2, 4, 8, 16, and 24 messages have been multiplexed. In designing frequency-division multiplexing circuits, linearity becomes of extreme importance because lack of linearity produces crosstalk among channels. Moreover, extreme care must be used in adjusting the levels of these signals to provide appropriate signal levels, particularly in those cases in which a large number of repeaters is required.

The assignment of frequencies to radio transmitters is a further example of the use of frequency-division multiplex. The tuning of a radio receiver to an appropriate frequency band is the means for demultiplexing in this case.

Time-division Multiplexing. In time-division multiplexing, continuous functions are not sent. Instead, a set of samples of the function to be transmitted is taken, and the signal transmitted is some function of the set of samples. A variety of functions of the samples can be used. For example, if pulses whose amplitude is proportional to the amplitude of the sample are sent, one has pulse amplitude modulation; if a pulse whose time of occurrence depends on the amplitude of the sample is sent, one has pulse time modulation; if a sequence of pulses which forms the binary representation of the sample amplitude is sent, one has pulse code modulation. Other types of modulation that might be used are pulse width modulation and pulse frequency modulation. In the former, the width of the pulse is modulated in accordance with the sample value; in the latter case, the number of pulses per second is proportional to the sample value. In each case, the interval between samples can be long compared with the sample duration, and in transmitting a message no signal is sent between samples. Thus one is led to consider samples from a multiplicity of messages interleaved.

A time-division multiplexed signal, therefore, consists of a sequence of frames in which each message is sampled once in that frame. An appropriate operation is performed on the sample depending on the type of modulation, and the samples are ordered in time.

Sampling Theorem. Fundamental to time-division multiplexing is the sampling theorem, which states that a signal can be represented by a sequence of samples of the function, provided certain conditions are met. It can be shown that, if a function $f(t)$ has a bounded spectrum, i.e., if

$$|F(\omega)| = 0 \quad \text{for } |\omega| > \omega_0 \tag{12-1}$$

where $F(\omega)$ is the Fourier transform of $f(t)$, then $f(t)$ can be represented by the expression

$$f(t) = \sum_{n=-\infty}^{\infty} f(nT) \frac{\sin \pi(t - nT)/T}{\pi(t - nT)/T} \tag{12-2}$$

In this expression, T is the interval between samples and must satisfy the inequality

$$T < \frac{\pi}{\omega_0} \tag{12-3}$$

which states that the sampling rate must be at least twice the highest frequency contained in the spectrum of $f(t)$. Equation (12-2) states that, given a function $f(t)$ whose spectrum satisfies (12-1), one can reconstruct the function from a set of samples of a function, provided the samples are taken sufficiently often, and provided that in the reconstruction process each sample is used to generate a waveform of the form $(\sin x)/x$.

Other sampling theorems exist. The sampling theorem stated above is appropriate when the signal has a spectrum which extends from 0 to some maximum value $\omega_0/2\pi$. For the band-pass case, a sampling theorem can be stated for which a function whose spectrum extends from $\omega_1/2\pi$ to $\omega_2/2\pi$ can be reconstructed from its samples. There are differences in detail, but the same minimum number of samples per second are required, namely, a sampling rate equal to at least twice the bandwidth of the signals. Still other forms of the sampling theorem exist, in which samples of the function together with one or more of its derivatives can be used to reconstruct the function. Some of these sampling theorems were discussed in a paper by D. Linden.[2]

SYSTEM COMPONENTS

Implicit in the sampling theorem is the fact that the samples of the function can be delta functions, i.e., very narrow functions. This provides the opportunity to interleave samples from a number of messages to obtain a time-division multiplex signal.

Pulse Time Modulation. In pulse time modulation, the value of the sample is used to determine the time of occurrence of a pulse with respect to a reference time. For example, if the pulse occurs at the reference time, the sample value was 0; if the pulse occurs 2 units of time earlier than the reference time, the sample value was -2; if the pulse occurs 5 units of time later than the reference time, the sample value was $+5$. In pulse time modulation, a reference instant of time, and a time interval about this instant, is set aside for each message. To generate a pulse time modulation signal, a function is sampled. From this sample value the time of occurrence of the pulse is controlled.

In forming a complete time-division multiplex system in which pulse time modulation is used, a frame is divided into a number of subintervals. A specific communication equipment for transmitting 24 voice channels will be used as an illustration.

In this case, it was considered that the highest important frequency for telephone quality was 3,000 cps. A sampling rate of 8,000 cps (125 μsec between samples) was selected, which exceeds twice the highest frequency of interest. This 125-μsec interval between samples can be subdivided into 25 subintervals, each 5 μsec long. The first of these subintervals is used for synchronization purposes, and a readily recognizable pulse pattern, such as two pulses, is used to distinguish this "synch interval." Each of the subsequent 5-μsec intervals is associated with one of the 24 messages. The center of each interval corresponds to 0 amplitude; hence a maximum excursion of 2.5 μsec is used to represent the maximum amplitude of the sample function.

In building up one frame of pulse time modulation, the pulses forming the synchronization signals are formed. Then message 1 is sampled, and a pulse whose time of occurrence represents the sample is placed into the first 5-μsec interval following the synch pulse. Similar operations are performed, in turn, on messages 2 to 24 to generate the full 125-μsec frame interval. The frame is repeated in the same sequence in each subsequent 125-μsec interval. The signal thus obtained can be sent as a video waveform, or it can be used to modulate a carrier in any of the standard ways, e.g., amplitude modulation, frequency modulation, etc. In the receiver, time gates are employed such that the samples from message number 1 and subsequent messages are routed in sequence to the appropriate equipment for extracting the set of sample values from the time-modulated pulses. In the example under consideration, 24 such channels would be used. The synch pulses are used to generate the time base for sorting the signals.

To achieve pulse time modulation, the value of the sample is added to an appropriate bias level. Then it is compared with the value of a linearly increasing voltage. A pulse is generated at the time at which this linearly increasing voltage achieves equality with the sample plus bias value. Thus lower sample values cause earlier-occurring pulses than higher sample values. The linear ramp function must have appropriate slope and duration. To demodulate pulse-time-modulated signals, a similar linearly increasing voltage is generated. The occurrence of the pulse is used to read out the value of the linearly increasing voltage at the time of occurrence of the pulse. Thus an earlier-occurring pulse reads out a lower voltage than does a later-occurring pulse, thereby regenerating the sample value.

In some cases an attempt is made to generate the $(\sin x)/x$ waveform required by the sampling theorem from the set of samples. In other cases the sequence of pulses (now in pulse-amplitude form) is simply passed through a low-pass filter, and the signal distortion resulting therefrom is accepted.

A motivation for the use of pulse time modulation is the improvement in signal-to-noise ratio which occurs as a result of the use of this modulation. The signal-to-noise ratio at the output of a pulse-time-modulation system is improved over the signal-to-noise ratio before the demodulation in the ratio $(\delta/\tau)^2$, where δ is the pulse deviation and τ is the pulse rise time.

This result can be derived as follows:

COMMUNICATIONS EQUIPMENT

The output e_o of a pulse-time-demodulation equipment can be written as

$$e_o = E_o \frac{t}{\delta} \tag{12-4}$$

where E_o is the maximum signal amplitude, δ is the maximum pulse deviation, and t is time measured from the reference time. If a pulse has rise time τ, the shape of its leading edge can be approximated by

$$e = E_o \frac{t}{\tau} \tag{12-5}$$

The input noise voltage n_i can be considered as an error in e. This causes an error in t. By differentiation of Eq. (12-5) one obtains

$$n_i = de = \frac{E_o}{\tau} dt \tag{12-6}$$

The output noise n_o can be considered as the fluctuation de_o. Differentiation of (12-4) gives

$$n_o = de_o = \frac{E_o}{\delta} dt \tag{12-7}$$

Combination of Eqs. (12-6) and (12-7) gives for the output signal-to-noise voltage ratio

$$\frac{E_o}{n_o} = \frac{E_o}{n_i} \frac{\delta}{\tau} \tag{12-8}$$

Since signal-to-noise ratios are usually expressed as power ratios, one obtains

$$\left(\frac{S}{N}\right)_{\text{out}} = \left(\frac{E_o}{n_o}\right)^2 = \left(\frac{\delta}{\tau}\right)^2 \left(\frac{E_o}{n_i}\right)^2 = \left(\frac{\delta}{\tau}\right)^2 \left(\frac{S}{N}\right)_{\text{in}} \tag{12-9}$$

For the parameters of the illustrative system described above, δ has a maximum value of 2.5 μsec. The system described can reasonably have a pulse rise time of $\tau = 0.1$ μsec. Hence a signal-to-noise improvement of up to 625:1 is possible. This increase in signal-to-noise ratio is, of course, achieved at the expense of wider bandwidth. In the example above, a bandwidth of 10 Mc was utilized; the total signal bandwidth transmitted is 24 × 3,000 = 72,000 cps. Thus a bandwidth about 140 times wider than required was used. This is quite typical of most signal-to-noise improvement techniques, namely, that a wider spectrum is used to transmit the message.

Other motivations for the use of pulse time modulation include the fact that the signals can be regenerated and amplified for retransmission without affecting the signal-to-noise ratio appreciably. This is important for signals transmitted over microwave links, in which case this regeneration of signals and amplification must be performed at intervals of 20 to 30 miles. Another reason for the use of pulse time modulation is the fact that crosstalk among channels is minimized, provided the pulse excursion has been reduced to sufficient values as not to extend into adjacent time intervals.

Pulse Code Modulation. As an alternative to pulse time modulation, the signals representing a sample value can be sent as a sequence of pulses or no-pulses (zero's or one's) which represent, in binary form, the amplitude of the sample.

In this case, a message is sampled and a sequence of pulses, which represents the sample value in binary form, is generated.

For the system of the previous section, for example, it might be considered sufficient to use 1,024 levels to represent the possible amplitudes of the functions being sampled. In this case, a sequence of 10 pulses (some of which may be 0) is required to represent this number of levels.

12-8   SYSTEM COMPONENTS

Using pulse code modulation, therefore, one frame having 125 μsec duration would consist of 25 5-μsec subintervals, the first 5-μsec subinterval being used, as before, for synchronization purposes. The next 5-μsec interval would consist of a set of pulses representing, in binary form, the amplitude of message number 1, etc., through message number 24. This frame is then repeated 8,000 times per second, or at intervals of 125 μsec.

Since 5 μsec is allotted to each channel, and 10 pulses must be allowed for, pulse widths of 0.5 μsec are the maximum appropriate for the system under consideration.

To achieve pulse code modulation from a sample, numerous techniques are available. Many of these have been developed for digital-computer purposes and are known as analog-to-digital converters. These devices often operate by generating signals which are weighted as powers of 2. These quantities are then added to give an analog value corresponding to the binary number. The sum thus formed is compared with the sample value, and the binary number is modified until equality results. The set of pulses as generated is the binary representation of the sample.

Alternatively, a coding device having a special aperture plate in a cathode-ray-tube-like device can be used. The aperture plate contains a hole wherever a 1 is required in the binary representation of the number. The binary representation for numbers from 0 to 1,023 is written on separate lines on this aperture plate. Behind the aperture plate are placed 10 electrodes, one for each of the possible 10 pulses, to represent the sample value. The sampled pulse amplitude is used to select the level at which the binary representation of that number is written. The beam is then swept across the aperture plate at this level. The passage of the electron beam through the holes in the aperture plate at this level activates the appropriate electrodes. A sequence of pulses, in which the numbers 1 are generated corresponding to the electrodes which are so energized, then gives the binary representation for the sample value.

To recover the sample value from the pulse code or binary representation, digital-to-analog converters are used. These devices develop voltages proportionate to powers of 2 and sum them. Signals with these weights are selected when there is a 1 in the binary representation of the number.

As with pulse time modulation, much of the motivation for pulse code modulation results from the output signal-to-noise improvement characteristics which are achievable with this type of modulation. Pulse code modulation comes closer to achieving the ideal performance predicted by Shannon than any other type of communications system thus far in general use. That this should be so can be made plausible as follows: In a pulse-code-modulation system, perfect demodulation of the signals will occur, provided all pulses are correctly decoded as one's and absences of pulses are properly decoded as zero's. Errors occur when noises are of such magnitude and polarity as to cancel a pulse, in which case the equipment will record a 0 when actually a 1 was transmitted. Conversely, if noise appears as a pulse, a 1 will be recorded whereas a 0 was sent. The probability of cancellation of a pulse or of creation of a false pulse decreases very rapidly as the signal-to-noise ratio at the input to a pulse code demodulator is increased. Thus pulse code modulation has a threshold. When the input signal-to-noise ratio exceeds this threshold value by a few decibels, the error rate decreases very rapidly.

The signal-to-noise improvement possible by the use of pulse code modulation can be attributed to the fact that at the reception of each pulse the receiver is required to determine solely the presence or absence of a pulse. If this is done correctly, the received message has no errors. Errors arise if pulses are not recognized or if noise is mistaken for a pulse. There are thus two kinds of errors, namely, missed pulses and false pulses. If the noise has a Gaussian distribution of amplitudes, then the probability density $P(n)$ for the noise amplitude being between n and $n + dn$ is

$$P(n)dn = \frac{1}{\sqrt{2\pi}\,\sigma} \exp\left(\frac{-n^2}{2\sigma^2}\right) dn \qquad (12\text{-}10)$$

where σ is the rms noise voltage.

If pulses of amplitude E are received, the probability density of signal plus noise is given by

$$P(E + n)dn = \frac{1}{\sqrt{2\pi}\,\sigma} \exp\left[\frac{-(E + n)^2}{2\sigma^2}\right] dn \qquad (12\text{-}11)$$

Let $R(a|b)$ be the probability that the receiver indicates the result a, given that b was transmitted. In pulse code modulation, a and b can assume only values 1 or 0.

Let a receiver be assumed which indicates pulse or no pulse depending upon whether or not a threshold Q has been exceeded. Thus four cases need to be considered. Two of these cases are for correct decoding, in which a pulse is sent and the receiver correctly indicates the pulse, and no pulse is sent and the receiver correctly indicates the absence of a pulse. The probability for these cases is given by Eqs. (12-12) and (12-13), respectively.

$$R(1|1) = \int_Q^\infty \frac{1}{\sqrt{2\pi}\,\sigma} \exp\left[\frac{-(E + n)^2}{2\sigma^2}\right] dn \qquad (12\text{-}12)$$

$$R(0|0) = \int_{-\infty}^Q \frac{1}{\sqrt{2\pi}\,\sigma} \exp\left(\frac{-n^2}{2\sigma^2}\right) dn \qquad (12\text{-}13)$$

Incorrect decoding results when a pulse is sent and the receiver indicates a 0, or when no pulse is sent and the receiver indicates a 1. The probabilities of this occurrence are given by Eqs. (12-14) and (12-15), respectively.

$$R(0|1) = \int_{-\infty}^Q \frac{1}{\sqrt{2\pi}\,\sigma} \exp\left[\frac{-(E + n)^2}{2\sigma^2}\right] dn \qquad (12\text{-}14)$$

$$R(1|0) = \int_Q^\infty \frac{1}{\sqrt{2\pi}\,\sigma} \exp\left(\frac{-n^2}{2\sigma^2}\right) dn \qquad (12\text{-}15)$$

By computation of the probabilities expressed by Eqs. (12-12) through (12-15), it can be seen that the probability of error decreases rapidly as the signal-to-noise ratio E^2/σ^2 is increased, and that the probability of correct decoding approaches 1 as the signal-to-noise ratio exceeds values of the order of 6 to 10 db.

It is because of this signal-to-noise improvement characteristic of pulse code modulation that use of this form of modulation has greatly increased during recent years.

Pulse code modulation can be used without generating much additional noise or crosstalk when messages are sent through a number of repeaters. This follows from the fact that the signals can be regenerated at each repeater. As with pulse time modulation, the increase in signal-to-noise ratio is accompanied by the use of a wider-than-necessary bandwidth. For the example considered herein, the signal band is, as before, 72,000 cps. Since 0.5 μsec pulses are used to convey information, a rise time of about 0.25 μsec can be used, in which case a bandwidth of 4 to 5 Mc would be adequate.

Summary Remarks. The earlier multiplexing schemes used in communications systems employed frequency-division multiplexing. In such cases, maintaining linearity and uniform characteristics was a problem of extreme importance. Failure to maintain these quantities with sufficient accuracy resulted in crosstalk among channels and wide variation in signal quality.

The use of time-division multiplexing is a more recent development. The processing, in this case, is deliberately nonlinear, particularly in that the pulses are regenerated at each repeater so that a clean pulse is transmitted at each new hop. Crosstalk and level problems are avoided in this case by circuit design which ensures that the pulses have decayed sufficiently so that the signals in the time allotted to one channel do not contribute appreciably to the signals in adjacent channels. Moreover, the time-division multiplexing signals have bandwidths and characteristics that suit them quite well for transmission over microwave links.

12-4. Communication Channels

It is the function of the communication channel to transmit signals, usually as electromagnetic waves, from one point to another. Sometimes the signals are guided by wire, transmission lines, waveguides, etc. Sometimes they are launched into space by an antenna and recovered at the receiving end by another antenna. While there are many similarities, there are also many differences in the characteristics of wire lines as compared with radio transmission.

When wire is used, the signals can be sent from point to point with somewhat greater assurance of privacy than when radio is used. The choices dictating the use of wire or radio depend in part upon economic considerations and in part upon the transmission characteristics. Generally, wire lines, including transmission lines and coaxial cables, are appropriate for the lower frequencies (not over a few megacycles), with waveguides also used in the microwave region; the radio spectrum, on the other hand, starts at a few kilocycles and, at present, extends with some gaps up to the optical region.

Most of the frequency-division multiplexing equipments were used with wire equipments, although largely for historical reasons. The time-division multiplexing signals are used primarily on microwave radio links. This is due in large part to the wide bandwidth of signals. In the case of wire links it is relatively easy to switch messages from various origination points to various destinations. In the case of radio, depending upon the antenna system and the frequencies, one can either transmit signals to cover a large area ("broadcast") or transmit them essentially as point-to-point messages.

At the lower frequencies, antenna gain is more difficult to achieve, and generally broad beams and wide coverage result. At the higher frequencies, particularly in the microwave regions, very directive beams can be achieved by the use of antennas having dimensions of many feet. Wide bandwidths are readily available in radio links.

In either case the characteristics of the medium affect the signals received in a number of ways. One of these effects is the addition of noise. Thus the signal arriving at the receiver by either path is a contaminated version of the signal transmitted. It is then a part of the receiver (and the transmitter) design to attempt to minimize the effect of noise.

In the case of radio, multipath is a problem. In this case, signals arrive at the receiving antenna by a variety of paths. These paths not only differ in time of arrival, but they have different distortion characteristics, particularly if the signals transmitted are wide-band. Space-diversity antennas are sometimes used to help combat the interference effects of multipath transmission.

12-5. The Coding-Decoding Problem

Coding and decoding are employed in communications systems for a variety of purposes. One reason has been mentioned in Sec. 12-3, namely, the improvement of signal-to-noise ratio. It was there stated that pulse-time-modulation systems and pulse-code-modulation systems are capable of yielding a higher signal-to-noise ratio at their output than is present at their input. The same is true of frequency-modulation systems.

Coding is also resorted to for purposes of error correction. In such cases, for example, in the transmission of data, binary numbers are sent, but a longer-than-necessary binary word is used. In this case, the extra bits are functions of the information being transmitted, and depending upon the sophistication of the code and on the number of extra bits used, a given number of errors can be detected and corrected.

Other reasons for using codes depend upon the intended purposes of the communication. For example, one may wish to minimize the transmitted energy required per unit of information transmitted. As another example, one may wish to remove or increase the redundancy of a message.

The coding of a communications system affects the design of both the transmitter and the receiver. The transmitter plays its role in that the required waveforms to be transmitted are generated in the transmitter. The designer must face the problem of designing his codes to achieve the desired objective. At the receiver, the code used by the transmitter must be known, and the receiver design operates upon the signals in such a manner as to optimize the performance characteristic or characteristics desired. While much has been done in theory of coding, many problems still remain to be solved in this area; see Chapter 22.

REFERENCES

1. Shannon, Claude E.: The Mathematical Theory of Communication, *Bell System Tech. J.*, July and October, 1948.
2. Linden, D. A.: A Discussion of Sampling Theorems, *Proc. IRE*, vol. 47, pp. 1219–1226, July, 1959.

Chapter 13

COMMUNICATION SYSTEMS

ARTHUR D. HALL, *Head, Television Engineering Department, Bell Telephone Laboratories.*

CONTENTS

13-1. Preview	13-2
13-2. Possible Objectives for Communication Networks	13-2
13-3. General Features of Communication Systems	13-3
End Instruments for the System	13-3
The Two Broad Functional Divisions of a System: Switching and Transmission	13-3
Circuit Switching and Message Switching	13-3
Auxiliary Messages and Signals	13-3
Multichannel Transmission Systems	13-4
13-4. The Properties of Messages and Signals	13-4
Space or Geographical Properties	13-5
Time Properties	13-6
Information Properties	13-6
Physical Properties	13-6
13-5. Channel Properties	13-7
Transmission Properties	13-7
13-6. Transmission Media	13-9
The Radio Medium	13-9
Choice between Wire and Radio	13-9
13-7. Functions of a One-way Transmission System	13-12
Model of a One-way Transmission System	13-12
Transmission Characteristics of System Channels	13-13
The Line	13-13
13-8. Functions of a Two-way Transmission System	13-15
13-9. Introduction to Network Aspects	13-17
13-10. Switching Plans	13-17
13-11. Traffic-handling Capacity of Trunk Groups	13-19
The Volume of Traffic	13-20
Service Standards and the Class of Trunk	13-20
Engineering Trunk Groups for Alternate Routing	13-21
13-12. A Major Transmission Aspect of Networks	13-23
13-13. Conclusion	13-24

13-1. Preview

This chapter on communication systems is meant to convey flavor for certain aspects of the field of communication-system engineering. This will be done by stressing functional design, which begins by designing a good set of objectives, proceeds to a detailed listing of system inputs and outputs which imply the functions required, and continues with a space-time ordering of the functions that will do the job specified by the objectives. Wherever possible, communication networks, as contrasted with a single link between two points, will be stressed, because the network aspects contribute perspective in dealing with all types of information-handling systems.

In preview, the general objectives for communication networks will be reviewed. Then some features shared by all communication systems will be introduced. Then we shall study the means of specifying the properties of messages and signals, which comprise the information to be processed by the system. Following this we shall survey the functions required in both one-way and two-way transmission systems. Last, we shall study certain problems that arise because of the network aspects of communication systems.

13-2. Possible Objectives for Communication Networks

Communication systems, like all man-made systems, are developed to satisfy certain needs. These needs both provide the stimulus for their inception and mold their objectives. The objectives, in turn, set the pattern of functions that must be built into the system by its designer.

The general objective or purpose of any communication system is to provide means for communication between people and/or machines that are in different locations in space and/or time. Communication signifies the conveyance of information. It includes both one-way distribution of information from a sending point to one or more receiving points, and two-way interchange of information between two or more places. The information communicated (the message) may originally exist in oral, written, pictorial, numerical, mechanical, or other forms. Though there are many methods of communication, we shall be concerned with systems in which the messages are processed in electrical form.

A number of specific objectives guide the planning and design of an electrical communication system. A particular set of objectives may include any or all of the following:

1. The kinds of messages to be processed are usually specified (numerical data, telephone, high-quality audio, telegraph, facsimile, television). A communication system may be required to handle any or all of these, as the demand arises. If it handles more than one kind of message, it may be called a *multiservice communication* system.

2. The system must be able to establish paths over which these messages can be sent between specified locations and at the desired times. This implies a *switched network*. If only permanent paths exist between all subsets of end points, the communication system is said to be a *nonswitched network*.

3. The system must have the capacity for setting up as many simultaneous paths between different sets of terminal locations as the traffic demands, while preserving the independence and privacy of the communications. The system may be required to expand or contract with changes in traffic demand.

4. The system must be so designed as to ensure that the messages reach their destinations without losing any serious amount of information. This leads to detailed requirements covering signal-to-noise ratios, bandwidth, distortion, etc. The satisfaction of the users is usually the final criterion of how much degradation can be tolerated.

5. The system must be reliable in service and easy to install, operate, and maintain, to such degree as is consistent with other requirements. Survivability, continuity of

service, and restoral time all are aspects of reliability, and these are particularly important in military networks.

6. The system must preserve privacy to the degree required by the application. In radio broadcasting the degree is zero; in secret military communications, the need for privacy may be absolute.

7. The system must be designed to be competitive in the marketplace, if it is a commercial system, or to have prescribed effectiveness, if it is a military system. Related objectives covering market, cost, or service features may need to be formulated.

8. The system must accomplish all these things with an expenditure in time, economic, or other values that is satisfactory to the users, designers, and other decision makers.

13-3. General Features of Communication Systems

End Instruments for the System. The boundary between any system and its environment is often an arbitrary matter. A complete communication system can, in the broadest sense, be considered to include the functions of generating and utilizing messages, as well as transmitting them. It is equally logical, however, and also more consistent with engineering practice, to treat the messages as a commodity handled by the system. With this view, the ultimate sources and destinations of the messages lie in the environment, external to the system. The system, then, begins and ends at certain transducers, called *end instruments*, or sometimes stations, through which the messages enter and leave it. Examples of end instruments are a telephone set, a television camera, a loudspeaker, or a radar display.

The Two Broad Functional Divisions of a System: Switching and Transmission. To provide communication when wanted between particular terminals, the system must perform two distinct sets of operations. It must first establish a path between the specified terminals for the wanted period, and then it must send the messages over the path thus set up. This indicates the two major functional divisions of a communication system. Switching functions are all those functions which establish the path, disestablish it when no longer needed, and perform logical operations upon the message such as computing and automatic quality control. All the functions that enable messages to be transmitted satisfactorily over the path are classified under transmission. There are, of course, many useful nonswitched communication systems, in which the channels between the end points are always available, e.g., the local-area TV broadcasting system.

Circuit Switching and Message Switching. There are two entirely different methods of switching. In one method a direct electrical path is established between the system terminals by switching circuits together. This is called *circuit switching* and is the method used, for example, in telephone systems and in teletypewriter exchange (TWX) systems. In the other method the messages are forwarded to their destinations over a succession of unconnected paths. After traversing each path, the message is recorded and later sent into the next path. The messages rather than the circuits are switched, and the method is called *message switching*, or sometimes *store and forward* systems. This is done at the expense of intermediate terminal and storage functions, but the saving is through more uniform loading of the network in time.

The radio and television broadcast services supply examples of message switching. A program originating in the East may be recorded and sent over the western networks at a later time to accommodate the time differences between the two parts of the country.

Auxiliary Messages and Signals. The main communication messages that are sent through the system from terminal to terminal may not be the only messages that must be handled by the system. Other auxiliary messages or signals* may be neces-

---

* Signals are the physical embodiment of a message. A message is the information conveyed by a set of signals.

sary to control and supervise the switching functions, and sometimes for other purposes. In telephony, for example, dial tone, dial pulsing, and dozens of other unique auxiliary messages are used between customers and switching centers or between one switching center and another. The generation, transmission, and utilization of the signals directly associated with switching will be classed here as part of the switching function.

Multichannel Transmission Systems. A transmission system may provide from one to thousands of transmission channels. In the multichannel systems, all the transmission channels are combined (by a multiplexing function—Sec. 12-3) in the terminal equipment and sent over one common line channel, which must have a frequency band wide enough for this purpose. This affords great flexibility, for such systems can be utilized for the transmission of signals of a variety of bandwidths. Signals of greater bandwidth than that of one of the basic transmission channels may be handled either by connecting directly to the line channel through special terminal equipment, or by modifying the normal terminal equipment so as to combine several message channels into one broader-band transmission channel.

There are many kinds of transmission systems. These differ in complexity, in the transmission medium they utilize, in the number of channels they furnish, and in the distance over which they can satisfactorily be used. In engineering the system, it is necessary to select for each application the type of facility which is technically and economically suitable. However, from the system standpoint these differences are not functionally important. To the switching mechanisms it makes little difference what kinds of facilities provide the circuits that are switched together, provided the auxiliary signals can be exchanged. Each circuit is merely a path to a desired distant destination that can be relied upon to have certain necessary transmission properties.

However, it is required that all circuits at a given switching center have similar terminal properties so that any two of them may be switched in tandem. All must accept and deliver signals which have the same frequency or time-slot allocation, bandwidth, and code. The received (power) levels of incoming signals must be suitable sending levels for the outgoing circuits (or must be so adjusted by the switching mechanisms). Impedances must be compatible. All circuits must terminate similarly as either two-wire or four-wire circuits.

13-4. The Properties of Messages and Signals*

This section aims to organize all the significant properties of messages and signals and to put them into the kind of perspective needed for the planning and design of practical communication systems. Part of this perspective will be obtained by comparing the properties of one message and its corresponding signals in an elementary communication system (Table 13-1) with the aggregate properties of a class of messages and signals (Table 13-2) which a communication network may be engineered to handle.

Inputs and outputs have four basic classes of properties, which are space, time, information, and physical.

A complete description of these properties is sufficient to describe the inputs and outputs of any information-handling system.

Presenting a complete list of message and signal properties does not mean that every system-design problem calls for complete description, or even mention, of each property. The input and output properties that need to be described depend strongly on the type of system under study, the analytical tools available to the designer, and other factors. In some problems the information properties may be of little consequence, while in other problems they play a leading role. Also, some properties may be described in alternate ways; e.g., signals may be described in the time domain by specifying their waveshapes or in the frequency domain (Chapter 39) by specifying their spectra. Which method is chosen may depend on relative convenience, ability to make measurements, or other factors.

* Adapted from Ref. 1, chap. 15, Sec. 15.5, with permission.

COMMUNICATION SYSTEMS 13-5

Table 13-1. Properties of One Message and Its Signals for a System with One Input and One Output

Space properties of message and signals:
  Location of the origin and destination
  Distance and path location between the origin and destination
  Physical dimensions of the origin and destination
Time properties of message and signals:
  Start, duration, and end
  Waveshape (see Physical properties)
Information properties of message and signals:
  Language of the message, or code of the signals
  Information content, both statistical and semantical
  Efficiency or redundancy
  Information rate
Physical properties of signals:
  Kind of wave phenomenon (electric, acoustic, light, heat, etc.)
  Wave property used to encode (amplitude, frequency, phase)
  Method of transmission (serial or parallel in space, time, or frequency)
  Waveshape (or appropriate measures thereof, such as distribution of amplitudes or autocorrelation)
  Frequency-band or time-slot location of the signals
  Frequency spectrum: amplitude, phase, and power density

Table 13-2. Properties of the Messages and Signals for a System with Multiple Inputs and Multiple Outputs

Space properties of messages and signals:
  Number and density of origins and destinations
  Distribution of distances between origins and destinations
  Spatial relations between channels
  Distribution of similar physical dimensions of messages
Time properties of messages and signals:
  Distribution of start times in all time units of interest
  Distribution of message durations
  Distribution of waveshapes (see Physical properties)
  Distribution of messages with different information properties
Information properties of messages and signals:
  Languages and codes used
  Information contents, both statistical and semantical
  Efficiencies
  Information rates
Physical properties of signals:
  Kinds of wave phenomena
  Wave properties used for encoding different inputs
  Methods of transmission, switching, and handling queues
  Waveshapes
  Cross correlation between inputs and outputs
  Frequency-band or time-slot locations of signals
  Frequency spectra: amplitude, phase, and power density

Space or Geographical Properties. The space coordinates of origins and destinations must nearly always be specified. Sometimes, as in systems which handle teletypewriter and telephone services, the sources are uniquely identified by assigning a different number to each source; in these systems definite paths between pairs of numbers are wanted and are set up on demand to provide transmission for the signals. The design of a good numbering plan for station identification depends on many factors including the capacity of the system, growth requirements, the need for freedom of numbering from routing and charging considerations, and others. In radio broadcasting the destination coordinates are not explicitly known; here it is important only to know the distances at which a specified signal power will be received.

Unifunctional devices called selectors (sometimes detectors) can be designed to select one or more originating signals from a plurality of sources, or to select a particular destination from a group of similar destinations. Also, unifunctional devices called connectors can be designed to set up a channel between a particular origin and a particular destination. Such a device can be made by a tandem connection of a many-to-one selector and a one-to-many selector, or by a cross-point (or channel-matching) array.

Both messages and signals may at times be stored in recorded form at various places in the information system. A facsimile message starts and ends in the recorded form of a picture, drawing, or map. Thus the original message may have important space properties, such as the size of the smallest detail to be transmitted, or the number of edges in the picture. Signals may be recorded on moving-picture film, magnetic tapes, holes in a card, a storage tube, relay contacts, etc. However, they can be transmitted over the usual system channels only as wave phenomena in the form of discrete or continuous time sequences of signals. The properties given in Tables 13-1 and 13-2 apply to such messages and signals.

Note that the message information from a single source is sometimes divided into segments that are sent simultaneously over two or more parallel channels. Thus a data-transmission system may use two channels to transmit a coordinate, one for the coarse values and the other for vernier interpolations between the coarse values. The portion of the message handled by a given channel is treated in Table 13-1 as "the message" for that channel. In such a system the space separation between channels may be an important property when consideration is given to signal interference between the two channels.

When engineering a multiple-input system, interest usually focuses upon the aggregate time and space properties of the whole class or subclass of sources that generate the same kind of messages. An example is the class of sources that generates telephone messages. Table 13-2 lists the properties for which statistical descriptions may be needed, including the average rate of messages from each source, the average message length, etc. The parameters of the corresponding distributions are not usually time-invariant; i.e., the underlying process is said to be nonstationary. An adequate description of such processes calls for autocorrelation studies to detect periodicities and trends.

Time Properties. The start and end of messages and signals are instants in time. In many cases these instants are important and are marked by special start and stop signals. The duration of a message or signal is the time difference between the start and the end. Often, in signal analysis, the terms epoch and era are used for the start and duration. The start, stop, and duration need not be the same for the message as for the signal. The start and stop of the signal must always be later than the corresponding instants of the message because some time delay is unavoidable in physically realizable encoders. The signal may be more efficiently coded than the message to realize a shorter signal duration, and it may be sent over a fast transmission system. Any such difference in the duration of the message and signal requires signal storage in the coding device.

The space and time properties together provide the logical basis for the topological design of the network. Such design involves the choice of location and classification of the switching centers, transmission branching points, the routes between each pair of points, and the number of channels between each node in the network. These choices are made to satisfy minimum cost, service continuity, route or facility dispersion for survivability enhancement, or other network objectives.

Information Properties. The need for information properties as part of a complete characterization of inputs and outputs is made clear in Chapter 22 on information theory. Furthermore, the reasons for a statistical description of these properties for multiple-input systems are easy to see. For example, the number and the number of types of signal encoders that must be provided in a communication system depend on the frequency of occurrence of different languages and the volume of messages in each language. As another example, the information rate, in bits per second, helps to determine the bandwidth required in the transmission paths of the system.

Physical Properties. The physical properties of the signals are particularly important, as most of them must be compatible with the signal channel or other element of a communication system on which they are impressed. A given signal channel, for example, can accept and transmit signals in any code, if the encoding is in the right kind of wave phenomenon and the resulting frequency spectra match the transmission properties of the channel within the tolerances established by technical standards. The physical properties of signals may need to be altered by suitable transducers to make this possible, however.

Tables 13-1 and 13-2 were arbitrarily limited to signals transmitted by wave phenomena. The message signals are encoded by modifying some property of the wave to be propagated, such as amplitude, phase, or frequency. The property modulated is the one that must be examined by the receiver to decode the message.

Stating the waveshape of a signal includes distinguishing among the various kinds of pulse modulation (Sec. 12-3) and between these as a class and continuous modulation. It also includes the distinction between signals that occupy periodic time slots and those active continuously.

Important properties of the encoded signal wave are its amplitude and phase spectra. In the case of the originating signal, the amplitude spectrum may extend between very wide limits, but usually most of the signal energy will be concentrated in a narrower portion of this wide range. The terms bandwidth and frequency location are engineering approximations to a description of the significant characteristics of the spectrum that must match the signal channel. The bandwidth is the frequency interval containing the principal part of the spectrum necessary to convey practically all of the message. The frequency location is the position of this band on the frequency scale. For example, the spectrum of a speech wave extends from about 50 to over 8,000 cycles, but for telephone purposes it is necessary to transmit only that portion of the spectrum lying between about 200 and 3,500 cycles. In actual systems, this band may be located anywhere on the frequency scale, limited only by the state of technology. At the present time microwave technology is limited to less than about 150 gigacycles (150,000 Mc), but laser technology may open up much higher ranges.

13-5. Channel Properties*

There are two bases for presenting the channel properties. One is in terms of readily measured transmission properties. The other is in terms of information theory. The transmission properties are the more directly useful for design purposes. The properties derived from information theory are in some respects the more fundamental. The two sets of properties are related, hence complementary.

Transmission Properties. These properties can be measured by standard engineering techniques from the terminals of a physical channel without knowledge of what lies between the terminals. Most of these properties are usually considered to be (and are sometimes called) *transmission impairments.* Table 13-3 lists the transmission properties.

The transmission properties differ in the extent to which they limit the transmission potential of the channel. In one class are those properties that can, at a cost, be compensated to any needed degree, so that the residual effects do not constitute basic limitations on the serviceability of the channel. For example, loss can be compensated by gain, attenuation and phase distortion by equalizers, and instabilities by regulating techniques, devices, or systems.

Flat delay is in a special class, for in the absence of physical means to produce negative delay it cannot be reduced. However, so long as consideration is restricted to one-way channels, delays of the magnitudes usually encountered are ordinarily not important. There are exceptions; for example, in data systems used to transmit signals to direct the aim of antimissile missiles, time is an important coordinate, and the delay of the channel must be allowed for by the missile-directing computers. Delay is important in two-way communication, both for its effect on echoes (or reflections) and for the interference it injects in rapid to-and-fro exchange of information. In modern high-speed carrier systems, delay problems are minimized. Satellite systems and space-to-space systems may introduce more serious delay problems.

* Adapted from Ref. 1, chap. 15, Sec. 15.7, with permission.

Table 13-3. Transmission Properties of a Signal Channel

Propagation effects (functions of frequency)
  Attenuation:
    Flat loss. A uniform reduction in amplitude
    Limited frequency band. The attenuation approaches infinity outside of certain frequency limits
    Attenuation distortion. A change in the relative amplitudes of the frequency components of a complex wave within the transmission band
    Attenuation stability with time
  Phase, Delay, Velocity:
    Flat delay. Average transmission time
    Delay (or phase) distortion. Change in the relative time of propagation (or relative phase) of the frequency components of a complex wave
    Phase or delay stability with time
    Impedance irregularities. Reflection (or echoes)
Nonlinear distortion (functions of amplitude)
  Amplitude distortion. The larger amplitudes are usually attenuated more than the smaller amplitudes (an overload effect)
  Modulation distortion. The creation of new frequencies that are harmonics of signal-frequency components, harmonic distortion, or that are modulation products due to the interaction of two or more components, i.e., intermodulation distortion
Interference
  A spurious wave added to the signal. Interference is often loosely called noise, regardless of its source. However, more strictly speaking, interference is *crosstalk* when its source is signals on one or more other communication channels, and *noise* when from any other source

In another class is a group of three properties that do fundamentally limit the transmission capabilities of the channel. These are bandwidth, signal power, and noise (interference). The maximum allowable signal power is closely related to the nonlinear effects in Table 13-3; it must be limited to a value that keeps these nonlinear effects within tolerable bounds. Sometimes the signal power is limited by other factors, such as the need to prevent crosstalk into other channels or a ruling of a regulating body (e.g., the FCC). Once the maximum power is set by one of these considerations, the relative levels of the signal and the noise become controlling. Therefore, the three fundamental properties in this group reduce to two: bandwidth and signal-to-noise ratio. A new and basic property, *channel capacity*, is derived from these two. The best-known formula for this is $C = W \log_2 (1 + S/N)$ bits per second where W = bandwidth and S/N = the average power ratio of signal to noise. One should always remember that the "terms of trade" implied by this relation also includes transmission time T; the total information transmitted is $h = CT$ bits.

The practical effects on the system of the different kinds of distortion and interference depend upon the kind of message. For example, in telephony the human ear that receives the message is a decoder that responds only to the relative amplitudes of the frequency components of the speech wave, substantially disregarding the relative phase of the components (unless the phase distortion is great). The converse is true in facsimile and television. With the customary methods of encoding, delay distortion displaces the picture elements defining the image from their true positions and blurs picture outlines, whereas moderate attenuation distortion merely produces unnoticeable errors in picture brightness. (Of course, if the channel employs phase or frequency modulation, phase distortion in the medium may be converted to amplitude distortion in the decoded signal.)

The different kinds of information signals also vary in their sensitivity to transmission instabilities. For example, amplitude-modulation facsimile systems are particularly sensitive to attenuation variations that are rapid, sudden, or periodic, as these produce visible patterns on the received picture, but are much less sensitive to gradual and nonperiodic changes. Frequency-modulation systems of any kind are insensitive to attenuation variations but are sensitive to phase variations.

Again, some signals tolerate more noise than others. For example, an important advantage of frequency modulation and many of the pulse-modulation systems is that they can operate through comparatively poor signal-to-noise ratios. High-quality audio channels are least tolerant of noise.

13-6. Transmission Media

Channels require a physical medium through which signals may be propagated. All electrical media fall into two classes—those furnishing *guided* transmission, and those furnishing *unguided* transmission in which the medium itself does not restrict the transmission path. In the first category are wire lines and the various structures conventionally called waveguides. In the second class is the radio medium. The most important types are listed in Table 13-4.

Table 13-4. Types of Transmission Media

Guided media:
  Single wire
  Paired wires: open wire and cable
  Coaxial cable: land and undersea
  Waveguides: rectangular and circular
  Others in experimental stage
Unguided media:
  Acoustical or underwater sound
  Ground-wave radio
  Line-of-sight radio
  Tropospheric and ionospheric scatter radio
  Ionized gas plasmas (for space communication)
  Propagation by ionospheric refraction (high-frequency radio)

An early and major problem in the design of any transmission system is to characterize the medium fully in terms of the properties already mentioned.

The Radio Medium. The radio medium, unlike wire lines, is provided by nature for the free use of man. Transmission occurs through the unconfined space above the earth's surface. The transmission properties of the medium are greatly affected by natural phenomena. The effects of the natural phenomena are different in various portions of the radio-frequency spectrum and are often erratic and uncontrollable. For this reason, the transmission behavior of the medium is a much more complex matter than the definite and fairly simple laws that govern transmission on wire lines. In spite of this, present techniques and knowledge permit the use of the radio medium to provide reliable circuits for communication networks. The important transmission effects are summarized qualitatively by Fig. 13-1.

Choice between Wire and Radio. It is technically feasible to provide channels by either wire or radio over any desired distance, singly or in groups, that have adequate reliability and transmission quality for communication systems. The choice between the two media therefore depends chiefly on other considerations than the mere ability to provide a needed service over a specified distance. However, the two media have markedly different properties, some of which favor the use of radio and others the use of wire in particular situations. In general it may be said that wire and radio are complementary to each other, with a large area in which they overlap in usefulness.

Radio is the natural choice of medium in certain fields.

1. Radio is useful for spanning territory where line construction is difficult or impossible, such as over large bodies of water, mountains, jungles, and enemy-held territory (in military communications).

2. Radio is the most widely used medium when one or both terminals are mobile.

3. The simpler types of point-to-point radio systems require less manpower and time to install and place in operation than corresponding wire systems (except for very short distances). They can also be readily taken down and moved to new locations. For this reason radio is often the natural choice for temporary service.

4. For similar reasons, radio is useful in emergencies for quick restoration of service when wire or other systems have failed because of a catastrophe.

5. Closely connected with the last two points is the fact that the volume and weight of material required to establish long communication channels by radio is generally

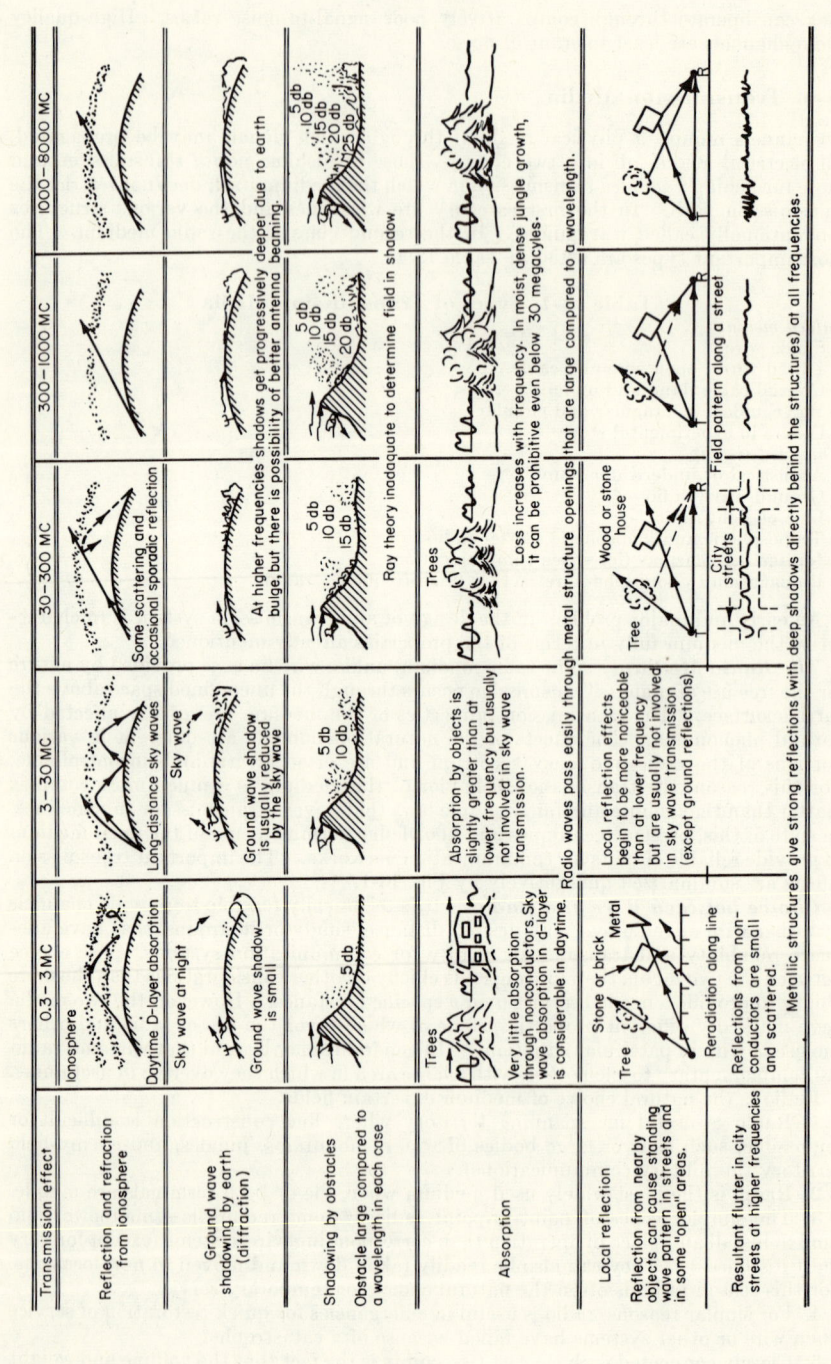

Transmission effect	0.3–3 MC	3–30 MC	30–300 MC	300–1000 MC	1000–8000 MC
Reflection and refraction from ionosphere	Ionosphere. Daytime D-layer absorption. Sky wave at night.	Long-distance sky waves	Some scattering and occasional sporadic reflection		
Ground wave shadowing by earth (diffraction)	Ground wave shadow is small	Ground wave shadow is usually reduced by the sky wave			
Shadowing by obstacles. Obstacle large compared to a wavelength in each case	5 db	5 db, 10 db	5 db, 10 db, 15 db	5 db, 10 db, 15 db, 20 db	5 db, 10 db, 15 db, 20 db, 25 db
			At higher frequencies shadows get progressively deeper due to earth bulge, but there is possibility of better antenna beaming.		
			Ray theory inadequate to determine field in shadow		
Absorption	Trees. Very little absorption through nonconductors. Sky wave absorption in d-layer is considerable in daytime.	Absorption by objects is slightly greater than at lower frequency but usually not involved in sky wave transmission.	Trees. Loss increases with frequency. In moist, dense jungle growth, it can be prohibitive even below 30 megacycles.		
		Radio waves pass easily through metal structure openings that are large compared to a wavelength.			
Local reflection. Reflection from nearby objects can cause standing wave pattern in streets and in some "open" areas.	Stone or brick. Metal. Tree. Reradiation along line. Reflections from nonconductors are small and scattered.	Local reflection effects begin to be more noticeable than at lower frequency but are usually not involved in sky wave transmission (except ground reflections).	Wood or stone house. Tree.		
Resultant flutter in city streets at higher frequencies			City streets. Field pattern along a street		
	Metallic structures give strong reflections (with deep shadows directly behind the structures) at all frequencies.				

13–10

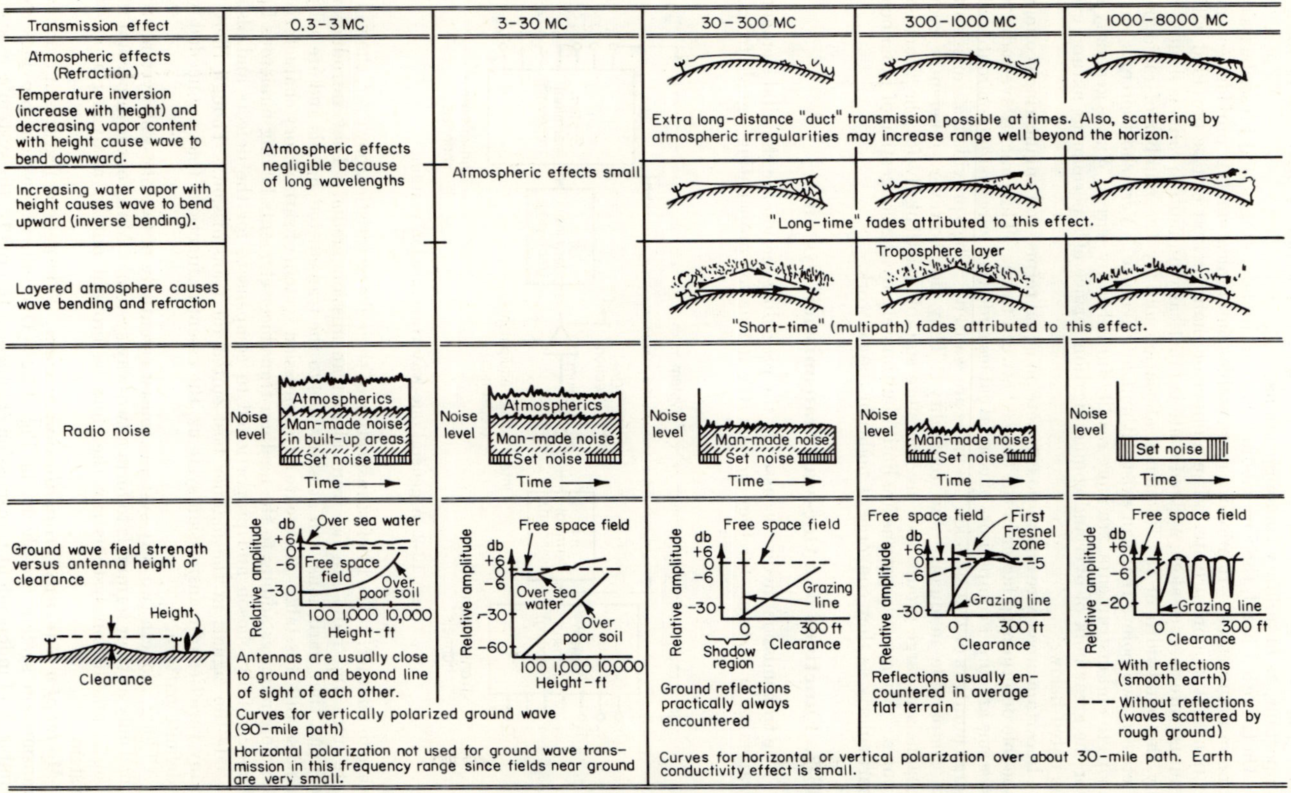

FIG. 13-1. General radio-propagation characteristics over the frequency spectrum.

much less than by wire. The factor of shipping space and weight is sometimes very important, as in military communications.

On the other hand, the wire medium has certain advantages that prescribe its selection in many cases. It provides more stable communication. It provides communication that is more private. The number of channels that can be provided without mutual interference along a given route is practically unlimited; radio is limited both by the bandwidth of the frequency allocation and by the number of radio beams between two points that can exist without interference. Very short circuits, such as loops to end instruments, are more readily furnished by wire. Wire is generally used for the great bulk of overland communication service of a permanent nature, except as noted below.

In certain areas wire and radio compete on equal terms, and engineering economy is the sole determining factor in the choice between them. Thus microwave-radio relay systems are direct competitors of coaxial cables and other wire systems on routes where large numbers of transmission channels are wanted. They provide transmission that is comparable in quality and reliability. They do, however, inherently provide channels in large groups and are therefore suitable particularly for the "backbone" routes. Microwave systems are limited ultimately by the congestion of radio beams, particularly as they converge at nodes in the network.

13-7. Functions of a One-way Transmission System

The transmission media discussed in the preceding section are the vital elements of transmission systems, for they provide the means for propagating signals from one

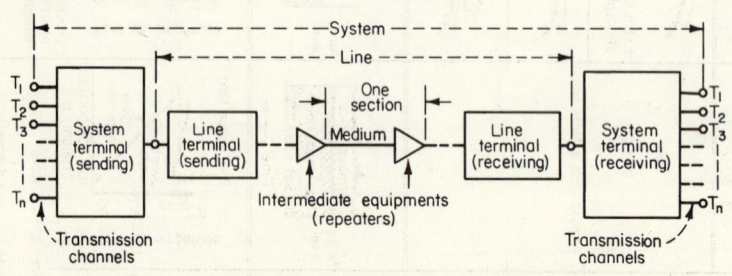

Fig. 13-2. A general one-way transmission system.

place to another. The briefest inspection of their characteristics, however, shows that wire lines and radio are far from ideal transmission media. They introduce distortions, time variations, and attenuations to the transmitted signals and are subject to noise and crosstalk. To create a transmission system of satisfactory characteristics therefore requires adding to the medium transducer elements having numerous functions carefully chosen and proportioned to compensate for the defects and also to encode the signals in the manner best suited for transmission. Though there are many varieties of transmission systems, the essential functions that are needed form a definite pattern.

All long, modern two-way systems consist essentially of two one-way systems combined only at their terminals to form a two-way system. Many of the important facts about transmission systems can therefore be brought out by analyzing a one-way system.

Model of a One-way Transmission System. The general features of a one-way transmission system are shown in Fig. 13-2. Over-all, the system provides some number N of independent transmission channels. Internally, the system consists of a single common-line channel throughout its length for transmitting all the signals—an arrangement dictated by economy. Considered by itself, the line is a single-channel transmission system consisting of line terminals and one or more sections of wire or radio media connected together through intermediate repeater equipments. The complete transmission system consists of the line plus system terminals that include the modulating and multiplexing arrangements for combining the N transmission-channel signals into one line signal at the sending end and for recovering the N signals at the receiving end. Such a system can be duplicated many times along a particular route; e.g., multiunit coaxial cable may provide the medium for this. These are the elements of a general one-way transmission system.

All transmission systems conform basically to the pattern shown by this model, though in some elementary systems the pattern is so simplified as to bear only a rudimentary resemblance to the diagram. All the elements of the model can usually be identified in carrier and radio systems.

The functions needed in the system are closely related to the transmission characteristics that are wanted in its channels.

Transmission Characteristics of System Channels. The transmission properties of a system channel can be measured from its input and output terminals without knowledge of the internal structure of the channel. They define the transmission performance of the channel. The transmission-performance objectives of the system are stated in terms of the required or expected transmission properties of its transmission and line channels. The properties already listed (in Table 13-3) are those to which specific numbers must be attached. Deriving such objectives requires extensive subjective testing, statistical analysis, and testing for economic and technical feasibility. Such work is an important branch of system engineering in the communications field.

The characteristics of the transmission channels of a system are directly set by the requirements of the communication system, for they are the links in the system. The requirements of the line channel are in turn determined by the characteristics of the transmission channels and also by their number and the kind of modulation to be employed.

In general, the greater the number of transmission channels the more exacting are the transmission requirements of the line. Nonlinear (amplitude) distortions in the common line tend to produce cross effects between the transmission channels so that the transmission-channel signals cannot be completely separated from one another. The precise interpretation of this in terms of line-transmission requirements depends upon the kinds of multiplexing and modulation employed. If frequency-division multiplexing is used, severe requirements are placed on the linearity of the line channel to avoid cross modulation between channels. If time-division multiplexing is used, strict limits must be placed on the phase distortion to keep the signals in their assigned time slots.

The Line. *Functions Needed in the Line.* The line provides a channel that accepts at its sending end a signal having a certain frequency allocation, bandwidth, and level. It is desired that this channel deliver at its receiving end a faithful replica of this signal, at a prescribed, stable level and at a prescribed frequency allocation that is usually the same as at the sending end. Distortion and noise must be kept below specified limits. The signal may be translated to other frequency allocations on the medium and may be sent directly or recoded and sent by any method of modulation. The kinds of function that must be present in the line to accomplish all these things are as follows:

1. Gain, to compensate for the loss in the medium.
2. Equalization (attenuation and phase) to compensate for the distortion in the medium.
3. Regulation, to compensate for the time variations of the propagation characteristics in the medium.
4. Frequency or time translation, to change the frequency or time-slot assignment of the signals.
5. Band limitation (filters).
6. Encoders, if the signals are to be recoded or sent by a different kind of modulation.
7. Other functions, depending on the type of service and the overall system of which the line is one component. Examples are line balancing, crosstalk-control measures such as shielding, fault location, and channel-protection switching.

The first three functions compensate for the adverse characteristics of the medium. It should be noted that if the attenuation of a system could be perfectly equalized over an infinite frequency range, this would automatically result also in perfect phase equalization. Phase equalization is required in an actual system only because (1) the attenuation is equalized over only a finite band and is sharply cut off by filters outside that band or (2) the attenuation equalization is not perfect within the band. Even then it is needed only if the transmitted signals are sensitive to phase distortion.

Regulation, when required, may be one of three types: manually controlled, temperature-controlled, or signal-controlled. Manually controlled regulation is satisfactory for some short systems and consists merely of a hand-adjusted gain or loss. Temperature-controlled regulation is an automatic type of regulation employed in some cable systems where temperature is the main source of the transmission variations. It is necessary that some form of thermometer for measuring the temperature be provided to serve as the control element. In the pilot-wire regulating system used on long voice-frequency cable circuits, the d-c resistance of a pair in the cable serves as the thermometer. Long carrier and radio systems usually employ signal-controlled regulation. The regulator measures the received level of a signal, which may be the main signal or one or more special pilot signals consisting of single frequencies sent over the line along with the main signal. The regulator then adjusts the loss of regulating networks as needed to keep the received signal at the correct level. In broadband systems the required amount of regulation may be different at different frequencies. In such cases several pilot frequencies are used to control the regulating networks. In one arrangement, three pilot frequencies are employed. One controls flat loss, another the slope of the attenuation-frequency characteristic, and the third the bulge or curvature of the characteristic. Signal-controlled regulation is the most accurate type because it compensates for all variations ahead of the regulating point due to any cause. However, attention must be paid to the time constants of the devices, when many regulators are used in tandem, to avoid large transient errors when sudden changes occur.

The fourth function, frequency translation, needs little comment. It is simply the application of the well-known techniques, for example, amplitude modulation, to translate a signal band from one position in the frequency spectrum to another.

Band limitation serves two purposes. It is applied at the sending end to restrict the transmitted signal to its essential frequency band. This minimizes the frequency space occupied by the channel and permits operating other channels on closely adjacent frequencies on the same or nearby wire or radio paths. Band limitation is applied at the receiving end to exclude from the received signal all interference energy (noise or crosstalk) that lies outside the essential signal-frequency band.

The sixth class of functional elements, encoders, refers to frequency modulators and demodulators, pulse modulators, etc., which may be introduced into the line terminals to encode the signals in the form in which it is desired to transmit them.

Two examples of "other functions" are alarm or protective features to guard against transmission troubles, and predistorters and restorers. Alarm features may merely signal an attendant, or they may automatically switch spare lines or equipment in place of those that have failed. For example, such functions are often added to automatic-transmission regulators. If a regulator reaches the limit of its range of adjustment, a trouble is indicated; this is made to actuate an alarm or a switch. Predistorters are used for altering the spectrum of a signal on the line to a more favorable relation with the noise spectrum for the purpose of improving the signal-to-noise ratio. The restorers return the spectrum to normal.

Model of a Single-line Section; Application to Practical Lines. Figure 13-3 shows a model of a line section incorporating the functions that in general are needed as discussed earlier. This, of course, is not a block diagram of any particular system but a basic pattern for any system. The functions may actually appear in different order or form than shown. For example, equalization and regulation are often incorporated in the feedback circuit of the amplifier rather than as tandem elements.

To make the model more general, a frequency translator (assumed to have no loss or gain) is included in the dotted box when it is desired to change the frequency loca-

COMMUNICATION SYSTEMS   13-15

tion of the channel on successive sections. The dashed lines indicate how the regulating controls will appear if signal-controlled regulation is employed.

Practically, the perfection of an ideal line is not possible, not economically justified, and not necessary. Many departures from the above pattern occur in practice. In the simplest lines, the functions of regulation, equalization, and even of amplification may be altogether omitted. In other cases, equalization may be rough or approximate; regulation may be done manually or, if automatic, may be incorporated in only, say, every second or third line section. In long multisection lines, a common engineering procedure is to use simple, rough, or approximate methods of equalization and regulation in most sections, with more refined methods in occasional sections to remove the accumulated errors before they become too large. Perfect equalization and regulation are of course impossible in any event. The designer's problem is, first, to determine how near perfection it is necessary to come and, second, to find the most practical and economical method of achieving the needed results.

The Line Terminals. The receiving line terminal (see Fig. 13-2) is no different functionally from an intermediate line section. Usually it is required that the signals be delivered to the receiving-system terminal at a different frequency from that used

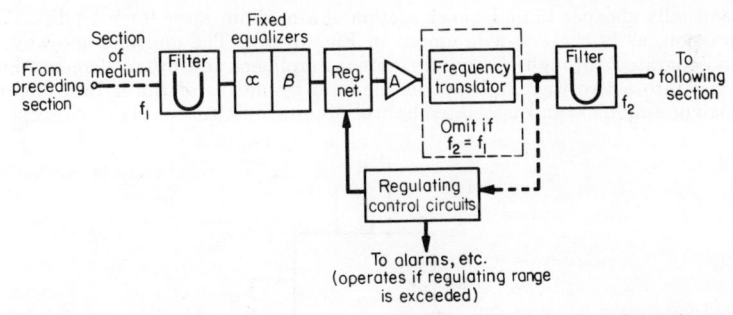

FIG. 13-3. General functional diagram of a line section. Note 1: A = flat-gain amplifier or regenerator. Note 2: Filters may be omitted if $f_1 = f_2$. Note 3: α = attenuation equalizer. Note 4: β = phase equalizer

on the line sections. Therefore (see Fig. 13-3), f_2 is ordinarily different from f_1, and the frequency translator is normally required. Also, it is generally desired to deliver the signals to the system terminal at a lower level than that used on the line. This is equivalent to adding a flat loss in the output of the line. Finally, since this is the final element of the line, the regulation of the entire line rests upon that incorporated in the terminal, and this is the last chance to improve the equalization. Therefore, the most refined methods of equalization and regulation used anywhere in the line will usually be found here.

The sending line terminal will include a frequency translator to shift the frequencies of the signals delivered by the sending terminal of the system to the frequencies wanted on the first line section. It will include a band-pass filter at the input to the line section. It will include an amplifier to raise the signal level to that wanted on the line.

The sending terminal will include the functions of encoders, frequency translators, frequency or time-division multiplexing, and amplification. Encoding may be done both before and after the signals of the separate input channels are combined into one signal by the multiplexing arrangements. The encoding may consist of amplitude modulation, frequency modulation, or some form of pulse modulation combined with AM or FM. If pilots are used for regulation, the sending terminal includes the function of generating them and regulating their outgoing levels.

13-8. Functions of a Two-way Transmission System

Some communication services, such as teletypewriter and telephone services, are two-way services. It is necessary on every such connection that two signal channels, one transmitting in each direction, be established from one end to the other through all the intervening transmission and switching systems. Two-way transmission introduces a new function: a transducer which combines two one-way channels in opposite directions into one two-way channel. For telephone service, this transducer is called a *hybrid circuit*, and its essential properties are given by Figs. 13-4 and 13-5. In teletype and radar systems, functionally similar transducers exist.

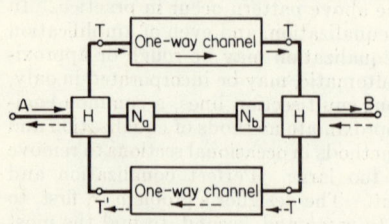

FIG. 13-4. Method of combining two one-way channels to form a two-way channel.

A conceivable method of providing two-way communication is to set up two completely separate one-way channels, as in Fig. 13-6a. This arrangement is four-wire throughout. It is technically an ideal arrangement, for the full independence of the two paths ensures that there can be no echoes or singing (oscillation). The only objection is that of cost of the extra path and of four-wire switching.

It is usually cheaper to make each section of a medium serve for both directions of transmission, as in the two-wire circuit in Fig. 13-6b. The opposite one-way paths remain separated through the unidirectional amplifiers and other elements but are combined into a two-way path on the medium by means of the hybrid circuits H. Each pair of amplifiers and associated hybrid circuits constitutes a "two-wire repeater."

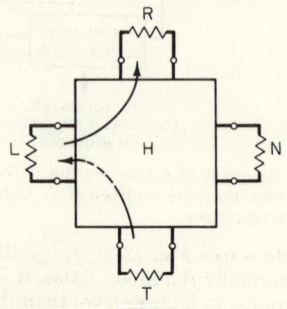

FIG. 13-5. Some properties of biconjugate circuits. L = impedance of two-way line. T = impedance of one-way line that transmits to L. R = impedance of one-way line that receives from L. N = impedance of balancing network. Loss T to L = n_1 = 3 db in symmetrical coil-type hybrid circuit. Loss L to R = n_2 = 3 db in symmetrical coil-type hybrid circuit. Loss T to R = return loss + n_1 + n_2, where return loss = $-20 \log_{10} (L - N)/(L + N)$.

Note: The above ignores some minute second-order effects, assuming N approximately balances L, and R balances T.

3 db is the theoretical value for ideal coils. The value is often taken as 4 db for actual coils. The hybrid circuit need not be symmetrical, in which case n_1 and n_2 are different. However, these are always so related that the sum of the power ratios corresponding to n_1 and n_2 is unity (for ideal coils). In symmetrical bridge-type hybrid circuits, n_1 and n_2 are 6 db or more.

Because of echo and singing limitations, long circuits are made four-wire, as indicated in Fig. 13-6c and d. Figure 13-6d is better from the transmission standpoint than c because it eliminates the two-wire path through switching center PC as a source of echoes. This is at the expense of four-wire switching at PC.

In a four-wire transmission system, the two signal channels are physically or electrically separated to such a degree that the couplings between them are small. In a two-wire system, however, the oppositely directed channels are more closely associated through the hybrid circuits. Gains in a two-wire system amplify the ill effects of the couplings, creating potential singing and echoes. This restricts the amount of gain that can be employed in a two-wire system and limits its length for satisfactory operation.

(a) Four-wire circuit throughout

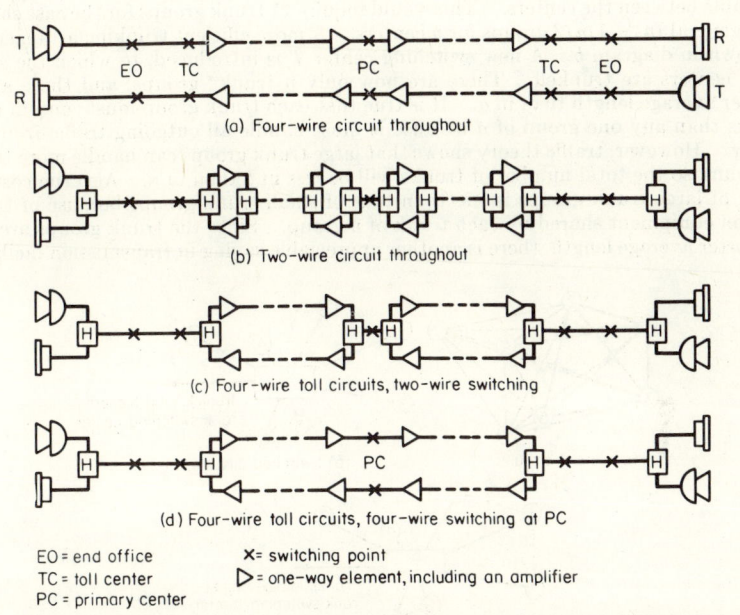

(b) Two-wire circuit throughout

(c) Four-wire toll circuits, two-wire switching

(d) Four-wire toll circuits, four-wire switching at PC

EO = end office
TC = toll center
PC = primary center
X = switching point
▷ = one-way element, including an amplifier

Fig. 13-6. Several arrangements for providing two-way telephone communication.

For much the same reasons, four-wire switching has considerable transmission advantages over two-wire switching. By not bringing the two signal channels together through the switching center, important sources of echoes are eliminated.

13-9. Introduction to Network Aspects

A complete communications network may be a complex aggregate of end instruments, transmission systems, and switching systems which are organized in a hierarchy of some sort. For this aggregate of elements to function as a communication network that gives fast service with acceptable transmission at low cost and high reliability, it must be carefully engineered as a whole. This involves transmission, switching, and traffic-engineering phases of the plan. One cannot assign losses and other transmission standards to particular circuits without knowing how many other circuits may be connected in tandem with these to complete a connection.

The overall engineering plan for the network may be called a switching plan. Switching plans are predicated on certain switching and operating criteria and methods and specify the pattern of grades and locations of switching centers, the routes of calls, and the number and lengths of circuits that may be switched in tandem. They include specification of rules for the losses that different classes of circuits called for by the plan must have. The application of the switching plan to the plant involves traffic-engineering considerations, for the sizes of trunk groups needed to handle given volumes of traffic are important economic factors in determining the locations of switching centers and the trunk groups to be provided.

13-10. Switching Plans

The general pattern of a switching plan may be described by using the elementary diagrams of Fig. 13-7. The three diagrams show three ways of interconnecting seven switching centers, represented by the circles. The switching centers may be local exchanges or they may be long-distance switching centers.

Diagram *a* shows the arrangement that would exist if only direct trunk groups were available between the centers. This would require 21 trunk groups for the case shown, or in general $(n - 1)n/2$ groups for n centers. A more efficient trunking arrangement is shown in diagram *b*. A new switching center T is introduced, to which the seven other centers are trunked. There are now only n trunk* groups, and these are of shorter average length than in *a*. It is true that each trunk group must contain more trunks than any one group of *a* because it must handle all outgoing traffic from the center. However, traffic theory shows that large trunk groups can handle more traffic per trunk so the total number of trunks will be less in *b* than in *a*. Also the cost per trunk of large trunk groups is less than that of small trunk groups because of transmission equipment shared by each trunk of a group. Since the trunk groups are also of shorter average length, there is another appreciable saving in transmission facilities.

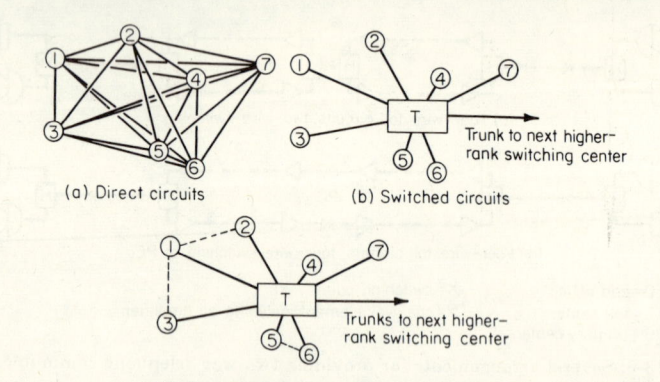

(a) Direct circuits

(b) Switched circuits

(c) Combined switched and direct circuits

Fig. 13-7. Elementary switching plans.

On the other side of the ledger are the switching penalty of requiring an additional switching operation to complete a connection, and the transmission penalty of requiring each circuit to be of higher transmission quality if the same overall quality of transmission is to be furnished, because each call now passes over two trunks instead of one.

Now there may be enough traffic between certain of the unit centers to warrant adding some direct trunks, as suggested by the dotted lines of diagram *c*. This heavy traffic can be handled over these trunks without the additional switching operation, reducing the load on T and improving the transmission for these calls. This configuration introduces the possibility of alternate routes for many of the calls. A call can reach (2) from (1), for example, either over the dotted trunk group or via the tandem center T. Therefore, the preferred dotted group does not have to be large enough to handle all the calls offered; the overflow can be routed via T. The dotted trunks are called *high-usage* trunks. The alternate routes to T which carry the overflow traffic from the high-usage trunks are called the *final* trunk groups. Note that the final trunk group from (1) to T handles the overflow traffic from two high-usage trunk groups: (1) to (2) and (1) to (3).

A third class of trunks must be distinguished: trunk groups such as those from (4) to T and (7) to T, for which there is no alternate route. These will be referred to as *direct* trunks in this section.

* This assumes that the trunks are two-way from the switching standpoint so that connections can be built up in either direction through them. If the trunks are one-way, as is common in the local plant, each line of the figure represents two trunk groups, one for each direction of calls.

COMMUNICATION SYSTEMS 13-19

Figure 13-8 is typical of a telephone network in a sizable city, or of a teletypewriter network in a much larger geographical area.

A nationwide switching plan can be synthesized from elementary plans like that just discussed. Imagine that tandem center T of Fig. 13-7c is itself one of several centers whose central switching center is one of higher rank in a larger-scale arrangement. This in turn can be a unit in a still larger scale arrangement, and so on until we finally reach the highest-ranking switching center in the whole system. Additional

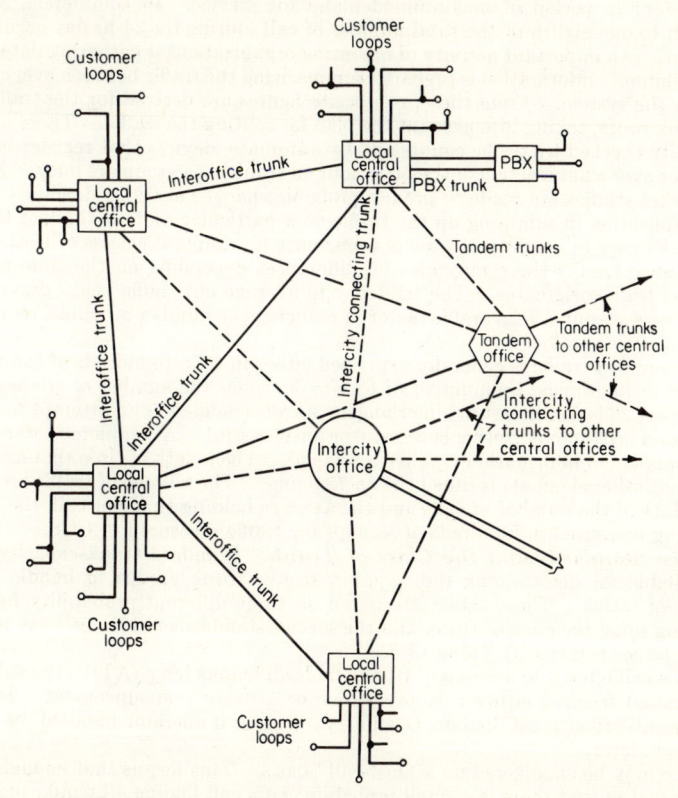

Fig. 13-8. Possible layout of loops and trunks in part of an exchange area.

high-usage trunks would be added as needed between members of different parts of this network of centers to take care of particular traffic demands and to provide alternate routes between more distant parts of the network.

13-11. Traffic-handling Capacity of Trunk Groups

This section discusses briefly the traffic-engineering principles that are applied in practice to determine the sizes of trunk groups needed in different parts of a switching network. A more complete discussion of this subject will be found in Ref. 2.

The general traffic objective is to provide sufficient facilities so that calls can be completed with reasonably small delay during the hour of heaviest daily traffic. The problem is to determine what constitutes "sufficient facilities" when planning trunk groups. The answer depends upon a number of factors:

1. The volume and time distribution of the traffic that is offered
2. Whether the trunk group is a high-usage group whose overflow is to be submitted to other routes, or a final trunk group which accepts the overflow traffic from high-usage groups
3. The standards as to what constitutes "reasonably small delay"

The Volume of Traffic. Each type of communication service has its own traffic properties. Telephone traffic, for example, occurs in daily cycles with one to three peaks each day. Engineering usually is on the basis of the busy hour, i.e., the consecutive 60-min period of maximum demand for service. In commercial systems one-tenth to one-eighth of the total number of calls during the 24-hr day occur in the busy hour. An important activity of operating organizations is gathering data on the traffic volume. Information is prepared summarizing the traffic between every pair of points in the system. From these, composite figures are derived for the traffic over each trunk route, taking into account the plan for routing the traffic. These data are continually checked by traffic counts and by automatic devices that register how frequently or over what fraction of the busy hour all trunks in a group are busy. Beyond this, market studies are made to predict probable changes in traffic demands.

A complication in summing up the traffic on a particular route is the fact that the daily peaks vary in height and time of occurrence for different classes of traffic. For long-distance trunks there may also be differences depending on the time zones in which the traffic originates. The tendency to average out traffic peaks due to these differences is of course a favorable factor in reducing the number of trunks required to handle the day's traffic.

The volume of traffic is generally expressed either in CCS (hundreds of call-seconds per hour) or in erlangs; the number of CCS = 36 times the number of erlangs. The ρ of Chapter 28 is also measured in erlangs. A *call* is defined as an attempt to reach a station (end instrument), either successful or unsuccessful. A completed call resulting in transmission of information is called a *message*. The length of time that a call uses a trunk (or other element) is called the *holding time*. The volume of traffic is equal to the product of the number of calls and the average holding time per call, the holding time being measured in hundreds of seconds for traffic measured in CCS.

Service Standards and the Class of Trunk. Standard trunk-capacity tables are available for determining the required size of trunk groups to handle various volumes of traffic. These tables are based on three different probability formulas, depending upon the class of trunk and the service standards to be met. An example of these tables is given by Table 13-5.

When a call offered to a group of trunks finds all trunks busy (ATB), the subscriber (or operator) receives either a suitable tone or a delay pronouncement. In either case, the subscriber must disconnect and try again, or if operator-handled, he may be called back later.

Service may be engineered on a "lost-call" basis. This means that enough trunks are provided so that there is a small probability of a call finding all trunks of a group busy. When thus engineered, the subscriber generally receives an all-trunks busy signal and must disconnect. The trunk-capacity tables used in the Bell System for engineering trunk groups on this basis (no alternate routes) are called P tables. They are based on the Poisson formula* and assume that the traffic is random in occurrence and must all be handled by the particular trunk group. Table 13-5 is for $P.01$ and represents probabilities of one in a hundred that a call will find all trunks busy.

The very great increase in the efficiency of large over small trunk groups is apparent from the table. For example, in Table 13-5 a group of two trunks can handle only 2.7 CCS per trunk per busy hour, while a group of 16 trunks can handle more than 18 CCS per trunk per hour—about seven times as much traffic per trunk.

A quite different situation exists when each call has a choice of routes to its destination. Alternate routing is a procedure whereby a parcel of traffic is provided a second-choice path through an intermediate switching office, which is also the first-, second-,

---

* The well-known traffic formulas referred to in this discussion are given without explanation in Table 13-6; the Poisson distribution is discussed in Chapters 28 and 38.

COMMUNICATION SYSTEMS

**Table 13-5. Trunk Capacity for $P.01$—Poisson Formula
(ATB* Probability of 0.01)**

Trunks per group	Capacity, CCS	% use	Trunks per group	Capacity, CCS	% use
1	0.4	1.1	26	562	60.1
2	5.4	7.5	27	590	60.7
3	16	14.5	28	618	61.3
4	29	20.1	29	647	62.0
5	46	25.6	30	675	62.5
6	64	29.8	31	703	63.0
7	84	33.3	32	732	63.6
8	105	36.5	33	760	64.0
9	126	38.9	34	789	64.5
10	149	41.4	35	818	64.9
11	172	43.4	36	847	65.4
12	195	45.1	37	876	65.8
13	220	47.0	38	905	66.2
14	244	48.4	39	935	66.6
15	269	49.8	40	964	66.9
16	294	51.1	41	993	67.3
17	320	52.3	42	1,023	67.7
18	346	53.4	43	1,052	68.0
19	373	54.5	44	1,082	68.3
20	399	55.4	45	1,112	68.7
21	426	56.4	46	1,142	69.0
22	453	57.2	47	1,171	69.2
23	480	58.0	48	1,201	69.5
24	507	58.7	49	1,231	69.8
25	535	59.4	50	1,261	70.1

* All trunks busy.

or third-choice path for other parcels of traffic. The first-choice trunk groups, the high-usage groups, can be tightly engineered since they do not have to handle all the offered traffic. The second-choice (or final-choice, if there are more than two choices) path, since it is used in common for the overflow from several first-choice paths, becomes relatively efficient because of its larger size. The overall result is that the total number of trunk paths to a particular destination is less than would be the case if each parcel had sole access to its own group of paths, and service is better. With automatic switching, this is accomplished without impairment of the average speed with which all the traffic is handled.

The trunk-capacity tables used for engineering high-usage trunk groups are based on the Erlang B formula and are called high-usage trunk-capacity tables. The Erlang B formula assumes that traffic is offered at random with "lost calls cleared," that is, not reoffered to the high-usage trunks (see Table 13-6 for formula).

Engineering Trunk Groups for Alternate Routing. The principles affecting the engineering of an alternate-routing trunk arrangement can be demonstrated from the typical Erlang B distribution shown in Fig. 13-9 for an offered load of 240 CCS. Curve a shows the load that would be carried by each of 14 trunks, assuming the calls are offered in succession to trunks 1, 2, 3, etc., in that order. Curve b shows the total load carried by n trunks. The remainder of the 240 CCS is cleared over other trunk routes. (The total load carried by a group of trunks will be the same regardless of the order in which the load is presented to the trunks.)

Table 13-6. Telephone-traffic Formulas, Assuming Infinite Sources

	Used for direct trunks	Used for high-usage trunks	Used for manually operated trunks, e.g., $T = 30$ tables are for probable delays of 30 sec
	Poisson	Erlang B	Erlang C
	Lost calls held; no alternate route, no delay	Lost calls cleared through an alternate route	Lost calls delayed
Frequency distributions $x \leq c$	$\dfrac{a^x e^{-a}}{x!}$	$\dfrac{a^x e^{-a}/x!}{1 - P(c+1, a)}$	$\dfrac{a^x e^{-a}/x!}{1 - P(c,a) + (a^c e^{-a}/c!)[c/(c-a)]}$
$x \geq c$	$\dfrac{a^x e^{-a}}{x!}$	0	$\dfrac{a^x e^{-a}/c! c^{x-c}}{1 - P(c,a) + (a^c e^{-a}/c!)[c/(c-a)]}$
Probability of delay	$P(c,a) = \displaystyle\sum_{x=c}^{\infty} \dfrac{a^x e^{-a}}{x!}$	$B(c,a) = \dfrac{a^c e^{-a}/c!}{1 - P(c+1, a)}$	$C(c,a) = \dfrac{(a^c e^{-a}/c!) \, [c/(c-a)]}{1 - P(c,a) + (a^c e^{-a}/c!)[c/(c-a)]}$
Load carried	$a[1 - P(c-1, a)] + cP(c,a)$	$a[1 - B(c,a)]$	a

x = number of trunks in use
c = number of full-access trunks
a = load submitted in average simultaneous calls
  = $\dfrac{\text{(number of calls per hour) (average holding time, sec)}}{3,600}$

COMMUNICATION SYSTEMS 13-23

Let us assume that the busy-hour traffic from exchange M to exchange N is 240 CCS and that this is offered first to a high-usage trunk group between the two exchanges and then, if no trunk is free, to a second (final) route via a tandem office T. What is the economic balance between the sizes of the high-usage and final trunk groups? Tandem trunk groups range in size from 20 trunks upward. From Table 13-5, adding a trunk to a 20-trunk group increases its capacity by 27 CCS, and adding one to a 40-trunk group increases its capacity by 29 CCS. Therefore, over a considerable range each additional trunk increases the traffic-handling capacity by a substantially constant figure of about 28 CCS.

The final route via the tandem involves two trunks M to T and T to N. Let us assume that an incremental path over the tandem route costs, say, 1.4 times as much as an incremental path over the M to N route. It follows that the economic size of the high-usage M to N trunk group is such that the traffic handled by an incremental path in that group is 28/1.4, or 20 CCS. From Fig. 13-9, curve a, this is the traffic handled by the seventh trunk of a seven-trunk group. The economic size of the M to N trunk group is therefore seven trunks. From curve b, the seven trunks will carry

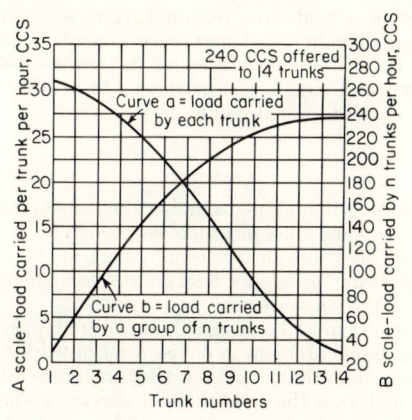

FIG. 13-9. Distribution of offered load according to the Erlang B assumption, lost calls cleared.

185 CCS, so that 240 minus 185, or 55 CCS, will overflow into the tandem trunks. To handle this overflow $^{55}\!/_{28}$, or 1.96 trunks, must be added to the tandem groups. Significantly, the service from M to N is now only 0.0023 instead of 0.01.

The above example illustrates the basic principles underlying the optimum size of high-usage trunk groups with a very simple case. The application of the principles becomes more complicated in many actual cases. This is particularly true of large networks where the switching pattern is more complex and circuits are more varied as to lengths, types, and costs.

13-12. A Major Transmission Aspect of Networks

Most switching plans for large networks impose strict transmission requirements on the individual trunks. For example, the loss (overall signal attenuation) must not be excessive when calls are routed over the maximum number of trunks in tandem, and there should not be too great a contrast between the transmission afforded over the different possible routes that a call may take. The general requirement for the switching plan to be successful is that every trunk must be operated at the least possible loss. This section discusses the problem of how the individual trunk loss is determined. There are other important network transmission problems, but this one arises in every network, and it is a good example of a whole class of communication-network problems.

The losses of two-wire trunks cannot be made zero because of echoes, crosstalk, noise, and sometimes because of singing limitations. Moreover, the plant may contain a wide variety of facilities, old and new, which are limited in different degrees by these transmission factors. The problem is to determine how low the trunk losses can be made, and to state in easily applied rules the losses that different classes of trunks called for by the switching plan should have.

The general plan is to operate every trunk at the lowest practicable loss, considering the length and type of facility, and to assign trunks to positions in the network in accordance with their transmission capabilities. The poorer trunks are assigned to short-trunk groups or to ones whose position in the network ensures that they will never be switched in tandem with more than one or two other trunks.

Ideally, each trunk should have a different loss in each connection in which it is used. This is not practicable, and a compromise is necessary. The general plan is to operate the intermediate trunk of a multitrunk connection at a low loss called the via net loss (VNL) and to add an extra loss S at both ends of the connection.

In one version of this scheme, each end of an intertoll trunk is equipped with a switched pad (an S pad) whose loss is 2 db in commercial telephone systems. The pad is left in when connection is made to a terminal trunk with a local exchange. It is switched out, affording 2-db gain, when connection is made to another intertoll trunk. The loss of a circuit with both S pads in is called its terminal net loss (TNL), and the loss with both pads switched out, its via net loss. In another version, the S loss is permanently included in the terminal trunk.

13-13. Conclusion

The above discussion of switching plans and of the transmission standards that the circuits involved in these plans must meet has necessarily been somewhat condensed. However, the main principles, and the philosophy back of the establishment of transmission standards, are meant to supply background for understanding the network aspects of large communication systems.

Communication networks must be designed, engineered, operated, and maintained part by part. It is no simple problem to devise a set of working standards and rules for dealing with these parts so that, together, they integrate into a system that gives customer satisfaction under all the conditions of service. These matters are under continual study, and the standards and rules may be expected to change with time, particularly as new systems with new capabilities emerge from system engineering and development to use.

REFERENCES

1. Hall, A. D.: *A Methodology for Systems Engineering*, D. Van Nostrand Company, Inc., Princeton, N.J., 1962.
2. Truitt, C. J.: Traffic Engineering Techniques for Determining Trunk Requirements in Alternate Routing Trunk Networks, *Bell System Tech. J.*, vol. 23, pp. 277–302, March, 1954.

Chapter 14

RADAR

JEROME FREEDMAN, *Lincoln Laboratory,* Massachusetts Institute of Technology.*

CONTENTS

14-1. Introduction	14-3
14-2. Detection of Radar Signals	14-5
Elements of the Detection Process	14-5
The Matched Filter	14-5
Decision Mechanisms	14-6
Signal-to-Noise Ratios in Coherent Systems	14-7
Signal-to-Noise Ratios in Noncoherent Systems	14-7
14-3. Design of Radar Signals	14-8
The Significance of Resolution	14-8
The Ambiguity Function	14-10
Relation between CW and Pulsed Systems	14-10
Large Time-Bandwidth Product Signals	14-10
Pulse Groups	14-12
Clutter Rejection of Pulse Groups	14-13
"Pulse Doppler"	14-14
Jittered Repetition Rates	14-14
14-4. Synthesis of Matched Filters	14-14
Matched Filter for a Pulsed Sinusoid	14-14
Ensembles of Matched Filters	14-15
Matched Filters for Pulse Groups	14-15
Clutter Rejection in Matched Filters	14-16
Postdetection "Matched" Filters	14-17
Sensitivity to Mismatch	14-18
14-5. Transmitters	14-18
Stable Oscillators	14-18
High-power Microwave Generators	14-19
Grid-controlled Tubes	14-19
Linear-beam Tubes	14-20
Crossed-field Tubes	14-21

* Operated with support of the U.S. Air Force.

14-2   SYSTEM COMPONENTS

Magnetron	14-22
Amplitron	14-22
Cathodes	14-23
Windows	14-24
Cooling	14-24
X Rays	14-24
Available High-power Tubes	14-25
Radar Pulse Modulators	14-26
Line-type Modulators	14-26
The Hard-tube Modulator	14-28
Modulating-anode Operation of High-power Tubes	14-28
Fault Protection	14-28
Special Microwave Components	14-29
Duplexers	14-29
The Rotating Joint	14-30
14-6. Antennas for Radar Systems	14-30
Reflector-type Antennas	14-31
Special Beam Shapes	14-34
Feed-displacement Scanning Systems	14-34
Tracking Feed Systems	14-35
Conical Scan	14-35
Four-horn Monopulse	14-35
Radiated Field Zones	14-35
Array Antennas and Scanning Systems	14-37
14-7. Receiving Systems	14-40
Functional Parts of the Receiver	14-40
Sources of Receiving-system Noise	14-41
Low-noise Amplifiers	14-44
The TWT	14-45
The Parametric Amplifier	14-45
Tunnel-diode Amplifier	14-46
Masers	14-46
14-8. Radar Targets	14-47
Scattering of Electromagnetic Energy	14-47
The Scattering Matrix	14-48
Methods for Computing Cross Section	14-48
Aircraft Targets	14-50
Rain	14-52
Reflection from the Earth	14-53
Reflection from a Smooth Earth	14-55
14-9. Influence of the Atmospheric Medium	14-55
Absorption and Backscatter in the Lower Atmosphere	14-55
Refraction in the Lower Atmosphere	14-56
Propagation through the Ionosphere at Microwave Frequencies	14-57
Other Atmospheric Effects	14-58
14-10. Utilization of Data	14-58
Search Systems	14-58
Tracking Systems	14-60
14-11. Radar-system Sensitivity	14-60
Derivation of the Canonic Search Equation	14-60
Limitation on Scanning Rate	14-62
System Losses	14-63
The Tracking or Nonscanning Form of the Search Equation	14-63
14-12. Typical Radar Systems	14-64
Surveillance Radar	14-64
Tracking Radar	14-65
Mapping Radar	14-66
Acknowledgment	14-70

14-1. Introduction

Man has always been interested in amplifying the capability of his principal sensory system, the visual system. Our history is full of recorded attempts to see at a distance. None of the devices which apply to seeing at a distance has ever come close to amplifying all the characteristics of our visual system. We have as a result been willing to accept devices which will amplify one characteristic at the expense of others. The telescope is an example of such a device; radar is another. The development of radar was motivated by our dissatisfaction with the sensitivity and reliability of our visual system for the detection of aircraft at great distances. We sought a system which could do this at distances of hundreds of miles, night or day, in good weather or bad.

Detection at a distance requires the detection of some form of energy generated or reflected by the "target." There are many forms of energy which might be useful to a detection system. However, because of the propagation limitations imposed by our terrestrial atmosphere, the most useful of these is electromagnetic energy at wavelengths greater than 1 centimeter; wavelengths are limited to less than 10 meters for other reasons.

While considerable energy is generated by such targets as aircraft in flight, in general we cannot depend reliably on the generating mechanisms and the radiation of this energy. In order to provide assured detection, radar has developed principally around the form in which the source of energy is supplied by the system and the detected energy is reflected from the target.

The basic elements of such a system are:

1. A source of electromagnetic energy, commonly called a transmitter
2. An antenna to radiate this energy into space
3. An antenna for receiving the energy reflected from the target in space
4. A receiver for processing the received energy
5. An output system for display or utilization of the energy

In early systems, the transmitting system (consisting of the transmitter and transmitting antenna) and the receiving system (consisting of the receiving antenna and receiver) were separated by a considerable distance. Such a system is known as a bistatic system. When the transmitted signal is continuous, this separation may be necessary in order to ensure that the receiving system will receive energy reflected from the target rather than directly from the transmitting system. The early radar experiments were conducted at low frequencies (less than 50 Mc) where it was difficult to obtain isolation by means of antenna directivity alone. Adequate spatial isolation was essential. Because of the poor angular resolution and the continuous transmission, very little information about a target was available except that it had come within the range of the system.

In further developments in the years between 1935 and 1940, the notion of a pulsed transmitter was introduced. The use of a pulsed signal also permitted the receiver to be connected to the same antenna as the transmitter through a time-sharing (gaseous discharge) switch activated by the transmitter pulse. This switch is called a duplexer. Systems using a common antenna are called monostatic radar systems; a block diagram is shown in Fig. 14-1.

The word *radar* is an acronym coined from the combination of the ability to measure target range (distance to the target) and the ability to detect; hence RAdio Detection And Ranging, or radar. Implicitly associated with the detection and ranging functions are the auxiliary functions of tracking (following the target) and determining the identity and nature of the target. At the time of the development of radar, the number of aircraft in flight at any given time was not large, and it was therefore possible to emphasize the problem of detection sensitivity and minimize the problem of target discrimination and recognition.

As much as the development of new components permitted, radar systems were constructed at higher and higher frequencies, permitting angular resolution with

SYSTEM COMPONENTS

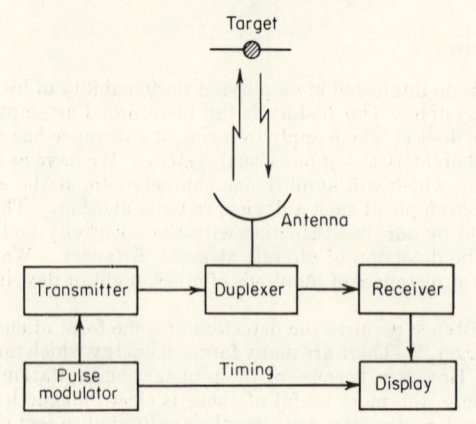

Fig. 14-1. Elementary monostatic pulse radar system.

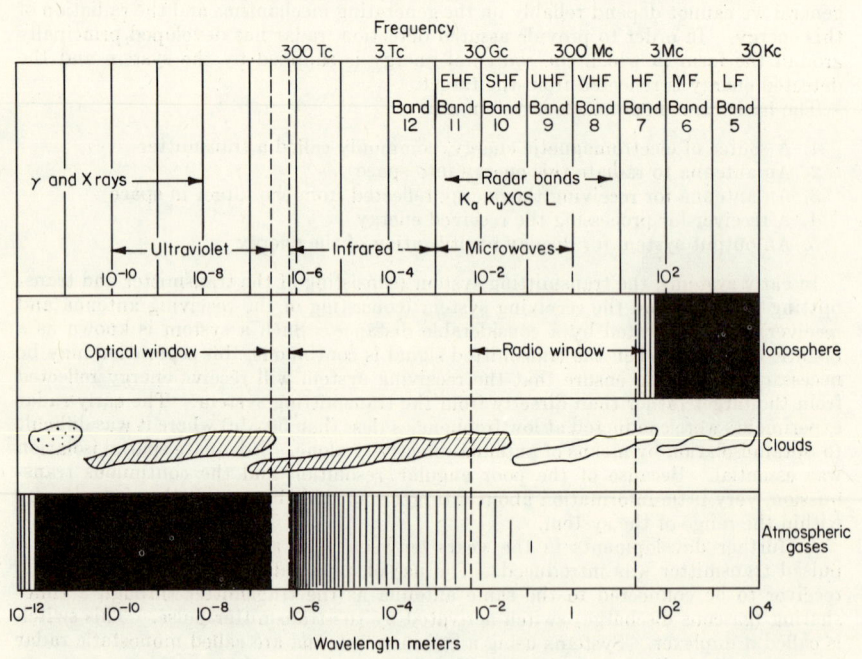

Fig. 14-2. Electromagnetic spectrum usage and atmospheric obstructions. (*Taken in part from J. L. Pawsey and R. N. Bracewell, Radio Astronomy, Oxford University Press, Fair Lawn, N.J., 1955.*)

modest antenna sizes and greater bandwidths for higher range resolution. By the end of World War II, portions of the frequency spectrum between 25 Mc and 25 Gc had been employed. The development had occurred in discrete bands which, during the war, had been given letter designations for security's sake. These designations are now in common usage. The occupancy of the microwave spectrum by radar is shown in Fig. 14-2.

Because of the great difficulty involved in constructing an all-purpose radar system, radars have taken on many specialized forms. Most radars can be classified into three most common functional categories:

1. Search systems, whose purpose is to search a solid angle for unknown point* targets. The output of a search system is the statement that targets are present or not present in the resolvable elements of search volume.

2. Tracking systems, whose purpose is to follow continuously the motion of a point target. The outputs of such a system are the coordinates of position and the motion parameters of the target.

3. Mapping systems, whose purpose is to provide a radar map of targets having considerable area.

The dividing line between categories is not sharp. Search systems frequently provide track data, but the data rate and accuracy may be much less than a tracking radar will provide. On the other hand, tracking systems frequently must perform a search function through a solid angle, albeit of small extent, in order to acquire the target to be tracked. Sometimes systems are deliberately built in combination. Airborne intercept systems have been built to perform a wide-angle search function and then to change to a tracking mode after target acquisition. The radars in each of the categories have the same basic elements previously listed. The designs may differ considerably in detail because of their application to a special function.

Radar systems may have a variety of locations as well as a variety of purposes. For example, they can be located on the ground, on a ship, in an aircraft, or in a satellite. They may be assigned the task of searching for an aircraft on an airport taxi strip or assisting in the rendezvous of two satellites. The most widely used radar system is the pulsed monostatic system. Bistatic systems are less frequently used. One application for the bistatic system, however, is in so-called semiactive missile guidance, where the missile-borne receiver homes on the energy reradiated from a target which is illuminated by a separate transmitter, usually ground-based.

14-2. Detection of Radar Signals

Elements of the Detection Process. One of the primary functions of a radar system is to detect the presence of targets in the space around it. There are many factors which interfere with the detection process. The radar signals may pass through a medium that introduces absorption which is selective in frequency and polarization. The medium may also introduce partial reflection, refraction, and multipath problems. The targets may have reflective properties which are selective in frequency and polarization. The reflective properties may also vary rapidly with time as a result of change in target orientation or other reasons. Random fluctuations of propagation-path characteristics and reflective properties result in random variations in the phase and amplitude of the reflected signal. Finally, there is the problem of the noise introduced by the transmission medium and the receiving system.

The presence of noise would not disturb the detection process if infinite observation time were available. But because either the target or the radar is in motion, the observation time must be finite. Conflict exists between the requirements for a short observation time to reduce the errors introduced by target motion, and a long observation time to diminish the errors introduced by noise. It is possible to establish by methods of decision theory that the optimum detection process in the presence of white Gaussian noise consists, first, of maximizing the signal-to-noise ratio and, second, establishing a decision mechanism for determining the presence or absence of targets in the processed signal (see Chapter 24).

The Matched Filter. It has been shown that, when noise is "white" and Gaussian, maximum signal-to-noise ratio at the output of the receiver is attained when the

---

* Point targets are defined as targets which are smaller in dimension than the range and angular resolution of the radar. Area targets are defined as having dimensions large compared with the range and angular resolution of the radar.

receiver has a filter characteristic such that

$$h(t) = s(T - t) \tag{14-1}$$
or
$$H(\omega) = S^*(\omega)e^{-j\omega T} \tag{14-2}$$

where $h(t)$ = impulse response of the filter
   $s(t)$ = received signal of duration T
   $H(\omega)$ = frequency response of the filter
   $S(\omega) = \int_{-\infty}^{+\infty} s(t)e^{-j\omega t}\,dt$ = Fourier transform of $s(t)$
   $S^*(\omega)$ = complex conjugate of $S(\omega)$

Such a filter has been given the title "matched filter" or "ideal receiver." Its impulse response is the reversed time image of the received signal, and its frequency response is the complex conjugate of the Fourier transform of the received signal.

The value of the signal-to-noise ratio at the output of the matched filter is

$$\left(\frac{S}{N}\right)_{\text{m.f.}} = \frac{E}{2N_0} \tag{14-3}$$

where E = received signal energy
   N_0 = noise power per cycle per second ($\tfrac{1}{2} KT_0$)
   K = Boltzmann's constant
   T_0 = absolute noise temperature
   $\dfrac{S}{N} = \dfrac{\text{rms signal power}}{\text{rms noise power}}$

This last equation (14-3) is extremely significant since it states that the output signal-to-noise ratio is independent of the shape of the signal and dependent only on the energy of the signal. The full significance of this statement begins to be realized when it is recognized that we now have a mechanism for the comparison, on the basis of sensitivity, of a wide variety of signal forms. For example, a continuous (CW) signal, a single-pulse signal, or a train of pulses can all be compared on the basis of energy content for a given observation time. Provided that in each case the receiver is a matched filter, the system sensitivity is the same for equal-energy-content signals. The matched filter for each signal is, of course, vastly different. Various types of matched filters are discussed below.

Decision Mechanisms. The second step in the detection process, establishment of a decision mechanism, logically follows the matched filter. In one form of detection system, an ideal observer is postulated who sets a threshold in his observation system, and any signal which exceeds this threshold is labeled as a target. This threshold may be exceeded by a target signal plus noise, or it may be exceeded by noise alone. In either case, the observer will label the threshold crossing as a target. This is illustrated in Fig. 14-3. The observer can control the rate of noise crossings of the threshold, or "false alarms," by the setting of the threshold. The higher the threshold, the less the false-alarm rate. The number of time intervals (the correlation interval is established by the filter bandwidth) which elapse between noise crossings

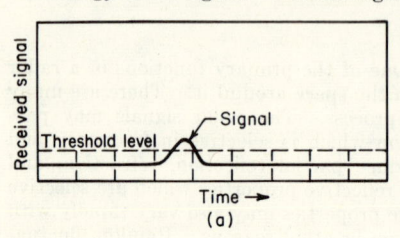

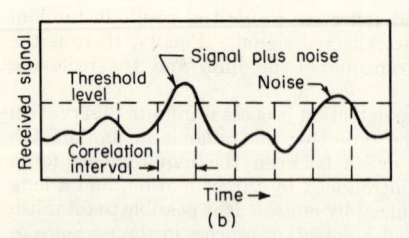

Fig. 14-3. Illustration of threshold decision criterion: (a) when the signal is very much greater than noise; (b) when the signal is not much greater than noise.

is called the false-alarm number n. The probability of a false crossing is called P_F. The conditional probability that the signal plus noise will cross the threshold if a target is present is called P_D, the probability of detection.

The relationships between P_D, n, and S/N are given in Fig. 14-4 for an ideal fixed nonfluctuating target. If the S/N is known or can be computed for a given system, then the permissible probability of detection and false-alarm rate can be determined from this figure. Conversely, the S/N required for a given combination of P_D and n can also be obtained from this figure.

Signal-to-Noise Ratios in Coherent Systems. The "matched filter" has been defined as a filter whose impulse response is the reversed time image of the received signal. The matched-filter designer must be able to describe the received signal and hence must be able to predict its amplitude and phase. A system which fully utilizes this information in the design of the filter is known as a coherent system. The computation of S/N in such a system with known or predictable signal phase and amplitude is straightforward. As previously stated, the S/N is proportional to the total amount of energy in the received signal.

However, there are many possible situations where the received signal amplitude and phase fluctuate. In the simplest case, the phase of the received signal may be unknown but quite regular in behavior. Only a small loss in S/N occurs for this situation, and it may be neglected. In radar systems dealing with simple moving targets having no acceleration, a systematic change of phase occurs. The progression of phase (Doppler shift) is proportional to the radial component of target velocity. The expected extent of Doppler shifts may be large compared with the spectrum of the received signal. In this case an ensemble of matched filters is required, giving overlapping coverage of the entire expected Doppler-frequency-shift region. A small loss in S/N is incurred in the combining of the outputs of the filter ensemble if P_D and n are large, and a decision is made at the output of each filter before combining.

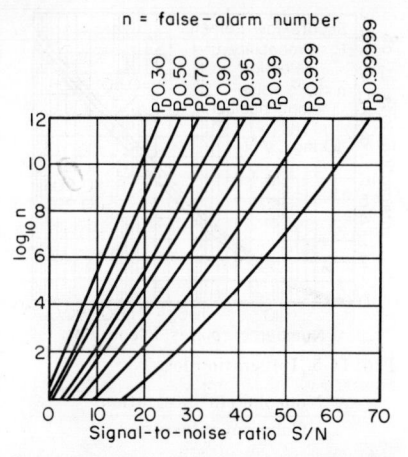

FIG. 14-4. Probability of detection with no integration. (J. I. Marcum, *A Statistical Theory of Target Detection by Pulsed Radar—Mathematical Appendix*, IRE Trans. Inform. Theory, vol. IT-6, no. 2, April, 1960.)

A more complex situation occurs when the received signal phase and amplitude fluctuate in an unpredictable or random manner. These fluctuations may be due to unpredictable target motions or oscillations. Fluctuations may also be introduced by changes in the propagation path or by instabilities in the transmitter or receiver equipment. These fluctuations set a time limit for which the behavior of the received signal is predictable.

The time constant of the matched filter cannot exceed the period for which the signal can be predicted and described (i.e., its period of "coherence"), although this may be considerably less than the period of observation.

Further processing of the received signal may continue beyond this time period, but it must be done by another process known as noncoherent integration or autocorrelation.

Signal-to-Noise Ratios in Noncoherent Systems. The equipment stability required for coherent matched-filter processing is sometimes difficult to attain when the period is long, such as for a train of pulses. Usually the "matched filter" is matched only to a single pulse of the train. The filter is then followed by a rectifier which eliminates phase information by detecting only the amplitude of the received signal. The signal processing is completed by a filter matched to the resulting video pulse train. This process is called incoherent integration or autocorrelation. The insertion of the rectifier in the process introduces nonlinearity into the system which causes the S/N to be proportional to the total signal energy only when the S/N is quite large (i.e., for large P_D and n). For small signal-to-noise ratio, the process is inefficient and a loss is incurred.

This loss has been computed for the case where the received signal amplitude does not fluctuate. In Fig. 14-5 this loss is plotted as a function of the number of pulses in the train.

The received signal may fluctuate in amplitude for reasons previously given, even though the phase information has been eliminated by rectification. This fluctuation rate may be so slow that the amplitudes of the returns are constant for the entire duration of a received pulse train, but it may have changed when the radar next observes the target. In this case additional sensitivity is required of the radar sys-

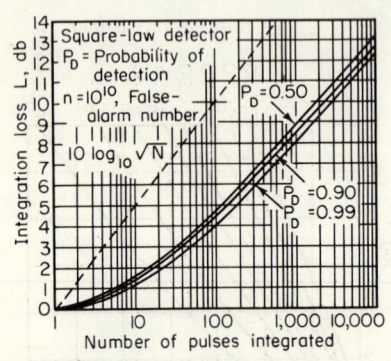

Fig. 14-5. Integration loss

$$\frac{\text{Noncoherent integration}}{\text{Coherent integration}}$$

for a nonfluctuating target. (*J. I. Marcum, A Statistical Theory of Target Detection by Pulsed Radar—Mathematical Appendix, IRE Trans. Inform. Theory, vol. IT-6, no. 2, April, 1960.*)

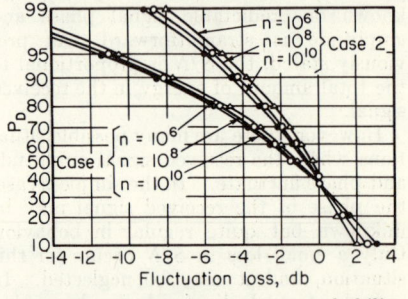

Fig. 14-6. Fluctuation loss vs. probability of detection. The increase in signal-to-noise ratio over that required to detect nonfluctuating targets is plotted as a function of the required probability of detection P_D and the false-alarm number n. Targets are correlated for T_T but fluctuate from scan to scan. Case 1 is an exponential target distribution (several independently fluctuating reflectors). Case 2 is a one-dominant target distribution (one large reflector with other small independently fluctuating reflectors).

tem to maintain a high probability of detection when the target is positioned such that there is a small return. An example of the losses for two different kinds of target-fluctuation statistics is plotted in Fig. 14-6.

If the fluctuation rate is fast, so that the correlation from pulse to pulse in a train is small, then not much loss is encountered if more than a few uncorrelated samples are obtained. In the case of slow fluctuation (high correlation from pulse to pulse), if a small number (three to five) of uncorrelated signals can be obtained simultaneously the fluctuation loss can be greatly diminished. These uncorrelated signals can be obtained through polarization diversity, frequency diversity, or both.

14-3. Design of Radar Signals

The Significance of Resolution. As pointed out by Woodward,[1] it is a peculiarity of radar systems that pure existence information (detection) cannot be divorced from location information. Thus we shall always face two problems, the detection of the desired signal in the presence of noise and the measurement of some parameters of the signal in the presence of noise. Examples of such parameters are the range of the target, the azimuth angle of the target, and the velocity of the target. Table 14-1 lists the relevant criteria for detection and for measurement of range, velocity, angle, and acceleration. As can be seen in this table, only for the case of simple detection is the result dependent only on total signal energy. The need for parameter estimation introduces additional constraints on the signal characteristics, and from needs such as these are created the problems of signal design.

Table 14-1. Detection and Parameter-estimation Criteria

1. Detection criterion:

$$P_D \text{ for a given } P_F \text{ is a function of } E_s/N_0$$

where P_D = probability of detection
P_F = probability of false alarm
E_s = total signal energy received
N_0 = noise power per cycle ($N_0 = KT/2$)
E_s/N_0 = [peak signal/rms noise]^2

$$\frac{S}{N} = \frac{E_s}{2N_0} = \frac{\text{rms signal power}}{\text{rms noise power}}$$

Maximum P_D (for fixed P_F) occurs when E_s/N_0 is maximized at the output of a matched filter.

2. Parameter-estimation criteria:

Standard deviation of error	Measurement based on pulse envelope	Measurement based on carrier phase
Range ΔR	$\dfrac{1}{4\pi} \dfrac{c/B}{\sqrt{E_s/N_0}}$	$\dfrac{1}{4\pi} \dfrac{\lambda_0}{\sqrt{E_s/N_0}}$
Angle $\Delta \theta$	$\dfrac{\theta}{\sqrt{E_s/N_0}}$	$\dfrac{2}{\pi} \dfrac{\lambda_0}{D \sqrt{E_s/N_0}}$ *
Velocity ΔV 2 pulses T_0 sec apart	$\dfrac{1}{2\pi} \dfrac{c/B}{\sqrt{E_s/N_0}\, T_0}$	$\dfrac{1}{2\pi} \dfrac{\lambda_0}{\sqrt{E_s/N_0}\, T_0}$
Acceleration $\Delta \alpha$ 3 equispaced pulses over a total of T_0 sec	$\dfrac{3\sqrt{2}}{\pi} \dfrac{c/B}{\sqrt{E_s/N_0}\, T_0^2}$	$\dfrac{3\sqrt{2}}{\pi} \dfrac{\lambda_0}{\sqrt{E_s/N_0}\, T_0^2}$

B = bandwidth
θ = beamwidth of the antenna
λ_0 = wavelength of the carrier frequency
D = diameter of a circular-aperture antenna
E_s = *total* signal energy
c = speed of light

* R. Manasse, Summary of Maximum Theoretical Accuracy of Radar Measurements, *Mitre Tech. Series Rept. 2.*

Until now, our discussion has been confined to the problem of separation of signals from noise. An extremely important requirement for many radar systems is the ability to separate two target returns from each other. For example, if two aircraft are present in a given volume of space, it may be important to be able to recognize the signals from these aircraft as two distinct targets without doubt or ambiguity. A further problem may arise from the need to separate desirable target returns from undesirable ones. Undesired target returns are frequently referred to as clutter. Examples of clutter are backscatter from clouds, rain and snow (except in weather radars), and trees and mountains (except in terrain-mapping radars).

An important asset in the problems of separation is high resolving power. In radar systems, the ability exists to resolve targets in range, Doppler frequency, or angle. The discussion here will be confined to the problem of designing signals for fine range and Doppler resolution; angular resolution is discussed in other sections.

The Ambiguity Function. There is an interrelationship between range and Doppler resolution, and this has been elegantly and compactly expressed by Woodward[1] in his ambiguity function, defined as the squared modulus of the two-dimensional χ function.

$$|\chi|^2 = \theta^2(\tau,f) = \left| \int_{-\infty}^{+\infty} u(t) u^*(t+\tau) e^{-j2\pi ft} \, dt \right|^2 \quad (14\text{-}4)$$

where $u(t) = E(t) e^{j\varphi t}$

$$\int_{-\infty}^{+\infty} \theta^2(\tau,f) \, df \, d\tau = (\text{signal energy})^2$$

$$= 1 \text{ if } \int_{-\infty}^{\infty} u^2(t) \, dt = 1$$

This is a mapping of the average signal-power output from a matched filter as a function of the delay and Doppler misalignment, τ and f, respectively.

In Fig. 14-7 the modulus of the χ function has been mapped for a rectangular constant-frequency pulse. It is to be noted that the range resolution (i.e., the half width of the major time lobe $\Delta\tau$) and the Doppler resolution (i.e., the half width of the frequency lobe Δf) are very simply related, namely,

$$(\Delta\tau)(\Delta f) = 1 \quad (14\text{-}5)$$

Because the volume of the mapped function is constrained to be equal to a constant, the range resolution can be improved only at the expense of Doppler resolution and vice versa.

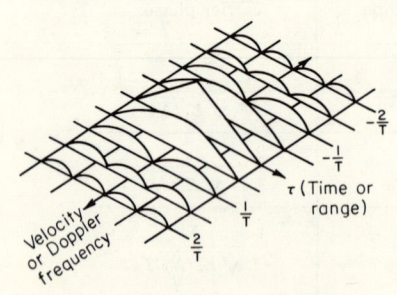

Fig. 14-7. Sketch of the $|\chi|$ function of a rectangular constant-frequency signal. Profiles are shown vs. time at intervals of $\tfrac{1}{2}T$ in Doppler frequency, where T is the signal duration. (*S. Applebaum and P. W. Howells, Waveform Design for Tomorrow's Radars, Space/Aeronautics, October, 1959.*)

Relation between CW and Pulsed Systems. It can be seen that, in a CW or a very long pulse system, range resolution is sacrificed to obtain Doppler resolution; in the very short pulse systems, Doppler resolution is sacrificed to obtain range resolution. In terms of some special set of requirements, one or the other may be more significant. Reasonable combinations of range and Doppler resolution cannot be achieved in the microwave region for simply shaped pulses. To illustrate this point, (14-5) can be rewritten as follows:

$$\lambda = \frac{4(\Delta R)(\Delta v)}{c} \quad (14\text{-}6)$$

where ΔR = range resolution
Δv = Doppler resolution
λ = wavelength of the carrier frequency
c = velocity of light

If $\Delta R = 10$ m and $\Delta v = 10$ m/sec, then $\lambda \approx 10^{-6}$ m. Thus this combination of range and Doppler resolution will be obtained only in the optical region.

Large Time-Bandwidth Product Signals. In order to obtain good range resolution while maintaining the long transmission time required to obtain Doppler resolution, it is necessary to superimpose a modulation characteristic on the waveform. The characteristic should be such that, when the signal is processed through its matched filter, a peak response will occur only for a limited interval of the total transmission. The modulation mechanism for accomplishing this can be frequency modulation, phase modulation, or amplitude modulation.

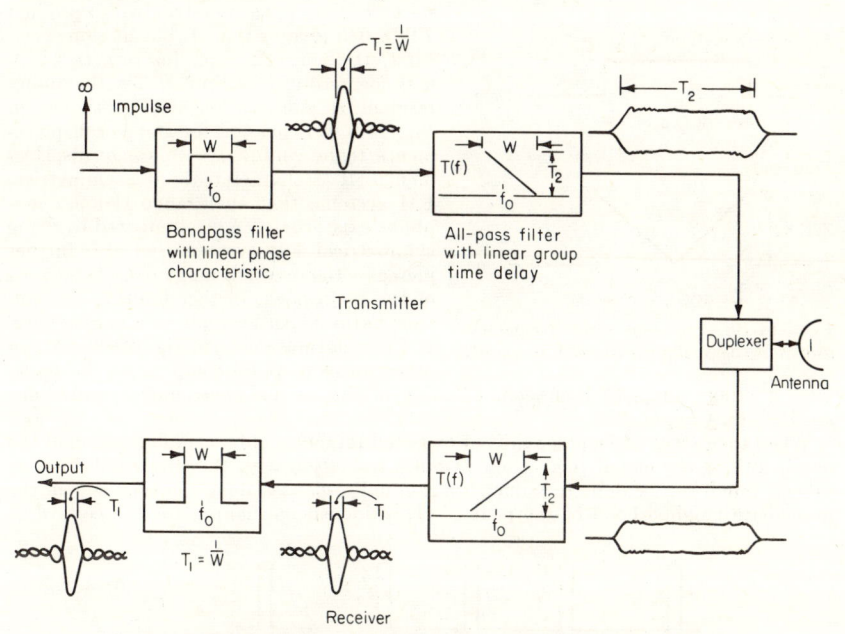

Fig. 14-8. Block diagram of a linear FM pulse-compression system.

Fig. 14-9. $|\chi|$ function of a signal with a rectangular envelope and linear frequency sweep. $TW = 10$. Profiles are vs. time at intervals in Doppler of one-tenth of the signal bandwidth. (*Photograph courtesy of C. E. Cook, Sperry-Rand Corp.*)

The most widely used system is linear frequency modulation. The elements of such a system are shown in Fig. 14-8. The transmitted signal is generated by exciting with a band of frequencies a dispersive filter having a uniform amplitude response over a band W and a time-delay characteristic proportional to frequency. A receiver consisting of a matched filter, for zero Doppler frequency, is shown in the figure. It is characteristic of such systems that the time response is not "clean." Note that there are some small residual responses over the entire interval of transmission $2T^2$ even though the peak region has a time duration approximately equal to $1/W$. Thus we have achieved a time-bandwidth product TW much greater than 1, but at some cost. First, the contrast ratio for targets which may be resolved is limited by the many residual responses. Second, as shown in Fig. 14-9, it is possible for Doppler displacements to be confused with range displacement. It is characteristic of asymmetrical FM systems that such range Doppler couplings exist; they can be eliminated by using symmetrical FM systems, but this further increases the residual energy outside the area of the main peak, further limiting the contrast ratios. An example of a symmetrical FM waveform is shown in Fig. 14-10. Many other forms of pulse compression are possible; for a more complete discussion, including amplitude- and phase-modulation forms, see Ref. 18, Chapter 3.

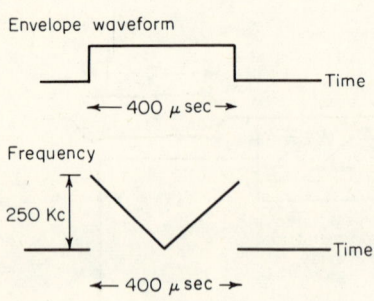

Fig. 14-10. Envelope and frequency sweep of the V-shaped linear FM signal.

When the extent of Doppler shift of expected targets is only a small fraction of the bandwidth of the signal (i.e., when Doppler resolution with the single pulse is forsaken), much can be done to reduce the residual responses along the time axis in the immediate neighborhood of the peak. The principal mechanism for the control of

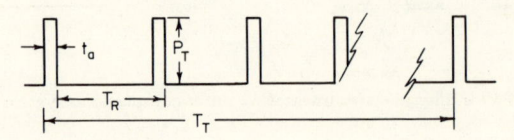

(a) Received signal in a pulsed system. $T_T = \dfrac{n}{F_R} = nT_R$

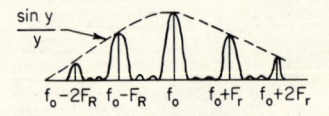

(b) Matched filter frequency characteristic for pulse train of n pulses. Noise bandwidth = $1/nt_a$

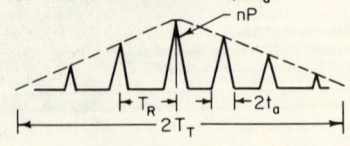

$$H(\omega) = \dfrac{\sin y}{y} \cdot \dfrac{\sin nx}{\sin x} e^{jx(n-1)}$$
$$x = \dfrac{\omega T_R}{2}$$
$$y = \dfrac{\omega t_a}{2}$$

(c) Output of matched filter for a pulse train of n pulses

Fig. 14-11. Essential characteristics of pulse-group signal.

these residues is shaping of the amplitude of the FM signal, in either transmission or reception.

Pulse Groups. One form of modulation consists of a group of pulses. The pulses are separated by intervals long compared with their duration. Ordinarily the group of pulses is not thought of as one long modulated signal. In Fig. 14-11 are shown a pulse group, its spectrum, and its autocorrelation function (i.e., the output of the matched filter). With such a signal, it is possible to achieve the range resolution

RADAR   14-13

of the short pulse t_a and Doppler resolution corresponding to the time duration of the entire pulse train nT_R. The Doppler resolution obtained from this signal is improved over that of a single pulse t_a by the ratio (nT_R/t_a). This ratio can be quite large, on the order of 10,000.

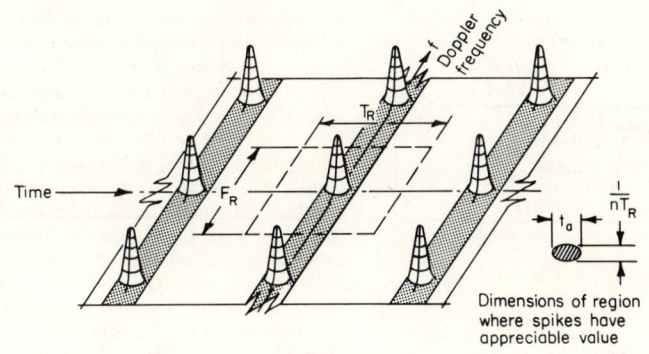

FIG. 14-12. Sketch of the $|\chi|$ function of a pulse-train signal.

The price paid for this vast improvement in resolution is the introduction of ambiguities in the range and Doppler measurements. The introduction of ambiguities is apparent from Fig. 14-11, and further emphasized in the contour of $|\chi|$ for a pulse group shown in Fig. 14-12. The location of targets in range can be determined unambiguously only if the targets are confined to a range less than $cT_R/2$. The Doppler frequency can be determined unambiguously only if the target velocities are confined to velocities less than $F_R\lambda/2$. These ambiguity regions are interrelated in the same way that transmission length and Doppler resolution are interrelated in the single-pulse case. Expressed in terms of velocity and range

$$v_0 R_0 = \frac{c^2}{4f_0} \qquad (14\text{-}7)$$

where v_0 = maximum unambiguous velocity
R_0 = maximum unambiguous range
f_0 = carrier frequency

This simple equation is plotted in Fig. 14-13, where f_0 is expressed in gigacycles, R_0 in nautical miles, and v_0 in knots ($v_0 f_0 R_0 = 2.3 \times 10^4$).

FIG. 14-13. Unambiguous range vs. unambiguous velocity for a pulse-group radar.

Clutter Rejection of Pulse Groups. The conventional pulse radar has its repetition frequency chosen to provide unambiguous range information on the target group of interest to the system. A typical choice of parameters from Fig. 14-13 for an aircraft search system might be

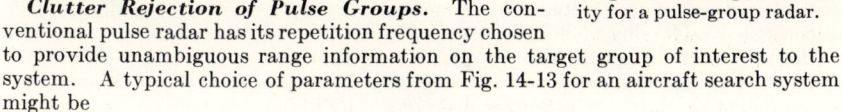

$$R_0 = 230 \text{ nm}$$
$$f_0 = 1 \text{ Gc}$$
$$v_0 = 100 \text{ knots}$$

If 10 pulses are processed in the matched-filter ensemble, then the Doppler resolution will be approximately 10 knots. The range of target velocities certainly exceeds 100 knots for aircraft targets. Therefore, the measurement of velocity would be highly ambiguous.

Nevertheless, such a system can have considerable utility in the mass processing of clutter that is confined to a narrow spectrum. This is illustrated in Fig. 14-14.

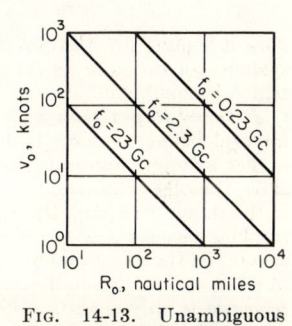

14-14 SYSTEM COMPONENTS

When the clutter is confined to a narrow region in the vicinity of zero velocity, then a filter can be devised (or filters omitted from the required ensemble) which rejects most of the clutter and accepts most of the targets of interest. For the situation illustrated in Fig. 14-14, radial target velocities between −5 and +5 knots, 95 and 105 knots, etc., will be excluded by the filter, as is the clutter; but sooner or later these targets, if they are not on radial paths, will attain radial velocities which are accepted by the system. Such a clutter-rejection system can be very useful in rejecting reflections from hills, buildings, trees, etc. If, however, the requirement is to reject broadband clutter, such as backscatter from rainclouds which may occupy spectrum widths as much as 30 knots, then for the case illustrated in Fig. 14-14, the usefulness of such a clutter-rejection system is doubtful.

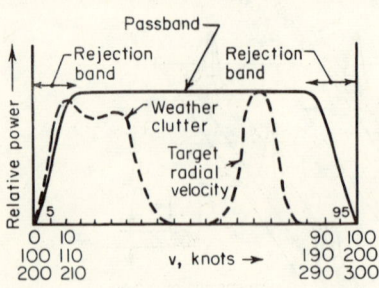

FIG. 14-14. Clutter rejection for pulse groups: $f_0 = 1$ Gc; $R_0 = 230$ nautical miles; $v_0 = 100$ knots; $N = 10$.

"Pulse Doppler." The so-called "pulse Doppler" radar systems are in principle no different from the systems just discussed. However, in this class of systems the choice of repetition period is made to favor unambiguous Doppler measurement, accepting the consequence of highly ambiguous range information. An example of such a system might be one where the unambiguous velocity is

If
then
$$v_0 = 1{,}000 \text{ knots}$$
$$f_0 = 1 \text{ Gc}$$
$$R_0 = 23 \text{ nautical miles}$$

Now it is quite practical to extend the filter design to reject weather clutter. Such a system can be quite useful where the density of aircraft targets at any azimuthal region is small, and where a considerable problem with wide-spectrum clutter exists.

Jittered Repetition Rates. It is possible to design pulse-group signals where the interval between pulses is deliberately varied to reduce range ambiguity. Thus a higher average pulse-repetition rate may be used than would ordinarily be permissible, thereby achieving better Doppler resolution. In general, if n pulses are used in the staggered group, the ratio of the main or desired amplitude response to all other residual responses is n. Thus a limitation in contrast ratio is imposed in such a system in the same manner as for pulse-compression systems. When the variation in pulse spacing is small compared with the average pulse spacing, then the Doppler response is only slightly different from the response of a group with uniform pulse spacing. When the variation in timing is spread over a small number of pulses, such as the two-pulse stagger systems sometimes used in MTI systems,[2] then the Doppler response is considerably modified.

Many variants of the jittered-repetition-rate approach are possible. There are systems where no attempt is made to obtain Doppler information for either target measurement or clutter rejection, and there are hybrid systems which use "coded" pulses for the individual pulses in a pulse group.

14-4. Synthesis of Matched Filters

Matched Filter for a Pulsed Sinusoid. The basis for the synthesis of matched filters is the requirement that the matched filter should have an impulse response which is the reversed time image of the received signal. For the case of a perfect transmission medium and point targets (nonfluctuating), the matched filter is the reversed time image of the transmitted signal corrected for the relative radial velocity (Doppler shift) between the radar system and the target.

RADAR 14-15

In Fig. 14-15 the matched filter is shown for a sinusoid of duration t_a and frequency f_0. Since in this example the received signal is symmetrical, the reversed image is identical with the signal. The integrator shown is essentially a resonant circuit of center frequency f_0, having very low losses and a long memory. The impulse response

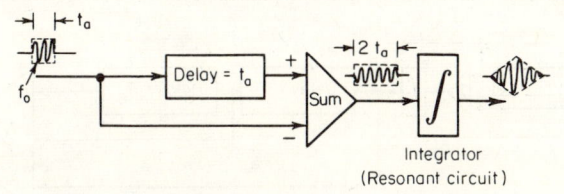

FIG. 14-15. Matched filter for a pulsed sinusoid.

will be a rectangularly shaped pulsed sinusoid if the delay line has a length precisely equal to the length of the sinusoid. If such a device is practical to construct at all, it would be at an intermediate frequency, which is large compared with the spectral width of the received signal. This filter is "matched" for only one Doppler frequency.

Ensembles of Matched Filters. If the anticipated Doppler frequency shifts are large compared with the spectral width of the received signal, then an ensemble of matched filters is required as shown in Fig. 14-16. Each filter has its integrator set for a different frequency, namely,

$$f = f_0 \pm \Delta f \qquad (14\text{-}8)$$

where f = resonant frequency of the filter
f_0 = carrier frequency

$$\Delta f = \frac{K}{t_a}, K = 1, 2, \ldots, n$$

$\dfrac{n}{t_a}$ = maximum Doppler frequency

FIG. 14-16. Ensemble of matched filters for a pulsed sinusoid of duration t_a and frequency f_0.

The output of each filter is combined with that of the other filters, following a threshold decision device such as discussed previously. The number of filters in the ensemble is approximately

$$N_f \approx F_d t_a \qquad (14\text{-}9)$$

where N_f = number of filters in the ensemble
F_d = expected range of Doppler frequencies (both + and −)
t_a = pulse width

In practice, filters such as described are rarely, if ever, constructed. The difference between the S/N obtained through use of an ideal filter and a simple approximation such as a single resonant circuit is small.

Matched Filters for Pulse Groups. Another example of a matched filter is shown in Fig. 14-17 for a pulse train of three sinusoids as shown in waveform A of part b of this figure. The matched filter is composed of two sections in cascade. The first section is similar to the matched filter shown in Fig. 14-15. The second section generates the repetition. This filter can be extended to match a train of n pulses by using $(n - 1)$ delay sections. The output train will have a length of $(2n - 1)$ pulses.

The frequency response of a matched filter for a pulse group was shown in Fig. 14-11b. The shape of the response indicates why these filters have been referred to as "comb" filters. When the expected range of Doppler frequencies exceeds the spectral width of one of the "teeth" of the "comb" (i.e., when it is greater than $1/nT_R$), then an ensemble of matched filters must be constructed which encompasses the expected range of Doppler frequencies. A compact method for constructing the ensemble using a single set of delay lines and a compensating phase matrix is shown in Fig. 14-18.

Another approximation to a matched filter used for pulse trains is shown in Fig. 14-19. A gate (electronic switch) must be provided for each range-resolution element.

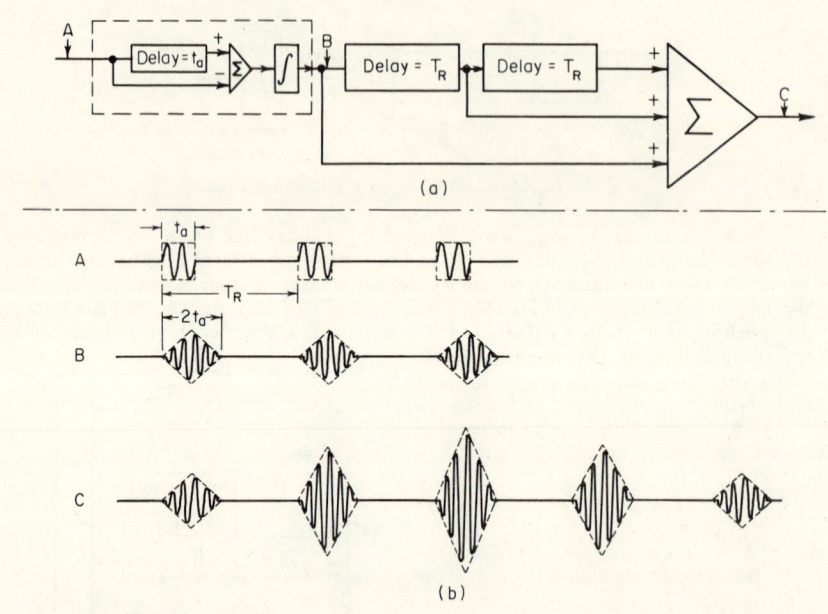

FIG. 14-17. Matched filter for a pulse train of three sinusoids. (a) The matched-filter block diagram. (b) The waveforms at points A, B, and C.

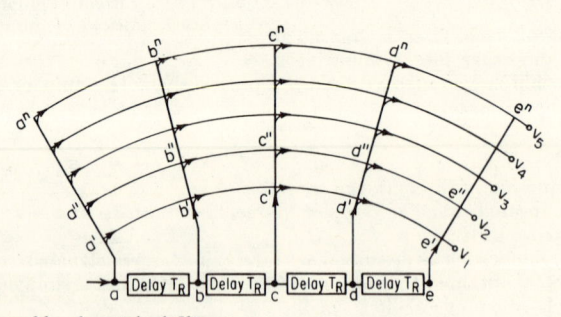

FIG. 14-18. Ensemble of matched filters.
T_R = interval between pulses
N = number of pulses (five for case shown above)
λ = wavelength
length $aa'b'c'd'e'$ = length $bb'c'd'e'$ = length $cc'd'e'$ = length $dd'e'$ = length ee'
length $a''b''c''d''e''$ = $a'b'c'd'e' + \lambda$
length $a^n b^n c^n d^n e^n$ = $a'b'c'd'e' + (n-1)\lambda$

Each filter in the filter bank shown at the output of each gate can be simple, but one must be provided for each Doppler-resolution element. Consequently the number of filters required is quite large.

Clutter Rejection in Matched Filters. Frequently, coherent processing is confined to rejection of bands of Doppler frequency occupied by clutter, as discussed in

Sec. 14-3 (see Fig. 14-14). The rejection of a narrow band of Doppler frequencies may be achieved by omitting from the ensembles previously discussed those filters which would perform the processing in the undesired Doppler region. In order to obtain a simpler system, the matched-filter ensemble could be eliminated and replaced with a rejection filter designed to provide high attenuation for the undesired Doppler region and low attenuation for all other regions. A simple rejection filter and its spectral response are shown in Fig. 14-20. The processing-time constant of the filter is equal to one repetition period. It is possible to obtain more desirable filter characteristics by cascading several delay lines and providing feedback loops to increase the filter response further.

Postdetection "Matched" Filters.
After rejection of undesired Doppler regions, there still remains the task of accumulating the energy of all the pulses in the train. This may be accomplished without a filter ensemble if the Doppler shifts are eliminated by using a rectifier or detector, which follows the envelope of the signal. This completion of the integration process subsequent to elimination of the Doppler or phase information is called noncoherent processing, noncoherent integration, or postdetection integration.

A loss is involved in resorting to this simplification, but when high probabilities of detection and low false-alarm rates are required, the S/N ratio preceding this nonlinear processor is generally high and hence the signal suppression is not very large. The losses for this situation are given in Fig. 14-5.

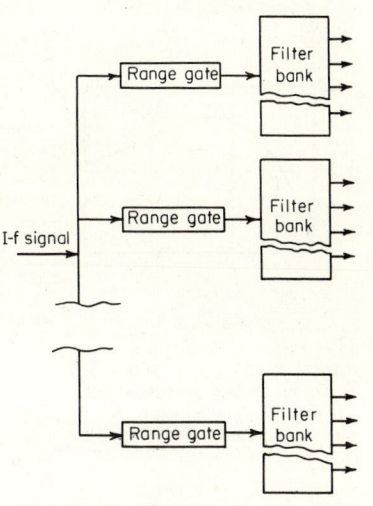

Fig. 14-19. Range gate and filter ensemble for a pulse group of n pulses of duration t_a and spacing T_R.

Number of filters in the bank:
$$N_f = F d n T_R \qquad N_f \leq n$$
Number of range gates:
$$N_R \leq T_R/t_a$$
Maximum number of filters:
$$N_m = n(T_R/t_a)$$

The optimum processor for the postdetection processing is in some ways similar to that for predetection processing, except that it is the video impulse response which must now be matched. The matched or approximately matched filters are now the

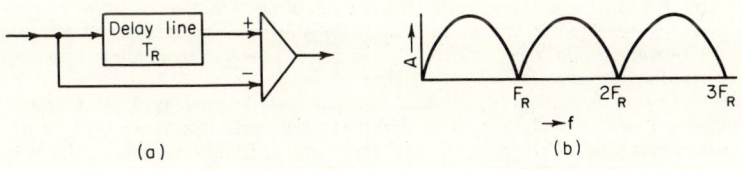

Fig. 14-20. (a) Simple clutter-rejection filter. (b) Spectral response of filter.

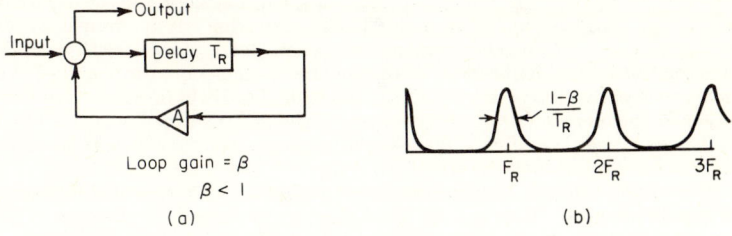

Fig. 14-21. (a) Delay-line integrator. (b) Spectral response of integrator.

low-pass equivalents of the filters previously discussed. The requirements are now much simpler, since the delays must be accurate only to within a fraction of the pulse length rather than to within a fraction of an i-f cycle.

A single delay line may be used with the output fed back into the input, as shown in Fig. 14-21, to generate an impulse response which is an approximation to the desired impulse response. In order for the system to be stable, the loop gain must be less than 1.[3] When loop gains greater than 0.85 are used, the impulse response of the system is closely approximated by a simple exponential. The mismatch loss (for proper adjustment of the loop gain) is small and quite analogous to the use of a simple resonant circuit for a rectangular pulse.

Sensitivity to Mismatch. An interesting question concerns the sensitivity of the signal-to-noise ratio to mismatch of the "matched" filter. For example, in the single-pulse case, what effect will the substitution of a simple resonant circuit of appropriate time constant have on the optimum S/N? As may be seen from Fig. 14-22, the value achieved by this approximation is within 1 db of the optimum, even without precise matching of the time constant. Similar results with many different shapes of filters lead to the conclusion that approximations to the matched filter which are simpler to realize may frequently be used with only a small sacrifice in sensitivity.

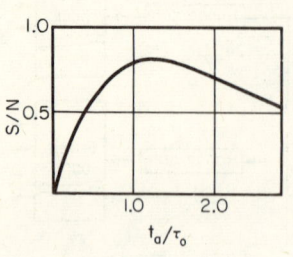

FIG. 14-22. Relative signal-to-noise ratio vs. filter time constant. The input to filter is a single pulsed sinusoid of duration t_a. The filter is a simple resonant circuit with time constant τ_0.

14-5. Transmitters

The function of the transmitter in a radar system is to convert energy, usually from a d-c source, to microwave energy. It is desirable to transmit as large an amount of energy, or average power, as the size, weight, and cost limitations of the system will permit. Since frequency and amplitude instability in the transmitter limits the ability to achieve a proper match in the receiving system, it is also desirable to obtain as much transmitter stability as size, weight, and cost limitations will permit.

Two basic forms of transmitters are in use: the high-power self-excited oscillator and the master-oscillator power amplifier. The self-excited high-power oscillator achieves compactness and economy at the expense of stability, and has applications where high stability is not essential. Separation of the function of frequency control from power generation is the basic reason for the greater stability of the master-oscillator power-amplifier transmitter. Also, it is possible to achieve higher power in a device used as an amplifier rather than an oscillator.

The most common signals transmitted in radar systems are CW, FMCW, pulse, and FM pulse. The CW and FMCW transmitters are quite like those utilized in communication systems. The pulse transmitter of low or moderate duty cycle, however, is unique to radar. It requires a mechanism for rapidly pulsing the r-f energy and, in efficient systems, rapidly switching the power generator on and off.

Stable Oscillators. The stable oscillator in a radar system may be a microwave oscillator (e.g., magnetron, klystron, or TWT) operating at the frequency of the radar transmitter. To prevent frequency variations, care is exercised in the design and construction of the oscillator to minimize circuit-element microphonics (acoustic isolation), to suppress power-supply ripple which might create frequency modulation, and to shield the oscillator from stray magnetic and electrostatic fields. If the oscillator tube has integral cavities of inadequate frequency stability, it may be locked to external cavities to improve the stability.

Another common form of stable oscillator utilizes the inherent stability and low loss of quartz crystals. If extremely good stability is desired, application of all the measures listed above will be necessary, and also temperature control of the crystal.

RADAR 14–19

The physical limitations of quartz limit the frequency of these oscillators to less than 100 Mc and preferably less than 50 Mc. It is therefore necessary to have a frequency-multiplier chain with a high multiplication ratio to obtain energy at microwave frequencies. Extremely stable multiplier chains can be constructed using solid-state elements.

High-power Microwave Generators. All microwave tubes are energy converters in which some of the input d-c power is converted to r-f output power. The conversion requires an input r-f control signal, applied externally (as in amplifiers) or internally (as in oscillators). The energy converters all have a source of electrons, a means of accelerating the electrons through an interaction space, a means of collecting the spent electrons, and a means of coupling in and out of the electron flow within

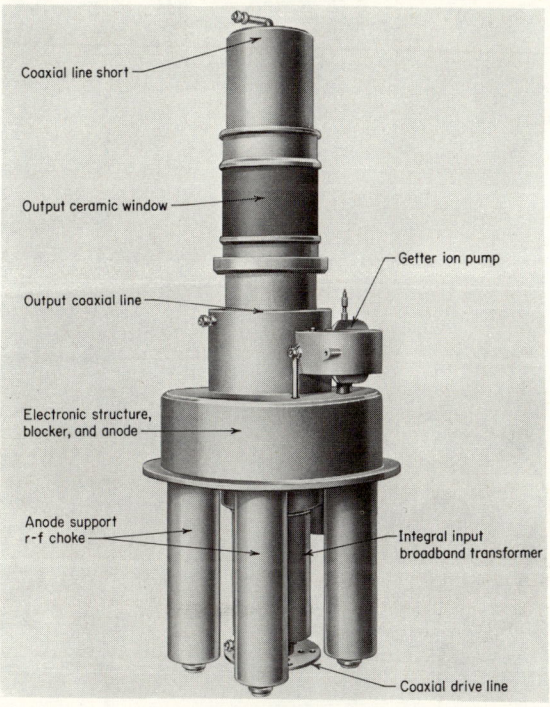

FIG. 14-23. The model C coaxitron; overall length is approximately 3½ ft. (*Courtesy of Radio Corporation of America.*)

the interaction space. The manner in which the d-c energy is converted into r-f power is the distinguishing feature of microwave tubes. The three basic forms are grid-control devices, linear-beam devices, and crossed-field devices.

Grid-controlled Tubes. In conventional grid-controlled tubes, the electron transit time must be small compared with the reciprocal of the r-f frequency. The spacing of the tube elements determines the transit time, and this spacing must exceed a certain minimum distance in order to accommodate the high voltages and power dissipation associated with a high-power tube. Consequently, these tubes are limited to the lower r-f frequencies.

No practical grid-controlled tubes are available now which can generate megawatts of peak power and kilowatts of average power above 1,000 Mc. However, in the region between 400 and 1,000 Mc, many types of tubes may be encountered; below about 400 Mc, grid-controlled tubes are used almost exclusively.

The current trend in design of high-power grid-controlled tubes for use in the uhf region is toward integrating the r-f circuitry as part of the structure of the tube.

14-20 SYSTEM COMPONENTS

In addition, since transit-time effects limit the size and hence the power capacity, the current design trends employ a plurality of unit electron-focusing systems in a coaxial cylindrical configuration.

An example of such a design is the RCA A15038 coaxitron (Fig. 14-23) which employs a circular array of 96 identical triode units. This tube has uniform (broadband) response from 400 to 450 Mc, with peak power in excess of 5 Mw.

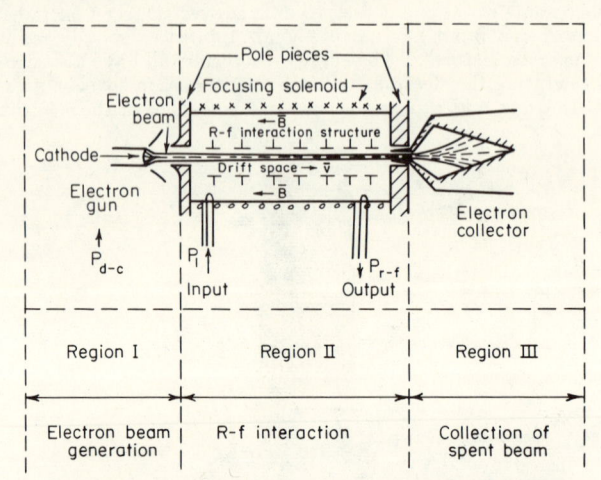

FIG. 14-24. Functional regions of a linear-beam amplifier.

FIG. 14-25. Cutaway view of VA-87. (*Courtesy of Varian Associates.*)

Linear-beam Tubes. In linear-beam tubes, the conversion of d-c energy to electron-beam energy takes place in a region separated from the conversion of the beam energy to r-f energy. As shown in Fig. 14-24, the beam is formed by a combination of cathode focusing and electrostatic focusing in region I. By the time the beam reaches region II, the electrons have been accelerated to their full kinetic energy of motion. The shape of the beam is maintained in region II by an external d-c magnetic field. The interaction with the beam to convert some of its kinetic energy into r-f energy takes place in region II. Collection of spent electrons takes place in region III.

In region II, the interaction circuit may be either a series of isolated resonant circuits, in which case the tube is a klystron, or it may be a propagating structure, in which case it is a traveling-wave tube. The interaction in the klystron is due to velocity bunching such as with gridded klystrons developed during World War II. The TWT circuit, instead of interacting through discrete gaps, as with the klystron, interacts with the beam continuously. As might be expected, broader-band operation is possible with a TWT than with a klystron because of the broader-band characteristic of a distributed structure as compared with a stagger tuning of isolated structures.

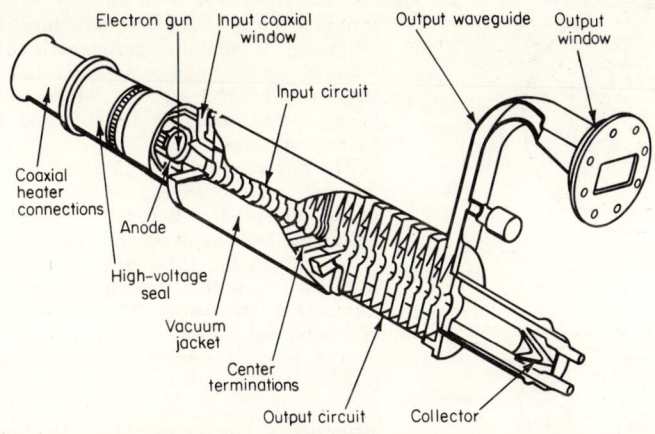

Fig. 14-26. A cutaway view of the VA-125 showing the internal construction of the tube. (*Courtesy of Varian Associates.*)

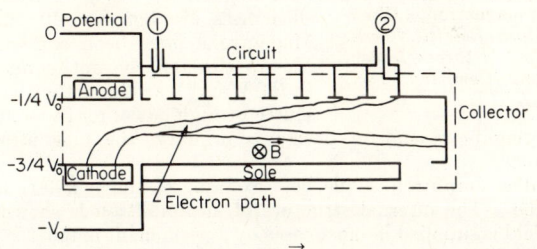

Fig. 14-27. Elementary crossed-field tube. $\otimes \vec{B}$ = uniform magnetic field perpendicular and into the plane of the paper within the dashed lines.

Bandwidths of approximately 5 per cent (of the nominal or center frequency) have been achieved with high-power klystrons at efficiencies of approximately 35 per cent and gains of 35 db. Greater bandwidths can be achieved with a high-power TWT. At present, higher peak and average power has been achieved with klystrons than with TWT's. Figure 14-25 illustrates a typical klystron, the VA-87, and Fig. 14-26 a typical TWT, the VA-125.

Crossed-field Tubes. In crossed-field tubes, the three regions are not clearly separable as in linear-beam tubes. The magnetic field not only serves for focusing but is an indispensable part of the energy-conversion process. The electric and magnetic fields are applied orthogonally, and the interaction region is orthogonal to both. Electrons perform work on the r-f field as they drift through the interaction region. The manner of motion and field interaction is more advantageous for the conversion of energy than in a linear-beam tube. An elementary form of crossed-field tube is shown in Fig. 14-27.

Magnetron. In the two most widely used high-power crossed-field tubes in this country, the magnetron and the amplitron, the sole shown in the elementary version (Fig. 14-27) becomes the cathode. The cathode is closed on itself in a cylindrical shape, giving a continuous emitting surface.

In the magnetron, the interaction circuit consists of a series of resonant cavities and slots arrayed periodically on the anode circumference and strapped. The beam interaction is with a standing-wave circuit. A typical internal-structure cross section is shown in Fig. 14-28. The magnetic field is usually supplied by a permanent magnet bolted to the external structure of the tube. The magnetron achieves the utmost in simplicity, since it is an entirely integral two-terminal device. All that is required to obtain r-f energy from the output port is to apply a filament voltage and a d-c potential between the cathode and anode. Magnetrons have been manufactured in tunable and fixed-frequency versions.

The introduction of tuning elements into the interaction structure can be a limitation on the power generated. Recently a coupled-cavity design has been evolved wherein a cavity which is integral in the tube structure but separate from the interaction structure is coupled to the interaction structure. Two beneficial effects are achieved. First, the frequency stability of the tube is improved, and second, it is possible to tune the tube by changing the resonant frequency of the cavity. The tuning structure is thus removed from the limited and complex interaction space.

Efficiencies of 60 per cent have been achieved with magnetrons, but more usually encountered efficiencies are between 40 and 50 per cent. Magnetrons have been constructed from 400 Mc up through the entire microwave region. In the World War II era, they were the principal source of high power for radar systems. When compact, lightweight transmitter design is primary, the magnetron remains the outstanding choice.

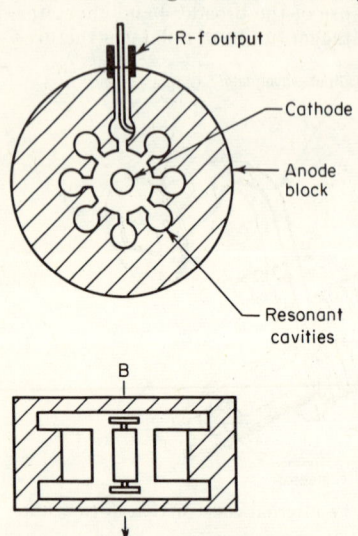

FIG. 14-28. Typical internal cross section of a magnetron. [*From Grayson Merrill (ed.), Povejsil, Raven, and Waterman's Airborne Radar, D. Van Nostrand Company, Inc., Princeton, N.J., 1961.*]

Amplitron. The amplitron is the best example of the four-terminal high-power crossed-field tube. The internal structure of an amplitron is shown in Fig. 14-29. The magnetic field is supplied in most cases by a permanent magnet. Externally, the device looks much like a magnetron except for the presence of an input port. At high power the amplitron behaves as an oscillator locked to the input frequency.

Considerably higher efficiency has been obtained with amplitrons than with any other microwave energy generator. Efficiencies as high as 80 per cent have been reported. More frequently, efficiencies between 60 and 70 per cent are encountered. Bandwidths higher than 10 per cent have also been obtained. Were it not for the low gain, on the order of 10 db, the amplitron would be a truly outstanding device. Because it is a backward-wave reentrant-beam device, power incident at the output terminal is not attenuated in its path to the input terminal. This has some advantage from the viewpoint of receiver protection since the receiver may be duplexed at the input to the amplitron, thus not encountering the peak power at the output. This, however, is a doubtful advantage for low-noise receivers. A disadvantage occurs when poor matching is achieved at the antenna terminals and considerable power is reflected to the input driver. Good matching and an absorption device between the amplitron and its driver are required.

Amplitrons have been constructed from 400 Mc up through the microwave spec-

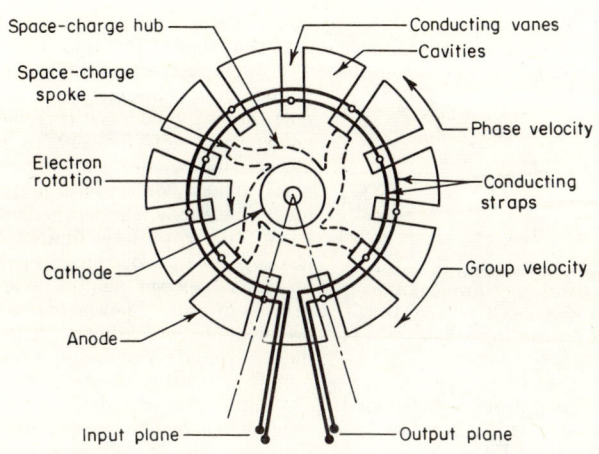

FIG. 14-29. Basic structure of the amplitron. A magnetic field is applied parallel to the axis of the tube. (*Courtesy of Raytheon Company.*)

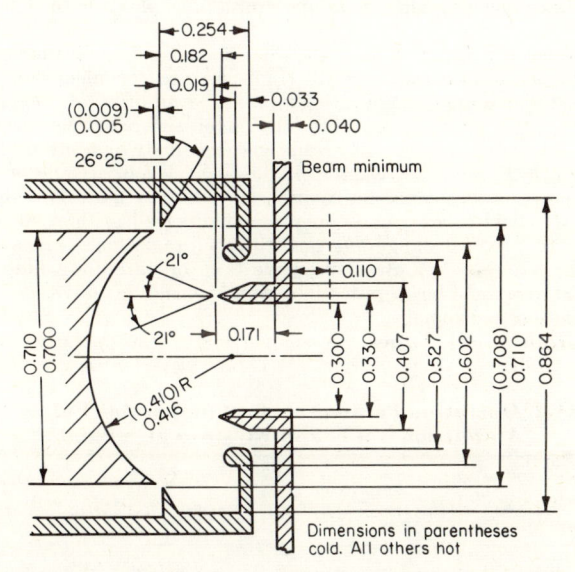

FIG. 14-30. Configuration of the perveance 2.2×10^{-4} (convergence 300:1) gun (type 5B). This gun was developed using multielectrodes to optimize the beam-boundary potential after the cathode was reshaped. Multielectrodes were then replaced by single electrode shown. (*R. D. Frost, O. T. Purl, H. R. Johnson, Electron Guns for Forming Solid Beams for High Perveance and High Convergence, Proc. IRE, vol. 50, pp. 1800–1807, 1962.*)

trum. Occasionally the high driver power required by the amplitron has been supplied by a magnetron.

Cathodes. Most high-power microwave tubes use space-charge limited emission from oxide-coated and impregnated-matrix cathodes. Emission densities vary from 0.2 to 4.0 amp/sq cm for CW tubes and 2 to 10 amp/sq cm for pulse tubes. The cathode shapes take two general forms: the spherical coated sections as shown in Fig. 14-30, and the cylindrical as shown in Fig. 14-31.

14-24   SYSTEM COMPONENTS

Windows. The conversion of energy from direct current to radio frequency in microwave devices takes place in a vacuum of the order of 10^{-7} mm. Coupling in and out of these devices must be accomplished through ports which maintain the vacuum. Under some circumstances the r-f output port, or "window" as it is termed, limits the power that can be sent to the antenna. These windows are generally constructed of glass, quartz, or ceramic material. Under ideal conditions, the power-handling capacity of the window is limited by dielectric heating because of the dissipation of some of the r-f power in the window. Other sources of limited power-handling capability are voltage breakdown, X-ray bombardment, secondary-emission breakdown, and nonlinear heating effects.

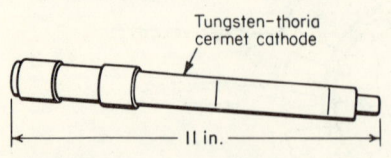

FIG. 14-31. QK 622 amplitron cathode. (*William A. Smith and Frank Zawada, A 3-megawatt, 15-kilowatt S-Band Amplitron, The Microwave Journal, vol. 2, no. 10, October, 1959.*)

Cooling. The difference between the d-c input power and the r-f output power appears as heat. In high-power tubes, where unconverted power on the order of tens or hundreds of kilowatts must be dissipated, liquid cooling is most frequently used. In linear-beam tubes, the cooling is applied to the collector; in crossed-field tubes, it is applied to the circuit structure immersed in the interaction field. Other parts of the tubes which develop high temperatures must also be cooled, by either liquid or air flow.

Common values of dissipation for water-cooled surfaces are on the order of several hundred watts per square centimeter. High-pressure turbulent-flow methods are being developed which may increase this number by an order of magnitude. Unintentional interception of even a small fraction of the beam by circuit elements, particularly in linear-beam tubes, may result in exceedingly high surface temperatures.

X Rays. Since X radiation can be produced at levels of tens of kilovolts, all existing high-power tubes must be suspect of radiation hazards. Grid-controlled tubes and crossed-field tubes usually operate at voltages less than 50 Kv and hence are less likely to produce dangerous quantities of radiation. Linear-beam tubes, particularly the higher-power klystrons, have been operated at voltages as high as 400 Kv. Great care must be exercised in design of shielding and other precautions to provide for safety of personnel.

Available High-power Tubes. A comparison of the operating parameters of three S-band amplifiers—an amplitron, TWT, and klystron—is given in Table 14-2.

Table 14-2. Operating Parameters for Comparable Pulsed Power: Amplitron, TWT, and Klystron at S Band

Parameter	Amplitron (Raytheon QKS622)	TWT (Varian VA-125B)	Klystron (Varian VA-839)
Frequency, Mc	2,900–3,100	2,900–3,250	2,730–2,870
Power output, peak, Mw	3.0	3.0	5.0
Power output, avg, Kw	15	4.0	10
Gain, db	7.8	35	45
Bandwidth, Mc		350	140
Pulse duration, μsec	10	10	7
Beam voltage, peak, Kv	50–54	120	125
Beam current, peak, amp	66	75	93
Dimensions, in	20 × 16.8 × 16	10 × 15¾ × 42	8 × 10 × 45
Weight, lb	110	120*	70*
Cooling	Liquid	Liquid	Liquid
Focusing	Permanent magnet	Electromagnet	Electromagnet

* Does not include electromagnet.

RADAR **14-25**

The available peak power of a number of current tubes has been plotted vs. frequency in Fig. 14-32. The average power of these pulse tubes, as well as average power of CW tubes, has been plotted vs. frequency in Fig. 14-33.

Radar Pulse Modulators. Three types of modulating methods are in current usage for pulsing high-power microwave tubes.

1. The line modulator, in which a switch (usually a thyratron) is used to discharge the total stored energy in a pulse-forming network (PFN) each time the switch is closed. The length of the pulse is determined by the PFN.

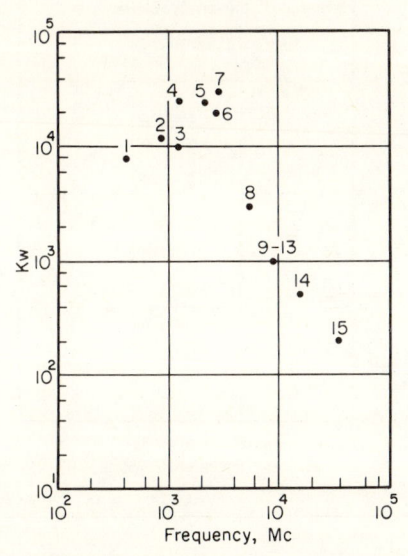

FIG. 14-32. Available peak power vs. frequency.

1. Varian Associates....................	VA 812C	KLY
2. Eitel-McCullough....................	X 831	KLY 6
3. CSF, France.........................		TWTM
4. Litton Industries.....................	L 3287	KLY
5. Eitel-McCullough....................	X 769	KLY
6. Stanford University..................	Acc. tube	KLY
7. Thomson-Houston, France............	TH 2011	KLY
8. Varian Associates....................	VA 126	TWT
9. Stanford University..................	273	KLY 3
10. Sperry Gyroscope, Great Neck, L.I......	SAX 151B	KLY 5
11. Raytheon Mfg. Co., Waltham..........	7718	MAG
12. SFD...............................	SFD 303	MAGC
13. Philips Industries....................	EAX 1	MAG
14. Bell Telephone Labs..................	BELL 4020	MAG 5
15. SFD...............................	SFD 302	MAGC

2. The hard-tube modulator in which a charged capacitor is connected to the load for the desired pulse duration by conduction of current through a hard tube. The energy stored in the capacitor must be large compared with the pulse energy for good pulse shape.

3. The modulating anode modulator in which the switching device has been integrated into the r-f power amplifier in the form of a control electrode. The voltage on the control electrode may be turned on and off by either of the above devices. The circuits in use are neither novel nor complicated. The majority of improvements in

modulators have taken place as a result of improved components rather than innovations in circuitry.

Line-type Modulators. The diagram for a basic line-type modulator is given in Fig. 14-34. Between pulses, the PFN capacitors are charged through the charging choke L_c and the diode D_1. Usually L_c is very large compared with the inductance of the pulse-forming network and the pulse transformer. The network capacitors charge

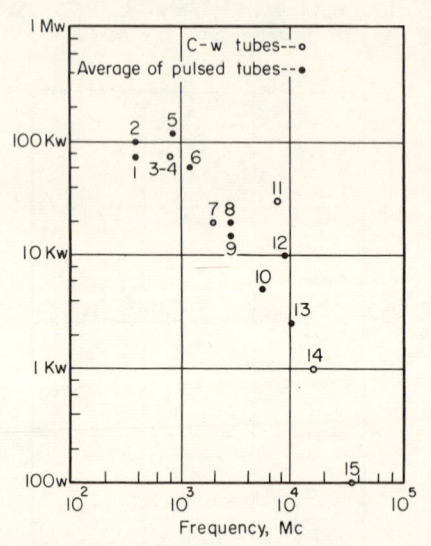

FIG. 14-33. Available average power vs. frequency.

1. Varian Associates..................	VA 842	KLY 4
2. Eitel-McCullough...................	X 626	KLY 3
3. Eitel-McCullough...................	X 768	KLY 3
4. Varian Associates..................	VA 853	KLY
5. Eitel-McCullough...................	X 831	KLY 6
6. Litton Industries....................	L 3287	KLY
7. Eitel-McCullough...................	X 812	KLY
8. Thomson-Houston, France..........	TH 2011	KLY
9. Raytheon Mfg. Co., Waltham.......	QK 622	AMP
10. Varian Associates..................	VA 126	TWT
11. Varian Associates..................	VA 849	KLY
12. Sperry Gyroscope, Great Neck, L.I.	SAX 151B	KLY 5
13. Raytheon Mfg. Co., Waltham.......		KLY 2
14. Varian Associates..................		TWT
15. SFD...............................	SFD 302	MAGC

in a resonant fashion, in one half of a resonant cycle, to twice the power-supply voltage. The charging time

$$T_c \approx \pi \sqrt{L_c C_n} \leq T_R \tag{14-10}$$

where L_c = charging inductance
C_n = total capacity of PFN
T_c = charging time
T_R = repetition period

When the thyratron is triggered, the network is connected to the load (transmitter tube) through the pulse transformer. The discharge forms a pulse of length T where

$$T = 2\sqrt{L_n C_n} = 2C_n Z_n \tag{14-11}$$

RADAR

where L_n = total PFN inductance
 C_n = total PFN capacitance
 Z_n = network impedance

The network is chosen to have an impedance approximately equal to the load impedance; hence the voltage of the pulse is equal to the power-supply voltage.

The charging diode D_1 is not essential if the resonant charging time of the network is set to be exactly $T_R/2$. Its use, however, provides flexibility in choosing the radar repetition period. Inverse voltage appears at the thyratron after PFN discharge as a result of mismatch and energy release from the pulse transformer, and the function

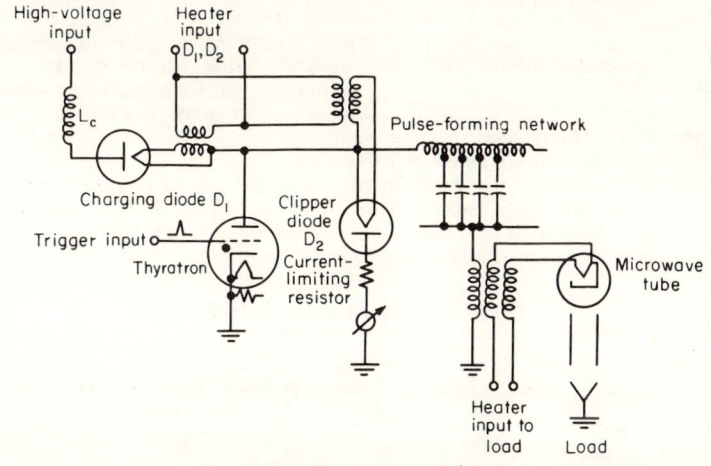

FIG. 14-34. Line-type modulator.

of the clipper diode is to limit this value. Pulse-forming network designs in use differ very little from those developed by Guillemin.[4]

The need for the pulse transformer is dictated by lack of switch tubes of sufficiently high voltage capability. Commonly available thyratrons have maximum voltage-hold-off capabilities on the order of 30 Kv. The maximum voltage available for the load is then 15 Kv. However, many transmitting tubes require voltages in excess of 100 Kv; so a step-up transformer is needed.

The major limitation in the power-handling capability of a line modulator is the switching thyratron. Characteristics of several recently developed types are given in Table 14-3. In order to achieve more power, several thyratrons may be used in parallel systems. The tubes themselves may not be paralleled because of the difficulty

Table 14-3. Hydrogen Thyratron Characteristics

	KU74	Z5101	GL7390
Max peak anode voltage. Kv:			
Forward.................	33	40	33
Inverse.................	33	40	33
Max cathode current, amp:			
Peak....................	2,000	2,400	2,000
Avg.....................	4.0	2.6	4.0
rms.....................	90	75	
Required trigger...........	.....	700 volts	1,300–2,500 volts
	.....	2 μsec	2 μsec
	.....	50 ohms	10–25 ohms

in sharing the load. The best arrangement is one in which each thyratron operates into a separate PFN. The network outputs are then paralleled. Series arrangements are also possible with special triggering provisions. Ignitrons are being considered to replace hydrogen thyratrons.

The Hard-tube Modulator. A typical hard-tube modulator circuit is shown in Fig. 14-35. It is simply a pulse amplifier capable of generating a pulse having sufficient power to drive the transmitter tube. Its use has been limited to low-power systems because of the lack of adequate pulse amplifiers. It has great flexibility, since the repetition period and pulse width may be varied over wide ranges provided that adequate energy storage is provided in capacitor bank C_s. Some tubes now in production are listed in Table 14-4.

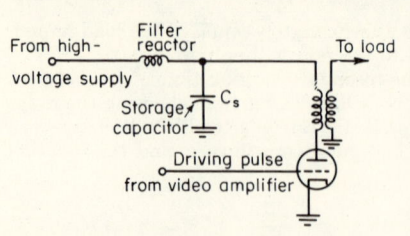

Fig. 14-35. Simplified output section; hard-tube modulator.

If amplitude and phase variations in the r-f output are to remain small, the d-c voltage must not vary from pulse to pulse, and the voltage droop during the pulse must be small. Instead of making C_s larger, which can be very costly, pulse-shape compensation in the pulse-amplifier drive can be used.

Table 14-4. Hard-tube Switches

Manufacturer	I.T.T.	Machlett	RCA	Litton
Type	Triode D1037	Triode ML7500	Beam triode 15030	Linear beam L3401
Typical operation:				
D-C plate voltage, Kv	50	46	40	150
Pulse plate current, amp	450	360	600	20
D-C grid bias, Kv	5	1.3	1.7	1
Pulse grid current, amp	90	70	22	0.1
Peak positive grid voltage, Kv	3.2	1.1	1.2	10
Plate output voltage, Kv	44	40	37	130
Pulse power output, Mw	19.8	14.4	22	2.6

Modulating-anode Operation of High-power Tubes. The modulating anode in a klystron or TWT or other linear-beam tube is an electrode which acts as a control grid of very low gain. This control anode is inserted in an insulating space between the beam-forming section and the body of the tube. When this anode is held at cathode potential, no current will flow from the cathode. When it is pulsed to a voltage dictated by the design of the tube, full current flows from the cathode. Present designs require the modulating anode to be pulsed to almost full anode voltage. A typical circuit, in simplified form, is shown in Fig. 14-36.

One important advantage of the modulating-anode device is that regulation of the modulating-pulse voltage need not be so exact as in a cathode pulsing system, since this voltage does not have nearly so much effect on phase. Another advantage is that less power is expended in modulation, especially in long-pulse operation, where rise times can be long and the voltage across the stray shunt capacity need not be rapidly changed.

Fault Protection. Transmitter tubes will spark occasionally for a large variety of reasons. When they are connected to an energy-storage system of high capacity, the entire energy of the system will then discharge through the tube. To protect the tube from destruction during a sparking interval, the storage bank is discharged through an alternate load capable of dissipating the stored energy. Circuits used to accomplish this are known as "crowbar" or "dump" circuits, and the dissipation element may be an ignitron.

Special Microwave Components. *Duplexers.* Monostatic radar systems can use one antenna for both transmission and reception if a switch can be constructed which will transfer the antenna from the transmitter to the receiver at the appropriate time. For most radars the basic requirements are that the switch must be fast, able to handle high power, and have small leakage through the switch from the transmitter

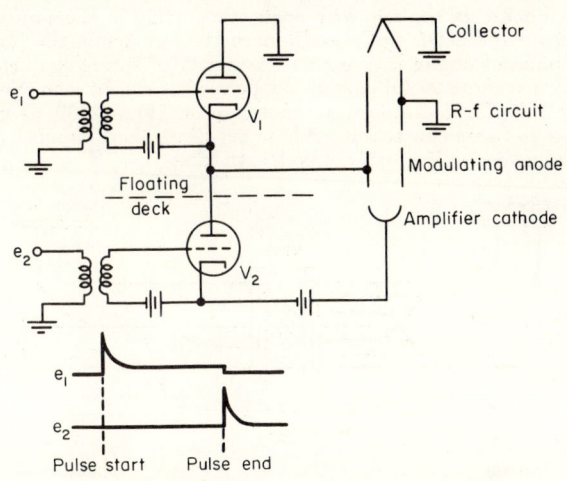

FIG. 14-36. Simplified modulating-anode switch.

to the receiver during transmission and low path loss (insertion loss) from the receiver to the antenna during reception. Also the path loss from the transmitter to the antenna must be low during the transmission period and high during the reception period. Such devices, which permit dual usage of the antenna, are called duplexers.

The most commonly used duplexer is a self-triggering gaseous discharge suitably engineered into the microwave circuitry. An example of an early form of duplexer is shown in Fig. 14-37. When the transmitter is pulsed, the switch tubes are "fired" and high conduction occurs. Because of their placement in the microwave circuit ($\lambda/4$ from the main transmission line), the path to the antenna has low loss. The TR tube opens the path to the receiver when it is fired. Additional protection in the form of another gas switch device may also be necessary to protect the receiver. The purpose of the ATR tube is to open the path to the transmitter when the ATR tube is not "fired."

Modern duplexers use inert gases enclosed in a chamber which may be an integral part of the microwave switching circuit. A typical arrangement, shown in Fig. 14-38, makes use of a balanced arrangement of directional couplers and switch tubes to achieve the same objective.

FIG. 14-37. Principle of branch-type radar duplexer. The TR provides receiver protection and switching; the ATR channels the echo power into the receiver. (*From M. I. Skolnik, Introduction to Radar Systems, McGraw-Hill Book Company, New York, 1962.*)

Ferrite devices such as isolators, circulators, and Faraday rotators can also be used in duplexer applications. They have shorter recovery times, longer life, and wider bandwidth operation than gaseous-discharge duplexers. However, they are heavier, bulkier, and have less power-handling capability than similar gaseous-discharge devices, which becomes especially significant at the lower radar frequencies.

14-30 SYSTEM COMPONENTS

The Rotating Joint. Most radar systems require a mechanical rotation of the complete antenna system, but it is undesirable to move the transmitter and receiver. Hence a microwave device is required which transmits energy efficiently while one terminus is fixed and the other rotates. Such a device is known as a rotating joint.

One of the simplest configurations is a section of circular waveguide which is divided on a plane orthogonal to the axis of the cylinder. The two adjacent edges of the cylinder must remain in contact with each other while in the process of rotation. Ordinary pressure contact is not enough since the loss across the junction may be high, and sparking or arcing may occur. A variety of spring-contact systems have been devised. These are useful only at low power but can be made very broadband. In order to avoid sliding contacts, a "choke" joint (Fig. 14-39) is generally used. The open space between contacts is made to act like a short circuit by connecting it to a shorted transmission line half a wavelength long.

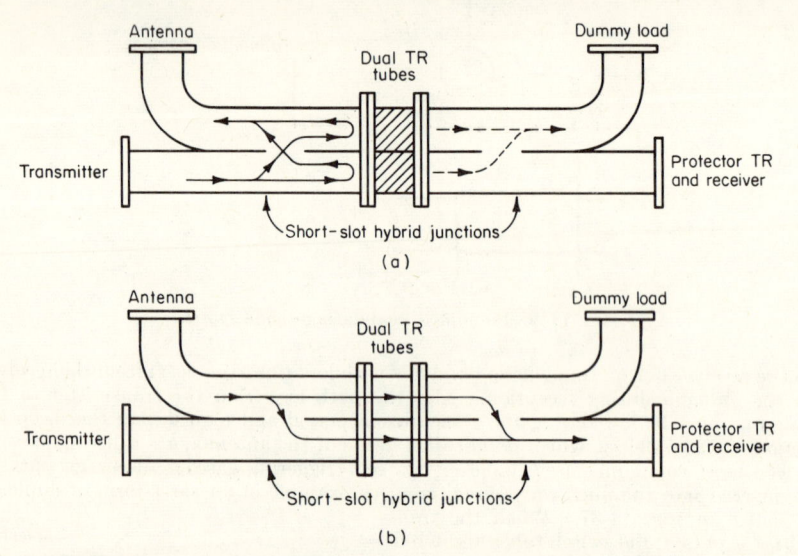

Fig. 14-38. Balanced duplexer using dual TR tubes and two short-slot hybrid junctions. (a) Transmit condition. (b) Receive condition. (*From M. I. Skolnik, Introduction to Radar Systems, McGraw-Hill Book Company, New York, 1962.*)

The rotating joint requires a symmetry, and so most joints are constructed from cylindrical waveguide or coaxial line sections. A constraint exists on the outer diameter of a coaxial line for transmission of power at a given wavelength. Also the outer diameter in general must be somewhat less than one-half wavelength to prevent transmission of higher-order modes.

Many designs have been proposed which would permit the assembly of a multiplicity of joints (greater than three). Most of these lack symmetry and cannot be rotated through 360° without a "blind angle" where transmission is unsatisfactory.

14-6. Antennas for Radar Systems

Antennas are used both to transmit and to receive energy in radar systems. As a transmitting device, the antenna accepts energy from the transmitter via a transmission line and radiates it into space. Generally the radiation pattern is far from isotropic, so that the bulk of the transmitted energy is selectively directed into a relatively small solid angle—i.e., into a "beam." As a receiving device, the antenna intercepts some of the reflected energy from the target and transfers it through a transmission line to the receiving system. According to the "reciprocity theorem,"

RADAR   14-31

many of the characteristics of an antenna are the same whether it is used for transmitting or receiving; one important difference is that a transmitting antenna must be able to handle high peak power.

Most antennas, especially those with narrow beams, are mounted on a supporting structure called a pedestal which permits the beam to be accurately pointed. A drive system and a servomechanism control the pointing angle to follow the motion of a

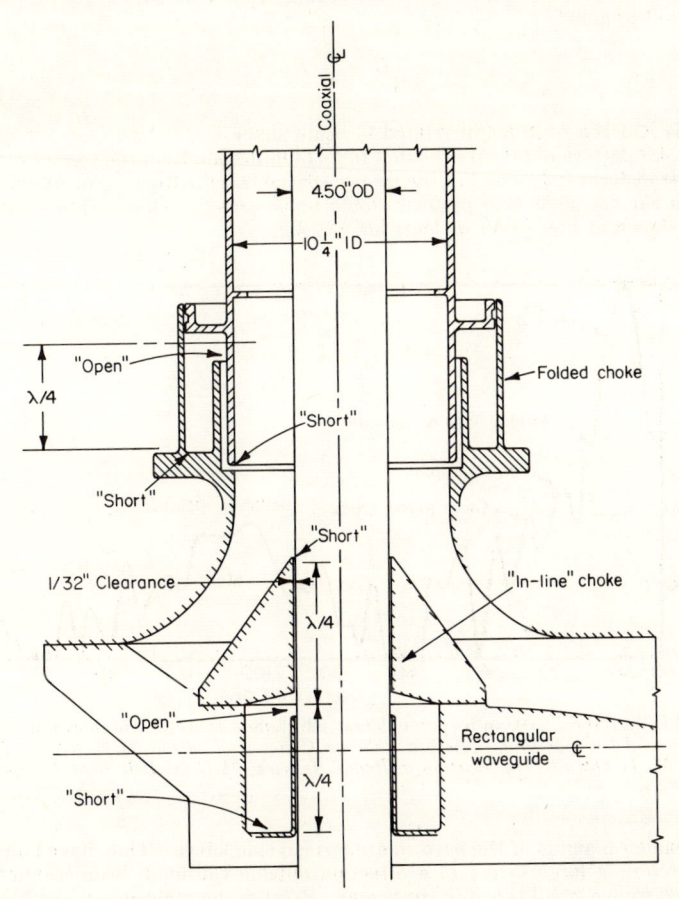

Fig. 14-39. A "choke" joint.

specific target (track) or to execute a predetermined angular pattern (search). In another class of antennas, the direction, and sometimes also the shape, of the beam are controlled electronically; these are discussed separately below. Moving the beam mechanically or electronically is called "scanning."

Reflector-type Antennas. In general concept the reflector-type antenna is similar to the searchlight. A "point" source of energy is placed at the focal point of a paraboloid reflector which, ideally, confines all the radiated energy within a specified angle. The gain of such a system, if it were feasible, would be

$$G = \frac{4\pi}{\Omega} \qquad (14\text{-}12)$$

where G = antenna gain (ratio of power averaged over main beam to power from an isotropic radiator)

Ω = solid angle of the antenna beam

Unfortunately, appreciable power is radiated in almost all directions; a typical polar plot of the radiated energy is shown in Fig. 14-40. The "beamwidth" θ is then arbitrarily defined as the angle between points on the polar plot where the power is one-half of the power at the center of the beam. Hence the gain of the antenna is less than the ideal gain:

$$G = \frac{4\pi}{\Omega} K_1 \qquad (14\text{-}13)$$

where $K_1 < 1$ is a form factor related to beam shape.

The polar pattern obtained is related to its illumination from the feed system by the Fourier transform (expressed in the proper spatial coordinates). For uniform illumination, a $\sin x/x$ polar field pattern should be obtained. This is approximately the pattern shown in Fig. 14-40, at least for $\theta < 40°$.

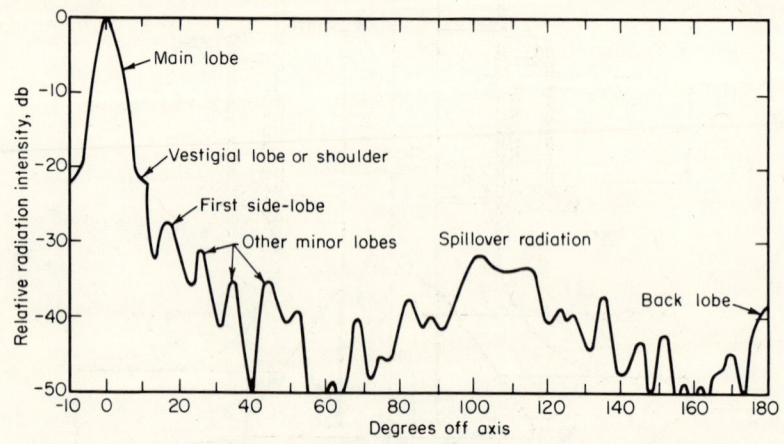

Fig. 14-40. Radiation pattern for a particular paraboloid reflector antenna illustrating the main-lobe and the side-lobe radiation. (*After Cutler et al., Proc. IRE, vol. 35, December, 1947, in M. I. Skolnik, Introduction to Radar Systems, McGraw-Hill Book Company, New York, 1962.*)

The smaller maxima of the pattern are termed side lobes. They have a finite gain, and therefore a large target in a direction outside the main beam cannot be distinguished from a small target in the beam. Because the same gain is also obtained in the receive pattern, the effective system field pattern for uniform illumination will have a $(\sin x/x)^2$ shape. The difference or contrast between the first side lobe and the main lobe in this case is approximately 27 db (two-way). This side-lobe level is frequently inadequate, and so uniform illumination is rarely used, although it is referred to as a standard of gain. By "tapering" the illumination, so that more energy is concentrated toward the center of the paraboloid, lower side lobes can be achieved, at the expense of less gain and greater beamwidth. A typical illumination taper is shown in Fig. 14-41 together with the effect of varying the taper on antenna gain, beamwidth, and first side-lobe level.

The beamwidth of such an antenna is given by

$$\theta = K \frac{\lambda}{D} \qquad (14\text{-}14)$$

where θ = half-power beamwidth, radians
 λ = wavelength
 D = aperture size
 K = illumination factor

The relation between the area (measured perpendicular to the beam axis) and the beamwidth of a circular-reflector antenna is

$$A = \frac{\pi}{4} K^2 \frac{\lambda^2}{\theta^2} \approx \frac{\pi K^2}{4} \frac{\lambda^2}{\Omega} \quad \text{since} \quad \Omega \approx \frac{\pi \theta^2}{4} \quad \text{for } \theta \ll 1 \quad (14\text{-}15)$$

The relation between area and gain is

$$G = \frac{4\pi A}{\lambda^2} \epsilon \quad (14\text{-}16)$$

where $\epsilon \approx \dfrac{4K_1}{\pi K^2} \approx 0.7$

There is no theoretical limit to the reduction in side-lobe level which may be obtained. A lower limit is set in practice by tolerances which can be maintained on the reflector shape and illumination.

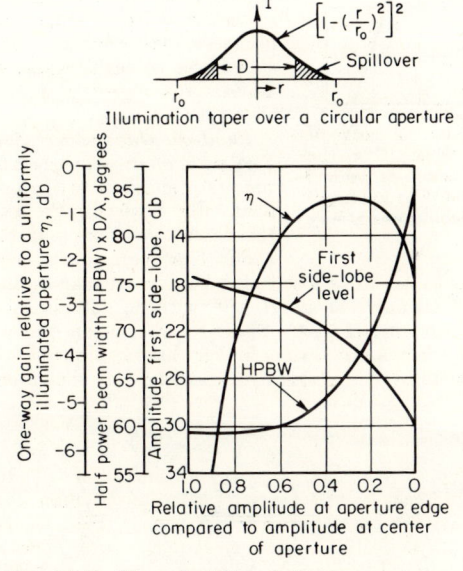

Fig. 14-41. Characteristics of circular apertures.

Usually antennas are designed to radiate either horizontal or vertical polarization (the direction of polarization is determined by convention by the direction of the electric-field vector). Occasionally antennas have been designed to radiate circularly polarized waves, a special case of elliptical polarization.

The reflecting surfaces of a paraboloidal antenna must be good electrical conductors at the radiated frequency. They need not be solid, but any openings or perforations must be small compared with a wavelength. A surface tolerance of 0.05 to 0.1 wavelength should be maintained; as the error increases, the side-lobe level increases, and finally the shape of the main beam is affected. The relation between tolerance and gain is shown in Fig. 14-42.

The paraboloid is illuminated by a feed placed at the focal point. Because of mechanical considerations, the ratio of focal length to diameter is usually in the range $0.25 < f/D < 1$. Several methods for transmitting energy to the feed are shown in Fig. 14-43. The method shown in Fig. 14-43c employs an intermediate reflector and is derived from Cassegrainian optics. In this system the feed position is approximately coincident with the reflector surface. This is a particular advantage in the four-horn monopulse (see below), when the feed system is bulky and complex.

Special Beam Shapes. The circular paraboloid will produce a symmetrical beam. Sometimes it is necessary to have a wide vertical beam combined with a narrow horizontal beam. This can be achieved with an elliptical paraboloid, the beamwidth in each direction being inversely proportional to the aperture dimension [see Eq. (14-14)]. Many antennas of this type have been used in search systems and height-finding systems, for ratios of vertical to horizontal beamwidth as great as 10:1. For greater ratios, array antennas must be used.

Distorted beam shapes have been employed for special applications. The most widely used, called the cosecant squared, is shown in Fig. 14-44. Its objective is to achieve a uniform elevation coverage.

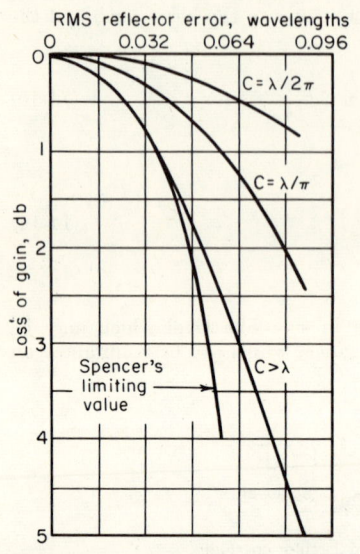

Fig. 14-42. Reduction of gain of parabolic reflector, where C is the correlation interval which is defined as that distance "on the average" where the errors become essentially independent.

Feed-displacement Scanning Systems. The paraboloid is a perfect focusing system and therefore sensitive to the proper location of the feed. By displacing the feed, one can displace the beam, and thus scan through angles up to about five beamwidths before serious pattern deterioration begins to set in. A reflector having circular symmetry, however, can be scanned through an angle of about 120° with no deterioration beyond that of a spherical reflector with central feed. Spherical aberration prevents perfect focusing, so that the side-lobe levels are always higher than in a paraboloidal reflector. The locus of the focal arc and the limitation on scan angle are shown in Fig. 14-45.

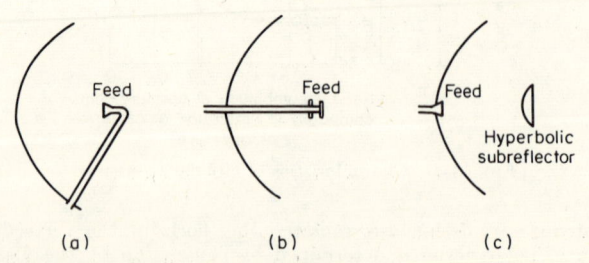

Fig. 14-43. Methods of feeding parabolic reflectors. (a) Horn fed by waveguide. (b) Half-wave dipole fed by waveguide. (c) Cassegrain system.

To avoid moving the feed along the focal arc in a large antenna a considerable linear distance, a continuum of feed horns may be provided along the focal arc (Fig. 14-46). These horns are connected by waveguides to a smaller circular arc. A feed horn is rotated past the waveguide openings in this smaller arc, coupling power into the openings it overlaps.

Tracking Feed Systems. Radar tracking systems obtain an error signal which is related to the displacement of the antenna pointing angle from the angular direction to the target. This error signal is fed to a servomechanism which operates to minimize the error and, hence, track the target.

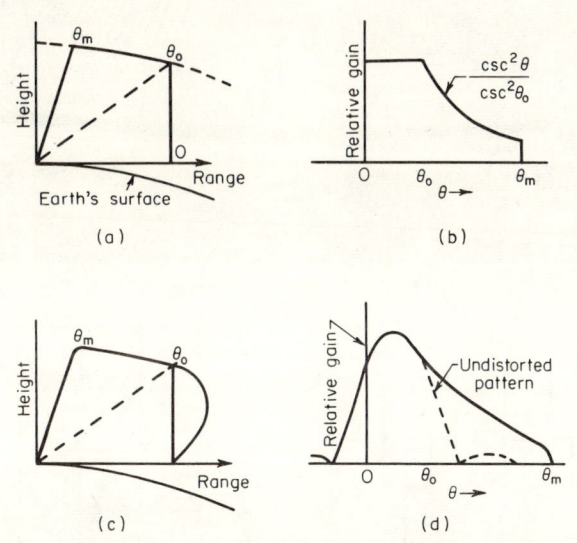

Fig. 14-44. Elevation coverage of a "cosecant-squared" antenna. (a) Desired elevation coverage from 0 to θ_0 to θ_m. (b) Corresponding antenna pattern desired. (c) Realizable elevation coverage with pattern shown in d. (d) Actual cosecant-squared antenna pattern.

The three most common methods of developing the error signal are:

1. Conical scan: The beam is offset one-half beamwidth from boresight by displacement of the feed. The feed is rotated to generate a conical surface. This system develops the angular error serially.
2. Sequential lobing: A special case of conical scan where only four beam positions are used on the conical surface in sequence.
3. Monopulse: The four beam positions are developed simultaneously, permitting angular error to be determined on each pulse. This minimizes errors which are due to time fluctuations of the target characteristics.

Conical Scan. Conical scan is simple in concept and requires only some mechanical ingenuity to produce an offset feed which can be mechanically rotated. A typical example uses a circular-waveguide feed which is offset from the center axis with a rotating joint at the offset.

Four-horn Monopulse. Many variations of the monopulse technique have been evolved. In one simple form, four horns are connected by a microwave network, as shown in Fig. 14-47. The circled junctions indicate a hybrid junction with sum (Σ) and difference (Δ) outputs. Four such junctions are used to provide an azimuth-error output, an elevation-error output, and an output known as the sum channel, which is the summation in equiphase of all four horns.

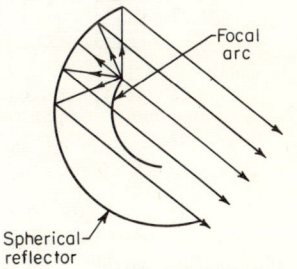

Fig. 14-45. Maximum scan angle of spherical reflector.

Radiated Field Zones. The polar radiation patterns previously discussed are actually achieved only at an infinite distance from the antenna. Near the antenna, in the so-called near field, or Fresnel zone, a polar plot of the field tends to resemble the

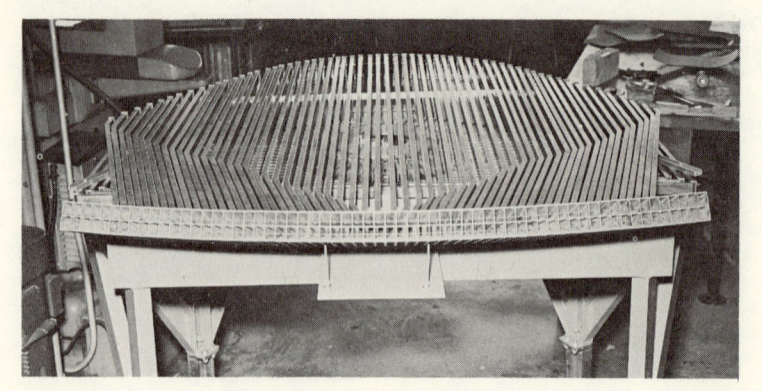

FIG. 14-46. Organ-pipe scanner.

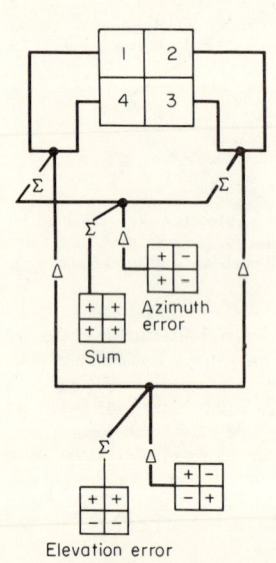

FIG. 14-47. Four-horn monopulse comparator circuit.

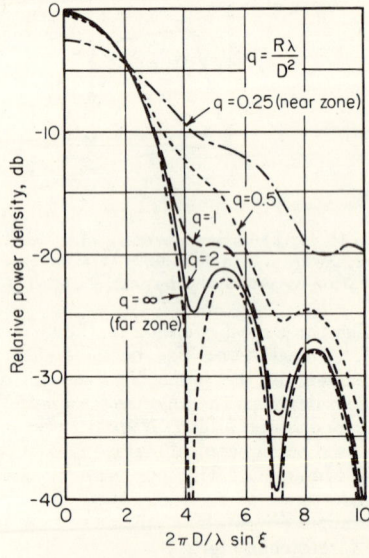

FIG. 14-48. Relative power densities vs. normalized beam angle for typical antenna. (*R. B. Barrar and C. H. Wilcox, On the Fresnel Approximation, IRE Trans. Antennas and Propagation, vol. AP-6, no. 1, pp. 43–49, January, 1958.*)

illumination pattern. As the distance from the antenna is increased, a transition gradually occurs into the far field, or Fraunhofer zone, where the pattern resembles the pattern at infinity. The pattern at distance R from the antenna is usually a good approximation to the far-zone pattern when

$$\frac{R\lambda}{D^2} > 2 \tag{14-17}$$

Figure 14-48 indicates the nature of the transition for a typical parabolic antenna.

An appreciation of this phenomenon is important for two reasons. In measuring antenna patterns, it is important that the measurement be made in the far field, or that appropriate corrections be made, lest serious errors occur in extrapolating to far-field conditions, where the radar system will most frequently employ the antenna. Knowledge of the near field in some detail is important in determining personnel hazards, which may be very serious for systems using high-power transmitters.

Array Antennas and Scanning Systems. Reflector-type antenna systems have the virture of simplicity, but they have a number of shortcomings.

1. The mechanical motion, which is required to scan, limits the scan techniques and the rapidity of scanning.

2. It is difficult to construct systems which provide for a multiplicity of simultaneous beams.

Both these limitations have led to a revival of interest in array antennas and lens systems.

An array antenna consists of a number of radiating sources. The shape and direction of the beam are a function of the number of sources, their spacing, and the relative

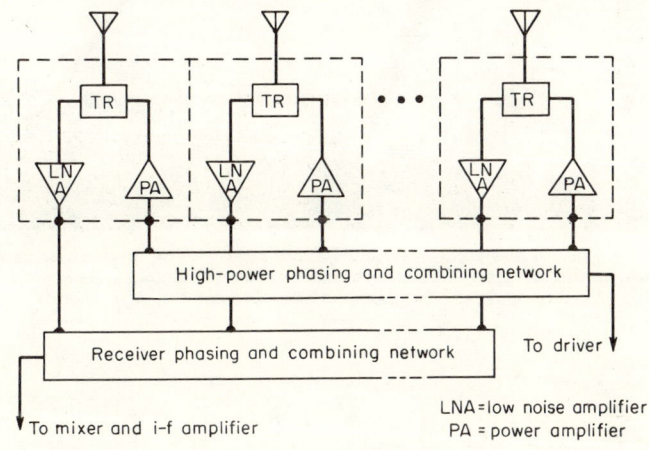

Fig. 14-49. Active element array using minimal element modules. TR = transmit-receive switch. (*J. L. Allen, Array Radars, A Survey of Their Potential and Their Limitations, Microwave J., vol. 5, no. 5, May, 1962.*)

amplitude and phase of the excitation to each source. The array is called a linear array or a planar array, respectively, if the elements are arranged in a straight line, or are distributed over a plane.

When the excitation of each element of such an array is held constant in amplitude, and the phase is varied, the direction of the beam is changed. Such a system is referred to as a "phased array." The phase can be controlled by electronic phase shifters yielding so-called "inertialess" scanning. The beam can then be positioned in any direction almost instantaneously (i.e., in microseconds).

Another advantage of arrays is that, for high-power radar applications, a large amount of overall system power can be radiated using only low-power sources to drive each element. Also it is possible to transmit or receive in many directions simultaneously. These advantages, however, can be obtained only at the expense of providing an individual transmitter and receiver for each element of the array. Such a system is referred to as an "active-element array" and is shown in Figs. 14-49 and 14-50.

In the simple case where the beam is to be perpendicular to the plane of the array, the considerations previously stated concerning gain, illumination, and patterns of reflectors are quite similar for arrays, and Eqs. (14-12) through (14-16) are applicable except for minor variations in the shape and efficiency factors. By appropriate element spacing, and control of phase and amplitude, arrays can be designed with small side-lobe levels.

The elements of the array may be dipoles, horns, slots in a waveguide, or other radiating devices. The method of feeding each element can be derived from a branching feed system, where each element is fed by a separate transmission line from the r-f power source, or a serial system, where the feed for each element is tapped consecutively off one long transmission line. One of the reasons reflector systems have

Fig. 14-50. ESAR—an electronic scanning L-band radar. The rectangular panels on the sloping face contain over eight thousand antennas embedded in plastic. (*J. L. Allen, Array Radars, A Survey of Their Potential and Their Limitations, Microwave J., vol. 5, no. 5, May, 1962.*)

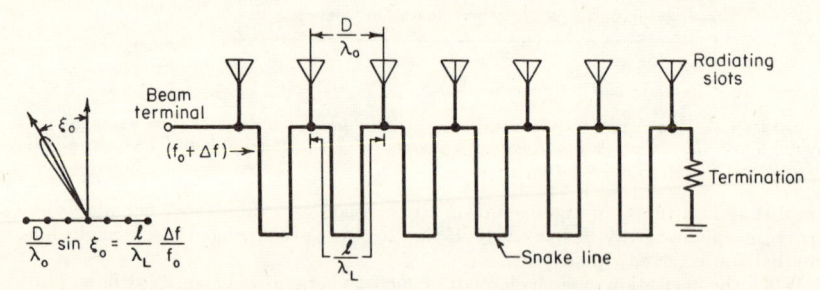

Fig. 14-51. Frequency-scanned array. (*J. L. Allen, Array Radars, A Survey of Their Potential and Their Limitations, Microwave J., vol. 5, no. 5, May, 1962.*)

been favored over array systems is the complexity of this feed and radiator system. To illustrate this point, consider the number of elements required in a square array. Since the dipoles are generally spaced by approximately $\lambda/2$, the number of elements in a square array is given approximately by

$$n \approx \frac{10^4}{\theta_h \theta_v} \tag{14-18}$$

where θ = beam angle, degrees

For $\theta_h = \theta_v = 1°$, $n = 10^4$ elements.

The direction of the beam in passive-element arrays is most often controlled by the phase shifters. These may be in the main transmission line, where the phase shifts are cumulative, or in the branch lines to the individual elements. The former is simpler

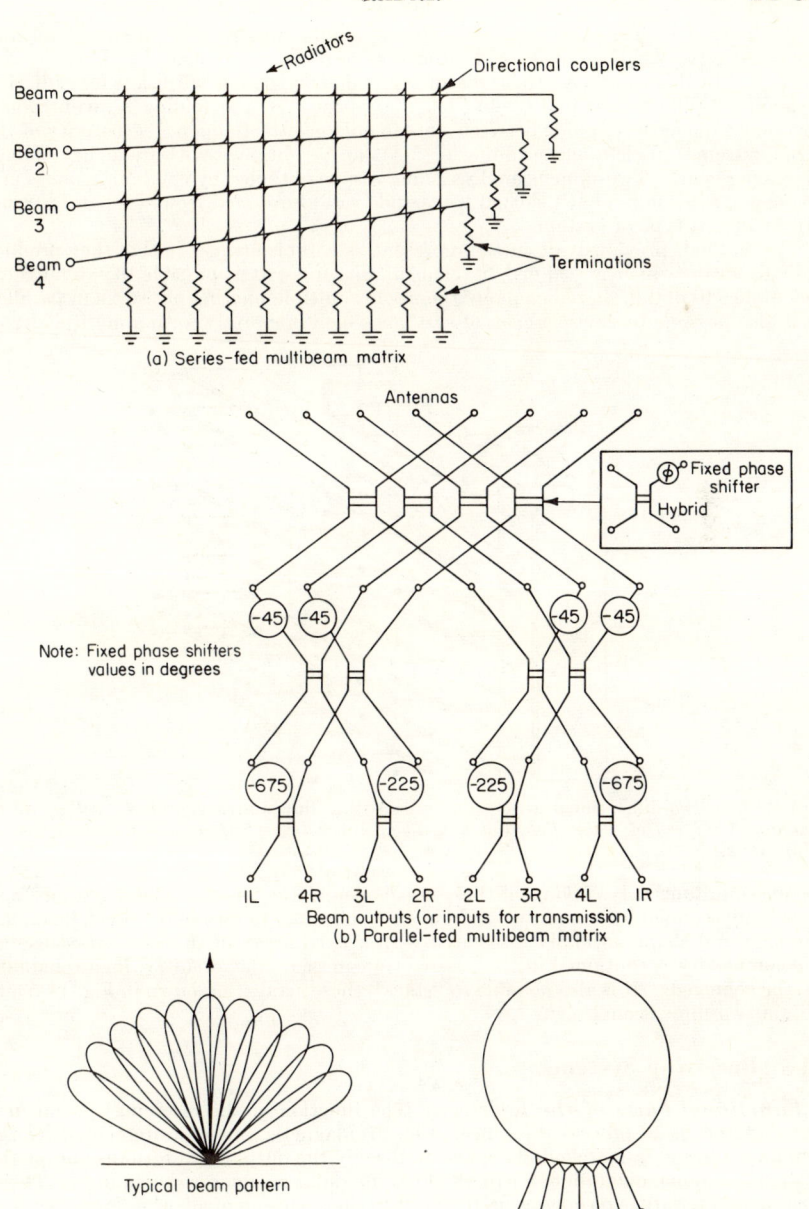

FIG. 14-52. Some multibeam array techniques. (*J. L. Allen, Array Radars, A Survey of Their Potential and Their Limitations, Microwave J.*, vol. 5, no. 5, May, 1962.)

to control but requires that the phase shifters handle high power. A special form of the serial array has been developed for control of phase (i.e., angle) by change of frequency. In this technique, the radiator is a folded waveguide with slots for radiating elements, as illustrated in Fig. 14-51. The purpose of the folding is to introduce several wavelengths of guide between slots to enhance the frequency sensitivity of the serial system. Wide-angle scanning ($\pm 50°$) can be obtained with frequency change of ± 5 per cent. Two-dimensional scanning can be obtained by combining phase and frequency scanning. Each slotted waveguide can also be fed from a separate transmitter in this type of system.

The methods previously discussed represent "steerable arrays" in that they produce a single beam which is scanned by element phasing. It is possible to extend these techniques to obtain more beams by the use of multiple phasing networks in parallel. It is also possible to devise a class of systems which inherently form a multiplicity of

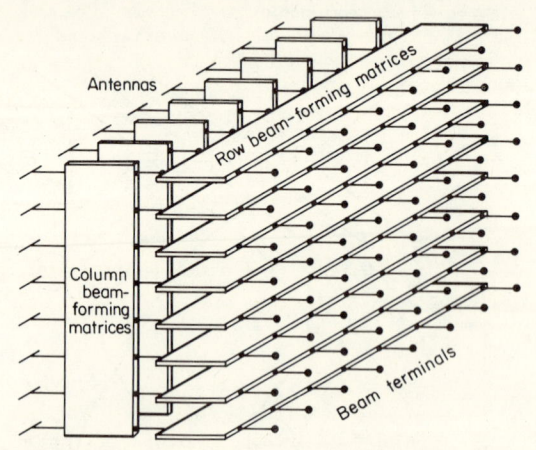

Fig. 14-53. Two-dimensional array synthesized from linear arrays. (*J. L. Allen, Array Radars, A Survey of Their Potential and Their Limitations, Microwave J., vol. 5, no. 5, May, 1962.*)

beams simultaneously. Some of these are shown in Fig. 14-52. These systems may be serially scanned by commutating from beam terminal to terminal. Each beam has the gain and shape normally associated with a fixed array of this size. Because all the complexity is contained in the passive transmission-line circuitry, high reliability can be achieved. It is also possible to "stack" these arrays, as shown in Fig. 14-53, to obtain two-dimensional scan.

14-7. Receiving Systems

Functional Parts of the Receiver. The function of the receiving system in a radar system is to process the reflected electromagnetic energy (intercepted by the antenna) through a matched filter, and to display the output in a form useful for the system. A block diagram of a superheterodyne radar receiver is shown in Fig. 14-54. The superheterodyne receiver is used almost to the exclusion of all other forms because of its convenience in permitting the separation of the functions of controlling the noise temperature and providing the "matched filter" necessary for optimum detection performance. An r-f amplifier is not essential to the superheterodyne circuit, although its use has become more common in radar for reasons discussed below. Other types of receivers possible, but rarely used, are r-f receivers (the mixer, local oscillator, and i-f amplifier are omitted), superregenerative receivers, and crystal video receivers.

The detectability of a received signal is determined by its signal-to-noise ratio, and as indicated in Sec. 14-3, the optimum signal-to-noise ratio is obtained when a matched-filter receiver is employed. Because the energy level of the received signal is usually very low, it is essential to introduce amplification in the receiver, and this amplifies the internal noise as well as the signal. Hence it is essential to keep this internal noise to a minimum. In a superheterodyne receiver, the principal source of noise is the r-f amplifier, if one is used, or the mixer and first i-f stage. Considerable attention has therefore been devoted to developing r-f amplifiers and mixer-preamplifier combinations with minimum noise characteristics (see below under Low-noise Amplifiers).

The greater part of the receiver gain is contributed by the i-f amplifier. The matched filter is "formed" in this section of the receiver because it is generally more convenient to achieve the required filter characteristics in this frequency region (usually between 10 and 100 Mc). When part of the matched filter is a postdetection device, it is incorporated in the video section of the receiver.

The local oscillator is usually a microwave oscillator when the transmitter itself is an oscillator. In this case an attempt is usually made to lock the frequency of the local oscillator to the transmitter frequency, offset by a frequency equal to the i-f frequency. This servomechanism is termed an automatic frequency control. If the transmitter is an amplifier, the frequency is usually derived from a crystal oscillator and multiplier chain. In this case the local oscillator frequency is also derived from this chain.

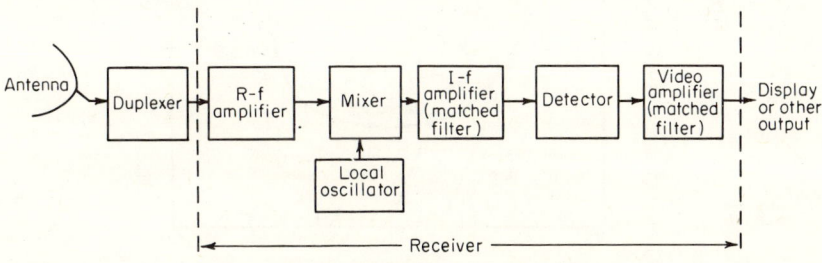

FIG. 14-54. A typical radar receiving system.

Modern r-f amplifiers and mixers are rarely "tuned" in synchronism with the transmitter frequency. Usually they are designed to have a 10 per cent bandwidth centered on the mean frequency of operation. Only the local oscillator is tuned in synchronism with the transmitter.

One of the problems of the receiver is to achieve amplification over a wide dynamic range without saturation or overload. A number of techniques are available for accomplishing this. Some of these techniques involve gain control of the i-f stages by a servomechanism action controlled by the signal itself. Examples of these types are the IAGC (instantaneous automatic gain control) and DBB (delayed back bias). Another useful technique for achieving wide dynamic range is the logarithmic i-f amplifier.

It is possible to avoid the problem of wide dynamic range by use of a class known as constant false-alarm rate (CFAR) receivers. These receivers do not depend on the signal- and noise-amplitude statistics but utilize inherent phase or frequency statistical differences between the signal and noise. The noise output of such devices is constant and independent of noise input, hence the name CFAR.

Sources of Receiving-system Noise. Until recently, at frequencies above 100 Mc, the greatest contribution to receiving-system noise came from the receiver itself. Therefore, in order to determine the noise power of the receiving system, it was only necessary to determine the noise power of the receiver. The recent introduction of low-noise r-f amplifiers and mixers has considerably complicated the determination of receiving-system noise, since it is now necessary to consider the contributions to system noise from the external environment and from the lossy components linking the receiver to the atmosphere. (It will be assumed in further discussion that all the sources of noise are Gaussian and uncorrelated.)

14-42   SYSTEM COMPONENTS

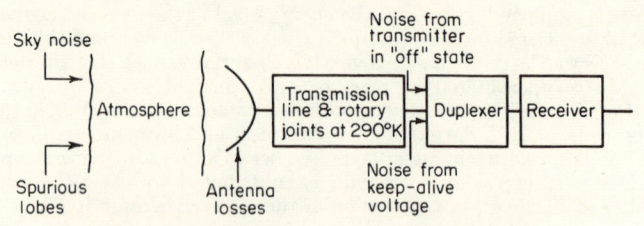

Fig. 14-55. Sources of receiving-system noise.

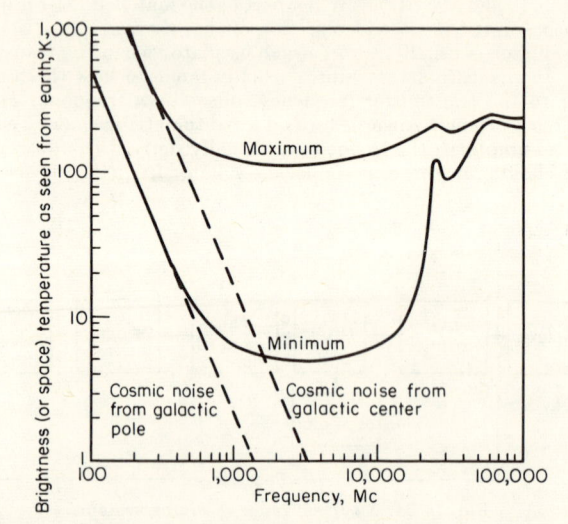

Fig. 14-56. Maximum and minimum brightness (or space) temperatures of the sky as seen by an ideal single-polarization antenna on earth. Dashed curves apply to cosmic noise. Cosmic noise predominates at lower frequencies, atmospheric-absorption noise at higher frequencies. Maximum and minimum cosmic noise correspond to the directions of the galactic center and the galactic pole, respectively. Maximum atmospheric absorption corresponds to the antenna beam pointing along the horizon, while minimum absorption corresponds to antenna pointed at the zenith. (*After Greene and Lebenbaum, Microwave J., vol. 2, pp. 13-14, October, 1959, in M. I. Skolnik, Introduction to Radar Systems, McGraw-Hill Book Company, New York, 1962.*)

IEEE definitions establish a "noise factor" which has been used extensively as an indicator of receiver or receiver-system sensitivity.[16,17] This noise factor is the ratio of the total available noise power to that portion attributable to the input termination at a temperature of 290°K.*

For a blackbody, at microwave frequencies, noise power is related to noise temperature by

$$P = kT\,\Delta f \qquad (14\text{-}19)$$

where P = noise power, watts
T = absolute temperature, °K
Δf = noise bandwidth, cps
k = Boltzmann's constant (1.38 × 10⁻²³ joules/°K)

The noise factor is then given by

$$F = \frac{(k)(290)(\Delta f) + P_e}{(k)(290)(\Delta f)} = 1 + \frac{P_e}{(k)(290)(\Delta f)} \qquad (14\text{-}20)$$

* 290°K is a rather chilly "room" temperature (63°F) but has the advantage that kT is then 4 × 10⁻²⁰, an "easy" number to remember.

where P_e = effective noise-power contribution of the receiving system
This may also be written as

$$F = 1 + \frac{T_e}{290} \qquad (14\text{-}21)$$

where T_e = effective input noise temperature of the receiving system

When T_e is large compared with 290°K, as was the case until recent years, the noise factor is a good indication of receiving-system sensitivity. However, F becomes insensitive to changes in T_e when $T_e < 290°$. It is therefore recommended that the effective input noise temperature T_e of a receiving system be used as a direct indication of the receiver's sensitivity.

The sources of noise for a receiving system are shown diagrammatically in Fig. 14-55. The antenna may be pointed at the sky, in which case it will receive cosmic or sky noise through the atmosphere. The atmosphere, being "lossy," will attenuate the cosmic noise somewhat but will provide an additional noise because of its lossiness. The sources of cosmic noise are a function of direction angle, tending to be much higher on the galactic plane. The effective temperature of the atmosphere is a function of altitude, location, and time. In addition to the main lobe noise, side lobes of the antenna will be receiving noise from other directions, some from the sky and some from the earth. The effective temperature of the earth is also variable, being low for highly reflective surfaces such as water (unless they reflect the sun or a radio star) and close to 290°K for highly absorbent surfaces. The antenna directivity pattern is a complex function when the details of the side lobes are involved, and as a result, the computation of antenna temperature is quite complex. A composite plot showing the combined effects of attenuation and cosmic noise is shown in Fig. 14-56. It may be seen that atmospheric attenuation causes a sharp increase in external noise (as well as introducing signal attenuation) at frequencies above 10 Gc. Cosmic noise causes a sharp increase in external noise at frequencies below several hundred megacycles. As a result, long-range systems which are severely dependent on low system temperature tend to be located in this "window" between 200 and 20,000 Mc. A typical experimentally derived result giving the variation of receiving-system temperature with zenith angle for a C-band antenna is shown in Fig. 14-57.

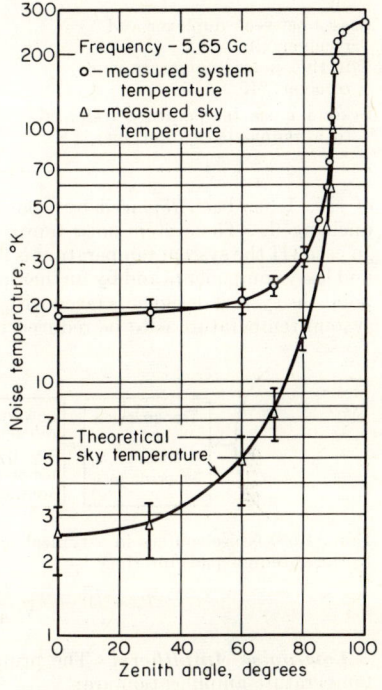

FIG. 14-57. Noise temperature vs. zenith angle. (*Degrasse, Hogg, Ohn, and Scovil, Ultra-low Noise Measurements Using a Horn Reflector Antenna and a Traveling Wave Maser, J. Appl. Phys., vol. 30, p. 2013, December, 1959.*)

In computing system temperature, it is desirable to refer all noise-power contributions to a common point within the system such as the antenna and space interface or the receiver input terminals. In Fig. 14-58 a method for accomplishing this is shown.

A tabulation of three representative cases computed in this manner is shown in Table 14-5.

In case I the dominant source of noise is the receiver. All the other contributions only result in increasing the temperature by a factor of 2. In case II the receiver

Table 14-5. Comparison of Noise Temperatures in Three Systems

Source of noise	Case I. Airport surveillance radar	Case II. Long-range tracking radar	Case III. Advanced deep-space surveillance radar
Sky temperature, °K	20	20	20
Antenna losses and spurious lobes, °K	100	100	5
Losses between antenna and duplexer, db	1.5	1.5	0.4
Noise from transmitter in "off" state, °K	100	100	0
Noise from duplexer and keep-alive, °K	200	200	0
Losses between duplexer and receiver, db	1.0	1.0	0.2
Effective noise temperature of receiver, °K	2500	150	10
System noise temperature (referred to antenna-space interface), °K	5220	1040	80

of case I has been replaced by a low-noise receiver; all other noise sources remain unchanged. The system noise temperature as a result was reduced by a factor of 5. In case III the system temperature is further reduced by eliminating or reducing noisy and lossy components and by further lowering the receiver noise temperature. At this point the system designer is faced with increasingly difficult component problems if the system temperature is to be reduced to the irreducible limit of the sky temperature.

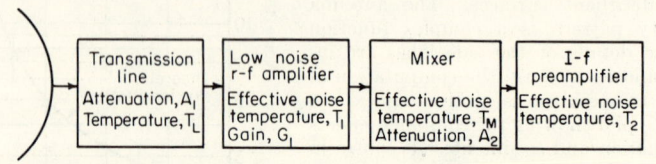

FIG. 14-58. Noise sources in a typical receiver. The effective noise temperature referred to the antenna-space interface is

$$T_e = (1 - A_1)\frac{T_L}{A_1} + \frac{T_1}{A_1} + \frac{T_M}{A_1 G_1} + \frac{T_2}{A_1 G_1 A_2}$$

Low-noise Amplifiers. The principal devices in current usage to achieve low-temperature amplification are:

1. Triodes
2. Traveling-wave tubes and backward-wave amplifiers
3. Parametric amplifiers
    a. Lumped varactor diode amplifiers
    b. Distributed varactor diode amplifiers
    c. Electron-beam amplifiers
4. Tunnel-diode amplifiers
5. Masers
    a. Cavity type
    b. Traveling-wave type

A rough comparison of noise figures achievable with these devices is shown in Fig. 14-59. Also shown is the noise figure achievable with modern crystal mixers. The trend toward the use of r-f amplifiers, even though this introduces additional complications, can be explained by the improved performance as indicated in this figure.

The comparison on the basis of noise temperature or noise figure is not in itself adequate since other factors enter into the performance requirements, such as achievable bandwidth and dynamic range, tunability, stability, complexity, size, weight, and reliability. If the choice were to depend solely on system temperature, then the maser would be the outstanding choice. Yet the maser is used only infrequently because of the complexity currently involved in achieving the required liquid helium temperature (4°K) for operation, and because of the limitation on achievable bandwidth with useful gain.

The TWT. A space-charge TWT consists of an electron gun for forming a pencil-shaped beam, a longitudinal magnetic field to maintain the beam, a coupling structure to provide an interaction region between an external circuit and the beam, and a collector to absorb the beam. Two longitudinal waves propagate, having phase velocities slightly higher and lower than the velocity of the electrons in the beam. The slow

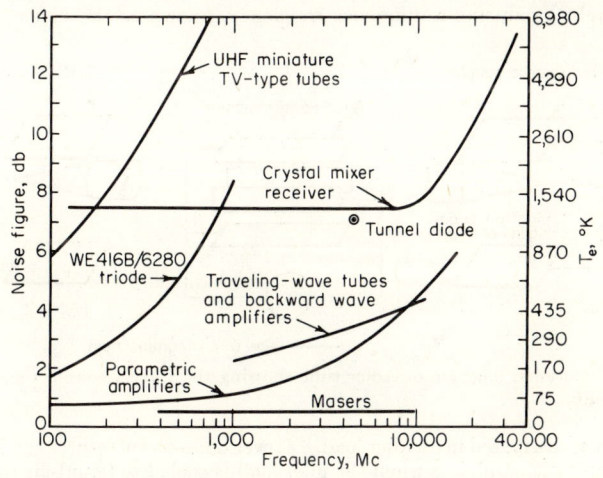

Fig. 14-59. Comparison of the noise figures of various receivers as a function of frequency. (*From M. I. Skolnik, Introduction to Radar Systems, McGraw-Hill Book Company, New York, 1962.*)

wave is excited when energy is removed from the beam and is therefore useful for amplification. A number of slow-wave structures have phase-velocity characteristics that match the phase velocity of the slow wave. These can be used as the coupling structure. The most commonly used structure is a wound helix.

The internal noise of the TWT is generated almost entirely by noise on the beam. This noise is due to the random velocity of the electrons and the shot noise of the beam. A considerable amount of progress has been achieved in reducing the noise coupled into the slow wave by properly shaping the potential contours near the cathode of the electron gun. Guns have been constructed with noise temperatures less than 300°K.

The Parametric Amplifier. Parametric amplification is not new, although it is newly applied to the problem of microwave amplification, and its principles are more clearly understood as a result of the work of Manley and Rowe.[5] The simplest and most widely known form of parametric amplification is the amplification of swinging achieved by children through "pumping." The parameter synchronously varied in this activity is radius of gyration.

Parametric amplification is derived from the conversion of energy from an r-f source called a pump to r-f energy at another frequency. The transfer of energy requires a nonlinear lossless element in the amplifier.

The most common form of microwave parametric amplifier uses a varactor diode, which is a back-biased junction diode. These diodes may be crudely compared with a parallel-plate capacitor in which the separation between the plates is a function of the applied voltage. This provides the nonlinear, lossless element (i.e., a nonlinear capacitance). The varactor diode may be used in a number of different circuits to achieve parametric amplification. The simplest circuit is the one-port amplifier using a circulator to separate the input from the output.

The input resistance of this amplifier is negative, and therefore the major problem is that of maintaining stable gain. An additional problem imposed by the negative resistance operation is the narrow bandwidth (a few per cent) that can be achieved.

Varactor diodes may be used in distributed circuits which in effect cascade the action of a number of varactors, each operating at a low gain. These circuits permit wider bandwidth and more stable operation. In the simplest form of parametric amplifier, the pump is operated at exactly twice the signal frequency. This is known as the degenerate form and is very sensitive to the relative phase between the signal and pump. To avoid this, the pump frequency is made higher than twice the signal frequency, thereby producing a difference frequency, known as the "idler" frequency,

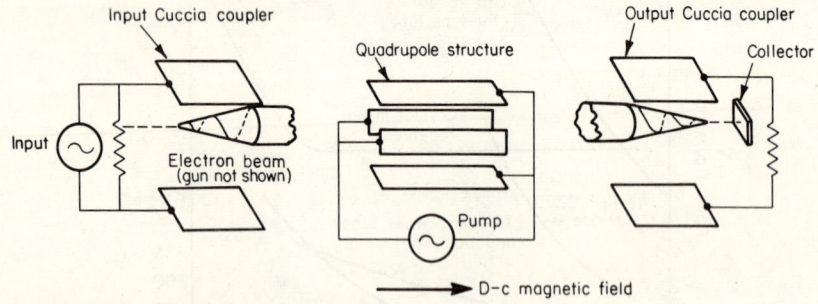

Fig. 14-60. Functional diagram of Adler tube showing the Cuccia couplers and the quadrupole structure.

whose energy is dissipated in an idler load. Lower noise temperatures can be achieved in some parametric amplifiers when the idler load is cooled to liquid-air temperature.

Parametric amplification has been achieved very successfully with electron-beam devices. The most widely known device of this type is the Adler tube,[6] which provides the necessary mechanism for nonlinear interaction of the pump and signal frequency sources with a filamentary beam of electrons. Figure 14-60 shows a functional sketch of the Adler tube. It consists of an electron beam-forming device, an input coupler, an output coupler (to produce and remove the fast cyclotron wave), an electric quadrupole coupler between them to pump the fast cyclotron wave parametrically, and a collector.

Although the Adler tube and its power supply represent more equipment than the varactor diode amplifiers, it has some advantages. Its input circuit is rugged and requires minimum protection from the transmitted pulses. Wide bandwidths (10 per cent or more) are easily obtained, and the gain is unconditionally stable.

Tunnel-diode Amplifier. The current vs. voltage characteristic of the tunnel diode has a region of negative slope. This region of negative resistance can be used to achieve amplification in the same manner as was achieved with some of the early tetrodes (dynatrons). The device is simpler to use than the varactor since no pumping source is required and no idler frequencies are present. To date, noise temperatures achievable are not as low as for parametric amplifiers.

Masers. Maser is an acronym coined by C. H. Townes:[7] Microwave Amplification by Stimulated Emission of Radiation. The amplification is obtained through the net emission of microwave energy when a substance is stimulated by a pumping source.

In the usual equilibrium conditions which exist in a solid, the absorption of energy exceeds the emission. To obtain emission of microwave energy from a solid, the usual relation between population and energy levels must be changed. The higher population which ordinarily exists at low energy levels must be transferred to a high energy level. In a maser using a paramagnetic solid (such as ruby) the types of energy levels employed are those corresponding to the several possible orientations of the magnetic moment of a spinning electron in a magnetic field. For this kind of level, alignment with the magnetic field represents a low level and misalignment corresponds to a high level. If the pumping energy is to be obtained from a microwave source, the difference in the two energy levels must correspond to the microwave source frequency.

Thermal vibrations at normal temperatures prevent good spin alignment except with very high magnetic fields or very high pump frequencies (above the microwave region). It is therefore necessary to operate the maser at very low temperatures (-270 to $-200°C$). The energy stored by this population inversion is released by the action of the incoming signal, and a net gain in energy of the signal is obtained. The ordinary types of maser are single-port devices, and a circulator is required to separate the output signal from the input.

Extremely low noise temperatures can be obtained. They are usually limited by the losses in the transmission-line components. The maser's disadvantages are the small gain-bandwidth product, susceptibility to saturation, and need for an exceedingly low temperature. This currently is an important limitation, since to date no suitable closed-cycle refrigerator unit is available. The low temperature required is achieved by immersing the paramagnetic solid in liquid helium together with its resonant circuit (Fig. 14-61).

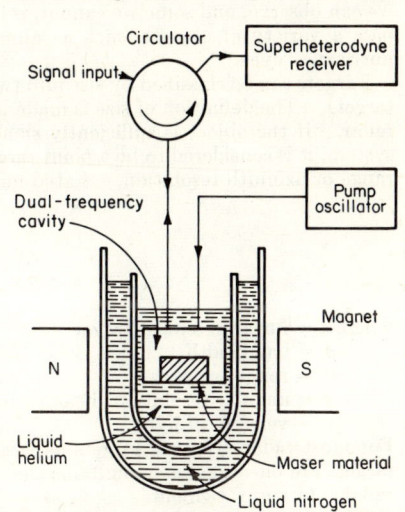

FIG. 14-61. Diagram of a three-level cavity-type maser. (*From M. I. Skolnik, Introduction to Radar Systems, McGraw-Hill Book Company, New York, 1962.*)

14-8. Radar Targets

Scattering of Electromagnetic Energy. The principles of radar are based on the ability of metallic and other objects to "reflect" electromagnetic energy, a phenomenon first noted by Hertz.[8] When a plane electromagnetic wave is intercepted by an object in free space, currents flow on the surface if it is a conductor. If it is a dielectric, atomic and molecular polarization occurs. As a result, electromagnetic energy is reradiated. It is the detection of this reradiated energy upon which radar depends. It is important to note that the reradiation may be directed not only toward the source of energy but in other directions as well.

An important function to define is the ratio of the power flux density P_r scattered from the object (target) to the incident power flux density P_2. This ratio leads to the following definition:

$$\sigma = \lim_{R \to \infty} \left(4\pi R^2 \frac{P_r}{P_2} \right) \tag{14-22}$$

where R = distance to object

The parameter σ has the dimensions of an area and is termed the "radar cross section" or "radar echoing area." The value of σ is dependent on a number of factors. Obviously, the material and its size and shape will have an effect on this quantity. The polarization and wavelength of the incident wave and the orientation of the object with respect to the source of radiation will also affect σ. In the case of bistatic observation, the value of σ is a more complicated function than in the case of monostatic observation, since the cross section may also change as a function of the angle between the transmitter and receiver. Because the orientation may be a function of time, if the target or the radar system is not stationary, σ may also be a function of time.

It should be obvious by now that any discussion of σ is apt to be a complex affair even if the shapes to be considered as targets are simple. The discussion is further complicated when a variety of shapes and materials is considered. Any object which we can observe, and some we cannot, is fair game, and therefore radar targets encompass a variety of objects, such as aircraft, men, electrons, raindrops, terrain, and hurricane "eyes."

Targets can be classified by size into two categories: point targets and extended-area targets. The definition of size is made in comparison with the resolving power of the radar. If the object is sufficiently small so as to be unresolvable by a given radar system, it is considered to be a point target. That is, its linear extent is less than the range or azimuth resolution. Stated more formally:

$$l < \frac{c}{2\beta} \quad \text{(range)}$$

$$l < R\theta \quad \text{(azimuth)}$$

where l = linear extent of target
β = bandwidth
R = range
θ = angular resolution (beamwidth)
c = velocity of light

For most radar systems, targets as large as jet liners are point targets. Extended targets are ones such as mountains that are much larger and can be resolved by the radar into many elements.

The Scattering Matrix. Except for some special cases, the reradiated energy from a target will be elliptically polarized even though the incident energy is linearly polarized. Thus it is possible to receive two orthogonal components even though one linear component is transmitted. In order to give a complete "radar" description of the target, a matrix equation is required which will state all the interrelationships. This can be written as

$$\mathbf{E}^r = \mathbf{A}\mathbf{E}^t \qquad (14\text{-}23)$$

where $\mathbf{E}^r = \begin{pmatrix} E_x^r \\ E_y^r \end{pmatrix} \quad \mathbf{E}^t = \begin{pmatrix} E_x^t \\ E_y^t \end{pmatrix}$ and $\mathbf{A} = \begin{pmatrix} a_{11} & a_{12} \\ a_{21} & a_{22} \end{pmatrix}$

$\mathbf{A}$ is called the scattering matrix. The superscripts r and t refer to reflected and transmitted energy, respectively. The subscripts x and y refer to two orthogonal polarizations. If the scattering matrix is determined for a given target as a function of aspect, then the target properties are completely described at the wavelength at which the matrix was obtained. Frequently, only the a_{11} or a_{22} coefficient is known.

Methods for Computing Cross Section. In principle, it should be possible to obtain the scattering cross section of any object through the solution of Maxwell's equations. In practice, such solutions are beset by many difficulties, not the least of these being the ability to state the boundary conditions at the surface of the scatterer. As a result of these difficulties, complete solutions have been obtained for only a few simple three-dimensional shapes such as the sphere and the cylinder. Other mathematical approximations have been used. The methods of physical optics lead to a somewhat simpler rather than a complete solution. Geometrical optics is also used, but this method, of course, gives only a "power" solution (phase is neglected). Hybrid methods have been developed to deal with complex shapes (anything more involved than a sphere is "complex"). The pioneering work in this area was accomplished by a group at the University of Michigan led by K. M. Siegel.[9]

The results obtained for backscatter from a sphere are shown in Fig. 14-62. The coordinates have been normalized. The backscattering characteristics can be con-

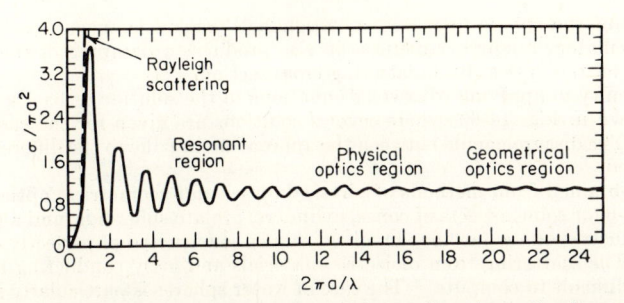

Fig. 14-62. Radar backscatter cross section σ of a sphere of radius a.

Table 14-6. Cross Sections of Simple Shapes (Size Large Compared with Wavelength) Obtained by Methods of Physical and Geometrical Optics

	Physical optics	Geometrical optics
Sphere	πa^2	πa^2
Cylinder (broadside)	$\dfrac{2\pi l^2 a}{\lambda}$	
Wire loop (normal incidence)	$\pi a^2 \dfrac{(\pi/2)^2 + [\ln(85/\gamma\pi)]^2}{(\pi/2)^2 + [\ln(\lambda/\gamma\pi b)]^2}$	a = loop radius $\gamma = 1.78$ b = wire radius
Thin circular flat plate. Incidence at angle θ to normal. a = radius of plate	$\dfrac{\pi a^2}{\tan^2 \theta}[J_1(x)]^2$ where $x = \dfrac{4\pi a}{\lambda}\sin\theta$	
Cone sphere (ice-cream cone) with continuous second derivative (axial aspect) (tip-scattering contribution only)	$\dfrac{\lambda^2}{16\pi}\tan^4 \theta_0$ $\theta_0 = \tfrac{1}{2}$ cone angle	
Prolate spheroid (axial aspect)	$\pi \dfrac{B^4}{A^2}\left[1 - \dfrac{\sin 2KA}{KA} + \dfrac{1-\cos 2KA}{2(KA)^2}\right]$ where $K = \dfrac{2\pi}{\lambda}$	$\pi\dfrac{B^4}{A^2}$ where B = semiminor axis A = semimajor axis
Paraboloid (axial aspect)		$4\pi\rho^2$ where ρ = radius of curvature of apex

veniently divided into three regions. In the first or Rayleigh region where

$$\frac{2\pi a}{\lambda} < 1$$

the cross-section function is given by

$$\sigma = \frac{144\pi^5 a^6}{\lambda^4}$$

SYSTEM COMPONENTS

As the size of the sphere increases, a "resonance" region is encountered where the result is oscillatory. Further increase in size produces a damping of the oscillation convergent on the optical backscattering cross section.

The difficulty in applying Maxwell's equations to the solution of backscatter problems is shown in Fig. 14-63, where several solutions are given for the case of a thin cylinder. The disagreements between the approaches are due to the different approximations used.

Results obtained from methods of physical optics and geometrical optics are given in Table 14-6 for some aspects of cones, cylinders, prolate spheroids, and a paraboloid.

All the illustrations so far have involved shapes constructed of perfectly conducting surfaces. The scattering from dielectric materials and lossy conducting materials is even more difficult to compute. The case of water spheres is particularly interesting, since raindrops are nearly spherical. The results obtained are shown in Fig. 14-64.

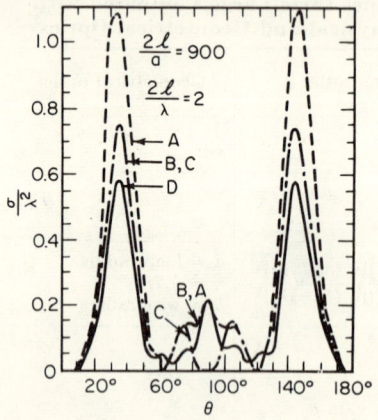

Fig. 14-63. Cylinder scattering. (a) Integral-equation method. (*von Vleck et al.*). (b) EMF method. (*von Vleck et al.*). (c) Chu resonance formula. (d) Variational method. (*Tai.*) (*von Vleck, Block, and Hamermesh, Theory of Radar Reflections from Wires or Thin Metallic Strips, J. Appl. Phys., vol. 19, p. 274, 1947. C. T. Tai, Electromagnetic Backscattering from Cylindrical Wires, J. Appl. Phys., vol. 23, p. 909, 1952.*)

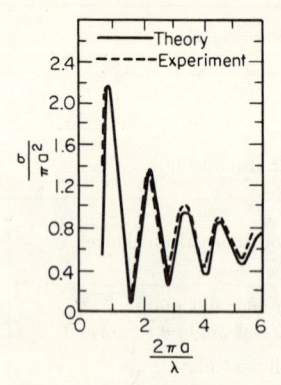

Fig. 14-64. Backscattering from water spheres. (*A. L. Aden, Electromagnetic Scattering from Spheres with Sizes Comparable to the Wavelength, J. Appl. Phys., vol. 22, pp. 601–605, May, 1951.*)

The application of coatings to surfaces to reduce their backscatter has been a subject of investigation since the early days of radar. The effort in Germany during World War II to produce radar camouflage for U-boats is mentioned in Ridenour.[10] Recently, efficient thin absorbers have been developed from ferrimagnetic insulators (ceramic ferrites).

Aircraft Targets. The aircraft is a good example of a complex radar target constructed in the main from a smooth joining of simple geometric shapes. A typical aircraft pattern is shown in Fig. 14-65. The frequency of the occurrence of maxima and minima should come as no surprise considering the overall size and the wavelength of measurement. Siegel et al.[9] have been able to obtain reasonable approximations to the envelope of such a scatterer. An example of their results is shown in Fig. 14-66. The detailed lobe structure of such patterns is sensitive to changes in wavelength and polarization. As the wavelength is increased, the fine lobe structure tends to become coarser. If the wavelength is increased sufficiently, resonance effects can be noted.

Aircraft in flight do not present a constant aspect to a radar, for either short-term or long-term observation. Relatively rapid changes in aspect occur as a result of

RADAR 14-51

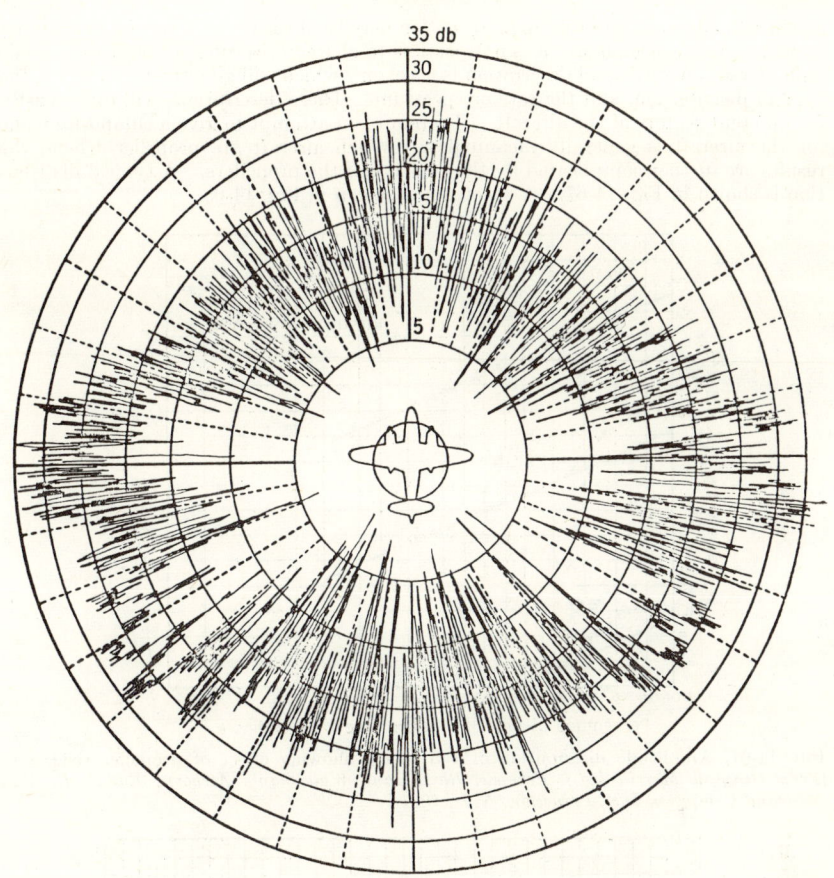

FIG. 14-65. Radar cross section of a B-26 bomber at 10-cm wavelength as a function of azimuth angle. (*From L. N. Ridenour, Radar System Engineering, McGraw-Hill Book Company, New York, 1947.*)

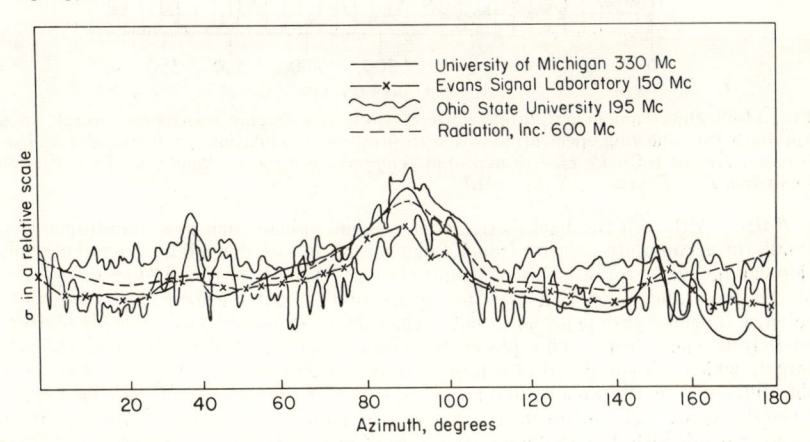

FIG. 14-66. Comparison of the theoretical and experimental cross section of the B-47 jet aircraft as a function of aspect and frequency as obtained by various investigators. (*Courtesy of K. Siegel, University of Michigan.*)

pitch, roll, and yaw, and this angular motion may be of the order of several degrees per second. As a consequence, it is not possible to characterize such a target by a single value for σ. A statistical description is necessary which will give the mean value, the type of distribution, and the rate or spectrum. These descriptions will differ vastly for different sectors of the aircraft. Therefore, an attempt to give a composite value for the aircraft is generally meaningless. When aircraft are propeller-driven, the results are further complicated by the rotation of the propellers. A typical distribution is shown in Fig. 14-67. A spectrum is shown in Fig. 14-68.

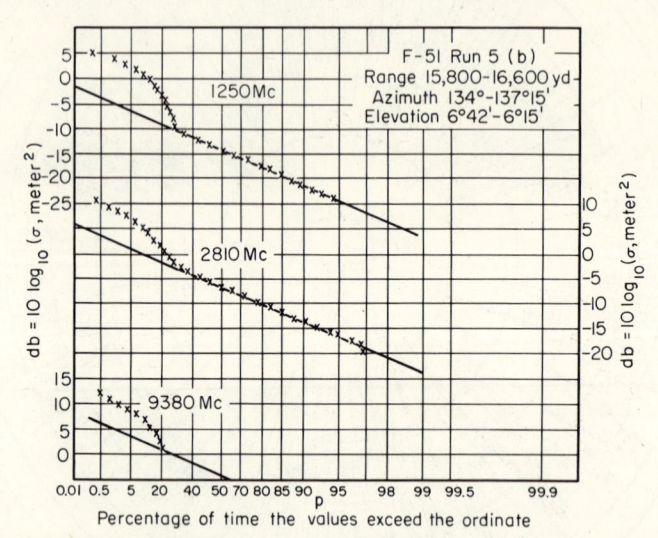

Fig. 14-67. Amplitude distribution of F-51 echo, showing effect of propeller reflection. [*From Grayson Merrill (ed.), Povejsil, Raven, and Waterman's Airborne Radar, D. Van Nostrand Company, Inc., Princeton, N.J., 1961.*]

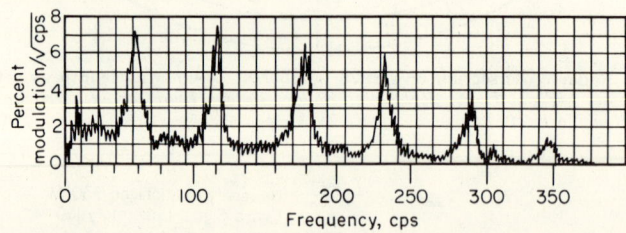

Fig. 14-68. Spectrum of amplitude noise of SNB (two-engine transport) aircraft in an approach run, showing spectral lines due to propeller modulation ($\approx$ 9,400 Mc). [*From Grayson Merrill (ed.), Povejsil, Raven, and Waterman's Airborne Radar, D. Van Nostrand Company, Inc., Princeton, N.J., 1961.*]

Rain. Although the backscatter from a small sphere such as a raindrop is very small, the accumulated scatter from a large volume of rain defined by the radar resolution can make the total backscatter quite large. Because of the random distribution of raindrops and, hence, the random phase of the contributions of the individual spheres, the net return is proportional to the sum of the power (not the field) scattered back from each drop. This power summation is characteristic of most extended targets which are composed of a large number of individual scatterers. Except for the difference in the mean fluctuation rate, the characteristics of rain backscatter resemble noise. The mean fluctuation rate of noise is generally determined by the receiver bandwidth, but the fluctuation rate of rain clutter is much slower and therefore presents more difficulty in that the ratio of signal to clutter cannot be improved by just increasing the observation time (i.e., integration).

The size of the raindrop sphere is in the Rayleigh region for most microwave radars where

$$\sigma_r \propto \frac{1}{\lambda^4}$$

Consequently, radar systems which are required to observe weather will choose short wavelengths. Radar systems which are required to avoid weather clutter will choose as long a wavelength as other factors permit. Figure 14-69 shows the cross section as a function of rainfall rate for various wavelengths.

Because of the spherical shape of a raindrop, circularly polarized transmission will have the reverse sense (or orthogonal polarization) reflected. Thus transmission of circular polarization and reception of the same sense will provide a means of rejecting rain clutter. The rejection is by no means complete, because raindrops are not perfectly spherical and because it is impossible to achieve perfect circular polarization in all sectors of the antenna beam. Because of the slow rate of fluctuation of rain clutter, attempts have been made to reject it by use of MTI techniques. Unfortunately the spectrum is sufficiently wide so the benefit from this type of filtering is small.

Reflection from the Earth. While backscatter from rain, snow, and sleet have the characteristics of volume targets, backscatter from the surface of the earth has the characteristic of an area target. The extent of this area is determined by the azimuth resolution (beam width × range) and the range resolution.

The reflection from the earth's surface is a complex function of the wavelength, polarization, angle of incidence, type of surface, and condition of the surface. It may be almost specular, as from a calm water surface, or quite diffuse, as from vegetation. However, the surface of the sea during high sea states appears as a diffuse reflector to a system using short wavelengths, and diffuse reflection from areas of heavy vegetation during the summer may turn to specular reflection during the winter.

The number of different surface reflectors is almost too large to catalog. Yet independent investigators have attacked parts of the problem. The state of knowledge of reflection from the sea's surface under a variety of circumstances has rapidly increased. Some results for backscatter as a function of wavelength and sea state are shown in Fig. 14-70. Also shown are some spectral distributions obtained in these experiments.

It may be the function of some radar systems to minimize reflection from the earth's surface (aircraft-detection systems) and the function of others to observe the surface reflection in considerable detail (bombing and navigation radar). The most widely used system for rejecting backscatter from the earth is MTI, where the earth is presumed to be motionless and the targets of interest are in motion. Unfortunately the earth's surface is not always motionless since trees wave in the breeze, water towers sway, and ocean waves propagate.

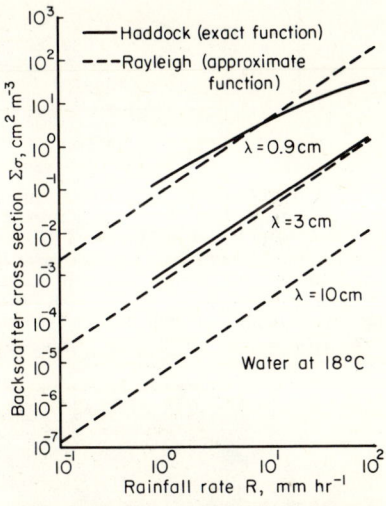

FIG. 14-69. Exact and Rayleigh approximate backscattering cross sections per unit volume of rain-filled space, plotted against rainfall intensity. (*D. Atlas et al., Air Force Surveys in Geophysics, no. 23, Geophysics Research Directorate, Cambridge, Mass., 1952.*)

14-54 SYSTEM COMPONENTS

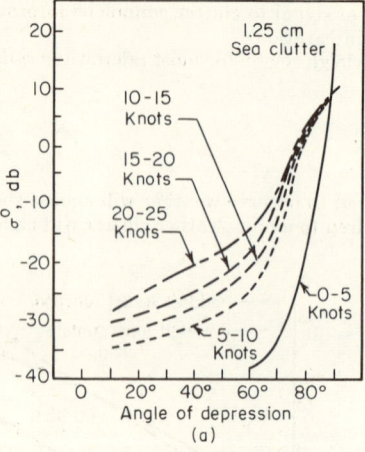

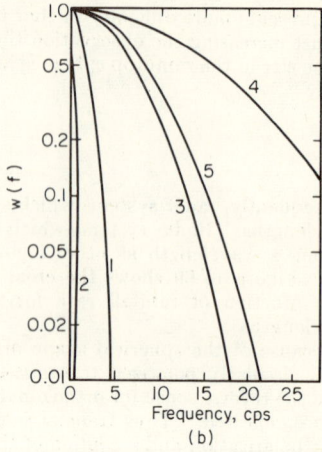

Fig. 14-70. (a) $\sigma°$ (radar cross section per unit area) as a function of wind velocity, $\lambda = 1.25$ cm. [*After Grant and Yaplee, Backscattering from Water and Land at Centimeter and Millimeter Wavelengths, Proc. IRE, vol. 45, p. 978, July, 1957, in Grayson Merrill (ed.), Povejsil, Raven, and Waterman's Airborne Radar, D. Van Nostrand Company, Inc., Princeton, N.J., 1961.*] (b) Power spectra of various clutter targets. (1) Heavily wooded hills, 20 mph wind blowing. (2) Sparsely wooded hills, calm day. (3) Sea echo, windy day. (4) Rain clouds. (5) Chaff. $w(f)$ = clutter echo power. (*E. J. Barlow, Doppler Radar, Proc. IRE, vol. 37, pp. 340–355, April, 1949.*)

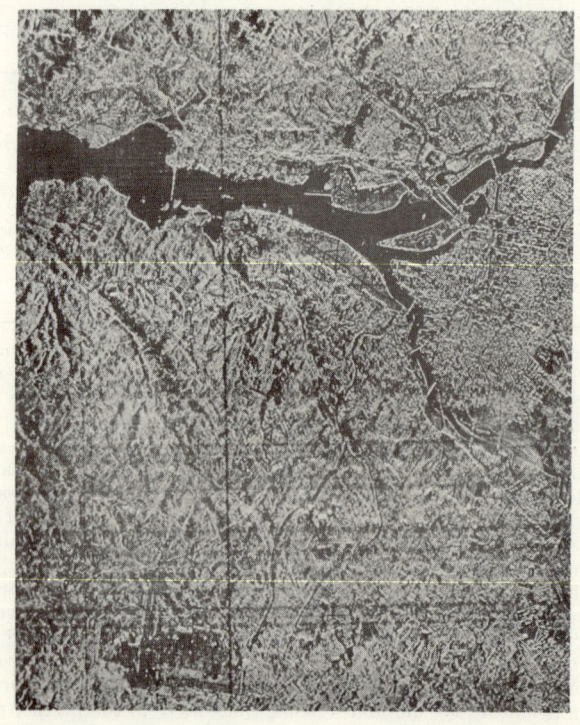

Fig. 14-71. AN/UPD-1 map of Washington. (*L. J. Cutrona, W. E. Vivian, E. N. Leith, and G. O. Hall, A High-resolution Radar Combat-surveillance System, IRE Trans. Military Electron., MIL-5, no. 2, April, 1961.*)

Reflections from the earth's surface do have recognizable characteristics because of their very complexity. As was demonstrated during World War II, observation of the ground by airborne radar can produce very useful maps of the earth surface which trained observers can interpret almost as well as photographic images. As the resolution of the mapping radar is improved, the results become more nearly photographic, with the added advantage that they may be obtained in night or day and, depending on the wavelength of operation, even during poor weather. A great deal of pioneering work in fine-resolution mapping systems has been done at the University of Michigan.[11] A typical result, obtained by a fine-resolution side-looking system, is shown in Fig. 14-71.

Reflection from a Smooth Earth. Some parts of the earth's surface behave almost like perfect reflectors under some conditions of weather and radar wavelength. Typical instances are flat grass meadows, stretches of sandy beach, and calm water. Propagation over such surfaces can have an important effect on the radar performance. A detailed treatment of such effects, taking into account variation of conductivity with grazing angle and dispersion due to the curvature of the earth's surface, is given in Ref. 12.

The results from the simple case of propagation over a flat perfectly reflecting plane are given here. This simple case lets the reflecting plane be replaced by an image point as far below the earth surface as the radar transmitter is above. When the vertical directivity of the system is low, the familiar interference pattern will be obtained.

A typical elevation coverage obtained over a smooth earth is shown in Fig. 14-72. The regions of constructive and destructive interference may be noted. This phenomenon has been used to advantage with shipborne detection radar at wavelengths of about 1 m. The first maximum occurs at an angle of approximately $\lambda/4h$ and having a width of approximately $\lambda/4h$. Thus a lower lobe of about 0.4° width and elevation is obtained which is suitable for detection of targets. Improvement in range performance by a factor of 2 is obtained because of the constructive interference. The successive maxima and minima, however, provide interrupted tracking which can be harmful to surveillance systems.

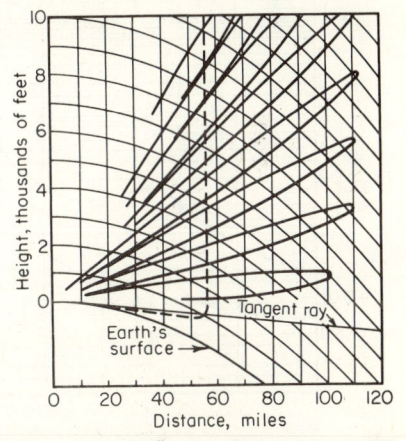

Fig. 14-72. Coverage diagram for 2,600 Mc/sec transmitter (height 120 ft). Solid curve for totally reflecting earth. Dotted curve for nonreflecting earth. (*From L. N. Ridenour, Radar System Engineering, McGraw-Hill Book Company, New York, 1947.*)

The use of higher frequencies reduces the reflection coefficient of the earth as does using vertical polarization. The directivity possible with microwave antennas permits the vertical pattern to be pointed upward sufficiently to reduce the earth's interference without seriously affecting the energy directed toward the airborne target. Search systems usually operate with the vertical beam tilted upward from the horizon by one-half beamwidth. Tracking systems generally do not attempt to operate within a beamwidth of the earth's surface.

14-9. Influence of the Atmospheric Medium

On their round trip between the radar and the target, electromagnetic waves are modified in a number of ways by the atmospheric path, especially if the path passes through the ionized regions of the upper atmosphere. The principal effects are due to absorption, refraction, and decoherence.

Absorption and Backscatter in the Lower Atmosphere. The absorptive effect of various atmospheric obstructions is shown in Fig. 14-73 in a rough, qualitative way. The lower bound at about 1 cm is imposed by atmospheric gases, mainly oxygen and water vapor. The density of these gases is a function of altitude; so the attenuation is a function not only of wavelength but also of the particular path through the atmosphere. This is illustrated in Fig. 14-73. Blake[13] has published a set of curves which are quite useful in computing the total path attenuation as a function of wavelength, elevation angle, and range.

In addition to attenuation by the normal atmosphere, additional attenuation and backscattering are introduced by rain and snow. Figure 14-74 shows the effect of

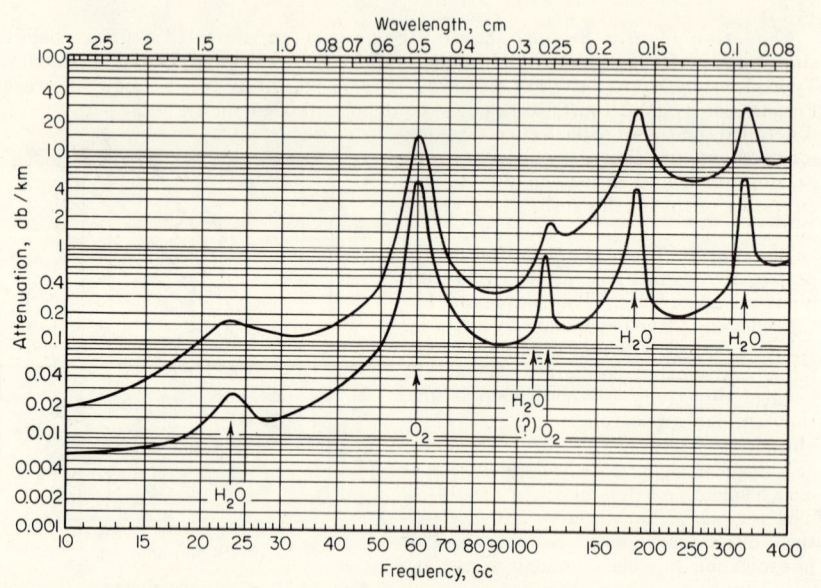

FIG. 14-73. Attenuation per kilometer for horizontal propagation.
Upper curve: "Sea level"

$$P = 760 \text{ mm Hg}$$
$$T = 20°C$$
$$\rho H_2O = 7.5 \text{ g/cu m}$$

Lower curve: "4 Km"

$$T = 0°C$$
$$\rho H_2O = 1.0 \text{ g/cu m}$$

(E. S. Rosenblum, Atmospheric Absorption of 10–400 Gc Radiation: Summary and Bibliography to 1961, Microwave J., March, 1961.)

attenuation vs. wavelength. The severe backscatter and attenuation effects of rainfall at wavelengths below 3 cm have limited the employment of these frequencies (within the atmosphere) to short path lengths, except in the special case of weather-detection devices.

Refraction in the Lower Atmosphere. The effect of refraction in the lower atmosphere is to produce retardation and bending so as to produce errors in measured directions and ranges of targets. The presence of small-scale turbulences and other inhomogeneities produces relatively rapid amplitude, phase, and angular scintillations which result in "tracking noise."

In order to provide a useful correction for the effects of atmospheric refraction on radio-wave propagation, models of the mean variation of the refractive index as a function of altitude have been proposed. In 1933, Schelleng, Burrows, and Ferrell[14] proposed a model in which the earth with its real radius a_0 and surrounding atmosphere is replaced by a fictitious earth of radius $\frac{4}{3}a_0$ surrounded by a vacuum. With this simplification, simple geometric methods may be used to calculate ranges and elevation angles.

The "linear" refractive index proposed by Schelleng et al. is correct only at lower altitudes and even then depends strongly on weather conditions—especially temperature. An atmospheric model proposed by Bauer et al.,[15] which is known as the exponential model, seems to fit over a greater range. The bending effects produced by this model are shown in Fig. 14-75 and the corresponding retardation effects in Fig. 14-76. Both figures are for the month of April. Because of the variability of average index profiles from one region to another, no single model may be expected to yield accurate predictions for all seasons and locations.

Sharp changes in temperature or in water-vapor content sometimes produce an unusually rapid decrease of the refractive index with height. Excessive bending of the rays then occurs, and it is possible for a ray to bend around the earth without leaving the surface for considerable distances. Such phenomena lead to propagation anomalies or "ducting."

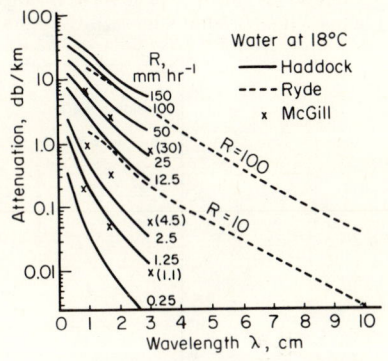

Fig. 14-74. Attenuation vs. wavelength for rainfall of various rates. (*D. Atlas et al., Air Force Surveys in Geophysics, no. 23, Geophysics Research Directorate, Cambridge, Mass., 1952.*)

Propagation through the Ionosphere at Microwave Frequencies. The major effects on radio propagation through the ionosphere occur at frequencies below 1,000 Mc. Although the ionosphere does not act as a reflecting layer at frequencies above roughly 30 Mc, it does refract, retard, and otherwise modify the electromagnetic waves propagated through it.

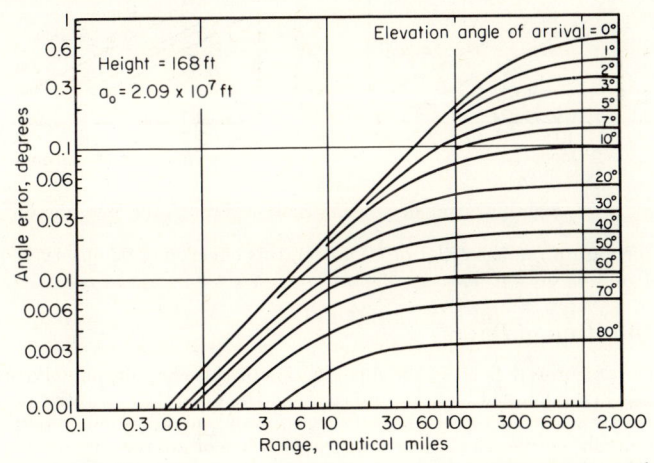

Fig. 14-75. Elevation-angle correction vs. range for exponential model.

For frequencies above 100 Mc, the attenuation due to passage through the ionosphere is small, even for long paths at low elevation angles. Above a few hundred megacycles, ionosphere attenuation is negligible, but there can be appreciable bending and retardation. These effects are shown in Fig. 14-77 as a function of frequency for daytime conditions and a horizontal beam (these conditions give a maximum error).

Dispersion and other phase effects can lead to severe problems with wide-band coherent radars which must operate through the ionosphere, even at gigacycle frequencies.

Other Atmospheric Effects. The Faraday effect produces a rotation of the polarization of the electromagnetic wave which diminishes with increasing frequency. This rotation is additive for the two passages of a radar signal through the ionosphere.

Significant auroral backscatter can occur at frequencies below 1,000 Mc. This phenomenon becomes more intense as the frequency is decreased. Because of the

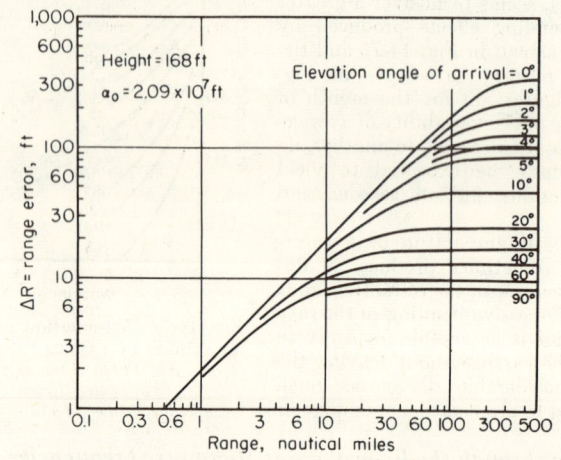

FIG. 14-76. Range correction vs. range for exponential model.

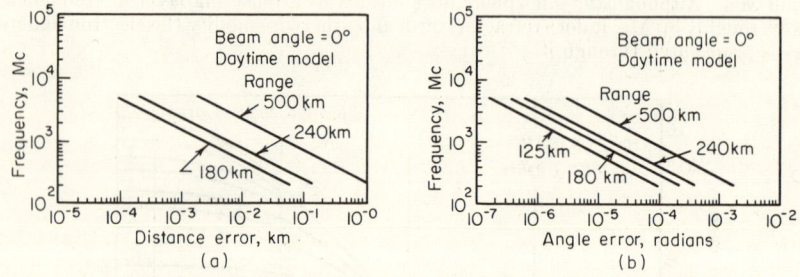

FIG. 14-77. Sighting errors due to propagation through the ionosphere.

height of the aurora (on the order of 60 miles), this effect may be observed over considerable distances (on the order of 500 miles).

14-10. Utilization of Data

The basic categories defined in the introduction of this chapter, namely, searching, tracking, and mapping, will be used in discussing the utilization of radar data.

Search Systems. The most rudimentary form of search system is an alarm system which merely notifies the observer of the presence of a target in a sector. If the system covers an extensive volume of space, unless the traffic is exceedingly small, some means must be provided to distinguish newly entering targets from old. This implies maintenance of some form of position identification associated with a track (radar parlance for a plot of past and current position of a target). This leads to a more advanced surveillance system in which the detection of a new target is followed by maintenance of a complete track history while the target is in view of the radar.

In the general form of a search system, a solid angle of space is scanned with a beam whose angular dimensions may be considerably smaller than the search angle. An interruption occurs in the track history of each target for a duration equal to the time necessary to scan the complete solid angle. Memory and association processes are necessary to bridge this gap.

There are three important time periods in a pulse radar system, each requiring a memory of different duration. First is the duration of the pulse length, often on the order of microseconds. The simple matched filter supplies this memory. Second is the time the antenna beam moves an angular distance of one beamwidth, often on the order of milliseconds, and usually allowing several "hits" on the target (i.e., pulses). This memory is supplied again by a more sophisticated matched filter of the type described in Sec. 14-4. Third is the time interval between successive scans, often on the order of seconds. Because of the length of time and the need for handling a large number of targets, bridging this interval has been difficult. For this reason the human operator performed this "track-while-scan" function for many years until the necessary electronic techniques were developed.

If target motion is small during the interval between scans and, more specifically, if target acceleration can produce only minor deviation from a linear path during this scan interval, then the process of association of position data from one scan to the next to form a continuous track history is simple. For this condition, a velocity estimate based on considerable past history (i.e., heavily smoothed linear prediction) can be used to predict future position. Several factors can combine to increase the difficulty of maintaining track history. Poor signal-to-noise ratio, target scintillation, or masking by clutter can cause loss of position data. For linear tracks, long-term prediction can bridge these gaps. If the target is capable of acceleration which can produce considerable deviation from a linear course during this interval, maintenance of track is more difficult. If, moreover, a number of other targets are in the vicinity, then confusion of tracks may occur.

In the earlier days of radar history, target speeds were slow and traffic density was low. The data output of the radar was presented on an A scope (an oscilloscope presentation where the horizontal deflection represents range and vertical deflection represents amplitude). A great deal of information is presented for operator judgment, making this a very sensitive device for detection and discrimination against various forms of interference. Track plots were maintained on a plotting board by a second operator who received the range data from the A-scope observer and coordinated this with antenna azimuth angle to produce a position plot.

This technique did not long survive the increasing speed and acceleration of aircraft. Two displays were developed to cope with this. The PPI (plan position indicator) displays a polar plot of range and azimuth on the face of a cathode-ray tube. Signal amplitude is used as an intensity modulation; i.e., bright spots appear at the location of targets. The B-scan display is similar except that range is displayed on the vertical axis and azimuth on the horizontal axis.

As air-traffic capacities increased, it became necessary to increase the capacity beyond the ability of single operators. The two-dimensional space (range and azimuth) was broken into sectors, each assigned to an operating team. The data were consolidated on a single plotting board.

Automatic maintenance of track history without human intervention obviously was desirable. One difficulty was that the data output was, on occasion, considerably corrupted by noise, interference, or clutter and required considerable interpretive judgment. Another was that long-term memories were difficult to achieve by available techniques.

With the development of low-noise systems and sophisticated filters, noise, clutter, and interference are not such a serious problem. The digital computer is an excellent device for position memory and association and can perform some of the adaptive tasks of human operators. As a result, many radars have been designed in recent years which no longer use human operators, except as supervisors having veto or override capability. The computer display also permits insertion of other data which may be associated with each track, such as altitude and target identification, permitting a rather complete target record.

In the meantime, improvement has also taken place in operator displays. The flicker rate on a PPI or B scope is troublesome to a human observer at rotation rates less than 30 rpm because of the short memory time. A longer-persistence phosphor and a storage-tube memory device preceding the display have helped. Both fall short of the desired result because they accumulate noise and clutter history as well. If the output of a filter matched for the entire time on target is presented on the display, then "noise" appears only when a false report comes through the decision mechanism. If the decision level is set for a low false-alarm rate, then a "clean" display is obtained consisting mainly of true target positions. Long-term memory can now be introduced with considerable success to create a track history equivalent to that produced by a computer display.

Tracking Systems. The data in a tracking system are generally used for two purposes: reinsertion in the tracking servomechanism loop to maintain the track, and insertion into a computer for control (of an interceptor or other device). Because a tracking radar devotes its attention to one target, the data rate is very high, and very good tracking performance can be achieved giving highly accurate position, velocity, and acceleration data.

Three servo loops may be used in a tracking system: an angle-tracking loop, a range-tracking loop, and a velocity-tracking loop. The memories from all three loops are combined to produce the accurate data mentioned above. In some systems, such as a homing missile, no display of the data is made at all. The data are supplied only to the computer in the missile. In any event, the data displays are relatively simple since data on only one target are displayed.

14-11. Radar-system Sensitivity

Derivation of the Canonic Search Equation. It has previously been shown that the factor which uniquely determines the probability of detection and the false-alarm rate is the ratio of the received signal energy to the noise power per cycle ($E/2N_0$). A solution for signal-to-noise ratio ($E/2N_0$) in terms of basic radar-system parameters will provide a measure of system sensitivity which is general and applicable for any of the wide variety of signal forms. As has been stated in Sec. 14-2, one of the numerous limitations on system sensitivity is imposed by the limited time available for observation. The solution for the sensitivity of a radar search system illustrates this aspect of the detection problem very well. The solution also may easily be reduced to other radar-system-sensitivity cases because of its generality.

In order to obtain a basic form, the transmission path will be assumed to be ideal and the target will be assumed to be small compared with the resolution, large compared with a wavelength, stationary, and nonfluctuating. A simple form of search system will be considered. Using a spherical coordinate system to define the geometry, a volume of space is defined by a range interval R and a solid angle Ω_s. The radar is located at the center of the coordinate system and is required to search this volume in a time T_s. It scans through this volume of space with a beam of solid angle Ω whose dimension in steradians is small compared with the angle to be searched Ω_s. The radar will search all ranges from 0 to R. The geometry of the problem is given in Fig. 14-78. Simplifying assumptions will also be made about the beam shape and beam motion, to eliminate the necessity for considering antenna scanning losses. A uniform distribution of power over the beam angle Ω is assumed, and the shape of the beam is assumed to be rectangular. The motion of the beam in scanning is assumed to take place in azimuth or elevation in the form of discrete, instantaneous angular displacements of one beamwidth. Later we shall consider how to expand this form to include some of the problems engendered by more complex situations.

The simplified radar located at the center of the coordinate system is assumed to have its antenna connected to a source of power $P_T(t)$. During the interval T_T that the beam remains stationary, the energy generated by this source of power is

$$E_T = \int_0^{T_T} P_T(t)\,dt \qquad (14\text{-}24)$$

RADAR

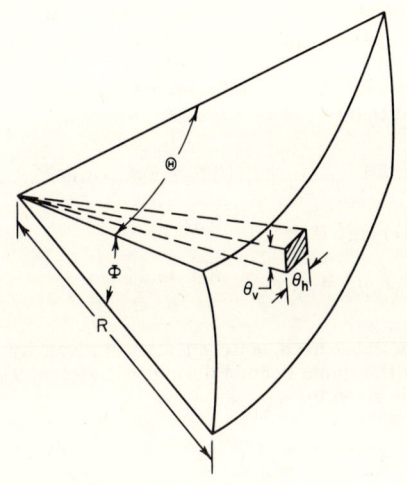

FIG. 14-78. The search-radar geometry.

Ω_s = solid angle to be scanned
$\Omega_s = \Theta \sin \phi$

where Θ = azimuth angle to be scanned
where ϕ = elevation angle to be scanned

$\Omega_s \approx \Theta\phi$, when $\phi \leq \pi/10$
R = distance to the target
Ω = solid angle of the scanning beam
$\Omega = \theta_h \sin \theta_v$

where θ_h = azimuth beamwidth
where θ_v = elevation beamwidth

$\Omega \approx \theta_h \theta_v$ when $\theta_v \leq \pi/10$
$\theta_h \leq \Theta, \theta_v \leq \phi$

The area A_R illuminated by the transmitting beam at range R is

$$A_R = R^2\Omega \tag{14-25}$$

The energy density at target range R is

$$\text{Energy density} = \frac{K_1 \int_0^{T_T} P_T(t)dt}{R^2\Omega} \tag{14-26}$$

where K_1 = unity for a perfect antenna beam
$K_1 < 1$ when side lobes exist

For the simple target assumed, the scattering cross-section area can be defined by a single value σ. The energy reflected from the target is then

$$\text{Reflected energy} = \frac{K_1\sigma}{R^2\Omega} \int_0^{T_T} P_T(t)dt \tag{14-27}$$

In the simple case being treated, the receiving antenna is assumed to be co-located with the transmitting antenna and to have the same effective area $A\epsilon$. (ϵ is an aperture-utilization factor.) The total energy received is then given by

$$E_R = \frac{K_1\epsilon}{4\pi} \frac{A\sigma}{R^4} \frac{1}{\Omega} \int_0^{T_T} P_T(t)dt \tag{14-28}$$

Returning to the geometry of Fig. 14-78, it is noted that

$$\frac{1}{\Omega} = \frac{T_s}{\Omega_s}\frac{1}{T_T} \qquad (14\text{-}29)$$

It is also to be noted that

$$\frac{1}{T_T}\int_0^{T_T} P_T(t)dt = \text{avg power } P_{\text{av}} \qquad (14\text{-}30)$$

so that the received energy is

$$E_R = \frac{K_1\epsilon}{4\pi}\frac{A\sigma}{R^4}\frac{T_s}{\Omega_s}P_{\text{av}} \qquad (14\text{-}31)$$

The signal-to-noise ratio for a matched filter is given by $E/2N_0$ where the filter must be matched to the signal defined during the interval T_T. In the radar system being evaluated N_0 is given by

$$N_0 = k\frac{T}{2} \qquad (14\text{-}32)$$

where k = Boltzmann's constant (1.38 × 10^{-23} joules/°K)
T = system noise temperature, °K
Finally,

$$\frac{E_R}{2N_0} = \frac{S}{N} = \frac{K_1\epsilon}{4\pi k}\frac{P_{\text{av}}A}{T}\frac{\sigma}{R^4}\frac{T_s}{\Omega_s} \qquad (14\text{-}33)$$

This basic search-equation form gives the S/N (on which the probability of detection P_D and false alarm P_F are based) in terms of some very simple performance requirements and some very simple system-component requirements.

In this simple form of the search equation, neither angular resolution nor range resolution appears explicitly. The sensitivity of the search system is independent of these parameters. It is also important to note that no wavelength preference is expressed explicitly either.

Limitation on Scanning Rate. For the simple single-beam scanning-pulse system being considered, the time between pulses T_r must be equal to or greater than the time required for the transmitted signal to travel to a target at the maximum range of interest and to return to the receiver.

$$T_R \gtrsim \frac{2R_m}{c} + \tau \qquad (14\text{-}34)$$

where R_m = range to most distant target
c = velocity of light
τ = duration of signal

Generally the duration τ is small compared with $2R_m/c$; so it can be neglected. It is frequently desirable to obtain several observations in a particular direction before proceeding to the next angular position. The total time on target T_T for a multiplicity of observations is

$$T_T = NT_R = N\frac{2R_m}{c} \quad \text{if} \quad T_R = \frac{2R_m}{c} \qquad (14\text{-}35)$$

where N = number of hits
The time T_s taken to scan the solid angle Ω_s is

$$T_s = NT_R\frac{\Omega_s}{\Omega} \qquad (14\text{-}36)$$

RADAR

There is a constraint on the minimum value of the solid beam angle Ω in order to obtain at least N hits per scan at a given scanning rate Ω_s/T_s. This is

$$\Omega \geq NT_R \frac{\Omega_s}{T_s} = \frac{2NR}{c}\frac{\Omega_s}{T_s} \tag{14-37}$$

There is a wide latitude in the beam shape since the constraint is imposed only on the solid angle (which for small angles is the product $\theta_h \theta_v$).

If a scanning rate or angular resolution is required greater than permissible in terms of the above constraint, then parallel systems may be employed. The "stacked" beam system and the "pincushion" systems are variants of the parallel-beam approach. The sensitivity of a stacked beam system is the same as that of a single-beam system for the same total system power and *receiving* aperture.

System Losses. The sensitivity of the search system was determined with the deliberate assumption of idealized conditions. The difference between the ideal condition and the practically achievable condition must be inserted in the sensitivity determination in order to obtain an accurate estimate of performance. This difference is attributed to a "loss," and the insertion in the calculation is made in the form of a multiplying "loss factor." These "losses" derive from a number of sources. Typically, these losses may be due to:

1. Imperfect transmission media (with attenuation) (Sec. 14-9)
2. Imperfectly matched filters (Sec. 14-4)
3. Fluctuating targets (Sec. 14-8)
4. Imperfect transmission lines (with attenuation)
5. Imperfect antennas
6. Scanning motion other than step scan

The losses due to the first three items have been treated above. The losses due to attenuation in the transmission-line path are simply accounted for by using the actual power delivered to the antenna by the transmitter rather than the power output of the transmitting tube. The losses due to the attenuation in the transmission line from the antenna to the receiver are simply accounted for by computing the reduction in signal power delivered to the receiver terminals.

Unless an antenna aperture of infinite length is used, it is impossible to achieve the perfect beam shape previously assumed. Apertures of ordinary extent produce beam shapes which are closely approximated in the major lobe by a Gaussian shape. When such a beam is scanned across a target, a sequence of signals which are not all of the same amplitude is received. The

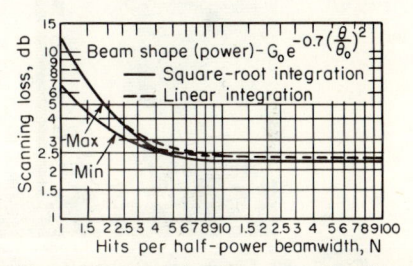

FIG. 14-79. Scanning losses as a function of N at maximum range. (*Courtesy of Sperry Gyroscope Company.*)

corresponding loss in sensitivity is a function of the shape of the pattern and the rate of scanning. It also is dependent on the time phasing between the transmitted pulse and the relative target angular direction. A plot of scanning losses is shown in Fig. 14-79.

The Tracking or Nonscanning Form of the Search Equation. In the tracking form of the search equation, which is a limiting form, the search angle Ω_s becomes equal to the beam angle Ω. The time of scan T_s becomes equal to the time on target T_T. This particular time must be given special interpretation, since it might otherwise appear that it is infinite. For the tracking case, T_T represents the integration time or memory time of the tracking system. Consequently, the tracking form of the search equation can be shown to be

$$\frac{S}{N} = \frac{\epsilon^2}{4\pi k}\frac{P_{av}}{T}\frac{A^2}{\lambda^2}\frac{\sigma}{R^4}T_T \tag{14-38}$$

14-12. Typical Radar Systems

Surveillance Radar. An experimental radar for search and tracking of aircraft was developed at Lincoln Laboratory in 1956. In order to obtain high sensitivity, a large average power (30 Kw) and aperture were employed. The operating frequency was chosen in the uhf region to avoid sensitivity to weather clutter and because it is easier to generate the requisite high average power at uhf than at shorter wavelengths. It was also possible to obtain elevation coverage with a simple cosecant-squared beam up to an angle of 30° and yet utilize a large vertical aperture. The

Fig. 14-80. Experimental aircraft-surveillance radar. (*Lincoln Laboratory, MIT.*)

dimensions of the antenna were 120 by 30 ft, and it rotated at 5 rpm. The azimuth beamwidth was 1.5° and the antenna gain 32 db. The antenna pedestal was mounted on a concrete tower 100 ft high which served the dual purpose of supporting the antenna and providing a building to house the radar equipment. A picture of this structure is shown in Fig. 14-80.

A block diagram of the transmitter-receiver system is shown in Fig. 14-81. The controlling frequencies for the transmitter and receiver were generated in a common source. The multiplier chain, used to derive the transmitter and local oscillator frequencies, was broadband so that by simply switching the master crystal-controlled oscillators, it was possible to change frequencies rapidly without retuning. The final power amplifier developed for this purpose, the VA-812, was a multicavity klystron capable of generating 8 Mw peak power at a pulse width of 6 μsec. Since the radar was constructed for experimental purposes, many varieties of MTI systems and data-processing systems were tested at this site. The size of the antenna system precluded the possibility of measuring antenna patterns in the usual fashion; so techniques were developed at this site to measure elevation patterns using the sun as a noise source.

RADAR **14-65**

In contrast to this large, sensitive but simple search radar, it is interesting to compare the ASDE system, a surveillance radar for taxi control at airports. Since the range required is short, the operating wavelength can be in the millimeter region, permitting fine angular resolution to be obtained with a relatively small antenna. A very short pulse is used to give fine range resolution. A picture of the radar is shown in Fig. 14-82, and a typical display is shown in Fig. 14-83.

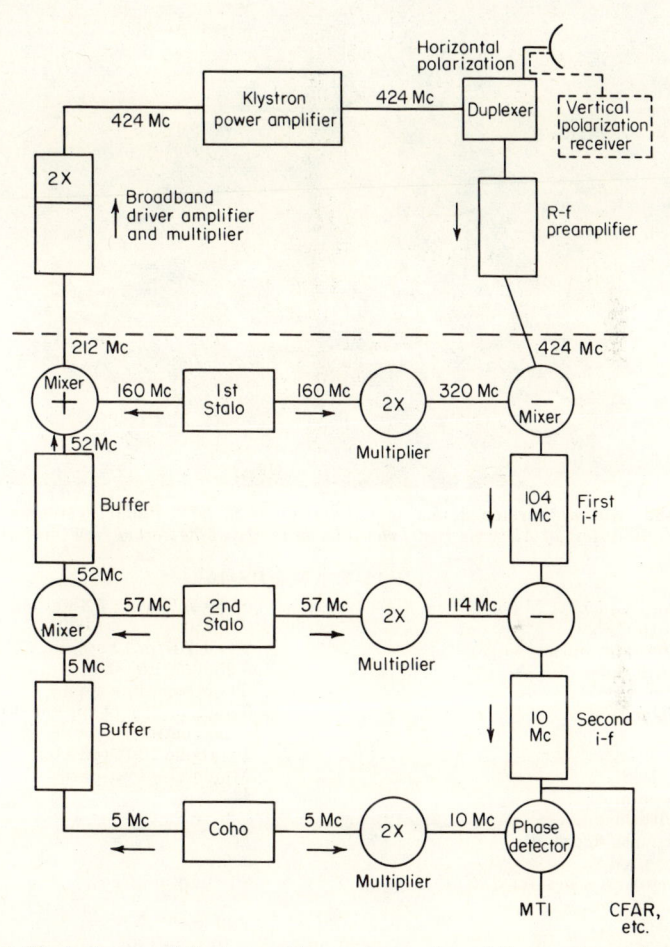

FIG. 14-81. Block diagram of the transmitter-receiver system of the experimental uhf radar.

Tracking Radar. An example of current design practice in tracking radar systems is illustrated by the AN/FPS-16 radar set. It was developed primarily for instrumentation usage on test ranges. It can track on either target backscatter or beacon returns, and the transmitter has a number of possible combinations of pulse width and pulse repetition rate. The salient characteristics of the system are given in Table 14-7, and a block diagram of the system is shown in Fig. 14-84. The antenna is a 12-ft parabola fed by a four-horn monopulse feed. High angular accuracy and tracking may be achieved, as indicated in the table. A picture of the antenna and pedestal is shown in Fig. 14-85. The many modes of operation are controlled from a console pictured in Fig. 14-86.

14-66 SYSTEM COMPONENTS

Fig. 14-82. Airport surface detection equipment (ASDE). Radar system at Newark Airport. (*Courtesy of Airborne Instruments Laboratory and the Port of New York Authority.*)

Table 14-7. AN/FPS-16

Operating frequency, Mc	5,450–5,825 5,480
Power output, Kw	250 1,000
Repetition rate, pulses/sec	285–1,707 in 12 steps
Pulse width, μsec	0.25, 0.5, 1.0
Modulator type	Thyratron, line type
Transmitter type	Tunable magnetron and fixed tuned magnetron
Antenna	Parabolic 12-ft reflector
Antenna feed	Monopulse four-horn
Antenna gain, db	44
Beamwidth, deg	1.2
Receiver noise figure, db	11
Max range, yd	520,000
Range, nautical miles, for S/N of 12 db	80 (1-sq-m target)
Angle limits, deg:	
Azimuth	Full 360°
Elevation	-10 to $+190°$
Angle tracking rates, mils/sec:	
Azimuth	750
Elevation	400
Angle accuracy, mil	0.1
Range accuracy, yd	5
Data outputs for range, azimuth, and elevation	Digital, potentiometer, and synchro

Mapping Radar. Many of the mapping-radar systems differ but little from surveillance-radar systems in principle. The primary concern of mapping radars is to develop a maplike or photographlike picture of a terrain, seacoast, clouds, etc. The backscatter from extended targets will vary in its magnitude and polarization depending on the nature of the surface. The contrast between types of surfaces will frequently give a maplike appearance to a radar picture of the backscatter if the radar system has sufficient resolution. In order to obtain high resolution, mapping radars are generally designed to operate at the higher frequencies (i.e., at 8,000 Mc and above) and with short pulses (0.1 μsec and less).

One form of mapping radar has been used as an adjunct to aerial photography in developing strip maps. In this method, radar observation is made of only a thin angular sector orthogonal to the path of the aircraft motion. The return from each sector is recorded by photographic techniques. The sectors can be connected to form a strip-map record of the terrain on each side of the aircraft's path. If an antenna is

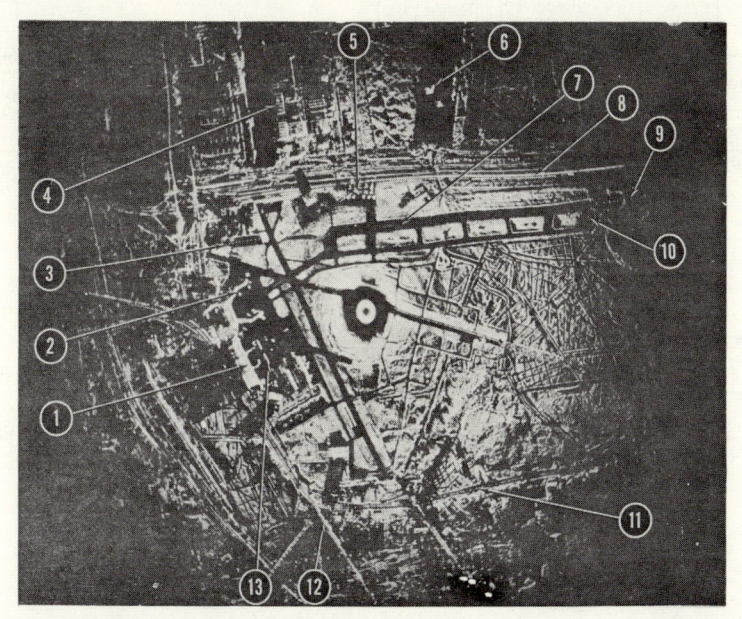

FIG. 14-83. Detailed radar image of a circle $3\frac{1}{2}$ miles in diameter at Newark Airport visible on the control-tower screen shows how the new system clearly outlines the surface features of an airfield and the area around it. The operator can see (1) terminal building; (2) parked commercial aircraft; (3) runway approach lights; (4) Port Newark berths; (5) parked military aircraft; (6) barge under tow; (7) track of landing plane; (8) New Jersey Turnpike; (9) runway approach lights; (10) aircraft near end of runway awaiting take-off; (11) U.S. Routes 1 and 9 with roadway lights showing as dots; (12) U.S. Route 22; (13) tracks of vehicles moving on the field apron. (*Courtesy of Airborne Instruments Laboratory and the Port of New York Authority.*)

constructed along the fuselage of an aircraft, the beam will be orthogonal to the path of aircraft flight. The range resolution can be made quite good by using very short pulses or pulse-compression techniques ("chirp"). A beam whose linear width is small and independent of range would be most desirable for mapping.

Radar set AN/UPD-1 (XPM-1)* is an airborne-radar mapping system. This system, by synthesizing an extremely long antenna which expands in length in direct proportion to radar range, provides a linear resolution in the azimuth direction that is constant for all radar ranges. The fundamental idea is illustrated in Fig. 14-87. In part *a* of this figure an ordinary linear array is illustrated. The transmission lines are designed so that the signal from each dipole arrives in phase at the main transmission line producing a narrow beam. In part *b* of this figure, the genera-

---

* This radar set was developed at the Institute of Science and Technology of the University of Michigan and is described in L. S. Cutrona, W. E. Vivian, E. N. Leith, and G. O. Hall, A High-resolution Radar Combat-surveillance System, *IRE Trans. Military Electron.*, April, 1961.[11] The information covering this system was obtained from this paper.

14-68 SYSTEM COMPONENTS

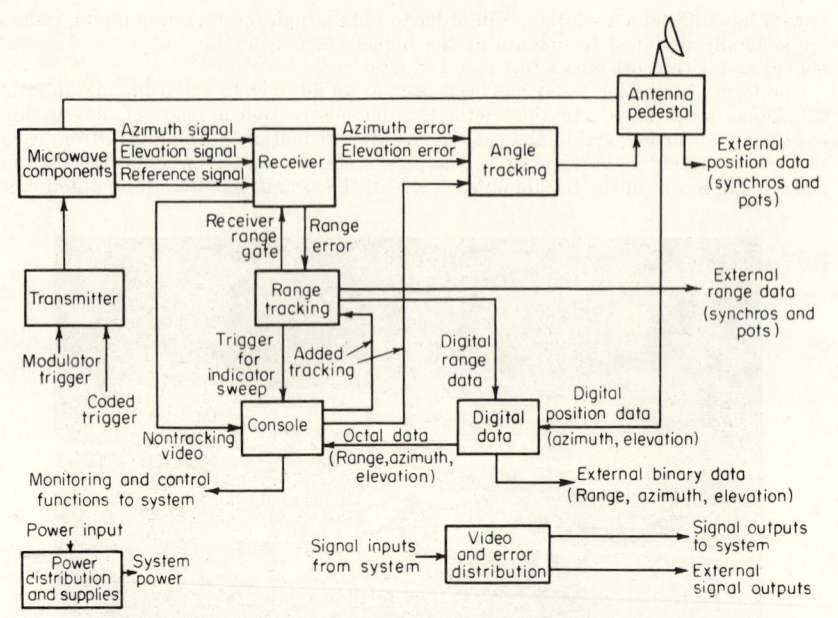

Fig. 14-84. Radar set AN/FPS-16 block diagram. [*Courtesy of Radio Corporation of America, Missile and Service Radar Division, Moorestown, N.J.*]

Fig. 14-85. Antenna pedestal AN/FPS-16. [*Courtesy of Radio Corporation of America, Missile and Service Radar Division, Moorestown, N.J.*]

RADAR 14-69

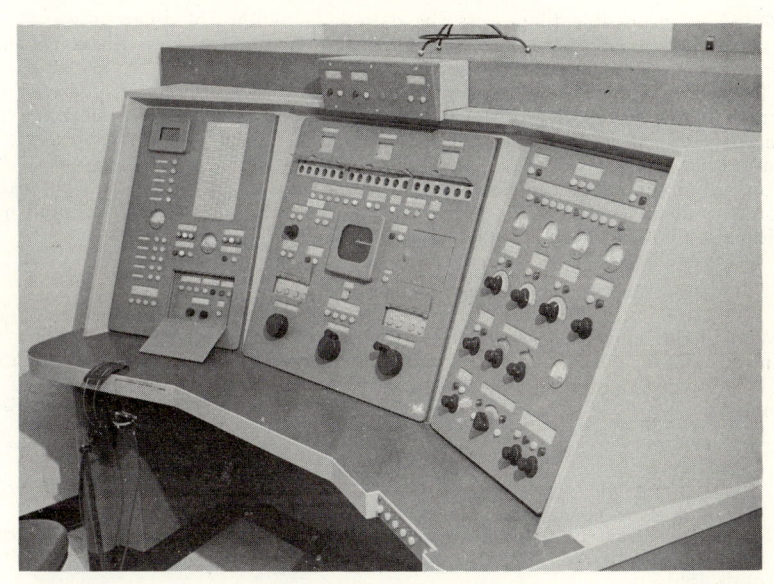

Fig. 14-86. Radar operator console. [*Courtesy of Radio Corporation of America, Missile and Service Radar Division, Moorestown, N.J.*]

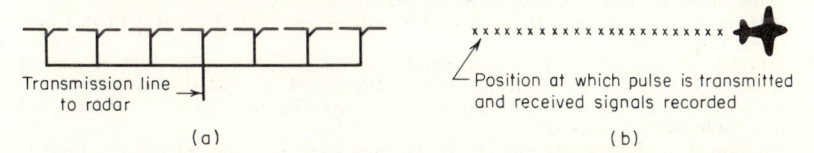

(a) Transmission line to radar

(b) Position at which pulse is transmitted and received signals recorded

Fig. 14-87. (a) Physical array of dipoles. (b) Synthetic-array generation. (*L. J. Cutrona, W. E. Vivian, E. N. Leith, and G. O. Hall, A High-resolution Radar Combat-surveillance System, IRE Trans. Military Electron., MIL-5, No. 2, April 1961.*)

Fig. 14-88. Experimental aircraft used by University of Michigan in mapping radar experiments. (*Courtesy of University of Michigan.*)

tion of the synthetic array is illustrated. The elements of the array do not exist simultaneously but are generated serially by the aircraft motion. It is necessary to record the output (amplitude and phase) as a function of aircraft position. By appropriate processing, it is possible to combine the data recorded for each aircraft position and produce the desired effect of a narrow-beam antenna. But even more important, since the data from each radiating element (position) are separately recorded, it is possible to process the data in a number of different ways. One obviously convenient method is to produce the effect of constant linear resolution rather than constant angular resolution.

The physical size of the radiator need not be very large to produce high-resolution pictures. An external view of the aircraft used in the University of Michigan experiments is shown in Fig. 14-88. The radome houses the side-looking antennas. The radar must be coherent to provide the phase data and must have high stability. The processing of the data need not be done in the aircraft in "real time" but can be recorded and processed on the ground. A high-resolution radar photograph made with the AN/UPD-1 system is shown in Fig. 14-71.

ACKNOWLEDGMENT

Much of the information presented in this chapter is derived from Ref. 18.

In addition, the author wishes to express his sincere appreciation to C. M. Steinmetz for his valuable suggestions and constructive criticism in the final revision of this manuscript.

REFERENCES

1. Woodward, P. M.: *Probability and Information Theory, with Applications to Radar*, Pergamon Press, London, 1957; McGraw-Hill Book Company, New York, 1953.
2. Fowler, C. A., A. P. Uzzo, Jr., and A. E. Ruvin: Signal Processing Techniques for Surveillance Radar Sets, *IRE Trans. Military Electron.*, vol. MIL-5, no. 2, pp. 103–108, April, 1961.
3. Freedman, J., and J. Margolin: Signal-to-Noise Improvement through Integration in a Delay-line Filter System, *Massachusetts Institute of Technology, Lincoln Laboratory Tech. Rept.* 22, May 13, 1953 (ASTIA 19279).
4. Glasoe, G. N., and Jean V. Lebacqz: *Pulse Generators*, McGraw-Hill Book Company, New York, 1948.
5. Manley, J. M., and H. E. Rowe: Some General Properties of Nonlinear Elements—Part I, General Energy Relations, *Proc. IRE*, vol. 44, pp. 904–913, July, 1956.
6. Adler, R.: Parametric Amplification of the Fast Electron Wave, *Proc. IRE*, vol. 46, pp. 1300–1301, June, 1958.
7. Grodon, J. P., H. J. Zeiger, and C. H. Townes: *Phys. Rev.*, vol. 99, p. 1264, 1955.
8. Hertz, Heinrich: Ueber electrodynamische Wellen in Luftraume und deren Reflexion, *Widemann's Annalen der Physik*, vol. 34, pp. 610–623, 1888. Reprinted in H. Hertz, *Untersuchungen ueber die Ausbreitung der Elektrischen Kraft*, pp. 133–146, Johann Ambrosius Barth, Leipiz, 1892; and reprinted in translation as *Electric Waves*, The Macmillan Company, New York, 1893, 2d ed., 1900.
9. Crispin, J. W., Jr., R. F. Goodrich, and K. M. Siegel: A Theoretical Method for the Calculation of the Radar Cross Sections of Aircraft and Missiles, *Univ. Mich. Radiation Lab. Rept.* 2591-1-H on Contract AF 19(604)-1949, July, 1959.
10. Ridenour, L. N.: *Radar System Engineering*, McGraw-Hill Book Company, New York, 1947.
11. Cutrona, L. J., W. E. Vivian, E. N. Leith, and G. O. Hall: A High-resolution Radar Combat-surveillance System, *IRE Trans. Military Electron.*, vol. MIL-5, no. 2, April, 1961.
12. Durlach, N., A. Carpenter, and M. Herlin: Curved-earth Computations for Airborne Early Warning and Control, *Massachusetts Institute of Technology, Lincoln Laboratory Tech. Rept.* 194, Jan. 13, 1959 (ASTIA 210006).
13. Blake, L. V.: Recent Advancements in Radar Range Calculation Technique, *IRE Trans. Military Electron.*, vol. MIL-5, no. 2, pp. 154–164, April, 1961.
14. Schelleng, J. C., C. R. Burrows, and E. B. Ferrell: *Proc. IRE*, vol. 21, p. 427, 1933.
15. Bauer, J. R., W. C. Mason, and F. A. Wilson: Radio Refraction in a Cool Exponential Atmosphere, *Massachusetts Institute of Technology, Lincoln Laboratory Tech. Rept.* 186, Aug. 27, 1958 (ASTIA 202331).
16. IRE Standards on Electron Tubes: Definitions of Terms, *Proc. IRE*, vol. 45, no. 7, pp. 983–1010, July, 1957.
17. IRE Standards on Electron Tubes: Definitions of Terms, *Proc. IEEE*, vol. 51, no. 3, pp. 434–435, March, 1963.
18. Freedman, J., and L. D. Smullin: *Elements of High-power Radar*, McGraw-Hill Book Company, New York, to be published.

Chapter 15

INFRARED

LLOYD G. MUNDIE, *Lockheed-California Company, Burbank, California.*

CONTENTS

15-1. Introduction	15-1
Advantages and Disadvantages of the Infrared Technique	15-2
15-2. Infrared Sources	15-3
Emission from Solids and Liquids	15-3
Emission from Gases	15-4
Practical Sources	15-4
Background Radiation	15-5
15-3. Atmospheric Transmission of Infrared Radiation	15-6
Absorption	15-6
Scattering	15-9
15-4. Infrared Optical Systems	15-9
Optical Materials for the Infrared	15-12
Infrared-optical-system Design	15-13
15-5. Infrared Detectors	15-16
15-6. Infrared-system-design Consideration	15-27
15-7. Applications of Infrared Systems	15-27

15-1. Introduction

Infrared radiation is electromagnetic in nature, falling in wavelength between the visible and microwave spectral regions. It is generally assigned the wavelength region extending from 0.75 to 1,000 microns (1 micron = 10^{-6} m). Details concerning the basic elements of modern infrared technology are available in a number of excellent treatises on the subject.[1-6,27]

Infrared systems may be used typically for chemical analysis, for communications, for the detection and tracking of targets, as horizon sensors, or as a ground reconnaissance technique. The basic elements of any infrared system include the source or target; the radiation background against which the target must be detected; the atmosphere, which attenuates the infrared signal; the optical system used to collect the radiation; the detector, which usually converts the radiant signal into an electrical one; and finally the electronic system used to amplify and display (or otherwise utilize) the signal. Such a generalized infrared system is illustrated schematically in Fig. 15-1. With these elements as our building blocks, the basic infrared range equation will be developed and used to predict the performance of the system. This range equation involves the integration over the wavelength region involved of the product of the target emission (relative to adjacent background), atmospheric transmission, transmission of the optical system, and detector response. A number of "trade-offs" must be considered in the design of a typical infrared system. Thus, for example, a high information rate, long detection range, high tracking accuracy, and low false-alarm rate may be acquired only at the cost of large size, weight, cost, and complexity.

Infrared systems may be either active or passive in nature. In active systems the target area is flooded with radiation from an infrared searchlight, and the reflected radiation is utilized for target location and identification. Active systems usually possess superior identification capability, since the target registration more closely resembles its visual appearance. This advantage is, however, counterbalanced by loss of security, greater power requirement, and the circumstance that the signal strength falls off as the inverse fourth power of range rather than the inverse square, as with passive systems.

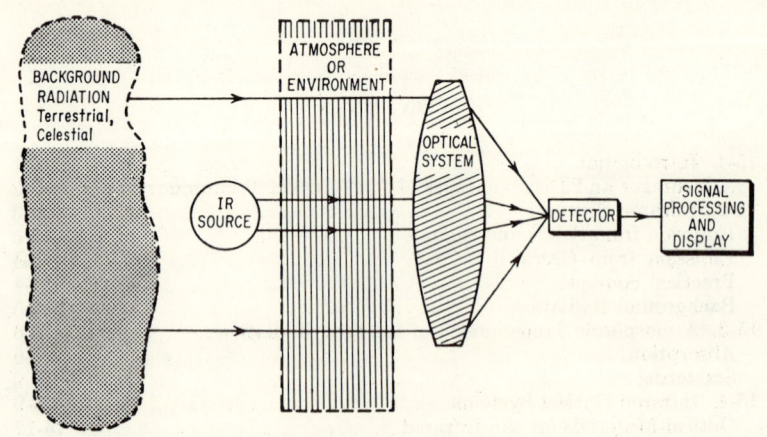

FIG. 15-1. Generalized infrared system. (*By permission from H. L. Hackforth, Infrared Radiation, McGraw-Hill Book Company, New York, 1960.*)

Advantages and Disadvantages of the Infrared Technique. Both inorganic and complex organic substances can be "fingerprinted" by their infrared-absorption spectra. This arises from the circumstance that the *natural frequencies* of vibration and rotation of these molecules lie in the infrared spectral region; so resonance absorption occurs at these wavelengths.

An advantage of infrared over radar (Chapter 14) in military applications is the added security associated with the fact that it may be used passively; targets may be detected by virtue of the radiation they emit naturally. The potentialities of the technique are evident when one considers that roughly 460 watts/sq m are radiated by most normally encountered surfaces at ambient temperatures and that modern detectors (with response times of microseconds) can easily detect 10^{-12} watts of this radiation. Modern aircraft and missiles at launch radiate large quantities of infrared radiation. Passive infrared systems experience only an inverse-square fall-off of signal with range, as compared with an inverse-fourth-power dependence in the case of radar. Since no transmitter is required, the power and weight requirements of passive infrared equipments are relatively modest. Since resonant high-frequency circuits are not involved, infrared equipments are usually considerably less complex.

Another advantage of infrared over radar is its higher potential resolution capabilities in relation to mirror (or antenna) size, as limited by diffraction. Thus, for example, a potential advantage by a factor of 10,000 is enjoyed by an infrared system operating at 3 microns over an X-band (3-cm wavelength) radar system with the same collector diameter. Conversely, for the same resolution, smaller antennas may be used in the infrared. No side lobes of sensitivity exist in infrared systems; they can be used to close (essentially zero) range, and no ground-clutter problem arises at low angles as with radar.

Infrared detection systems are difficult to jam and countermeasure because of their high resolution and passive nature. False decoy targets, such as flares, can usually be discriminated against on the basis of wavelength distribution. Low-temperature decoys which might escape this discrimination require a great deal of power.

In comparison with visible light (Chapter 16), infrared has the advantage of "round-the-clock" (day-and-night) operation. In common with visual techniques, however, infrared has the disadvantage, when used inside the earth's atmosphere, of being unable to penetrate fog and clouds, although infrared is superior to visible light in haze penetration.

15-2. Infrared Sources

All objects at temperatures above absolute zero emit electromagnetic radiation. This fact is not surprising when it is remembered that matter is made up of electrically charged particles in constant thermal agitation. Radiation from solids and liquids usually contains a broad continuous band of wavelengths, while that from gases generally consists of discrete wavelengths (spectral lines) which may be grouped in bands. For further discussion of radiometry, see Sec. 16-2.

Emission from Solids and Liquids. A hypothetical surface which absorbs all radiation incident upon it, a perfect absorber, is called a "blackbody." The radiant emittance W of such a surface at temperature T, defined as the radiation it emits into a hemisphere, is given by the *Stefan-Boltzmann law:*

$$W = \sigma T^4 \quad \text{watts/sq cm} \tag{15-1}$$

where $\sigma = 5.673 \times 10^{-12}$ watts/(sq cm) (deg^4)

As mentioned above, a blackbody at room temperature (300°K) emits 460 watts/sq m; for every degree of temperature increase in this temperature range, an extra 6.2 watts is emitted.

The spectral radiant emittance W_λ of a blackbody is given by *Planck's law:*

$$W_\lambda = C_1 \lambda^{-5} \exp\left(\frac{C_2}{\lambda T} - 1\right)^{-1} \quad \text{watts/(sq cm)(cm)}* \tag{15-2}$$

where $C_1 = 3.7402 \times 10^{-12}$ watts/sq cm
$C_2 = 1.43848$ cm deg

The spectral radiant emittance of blackbodies at a number of temperatures is plotted against wavelength in Fig. 15-2. The spectral radiant emittance of a perfectly diffusely reflecting surface illuminated by sunlight is included because of the importance of scattered sunlight as a background element.

Radiation slide rules giving W and W_λ for any temperature with sufficient accuracy for most practical applications are now available commercially.†

The wavelength λ_m at which W_λ is a maximum is given by *Wien's displacement law:*

$$\lambda_m T = 2{,}897 \text{ micron deg} \tag{15-3}$$

Actual surfaces are not perfectly absorbing and hence radiate less than blackbodies. In accordance with *Kirchhoff's law*, the absorptance and emissivity are equal at any wavelength. Most naturally occurring surfaces have emissivities in excess of 0.7 in the infrared; for highly polished metal surfaces the emissivity ranges from 0.02 to 0.25.

---

* Watts/sq cm of surface area/cm wavelength.
† From the General Electric Company, Schenectady, N.Y., or (in more elaborate form) from the Jarrell-Ash Company, Boston, Mass.

The radiation received from a plane diffuse radiator varies as the projection of its area normal to the line of sight, and hence as the cosine of the angle between the line of sight and the normal to the surface. This is *Lambert's cosine law*. Since the average value of the cosine function over the hemisphere is 0.5, the radiance N (watts/sq cm/steradian normal to the surface) is found by dividing the radiant emittance W (watts/sq cm into the hemisphere) by π (not 2π!).

Since the instantaneous rate at which photons are emitted by a source is the result of many individual processes, it fluctuates with time. This fluctuation in signal intensity causes "radiation noise" in an otherwise perfect radiation detector, and provides the ultimate limitation to the performance of detectors.[7]

Emission from Gases. As mentioned above, gases generally emit (and absorb) radiation at discrete frequencies (spectral lines), which may be grouped together in bands. At high temperatures and pressures, however, the lines become broadened.

Each line of a spectrum corresponds to a transition of the emitting atom or molecule from one discrete energy state to another. Atomic lines correspond to changes in electronic energy, associated with transitions of orbital electrons from one orbit to another; molecular lines correspond to some combination of changes in rotational, vibrational, and electronic energy. Changes only in rotational energy give rise to lines in the 25- to 1,000-micron wavelength region; if changes in vibrational energy are included, groups of lines (bands) will occur in the 2.25- to 25-micron region. From 1.3 to 2.5 microns, harmonics of the fundamental molecular vibrations are involved. When electronic transitions are involved, bands will appear in the visible and ultraviolet spectral regions.

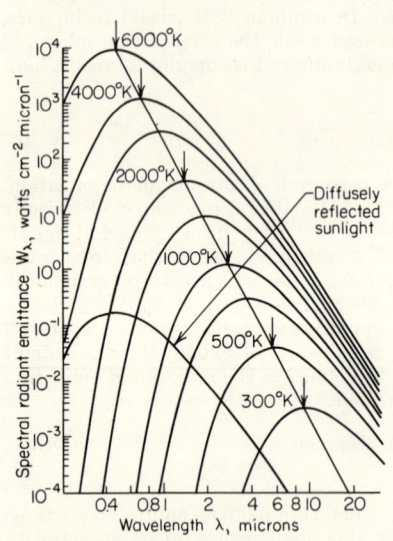

Fig. 15-2. Spectral-radiant-emittance curves for blackbodies at several absolute temperatures and for diffusely reflected sunlight.

Practical Sources. Practical sources include those designed for such things as spectroscopy, communications, photography, night driving, microscopy, general laboratory measurements, searchlights, beacons, and radiation standards. Targets such as aircraft and missiles must also be included in this category. The radiant emission may arise from a heated solid, a flame, or an arc discharge.

Sources involving heated solids include (1) the globar, an electrically heated silicon carbide rod, which may be operated at temperatures as high as 1400°K; (2) the tungsten filament, which can be operated up to 2900°K; (3) the Nernst glower, a tube composed of a mixture of zirconium and yttrium oxides, and heated electrically as high as 2000°K; (4) the carbon arc, which has an effective crater temperature of approximately 4000°K; and (5) the zirconium oxide concentrated arc which is especially designed to provide a very small bright source.

Sources which approximate blackbodies and are used as radiation standards may be fabricated by drilling a conical hole in the end of a heated cylindrical steel slug. The emissivity of the conical surface is raised to a value of 0.8 by oxidation, whereupon the effective emissivity of the source, for a 15° cone angle, becomes 0.99 through a multiple-reflection process.

Gaseous sources, used primarily for infrared communication, include arc lamps containing (1) a mixture of mercury and xenon or (2) cesium vapor. The latter can be modulated conveniently at voice frequency.

The primary sources of infrared radiation on aircraft are the hot metal of a jet tailpipe or an engine exhaust manifold, and the jet plume, which has a higher tem-

INFRARED

perature but lower emissivity. At supersonic speeds, the aircraft's skin becomes a strong infrared source because of aerodynamic heating. Infrared radiation from aircraft and rocket exhausts usually consists of continuous blackbody-type background radiation on which are superimposed emission bands due to water vapor, carbon dioxide, and other chemicals. Large aircraft emit several kilowatts of infrared energy. Many military land vehicles are also strong emitters of infrared.

Background Radiation. Targets can be detected only by virtue of their *contrast* against their background. During the day, the signal from natural backgrounds at wavelengths below 3 microns arises mainly from scattered sunlight. Above 3 microns the background signal is of primarily thermal origin, and very similar day and night. Figures 15-3 and 15-4 present short- and long-wavelength portions, respectively, of infrared sky spectra, together with corresponding solar and lunar spectra. Note that atmospheric emission bands coincide with absorption bands in the solar and lunar spectra, in accordance with Kirchhoff's law.

Celestial sources including the sun, moon, planets, and stars[10] may constitute false background targets. The sun closely resembles a blackbody source at 6000°K. Direct sunlight is sufficiently intense to damage detectors on which it is imaged, so protective measures must usually be taken.

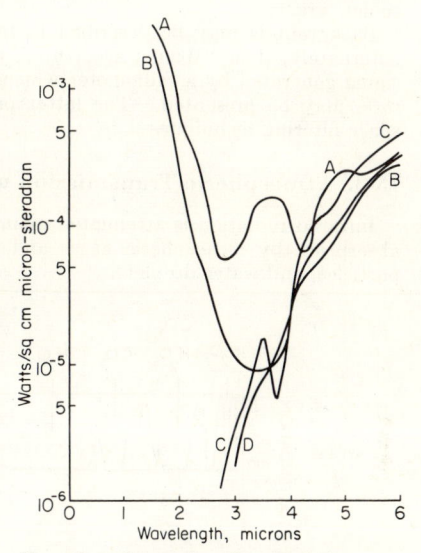

FIG. 15-3. Spectral radiance of the moon and sky. (A) Moon in clear sky, sun about 15° above horizon. (B) Clear sky, also early evening. (C) Summer night sky, clear. (D) Summer night sky, overcast. (*By permission from N. Ginsburg, W. R. Fredrickson, and R. Paulson, J. Opt. Soc. Am., vol. 50, p. 1176, 1960.*)

Discrimination against such background elements may be carried out in two ways: space-filtering techniques, which reject extended targets in favor of point targets, may be used, or spectral discrimination may be used to reject either direct or scattered

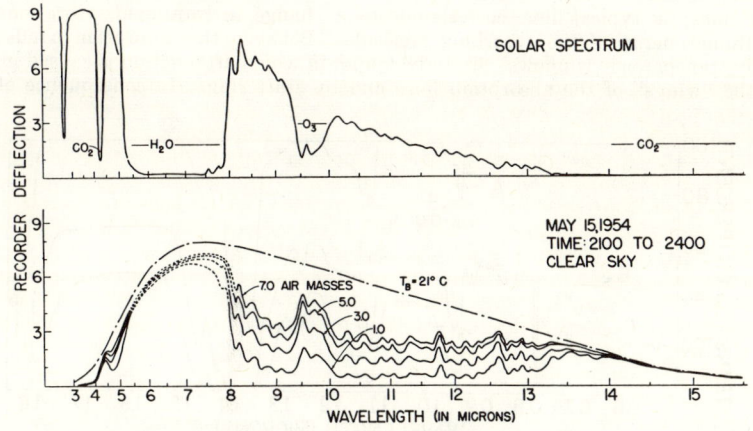

FIG. 15-4. Typical zenith sky spectra with solar and 21°C blackbody spectra for comparison. (*By permission from R. Sloan, J. H. Shaw, and D. Williams, J. Opt. Soc. Am., vol. 45, p. 455, 1955.*)

solar radiation; the latter technique capitalizes on the circumstance that the spectral distribution of solar background radiation usually differs markedly from that of the cooler target.

Backgrounds may be described in terms of their spatial and spectral radiance. Alternately, their "Wiener spectrum," derived from the frequency spectrum of the signal generated by a radiometer which scans the background at a uniform angular rate, may be presented. The latter presentation is useful in the development of space-filtering techniques.[22]

15-3. Atmospheric Transmission of Infrared Radiation

Infrared radiation is attenuated upon passing through the atmosphere because of absorption by atmospheric gases and scattering by molecules, dust, smoke, salt particles, and water droplets.

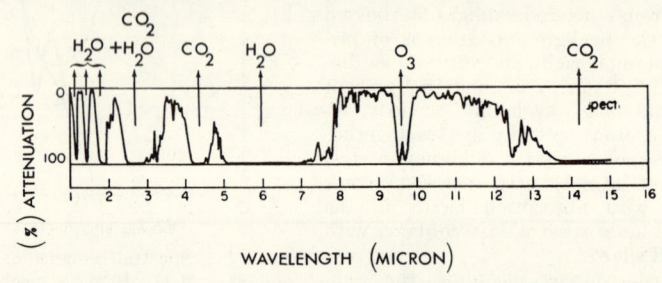

FIG. 15-5. The absorption spectrum of the atmosphere for solar radiation. (*By permission from J. N. Howard and J. S. Garing, Infrared Phys., vol. 2, p. 155, 1962.*)

Absorption. Water vapor, carbon dioxide, and ozone are the principal absorbers of infrared in the atmosphere, with minor additional contributions by CH_4, CO, and N_2O. Water vapor is concentrated mainly at lower levels, and ozone at altitudes between 10 and 40 km.[11] The other gases are rather uniformly distributed throughout the atmosphere (see Sec. 6-4). Figure 15-5 indicates the attenuation along a vertical path through the entire atmosphere, with the chief absorbing molecules identified. Absorption is seen to occur strongly in several bands. When observed with high spectral resolution, each band is found to be composed of a large number of absorption lines; a typical line corresponds to a change in rotational, vibrational, or electronic energy of the absorbing molecule. Between the absorption bands relatively transparent "windows" are to be found, in which attenuation is caused mainly by the "wings" of the absorption lines and by scattering. In consequence of this

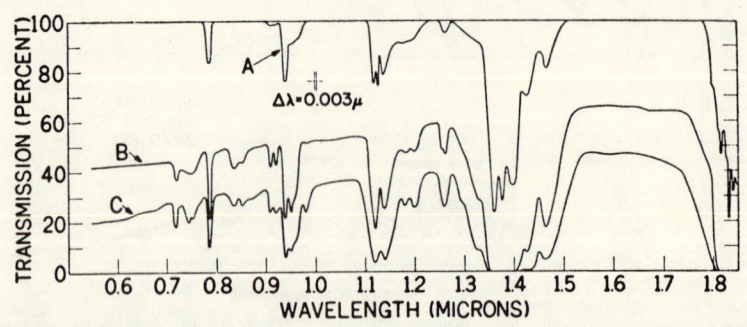

FIG. 15-6. Atmospheric transmission. (*By permission from J. H. Taylor and H. W. Yates, J. Opt. Soc. Am., vol. 47, p. 223, 1957.*)

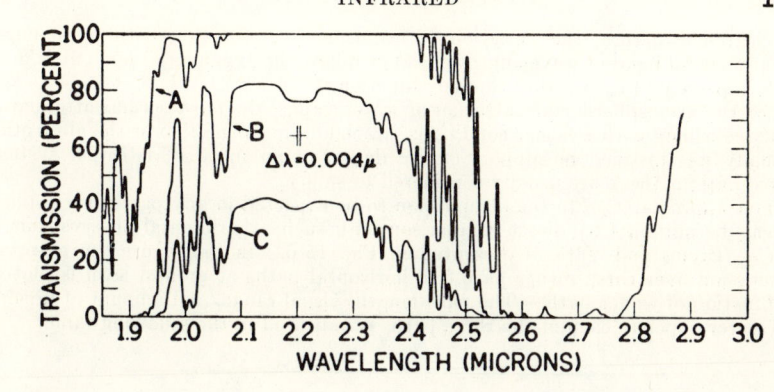

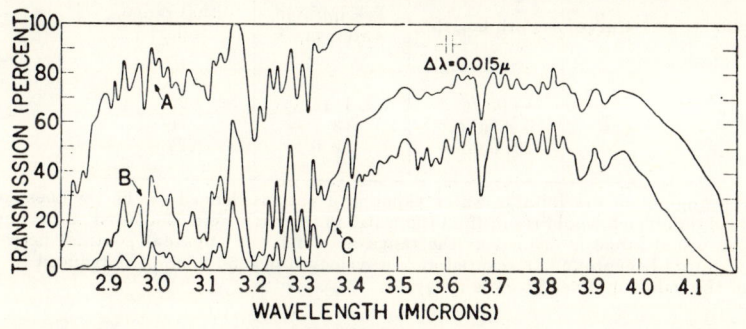

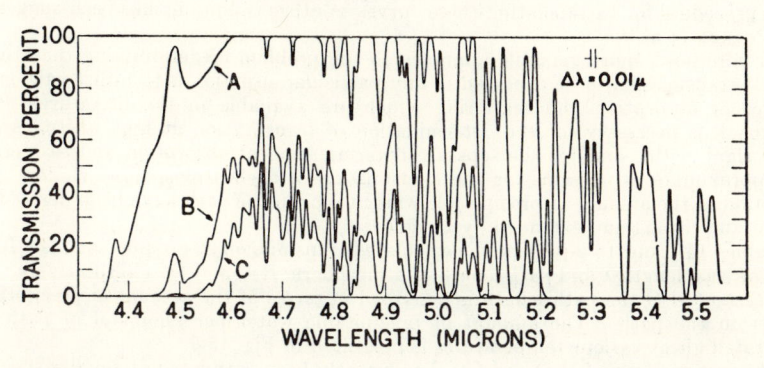

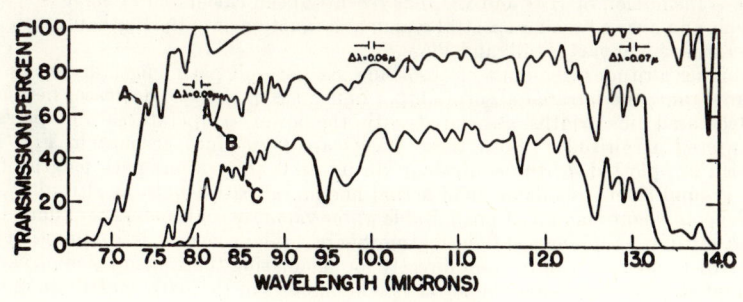

FIG. 15-6. (*Continued*)

complicated structure, the variation of atmospheric absorption with path length over any useful band of wavelengths does not follow the exponential law but is more closely represented by a rather complex function.[12]

Since the atmospheric concentration of water vapor, the most serious attenuator, fluctuates widely with weather conditions, an elaborate calculation of the absorption is usually not justified in support of the design of an infrared detection system; approximate methods are usually considered to suffice.

A first approximation to the attenuation to be expected in any particular application can be obtained by direct comparison with such experimental observations as those of Taylor and Yates,[13] presented in Fig. 15-6. In this figure the observed transmission over three rather haze-free horizontal paths at ground level is plotted as a function of wavelength. The path length, visual range, and amount of precipitable water vapor in each of the three paths is indicated in the following table.

Curve	Path length	Precipitable water, mm*	Visual range, miles†
A	1,000 ft	1.1	22
B	3.4 miles	13.7	16
C	10.1 miles	52.0	24

* The amount of precipitable water vapor in a path is defined as the thickness of the liquid column which would result if all the water in the path were condensed at one end.

† The visual range is defined as the range at which the apparent contrast between a target and its background (measured at a wavelength of 0.55 micron) is reduced to 2 per cent of the value obtained at close range.

A procedure for extrapolating these curves to other conditions has been suggested by Prostak et al.[14]

An alternate, more versatile approach to the problem of determining the atmospheric transmission to be expected in any particular situation is to utilize calculated values of absorption and scattering which are available in the literature. This approach is necessary for the determination of transmission at high altitudes and over slant paths, and for the separate determination of absorption and scattering.

Approximate expressions for the transmission of each atmospheric window as a function of the amount of precipitable water vapor in the path have been derived by Elder and Strong and extended by Langer.[15]

Tables 15-1 and 15-2 present the calculated transmission at ground level as a function of wavelength λ and path length of atmospheric H_2O and CO_2, respectively. In the case of H_2O the path length is described in terms of the amount of precipitable water in the path. The amount of precipitable water per kilometer of path for saturated air at various temperatures is presented in Fig. 15-7.

The transmission of H_2O and CO_2 has recently been calculated at smaller spectral intervals and over a broader spectral region as a function of path length, pressure, and temperature by Wyatt, Stull, and Plass.[17]

At higher altitudes, the mass of absorbing gas per unit path length decreases. At the same time, the infrared absorption per unit mass of gas also decreases because of the decreased line widths associated with the lower temperatures and pressures encountered at altitude. Both these effects are taken into account in Fig. 15-8, in which are plotted d_h, the equivalent (in infrared absorption) path length of dry air at ground level per kilometer of actual horizontal path length at altitude h, and W_h, the equivalent amount of precipitable water vapor, in centimeters per kilometer of path at altitude assuming a standard atmosphere. These curves, in conjunction with Tables 15-1 and 15-2, can be used directly for calculating the transmission of vertical and slant paths. Note that the lower few kilometers of the path contribute most of the absorption, particularly in the case of water vapor.

INFRARED

Scattering. Superimposed on molecular absorption, scattering by molecules, dust, smoke, salt particles, and water droplets also limits atmospheric transmission. Since detailed information concerning the abundance, type, and size distribution of scattering particles is usually not available, approximate methods, in which the visual range is used as a rough indicator of these parameters, are generally used for determining the transmittance. Attenuation by scattering is found to vary nearly exponentially with range. Figure 15-9, presents the best available, but admittedly only approximate, transmittance per kilometer of path length as limited by scattering as a function of wavelength for various visual ranges. The index at the right-hand side

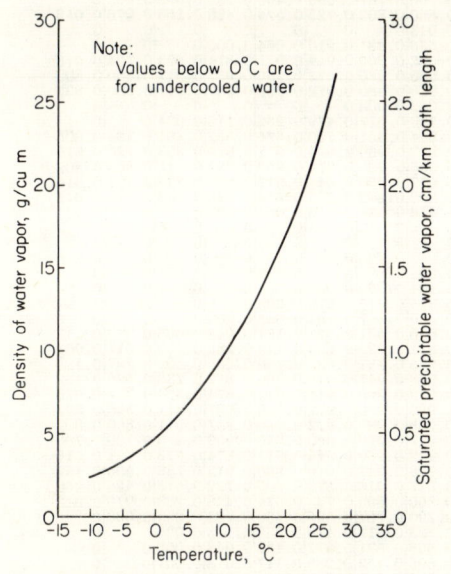

Fig. 15-7. Density of saturated water vapor.

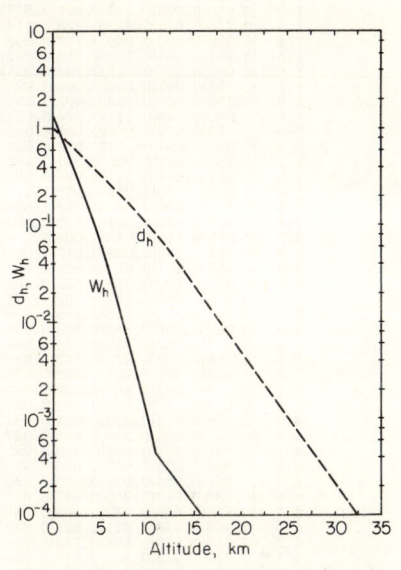

Fig. 15-8. Equivalent horizontal paths of atmospheric gases vs. altitude. (*By permission from T. L. Altshuler, Infrared Transmission and Background Radiation by Clear Atmospheres, General Electric Missile and Space Vehicle Department, Valley Forge, Pa., Document 61SD199, December, 1961.*)

can be used to determine the transmittance for other ranges. To do this, one first notes the linear distance on the index between the "1" mark and that indicating the range in question (use of dividers will facilitate this process). This distance is then stepped off vertically (in magnitude and direction) from the appropriate curve in the figure at the appropriate wavelength. The desired transmittance can then be read directly at the new ordinate.

15-4. Infrared Optical Systems

Infrared optical systems are used to collect radiation at an entrance aperture, process it spectrally and spatially, and direct it to the detector. High transmission and good image quality are usually imperative, requiring good-quality surfaces and mirrors, and homogeneous transmission components. When the optical system "looks" forward from a high-speed carrier in the atmosphere, a convex entrance window (irdome) is used. The irdome must withstand atmospheric moisture, the shocks associated with high acceleration, abrasion from dust and water particles, and aerodynamic

Table 15-1. Atmospheric Transmittance of Water Vapor at Sea Level*

μ \ w, cm	0.01	0.02	0.05	0.1	0.2	0.5	1	2	5	10	20	50	100
0.3	0.980	0.972	0.955	0.937	0.911	0.860	0.802	0.723	0.574	0.428	0.263	0.076	0.012
0.4	0.980	0.972	0.955	0.937	0.911	0.860	0.802	0.723	0.574	0.428	0.263	0.076	0.012
0.5	0.986	0.980	0.968	0.956	0.937	0.901	0.861	0.804	0.695	0.579	0.433	0.215	0.079
0.6	0.990	0.986	0.977	0.968	0.955	0.929	0.900	0.860	0.779	0.692	0.575	0.375	0.210
0.7	0.991	0.987	0.980	0.972	0.960	0.937	0.910	0.873	0.800	0.722	0.615	0.425	0.260
0.8	0.989	0.984	0.975	0.965	0.950	0.922	0.891	0.845	0.758	0.663	0.539	0.330	0.168
0.9	0.965	0.951	0.922	0.890	0.844	0.757	0.661	0.535	0.326	0.165	0.050	0.002	0
1.0	0.990	0.986	0.977	0.968	0.955	0.929	0.900	0.860	0.779	0.692	0.575	0.375	0.210
1.1	0.970	0.958	0.932	0.905	0.866	0.790	0.707	0.595	0.406	0.235	0.093	0.008	0
1.2	0.980	0.972	0.955	0.937	0.911	0.860	0.802	0.723	0.574	0.428	0.263	0.076	0.012
1.3	0.726	0.611	0.432	0.268	0.116	0.013	0	0	0	0	0	0	0
1.4	0.930	0.902	0.844	0.782	0.695	0.536	0.381	0.216	0.064	0.005	0	0	0
1.5	0.997	0.994	0.991	0.988	0.982	0.972	0.960	0.944	0.911	0.874	0.823	0.724	0.616
1.6	0.998	0.997	0.996	0.994	0.991	0.986	0.980	0.972	0.956	0.937	0.911	0.860	0.802
1.7	0.998	0.997	0.996	0.994	0.991	0.986	0.980	0.972	0.956	0.937	0.911	0.860	0.802
1.8	0.792	0.707	0.555	0.406	0.239	0.062	0.008	0	0	0	0	0	0
1.9	0.960	0.943	0.911	0.874	0.822	0.723	0.617	0.479	0.262	0.113	0.024	0	0
2.0	0.985	0.979	0.966	0.953	0.933	0.894	0.851	0.790	0.674	0.552	0.401	0.184	0.006
2.1	0.997	0.997	0.991	0.988	0.982	0.972	0.960	0.944	0.911	0.874	0.823	0.724	0.616
2.2	0.998	0.997	0.996	0.994	0.991	0.986	0.980	0.972	0.956	0.937	0.911	0.860	0.802
2.3	0.997	0.994	0.991	0.988	0.982	0.972	0.960	0.944	0.911	0.874	0.823	0.724	0.616
2.4	0.980	0.972	0.955	0.937	0.911	0.860	0.802	0.723	0.574	0.428	0.263	0.076	0.012
2.5	0.930	0.902	0.844	0.782	0.695	0.536	0.381	0.216	0.064	0.005	0	0	0
2.6	0.617	0.479	0.261	0.110	0.002	0	0	0	0	0	0	0	0
2.7	0.361	0.196	0.040	0.004	0	0	0	0	0	0	0	0	0
2.8	0.453	0.289	0.092	0.017	0.001	0	0	0	0	0	0	0	0
2.9	0.689	0.571	0.369	0.205	0.073	0.005	0	0	0	0	0	0	0
3.0	0.851	0.790	0.673	0.552	0.401	0.184	0.060	0.008	0	0	0	0	0
3.1	0.900	0.860	0.779	0.692	0.574	0.375	0.210	0.076	0.005	0	0	0	0
3.2	0.925	0.894	0.833	0.766	0.674	0.506	0.347	0.184	0.035	0.003	0	0	0
3.3	0.950	0.930	0.888	0.843	0.779	0.658	0.531	0.377	0.161	0.048	0.005	0	0
3.4	0.973	0.962	0.939	0.914	0.880	0.811	0.735	0.633	0.448	0.285	0.130	0.017	0.001
3.5	0.988	0.983	0.973	0.962	0.946	0.915	0.881	0.832	0.736	0.635	0.502	0.287	0.133
3.6	0.994	0.992	0.987	0.982	0.973	0.958	0.947	0.916	0.866	0.812	0.738	0.596	0.452
3.7	0.997	0.994	0.991	0.988	0.982	0.972	0.960	0.944	0.911	0.874	0.823	0.724	0.616
3.8	0.998	0.997	0.995	0.994	0.991	0.986	0.980	0.972	0.956	0.937	0.911	0.860	0.802
3.9	0.998	0.997	0.995	0.994	0.991	0.986	0.980	0.972	0.956	0.937	0.911	0.860	0.802
4.0	0.997	0.995	0.993	0.990	0.987	0.977	0.970	0.960	0.930	0.900	0.870	0.790	0.700
4.1	0.997	0.994	0.991	0.988	0.982	0.972	0.960	0.944	0.911	0.874	0.823	0.724	0.616
4.2	0.994	0.992	0.987	0.982	0.973	0.958	0.947	0.916	0.866	0.812	0.738	0.596	0.452
4.3	0.991	0.984	0.975	0.972	0.950	0.937	0.910	0.873	0.800	0.722	0.615	0.425	0.260
4.4	0.980	0.972	0.955	0.937	0.911	0.860	0.802	0.723	0.574	0.428	0.263	0.076	0.012
4.5	0.970	0.958	0.932	0.905	0.866	0.790	0.707	0.595	0.400	0.235	0.093	0.008	0
4.6	0.966	0.943	0.911	0.874	0.822	0.723	0.617	0.478	0.262	0.113	0.024	0	0
4.7	0.950	0.930	0.888	0.743	0.779	0.658	0.531	0.377	0.161	0.048	0.005	0	0
4.8	0.940	0.915	0.866	0.812	0.736	0.595	0.452	0.289	0.117	0.018	0.001	0	0
4.9	0.930	0.902	0.844	0.782	0.695	0.536	0.381	0.216	0.064	0.005	0	0	0
5.0	0.915	0.880	0.811	0.736	0.634	0.451	0.286	0.132	0.017	0	0	0	0
5.1	0.885	0.839	0.747	0.649	0.519	0.308	0.149	0.041	0.001	0	0	0	0
5.2	0.846	0.784	0.664	0.539	0.385	0.168	0.052	0.006	0	0	0	0	0
5.3	0.792	0.707	0.555	0.406	0.239	0.062	0.008	0	0	0	0	0	0
5.4	0.726	0.611	0.432	0.268	0.116	0.013	0	0	0	0	0	0	0
5.5	0.617	0.479	0.261	0.110	0.035	0	0	0	0	0	0	0	0
5.6	0.491	0.331	0.121	0.029	0.002	0	0	0	0	0	0	0	0
5.7	0.361	0.196	0.040	0.004	0	0	0	0	0	0	0	0	0
5.8	0.141	0.004	0.001	0	0	0	0	0	0	0	0	0	0
5.9	0.141	0.044	0.001	0	0	0	0	0	0	0	0	0	0
6.0	0.180	0.058	0.003	0	0	0	0	0	0	0	0	0	0
6.1	0.260	0.112	0.012	0	0	0	0	0	0	0	0	0	0
6.2	0.652	0.524	0.313	0.153	0.043	0.001	0	0	0	0	0	0	0
6.3	0.552	0.401	0.182	0.060	0.008	0	0	0	0	0	0	0	0
6.4	0.317	0.157	0.025	0.002	0	0	0	0	0	0	0	0	0
6.5	0.164	0.049	0.002	0	0	0	0	0	0	0	0	0	0
6.6	0.138	0.042	0.001	0	0	0	0	0	0	0	0	0	0
6.7	0.322	0.162	0.037	0.002	0	0	0	0	0	0	0	0	0
6.8	0.361	0.196	0.040	0.004	0	0	0	0	0	0	0	0	0
6.9	0.416	0.250	0.068	0.010	0	0	0	0	0	0	0	0	0
7.0	0.532	0.377	0.161	0.048	0.005	0	0	0	0	0	0	0	0

* By permission from L. Larmore, Transmission of Infrared Radiation through the Atmosphere, *Proc. IRIS*, vol. 1, no. 1, June, 1956.

Table 15-2. Atmospheric Transmittance of CO_2 at Sea Level*

d = path length, km

λ, μ	0.3	0.4	0.5	0.6	0.7	0.8	0.9	1.0	1.1	1.2	1.3	1.4	1.5	1.6	1.7	1.8	1.9	2.0
d = 0.1	1.000	1.000	1.000	1.000	1.000	1.000	1.000	1.000	1.000	1.000	1.000	0.996	0.999	0.996	1.000	1.000	1.000	0.978
d = 0.2	1.000	1.000	1.000	1.000	1.000	1.000	1.000	1.000	1.000	1.000	1.000	0.995	0.999	0.995	1.000	1.000	1.000	0.969
d = 0.5	1.000	1.000	1.000	1.000	1.000	1.000	1.000	1.000	1.000	1.000	1.000	0.992	0.998	0.992	1.000	1.000	1.000	0.951
d = 1	1.000	1.000	1.000	1.000	1.000	1.000	1.000	1.000	1.000	1.000	0.999	0.988	0.998	0.988	0.999	1.000	0.999	0.931
d = 2	1.000	1.000	1.000	1.000	1.000	1.000	1.000	1.000	1.000	1.000	0.999	0.984	0.997	0.984	0.999	1.000	0.999	0.903
d = 5	1.000	1.000	1.000	1.000	1.000	1.000	1.000	1.000	1.000	1.000	0.999	0.975	0.995	0.975	0.999	1.000	0.999	0.847
d = 10	1.000	1.000	1.000	1.000	1.000	1.000	1.000	1.000	1.000	1.000	0.998	0.964	0.993	0.964	0.998	1.000	0.998	0.785
d = 20	1.000	1.000	1.000	1.000	1.000	1.000	1.000	1.000	1.000	1.000	0.997	0.949	0.990	0.949	0.997	1.000	0.997	0.699
d = 50	1.000	1.000	1.000	1.000	1.000	1.000	1.000	1.000	1.000	1.000	0.996	0.919	0.984	0.919	0.996	1.000	0.996	0.541
d = 100	1.000	1.000	1.000	1.000	1.000	1.000	1.000	1.000	1.000	1.000	0.994	0.885	0.976	0.885	0.994	1.000	0.994	0.387
d = 200	1.000	1.000	1.000	1.000	1.000	1.000	1.000	1.000	1.000	1.000	0.992	0.838	0.967	0.838	0.992	1.000	0.992	0.221
d = 500	1.000	1.000	1.000	1.000	1.000	1.000	1.000	1.000	1.000	1.000	0.987	0.747	0.949	0.747	0.987	1.000	0.987	0.053
d = 1,000	1.000	1.000	1.000	1.000	1.000	1.000	1.000	1.000	1.000	1.000	0.982	0.649	0.927	0.649	0.982	1.000	0.982	0.006

λ, μ	2.1	2.2	2.3	2.4	2.5	2.6	2.7	2.8	2.9	3.0	3.1	3.2	3.3	3.4	3.5	3.6	3.7
d = 0.1	0.998	1.000	1.000	1.000	1.000	1.000	0.799	0.871	0.997	1.000	1.000	1.000	1.000	1.000	1.000	1.000	1.000
d = 0.2	0.997	1.000	1.000	1.000	1.000	1.000	0.718	0.804	0.995	1.000	1.000	1.000	1.000	1.000	1.000	1.000	1.000
d = 0.5	0.996	1.000	1.000	1.000	1.000	1.000	0.569	0.695	0.993	1.000	1.000	1.000	1.000	1.000	1.000	1.000	1.000
d = 1	0.994	1.000	1.000	1.000	1.000	1.000	0.419	0.578	0.990	1.000	1.000	1.000	1.000	1.000	1.000	1.000	1.000
d = 2	0.992	1.000	1.000	1.000	1.000	1.000	0.253	0.432	0.985	1.000	1.000	1.000	1.000	1.000	1.000	1.000	1.000
d = 5	0.987	1.000	1.000	1.000	1.000	1.000	0.071	0.215	0.977	1.000	1.000	1.000	1.000	1.000	1.000	1.000	1.000
d = 10	0.982	1.000	1.000	1.000	1.000	1.000	0.011	0.079	0.968	1.000	1.000	1.000	1.000	1.000	1.000	1.000	1.000
d = 20	0.974	1.000	1.000	1.000	1.000	1.000	0	0.013	0.954	1.000	1.000	1.000	1.000	1.000	1.000	1.000	1.000
d = 50	0.959	1.000	1.000	1.000	1.000	1.000	0	0	0.927	1.000	1.000	1.000	1.000	1.000	1.000	1.000	1.000
d = 100	0.942	1.000	1.000	1.000	1.000	1.000	0	0	0.898	1.000	1.000	1.000	1.000	1.000	1.000	1.000	1.000
d = 200	0.919	1.000	1.000	1.000	1.000	1.000	0	0	0.855	1.000	1.000	1.000	1.000	1.000	1.000	1.000	1.000
d = 500	0.872	1.000	1.000	1.000	1.000	1.000	0	0	0.772	1.000	1.000	1.000	1.000	1.000	1.000	1.000	1.000
d = 1,000	0.820	1.000	1.000	1.000	1.000	1.000	0	0.683	1.000	1.000	1.000	1.000	1.000	1.000	1.000	1.000	1.000

λ, μ	3.8	3.9	4.0	4.1	4.2	4.3	4.4	4.5	4.6	4.7	4.8	4.9	5.0	5.1	5.2	5.3	5.4
d = 0.1	1.000	1.000	0.998	0.983	0.673	0.098	0.481	0.057	0.995	0.995	0.976	0.975	0.999	1.000	0.986	0.997	1.000
d = 0.2	1.000	1.000	0.997	0.975	0.551	0.016	0.319	0.949	0.993	0.993	0.966	0.064	0.998	0.999	0.980	0.995	1.000
d = 0.5	1.000	1.000	0.996	0.961	0.445	0	0.115	0.903	0.989	0.989	0.945	0.943	0.997	0.999	0.968	0.993	1.000
d = 1	1.000	1.000	0.994	0.944	0.182	0	0.026	0.863	0.985	0.985	0.922	0.920	0.995	0.998	0.955	0.989	1.000
d = 2	1.000	1.000	0.991	0.921	0.059	0	0.002	0.807	0.978	0.978	0.891	0.886	0.994	0.998	0.936	0.984	1.000
d = 5	1.000	1.000	0.986	0.876	0.003	0	0	0.699	0.966	0.966	0.828	0.822	0.990	0.996	0.899	0.976	1.000
d = 10	1.000	1.000	0.980	0.825	0	0	0	0.585	0.951	0.951	0.759	0.750	0.986	0.994	0.857	0.966	1.000
d = 20	1.000	1.000	0.971	0.755	0	0	0	0.439	0.931	0.931	0.664	0.652	0.979	0.992	0.799	0.951	1.000
d = 50	1.000	1.000	0.955	0.622	0	0	0	0.222	0.891	0.891	0.492	0.468	0.968	0.988	0.687	0.923	1.000
d = 100	1.000	1.000	0.937	0.485	0	0	0	0.084	0.845	0.845	0.331	0.313	0.954	0.984	0.569	0.891	1.000
d = 200	1.000	1.000	0.911	0.322	0	0	0	0.014	0.783	0.783	0.169	0.153	0.935	0.976	0.420	0.846	1.000
d = 500	1.000	1.000	0.859	0.118	0	0	0	0	0.663	0.663	0.030	0.024	0.897	0.961	0.203	0.760	1.000
d = 1,000	1.000	1.000	0.802	0.027	0	0	0	0	0.539	0.539	0.002	0.001	0.855	0.946	0.072	0.066	1.000

λ, μ	5.5	5.6	5.7	5.8	5.9	6.0	6.1	6.2	6.3	6.4	6.5	6.6	6.7	6.8	6.9	7.0
d = 0.1	1.000	1.000	1.000	1.000	1.000	1.000	1.000	1.000	1.000	1.000	1.000	1.000	1.000	1.000	1.000	1.000
d = 0.2	1.000	1.000	1.000	1.000	1.000	1.000	1.000	1.000	1.000	1.000	1.000	1.000	1.000	1.000	1.000	1.000
d = 0.5	1.000	1.000	1.000	1.000	1.000	1.000	1.000	1.000	1.000	1.000	1.000	1.000	1.000	1.000	1.000	1.000
d = 1	1.000	1.000	1.000	1.000	1.000	1.000	1.000	1.000	1.000	1.000	1.000	1.000	1.000	1.000	1.000	1.000
d = 2	1.000	1.000	1.000	1.000	1.000	1.000	1.000	1.000	1.000	1.000	1.000	1.000	1.000	1.000	1.000	1.000
d = 5	1.000	1.000	1.000	1.000	1.000	1.000	1.000	1.000	1.000	1.000	1.000	1.000	1.000	1.000	1.000	1.000
d = 10	1.000	1.000	1.000	1.000	1.000	1.000	1.000	1.000	1.000	1.000	1.000	1.000	1.000	1.000	1.000	1.000
d = 20	1.000	1.000	1.000	1.000	1.000	1.000	1.000	1.000	1.000	1.000	1.000	1.000	1.000	1.000	1.000	1.000
d = 50	1.000	1.000	1.000	1.000	1.000	1.000	1.000	1.000	1.000	1.000	1.000	1.000	1.000	1.000	1.000	1.000
d = 100	1.000	1.000	1.000	1.000	1.000	1.000	1.000	1.000	1.000	1.000	1.000	1.000	1.000	1.000	1.000	1.000
d = 200	1.000	1.000	1.000	1.000	1.000	1.000	1.000	1.000	1.000	1.000	1.000	1.000	1.000	1.000	1.000	1.000
d = 500	1.000	1.000	1.000	1.000	1.000	1.000	1.000	1.000	1.000	1.000	1.000	1.000	1.000	1.000	1.000	1.000
d = 1,000	1.000	1.000	1.000	1.000	1.000	1.000	1.000	1.000	1.000	1.000	1.000	1.000	1.000	1.000	1.000	1.000

* By permission from L. Larmore, Transmission of Infrared Radiation through the Atmosphere, *Proc. IRIS*, vol. 1, no. 1, June, 1956.

heating at supersonic speeds. It must remain transparent at the high temperatures reached, and must itself emit a minimum amount of infrared which might saturate the detector and provide false targets by random reflections within the optical system.

Optical Materials for the Infrared. In selecting optical materials, the infrared-system designer must consider their mechanical, chemical, and thermal properties as well as their optical properties.

Optical properties to be considered include, first of all, high transmission in the spectral region of interest. For filters, opacity in adjacent regions with sharp cut-off may be highly desirable.

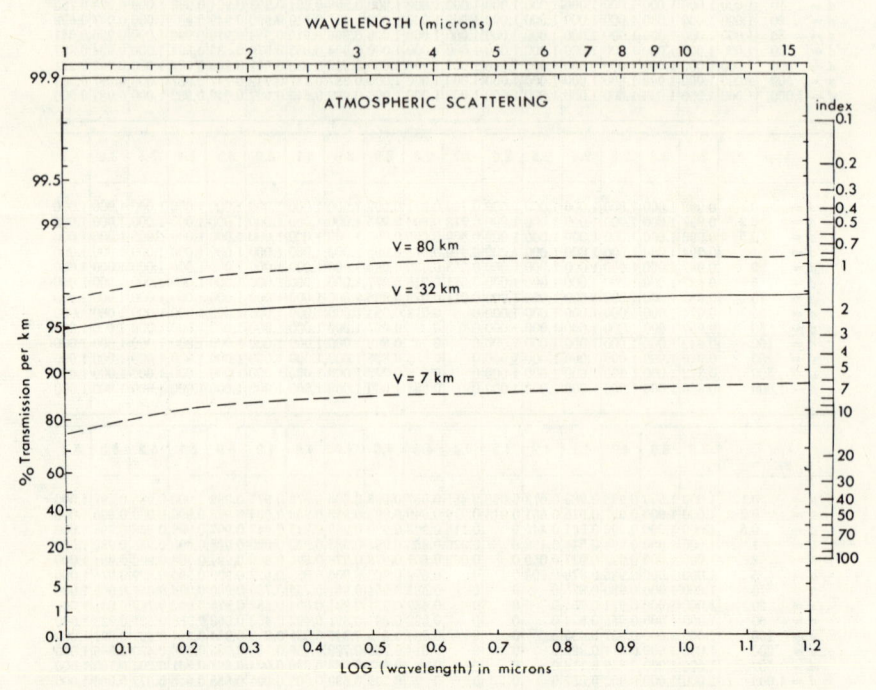

Fig. 15-9. Atmospheric transmittance at sea level, as limited by scattering, for various values of visual range V. (*By permission from T. L. Altshuler, Infrared Transmission and Background Radiation by Clear Atmospheres, General Electric Missile and Space Vehicle Department, Valley Forge, Pa., Document 61SD199, December, 1961.*)

The refractive index of the material is another important consideration. High-index materials, unless coated, are subject to high reflection losses, in accordance with the *Fresnel law:*

$$\text{Reflectance} = \frac{(n-1)^2}{(n+1)^2}$$

where n is the index of refraction. From this equation we see that, upon going from glass ($n = 1.5$) to germanium ($n = 4$), the surface reflection (at normal incidence in air) increases from 4 to 36 per cent. This reflection loss, however, can be largely overcome over a restricted wavelength region, and for a narrow range of angles of incidence, by coating one-quarter wavelength thick with a material whose index is equal to the square root of the substrate index. Lenses with high refractive indices possess less curvature for the same optical speed than low-index lenses and so are lighter in weight and less afflicted with spherical aberration. Achromatic doublets may be fashioned by combining two or more elements with different indices.

INFRARED

An excellent discussion of optical materials suitable for infrared instrumentation has been prepared by Ballard et al.[19] The properties of a number of infrared optical materials are summarized in Table 15-3.

The most commonly used optical materials in infrared systems include fused silica, sapphire, KRS-5 (a mixture of thallium iodide and bromium iodide), germanium, silicon, arsenic trisulfide, and Irtran-2 (compacted zinc sulfide).

Several types of spectral filters are available for use in the infrared. These include absorption-edge filters, which utilize selective absorption of certain materials; interference filters,[20] having single or multiple layers, in which radiation is attenuated by interference; reflectance filters, which selectively reflect certain wavelength bands; and scattering filters, consisting of fine particles embedded in a transparent binder, which scatter shorter wavelengths out of the radiation path in accordance with the Rayleigh law, while transmitting longer wavelengths.

Infrared-optical-system Design. Radiation from the target, having traversed the irdome, falls on the entrance optics, which usually image the target on the detector, reticle, or field lens. The collecting optics may utilize lenses, mirrors, or a combination

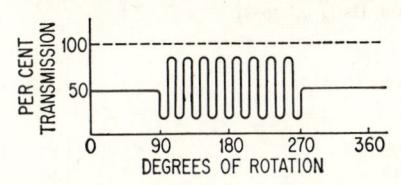

FIG. 15-10. An elementary tracking reticle. (*By permission from H. L. Hackforth, Infrared Radiation, McGraw-Hill Book Company, New York, 1960.*)

of both to accomplish the imaging function; the basic design principles are identical with those used with visible light.

Reflecting systems have the advantages of high optical efficiency at all wavelengths and freedom from chromatic aberrations. They also have less spherical aberration than would a single refracting element. Use of "folded" reflecting systems permits long focal lengths in a compact optical system. Reflecting optical systems, on the other hand, have the disadvantages that their focal surface is curved, and the field of view is limited by off-axis aberrations. These aberrations can, however, be corrected by introduction of refracting elements. The optical gain of reflecting systems when used on-axis is reduced by the blocking effect of either the detector assembly or the secondary mirror.

Refracting systems have the advantage that they possess more control of aberrations and there exists greater freedom of design. They have the disadvantages, however, of greater reflectance losses (since more surfaces must be traversed), limitations in spectral transmission because of lack of available materials which are free from selective absorption, and presence of chromatic aberrations. Combined reflective-refracting systems, known as catadioptric systems, have been developed[21] which possess advantages over pure reflecting or refracting systems for many applications.

In the process of guiding it to the detector, the optical system is often required to process the radiation either spectrally or spatially. Spectral processing may, for example, be used to discriminate against background elements on a wavelength basis, and is performed by a selective optical filter. Spatial processing may be accomplished through the use of a reticle, and is used both to discriminate against background elements on the basis of geometrical shape and to provide information concerning the direction of the line of sight to the target.

An elementary tracking reticle which performs these functions is shown in Fig. 15-10. This reticle rotates about its axis and is divided into two equal halves, one of which contains a set of radial spokes, while the other is divided into annular rings. A point image formed on such a rotating reticle will generate an electrical signal similar to that indicated in the right-hand portion of this figure. An extended image such as a large cloud, on the other hand, will produce relatively little a-c signal. The

Table 15-3. Properties

Material	Trans. Limit (microns)	Index of Refraction 2.2 μ	Index of Refraction 4.3 μ	Young's Modulus (10⁶ psi)	Knoop Hardness	Density (gms/cm³)
Optical glasses	2.7	1.5	1.7	7–10	300–600	2.3–4.6
Fused silica	4.5	1.43	1.37	10.1	470	2.20
RIR 2	4.7	1.75	—	15.2	∼600	∼3
850324 (thorium glass) Bu. Std. Desig. F-158	4.7	1.80	—	—	—	4.62
915213 (flint glass) Bu. Std. Desig. A-2059	4.8	1.85	—	7	—	6.01
Sapphire (Al_2O_3)	5.5	1.73	1.68	53	1370	3.98
RIR 20	5.5	1.82	1.79	12–14	542	5.18
RIR 12	5.7	1.62	—	15.2	594	3.07
Rutile (TiO_2)	6	—	2.45	—	880	4.26
Lithium fluoride (LiF)	6	1.38	1.34	11	110	2.6
Periclase (MgO)	6.8	1.71	1.66	36	690	3.59
Strontium titanate ($SrTiO_3$)	7.0	2.23	2.19	—	620	5.13
Irtran-1 (MgF_2)	8	1.37	1.35	—	576	3.18
Tellurium (Te)	—	—	4.93 ⊥ to c; 6.37 ∥ to c	—	—	6.24
Calcium fluoride (CaF_2)	9	1.42	1.41	15	158	3.18
Arsenic trisulfide (As_2S_3)	12	2.38	2.35	2.3	109	3.20
Barium fluoride (BaF_2)	13.5	1.46	1.45	8	82	4.89
Silicon (Si)	15*	3.44	3.42	19	1150	2.33
Irtran-2 (ZnS)	15	2.26	2.25	14	354	4.11

* By permission from P. W. Kruse, L. D. McGlaughlin, and R. B. McQuistan, *Elements*

of Optical Materials*

Melting Point (°C)	Thermal Expansion Coefficient (10^{-6}/°C)	Solubility gr/100 ml of H_2O at 20°C	Soluble in	Comments
700	4–10	0.00	HF	Easily cut, polished; nontoxic, nonhygroscopic.
1667	0.55	0.00	HF	Excellent mechanical and thermal properties; nontoxic.
~900	8.3	0.00	—	Available in production quantities. Preliminary studies indicate that no surface protection is needed.
780	8.3	—	Slightly in HNO_3	Contains thorium. Conventional surfacing techniques apply.
430	9.8	—	Dissolves in 1% HNO_3	Conventional surfacing techniques apply. Available in random slabs of the order of 2" × 2" × 1½".
2030	5.0 ⊥ to c; 6.7 ∥ to c	0.00	NH_4 salts	Excellent mechanical and thermal properties; nontoxic. Can be fused directly to 7052 and 7520 glass.
760	9.6	—	1% HNO_3	Good mechanical and optical properties; nontoxic.
~900	8.3	—		
1825	9	0.00	H_2SO_4	Nonhygroscopic; nontoxic.
870	36	0.27	HF	Scratches easily; nontoxic; noncorrosive.
2800	13	0.00	NH_4 salts	Nonhygroscopic; nontoxic; noncorrosive. A surface scum forms if stored in air.
2080	9.4	—	—	Of special interest because its refractive index $\approx \sqrt{5}$.
1396	16	Small	—	Emissivity very low even at 800°C.
450	16.8	0.00	—	Poisonous. ⊥ denotes E vector normal to c-axis. ∥ denotes E vector parallel to c-axis.
1403	25	0.002	NH_4 salts	Scratches easily; not resistant to thermal or mechanical shock; nontoxic.
196	26	0.00	—	Nonhygroscopic; noncorrosive.
1280	—	0.17	NH_4Cl	Slightly hygroscopic; nontoxic.
1420	4.2	0	HF + HNO_3	Resistant to corrosion. *Long wavelength limit depends upon impurity concentration as well as thickness and temperature. Very pure specimens are transparent even into the microwave region. Must be highly polished to reduce scattering losses at the surfaces.
800	7.9	0	Slightly in HNO_3, H_2SO_4	

of Infrared Technology, John Wiley & Sons, Inc., New York, 1962.

15-16   SYSTEM COMPONENTS

Table 15-3. Properties of

Material	Trans. Limits (microns)	Index of Refraction 2.2μ	Index of Refraction 4.3μ	Young's Modulus (10^6 psi)	Knoop Hardness	Density (gms/cm³)
Cadmium telluride (CdTe)	15	—	2.56**	—	—	—
Indium antimonide (InSb)	16	—	3.99***	6.21	—	5.78
Germanium (Ge)	25*	4.09	4.02	14.9	—	5.33
Sodium chloride (NaCl)	25	1.53	1.52	5.8	17	3.16
Silver chloride (AgCl)	30	2.01	2.00	2.9	9.5	5.53
KRS-6 (Thallium bromide thallium chloride)	30	2.20	2.19	3.0	35	7.19
KRS-5 (Thallium bromide thallium iodide)	45	2.62	—	—	—	—
Cesium bromide (CsBr)	48	1.66	1.66	2.3	—	4.44
Cesium iodide (CsI)	60	1.75	1.73	0.8	—	4.53
Kel-F	4	—	—	—	—	—
Lucite	5.5	—	—	—	—	—

phase of the modulated signal from the point target provides information concerning the angular position of the target. This information can subsequently be used for tracking, homing, or fire control. The intelligent design of such a reticle requires a knowledge of the spatial distribution of background clutter such as clouds, so that most efficient cloud rejection can be accomplished.[22] For the detection of small-sized targets, best discrimination is usually accomplished by making the reticle grids as fine as the state of the art will permit. Aroyan presents an excellent discussion of the spatial filtering problem.[23]

15-5. Infrared Detectors

Infrared detectors fall into two general classes: thermal detectors and photodetectors. Thermal detectors usually have an active element of low thermal capacity coated with an absorbing black material. Detection relies on the change of some physical property of the element associated with the temperature elevation caused by the absorption of radiant power. The bolometer, thermocouple, pneumatic detector,[24] evaporograph,[25] and absorption-edge image tube[26] fall in this category.

Photodetectors utilize the action of individual photons upon electrons within a sensitive film or crystal either to eject the electron (photoemission) or to raise the electron to an excited state, as in photoconductive (PC), photovoltaic (PV), and photoelectromagnetic (PEM) detectors (see Sec. 16-3). Photodetectors generally have short response times (since no temperature rise is necessary), and are characterized by a sharp long-wavelength cut-off in sensitivity, since photons with energy below a certain cut-off value are unable to induce the electronic excitation on which the detection process depends. Photoconductors which require relatively little energy for excitation are sensitive to longer wavelengths but must be cooled to lower temperatures in order to avoid saturation by thermal energy. Photodetectors are

INFRARED 15-17

Optical Materials. (*Continued*).

Melting Point (°C)	Thermal Expansion Coefficient (10^{-6}/°C)	Solubility gr/100 ml of H_2O at 20°C	Soluble in	Comments
1045	4.5	0.00	—	**At 10 μ. Transmission is 38% in the range 1 to 10 μ.
523	4.9	0.00	CP4	***At 8 μ.
940	6.1	0.00	Hot H_2SO_4, Aqua regia	*Same note as for silicon above.
803	44	35.7	Glycerin, H_2O	Corrosive; hygroscopic.
458	30	0.00	NH_4OH, KCN	Corrosive; nonhygroscopic.
424	51	0.01	HNO_3, Aqua regia	Toxic; cold flows; nonhygroscopic.
415	60	0.02	HNO_3, Aqua regia	Toxic; cold flows; nonhygroscopic; high reflection loss.
636	48	124.3	—	Soft, scratches easily; hygroscopic.
621	50	44	—	Soft, scratches easily; hygroscopic.
—	—	—	—	Soft, easily scratched; becomes milky if overheated.
Distorts at 72	110–140	—	—	Soft, easily scratched.

termed intrinsic, *N*-type, or *P*-type depending upon whether they rely for their sensitivity upon electron excitation from the valence to the conduction band, from an impurity level to the conduction band, or from the valence band to an impurity level, respectively.

The logarithm of the mean-square noise current i_N in photoconductive detectors is plotted as a function of the frequency f in Fig. 15-11. At low frequencies (below 100 cps) the dominating noise power, termed "$1/f$" noise, is seen to be inversely proportional to frequency and appears to be due to surface effects (and possibly contact noise). At intermediate frequencies the noise follows the detector-response curve and

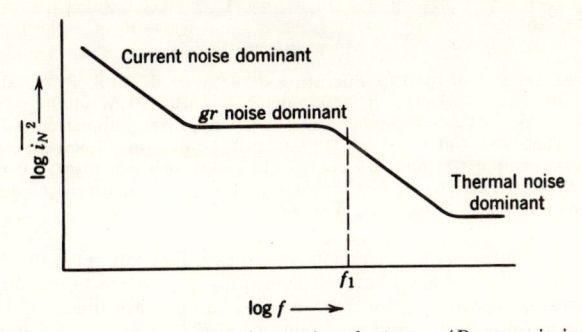

FIG. 15-11. Spectrum of noise current in semiconductors. (*By permission from P. W. Kruse, L. D. McGlaughlin, and R. B. McQuistan, Elements of Infrared Technology, John Wiley & Sons, Inc., New York, 1962.*)

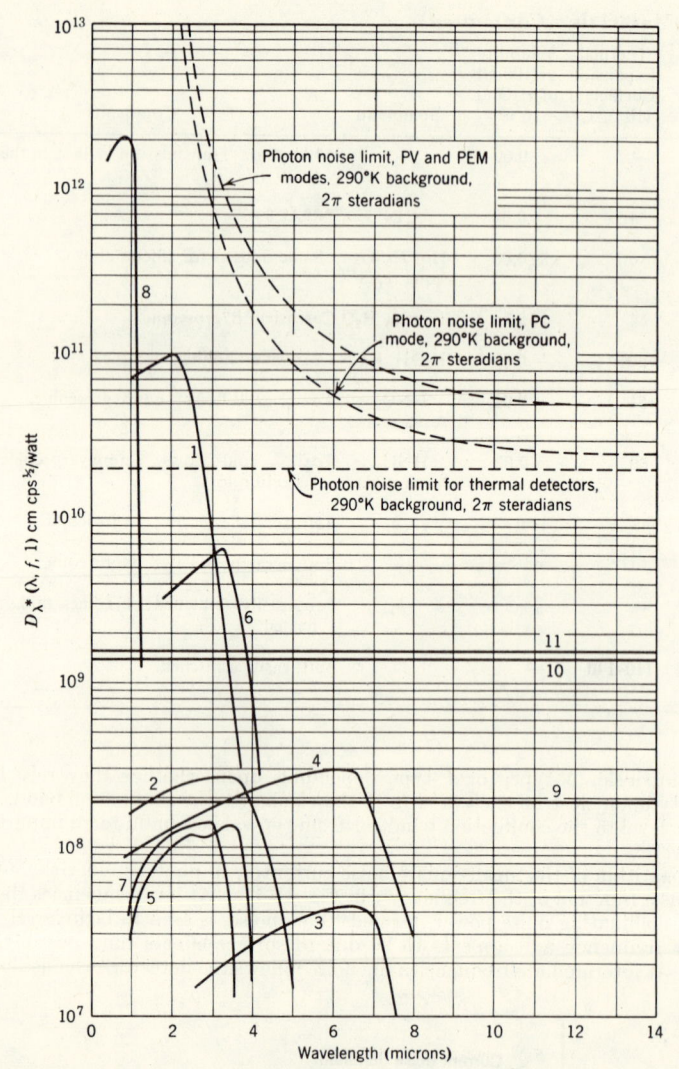

FIG. 15-12. Spectral D_λ^* of room-temperature detectors. 1. PbS, PC (250 μsec)(90 cps). 2. PbSe, PC (90 cps). 3. InSb, PC (800 cps). 4. InSb, PEM (400 cps). 5. InAs, PC (90 cps). 6. InAs, PV (frequency unknown)(sapphire immersed). 7. InAs, PEM (90 cps). 8. Tl$_2$S, PC (90 cps). 9. Thermistor bolometer (1,500 μsec)(10 cps). 10. Radiation thermocouple (36 msec)(5 cps). 11. Golay cell (20 msec)(10 cps). (*By permission from P. W. Kruse, L. D. McGlaughlin, and R. B. McQuistan, Elements of Infrared Technology, John Wiley & Sons, Inc., New York, 1962.*)

is caused by fluctuations in generation and recombination rates of charge carriers. When the charge carriers are liberated thermally by lattice vibrations this noise is termed "generation-recombination" noise. When they are liberated by background radiation, this noise is designated as "photon noise." Since, in PV and PEM detectors, these fluctuations occur only in the generation process, this type of noise is less by $\sqrt{2}$ in these detectors than in PC detectors. In this case the noise can be reduced

INFRARED 15-19

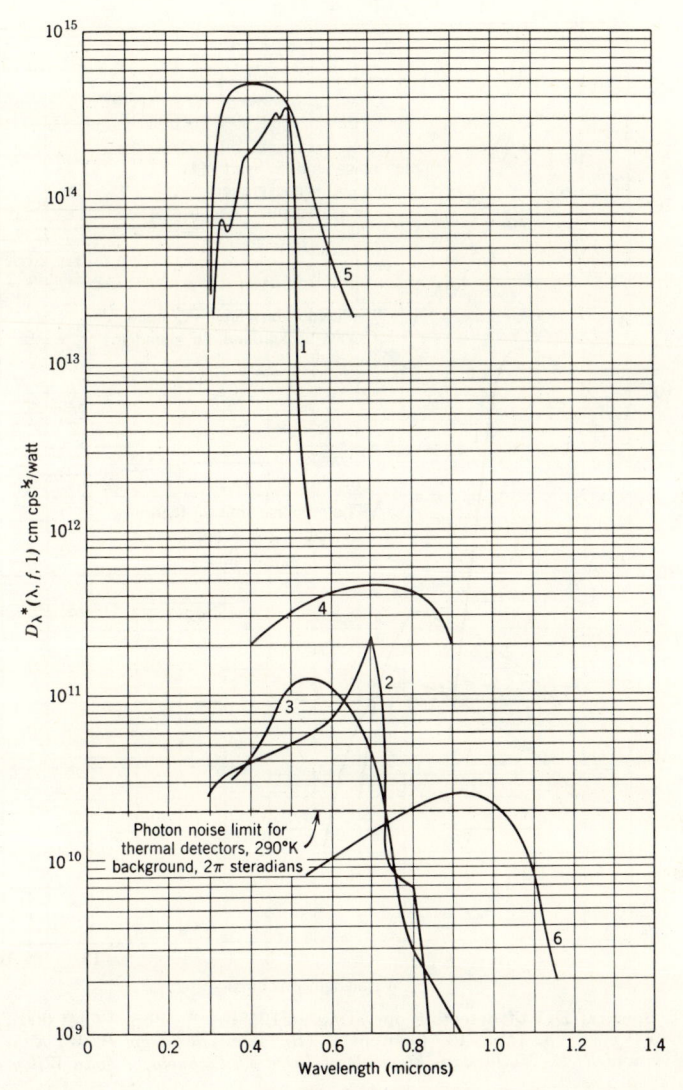

FIG. 15-13. Spectral D_λ^* of room-temperature detectors responding in the visible spectrum. 1. CdS, PC (90 cps). 2. CdSe, PC (90 cps). 3. Se-SeO, PV (90 cps). 4. GaAs, PV (90 cps). 5. 1P21 photomultiplier (measuring frequency unknown). 6. 1N2175 Si photo-duo-diode, PV (400 cps). (*By permission from P. W. Kruse, L. D. McGlaughlin, and R. B. McQuistan, Elements of Infrared Technology, John Wiley & Sons, Inc., New York, 1962.*)

by reducing the solid angle over which the detector "sees" the background. At higher frequencies thermal "white" noise is dominant.

A commonly used figure of merit for infrared detectors is the noise-equivalent power (NEP), usually measured in watts. It is defined as the rms value of the sinusoidally modulated radiant power falling on the detector which will give rise to an rms signal voltage equal to the rms noise voltage from the detector.

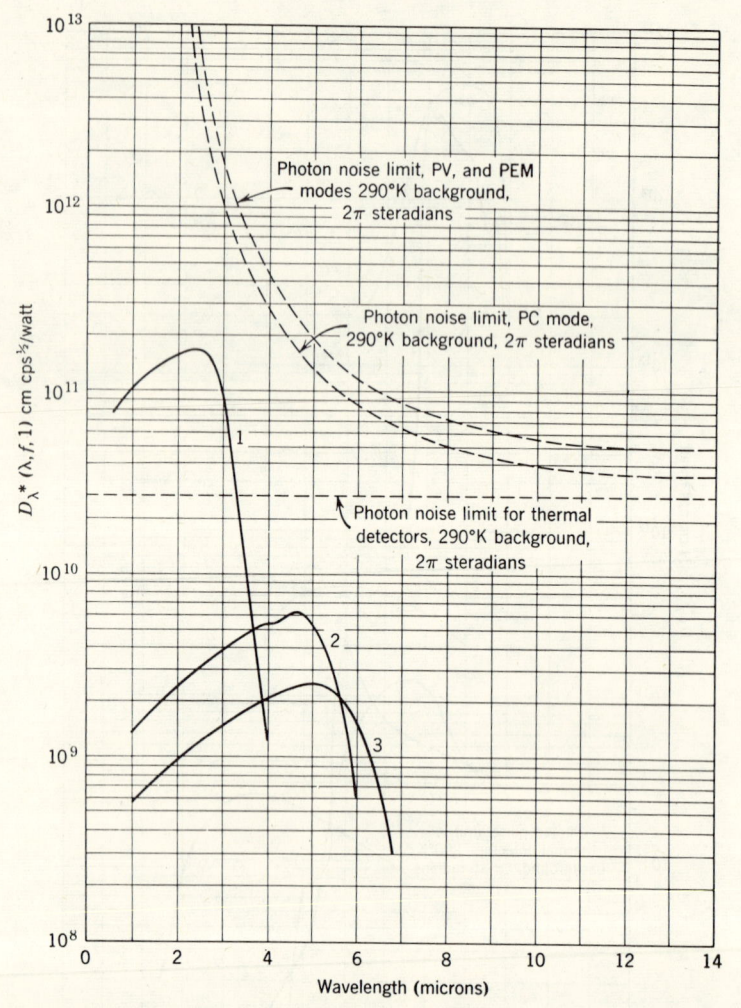

FIG. 15-14. Spectral D_λ^* of detectors operating at 195°K. 1. PbS, PC (1,000 cps). 2. PbSe, PC (900 cps). 3. InSb, PC (900 cps). (*By permission from P. W. Kruse, L. D. McGlaughlin, and R. B. McQuistan, Elements of Infrared Technology, John Wiley & Sons, Inc., New York, 1962.*)

A more generally useful figure of merit, the normalized detectivity D^* is defined by the relation

$$D^*(T°K, f \text{ cps}, 1 \text{ cps}) = \frac{(A \, \Delta f)^{1/2}}{\text{NEP}} \quad \text{cm cps}^{1/2}/\text{watt} \quad (15\text{-}4)$$

where A is the detector area and the NEP is measured using amplifier bandwidth of Δf cps, and blackbody radiation from a source at temperature $T°K$, modulated at a frequency of f cps. D^* is thus normalized for both detector area and bandwidth; it is, actually, the signal-to-noise ratio arising from a hypothetical detector of unit area upon irradiation by 1 watt of radiant power with a unit-bandwidth amplifier.

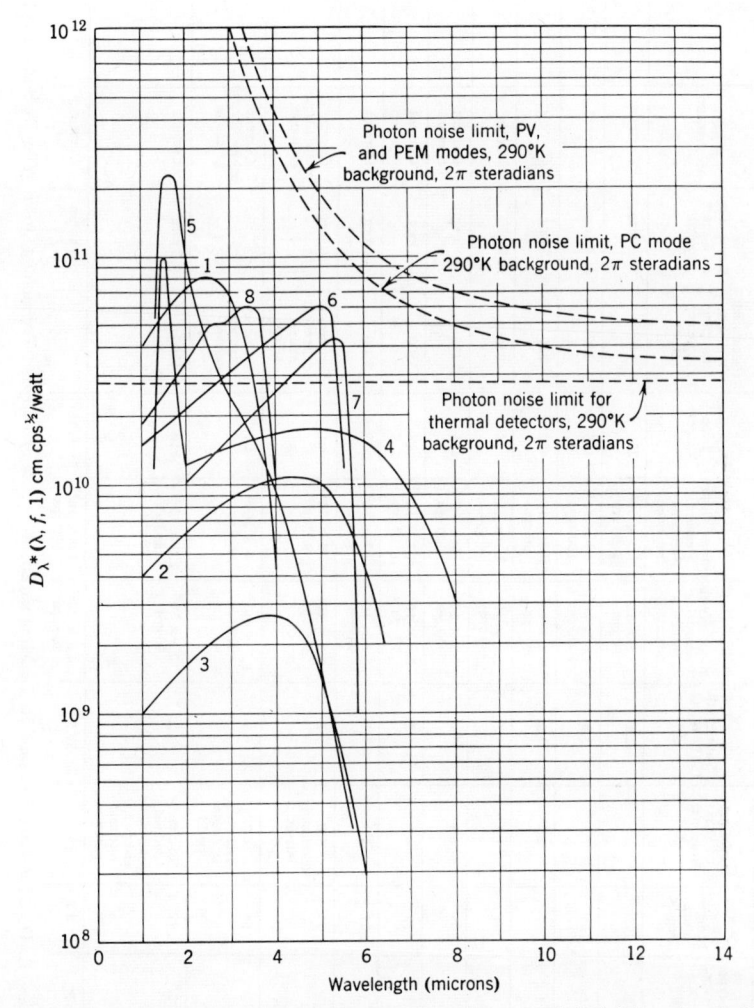

FIG. 15-15. Spectral D_λ^* of detectors operating at 77°K. 1. PbS, PC (90 cps). 2. PbSe, PC (90 cps). 3. PbTe, PC (90 cps). 4. Ge:Au, PC (900 cps). 5. Ge:Au,Sb, PC (90 cps). 6. InSb, PC (900 cps) (60° field of view). 7. InSb, PV (900 cps). 8. Te, PC (900 cps). (*By permission from P. W. Kruse, L. D. McGlaughlin, and R. B. McQuistan, Elements of Infrared Technology, John Wiley & Sons, Inc., New York, 1962.*)

When the NEP is measured using monochromatic radiation of wavelength λ, the corresponding normalized detectivity is characterized: D_λ^* (λ, f cps, 1 cps).

Table 15-4 presents the characteristics of a number of infrared detectors. Their spectral responses are presented in Figs. 15-12 to 15-16 together with the photon noise limit at the spectral peak for an ideal detector as a function of its threshold wavelength.

Until recently, the thought of taking a liquid nitrogen-, hydrogen-, or helium-cooled detector into the field was appalling. Miniature cryostats have been developed over the past few years, however, which make this procedure completely feasible. Cooled detectors can now be used even on small rockets. These minature cooling units may utilize gas from a bottle under pressure or from a compressor in a closed recirculating system.

Table 15-4. Performance of Elemental Detectors*

Material	Photon or Thermal	Mode of Operation	Film or Single Crystal	N-type, P-type, or Intrinsic	Operating Temperature (Deg. K)	Wavelength of Peak Response $\lambda_p(\mu)$	Cutoff Wavelength (50% value) $\lambda_0(\mu)$	D^* (500° K, f, 1) cm cps$^{1/2}$/watt (Measuring Frequency Indicated)	$D^*_{\lambda_p}$ ($\lambda_p, f, 1$) cm cps$^{1/2}$/watt (Measuring Frequency Indicated)	Response Time (μsec)	Calculated Optimum Chopping Frequency (cps)	Resistance per Square (ohm)	Noise Mechanism
1 PbS	P	PC	F	I	295	2.1	2.5	4.5×10^8 90 cps	1.0×10^{11} 90 cps	250	640	1.47 meg	Current
2 PbS	P	PC	F	I	195	2.5	3.0	4.0×10^9 1000 cps	1.7×10^{11} 1000 cps	455	350	4 meg	Current
3 PbS	P	PC	F	I	77	2.5	3.3	4.0×10^9 90 cps	8.0×10^{10} 90 cps	455	350	5 meg	Current
4 PbSe	P	PC	F	I	295	3.4	4.2	3.0×10^7 90 cps	2.7×10^8 90 cps	4	40 kc	50 kohm	Current
5 PbSe	P	PC	F	I	195	4.6	5.4	7.5×10^8 900 cps	6×10^9 900 cps	125	1270	40 meg	Current below 6 kc
6 PbSe	P	PC	F	I	77	4.5	5.8	2.2×10^9 90 cps	1.1×10^{10} 90 cps	48	3300	5 meg	Current
7 PbTe	P	PC	F	I	77	4.0	5.1	3.8×10^8 90 cps	2.7×10^9 90cps	25	6500	32 meg	Current
8 Ge:Au	P	PC	SC	P	77	5.0 (excluding intrinsic peak)	7.1	7.5×10^9 900 cps	1.75×10^{10} 900 cps	<1	Frequency independent above 40 cps	1.0 meg	Current below 40 cps, gr above
9 Ge:Au	P	PC	SC	P	65	4.7 (excluding intrinsic peak)	6.9	1.7×10^{10} 900 cps	4×10^{10} 900 cps	<1	Frequency independent above 40 cps		Current below 40 cps, gr above
10 Ge:Au,Sb	P	PC	SC	N	77	No clearly defined peak exists except for intrinsic excitation.		2.9×10^9 90 cps	2.5×10^{10} at 3μ 90 cps	110	1500	1.0 meg	Current
11 Ge:Zn (Zip)	P	PC	SC	P	4.2	36	39.5	4.0×10^9 800 cps	1.0×10^{10} 800 cps	<0.01		300 kohm	Current

* By permission from P. W. Kruse, L. D. McGlaughlin, and R. B. McQuistan, *Elements of Infrared Technology*, John Wiley & Sons, Inc., New York, 1962.

12 Ge:Zn,Sb	P	PC	SC	N	50	12	15	2 × 10⁹ 900 cps	3 × 10⁹ 900 cps		0.1 meg	Current below 1 kc; gr above 1 kc
13 Ge:Cu	P	PC	SC	P	<20	20	27	1 × 10¹⁰ (60° field of view) 900 cps	2.5 × 10¹⁰ (60° field of view) 900 cps			Current below 500 cps; gr above 500 cps
14 Ge:Cd	P	PC	SC	P	<25	16	21.5	7 × 10⁹ (60° field of view) 500 cps	1.8 × 10¹⁰ (60° field of view) 500 cps			
15 Ge-Si:Au	P	PC	SC	P	50	7.3	10.1	3.1 × 10⁹ 90 cps	7.0 × 10⁹ 90 cps	0.1	Frequency independent to approximately 1 Mc	10 meg	gr
16 Ge-Si:Zn,Sb	P	PC	SC	P	50	10	13.3	4.0 × 10⁹ 100 cps	1.0 × 10¹⁰ 100 cps	0.1	Frequency independent below approximately 1 Mc	20 meg	gr
17 InSb	P	PC	SC	I	295	6.5	7.3	1.4 × 10⁷ 800 cps	4.3 × 10⁷ 800 cps	0.2	Frequency independent to 500 kc	20	Thermal
18 InSb	P	PC	SC	I	195	5.0	6.1	5 × 10⁸ 900 cps	2.5 × 10⁹ 900 cps	<1	Frequency independent above 500 cps	60	Current below 400 cps
19 InSb	P	PC	SC	P	77	5.0	5.4	1.2 × 10¹⁰ (60° field of view) 900 cps	6 × 10¹⁰ (60° field of view) 900 cps	<2		10 kohm	Current
20 InSb	P	PV	SC	PN	77	5.3	5.6	8.6 × 10⁹ 900 cps	4.3 × 10¹⁰ 900 cps	<1	Frequency independent above 500 cps	1 kohm	Current below 100 cps; gr above
21 InSb	P	PEM	SC	I	295	6.2	7.0	1.0 × 10⁸ 400 cps	3.0 × 10⁸ 400 cps	0.2	Frequency independent below 100 kc	20	Thermal

15–23

Table 15-4. Performance of Elemental Detectors. (Continued).

Material	Photon or Thermal	Mode of Operation	Film or Single Crystal	N-type, P-type, or Intrinsic	Operating Temperature (Deg. K)	Wavelength of Peak Response $\lambda_p(\mu)$	Cutoff Wavelength (50% value) $\lambda_0(\mu)$	D^* (500°K, f, 1) cm cps$^{1/2}$/watt (Measuring Frequency Indicated)	$D\lambda_p^*$ (λ_p, f, 1) cm cps$^{1/2}$/watt (Measuring Frequency Indicated)	Response Time (μsec)	Calculated Optimum Chopping Frequency (cps)	Resistance per Square (ohm)	Noise Mechanism
22 InAs	P	PC	SC	N	295	3.6	3.8	1.4×10^7 90 cps	1.4×10^8 90 cps	0.2	Frequency independent below 100 kc		
23 InAs	P	PV	SC	PN	295	3.4	3.7	2.5×10^8 90 cps	2.5×10^9 750 cps	2		50	Assumed thermal
24 InAs	P	PEM	SC	N	295	2.5	3.4	1.4×10^7 90 cps	1.4×10^8 90 cps	0.2	Frequency independent below 100 kc		Assumed thermal
25 Te	P	PC	SC	P	77	3.5	3.8	4.0×10^9 900 cps	6.0×10^{10} 900 cps	60	2700	2 kohm	Current
26 Tl$_2$S	P	PC	F	I	295	0.9	1.1		2.2×10^{12} 90 cps	530	300	5 meg	Current
27 86% HgTe 14% CdTe	P	PC	SC	I	295	6	6.5	5×10^6	1.5×10^7			~1	
28 Thermistor Bolometer	T	Bolometer			295			1.95×10^8 (1.5 millisec) 10 cps	1.95×10^8 (1.5 millisec) 10 cps	1500	Frequency independent below 30 cps	2.4 meg	Thermal
29 Radiation Thermocouple	T	Thermoelectric effect			295			1.4×10^9 5 cps	1.4×10^9 5 cps	3.6×10^4	<5	5	Thermal
30 Golay Cell	T	Expansion of air			295			1.67×10^9 10 cps	1.67×10^9 10 cps	2×10^4	<5		Temperature
31 NbN Bolometer	T	Superconducting Bolometer			15			4.8×10^9 360 cps	4.8×10^9 360 cps	500		0.2	Unknown

32 Carbon Bolometer	T	Bolometer		2.1			4.25×10^{10} 13 cps	10^4	16	0.12 meg	Current
33 CdS	P	PC	SC F Sintered		295	0.5	0.51	4.25×10^{10} 13 cps			
								3.5×10^{14} 90 cps	5.3×10^4	3	5×10^{11}	Current
34 CdSe	P	PC	SC Sintered		295	0.7	0.72	2.1×10^{11} 90 cps	1.2×10^4	13	1.5×10^{11}	Current
35 Se-SeO	P	PV	SC	PN	295	0.55	0.69	1.2×10^{11} 90 cps	910	160	3 kohm Area dependent	
36 GaAs	P	PV	SC	PN	295	0.8	0.89	4.5×10^{11} 400 cps	1000	160	4.6 meg Area dependent	Current
37 IN 2175 Photo-duodiode	P	PC	SC	PN	295	0.95	1.07	2.5×10^{10} 400 cps	8	20 kc	4×10^9	
38 1P21 Photo-multiplier	P	PE	F		295	0.40	0.53	5×10^{14} 1000 cps	<0.01	Frequency independent to about 100 Mc		Shot

Notes (Numbers correspond to those in column 1)

1,2,3 Detectors with time constants ranging from about 1 μsec to 10,000 μsec are available. The detectivity will vary with time constant according to the McAllister relation. The cutoff wavelength, may also be shifted to greater values with a sacrifice in detectivity. Detectors operating at 77°K may exhibit double time constants.
6 May exhibit double time constant.
7 Resistance may be reduced by grid type electrodes. Detector is background limited. Performance at 90°K same as at 77°K. May have second time constant for 1.5 μ radiation.
8 Exhibits long time constant for intrinsic excitation (less than 2 μ). See below.
9 Exhibits long time constant for intrinsic excitation (less than 2μ).
10 Detectivity at 90°K equals that at 77°K. Exhibits wavelength dependent time constant.
12 Not readily available.
15 Spectral response can be changed by varying alloy composition. Frequency response may be limited by RC time constant.
16 Spectral response can be changed by varying alloy composition. Frequency response may be limited by time RC constant.
19 Responsivity is superior to InSb PV, 77°K.
20 May be either broad area diffused junction or line type of grown junction. May be operated with or without bias voltage.

21 Maximum dimensions approx. 2 × 10 mm. Should be transformer coupled to amplifier. Sensitive to magnetic pickup from electrical mains.
22 Not readily available.
23 Detector is sapphire immersed.
24 Not readily available.
25 Peak detectivity in solar and earth background minimum.
26 Not readily available.
27 Not readily available.
28 Detectors with time constants ranging from about 1 msec to 50 msec are available.
29 Widely used in infrared spectroscopy.
30 Fragile, microphonic.
31 Not readily available. Noise appears to arise from some unknown mechanism associated with superconductivity.
32 Not readily available. Use of quartz and paraffin filters cut out response at wavelengths shorter than 40 μ.
33 Highest responsivity of any photoconductor.
34 Responds to longer wavelengths and is faster than CdS.
35 Used in exposure meters.
36 Useful for star tracking.
37 Very small over-all size.

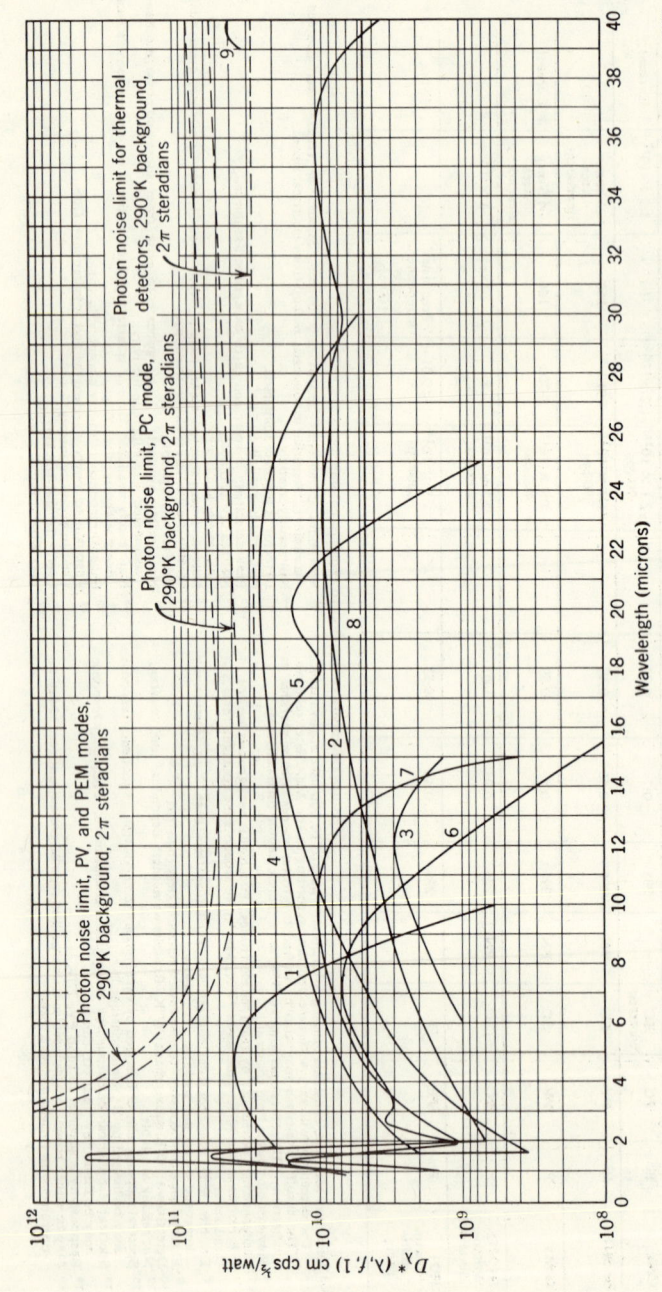

FIG. 15-16. Spectral D_λ^* of detectors operating at temperatures below 77°K. 1. Ge:Au, 65°K, PC (900 cps). 2. Ge:Zn, 4.2°K, PC (800 cps). 3. Ge:Zn,Sb, 50°K, PC (900 cps). 4. Ge:Cu, 4.2°K, PC (900 cps, 60° field of view). 5. Ge:Cd, 4.2°K, PC (500 cps, 60° field of view). 6. Ge-Si: Au, 50°K, PC (90 cps). 7. Ge-Si: Zn,Sb, 50°K, PC (100 cps). 8. NbN superconducting bolometer, 15°K (360 cps). 9. Carbon bolometer, 2.1°K (13 cps). (*By permission from P. W. Kruse, L. D. McGlauphlin, and R. B. McQuistan, Elements of Infrared Technology, John Wiley & Sons, Inc., New York, 1962.*)

15-6. Infrared-system-design Consideration

The design of an infrared system is an iterative process. Thus a tentative set of components and design parameters is first selected, and the performance and shortcomings of the hypothetical system so conceived are determined. An improved system is then tried, and so on, in an attempt to develop the system which optimally fulfills the prevailing boundary conditions.

The performance of a detection system is usually described in terms of angular coverage, probability of target detection at the required range, false-alarm rate, and mean time to failure in relation to size, weight, power requirement, and cost. These performance requirements usually dictate the signal-to-noise ratio required in any specified target configuration as it relates to angular resolution, field of view, frame rate, collector area, cooling requirements, etc. The angular resolution, field of view, and frame rate usually determine the dwell time of the target image on the detector, and hence the required amplifier bandwidth.

As a culmination, in effect, of the discussion in the preceding sections of this chapter, the signal-to-noise output of an infrared system s/n can be written

$$\frac{s}{n} = \frac{A_0 E_s}{r^2 (A_d \, \Delta f)^{1/2}} \int_{\lambda_1}^{\lambda_2} J_\lambda T_a T_0 D_\lambda^* \, d\lambda \quad (15\text{-}5)$$

where A_0 = collector area, sq cm
E_s = efficiency of the scanning pattern employed
r = range to target, cm
A_d = detector area, sq cm
Δf = amplifier bandwidth, cps
J_λ = spectral radiant intensity of the target relative to its background, watts/(steradian) (micron)
T_a, T_0 = transmission of atmosphere and optical system, respectively
D_λ^* = normalized detectivity of detector, cm cps$^{1/2}$/watt
λ_1, λ_2 = define the useful spectral range, microns

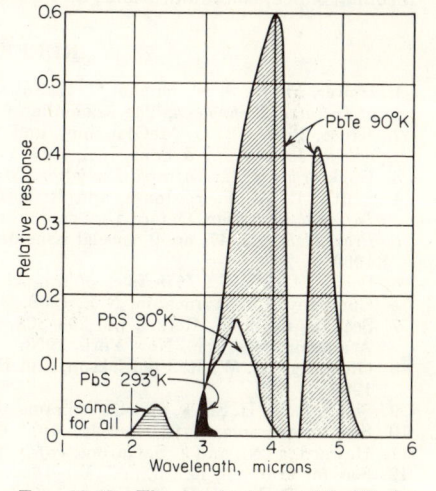

FIG. 15-17. The products $J_\lambda \, D^*_\lambda \, T_a$ for PbS at 293°K and 90°K, and PbTe at 90°K for 500°K source and 17 mm precipitable water vapor. (*By permission from Arthur S. Locke, Guidance, D. Van Nostrand Company, Inc., Princeton, N.J., 1955.*)

Since T_a is a function of range, an explicit solution of this equation for r does not yield a convenient function.

An example of the effects of the factors J_λ, D_λ^*, and T_a is presented in Fig. 15-17 in which the product of these factors is plotted against λ for a 500°K blackbody, detected through 17 mm of precipitable water vapor, by PbS detectors at 293 and 90°K, and by a PbTe detector at 90°K.

15-7. Applications of Infrared Systems

Infrared equipments are used in a wide variety of applications. In the laboratory, infrared spectrometers are used for chemical analysis. The atmospheres of the earth, sun, planets, and stars have been investigated using infrared techniques. The temperatures of celestial bodies have been calculated on the basis of infrared measurements. Satellite and star tracking has been accomplished in this way. Information concerning the earth's cloud cover, both day and night, has been acquired from artificial satellites carrying scanning radiometers. The "heat budget" of the atmosphere (Sec. 5-2) is determined using infrared techniques.

Chemical analysis by infrared means has proved useful in the field of agriculture, and in the rubber, petroleum, printing, and cement industries. In the railroad industry, infrared techniques have been used to detect hot journal boxes, using trackside infrared detection systems. In the aircraft and missile industries, fuel studies and aerodynamic-heating effects have been investigated using infrared techniques. Infrared communications systems provide a high degree of security.

The infrared technique competes most successfully with such other techniques as radar in space applications, where the problems of atmospheric attenuation and refraction are avoided and background radiance is low. Missiles, rockets, and satellites may be detected using this technique, and weather observations, both day and night, can be made effectively from a satellite. Infrared horizon sensors are used in earth-oriented satellites.

Military applications of infrared include the detection and tracking of targets, reconnaissance, communications, and gas analysis.

REFERENCES

1. Holter, M. R., S. Nudelman, G. H. Suits, W. L. Wolfe, and G. J. Zissis: *Fundamentals of Infrared Technology*, The Macmillan Company, New York, 1962.
2. Kruse, P. W., L. D. McGlaughlin, and R. B. McQuistan: *Elements of Infrared Technology*, John Wiley & Sons, Inc., New York, 1962.
3. Hackforth, H. L.: *Infrared Radiation*, McGraw-Hill Book Company, New York, 1960.
4. Smith, R. A., F. E. Jones, and R. P. Chasmar: *The Detection and Measurement of Infrared Radiation*, Oxford University Press, Fair Lawn, N.J., 1957.
5. *Proc. IRE*, vol. 47, no. 9, special issue on infrared physics and technology, September, 1959.
6. Locke, Arthur S.: *Principles of Guided Missile Design*, chap. 5, D. Van Nostrand Company, Inc., Princeton, N.J., 1958.
7. See, for example, Ref. 4, pp. 293–308, and R. C. Jones, *Advances in Electronics*, V, Academic Press Inc., New York, 1953.
8. Ginsburg, N., W. R. Fredrickson, and R. Paulson: *J. Opt. Soc. Am.*, vol. 50, p. 1176, 1960.
9. Sloan, R., J. H. Shaw, and D. Williams: *J. Opt. Soc. Am.*, vol. 45, p. 455, 1955.
10. See, for example, R. C. Ramsey, *Appl. Opt.*, vol. 1, p. 465, 1962.
11. Howard, J. N., and J. S. Garing: *Infrared Phys.*, vol. 2, p. 155, 1962.
12. See, for example, Ref. 5, pp. 1451–1457, and Ref. 11.
13. Taylor, J. H., and H. W. Yates, *J. Opt. Soc. Am.*, vol. 47, p. 223, 1957.
14. Prostak, A., A. J. LaRocca, J. Livisay, and R. Nichols: *Proc. IRIS*, vol. 5, no. 2, 1960.
15. See, for example, Ref. 2, pp. 172–181.
16. Larmore, L.: Transmission of Infrared Radiation through the Atmosphere, *Proc. IRIS*, vol. 1, no. 1, June, 1956 (see also Ref. 3, pp. 54, 59).
17. Wyatt, P. J., V. R. Stull, and G. N. Plass: *Appl. Opt.*, vol. 3, p. 229, 1964.
18. Altshuler, T. L.: *Infrared Transmission and Background Radiation by Clear Atmospheres*, General Electric Missile and Space Vehicle Department, Valley Forge, Pa., Document 61SD199, December, 1961.
19. Ballard, S. S., K. A. McCarthy, and W. L. Wolfe: IRIA State-of-the-art Report on Optical Materials for Infrared Instrumentation, *Univ. Michigan Rept.* 2389-11-S, 1959.
20. See, for example, J. Grant, E. Michel, and A. Thelen, *Infrared Phys.*, vol. 2, p. 123, 1962.
21. See, for example, A. Bouwers, *Achievements in Optics*, American Elsevier Publishing Company, New York, 1950.
22. For an excellent discussion of the methods of describing backgrounds, see the article by D. Z. Robinson in Ref. 5, pp. 1554–1561.
23. See Ref. 5, pp. 1561–1568.
24. An excellent discussion of thermal detectors may be found in Ref. 4.
25. McDaniel, G. W., and D. Z. Robinson: *Appl. Opt.*, vol. 1, p. 311, 1962.
26. Hilsum, C., and W. R. Harding: *Infrared Phys.*, vol. 1, p. 67, 1961.
27. Jamieson, J. A., G. N. Plass, R. H. McFee, R. H. Grube, and R. G. Richards: *Infrared Physics and Engineering*, McGraw-Hill Book Company, New York, 1963.

Chapter 16

OPTICAL SENSORS

DUNCAN E. MACDONALD, *Itek Corporation.*
RAYMOND L. DUSSAULT, *Itek Corporation.*

CONTENTS

16-1. The Nature of Light	16-2
16-2. Energy Sources	16-2
Thermal Radiation and Radiometry	16-3
Photometry: Luminous Flux, Efficiency, and Intensity	16-4
Luminance	16-5
Illuminance	16-6
Optical Filters	16-7
16-3. Radiation Sensors	16-7
Classification of Radiation Detectors	16-7
The Human Eye	16-8
Photographic Emulsions	16-9
Photoemissive Tubes	16-10
Photoconductive Cells	16-10
Photovoltaic Cells	16-11
16-4. Practical Considerations	16-12
16-5. Resolution	16-13

This chapter, which parallels and supplements the preceding chapter on infrared, discusses detectors of electromagnetic radiation which are responsive to the same region of the spectrum to which the human eye responds, i.e., "visible light." Optical sensors, so defined, are of three basic types: photochemical, photoelectric, and photothermal. These classifications are derived from the process by which radiation is converted to a usable or measurable display, e.g., a photograph or a change in electrical or thermal property of the sensor. The following pages consider the nature of electromagnetic radiation in the visible region of the spectrum, the emittance (spectral quality and quantity) of energy sources in this region, the qualitative and quantitative response of sensors, and the method of conversion from detected energy to useful information.

The system engineer concerned with optical sensors must optimize the following: light source, energy detector, method of energy conversion or amplification, and finally, secondary detection and/or visual display.

16-1. The Nature of Light

Because light is an electromagnetic wave, it can be described by the basic relationship

$$c = \lambda \nu$$

where c is the velocity, λ the wavelength, and ν the frequency. In a vacuum, light has a constant velocity of approximately 3×10^{10} cm/sec. The normal eye sees over a range from violet to red, which may be expressed in terms of frequencies but is more commonly described by wavelengths in angstrom units (1 A = 10^{-8} cm), millimicrons (1 mμ = 10^{-7} cm), or microns (1 micron = 10^{-4} cm), and sometimes by wave number where WN = $10^8/\lambda$ for λ in angstroms.

The interaction of this radiant energy with matter is usually best considered from the viewpoint of quantum theory, where the photon is the basic unit of light energy. Planck's relationship for the energy of a photon and its frequency is

$$E = h\nu$$

where E is the energy in ergs and h is Planck's constant (6.624×10^{-27} erg-sec). From this it follows that the momentum of a photon is given by $h\nu/c$ and its mass equivalent by $h\nu/c^2$. Table 16-1 lists the magnitudes of some of these quantities. In addition, it can be determined that a near-infrared photon has a momentum of 6×10^{-23} g cm/sec and a mass equivalent of 2×10^{-33} g/photon.

Table 16-1. Wave and Energy Characteristics of Radiation in the Near and Visible Portions of the Spectrum

Spectral region	λ, A	λ, microns	WN $\times$ 10^4/cm	$\nu \times 10^{-12}$/sec	Energy of photon $\times 10^{-12}$ ergs
Near infrared.........	10,000	1.00	1.00	300	1.99
Visible red limit......	7,200	0.72	1.39	417	2.76
Visible violet limit....	4,000	0.40	2.50	750	4.97
Ultraviolet...........	1,000	0.10	10.0	3,000	19.87

While these relationships are seldom employed near the visible portion of the spectrum in engineering aspects, they do enable us to understand better the transition to the extremely high-energy end of the electromagnetic spectrum, i.e., gamma rays and secondary atomic radiations. For where $mv = h\nu/c$ we can write instead $\lambda = h/mv$ and then consider de Broglie's fundamental postulate of wave mechanics for the wavelength of a moving electron. In this high-energy region the energy is measured in units of electron volts (the electron charge of 4.803×10^{-10} esu being accelerated through a potential difference of 1 volt; therefore, 1 electron volt (ev) equals 1.60×10^{-12} ergs). To a good approximation the wavelength in angstroms of a moving electron is given by $\lambda = \sqrt{150/\text{ev}}$. For example, an electron of 600 volts has a wavelength of 0.5 angstrom.

16-2. Energy Sources

In the broadest sense there are two classes of radiators, those which radiate in discrete wavelengths (line spectra) and those which radiate in all wavelengths (continuous spectra). Materials in the gaseous state through which an electrical discharge is passing, such as a neon tube or a sodium-vapor lamp, are typical of line-spectra sources. The line spectrum emitted by an electrically excited gas is characteristic of the atoms composing it. For these types of sources, Bohr's second postulate is applicable. This postulate states that, when an atom is excited, i.e., absorbs energy, an electron is raised to a higher-energy state. This excited state is inherently unstable, and the electron at some time will probably "jump" back to a lower-energy state, emitting radiant energy equal to the energy difference between the two states. When this occurs, a single photon of frequency ν is emitted whose energy is given by the relationship

$$E_1 - E_2 = h\nu$$

where h is Planck's constant and E_1 and E_2 are the initial and final energies.

For most continuous sources, the spectral distribution of energy is determined chiefly by the temperature of the emitting surface and is not characteristic of the material of which it is composed. Solids and liquids radiate visible light at temperatures of 500°C or more. Examples of such continuous sources are the sun, the tungsten lamp, and the flame. The phosphor is another type of continuous source where the characteristics of the radiation are dependent on the characteristics of the material. In this case, the absorption, the storage, and the emission of energy are each controlled by the nature of the crystal. An example of the spectral-energy emission of a typical phosphor (P-17) is shown in Fig. 16-1.

Thermal Radiation and Radiometry. The radiant energy striking a surface or emitted by a source per unit time is called the radiant flux P and is expressed in watts in the mks system. Radiant flux incident on a surface per unit area is called irradiance H, expressed in watts per square meter. A fraction of the incident flux is reflected, the remainder absorbed or transmitted. The fraction reflected is called reflectance r, that absorbed is called the absorptance a, and that transmitted the transmittance t. For an opaque surface, $r + a = 1$.

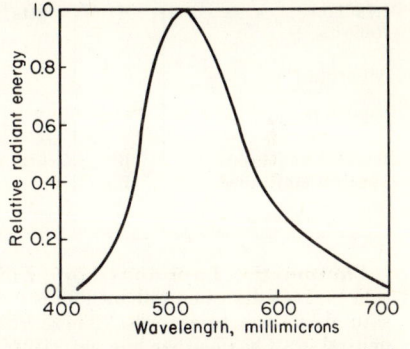

Fig. 16-1. Spectral emittance of a typical phosphor (P-17).

The rate at which radiant energy is emitted per unit area is called radiant emittance W, in watts per square meter. Kirchhoff's law states that the ratio of radiant emittance to absorptance is the same for all surfaces at the same temperature. This is expressed as $W = aH$. A surface which absorbs all the radiant energy incident on it would appear black (assuming its temperature is not so high that it is self-luminous) and it is called a blackbody. From Kirchhoff's law, since $a = 1$ for a blackbody,

$$W = aW_{bb}$$

where W_{bb} is the radiant emittance of a blackbody. The spectral emittance W_λ of a blackbody is defined by $dW_{bb}/d\lambda$, the radiant emittance per unit wavelength at a given temperature. The units of spectral emittance are watts/(sq m) (mμ). For this case, Wien's displacement law states that the wavelength of maximum spectral emittance λ_m is inversely proportional to absolute temperature. This is given by the relationship

$$\lambda_m T = 2.8971 \times 10^6$$

where λ_m is in millimicrons, T in degrees Kelvin.

The radiant emittance W_{bb} of a blackbody can be found by integration and is given by

$$W_{bb} = \sigma T^4$$

the Stefan-Boltzmann equation, where T is absolute temperature and σ is the Stefan-Boltzmann constant equal to 5.672×10^{-8} when T is in degrees Kelvin and W_{bb} in watts per square meter. The emissivity e of a surface is defined as the ratio of the radiant emittance of the surface to that of a blackbody at the same temperature.

The Stefan-Boltzmann equation then becomes

$$W = e\sigma T^4$$

The spectral emittance of a blackbody showing λ_m for increasing temperature is shown in Fig. 16-2. A summary of radiometric quantities and definitions is shown in Table 16-2; see also Sec. 15-2.

Table 16-2. Summary of Radiometric Quantities and Definitions

Radiometric quantity	Symbol	Unit	Definition
Radiant flux	P	watts	Radiant energy crossing or striking a surface per unit time
Irradiance	H	watts/sq m	Radiant flux incident per unit area
Reflectance	r	. . . .	Fraction of incident radiant flux reflected
Absorptance	a	. . . .	Fraction of incident radiant flux absorbed
Emissivity	e	. . . .	Ratio of radiant emittance of a body to that of a blackbody
Radiant emittance	W	watts/sq m	Radiant flux emitted per unit area
Spectral emittance	W_λ	watts/(sq m) (mμ)	Radiant emittance per unit range of wavelength

Photometry: Luminous Flux, Efficiency, and Intensity. The visual sensation that results when radiant flux is incident on the retina has three attributes: hue, saturation, and brightness. These attributes make up the sensation of color. A neutral gray has neither hue nor saturation and is called achromatic. Brightness is that attribute of any color sensation which permits it to be classified as equivalent to

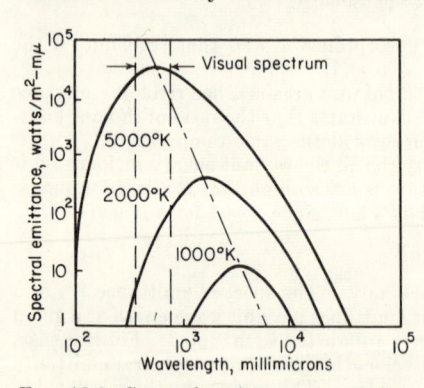

Fig. 16-2. Spectral emittance for various temperatures.

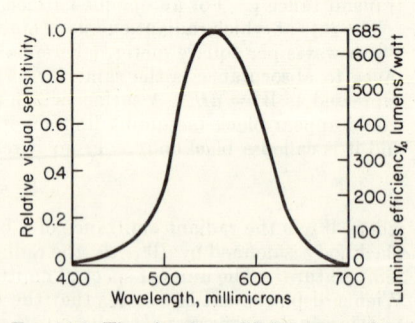

Fig. 16-3. The visual-sensitivity curve of the human eye.

the sensation produced by some member of a series of neutral (achromatic) grays. Equal amounts of radiant flux of different wavelengths do not produce visual sensations of equal brightness. For purposes of standardization of photometric data, the curve shown in Fig. 16-3 has been adopted by the International Commission on Illumination as the standard luminosity curve for the human eye; see also Fig. 31-3.

With the aid of the standard relative-luminosity curve, a given sample of radiant flux can be evaluated with respect to its brightness-producing capacity. Radiant flux, evaluated with respect to its capacity to evoke the sensation of brightness, is called luminous flux F. The unit of luminous flux is the lumen. One watt of radiant

OPTICAL SENSORS 16-5

flux of wavelength 555 mμ is equal to a luminous flux of 685 lumens. One lumen, therefore, is equal to 1.46×10^{-3} watts.

The luminous efficiency of any sample of flux is given by

$$\text{Luminous efficiency} = \frac{\text{luminous flux}}{\text{radiant flux}} = \frac{F}{P}$$

For a radiating blackbody the luminous efficiency must be found by integration methods since the radiant flux is not monochromatic. For a blackbody

$$\frac{F}{P} = 685 \frac{\int_0^\infty vf(\lambda)d\lambda}{\int_0^\infty f(\lambda)d\lambda}$$

where v is the relative luminosity at any wavelength λ and $f(\lambda)$ is Planck's equation for a radiating blackbody. Figure 16-4 shows the luminuous efficiency of a blackbody as a function of absolute temperature.

The luminous intensity I of a point source, in a given direction, is defined as the ratio of the luminous flux dF to the solid angle $d\omega$ or flux emitted per unit solid angle

$$I = \frac{dF}{d\omega}$$

The unit of luminous intensity is lumens per steradian. This unit is called the candle. A source whose intensity in all directions is I candles emits, therefore, a total of $4\pi I$ lumens.

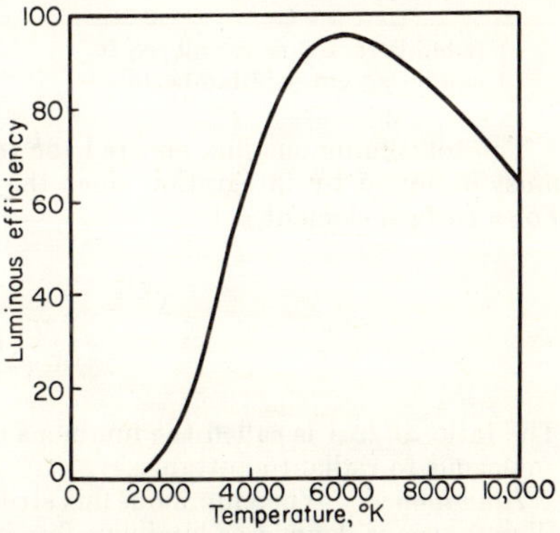

FIG. 16-4. Luminous efficiency of a blackbody's radiant flux.

Luminance. For extended light sources, which may be self-luminous, diffusely reflecting, or transmitting, the intensity is found to be different in different directions, and for many surfaces it varies in accordance with Lambert's law:

$$\Delta I_\theta = \Delta I_m \cos \theta$$

where ΔI_θ is the intensity in a direction making an angle θ with the normal to the surface and ΔI_m is the intensity along the normal. A surface element (self-luminous, transmitting, or reflecting) which emits (transmits or reflects) in this manner is said to be perfectly diffuse. Specular sources are defined as those which emit (reflect or transmit) unidirectionally.

The luminance of a surface element in any direction is defined as the ratio of the intensity of the element in that direction ΔI_θ to the area of the element projected on a plane perpendicular to the direction. The luminance in the direction θ is given by

$$B_\theta = \frac{\Delta I_\theta}{\Delta A \cos \theta}$$

where ΔA is the unit of area. Luminance is, therefore, intensity per unit projected area of a source. If the intensity is in candles, the luminance of the source is in candles/sq m or candles/sq ft. The standard candle, as such, is defined as the unit of luminous intensity such that the luminance of a blackbody at the temperature of freezing platinum is 600,000 candles/sq m. Table 16-3 gives typical values of luminance of some light sources.

Table 16-3. Typical Values of Luminance of Some Light Sources

Light source	Ft-lamberts*	Candles/sq ft	Candles/sq cm†
Sun (seen without earth atmosphere)...	6.57×10^8	2.09×10^8	2.25×10^5
Moon............................	1.67×10^3	5.33×10^2	5.74×10^{-1}
Blue sky.........................	1.81×10^2	5.76×10	6.20×10^{-2}
Snow (in sunlight)................	4.97×10^3	1.58×10^3	1.71
Candle flame.....................	2.36×10^3	7.50×10^2	8.07×10^{-1}
Blackbody, 4000°K................	7.19×10^7	2.29×10^7	2.46×10^4
Tungsten, 3000°K.................	3.68×10^5	1.17×10^5	1.26×10^2
Filament lamp, 500-watt...........	9.50×10^4	3.02×10^4	3.26×10
Mercury arc......................	9.00×10^7	2.88×10^7	3.10×10^4
Mercury-vapor lamp...............	6.80×10^3	2.16×10^3	2.33
Fluorescent lamp, 40-watt.........	9.00×10^2	2.28×10^2	3.10×10^{-1}
Carbon arc.......................	3.79×10^7	1.21×10^7	1.30×10^4
High-intensity arc................	2.48×10^8	7.90×10^7	8.50×10^4

* 1 ft-lambert = $1/\pi$ candle/sq ft
† 1 candle/sq cm = π lamberts

The total luminous flux emitted (or transmitted or reflected) by a diffuse surface must be found by integration since the intensity is not the same in all directions. For a surface element ΔA,

$$\Delta F = \frac{2\pi \Delta A}{\int_0^{\pi/2} B_\theta \sin \theta \cos \theta \, d\theta}$$

The ratio $\Delta F/\Delta A$ is called the luminous emittance, is designated by L, and is exactly analogous to radiant emittance.

Illuminance. When luminous flux strikes a surface we say the surface is illuminated. Illuminance is defined as luminous flux incident per unit area

$$E = \frac{F}{A}$$

and is expressed in lumens/sq m (lux) or lumens/sq ft (foot-candle). As such, illuminance is exactly analogous to irradiance previously defined. Table 16-4 lists values

Table 16-4. Typical Values of Illuminance

Type of illumination	Lumens/sq m (lux)	Lumens/sq ft (foot-candles)
Sunlight plus skylight (max).........	100,000	10,000
Sunlight plus skylight (dull day).....	1,000	100
Interiors–daylight...................	200	20
Interiors—artificial light.............	100	10
Full moonlight......................	0.2	0.02

of illuminance for some typical modes of illumination. For point sources of illumination, the illuminance is given by

$$E = \frac{I \cos \theta}{r^2}$$

OPTICAL SENSORS

Table 16-5. Summary of Photometric Quantities and Definitions

Photometric quantity	Symbol	Unit	Definition
Luminous flux.............	F	Lumen	1 lumen = 0.00146 watt of radiant flux of wavelength 555 mμ or its equivalent in evoking the sensation of brightness
Luminous efficiency........	F/P	Lumens/watt	Ratio of luminous flux to radiant flux
Luminous intensity........	I	Lumens/steradian (candles)	Luminous flux emitted per steradian
Luminous emittance.......	L	Lumens/sq m	Luminous flux emitted per unit area
Illuminance...............	E	Lumens/sq m	Luminous flux incident per unit area
Luminance................	B	Candles/sq m	Luminous intensity per unit projected area

where θ is the angle which the illuminated surface makes with a point source at distance r. For extended sources, e.g., a disk of radius a and luminance B, the illuminance on a surface at a distance b is given by

$$E = \frac{BA}{a^2 + b^2}$$

where A is the area of the source.

A summary of photometric quantities and definitions is given in Table 16-5.

Optical Filters. Optical filters help optimize the spectral quality and the total flux of a radiant source to the optical transducer, e.g., a lens, and the response of the radiation sensor. Optical filters provide a means of selective spectrophotometric absorption (filtering) of the radiant energy as well as a means of flux-level control.

Optical filters can be broadly classified in three categories: neutral-density, narrow-band, and broad-band filters. Neutral-density (density = − log transmittance) filters absorb radiation equally at all wavelengths of the visible spectrum. Fine-grained photographic (silver halide) emulsions, uniformly exposed and developed, are good examples of neutral filters. Narrow-band (interference) filters are made primarily of thin metallic films evaporated on a transparent base. They transmit light in a very narrow band of the spectrum, and as such are primarily useful in spectroscopic applications. The broad-band spectrophotometric filters are made of organic dyes which absorb radiation in selective broad regions of the spectrum and are the most-used optical filters to balance the spectral output of a radiation source to the spectral response of a detector through an optical transducer. Figure 16-5 shows the spectral transmittance of a typical Kodak Wratten filter.

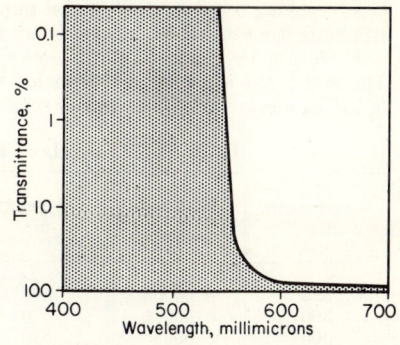

FIG. 16-5. Typical spectral-transmittance curve of a Kodak Wratten filter.

16-3. Radiation Sensors

Classification of Radiation Detectors. Radiation detectors may be broadly classified into two categories, thermal detectors and quantum detectors. In the thermal detector, radiation is absorbed and transformed into heat. Some physical characteristic of the detector changes as a function of temperature, and this change can be measured to determine the quantity of radiation striking the detector. The quantity actually measured is the temperature change, the incident radiation being measured indirectly. Examples of thermal detectors include thermistor bolometers, thermocouples, and pneumatic detectors. Since thermal detectors are used primarily for radiation outside the visible spectrum, they will not be discussed in this chapter.

In quantum detectors (optical sensors), the incident photons change the characteristics of the detector directly. Quantum detectors can be further classified as photochemical and photoelectric. Table 16-6 summarizes the optical sensors and their response characteristics.

Table 16-6. Classification and Measured Characteristics of Optical Sensors

Sensor	Measured characteristics
1. Photochemical	
a. Human eye	A chemical change induced by photon absorption
b. Photoemulsion	
2. Photoelectric	
a. Photoemissive	The emission of electrons from a surface incident to sufficiently energetic photons
b. Photoconductive	The resistance of a cell changes directly with photon absorption
c. Photovoltaic	A voltage is generated directly as the result of the absorption of a photon

The Human Eye. The human eye is a complex optical system which includes a sensor (retina), transducer (lens), and filter (iris). The response of the eye to radiation cannot, therefore, be described simply in terms of the basic detector (retina), since the other elements are continually optimizing for maximum retinal response.

The retina consists of a large number of elements (1.2×10^8) termed rods and cones. The rods contain a photochemically sensitive substance (visual purple) which is rapidly bleached on exposure to light. The physiological mechanism by which this bleaching triggers an electrical impulse through the optic nerves to the brain to stimulate the sensation of vision is not yet clearly understood.

There are two aspects of the eye's response which concern us in this chapter. The first is the eye's spectral response to radiation, the second its self-adaptation to total flux level. The human eye's sensitivity to radiation as a function of wavelength

Table 16-7. Standard Relative-luminosity Factors

Wavelength, mμ	Relative luminosity	Wavelength, mμ	Relative luminosity	Wavelength, mμ	Relative luminosity
380	0.00004	510	0.503	640	0.175
390	0.00012	520	0.710	650	0.107
400	0.0004	530	0.862	660	0.061
410	0.0012	540	0.954	670	0.032
420	0.0040	550	0.995	680	0.017
430	0.0116	560	0.995	690	0.0082
440	0.0230	570	0.952	700	0.0041
450	0.0380	580	0.870	710	0.0021
460	0.0600	590	0.757	720	0.00105
470	0.0910	600	0.631	730	0.00052
480	0.1390	610	0.503	740	0.00025
490	0.2080	620	0.381	750	0.00012
500	0.3230	630	0.265	760	0.00006

OPTICAL SENSORS 16-9

is shown in Figs. 16-3 and 31-3. Based on the eye's response, standard luminosity factors have been established (Table 16-7) to standardize photometric calculations. In this context the term luminance is preferred to brightness when referring to the physical stimulus rather than the sensation.

In front of the eye's lens is the iris, at the center of which is an opening, the pupil. The pupil regulates the quantity of light entering the eye by a process of adaptation by variation of pupil diameter. Figure 16-6 shows pupillary diameter as a function of luminance. It should be noted that the eye's range of flux-level control is relatively small, a ratio of area, according to our graph, of 5.2:1 to cover a luminance range of 100,000.

Photographic Emulsions. A photographic emulsion is composed of a large number of silver halide (e.g., silver bromide) crystals embedded in a layer of gelatin. The spectral and quantitative responses of photographic emulsions depend primarily on the type and size of these crystals and on the absorbing or sensitizing dyes added to the emulsion. Radiant energy absorbed by a silver halide crystal causes the formation of a silver "latent" image on the surface of, and in, the crystal. By subsequent chemical development, those crystals exposed are completely converted to

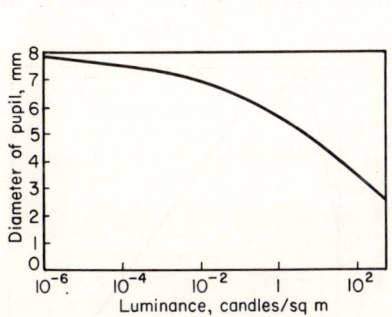

FIG. 16-6. Pupillary diameter as a function of incident luminance.

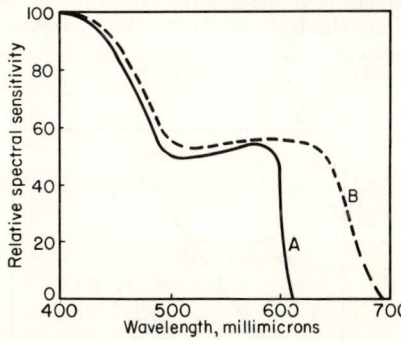

FIG. 16-7. Typical spectral-sensitivity curves for two photographic emulsions. (A) Orthochromatic; (B) panchromatic.

the developed silver image. Photographic emulsions are therefore photochemical optical sensors.

The silver halide, unsensitized, is blue-sensitive, responding to wavelengths shorter than 5,000 A. Through the incorporation of dyes which absorb in the longer wavelengths, the sensitivity of the silver halide receptor has been extended into the red region of the visible spectrum, and even into the infrared. Figure 16-7 shows the spectral sensitivity of two typical early emulsions, the green-sensitive "orthochromatic," and the dye-sensitized "panchromatic." Today, sensitizing dyes are used in spectrographic emulsions which provide selected regions of spectral sensitivity that extend beyond 1 micron.

The basic response of a photographic emulsion is expressed in terms of optical density produced as a function of incident-radiant-energy exposure. The resultant characteristic response curve is normally described as follows (Fig. 16-8):

$$\text{Density} = -\log(\text{transmittance})$$
$$\text{Exposure} = \text{incident exposure radiation, m-candle sec}$$

The characteristic curve has three regions of response which are described as (1) toe region, which includes response caused by extraneous radiation (fog), (2) straight-line portion, the slope of which is the contrast (gamma) of the sensor, and (3) the shoulder, a nonlinear region reaching saturation response. It is obvious that the optimum response of a photographic emulsion is within the straight-line portion. Imaging quality (resolution, sharpness, etc.) characteristics of emulsions are usually measured in this region. Figure 16-8 shows the characteristic curves of a typical photographic emulsion for two degrees of chemical development. This illustrates the manner by which gamma and the effective speed are influenced by processing, and points up the need for good process control in laboratories.

Photoemissive Tubes. Photoemissive tubes are those whose light-sensitive surfaces (e.g., cesium-magnesium) emit electrons when exposed to sufficiently energetic photons. There are two basic laws which describe the response of photoemissive materials. The first states that the number of electrons released per unit time at a photoelectric surface is directly proportional to the intensity of the incident light. The second states that the maximum energy of the photoelectrons released is independent of the intensity of the incident light but increases linearly with the frequency of the radiation.

The second law is expressed by Einstein's equation for the kinetic energy of a photoelectron as

$$\frac{mv^2}{2} = h\nu - W$$

where W represents the excess energy which the electron has to expend to free itself from the body of the emitting substance after its reaction with a quantum of frequency ν.

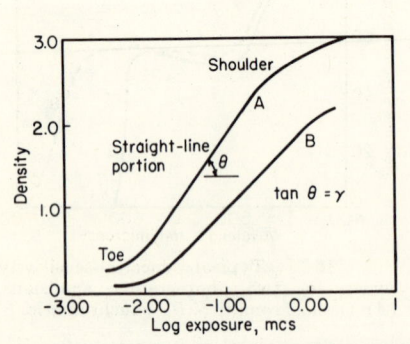

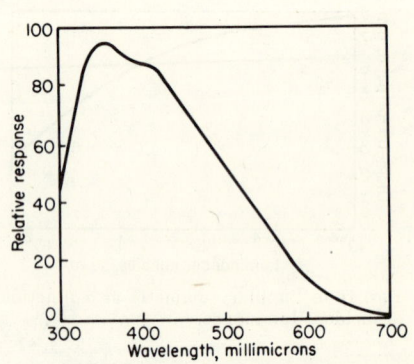

FIG. 16-8. Characteristic curve of a photographic emulsion. (A) Full development; (B) underdevelopment.

FIG. 16-9. Spectral response of a typical photoemissive (cesium-magnesium) surface.

The Radio Manufacturers' Association and the National Electrical Manufacturers' Association have jointly adopted a classification system of phototubes based on their spectral-response characteristics. The designations are S-1, S-3, S-4, etc. For example, S-3 is a silver-rubidium, oxide-rubidium surface with a maximum response near 4,200 A in limeglass. It has high sensitivity throughout the visible spectrum, approximating the response of the human eye. Figure 16-9 shows the relative spectral response of a typical photoemissive (cesium-magnesium) phototube. The absolute luminous sensitivity is generally given in microamperes per lumen, an incandescent tungsten filament at a color temperature of 2870°K serving as the energy source. On this basis the sensitivity of the cesium-magnesium phototube shown in Fig. 16-9 is about 2 microamperes/lumen.

The current output of a photoemissive tube is only approximately a linear function of the light flux falling on the photo surface. This nonlinearity is primarily dependent on the geometrical structure of the tube. Figure 16-10 shows the output current as a function of incident light flux for a typical gas-filled photoemissive tube.

Photoconductive Cells. Photoconductive cells (e.g., selenium) contain materials whose electrical conductance (ohmic resistance) changes as a function of the intensity and spectral quality of radiation. All practical photoconductors have the

OPTICAL SENSORS

common property of being semiconductors, that is, of passing a small current increasing with temperature when subjected to an electric field in the dark.

When a semiconductor of this type is exposed to light, conduction may be initiated by the photoexcitation of an impurity or of an electron in the top filled energy band of the material. The threshold wavelength λ_0 will be determined by the energy difference ΔE between the original and excited (illuminated) state of the material. If ΔE is measured in electron volts then

$$\lambda_0 = \frac{hc}{\Delta E} = \frac{12{,}395}{\Delta E} \quad \text{angstrom units}$$

The spectral response of a typical photoconductive (selenium) cell is shown in Fig. 16-11. As can be seen, the response is generally broad with a fairly pronounced maximum near 7,000 A. As expected, the relation between photocurrent and illumination is nonlinear. Figure 16-12 shows net current as a function of incident light for a typical photoconductive cell.

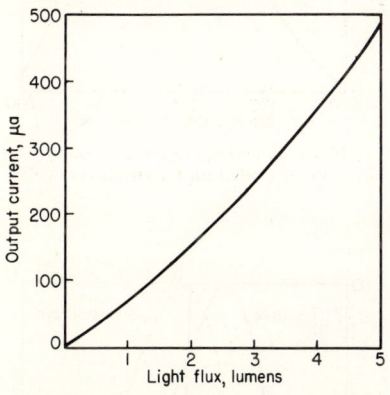

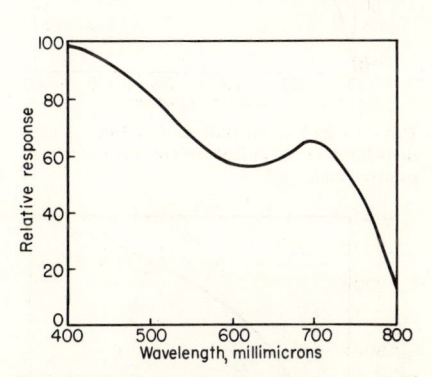

FIG. 16-10. Output current as function of incident light for a typical gas-filled photoemissive tube.

FIG. 16-11. Spectral response of a typical photoconductive (selenium) cell.

A good selenium cell yields about one electron per incident quantum of visible radiation. The dark resistances of such cells range from 50,000 ohms to 10 megohms with corresponding operating voltages from 10 to 500 volts. A luminous flux of 0.1 lumen may reduce the effective resistance of the cell by a factor of 3.

Other types of photoconductive cells include thallous sulfide, lead sulfide, cadmium sulfide, and silicon.

Photovoltaic Cells. Photovoltaic cells are self-contained current and voltage generators, setting up a potential difference between their terminals when exposed to light. They may be divided into "wet" and "dry" cells. Wet photovoltaic cells consist of two electrodes immersed in an electrolyte; dry cells consist of a semiconducting layer between two metal electrodes. The latter has almost completely displaced the less convenient and less stable wet cells.

Photovoltaic cells are made from cuprous oxide, silver sulfide, zinc sulfide, lead sulfide, and a number of other materials including selenium. The spectral response of a typical selenium barrier-layer photovoltaic cell is shown in Fig. 16-13.

Electrically, a barrier-layer cell can conveniently be represented as a current generator, shunted by the capacity C and the inner resistance R_i in series with a small resistance R_s. If the primary photoelectric current is I_p and the load resistance R, the current through the latter becomes simply

$$I = \frac{I_p + R_i}{R_i + R_s + R}$$

It follows that, for finite values of the load resistance R, the variation of current becomes increasingly nonlinear as the internal resistance R_i decreases. Hence the characteristics for the more sensitive cells are less linear (for large values of load resistance) than those for less sensitive cells. Figure 16-14 shows photocurrent as a function of incident light for a typical selenium barrier-layer photovoltaic cell.

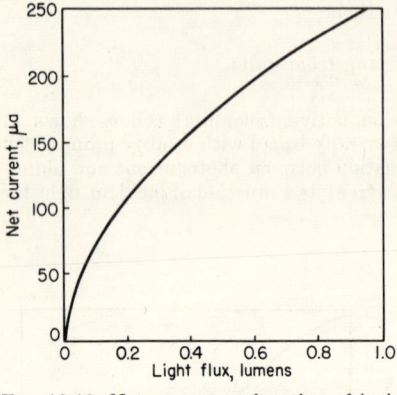

FIG. 16-12. Net current as function of incident light for a typical (selenium) photoconductive cell.

FIG. 16-13. Spectral response of a typical photovoltaic (selenium barrier-layer) cell.

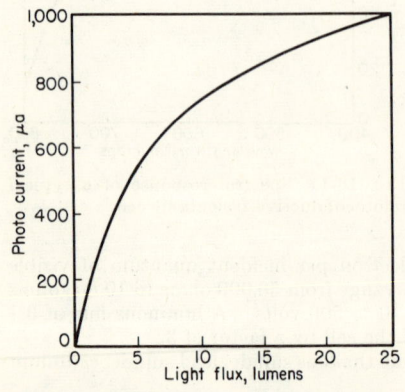

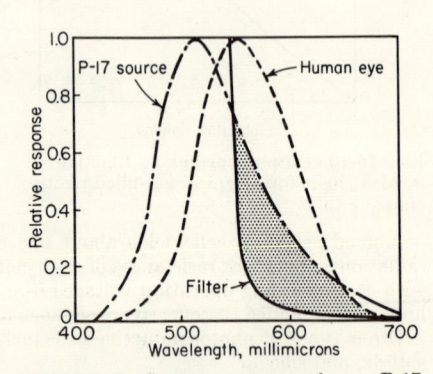

FIG. 16-14. Photocurrent as function of incident light for a typical selenium barrier-layer photovoltaic cell.

FIG. 16-15. System response for a P-17 phosphor source and the human eye.

Selenium barrier-layer cells are used principally in exposure meters and for relay operation. Short-circuit current sensitivities range from about 150 to 600 μa/lumen, the less sensitive cells being designed for high light levels. Power-conversion ratios from 40 to 60 μw/lumen may be reached at a light level of 100 ft-candles.

16-4. Practical Considerations

In practical engineering applications, the optimum use of an optical sensor must consider five elements of the system: source, filter, transducer, sensor, and display. Display is used here to include utilization of sensor output as well as visual display. In this sense a display can be considered as a secondary radiant source in the system. The important characteristics of the transducer element in the system are the spectral-energy response and total radiant flux control.

As a practical example, as lenses are designed for maximum image quality in a specified spectral region, the system engineer may first decide on the spectral region he will utilize. Having made this decision he then can select the lens that best performs over this region. After these decisions his problem becomes one of selecting the appropriate combination of source and filter to restrict the radiation to this selected spectral region. At this point his problem is reduced to that of controlling flux by aperture setting.

The system engineer's ability to optimize the spectral quality and flux-level characteristics of the source, filter, and sensor is the important consideration in the design of any system utilizing optical sensors. Since radiant flux-level control is usually independent of spectral considerations, the optimum balance is primarily achieved through consideration of the spectral emittance, transmittance, and response of these three elements.

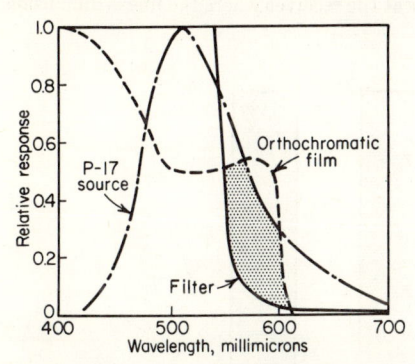

FIG. 16-16. System response for a P-17 phosphor source and a typical orthochromatic emulsion.

FIG. 16-17. System response for a P-17 phosphor source and a photoconductive cell sensor.

It has been mentioned that the spectral emittance, transmittance, and response of a source, filter, and sensor cannot be described analytically. Therefore, the relative efficiency of any such combination must be determined by graphical or numerical integration. Since the efficiency (total relative flux response) of any source-filter-sensor combination for a given wavelength is given by

$$(\text{Efficiency})_\lambda = f[(\text{source})_\lambda, (\text{filter})_\lambda, (\text{sensor})_\lambda]$$

the relative efficiency for a spectral range λ_1 to λ_2 is given by

$$\text{Rel. eff.} = \int_{\lambda_2}^{\lambda_1} E_\lambda \, d\lambda$$

where E_λ is defined as the overall spectral response of source, filter, and sensor.

In Figs. 16-15 to 16-17, the shaded areas of the curves show the relative efficiency of three typical source-filter-sensor combinations. The spatial response of the transducer and the basic response of the display must finally be superimposed to determine the overall performance of an optical sensor.

16-5. Resolution

In many applications of optical engineering, the ability of the system to operate on or to see fine detail becomes an important consideration. Because of the wave nature of light, each point of the object space is imaged, not as a point, but as a diffraction disk. The size and shape of the disk are a function of the angular subtense and shape of the exit pupil as viewed from that point in the focal surface, and the residual aberration in the lens. The classical limit of resolving power, known as the Rayleigh

16-14 SYSTEM COMPONENTS

limit, is given by

$$\alpha = \frac{1.22\lambda}{D}$$

where α is the angular separation between two points that can just be separated, λ the wavelength of the light, and D the diameter of the entrance pupil.

In practice, the resolution has been determined by the use of resolving-power targets to test the system's performance. Over the years, these targets have assumed many sizes and shapes. Most common in the United States today is the U.S. Air Force Military Standard, shown in Fig. 16-18.

As one moves to finer patterns, as long as the intervening spaces can still be seen or detected, the pattern is said to be resolved. Thus in this test the resolution limit is that point where the contrast in the image approaches zero. This practical resolution limit is expressed in lines per millimeter at the receiver where the line is dimension

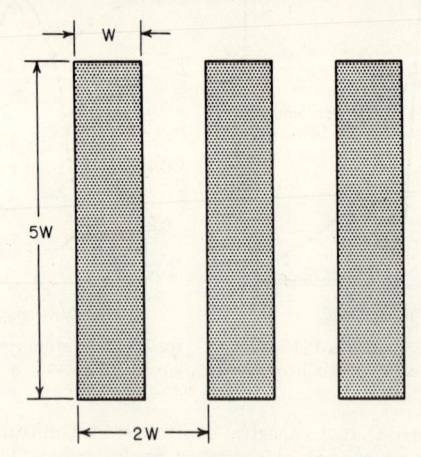

FIG. 16-18. U.S. Air Force Military Standard test pattern.

2W. As a rule of thumb, for the well-corrected lens at about the center of visual efficiency (550 mμ) resolution (Res.) is given approximately by

$$\text{Res.} = \frac{1,500}{f \text{ number}}$$

In reality the problem is considerably more involved, for resolution of objects depends upon their innate contrast, shape, size, etc., and the operation of each element of the system on each of these factors. Coupled with the fact that the systems utilize nonlinear detectors, this suggests that there can be no general rule for evaluation and that each system or application should be examined on the basis of its unique function.

Each element of the system (the receptor, the lens, the filter, etc.) has a limiting resolving power (the finest possible detail that can be passed). At each dimension, i.e., spatial frequency (lines/mm or cycles/mm), over the range from zero to the resolving-power limit, each element operates differently upon the contrast. As we are concerned, at the end point, with where the contrast is reduced to zero, we should think of the role of each element of our system in terms of its ability to transfer contrast to the next element with the minimum degradation. Much laboratory work is now in progress utilizing the transfer function T given by

$$T(N) = \frac{C_i(N)}{C_o(N)}$$

where N is spatial frequency, C is contrast, and the subscripts i and o denote image and object, respectively. For the engineer, the important point is that the system's performance is the product of the transfer functions of each of its components, irrespective of the method of test or analysis of optical systems.

BIBLIOGRAPHY

Gray, Dwight E. (ed.): *American Institute of Physics Handbook*, 2d ed., McGraw-Hill Book Company, New York, 1963.
General Electric Company, *Design Considerations in Selecting Photoconductive Cells*, 1962.
Allen B. Dumont Laboratories, Inc., *Dumont Multiplier Photo Tubes*, 1959.
Condon, E. U., and Hugh Odishaw (eds.): *Handbook of Physics*, McGraw-Hill Book Company, New York, 1958.
Eastman Kodak Company: *Kodak Plates and Films*, 1962.
Eastman Kodak Company: *Kodak Wratten Filters*, 1962.
Barros, William E.: *Light, Photometry, and Illuminating Engineering*, McGraw-Hill Book Company, New York, 1951.
Mauro, J. A., (ed.): *Optical Engineering Handbook*, General Electric Company, 1957.
Sears, Francis Weston: *Optics*, Addison Wesley Publishing Company, Inc., Reading, Mass., 1956.
Zworykin, V. K., and E. G. Ramberg: *Photoelectricity*, John Wiley & Sons, Inc., New York, 1949.
Walsh, John W. T.: *Photometry*, Constable and Co., Ltd., London, 1953.
Radio Corporation of America: *RCA Photosensitive Devices and Cathode Ray Tubes*, 1960.
Forsythe, W. E.: *Smithsonian Physical Tables*, The Smithsonian Institute, Washington, D.C., 1954.
Sasuga, John: *Solar Cell and Photocell Handbook*, International Rectifier Corporation, 1960.
Mees, C. E. Kenneth: *The Theory of the Photographic Process*, The Macmillan Company, New York, 1954.

Chapter 17

SATELLITE SYSTEMS*

SIDNEY STERNBERG, *Electro-Optical Systems, Inc., Pasadena, California.*

VERNON D. LANDON, *Radio Corporation of America, Defense Electronic Products, Astro-Electronics Division, Princeton, New Jersey.*

CONTENTS

17-1. Introduction	17-2
17-2. Astrodynamics	17-2
Glossary of Astronomical Terms	17-2
The Elements of Orbital Mechanics	17-3
Approaches to Orbit Calculation	17-6
Special Perturbation Theory	17-6
General Perturbation Theory	17-7
First-order Effects of Particular Perturbations	17-7
Ballistic-missile Orbits	17-10
17-3. Spinning-body Dynamics	17-10
Introduction	17-10
General Conditions for the Motion of a Spinning Rigid Body	17-10
Precession Damping	17-11
Damper-design Considerations	17-12
Control of Spin-axis Drift	17-12
Differential Gravity	17-13
Eddy Currents	17-13
Magnetic Dipole	17-13
Controlled Magnetic Torquing	17-13
Yo-yo Speed Reduction	17-13
17-4. Design Considerations	17-14
Introduction	17-14
The Launching System	17-14
Arrangement of Component Systems	17-14

* For the material presented in this chapter, the authors gratefully acknowledge the assistance of D. Almy, W. Cable, J. Newman, R. Northrup, C. Osgood, J. Owens, H. Perkel, and K. Robinson.

The Tiros and Relay satellites described herein were built by the Radio Corporation of America under contract with the National Aeronautics and Space Administration. The Ranger structure described was designed for the Ranger TV subsystem, which was built by the Radio Corporation of America under contract with the Jet Propulsion Laboratories of the California Institute of Technology under the sponsorship of the National Aeronautics and Space Administration.

17-2   SYSTEM COMPONENTS

       Internal Heat Control... 17-15
       Types of Structures.. 17-15
       Structural Testing and Analysis................................ 17-16
       Fabrication and Assembly...................................... 17-17
       Thermal Analysis.. 17-18
          Passive Thermal Control................................... 17-18
          Active Thermal Control.................................... 17-21
    17-5. Satellite Power Supplies.................................. 17-22
       Introduction.. 17-22
       Silicon Solar Cells... 17-22
       The Storage Battery... 17-23
       Overall Power-supply Considerations........................... 17-25
       The Voltage Regulator... 17-25
    17-6. Image Sensing and Storage................................ 17-25
       Introduction.. 17-25
       The Vidicon Camera... 17-25
       The Optical System.. 17-26
       Ground Resolution of the Vidicon............................... 17-26
       Bandwidth Requirements...................................... 17-27
       The Use of Infrared Sensors.................................... 17-27
       Satellite Tape Recorders....................................... 17-28
    17-8. Conclusions... 17-28

17-1. Introduction

    The first Russian satellite was placed in orbit on Oct. 4, 1957, and the first American satellite, Explorer 1, on Jan. 31, 1958. Within the next seven years, over 300 satellites had been successfully placed in orbit.

    These early spacecraft have performed a wide variety of functions, usually involving the sensing of scientific information over a wide band of power spectra, from x-rays and cosmic rays down through infrared (see Chapter 7). Because of a requirement for reliability, simplicity became a primary goal. As a consequence, the first generation of satellites were spin-stabilized (that being the simplest known method of stabilization), including those used for picture taking. As bigger boosters become available and weight limitations become less stringent, more complicated spacecraft will become practical. Some of these will undoubtedly utilize other attitude-control systems, but the simplicity and reliability of spin-stabilized systems ensures their utilization for special purposes for many years to come.

    For the reason that more complicated satellites will owe their evolution to these early spinning systems, and the fact that spacecraft using spin or angular momentum for stabilization are generic from a design viewpoint, these types of satellites have been chosen for the exposition of design information in this chapter. The principles developed herein will prove useful in the future for many other types of spacecraft systems.

17-2. Astrodynamics

    Glossary of Astronomical Terms. The design of a satellite system involves lunar, solar, and stellar considerations. For example, initial orbital considerations must account for the fact that the satellite will be orbiting a planet (earth) which rotates on its axis once every 24 hr. At the same time, both the planet and the satellite, considered as a system, will be revolving around the sun in an elliptical orbit—the plane of which is inclined 23.5° to the earth's equatorial plane—with a period of approximately 365 days. A closer examination of these and other orbital considerations requires the introduction of certain astronomical terms. A few of the most important of these terms (from the standpoint of orbital mechanics) are listed below, together with their definitions. Figure 17-1 further illustrates the relationships among these terms. Further discussion of most of these terms is given in Sec. 7-4.

Celestial Sphere. A sphere of infinite radius whose center is the center of the earth, and on which appear projected the stars and other astronomical bodies fixed in space. Although the celestial sphere is defined as a sphere of infinite radius, Fig. 17-1 shows it as viewed from the outside. To make this view at all rational, the celestial sphere must be thought of as very large but less than infinite.

Celestial Equator. The projection of the earth's equator on the celestial sphere.

Ecliptic. The projection of the earth's orbital plane around the sun on the celestial sphere. It is the apparent annual path of the sun among the stars, and inscribes a great circle on the celestial sphere inclined 23.5° to the celestial equator.

Vernal Equinox. The intersection of the ecliptic with the celestial equator when the apparent motion of the sun is from south to north. It occurs about Mar. 21.

Autumnal Equinox. The intersection of the ecliptic with the celestial equator when the apparent motion of the sun is from north to south. It occurs about Sept. 23.

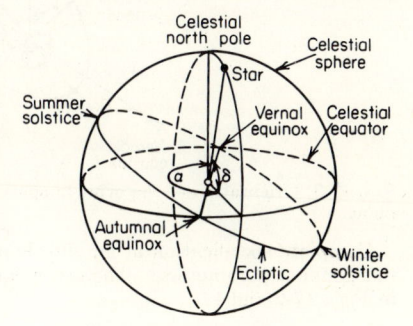

FIG. 17-1. The celestial sphere.

Summer Solstice. The point on the ecliptic which is farthest north. The sun arrives there about June 22.

Winter Solstice. The point on the ecliptic which is farthest south. The sun arrives there about Dec. 22.

Declination δ. The angular distance of a star north or south from the celestial equator, along a celestial great circle containing the star and the celestial north pole.

Right Ascension α. The angle measured eastward along the celestial equator from the vernal equinox to the celestial great circle containing the star and the celestial north pole.

The Elements of Orbital Mechanics. The classical approach to orbit calculations applies the methods of Newton to the case of a spherical mass exerting a central gravitational attraction on a small orbiting mass. The result of these calculations is an elliptic orbit which, in order to obtain a more accurate result, must then be modified to account for perturbations caused by (1) the varying geopotential of a pear-shaped earth; (2) the presence of a rotating atmosphere which is subject to

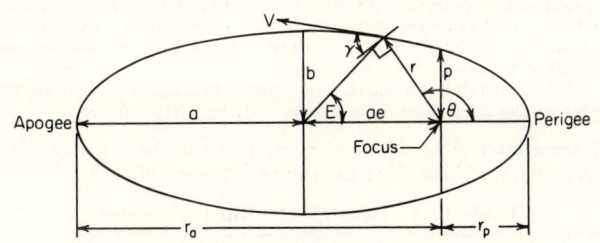

FIG. 17-2. The orbit plane and related quantities.

seasonal and diurnal variations, density fluctuations (due to solar flare activity), and tidal motions; (3) electromagnetic drag; and (4) the gravitational gradients of the moon, sun, and planets. The relative magnitudes of each of these perturbations depend markedly upon the proximity of the satellite to the earth.

Another approach (general perturbation theory) incorporates the general analytical expressions for the perturbative forces in the basic equations of motion of the satellite. Analytical integration of these equations leads to the desired orbital information.

Still another approach (special perturbation theory) utilizes numerical methods of calculating the required orbital data on the basis of specific numerical values for the various parameters.

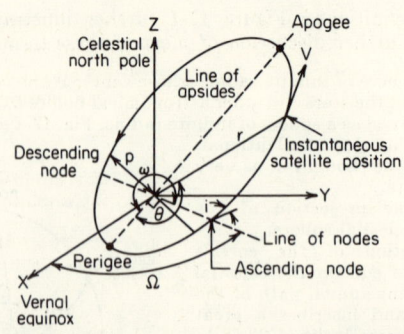

FIG. 17-3. Orientation of the orbit in space. xy plane is equatorial; ω is measured in orbit plane.

Complete specification of an elliptic orbit and the position in orbit requires seven independent parameters. The seven commonly used are listed below and illustrated in Figs. 17-2 and 17-3.

Symbol   *Parameter*
- a The semimajor axis of the orbital ellipse. It is equal to one half the distance from the apogee (high point) to the perigee (low point) of the orbit
- b The semiminor axis of the orbital ellipse. It is equal to one half the width of the ellipse measured at right angles to the major axis midway between the foci. (The semiminor axis is not one of the seven independent parameters required for specification of the orbit, since it may be derived from the semimajor axis a and the eccentricity e)
- e The eccentricity of the orbital ellipse, defined by

$$e = \frac{\sqrt{a^2 - b^2}}{a}$$

- i The inclination of the orbital ellipse. It is the angle between the orbital plane and the equatorial plane
- Ω The right ascension of the ascending node (see the Glossary of Astronomical Terms in Chapter 7). The ascending node is the intersection of the orbit with the equatorial plane when the satellite is moving generally northward
- ω The argument of perigee. This is the angle in the ellipse, as measured from the ascending node to perigee, with the (force center) focus as the origin
- T The time of passage through perigee. All other parameters are commonly measured at this time
- P The orbital period. For earth satellites, this parameter can be expressed as a function of the orbital semimajor axis and the earth's mass

It may be of some interest to see what typical values these parameters may have. Table 17-1 gives a list for Tiros IV of the values of the seven parameters as of June 13,

Table 17-1. Tiros IV Orbital Parameters

Parameter	Value	Change per day
Semimajor axis, statute miles	4,446.15	−0.0002
Eccentricity	0.00933	Negligible
Orbital inclination, deg	48.301	Negligible
Right ascension of ascending node, deg	26.657	−4.4324
Argument of perigee, deg*	264.230	4.0397
Period, min	100.397	−0.000005
Time of perigee (June 13, 1962) (time: AM, EST)	7:45:25.14	Not applicable

* Values recorded at the indicated time of perigee.

SATELLITE SYSTEMS   17–5

1962, together with their rates of change. (Tiros is a meteorological satellite used for taking pictures of cloud cover from outside the atmosphere. It was designed and built by RCA under the sponsorhsip of NASA.) In addition to the seven fundamental parameters defined above, the following are also commonly employed:

Symbol	Parameter
p	The semilatus rectum
r	The distance (radius) from the force center to the satellite at any time
r_a, r_p	The distances (radii) from the force center to the satellite at apogee and perigee, respectively
n	The mean motion rate, or average angular velocity of the satellite around the orbit as measured from the force center; equal to $2\pi/P$
t	Any time increment, typically measured from perigee passage T
M	The mean anomaly—an angle, measured from the force center, which increases uniformly with time from perigee passage T, passing through 360° in one orbital period; equal to nt
m	The mass of the orbiting body
V	The tangential velocity of the satellite at any time
V_a, V_p	The tangential velocity of the satellite at apogee and perigee, respectively
θ	The orbit central angle between perigee and satellite position as measured from the force center
E	The eccentric anomaly—the angle between perigee and satellite position as measured from the intersection of the semimajor and semiminor axes
γ	The flight-path angle—the angle between the flight path at any time and the perpendicular to the radius vector from the force center
μ	The gravitational constant for the appropriate central (force center) body, such that, if the force of gravitational attraction between the force center and the satellite is F, then

$$F = \mu \frac{m}{r^2} = G \frac{Mm}{r^2}$$

where G is the universal gravitational constant, and M is the mass of the central body. Hence

$$\mu = GM$$

The distance r between the force center and the satellite is given as a function of the orbit central angle θ by

$$r = \frac{a(1 - e^2)}{1 + e \cos \theta} = a(1 - e \cos E)$$

where θ is related to the eccentric anomaly E according to

$$\tan \frac{\theta}{2} = \frac{(1 + e)^{1/2}}{(1 - e)^{1/2}} \tan \frac{E}{2}$$

But from Kepler's equation, it is seen that E is a nonuniform function of time:

$$M = nt = E - e \sin E$$

Hence, given only the orbital elements a, e, and n, the satellite position in orbit can be found as a unique function of time.

Convenient relationships among some of the listed parameters are:

(1) $$P = 2\pi \left(\frac{a^3}{\mu}\right)^{1/2}$$

(2) $$p = \frac{b^2}{a} = a(1 - e^2)$$

(3) $$a = \frac{r_a + r_p}{2} = \frac{\mu}{V_a V_p}$$

(4) $$b = a(1 - e^2)^{1/2}$$

17-6 SYSTEM COMPONENTS

(5) $$e = \frac{r_a - r_p}{r_a + r_p} = \frac{V_p - V_a}{V_p + V_a}$$

(6) $$r_a = a(1 + e)$$

(7) $$r_p = a(1 - e) = r_a \frac{1 - e}{1 + e}$$

(8) $$V_a = \sqrt{\frac{\mu}{a} \frac{1 - e}{1 + e}}$$

(9) $$V_p = \sqrt{\frac{\mu}{a} \frac{1 + e}{1 - e}}$$

(10) $$V = \sqrt{\mu \left(\frac{2}{r} - \frac{1}{a}\right)}$$

(11) $$\gamma = \tan^{-1}\left(\frac{e}{\sqrt{1 - e^2}} \sin E\right)$$
$$= \tan^{-1}\left\{\left[1 - \frac{r}{a(1 - e^2)}\right] \tan \theta\right\}$$

(12) $$n = \left(\frac{\mu}{a^3}\right)^{1/2}$$

Accurate orbit computation requires consideration of the perturbing factors, since they cause the orbit to deviate from pure conic form and to become nonstationary in space.

Approaches to Orbit Calculation. *Special Perturbation Theory.* Special perturbation methods implement the calculation of orbits by direct numerical integration of the equations of motion. No attempt is made to find the equations of the curve traced, and usually no information on long-term behavior is secured. These techniques are especially useful for short-term descriptions of orbits.

The equations of motion may be written in the form

$$\ddot{x} = -\mu \frac{x}{r^3} + \sum_i \ddot{x}_i$$

$$\ddot{y} = -\mu \frac{y}{r^3} + \sum_i \ddot{y}_i$$

$$\ddot{z} = -\mu \frac{z}{r^3} + \sum_i \ddot{z}_i$$

where $\ddot{x}_i$, $\ddot{y}_i$, and $\ddot{z}_i$ represent the various perturbative accelerations in the x, y, and z directions, respectively.

Any one of three well-known approaches may be used for the integration. Cowell's method[1,2] integrates the entire right side (total acceleration) and is especially useful where large perturbations—thrust, for example—are encountered, but it has a tendency to accumulate round-off error rapidly.

In contrast, Encke's method[1,2] integrates only the perturbative accelerations to determine the deviations in path from a Keplerian reference orbit. This method is well suited to computations involving perturbations which are small with respect to the central-force term, and have a limited influence on the orbit.

Finally, there is the variation-of-parameters method,[3] in which the instantaneous orbital position of the satellite may be represented by a conventional set of conic elements expressed as functions of time. Planetary equations, giving the rates of change of the orbital elements in terms of the perturbing functions, are integrated to yield the instantaneous orbit, from which the position and velocity of the satellite may be found. This method is useful for trajectories where small perturbations are present throughout the orbit. It has been applied, for example, to missions with continuous low-thrust propulsion, and to missions where the central body possesses distinct gravitational asymmetries.

Listed below are Lagrange's planetary equations. In the forms shown, they are typical for the approach, although frequently other forms—involving different sets of orbital elements—will offer computational advantages, depending on the problem. In these equations, F represents the perturbing potential function. The partial derivatives of F may be replaced by forces not derivable from a potential.

$$\frac{da}{dt} = \frac{2}{na} \frac{\partial F}{\partial M}$$

$$\frac{de}{dt} = \frac{1-e^2}{na^2 e} \frac{\partial F}{\partial M} - \frac{\sqrt{1-e^2}}{na^2 e} \frac{\partial F}{\partial \omega}$$

$$\frac{d\omega}{dt} = -\frac{\cos i}{na^2 \sin i \sqrt{1-e^2}} \frac{\partial F}{\partial i} + \frac{\sqrt{1-e^2}}{na^2 e} \frac{\partial F}{\partial e}$$

$$\frac{di}{dt} = \frac{\cos i}{na^2 \sin i \sqrt{1-e^2}} \frac{\partial F}{\partial \omega} - \frac{1}{na^2 \sin i \sqrt{1-e^2}} \frac{\partial F}{\partial \Omega}$$

$$\frac{d\Omega}{dt} = \frac{1}{na^2 \sin i \sqrt{1-e^2}} \frac{\partial F}{\partial i}$$

$$\frac{dM}{dt} = n - \frac{1-e^2}{na^2 e} \frac{\partial F}{\partial e} - \frac{2}{na} \frac{\partial F}{\partial a}$$

General Perturbation Theory. Because of the limited long-term accuracy of results obtained from special perturbation methods, general perturbation methods, i.e., analytic approximations, have been developed. Such methods lead to expressions for the perturbed motion through series expansions of analytic functions, and therefore do not accumulate the degree of error characteristic of special perturbation methods.

Orbital perturbations consequent to the presence of a third body, the oblateness of the earth, and the presence of an atmosphere have been extensively treated by many prominent mathematicians and astronomers making use of general perturbation methods.[4,5]

First-order Effects of Particular Perturbations. In order to obtain a physical understanding of the orbital motion of a satellite, it is useful to discuss the first-order effects of the principal perturbative forces acting on the satellite.

For an earth satellite orbiting at an altitude of 250 miles or less, atmospheric drag becomes a perturbation of importance. In effect it controls the orbital lifetime of the satellite by acting as a constant drain on the orbital energy. This energy drain is principally manifested by a steady reduction in the orbital elements r_a (apogee distance) and e (eccentricity of the orbital ellipse). Most of the drag, of course, occurs at the lowest altitudes. After the initial injection, as the orbit becomes more nearly circular (i.e., as the eccentricity approaches zero), the altitude is gradually reduced until the satellite enters the high-drag region below 100 miles. Once in the high-drag region, the satellite rapidly loses altitude and finally reenters the atmosphere.

The inclusion of an atmospheric drag force in the equations of motion for a mass orbiting in a central-force field leads to a set of simultaneous nonlinear differential equations. The solution of these equations by analytical or numerical methods provides an orbit-decay history. The rate of change of the radius of a circular orbit and the rate of change of its period are part of this decay history and are given by the following equations:

$$\frac{da}{dt} = -2B\rho(\mu a)^{1/2} \quad \text{ft/sec}$$

$$\frac{dP}{dR} = -6\pi B\rho a P \quad \text{sec/rev}$$

where B = ballistic coefficient ($C_d A/2m$), sq ft/slug
  ρ = atmospheric density, slugs/cu ft
  A = projected area of the satellite
  C_d = drag coefficient
  $\dfrac{dP}{dR}$ = increment of period per revolution or orbit

The lifetime of a satellite in an elliptic orbit with a ballistic coefficient of 1.0 sq ft/slug (equivalent to a 25-lb sphere of 1.0 ft diameter) is illustrated in Fig. 17-4 as a function of initial perigee altitude.

The rotation of the atmosphere has some effect on the inclination of the orbit, most noticeably for low-density satellites such as the Echo balloon. For direct (eastward) orbits, the inclination tends to be decreased; i.e., the rotating atmosphere tends to align the plane of the orbit with the plane of atmospheric rotation; for retrograde (westward) orbits, the inclination tends to be increased.

The principal perturbative effects of the oblateness of the earth are continuous changes in the right ascension of the ascending node, in the argument of the perigee, and in the mean angular rate. Since these effects range up to several degrees per day, they must often be taken into account even for missions requiring relatively little precision of orbit.

Figure 17-5 is a graph of the period of the orbit of an earth satellite as a function of semimajor axis. Figure 17-6 shows the effect of semimajor axis on $\dot\Omega$ and $\dot\omega$, while Fig. 17-7 shows the dependence of these parameters on inclination. The data contained in these figures are based upon the following equations:

$$P = 2\pi \left(\frac{a^3}{\mu}\right)^{1/2}$$

$$\dot\Omega = -\frac{3}{2a}\sqrt{\frac{\mu}{a}} J_2 \left(\frac{R}{p}\right)^2 \cos i$$

$$\dot\omega = \frac{3}{2a}\sqrt{\frac{\mu}{a}} J_2 \left(\frac{R}{p}\right)^2 \frac{-1 + 5\cos^2 i}{2}$$

$$\dot M = \sqrt{\frac{\mu}{a^3}} + \frac{3}{2a}\sqrt{\frac{\mu}{a}} J_2 \left(\frac{R}{p}\right)^2 \frac{1 - e^2(-1 + 3\cos^2 i)}{2} = n + \Delta\dot M$$

where R = radius of the earth
  J_2 = oblateness coefficient of the zonal harmonic expansion of the earth's gravitational potential[10]

With the insertion of known values, these equations may be transformed into

$$\dot\Omega = -C \frac{\cos i}{a^{7/2}(1-e^2)^2}$$

$$\dot\omega = C \frac{-1 + 5\cos^2 i}{2a^{7/2}(1-e^2)^2}$$

$$\Delta\dot M = C \frac{-1 + 3\cos^2 i}{a^{7/2}(1-e^2)^{3/2}}$$

where $C = 3.939 \times 10^{13}$ deg miles$^{7/2}$/day

For retrograde orbits below approximately 3,500 miles altitude, it is possible to choose an inclination i such that the node moves in right ascension at the same rate as the annual motion of the sun, thus permitting the satellite to have constant solar "look angles" and fraction of time in the sun, and to view the ground area with an essentially unchanging illumination profile.

Third-body perturbations have recently been the subject of renewed investigation, particularly for satellites of the moon. It has been shown that, in the presence of a third body, changes may be induced in Ω, ω, and i, and that satellite lifetime may be

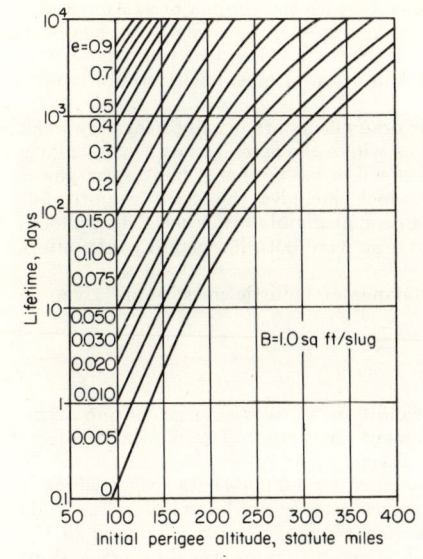

FIG. 17-4. Satellite lifetime as a function of initial perigee altitude and eccentricity.

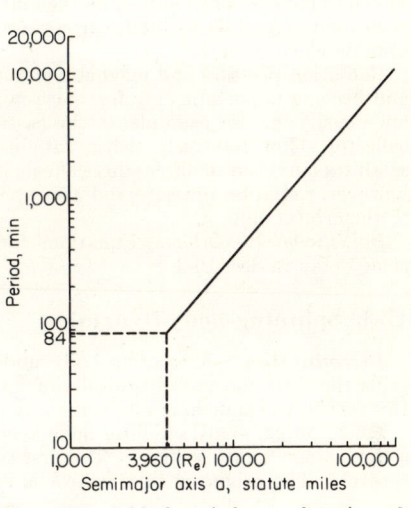

FIG. 17-5. Orbital period as a function of semimajor axis.

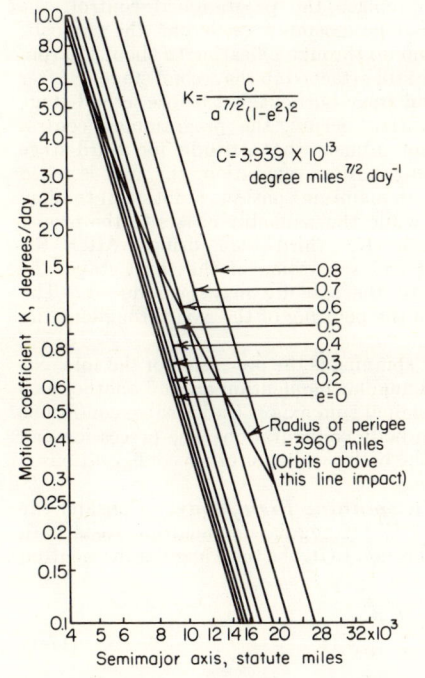

FIG. 17-6. Motion coefficient k for motion rates of line of nodes $\dot{\Omega}$ and of line of apsides $\dot{\omega}$ as a function of semimajor axis and eccentricity.

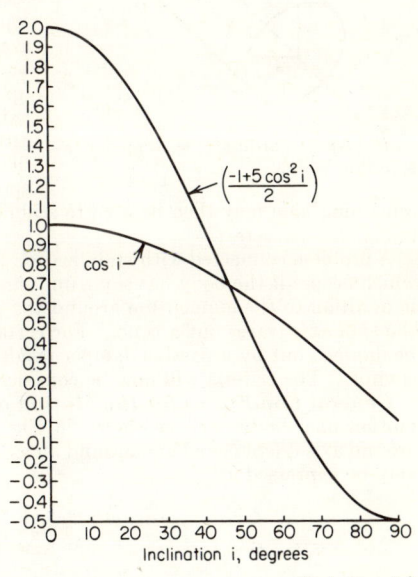

FIG. 17-7. Values of $\cos i$ and $\dfrac{-1 + 5 \cos^2 i}{2}$ for the calculation of $\dot{\Omega}$ and $\dot{\omega}$.

limited by secular drift induced in the eccentricity e with the semimajor axis a remaining essentially constant. For certain orbit configurations, the eccentricity has been shown to increase until impact on the lunar surface occurs. For nearly circular orbits, even relatively close to the lunar surface, this increase in eccentricity would require some weeks or months.

Radiation pressure and magnetic and ionic drag effects are usually relatively weak and become important only for those missions where the satellite has a particularly low density, as, for example, in the cases of a sail, a balloon, or a large solar-power collector. For relatively dense satellites—which includes the great majority of satellites launched to date—these effects have been negligible. For the Echo balloon, however, radiation pressure and atmospheric drag were both important contributors to the orbital motion.

Ballistic-missile Orbits. Equations for guidance of ballistic missiles are given in some detail in Sec. 19-3.

17-3. Spinning-body Dynamics

Introduction. A rotating body under conditions of zero exterior torque maintains the direction and magnitude of its momentum vector. Use is often made of this fact in the launching and control of satellites.

For example, many satellites have been launched by a Thor Delta rocket which is a three-stage launch vehicle. The first two stages make use of jet-stabilized attitude control. The origin of this control is an automatic programmer contained in the second stage. To save weight in the third stage, the control system is thrown away with the second stage after second-stage burnout.

To replace the programmed control, the third-stage assembly (including the payload) is spun up through a bearing to about 120 rpm while still attached to the second stage. After second-stage burnout, but before second-stage separation occurs, the programmed control system adjusts the attitude for third-stage burning. After separation, the spin is relied upon to maintain constant pointing of the spin axis while the assembly coasts to the proper altitude for third-stage firing. After the firing and separation of the third stage, the spin of the satellite may be reduced. The remaining spin may then be used to stabilize the pointing of the axis throughout the life of the spacecraft.

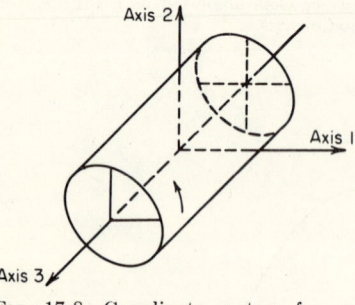

FIG. 17-8. Coordinate system for a spinning body.

A problem connected with stabilization by spinning is the precession of the spin axis which occurs if the body has some undesired angular momentum around another axis in addition to the momentum around the nominal spin axis. Under these conditions the spin axis traces out a cone. For certain body configurations, the precession can be damped out by a passive damper inside the body, while for other configurations it cannot. The criteria will now be considered.

General Conditions for the Motion of a Spinning Rigid Body. Consider the motion of a body such as shown in Fig. 17-8. If angular momentum exists both around axis 3 (spin axis) and around an axis normal to it, the total angular momentum may be expressed as

$$h_T = \sqrt{h_3^2 + h_{12}^2} \\ = \sqrt{\omega_3^2 C^2 + \omega_{12}^2 A^2} \qquad (17\text{-}1)$$

where the symbols are as defined in Table 17-2.

The geometric interpretation of Eq. (17-1) is shown in Fig. 17-9. In the figure, axis 3 is coincident with h_3 and traces out a precession cone as indicated. Note that

SATELLITE SYSTEMS   17–11

Table 17-2. Definition of Symbols for the Motion of a Spinning Rigid Body

Symbol	Definition
ω_1	Angular velocity about axis 1
ω_2	Angular velocity about axis 2
ω_{12}	$\sqrt{\omega_1^2 + \omega_2^2}$ (in the absence of damping, this is a constant)
ω_3	Angular velocity about axis 3
A	Moment of inertia about axes 1 and 2
C	Moment of inertia about axis 3
h_{12}	Angular momentum at right angles to axis 3; equal to $\omega_{12}A$
h_3	Angular momentum about axis 3; equal to $\omega_3 C$
$\dot{\psi}$	Angular velocity of precession
θ	Half angle of the precession cone
h_T	Total angular momentum

the half angle of the cone is θ and that

$$\tan \theta = \frac{h_{12}}{h_3} = \frac{\omega_{12} A}{\omega_3 C} \tag{17-2}$$

The angular velocity of the precession around the cone is $\dot{\psi}$. To evaluate $\dot{\psi}$, consider the precession motion of a point P on axis 3, as shown in Fig. 17-10. Let the linear velocity of the point P as it moves around the circle be v. Then

$$\omega_{12} = \frac{v}{r}$$

$$\dot{\psi} = \frac{v}{a}$$

so that

$$\dot{\psi} = \omega_{12} \frac{r}{a} = \frac{\omega_{12}}{\sin \theta}$$

$$= \omega_{12} \frac{1}{\tan \theta \cos \theta}$$

Substituting this expression in (17-2), we obtain

$$\dot{\psi} = \frac{\omega_3}{\cos \theta} \frac{C}{A} \tag{17-3}$$

It will be shown that $\dot{\psi}$ must exceed ω_3, and hence that C must exceed $A \cos \theta$ for precession damping to be possible.

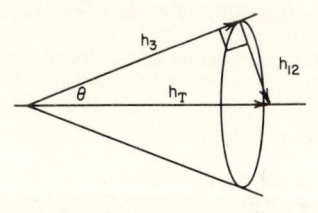

FIG. 17-9. Angular-momentum diagram for a spinning rigid body.

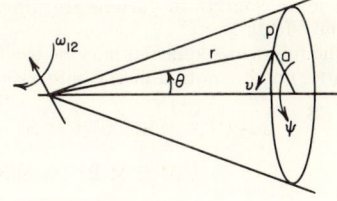

FIG. 17-10. Precession diagram for a spinning rigid body.

Precession Damping. When a body is spinning smoothly about its axis of rotational symmetry, no energy can be extracted by a self-contained damper, because all centrifugal forces are constant and directed away from the axis. However, in a precessing body the forces are variable, and this variability presents the possibility of damping. A type of damper which is especially easy to demonstrate is shown in Fig. 17-11. A doughnut-shaped cavity is installed as part of the spinning body. As shown, the axis of the cavity coincides with the axis of symmetry of the body. The cavity is not at the center of mass, however, but well away from it, so that the motion of precession moves the center of the cavity in a circular path.

The cavity is assumed to be partially filled with a liquid. As shown in the figure (shaded area), the liquid is thrown by centrifugal force to the point in the cavity farthest from the axis of the cone. As a result, the liquid rotates around the axis of the cone (and also around axis 3) at the angular velocity ψ.

If ψ is greater than ω_3, the friction of the liquid on the sides of the cavity will tend to increase ω_3 at the expense of the energy of precession. This condition corresponds to disk-shaped bodies, since, from (17-3),

$$\psi > \omega_3 \text{ only if } C > A$$

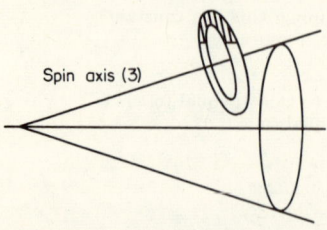

FIG. 17-11. Doughnut-type precession damper.

(assuming that θ is small, so that cosine θ is nearly unity).

If ψ is less than ω_3, the friction of the liquid decreases ω_3, energy is fed into the propeller mode, and the precession increases (i.e., θ increases). This condition (C less than A) corresponds to rod-shaped bodies.

It is clear that, for precession to be damped out, the body must be disk-shaped (i.e., C must exceed A).

Damper-design Considerations. In general, the type of damper used has very little effect on the result. Anything that is caused by the precession to move relative to the body will serve the purpose. A defect common to most dampers is a tendency to freeze up and stop functioning when θ becomes small. The chief advantage of the doughnut-shaped damper shown in Fig. 17-11 is the ease with which its operation can be understood. It is, in addition, a very effective damper for large amplitudes of precession. For small values of θ, however, the liquid will fail to follow the precession and the damping will be reduced. For very thin doughnut-shaped cavities, the effects of adhesion, surface tension, and friction may cause the liquid to stop all motion relative to the body. For complete damping, then, the doughnut shape should perhaps be replaced by a small sphere. However, a disadvantage of any damper involving a liquid is the fact that such a damper is very difficult to clamp. It is usually necessary to clamp the damper up to the time of third-stage separation (when spin is used to stabilize the third-stage rocket). Damping when the third stage is present would cause precession to increase, since the assembly is rod-shaped at that time.

Control of Spin-axis Drift. Spin stabilization is quite effective in maintaining constant axis pointing for short periods of time. For longer periods, the very small torques to which the satellite is subjected may cause a gradual drift of the axis angles. The main sources of torque are differential gravity, eddy currents, and magnetic-dipole effects.

The following equation may be used for accurate calculation of the effects of these torques on the motion of the spin axis:

$$C\omega_3 \frac{d\mathbf{S}}{dt} = 3\omega_0^2 (C - A)(\mathbf{r} \cdot \mathbf{S})(\mathbf{r} \times \mathbf{S}) \text{ (differential gravity)}$$

$$+ M(\mathbf{S} \times \mathbf{B}) \text{ (magnetic dipole along the spin axis)}$$

$$+ k[(\mathbf{S} \cdot \mathbf{B})\mathbf{B} - (\mathbf{S} \cdot \mathbf{B})^2 \mathbf{S}] \text{ (eddy currents)}$$

where $\mathbf{S}$ = unit vector in direction of the spin axis
 ω_3 = spin rate of the satellite
 ω_0 = orbital frequency, equal to $2\pi/T$, where T is the orbital period
 C = moment of inertia about the spin axis
 A = moment of inertia transverse to the spin axis
 $\mathbf{r}$ = unit vector from the center of the earth to the satellite
 M = magnetic-dipole moment along the spin axis of the satellite
 $\mathbf{B}$ = earth's magnetic-field vector
 k = eddy-current constant (a function of the geometry of the satellite)

Differential Gravity. The differential-gravity torque is caused by the fact that the pull of gravity varies as the inverse square of the distance. Thus the pull on the near edge of a satellite is greater than the pull on the far edge.

A nonsymmetrical rigid body in a gravitational field will experience a measurable torque if its principal moments of inertia are unequal. If the satellite is spin-stabilized, the moment of inertia about the satellite spin axis should be larger than the other principal moments of inertia. Although the effect of differential gravity is small, it must be included in the analysis in order to preserve predictive capabilities during periods of zero or small impressed fields.

Eddy Currents. The rotating satellite is immersed in a varying magnetic field which induces an oscillating electric field on the surface of the vehicle. This in turn causes currents to circulate, thereby resulting in an induced magnetic field. The torque due to eddy currents has a negligible effect on the spin-axis orientation. It does, however, cause an exponential decay of the spin rate. If the spacecraft contains magnetic materials, there will also be spin-rate decay resulting from magnetic hysteresis.

Magnetic Dipole. The primary perturbation of spin-axis direction is magnetic in origin, namely, a satellite-dipole interaction with the earth's magnetic field. Such a dipole is the result of closed current loops and permanent magnetism of ferromagnetic materials in the vehicle.

A simple dipole model for the earth's magnetic field is chosen for economy in computer time, and results in a high degree of predictive accuracy.

In a typical spin-stabilized satellite there may be a magnetic-dipole moment of the order of 1.0 amp-turns/sq m.

In a constant magnetic field of 0.2 gauss (approximate value of the earth's magnetic field), the dipole moment might cause a tilt of the axis of about 6°/day. However, the field of the earth varies in magnitude and direction as the satellite moves around the orbit. The tiltings at various parts of the orbit would partially cancel each other out, but there is usually a residual effect. This residual effect might, at times, amount to as much as 3°/day for a satellite in such an orbit.

Controlled Magnetic Torquing. In some existing spin-stabilized satellites, magnetic torquing has been employed to change the spin-axis attitude in a desired way. This is accomplished by the use of a coil of wire wrapped around the perimeter of the satellite in a plane perpendicular to the spin axis. The circuit should be so arranged that currents of various magnitudes and of either polarity can be sent through this coil for any desired period of time. This arrangement provides good control over axis pointing, and greatly enhances the usefulness of spin stabilization.

Yo-yo Speed Reduction. Another well-known device of some interest is the equipment used for spin-rate reduction. As mentioned previously, the Thor Delta third-stage assembly is spun up to 120 rpm for spin stabilization during third-stage burning. After third-stage burning and separation, it may be necessary to slow down the spin of the satellite to perhaps 12 rpm. A slow-down device which is sometimes used is called a "yo-yo" for obvious reasons. It consists of two inert weights attached to separate cables which are wrapped around the perimeter of the spacecraft. At a preprogrammed signal, the weights are released from the spacecraft (but still fastened to the cables). The centrifugal force from the spin forces the weights away from the vehicle, and as the radius increases, the centrifugal force increases. The tangential pull on the two cables provides a retarding force which reduces the spin rate. When the cable is completely unwound it is released. The spin-rate reduction can be accurately predicted and adjusted by the length of the cables and the size of the weights.

The following equation is a formula for the ratio by which the spin rate is reduced by the use of a yo-yo.

Frictionless-design formula:

$$\frac{\omega_3}{\omega_3'} = \frac{D}{F} \left\{ 1 - \left[\left(\frac{F}{C} - 1 \right) \left(\frac{FH}{G^2} - 1 \right) \right]^{\frac{1}{2}} \right\}$$

where $F = E + 2mL^2 + \tfrac{2}{3}\rho L^3 + 2aL(2m + \rho L) \sin Q$
$G = 2mL^2 + \tfrac{2}{3}\rho L^3 + aL(2m + \rho L) \sin Q$
$H = 2mL^2 + \tfrac{2}{3}\rho L^3$
$E = C + 2ma^2 + 2\rho La^2$
$D = E$ (for all practical purposes)
m = mass of one despin weight, lb-sec^2/in.
L = length of cable from hook to center of mass of despin weight, in.
d = distance from spin axis to pivot point of cable hook, in.
ρ = mass per unit length of cable, lb-sec^2/sq in.
C = mass moment of spin inertia without yo-yo, lb-sec^2/in.
D = mass moment of spin inertia with yo-yo, lb-sec^2/in.
Q = angle between tangential position and extended cable (for radial release, $Q = 90° = \pi/2$ radians)
ω_3' = initial spin speed
ω_3 = final spin speed

17-4. Design Considerations

Introduction. The design of a satellite structure is primarily determined through the effective integration of the system components. The system must be considered not only from the standpoint of the physical characteristics and performance of the satellite itself but from the standpoint of the launching system as well. The structure cannot be designed as an independent subsystem, since its capabilities, configuration, and in many cases, design philosophy are governed by the system elements and mission of the satellite. The principal influences which usually affect the structural design are:

1. Volume restrictions
2. Total payload weight
3. Dynamic load capability under launch conditions
4. Subsystem performance and configuration
5. Thermal considerations
6. Inertia, balance, and guidance characteristics

System flexibility increases with decreasing structure-to-payload weight ratio. A much greater degree of design flexibility is evident in ratios in the 0.15 to 0.20 range that was realized in Tiros than in the 0.08 range of Relay. However, relatively flexible structures of 0.06 to 0.10 are possible through extremely efficient integration of the subsystem components (which then become part of the structure), including thermal mass and gradient requirements, and the overall system hardware.

The Launching System. In addition to the influences discussed above, the geometry of the structure is governed by the capabilities, configuration, and requirements of the launching system. The separation mechanism is normally established by the launching requirements. The separation device is generally a highly reliable single coupling, of either the Marmon clamp or explosive tie-bolt type. The nose-cone configuration, volume, and the necessary thermal shrouds also govern the payload envelope. Further, the satellite vibration environment, established by the rocket performance, is probably one of the most important influences on the design of the payload. Thus the evolution of the basic payload design is strongly guided by the characteristics of the launching system.

Arrangement of Component Systems. The design of the structure is further affected by the weights and shapes of the numerous subsystem components. Proper arrangement of these components facilitates structure design. A general symmetry should be sought in order to minimize structure bending moments. Component location may be used to modify mass moments of inertia, rotational stability, drag, and center of mass. Some missions have a requirement for the unfolding of a mechanism after the protective shroud is discarded. Such deployment mechanisms must be analyzed for their attendant changes in inertia and their effects on the stabilization of the satellite in orbit.

SATELLITE SYSTEMS 17–15

The life of certain solid-state components may be adversely affected by radiation, particularly that in the Van Allen belt (Sec. 7-1). The resultant deterioration may be reduced by the use of shielding, and the configuration and weight of this shielding must be considered in the structural design.

Internal Heat Control. Local masses, finishes, materials, and active heat-controlling devices are frequently made use of for thermal control. The structure itself must also be utilized for conductive and radiation paths necessary to the thermal performance. It is evident that, for optimized designs in the 0.1 weight-ratio range, the thermal requirements must be analyzed to avoid canceling the weight advantages gained in the design of the structure. This aspect again demonstrates the importance

Fig. 17-12. View of the Tiros structure.

of efficiently integrating all elements of the payload in the design or prehardware phases.

Types of Structures. Several distinct structural-fabrication techniques have been utilized successfully in satellites up to the present time. These include skeletal framing, stressed skin, casting or forging, and combinations of these. Although aluminum has been used in the majority of our existing satellite structures, the use of magnesium and beryllium is increasing in current designs. Magnesium is generally used in the form of forged or cast elements which present distinct weight, machinability, and damping advantages. Some structural weights in the range from 150 to 400 lb have been of the combination type of construction. Examples of this are the Tiros and Ranger structures depicted in Figs. 17-12 and 17-13. The skeletal concept has been used by RCA in the design of the Relay communications satellite to provide a cruciform or framed structure. The inherent strengths of a number of the subsystem components are utilized in this overall structure. The Relay structure, shown in Fig. 17-14, also provides for "soft" mounting of the honeycomb solar panels to provide isolation under dynamic load conditions. Riveted skeletal frames may be combined with the wrought or cast elements to complete the total structure.

Structural Testing and Analysis. The test and development phases of the structure and payload also affect the overall design. Once again, the importance of development alterations is much greater in the low weight-ratio structures, the nature of such alterations being a direct function of permissible schedule times and design latitude. Such alterations should be treated with the same degree of regard as the basic design in order to avoid the introduction of latent problems relative to weight, stress, stabilization, or dynamic loads.

FIG. 17-13. View of the Ranger structure.

Once the physical configuration of the total payload has been established through consideration of the influences discussed above, the structure must be analyzed for the stress levels to which it will be subjected under launch conditions. Further analysis is made to determine the rigidity necessary to obtain desired system and component performance aloft. When a satellite is launched into orbit, the structure has to withstand not only the vertical thrust of the rocket (of perhaps 3 to $8g$) but also a high level of mechanical vibration. The vibratory acceleration is also measured in g levels. To make sure that the structure will withstand this vibration without failure, the entire structure is tested on a vibration table. The vibration table is driven sinusoidally at varying frequencies, or by random noise containing a limited band of frequencies. In a typical case, the prototype model was tested by applying noise vibration over the frequency range from 20 to 2,000 cps, with $35g$ (rms) applied along the spin axis, and $20g$ applied laterally. To avoid unnecessary damage in the tests, flight models were tested at a more realistic level, i.e., $10g$ on all three axes.

SATELLITE SYSTEMS 17–17

Usually, steady-state load levels are about 10g, with vibration inputs simultaneously applied in combinations such as 7.0g (rms) sinusoidal plus "white noise" at densities of approximately 0.05 g^2/cps over the range from 20 to 500 cps.

Yielding (inelastic distortion) of the basic structure during the launch period must be avoided or else alignment problems will arise. Therefore, an analysis of the combined loads should be made in conjunction with an investigation of the transmissibility of structural and component natural frequencies.

Coincidence of the natural frequencies of the structure and of the components must be avoided. Each structural member must be analyzed to determine the peak load for each condition to verify that the stress is not above the yield point or some preselected limit such as 95 per cent of the yield point. The structure must also be

Fig. 17-14. View of the Relay structure.

analyzed for failure due to cumulative fatigue. The peak stress for each structural member under the load condition must be evaluated as a function of the number of cycles accumulated in the life expectancy of the payload.

Stress analysis for fatigue is generally based upon a total number of cycles equivalent to three times the number of stress cycles which occur in prototype qualification testing. Thermal stresses are also analyzed where deemed critical to the satellite's orbital mission; however, these generally are negligible compared with the overall load conditions at launch. Combined thermal and structural deflections must be evaluated wherever precise positioning is required. For example, high-performance optical systems may be misaligned or defocused by these deflections.

Fabrication and Assembly. The preceding section described the various aspects generally involved in the design of the structure. In order to produce structures of high reliability and consistent performance, the manufacturing, inspection, and assembly operations must be conducted with an appropriate degree of precision and control. Obviously, structures which have been optimized with respect to volume, weight, safety factors, and fatigue must incorporate a high degree of process control. The trend toward integral packaging on a module basis is increasing, and it is possible that the structure itself may replace the numerous covers, plates, brackets, etc., which are common to present-day methods of component and subsystem integration.

Thermal Analysis. *Passive Thermal Control.* An important factor in the design of a satellite is the control of component and vehicle temperatures. This control has the effect of restricting the performance variation of temperature-sensitive components as well as permitting the utilization of components in a range around room temperature (0 to 40°C) or a narrower range when required by certain components.

The steady-state temperature of the orbiting craft is a result of equilibrium between the external energy received, the internal heat dissipated, and the energy radiated from the exterior surfaces. Neglecting the interior heat dissipation, the resulting temperatures are governed by the solar absorptivity α and the total emissivity ϵ of the exterior satellite surface. At any given wavelength $\alpha(\lambda)$ and $\epsilon(\lambda)$, the spectral absorptivity and spectral emissivity, are equal and insensitive to the surface temperature. However, the solar-radiation energy absorbed peaks in the visible region at a wavelength of about 470 millimicrons while the radiation from a body at about room temperature has an energy peak at about 10 microns. Since the wavelengths involved are so far different, the value of α/ϵ may be varied considerably by the choice of the type of satellite surface material. When considering a satellite in space that is receiving only direct solar energy with negligible internal heat dissipation, the conditions of radiative equilibrium can be stated as a simple steady-state energy balance. If the satellite is assumed to be an isothermal sphere, it may be described by the following expression (see also Sec. 15-2):

$$\alpha a S = \epsilon A \sigma T^4$$

or

$$T = \left(\frac{\alpha a S}{\epsilon A \sigma}\right)^{1/4} \quad (17\text{-}4)$$

where α = solar absorptivity of the surface
a = projected satellite area as viewed from the sun, sq ft
S = solar constant, equal to 130 watts/sq ft
ϵ = total emissivity of the surface at the temperature T
A = total satellite area, sq ft
σ = Stefan-Boltzmann constant, equal to 5.26×10^{-9} watts/(sq ft)(°K^4)
T = satellite temperature, °K

For a sphere,

$$\frac{a}{A} = \frac{\pi r^2}{4\pi r^2} = \frac{1}{4}$$

so that

$$T = 280 \left(\frac{\alpha}{\epsilon}\right)^{1/4} \quad (17\text{-}5)$$

Table 17-3 shows the ratio of α/ϵ and the corresponding operating temperature for various materials as derived from (17-5).

Table 17-3. Theoretical Operating Temperatures of Various Materials When Exposed to Solar Radiation

Surface material	α/ϵ	$t(°C) = T - 273$
White paints	0.3	−66°C
Black paints	1.0	+7°C
Bare aluminum	2.0	+57°C
Bare copper	5.0	+145°C

SATELLITE SYSTEMS   17–19

For satellite shapes other than spherical, the temperatures indicated in Table 17-3 will vary somewhat depending on the value of a/A. For example, the a/A ratio can vary between $1/6$ and $\sqrt{3}/6$ for a cubical-shaped satellite, depending on whether the side or the corner of the cube is facing the sun. This effect will cause a temperature difference that can be easily calculated if the cube is assumed to be isothermal. It can be shown that the a/A ratio for a convex solid that takes up all orientations randomly is $1:4$. This is often a useful approximation for the preliminary analysis of a spinning vehicle.

The earth, aside from the effects of shadowing the satellite during a part of its orbit, has an additional effect on the thermal balance of satellites orbiting at altitudes less than a few thousand miles. The earth emits infrared energy as a function of its own temperature and, in addition, reflects solar energy. Both these sources of radiant energy constitute part of the total external energy input to the satellite. If the earth is considered to be a Lambert surface (see Sec. 16-2, *Luminance*) for both emission and reflection, the following expressions can be derived for the energy inputs (averaged over one orbit) to a spherical satellite due to the earth's acting as both a radiating and a reflecting source of energy:

Average earth radiation:

$$\bar{\mu} = \frac{S(1-\xi)}{8}\left[1 - \sqrt{1 - \frac{R^2}{(R+h)^2}}\right]$$

Average earth-reflected solar energy:

$$\bar{\rho} = \frac{S\xi}{8}\left[1 - \sqrt{1 - \frac{R^2}{(R+h)^2}}\right]$$

where ξ = average reflectivity of the earth, equal to 0.38
  R = radius of the earth, equal to 3,960 statute miles
  h = mean satellite altitude, statute miles
  S = solar constant, equal to 130 watts/sq ft

Considerable investigation has been carried out relating earth radiation and earth-reflected solar radiation to variously shaped bodies at various altitudes.[20,21]

An isothermal spherical or tumbling satellite operating at a steady-state temperature is described by the following energy-balance equations when in the sunshine and in the earth's shadow, respectively:

Sunshine: $\alpha a S + \bar{\rho}\alpha A + \bar{\mu}\epsilon A + Q = \epsilon A \sigma T_D^4$
Shadow: $\bar{\mu}\epsilon A + Q = \epsilon A \sigma T_S^4$

where Q = internal heat dissipation
  T_D = daytime equilibrium temperature
  T_S = shadow-time equilibrium temperature
  the other symbols are as defined earlier

It should be noted that most satellites do not achieve an equilibrium temperature during the shadow-time portion of an orbit unless they have an extremely large ratio of surface area to mass, as the Echo balloons did. Similarly, most satellites do not achieve an equilibrium temperature during the sunshine portion of the orbit for the same reason, unless the orbit is a 100 per cent sun-time orbit (i.e., when the normal to the orbit points approximately to the sun).

Usually the temperature of the satellite is in a transient state, warming up during the sunshine portion of the orbit and cooling down during the shadow portion. Under the transient thermal conditions, the rate of change of stored thermal energy is given by the following equation for an isothermal body:

Rate of change of energy stored = power input − power output which actually must be stated as two equations, one for the sunshine portion of the orbit and one for the shadow portion:

Sunshine: $$mc\frac{dT}{dt} = \alpha aS + \bar{p}\alpha A + \bar{\mu}\epsilon A + Q - \epsilon A\sigma T$$

Shadow: $$mc\frac{dT}{dt} = \bar{\mu}\epsilon A + Q - \epsilon A\sigma T^4$$

where m = weight of the satellite
c = average specific heat of the satellite
$\frac{dT}{dt}$ = change of temperature with respect to time

The solutions of these nonlinear differential equations (when Q is a constant) are

Sunshine: $$\frac{t}{2\tau_D} = \tanh^{-1}\frac{T}{T_D} + \tan^{-1}\frac{T}{T_D}\Big|_{T_1}^{T} \qquad (17\text{-}6)$$

Shadow: $$\frac{t}{2\tau_S} = \coth^{-1}\frac{T}{T_S} - \cot^{-1}\frac{T}{T_S}\Big|_{T_2}^{T} \qquad (17\text{-}7)$$

where t = time, sec
$$\tau_D = \frac{mc}{4\epsilon A\sigma T_D^3}$$
$$\tau_S = \frac{mc}{4\epsilon A\sigma T_S^3}$$
T_1 = temperature at the end of the shadow period
T_2 = temperature at the end of the sunshine period

and the remainder of the symbols are as previously defined. The τ_i symbols are defined as the thermal time constants of the satellite. A typical value (Relay satellite) of the thermal time constant is 10 hr. When $t = \psi P$ in (17-6), $T = T_2$; when $t = \psi P$ in (17-7), $T = T_1$. ψ is the fractional sun time.

Exponential approximations to the above equations are much easier to work with:

Sunshine: $\quad T = T_D + (T_1 - T_D)e^{-t/\tau_D}$
Shadow: $\quad T = T_S + (T_2 - T_S)e^{-t/\tau_S}$

where T_i = an initial temperature
e = base of the natural logarithms

The length of time that the satellite spends in the sunshine and shadow has a major effect on the temperature response. The proportions of sunlight and shadow time can be determined from the intersection of the orbit path and the shadow region of the earth. Because of the orbital changes such as precession and altitude variation, these times are variable during the orbit life, and design should include consideration of the most severe conditions of temperature and temperature gradients. The fraction of a circular orbit that is spent in sunlight can be determined from

$$\psi = \tfrac{1}{2} + \frac{1}{\pi}\cos^{-1}\frac{(1-B^2)^{1/2}}{\sin\theta} \qquad B \geq \cos\theta$$
$$\psi = 1 \qquad\qquad\qquad\qquad\qquad\qquad B \leq \cos\theta$$

where ψ = fractional sun time
θ = angle from the sun's rays to the orbit normal
R = earth's radius
h = altitude (assumed to be small enough that the earth's shadow region is a cylinder)
$B = R/(R + h)$

SATELLITE SYSTEMS

The isothermal sphere is not usually encountered. For actual satellite analysis the craft is divided up into a number of thermal bodies or regions, each of which is considered to be isothermal. Each of these bodies is then joined together by conduction- and radiation-coupling factors. This essentially sets up an analytical model to forecast the actual satellite temperature.

For example, consider a cubical satellite where each of the six sides is considered to be an isothermal body. Thus, in general terms, the temperature of the ith body can be calculated from

$$\frac{dT_i}{dt} = \frac{1}{m_i c_i} \left[(\bar{p}_i \alpha_i A_i + S_i \alpha_i a_i)\delta + \bar{\mu}_i \epsilon_i A_i + Q_i - \sigma \sum_{j=1}^{6} R_{ij}(T_i^4 - T_j^4) - \sum_{j=1}^{6} K_{ij}(T_i - T_j) - \epsilon_i A_i \sigma T_i^4 \right]$$

where K_{ij} = conduction-coupling factor, sq in., between bodies i and j
R_{ij} = radiation-coupling factor, sq in., between bodies i and j
$\delta = 1$ (sunshine portion of orbit)
$\delta = 0$ (shadow portion of orbit)

For complex satellite shapes, these equations are solved on computers. The conduction- and radiation-coupling factors are difficult to calculate exactly, and the predictions of analytical models are verified by a thermal-vacuum test. In this test, various temperature heat sinks are set up in the vacuum chamber to allow each of the external surfaces on the satellite to absorb the same amount of radiant energy that it would absorb in space. Thus the predicted orbital temperature should be verified in the vacuum chamber.

Active Thermal Control. As pointed out in the foregoing section, a certain amount of thermal control may be obtained by a suitable choice of the surface parameters α and ϵ of the satellite. However, a purely passive control of this sort may not be adequate in view of varying times of sun eclipse, varying sun angles, and different modes of operation. When closer thermal control is required, an active thermal controller may be used.

An active controller may consist of a shutter, a temperature sensor, and an actuator. The shutter is a device for changing the effective value of α/ϵ as required. It may be designed in a number of ways. In one form it might be made as follows: A circular area of the satellite skin is chosen. This circle is divided into an even number of pie-shaped segments, and alternate segments are given surface finishes having high and low values of α/ϵ. The circular area is covered by a rotating disk. The disk has alternate pie-shaped segments cut away. As the disk is rotated, the underlying areas of high and low α/ϵ are exposed alternately, thus varying the overall effective α/ϵ.

A second type of controller utilizes the two disks as described, but the alternate sections of both disks are covered with aluminized mylar, a low-emissivity material. These disks are located internally in the satellite, between the electronic components and an external surface. The external surface is chosen as a surface that is always cold. Thus, when these two disks are aligned, the hot electronic components can radiate to the cold surface, and when the one disk rotates, the alternate pie-shaped sections are covered over, thus introducing a very effective radiation barrier between the electronic components and the cold surface. This type of controller changes the effective radiation-coupling factor between the components and the cold heat sink of space.

The temperature sensor may be any device for sensing deviations from the desired temperature. It may take the form, for example, of a bimetallic strip or a fluid-filled bellows. The actuator is a device for converting the sensed temperature deviation into a mechanical motion of the shutter.

An active controller of this nature can be used to reduce considerably the temperature fluctuations of the satellite structure. An example of an active controller of the aluminized-mylar type is shown in Fig. 17-15. The controller is 24 in. in diameter, weighs 2.8 lb, and provides over 150 sq in. of controllable area. It is fully closed at 7°C (44°F) and fully opened at 20°C (67°F). The effective coupling from the internal package to outer space is changed by about 70 sq in.

Fig. 17-15. Active thermal controller of the aluminized-mylar type.

17-5. Satellite Power Supplies

Introduction. Research and development work is currently being carried out on satellite power supplies using several primary sources of power, namely, solar power, nuclear reactors, and nuclear isotopes. The power converters may be thermionic or thermoelectric (described in detail in Chapter 21) or, in the case of solar power, photovoltaic cells. In spite of the wide diversity of possible power-supply types, a great majority of the power supplies which have been placed in orbit have been of the silicon photovoltaic type of solar cell, and this is the only type which will be considered here.

Silicon Solar Cells. A silicon solar cell is usually a thin plate (0.025 in.) of arsenic-doped (N-type) silicon, 1 by 2 cm. (Some later cells are 2 by 2 cm.) One side of the silicon plate is coated with a very thin layer of nickel and then a layer of solder which produces an ohmic contact. This forms the negative terminal of the solar cell. The positive terminal is formed by doping the other side of the silicon plate with boron, which penetrates very slightly to give a very thin layer of P-type silicon (thus giving the name to this type of solar cell: "P-on-N"). The impedance of this positive terminal is reduced by running a very narrow strip of tinned nickel plating down one edge. Some models of solar cells also have three or more additional thin strips across the plate at right angles to the first.

The solar cells designed for use in space are usually protected against damage from high-speed electrons and micrometeorites by the use of a thin (0.006 to 0.06 in.) layer of silica glass which is transparent to ultraviolet radiation down to about 200 millimicrons.

A light filter also is usually used to reduce the temperature of the cells by radiating the infrared portion of the spectrum, because the available power output drops approximtely 0.5 per cent/°C.

The operating voltage of a solar cell is of the order of 0.4 volt. The efficiency with optimum load runs from about 8 to 12 per cent. The current and voltage characteristics of a typical cell are shown by the dotted curves in Fig. 17-16 for operating temperatures of 0, 30, and 60°C. The solid curves are rectangular hyperbolas of constant conversion efficiency (or constant power: $P = IV$). For the cell at 0°C, efficiencies better than 11 per cent may be obtained, while at 60°C the maximum efficiency obtainable is about 8.5 per cent. The cells are usually connected in a series-parallel network designed to produce the desired output voltage and power.

Considerable study has been made of the effect of Van Allen belt radiation on the efficiency of silicon solar cells. The importance of such study has been increased by the July 1, 1962, high-altitude thermonuclear explosion which resulted in an artificial increase in the intensity of the Van Allen belt. This increase in intensity amounts to about one order of magnitude for high-energy electrons ($E > 600$ kev), but only about 50 per cent for low-energy electrons ($150 < E < 600$ kev); see Sec. 7-1. The

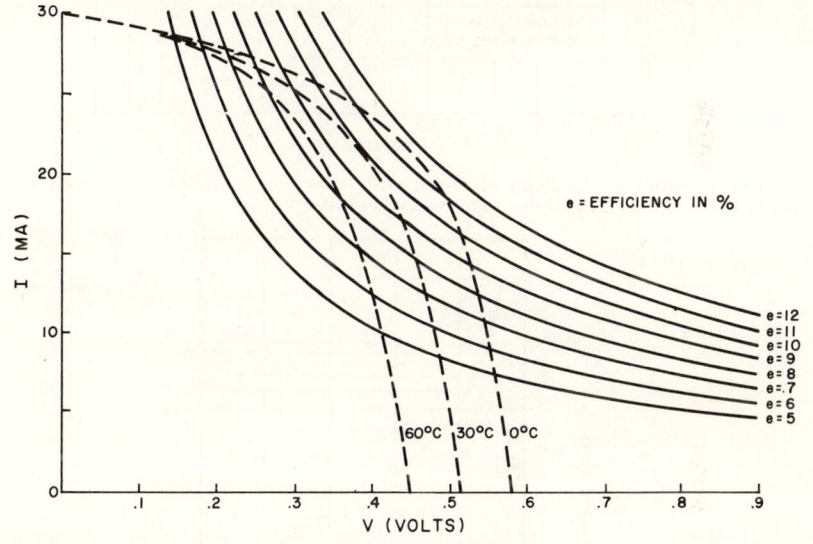

Fig. 17-16. Solar-cell characteristic curves.

deterioration of a solar-cell power supply with time has been calculated for the Relay satellite on the basis of operation in the combined natural and artificial fields. For the calculations, the orbit of the satellite was assumed to have a perigee of 800 nautical miles and an apogee of 3,000 nautical miles. The estimated and measured results are plotted in Fig. 17-17, where the abscissa is time in days, and the ordinate is the power output in per cent of that required for proper operation of the Relay satellite. The data have been plotted for the P-on-N cells described above, and for the newer N-on-P cells. The lifetime of the N-on-P cells is approximately 25 times as great as that of the P-on-N cell. For this reason the N-on-P type cells may gradually replace the P-on-N type as they become available. Figure 17-17 also shows the effect of a change in thickness of the fused-silica cover glass.

The Storage Battery. A storage battery is usually connected in parallel with the array of solar cells. There are two reasons for doing this. First, it allows a peak power to be drawn (for short intervals) which is greater than that which the solar cells can supply directly. Second, it supplies a source of power during the dark portion of the orbit.

The storage battery is usually composed of nickel-cadmium cells, but silver-cadmium cells are being investigated. Such wet cells must be hermetically sealed to protect them from the vacuum of space. Battery life expectancy is dependent upon the depth of discharge, as shown in Fig. 17-18. As a consequence, the battery is often so operated that it is never discharged by more than 20 per cent of total capacity.

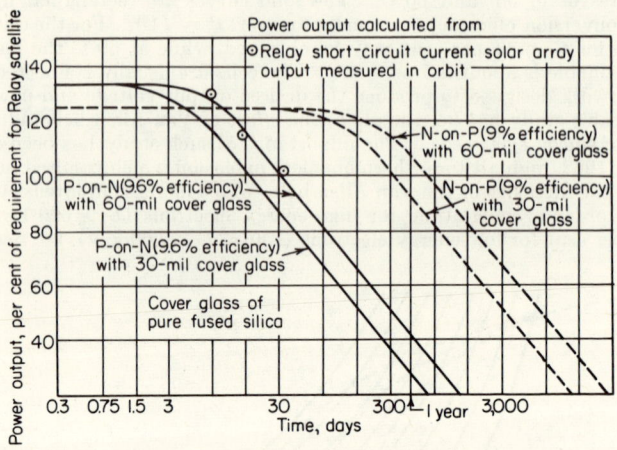

Fig. 17-17. Output of the Relay solar-cell array as a function of operating time in the natural and artificial (after July, 1962) radiation belts.

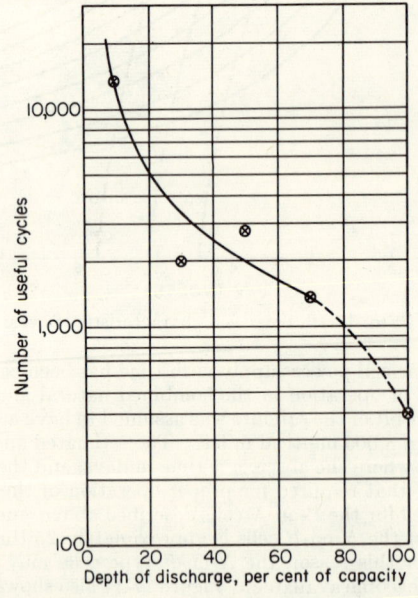

Fig. 17-18. Battery life in cycles vs. depth of discharge (nickel-cadmium cells).

Another reason for restricting the depth of discharge is the limited charge rate which may be used if excessive gassing is to be avoided.

The storage battery may comprise several parallel branches, each branch having a number of cells in series arrangement. Isolating diodes are usually placed in the load lines of each branch so that a branch cannot act as a load on the other branches. Also desirable is the use of a current limiter in the charging circuit of each branch to limit

SATELLITE SYSTEMS

the current to a suitable maximum value consistent with the characteristics of the cells.

Overall Power-supply Considerations. Above the earth's atmosphere, a surface is exposed to a solar irradiance of 130 watts/sq ft (solar constant). Since the solar cells have an average efficiency of about 10 per cent, this means that the nominal power output of a solar-cell array is about 13 watts/sq ft. There are, however, a number of degradation factors which make it necessary to use a somewhat lower figure. These factors, with approximate values, are as follows:

1. Solar-cell packing factor 0.90
2. Allowance for storage-battery efficiency 0.80
3. Series-parallel degradation 0.94
4. Solar-cell operating efficiency at +60°C 0.85
5. Transmission of glass cover 0.90
6. Efficiency of blocking diode 0.98
7. Solar-cell array illuminated 65% of the time 0.65
   Product of factors 1 through 7 .... 0.33

Thus the effective output figure is about $(0.33)(13) = 4.3$ watts/sq ft assuming the above degradation factors. This figure assumes that the array is oriented in a direction normal to the sun. If the array under consideration is a spinning satellite covered with solar cells, then the area to be used in calculations is the projected area as viewed by the sun.

Where some of the cells are in the dark at a given instant, it is desirable to disconnect effectively the unilluminated cells. This can be accomplished by dividing the cells into narrow-angle sectors and connecting each sector into the circuit through a diode. The diode allows current to flow only in the forward direction, thus preventing the impedance of the dark cells from acting as a load on the power supply.

The Voltage Regulator. In many types of satellites, the power-supply specification requires a regulated voltage for some portions of the load. When this specification is present, a voltage regulator is included only for the critical items of the load. In a typical satellite much of the electronic equipment utilizes a regulated voltage. Exceptions are shutter solenoids, telemetry switches, and the power converter for the tape recorder. Isolating such items from the regulators may reduce the required capacity of the regulators by as much as 50 per cent.

Several separate regulators may be used, for redundancy (thus giving higher reliability) or for obtaining different useful regulated voltages. The regulators usually contain transistors which act as series arms to control the output voltage. The error voltage, which is used for control, is obtained by comparing the regulated voltage with a zener-diode-controlled reference voltage.

17-6. Image Sensing and Storage

Introduction. There are a number of satellite missions which involve image sensing, storage, and retrieval. These missions include astronomical observations, solar observations, meteorological surveillance, and military reconnaissance. The images may be obtained in visible light, infrared, or other portions of the spectrum.

The Vidicon Camera. When images are to be obtained with visible light, the most commonly used sensor is the vidicon camera. The vidicon was originally designed for commercial television where a low-cost standard-rate (30 frames/sec) camera was desired. In satellite applications it is usually used to generate separate still pictures with read-out times per frame measured in seconds. Since each frame is a completely new picture, the system probably should not be called television, but the term is still frequently used.

In the $\frac{1}{2}$-in. vidicon-camera tube, the optical image of the picture is focused on an area 6.3 by 6.3 mm on the end of the tube. A 1-in. vidicon has a sensitive area 11.2 by 11.2 mm. This area, called the target, is a photoconductor. By photoconductor action, the light pattern is converted to an analogous charge pattern, which is then read off by a scanning electron beam.

For some applications, the pictures are taken with the cameras pointing along the spin axis of the satellite. In this case the spin motion would cause a blur in the peripheral portions of the picture if the optical shutter were opened for too long an exposure interval. The maximum exposure time which should be used is that which produces half a picture element (TV line) of smear in the corners of the picture.

The picture is stored for several seconds on the vidicon target in the form of a charge pattern. Since the photoconductor has some conductivity, even in the dark, the charge pattern gradually leaks off. To minimize this effect the pattern is usually read off by a scanning electron beam within 10 sec of exposure. The sensitivity of a typical 1-in. vidicon is illustrated graphically in Fig. 17-19. The response as a function of resolution is shown in Fig. 17-20.

In order to store the picture until the satellite reaches a point in the vicinity of a ground station, storage times of the order of 1 hr or more are necessary. A magnetic-tape recording may be used to accomplish this.

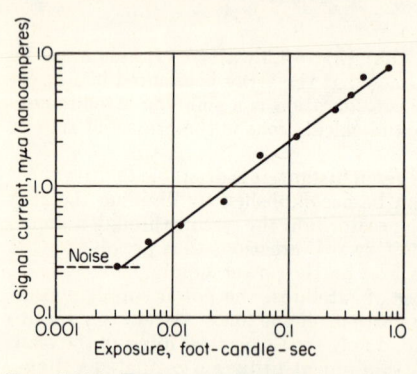

FIG. 17-19. Transfer characteristics of the 1-in. vidicon with 2-sec read-out time.

FIG. 17-20. Aperture characteristic of the 1-in. vidicon.

The Optical System. In order to minimize the picture smear due to image motion (caused by spin, vehicle velocity, etc.) the open-shutter interval (exposure time) must be minimized. To obtain adequate exposure of the vidicon target, the lens aperture must be made as large as practicable (i.e., the f number—focal length divided by diameter—must be small).

The size of the object area actually photographed is an inverse function of the focal length of the lens. If S is the length of one side of the square to be photographed, then

$$S = \frac{hw}{f}$$

where h = altitude of the camera
  w = width of the vidicon target
  f = focal length of the lens
For example, in Tiros IV the values are

h = 400 miles
w = 6.33 mm
and f = 5 mm
Hence S = 500 miles

Ground Resolution of the Vidicon. In a TV camera using the ½-in. vidicon, the resolution is about 350 lines per picture, regardless of the size of the area photographed. This is a consequence of the fact that a resolution of better than 350 lines cannot be obtained on the vidicon target regardless of what optical system is used to focus an image upon it.

Another method of expressing the resolution is the width of one resolution line, which is $S/350$ for the ½-in. vidicon. Thus the ground resolution of Tiros IV is 1.4 miles. For some purposes, much finer resolution is required. Finer subject resolution requires the use of a lens with both a larger focal length and a larger diameter. If the f number is to remain small, optical systems with very large apertures are required.

Resolution can be improved, without increasing the focal length, if the resolution on the photoconductor target can be improved. The resolution on the ½-in. vidicon target is 350 lines in 6.3 mm, or 56 lines/mm. Research now in progress indicates that a resolution of 200 lines/mm is feasible and will soon be available.

Bandwidth Requirements. The frequency bandwidth required to transmit a picture to ground is an important consideration, because it affects the signal-to-noise ratio, which in turn affects the power required of the picture transmitter.

The minimum bandwidth required to transmit picture information may be calculated from the following formula:

$$B = \frac{hv}{2T}$$

where h = horizontal resolution in lines per picture
v = number of scanning lines
T = frame time

Usually an attempt is made to keep the resolution the same in the horizontal and vertical directions. The vertical resolution is not the same as the number of scanning lines but is reduced because of beating effects. The reduction factor (also called the Kell factor) has a value of about 0.7. Thus, for a square picture with equal horizontal and vertical resolution,

$$h = 0.7v$$

For example, in the Tiros satellite,

$$v = 500$$
$$h = 0.7v = 350$$
and $\quad T = 2$ sec
Hence $\quad B = 43$ Kc

It is desirable to get the best resolution possible within the limitations of the ½-in. vidicon. To make certain that the bandwidth is not the limiting parameter, the nominal bandwidth in Tiros was chosen somewhat greater than that calculated above, namely, 62.5 Kc. This figure corresponds to a theoretical horizontal resolution of 500 lines per picture. The actual horizontal resolution in some cameras is 400 lines per picture.

The Use of Infrared Sensors. The use of infrared sensors is increasing in space technology. Infrared technology (Chapter 15) differs from that of visible light (Chapter 16) in that the light being sensed is emitted by the target rather than being reflected from it. Thus infrared is usually utilized to detect and record differences in temperature rather than differences in reflectivity.

A detector which is frequently used for sensing and measuring infrared radiation is the thermistor bolometer. These thermistor bolometers are tiny flakes of a material whose resistance changes rapidly with a change in temperature.

In a meteorological-satellite program, infrared detectors have been used to provide information about the thermal budget of the earth. Data obtained from these detectors furnished information about the thermal-radiation exchange between the various portions of the earth and space, and facilitated calculations of the temperatures of the oceans, land masses, and clouds. The extent of cloud cover on the "night" side of the earth can be inferred to some extent from these data. In deep-space probes, the temperatures of the planets can be determined by using infrared sensing.

A filter is often used in front of an infrared detector to isolate the preselected band of wavelengths to be received. Certain wavelength regions are especially useful. For example, the range from 8 to 12 microns is the best atmospheric window. It may be used to measure earth-surface temperatures and cloud temperatures. The band from 5.6 to 7.0 microns is a strong water-vapor-absorption band. It may be used to measure the temperature at the top of the troposphere.

Satellite Tape Recorders. In many satellite imaging systems, the image is sensed at a point in the orbit which is far removed from any ground station. When this is the case, it becomes necessary to store the image in the spacecraft until it reaches a point in its orbit sufficiently close to a ground station that the picture signal can be transmitted to ground. It is sometimes convenient to accomplish this storage by the use of a suitable magnetic-tape recorder. A photodielectric tape recorder has been developed which performs both the operations of image sensing and storage. This tape system offers increased capability along with increased simplicity, as compared with both the magnetic and electrostatic systems, in that it requires one less image-processing step.

The requirements for image-storing tape recorders are so stringent that it is usual to design for the specific purpose. To prevent jitter in the display process, the tape speed must not vary by more than 0.2 per cent over a short time interval. To prevent linearity distortion, the picture signal is usually recorded on the tape as a frequency-modulated carrier wave. To minimize weight, densities of 2,000 cycles/in. or higher are commonly employed. System considerations sometimes require that the read-in and read-out tape speeds be different, and several tape speeds are sometimes provided. A number of signal channels may be utilized on the tape, depending upon mission requirements, with a minimum of weight and power always required.

17-7. Conclusions

The preceding sections discuss in some detail the design of various parts of a satellite. In general, however, the missions for which satellite systems are intended are satisfied only by the operation of the satellite as an integrated system. It is therefore essential that the design of the separate parts be optimized in order to fulfill the requirements of the overall system. A working knowledge of the design factors in the parts and subsystems as presented in this chapter will enable the system engineer to understand better the relative importance of the many engineering decisions required of him in the design of a satellite system. With this knowledge he will be able to effect a proper match between satellite technology and mission requirements.

REFERENCES

1. Brouwer, D., and G. M. Clemence: *Methods of Celestial Mechanics*, Academic Press Inc., New York, 1961.
2. Jensen, J., G. E. Townsend, J. Kork, and J. D. Kraft: *Design Guide to Orbital Flight*, McGraw-Hill Book Company, New York, 1962.
3. Smart, W.: *Celestial Mechanics*, Longmans, Green & Co. Inc., New York, 1953.
4. Anderson, J. D.: *Long Term Perturbations of a Moon Satellite by the Earth and Sun*, JPL-TM 312-162, Jet Propulsion Laboratory, California Institute of Technology, Pasadena, Calif., Feb. 8, 1962.
5. Lorell, J.: *Precession Rates of an Artificial Satellite Due to a Third Body Perturbation*, JPL-TM 312-175, Jet Propulsion Laboratory, California Institute of Technology, Pasadena, Calif., Feb. 15, 1962.
6. Kork, J.: Satellite Lifetimes in Ecliptic Orbits, *J. Aerospace Sci.*, vol. 29, no. 11, November, 1962.
7. Billik, B. H.: Survey of Current Literature on Satellite Lifetimes, *Am. Rocket Soc. J.*, vol. 32, no. 12, December, 1962.
8. Baker, R. H.: *An Introduction to Astronomy*, D. Van Nostrand Company, Inc., Princeton, N.J., 1961.
9. Van de Hulst, H. C., C. de Jager, and A. F. Moore: *Space Research* II, North Holland Publishing Company, Amsterdam, 1961.
10. Vinti, J.: New Method of Solution for Unretarded Satellite Orbits, *J. Res. Natl. Bur. Std.*, Sec. B, vol. 63B, no. 2, 1959.

11. Routh, E. J.: *Dynamics of a System of Rigid Bodies*, 6th ed., Dover Publications, Inc., New York, 1955.
12. Thomson, W. T.: *Introduction to Space Dynamics*, John Wiley & Sons, Inc., New York, 1961.
13. Timoshenko, S., and D. H. Young: *Advanced Dynamics*, McGraw-Hill Book Company, New York, 1948.
14. Shanley, F. R.: *Weight-Strength Analysis of Aircraft Structures*, 2d ed., Dover Publications, Inc., New York.
15. Timoshenko, S., and D. H. Young: *Vibration Problems in Engineering*, 3d ed., D. Van Nostrand Company, Inc., Princeton, N.J., 1961.
16. Crandall, S. H.: *Random Vibrations*, 2d ed., The Technology Press of the Massachusetts Institute of Technology, Cambridge, Mass., 1959.
17. Freberg, C. R., and E. N. Kemler: *Elements of Mechanical Vibration*, 2d ed., John Wiley & Sons, Inc., New York, 1949.
18. Spence, H. R., and H. N. Luhrs: *Structural Fatigue under Combined Sinusoidal and Random Vibration*, Space Technology Laboratories Rept. STL/TN-60-0000-19213, Los Angeles, November, 1960.
19. Lipson, C., and D. H. Thiel: *Applications of Stress Analysis*, University of Michigan Summer Course, 1959.
20. Cunningham, F. G.: *Power Input to a Small Flat Plate from a Diffusely Radiating Sphere with Application to Earth Satellites*, NASA TNO 710, July, 1961.
21. Cunningham, F. G.: *Earth Reflected Radiation Input to Spherical Satellites*, NASA TNO 1099, October, 1961.
22. Fink, D. G. (ed.): *Television Engineering Handbook*, McGraw-Hill Book Company, New York, 1957.
23. Zworykin, V. K., and G. A. Morton: *Television*, 2d ed., John Wiley & Sons, Inc., New York, 1954.
24. *Air Navigation*, U.S. Navy Hydrographic Office Publication 216, Government Printing Office, Washington, D.C., 1955.

Chapter 18

AERODYNAMIC SYSTEMS

JACK N. NIELSEN, *Director of Research and Development, Vidya Division, Itek Corporation.*

CONTENTS

18-1. Introduction	18-1
18-2. Physical Characteristics of Air	18-2
18-3. Components of Airframes	18-3
18-4. Axes, Sign Convention	18-4
18-5. Low-speed Configurations	18-5
Airfoils and Bodies of Revolution	18-6
Wing and Wing-Body Combinations	18-7
Boundary Layer, Skin Friction	18-8
Low-speed Stability and Control	18-9
18-6. Compressibility; Airframes at Transonic Speeds	18-10
Airfoils and Wings	18-10
Slender Airframes; Whitcomb Transonic Area Rule	18-11
18-7. Airframes at Supersonic Speeds	18-13
Airfoils	18-13
Supersonic Wings	18-14
Supersonic Wing-Body Combinations	18-15
18-8. Hypersonic Configurations	18-16
Reentry Configurations	18-16
Hypersonic Airframes	18-16
Newtonian Theory	18-17

18-1. Introduction

The principal purpose of this chapter is to summarize certain essential results widely used in the design of aerodynamic configurations in a form readily understandable to one not schooled in aeronautical engineering. For this purpose we might define aerodynamics as a special branch of fluid mechanics which deals with the flow of air about bodies, and which is particularly concerned with the force and moment reactions of the bodies to such air flow. While defined solely as involving the flow of air, aerodynamics of low-speed flows is identical with the theory of the flow of incompressible inviscid fluids to the extent that the effects of viscosity are negligible. Thus many of the results of low-speed aerodynamics apply to ships, torpedoes, submarines, hydrofoils, propellers, etc., operating in water as well as they do to comparable shapes operating in air.

In an aerodynamic system we can include airframes, propulsion, instrumentation, etc. The emphasis here will be on airframes, be they missiles, airplanes, torpedoes, submarines, etc. In the design of such configurations, the first step is usually the preliminary design, which is based on existing information, both experimental and theoretical. In the detailed-design phase, extensive theoretical analyses and wind-tunnel testing will usually occur, culminating in a prototype design. The next phases include prototype construction and flight testing. The formulas herein are intended to be useful in the preliminary- and detailed-design phases of airframes.

In order to achieve the maximum possible generality within the present limited treatment, it seems best to approach aerodynamic configurations as combinations of various simple geometric components such as bodies of revolution, wing surfaces, and nozzles. We shall first consider the aerodynamic characteristics of the isolated components. These characteristics will be found to vary with the speed range in which the component operates. The Mach number, the ratio of the speed of the configuration to the speed of sound, is the usual parameter used to specify the speed range and is probably the most important parameter discussed herein. Viscosity, which modifies some of these characteristics, will also be considered. The formation of configurations by physical addition of the single geometric components will be discussed.

18-2. Physical Characteristics of Air

For the purposes of this chapter, we may consider air to be a perfect gas, obeying the gas law

$$p = R\rho T \tag{18-1}$$

where p = pressure
ρ = mass density
T = absolute temperature
R = gas constant for air

(See List of Common Symbols at the end of the chapter.) The speed of sound in air depends on the absolute temperature T

$$a = \sqrt{\gamma R T} \tag{18-2}$$

where $\gamma = c_p/c_v = 1.4$ at ordinary temperatures

The absolute viscosity of air over the temperature range $180°\text{R} < T < 3400°\text{R}$ is given closely by the formula of Sutherland

$$\mu = 2.270 \frac{T^{3/2}}{T + 198.0} 10^{-8} \quad \text{lb-sec/ft} \tag{18-3}$$

The following dimensionless quantities are widely encountered in aerodynamics:

$$M = \text{Mach number} = \frac{V}{a} \tag{18-4}$$

$$\text{Re} = \text{Reynolds number} = \frac{V l \rho}{\mu} \tag{18-5}$$

The local Mach number is the ratio of the flow velocity V to the local speed of sound at the point where V is measured. The quantities V and a may be specified at different locations in some applications. A flow is said to be locally subsonic, supersonic, or hypersonic, in accordance with the following listing:

$0 < M < 1$, subsonic flow
$1 < M < 5$, supersonic flow
$5 < M$, hypersonic flow

AERODYNAMIC SYSTEMS   18-3

The boundary between supersonic and hypersonic is a nominal one and is to be used only for general purposes. For a small range of M about $M = 1$, the flow is said to be transonic. A nominal range might be $0.8 < M < 1.2$.

The Reynolds number is the ratio of the inertial force to the viscous force acting on an elementary fluid mass. At low values of the Reynolds number, flows in pipes, for instance, tend to be laminar; at high Reynolds numbers they tend to be turbulent. The flow in the boundary layer next to the surface of a fuselage or a wing tends to be laminar or turbulent in the same manner. Any dimension which characterizes the scale of geometrically similar bodies can be used for the reference length l. For a pipe l might be the diameter; for a wing it might be the span or chord. For each set of geometrically similar objects we have a different Reynolds number. Comparisons of geometrically dissimilar objects on the basis of Reynolds number are meaningless, since the magnitude of the Reynolds number depends to some extent on the arbitrary choice of reference length used in its definition.

The science of hydrodynamics as the theory of incompressible inviscid fluid flow represents aerodynamic theory for the limiting case as M approaches 0 independent of Reynolds number. Any systematic variation of a flow quantity with Re is probably caused by deviations of the flow from hydrodynamics theory which are due to viscosity.

18-3. Components of Airframes

Consider an aerodynamic configuration consisting of a body (fuselage), wing, and tail as shown in Fig. 18-1. The body will probably be a streamline shape with a blunt nose for subsonic operation, or a pointed nose for supersonic operation, and will probably have a blunt nose again at hypersonic speeds where aerodynamic heating is usually important. Some aerodynamic bodies of interest include the following:

Subsonic: ellipsoids, circular cylinders
Supersonic: tangent ogives, Sears-Haack bodies, cones, Karman ogives
Hypersonic: hemispheres, paraboloids, hemispherically tipped cones

Wings and tails are generally termed lifting surfaces if they are very thin. The wing is used to provide the main lift necessary to maintain flight, and the tail is used for stability and control. If the tail precedes the wing, as in the airplane of the Wright brothers, the airplane is termed a canard. Generally speaking, most airplanes or missiles are wing-body-tail combinations equipped with aerodynamic controls such as ailerons, elevators, and stabilizers to trim the airframe in equilibrium flight, that is,

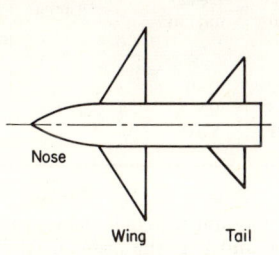

Fig. 18-1. Wing-body-tail combination.

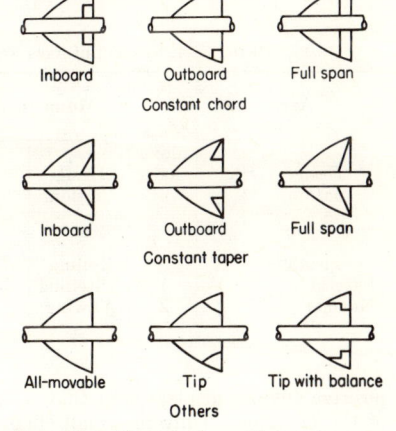

Fig. 18-2. Types of aerodynamic controls.

to reduce the moments acting about its center of gravity to zero. They also serve the purposes of providing untrimmed moments for changing the flight path of the airplane. There are many types of aerodynamic controls, a number of which are shown in Fig. 18-2.

18-4. Axes, Sign Convention

A large number of different axis systems are used in the design of aeronautical systems. We shall consider only *body axes* and *wind axes* in this chapter. A set of body axes is a set fixed in an aerodynamic body and is usually a cartesian set with one or more of the axes coinciding with intersections of planes of symmetry, as shown in Fig. 18-3. This set is widely used in theoretical aerodynamics. A set of axes usually used in published reports of the National Advisory Committee for Aeronautics, now the National Aeronautics and Space Administration, is shown in Fig. 18-4.

Consider the general notation for positive directions of forces, moments, rotations, linear velocities, and angular velocities. With regard to forces, if the X force points along the direction of translation of the center of gravity of the airframe, then $-X$ is the *drag* D and $-Z$ is the *lift force* L. If X is along a longitudinal axis of the missile, $-X$ is the *axial force* and $-Z$ is the normal force N.

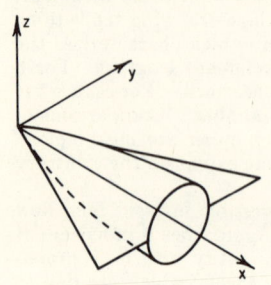

Fig. 18-3. Body axes.

With regard to moments, the positively indicated direction of ϕ, θ, and ψ are the positive directions for rolling, pitching, and yawing moments, respectively. Positive components of the airframe linear velocity u, v, and w are in the positive directions of X, Y, and Z, respectively; and a similar convention exists for the components p, q, and r of the airframe angular velocity. The angles ϕ, θ, and ψ represent directions of

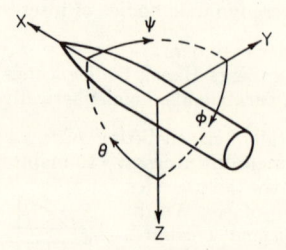

Fig. 18-4. System of axes and symbols used in NACA reports.

Axes Designation	Force along axis	Moments about axis Designation		Positive direction	Angles Designation		Velocities Linear	Angular
Longitudinal	X	X	Rolling	L	$Y \to Z$	Roll	ϕ	u	p
Lateral	Y	Y	Pitching	M	$Z \to X$	Pitch	θ	v	q
Normal	Z	Z	Yawing	N	$X \to Y$	Yaw	ψ	w	r

positive angular displacement that X, Y, and Z may undergo in a change in attitude of the airframe. Only for small changes in ϕ, θ, and ψ is the order of the angular displacements commutative. For large displacements, the cyclic order of ϕ, θ, and ψ determines the final position.

In practical airframe design, one of the most important problems is the calculation of the aerodynamic forces and moments acting on the airframe. For steady flow or steady flight, the forces depend on the attitude of the airframe with respect to the

AERODYNAMIC SYSTEMS

flow velocity, as well as the air density and the airspeed. In order to remove this dependence, the forces and moments usually defined are in coefficient form as follows:

$$C_D = \frac{D}{q_\infty S} = \text{drag coefficient} \tag{18-6a}$$

$$C_L = \frac{L}{q_\infty S} = \text{lift coefficient} \tag{18-6b}$$

$$C_m = \frac{M}{q_\infty S l} = \text{pitching-moment coefficient} \tag{18-6c}$$

where
$$q_\infty = \frac{1}{2}\rho_\infty V_\infty^2 = \text{dynamic pressure} \tag{18-6d}$$

The subscript ∞ is used to designate flow quantities in a parallel flow at any point out of the range of influence of a body immersed in the flow. The quantities S and l are

S = arbitrary reference area
l = arbitrary reference length

The area S might be the base area for a body of revolution or it might be the planform area for a wing.

One of the principal purposes of wind-tunnel testing is to measure the coefficients for various aerodynamic configurations. It is found that the coefficients depend on a number of parameters, the principal ones of which are angle of attack α, Mach number M, Reynolds number Re, and angle of sideslip β. Thus,

$$\begin{aligned}C_D &= C_D(\alpha,\beta,M,\text{Re})\\ C_L &= C_L(\alpha,\beta,M,\text{Re})\end{aligned} \tag{18-7}$$

In the definition of α and β consider the body axes illustrated in Fig. 18-3. Let the velocity of the airflow relative to the x, y, z system be V_∞, and let the velocity components along x, y, and z be u, v, and w. It is understood here that u, v, and w are airspeed velocity components, not velocity components of the center of gravity as in Fig. 18-4. Then the angles of attack and sideslip for small values of α and β are sometimes defined by

$$\alpha = \frac{w}{V_\infty} \qquad \beta = -\frac{v}{V_\infty} \tag{18-8}$$

For large values of α and β, use is sometimes made of the sine definitions indicated by the subscript s

$$\sin \alpha_s = \frac{w}{V_\infty} \qquad \sin \beta_s = -\frac{v}{V_\infty} \tag{18-9}$$

or the tangent definitions indicated by the subscript t

$$\tan \alpha_t = \frac{w}{V_\infty} \qquad \tan \beta_t = -\frac{v}{V_\infty} \tag{18-10}$$

For data taken at large angles of attack or sideslip, it is important to specify which definitions have been adopted.

18-5. Low-speed Configurations

By low-speed aerodynamics, we mean aerodynamics for which changes in air density due to compressibility do not have a significant influence on the airflow pattern (possibly excepting free convection flows). The *compressibility effects* in question are those due to density changes associated with speed changes. Increase in the speed of air without addition of energy results in lowered density, pressure, and temperature in accordance with the adiabatic gas law and the law of Bernoulli. For airflow over a body, the free-stream Mach number is a simple measure of the importance of compressibility. For instance, for a parallel flow of Mach number 0.4 past a circular cylinder, sonic flow will appear at the extremities of the diameter normal to the flow. In this instance, 0.4 is said to be the *critical Mach number* of the cylinder; namely, the Mach number at which locally sonic flow first appears in the flow pattern. For Mach numbers substantially less than the critical Mach number, we have the realm of low-speed aerodynamics. For a thin airfoil at zero angle of attack, the critical Mach number might be 0.7; for the same airfoil near maximum lift, the critical Mach number might be 0.2.

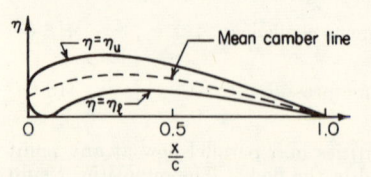

FIG. 18-5. Subsonic airfoil section.

Airfoils and Bodies of Revolution. Consider now the low-speed characteristics of two-dimensional wings, or airfoils. The aspect ratio A of a wing is the span squared divided by its planform area. Alternately, the aspect ratio is the ratio of the span to the mean chord. An airfoil is a wing of infinite aspect ratio.

An airfoil can be defined by the ordinates of its upper and lower surfaces η_u and η_l measured from some chord line as shown in Fig. 18-5. The thickness of the airfoil is taken to be

$$t = \eta_u - \eta_l \tag{18-11}$$

and the ordinate of the so-called mean camber line is given by

$$\eta_c = \tfrac{1}{2}(\eta_u + \eta_l) \tag{18-12}$$

The thickness and camber are usually expressed as a percentage of the chord length. The shape of the camber line and the thickness distribution thus define an airfoil shape.

Consider now the lift and drag characteristics of airfoils. The plot of lift coefficient c_l against angle of attack is usually quite linear and remains so to values of c_l of unity or thereabouts. Over the linear range an airfoil has a lift-curve slope of 2π per radian; that is,

$$c_l = \frac{2\pi}{57.3}\, \alpha \qquad \alpha \text{ in degrees} \tag{18-13}$$

The term *section lift coefficient* refers always to an airfoil or to a two-dimensional strip of a three-dimensional wing and is designated c_l.

With regard to drag, an airfoil can have drag as a result of the pressures acting on it or as a result of viscous drag (shearing forces) at its surface. The famous *paradox of d'Alembert* says that in an inviscid fluid an airfoil can have no drag. Since air is a real viscous fluid, practical airfoils do have some drag. The viscosity even permits the existence of some

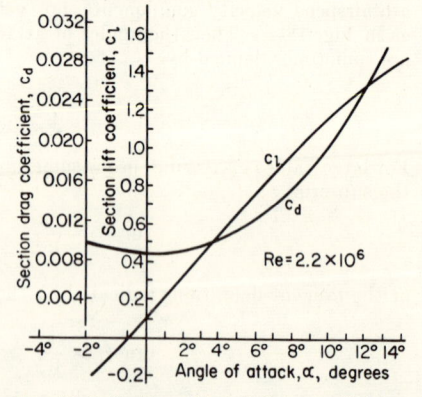

FIG. 18-6. Section lift and drag coefficients of NACA 23012 airfoil. $Re = 2.2 \times 10^6$.

pressure drag contrary to d'Alembert's paradox. It is possible to reduce the drag of an airfoil by maintaining laminar flow over its surface. Much effort has been expended in developing laminar-flow airfoils. While airfoil lift can be calculated, airfoil drag must be measured.

A typical plot of c_d and c_l vs. α for a low-speed airfoil is shown in Fig. 18-6 for an airfoil section designated as NACA 23012. It is noted that the Reynolds number based on the chord length is included. While not much influence of Reynolds number on the linear part of the lift curve is usually found, it can have large effects on the nonlinear part of the lift curve near maximum lift and on the drag curve everywhere. The value of α for $c_l = 0$ is termed the *angle of zero lift*, and the least value of c_d is the *minimum drag coefficient*. Experimental lift and drag data are to be found in Refs. 1 to 3. For a symmetrical airfoil, the center of lift is at the 25 per cent chord. For a cambered airfoil the lift at zero angle of attack will usually not act at the quarter-chord location, but the additional lift due to angle of attack will act near the quarter-chord location.

Bodies of revolution at low speeds have no drag in an inviscid fluid in accordance with d'Alembert's paradox. For bodies which have zero radius at each end such as submarines and dirigibles, the lift is also zero in an inviscid fluid, although some lift can occur in a viscous fluid because of boundary-layer effects. For a slender body with an aft end (base) of radius r, the lift force in an inviscid fluid on the basis of slender-body theory[4] is

$$L = 2q_\infty (\pi r^2) \alpha \qquad (18\text{-}14)$$

Wing and Wing-Body Combinations. Airfoils are lifting surfaces of infinite aspect ratio; wings are lifting surfaces of finite aspect ratio. As a consequence of finite aspect ratio, d'Alembert's paradox is not applicable to wings. Thus, unlike airfoils, wings have pressure drag associated with their loading distribution, so-called *induced drag*. Also, the lift-curve slope of a wing is generally less than that of an airfoil. For wings of low aspect ratio, the lift-curve slope is given by slender-wing theory as

$$\frac{dC_L}{d\alpha} = \frac{\pi}{2} A \qquad (18\text{-}15)$$

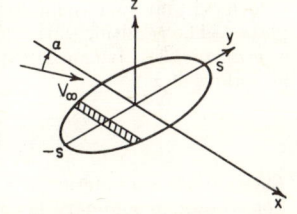

FIG. 18-7. Chordwise strip of wing.

With regard to subsonic wing drag, in addition to friction drag which exists at all angles of attack (and varies with α), we have "induced drag" due to pressure. The total drag coefficient can thus be written

$$C_D = C_{D_f} + C_{D_i}$$

where C_{D_f} = friction drag due to viscosity
C_{D_i} = induced drag

The induced drag can be thought of as drag which necessarily exists whenever wings of finite span are used to induce downward momentum to air to generate lift. Consider a wing as shown in Fig. 18-7 with a chordwise strip shown hatched. Let c_l be the section lift coefficient on this strip and let c be its chord. Then the wing *span-load distribution* is the variation of cc_l across the wing span. An elliptical span-load distribution is given by

$$cc_l = (cc_l)_{y=0} \sqrt{1 - \frac{y^2}{s^2}} \qquad (18\text{-}16)$$

For an elliptical span-load distribution, the induced drag is a minimum for a given lift and a given span. The induced-drag coefficient is then given by

$$C_{D_i} = \frac{1}{\pi A} C_L^2 \qquad (18\text{-}17)$$

Any low-speed aerodynamic configuration can minimize its induced drag by attempting to achieve an elliptical span-load distribution. In general, the greater the departure of the span loading from elliptical, the greater the drag penalty. A precise analysis of this matter based on Fourier series is included in Ref. 6.

The aerodynamic characteristics of low-speed wing-body combinations are influenced by *interference* between body and wing. A rule of thumb long used in subsonic airplane design has been that the lift generated by a wing mounted on a body is the same as if the body were not there. For aircraft for which the wing aspect ratio is 6 or greater, and for which the body diameter is a small fraction of the wing span, this rule of thumb is generally valid. However, a more precise way of calculating lift interference applicable to all speed ranges is described in Sec. 18-7 under Supersonic Wing-Body Combinations.

Boundary Layer, Skin Friction. A principal cause of the drag of a low-speed aerodynamic configuration is the skin friction associated with the viscosity of air or water. Calculation of skin friction is possible only for geometrically simple bodies. The friction drag of an airfoil, for instance, might be approximated by that for a flat plate. The general boundary-layer characteristics for a flat plate are illustrated in Fig. 18-8. At the surface of a flat plate, in continuum viscous flow, the local fluid velocity tangent to the body is zero. For parallel airflow over the flat plate, the flow velocity u decreases from its velocity V_∞ at some point far from the plate to a value of zero at the surface of the plate. The resulting variation in velocity from V_∞ to zero is called the *boundary-layer velocity profile*. It can be expressed in the following approximate functional form for all values of x:

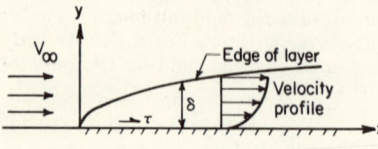

FIG. 18-8. Boundary layer on flat plate.

$$\frac{u}{V_\infty} = f\left(\frac{y}{\delta}\right) \approx \left(\frac{y}{\delta}\right)^{1/n} \tag{18-18}$$

The velocity is a function of x and y since the boundary-layer thickness δ varies with x. For a laminar boundary layer the value of n is close to 2, while for a turbulent layer it increases from 5 to 9 as the Reynolds number increases. At the wall the equation is not valid for calculating the derivative $\partial u/\partial y$ which is singular there.

The skin friction at the wall is a shearing stress τ given by

$$\tau = \mu \frac{\partial u}{\partial y}\bigg|_{y=0} \tag{18-19}$$

It is customary to express τ in terms of a skin-friction coefficient C_f:

$$C_f = \frac{\tau}{q_\infty} \tag{18-20}$$

It is noted that C_f is a local quantity and varies with x. The usual formulas for skin-friction coefficient are
Purely laminar flow:

$$C_f = \frac{0.664}{(R_x)^{1/2}} \tag{18-21}$$

Purely turbulent flow:

$$C_f = \frac{0.370}{(\log_{10} R_x)^{2.584}} \tag{18-22}$$

$$R_x = \frac{\rho_\infty V_\infty x}{\mu_\infty} \tag{18-23}$$

It is to be noted that the value of C_f depends not only on Reynolds number, Mach number, and heating or cooling of the plate, but also on whether the flow is two-dimensional or three-dimensional. For an extensive treatment of boundary layers, see Ref. 8.

Low-speed Stability and Control. For successful flight of an airframe (airplane or missile), it is necessary to control the flight path of the airframe and also to stabilize it. For the flight of an airplane in a vertical plane we have the case of *longitudinal stability and control,* and for motions in yaw or roll we encounter *lateral stability and control.*

The longitudinal motions of an aircraft are accompanied by changes in angle of attack, speed, and altitude with fixed controls. Both short-term and long-term periodic variations in these quantities are possible. The long-term motions involving gravity are not usually so important as the short-period oscillations. The short-period oscillations correspond to those of a spring-mass system in which the spring constant is the slope of the pitching-moment curve C_{m_α}. The natural frequency of an airframe is usually taken to be that of its short-period oscillations and is given by

$$\omega_n = \sqrt{\frac{(-C_{m_\alpha})q_\infty S l}{I_y}} \tag{18-24}$$

It is important to note that we can talk about *stability* only when considering an *equilibrium* state. Consider, for example, the moment curve for a complete airframe as shown in Fig. 18-9. There are two points A and B where the moment is zero so that an equilibrium condition can exist. Points A and B are called *trim points.* At point A, $dC_m/d\alpha$ is a negative quantity, and Eq. (18-24) yields a real natural frequency for small oscillations about A. Thus A is a *stable trim point.* At point B, $dC_m/d\alpha$ is positive and we have an *unstable trim point;* that is, the moment developed by small departures from this point will tend to increase the departure. The criteria for static longitudinal stability are thus

1. Equilibrium: $C_m = 0$
2. Stability: $C_{m_\alpha} < 0$, stable
                           $C_{m_\alpha} = 0$, neutrally stable
                           $C_{m_\alpha} > 0$, unstable

These criteria refer to small departures from a trim point.

Another term sometimes used to describe the longitudinal static stability is that of *static margin,* the distance measured rearward from the center of gravity of an airframe to the intersection of the resultant of the aerodynamic forces acting on the airframe with the longitudinal axis. If this distance is positive, the airframe is statically stable and if negative, statically unstable. Airframes with positive static margin with the controls fixed are said to have inherent static stability. For an airplane with a static margin of 1 or 2 per cent of the wing chord, the degree of static stability is small, while if the static margin is 10 per cent, it would be considered large. Small static margins usually make for increased maneuverability. For negative static margins, artificial stabilization is introduced into the control system using various methods of automatic stabilization.

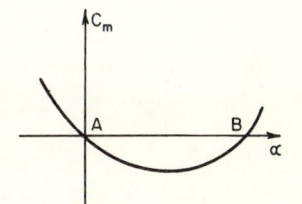

Fig. 18-9. Nonlinear pitching-moment curve.

With regard to lateral stability, it is desired that if the airframe sideslips with fixed controls so that β is no longer zero, the yawing moment should be such that the airframe tends to return to $\beta = 0$. Thus C_{n_β} becomes a measure of weathercock stability just as C_{m_α} is a measure of static longitudinal stability. Because of sign conventions, positive C_{n_β} is stable.

With regard to dynamic stability, it should be noted that an aft horizontal tail usually provides positive damping for most airframes. References 9 and 10 treat stability and control of airframes including dynamic stability.

18-6. Compressibility; Airframes at Transonic Speeds

We now take up the subject of transonic flow. The subject traditionally has been approached from the low-speed side and has been associated with the myth of the "sonic barrier." Transonic aerodynamics is made difficult by the fact that the basic differential equation governing transonic flows is nonlinear and must possess solutions containing basic elements sometimes of subsonic flow and sometimes of supersonic flow. Its engineering aspects are complicated by the fact that in this narrow region the aerodynamic characteristics of an airframe must change in nature from subsonic to supersonic, a change which is usually large. Since the behavior of wings can differ basically from that of bodies in this region, we shall examine each individually.

Airfoils and Wings. A fruitful approach to the characteristics of airfoils at high subsonic Mach number where compressibility effects are of some concern is through *velocity correction factors*. In this approach, suggested by the differential equation for linearized subsonic compressible inviscid flow, the velocity, potential, or pressure at any point on an airfoil at low speed are related to those at the same point at some other subsonic Mach number by a factor depending only on the Mach number and independent of the actual position of the point. There thus results the Glauert-Prandtl relationship between the pressure coefficient C_{P_0} for zero Mach number and C_P at the same point and angle of attack at Mach number M; that is,

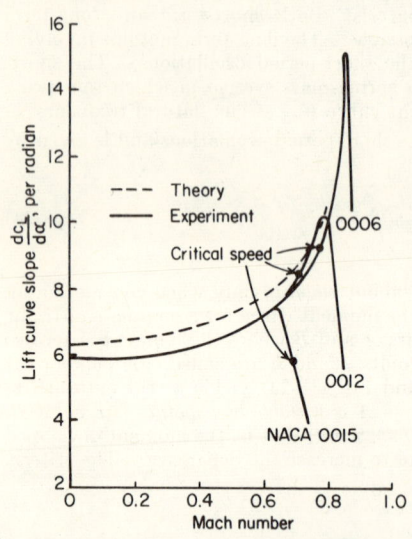

Fig. 18-10. Effect of compressibility on lift-curve slopes of symmetric airfoils.

$$C_P = \frac{C_{P_0}}{\sqrt{1 - M_\infty^2}} \qquad M_\infty < 1 \tag{18-25}$$

where the pressure coefficient is defined as

$$C_P = \frac{p - p_\infty}{q_\infty} \tag{18-26}$$

Since (18-25) is valid for every point,

$$c_l = \frac{(c_l)_0}{\sqrt{1 - M_\infty^2}} \qquad M_\infty < 1 \tag{18-27}$$

and

$$c_m = \frac{(c_m)_0}{\sqrt{1 - M_\infty^2}} \qquad M_\infty < 1 \tag{18-28}$$

A more sophisticated approach due to von Karman and Tsien[12] yields the following relationship for pressure coefficients:

$$C_P = \frac{C_{P_0}}{\sqrt{1 - M_\infty^2} + (C_{P_0}/2)(1 - \sqrt{1 - M_\infty^2})} \tag{18-29}$$

To see the transonic effect consider Fig. 18-10, which shows the actual variation in $dC_L/d\alpha$ with M_∞ at zero angle of attack for symmetrical airfoils of various thicknesses. The airfoil section NACA 0006 is a symmetrical one with a maximum thickness of 6 per cent of the chord, and NACA 0012 and NACA 0015 profiles have analogous properties.

The Glauert-Prandtl rule for lift-curve slope works well up to the critical speed for airfoils which are not too thick. For thick sections it is not successful. By the same token, for thin sections it becomes less successful as the angle of attack increases. The reason for the lack of success is the appearance of shock waves in the local regions of supersonic flow which develop as the critical Mach number is exceeded. These shock waves themselves cause large sudden changes in pressure coefficient at the airfoil surface, which, in turn, cause large changes in its center of pressure and cause the airfoil to lose lift. The shock waves also cause thickening or separation of the boundary layer, which is usually attended by large increases in drag. In fact, the drag can start to rise so rapidly that the term *drag divergence* has been applied to the sudden increase in drag with increase in Mach number which can occur near the critical speed. Experimental data are almost indispensable for engineering use when transonic aerodynamic characteristics are dominated by shock-wave effects.

For complete airframes, very significant and usually undesirable aerodynamic effects accompany the appearance of so-called *shock stall* which appears in flow past a wing at transonic speeds. Not only do large increases in drag occur, but deleterious changes in stability and control usually result. Means for alleviating the effects include the use of thin wings, small angles of attack, symmetrical airfoil sections, sweepback, and area-rule contouring. The first three means are clear, the fourth will now be discussed, and the fifth will be described subsequently.

One of the practical means for alleviating adverse transonic effects on wings is the use of *sweepback* advocated in the United States by Jones and described by Jones and Cohen.[13] The theory of a simple sweepback in essence says that the aerodynamic forces on an airfoil section viewed perpendicular to the leading edge of a swept wing are the same as those of an unswept wing at the normal Mach number M_N as shown in Fig. 18-11. We can eliminate the parallel component of the Mach number without change in the aerodynamic characteristics. It is thus seen that even for Mach number M_∞ exceeding unity, the swept wing can still act subsonically as if at M_N.

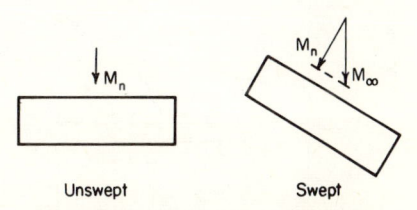

FIG. 18-11. Simple sweep principle.

It is worthwhile noting that the Glauert-Prandtl rule exhibits infinite lift-curve slope at $M = 1$, a Mach number well out of its range of application. For $M = 1$ slender-body theory is usually quite good and yields a lift-curve slope of 2 based on a circular reference area having a diameter equal to the wing span.

Slender Airframes; Whitcomb Transonic Area Rule. A body of revolution at subsonic speed possesses no drag in an inviscid fluid, but the existence of viscosity causes friction drag in a real fluid. Also, if shock waves appear in the airflow as they do at transonic speeds, there can then be ascribed a certain *wave drag* to the body. At Mach numbers slightly in excess of unity a definite bow wave occurs on such a body and accounts for part of the wave drag; while near the critical speed, shock waves appear near the position of maximum diameter and contribute to the wave drag. A remarkable formula exists for the wave drag of a slender body in the transonic range. Define a ξ axis along the body longitudinal axis such that $\xi = 0$ corresponds to the front end of the body and $\xi = 1$ corresponds to the back end. Let the body cross-sectional area be $S(\xi)$ and indicate ξ derivatives of S by S', S'', etc. Then for a body with a base such that $S'(1) = 0$ (pointed or cylindrical base) we have for the wave drag[14]

$$\frac{D}{q_\infty} = \frac{1}{2\pi} \int_0^1 \int_0^1 \log \frac{1}{|s - \xi|} S''(s) S''(\xi) \, d\xi \, ds - C_{P_B} S(1) \qquad (18\text{-}30)$$

where C_{P_B} is the base pressure coefficient and s is a dummy variable of integration. This represents an analytical result for which several aerodynamicists are variously given credit, and which Whitcomb[15] verified experimentally.

What is important about Eq. (18-30) is that the drag depends only on the axial distribution of the cross-sectional area and not on whether the area is deployed as a body or a wing. Although strictly valid only at sonic speed, it is nevertheless quite accurate for a small range of Mach numbers above and below $M = 1$. Whitcomb measured the drag of a body alone and a wing-body of equal area distribution through the transonic-speed range, and found that their drags were nearly equal and differed only by a small amount attributable to skin friction as shown in Fig. 18-12. In fact, if the minimum drag coefficient C_{D_0} is reduced by the constant value at $M_\infty = 0.84$ to form ΔC_{D_0}, it is seen that the drag of the wing-body combination is close to that of the equivalent body. This result leads to the so-called "Marilyn Monroe" shape in the fuselages of aircraft which fly in or through the transonic-speed regime. If a body of revolution has a small wave drag through the transonic-speed range, then a wing-body configuration of the same

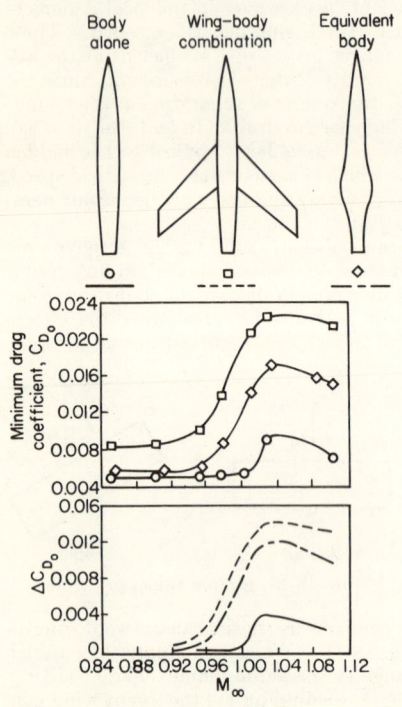

Fig. 18-12. Drag coefficients of equivalent bodies in transonic flow.

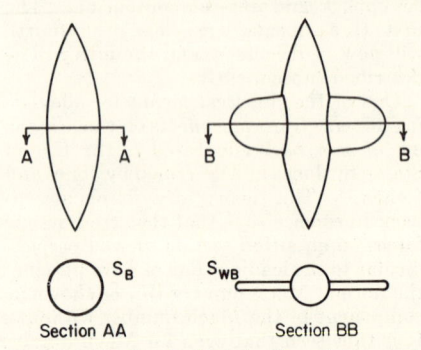

Fig. 18-13. Equivalent bodies at transonic speed.

drag can be found by necking in the fuselage to account for the wing cross-section area. Figure 18-13 illustrates the principle. Thus the area S_B of the section through the body equals S_{WB} for the wing-body combination.

Another advantage of the drag formula is that it is possible to find the area distributions yielding minimum values of the wave drag under various constraints. This leads to efficient transonic airframes. As an example, if we calculate the body of minimum wave drag having zero base area and a given length l and volume, we find the *Sears-Haack* body given by

$$S(\xi) = \frac{16 \text{ vol.}}{3\pi l} \left[1 - \left(1 - \frac{2\xi}{l}\right)^2 \right]^{3/2} \qquad (18\text{-}31)$$

with a drag given by

$$\frac{D}{q_\infty} = \frac{128 \, (\text{vol.})^2}{\pi l^4} \qquad (18\text{-}32)$$

If we calculate the shape of a body of minimum wave drag having a cylindrical base of given area, we find the *Karman ogive*.

18-7. Airframes at Supersonic Speeds

At supersonic speeds aerodynamic theory is perhaps simpler than at subsonic speeds, and for most configurations fairly good analytical estimates can be made of lift, moment, and even drag. Let us start with airfoils, then consider wings, and finish with wing-body configurations.

Airfoils. It is well to consider the type of flow which prevails over a single airfoil at supersonic speeds. Various types of airfoil sections become of interest in this regime including parabolic-arc sections and wedge sections. Consider first the supersonic flow past a parabolic-arc airfoil section at zero angle of attack in Fig. 18-14 as described by the theory of Ackeret.[17] The flow approaching the sharp leading edge is suddenly deflected by two oblique shocks. These shocks have the property that they are of constant strength along their length and suddenly turn the flow along them through a constant angle all along their length. In so doing, the static pressure of the flow is increased and its velocity is decreased. Immediately behind the oblique shocks (bow wave) the flow starts to follow streamlines which are parallel to the body contour but displaced downstream by an amount proportional to the distance from the airfoil. For weak oblique shocks, Λ is given by the condition that the component of the free-stream Mach number normal to the bow wave be unity.

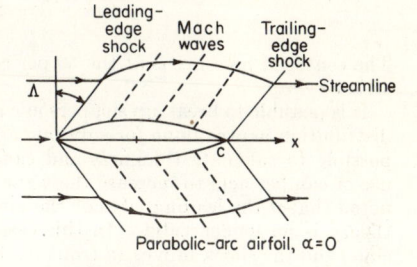

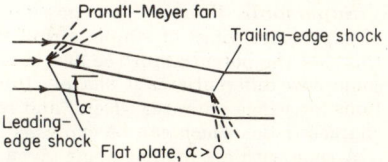

FIG. 18-14. Supersonic flow past airfoils.

$$\cos \Lambda = \frac{1}{M_\infty} \tag{18-33}$$

As the air flows around the airfoil in streamlines parallel to the airfoil contour, its velocity, pressure, and density remain constant along the Mach waves shown in Fig. 18-14. At the trailing-edge oblique waves, the flow undergoes a finite shock which deflects it parallel to the free-stream direction again.

This simple flow pattern for a two-dimensional wing gives rise to a very simple pressure field described by the Ackeret formula.

$$C_P = \frac{p - p_\infty}{q_\infty} = \frac{2\delta}{\sqrt{M_\infty^2 - 1}} \tag{18-34}$$

where δ = slope of airfoil surface with respect to x axis

Through an oblique shock wave δ is positive and through a Prandtl-Meyer fan (a series of expansion waves emanating from a point as shown in Fig. 18-14) δ is negative. For the flat plate at angle of attack α, $\delta = +\alpha$ for the lower surface and $\delta = -\alpha$ for the upper. A parabolic-arc airfoil has a thickness distribution given by

$$t = t_{\max} \frac{x}{c} \left(1 - \frac{x}{c}\right) \tag{18-35}$$

and when the airfoil is at zero angle of attack as in Fig. 18-14, the angle δ is given by

$$\delta = \frac{dt}{dx} = \frac{t_{\max}}{c} \left(1 - \frac{2x}{c}\right) \tag{18-36}$$

Thus, from Eq. (18-34) the pressure coefficient is a linear function of x, and by symmetry is the same on top and bottom surfaces.

From such a simple formula it is possible to determine c_l and c_d for any supersonic airfoil for which δ is small. For a flat plate we obtain directly a uniform negative pressure on the upper surface since $\delta = -\alpha$ in (18-34). With $\delta = +\alpha$ for the lower surface, we find for the lift and drag coefficients

$$c_l = \frac{4\alpha}{\sqrt{M_\infty^2 - 1}} \qquad (18\text{-}37)$$

$$c_d = \frac{4\alpha^2}{\sqrt{M_\infty^2 - 1}} \qquad (18\text{-}38)$$

The center of pressure is at the 50 per cent chord point. The drag in this instance is all *wave drag*.

It is possible to break up a supersonic airfoil into a mean camber line plus a thickness distribution as was done for subsonic airfoils in Eqs. (18-11) and (18-12). It is then possible to calculate thickness and camber effects separately. It is noted that the use of camber acts to increase the wave drag over that of a symmetrical airfoil. It is noted that if the leading edge of the airfoil is blunt so that $\delta \to \infty$, then the Ackeret theory is no longer valid. In this case the bow shock cannot attach to the leading edge, and the shock moves in front of the airfoil. We then have increased drag.

Supersonic Wings. It is possible to determine fully the lift, drag, and moment of supersonic wings of simple airfoil section using linearized supersonic theory. In this case the potential for the flow satisfies the wave equation. This aspect of supersonic flow differentiates it sharply from subsonic flow where precise linearized solutions for wings are rare. Delta and rectangular wings have supersonic aerodynamic characteristics which can be expressed in fairly simple form.

A cone with a semiapex angle given by

$$\sin \theta = \frac{1}{M_\infty} \qquad (18\text{-}39)$$

is termed the *Mach cone*. It is the surface created by a weak point disturbance in a supersonic stream. If the leading edge of a delta wing is swept back within the Mach cone from the wing apex, it is termed a *subsonic leading edge;* otherwise it is a *supersonic leading edge*. For a flat delta wing with supersonic leading edges, the lift-curve slope for all aspect ratios is

$$\frac{dC_L}{d\alpha} = \frac{4}{\sqrt{M_\infty^2 - 1}} = \frac{4}{B} \qquad (18\text{-}40)$$

For a delta wing with subsonic leading edges, the life-curve slope is given by

$$\frac{dC_L}{d\alpha} = \frac{2\pi \tan \omega}{E(k)} \qquad (18\text{-}41)$$

where ω = semiapex angle of delta wing
E = complete elliptic integral of second kind
k = modulus of E, $\sqrt{1 - B^2 \tan^2 \omega}$

The span loading is elliptical.

All delta wings have their centers of pressure at the two-thirds root chord because the pressure is constant along lines through the apex.

For rectangular wings, the lift-curve slope depends on the aspect ratio and Mach number, specifically on the product BA, the so-called *effective aspect ratio*. Except within the Mach cones emanating from the leading edges of the wing tips, the pressure distribution is the same as for an airfoil. The pressure distribution within the tip Mach cones is constant along lines from the leading edge of the tips. The lift-curve slope is given by

$$\frac{dC_L}{d\alpha} = \frac{4}{B}\left(1 - \frac{1}{2BA}\right) \qquad BA \geq 1 \qquad (18\text{-}42)$$

The center of pressure lies forward of the mid-chord at a position given by

$$\frac{\bar{x}}{c} = \frac{1}{2}\left[\frac{1 - (2/3BA)}{1 - (1/2BA)}\right] \qquad BA \geq 1 \qquad (18\text{-}43)$$

Supersonic Wing-Body Combinations. At supersonic speeds the aspect ratio of wings tends to be significantly less than that for subsonic aircraft, and the fuselages tend to be of significantly larger diameter compared with the wing span. As a result, it can no longer be assumed as for subsonic airframes that the part of the wing blanketed by the body still acts as if the body were not there. Two significant effects tend to increase the lift. First, consider the body alone as shown in Fig. 18-15.

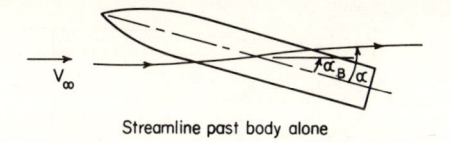

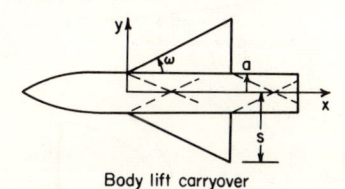

Fig. 18-15. Body and wing-body combination.

The streamlines in passing the location of a hypothetical mid-wing have an upward angle of attack α greater than the body angle of attack α_B. The relationship is given by

$$\alpha = \alpha_B \left(1 + \frac{a^2}{y^2}\right) \qquad y \geq a \qquad (18\text{-}44)$$

where a is the body radius. As a result the wing panels have a greater lift than if they were joined together and acted at angle of attack α_B. The resulting lift ratio is given the notation K_W.[18]

A second effect tends to increase the lift of the wing-body combination. This lift is that associated with the lift "carried over" onto the body within the Mach cones from the leading edges of the wing-body junction as shown in Fig. 18-15. The ratio of this lift to that of the wing alone (panels joined together) is called K_B. As a result, the lift-curve slope of the wing-body combination (exclusive of lift on the nose which is given closely by slender-body theory) is

$$\left(\frac{dC_L}{d\alpha}\right)_{WB} = \left(\frac{dC_L}{d\alpha}\right)_W (K_W + K_B) \qquad (18\text{-}45)$$

For the speed range from subsonic to supersonic and for a wide variety of wing planforms, the values of K_W and K_B are closely given by

$$K_W = 1 + \frac{a}{s} \qquad (18\text{-}46a)$$

$$K_B = \left(1 + \frac{a}{s}\right)^2 - K_W \qquad (18\text{-}46b)$$

where a/s is the ratio of body radius to wing-panel semispan.

In case of a wing-body-tail combination, there is interference between the wing wake and the tail surfaces. Methods for calculating this interference are given in Ref. 18.

18-8. Hypersonic Configurations

At hypersonic speeds almost all configurations are subject to very high rates of heating because of viscous friction in the boundary layers next to them. This is the phenomenon of *aerodynamic heating* which is illustrated by the heating of reentry capsules and the ablation of meteors.

Reentry Configurations. One of the distinctive features of a hypersonic configuration is the bluntness of its nose. This feature results from the fact that aerodynamic heating burns off sharp noses more easily than blunt noses. Thus the blunt nose of subsonic airframe design is again used at hypersonic speeds.

The use of nose bluntness has particular advantages for vehicles which must reenter the atmosphere from satellite or planetary trajectories. These advantages have

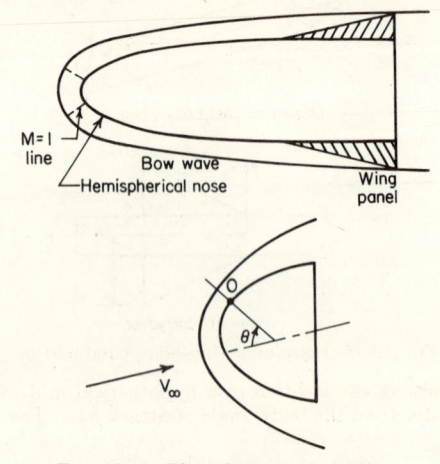

FIG. 18-16. Blunt hypersonic bodies.

been pointed out by Allen.[19] Here it is pointed out that aerodynamic heating can be greatly reduced in increasing the ratio of the drag force on the body to its weight. Such a reduction results in a dissipation of the energy of the body more through wave drag than by boundary-layer friction. The parameter of importance in this connection is the ratio of weight to drag $W/C_D S$, where W is the reentry-configuration weight and $C_D S$ is its equivalent drag area for $C_D = 1$. For a balloon, $W/C_D S$ may be of the order of 1 lb/sq ft or less; for a manned reentry capsule, it may be of the order of 10 to 100; while for military applications, it may be still greater.

Hypersonic Airframes. Hypersonic aircraft, such as the X-15, are not many in number, and their design has not reached the highly developed art of subsonic and supersonic aircraft. A number of hypersonic missiles exist, however. Hypersonic airframe aerodynamic design is dominated by the blunt nose. The use of a blunt nose causes a bow wave to stand off the nose at a certain stand-off distance, as shown in Fig. 18-16. For high Mach numbers, this bow wave tends to "wrap around" the nose and lie close to the downstream sides of the configuration. The calculation of the position and shape of the bow shock is the starting point in the calculation of the flow field around the vehicle. At the present time, a number of procedures have been contrived for use with computers to calculate the flow field.

It is clear that the closeness of the shock wave to the wing panels of the airframe can have significant interference effects on the panels. Design procedures for such interference effects are not so well developed as those for subsonic or supersonic speeds.

AERODYNAMIC SYSTEMS 18-17

To keep the wing-panel leading edges subsonic, it is necessary to sweep the edges back almost parallel to the flow direction. There is thus a trend in hypersonic airframe design to use wing or tail panels of very low aspect ratio. In fact, the panels of low aspect ratio tend to merge into the body at hypersonic speeds. Thus the distinction between wing and body loses its sharpness, and there is a trend to make the body itself the lifting surface.

Newtonian Theory. A very simple and useful theory exists for the calculation of pressure distribution on blunt hypersonic configurations in Newtonian theory. This theory says that the static pressure is positive on any surface inclined to the general flow direction in such a way that the component of the free-stream velocity *normal* to the surface is inward. In such case the static pressure rise above the free-stream static pressure is $\rho_\infty V_n^2$, where V_n is the velocity normal to the surface. With respect to Fig. 18-16, the static pressure at point 0 would be

$$p = p_\infty + \rho_\infty V_n^2 = p_\infty + \rho_\infty V_\infty^2 \cos^2 \theta \tag{18-47}$$

The pressure coefficient is thus

$$\frac{p - p_\infty}{q_\infty} = 2 \cos^2 \theta \tag{18-48}$$

This simple formula allows a good first approximation to the lift and drag of a blunt hypersonic body when coupled with the assumption that the static pressure on leeward surfaces is zero.

List of Common Symbols

a	= local speed of sound, ft/sec; also body radius, ft
A	= wing aspect ratio, (span)^2/area
B	= $\sqrt{M_\infty^2 - 1}$; $M_\infty > 1$
c	= chord of airfoil or wing, ft
c_d	= drag coefficient of airfoil
C_D	= drag coefficient, Eq. (18-6a)
C_{D_i}	= induced-drag coefficient
c_l	= lift coefficient of airfoil section
C_L	= lift coefficient, Eq. (18-6b)
$\dfrac{dC_L}{d\alpha} = C_{L\alpha}$	= lift-curve slope, per radian
C_m	= pitching-moment coefficient, Eq. (18-6c)
$\dfrac{\partial C_m}{\partial \alpha} = C_{m\alpha}$	= moment-curve slope, per radian
c_p	= specific heat at constant pressure for air, Btu/(lb)(°F)
C_P	= pressure coefficient $\dfrac{p - p_\infty}{q_\infty}$
C_{P_0}	= value of C_P for incompressible flow
c_v	= specific heat at constant volume for air, Btu/(lb)(°F)
D	= drag force, lb
I_y	= moment of inertia of airframe about lateral axis through center of gravity
l	= arbitrary reference length, ft
L	= lift force, lb
M	= local Mach number, Eq. (18-4)
M_∞	= Mach number of parallel flow
p	= local static pressure, lb/sq ft
p_∞	= value of p for parallel flow
q_∞	= free-stream dynamic pressure, $\frac{1}{2}\rho_\infty V_\infty^2$
R	= gas constant for air
Re	= Reynolds number, Eq. (18-5)
s	= semispan of wing or wing-body combination, ft
S	= arbitrary reference area, sq ft
T	= absolute temperature, °R, °F + 460

u, v, w	= components of V_∞ along x, y, z
V	= local velocity of air, ft/sec
V_∞	= value of V for parallel stream
W	= weight of reentry vehicle, lb
x, y, z	= set of body axes, Fig. 18-3
$\bar{x}$	= distance of center of pressure behind leading edge, ft
α	= angle of attack, radians unless otherwise noted
β	= angle of sideslip, radians unless otherwise noted
γ	= ratio of specific heats, c_p/c_v
δ	= slope of airfoil surface with respect to x axis
Λ	= sweepback angle, deg
ρ	= mass density, lb/cu ft
τ	= wall shearing stress, lb/sq ft

Subscripts:

W	= wing alone
WB	= wing body
∞	= parallel stream
0	= $M_\infty = 0$

REFERENCES

1. Abbott, I. H., A. E. Von Doenhoff, and L. S. Stivers: Summary of Airfoil Data, *NACA Rept.* 824, 1945.
2. Riegels, F. W.: *Aerofoil Sections*, Butterworth & Co. (Publishers), Ltd., London, 1961.
3. Hoerner, Sighard F.: *Fluid-Dynamic Drag*, published by the author, Midland Park, N.J., 1958.
4. Ward, G. N.: Supersonic Flow Past Slender Pointed Bodies, *Quart. J. Mech. Appl. Math.*, vol. 2, part I, 1949.
5. DeYoung, John, and Charles W. Harper: Theoretical Symmetric Span Loading at Subsonic Speeds for Wings Having Arbitrary Planform, *NACA Rept.* 921, 1948.
6. Glauert, H.: *The Elements of Aerofoil and Airscrew Theory*, pp. 142-143, The Macmillan Company, New York, 1943.
7. Hinze, J. O.: *Turbulence*, McGraw-Hill Book Company, New York, 1959.
8. Schlichting, Hermann: *Boundary Layer Theory*, 4th ed., McGraw-Hill Book Company, New York, 1960.
9. Etkin, Bernard: *Dynamics of Flight*, John Wiley & Sons, Inc., New York, 1959.
10. Perkins, C. D., and R. E. Hage: *Airplane Performance, Stability and Control*, John Wiley & Sons, Inc., New York, 1949.
11. Glauert, H.: The Effect of Compressibility on the Lift of an Airfoil, *Aero. Research Council Repts. and Mem.* 1135, 1927.
12. von Karman, T., and H. S. Tsien: Compressibility Effects in Aeronautics, *J. Aeronaut. Sci.*, vol. 8, p. 351, 1941.
13. Jones, Robert T., and Doris Cohen: *Aerodynamic Components of Aircraft at High Speeds*, Sec. A, *Aerodynamics of Wings at High Speeds. High Speed Aerodynamics and Jet Propulsion*, vol. VII, Princeton University Press, Princeton, N.J., 1957.
14. Nielsen, J. N.: *Missile Aerodynamics*, McGraw-Hill Book Company, New York, 1961.
15. Whitcomb, R. T.: A Study of the Zero-lift Drag-rise Characteristics Near the Speed of Sound, *NACA Research Memo* L52H08, 1952.
16. Jones, R. T.: Theory of Wing Body Drag at Supersonic Speeds, *NACA Tech. Rept.* 1284, 1956.
17. Ames Research Staff: Equations, Tables, and Charts for Compressible Flow, *NACA Tech. Rept.* 1135, 1953.
18. Pitts, W. C., Jack N. Nielsen, and George E. Kaattari: Lift and Center of Pressure of Wing-body-tail Combinations at Subsonic, Transonic, and Supersonic Speeds, *NACA Tech. Rept.*, 1307, 1957.
19. Allen, H. Julian: Hypersonic Flight and the Reentry Problem, The Twenty-first Wright Brothers Lecture, *J. Aeronaut. Sci.*, vol. 15, no. 4, pp. 217-227, April, 1958.

Chapter 19

GUIDANCE

R. M. HOWE, *University of Michigan.*

CONTENTS

19-1. Introduction	19-1
19-2. Navigational Homing Systems	19-2
Pursuit Navigation	19-3
Fixed-lead Navigation	19-4
Constant-bearing Navigation	19-4
Proportional Navigation	19-5
The Linearized Proportional-navigation Equations	19-5
Use of Adjoint Functions for Computing Miss Distances	19-8
Miss Due to a Launch Error	19-10
Miss Due to Step Target Acceleration	19-11
Miss Due to Target Sinusoidal Weave	19-11
Miss Due to Noise	19-12
Beam-rider Navigation	19-13
19-3. Ballistic Guidance Systems	19-14
Trajectory of a Point Mass in a Central-force Gravity Field	19-14
The Two-dimensional Hit Equation	19-16
Ballistic Guidance in Three Dimensions	19-19
Coordinate Determination by Inertial Means	19-24

19-1. Introduction

The word "guidance," as interpreted in technology, is generally associated with the navigation of missiles such that they hit or come close to some preselected target. The mechanization of a missile guidance system can be viewed as the implementation of a final-value control system, that is, a control system where the variables being controlled, e.g., the missile position, are required to attain some final value, e.g., the target position. Although the guidance-mechanization techniques which we shall consider in this chapter will be drawn specifically from missile examples, it should be emphasized at the outset that the techniques discussed will be applicable to final-value control systems in general.

There would appear to be two basic categories of guidance systems for missiles. The first category involves guidance systems where the missile path can be controlled more or less continuously until it reaches the target or some nearby point; a good example is a ground-to-air missile which homes on an airborne target. We shall designate guidance systems of this type *navigational homing guidance systems*. The second category involves guidance systems where the vehicle path cannot be controlled over a major portion of the trajectory; typical examples include both ballistic-missile and space-vehicle guidance systems, where the vehicle spends long periods coasting under no influence from external forces except gravity. We shall designate guidance systems of this type *ballistic guidance systems*. The detailed mechanization of ballistic guidance systems is considerably different from that of navigational homing systems, but many of the analysis techniques are common. These include procedures of linearization of the trajectories with respect to nonlinear reference trajectories. In this chapter we shall consider, successively, navigational homing systems and ballistic systems. The discussion will be elementary and nonexhaustive, and will not include the problem of sensing position and velocity information as needed to implement guidance (except for a brief discussion of errors in inertial systems), or the missile control systems (autopilots) needed to carry out the commands of the guidance computer. Some of these problems are discussed in Chapters 17 and 29. Here, attention will be focused on the guidance system itself, the mathematical basis for the various guidance laws, the required computational methods, and in the case of navigational homing systems, the influence of control-system dynamics and sensor noise on guidance-system performance.

19-2. Navigational Homing Systems

In this type of guidance system the missile path is considered to be under continuous control throughout the trajectory. To simplify the discussion, let us restrict the problem to two dimensions, as shown in Fig. 19-1. Here we designate target and missile positions with respect to a two-dimensional inertial frame with coordinates x, y. The target has velocity of magnitude v_T and direction θ_T, as measured with respect to the x axis reference direction. Similarly, the missile has velocity of magnitude v_M and direction θ. From now on we shall refer to θ as the missile heading angle, although this is true in fact only if the missile has zero angle of attack, which for simplicity we shall assume. The line-of-sight angle in Fig. 19-1 is β, and the distance separating target and missile is the range R. Target and missile have velocity components $v_{\beta T}$ and $v_{\beta M}$, respectively, along the line of sight and $v_{\alpha T}$ and $v_{\alpha M}$, respectively, perpendicular to the line of sight.

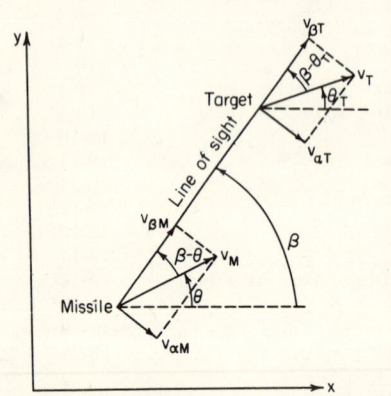

Fig. 19-1. Geometry of two-dimensional homing problem.

The purpose of the guidance system is to provide steering commands to the missile autopilot such that the missile homes on the target regardless of target maneuvers. We assume that the guidance system has as an effective input the line-of-sight angle β (or the time rate of change $\dot{\beta}$). With this as an input there are several different homing laws which can be used for guidance. These include

1. Pursuit
2. Fixed-lead
3. Constant-bearing
4. Proportional-navigation
5. Beam-rider

In type 1 (pursuit) the missile always heads toward the target (i.e., $\theta = \beta$). In type 2 (fixed-lead) the missile is always headed a fixed angular distance ahead of the target ($\theta = \beta - \theta_0$). In type 3 (constant-bearing) the missile is steered such that the

GUIDANCE

19-3

line-of-sight angle does not change ($\dot{\beta} = 0$). For a nonmaneuvering target this means that a missile with constant velocity will ideally be directed onto a straight-line collision course. In type 4 (proportional-navigation) the missile turning rate is made proportional to the line-of-sight angular rate ($\dot{\theta} = \lambda\dot{\beta}$). Here λ is designated as the navigation constant. In type 5 the target is tracked from a location other than the missile, and the missile flies along the beam from the tracking station to the target.

Reference to Fig. 19-1 shows that the range rate $\dot{R}$ is given by

$$\dot{R} = v_{\beta T} - v_{\alpha T} = v_T \cos(\beta - \theta_T) - v_M \cos(\beta - \theta) \tag{19-1}$$

The angular rate of the line of sight is given by

$$\dot{\beta} = -\frac{v_{\alpha T} - v_{\alpha M}}{R} = -\frac{v_T \sin(\beta - \theta_T) - v_M \sin(\beta - \theta)}{R} \tag{19-2}$$

Pursuit Navigation. Because it represents the simplest system, let us analyze first the case of pursuit navigation. For simplicity assume a nonmaneuvering target with v_T = constant. Furthermore, we let $\theta_T = 0$ since we can always reorient the x, y reference system of Fig. 19-1 so that this is true. For ideal pursuit navigation $\theta = \beta$, and for $\theta_T = 0$, (19-1) and (19-2) become

$$\dot{R} = \frac{dR}{dt} = v_T \cos \beta - v_M \tag{19-3}$$

$$\dot{\beta} = \frac{d\beta}{dt} = -\frac{v_T \sin \beta}{R} \tag{19-4}$$

Note that $\dot{\beta}$ is not zero unless $\beta = 0$ or π, i.e., unless the attack is head on or tail on. Since $\dot{\theta} = \dot{\beta}$ in pursuit navigation, the missile will always have to turn during the attack unless it is head on or tail on.

By eliminating time from (19-3) and (19-4) we can solve for β and hence the missile heading θ as a function of the range R. Dividing Eq. (19-3) by (19-4), we obtain

$$\frac{dR}{d\beta} = \left(-\cot\beta + \frac{v_M}{v_T}\csc\beta\right) R$$

or

$$\frac{dR}{R} = (-\cot\beta + \gamma\csc\beta)\, d\beta \tag{19-5}$$

where $\gamma = v_M/v_T$, the ratio of missile to target velocity. After integration, (19-5) becomes

$$\ln R = -\ln|\sin\beta| + \gamma \ln\left|\tan\frac{\beta}{2}\right| + \text{const}$$

Assuming that $0 \leq \beta \leq \pi$, we can drop the absolute-value signs and write

$$\ln \frac{R \sin\beta}{(\tan\beta/2)^\gamma} = \text{const}$$

or

$$\frac{R \sin\beta}{(\tan\beta/2)^\gamma} = \frac{R_0 \sin\beta_0}{(\tan\beta_0/2)^\gamma} = K \tag{19-6}$$

where R_0 and β_0 are initial values of range and line-of-sight angle, respectively. As the missile approaches the kill, R approaches zero, and from (19-6) it is evident that β must also approach zero for the left-hand side of (19-6) to remain constant. Thus we are led to the important conclusion that in ideal pursuit navigation the trajectory always terminates in a tail chase with $\theta = \beta = 0$ (except for the case of an exact head-on attack).

Substituting R from (19-6) into (19-4), we obtain

$$\dot{\beta} = -\frac{v_T}{K} \frac{\sin^2\beta}{(\tan\beta/2)^\gamma} \tag{19-7}$$

Near the trajectory termination, when $\beta \ll 1$ and $\sin \beta \cong \beta$, $\tan \beta/2 \cong \beta/2$, Eq. (19-7) becomes

$$\dot{\beta} \cong -\frac{v_T 2^\gamma}{K} \beta^{2-\gamma} \qquad (19-8)$$

Since the missile turning rate $\theta = \dot{\beta}$ in ideal pursuit navigation and since $\beta \to 0$ at the termination of the trajectory, (19-8) leads to the interesting result that the required missile turning rate becomes infinite at the instant of kill if the ratio γ of missile to target velocity is greater than 2. For $\gamma < 2$ the turning rate becomes zero at kill.

By differentiating (19-8) again, it is easy to show that the required turning acceleration $\ddot{\theta}$ becomes infinite at kill if the ratio γ of missile to target velocity is greater than 1.5. For $\gamma < 1.5$ the required acceleration $\ddot{\theta}$ becomes zero at kill.

Thus one concludes that for pursuit navigation for finite target turning rates $\dot{\theta}$ or angular acceleration $\ddot{\theta}$ the missile velocity cannot exceed 2 and 1.5 times the target velocity, respectively, for a nonmaneuvering target. Some typical pursuit trajectories are shown in Fig. 19-2. Because of the high maneuver requirements to end the attack on a tail chase, pursuit navigation is not considered practical as a missile guidance law for homing against moving targets, in spite of its simple implementation. One

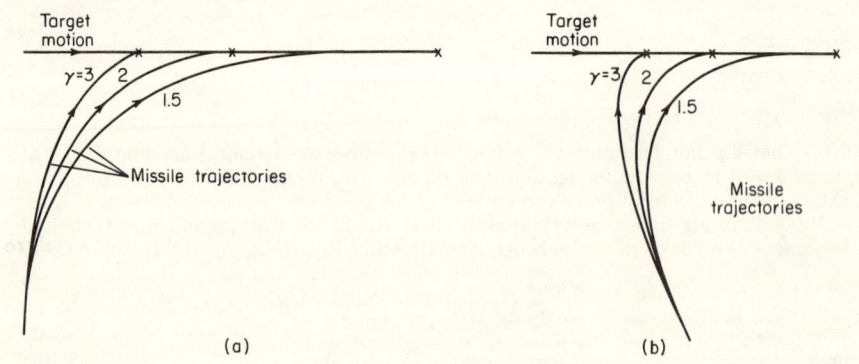

Fig. 19-2. Typical pursuit trajectories for a nonmaneuvering target. (a) Beam attack. (b) Front-quarter attack.

exception would be for a target moving very slowly compared with the missile (e.g., a ship being attacked by an airborne missile), in which case the small miss caused by the intense last-instant maneuver requirement of pursuit navigation would still allow the missile to hit a large target. Another exception might be for a missile system which always attacks almost directly from the rear of a moving target.

Because pursuit navigation is a fairly impractical guidance law in most cases, we shall not consider the effect of missile control-system dynamics in this case.

Fixed-lead Navigation. For fixed-lead navigation the guidance law is $\theta = \beta - \theta_0$; i.e., the missile flies at a fixed angle ahead of the line of sight. For a given initial line-of-sight angle β_0 and a nonmaneuvering target, it is apparent that there is a specific lead angle $\beta - \theta$ (see Fig. 19-1) which represents a collision course. In this case a constant-velocity missile will fly a nonmaneuvering course directly to the target. However, this fixed lead angle will not provide a constant-bearing intercept for any other line-of-sight angle β_0. Thus fixed-lead navigation is useful compared with pursuit only when one knows ahead of time the missile-target velocity ratio and from which relative direction the attack will take place. For this reason it is of doubtful value as a missile guidance law.

Constant-bearing Navigation. In the case of constant-bearing navigation, the missile is steered such that the line of sight does not rotate ($\dot{\beta} = 0$). Reference to Eq. (19-2) shows that this will be achieved if the missile velocity component $v_{\alpha T}$ perpendicular to the line of sight matches the target component $v_{\beta T}$ perpendicular to

GUIDANCE 19-5

the line of sight. This in turn will be maintained, once it has been achieved, if the missile acceleration $\dot{v}_{\alpha T}$ perpendicular to the line of sight matches the target acceleration $\dot{v}_{\beta T}$. Thus ideal constant-bearing navigation does not require the missile acceleration to exceed the acceleration of a maneuvering target.

It turns out to be difficult to implement ideal constant-bearing navigation. One obvious way to attempt implementation is to consider $\dot{\beta}$ an error signal which the missile system drives toward zero by commanding a missile turning rate proportional to the error signal, i.e., $\dot{\theta} = \lambda \dot{\beta}$. But this is simply proportional navigation. Thus proportional navigation actually represents a simple mechanization of constant-bearing navigation. For this reason we turn directly to a consideration of proportional navigation, which is the most widely used guidance law for missile homing navigation.

Proportional Navigation. The equation for ideal proportional navigation is given by

$$\dot{\theta} = \lambda \dot{\beta} \tag{19-9}$$

where λ is called the navigation constant. Integrating (19-9), we obtain

$$\theta = \lambda \beta + \theta_0 \tag{19-10}$$

where θ_0 is the integration constant. It is immediately apparent that our previous guidance schemes are special cases of proportional navigation. Consider the following examples:

1. $\lambda = 1$, $\theta_0 = 0$; this is pursuit navigation.
2. $\lambda = 1$, $\theta_0 \neq 0$; this is fixed-lead navigation.
3. $\lambda = \infty$; from Eq. (19-9) for finite $\dot{\theta}$ this corresponds to $\dot{\beta} = 0$, or constant-bearing navigation.

For a missile using proportional navigation and launched on a perfect collision course ($\dot{\beta} = 0$), Eq. (19-9) shows that $\dot{\theta} = 0$ initially. If the target fails to maneuver and missile and target velocities remain constant, the missile trajectory will remain on a collision course. Thus constant-bearing interception is achieved. From Eq. (19-2) it is apparent that the initial target launch angle θ_0 needed for a collision course ($\dot{\beta} = 0$) is given by

$$\sin (\beta_0 - \theta_0) = \frac{v_T}{v_M} \sin (\beta_0 - \theta_T) \tag{19-11}$$

If the missile is launched off of the collision course, then the resulting trajectories have curvature, depending on the value of the navigation constant λ. Some typical trajectories for ideal proportional navigation for constant-velocity target and $v_M/v_T = 2$ are shown in Fig. 19-3.

In the following sections we shall study the effect of the navigation constant λ on typical proportional-navigation trajectories where we consider missile autopilot dynamics as well as target maneuvers. In addition, the effect of random noise superimposed on the target-bearing signal as received by the missile will be considered. All these effects will influence the final choice of λ and the autopilot time constants.

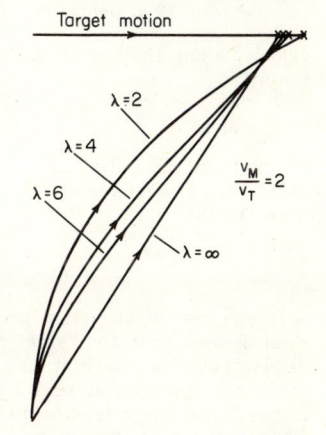

FIG. 19-3. Typical proportional-navigation trajectories for a beam attack with 30° initial launch error.

The Linearized Proportional-navigation Equations. Equations (19-1), (19-2), and (19-9) which the ideal proportional-navigation trajectory obeys are highly nonlinear and hence difficult to solve directly except using computing machines. If one includes missile control-system dynamics in (19-9), the overall trajectory equations become impossible to solve directly; and when solved by computing machines, the equations must be solved completely for every initial condition, navigation constant, missile-to-target velocity ratio, target maneuver, etc. In order to simplify this problem and produce more general results, it is useful to linearize the equations about a given constant-bearing reference trajectory, as shown in Fig. 19-4. Here we assume for the reference-trajectory case that the missile is launched at time $t = 0$ with velocity $v_M{}^*$ and lead angle Θ^* such that it follows an exact collision course to the target. If $v_{\beta M}{}^*$ is the missile velocity component along the line of sight and $v_{\beta T}{}^*$ is the target

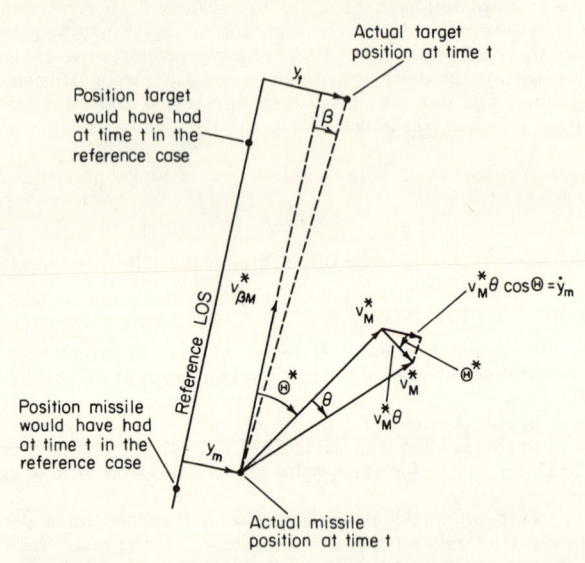

FIG. 19-4. Geometry for linearized proportional-navigation coordinates.

velocity along the line of sight, then the reference-trajectory range rate $\dot{R}^*$ is given simply by

$$\dot{R}^* = v_{\beta T}{}^* - v_{\beta M}{}^* = \text{const} \tag{19-12}$$

from which
$$R^* = R_0 - |v_{\beta T}{}^* - v_{\beta M}{}^*|t \tag{19-13}$$

where R_0 is the initial range and where we assume $v_{\beta T}{}^* - v_{\beta M}{}^*$, the range rate, is negative. The reference-trajectory impact time $t_f{}^*$ is achieved when $R^* = 0$. Thus from (19-13)

$$t_f{}^* = \frac{R_0}{|v_{\beta T}{}^* - v_{\beta M}{}^*|} \tag{19-14}$$

In the case of an actual trajectory, as distinct from the reference trajectory, we shall assume that the target position at time t is designated by the perpendicular displacement y_t which it has from the reference LOS (line of sight). Similarly, the missile position at time t is designated by the perpendicular distance y_m which it has from the reference LOS. The actual-trajectory LOS β will be measured from the reference-trajectory LOS, as shown in Fig. 19-4. Similarly, the actual missile heading θ will be measured from the reference-trajectory missile heading. Missile velocity $\dot{y}_m$ perpendicular to the reference LOS will result from heading changes θ. In Fig. 19-4 it is apparent that this produces a velocity change $v_M{}^*\theta$ with respect to the reference-trajectory missile velocity, assuming that $\theta \ll 1$. The projection of this velocity increment along the direction perpendicular to the reference LOS is just $\dot{y}_m$. Thus

$$\dot{y}_m = v_M{}^*\theta \cos \Theta^* = v_{\beta M}{}^*\theta \tag{19-15}$$

Consider now the equation for proportional navigation, ideally $\theta = \lambda\beta$. In actuality the implementation of this equation will be with a rate autopilot with dynamic lags. If we assume that the rate autopilot can be represented as a linear system with

GUIDANCE

transfer operator $\lambda Y(p)$, where p is the operator d/dt and $Y(p)$ is the ratio of two polynomials in p, then the actual equation for our proportional-navigation system becomes

$$\dot{\theta} = \lambda Y(p)\dot{\beta} \tag{19-16}$$

Integrating both sides, we obtain

$$\theta = \lambda Y(p)\beta \tag{19-17}$$

where we have assumed $\beta(0) = \theta(0) = 0$, i.e., a perfect collision-course launch with target at the reference position. Later we shall take into account imperfect launches by allowing the target to undergo a step velocity an instant after launch.

From Fig. 19-4 it is evident that $\beta \cong (y_t - y_m)/R$. If we assume that the actual range R at time t does not differ appreciably from the reference-trajectory range R^* at time t, then we can write

$$\beta \cong \frac{y_t - y_m}{R^*} \qquad |y_t - y_m| \ll R^* \tag{19-18}$$

From (19-15) we have

$$\theta \cong \frac{\dot{y}_m}{v_{\beta M}{}^*} \qquad \theta \ll 1 \tag{19-19}$$

Finally, if we let the autopilot transfer operator $Y(p)$ be the ratio of two polynomials $N(p)/D(p)$, (19-17) can be written

$$D(p)\theta = \lambda N(p)\beta \tag{19-20}$$

Substituting (19-18) and (19-19) in (19-20) we obtain

$$D(p)\dot{y}_m = \frac{v_{\beta M}{}^* \lambda N(p)}{R^*(t)}(y_t - y_M) \tag{19-21}$$

From (19-13) and (19-14) this can be rewritten as

$$D(p)\dot{y}_m = \frac{\Lambda}{t_f{}^* - t} N(p)(y_t - y_m) \tag{19-22}$$

where the parameter Λ is related to the navigation constant λ and is given by

$$\Lambda = \frac{v_{\beta M}{}^* \lambda}{|v_{\beta T}{}^* - v_{\beta M}{}^*|} \tag{19-23}$$

Equation (19-22) represents a time-varying linear system with transverse target displacement y_t as the input or forcing function and missile transverse displacement y_m as the response function. Assume that the time-varying weighting function for the system is $W(t,\tau)$; i.e., the response y_m at time t to a unit impulse in y_t applied at time τ is given by $W(t,\tau)$. Then we can write the response at time t to any input y_t by means of the superposition integral. Thus

$$y_m(t) = \int_0^t W(t,\tau) y_t(\tau)\, d\tau \tag{19-24}$$

Let us now assume that the actual miss distance can be calculated as the difference between lateral target and missile displacements at the reference-trajectory impact time $t_f{}^*$. Thus let

$$\text{Miss distance} = M = y_t(t_f{}^*) - y_m(t_f{}^*) \tag{19-25}$$

Although the actual time t_f of closest approach will differ slightly from $t_f{}^*$, it is clear that this can make only second-order changes in $y_t - y_m$ over the interval $t_f - t_f{}^*$.

Combining (19-24) and (19-25), we obtain

$$M = y_t(t_f^*) - \int_0^{t_f^*} W(t_f^*,\tau)y_t(\tau)\,d\tau \qquad (19\text{-}26)$$

as the expression for miss distance for any arbitrary input target displacement $y_t(t)$.

Use of Adjoint Functions for Computing Miss Distances. To obtain the weighting function $W(t,\tau)$ it is necessary to solve the time-varying proportional-navigation equation (19-21) on a computing machine. In particular the response $y_m(t)$ is obtained for a unit impulse at τ. Computer runs are made for a large number of different impulse-application times τ ranging from launch ($\tau = 0$) to reference impact ($\tau = t_f^*$). For each computer run, response $y_m(t_f^*)$ is noted; a plot of $y_m(t_f^*)$ vs. τ represents $W(t_f^*,\tau)$, the system response at time t_f^* as a continuous function of impulse-application time τ.

By considering the solutions to the equation adjoint (defined below) to (19-22) we can obtain $W(t_f^*,\tau)$ with a single computer run. As a specific example consider the case of a missile with a rate autopilot with a simple time lag T. Thus let

$$Y(p) = \frac{1}{Tp + 1} \qquad (19\text{-}27)$$

i.e., $N(p) = 1$, $D(p) = Tp + 1$ in (19-22), which becomes

$$(Tp + 1)\dot{y}_m = \frac{\Lambda}{t_f^* - t}(y_t - y_m)$$

or

$$\ddot{y}_m = -\frac{\Lambda}{T(t_f^* - t)}y_m - \frac{1}{T}\dot{y}_m + \frac{\Lambda}{T(t_f^* - t)}y_t \qquad (19\text{-}28)$$

We next write this second-order equation as two first-order equations. Thus let

$$y_1 = y_m \qquad y_2 = \dot{y}_m \qquad (19\text{-}29)$$

from which

$$\begin{aligned}\dot{y}_1 &= y_2 \\ \dot{y}_2 &= -\frac{\Lambda}{T(t_f^* - t)}y_1 - \frac{1}{T}y_2 + \frac{\Lambda}{T(t_f^* - t)}y_t(t)\end{aligned} \qquad (19\text{-}30)$$

For an nth-order system, (19-30) will have the form

$$\dot{y}_i = \sum_{j=1}^n a_{ij}(t)y_j + Y_i(t) \qquad i = 1, 2, \ldots, n \qquad (19\text{-}31)$$

The adjoint to this equation is defined as follows:

$$\dot{\lambda}_i = -\sum_{j=1}^n a_{ji}(t)\lambda_j \qquad i = 1, 2, \ldots, n \qquad (19\text{-}32)$$

Here λ_i is the ith adjoint coordinate and should not be confused with the navigation constant λ. If we multiply (19-31) by λ_i and (19-32) by y_i, add the two equations, and sum over i, we obtain

$$\sum_{i=1}^n (\lambda_i \dot{y}_i + \dot{\lambda}_i y_i) = \sum_{i=1}^n \sum_{j=1}^n [a_{ij}(t)y_j\lambda_i - a_{ji}(t)y_i\lambda_j] + \sum_{i=1}^n \lambda_i Y_i(t) \qquad (19\text{-}33)$$

The left-hand side of (19-33) is the perfect differential $(d/dt)\sum_{i=1}^n \lambda_i y_i$. The double summation on the right side vanishes. Integrating both sides from $t = 0$ to $t = t_f^*$, we obtain

$$\sum_{i=1}^{n} \lambda_i(t_f^*) y_i(t_f^*) = \sum_{i=1}^{n} \lambda_i(0) y_i(0) + \int_0^{t_f^*} \sum_{i=1}^{n} \lambda_i(t_0) Y_i(t_0) \, dt_0 \qquad (19\text{-}34)$$

Let us assume $y_i(0) = 0$, corresponding to a perfect collision-course launch [we made this assumption earlier in integrating (19-16) to obtain (19-17)]. Furthermore, let us choose initital conditions on the adjoint functions λ_i at time t_f^* as follows:

$$\lambda_1(t_f^*) = 1 \qquad \lambda_2(t_f^*) = \lambda_3(t_f^*) = \cdots = \lambda_n(t_f^*) = 0 \qquad (19\text{-}35)$$

Then the left-hand side of (19-34) becomes $y_1(t_f^*) = y_m(t_f^*)$, which is the desired solution. For the specific case represented in (19-30), $Y_1(t) = 0$ and

$$Y_2(t) = \frac{\Lambda}{T(t_f^* - t)]} y_t(t)$$

in which case (19-34) becomes

$$y_1(t_f^*) = y_m(t_f^*) = \frac{\Lambda}{T} \int_0^{t_f^*} \frac{\lambda_2(t_0)}{(t_f^* - t_0)} y_t(t_0) \, dt_0 \qquad (19\text{-}36)$$

For a unit impulse in target displacement at time τ, $y_t(t_0) = \delta(t_0 - \tau)$, and the right-hand side of (19-36) is equal to the factor in the integrand which multiplies the delta function $\delta(t_0 - \tau)$ evaluated at the time τ at which the impulse occurs. Thus

$$y_m(t_f^*) = \frac{\Lambda \lambda_2(\tau)}{T(t_f^* - \tau)} = W(t_f^*, \tau) \qquad (19\text{-}37)$$

Thus the weighting function $W(t_f^*, \tau)$ can be determined as a function of τ in a single computer run by computing the adjoint function $\lambda_2(\tau)$ which satisfies (19-32) subject to the initial conditions given in (19-35). Although (19-37) applies specifically to the case where the order of the original Eq. (19-30) is only $n = 2$, the

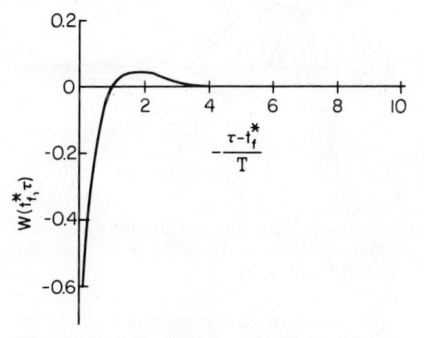

Fig. 19-5. Miss due to a unit impulse in target displacement at time τ for $\Lambda = 4$. $Y(p) = 1/(1 + Tp)$.

generalization to higher-order systems is obvious. For example, if the autopilot transfer operator were given by

$$Y(p) = \frac{1}{a_m p^m + a_{m-1} p^{m-1} + \cdots + a_1 p + 1} \qquad (19\text{-}38)$$

then it is easy to show that Eq. (19-37) becomes

$$W(t_f^*, \tau) = \frac{\Lambda \lambda_{m+1}(\tau)}{a_m (t_f^* - \tau)} \qquad (19\text{-}39)$$

where the order n of the linearized equations is equal to $m + 1$.

For $Y(p) = 1/(Tp + 1)$ as assumed earlier (a simple time-lag autopilot) the weighting function $W(t_f^*, \tau)$ is shown in Fig. 19-5 for $\Lambda = 4$. Note that the abscissa time scale is made dimensionless by dividing by the missile autopilot time constant T, and that the time scale is given by $(\tau - t_f^*)/T$ rather than τ/T; this makes the plot independent of the reference time of flight t_f^*. Note also that the maximum miss results from a target impulse at the last instant before reference-trajectory impact, i.e., at

$\tau = t_f^*$ or $\tau - t_f^* = 0$. For target impulses applied earlier along the time of flight ($\tau < t_f^*$) the resulting miss decreases but with the oscillatory characteristic shown. For $|\tau - t_f^*| \gg T$ the resulting miss approaches zero, since here the impulse in target displacement occurs many autopilot time constants prior to impact and the guidance system has time to correct for the transient offset caused by the impulse maneuver.

Miss Due to a Launch Error. As stated earlier, we have assumed up to now that the missile is launched on a perfect collision course. The effect of a launch error θ_0 can be determined by calculating the miss resulting from a step target velocity occurring just after launch and given by

$$\dot{y}_t = -v_M^*\theta_0 \cos \Theta^* = -v_{\beta M}^*\theta_0 \qquad (19\text{-}40)$$

as is evident in Fig. 19-4. This can be computed in turn from Eq. (19-26) by letting $y_t(\tau) = -v_{\beta M}^*\theta_0 t$. An alternative method for computing the miss due to a launch

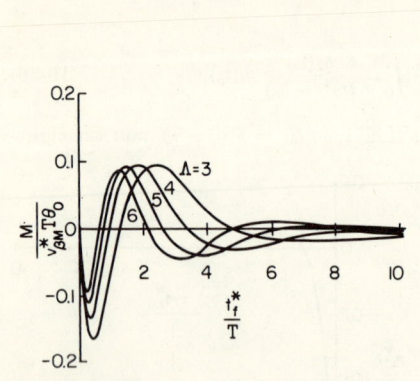

FIG. 19-6. Miss due to launch error of θ_0 radians for $\Lambda = 4$, $Y(p) = 1/(1 + Tp)$, time of flight = t_f^*.

FIG. 19-7. Miss due to step target acceleration a_t at time t_0 for $\Lambda = 4$, $Y(p) = 1/(1 + Tp)$.

error is obtained by setting $y_t = 0$ and letting the initial condition on the missile velocity $\dot{y}_m(0)$ be given by $v_{\beta M}^*\theta_0$. If we let $y_i(0) = 0$, $i \neq 2$, and

$$y_2(0) = \dot{y}_m(0) = v_{\beta M}^*\theta_0$$

in Eq. (19-34), then we can write

$$\sum_{i=1}^{n} \lambda_i(t_f^*) y_i(t_f^*) = \lambda_2(0) y_2(0) \qquad (19\text{-}41)$$

Letting $\lambda_1(t_f^*) = 1$ and $\lambda_2(t_f^*) = \lambda_3(t_f^*) = \cdots = \lambda_n(t_f^*) = 0$ as before, the left side of (19-41) becomes $y_1(t_f^*) = M$ = miss. Thus for $y_2(0) = v_{\beta M}^*\theta_0$ we have

$$M = v_{\beta M}^*\lambda_2(0)\theta_0 \qquad (19\text{-}42)$$

where θ_0 is the launch-error angle in radians. By starting the adjoint-solution computation at $t = t_f^*$ and integrating backward to $t = 0$, we obtain from $\lambda_2(0)$ the required miss coefficient. This can be recorded in a single computer run for all time-of-flight values t_f^* by plotting $\lambda_2(t_f^* - t)$ which, actually, is equal to $\lambda_2(t_f^*)$ for $t = 0$. The dimensionless miss coefficient $M/v_{\beta M}^*T\theta_0$ is plotted vs. dimensionless time of flight t_f^*/T for a simple time-lag system with time constant T in Fig. 19-6 for $\Lambda = 3, 4, 5,$ and 6.

GUIDANCE   19-11

Miss Due to Step Target Acceleration. For a step target acceleration of magnitude a_t at t_0 sec, y_t is given by

$$y_t = 0 \qquad\qquad 0 \leq t \leq t_0 \qquad (19\text{-}43)$$

$$= \frac{a_t}{2}(t - t_0)^2 \qquad t > t_0$$

The resulting miss can be computed from Eq. (19-26). However, an easier procedure is evident when one recalls that the curves of Fig. 19-6 represent, essentially, the miss resulting from a negative unit-step target velocity at time $t_0 = t_f{}^*$. By integrating each curve with respect to $t_f{}^*$ the resulting unit-acceleration miss curves are obtained, as shown in Fig. 19-7 for $Y(p) = 1/(1 + Tp)$ as before.

For a target with a maximum acceleration capability perpendicular to the line of sight given by a_t, it is of interest to calculate the optimum target maneuver to maximize the miss. Reference to Fig. 19-7 shows that the maximum miss will be obtained if the target starts initially with full positive acceleration a_t which is switched to full negative acceleration when the miss error goes through a maximum (at $t_0 - t_f{}^* = -4.8T$ in Fig. 19-7, or 4.8 autopilot time constants before impact) and back to full positive

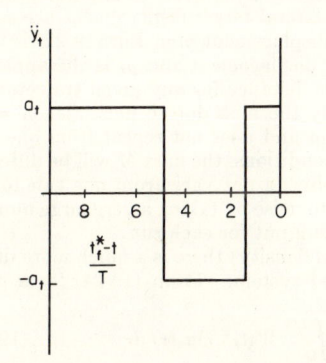

Fig. 19-8. Optimum target-acceleration program to maximize the miss distance for $\Lambda = 4$, $Y(p) = 1/(1 + Tp)$.

Fig. 19-9. Rms miss due to target sinusoidal weave of frequency ω and amplitude A for $\Lambda = 4$, $Y(p) = 1/(1 + Tp)$.

acceleration at the next maximum (1.4 autopilot time constants before impact in Fig. 19-7). The resulting target-acceleration program is shown in Fig. 19-8.

As a specific example, consider a target with maximum acceleration capability perpendicular to the line of sight equal to $\pm 2g$ (± 64 ft/sec^2). Assume that the missile autopilot time constant $T = 0.5$ sec. From Fig. 19-7 the miss resulting from the step change in target acceleration of $-4g$ at $(t_f{}^* - t)/T = 4.8$ is equal to

$$0.06T^2(4g) = -0.06(0.25)(-128) = 1.92 \text{ ft}$$

The miss resulting from the next step change in target acceleration of $+4g$ at

$$(t_f{}^* - t)/T = 1.4$$

is equal to $0.14(0.25)(128) = 4.48$ ft. Since we are dealing with a linearized system, the total miss is the sum of the individual miss distances, or $1.92 + 4.48 = 6.40$ ft. This is not impressively large in view of the fact that the target must know the geometry of the missile attack, the exact timing, the missile speed, the missile navigation constant, and the time constant of the missile autopilot in order to implement this optimum maneuver of $\pm 4g$ step accelerations.

Miss Due to Target Sinusoidal Weave. Next consider the miss resulting from a sinusoidal target weaving motion given by

$$y_t = A \sin(\omega t + \phi) \qquad (19\text{-}44)$$

From Eq. (19-26) we can calculate the resulting miss M, which will be a function of the phase angle ϕ of the sinusoid. Thus

$$M(\phi) = A \sin(\omega t_f^* + \phi) - A \int_0^{t_f^*} W(t_f^*,\tau) \sin(\omega\tau + \phi)\, d\tau \qquad (19\text{-}45)$$

If the phase shift has equal probability of falling between 0 and 2π, then the rms miss is given by the formula

$$\text{rms miss} = \sqrt{\frac{1}{2\pi} \int_0^{2\pi} [M(\phi)]^2\, d\phi} \qquad (19\text{-}46)$$

In Fig. 19-9 the rms miss parameter is plotted as a function of weave frequency ω for $\Lambda = 4$ and $Y(p) = 1/(1 + Tp)$. Note that for $\omega \gg 1/T$, the rms miss approaches $0.707A$, where A is the peak amplitude. This is not unexpected since, at frequencies high compared with the reciprocal of the autopilot time constant, the missile is unable to follow the target and simply guides to its mean position of $y_t = 0$, resulting in a miss equal to the rms value of a sine wave of amplitude A.

Miss Due to Noise. Let us now consider the effect of random noise in the line-of-sight angle β as measured by the guidance system. This noise in angular position can be translated into an equivalent noise y_n in lateral target displacement according to the formula $y_n(t) = \beta_n(t)R^*(t)$. The target displacement y_t in Eq. (19-22) is then replaced by $y_t + y_n$, where y_t is the true target displacement and y_n is the apparent target displacement due to the noise. The miss distance for any given trajectory is still $M = y_t(t_f^*) - y_m(t_f^*)$. If we consider only the miss due to noise (let $y_t = 0$), then $M = -y_m(t_f^*)$. Since the noise is random and does not repeat from one trajectory to the next, even under the same initial conditions, the miss M will be different for each trajectory. If the statistics of the noise do not vary from one run to the next, however, we can calculate an rms miss due to noise by taking a very large number of computer runs with a simulated random-noise input for each run.

For the case of white noise (flat power spectral density) there is a much more direct way to calculate the rms miss for our linearized system. From (19-24) for a noise input y_n,

$$-\text{Miss distance} = y_m(t_f^*) = \int_0^{t_f^*} W(t_f^*,\tau) y_n(\tau)\, d\tau \qquad (19\text{-}47)$$

The square of the miss distance can be written as

$$y_m^2(t_f^*) = \int_0^{t_f^*} \int_0^{t_f^*} W(t_f^*,\tau_1) W(t_f^*,\tau_2) y_n(\tau_1) y_n(\tau_2)\, d\tau_1\, d\tau_2 \qquad (19\text{-}48)$$

The mean-square miss distance is the expected value of $y_m^2(t_f^*)$, which can be written as

$$E[y_m^2(t_f^*)] = \int_0^{t_f^*} \int_0^{t_f^*} W(t_f^*,\tau_1) W(t_f^*,\tau_2) \phi_{y_n y_n}(\tau_1,\tau_2)\, d\tau_1\, d\tau_2 \qquad (19\text{-}49)$$

where $\phi_{y_n y_n}(\tau_1,\tau_2)$ is the joint distribution between $y_n(\tau_1)$ and $y_n(\tau_2)$. For an ergodic process [time average equivalent to ensemble average of $y_n(t)$] the joint distribution is equal to the autocorrelation function $\phi(\tau_1 - \tau_2)$, i.e., the time average of $y_n(t)y_n(t - \tau_1 + \tau_2)$. Let us assume that the statistics of the noise y_n are independent of time t following launch (later we shall see that this restriction can be removed) so that the ergodic assumption holds. For white noise with spectral density Φ_0 sq ft/cps the autocorrelation function $\phi(\tau_1 - \tau_2) = \Phi_0 \delta(\tau_1 - \tau_2)$. Then (19-49) becomes

$$E[y_m^2(t_f^*)] = \Phi_0 \int_0^{t_f^*} \int_0^{t_f^*} W(t_f^*,\tau_1) W(t_f^*,\tau_2) \delta(\tau_1 - \tau_2)\, d\tau_1\, d\tau_2$$

$$= \Phi_0 \int_0^{t_f^*} [W(t_f^*,\tau_1)]^2\, d\tau_1 \qquad (19\text{-}50)$$

and

$$\text{rms miss} = \sqrt{\Phi_0 \int_0^{t_f^*} [W(t_f^*,\tau_1)]^2\, d\tau_1} \qquad (19\text{-}51)$$

GUIDANCE

Equation (19-51) can be solved with a single computer run where the weighting function $W(t_f^*,\tau)$ is calculated from (19-39) from the adjoint function as before.

Figure 19-10 shows a plot of rms miss parameter vs. Λ for white noise with

$$Y(p) = 1/(1 + Tp)$$

In many cases the spectral density of the noise is dependent on the time t from launch at $t = 0$ to impact (or near impact) at $t = t_f^*$. In this case we can write the noise $y_n(t)$ as

$$y_n(t) = f(t)\bar{y}_n(t) \tag{19-52}$$

where $\bar{y}_n(t)$ represents an ergodic process with constant spectral density Φ_0. It is easy to show in this case that

$$\text{rms miss} = \sqrt{\Phi_0 \int_0^{t_f^*} [W(t_f^*,\tau_1)f(\tau_1)]^2 \, d\tau_1} \tag{19-53}$$

If the noise is not white (spectral density a function of frequency ω), it can be viewed as white noise passed through a linear filter with the appropriate transfer function $Y_F(j\omega)$ such that the spectral density of the filter output $\Phi(\omega) = |Y_F(j\omega)|^2 \Phi_0$ and matches the given noise spectral density. The filter transfer operator $Y_F(p)$ is then included in Eq. (19-22); i.e., y is replaced by $Y_F(p)y_n$. The analysis then proceeds, as before.

In summary it should be noted that the results of this and the previous sections on proportional navigation are based on a linearized system, including a linear auto-

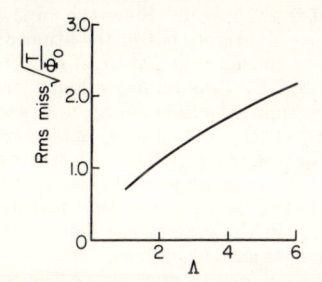

Fig. 19-10. Rms miss vs. navigation parameter Λ for white noise of spectral density Φ_0 sq ft/cps; $Y(p) = 1/(1 + Tp)$.

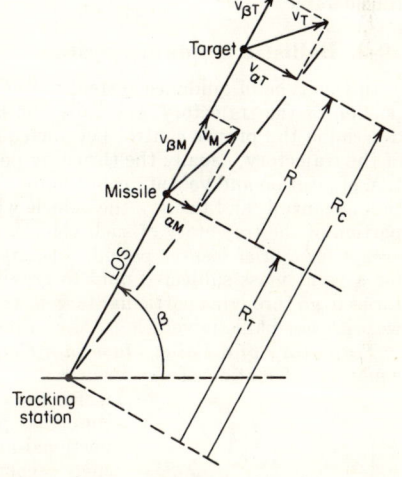

Fig. 19-11. Geometry for beam-rider navigation.

pilot. The analysis is much more complicated if important nonlinearities are present, since superposition no longer applies. For example, to study the influence of noise, hundreds of computer trajectory runs must be made to obtain the rms miss with reasonable confidence as opposed to the single run represented by Eq. (19-51) in the linear case. It seems apparent that there are enormous advantages to be gained if the system can be linearized without undue loss of accuracy.

Beam-rider Navigation. As a final example of homing navigation let us consider beam-rider or line-of-sight navigation. Here the missile flies along the line of sight between some remote tracking station and the target. The remote station may be moving or stationary. An example of a system of this type might consist of a surface radar which tracks target aircraft and a missile which flies along the radar beam from surface to target.

For a fixed tracking station the geometry of beam-rider navigation is shown in Fig. 19-11. In order to fly along the line of sight the missile must have a velocity component $v_{\alpha M}$ perpendicular to the line of sight which is equal to the line-of-sight velocity $R_T\dot\beta$ at the same point, where R_T is the distance from missile to tracking station. Thus

$$v_{\alpha M} = R_T\dot\beta \tag{19-54}$$

But $\dot\beta$ is also given by

$$\dot\beta = \frac{v_{\alpha T}}{R_C} \tag{19-55}$$

where R_C is the range from tracking station to target. From (19-54) and (19-55),

$$v_{\alpha M} = \frac{R_T}{R_C} v_{\alpha T} \tag{19-56}$$

When the beam-rider missile is launched into the beam, $R_T \cong 0$ and $v_{\alpha M} = 0$. Hence the missile flight path is directed along the line of sight and the beam-rider trajectory begins as pursuit navigation.

On the other hand, just before impact with the target, $R_T \cong R_C$ and $v_{\alpha M} \cong v_{\alpha T}$. Thus the beam-rider trajectory ends approximately on a constant-bearing trajectory.

The main advantage of a beam-rider system is its simplicity. Its principal disadvantages include inaccuracy at long ranges and inefficiency because much of the missile flight path lies in the lower atmosphere where aerodynamic drag losses are considerable.

19-3. Ballistic Guidance Systems

In this type of guidance system the vehicle path cannot be controlled over a large portion of the trajectory, generally the latter portion. The guidance problem is to determine the proper control laws or equations so that during the controllable part of the trajectory, usually the thrusting portion, the vehicle reaches a correct combination of position and velocity coordinates. If this is correctly done, then after termination of control (and thrust), the vehicle will coast to the target. Since the unpowered portion of the trajectory of such vehicles is almost invariably out of the atmosphere, except for a brief reentry period, it is important to consider the nature of trajectories for a point mass subjected only to gravity forces. By constraining all such trajectories to go through a particular target, we can determine the class of trajectories which we wish our ballistic vehicle to have when thrust (and hence control) is terminated.

Trajectory of a Point Mass in a Central-force Gravity Field. In writing the equations of motion of any vehicle it is important to choose a coordinate system in which the equations can be written as conveniently and simply as possible. In the case of the linearized proportional-navigation equations of Sec. 19-2 we chose such a coordinate system. In considering the trajectory of a ballistic missile or space vehicle outside the atmosphere and subjected only to gravity forces, we shall assume at the outset that the only significant force results from a single inverse-square gravity-force field. This would be the case for a vehicle in free fall, out of the atmosphere, in the vicinity of a spherical earth, where we neglect the gravitational effects of moon, sun, or any other attractive bodies. The effect of neglecting the acceleration due to the curvilinear motion of the earth about the sun is almost exactly canceled by the effect of neglecting the sun's gravity, provided the vehicle remains in the vicinity of the earth. We shall also neglect the noncentral-force-field gravity terms due to the fact that the earth is somewhat oblate. These neglected gravity forces put in only small corrective terms (Sec. 17-2) which often must be considered in practice but which would complicate our discussion needlessly.

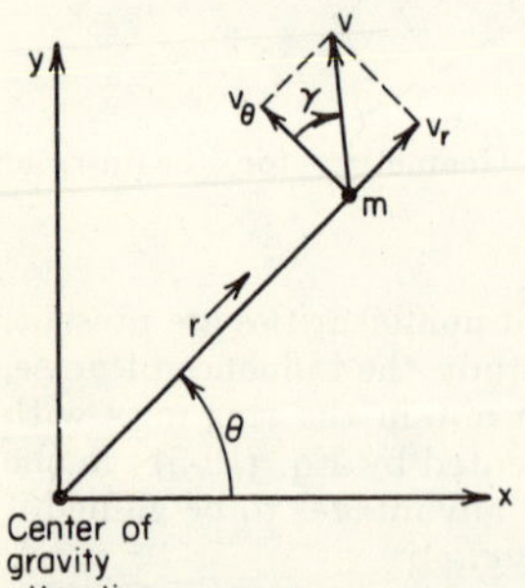

Fig. 19-12. Coordinates for two-dimensional trajectory.

Assuming, then, that the only force acting on our vehicle is the inverse-square gravity force, it is easy to show that the motion of the vehicle, hereafter referred to as

GUIDANCE

the point mass m, will remain in a plane which includes the center of gravitational attraction. If we write the coordinates of the point mass in this plane, then the problem becomes two-dimensional, which is already a significant simplification. Further, let us use polar coordinates r, θ with origin at the center of gravitational attraction to represent the position in the plane, as shown in Fig. 19-12. This simplifies the expression of the gravity force. Here x is a fixed reference direction in the plane of the motion. We assume that the point mass m has a velocity v with components v_r and v_θ along the direction of increasing r and θ, respectively.

Summing inertial and gravity forces along the v_r and v_θ direction, we obtain the following familiar equations:

$$m(\ddot{r} - r\dot{\theta}^2) = -m \frac{g_0 r_0^2}{r^2} \tag{19-57}$$

$$m(r\ddot{\theta} + 2\dot{r}\dot{\theta}) = 0 \tag{19-58}$$

Here g_0 is the acceleration due to gravity at a prescribed radial distance r_0 (e.g., the radius of the earth). In (19-57) $-r\dot{\theta}^2$ is the familiar centrifugal acceleration term; $2\dot{r}\dot{\theta}$ in (19-58) is the Coriolis acceleration.

Equation (19-58) can be integrated directly by noting that

$$\frac{d}{dt}(mr^2\dot{\theta}) = mr(r\ddot{\theta} + 2\dot{r}\dot{\theta})$$

Thus (19-58) can be written

$$\frac{d}{dt}(mr^2\dot{\theta}) = 0$$

or
$$mr^2\dot{\theta} = p = \text{const} \tag{19-59}$$

where p is the angular momentum. From (19-59) we can write

$$\dot{\theta} = \frac{p}{mr^2} \tag{19-60}$$

which can be substituted into (19-57) to obtain

$$\ddot{r} - \frac{p^2}{m^2}\frac{1}{r^3} + \frac{g_0 r_0^2}{r^2} = 0 \tag{19-61}$$

Thus we now have a single second-order equation in r which is, unfortunately, nonlinear. However, in the usual fashion we define a new variable $u = 1/r$, i.e., an inverse radius. Then

$$r = \frac{1}{u} \qquad \dot{r} = -\frac{1}{u^2}\frac{du}{d\theta}\dot{\theta} = -\frac{p}{m}\frac{du}{d\theta}$$

Similarly,

$$\ddot{r} = -\frac{p}{m}\frac{d^2u}{d\theta^2}\dot{\theta} = -\left(\frac{p}{m}\right)^2 u^2 \frac{d^2u}{d\theta^2} \tag{19-62}$$

Substituting (19-62) into (19-61) and letting $r = 1/u$, we obtain

$$\frac{d^2u}{d\theta^2} + u = \frac{m^2 g_0 r_0^2}{p^2} \tag{19-63}$$

This is now a linear equation which has a general solution

$$\frac{1}{r} = u = A\cos\theta + B\sin\theta + \frac{m^2 g_0 r_0^2}{p^2} \tag{19-64}$$

where A and B are arbitrary constants which depend on initial conditions. It is well known that the solution (19-64), which represents the trajectory of our point mass m in free fall, is a conic section which may be an ellipse, parabola, or hyperbola depending on the initial conditions. For a ballistic missile the trajectory is always an ellipse which intersects the surface of the earth.

The Two-dimensional Hit Equation. If we differentiate (19-64) with respect to time, we obtain

$$-\frac{\dot{r}}{r^2} = -A\dot{\theta}\sin\theta + B\dot{\theta}\cos\theta \tag{19-65}$$

Let us now assume as initial conditions at some time t_1 the following:

$$r(t_1) = r_1 \qquad \theta(t_1) = \theta_1 \qquad \dot{r}(t_1) = v_{r_1} \qquad r\dot{\theta}(t_1) = v_{\theta_1} \tag{19-66}$$

Noting that $m^2/p^2 = 1/r_1^2 v_{\theta_1}^2$, we can rewrite (19-64) and (19-65) at $t = t_1$ as

$$\frac{1}{r_1} = A\cos\theta_1 + B\sin\theta_1 + \frac{g_0 r_0^2}{r_1^2 v_{\theta_1}^2} \tag{19-67}$$

and

$$-\frac{v_{r_1}}{r_1 v_{\theta_1}} = -A\sin\theta_1 + B\cos\theta_1 \tag{19-68}$$

Solving for A and B from (19-67) and (19-68) and substituting into (19-64) and (19-65) we obtain

$$\frac{1}{r} = \left[\left(\frac{1}{r_1} - \frac{g_0 r_0^2}{r_1^2 v_{\theta_1}^2}\right)\cos\theta_1 + \frac{v_{r_1}}{r_1 v_{\theta_1}}\sin\theta_1\right]\cos\theta + \left[\left(\frac{1}{r_1} - \frac{g_0 r_0^2}{r_1^2 v_{\theta_1}^2}\right)\sin\theta_1 - \frac{v_{r_1}}{r_1 v_{\theta_1}}\cos\theta_1\right]\sin\theta + \frac{g_0 r_0^2}{r_1^2 v_{\theta_1}^2} \tag{19-69}$$

This is now the equation for the trajectory of our point mass with initial conditions r_1, θ_1, v_{r_1}, and v_{θ_1}. Let us now assume that the point mass is a ballistic missile with a target on the surface of the earth and located in the plane of motion of the point mass. Furthermore, assume that we have chosen our x reference direction in Fig. 19-12 to pass through the target, so that the target coordinates are $r = r_0$, $\theta = 0$. Obviously, if the trajectory given by (19-69) is to pass through the target, then $r = r_0$ when $\theta = 0$. Making this substitution, multiplying by r_1 and rearranging terms we obtain

$$\frac{r_1}{r_0} - \left[\left(1 - \frac{g_0 r_0^2}{r_1 v_{\theta_1}^2}\right)\cos\theta_1 + \frac{v_{r_1}}{v_{\theta_1}}\sin\theta_1\right] - \frac{g_0 r_0^2}{r_1 v_{\theta_1}^2} = F(r_1,\theta_1,v_{r_1},v_{\theta_1}) = 0 \tag{19-70}$$

This is called the hit equation, since it establishes the combination of position coordinates r_1, θ_1 and velocity coordinates v_{r_1}, v_{θ_1} which will cause the resulting trajectory to pass through the target at $r = r_0$, $\theta = 0$. There are, of course, an infinite number of such combinations. The guidance problem is to steer the ballistic missile during its powered flight so that its trajectory in the four-dimensional r, θ, v_r, v_θ phase space intersects the four-dimensional hypersurface represented by the hit equation,

$$F(r_1,\theta_1,v_{r_1},v_{\theta_1}) = 0$$

When the equation is satisfied, engine thrust is terminated. The missile will then coast in free fall on a trajectory which intersects the target at $r = r_0$, $\theta = 0$.

Since Eq. (19-70) is highly nonlinear, let us introduce the concept of a reference trajectory which satisfies (19-70) and represents a trajectory which our ballistic missile could achieve with nominal engine thrusting. Let r_1^*, θ_1^*, $v_{r_1}^*$, and $v_{\theta_1}^*$ be the position and velocity coordinates for the reference trajectory at the thrust termination time t_1^*. Let us now expand $F(r_1,\theta_1,v_{r_1},v_{\theta_1})$ in a Taylor series about the reference-trajectory cut-off values. Thus let

$$F(r_1,\theta_1,v_{r_1},v_{\theta_1}) = F(r_1{}^*,\theta_1{}^*,v_{r_1}{}^*,v_{\theta_1}{}^*) + \left.\frac{\partial F}{\partial r_1}\right|^* (r_1 - r_1{}^*) + \left.\frac{\partial F}{\partial \theta_1}\right|^* (\theta_1 - \theta_1{}^*)$$
$$+ \left.\frac{\partial F}{\partial v_{r_1}}\right|^* (v_{r_1} - v_{r_1}{}^*) + \left.\frac{\partial F}{\partial v_{\theta_1}}\right|^* (v_{\theta_1} - v_{\theta_1}{}^*) + \cdots \quad (19\text{-}71)$$

(Here the asterisk on the partial derivatives indicates that they are evaluated along the reference trajectory.) But $F(r_1{}^*,\theta_1{}^*,v_{r_1}{}^*,v_{\theta_1}{}^*) = 0$ because the reference trajectory satisfies the hit equation. If we now assume that the actual trajectory thrust termination coordinates r_1, θ_1, v_{r_1}, v_{θ_1} are close to the reference-trajectory values, we can neglect terms in (19-71) of order higher than first. The linearized hit equation then becomes

$$\left.\frac{\partial F}{\partial r_1}\right|^* (r_1 - r_1{}^*) + \left.\frac{\partial F}{\partial \theta_1}\right|^* (\theta_1 - {}_1\theta^*) + \left.\frac{\partial F}{\partial v_{r_1}}\right|^* (v_{r_1} - v_{r_1}{}^*) + \left.\frac{\partial F}{\partial v_{\theta_1}}\right|^* (v_{\theta_1} - v_{\theta_1}{}^*) = 0 \quad (19\text{-}72)$$

It is obvious that a deviation $(\theta_1 - \theta_1{}^*)$ in polar angle at thrust termination as compared with the reference value will simply rotate the whole free-fall trajectory through the angle $\theta_1 - \theta_1{}^*$ and will produce a downrange miss at the target which is $r_0(\theta_1 - \theta_1{}^*)$, where r_0 is the radius of the earth. For this reason we multiply (19-72) by $r_0 \Big/ \left.\dfrac{\partial F}{\partial \theta_1}\right|^*$ and rewrite it as

$$r_0 \frac{\left.\frac{\partial F}{\partial r_1}\right|^*}{\left.\frac{\partial F}{\partial \theta_1}\right|^*} (r_1 - r_1{}^*) + r_0(\theta_1 - \theta_1{}^*) + r_0 \frac{\left.\frac{\partial F}{\partial v_{r_1}}\right|^*}{\left.\frac{\partial F}{\partial \theta_1}\right|^*} (v_{r_1} - v_{r_1}{}^*) + r_0 \frac{\left.\frac{\partial F}{\partial v_{\theta_1}}\right|^*}{\left.\frac{\partial F}{\partial \theta_1}\right|^*} (v_{\theta_1} - v_{\theta_1}{}^*) = 0$$
$$(19\text{-}73)$$

Since the second term in this equation represents the downrange miss due to the deviation $(\theta_1 - \theta_1{}^*)$ in polar angle, it is obvious that the other terms represent downrange misses due to $(r_1 - r_1{}^*)$, $(v_{r_1} - v_{r_1}{}^*)$, and $(v_{\theta_1} - v_{\theta_1}{}^*)$, respectively. To hit the target the sum of the individual downrange misses must add up to zero. In terms of individual miss coefficients we can write the overall downrange miss M_{DR} as

$$M_{DR} = M_r(r_1 - r_1{}^*) + M_\theta(\theta_1 - \theta_1{}^*) + M_{v_r}(v_{r_1} - v_{r_1}{}^*) + M_{v_{\theta_1}}(v_{\theta_1} - v_{\theta_1}{}^*) \quad (19\text{-}74)$$

where

$$M_r = r_0 \frac{\left.\frac{\partial F}{\partial r_1}\right|^*}{\left.\frac{\partial F}{\partial \theta_1}\right|^*} \qquad M_\theta = r_0$$
$$M_{v_r} = r_0 \frac{\left.\frac{\partial F}{\partial v_{r_1}}\right|^*}{\left.\frac{\partial F}{\partial \theta_1}\right|^*} \qquad M_{v_\theta} = r_0 \frac{\left.\frac{\partial F}{\partial v_{\theta_1}}\right|^*}{\left.\frac{\partial F}{\partial \theta_1}\right|^*}$$
$$(19\text{-}75)$$

Referring to Eq. (19-70) for $F(r_1,\theta_1,v_{r_1},v_{\theta_1})$ we find that

$$\frac{\partial F}{\partial r_1} = \frac{1}{r_0} + \frac{g_0{}^2 r_0{}^2}{r_1{}^2 v_{\theta_1}{}^2}(1 - \cos\theta_1)$$
$$\frac{\partial F}{\partial \theta_1} = \left(1 - \frac{g_0 r_0{}^2}{r_1 v_{\theta_1}{}^2}\right) \sin\theta_1 - \frac{v_{r_1}}{v_{\theta_1}} \cos\theta_1$$
$$\frac{\partial F}{\partial v_{r_1}} = -\frac{\sin\theta_1}{v_{\theta_1}} \quad (19\text{-}76)$$
$$\frac{\partial F}{\partial v_{\theta_1}} = \frac{2 g_0 r_0{}^2}{r_1 v_{\theta_1}{}^3}(1 - \cos\theta_1) + \frac{v_{r_1}}{v_{\theta_1}{}^2} \sin\theta_1$$

As a specific example, let us consider an intercontinental ballistic missile with a reference trajectory for which θ_1, the polar angle at cut-off, is $-90°$. This represents a free-fall trajectory of one-quarter of the earth's circumference, since $\theta_{\text{target}} = 0$ by definition. For simplification, let us assume that the radial distance at thrust cut-off $r_1 = r_0$, the earth's radius (actually, cut-off would typically occur at $r \approx 1.02 r_0$). We would like to calculate the miss coefficients at thrust termination as given by (19-75) and (19-76).

Our first task is to establish a reference trajectory. This can be done by setting $r_1 = r_0$ and $\theta_1 = -\pi/2$ in (19-70) and by determining through trial and error a pair of velocity components v_{r_1}, v_{θ_1} which satisfies the hit equation. To make the problem more definitive, however, let us assume that the reference trajectory is the minimum-energy trajectory, i.e., the one which requires the least total velocity v_1 at thrust cut-off and hence the least kinetic energy. The kinetic energy T is given by

$$T = \tfrac{1}{2} m v_1^2 = \tfrac{1}{2} m (v_{r_1}^2 + v_{\theta_1}^2) \tag{19-77}$$

Let us differentiate T with respect to v_{θ_1} and set $\partial T/\partial v_{\theta_1} = 0$, remembering that v_{r_1} is a function of v_{θ_1} in accordance with the hit equation (19-70). Then

$$\frac{\partial T}{\partial v_{\theta_1}} = m \left(v_{\theta_1} + v_{r_1} \frac{dv_{r_1}}{dv_{\theta_1}} \right) = 0$$

and

$$\frac{dv_{r_1}}{dv_{\theta_1}} = -\frac{v_{\theta_1}}{v_{r_1}} \tag{19-78}$$

If we now solve (19-70) for v_{r_1} in terms of v_{θ_1}, differentiate, and set $dv_{r_1}/dv_{\theta_1} = -v_{\theta_1}/v_{r_1}$ in accordance with Eq. (19-78), we shall obtain a second equation relating v_{θ_1} and v_{r_1}. Along with (19-70) this allows us to solve for v_{θ_1} for the minimum-energy trajectory, obtaining

$$v_{\theta_1} = \left[\frac{g_0 r_0^3 (1 - \cos \theta_1)}{r_1 \sqrt{r_1^2 - 2 r_1 r_0 \cos \theta_1 + r_0^2}} \right]^{1/2} \tag{19-79}$$

For our example ICBM problem with $r_1 = r_0$, $\theta_1 = -\pi/2$, (19-79) gives

$$v_{\theta_1} = 21{,}800 \text{ ft/sec}$$

and (19-70) gives $v_{r_1} = 9{,}000$ ft/sec. Using these as reference-trajectory cut-off values and substituting into (19-75) and (19-76) we obtain the following miss coefficients:

$$M_{r_1} = 5.85 \text{ ft/ft} \qquad M_{\theta_1} = r_0 = \text{radius of earth}$$
$$M_{v_{r_1}} = 2310 \text{ ft/ft/sec} \qquad M_{v_{\theta_1}} = 5{,}590 \text{ ft/ft/sec}$$

The velocity miss coefficients illustrate the guidance accuracy needed for an ICBM; velocity errors at thrust cut-off of the order of 1 ft/sec produce miss distances at the target of the order of 1 mile.

Returning to the linearized downrange miss equation (19-74), we can see a method of controlling the ballistic missile during thrusting to ensure that the hit equation is satisfied at thrust cut-off. The missile attitude is completely preprogrammed during thrusting in accordance with the reference trajectory. At the same time, during thrusting, the guidance computer accepts calculated position and velocity coordinates for the missile, either by on-board inertial means or from a ground track of the missile. The deviations $(r_1 - r_1^*)$, $(\theta_1 - \theta_1^*)$, $(v_{r_1} - v_{r_1}^*)$, and $(v_{\theta_1}/v_{\theta_1}^*)$ of the instantaneous missile position and velocity coordinates from the reference-trajectory cut-off values are computed simultaneously and the right side of (19-74) is computed. At any instant this calculation represents the downrange miss M_{DR} which would occur if the engine thrust were terminated at that instant. Note that M_{DR} in (19-74) is based on a linearized formula good only for small coordinate deviations. This means that when the deviations are large the formula is very much in error. However, as thrusting continues and the deviations become smaller and smaller, M_{DR}, which starts out very negative, will increase monotonically toward zero. At the same time the accuracy of

GUIDANCE

the linearized formula improves. Finally, when M_{DR} goes through zero, engine thrust is terminated and, to within the accuracy of the linearized equation (19-74), the ballistic missile should be on a hit trajectory with $M_{DR} = 0$.

If the deviations between actual and reference-trajectory coordinates at cut-off are too large to make (19-74) sufficiently accurate, then the necessary second-order terms in the Taylor-series expansion (19-71) of the exact hit equation function in (19-70) can be used. Also, in writing (19-74) for downrange miss M_{DR} we have assumed that the target remains fixed in our inertial frame of Fig. 19-12. Actually, a target on the surface of the earth will be moving with respect to this frame because of the earth's rotation. We shall show how the linearized hit equation can be modified to take this into account in the next section. Another way to solve this problem is to constrain the time of flight to that of the reference trajectory. This modifies the hit equation such that a relationship between v_{r_1} and v_{θ_1} is enforced.

In addition, we have ignored the trajectory modification due to missile-warhead reentry. This can be taken care of by displacing the target to a new fictitious position which it would need to have for the vacuum trajectory to be the correct trajectory in view of reentry considerations. This same means can be used to take care of slight trajectory modifications due to the noncentral-force-field gravity forces, which can influence an ICBM trajectory impact point by up to 20 miles.

Ballistic Guidance in Three Dimensions. In the previous section we considered the equations describing the free-fall trajectory of a ballistic missile in two dimensions. Thus the motion was assumed to take place in a plane, e.g., the plane of the reference trajectory. In practical cases this restriction is unrealistic, and for this reason we shall consider the three-dimensional problem in this section. A reference-trajectory plane is still useful in this case, and the projection of the vehicle point mass onto the plane is located using polar coordinates as before, as shown in Fig. 19-13, where the xy plane is the reference plane. A third coordinate z denotes the perpendicular distance of the point mass m from the plane. Thus cylindrical coordinates r, θ, and z locate the vehicle position in three dimensions, while v_r, v_θ, and v_z denote the vehicle velocity components along the direction of increasing r, θ, and z, respectively (see Fig. 19-13).

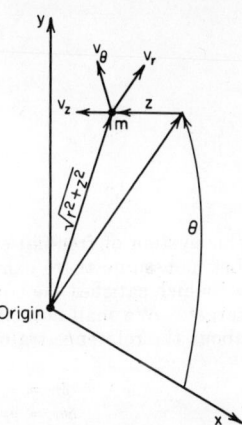

FIG. 19-13. Cylindrical coordinates for three-dimensional trajectory.

Summing accelerations along the three coordinate directions, we obtain the following equations of motion:

$$\ddot{r} - r\dot{\theta}^2 = -\frac{g_0 r_0^2 r}{(r^2 + z^2)^{3/2}} + \frac{F_r}{m}$$

$$r\ddot{\theta} + 2\dot{r}\dot{\theta} = \frac{F_\theta}{m} \qquad (19\text{-}80)$$

$$\ddot{z} = -\frac{g_0 r_0^2 z}{(r^2 + z^2)^{3/2}} + \frac{F_z}{m}$$

where F_r, F_θ, and F_z are external forces along the direction of v_r, v_θ, and v_z, respectively. Note that the distance of m from the origin in Fig. 19-13 is $\sqrt{r^2 + z^2}$ and that the gravity acceleration is $g_0 r_0^2/(r^2 + z^2)$ and is directed from m toward the origin. This acceleration is multiplied by $-r/\sqrt{r^2 + z^2}$ to obtain the component along v_r in the first of these three equations and by $-z/\sqrt{r^2 + z^2}$ to obtain the component along v_z in the third. For a pure inverse-square gravity field with no other external forces, $F_r = F_\theta = F_z = 0$ in (19-80). Any gravity perturbation forces make F_r, F_θ, and F_z functions of r, θ, and z. Also, thrusting and aerodynamic forces contribute to F_r, F_θ, F_z.

19-20 SYSTEM COMPONENTS

Let us now simplify our problem by assuming no external forces except gravity, so that the vehicle is assumed to be in free fall in a vacuum. This makes F_r, F_θ, and F_z only a function of r, θ, and z. Furthermore, let us neglect z^2 compared with r^2 on the assumption that the vehicle will not move too far away from the reference plane. Finally, let us denote $\dot{r}$ by v_r, $r\dot{\theta}$ by v_θ, and $\dot{z}$ by v_z. Then (19-80) can be written as follows:

$$\frac{dr}{dt} = v_r$$
$$\frac{d\theta}{dt} = \frac{v_\theta}{r}$$
$$\frac{dz}{dt} = v_z$$
$$\frac{dv_r}{dt} = \frac{v_\theta^2}{r} - \frac{g_0 r_0^2}{r^2} + \frac{F_r}{m} \quad (19\text{-}81)$$
$$\frac{dv_\theta}{dt} = -\frac{v_r v_\theta}{r} + \frac{F_\theta}{m}$$
$$\frac{dv_z}{dt} = -\frac{g_0 r_0^2 z}{r^3} + \frac{F_z}{m}$$

Our system of free-fall equations now consists of six first-order nonlinear equations. Let us assume there exists a reference trajectory with coordinates r^*, θ^*, z^*, v_r^*, v_θ^*, v_z^* which satisfies the equations and the terminal condition, namely, impact with the target. We shall now linearize the equations of motion by considering perturbations about the reference trajectory. Thus let

$$\begin{array}{lll} \delta r = r - r^* & \delta \theta = \theta - \theta^* & \delta z = z - z^* \\ \delta v_r = v_r - v_r^* & \delta v_\theta = v_\theta - v_\theta^* & \delta v_z = v_z - v_z^* \end{array} \quad (19\text{-}82)$$

The right side of each of Eqs. (19-81) is now expanded in a Taylor series about the reference-trajectory coordinates and terms of order higher than first are neglected. Recognizing that the zeroth-order terms on both sides of each equation cancel each other, we obtain the following set of linearized equations:

$$\frac{d\delta r}{dt} = \delta v_r$$
$$\frac{d\delta \theta}{dt} = -\frac{v_\theta^*}{r^{*2}} \delta r + \frac{1}{r^*} \delta v_\theta$$
$$\frac{d\delta z}{dt} = \delta v_z$$

$$\frac{d\delta v_r}{dt} = \left(-\frac{v_\theta^{*2}}{r^{*2}} + \frac{2 g_0 r_0^2}{r^{*3}} + \frac{1}{m} \frac{\partial F_r}{\partial r} \bigg|^* \right) \delta r + \frac{1}{m} \frac{\partial F_r}{\partial \theta} \bigg|^* \delta \theta$$
$$+ \frac{1}{m} \frac{\partial F_r}{\partial z} \bigg|^* \delta z + \frac{2 v_\theta^*}{r^*} \delta v_\theta \quad (19\text{-}83)$$

$$\frac{d\delta v_\theta}{dt} = \left(\frac{v_r^* v_\theta^*}{r^{*2}} + \frac{1}{m} \frac{\partial F_\theta}{\partial r} \bigg|^* \right) \delta r + \frac{1}{m} \frac{\partial F_\theta}{\partial \theta} \bigg|^* \delta \theta + \frac{1}{m} \frac{\partial F_\theta}{\partial z} \bigg|^* \delta z$$
$$- \frac{v_\theta^*}{r^*} \delta v_r - \frac{v_r^*}{r^*} \delta v_\theta$$

$$\frac{d\delta v_z}{dt} = \left(\frac{3 g_0 r_0^2 z^*}{r^{*4}} + \frac{1}{m} \frac{\partial F_z}{\partial r} \bigg|^* \right) \delta r + \frac{1}{m} \frac{\partial F_z}{\partial \theta} \bigg|^* \delta \theta$$
$$+ \left(-\frac{g_0 r_0^2}{r^{*3}} + \frac{1}{m} \frac{\partial F_z}{\partial z} \right) \delta z$$

GUIDANCE

Equations (19-83) can be rewritten in more compact form by letting the perturbation variables be represented as follows:

$$y_1 = \delta r \quad y_2 = \delta\theta \quad y_3 = \delta z$$
$$y_4 = \delta v_r \quad y_5 = \delta v_\theta \quad y_6 = \delta v_z \tag{19-84}$$

In this case the linearized equations can be written

$$\dot{y}_i = \sum_{j=1}^{6} a_{ij}(t) y_j \quad i = 1, 2, \ldots, 6 \tag{19-85}$$

where the time-varying coefficients $a_{ij}(t)$ are functions of the reference-trajectory coordinates as given in (19-83). For example,

$$a_{55}(t) = -\frac{v_r^*(t)}{r^*(t)}$$

We shall now show how the adjoint functions can be used to obtain the miss coefficients. As introduced earlier in Eq. (19-32), the adjoint functions satisfy the equation

$$\dot{\lambda}_i = -\sum_{j=1}^{6} a_{ji}(t) \lambda_j \quad i = 1, 2, 3, 4, 5, 6 \tag{19-86}$$

Multiplying (19-85) by λ_i, (19-86) by y_i, summing over i, and integrating from $t = t_1$ to t_f^*, we obtain the following equation:

$$\sum_{i=1}^{n} \lambda_i(t_f^*) y_i(t_f^*) = \sum_{i=1}^{n} \lambda_i(t_1) y_i(t_1) \tag{19-87}$$

The proof follows that given in Eqs. (19-33) and (19-34). Let us denote t_f^* as the time of impact of the missile on the target for the reference trajectory.

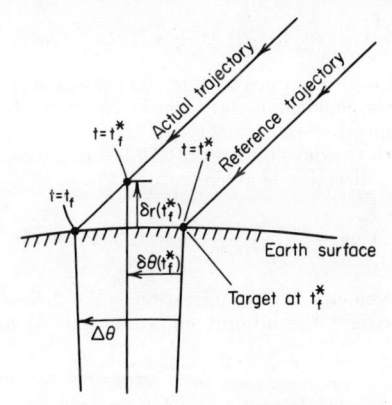

Fig. 19-14. Terminal portion of the trajectory projected into the reference plane.

Next consider the terminal portion of the trajectory projected into the reference plane, as shown in Fig. 19-14. Here we assume the target is located on the surface of the earth and at time t_f^* is intersected by the reference trajectory. From the figure it is evident that the actual trajectory impact time t_f is given by

$$t_f = t_f^* - \frac{\delta r(t_f^*)}{v_r^*(t_f^*)} \tag{19-88}$$

where $v_r^*(t_f^*)$ is negative. Also from the figure it is evident that the actual-trajectory impact will be downrange from the reference-trajectory impact by a polar angle $\Delta\theta$ and hence by a distance $r_0 \Delta\theta$ given by

$$r_0 \Delta\theta = r_0 \,\delta\theta(t_f^*) + v_\theta^*(t_f^*)(t_f - t_f^*)$$

or from (19-88)

$$r_0 \Delta\theta = -\frac{v_\theta^*(t_f^*)}{v_r^*(t_f^*)} \,\delta r(t_f^*) + r_0 \,\delta\theta(t_f^*) \qquad (19\text{-}89)$$

Next assume that, because of rotation of the earth's surface, the target itself has a downrange velocity given by v_{DR}. Then in the time interval $(t_f - t_f^*)$ the target will move downrange a distance equal to $v_{DR}(t_f - t_f^*)$. The actual downrange miss distance M_{DR} will then be the difference between the missile distance downrange at impact and the target distance. Thus

$$M_{DR} = r_0 \Delta\theta - v_{DR}(t_f - t_f^*)$$

or from (19-88) and (19-89)

$$M_{DR} = \frac{v_{DR} - v_\theta^*(t_f^*)}{v_r^*(t_f^*)} \,\delta r(t_f^*) + r_0 \,\delta\theta(t_f^*) \qquad (19\text{-}90)$$

We shall now choose the initial conditions at t_f^* for the six adjoint functions so that the left side of (19-87) represents the downrange miss M_{DR} given in (19-90). Remembering that $y_1 = \delta r$ and $y_2 = \delta\theta$, we let

$$\begin{aligned}\lambda_1(t_f^*) &= \frac{v_{DR} - v_\theta^*(t_f^*)}{v_r^*(t_f^*)} \qquad \lambda_2(t_f^*) = r_0 \\ \lambda_3(t_f^*) &= \lambda_4(t_f^*) = \lambda_5(t_f^*) = \lambda_6(t_f^*) = 0 \end{aligned} \qquad (19\text{-}91)$$

so that from (19-90) we can write (19-87) as

$$\begin{aligned}M_{DR} &= \sum_{i=1}^{n} \lambda_i(t) y_i(t) \\ &= \lambda_1(t)\,\delta r + \lambda_2(t)\,\delta\theta + \lambda_3(t)\,\delta z + \lambda_4(t)\,\delta v_r + \lambda_5(t)\,\delta v_\theta + \lambda_6(t)\,\delta v_z \end{aligned} \qquad (19\text{-}92)$$

Thus we see that the adjoint functions with initial conditions of (19-91) are the instantaneous downrange miss coefficients for coordinate perturbations in position and velocity from the reference free-fall trajectory.

By direct analogy with the development of (19-90) it is easy to show that the crossrange miss M_{CR} in the z direction is given by

$$M_{CR} = \frac{v_{CR} - v_z(t_f^*)}{v_r^*(t_f^*)} \,\delta r(t_f^*) + \delta z(t_f^*) \qquad (19\text{-}93)$$

where v_{CR} is the target velocity in the z direction. We define a second set of adjoint functions $\mu_i(t)$ which satisfy the adjoint equation (19-86) and have the following initial conditions:

$$\mu_1(t_f^*) = \frac{v_{CR} - v_z(t_f^*)}{v_r^*(t_f^*)} \qquad \mu_2(t_f^*) = 0 \qquad \mu_3(t_f^*) = 1$$
$$\mu_4(t_f^*) = \mu_5(t_f^*) = \mu_6(t_f^*) = 0 \qquad (19\text{-}94)$$

Equation (19-87) then becomes

$$\begin{aligned}M_{CR} &= \sum_{i=1}^{6} \mu_i(t) y_i(t) \\ &= \mu_1(t)\,\delta r + \mu_2(t)\,\delta\theta + \mu_3(t)\,\delta z + \mu_4(t)\,\delta v_r + \mu_5(t)\,\delta v_\theta + \mu_6(t)\,\delta v_z \end{aligned} \qquad (19\text{-}95)$$

GUIDANCE

Here we see the adjoint functions with initial conditions of (19-94) are the instantaneous cross-range miss coefficients for coordinate perturbations in position and velocity from the reference trajectory.

Having developed formulas for continuous calculation of downrange miss M_{DR} and cross-range miss M_{CR} which will occur if engine thrust is terminated, we can consider the mechanization of a complete three-dimensional ballistic-missile guidance system. The block diagram is shown in Fig. 19-15. During thrusting of the missile the three position and three velocity coordinates, as computed from inertial measurements (see the next section) or a ground track, are continuously compared with the free-fall reference-trajectory coordinates which are stored in the guidance computer. The resulting deviations δr, $\delta \theta$, etc., are used to compute continuously the downrange miss M_{DR} and cross-range miss M_{CR} which would result if thrust were terminated. Again this involves stored adjoint functions $\lambda_i(t)$ and $\mu_i(t)$ in the guidance computer. M_{CR} is used continuously to provide steering commands to the missile autopilot such that M_{CR} is driven to zero and held there. M_{DR} is a monotonic function which

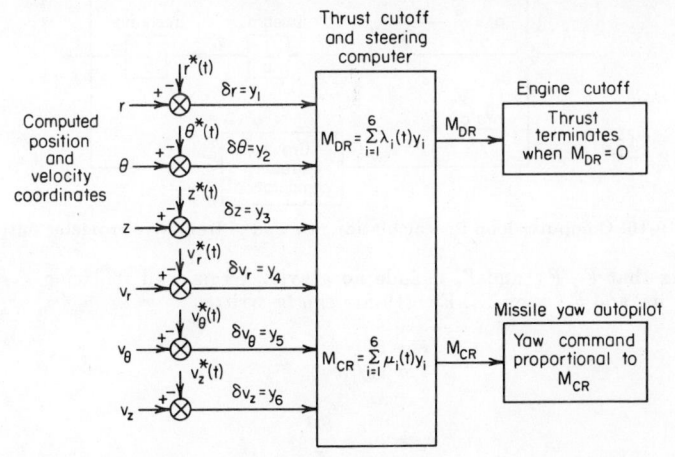

FIG. 19-15. Block diagram of ballistic-missile guidance system.

increases from negative values toward zero, at which time the engine thrust is terminated. The autopilot pitch-angle input is programmed in accordance with the reference-trajectory pitch angle throughout the thrusting portion of the trajectory.

An obvious simplification in the mechanization of Fig. 19-15 would result by using fixed values for the reference-trajectory coordinates, e.g., $r^*(t_1^*)$, $\theta^*(t_1^*)$, etc., where t_1^* is the reference-trajectory thrust cut-off time. Also, fixed values for the adjoint functions, e.g., $\lambda_1(t_1^*)$, $\lambda_2(t_1^*)$, etc., could be used. In this case the computation of the miss distance is essentially the same as that used earlier in Eq. (19-74), except that here we have accounted for target motion due to earth's rotation and have also included the third dimension. It might be remarked at this point that the z coordinate equations are at best only very lightly coupled to the r and θ equations, and become essentially decoupled in the absence of non-central-field gravity forces [$F_z = 0$ in last equation of (19-83) and $3g_0 r_0^2 z^*/r^{*4} \ll g_0 r_0^2/r^{*3}$].

The generalization of the adjoint method described in this section to space-vehicle trajectories where the target is not on the surface of the earth but is a moving target in space is quite obvious. It is only necessary to pick initial conditions on the adjoint functions such that the left side of (19-87) represents the appropriate miss distance. The adjoint equations are then solved backward in time using a computer in order to obtain the miss coefficients for subsequent use in the guidance computer.

As mentioned in the previous section, practical guidance considerations may require the inclusion of second-order perturbation terms in the miss equations if the perturbations become too large.

Coordinate Determination by Inertial Means. One of the most effective means of computing the necessary position and velocity coordinates in a missile guidance system is to measure the vehicle acceleration during thrusting. With proper corrections for gravity these accelerations can be integrated successively to obtain the required velocity and position information. Assume that accelerometers on board the missile are used to measure the acceleration components a_r, a_θ, and a_z in the v_r, v_θ, and v_z directions, respectively, in Fig. 19-13. Since these accelerations are measured by inertial means, they do not include the gravity acceleration, which must be added as a correction. For example, the radial accelerometer will measure F_r/m.

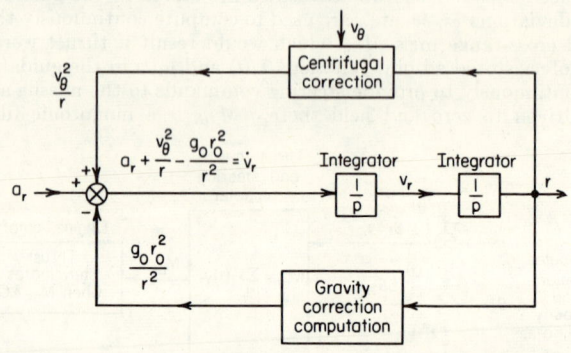

FIG. 19-16. Computer loop for calculation of r and v_r from accelerometer output a_r.

Assuming that F_r, F_θ, and F_z include no gravity terms, and replacing F_r/m by a_r, F_θ/m by a_θ, and F_z/m by A_z, Eq. (19-81) can be written

$$\dot{r} = v_r$$
$$\dot{\theta} = v_\theta/r$$
$$\dot{z} = v_z$$
$$\dot{v}_r = a_r + \frac{v_\theta^2}{r} - \frac{g_0 r_0^2}{r^2} \tag{19-96}$$
$$\dot{v}_\theta = a_\theta - \frac{v_r v_\theta}{r}$$
$$\dot{v}_z = a_z - \frac{g_0 r_0^2 z}{r^3}$$

The on-board guidance computer calculates $\dot{v}_r$, $\dot{v}_\theta$, and $\dot{v}_z$ from the measured acceleration components a_r, a_θ, and a_z in accordance with (19-96). $\dot{v}_r$, $\dot{v}_\theta$, and $\dot{v}_z$ are each integrated once to obtain v_r, v_θ, and v_z, and a second time to obtain r, θ, and z. Note that the velocity and position coordinates computed in this fashion must be used implicitly to mechanize (19-96).

Consider the mechanization of the computation of v_r and r from the radial accelerometer output a_r, as shown in the block diagram of Fig. 19-16. If a_r has no errors, if the initial-condition settings for r and v_r on the two integrators are perfect, and if the entire computation is errorless, then the values of r and v_r as obtained from the computer will continue to agree perfectly with the true missile trajectory values. On the other hand, suppose the measured acceleration a_r has a small error ϵ_{ar}, so that $a_r = a_{rt} + \epsilon_{ar}$, where a_{rt} is the true value the accelerometer should read. Suppose that this causes the computed radial distance r to have an error ϵ_r, so that $r = r_t + \epsilon_r$, where r_t is the true radial distance. Then

$$\dot{v}_r = \ddot{r} = \ddot{r}_t + \ddot{\epsilon}_r = a_{rt} + \epsilon_{ar} + \frac{v_\theta^2}{r_t + \epsilon_r} - \frac{g_0 r_0^2}{(r_t + \epsilon_r)^2} \tag{19-97}$$

GUIDANCE 19-25

Expanding those terms with $(r_t + \epsilon_r)$ in the denominator and neglecting terms in ϵ_r^2 and higher order, we obtain

$$\ddot{r}_t + \ddot{\epsilon}_r = a_{rt} + \epsilon_{ar} + \frac{v_\theta^2}{r_t} - \frac{v_\theta^2}{r_t^2}\epsilon_r - \frac{g_0 r_0^2}{r_t^2} + \frac{2g_0 r_0^2}{r_t^3}\epsilon_r \qquad (19\text{-}98)$$

But $\ddot{r}_t = a_{rt} + v_\theta^2/r_t - g_0 r_0^2/r_t^2$, since a_{rt} and r_t are true values. Hence we are left with an equation for the error ϵ_r. Noting that the velocity v_c for a circular orbit of radius r_0 is given by

$$v_c = \sqrt{g_0 r_c}$$

we can rewrite (19-98) as follows:

$$\ddot{\epsilon}_r - \frac{2g_0 r_0^2}{r_t^3}\left[1 - \frac{1}{2}\frac{r_t}{r_0}\left(\frac{v_\theta}{v_c}\right)^2\right]\epsilon_r = \epsilon_{ar} \qquad (19\text{-}99)$$

For a ballistic-missile trajectory the true radial distance r_t will not exceed the earth radius r_0 by more than perhaps 10 to 20 per cent, and the horizontal velocity v_θ remains less than circular orbit velocity v_c. Thus the coefficient of ϵ_r in (19-99) remains negative and the equation for ϵ_r is inherently unstable. For the simple case where $r_t \cong r_0$ and $v_\theta \ll v_c$, (19-99) becomes

$$\ddot{\epsilon}_r - \frac{2g_0}{r_0}\epsilon_r = \epsilon_{ar} \qquad (19\text{-}100)$$

which exhibits a transient solution $\exp(t\sqrt{2g_0/r_0})$ which grows exponentially with time. The time constant is $(\sqrt{2g_0/r_0})^{-1} = 571$ sec $\cong 9.5$ min. Thus the calculation of radial distance from inertial measurements is inherently unstable; any errors will tend to grow exponentially with time. Fortunately, the time of powered flight for ballistic missiles is usually only several minutes and the above errors do not have time to build up before the guidance task is completed (thrust terminated). However, this does show that pure inertial systems are not adequate for long-time space navigation. They must be supplemented with external observations of position.

Next consider the effect of an error ϵ_{az} in the z axis measured acceleration. As before we let $a_z = a_{zt} + \epsilon_z$ and $z = z_t + \epsilon_z$, in which case, from (19-96),

$$\dot{v}_z = \ddot{z} = \ddot{z}_t + \ddot{\epsilon}_z = a_{zt} + \epsilon_{az} - \frac{g_0 r_0^2}{r^3}(z_t + \epsilon_z) \qquad (19\text{-}101)$$

from which

$$\ddot{\epsilon}_z + \frac{g_0 r_0^2}{r^3}\epsilon_z = \epsilon_{az} \qquad (19\text{-}102)$$

Here the transient solution is an undamped sinusoid with angular frequency $\sqrt{g_0 r_0^2/r^3}$. For $r \cong r_0$ this corresponds to a period of 84.5 min. Thus we see that the computation of lateral displacement z from inertial measurements is neutrally stable. When the coordinate transformations necessary to obtain a_θ are considered, one can show that the computation of fore and aft position (i.e., the θ coordinate) from inertial measurements is also neutrally stable with a similar period. Although this does not present a serious problem for ballistic missiles because of their short powered flight, it can present a sizable problem in the inertial guidance of satellites; see Chapter 17.

Chapter 20

PROPULSION

ROBERT H. BODEN, *Rocketdyne Division of North American Aviation, Inc.*

CONTENTS

20-1. Introduction	20-1
20-2. Land Vehicles	20-2
20-3. Marine Vehicles	20-3
20-4. Airborne Vehicles	20-4
20-5. Space Vehicles	20-6
20-6. Internal-combustion-engine Systems	20-9
Reciprocating Engines	20-9
Propellants	20-15
Gas Turbines	20-15
Jet Engines	20-17
Rocket-engine Systems	20-24
20-7. Electrical Propulsion Systems	20-31
Energy Sources	20-32
Electrothermodynamic Thrust Devices	20-34
Electromagnetic Thrust Devices	20-35
Button Plasma Guns	20-35
Crossed-field Motors	20-36
Traveling-field Accelerators	20-36
Ion Rocket Engine	20-37
Ion Sources	20-39
Charged Colloid Sources	20-41
Propulsion-system Applications	20-41

20-1. Introduction

Propulsion is formally defined as the act of propelling a body. As an alternative definition for the system-oriented engineer or scientist, propulsion is the application of the basic sciences to generate, assess, and estimate the effectiveness of methods for achieving controlled motion of a vehicle. There are many abstract implications to these definitions; one is that motion is the result of human intervention, not free fall or drifting.

Propulsion, considering the alternative definition, has several points of view. The objective may be to establish the principles that underlie the design, application, and operation of the devices, to organize information about the host of details that the practicing propulsion engineer must have at call, or to present those significant data that permit the organization of a feasible propulsion system for a vehicle. The objective here is to present system data.

A general analysis of the power-plant system does not reduce to simple mathematical terms; an individual study is required for each type of vehicle. The important vehicle-performance parameters influenced by power-plant performance are the range X, the time duration of power-plant operation t_b, altitude h, and the mission velocity or characteristic velocity of the vehicle ΔV.

In general, the vehicle performance improves with decrease in specific fuel consumption, drag coefficient, and vehicle velocity. The performance improves with increases in the propeller or transmission efficiency η and the ratio of the lift coefficient to drag coefficient C_L/C_D. The wing loading W/A_w and specific power of the vehicle P/W may improve or diminish the performance, depending upon what is wanted.

20-2. Land Vehicles

For terrestrial vehicles, the changes in gross weights of the vehicle system are used to establish the factors that control the propulsion-system performance.[1] The basic relationships are for the rate of change of the gross weight of the vehicles in horizontal motion

$$\frac{dW}{dX} = \frac{w_f D}{375\eta} \qquad \text{lb/mile} \tag{20-1}$$

and the time rate of change of the gross weight

$$\frac{dW}{dt} = \frac{w_f DV}{375} \qquad \text{lb/hr} \tag{20-2}$$

The parameters of the integrated equations are

P = brake horsepower (bhp) for a given speed V
V = speed, mph
T = thrust = D = drag, lb
W = gross weight = lift = displacement
w_f = specific fuel consumption, lb fuel/bhp-hr
η = propeller or transmission efficiency
X = range, miles

The drag of automotive and rail vehicles on level roads[2,3] includes (1) rolling resistance, proportional to the vehicle gross weight W, and (2) air resistance, proportional to the vehicle speed V^2. Grade resistance is a factor on hills but is not considered here inasmuch as overall performance yardsticks for the power-plant systems are desired. The total drag D is expressed in terms of the rolling resistance per unit weight D_1 and the air resistance per unit frontal area D_2. Thus the total drag or resistance in pounds is $D = D_1 W + D_2 A V^2$.

The rolling resistance per unit vehicle weight depends upon road-surface characteristics, material, type, and condition of tires, and wheel-bearing friction. For automotive vehicles, the range of rolling friction is between 0.010 and 0.030 lb/lb gross weight on smooth roads. The upper limit can approach 0.10 to 0.20 lb/lb on rough roads.

The distance an automobile will travel under power is

$$X = \frac{375\eta}{D_1 w_f} \log_e \frac{D_1 W_1 + D_2 A V^2}{D_1 W_2 + D_2 A V^2} \qquad \text{miles} \tag{20-3}$$

When the weight of fuel carried is less than 5 per cent of the total vehicle weight,

$$X = \frac{375\eta}{w_f} \frac{W_f}{D_1 W_2 + D_2 A V^2} \qquad \text{miles} \tag{20-4}$$

PROPULSION

where η = efficiency of the power-transmission system
W_f = weight of fuel carried
W_1 = vehicle gross weight, including fuel
W_2 = dry weight of the vehicle
D_1 = rolling resistance, lb/lb gross weight
D_2 = air resistance, lb in./sq ft

The resistance factors for each axle of rail cars[2,3] are of the form

$$D_i = 1.3 + \frac{29}{w_i} + B_i V + \frac{C_i A_i V^2}{w_i n_i} \qquad (20\text{-}5)$$

where D_i = resistance from each axle, lb/ton
w_i = average load per axle, ton
n_i = number of axles of ith car
B_i, C_i = constants characteristic of equipment type

Typical values of B_i and C_i are tabulated for standard types of equipment in Table 20-1. Streamlined and lightweight units require modification.

Table 20-1. Constant Values for Standard Types of Vehicles

Equipment	B_i	C_i
Locomotive	0.03	0.0024
Freight cars	0.045	0.0005
Passenger cars	0.03	0.0004
Motor cars	0.09	0.0024

The drag D of a complete train is

$$D = \sum_{L}^{C} [(1.3 + B_i V) w_i n_i + 29 n_i + V^2 A_i C_i] \qquad (20\text{-}6)$$

in which the summation includes all the cars in the train, from the locomotive to the caboose.

The first two terms in (20-5) are obtained empirically from coasting and dynamometer tests on standard equipment, and are almost entirely from journal friction. These terms are a function of operating temperature. The values of the coefficients used in Table 20-1 are for normal temperatures. The third term is from flange friction and other factors proportional to speed. The last term is from air resistance. The coefficient C_i of V^2 in the last term should be modified to use the Totten coefficients when studying streamlined trains (for example, the diesel-electric streamliners such as the Santa Fe Super Chief).

The range or distance X the rail train will travel under its own power on tangent, level track is

$$X = \frac{375\eta}{w_f(1.03 + 0.03V)} \log_e \frac{D_1}{D_2} \quad \text{miles} \qquad (20\text{-}7)$$

where D_1 = drag of the train fully loaded with fuel
D_2 = drag of the train after using the fuel to travel X miles

When the ratio of fuel used to gross weight of the train is less than 1:20, the range becomes

$$X = \frac{375}{w_f} \eta \frac{W_f}{D_2} \quad \text{miles} \qquad (20\text{-}8)$$

where W_f = weight, tons, of fuel burned while traveling the X miles

20-3. Marine Vehicles

A marine propulsion system probably is governed by more variables than any other vehicle-propulsion system.[2,5,6] No attempt is made to generalize here, and if an accurate estimate of a ship's performance is required, it should be made by methods given in the references. Only a few of the vehicle-system variables are indicated here.

The performance of a completely submerged vessel is probably the most easily estimated. The drag is a function of the displacement, speed of the ship, a skin-friction term, and its wave resistance. The range of constant speed thus is proportional to the (speed)^2 and the logarithm of the mass ratio, similar in many respects to the function describing the range of aircraft.

The performance of a surface ship is complicated by the wave resistance; trim, roll, and pitch of the ship; and wind resistance. The power requirements of a displacement vessel can be estimated roughly from the Admiralty coefficient:

$$\text{Admiralty coefficient} = \frac{(\text{displacement})^{2/3}(\text{speed})^3}{\text{bhp}} \tag{20-9}$$

This coefficient is a function of the design speed, proportions of the ship, speed-to-length ratios, etc. It exhibits a consistency in value, ranging from 370 to 380 for single-screw vessels, and from 360 to 390 for twin-screw ships. Reliability is a prime factor in a marine system, as in aircraft. Reduction of weight per horsepower is accomplished by increased power-plant speeds but is limited by lowered propeller efficiencies, increased power requirements, fuel consumption, and dynamic effects which may affect other gains.

The performances of hydrofoils and vessels with planing hulls are specialized developments introducing variables different from displacement vessels and are not considered here.

In summarizing the marine systems, the important power-plant system variables are the shaft, or brake, horsepower; specific power (bhp/displacement); overall weight; mass ratio; and reliability.

20-4. Airborne Vehicles

Aircraft must satisfy three major performance criteria: (1) the rate of climb R_c, (2) the absolute ceiling H, and (3) the range X. The major power-plant parameters influencing these performance parameters are the net power P, the specific power of the vehicle $\eta P/W$ and the mass ratio W_1/W_2. The propeller efficiency η is a function of the parameter V/nD in which V is the flight speed, n is the propeller speed, and D is the propeller diameter.[1,7]

The rate of climb of an aircraft at sea level depends upon the reserve horsepower available:

$$R_c = \frac{\text{reserve hp} \times 33{,}000}{W} \quad \text{ft/min} \tag{20-10}$$

in which

Reserve hp = thrust hp less drag power losses

In terms of the equivalent monoplane span b, the flat-plate frontal area of the aircraft, and the specific power P/W,

$$R_c = \frac{33{,}000\eta P}{W} - \frac{11{,}000 W}{b^2 V} - \frac{0.288 A V^3}{W} \quad \text{ft/min} \tag{20-11}$$

Specific power of the vehicle is of prime importance here. The effect of high speed is not great, because the lift-to-drag ratio increases with speed and tends to reduce the effect of the last term.

PROPULSION

The maximum altitude, or ceiling, of an aircraft is estimated from the computed density ratio. Then the altitude is determined from a standard atmospheric-density table (Tables 6-3 and 6-4; see also Ref. 8).

$$\frac{\rho_0}{\rho} = 358 C_L \frac{A}{W} \left(\frac{L}{D}\right)^2 \frac{\eta' P'}{W} \tag{20-12}$$

where ρ_0 = sea-level density
ρ = altitude density
$\eta' P'$ = thrust horsepower available at altitude

This ratio should be as large as possible for altitude performance and is achieved by supercharging internal-combustion engines or high-performance compressors in gas turbines.

The range X of propeller-driven aircraft can be estimated from Bréguet's formula.[1,9]

$$X = 375 \frac{\eta}{w_f} \frac{C_L}{C_D} \log_e \frac{W_1}{W_2} \quad \text{miles} \tag{20-13}$$

in which η is the efficiency of the propeller, C_L is the lift coefficient, and C_D is the drag coefficient. The propeller efficiency η^{10} is a function of the parameter V/nD. V is the air speed, n is the propeller speed of rotation, and D is the propeller diameter. For a given flight speed, the product of the propeller efficiency and the lift-to-drag ratio is inversely proportional to the specific power of the aircraft. High maneuverability requires high specific power, and good range performance requires low specific power as long as the propeller efficiency remains high.

The helicopter obeys the same relationships for rate of climb, ceiling, and range. However, the efficiency is no longer that of the propeller alone, but is the product of propeller efficiency and transmission efficiency. Furthermore, typical lift-to-drag ratios of helicopters vary from 3 to 10, while the same ratio varies from 12 to 22 for contemporary winged aircraft. The net result is that the performance of the helicopter is substantially less than that of conventional aircraft, and a high-performance power-plant system contributes toward maintaining the performance of this type of aircraft.

The rate of climb of a jet aircraft depends upon the thrust-to-weight ratio of the aircraft, its airspeed, and the lift-to-drag ratio:

$$R_c = 1.467 \left(\frac{F}{W} - \frac{C_D}{C_L}\right) V \quad \text{ft/min} \tag{20-14}$$

The density ratio from which the ceiling of jet aircraft can be obtained from standard atmospheric-density tables is

$$\frac{\rho_0}{\rho} = \frac{C_L}{393} \frac{A}{W} \left(\frac{C_L}{C_D}\right)^2 \left(\frac{FV}{W}\right)^2 \tag{20-15}$$

in which A is the wing area and V is the maximum speed associated with the altitude lift coefficient and lift-to-drag ratio of the airplane.

The range X of the jet airplane can be estimated from an approach similar to that used by Bréguet. However, the specific fuel consumption w_f' becomes the ratio of the fuel rate per unit thrust.

$$X = \frac{V}{w_f'} \frac{C_L}{C_D} \log_e \frac{W_1}{W_2} \quad \text{miles} \tag{20-16}$$

$$w_f' = \frac{\text{lb fuel}}{(\text{lb thrust}) \text{ hr}} \tag{20-17}$$

The performance parameters for range, climb, and ceiling of rocket-powered winged aircraft are precisely the same as those for jet-powered vehicles [Eqs. (20-14) through (20-17)]. However, the specific fuel consumption w_f' becomes the specific propellant consumption, i.e., the total propellant flow in pounds of fuel and oxidizer per pounds thrust hour. The specific propellant consumption, expressed in the units of pounds and hours, is inversely proportional to the specific impulse I_s. In units of mile, pound, and hour, the specific propellant consumption is

$$w' = \frac{\dot{W}_0 + \dot{W}_f}{F} = \frac{3,600}{I_s} \quad \text{lb/lb-hr} \qquad (20\text{-}18)$$

in which the specific impulse is given in units of pounds and seconds. Frequently, the air specific impulse I_{sa} is used in assessing the performance of jet engines. This is directly related to the specific impulse based upon total propellant consumption and is derived by the equation

$$I_{sa} = I_s \left(1 + \frac{\dot{W}_f}{\dot{W}_0}\right) = \frac{3,600}{w_a'} \left(1 + \frac{\dot{W}_f}{\dot{W}_0}\right) \quad \text{sec} \qquad (20\text{-}19)$$

in which the ratio $\dot{W}_f/\dot{W}_0$ is the fuel-to-air ratio of the jet engine. w_a' is the specific air consumption. Fuel specific impulse can be expressed in a similar way.

Other aircraft-performance parameters that play a significant role are the cruising speed, which is established by the most efficient operation of the power plant and propeller; the high speed, which varies as the cube root of the specific power of the vehicle; the speed range, which also varies as the cube root of the specific power; and the endurance, which is the ratio of the range X to the average flight speed V.

20-5. Space Vehicles

Rocket-powered vehicles under steady motion obey the same basic laws of motion as do submersibles and aircraft.[11] The external forces acting are the rocket thrust, the aerodynamic and control forces, gravitational forces, centrifugal forces, and Coriolis forces. These forces vary greatly in magnitude, depending upon the ambient environmental conditions. Inherently, the rocket vehicle has very high terminal and cruise speeds, whether it is an aircraft, missile, projectile, space ship, or satellite. During the period of operation, or burning time, of the power plant, the motion is dynamic. Since many variables, including control forces, stability, lift coefficient, drag coefficient, altitude, flight angle, burning time, thrust level, and the environment, are usually different for each vehicle, no general solution can be given.

The velocity and altitude reached by a vertically ascending, rocket-powered vehicle are obtained from the basic equations of motion. These parameters indicate the significant performance factors to be looked for in rocket vehicles operating in the neighborhood of the earth. The velocity of the vehicle at burnout, or cut-off velocity V_c, is expressed by the equation

$$V_c = g I_s \log_e \frac{W_1}{W_2} - g t_c - \frac{B_1 C_D A}{W_1} + V_0 \qquad (20\text{-}20)$$

where g = acceleration of gravity, 32.2 ft/sec^2
t_c = time of end of burning, sec
A = equivalent flat-plate cross section of vehicle, sq ft
V_0 = initial velocity, ft/sec
$B_1 = \int_0^{t_c} \frac{\frac{1}{2}r^2}{1 - W_p t / W_1 t_c} dt$
I_s = specific impulse, sec
$W_p = W_0 + W_f$ = total propellant burned, lb

PROPULSION

In free space, where the gravitational losses are small and there is no atmosphere, the negative terms vanish, and an important performance parameter is generated,

$$\Delta V = V_c - V_0 = gI_s \log_e \frac{W_1}{W_2} \tag{20-21}$$

which is defined as the characteristic, or mission, velocity.

The height at end of burning is

$$h_c = gt_c I_s \left(1 - \frac{W_p}{W_2} \log_e \frac{W_1}{W_2}\right) - \tfrac{1}{2} gt_c^2 + V_0 t_c + h_0 - \frac{B_2 C_D A}{W_1} \tag{20-22}$$

where $W_p = W_0 + W_f$ = total weight of propellants, lb

The maximum altitude reached is

$$h = h_c + \frac{V_c^2}{2g} \quad \text{ft} \tag{20-23}$$

The kinetic energy of the vehicle at the end of burning is converted into potential energy; i.e., the vehicle coasts to its ceiling. The rocket power plant must generate sufficient thrust to make the initial thrust-to-weight ratio of the vehicle greater than unity. Power-plant operation in which time of burning is small in comparison with the flight time required to reach a desired range is defined as ballistic operation.

Space flight of a vehicle requires that the vehicle expend energy and undergo large changes in velocity to accomplish a mission. The energies and velocity changes are direct functions of the time of flight, thrust-to-weight ratio of the vehicle, drag losses, etc. The characteristic velocities required for various missions are summarized in Table 20-2. These are estimated from mission analyses[12-14] of ballistic rockets and

Table 20-2. Approximate Velocities for Typical Interplanetary Missions

Mission	Velocity, ft/sec Ballistic vehicle	Velocity, ft/sec Low-thrust vehicle*
Earth escape	37,000	25,000–27,000
Satellite orbit around earth	26,000–35,000	Not applicable
Earth to moon (no return)	35,000–37,000	25,000–36,000
Earth to Mars:		
No recovery	41,000–45,000	80,000–100,000
Return	53,000–65,000	
Earth to Venus:		
No recovery	43,000–44,000	80,000–100,000
Return	67,000–87,000	
Earth to Jupiter	80,000–90,000	

* From satellite orbit around the earth.

low-thrust rocket engines (such as ion rockets) that can generate a thrust-to-weight ratio less than 10^{-4} from power-supply systems that can be built within the limitations of current state of the art. The larger characteristic velocities required by the low-thrust-to-weight-ratio vehicles result from gravity losses in flight.

The performance regimes in which different vehicles work are quite different. The specific jet power, specific impulse, and thrust-to-weight ratio of the vehicles are functionally related by

$$\frac{P}{W} = 0.0292 G I_s \quad \text{hp/lb} \tag{20-24}$$

in which G is the thrust-to-weight ratio of the vehicle.

The range of variation for the chemical rocket, the aircraft, and the low-thrust-to-weight-ratio electrical propulsion systems are summarized in Fig. 20-1. The specific jet power for modern aircraft varies from 0.05 hp/lb for vehicles with propeller-driven internal-combustion engines to little more than 0.10 hp/lb for high-performance jet aircraft. The specific impulse of these engines seldom exceeds 100 sec.

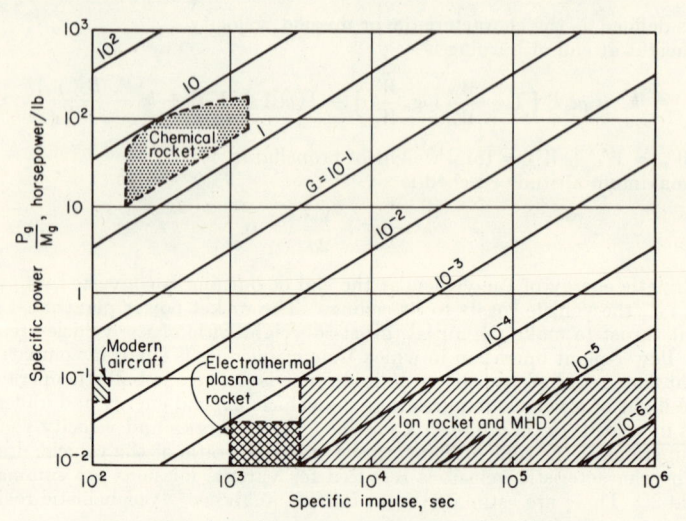

Fig. 20-1. Specific power requirements of vehicles.

Booster-rocket engines operate in the thrust-to-weight-ratio regime above unity to lift the vehicle from the ground. The terminal thrust-to-weight ratio near end of burning can exceed 10:1 in light low-payload vehicles.

Electrically powered vehicles are limited by current state of the art to maximum specific powers of approximately 0.10 jet hp/lb.

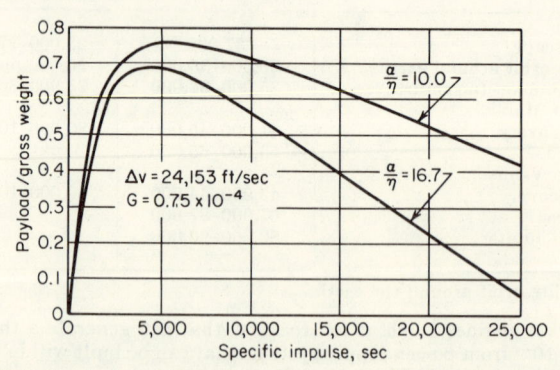

Fig. 20-2. Typical variation of payload vs. specific impulse, ion rocket.

As the specific impulse increases, the thrust-to-weight ratios must steadily decrease. This enforced low thrust-to-weight ratio results in an optimum specific impulse to obtain maximum payload for each mission. A typical estimate for a lunar vehicle using a low-thrust power plant is shown in Fig. 20-2.

The thrust and specific impulse of low-thrust rockets require programming to achieve the best payload and shortest flight time.[14,15] Because the propellant flow rate varies as I_s^{-2}, a mass flow variation of 100 is required when the specific impulse varies

PROPULSION   20-9

by a factor of 10. The estimated variations of specific impulse I_s, thrust-to-weight ratio G, mass ratio W_2/W_1, and trip times to Mars and return are summarized in Table 20-3.[14]

Table 20-3. Performance-parameter Variations for Round Trips to Mars

Trip time, days	Specific impulse, sec	Thrust/weight ratio	Mass ratio
500	2,500–37,000	0–3.5 × 10⁻⁴	1–0.23
800	5,000–38,000	0–2.4 × 10⁻⁴	1–0.45

Specific power = 0.10 Kw/lb

$$\frac{\text{Power-plant weight}}{\text{Vehicle weight}} = 0.20$$

20-6. Internal-combustion-engine Systems

Internal-combustion engines are arbitrarily divided into two categories, reciprocating systems[7] and gas turbines.[9] There are other specialized systems, but they are little used because of low efficiency, complexity, etc. The details of the systems are different, but since the principles are the same they are not considered further. The thermal efficiencies of these power plants are governed by the principles of the Carnot cycle (Sec. 21-2).

Reciprocating Engines. Four types of engines are available for application to different reciprocating-engine systems:

1. Liquid fuel (gasoline)/gaseous oxidizer (air), spark ignition
2. Gaseous fuel (H_2)/gaseous oxidizer (O_2), spark ignition
3. Liquid fuel (diesel fuel)/gaseous oxidizer (air), compression ignition
4. Liquid fuel/liquid oxidizer

The power output of a given engine cylinder is determined by six primary items:

1. The heating value Q_f of the fuel supplied to the engine, Btu/lb
2. The weight of fuel W_f supplied to the engine, lb/cycle
3. The mass of air (or fuel/oxidizer mixture) which is introduced into the cylinder during a complete cycle of the engine W_0, lb/cycle
4. The number of cycles per unit of time, N/min
5. The efficiency η_i with which the energy entering and released is transformed to work at the piston face
6. The efficiency η_m with which the power delivered to the piston face is delivered to the engine drive shaft

The brake horsepower is determined by the equation

$$\text{bhp} = J\frac{[Q_f(W_f/W_0)]W_0 N \eta_i \eta_m}{33,000} \quad \text{hp} \tag{20-25}$$

in which J = mechanical equivalent of heat
     = 778 ft-lb/Btu

The weight of air displaced per cycle is directly proportional to the volumetric efficiency η_v

$W_0 = \rho L A \eta_v$, lb/cycle
 ρ = air density, lb/cu ft
 L = piston stroke, ft
 A = piston area, sq ft
 η_v = volumetric efficiency
    $= \dfrac{\text{volume of air displaced (cu ft)/cycle}}{LA}$

The net thermal efficiency of the internal-combustion engine is the ratio of the brake horsepower produced to the power release of the fuel consumed.

$$\text{Net thermal efficiency} = \frac{2.544 \times 10^5}{w_f Q_f} \quad \text{per cent} \quad (20\text{-}26)$$

The relative efficiency of the engine is the ratio of the efficiency actually obtained to the air standard efficiency.

$$\text{Relative efficiency} = \frac{\text{net thermal efficiency}}{\text{air standard efficiency}} \times 100 \text{ per cent} \quad (20\text{-}27)$$

The air standard efficiency is the thermal efficiency in which the working medium is assumed to be air, and it has a constant specific heat throughout the cycle.

For reciprocating engines, the thermal efficiency is

$$\eta = 1 - \left(\frac{1}{r}\right)^{\gamma-1} \quad (20\text{-}28)$$

in which r is the compression ratio, the ratio of the maximum volume of the engine cylinder to the minimum, and γ is the average specific heat of the combusted mixture in the engine cylinder.

Maximum performance is achieved when there is a maximum flow of fuel and oxidizer through the system, i.e., a high volumetric efficiency, the ratio of the actual weight of air swept through the engine to the weight of air in the displacement volume. To achieve maximum power, the fuel-oxidizer ratio has a well-defined value for each fuel/oxidizer combination used; therefore, the volumetric efficiency is a close measure of the overall performance of the reciprocating engine when operating at the best fuel-air ratio.

For both spark- and compression-ignition engines, a high compression ratio is advantageous because of high thermal efficiency. The tendency of fuels to detonate, and ultimately preignite, places an upper limit on the ratio which can be employed. The optimum ratio for a particular engine depends upon design details and the amount of supercharging.

In addition to the brake horsepower, it has become customary to evaluate the performance of an engine on the basis of specific fuel consumption (sfc), indicated horsepower (ihp), the indicated mean effective pressure (imep), and the brake mean effective pressure (bmep). The brake horsepower can be measured directly on a torque dynamometer. The brake mean effective pressure is calculated from it:

$$\text{bmep} = \frac{229(\text{bhp})}{LAN} \quad \text{psi} \quad (20\text{-}29)$$

The indicated horsepower is the sum of the brake horsepower and the friction, or motoring, losses:

$$\text{ihp} = \frac{\text{bhp}}{\eta_m} \quad (20\text{-}30)$$

The indicated mean effective pressure, which gives a closer measure of the maximum pressures in the cylinder, is

$$\text{imep} = \frac{\text{bmep}}{\eta_m} \quad (20\text{-}31)$$

Comparison of engine cylinder-pressure records and imep tests indicates that the maximum pressures reached in internal-combustion-engine cylinders are 4.5 to 6 times the imep, depending upon the compression ratio and operating conditions.

PROPULSION

Data on the experimental performance of an internal-combustion engine (ihp, imep, bhp, bmep, and sfc) are usually presented as a function of the engine speed. Figure 20-3 is a typical presentation.

There are no basic differences in the pressure records, or indicator cards, of Otto-cycle and diesel-cycle engines. Because the Otto-cycle engine is spark-ignited, it can operate at lower compression ratios and pressures than the diesel cycle, which requires compression ignition of the fuel. The net result is that the diesel engine is substantially heavier than the gasoline engine because of increased dynamic stresses and, from a system standpoint, can be less desirable because of its lower specific power.

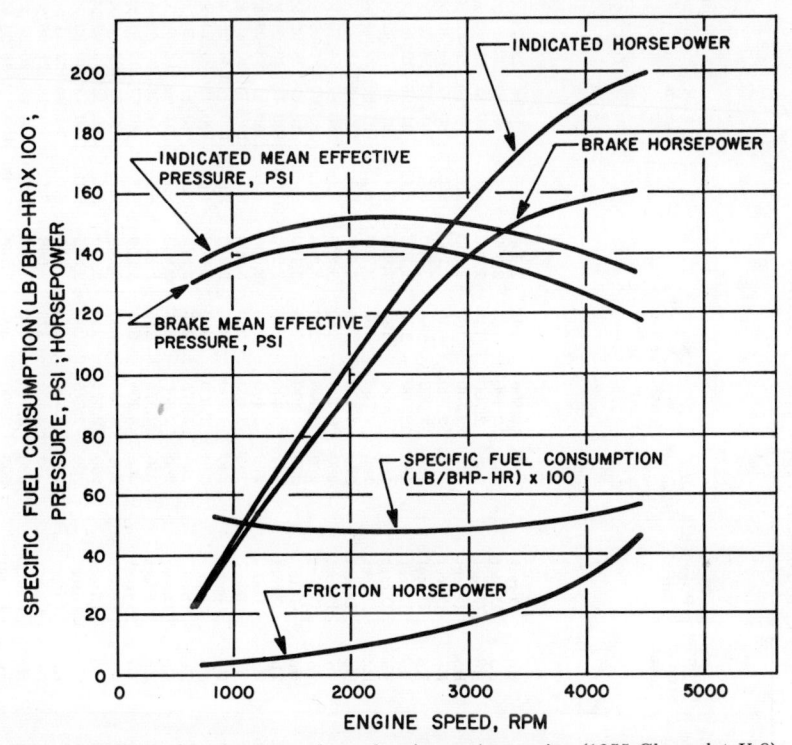

Fig. 20-3. Typical performance data of reciprocating engine (1955 Chevrolet V-8).

In practice it is found that the ideal efficiency of the compression-ignition, or diesel engine, is higher than that of the spark-ignition engine, but a maximum of approximately 75 per cent of the fuel is burned in the diesel cycle, so that the full calorific value of the fuel is not achieved.

The power-plant system for a flight vehicle includes the engine, gearing, and propeller.[7,9] Both the engine and propeller performance are affected by the ambient operating conditions. Because of the wide variation in engine systems and ambient conditions, the performance must be considered for each specific case. In general, the power output varies directly with the air density and inversely with the square root of the absolute temperature. Because the engine power required is constant if the airspeed is constant, the following conclusions hold:

1. The fuel consumption per mile of a given vehicle depends only upon the indicated airspeed and, at a specific indicated airspeed, is independent of altitude.

2. The minimum fuel consumption, which occurs at the indicated airspeed for which the power requirement is least, is independent of altitude except insofar as the specific fuel consumption w_f may vary with altitude.

Table 20-4. Reciprocating Engines*

Manufacturer, address, and designation	No. of cylinders	Cylinder arrangement	Propeller drive	Max take-off power at S.L., hp	Power ratings — At rpm	Normal rated power, hp	At rpm	At altitude, ft	Fuel grade	Diam or dimensions, in. (without cowling)	Blower ratio	Dry weight, lb
Avco Corp., Lycoming Div., Williamsport, Pa.:												
O-235-C1B	4	Ho	Direct	115	2,800	108	2,600	S.L.	80	22 × 53 × 32	...	240
O-290-D2C	4	Ho	Direct	140	2,800	135	2,600	S.L.	80/87	22.8 × 32.24	...	264
O-320-A2C	4	Ho	Direct	150	2,700	150	2,700	S.L.	80/87	23.12 × 32.24	...	271
O-320-B2C	4	Ho	Direct	160	2,700	160	2,700	S.L.	91/96	23.12 × 32.24	...	277
IO-320-A1A	4	Ho	Direct	150	2,700	150	2,700	S.L.	80/87	19.30 × 32.24	...	279
O-360-A1D	4	Ho	Direct	180	2,700	180	2,700	S.L.	91/96	24.72 × 33.37	...	284
HO-360-B1A	4	Ho	Direct	180	2,900	180	2,900	S.L.	91/96	19.68 × 33.37	...	288
VO-360-A1B	6	Vo	Direct	180	2,900	180	2,900	S.L.	91/96	21.42 × 33.37	...	298
IO-360-B1A	4	Ho	Direct	180	2,700	180	2,700	S.L.	91/96	22.47 × 33.37	...	291
IMO-360-B1B	4	Ho	Direct	225	3,400	225	3,400	S.L.	100/130	17.47 × 33.37	...	274
VO-435-A1E	6	Vo	Direct	260	3,400	250	3,200	S.L.	80/87	34.73 × 33.58	...	391
VO-435-A1F	6	Vo	Direct	260	3,400	250	3,200	S.L.	80/87	24.13 × 33.58	...	399
GO-435-C2B2-6	6	Ho	Gear	260	3,400	245	3,100	S.L.	80/87	28.02 × 33.12	...	430
TVO-435-A1A	6	Vo	Direct	260	3,200	220	3,200	20,000	100/130	36.51 × 33.58	...	468
GO-480-B1D	6	Ho	Gear	270	3,400	260	3,000	S.L.	80/87	28.02 × 33.12	...	432
GO-480-G1F6	6	Ho	Gear	295	3,400	280	3,000	S.L.	100/130	27.46 × 33.12	...	437
GSO-480-B1B6	6	Ho	Gear	340	3,400	320	3,200	8,000	100/130	33.26 × 33.12	...	500
IGSO-480-A1B6	6	Ho	Gear	340	3,400	320	3,200	11,500	100/130	23.29 × 33.12	...	497
O-540-B1A5	6	Ho	Direct	235	2,575	235	2,575	S.L.	80/87	24.56 × 33.37	...	395
O-540-F1B5	6	Ho	Direct	260	2,800	235	2,800	S.L.	91/96	24.56 × 33.37	...	398
O-540-A1D5	6	Ho	Direct	250	2,575	250	2,575	S.L.	91/96	24.56 × 38.42	...	397
VO-540-B1B	6	Vo	Direct	305	3,200	305	3,200	S.L.	80/87	24.57 × 34.14	...	429
VO-540-C1A	6	Vo	Direct	305	2,575	305	2,575	S.L.	100/130	24.58 × 34.70	...	439
IO-540-A1A5	6	Ho	Direct	290	2,575	290	2,575	S.L.	100/130	19.92 × 34.25	...	441
IO-540-B1A5	6	Ho	Direct	290	2,575	290	2,575	S.L.	100/130	19.60 × 34.25	...	437
IO-540-C1A5	6	Ho	Direct	250	2,575	250	2,575	S.L.	91/96	22.21 × 33.37	...	403
IGO-540-B1A	6	Ho	Gear	350	3,400	325	3,000	S.L.	100/130	21.66 × 34.25	...	504
IGSO-540-B1A	6	Ho	Gear	380	3,400	300	3,200	11,500	100/130	20.29 × 34.25	...	532
TVO-540-A1A	6	Vo	Direct	305	3,200	305	3,200	17,000	100/130	36.00 × 34.70	...	499
IO-720-A1A	8	Ho	Direct	400	2,650	380	2,650	S.L.	100/130	46.7 × 34.2	...	609

20-12

Engine	Cyl.									
Continental Motors Corp., Muskegon, Mich.:										
A-65-BF	4	Ho	Direct	65	2,300	65	2,300	S.L.	80/87		170
C-85-12F	4	Ho	Direct	85	2,575	85	2,575	S.L.	80/87		184
C-90-12F	4	Ho	Direct	90	2,475	90	2,475	S.L.	80/87		188
E-185-9	6	Ho	Direct	205	2,600	185	2,300	S.L.	80/87	31.0 × 31.56	352
O-200-A	4	Ho	Direct	100	2,750	100	2,750	S.L.	80/87	31.28 × 31.56	190
E-225-4	6	Ho	Direct	225	2,650	185	2,300	S.L.	80/87	31.25 × 31.56	355
O-300-A,-B,-C,-D	6	Ho	Direct	145	2,700	145	2,700	S.L.	80/87	47.62 × 33.38	268
GO-300-A,-B,-C,-D	6	Ho	G0.750:1	175	3,200	175	3,200	S.L.	80/87	30.53 × 31.56	314
O-470-4	6	Ho	Direct	225	2,600				80/87	47.62 × 33.38	415
O-470-15	6	Ho	Direct	213	2,600				80/87	38.38 × 31.50	405
O-470-J	6	Ho	Direct	225	2,550	225	2,550	S.L.	91/96	39.62 × 31.50	378
O-470-K,-L	6	Ho	Direct	230	2,600	230	2,600	S.L.	80/87	36.03 × 33.32	404
O-470-G,-M,-H†	6	Ho	Direct	240	2,600	240	2,600	S.L.	80/87	36.03 × 33.56	410
IO-470-J,-K	6	Ho	Direct	225	2,600	225	2,600	S.L.	91/96	37.56 × 33.56	309
IO-470-C,-G,-P†,-R	6	Ho	Direct	250	2,600	250	2,600	S.L.	80/87	37.93 × 33.56	432
IO-470-D,-E,-F,-H,-L,-M,-N	6	Ho	Direct	260	2,625	260	2,625	S.L.	91/96	37.93 × 33.56	426
TSIO-470-B	6	Ho	Direct	260	2,600	260	2,600	16,000	100/130	43.53 × 33.56	427§
FSO-526-A†	6	Ho	Direct	270	3,200	270	3,200	7,900	91/96	42.82 × 33.56	553
										50.33 × 36.64 10.04	
Franklin Engine Co., Inc., Syracuse, N.Y.:										
4A4-100-B3	4	Ho	Direct	100	2,550				80		200
6A4-150-B3	6	Ho	Direct	150	2,600				80		278
6A4-165-B3	6	Ho	Direct	165	3,800				80		278
6A64-185-B12	6	Ho	G0.632:1	185	3,100				80		232
6A4-200-C6	6	Ho	Direct	200	3,100				91		307
6V4-200-C32† C33 O335-5,-6	6	Vo	Direct	200	3,100				91		297
6V6-245-B16F† O 425-1	6	Vo	Direct	245	3,275				80		372
6V335-A-,B	6	Vo	Direct	210	3,100				91		295
6VS-335	6	Ho	Direct								
Pratt & Whitney Aircraft Div., United Aircraft Corp. East Hartford, Conn.:										
R2000-D5	14	Rad	G0.500	1,450	2,700	1,200	2,550	6,100	100/130	49.10 7.15	1,585
R2000-2SD13-G	14	Rad	G0.500	1,450	2,700	1,200	2,550	5,000	100/130	49.10 7.15 + 9.52	1,605
R2800-CB3	18	Rad	G0.450:1	2,400	2,800	1,800	2,600	8,500	100/130	52.80 7.29	2,357
R2800-CB4	18	Rad	G0.450:1	2,500	2,800	1,800	2,600	8,500	108/135	52.80 7.29	2,357
R2800-CB16	18	Rad	G0.450:1	2,400	2,800	1,800	2,600	8,500	100/130	52.80 7.29 + 8.58	2,390
R2800-CB17	18	Rad	G0.450:1	2,500	2,800	1,800	2,600	8,500	108/135	52.80 7.29 + 8.58	2,390
R2800-CB5,-54	18	Rad	Direct	2,100	2,700	1,900	2,600	7,000	115/145	52.80 7.29	2,330
R2800-CB15,-99W	18	Rad	G0.450:1	2,500	2,800	1,900	2,600	7,000	115/145	52.80 7.29 + 8.58	2,403

20–13

Table 20-4. Reciprocating Engines* (Continued)

| Manufacturer, address, and designation | No. of cylinders | Cylinder arrangement | Propeller drive | Power ratings ||||| Fuel grade | Diam or dimensions, in. (without cowling) | Blower ratio | Dry weight, lb |
				Max take-off power at S.L., hp	At rpm	Normal rated power, hp	At rpm	At altitude, ft				
Wright Aeronautical Div., Curtis-Wright Corp., Wood-Ridge, N.J.:												
867C9HE1,2 R1820-82-82A	9	Rad	G0.5625:1	1,525	2,800	1,275	2,500	3,500	115/145	55.74	7.21	1,469
878TC18EA1,2 R3350-32W-32WA	18	Rad	G0.4375:1	3,700	2,900	2,800	2,600	4,100	115/145	56.6	6.46	3,560
895C9HE6,7 R1820-84C,-84D	9	Rad	Direct	1,525	2,800	1,275	2,500	3,500	115/145	55.74	3.21	1,419
989C9HE1,2	9	Rad	Direct	1,525	2,800	1,275	2,500	3,500	115/145	55.74	7.21	1,419
990C7BA1	7	Rad	Direct	800	2,600	700	2,400	7,300	91/98	50.45	7.21	1,080
982C9HE1,2	9	Rad	G0.5625:1	1,525	2,800	1,275	2,500	3,500	115/145	54.95	7.21	1,470
998C9HE1,2	9	Rad	Direct	1,525	2,800	1,275	2,500	3,500	115/145	55.75	7.21	1,422
977C9HD3	9	Rad	Direct	1,425	2,700	1,275	2,500	3,300	100/130	54.95	7.21	1,363
988TC18EA6	18	Rad	G0.4375:1	3,400	2,900	2,800	2,600	4,700	115/145	56.59	6.46	3,712

G = geared
Ho = horizontally opposed
Rad = radial
S.L. = sea level
Vo = vertically opposed
* From Ref. 17.
† For helicopter installation.
‡ Extended propeller shaft.
§ Engine has a 42-lb turbocharger that is mounted separately.

PROPULSION   20-15

3. The time required to fly a given distance at any given fuel consumption per mile decreases with altitude in proportion to $\sqrt{\text{altitude density } \rho}$ because the flight speed varies inversely as $\sqrt{\rho}$.

Propellants. The propellants generally associated with internal-combustion engines are gasoline for spark-ignition-engine fuel, diesel fuel for compression-ignition engines, and air as the oxidizer. Industrial use of gaseous fuels such as coal gas, blast-furnace gas, natural gas, and others is fairly common where economy and availability outweigh fuel-storage space and high specific power. The behavior of hydrogen is unique in that it is possible to control the power from no load to full load by regulation of the hydrogen fuel supply alone, up to the stoichiometric mixture ratio. Auxiliary power units, which use cryogenic oxygen and hydrogen[16] or storable liquid rocket propellants (N_2O_4 and a 50-50 mixture of N_2H_4-UDMH), are currently proposed for space vehicles.

Detailed descriptions, performance data, and fuel requirements for reciprocating engines are summarized in Ref. 2 and Table 20-4.

Gas Turbines. A gas turbine is a heat engine in which the available energy of the working fluid is converted into kinetic energy by expansion through a nozzle from which the fluid issues as a jet.[9,18] The jet kinetic energy is partially converted into mechanical energy by directing the jet across contoured blades mounted in the periphery of the turbine wheel, or spool. Rotary motion is given to the turbine wheel by pressure on the turbine blades; this pressure is developed from the change in momentum of the working fluid jets as they pass through the blades.

The performance of this simple, ideal system depends upon the gas flow, the pressure ratio in the working fluid across the wheel, and the turbine inlet temperature. It has been confirmed, both analytically and experimentally, that at a specific gas flow, the following prevail:

1. The power output increases with pressure ratio (for each turbine inlet temperature) to a maximum value, then decreases at a lesser rate with any further pressure-ratio increase.

2. The power output increases markedly as the inlet temperature increases for a specific pressure ratio.

3. The maximum power outputs, as the inlet temperatures are increased, occur at increasing pressure ratios.

The limiting thermal efficiency η_a of this ideal system is

$$\eta_a = 1 - \left(\frac{1}{r}\right)^{(\gamma-1)/\gamma} \tag{20-32}$$

The experimentally determined thermal efficiency is a function of the output brake horsepower, the specific fuel consumption w_f, and the heat release of the fuel Q_f.

$$\text{Thermal efficiency (experimental)} = \frac{2.545 \times 10^{-5}}{Q_f w_f} \tag{20-33}$$

An analysis of the open-cycle, regenerative gas-turbine power-plant system (Fig. 20-4) is based on the assumptions that (1) the working fluid is a perfect gas, (2) there are no changes in the working fluid as the result of fuel combustion in the combustion chambers, and (3) there are no pressure drops. The first two assumptions have compensating errors. An accurate analysis requires that specific-heat variations and specific-heat ratios be considered. The changes referred to in assumption (2) are small because the gas turbine operates at air/fuel ratios of 60 to 70:1, in contrast to the internal-combustion engine value of 13:1 to 14:1.

$$\eta = \eta_m \eta_b \eta_a \frac{\eta_t T_4 - (T_r/\eta_c)^{(\gamma-1)/\gamma}}{y T_4 - Z T_r v^{(\gamma-1)/\gamma}} \tag{20-34}$$

in which $y = 1 - \eta_R(1 - \eta_t \eta_a)$

$$Z = (1 - \eta_R)\left(\frac{1}{r}\frac{\gamma-1}{\gamma} + \frac{\eta_a}{\eta_c}\right)$$

The overall efficiency of the engine system depends upon

1. The ideal cycle efficiency η_a
2. The combustion-chamber efficiency η_b
3. The mechanical efficiency of the turbines, compressor drives, gear train, and propeller η_m
4. The internal efficiency of the turbine η_t
5. The internal efficiency of the compressor η_c
6. The pressure ratio $r = P_2/P_1 = P_5/P_4$
7. The specific-heat ratio γ
8. The inlet temperature to the turbine T_4
9. The inlet temperature to the compressor T_1
10. The effectiveness of the regenerator η_R

$$R = \frac{\text{enthalpy change, cold side}}{\text{enthalpy change, hot side}}$$

Tables for computing the efficiency and specific examples are available in Ref. 9.

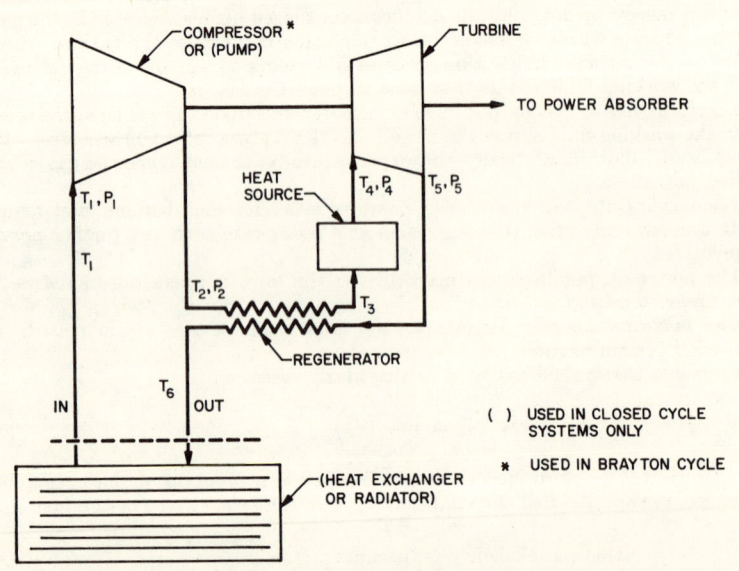

Fig. 20-4. Elements of a regenerative gas-turbine power plant.

The closed-cycle system, which is characterized by the regenerative steam turbine, can operate on the Brayton cycle in which the working fluids are gaseous in all parts of the engine system, or on the Rankine cycle in which the working fluid is condensed in the radiator or heat exchanger. The condensed liquid is pumped back through the system to be partially reheated in the regenerator, which recovers part of the waste heat from the cycle; then the temperature is brought to its working level in the heat source.

Open-cycle gas-turbine systems are applicable in land and marine vehicles and preferable in aircraft applications, because no radiator is required. The systems with no regeneration are characterized by simplicity, light weight, compactness, and low cost; however, their fuel consumption is high at full load and very high under variable- or partial-load operation. The regenerative systems have specific fuel consumptions ranging from 0.40 to 0.80 and can have net thermal efficiencies in the range from 20 to 35 per cent over a wide variation of loads. These systems are more costly and complex but are still relatively compact, lightweight, of high reliability, and easily maintained.

The working fluid of these power-plant systems is the air/fuel combustion product. Fuels are dictated by cost, availability, and the application. They include pulverized coal, kerosene, heavy fuel oil such as Bunker C, or fuel gas. Aircraft gas turbines use the JP series of fuels. These are distillation cuts taken in the manufacturing process to satisfy the operating-temperature requirements of the jet fuel but still produce the highest fraction of gasoline from the crude oil.

The closed-cycle systems show excellent fuel consumption under all conditions. This advantage is offset by the cooling requirements. The system weight includes the working fluid, the radiator or heat exchanger, and the associated piping. A system operating on the all-gas Brayton cycle tends to be heavier than one operating on the condensing Rankine cycle because of higher operating pressures.

The working fluids for the closed-cycle systems are summarized in Table 20-5.

Table 20-5. Working Fluids for Closed-cycle Gas-turbine Systems

Brayton cycle	Rankine cycle
Air	Water
Steam	Alkali metals (Na, Rb, K, Li)
Mercury vapor	Mercury
Helium	Ammonia
Hydrogen	Diphenyl
Other gases	Dowtherm A
	Sulfur
	Aluminum bromide
	Other hydrocarbons

Detailed specifications of many gas-turbine engines are summarized in Refs. 2, 3, and 17. A few selected gas turbines for different applications are summarized in Table 20-6.

Jet Engines. The propeller driven by a reciprocating engine or gas turbine, and the turbojet, ramjet, and pulsejet engines all obey the same physical laws. An idealized analysis[9] suffices to establish the general characteristics of each type of power plant and indicates the situation in which each is best applied to a vehicle system. In the cases of the propeller- and turbojet-powered vehicles, the thrust forces generated are used for take-off and to maintain the rate of climb in the flight path, to maintain the cruising speed of the vehicle by overcoming its drag, and to accelerate the vehicle to a higher velocity. Each action can occur in sequence or concurrently. Ramjet and pulsejet engines must be accelerated initially by external means to a sufficiently high velocity to develop sufficient pressures and air densities within the engine systems to achieve this objective.

Acceleration of an air-mass flow $\dot{W}_0$ from an initial velocity V, the vehicle flight velocity, to the jet velocity c generates a reaction force equal to the product of the mass flow and the velocity increment. This follows from the momentum theorem. If fuel is added in the combustion chamber (Fig. 20-5a), it is accelerated from rest to the exit velocity c and contributes a small amount of thrust, as does the pressure differential $P_j - P_0$ that may exist across the engine system.

The total thrust from the accelerated air and fuel mixture is shown by

$$F = \frac{\dot{W}_0}{g}(c - V) + \frac{\dot{W}_f}{g}c + A_j(P_j - P_0) \qquad (20\text{-}35)$$

In practice, the air/fuel ratio in jet engines is between 50:1 and 70:1, and P_j is little different from P_0. The first term describes the magnitude of the thrust with an accuracy of a few per cent. Therefore, in terms of the air specific impulse $I_{sa} = c/g$, the thrust becomes

$$F \cong \dot{W} I_{sa}\left(1 - \frac{V}{c}\right) \quad \text{lb} \qquad (20\text{-}36)$$

Table 20-6. Gas-turbine Engines*

Manufacturer and address	Military designation	Type	No. compressor stages	No. turbine stages	No. combustors	Max power at S.L.	Specific fuel consumption at max power	Compression ratio at max rpm	Max envelope diam, in.	Max envelope length, in.	Dry weight, less tailpipe, lb	Remarks
Avco Corp. Lycoming Div., Stratford, Conn.	T53-L-3 T55-L-7	AFP ACFS	5,1 7,1	1 2	1 1	960 shp 2,650 shp	0.68 0.61	6.0 6.0	23 24.25	58.40 44.2	524 580	Grumman OV-1 Vertol CH-47A (advanced)
The Boeing Company, Seattle, Wash.	T50-BO-8A T50-BO-10 550	CFS ACFS ACFS	1 1,1 1,1	2 2 2	2 2 2	300 shp 330 shp 440 shp	0.98 0.85 0.77	4.6 4.9 5.3	24 22.5 22.5	40 35.5 37.5	330 238 197	Navy DASH antisubmarine destroyer-carried drone-helicopter program (As above) Turboshaft
Continental Aviation & Engineering Corp., Detroit, Mich.	J69-T-25 J69-T-29 J69-T-39 T65-T-1 217-8B 217-7B 141 356-23	CFJ ACFJ ACFJ ACFS ACFS ACFP CFG ACFJ	1,1 1,1 3 2 1,1 1,1 1 3	1 1 1 3 3 3 1 2	1 1 1 1 1 1 1 1	1,025 lb t 1,700 lb t 2,550 lb t 250 shp 600 shp 623 eshp 191 shp 2,400 lb t	1.14 1.10 0.99 0.71 0.63 0.63 0.98	 6.1 5.4 6.1	24.9 22.3 18.3 19.4 19.4 22.3	50 46.3 44 38.7 42.5 42.5 47	364 335 385 130 220 245 197 410	Cessna T-37B Ryan Q-2C Target missile engine Turboshaft Turboprop For pressure-jet helicopter
Curtiss-Wright Corp., Wright Aeronautical Div., Wood-Ridge, N.J.	J65-W-16A J65-W-18	AFJ AFJ	13 13	2 2	1 1	7,700 lb t 10,500 lb t			37.5 37.5	113 181	2,757 3,485	North American AF-1E Grumman F-11A
Garrett Corp., AiResearch Mfg. Co., Phoenix, Ariz.	GTP30 GTCP85	CFS CFSG	2 2	1 1	1 1	60 shp 345 shp/215 shp	1.46 1.07	3.0 4.0	17.8 25.8	20.8 42.4	67.5 284	Portable generators On-board auxiliary power for B-727 BAC-111, C-141, Caravelle KC-135
General Electric Co., Flight Propulsion Div., Evendale, Ohio	J47-GE-25 J79-GE-2 J93-GE-3 X353-5B CJ805-23	AFJ AFJ AFJ AFLF AFF	12 17 8,1 17	1 3 2,1 3	8 10 1 10	7,200 lb t 16,150 lb t 30,000 lb t 7,430 lb lift 16,050 lb t	1.06 1.97 0.38 0.56	5.4 12.0 1.1 13.0	39.3 38.3 76.0 53	145 208 14.5 149	2,554 3,620 1,145 3,766	Boeing B-47E, water/alcohol injection McDonnell F-4B, North American A-5A For NAA RS-70 Lift-fan for XV-5A Aft-fan engine, Convair 990A
General Electric Co., Flight Propulsion Div.,	T64-GE-6 (was 1) J85-GE-5A	AFS AFJ	14 8	4 2	1 1	2,850 shp 3,850 lb t	0.50 2.20	12.6 6.6	30 22	83 109	713 559	XC-142A; CH-53A Northrop T-38A; Bell X-14A; Afterburner

Lynn, Mass.	CJ610-1	AFJ	8	2	1	2,850 lb t	0.99	6.8	18	41	381	Aero Commander 1121; Lear Jet; C-123H; Piaggio-Douglas PD-208; Commercial J85
General Motors Corp., Allison Div., Indianapolis, Ind.	T56-A-1 501-D13	AFP AFP	14 14	4 4	6 6	3,750 shp 3,750 shp	0.54 0.54	9.2 9.2	38 41	146 146	1,645 1,645	Lockheed C-130A Lockheed Electra
Solar, San Diego, Calif.	T62-S-2 T350 (Mfr. No.)	CFS CFS	1 1	1 1	1 1	80 shp 370 shp	1.10 1.0	3.4 4.1	15.6 26	24.7 38	61 195	Single shaft, constant-speed; APU Single shaft, air bleed, combination; APU
United Aircraft Corp., Pratt & Whitney Aircraft Division, E. Hartford, Conn.	TF33-P-9 J57-P-43 WB J60-P-3	AFF AFJ AFJ	15 16 9	4 3 2	8 8 8	18,000 lb t 13,750 lb t 3,000 lb t	0.54 0.95 0.96	13.0 13.0 6.5	53 38.9 21.9	167 76	4,170 3,870 436	JT3D-3A; Boeing KC-135R JT3C; B-52F & G JT12A-5; North American T-39; Lockheed C-140
	JT3C-6 JT8D-5 JT12A-6	AFJ AFF AFJ	16 13 9	3 4 2	8 9 8	13,500 lb t 12,000 lb t 3,000 lb t	0.78 0.57 0.96	13.0 14.0 6.5	38.8 44 21.9	76	4,234 2,994 436	Boeing 707-120; Douglas DC-8-10 Transport engine Lockheed Jet Star; NAA Sabreliner
Canada: Orenda Engines, Div., Toronto, Ontario	Orenda 10 Orenda 11 Orenda 14 J79-OEL-7 J79-11A J85-Can-40	AFJ AFJ AFJ AFJ AFJ AFJ	10 10 10 17 17 8	1 2 2 3 3 2	6 6 6 10 10 1	6,355 lb t 7,275 lb t 7,275 lb t 15,000 lb t 15,000 lb t 3,000 lb t	1.12 0.99 0.99 …… …… 0.99	5.3 6.1 6.1 12 12 7.1	42 45 45 38.31 38.31 17.7	123 121 121 204 204 40.5	2,515 2,425 2,425 3,200+ 3,200+ 355	CF104, built under license Canadair CT-114 Tudor
Great Britain: Hawker Siddeley Group, Bristol Siddeley Engines, Ltd., Filton	Astazou 2 Double Mamba 8 Nimbus 750 (series) Olympus 2001 Sapphire 209 (7R) Avon RA.29/6 Kk.531	ACFP APP ACFJ Ind AFJ AFJ	1-1 11 2 5,7 13 17	3 3 1 1,1 2 3	1 24 1 8 24 8	523 shp 3,600 shp 930 shp (1 hr) 23,600 lb t 12,230 lb t 12,200 lb t	0.632 0.66 0.68 0.52 1.2 0.724	6.0 5.8 6.0 …… 7.49 10.1	18.1 51.28 …… …… 39	69 103.35 62 312 107.7 134	271 2,500 390 38,000 3,180 3,491	Short Skyvan Gannet Westland Wasp and Scout 15-Mw generator, industrial Limited A/B Javelin
Rolls-Royce, Ltd., Derby.	Conway R. Co. 42 Dart R. Da. 6 Mk. 514	AFF CFPP	7 × 9 2	2 × 1 2	10 7	20,370 lb t 1,710 shp†	0.603 0.690	14.8 5.4	50 38	154 98	5,001 1,114	† Water-methanol restoration.
	Spey R. Sp. 1 Mk. 505	AFF	4 × 12	2 × 2	10	9,850 lb t	0.545	16.8	39	110	2,200	
	Tyne R. Ty. 20 Mk. 22	AFP	6 × 9	3 × 1	10	5,440 shp	0.446	13.5	43.2	108.7	2,218	

20-19

Table 20-6. Gas-turbine Engines* (Continued)

Manufacturer and address	Military designation	Type	No. compressor stages	No. turbine stages	No. combustors	Max power at S.L.	Specific fuel consumption at max power	Compression ratio at max rpm	Max envelope diam, in.	Max envelope length, in.	Dry weight, less tailpipe, lb	Remarks
France:												
Hispano-Suiza, Paris......	H.S. Verdon	CFJ	1	1	9	7,700 lb t	1.10	4.9	50	104	2,072	Alouette 2, Sud. Aviation helicopter
Societe Turbomeca, Paris.	Artouste 2C	CFS	1	2	A	500 eshp	0.87	3.68	1.8	4.72	253	Doenier DO 27-Beech SFERMA
	Astazou 2	ACFS	2	3	A	562 eshp	0.60	6.00	1.5	5.75	271	"Marquiz" Potez 840-Nord 1110-SIPA 251
	Bastan 4	ACFP	2	2	A	1,000 eshp	0.61	5.5	1.8	6.20	468	Morane Saulnier 1500-Sud Aviation "Voltigeur," "Diplomate"-SFERMA PD 185-Beech 18 S-Max-Holste "Super Broussard" GAM Dassault 415, 410
	Turmo 3C3	.....	2	2 + 2	A	1,500 eshp	0.61	5.50	2.15	5.5	414	Sud Aviation "Super Frelon" helicopter
Italy:												
Fiat-SMA, Turin........	4301.0000	CFP	1	1 + 1	A	670 shp	1.11	4.0	22.4	58.1	386	Free turbine moving the reduction gear

NOTE: Type of engine is designated by capital letters. First, or first two, designate compressor; A = axial, C = centrifugal. Next letter, F, stands for flow. Final letter or letters define output; J = jet; P = propeller; F = fan; S = shaft; G = gas; and LF = lift fan. For example, AFLF = axial-flow lift fan.
Abbreviations:
  ahp = air horsepower
  eshp = equivalent shaft horsepower
  lb t = pounds of thrust
  shp = shaft horsepower
* From Ref. 17.

which is a satisfactory formula to use in estimating the net thrust of the propeller and jet-engine systems under all operating conditions.

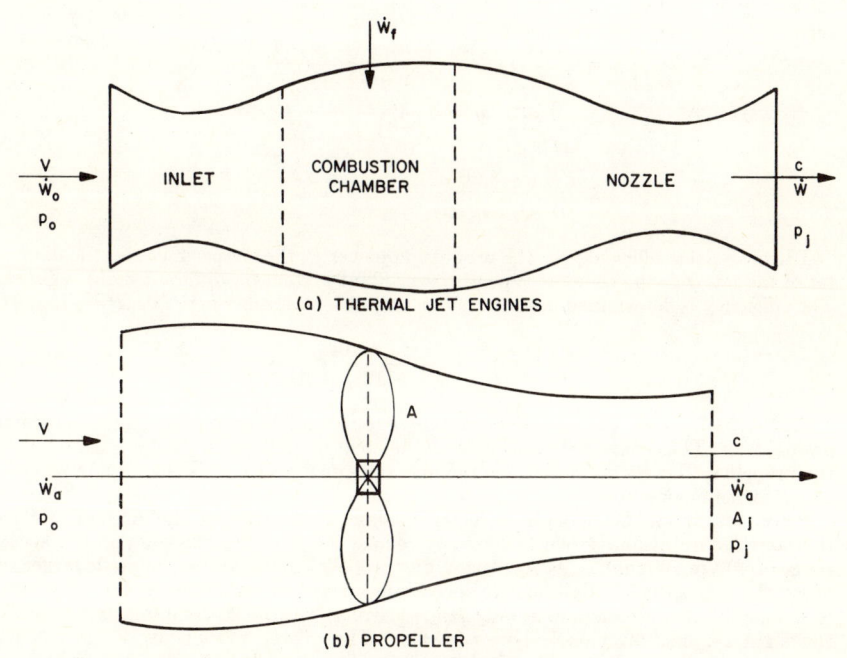

Fig. 20-5. Schematic of propeller and thermal-jet engine systems.

The internal efficiency η_i is a measure of the effectiveness with which the power (mechanical in the case of the propeller drive or thermal in the case of the jet) supplied to the engine system is transformed into propulsion power

$$\eta_i = \frac{P}{J\dot{W}_f Q_f} = \frac{c^2}{2gJ\dot{W}_f Q_f}\left[(\dot{W}_0 + \dot{W}_f)^2 - \dot{W}_f{}^2\left(\frac{V}{c}\right)^2\right]$$
$$\cong \frac{c^2 \dot{W}_0{}^2}{2gJ\dot{W}_f Q_f} \tag{20-37}$$

in which
$$P = \frac{FV}{375} + \frac{1}{750}\frac{\dot{W}_0 + \dot{W}_f}{g}c^2 - \frac{1}{750}\frac{\dot{W}_0 V^2}{g} \quad \text{hp} \tag{20-38}$$

The propulsive efficiency η_p of a jet-powered vehicle is the ratio of the power required by the vehicle and the total power (the sum of the vehicle power and the residual jet power)

$$\eta_p = \frac{2(V/c)}{1 + (V/c)^2} = \frac{2.933 V/gI_s}{1 + (2.933 V/gI_s)^2} \tag{20-39}$$

in which V, the absolute vehicle velocity, and c, the effective jet velocity, are in the same units, and I_s, the specific impulse, is in seconds. As shown in Fig. 20-6, the propulsive efficiency rapidly increases as the jet velocity approaches the vehicle velocity, reaches a maximum when the velocities are equal, and then slowly decreases as the jet velocity exceeds the vehicle velocity.

Comparison of propeller-driven and jet-powered engine systems is made by averaging the flow velocity over the propeller disk. In this way, an average value of the specific impulse can be assigned to a propeller-driven vehicle.

SYSTEM COMPONENTS

The total efficiency η of an air-jet propulsion system is the product of the internal and propulsive efficiencies. It is a measure of the effectiveness with which the chemical energy of the fuel is converted into propulsive power.

$$\eta = \frac{c^2}{gJ\dot{W}_f Q_f} \frac{(\dot{W}_0 + \dot{W}_f)^2 - \dot{W}_f^2 (V/c)^2}{1 + (V/c)^2}$$

$$\cong \frac{\dot{W}_0^2 C^2}{gJ\dot{W}_f Q_f} \left[1 - \left(\frac{V}{c}\right)^2\right] \quad \begin{array}{c} V < C \\ \dot{W}^2 > \dot{W}_f \end{array}$$

$$\cong \frac{F\dot{W}_0 c_\gamma}{j\dot{W}_f Q_f} \frac{3{,}600c}{J\dot{W}_f Q_f} \qquad (20\text{-}40)$$

The propulsion efficiency η_p, the propulsion power P, the airspeed V, and the diameter of the jet D define the best type of power-plant system for different flight regimes. The efficiency is determined in terms of the propulsion parameter $V(D^2\rho/P)^{1/3}$ (Ref. 9).

$$V\left(\frac{D^2\rho}{P}\right)^{1/3} = \frac{1.364\eta_p}{(8 - 12\eta_p + 4\eta_p^2)^{1/3}} \qquad (20\text{-}41)$$

Figure 20-7 shows the variation of the propulsion efficiency with the propulsion parameter. These data are applicable to the ideal propeller and the jet engine. For the propeller, D is its diameter, ρ is the atmospheric density, V is the airplane speed, and P is the engine power.

Power requirements for flight vehicles increase as the cube of the airspeed. Furthermore, compressibility effects seriously reduce the propeller efficiency in the speed range of 400 to 500 mph. At low speed, the propeller diameter can be made larger to offset the low speed. High vehicle speeds are limited to high-altitude flight because of compressibility effects and aerodynamic heating. Table 20-7 summarizes roughly the flight regimes for vehicle power plants. The overall efficiencies of various jet engines and a turboprop engine, the latter "weighted" to allow for excess weight over jet engines, are compared in Fig. 20-8.

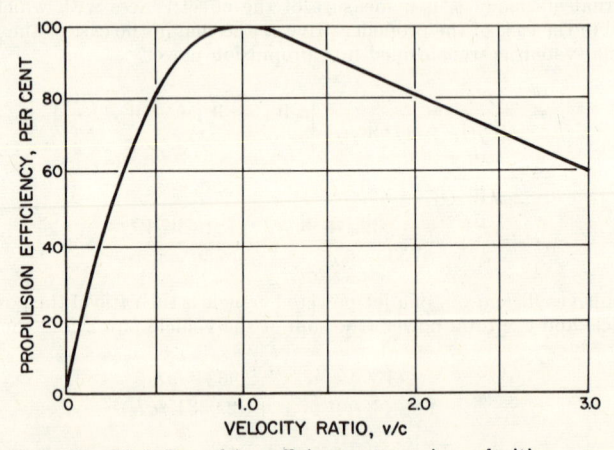

FIG. 20-6. Propulsion efficiency at varying velocities.

The ramjet engine is the simplest of the thermal-jet engines. It cannot produce static thrust and must rely on some other form of propulsion to bring it up to flight speed. At this point, the inlet duct and diffuser must collect and slow down a sufficient quantity of air to produce a pressure high enough for sustained combustion of the fuel. To accomplish this, the fuel is injected upstream of an array of flame holders that mix the fuel and air and supply a sheltered region in which the combustion

PROPULSION 20–23

can be maintained under relatively low velocity conditions. The combustion products then are expanded through a suitably designed nozzle to generate thrust. It is difficult to control the airflow and fuel/air ratio to maintain combustion, because of relatively low pressures and high velocities in the combustion chamber. The external drag of the

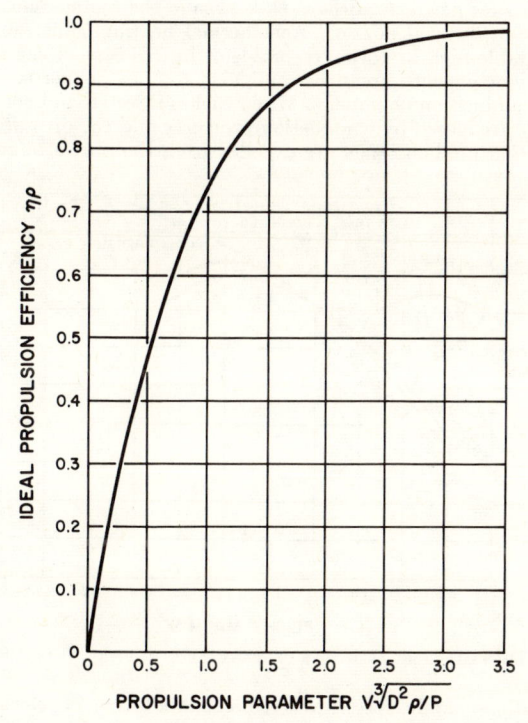

Fig. 20-7. Ideal propulsion efficiency vs. parameter for the ideal engine system.

engine must be considered carefully along with the specific impulse. The specific impulse tends to be less than 100 sec and shows a maximum at a specific flight speed. The range of specific fuel consumption is from 2 to 3 lb/hr-lb thrust for best operating conditions.

Table 20-7. Flight Regimes for Vehicle Power Plants

Engine type	Speed range, mph	Altitude ceiling, ft
Reciprocating, unsupercharged	0–250	15,000
Reciprocating, supercharged	150–450	30,000
Turboprop	150–600	45,000
Pulsejet	300–600	10,000
Turbojet	359 to Mach 3.5	60,000
Ramjet	Mach 1–5	45,000–150,000
Supersonic ramjet	Mach 5–25	50,000–150,000

In the nuclear ramjet, heat exchange from a nuclear reactor supplies energy to the air flowing through the duct. Because of the minimum critical mass of the reactor, and heat-exchanger and materials problems, this engine system is limited to large vehicles. Interest in the nuclear ramjet stems primarily from the potentially large supply of energy available from the nuclear-fission reaction, 1 g of fissioned uranium-235 producing 1 Mw-day of energy.

Recent work has been directed toward the supersonic and hypersonic ramjets in which combustion takes place at supersonic velocities external to aerodynamically shaped bodies that slow down the airflow, then expand the combustion products without benefit of a conventional nozzle. Aerothermal heating, supersonic combustion processes, and suitable fuels are primary problems in this type of engine system.

The pulsejet engine depends upon the ram effect to open flapper-type shutters and admit air to the combustion chamber. When fuel is injected and combustion takes place, the shutters are closed by combustion pressures and the pressure surges in the duct; the combustion products then are expelled from the nozzle to generate thrust.

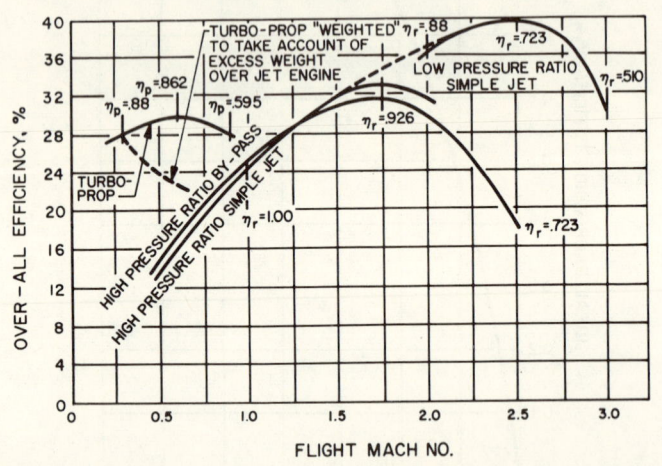

Fig. 20-8. Overall efficiency over a range of flight speed. (*From Ref. 19.*)

When the combustion-chamber pressure is reduced to a low value, the shutters reopen and the cycle is repeated. The pulsed thrust results in a low-efficiency system, and the resonant frequency of the duct strongly influences the performance.[20]

The turbojet engine includes a gas compressor system between the inlet and the combustors. Power is extracted from the combustion products by a turbine that drives the compressor. The residual power of the combustion products is used to generate thrust. The detailed design and arrangements of the components of this engine system are as varied, if not more so, as those of reciprocating engines. References 2, 3, 9, 18, and 19 present extensive descriptions and detailed design data. Specifications of currently available power plants are summarized in Table 20-6.

The specific impulses of these engine systems range from 50 to 155 sec. This is deliberately lowered by air augmentation or fluid injection (alcohol, water, or fuel) into the nozzle to obtain higher efficiency and thrust at low flight speed. An example of air augmentation is the fanjet engine, in which part of the inlet air is ducted around the combustion system to mix with the effluent combustion products before being expanded through the nozzle.

The effect of the specific weight of these power-plant systems on vehicle performance is illustrated in Fig. 20-9.

Rocket-engine Systems. Rocket-engine systems generate thrust by ejecting a high-energy jet from a closed thrust chamber in which a suitable oxidizer and fuel are reacted to release energy in the form of heat.[9,11] Basically no different in concept from air-breathing jet engines, for many applications they gain in flexibility from the use of many different propellant combinations, and the ability to operate in free space because they carry their own oxidizer. Propulsion is dependent upon internal conditions alone and not upon the effect of airflow through the system.

PROPULSION 20-25

The thrust of rocket engines depends upon the total propellant flow rate $\dot{W}$, the exhaust velocity c, the area of the exit plane of the nozzle A_e, the pressure at the exit plane of the nozzle P_e, and the ambient atmospheric pressure P_0.

$$F = \frac{\dot{W}}{g} c + (P_e - P_0) A_e \qquad (20\text{-}42)$$

The exhaust velocity depends upon the effectiveness of the combustion process

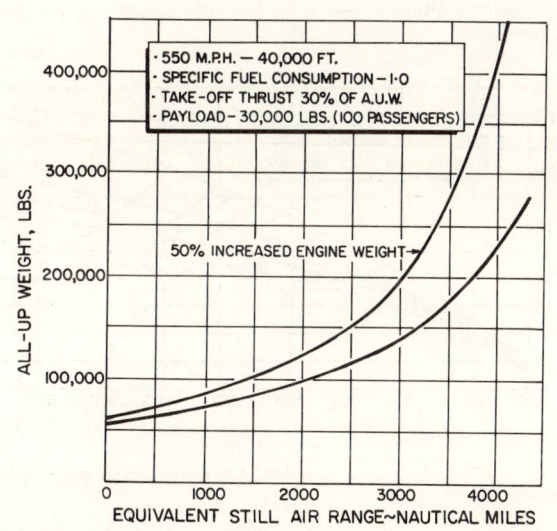

FIG. 20-9. Effect on aircraft weight of change in engine specific weight. (*From Ref. 19.*)

- 550 M.P.H. — 40,000 FT.
- SPECIFIC FUEL CONSUMPTION – 1·0
- TAKE-OFF THRUST 30% OF A.U.W.
- PAYLOAD – 30,000 LBS. (100 PASSENGERS)
- 50% INCREASED ENGINE WEIGHT

described by the characteristic velocity c^* and the behavior of the nozzle under differing ambient conditions, described by thrust coefficient C_F.

$$c = c^* C_F = g I_s \quad \text{ft/sec} \qquad (20\text{-}43)$$

The characteristic velocity of the thrust chamber is established from the thrust-chamber pressure P_c, the throat area A_t, and the propellant flow rate $\dot{W}$.

$$c^* = \frac{P_c A_t g}{\dot{W}} = \frac{g I_s}{C_F} \quad \text{ft/sec} \qquad (20\text{-}44)$$

$$= \frac{[g \gamma R(T_c/M)]^{1/2}}{\gamma [2/(\gamma + 1)]^{\frac{\gamma+1}{2(\gamma-1)}}}$$

where γ = specific-heat ratio of combustion products C_p/C_v
T_c = absolute temperature of combustion products, °R
R = gas constant
M = average molecular weight of combustion products, lb/mole volume
A_t = area of nozzle thrust, sq ft

The thrust coefficient depends upon the pressure ratio across the nozzle, nozzle dimensions, and combustion-product physical properties.

$$C_F = \frac{F}{P_c A_t} = \left\{ \frac{2\gamma^2}{\gamma - 1} \left(\frac{2}{\gamma + 1} \right)^{\frac{\gamma+1}{\gamma-1}} \left[1 - \left(\frac{P_e}{P_c} \right)^{\frac{\gamma-1}{\gamma}} \right] \right\}^{1/2} + \frac{P_c - P_0}{P_c} \frac{A_e}{A_t} \qquad (20\text{-}45)$$

where A_e = area of exit section of nozzle, sq ft

The variation of the thrust coefficient with pressure ratio, area ratio, $\epsilon = A_e/A_t$, and specific-heat ratio is summarized in Fig. 20-10.

These formulas are based upon the behavior of ideal gases and the assumption that the chemical equilibrium does not change at a specific operating mixture ratio. Good design practice permits experimentally determined c^* values within 90 to 95 per cent of these computed values, and experimental values of C_F approach within 95 to 98 per cent of theoretical values.

The thrust coefficient C_F is a function of altitude. As a result, the area ratio $\epsilon = A_e/A_t$ for booster engines is less than for upper-stage engines operating at high altitudes or in free space.

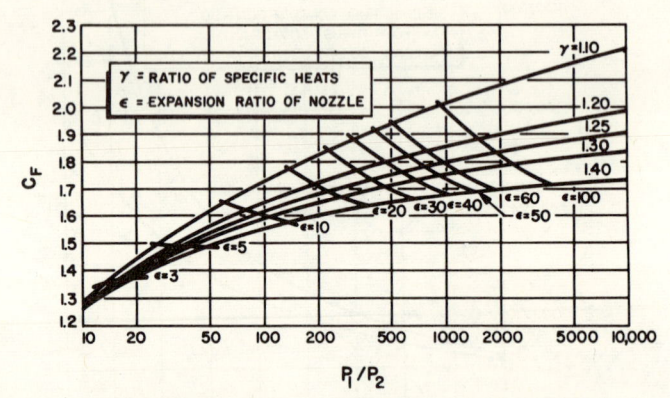

Fig. 20-10. Thrust coefficient C_F as a function of pressure ratio, and specific-heat ratio, for optimum expansion conditions.

Both the characteristic velocity c^* of the thrust chamber and the thrust coefficient are functions of the mixture ratio of the propellants; maximum performance occurs at slightly fuel-rich conditions. Therefore, the performance of the propellants is specified not only by the combination but also by the weight ratio of oxidizer to fuel.

A high-performance propellant combination does not always result in the most compact system (includes the propellants and their tanks, the propellant-feed system, and the thrust chamber). A measure of the compactness of the engine system is the density-impulse parameter I_{sd} defined by the relation

$$I_{sd} = I_s \delta_p \tag{20-46}$$

in which the propellant bulk specific gravity is defined by the mixture ratio r, the oxidizer density δ_o and the fuel density δ_f.

$$\delta_p = \frac{(1+r)\delta_f \delta_o}{\gamma \delta_f + \delta_o} \tag{20-47}$$

The chamber pressure P_c also influences the weight and overall dimensions of the rocket-engine system. As the chamber pressure rises to a few hundred psi, the overall size and weight of an engine system of given thrust level diminish to a minimum, then increase because of heavier structures required to withstand the higher stresses induced by increasing chamber pressures. The maximum size and weight and the optimum operating chamber pressure must be determined by a specific analysis of each given system.

The total impulse I_t is defined by the relation

$$I_t = \int_0^{t_b} F\, dt = \bar{F} t_b = I_s W_p \tag{20-48}$$

and is a parameter associated particularly with solid rocket-engine systems. By testing, the total impulse can be accurately determined from thrust F and burning time t_b; and from burned propellant weight W_p, the effective specific impulse can then be computed.

The equivalent power in the jet is

$$\text{hp}_{\text{jet}} = \frac{Fc}{550} \tag{20-49}$$

and the propulsive efficiency is the same as that of the air-breathing thermal jet and propeller-driven systems.

$$\eta_p = \frac{2V/c}{1 + (V/c)^2} = \frac{2gV/I_s}{1 + (gV/I_s)^2} \tag{20-50}$$

The liquid-propellant engine system includes the propellant tanks, propellant feed and control system, a power source to drive the feed system, and the thrust chamber.

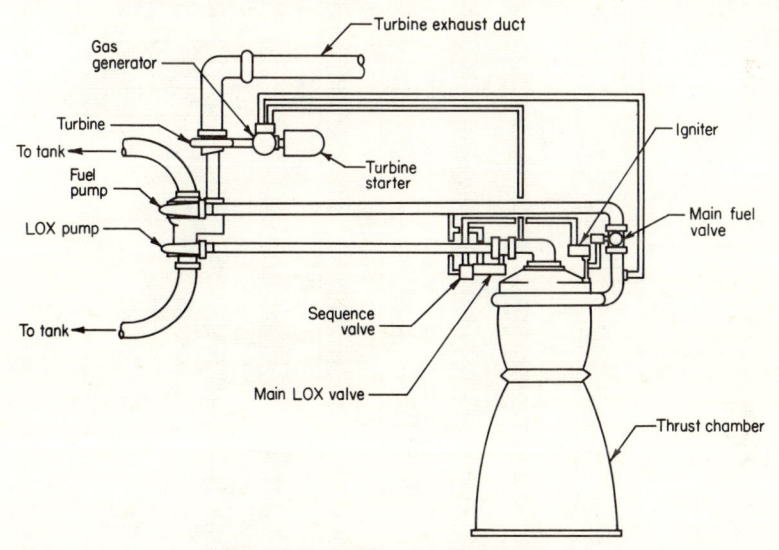

Fig. 20-11. Liquid-propellant engine.

There are major differences in propellant-feed systems. Pressure-fed systems are often used in smaller engines, control being obtained by balancing pressure drops in the feed lines. However, the pressure-fed system rapidly increases in weight as the thrust level increases. The favored system is the pump-fed type (Fig. 20-11) which is turbine-driven from a gas generator or bleed from the thrust chamber. Bipropellant engine systems are by far the most widely used because of the relative ease in handling and controlling the propellants. While the monopropellant systems are simpler, having only a single feed system and requiring no balancing, the propellants (hydrazine, hydrogen peroxide, and nitromethane) are sensitive to contaminants, heat, and impact.

The theoretical performance of a number of bipropellants is summarized in Table 20-8, and that of several monopropellants is shown in Table 20-9; other propellants are summarized in Ref. 21. A well-designed engine system can achieve a performance in the neighborhood of 95 per cent of the theoretical values, depending upon the specific design.

Currently, some attention is being given to hybrid engines that use a liquid oxidizer and solid fuel, or vice versa. No applications to vehicle systems have been made.

SYSTEM COMPONENTS

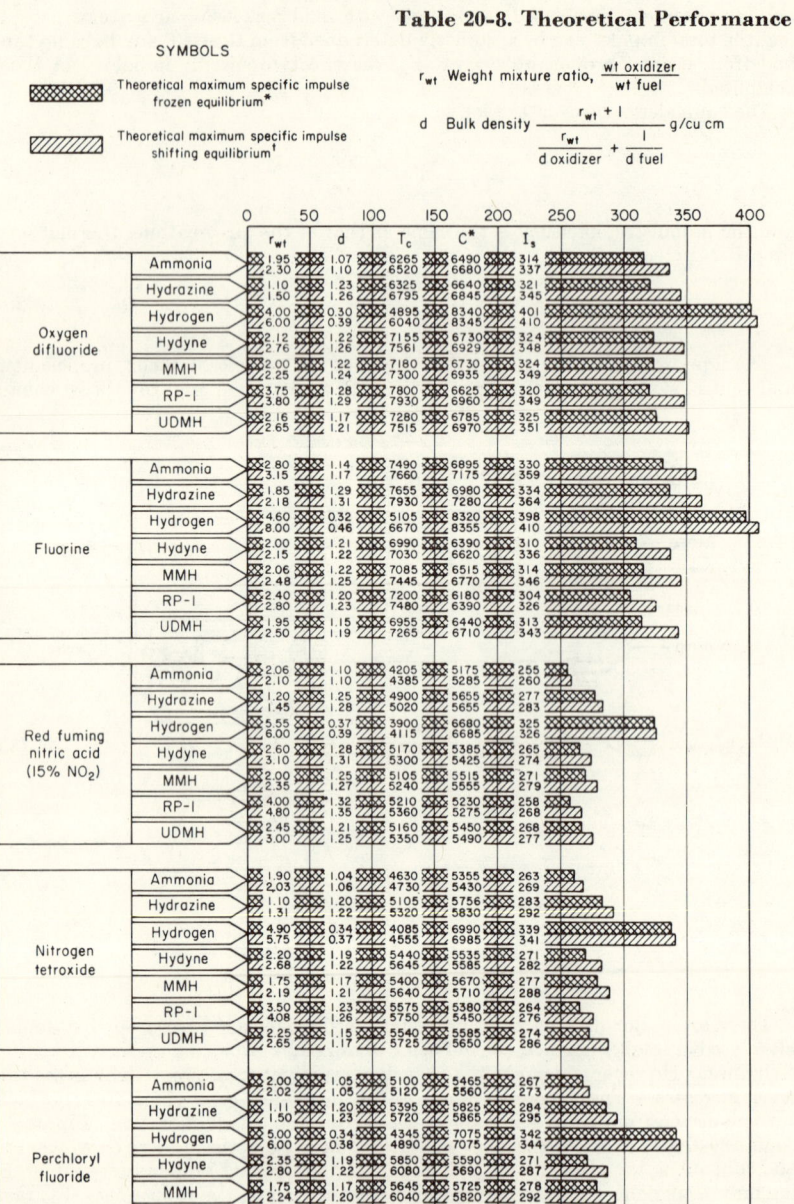

Table 20-8. Theoretical Performance

SYMBOLS

▓▓▓ Theoretical maximum specific impulse frozen equilibrium*

▨▨▨ Theoretical maximum specific impulse shifting equilibrium†

r_{wt} Weight mixture ratio, $\dfrac{\text{wt oxidizer}}{\text{wt fuel}}$

d Bulk density $\dfrac{r_{wt}+1}{\dfrac{r_{wt}}{d\,\text{oxidizer}} + \dfrac{1}{d\,\text{fuel}}}$ g/cu cm

Oxidizer	Fuel	r_{wt}	d	T_c	C^*	I_s
Oxygen difluoride	Ammonia	1.95 / 2.30	1.07 / 1.10	6265 / 6520	6490 / 6680	314 / 337
	Hydrazine	1.10 / 1.50	1.23 / 1.26	6325 / 6795	6640 / 6845	321 / 345
	Hydrogen	4.00 / 6.00	0.30 / 0.39	4895 / 6040	8340 / 8345	401 / 410
	Hydyne	2.12 / 2.76	1.22 / 1.26	7155 / 7561	6730 / 6929	324 / 348
	MMH	2.00 / 2.25	1.22 / 1.24	7180 / 7300	6730 / 6935	324 / 349
	RP-1	3.75 / 3.80	1.28 / 1.29	7800 / 7930	6625 / 6960	320 / 349
	UDMH	2.16 / 2.65	1.17 / 1.21	7280 / 7515	6785 / 6970	325 / 351
Fluorine	Ammonia	2.80 / 3.15	1.14 / 1.17	7490 / 7660	6895 / 7175	330 / 359
	Hydrazine	1.85 / 2.18	1.29 / 1.31	7655 / 7930	6980 / 7280	334 / 364
	Hydrogen	4.60 / 8.00	0.32 / 0.46	5105 / 6670	8320 / 8355	398 / 410
	Hydyne	2.00 / 2.15	1.21 / 1.22	6990 / 7030	6390 / 6620	310 / 336
	MMH	2.06 / 2.48	1.22 / 1.25	7085 / 7445	6515 / 6770	314 / 346
	RP-1	2.40 / 2.80	1.20 / 1.23	7200 / 7480	6180 / 6390	304 / 326
	UDMH	1.95 / 2.50	1.15 / 1.19	6955 / 7265	6440 / 6710	313 / 343
Red fuming nitric acid (15% NO₂)	Ammonia	2.06 / 2.10	1.10 / 1.10	4205 / 4385	5175 / 5285	255 / 260
	Hydrazine	1.20 / 1.45	1.25 / 1.28	4900 / 5020	5655 / 5655	277 / 283
	Hydrogen	5.55 / 6.00	0.37 / 0.39	3900 / 4115	6680 / 6685	325 / 326
	Hydyne	2.60 / 3.10	1.28 / 1.31	5170 / 5300	5385 / 5425	265 / 274
	MMH	2.00 / 2.35	1.25 / 1.27	5105 / 5240	5515 / 5555	271 / 279
	RP-1	4.00 / 4.80	1.32 / 1.35	5210 / 5360	5230 / 5275	258 / 268
	UDMH	2.45 / 3.00	1.21 / 1.25	5160 / 5350	5450 / 5490	268 / 277
Nitrogen tetroxide	Ammonia	1.90 / 2.03	1.04 / 1.06	4630 / 4730	5355 / 5430	263 / 269
	Hydrazine	1.10 / 1.31	1.20 / 1.22	5105 / 5320	5756 / 5830	283 / 292
	Hydrogen	4.90 / 5.75	0.34 / 0.37	4085 / 4555	6990 / 6985	339 / 341
	Hydyne	2.20 / 2.68	1.19 / 1.22	5440 / 5645	5535 / 5585	271 / 282
	MMH	1.75 / 2.19	1.17 / 1.21	5400 / 5640	5670 / 5710	277 / 288
	RP-1	3.50 / 4.08	1.23 / 1.26	5575 / 5750	5380 / 5450	264 / 276
	UDMH	2.25 / 2.65	1.15 / 1.17	5540 / 5725	5585 / 5650	274 / 286
Perchloryl fluoride	Ammonia	2.00 / 2.02	1.05 / 1.05	5100 / 5120	5465 / 5560	267 / 273
	Hydrazine	1.11 / 1.50	1.20 / 1.23	5395 / 5720	5825 / 5865	284 / 295
	Hydrogen	5.00 / 6.00	0.34 / 0.38	4345 / 4890	7075 / 7075	342 / 344
	Hydyne	2.35 / 2.80	1.19 / 1.22	5850 / 6080	5590 / 5690	271 / 287
	MMH	1.75 / 2.24	1.17 / 1.20	5645 / 6040	5725 / 5820	278 / 292
	RP-1	3.50 / 4.35	1.22 / 1.25	5965 / 6175	5460 / 5535	264 / 280
	UDMH	2.14 / 2.70	1.13 / 1.17	5765 / 6130	5670 / 5755	275 / 290

PROPULSION 20-29

of Rocket Propellant Combinations

T_c Chamber temperature, °F

C^* Characteristic velocity, ft/sec

* Frozen equilibrium: Condition of fixed chemical composition of the gaseous products throughout expansion in the nozzle

† Shifting equilibrium: Condition of changing chemical composition of the gaseous products throughout expansion in the nozzle

‡ The density at the boiling point was used for those oxidizers or fuels which boil below 68°F at one atmosphere pressure

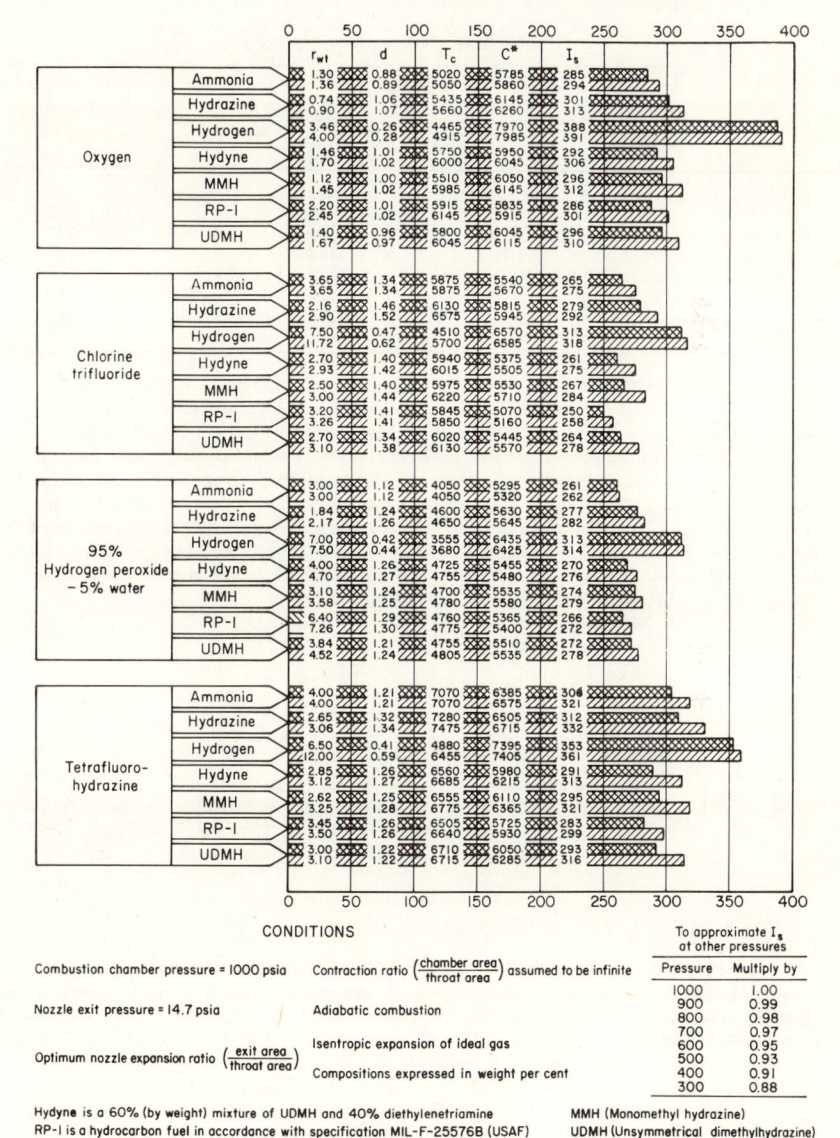

Oxidizer	Fuel	r_{wt}	d	T_c	C^*	I_s
Oxygen	Ammonia	1.30 / 1.36	0.88 / 0.89	5020 / 5050	5785 / 5860	285 / 294
	Hydrazine	0.74 / 0.90	1.06 / 1.07	5435 / 5660	6145 / 6260	301 / 313
	Hydrogen	3.46 / 4.00	0.26 / 0.28	4465 / 4915	7970 / 7985	388 / 391
	Hydyne	1.46 / 1.70	1.01 / 1.02	5750 / 6000	5950 / 6045	292 / 306
	MMH	1.12 / 1.45	1.00 / 1.02	5510 / 5985	6050 / 6145	296 / 312
	RP-1	2.20 / 2.45	1.01 / 1.02	5915 / 6145	5835 / 5915	286 / 301
	UDMH	1.40 / 1.67	0.96 / 0.97	5800 / 6045	6045 / 6115	296 / 310
Chlorine trifluoride	Ammonia	3.65 / 3.65	1.34 / 1.34	5875 / 5875	5540 / 5670	265 / 275
	Hydrazine	2.16 / 2.90	1.46 / 1.52	6130 / 6575	5815 / 5945	279 / 292
	Hydrogen	7.50 / 11.72	0.47 / 0.62	4510 / 5700	6570 / 6585	313 / 318
	Hydyne	2.70 / 2.93	1.40 / 1.42	5940 / 6015	5375 / 5505	261 / 275
	MMH	2.50 / 3.00	1.40 / 1.44	5975 / 6220	5530 / 5710	267 / 284
	RP-1	3.20 / 3.26	1.41 / 1.41	5845 / 5850	5070 / 5160	250 / 258
	UDMH	2.70 / 3.10	1.34 / 1.38	6020 / 6130	5445 / 5570	264 / 278
95% Hydrogen peroxide – 5% water	Ammonia	3.00 / 3.00	1.12 / 1.12	4050 / 4050	5295 / 5320	261 / 262
	Hydrazine	1.84 / 2.17	1.24 / 1.26	4600 / 4650	5630 / 5645	277 / 282
	Hydrogen	7.00 / 7.50	0.42 / 0.44	3555 / 3680	6435 / 6425	313 / 314
	Hydyne	4.00 / 4.70	1.26 / 1.27	4725 / 4755	5455 / 5480	270 / 276
	MMH	3.10 / 3.58	1.24 / 1.25	4700 / 4780	5535 / 5580	274 / 279
	RP-1	6.40 / 7.26	1.29 / 1.30	4760 / 4775	5365 / 5400	266 / 272
	UDMH	3.84 / 4.52	1.21 / 1.24	4755 / 4805	5510 / 5535	272 / 278
Tetrafluoro-hydrazine	Ammonia	4.00 / 4.00	1.21 / 1.21	7070 / 7070	6385 / 6575	304 / 321
	Hydrazine	2.65 / 3.06	1.32 / 1.34	7280 / 7475	6505 / 6715	312 / 332
	Hydrogen	6.50 / 12.00	0.41 / 0.59	4880 / 6455	7395 / 7405	353 / 361
	Hydyne	2.85 / 3.12	1.26 / 1.27	6560 / 6685	5980 / 6215	291 / 313
	MMH	2.62 / 3.25	1.25 / 1.28	6555 / 6775	6110 / 6365	295 / 321
	RP-1	3.45 / 3.50	1.26 / 1.26	6505 / 6640	5725 / 5930	283 / 299
	UDMH	3.00 / 3.10	1.22 / 1.22	6710 / 6715	6050 / 6285	293 / 316

CONDITIONS

Combustion chamber pressure = 1000 psia

Nozzle exit pressure = 14.7 psia

Optimum nozzle expansion ratio $\left(\frac{\text{exit area}}{\text{throat area}}\right)$

Contraction ratio $\left(\frac{\text{chamber area}}{\text{throat area}}\right)$ assumed to be infinite

Adiabatic combustion

Isentropic expansion of ideal gas

Compositions expressed in weight per cent

To approximate I_s at other pressures

Pressure	Multiply by
1000	1.00
900	0.99
800	0.98
700	0.97
600	0.95
500	0.93
400	0.91
300	0.88

Hydyne is a 60% (by weight) mixture of UDMH and 40% diethylenetriamine

RP-1 is a hydrocarbon fuel in accordance with specification MIL-F-25576B (USAF)

MMH (Monomethyl hydrazine)

UDMH (Unsymmetrical dimethylhydrazine)

SYSTEM COMPONENTS

The advantages of the liquid-propellant system include light weight, throttling capability, relative ease of handling the propellants for large systems, and the availability and low cost of a wide variety of propellant combinations.

Currently available liquid-propellant rocket engines range in thrust from a few pounds up to 1,500,000 lb. An example of the latter is the Rocketdyne F-1 engine, which burns liquid oxygen and kerosene (JP-4). The Saturn launch vehicle uses an engine system consisting of a cluster of eight Rocketdyne H-1 engines, each having a thrust of 188,000 lb.

Table 20-9. Monopropellants

Name	Formula	Specific impulse I_s at $P_c = 300$ psi	Specific gravity	Density impulse I_{sd}, sec
Ethylene oxide	C_2H_4O	166	0.887	147
Hydrazine	N_2H_4	174	1.0045	175
Nitromethane	CH_3NO_2	218	1.13	246
n-Propyl nitrate	$CH_3(CH_2)_2NO_3$	179	0.935	167
90% hydrogen peroxide	H_2O_2	133	1.387	184

Nuclear rocket engines (Fig. 20-12) are liquid monopropellant systems; the energy of the nuclear reactor utilizes heat-exchange processes to transfer the energy to the monopropellant, which then expands through the nozzle to generate jet power. The propellants used are preferably of low molecular weight to obtain high specific impulse at low operating temperatures. Typical propellants are hydrogen (H_2), or ammonia (NH_3). Currently in the early development phases, specific impulses of 600 to 900 sec are anticipated. The engines will be relatively large because of critical-mass requirements of the reactor. The specific weight of the engine system is expected to be an order of magnitude larger than that of the best chemical rocket engines, partially offsetting the advantage of high specific impulse.

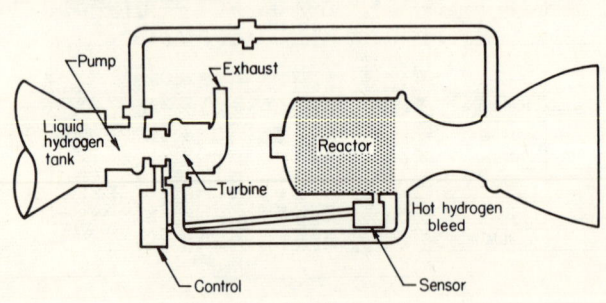

Fig. 20-12. Nuclear engine.

A solid-propellant rocket-engine system is an integrated assembly including an igniter, the engine case, the solid-propellant grain, and the nozzle.[11] There is no separate combustion chamber as in the liquid-propellant system. The propellant storage tank and the combustion chamber are combined in the engine case. After ignition, control of the propellant mass flow rate is accomplished by precise control of the exposed surface area of the solid propellant. The reacting grain generates combustion products at the exposed surface, which regresses normal to itself in parallel layers. The rate of regression is defined as the burning rate of the propellant. The burning rate of the grain must be matched to the flow rate of the nozzle to maintain a chamber pressure sufficient to give stable reaction, yet produce the combustion-product flow at a rate high enough to yield the required thrust level.

PROPULSION 20-31

A solid propellant is a homogeneous material, or a mixture of materials which, once ignited, reacts to evolve combustion products continuously at elevated temperatures without dependence on the atmosphere.[21] A grain, encased in the rocket chamber, must hold its shape over an extended temperature range and must withstand the mechanical stresses resulting from handling, igniting, and firing the rocket. To meet these requirements, a special class of plastic combinations of oxidizing and reducing agents has been developed.

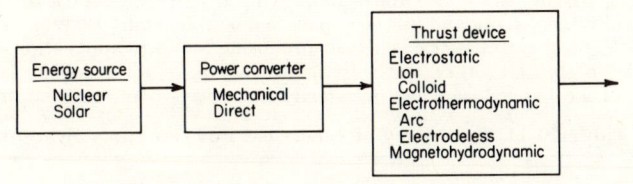

FIG. 20-13. Basic subsystems of electrical rocket engine.

```
Energy source        Power converter      Thrust device
Nuclear              Mechanical           Electrostatic
Solar                Direct               Ion
                                          Colloid
                                          Electrothermodynamic
                                          Arc
                                          Electrodeless
                                          Magnetohydrodynamic
```

The principal homogeneous propellant is nitrocellulose, which is plasticized with nitroglycerin and other organic materials.[21] The composite propellants include the oxidizers, potassium perchlorate, ammonium nitrate, ammonium picrate, etc. The fuels include the asphalts; rubber; plastic materials such as the polyisobutylenes, vinyl polyesters, polybutadienes, and polysulfides; and the metals aluminum and beryllium. The particular combination used depends upon the application and its requirements.

The sizes of solid-propellant engine systems range from small units, such as used in the Bazooka weapon, to large units of 154 and 260 in. diameter, currently in development. Maximum thrusts currently being achieved are in the million-pound class.

20-7. Electrical Propulsion Systems

Concepts of electrical power-plant systems are varied, both from the system standpoint and in the use of components. The electrical rocket engine has three important subsystems: the energy source, the power converter, and the electrical thrust devices (Fig. 20-13). The engine-system parameters for overall performance of the electrical propulsion systems are summarized in Table 20-10.

Table 20-10. Summary of Performance Parameters, Electrical Rocket Engines

Parameter	English system	CGS system
Specific power	$\dfrac{P}{W} = \dfrac{21.8 G I_s}{\eta}$ watts/lb $\dfrac{P}{W} = \dfrac{0.0293 G I_s}{\eta}$ hp/lb	$\dfrac{P}{W} = \dfrac{4.91 \times 10^{-5} a I_s}{\eta}$ watts/gm
Power per unit thrust	$\dfrac{P}{F} = \dfrac{21.8 I_s}{\eta}$ watts/lb $\dfrac{P}{F} = \dfrac{0.0293 I_s}{\eta}$ hp/lb	$\dfrac{P}{F} = \dfrac{4.91 \times 10^{-5} I_s}{\eta}$ watts/dyne
Exhaust velocity	C ft/sec	C cm/sec
Specific fuel consumption	$w_f = \dfrac{3{,}600}{I_s}$ hr^{-1}	$w_f = \dfrac{3.53 \times 10^6}{I_s}$ hr^{-1}
Specific impulse	$I_s = \dfrac{C}{32.2}$ sec	$I_s = \dfrac{C}{980}$ sec
Engine operating time	$t = \dfrac{1 - \exp(-\Delta v / 32.2 I_s)}{G} I_s$ sec	$t = \dfrac{1 - \exp(-\Delta v / 980 I_s)}{G} I_s$ sec

NOTE: G = thrust-to-weight ratio, a = acceleration, cm/sec^2, η = power-plant efficiency.

SYSTEM COMPONENTS

The key factor in the successful development of an electrical rocket engine is the energy source. Chemical sources known today produce specific energy too low to be used effectively. Solar energy, while it is relatively limitless in *amount*, is limited in *rate*, and the rate varies with distance from the sun. Nuclear energy, however, can produce the large amounts of energy at the rate required for extended space missions. It can be applied using present techniques and has future potential which will be further augmented by improvements in power-conversion systems.

Power conversion can be accomplished mechanically, thermoelectrically (Sec. 21-4), thermionically (Sec. 21-5), and by the application of magnetohydrodynamic principles (Sec. 21-6). Mechanical methods appear promising for early application in electrical power-plant systems. Other methods mentioned have not reached the technical perfection of mechanical systems, but steady progress is being made in many areas.

Table 20-11. Summary of Electrical Power-supply Systems

Type	Specific weight	Operating life	Status
Chemical:			
Batteries............	20 lb/Kwhr	50–200 days	Production
Gas turbine.........	7 to 10 lb/Kwhr	0.5–5 hr	Production
Fuel cell............	12 lb/Kwhr	30 days	Development
Radioisotope..........	1,000–3,000 lb/Kw	280 days	Development
Solar:			
Photovoltaic........	50–200 lb/Kw	Indefinite	Development
Gas turbine.........	10–20 lb/Kw	Unknown	Research
Thermoelectric......	100 lb/Kw	Unknown	Research
Nuclear:			
Turbine.............	8–100 lb/Kw	1 year	Development
Thermoelectric......	100 lb/Kw	Unknown	Research
Thermionic..........	Unknown	Unknown	Research
Fusion................	Unknown	Unknown	Research

As a result, direct conversion devices combined with an electrical thrust chamber may make early contributions to the accomplishment of continuously powered and controlled space flight. This section is supplemented by phenomenological and theoretical material in the following chapter.

Electrical thrust chambers have a variety of forms. The electrothermodynamic thrust chamber (see below), variously described as the arc or plasma jet, uses electrical energy for heating the propellant gas which, in turn, is expanded through a Laval nozzle to generate thrust in the same manner as a chemical rocket. Other devices generate a plasma that may be accelerated by external, or induced, electromagnetic fields (see below). Ions can be extracted from the plasma, or formed directly by surface-contact phenomena and then accelerated electrostatically; this is the ion rocket engine (see below). Substitution of charged colloid particles (see below) for ions offers the potential advantages of lower specific impulse ($1,000 < I_s < 5,000$ sec) to the electrostatic ion rocket. The concepts embodied in the electrical thrust chambers are as varied as those in the power-conversion systems.

Energy Sources. The energy sources for electrical propulsion include conventional chemical, radioisotope, the nuclear-fission reactor, the fusion reactor, and the sun. If the demands of electrical space propulsion are to be met, the specific power of the vehicle must be approximately $\frac{1}{10}$ hp/lb to achieve a thrust-to-weight ratio of 10^{-4} (Fig. 20-1) if the specific impulse is 10,000 sec. Since a moon mission requires a characteristic velocity of approximately 25,000 ft/sec, this implies an operating time from 1,700 to 1,800 hr. Similarly, a Mars mission, which requires a characteristic velocity approaching 100,000 ft/sec, demands nearly 9,000 hr of engine operation. These data allow an overall comparison for determining the capabilities of the different power-supply systems.

Table 20-11 is a summary of energy sources and power-conversion systems considered for electrical power-plant systems. Comparison of these data with the requirements presented above eliminates all systems except the nuclear and solar power supplies.

Among the chemical systems, batteries are reliable and inexpensive. These advantages are offset by low power, heavy weight, and short life. The turbines driven by a gas generator give a good power yield for short times, but the systems are heavy, have too short a life, and require too much fuel. The fuel cells apparently combine many of the disadvantages of the chemical systems, in addition to having low efficiency.

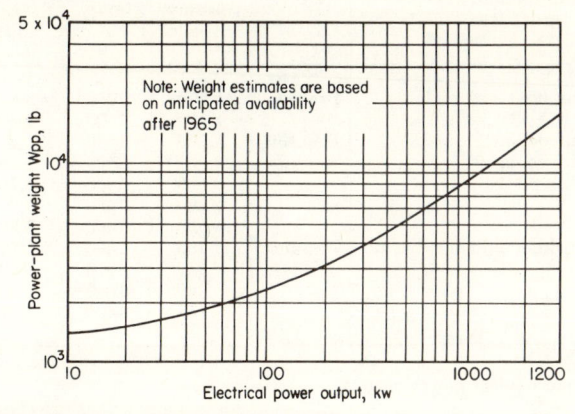

Fig. 20-14. Nuclear turboelectric power-plant weight including 1,000-lb shield.

The radioisotope sources have the advantage of being direct convertors, and the thermoelectric radioisotope type is reliable. However, the specific power is low, ranging from 0.0003 to 0.001 Kw/lb. The lifetime is limited by the half-life of the radioisotope.

Approximately 218 Kw of electrical power is required for a 1-lb-thrust electrical engine operating at a specific impulse of 10,000 sec and 100 per cent efficiency. From the data in Fig. 20-14, a nuclear turboelectric power plant with 1,000 lb of shielding is estimated to weigh approximately 3,200 lb. Allowing 500 lb for the nuclear reactor and controls, the mechanical power-conversion system weight is estimated to be 1,700 lb, approximately 8 lb/Kw. In contrast, currently available photovoltaic cells weigh approximately 72 lb/Kw. Assuming collector structural weights to be the same, these rough data indicate that a 1-lb-thrust photovoltaic engine has a weight penalty of 14,000 lb. This factor is offset by a 700 sq ft radiator area, increased mechanical complexity, and possible lesser reliability of the nuclear turboelectric system. The estimated performance of mechanical power-conversion systems using different working fluids and operating cycles is summarized in Table 20-12.

The choice between a solar-electric or a nuclear turboelectric system probably will be based upon the cost of placing in orbit a large-area heavy system of possibly higher reliability, as compared with the costs of the nuclear turboelectric system.

The overall picture of a low-powered system may be clarified by considering the SNAP II reactor, developed by the Atomics International Division of North American Aviation, Inc. This reactor has a power output of 3 Kw and a weight of 220 lb. An electric rocket using this power source would be capable of producing not more than 0.014 lb of thrust at a specific impulse of 10,000 sec. Arbitrarily allowing 60 lb for propellant and thrust device, the maximum thrust-to-weight ratio is 5×10^{-5}. A 3-Kw photovoltaic solar rocket would weigh approximately 360 lb and have a maximum area of 25 sq ft exposed to the sun, if operated in cislunar space. The thrust-to-weight ratio would be approximately 2.5×10^{-5}. Because of the possibility of

SYSTEM COMPONENTS

reducing the photovoltaic cell weights, the solar-electric and nuclear-electric systems are competitive at this power level. At lower power levels, the solar-electric rocket may have an advantage.

The fusion reactor is now an unknown quantity when considered for application in an electrical rocket-engine system. Because of the anticipated power levels and

Table 20-12. Operating Requirements for Mechanical Power-conversion Systems*

Working fluid	Cycle	Turbine inlet pressure, psia	Reactor outlet temp, F	Radiator† area, sq ft	Reactor† power, Mw
Helium (working fluid and coolant)...	Brayton	1,500	1900	8,800	4.00
		1,500	1500	12,300	5.26
Helium (sodium coolant).............	Brayton	1,500	1460	21,400	6.25
Mercury (sodium coolant)............	Rankine	5,000	2000	1,000	4.00
		2,500	1800	1,700	4.16
		1,200	1600	2,000	5.26
		500	1400	3,200	5.26
Alkali metal (working fluid and coolant)	Rankine	460	2000	1,000	3.57
		217	1800	1,300	3.57
		89	1600	1,500	4.16
		31	1400	3,400	7.14

* From Ref. 22.
† For 1-Mw mechanical power output.

the energy levels of the particles produced, direct use of the particles to generate reactive thrust appears to be more logical than converting energy to electrical power and then accelerating the same or other particles. There are so many unknown factors in this new development that the most that can be recommended at this time is to watch the progress of the basic research programs and, if possible, devise a sound approach to a propulsion system.[23]

The electrical thrust devices are dependent upon the formation of a plasma, which is an electrically neutral gaseous mixture of electrons, ions, and neutral particles.

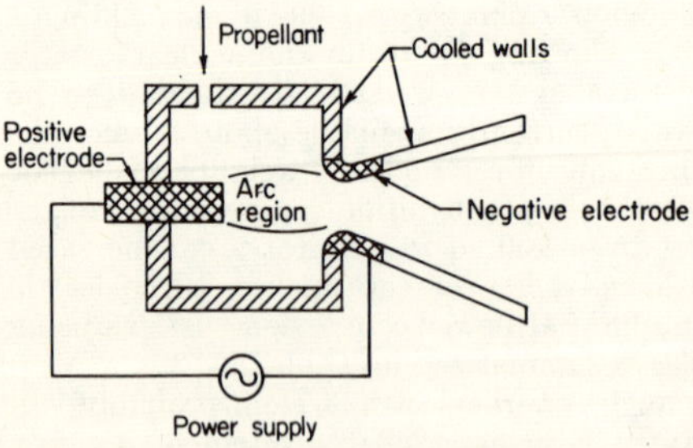

Electrothermodynamic Thrust Devices. The electrothermodynamic thrust device, or plasma jet, is characterized by high-temperature gas, high-temperature electrons, and high pressures.[24] Several techniques are used to inject propellant into the thrust device (Fig. 20-15). The propellant is heated by a low-voltage high-current arc, then expelled through the nozzle.

Table 20-13 summarizes gas temperatures measured experimentally within high-pressure arcs, and indicates operating temperatures for a number of promising propellants. At high specific impulses, these high temperatures, the relatively high pressures in the arc chamber, and the low propellant flow rates induce severe convective heat transfer; therefore, major material problems that will limit the maximum specific impulse can be expected with this device.

The lower limit for efficient use of this device is established by the power required to ionize and heat the large flow of propellants, which increases inversely with the specific impulse. For specific impulses less than 1,000, the reported power efficiencies

Fig. 20-15. Electrothermodynamic thrust device.

PROPULSION 20-35

are consistently less than 20 per cent. Because of these low efficiencies and current material limitations, and because the maximum payload for a given mission depends upon initial selection of the specific impulse of the rocket engine (Fig. 20-2) application of the electrothermodynamic engine in vehicles for space missions should be limited to those missions which require a specific impulse above 1,000 and below 2,000 sec.

Electromagnetic Thrust Devices. The acceleration of a plasma by electromagnetic fields has the objective of generating particle motion in an ordered linear flow with a minimum expenditure of energy. This objective has not been achieved to date and is reflected in low power efficiencies of the engine systems.

Two forces act on a charged particle of mass M and charge e when it moves through a region in which an electric field E and a magnetic field H exist.[25] If e is the charge

Table 20-13. Measured Temperatures at the Axis of Positive Column of Arcs Operating at 1 Atm Pressure

Gas	Electrode	Arc length, cm	Current, amp	Temp, °K
Air	Carbon	0.8	2	5,900
Air	Carbon	0.3–1.8	1–12	6,200–7,800
Air	C + Ca	Few	180	12,000
H_2	Tungsten	....	4	4,900
H_2	Tungsten	....	8–9	5,300–6,300
H_2	Carbon	....	Few	5,000
N_2	Copper	15	2	5,300
Air	Copper	15	2	4,700
Air	Copper	....	14	5,470
Air	Copper	....	10	6,100
H_2O	....	(Jet)	500	35,000
H_2O	....	(Jet)	1,500	50,000

on the particle, the product eE is the electrostatic force parallel to the field. The magnetic force is at right angles to the particle velocity v and to the magnetic field, and is the vectorial cross product of v and H.

$$M\frac{d\mathbf{v}}{dt} = e(\mathbf{E} + \mathbf{v} \times \mathbf{H})$$

When H vanishes, the charged particles move with constant acceleration $(e/M)E$, as in the electrostatic ion engine. When there is no electric field, the acceleration $(e/M)(\mathbf{v} \times \mathbf{H})$ is perpendicular to the particle velocity, curving its trajectory but not changing the scalar velocity v.

The motion of a discrete particle is complicated in a microscopic situation such as that which exists in the discharge of a high-intensity arc. When consideration is given to the electric- and magnetic-field distribution, the characteristics of the plasma, and transient discharges (see Sec. 21-6), the magnitude of fully understanding the phenomena emphasizes why magnetohydrodynamic systems are still in an early stage of development.

Button Plasma Guns. In button-type plasma guns (Fig. 20-16 and Ref. 24, p. 105) the discharge of a capacitor bank across the exposed ends of the conductors produces a plasma of metallic ions and electrons. The high discharge currents induce magnetic fields that project small plasma clouds at speeds up to 6.7×10^5 ft/sec. Figure 20-16c is a variation in which plasma can be injected into the gun from an external source.

Major problems associated with these units are low thrust per button, intermittent operation, coupling of the energy-storage system to obtain efficient energy transfer, and a wide-angle plasma pattern that results in a loss of thrust.

Crossed-field Motors. The rail-type motor (Fig 20-17a) depends upon the initiation of a discharge between the two rails. Once the discharge is started, the current in the system induces a magnetic field to drive the plasma along the rails to the ends. When no external magnetic field is applied perpendicular to the plane of the rails, the rail motor is analogous to a series-wound motor. With an external magnetic field, the system is analogous to a series shunt-wound motor. Substitution

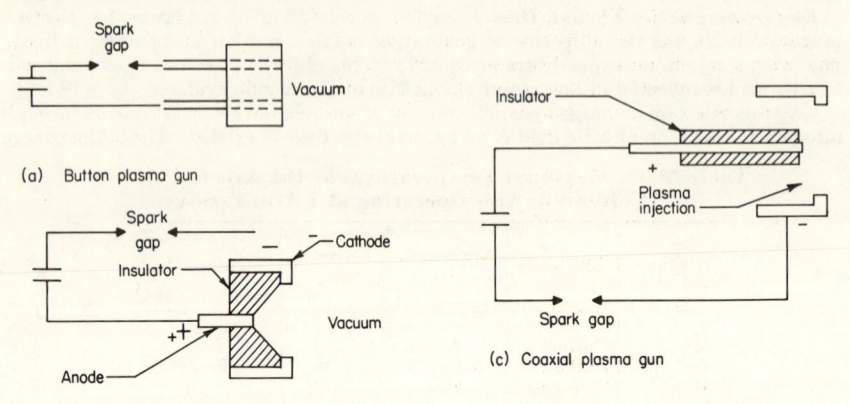

Fig. 20-16. Button plasma guns.

of flat plates for cylindrical rails provides better focusing of the plasma.[24] Plasma velocities up to 3.3×10^6 ft/sec have been obtained.

The rail-type motor presents fewer energy-source coupling difficulties than do button guns. Operation is necessarily intermittent, the flow pattern is wide-angled, and the thrust limited.

The Kolb T-strap motor (Fig. 20-17b) also applies the principles of crossed-field acceleration. The electric field across the arc gap and the induced magnetic field of the cathode lead, which is brought close to the arc gap, are at right angles. The plasma velocity is away from the cathode lead. The plasma generated includes metallic ions from the electrodes and electrons.

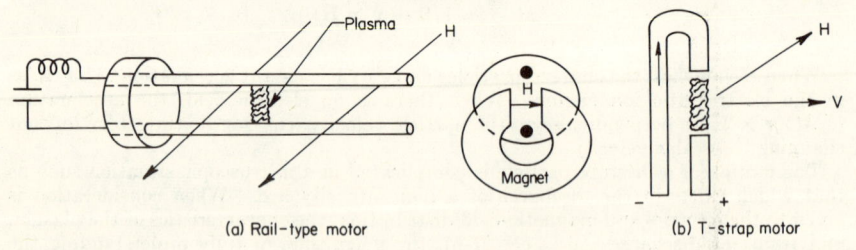

Fig. 20-17. Crossed-field motors.

In both the rail-type and Kolb guns, plasmas have been successfully produced by exploding wires across the arc gaps. Using 1-mil tungsten wire, specific impulses of 2,200 sec at efficiencies of 39 per cent have been obtained.[26]

Traveling-field Accelerators. Thrust can be generated by the interaction of a plasma ring and a moving magnetic field. The ring corresponds to the rotor bar of an induction motor, and is forced to move by the moving field. A schematic of such a system is shown in Fig. 20-18. The ionized gas leaves the plasma source and enters a cylindrical coil whose windings are arranged to produce an alternating radial field moving in the direction of flow. The coil arrangement moves the field with increasing velocity (constant acceleration). The initial field velocity is somewhat higher than that of the entering plasma. The final velocity is determined by the number of coils.

The force required to produce the constant acceleration of the plasma emanates from the interaction of the magnetic field and the induced current circulating around the ring through the conducting plasma. Because the ring is moving with a velocity slightly less than that of the field, the current is maintained. The current tends to maintain a high degree of ionization in the plasma and a stable configuration from pinch effect. These effects are sustained only if the acceleration is high; thus a high specific impulse is expected to result.

Present estimates of the obtainable specific impulse are from 3,000 to 15,000 sec.[27] Experimental values up to 25,000 sec also have been obtained.[28] The estimated efficiency of this type of device is 85 per cent. This figure does not include the power required to generate the plasma.

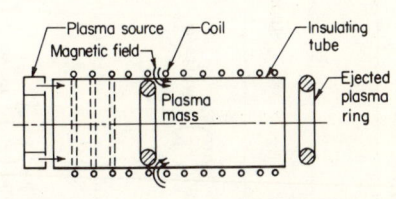

FIG. 20-18. Traveling-magnetic-field accelerator.

Ion Rocket Engine. The ion rocket engine is a system including three major subsystems: the energy source, the power convertor, and the ion thrust chamber (Fig. 20-19). The thrust chamber[30] includes an ion source, an array of electrodes to focus and accelerate them, and an electron emitter.

Vaporized propellants are fed to the ion source, where they are ionized either by contact with heated metals of high work function, or by being subjected to electron bombardment within electromagnetic fields. The latter method of ionization is commonly described as the arc source.

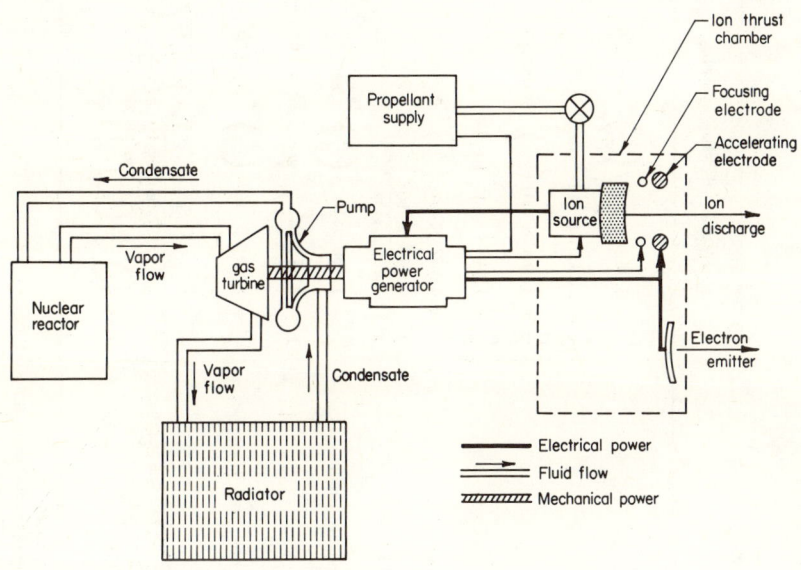

FIG. 20-19. Typical ion propulsion system.

The ionized propellant is directed from the ion source through focusing electrodes and accelerating electrodes, which impart kinetic energy to the ions. This results in the required changes of momentum of the particles to develop thrust for the vehicle. The major performance parameters of the ion engine[30] depend upon the rates of the accelerating potential V and the ion mass A, the gram-atomic or gram-molecular weight. These parameters[30] are summarized in Table 20-14.

Table 20-14. Summary of Performance Parameters, Ion Rocket Engines

Parameter	English system	Definition CGS system
Specific power	$\dfrac{P}{W} = 3.089 \times 10^4 G \left(\dfrac{V}{A}\right)^{1/2}$ watts/lb $\dfrac{P}{W} = 41.42 G \left(\dfrac{V}{A}\right)^{1/2}$ hp/lb	$\dfrac{P}{W} = 0.0695 \left(\dfrac{V}{A}\right)^{1/2}$ watts/g
Power per unit thrust	$\dfrac{P}{F} = 3.089 \times 10^4 \left(\dfrac{V}{A}\right)^{1/2}$ watts/lb $\dfrac{P}{F} = 41.42 \left(\dfrac{V}{A}\right)^{1/2}$ hp/lb	$\dfrac{P}{F} = 0.0695 \left(\dfrac{V}{A}\right)^{1/2}$ watts/dyne
Average exhaust velocity	$C = 4.557 \times 10^4 \left(\dfrac{V}{A}\right)^{1/2}$ ft/sec	$C = 13{,}890 \left(\dfrac{V}{A}\right)^{1/2}$ m/sec
Exhaust current per unit thrust	$\dfrac{I}{F} = \dfrac{3.089 \times 10^4}{A(V/A)^{1/2}}$ amp	$\dfrac{I}{F} = \dfrac{0.0695}{A(V/A)^{1/2}}$ amp
Propellant flow rate per unit thrust	$\dfrac{\dot{W}}{F} = 7.060 \times 10^{-4} \left(\dfrac{V}{A}\right)^{-1/2}$ sec^{-1}	$\dfrac{\dot{W}}{F} = 7.199 \times 10^{-7} \left(\dfrac{V}{A}\right)^{-1/2}$ sec^{-1}
Specific impulse	$I_s = 1{,}417 \left(\dfrac{V}{A}\right)^{1/2}$ sec	$I_s = 1{,}417 \left(\dfrac{V}{A}\right)^{1/2}$ sec
Minimum throat area per unit thrust (ion rocket only)	$\dfrac{A_t}{F} = 5.651 \times 10^{11} \left(\dfrac{V}{d}\right)^{-2}$ sq ft/lb	$\dfrac{A_t}{F} = 127.1 \left(\dfrac{A}{d}\right)^{-2}$ sq m/dyne
Engine operating time	$t = \dfrac{1 - \exp\dfrac{4.557 \times 10^4 (V/A)^{1/2}}{-\Delta v} \left(\dfrac{V}{A}\right)^{1/2}}{7.060 \times 10^{-4} G}$ sec	$t = \dfrac{1 - \exp\dfrac{1.386 \times 10^6 (V/A)^{1/2}}{-\Delta v} \left(\dfrac{V}{A}\right)^{1/2}}{7.199 \times 10^{-7} G}$ sec
Equivalent particle weight	$A = \left[\sum \dfrac{I_i}{I} \left(\dfrac{A_i}{h_i}\right)^{1/2}\right]^2$	$A = \left[\sum \dfrac{I_i}{I} \left(\dfrac{A_i}{h_i}\right)^{1/2}\right]^2$

NOTE: G = thrust/wt. ratio
A_i = atomic weight number of ejected ions
a = acceleration, cm/sec^2

20-38

PROPULSION

The ion discharge is balanced by an equal flow of oppositely charged particles; otherwise the surface of the vehicle develops a high induced charge, inhibiting the ion flow. An electron emitter is shown in Fig. 20-19. The electron discharge can be achieved by thermionic emission from a heated tungsten plate or by field emission of electrons from discharge points incorporated in the accelerating electrodes. An alternative method of obtaining a neutral beam is by charge-exchange techniques.

Power to drive the ion thrust chamber can be obtained from high-speed electrical power generators. Three types which may be applied to the ion rocket engine are a permanent-magnet generator, an a-c generator, or an electrostatic generator. The distribution of power to the various components of the system is shown by the solid lines in Fig. 20-19. Most of the power is absorbed in pumping the electrons, which are collected at the ion source, up to the potential of the accelerating electrode. Power required to generate ions in the ion source is a significant factor in determining the power efficiency of the thrust device.

Ion Sources. Several types of ion sources, including radio-frequency, thermionic arc, and surface contact, have been investigated for application in ion rocket engines.

What little information is available on radio-frequency ion sources indicates that the efficiency is low and the associated operating equipment heavy. These factors eliminate these sources for applications in the ion engine. Thermionic sources are based on the emission of ions from a heated mixture of chemical compounds. These also have proved heavy and inefficient.

The adoption of an arc-type ion source or a surface-contact source depends upon the power demands of the source. A brief preliminary analysis of the efficiency is carried out for both sources.

The power required for a surface-contact source includes that radiated from the heated ionizer surface, the heat required to vaporize the propellant from ambient conditions, and radiation from structural parts. The last is difficult to predict except by test. The efficiency of the surface-contact engine is shown by

$$\eta = \frac{P}{P + P_R + P_R'}$$
$$= \frac{1}{1 + \dfrac{1.04 \times 10^{-4}\epsilon T^4}{X^2(V/A)^{1/2}} + \dfrac{P_R'}{P}}$$

in which P_R' represents the power required to vaporize the propellant, beam loss from interception, and other external losses. Preliminary computations indicate that these power losses are small in comparison with the radiation loss if the ion engine is well designed.

The efficiency of the arc system becomes

$$\eta = \frac{P}{P + P_{\text{arc}} + P_R'}$$
$$= \frac{1}{1 + (V_a/V)\sqrt{1,836 A} + P_R'/P}$$

in which V_a is the operating voltage of the arc. Again, P_R' represents the power for vaporization of propellant, loss from beam interception, and other external losses.

The data of Fig. 20-20 summarize and compare the variation of the power efficiency η of an arc-type ion engine and a surface-contact engine. These preliminary data indicate how this parameter can be expected to vary, furnish additional information for selection of a propellant, and present areas of application for the two types of engines. Curve 1 of Fig. 20-21 shows the efficiency variation with accelerating voltage of an arc-type engine, with the arc operating at 20 volts and using cesium propellant. Curve 2 represents similar data for an engine incorporating a 10-volt arc, again with cesium metal as the propellant. Many possible propellant materials are most efficiently ionized for arc voltages between 100 and 150 volts. Curve 5 indicates the efficiency of a 100-volt arc, using a propellant of the same ion weight as cesium.

Curve 3 is the efficiency variation of a surface-contact engine using cesium that is ionized by a 2200°K tungsten plate. The strength of the accelerating field X is maintained at 10,000 volts/cm. As the field strength exceeds 100,000 volts/cm, the efficiency approaches 100 per cent, as shown in curve 4. In all computations, the total emissivity of the tungsten ionizer was assumed to be 0.28.

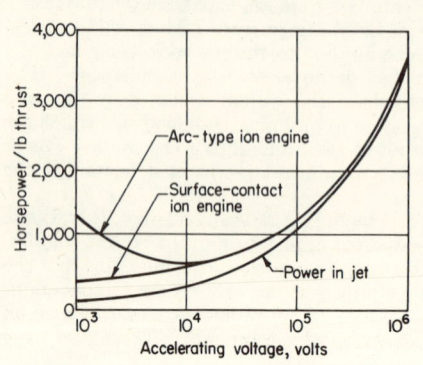

Fig. 20-20. Power requirements of ion engine.

These curves are not generally applicable, being computed for specific operating conditions that were selected to establish the minimum efficiencies and maximum power per unit thrust of the two types of engines. A significant factor is that in the specific impulse range (3,900 to 124,000 sec) covered by this analysis, the ion engine with an arc-type ion source can be expected to have a minimum total power demand for a specific impulse between 11,000 and 13,000 sec.

These qualitative data indicate that, within the limits of the analysis, cesium ionized on tungsten is superior for low accelerating voltages and low specific impulse. In the intermediate range of specific impulse, a cesium arc is competitive with a cesium surface-contact engine, and is superior to an arc-type engine using a propellant requiring a high operating voltage. For high accelerating voltage and specific impulse, the two types of sources are competitive.

The difference between the power requirements at lower specific impulse indicates a significant superiority for the surface-contact engine if the accelerating field strength

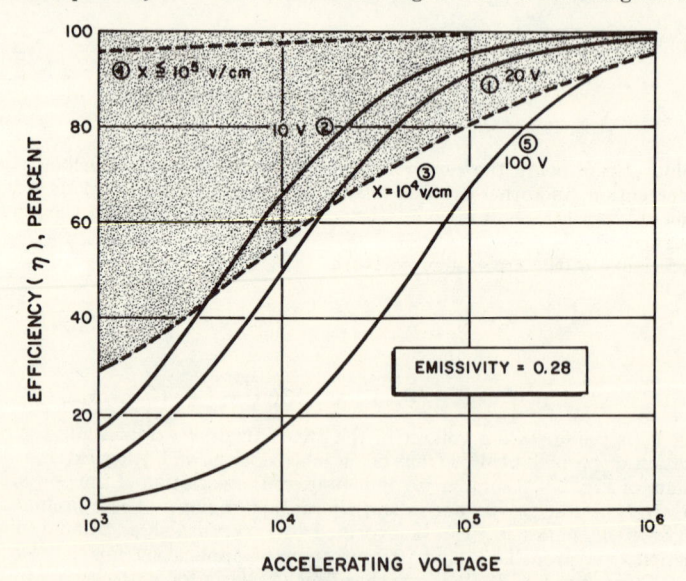

Fig. 20-21. Comparative efficiencies of ion rocket engine.

is high. An accurate evaluation can be made only from the experimental results from tests of the two different engine types.

Since the plasma, jet, and magnetohydrodynamic engines all depend upon generation of plasma, and the designs reviewed postulated arc-type sources, consideration of the power efficiency of the plasma sources is important.

PROPULSION 20–41

Charged Colloid Sources. The generation of charged colloid particles is a two-stage process. First, the propellant must be dispersed as colloidal-sized particles in a near vacuum. Then, the colloidal particles are charged to give the desired charge-to-mass ratio and ejected through an accelerating electrode system to generate thrust. Colloid propulsion is still in early stages of research.

Propulsion-system Applications. The propulsion-system weights and the operating times of electrical rocket engines also are convenient variables with which to study the operation of different low-thrust power plants. The results of a study performed at Rocketdyne to determine the competitive operating regimes for rocket engines of thrust levels between 0.1 and 5.0 lb[29] are summarized in Fig. 20-22. The minimum propulsion-system weight is shown as a function of the operating time of

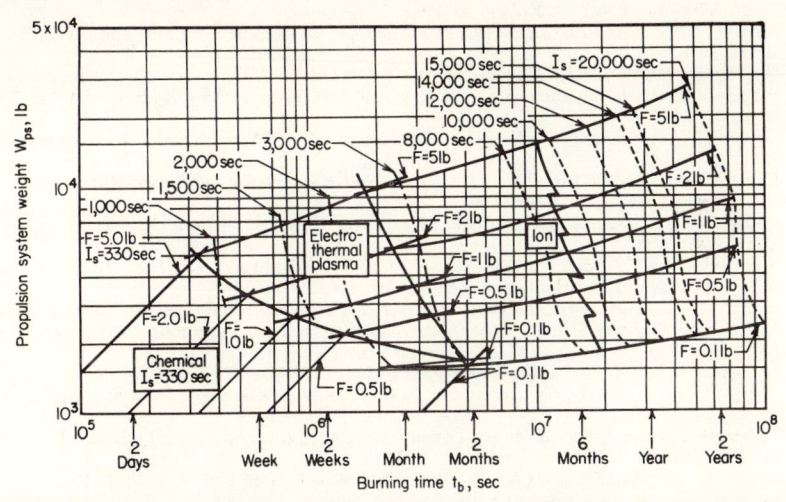

Fig. 20-22. Propulsion-system weight vs. burning time for optimum specific impulse.

the rocket engines between the two thrust levels. The specific impulse for a minimum-weight power plant establishes the operating conditions of the electrical engines. These preliminary data indicate that chemical, electrothermal plasma, and ion power plants for an upper-stage vehicle will be the lightest in the areas so labeled in Fig. 20-22.

Intermediate thrust levels and the specific impulse of the electrical thrust devices also are shown. The chemical rocket system is arbitrarily held at the fixed specific impulse of 330 sec, substantially all the propulsion-system weight being in the propellant. The operating regime of the colloid engine overlaps the electrothermal plasma engine and the ion engine; how much overlap exists is now difficult to determine, because the colloid engine is in the early research stage.

With the method of analysis used to obtain the data in Fig. 20-22, a working assumption was to neglect the thrust chamber, feed system, and propellant-tank weights of electrical propulsion systems that did not use heavy magnets or large capacitors (as in the magnetohydrodynamic engines).

The weight of the power-plant system considered includes the power-plant-section weight and the propellant.

If the thrust of the second-stage vehicle is less than 0.1 lb, an ion engine is superior to both the chemical and colloid rockets, provided the operating time for the vehicle mission is longer than 6 weeks. For shorter duration, the low-thrust chemical engine will be the lightest system.

REFERENCES

1. Durand, W. F.: *Aerodynamic Theory*, vol. V, Durand Reprinting Committee, California Institute of Technology, Pasadena, Calif. 1943.

2. Salisbury, J. K.: *Kent's Mechanical Engineers Handbook*, 12th ed., Power Volume, John Wiley & Sons, Inc., New York, 1960.
3. Marks, L. S. (ed.): *Mechanical Engineers' Handbook*, 6th ed., McGraw-Hill Book Company, New York, 1958.
4. Totten, A. I.: *Trans. ASME*, Paper RR59-3, May, 1937.
5. Chapman, L. B.: *The Marine Power Plant*, 2d ed., McGraw-Hill Book Company, New York, 1942.
6. Taylor, D. W.: *Speed and Power of Ships*, Ransdell, Inc., Washington, D.C., 1933.
7. Pye, D. R.: *The Internal Combustion Engine*, 2 vols. Oxford University Press, Fair Lawn, N.J., 1931.
8. Livingston, S. P., and W. Gravey: *Tables of Airspeed, Altitude and Mach Number Based on Latest International Values for Atmospheric Properties and Physical Constants*, National Aeronautics and Space Administration, Langley Research Center, Langley Field, Virginia, TN D-822, August, 1961.
9. Zucrow, M. J.: *Jet Propulsion and Gas Turbines*, John Wiley & Sons, Inc., New York, 1947.
10. Weick, F. E.: Working Charts for the Selection of Aluminum Alloy Propellers of Standard Form to Operate with Various Aircraft Engines and Bodies, *NACA Tech. Rept.* 350, 1935.
11. Sutton, G. P.: *Rocket Propulsion Elements*, 2d ed., John Wiley & Sons, Inc., New York, 1956.
12. Koelle, H. H. (ed.): *Handbook of Astronautical Engineering*, McGraw-Hill Book Company, New York, New York, 1961.
13. Clarke, H. C.: *Interplanetary Flight*, Harper & Row, Publishers, Incorporated, New York, 1951.
14. Moeckel, W. A.: *Fast Interplanetary Missions with Low-thrust Propulsion Systems*, NASA TR R-79, Lewis Research Center, Cleveland, Ohio, 1961.
15. Irving, J. H., and E. K. Blum: *Comparative Performance of Ballistic and Low-thrust Vehicles for Flight to Mars*, Vistas in Astronautics, vol. II, Pergamon Press, New York, 1959.
16. Morgan, N. E., W. D. Maroth, and R. C. Thomas: *Piston Engines Offer Advantages for Space Power Applications*, Aerospace Fluid Power Conference, Detroit, Mich., Oct. 29–30, 1962.
17. Thirtieth Inventory of Aerospace Power, *Aviation Week and Space Technology*, vol. 78, no. 10, Mar. 11, 1963.
18. Judge, A. W.: *Gas Turbines for Aircraft*, Chapman & Hall, Ltd., London, 1958.
19. Lombard, A. A.: *Low Consumption Turbine Engines*, pp. 482–503, Fifth International Aeronautical Congress, Los Angeles, Calif., June 20–25, 1955, Institute of the Aeronautical Sciences, Inc., New York.
20. Giles, C.: The V-1 Robot Bomb, *Astronautics*, September, 1944.
21. Warren, F. A.: *Rocket Propellants*, Reinhold Publishing Corporation, New York, 1958.
22. Huebner, A. L., and R. H. Boden: *Critical Power Supply Problems in Ion Propulsion*, Space Exploration Regional Meeting, Institute of the Aeronautical Sciences, American Rocket Society, San Diego, Calif., Aug. 5–6, 1958.
23. Post, Richard F.: Controlled Fusion Research—an Application of the Physics of High Temperature Plasmas, *Rev. Mod. Phys.*, vol. 28, no. 3, pp. 338–362, July, 1956.
24. Alperin, M., and G. P. Sutton: *Advanced Propulsion Systems*, Pergamon Press, New York, 1959.
25. Spitzer, L., Jr.: *The Physics of Fully Ionized Gases*, Interscience Publishers, Inc., New York, 1956.
26. Kash, S. W., and W. L. Starr: *Experimental Results with a Colinear Electrode Plasma Accelerator and a Comparison with Ion Accelerators*, Reprint 1008–1959, American Rocket Society Annual Meeting, Washington, D.C., Nov. 16–20, 1959.
27. Fonda-Bonardi, G.: *Final Report—Research Study on Plasma Acceleration*, AFOSR TR-59-170, Litton Industries, Beverly Hills, Calif., Sept. 30, 1959.
28. Patrick, R. M.: Description of a Propulsive Device Employing a Magnetic Field, *Research Rept.* 28, ASTIA No. AD 159614, Avco Research Lab., Everett, Mass., May, 1958, Contract AF49(638)-61.
29. Littman, T. M.: *A Critical Evaluation of the Ion Rocket*, Second Symposium on Advanced Propulsion Concepts, AFOSR, Boston, Mass., Oct. 1, 1959.
30. Boden, R. H.: *The Ion Rocket Engine*, AFOSR TN 57-573 (R-645), Rocketdyne, A Division of North American Aviation, Inc., Canoga Park, Calif., July 3, 1958.
31. Sanders, R. F.: The New Chevrolet V-8 Engine, *Trans. SAE*, vol. 63, pp. 401–420, 1955.

Chapter 21

SELECTED ENERGY-CONVERSION SUBSYSTEMS

W. G. DOW, *Professor and Chairman, Department of Electrical Engineering, University of Michigan.*

CONTENTS

21-1. Energy-conversion-subsystem Optimization Principles........	21-2
Underlying Measures of Value.................................	21-2
Limited and Nontraditional Scope of This Energy-conversion-subsystem Treatment.................................	21-2
Value Measures for Energy-conversion Subsystems...............	21-2
21-2. Principles Underlying Conversion from Thermal Sources......	21-3
Conceptual Comments as to the First Law of Thermodynamics...	21-3
Effectiveness or "Efficiency" of Thermodynamic Conversion Processes.................................	21-4
21-3. Voltage Profiles in Solid-state Materials.....................	21-5
Electrical Generation by Charged-particle Passage against Potential Steps or Slopes.................................	21-5
Steps or Barriers Existing in Contrasting Inside, Outside, and Electrochemical (Fermi-level) Potential Profiles around a Circuit...	21-6
21-4. Thermoelectric Conversion.................................	21-7
The Electrochemical or Fermi-level Potential as the Thermoelectric-conversion Random-energy-distribution Reference.............	21-7
Charge Transport by Majority-carrier Electrical Conduction and Minority-carrier Diffusion.................................	21-11
Power Generation; Junction Heating and Cooling Due to Recombination and Electron-hole Pair Generation.....................	21-11
The Thermoelectric Open-circuit Voltage........................	21-14
The Seebeck Coefficients for Thermoelectric Conversion..........	21-14
Thermoelectric-conversion-subsystem Geometry..................	21-16
Thermoelectric-conversion Efficiency; Materials Figure of Merit..	21-16
Properties of Thermoelectric Materials........................	21-19
Summary of Engineering Attributes of Thermoelectric Conversion	21-21
21-5. Thermionic Conversion.................................	21-21
The Surface Motive Barriers Surmounted by Electrons in Thermionic Emission.................................	21-21
The Internal-profile Barrier Potential and the External-profile Contact Potential for Metals.................................	21-24
The Thermionic-converter Plasma........................	21-25
Plasma Sheaths.................................	21-25
Motive Distribution.................................	21-26

Circuit Voltage and Current Relationships.................... 21-27
Thermionic-conversion Heat-power Relations.................. 21-28
Converter Effectiveness and Efficiency......................... 21-29
Engineering Considerations................................... 21-32
21-6. Magnetohydrodynamic Conversion........................ 21-32
Introduction; Approximate State of the Art...................... 21-32
Introductory Analysis of Electrical Aspects of MHD
Generation.. 21-34
Engineering Design and Operational Problems.................. 21-37
21-7. Thermonuclear-fusion Generation......................... 21-38
Introduction; Scientific Basis.................................. 21-38
State of the Art... 21-39

21-1. Energy-conversion-subsystem Optimization Principles

Underlying Measures of Value. The phrase "system engineering" implies optimization, which in turn implies attention either to a not-yet-existent system during its planning stage or to devising substantial changes in an existing system, as to either apparatus or operation.

Optimization requires measures of value. The traditional engineering measure of value in United States economy has been the dollar. In the present decade at least four other underlying contrasting and sometimes dominating measures must be added, in all including at least:

1. Values described in terms of dollars—this establishes the scale of dollar measure which threads in a subsidiary way through all the others.

2. Individual human values such as basic education, resources for enjoyment of life, a healthy physique, and survival into pleasant and useful later years.

3. The normal collective human values of good government, a stable economy, and wisdom and mutual good will among our own people.

4. Value toward success in political, social, and technological efforts to establish confidence in and gain adherence to the cause of the free world.

5. Values toward success, from military posture or in limited military actions, in the cold war, and toward victory and national survival if a hot war comes.

Limited and Nontraditional Scope of This Energy-conversion-subsystem Treatment. Attention here will be limited to conversion toward providing the electrical energy resources for more broadly purposeful engineering objectives. Thus the present treatment deals with energy-conversion *subsystems* that form parts of larger engineering systems. To narrow attention still further to meet space limitations, this treatment will deal, in detail, only with certain kinds of conversion from thermal to electrical energy which are not exhaustively dealt with in traditional engineering literature. These are:

Thermoelectric conversion, in some detail (Sec. 21-4)
Thermionic conversion, in some detail (Sec. 21-5)
Magnetohydrodynamic ("MHD") generation (Sec. 21-6)
Controlled thermonuclear-fusion generation—a glance into the future (Sec. 21-7)

Areas of applicability of the familiar Carnot-cycle efficiency will be discussed. There will be a few comments on fuel cells.

Value Measures for Energy-conversion Subsystems. Optimization must apply to a total system, or at least as large a portion as human intellectual resources can encompass in one package. Therefore, one must identify value measures—subsidiary to the underlying ones identified above—that can apply as well to other systemically adjacent subsystems, to permit measuring net gains for trade-offs.

The phrases "dollar or difficulty costs" and "expensive or difficult" will sometimes be used here. These are required because of a parallelism in the intersystem com-

SELECTED ENERGY-CONVERSION SUBSYSTEMS   21-3

parison values to the contrasts between the underlying value measures. The dollar cost in 1942 of getting operational fuel oil overland to a military radio-communication outpost 20 miles away through an enemy-infested South Pacific jungle was perhaps not great, but the "difficulty cost" involved risks to the lives of men who were precious to the success of the overall mission.

For interrelating energy-conversion subsystems and other systemically adjacent subsystems, as well as for comparisons internal to the subsystem, useful positive or negative value measures include:

1. Input-output energy efficiency (often of secondary importance, partly redundant with items 6, 7, 8 below)
2. Assurance that performance will continue as specified (i.e., reliability)
3. Weight (the apparatus must be supported and transported)
4. Geometric size and shape
5. Time required to design and build the apparatus
6. Dollar or difficulty costs of obtaining and putting in the input energy, whatever its embodiment
7. Dollar or difficulty costs of the specific energy-converting apparatus
8. Dollar or difficulty cost of apparatus for rejecting the nonconverted or misconverted energy (sometimes a dominating cost item)
9. Ease, flexibility, and convenience of control and use (that is, human factors)

As to item 1, it is only rarely desirable to use conversion that is less than 10 per cent efficient, and rarely worthwhile to pay for one that is more than 90 per cent efficient.

An *inefficient* converter may require relatively heavy and expensive or difficult-to-get waste-energy-rejection apparatus. Yet if this inefficient converter is itself much lighter in weight, is desirably flexible, and uses cheaper input energy than an efficient converter, the optimization may show substantial net subsystem value gain from using it, and accepting the processing of rather large amounts of power apparently wastefully. This is a very common situation.

Frequently item 9, because it involves human values, dwarfs all the others in importance. These values are hard to measure; they are intangible, therefore requiring experienced professional judgment in the decision. It may, in an individual case, require much more than an order-of-magnitude accumulated penalty from items 1 through 8 to negate a tentative intangibly based item-9 decision choice.

There are many parallels between the process of design optimization and the identification of the "critical path" for the actual building of a large-scale system (Chapter 36). In both, the resources of nature and of mankind do in fact contain an optimum, the initial engineering problem being to discover it. The "critical path" for many projected system-design optimization procedures leads through, is dominated by, and is sometimes driven to nonrealizibility by, the energy-conversion subsystem. It therefore follows that there exists a major present need, only partially now in progress, for imaginative applied-science research toward obtaining order-of-magnitude improvements in energy-conversion subsystems.

21-2. Principles Underlying Conversion from Thermal Sources

Conceptual Comments as to the First Law of Thermodynamics. The first law of thermodynamics provides a dominating principle governing conversion from thermal to any other kind of energy. Its basic meaning can be expressed as follows, for a subsystem portion (i.e., a gas) that is internally at a uniform temperature, that is, has its random-state energies internally in statistical equilibrium:

$$\begin{bmatrix} \text{Incremen-} \\ \text{tal heat-} \\ \text{energy} \\ \text{input} \end{bmatrix} = \begin{bmatrix} \text{incremental} \\ \text{increase in} \\ \text{energy of or-} \\ \text{dered states} \end{bmatrix} + \begin{bmatrix} \text{incremental increase} \\ \text{in "internal energy"} \\ \text{statistically distrib-} \\ \text{uted among random} \\ \text{states} \end{bmatrix} \quad (21\text{-}1)$$

Of course, "input" and "increase" can if negative signify "output" and "decrease." When partially converted to a mathematical format, this becomes

$$dH = \begin{bmatrix}\text{incremental increase} \\ \text{in energy of ordered} \\ \text{states}\end{bmatrix} + T\,dS \qquad (21\text{-}2)$$

"Conversion from thermal sources" implies a study of the division of energy between the two right-side terms. In this and later expressions:

dH = incremental heat energy transferred into the subsystem portion, in joule units

H = a cumulative measure in joules of this heat transfer; the symbol H has little meaning except as a measure of *transfer*, because *storage* of the energy before or after transfer is always described in terms of the two items on the right side of (21-1)

$T\,dS$ = the change, in joules, of the "internal energy" *storage*, that is, in the energy stored in the various random states or modes of the subsystem portion, being in statistical equilibrium among these states; $T\,dS$ is measurable

T = absolute temperature of the subsystem portion, in degrees Kelvin, a measurable quantity

dS = the increment in entropy associated with the change in internal energy $T\,dS$; it is thought of as $T\,dS/T$; units are joules per degree Kelvin

S = a cumulative measure of the entropy into the subsystem portion, in joules per degree Kelvin; in general, entropy is defined only within a constant

The "energy of ordered states" can consist of one or more kinds of energy storage measurable in macroscopic terms. Examples are kinetic energy of a moving piston and objects linked to it; energy storage in the magnetic field of a thermoelectrically generated current; gravitational potential energy of a weight lifted by gas expansion; the energy associated with permanent enthalpy changes of processed gas.

The fundamental engineering problem in conversion from thermal sources is that of finding means for causing the entering heat to appear in particular desired "ordered states," rather than as internal energy $T\,dS$.

Effectiveness or "Efficiency" of Thermodynamic Conversion Processes. One must not apply blindly what is sometimes called the "Carnot-cycle" limitation to achievable efficiency. Rather, it is necessary to understand clearly the principles that govern values in thermodynamically based conversion subsystems.

Regardless of the nature or processing of the working fluid, a very simple and fundamental energy-balance relationship applies, as follows:

$$\begin{bmatrix}\text{Heat energy trans-}\\ \text{ferred in from a}\\ \text{high-temperature}\\ \text{heat source}\end{bmatrix} = \begin{bmatrix}\text{heat energy trans-}\\ \text{ferred out to a}\\ \text{specific low-tem-}\\ \text{perature heat sink}\end{bmatrix} + \begin{bmatrix}\text{heat-energy}\\ \text{leakage, other-}\\ \text{wise than to}\\ \text{the heat sink}\end{bmatrix}$$

$$+ \begin{bmatrix}\text{energy miscon-}\\ \text{verted into use-}\\ \text{less nonthermal}\\ \text{ordered states}\end{bmatrix} + \begin{bmatrix}\text{energy converted}\\ \text{into the desired}\\ \text{ordered states}\end{bmatrix} \qquad (21\text{-}3)$$

Ignoring for the moment the "heat-leakage" term, a dimensionless-ratio description of conversion-process performance might be:

Converter effectiveness

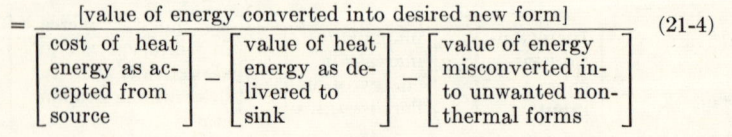

$$= \frac{[\text{value of energy converted into desired new form}]}{\begin{bmatrix}\text{cost of heat}\\ \text{energy as ac-}\\ \text{cepted from}\\ \text{source}\end{bmatrix} - \begin{bmatrix}\text{value of heat}\\ \text{energy as de-}\\ \text{livered to}\\ \text{sink}\end{bmatrix} - \begin{bmatrix}\text{value of energy}\\ \text{misconverted in-}\\ \text{to unwanted non-}\\ \text{thermal forms}\end{bmatrix}} \qquad (21\text{-}4)$$

SELECTED ENERGY-CONVERSION SUBSYSTEMS 21-5

In most thermally driven converters the heat from the source is expensive or difficult to obtain. The heat delivered to the sink and the misconverted energy may have positive or negative value, depending on whether they can be usefully employed or prove to be disposed of only with difficulty. The use of (21-3) to rearrange the numerator of (21-4) leads then to the following formal but sometimes rather uninformative description of "conversion efficiency," presuming *zero value* for energy misconverted, for that into the heat sink, and ignoring heat leakage:
Conversion efficiency

$$= \frac{\begin{bmatrix} \text{heat energy} \\ \text{accepted from} \\ \text{the source} \end{bmatrix} - \begin{bmatrix} \text{heat energy} \\ \text{delivered to} \\ \text{the sink} \end{bmatrix} - \begin{bmatrix} \text{misconverted} \\ \text{nonthermal} \\ \text{energy} \end{bmatrix}}{[\text{heat energy accepted from the source}]} \quad (21\text{-}5)$$

To make this take the form of the familiar "Carnot-cycle" efficiency limitation, two further assumptions are made, itemized here in combination with two already introduced:

1. The heat energies accepted from the source and delivered to the sink are in fact proportional to the temperatures at acceptance and delivery.
2. All energy conversion is into the desired form—there is no misconverted energy.
3. All rejected heat goes to the sink—none "leaks."
4. As presumed earlier, the heat rejected into the sink is presumed to have zero value.

Only by introducing these rather severe specifications does one obtain the familiar Carnot form

$$\text{Carnot efficiency} = \frac{T_h - T_c}{T_h} \quad (21\text{-}6)$$

Here and in many later equations, in degrees Kelvin,

T_h = the temperature at which heat energy is accepted from the heat source
T_c = the temperature at which heat energy is delivered to the heat sink

The essential point here is that, although the principles of the Carnot-cycle limitation apply to any conversion from a thermal source, the precise form which the limiting takes can have many variants, often exhibiting substantial contrasts from Eq. (21-6).

21-3. Voltage Profiles in Solid-state Materials

Electrical Generation by Charged-particle Passage against Potential Steps or Slopes. In all so-called "direct-conversion" subsystems the electric-power generation results from charged particles being impelled to surmount a slope or successive steps of electrical potential opposing their passage.

In an MHD (magnetohydrodynamic) generator the high-velocity forced passage of a conducting gas (plasma) through strong crossed d-c electric and magnetic fields uses the magnetic Lorentz force to drive the electrons of the plasma against a potential slope transverse both to the gas flow and to the magnetic field. The total potential along the slope may be substantial—perhaps several score or a hundred or more volts.

In all other familiar direct-conversion subsystems—thermoelectric and thermionic conversion, solar cells, and fuel cells—the electric-power generation results from particle-nature electric-charge carriers acquiring energies, from thermal, photoelectric, or chemical sources, or a combination, which cause them to surmount a step of electrical potential opposing their passage. This step occurs at or in the immediate neighborhood of an interface between unlike solid or liquid or semiliquid substances, or in some cases between a solid or liquid substance and a vacuum or plasma. In all cases individual voltage steps are small; useful power generation is obtained by having many steps electrically in series, while having the flow of input energy pass through them in parallel.

Two kinds of charged "particles" have primary importance as current carriers in conversion subsystems:

Electrons. Being unattached to atoms or molecules, the electrons comprise an "electron gas"; the electron gas is the primary charge-carrying medium within metallic conductors, within n-type semiconductors (this designation identifying the negatively charged nature of the principal charge carriers), and in the gaseous-conducting plasmas of magnetohydrodynamic and thermionic conversion subsystems.

Holes. Being electron deficiencies, the holes in a p-type semiconductor comprise a freely moving "hole gas," leading to the "p-type" identification that the principal carriers are positively charged but otherwise behave much like electrons; hole gases do not exist either in metallic conductors or in plasmas.

Positive ions, being atoms or molecules that have each a positive charge from having lost a valence-bond electron, or negative ions, being atoms or molecules that have each a negative charge from an electron attachment, may be very important because in fixed position, or in very slow motion, they establish a volume charge that neutralizes the volume charge of the current carriers. Positive or negative ions become the current carriers in fuel cells.

The thermal attributes of all these particles are very nearly those of an ideal gas, *wherever they exist at relatively low or moderate particle densities.* This low-density occurrence applies for electrons and holes in semiconductors used in thermoelectric subsystems and in transistors, and for electrons and positive and negative ions as they occur in plasmas and in the electrolytic environment of fuel cells; in all these cases the particles exhibit the familiar Maxwell-Boltzmann distribution of random energies and velocities.[3]

The thermal attributes of electrons are those of a degenerate gas when the electron density becomes extremely high, as in metallic conductors. The electrons then exhibit the Fermi-Dirac distribution[1] of random energies and velocities.

Steps or Barriers Existing in Contrasting Inside, Outside, and Electrochemical (Fermi-level) Potential Profiles around a Circuit. Four distinct types of potential profiles exhibit steps or barriers around a series-arranged multimaterial circuit. These are illustrated in Fig. 21-1, which represents a thermoelectric-conversion junction pair in the open-circuit condition, using an energy-level-diagram representation of the electrical properties of familiar semiconductors. To the left of the left junction is an n-type semiconductor, to the right a p-type. The vertical coordinate v is for some purposes an inverted measure of electrical potential in volts, an increasing v describing a declining potential; for other purposes v is a measure of kinetic or potential energy in electron volts, higher values of v describing greater values of random electron kinetic energy in the "conduction band," or lesser values of random hole energy in the "valence-bond band." Quantum-mechanical discreteness of random-energy values, or "energy states," is suggested by the regularly spaced lines in the two bands. Good electrical conductivity exists, because of the majority-carrier electrons in the conduction band for the n-type material, and because of majority-carrier holes in the valence-bond band for the p-type material.

The four kinds of profiles in Fig. 21-1 will be itemized:

1. *There is the V_c inside-the-bulk-material or "internal" electrical-potential profile.* This exhibits interface barriers that in the open-circuit condition in Fig. 21-1 equalize at each interface the two-way random-velocity flow of electrons between materials of greatly different electron densities. This is a profile of a true macroscopic electrical potential; it is the $v = V_c$ profile of the bottom of the conduction band, corresponding to a zero value of random thermal kinetic energy of electrons.

2. *There is the V_v inside-the-bulk-material "hole" potential-energy profile:* this exhibits interface potential-energy steps that in the open-circuit condition in Fig. 21-1 equalize the two-way random-energy-caused passage of holes between materials of greatly different hole densities. This profile's relation to hole energies and movement parallels, except for a difference between hole and electron effective mass, the effects of the (1) internal potential profile on electrons; it is the profile of the $v = V_v$ top of the valence-bond band in Fig. 21-1. Within a given semiconductor (germanium, silicon, cadmium telluride, etc.) V_v is lower than V_c by a fixed amount V_g, this being

the internal ionizing potential ("energy gap") of the semiconductor atoms, describing the electron-volt energy necessary to release an electron from an atom, to permit it to join the electron gas of conduction-band energies; at a junction between unlike semiconductors (germanium-silicon, etc.) the V_v barrier step has a different value than the V_c barrier step, because of the unlike values of the internal ionizing potentials for the two materials.

3. *There is the V_q just-outside-the-surface or "external" electrical-potential profile:* this exhibits electrostatically measurable steps or barriers, surmounted by electrons in electric-power generation by thermionic conversion. This along-the-conductor-surface profile is the line integral of the "contact potential" electric field.[3]

4. *There is the "electrochemical-potential" profile V_F, also called the Fermi-level profile, which is measured by ordinary current-responding d'Arsonval-movement voltmeters:* this is fundamentally the profile of the random-energy-distribution reference level (the Fermi level). This profile's *gradient* exhibits marked and important contrasts

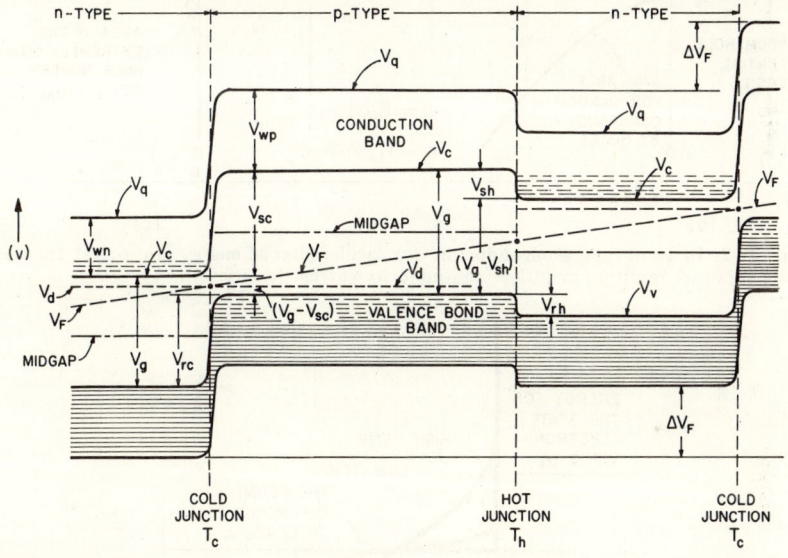

FIG. 21-1. Potential profiles and energy levels for a thermoelectric junction pair in the open-circuit condition.

with the electron-forcing and hole-forcing gradients in V_c and V_v; this profile exhibits barriers surmounted by ionic charge carriers during electric-power generation at the electrochemically active interfaces of fuel cells. The V_F profile is continuous (i.e., exhibits no steps) at interfaces between metallic conductors, or semiconductors, or from either to the other; for this reason the usual d'Arsonval-movement current-responding voltmeters measure steps in the electrochemical profile rather than in any of the other three profiles; thus such voltmeters fail completely to measure changes in the V_c internal potential which governs electron flow within the semiconductor, or in V_v, or in the V_q external profile.

Elaboration relative to properties of these profiles will appear in discussions of the kinds of direct conversion in which each is involved.

21-4. Thermoelectric Conversion

The Electrochemical or Fermi-level Potential as the Thermoelectric-conversion Random-energy-distribution Reference. Thermoelectric conversion occurs by virtue of random thermal energies of charged particles causing sur-

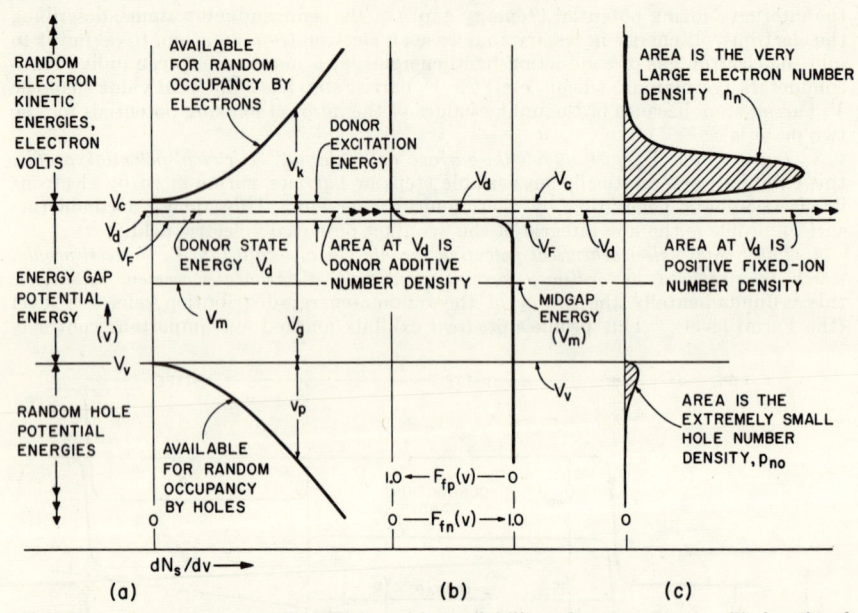

Fig. 21-2. In an *n*-type semiconductor: (*a*) availability of energy states; (*b*) fractional occupancy; (*c*) resulting majority- and minority-carrier densities.

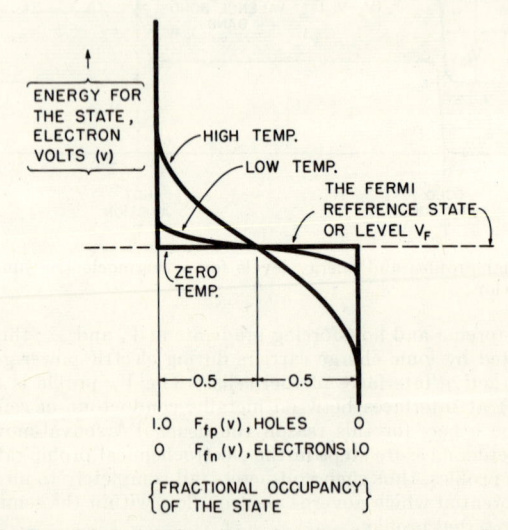

Fig. 21-3. The Fermi function describing fractional occupancy of available states.

mounting of internal voltage profiles. Quantum-mechanical statistics underlie all the electrical aspects of solid-material behavior, including particle-energy randomness. Two fundamental aspects of such statistics will be discussed relative to Figs. 21-2 and 21-3, the vertical coordinates having the same significance as in Fig. 21-1.

Availability of Energy States for Occupancy.[1] A simple particle-model demonstration based on action quantization and on the Pauli exclusion principle (that different

SELECTED ENERGY-CONVERSION SUBSYSTEMS 21–9

particles cannot have the same action quanta) leads directly to the following expression for the "density distribution in electron kinetic energy" for "available" kinetic-energy states, that is, in values of kinetic energy that it is possible for an electron to have:

$$\frac{dN_{sn}}{dv} = 4\pi \left(\frac{2m_e q_e}{h^2} \frac{m_n}{m_e}\right)^{3/2} (v - V_c)^{1/2} \qquad (21\text{-}7a)$$

A duality principle gives similarly, as applied to holes,

$$\frac{dN_{sp}}{dv} = 4\pi \left(\frac{2m_e q_e}{h^2} \frac{m_p}{m_e}\right)^{3/2} (V_v - v)^{1/2} \qquad (21\text{-}7b)$$

Symbolism used here or later is, in mks units:

m_e = mass of an electron, 9.11×10^{-31} Kg
q_e = absolute value of the charge carried by an electron, 1.60×10^{-19} coulomb
h = Planck's quantum of action, 6.62×10^{-34} joule-sec
k_b = Boltzmann's gas constant, 1.380×10^{-23} joule/°K
m_n/m_e, m_p/m_e = ratios, to the electron mass m_e, respectively, of the effective masses m_n and m_p of electrons and holes in the semiconductor interior environment; these are determined by experiment, and lie between perhaps 1.5 and 10 for different semiconductor materials (germanium, silicon, etc.)
dN_{sn}/dv = density distribution, in energy, of availability of kinetic-energy states for occupancy by electrons, being treated as a continuous function of energy because of the exceedingly small discrete energy differences between states
dN_{sp}/dv = similar distribution for holes
V_T = kinetic temperature, in electron volts, related as follows to temperature:

$$q_e V_T = k_b T = \begin{bmatrix} \text{energy per particle} \\ \text{characteristic of} \\ \text{the temperature } T \end{bmatrix} \qquad (21\text{-}8a)$$

$$V_T = \frac{T}{11{,}600} \qquad (21\text{-}8b)$$

where T = temperature, °K
V_{Th}, V_{Tc} = hot- and cold-junction kinetic temperatures, corresponding to the T_h, T_c, hot and cold degree-Kelvin temperatures

The upward-curving and downward-curving parabolas in Fig. 21-2a portray, respectively, Eqs. (21-7a) and (21-7b).

The Fermi Function $F_f(v)$ Describing Fractional State Occupancy.[1] A complicated quantum statistical analysis shows that the fractional occupancy of states depends on temperature and on a reference energy according to the following result, describing the *Fermi function $F_{fn}(v)$:*

$$\begin{bmatrix}\text{Fractional occupancy of random kinetic-}\\ \text{energy states available to electrons}\end{bmatrix} = F_{fn}(v) = \frac{1}{1 + \exp\left[(v - V_F)/V_T\right]} \qquad (21\text{-}9a)$$

A parallel relation is

$$\begin{bmatrix}\text{Fractional occupancy of random poten-}\\ \text{tial-energy states available to holes}\end{bmatrix} = F_{fp}(v)$$

$$= \frac{1}{1 + \exp\left[(V_F - v)/V_T\right]} \qquad (21\text{-}9b)$$

Here
$$F_{fp}(v) = 1 - F_{fn}(v) \qquad (21\text{-}10)$$

V_F is a reference energy, its placement governed by the requirement that the bulk material must be electrically neutral; that is, the total interior charge per unit volume due to positively charged particles is equal and opposite to that due to those negatively charged.

The dependence of (21-9) is illustrated in Fig. 21-3; Fig. 21-2b illustrates the placement of V_F for an n-type semiconductor.

The actual occupancy of states by electrons or holes, and so the "density distribution in random energy" for either kind of carrier, is the product of the availability (21-7) by the fractional occupancy (21-9). The density distributions so obtained are illustrated in Fig. 21-2c, for electrons above the $v = V_c$ zero for random kinetic energy, and for holes below the $v = V_v$ zero for random hole potential energy. The crosshatched area in each case describes the number density for the carrier.

The material in Fig. 21-2 has n-type attributes because of introduction, in processing, of donor additive atoms, as, for example, arsenic, to the extent of perhaps 1/100,000 to 1/1,000,000 atoms of additive per semiconductor atom.[2,8] A suitable donor atom requires only a very small amount of energy, perhaps 0.01 ev, to become ionized, thus releasing a valence electron into the electron gas and becoming itself a fixed-position ion. Because this energy is low, at all ordinary and useful temperatures the additive is to all intents and purposes fully ionized, making the number of additive ions essentially the number of additive atoms.

In Fig. 21-2 the energy level for the donor atoms is shown as $v = V_d$, down very slightly from $v = V_c$. It will be seen that at $v = V_d$ the electron occupancy $F_{fn}(v)$ is very nearly zero, corresponding to essentially complete release of electrons by the donor additives. The value of $F_{fn}(v)$ in the conduction band is still smaller, but the number of available states is so enormously great that the product of availability by occupancy gives the substantial density distribution and total electron number density illustrated in Fig. 21-2c. However, in the valence-bond band, below $v = V_v$, the exponential nature of $F_{fp}(v)$ makes it extremely small, so that the hole density distribution in Fig. 21-2c and the number density of holes are lower by many orders of magnitude than for the electrons. Since the hole density is extremely small, the electron number density is essentially that of the donor additive ions, shown above to be essentially that of the donor additive atoms.

A similar argument applies for the hole number density in a p-type material having acceptor additive atoms, placed as to energy at $v = V_a$, only very little above $v = V_v$.

However, although the neutrality requirement for placement of V_F results in the number density of minority carriers (holes in an n-type material) to be orders of magnitude less than that for majority carriers, the minority-carrier density is still far from zero, and diffusion transport of current by minority carriers is important in thermoelectric energy conversion.

The Fig. 21-2c density distributions in energy of the electrons and of the holes are of precisely the same mathematical form as the density distribution in energy for an ideal gas; hence, as stated earlier, electrons and holes as here dealt with behave like ideal-gas particles.

When the product of Eq. (21-7a) for state availability by (21-9a) for fractional occupancy is integrated from $v = V_c$ to $v = \infty$, the number density of majority carriers so obtained equals, nearly enough, the donor additive density. Before integrating it is recognized from Fig. 21-2b that $F_{fn}(v)$ is extremely small; this can be true only when the exponential term in the denominator of (21-9a) completely masks the unit-value term, which is dropped. The results of the integration rearrange to give the following forms identifying placement of the V_F electrochemical potential (the Fermi level):

$$V_c - V_F = V_T \ln \left[\frac{2}{N_d} \left(\frac{2\pi m_e q_e V_T}{h^2} \frac{m_n}{m_e} \right)^{3/2} \right] \qquad (21\text{-}11a)$$

$$V_F - V_v = V_T \ln \left[\frac{2}{N_a} \left(\frac{2\pi m_e q_e V_T}{h^2} \frac{m_p}{m_e} \right)^{3/2} \right] \qquad (21\text{-}11b)$$

These are for n-type and p-type materials, respectively. Here N_d, N_a, are, respectively, the number densities of donor and acceptor additive atoms.

SELECTED ENERGY-CONVERSION SUBSYSTEMS 21-11

Interesting constants of nature applicable here and later are

$$\left(\frac{2\pi m_e q_e}{h^2}\right)^{3/2} = 3.03 \times 10^{27} \qquad (21\text{-}12a)$$

$$\left(\frac{2\pi m_e k_b}{h^2}\right)^{3/2} = 2.43 \times 10^{21} \qquad (21\text{-}12b)$$

Equations (21-11) have been obtained by requiring bulk-material electrical neutrality. They show that a rise in temperature causes V_F to move away from the V_c or V_v profile toward midgap. This is accounted for in Fig. 21-1, by having the V_F line slope gradually upward, lying close to V_c and to V_v at the cold junction, but remote from them at the hot junction.

This temperature dependence of the electrochemical potential is the basic attribute semiconductor materials have for permitting thermoelectric conversion.

Charge Transport by Majority-carrier Electrical Conduction and Minority-carrier Diffusion. Figure 21-1 illustrates the internal potential and potential-energy profiles for a thermoelectric junction pair in the open-circuit condition, exhibiting the open-circuit thermoelectric electromotive force, but without current flow. Absence of current flow corresponds to zero gradient in the V_c and V_v profiles between the junctions—that is, in the bulk material. When the circuit is closed through a useful load, thermal energies of electrons and holes provide significant overcoming of the net of the junction barriers, resulting in a flow of generated current through the circuit.

In semiconductors generally, electric current can flow either because charge carriers are driven by the appropriate gentle profile gradient, this being the case with *majority* carriers; or, because the charge carriers are driven by a steep local concentration gradient, causing a diffusion current. This latter is the only way in which *minority* carriers can provide a substantial current density. To permit quantitative study, let

n_n, p_p = number densities (per cu m), respectively, of the majority-carrier electrons in an n-type material and the majority-carrier holes in a p-type material; n_n is the crosshatched area in the upper right of Fig. 21-2c

n_{p0}, p_{n0} = for an equilibrium region in the bulk material the number densities, respectively, of the minority-carrier electrons in a p-type material and minority-carrier holes in an n-type material; p_{n0} is the small crosshatched area in the lower right of Fig. 21-2c

n_p, p_n = for diffusion regions, the number densities, respectively, of the minority-carrier electrons and the minority-carrier holes

I = load current, amp, due to carrier flow

Power Generation; Junction Heating and Cooling Due to Recombination and Electron-hole Pair Generation.[9] Figure 21-4 illustrates a cold junction between n-type and p-type materials at which thermoelectric generation of electric power is occurring. The potential step occurs in a very thin "depletion layer" so called because it is depleted of majority carriers, thereby leaving a strong space-charge density due to donor or acceptor additive ions. The depletion exists because the thermal energies of all but a small proportion of the majority carriers are so small that they cannot climb appreciably into the step voltage region.

In Fig. 21-4b electrons "fall" downward and "climb" upward along $v = V_c$ potential steps (the figure is drawn with potentials inverted). In this power-generation condition, a small but very significant net number of majority-carrier electrons from the n-type material succeed because of their thermal energies (substantial even at the cold junction) in climbing up the $v = V_c$ potential hill toward the right, thereby generating electric power in the circuit, because this is a net movement of charge against the field force of the step. Similarly a net number of majority-carrier holes from the p-type material succeed in climbing down the $v = V_v$ potential-energy hill, also generating power. The generation occurs *in the depletion layer*, as being the location of the forcing gradients that are surmounted. The electric-power generation occurs at the expense of thermal energies of electrons and holes; therefore, cooling occurs (of second-

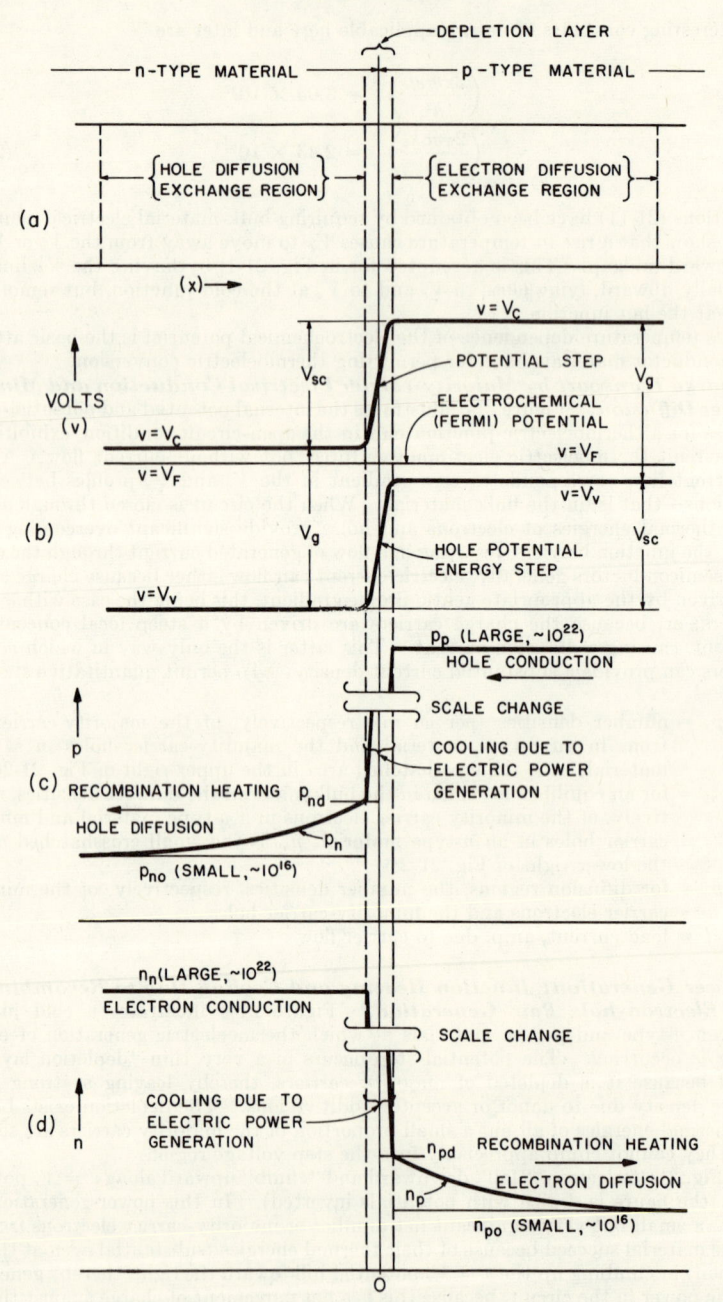

Fig. 21-4. An *np* cold junction generating electric power. (*a*) Region identification. (*b*) Potential steps surmounted by carriers. (*c*), (*d*) Recombination heating in exchange regions.

ary subsystem importance), equal in power rate to the power generation, in the depletion layer.

In Fig. 21-4d there appears, at the interface of the depletion layer with the p-type material, a steep *electron-concentration gradient*, causing the current-carrying electrons that have climbed up the depletion-layer voltage step to diffuse to the right away from the interface and into the bulk p-type material. Although short, the extent of this "diffusion region" is still much greater than the depletion-layer thickness. This diffusion region is also well called an "exchange region." It is here that the current flow represented by movement of electrons into it from the depletion layer, *diffusing* to the right, converts to current flow due to movements of holes into it from the bulk p-type material, flowing by *electrical conduction* toward the left. The flexion of the n_p curve in the exchange region signifies that this region is a "sink" for the electrons; this corresponds to disappearance of electrons by recombination with holes that enter from the right.

Part of the total load current flow is thus exchanged in the electron diffusion region of Fig. 21-4d, the rest of it in the hole diffusion region of Fig. 21-4c. Cold-junction recombinations each release, as heat energy per recombination event, the ionization energy V_g, and in addition an average V_{kc} electron volts comprising the mean random energy of the electron and hole that disappear. Here and later:

V_k, V_{kc}, V_{kh} = respectively in the general case and at hot and cold junctions, the mean random energy of holes and electrons converted to heat energy of the crystal lattice per recombination event, or appearing as the result of an extraction of heat from the crystal lattice per electron-and-hole pair generation; they are of the order of magnitude of V_{Tc} and V_{Th}, and a logical case can be made for V_{kc} being $3V_{Tc}$, and V_{kh} being $3V_{Th}$

V_{kj} = net cooling effect of these exchanges per pair of junctions; $V_{kj} \equiv V_{kh} - V_{kc}$, in the thermoelectric converter

The current I is a measure of the total number of recombination exchanges in the two exchange regions adjacent to a junction. Therefore, the total recombination heat power in the two cold-junction regions taken together is $(V_g + V_{kc})$, presuming a common value for V_g in the two regions. Passage of electrons through the cold-junction depletion layer results in removal from the n-type material of electron kinetic energy in the amount V_{sc} per electron, that appears as electric-circuit energy. Thus, $V_{sc}I$ describes both the electric-power generation and the cooling rates, due to carrier passage across the depletion-layer barrier. Therefore,

$$\begin{bmatrix} \text{Electric-power genera-} \\ \text{tion at this cold junction} \end{bmatrix} = V_{sc}I \quad \text{watts} \qquad (21\text{-}13)$$

$$\begin{bmatrix} \text{Net heating effect at} \\ \text{this cold junction} \end{bmatrix} = (V_g + V_{kc} - V_{sc})I \quad \text{watts} \qquad (21\text{-}14)$$

The behavior at the hot junction exhibits the following contrasts with Fig. 21-4:

1. The hot-junction depletion-layer voltage step V_{sh} is much less than the value V_{sc} at the cold junction, because from Eqs. (21-11) the electrochemical potential V_F must be much closer to midgap at the hot junction than at a cold junction.

2. The exchange regions have concentration gradients sloping downward toward the depletion layer, and represent conversion from hole flow to electron flow, requiring electron-hole pair *generation* rather than recombination. The V_g ev of energy required per pair generation come from the phonon vibration energy of the lattice. Therefore, the hot-junction exchange regions produce cooling, in a combined amount $(V_g + V_{kh})I$.

3. In the hot-junction depletion layer the electrons "fall down" and holes "fall up" the voltage step V_{sh}, both representing conversion of electric-circuit energy into thermal energy. The depletion-layer heat rate, also the electrical power rate, is $V_{sh}I$.

Thus,

$$\begin{bmatrix} \text{Electric-power absorp-} \\ \text{tion at the hot junction} \end{bmatrix} = V_{sh}I \quad \text{watts} \quad (21\text{-}15)$$

$$\begin{bmatrix} \text{Net cooling effect at} \\ \text{the hot junction} \end{bmatrix} = (V_g + V_{kh} - V_{sh})I \quad \text{watts} \quad (21\text{-}16)$$

Thus, in summary,

At the *cold* junction there occurs electric-power generation and a net heating effect.

At the *hot* junction there occurs electric-power absorption and a net cooling effect. More generally, pair-generation cooling and recombination heating, in exchange regions adjacent to junctions, are the dominating aspects of the thermal exchanges in thermoelectric conversion.

The Thermoelectric Open-circuit Voltage. From Fig. 21-1, the change in V_F per series junction pair is the net of V_{sc} and V_{sh}. Therefore,

$$\Delta V_F = V_{sc} - V_{sh} \quad (21\text{-}17)$$

where ΔV_F = open-circuit thermoelectric voltage per junction pair, being the change per pair in the electrochemical potential (the Fermi-level potential V_F)

From the graphical construction of Figs. 21-1 and 21-4,

$$V_g - V_{sc} = (V_c - V_F)_{\text{cold } n\text{-type}} + (V_F - V_v)_{\text{cold } p\text{-type}} \quad (21\text{-}18)$$
$$V_g - V_{sh} = (V_c - V_F)_{\text{hot } n\text{-type}} + (V_F - V_v)_{\text{hot } p\text{-type}} \quad (21\text{-}19)$$

Subtraction gives $V_{sc} - V_{sh}$, which from the preceding equation is ΔV_F. Thus with $(V_c - V_F)$, $(V_F - V_v)$, etc., evaluated from (21-11), magnitudes could be found for the open-circuit voltage. For materials here dealt with the cold V_F never quite rises to V_c or drops to V_v; so that the total net change per junction pair can never quite reach V_g. Thus

$$\begin{bmatrix} \text{Upper limit to the} \\ \text{open-circuit voltage} \end{bmatrix} = V_g \quad (21\text{-}20)$$

Actually, ΔV_F is always substantially less than this limiting value.

The Seebeck Coefficients for Thermoelectric Conversion.[4] The Seebeck coefficient is a parameter used to measure the junction heating and cooling properties of materials that make thermoelectric conversion possible.

Let S_c, S_h = Seebeck coefficients for the cold and hot junctions, respectively; this coefficient is a proportional measure of the heat power produced in a junction at which current flow is out of the p-type and into the n-type material; reverse current flow results in equal cooling; watts/(amp)(°K)

S = in the general case either S_c or S_h

S_n, S_p = the contributions, respectively, from the n-type and p-type material to the general-case Seebeck coefficient; thus

$$S = S_p - S_n = |S_p| + |S_n| \quad (21\text{-}21)$$

From these definitions and (21-13) and (21-14),

$$\begin{bmatrix} \text{Watts of heating} \\ \text{power at the cold} \\ \text{junction, Fig. 21-4} \end{bmatrix} = I(V_g + V_{kc} - V_{sc}) = S_c T_c I \quad (21\text{-}22)$$

$$\begin{bmatrix} \text{Watts of cooling} \\ \text{power at the hot} \\ \text{junction} \end{bmatrix} = I(V_g + V_{kh} - V_{sh}) = S_h T_h I \quad (21\text{-}23)$$

$$\begin{bmatrix} \text{Net watts of} \\ \text{cooling per} \\ \text{junction pair} \end{bmatrix} = S_h T_h I - S_c T_c I \quad (21\text{-}24a)$$

SELECTED ENERGY-CONVERSION SUBSYSTEMS 21-15

The net wattage of electric-power generation is, from (21-13) and (21-15), $I(V_{sc} - V_{sh})$; use of (21-22) and (21-23) to evaluate this gives

$$\begin{bmatrix} \text{Net watts of} \\ \text{electric-power} \\ \text{generation per} \\ \text{junction pair} \end{bmatrix} = S_h T_h I - S_c T_c I - V_{kj} I \qquad (21\text{-}24b)$$

The terms V_{kh}, V_{kc}, and V_{kj} in these equations are sometimes ignored in treatments and discussions of this subject, and are sometimes not clearly dealt with in reportings of the interpretations of data used to measure Seebeck coefficients. They can be important for operation extending into high ranges of temperature.

The open-circuit voltage is thus expressible in terms of the Seebeck coefficients:

$$\begin{bmatrix} \text{Open-circuit junction-pair} \\ \text{voltage } \Delta V_F = V_{sc} - V_{sh} \end{bmatrix} = S_h T_h - S_c T_c - V_{kj} \qquad (21\text{-}25)$$

Here, of course, $S_h = |S_p| + |S_n|$, with S_p and S_n evaluated at T_h, and $S_c = |S_p| + |S_n|$, with evaluation at T_c. Note that $(S_h T_h - S_c T_c)$ includes a contribution V_{kj}; see (21-26) and (21-27).

These equations establish the manner of using the Seebeck coefficients. Establishment of dependence of their magnitudes in relation to materials properties and temperature is initiated by introducing (21-18) and (21-19) into (21-22) and (21-23), and then rearranging by separation of contributions from n-type and p-type materials as suggested by (21-21). This gives, at any temperature T, if V_k is equally shared,[4]

$$-S_n - \tfrac{1}{2}(V_k/T) = \frac{(V_c - V_F)_{n\text{-type}}}{T} = \frac{k_b}{q_e} \frac{(V_c - V_F)_{n\text{-type}}}{V_T} \qquad (21\text{-}26)$$

$$S_p - \tfrac{1}{2}(V_k/T) = \frac{(V_F - V_v)_{p\text{-type}}}{T} = \frac{k_b}{q_e} \frac{(V_F - V_v)_{p\text{-type}}}{V_T} \qquad (21\text{-}27)$$

Introduction of (21-11) gives, nearly enough,

$$S_p - \tfrac{1}{2}(V_k/T) = \frac{k_b}{q_e} \ln \left[\frac{2}{N_a} \left(\frac{2\pi m_e k_b T}{h^2} \frac{m_p}{m_e} \right)^{3/2} \right] \qquad (21\text{-}28)$$

$$-S_n - \tfrac{1}{2}(V_k/T) = \frac{k_b}{q_e} \ln \left[\frac{2}{N_d} \left(\frac{2\pi m_e k_b T}{h^2} \frac{m_n}{m_e} \right)^{3/2} \right] \qquad (21\text{-}29)$$

$$S = S_p - S_n = \frac{k_b}{q_e} \ln \left[\frac{4}{N_a N_d} \left(\frac{2\pi m_e k_b T}{h^2} \right)^3 \sqrt{\frac{m_n}{m_e} \frac{m_p}{m_e}} \right] + \frac{V_k}{T} \qquad (21\text{-}30)$$

Numerically the first of these can be expressed as

$$S_p = \frac{1}{11{,}600} \ln \left[\frac{4.86 \times 10^{21}}{N_a} \left(\frac{m_p}{m_e} T \right)^{3/2} \right] + \tfrac{1}{2}(V_k/T) \qquad (21\text{-}31a)$$

$$S_p = 10^{-3} \left[4.32 + 0.129 \ln \left(\frac{m_p}{m_e} \right) + 0.129 \ln T - 0.086 \ln N_a \right] + \tfrac{1}{2}(V_k/T) \qquad (21\text{-}31b)$$

To obtain $-S_n$ numerically, replace m_p and N_a by m_n and N_d. As these equations are based on a gross-aspect theory, only qualitative agreement with experiment is to be expected.

Straightforward analysis using a semi-graphical study of Fig. 21-1 similar in method to that used in obtaining (21-18) and (21-19), followed by use of (21-26) and

(21-27), shows that (21-17) and (21-25) for ΔV_F are consistent with the thermodynamically-based relationship that

$$\Delta V_F = \int_{T_c}^{T_h} [S_p(T) - S_n(T)]dT \qquad (21\text{-}31c)$$

Limits to validity of the logic underlying these equations for the Seebeck coefficient are that

$$0 < (V_c - V_F) < \tfrac{1}{2} V_g \qquad (21\text{-}32)$$

There are two basic reasons for this:
1. *The cold-junction validity ceases if the electrochemical potential V_F reaches V_c.* This requires that $0 < (V_c - V_F)$; numerically,

$$\begin{bmatrix}\text{The equation for } S_n \\ \text{is valid at the cold} \\ \text{junction only if}\end{bmatrix} N_d < 4.86 \times 10^{21} \left(T_c \frac{m_n}{m_e}\right)^{3/2} \qquad (21\text{-}33a)$$

2. *The hot-junction validity fails as the electrochemical potential V_F approaches closely to midgap.* This requires that $(V_c - V_F) < \tfrac{1}{2} V_g$,

$$\begin{bmatrix}\text{The equation for } S_n \\ \text{can be acceptable} \\ \text{at the hot junction} \\ \text{only if}\end{bmatrix} \frac{2V_{Th}}{V_g} \ln\left[\frac{6.06 \times 10^{27} V_g^{3/2}}{N_d}\left(\frac{m_n}{m_e}\frac{V_{Th}}{V_g}\right)^{3/2}\right] < 1 \qquad (21\text{-}33b)$$

A similar limitation applies for S_p.

At sufficiently large values of the ratio V_{Th}/V_g, of the hot-junction kinetic temperature V_{Th} to V_g, the material approaches intrinsic semiconductor behavior, and the experimentally observed values of $-S_n$ cease to follow the rise with temperatures indicated by (21-29); they reach a maximum, and rapidly decline toward zero; similarly for S_p.

Thermoelectric-conversion-subsystem Geometry. Figure 21-5 illustrates the typical thermoelectric design plan of passing heat in parallel paths and electric current in series through a multielement structure, together with a temperature profile. A conversion-efficiency analysis uses junction temperatures T_h and T_c, which differ somewhat from the temperatures T_{c0} and T_{h0} of the actual heat source and sink.

Thermoelectric-conversion Efficiency; Materials Figure of Merit. Following the logic of Eq. (21-5), "conversion efficiency" is here given the following meaning:

$$\begin{bmatrix}\text{Conversion} \\ \text{efficiency}\end{bmatrix} = \frac{\begin{bmatrix}\text{watts delivered to the} \\ \text{electric circuit}\end{bmatrix}}{\begin{bmatrix}\text{heat-power watts accepted} \\ \text{from the source}\end{bmatrix}} \qquad (21\text{-}34)$$

As applied to a thermoelectric conversion subsystem this becomes

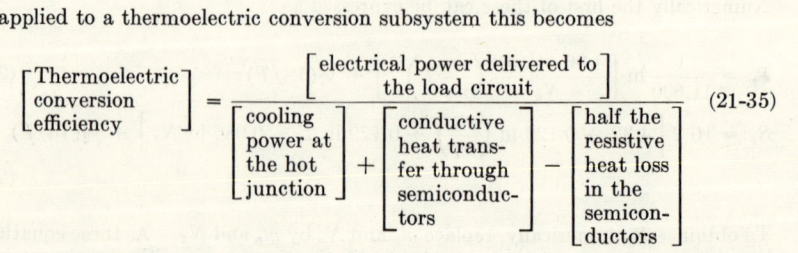

$$\begin{bmatrix}\text{Thermoelectric} \\ \text{conversion} \\ \text{efficiency}\end{bmatrix} = \frac{\begin{bmatrix}\text{electrical power delivered to} \\ \text{the load circuit}\end{bmatrix}}{\begin{bmatrix}\text{cooling} \\ \text{power at} \\ \text{the hot} \\ \text{junction}\end{bmatrix} + \begin{bmatrix}\text{conductive} \\ \text{heat trans-} \\ \text{fer through} \\ \text{semiconduc-} \\ \text{tors}\end{bmatrix} - \begin{bmatrix}\text{half the} \\ \text{resistive} \\ \text{heat loss} \\ \text{in the} \\ \text{semicon-} \\ \text{ductors}\end{bmatrix}} \qquad (21\text{-}35)$$

As to the last denominator term, imagine source and sink temperatures to be held equal, while current from an external source is caused to flow. The heat then produced, because of current passing through the resistance of the semiconductor, flows

SELECTED ENERGY-CONVERSION SUBSYSTEMS   **21**–17

out equally through the source and sink heat terminations. When hot and cold source and sink temperatures are restored, superposing their heat flow, the current now being generated by the subsystems, the internal resistive heat production, and two-way removal is not changed. Thus half of the internal resistive heat is still thought of as passing to the hot junction.

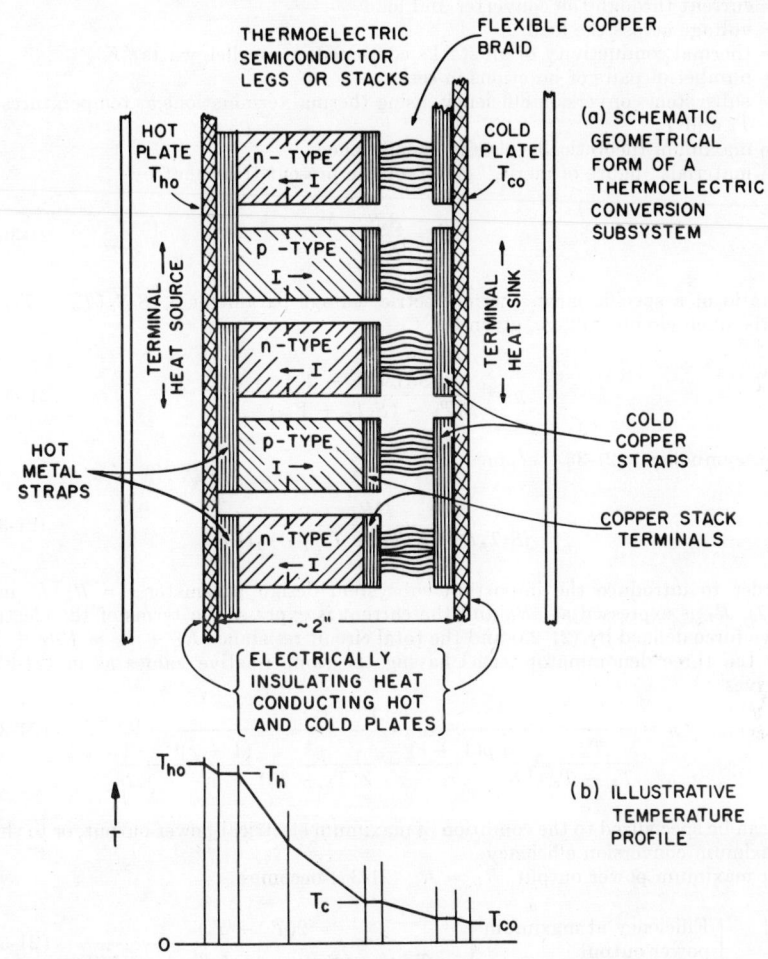

FIG. 21-5. Schematic physical arrangement and illustrative temperature profile for thermoelectric conversion.

The conversion efficiency analysis that follows goes beyond much current literature in the two following respects, that can both be important for thermoelectric conversion:

1. This analysis permits taking account of the very great differences between the hot-junction and cold-junction Seebeck coefficients S_h and S_c when there is a large difference between the temperatures T_h and T_c of the two junctions.

2. It accounts for the effect on the performance due to the random energy $V_{kj} = V_{kh} - V_{kc}$ that contributes at recombination and pair generation. This can be of importance when the hot junction is at a temperature greatly exceeding that of the cold junction.

For use in an efficiency analysis, let

R_L = resistive load, ohms
R_i = internal resistance of all semiconductor "stacks" combined in series
r = ratio R_L/R_i of load to internal resistance
r_{me} = value of r that maximizes efficiency
I = current through the converter and load
V_L = voltage across the load
K = thermal conductivity of all stacks combined in parallel, watts/°K
j = number of pairs of junctions in series
η = subsystem conversion efficiency, using thermal terminations at temperatures of T_h and T_c
η_m = maximum theoretically obtainable efficiency
Z = materials "figure of merit," given for the present treatment by

$$Z = \frac{S_h S_c}{K R_i} j^2 \tag{21-36a}$$

p = ratio of a specific mean thermoelectric voltage parameter $\sqrt{S_h S_c}\,(T_h - T_c)$ to the open-circuit voltage, given by

$$p = \frac{\sqrt{S_h S_c}\,(T_h - T_c)}{S_h T_h - (S_c T_c + V_{kj})} \tag{21-36b}$$

In this symbolism, (21-35) becomes

$$\eta = \frac{I^2 R_L}{j S_h T_h I + K(T_h - T_c) - \tfrac{1}{2} I^2 R_i} \tag{21-37}$$

In order to introduce the important subsystem design parameter $r = R_L/R_i$ into (21-37), R_L is expressed as rR_i, and the current is expressed in terms of the electromotive force defined by (21-25) and the total circuit resistance $R_L + R_i = R_i(r + 1)$. With the three denominator terms having the same relative values as in (21-37), this gives

$$\eta = \frac{1}{\dfrac{T_h}{T_h - T_c}\sqrt{\dfrac{S_h}{S_c}}\,\dfrac{p(1 + r)}{r} + \dfrac{p^2}{Z(T_h - T_c)}\,\dfrac{(1 + r)^2}{r} - \dfrac{1}{2r}} \tag{21-38}$$

This can be specialized to the condition of maximum electrical-power output, or to that of maximum conversion efficiency.

For maximum power output, $R_L = R_i$; (21-38) becomes

$$\left[\begin{array}{l}\text{Efficiency at maximum} \\ \text{power output}\end{array}\right] = \frac{2(T_h - T_c)}{T_h(4p\sqrt{S_h S_c} - 1) + T_c + (8p^2/Z)} \tag{21-39}$$

The maximum-efficiency value r_{me} of r is that at which $d\eta/dr = 0$, and can be expressed as

$$r_{me}^2 = 1 + \frac{Z}{2}\,\frac{S_h^2 T_h^2 - (S_c T_c + V_{kj})^2}{S_h S_c(T_h - T_c)} = 1 + \frac{1}{2KR_i}\,\frac{S_h^2 T_h^2 - (S_c T_c + V_{kj})^2}{T_h - T_c} \tag{21-40}$$

The corresponding form taken by (21-38) is

$$\eta_m = \frac{S_h T_h - (S_c T_c + V_{kj})}{S_h T_h}\,\frac{r_{me} - 1}{r_{me} + [(S_c T_c + V_{kj})/S_h T_h]} \tag{21-41a}$$

SELECTED ENERGY-CONVERSION SUBSYSTEMS 21-19

The two factors can be described as follows:

$$\begin{bmatrix} \text{A modified} \\ \text{Carnot} \\ \text{efficiency} \end{bmatrix} = \frac{S_h T_h - (S_c T_c + V_{kj})}{S_h T_h} = \frac{T_h - (S_c/S_h)T_c - (V_{kj}/S_h)}{T_h} \quad (21\text{-}41b)$$

$$\begin{bmatrix} \text{A device} \\ \text{efficiency} \end{bmatrix} = \frac{r_{me} - 1}{r_{me} + [(S_c T_c + V_{kj})/S_h T_h]} \quad (21\text{-}41c)$$

The quantity $S_h T_h - (S_c T_c + V_{kj})$ is from Eq. (21-25) numerically positive when $V_{sc} - V_{sh}$ is numerically positive, as is always true when the device is converting thermal to electrical energy; thus the numerator of (21-41b) is positive in thermoelectric converter operation, and the denominator of (21-41c) is greater than r_{me}.

To aid in evaluating the somewhat more broadly interesting "converter effectiveness" of Eq. (21-4), note that

$$\begin{bmatrix} \text{Heat-power} \\ \text{watts re-} \\ \text{moved by sink} \end{bmatrix} = jS_c T_c I + K(T_h - T_c) + \tfrac{1}{2} I^2 R_i \quad (21\text{-}42a)$$

$$\frac{[\text{Heat rejected into the sink}]}{[\text{Heat drawn from the source}]} = 1 - \eta \quad (21\text{-}42b)$$

Properties of Thermoelectric Materials.[4,10,11,12] The hot-junction and cold-junction Seebeck coefficients were analyzed separately; this is necessary because utility

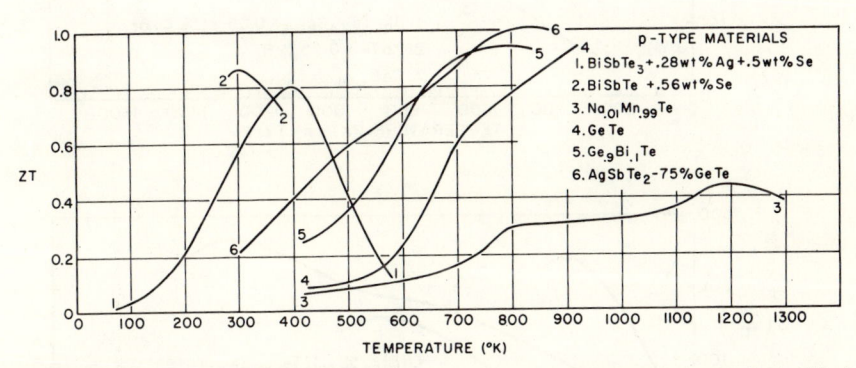

FIG. 21-6. For p-type materials, illustrative temperature variation of the product ZT of figure of merit by absolute temperature.

requires operation over a wide range of temperatures, and S_h and S_c are strongly dependent on temperature. At any junction, each of the two materials makes its own separable contribution, S_p or $-S_n$, to the S_h or the S_c, whether the design employs direct semiconductor interfaces as illustrated in Figs. 21-1 and 21-4 or employs metal terminal plates as illustrated in Fig. 21-5.

Temperature variations of the figure of merit Z occur broadly as illustrated in Figs. 21-6 and 21-7 for p-type and n-type materials, respectively, and Fig. 21-8 charts variations of the Seebeck coefficients for a few materials. Very intensive research effort is currently devoted to a search for better thermoelectric materials, so that any particular systems study must be based on materials known and available at the time of the study. It is not always clear whether or not values reported for S_p or S_n include the V_k contribution.

In order for a material to have a high figure of merit, its Seebeck coefficient must be high and its electrical resistivity and thermal conductivity low, within the temperature range of use. Part of the thermoelectric materials-research problem toward improving

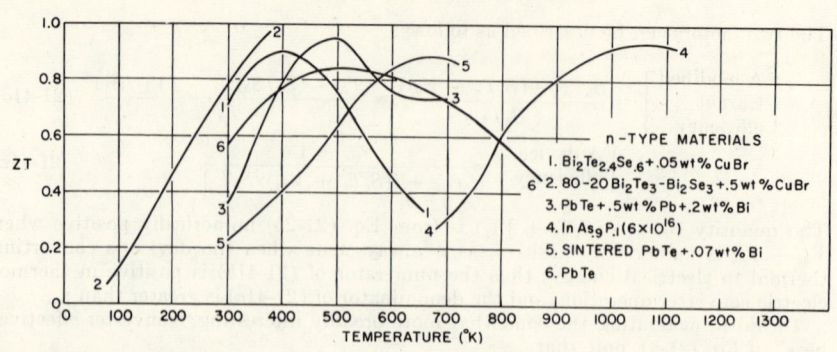

FIG. 21-7. For *n*-type materials, illustrative temperature variation of the ZT product.

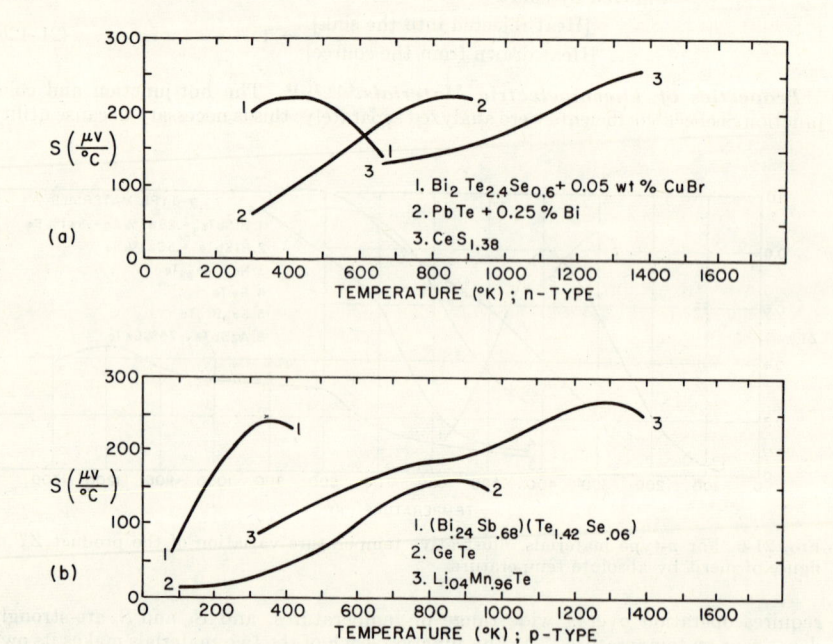

FIG. 21-8. Illustrative temperature variations for the Seebeck coefficient; units, microvolts per degree centigrade or Kelvin. (a) For *n*-type materials. (b) For *p*-type materials. (*From Ref. 12.*)

Z is that of reducing the thermal conductivity of the crystal lattice, which is not strongly dependent on the carrier thermal conductivity. The following choices tend toward decreasing the lattice thermal conductivity:

1. Use of semiconductor compounds rather than elemental semiconductors
2. Use of relatively large molecular weights
3. Use of increased atomic-weight ratios
4. Use of predominantly ionic rather than covalent bonding
5. Introduction of randomly distributed disturbances into the crystal structure, as, for example, by solid-solution alloying

SELECTED ENERGY-CONVERSION SUBSYSTEMS   21-21

Figure 21-1 and Eqs. (21-29) through (21-33), the comments after the latter, Eqs. (21-40) and (21-41), and the values of Fig. 21-8 suggest, as to selection of materials for their Seebeck coefficients, that

1. This coefficient rises rapidly with temperature over a substantial range of temperatures, reaches a maximum, and declines rapidly as the temperature becomes high enough to drive the electrochemical potential nearly to midgap.
2. The temperature at which the coefficient has its greatest value is high for materials with large energy gaps, low with low energy gaps.
3. Increasing the density of donor or acceptor additives N_d, N_a shifts the whole S_p or S_n curve downward; see Eq. (21-31).
4. A high value of the ratio S_h/S_c is generally favorable toward a high value of the maximum efficiency.

Materials can now be obtained with values of the Z figure of merit in the range 0.003 to 0.004. With T_h and T_c at illustrative values of 600° and 300°K (325°C and room temperature), and ignoring the change between S_h and S_c, the maximum theoretical efficiency comes out about 15 per cent. Substituting values from curves 2 and 2 in Fig. 21-8 into Eq. (21-25), ignoring V_{kj}, gives 0.29 volt at open circuit with $T_h = 850°K$ and $T_c = 300°K$.

It may not be feasible to cover a given temperature range, even 300 to 600°K, with a single semiconductor material. It is certainly necessary to use several materials in each leg to go from 300 to 1000°K. Use of multielement stacks introduces difficult design and processing problems, to keep contact resistances low and at the same time avoid poisoning of the interfacing materials by bonding substances. Research and development are needed and in progress on these and other design and techniques problems, as well as in materials research. The art will undoubtedly advance substantially, but probably not rapidly. A primary need is for the discovery of a low-cost source of tellurium, or else of a low-cost substance that can provide thermoelectric materials comparable with those based on tellurium.

Summary of Engineering Attributes of Thermoelectric Conversion. There are at present three primary demerits:

1. *Low efficiency.* Obtainable conversion efficiencies are low, 10 per cent being well on the high side for designs having engineering feasibility.
2. *Low voltage per junction pair.* Many units are required.
3. *High cost.* Thermoelectric converters are too costly for use except in specialty applications, as for military use and in space vehicles.

There are a number of engineering merits:

4. *Small units can be as efficient as large ones.*
5. *The direct-conversion attribute.* The power is converted directly from heat into electric power.
6. *Operability at high electric-circuit temperatures.* Operation appears feasible at source and sink temperatures up to perhaps 1500 and 1000°C, respectively; applications in prospect will need this.
7. *Flexibility as to heat-supply conditions.* Conversion can take place, for a given device, over a wide range of source and sink temperatures.
8. *Mechanical simplicity.* The structural design is simple; only the coolant system has moving parts.
9. *Quietness.* The converter itself operates silently.
10. *Maintenance deals with thermal rather than mechanical problems.* Expected maintenance problems should have to do with thermal stresses or fatigue, or with long-term deterioration of materials.
11. *Size and weight are moderate.*

It is too early to predict the degree of materials-research success that may ultimately come toward extending engineering feasibility by lowering equipment cost and increasing the efficiency.

21-5. Thermionic Conversion

The Surface Motive Barriers Surmounted by Electrons in Thermionic Emission. Thermionic conversion results from electrons being thermally driven out ("emitted") from a metal against a high surface "work-function" barrier in the "motive" energy of Fig. 21-9b, then returning through a lower similar barrier. The cooling due to thermionic emission outward through the high barrier is greater than the heating on reentry through the low one, the net cooling watts being converted into electrical power. Thermionic emission is also used in propulsion (Sec. 20-7).

Figure 21-9c illustrates an electron emerging from the metal equidistant from four fixed-location last-layer ions of the crystal lattice—in a metal all atoms are fixed-location ions, having given up valence electrons to form the electron gas. While still within a lattice spacing or two of the last layer, the electron is in a true potential field —inverted in the figure—because the ions maintain the electric field producing the close-in restraining force in Fig. 21-9a whether the electron is passing or not. For farther-out positions the restraining force is due to the electron's electrostatic image in the surface, now appearing to be an equipotential.[3] Because the image moves when the electron moves, the restraining force is not due to a fixed electric field; its outer-region line integral in Fig. 21-9b thus describes a changing potential energy but not a

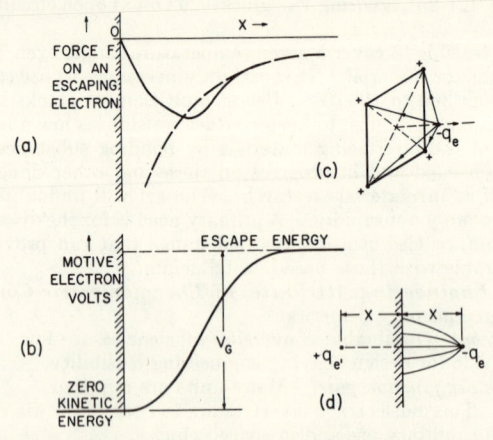

FIG. 21-9. Metal-surface motive barrier surmounted by emitted electrons. (a) Restraining force. (b) The "motive energy," being the line integral of the force. (c) Geometry of close-in emergence between ions. (d) The farther-out image force.

true potential. The wordage "motive energy" encompasses simultaneously as a volt measure the close-in potential, and as an electron-volt measure the farther-out potential energy, with complete merging of the two concepts. Electrons from the metal interior must "climb up" the motive barrier in Fig. 21-9b to escape, thus losing thermal kinetic energy (a cooling effect) at the expense of potential energy, which appears as electric-circuit energy.

A study of energy-state availability, and of placement of the electrochemical potential V_F relative to V_c with details much simpler than for a semiconductor, shows that, in a metal,[1,3]

$$V_F - V_c = \frac{n^2}{8 q_e m_e} \left(\frac{3 n_n}{\pi} \right)^{2/3} = 3.61 \times 10^{-19} \, n_n^{2/3} \qquad (21\text{-}43)$$

where n_n is the number density of electrons; values of $V_F - V_c$ range from perhaps 3 to 10 ev. The energy-level diagram in Fig. 21-10 employs at the surface the motive curve in Fig. 21-9b. At the $v = V_c$ internal "bottom-level" potential the electrons have zero kinetic energy. An electron that from any source accepts enough energy to bring it to the upper level of the motive barrier has the potentiality to escape from the metal. In Fig. 21-10,

V_e = the escape-level v; motive-barrier top
V_c = the internal-potential "bottom-level" v
V_G = "gross work function" (motive-barrier height)
V_W = "work function," the excess of V_G over $(V_F - V_c)$; adding kinetic energy V_W to an electron at V_F can just barely cause escape; thus

$$V_W = V_G - (V_F - V_c) \tag{21-44}$$

Table 21-1 gives experimentally determined values of V_W.

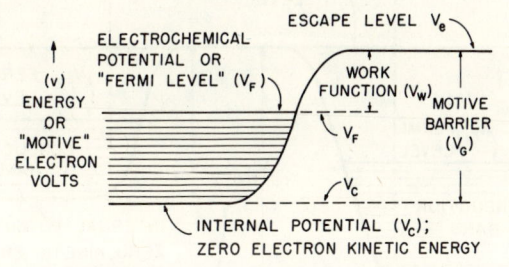

FIG. 21-10. The "work function," motive barrier, and the electrochemical potential.

Table 21-1. Work Functions V_W; Thermionic-emission Constants A_0*

Material	V_W	A_0	Material	V_W	A_0
Cs	1.81	162 × 10⁴	Pd	4.99	60 × 10⁴
Ba	2.11	60	C	4.60	46
Zr	4.00	120	Si	3.59	5–11
Mo	4.20	55	Ge	4.8	
Ta	4.19	55	As	4.72	
W	4.52	60	BaO	1.1	0.004†
Pt	5.32	32	SrO	1.4	0.01†
Th	3.36	70	CaO	1.9	0.4†
U	3.27	6	ThO₂	2.67	2.63

* From Ref. 3.
† Average values. Emission may be a factor of 10 higher or lower, depending on processing details and state of activation.

At a sufficiently high temperature, the slight state occupancy above $v = V_e$ permits significant electron escape, called thermionic emission. Straightforward physical logic based on the energy-state model gives[1]

$$J_{th} = A_0 T^2 \exp \frac{-V_W}{V_T} \tag{21-45a}$$

where J_{th} = thermionic emission, amp/sq m; when the potential gradient just outside the surface is positive, every electron comprising J_{th} is accelerated away from the surface

A_0 = a semiempirical constant; Table 21-1 gives experimentally found values; theory gives

$$A_0 = \frac{4\pi m_e k_b^2 q_e}{h^3} = 120 \times 10^4 \tag{21-45b}$$

in good agreement with experiment, considering the complexity of the theory.

The Internal-profile Barrier Potential and the External-profile Contact Potential for Metals.[3] Figure 21-11a shows two unlike uncharged metal objects moving together but not yet touching; note the alignment of the escape levels, implying zero electric field between them, because they are uncharged. Symbols V_{G1}, V_{G2}, V_{W1}, V_{W2}, etc., have obvious meanings carried over from Fig. 21-10. At the earliest contact moment electrons flow from right to left, because V_{F2} is higher than V_{F1}.

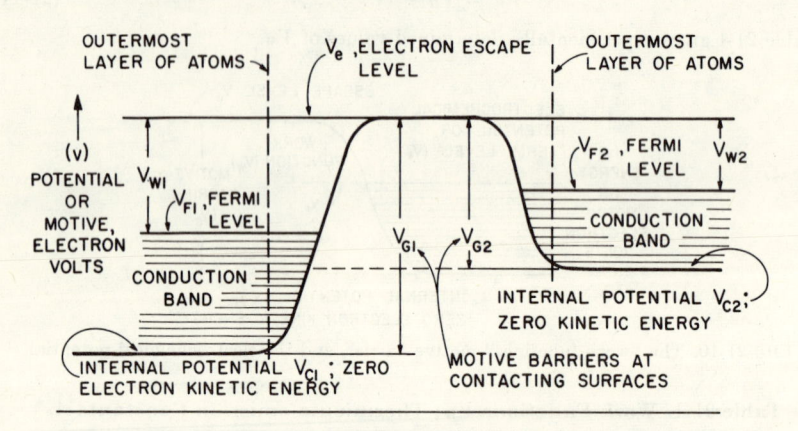

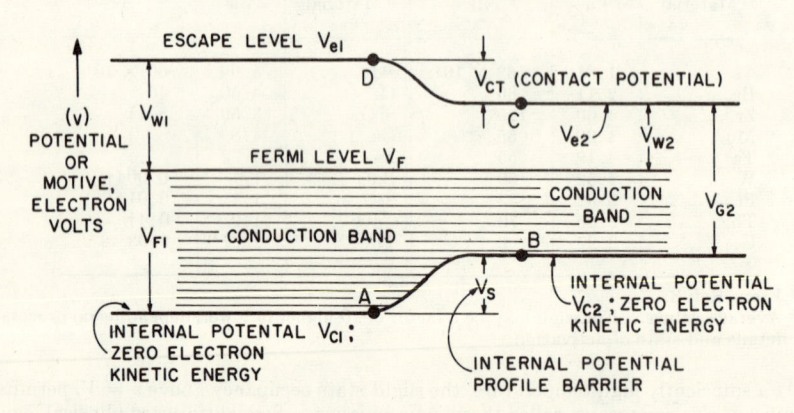

FIG. 21-11. Relation of surface work function to contact potential between metals.

This charges the left one negatively, raising its levels, and the right one positively, depressing its levels, until V_{F2} and V_{F1} become aligned as in Fig. 21-11b. Thus at this metal-to-metal contact, from Fig. 21-11b,

$$V_S = V_{F1} - V_{F2} \qquad (21\text{-}46)$$

This is the *barrier potential* at an intermetallic interface.

$$V_{CT} = V_{W1} - V_{W2} \qquad (21\text{-}47)$$

This is the *contact-potential* difference at an intermetallic interface.

SELECTED ENERGY-CONVERSION SUBSYSTEMS   21-25

Contact-potential difference thus results from transfer of charge and therefore establishes an electric field between noncontacting surfaces of these contacting objects. This field can be measured by an *electrostatic* voltmeter. The absence from Fig. 21-11b of a step in the electrochemical profile shows that contact-potential difference is not measured by ordinary current-responding voltmeters.

The Thermionic-converter Plasma. The plasma in a thermionic converter is of the low-density moderate-energy type, with properties as follows, in typical devices:

1. *Good electrical conductivity exists because of the presence of a great many free electrons.*
2. *The electron temperature is several thousands of degrees Kelvin, much higher than either cathode or plasma gas temperatures.*
3. *There exist significant-amplitude standing space-charge waves ("plasma oscillations") at or very near the "electron-density frequency."* Quantitatively,

$$\begin{bmatrix} \text{Electron-density} \\ \text{radian frequency} \end{bmatrix} = q_e \sqrt{\frac{n_n}{\epsilon_0 m_e}} = 9.05 \sqrt{n_n} \qquad (21\text{-}48)$$

where $\epsilon_0 = 8.85 \times 10^{-12}$, the permittivity of a vacuum, and n_n is the plasma electron density. The space-charge wave presence correlates experimentally with high electron temperatures; such waves are not measurable by conventional plasma diagnostic techniques.

4. *Thermal random-particle velocities greatly exceed current-carrying drift velocity.* The Maxwell-Boltzmann random thermal-velocity distribution gives rise to a "random electron-current density," carried equally in both directions across any small randomly oriented area increment interior to the plasma; this is enormously greater than the directed "drift-current density" that carries the circuit current. Electrons carry this random electron-current density up to plasma bounding regions.

5. *D-C space-charge forces due to the presence of the electrons are neutralized by opposing forces and so do not affect bulk-plasma electron motion.* In thermionic-converter plasmas the neutralization is from space-charge forces due to positive ions at a density equal to the electron density.

Electrons enter the plasma from a cathode, thermionically releasing electrons at the rate specified by (21-45a), and leave through an anode cool enough to have negligible thermionic emission. The cesium vapor commonly employed as the plasma gas can be ionized by either of two processes:[7,13]

1. By random-motion contact of neutral cesium-vapor atoms with the high-work-function cathode; they are ionized because the cathode work function exceeds the 3.87 ev required to ionize cesium; this process predominates.
2. By impact, on neutral cesium particles in the plasma, of electrons of the high-energy "tail" of the random-energy distribution.

Plasma Sheaths.[3] Adjacent to low-density plasma boundaries are "positive ion sheaths" (the counterparts of semiconductor depletion layers in *n*-type materials). A slow-rate positive-ion flow occurs from the plasma interior to the plasma face of a sheath, going on "downhill" through the sheath potential drop to the physically bounding surface. This ion-flow rate depends on plasma properties, not on sheath properties. The enormously larger random electron flow also arrives from the plasma interior at the plasma face of the sheath. The "barrier potential" the sheath presents to this flow prohibits passage onto the physical bounding surface of all but a very small exponentially determined portion of this electron flow.

At a thermionic converter's cathode the sheath barrier potential assumes the value necessary to cause the excess of thermionic-emission current, over electron penetration from the plasma through the sheath to the cathode, to just provide the current required by the circuit as a whole.[13] At the anode, the barrier potential limits the electron penetration to just that necessary to provide the circuit current. If, as can usually be assumed, the plasma electron density and temperature are uniform, a unique reference condition exists when the two sheaths are equal, so that the random electron penetration through from the plasma to the anode is the same as to the cathode, making the anode electron current just half the thermionic emission from the cathode specified by (21-45a).

21-26 SYSTEM COMPONENTS

Motive Distribution.[13] The motive and energy-level diagram in Fig. 21-12 illustrates the closed-circuit loop for a single-cell thermionic converter providing power to a load resistor. The AA right end is identically also the AA left end. From left to right there are: a piece of the high-work-function cathode; the essentially vertical motive barrier at the cathode surface; the plasma sheath adjacent to the cathode, through which emitted electrons "fall down" and plasma electrons "climb up"; the electrically conducting plasma; the sheath adjacent to the anode, through which some electrons "climb" by virtue of thermal energies and enter the anode to provide circuit

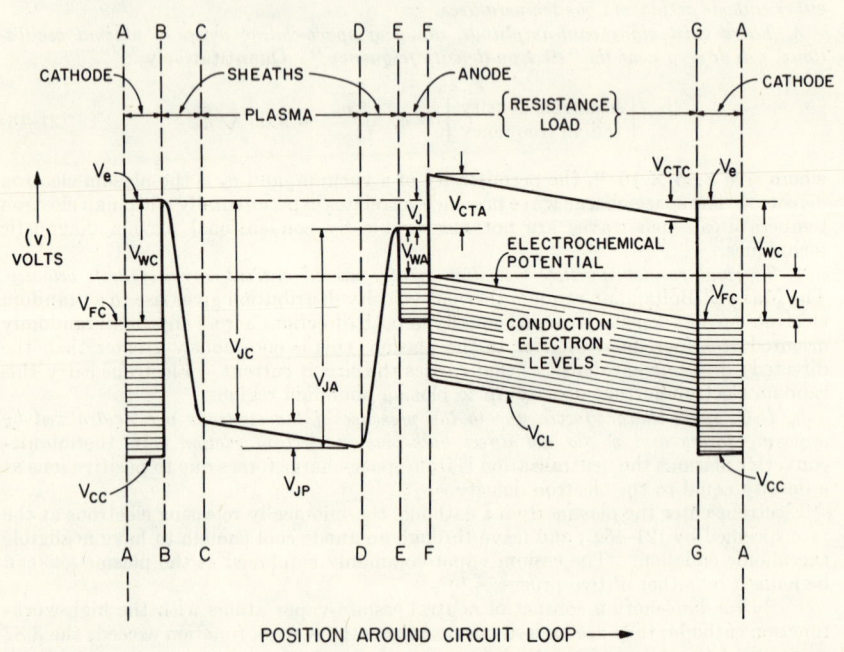

FIG. 21-12. Motive and energy-level diagram around a thermionic-converter circuit during electric-power generation.

current; the essentially vertical motive-barrier drop into the anode; the low-work-function anode; the resistive load; the other piece of the cathode. In addition to symbol meanings obvious from Figs. 21-10 and 21-11:

V_{JC} = "cathode drop" across the cathode sheath
V_{JA} = voltage across the anode sheath
V_{JP} = plasma voltage drop between sheaths
V_J = *electrostatically* measured voltage between cathode and anode, being the net voltage drop through the plasma and sheaths
V_{CT} = Eq. (21-47) contact-potential difference that would exist between cathode and anode materials if placed in contact (which they are not)

The electrochemical potential declines in the load, corresponding to delivering power as ohmic heat. The two sheaths are shown in Fig. 21-12 as being equal, corresponding to the reference condition. For an alternative situation see the paper "Electron space-charge-limited Operation of Cesium Thermionic Converters," by E. N. Carabateas in the *Journal of Applied Physics*, vol. 33, pp. 1445–1449, April, 1962.

SELECTED ENERGY-CONVERSION SUBSYSTEMS   21-27

Circuit Voltage and Current Relationships. Let

I_{th} = thermionic-emission current, at Eq. (21-45a) rate
I_{re} = random electron-current density arriving from plasma at plasma faces of both sheaths
I_i = ion current passing through the anode sheath to the anode
T_e, V_{Te} = the degree-Kelvin temperature and the electron-volt kinetic temperature of plasma electrons; $V_{Te} = T_e/11{,}600$
I = circuit current through converter and load
V_L = voltage across the load

Climbing-up penetration of plasma electrons through both sheaths is exponentially governed. The anode-sheath penetration minus the small ion current to the anode gives the circuit current, thus

$$I = I_{re} \exp \frac{-V_{JA}}{V_{Te}} - I_i \tag{21-49a}$$

At the cathode, the circuit current is I_{th} less the penetration, thus

$$I = I_{th} - I_{re} \exp \frac{-V_{JC}}{V_{Te}} \tag{21-49b}$$

As to voltages, tracing motive changes experienced by an electron starting from and

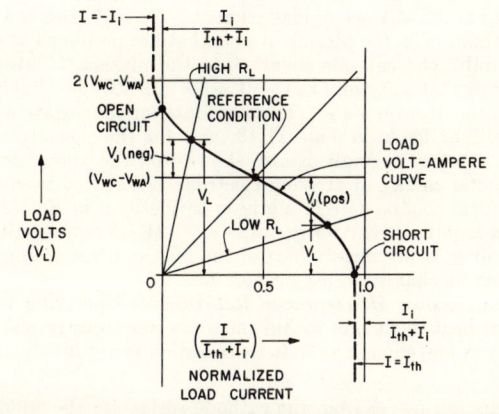

Fig. 21-13. Volt-ampere relationship and resistance lines for a thermionic converter; see Eq. (21-50c).

returning to the electrochemical potential of the cathode gives

$$V_L = (V_{WC} - V_{WA}) - (V_{JC} + V_{JP} - V_{JA}) \tag{21-50a}$$

The plasma resistive drop V_{JP} is usually very small; neglecting this, nearly enough,

$$V_L = (V_{WC} - V_{WA}) + (V_{JA} - V_{JC}) \tag{21-50b}$$

Combination of Eqs. (21-49) and (21-50b), with use of $V_{WC} - V_{WA} = V_{CT}$, gives the volt-ampere equation

$$\frac{I}{I_{th} + I_i} = \frac{1}{1 + \exp[(V_L - V_{CT})/V_{Te}]} - \frac{I_i}{I_{th} + I_i} \tag{21-50c}$$

This is illustrated in Fig. 21-13, for $V_{CT} = 2.2$ volts, and $V_{Te} = 0.55$ ev ($T_e = 6400°K$), treating I_i as a small constant fraction of I_{th}. The current and voltage for given load resistances are as indicated by the intersections in Fig. 21-13 of the resistance lines with the volt-ampere curve. The open-circuit voltage is that required to make the electron current equal $-I_i$; at short circuit the current is becoming rapidly asymptotic to a value less than I_{th} by I_i. At the reference resistance the sheath voltages are equal.

Power relationships are given by (21-51) with load current flowing, and noting the Fig. 21-12 similarity between cathode motive barrier and anode sheath, and between anode motive barrier and cathode sheath, neglecting bulk-plasma heating $V_{JP}I$,

$$\begin{bmatrix}\text{Cathode internal electric-power gener-} \\ \text{ation and electrode-cooling rate}\end{bmatrix} = V_{WC}I \quad (21\text{-}51a)$$

$$\begin{bmatrix}\text{Anode internal electric-power ab-} \\ \text{sorption and electrode-heating rate}\end{bmatrix} = V_{WA}I \quad (21\text{-}51b)$$

$$\begin{bmatrix}\text{Cathode-sheath electric-power ab-} \\ \text{sorption and plasma-electron heating rate}\end{bmatrix} = V_{JC}I \quad (21\text{-}51c)$$

$$\begin{bmatrix}\text{Anode-sheath electric-power gener-} \\ \text{ation and plasma-electron cooling rate}\end{bmatrix} = V_{JA}I \quad (21\text{-}51d)$$

$$[\text{Net electric-power generation}] = I[(V_{WC} - V_{WA}) + (V_{JA} - V_{JC})] \quad (21\text{-}51e)$$

Equations (21-51a) and (21-51b) result because electrons emitted or reentering may be treated, as to energies, as coming from or returning to the highest occupied level, that is, V_{FC} or V_{FA}. The cooling at electron emission and at "climbing" into the anode occurs because the high-energy electrons escape. The general agreement of Fig. 21-13 with experiment[13] indicates that the plasma electron temperature is maintained by an agency that persists in the absence of load current. Possibly this is a result, direct or indirect, of recombination in the plasma of ions that are produced at cathode contact and thermally "climb" the cathode sheath into the plasma. This or some similar energy source provides the varying $(V_{JA} - V_{JC})$ of (21-50b) as a "generated voltage" in addition to the fixed $V_{WC} - V_{WA}$. This accounts for the negative (power-generating) V_J values at light loads in Fig. 21-13, and for the power-exchange terms in Eqs. (21-51c) and (21-51d). Analysis also shows that the anode dissipation energy $I \cdot 2V_{Te}$ due to kinetic energy of arriving electrons is matched by an equal cathode-kinetic-energy-removal cooling rate in addition to the item in Eq. (21-51a), and that there is an electron kinetic-energy term $I_{th}(2V_{Te} - 2V_{TC})$ representing cathode heating and plasma cooling, being merely part of the plasma maintaining energy that the ion energy-transport mechanism must provide for.

Thermionic-conversion Heat-power Relations. Operating temperatures are high enough so that heat exchanges within the converter occur primarily by radiation, and the gas density is low enough so that its radiation is not involved (see Sec. 15-2). Let

W_R = heat-radiation power from the cathode surface to the anode surface, watts

T_C, T_A = respectively, the temperature of the cathode and anode (source and sink temperatures, °K)

γ_A, γ_B = respectively, the thermal-emissivity coefficients (fractional quantities) for the materials of these two surfaces at the respective temperatures; these are determinable for a number of substances from Fig. 21-14

S_R = area through which radiation flux passes, being for flat parallel facing electrodes the area of each; for this simple geometry, in watts,[5]

$$W_R = 5.73 \times 10^{-8} S_R (T_C^4 - T_A^4) \frac{1}{1/\gamma_C + 1/\gamma_A - 1} \quad (21\text{-}52)$$

This is obtained by analyzing the multiple diffuse reflections between the two surfaces. A design objective is to minimize W_R, for given current, as it goes to the heat sink. This involves obtaining low values for both γ's and would benefit from having a reasonably high anode temperature, not usually feasible. To give a useful $(V_{WC} - V_{WA})$ voltage in Eq. (21-50b), V_{WC} must substantially exceed V_{WA}; anode emission current must be substantially less than that from the cathode, to give a net current;

therefore, from Eq. (21-45) for emission-current density, the anode temperature must be very substantially less than that of the cathode. Therefore, $T_A^4 \ll T_C^4$ in (21-52), so that radiation loss is governed primarily by cathode temperature. Therefore, in most interesting cases, nearly enough,

$$W_R = 5.73 \times 10^{-8} S_R T_C^4 \frac{1}{1/\gamma_C + 1/\gamma_A - 1} \tag{21-53}$$

For the refractory metals that have the needed high cathode work functions, at useful emission temperatures of 2500°K and above, values of γ_C are near 0.25 or 0.30.

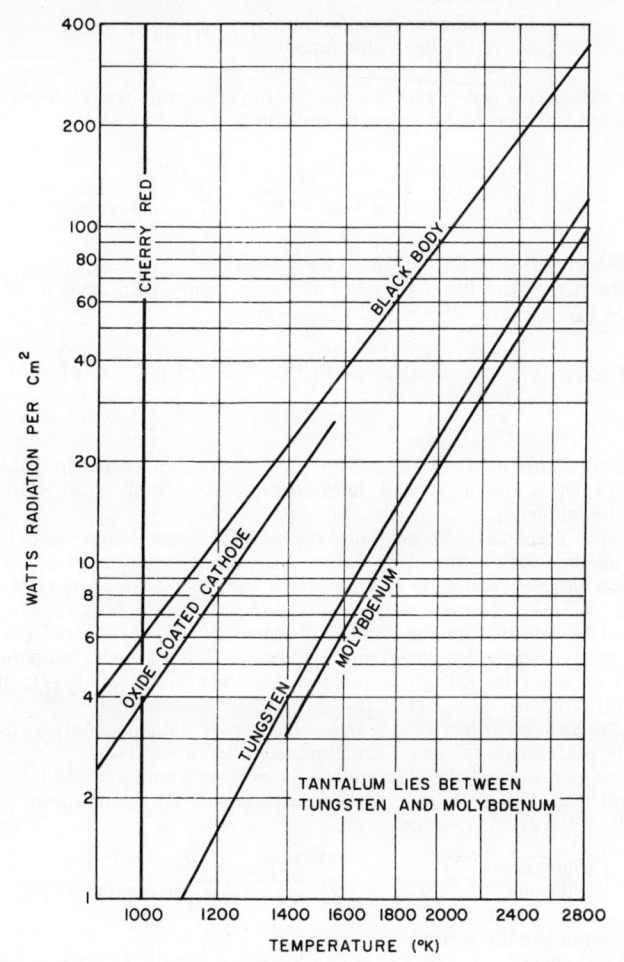

FIG. 21-14. Heat radiation from surfaces; emissivity coefficient γ is the ratio of radiation from the stated metal to that from a blackbody at the same temperature.

Converter Effectiveness and Efficiency.[14] Thermionic conversion is in too early a development state to predict expected ultimate performance. However, attention to an oversimplified model permits identifying expected limitations. In working with energy magnitudes, for use in Eq. (21-4) for conversion effectiveness, the plasma electron heating and cooling items in Eqs. (21-51c) and (21-51d) will be treated as basically

21-30 SYSTEM COMPONENTS

cathode heating and cooling items. This is under the presumption that at all load values the cathode is thermally linked to (but not in thermal equilibrium with) the plasma electron swarm, by an ion potential-energy-transport process that is reversible, permitting heat exchange either way. Thus, for simplicity neglecting heat leakage, ion-current electrode dissipation, and the $V_{JP}I$ resistive heat generation in the plasma, in watts,

$$\begin{bmatrix} \text{Energy converted} \\ \text{into electrical form} \end{bmatrix} = I[(V_{WC} - V_{WA}) + (V_{JA} - V_{JC})] \quad (21\text{-}54a)$$

$$\begin{bmatrix} \text{Thermal energy accepted} \\ \text{from heat source (cathode)} \end{bmatrix} = I[V_{WC} + (V_{JA} - V_{JC}) + 2V_{Te}] + W_R \quad (21\text{-}54b)$$

$$\begin{bmatrix} \text{Thermal energy disposed} \\ \text{of via the heat sink (anode)} \end{bmatrix} = I[V_{WA} + 2V_{Te}] \quad (21\text{-}54c)$$

Adaptation to describe conversion efficiency as in (21-5) employs (21-54) with the same items neglected. Thus, nearly enough, and using $V_J = V_{JC} - V_{JA}$,

$$\begin{bmatrix} \text{Thermionic-} \\ \text{conversion} \\ \text{efficiency} \end{bmatrix} = \frac{(V_{WC} - V_{WA}) - V_J}{V_{WC} - V_J + (W_R/I) + 2V_{Te}} \quad (21\text{-}55)$$

In the light-load region of Fig. 21-13, $-V_J$ is positive. The temperature dependence appears primarily in the W_R/I term. For the reference-point case in which, nearly enough, $I = \tfrac{1}{2}I_{th}$, use of (21-45) and (21-52) gives

$$\frac{W_R}{I} = 9.56 \times 10^{-14} \left(\frac{1}{1/\gamma_A + 1/\gamma_C - 1} \right) \left(\frac{120 \times 10^4}{A_{0C}} \right) \left(1 - \frac{T_A^4}{T_C^4} \right) \left(T_C^2 \exp \frac{+V_{WC}}{V_{TC}} \right) \quad (21\text{-}56a)$$

Here A_{0C} is the value of A_0 of (21-45) for the cathode, appearing in Table 21-1, and $V_{TC} = T_C/11{,}600$. For a typical refractory-metal cathode $(120 \times 10^4)/A_{0C} = 2$. Now note the following:

1. *Keeping W_R/I small is the dominant requirement for good conversion efficiency.* In (21-55), W_R/I is nearly always larger than V_{WC}, often much larger.
2. *It is usually not feasible to operate with a high anode temperature to gain benefit from the $[1 - (T_A/T_C)^4]$ term.*
3. *Raising the cathode temperature for given material very sharply reduces W_R/I, and correspondingly increases the conversion efficiency.* The cathode temperature factor $T_C^2 \exp(V_{WC}/V_{TC})$ in (21-56) dominates the magnitude of W_R/I; it decreases extremely rapidly with rising temperature.
4. *Use of a moderate rather than a high cathode work function sharply reduces W_R/I.* This and the statements in item 3 are illustrated in Table 21-2.

For the reference condition for which $V_J = 0$, presuming that $T_A^4 \ll T_C^4$, that $\gamma_A = \gamma_C$ and that $2/\gamma_C \gg 1$, that cathode area equals anode area, and that $2V_{Te} \ll [V_{WC} + (W_R/I)]$, (21-55) becomes

$$\begin{bmatrix} \text{Conversion} \\ \text{efficiency} \end{bmatrix} = \frac{V_{WC} - V_{WA}}{V_{WC} + 5.73 \times 10^{-8} \gamma_C A_{0C} T^2 \exp(V_{WC}/V_{TC})} \quad (21\text{-}56b)$$

which appears in the literature.[14]

Figure 21-15 based on Eq. (21-56b) illustrates clearly the advantage, for any selected cathode temperature, of having a moderate rather than a too high cathode work function.

5. *Temperatures that give good efficiency may severely limit cathode life.* Consider a design using a tungsten cathode, with $2V_{Te} = 1.0$ ev, $V_{JP} = 0$, with operation at the Fig. 21-13 reference point for which $V_J = 0$. The W_R/I and conversion-efficiency values appear in Table 21-3. Tungsten filaments in large commercial power triodes often operate at about 2500°K and have reasonable life before burnout because of

SELECTED ENERGY-CONVERSION SUBSYSTEMS 21–31

Table 21-2. Effects of Temperature and Work Function on Loss Term
Values of the Eq. (21-56) quantity $9.56 \times 10^{-14} T_C^2 \exp(V_{WC}/V_{TC})$

Cathode temp T_C	For $V_{WC} = 4.52$ (tungsten)	For $V_{WC} = 3.38$
1750°K	....	1500
2000°K	....	124
2250°K	7400	17.2
2500°K	860	3.82
2750°K	138	
3000°K	34	
3250°K	10.2	

Table 21-3. Conversion Efficiency vs. Temperature

Tungsten-cathode temp	W_R/I	Conversion efficiency, %
2500°K	504	0.53
2750°K	78	3.2
3000°K	19.2	10.5
3250°K	5.76	22

These use $V_J = 0$, $V_{WC} = 4.52$, $V_{WA} = 1.81$, $A_{0C} = 60 \times 10^4$, $\gamma_A = 0.6$, $\gamma_C = 0.33$, $T_A^4 \ll T_C^4$, $V_{Te} = 0.5$, in Eqs. (21-55) and (21-56).

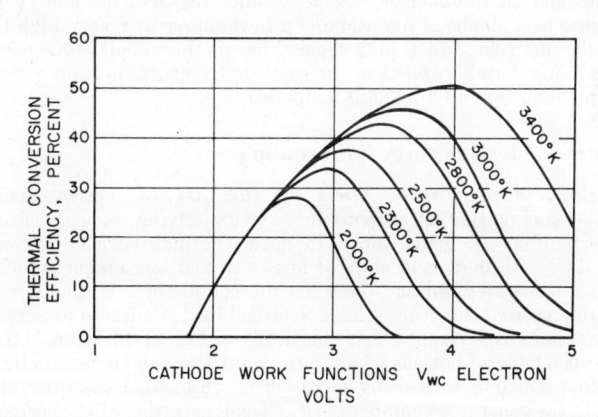

FIG. 21-15. Variation of thermionic conversion efficiency with cathode work function V_{wc}, with cathode temperature as the parameter, showing existence of an optimum work function for each temperature. *(From Ref. 14.)*

evaporation of the tungsten. At 2750° life expectancy is considerably reduced; 3000°K is beyond the range of useful engineering practice. Tungsten melts at about 3300°K. It is clear from Tables 21-2 and 21-3 that at the predictable near-term state of the art, conversion efficiency as defined by Eq. (21-5) is not likely to exceed 10 per cent; this may eventually improve substantially. Cathodes having properties like those in the right-hand column of Table 21-2, which might give efficiencies in the 15 to 20 per cent range, at temperatures that would not severely limit cathode life, may become usable. Voltage per cell would be lower than for the tungsten-cathode device. See also Ref. 14.

6. *The resistive plasma voltage drop must be small.* It is clear from the equations that the plasma voltage drop V_{JP} must be a small fraction of a volt, to have good efficiency; this is usually not difficult to accomplish.

7. *Carnot efficiency.* It is straightforward to rearrange (21-56) into the product of a "device efficiency," always less than unity, by a modified Carnot efficiency close to $(T_C - T_A)/T_C$.

8. *Slotted-cathode benefit.* An extension of the heat-flow analysis indicates that use of a deeply slotted cathode, having emitting area two or three times the projected area, might significantly increase the conversion efficiency. For given thermionic-emission current density, such slotting substantially increases the device current without an increase in heat-flux area.

Engineering Considerations. The thermionic conversion voltage per cell can be of the order of 2 volts, as, for example, by using at the anode a layer of cesium deposit on copper, giving $V_{WA} = 1.81$, and a tantalum cathode, $V_{WC} = 4.20$. Therefore, useful converters will employ many cells arranged thermally in parallel and electrically in series.

If the total system is such that heat rejected through the sink can be usefully employed, so having positive value, converter effectiveness is improved by operating the anode at a fairly high temperature, but not high enough to cause anode thermionic emission to be more than perhaps one-fourth or one-fifth of cathode thermionic emission. If the rejected heat has only negative value, the anode temperature should be chosen for optimum heat-rejection economy. In principle a space vehicle can employ solar heating of a thermionic-converter cathode to as high a temperature as engineering feasibility permits, related to survival of materials. Also, it can in principle use blackbody radiation to reject heat to outer space at the rate $5.73 \times 10^{-8} T_A^4$ watts/sq m. The principal costs of acceptance and rejection of heat in this case are in obtaining the area exposure; a high heat-rejection temperature favors a low rejection area.

When controlled thermonuclear *fusion* becomes feasible, its energy output will probably appear as a means of maintaining a heat source at a very high temperature with very little fuel cost, which may suggest use of thermionic conversion. On the nuclear *fission* side, bombardment of the cathode by neutrons from a fission reactor can in principle maintain an adequate temperature.

21-6. Magnetohydrodynamic Conversion

Introduction; Approximate State of the Art.[7,17] Direct magnetohydrodynamic conversion (MHD conversion) employs underlying concepts closely paralleling those for familiar d-c generators. In both, the high-velocity movement of an electrical conductor through and at right angles to a strong magnetic field results in production of a "generated voltage" between this conductor's terminals, with resulting power-delivering current flow through an electrical load connected externally between the generator terminals. As shown schematically in Fig. 21-16, in an MHD generator the conductor is a body of hot gas whose movement through the magnetic field can be caused by expansion due to heating either in a combustion chamber ahead of the generator, or in the generator chamber itself. Demonstration of the correctness of the principles has been made at moderate power levels, and experimental MHD generators have operated at substantial power levels. Both closed-returning-gas-cycle systems and open-cycle systems are feasible. MHD plasmas are also used in propulsion (Sec. 20-7).

The central requirement, that the hot gas must have an electrical conductivity high enough so that the internal voltage drop is perhaps half or less of the generated voltage, can in the presently achieved state of the art be met only for gas temperatures near or above about 2500°K (4000°F). To achieve even this requires "seeding" of the gas, by introducing perhaps 1 per cent or less (in particle-number density) of alkali-metal

SELECTED ENERGY-CONVERSION SUBSYSTEMS 21-33

vapors, chosen because of their low ionizing potentials, being 3.87 ev for cesium and 4.32 for potassium, the ones most commonly considered. Substantial and promising research effort is in progress toward employing non-thermal-equilibrium behavior to produce adequate electrical conductivity at gas temperatures below 2000°K, which would greatly enhance the economics.

The highest gas temperatures at present usable in familiar thermal-to-mechanical-to-electrical converters are very much lower than 2500°K—half or less of this absolute temperature. This makes necessary heat exchangers to reduce the temperature of the MHD exhaust gas to a temperature usable in such apparatus as gas turbines or turbo-electric steam generators. The MHD unit thus becomes a "topping" generator to convert power at a very high temperature.

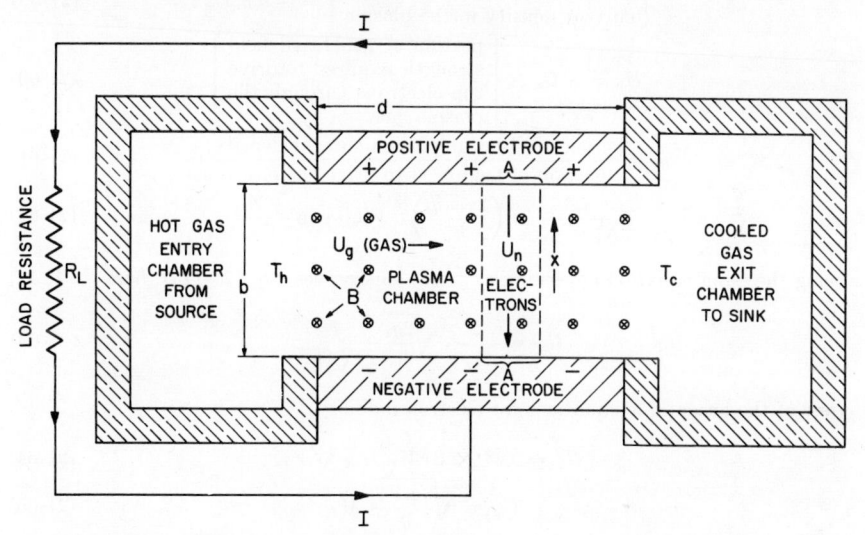

FIG. 21-16. Plasma electric-power-generation channel for magnetohydrodynamic conversion

Because MHD generation is thus primarily useful in systems giving major value to heat rejected into the sink, the formal thermal-conversion efficiency as stated in Eq. (21-5) is perhaps of less interest than the following:

$$\begin{bmatrix}\text{Net conversion}\\ \text{efficiency}\end{bmatrix} = \frac{[\text{electrical power delivered to load}]}{\begin{bmatrix}\text{heat power ac-}\\ \text{cepted from}\\ \text{the source}\end{bmatrix} - \begin{bmatrix}\text{heat power re-}\\ \text{jected } into\\ the\ sink\end{bmatrix}} \quad (21\text{-}57)$$

Being of the constant-generated-voltage type, an MHD machine delivers maximum power, from given heat input, when internal plasma resistive-heat generation equals the power into the load. Resistive heat exits via the sink and therefore does not reduce net conversion efficiency. Maximum power output is thus also the optimum for efficiency. Departure of this net conversion efficiency from unity results primarily from rejection of heat into the boundary-wall electrodes. Such heat has, in fact, serious negative value because of its destructive effect on the walls.

For a given gas, and a given seeding fraction and operating temperature, the elementary theory describes a useful electric-power output in watts per unit volume, which varies as the square of the magnetic-flux density. Operation usually takes place at or slightly above atmospheric pressure, at gas-flow velocities approaching the Mach 1 value; operating above Mach 1 may become feasible. For gases commonly used (air, argon, or combustion products) at temperatures in the neighborhood of

2500°K, and with magnetic fields around 10,000 gauss, theory predicts and operation achieves power-generation densities ranging up toward 100 watts/cu cm. Programs are in progress having as an ambition the achievement of 1,000 watts/cu cm at temperatures below 2000°K, by using nonequilibrium conditions.

Introductory Analysis of Electrical Aspects of MHD Generation. The logic of MHD generation will be discussed relative to an elemental portion of the generation or plasma chamber of Fig. 21-16, as at AA. Anywhere within any such section the following relations apply:

$$\begin{bmatrix}\text{Generated transverse}\\\text{electric-field strength}\end{bmatrix} = -BU_g \tag{21-58}$$

$$\begin{bmatrix}\text{Transverse electron-borne}\\\text{current density in the plasma}\end{bmatrix} = -q_e n_n U_n \tag{21-59}$$

$$U_n = -G_n \times \begin{bmatrix}\text{portion of generated field}\\\text{strength required to drive}\\\text{the electrons through the}\\\text{plasma gas}\end{bmatrix} \tag{21-60}$$

$$\sigma_n = \frac{[\text{electron-borne plasma current density}]}{[\text{portion of field used to drive the electrons}]} \tag{21-61}$$

$$\frac{n_n^2}{N_s - n_n} = \left(\frac{2\pi m_e q_e}{h^2}\right)^{3/2} V_{Te}^{3/2} \exp\frac{-V_{Is}}{V_{Te}} \tag{21-62}$$

being the Saha equation governing electron density; see Eq. (21-12a); here $n_n \ll N_s$;

$$G_n = \frac{2}{3\sqrt{\pi}}\sqrt{\frac{2q_e}{m_e}}\frac{1}{\xi_n \sqrt{V_{Te}}} \tag{21-63}$$

(one of the acceptable equations for electron mobility)

$$G_n = 9.44 \times 10^6 V_{Tg}/\xi_{n0}\sqrt{V_{Te}} \tag{21-64}$$

$$U_g = \sqrt{\frac{\gamma q_e V_{Tg}}{m_g}} M_f \tag{21-65}$$

(relating gas-flow rate to Mach number M_f)

$$U_g = 4.20 \times 10^5 \frac{\sqrt{\gamma}}{\sqrt{m_g/m_e}}\sqrt{V_{Tg}} M_f \tag{21-66}$$

See below Eq. (21-7) for values of q_e, m_e, k_b, h; also, as used here and later in relation to the Fig. 21-16 model,

- B = uniform magnetic-flux density, webers/sq m (1 weber/sq m = 10,000 gauss)
- U_g = longitudinal forward velocity of conducting gas
- U_n = transverse downward electron "drift velocity" carrying the transverse current density
- V_g = "generated voltage" within the chamber
- V_R = portion of V_g required to drive the electrons through the plasma, the counterpart of the armature "IR drop" in a d-c generator
- V_L = terminal voltage, appearing across the load resistance; for maximum power output $V_L = V_R$; note that

$$V_g = V_L + V_R \tag{21-67}$$

- G_n = electron mobility, *defined* by Eq. (21-60); units are meters per second per volt per meter

SELECTED ENERGY-CONVERSION SUBSYSTEMS 21-35

N_s = "seed-atom" number density in the plasma, being the density of the low-ionizing-potential metal vapor used to ensure conductivity; usually potassium or cesium

N_g = density of gas particles of the primary gas constituent of the plasma

N_{g0} = density of any gas at 1 atm and 273°K, being 2.687×10^{25} per cu m

f_s = seeding fraction N_s/N_g; usually $f_s \leq 0.01$

n_n = electron density, resulting from ionization of a portion of the seed atoms

T_g, T_e = temperatures in degrees Kelvin of the gas particles and electrons, respectively, in the plasma

V_{Tg}, V_{Te} = respective kinetic temperatures; see Eq. (21-8)

(W_m/volume) = power to the load, per unit volume of plasma chamber, under maximum power-output conditions for which $V_L = \frac{1}{2}V_g$

V_{Is} = electron-volt ionizing energy, or "ionizing potential," for the seed; 3.87 for cesium; 4.32 for potassium

σ_n = conductivity of the plasma, mhos/m

ϕ_{ng} = experimentally determined collision cross section (for momentum transfer) of an electron with a particle of the primary-content gas; $N_g\phi_{ng}$ is the reciprocal of the so-called electron "mean free path" for this gas; uncertainties as to values of ϕ_{ng} in operating equipment represent an important deficiency in information needed for design optimization; the "Ramsauer effect" refers to the sensitivity of ϕ_{ng} to electron energy, very pronounced for argon; ϕ_{ng} declines somewhat as the magnetic field becomes very intense

ϕ_{ns} = similar cross section for electron collision with a seed atom

ξ_n = cross section for electron collision with all gas particles in a cubic meter, gas density $N_g + N_s$

ξ_{n0} = corresponding cross section at standard conditions, 1 atm, and gas-particle temperature 273°K

$$\xi_n = \xi_{n0} \frac{273}{T_g} = 2.35 \times 10^{-2} \frac{\xi_{n0}}{V_{Tg}} \tag{21-68}$$

From the definition of ξ_{n0}, and noting that $f_s \ll 1$, nearly enough

$$\xi_{n0}(f_s) = N_g \phi_{ng} \left(1 + \frac{\phi_{ns}}{\phi_{ng}} f_s\right) \tag{21-69}$$

$$\frac{f_s}{\xi_{n0}(f_s)} = \frac{1}{N_{g0}\sqrt{\phi_{ng}\phi_{ns}}} \frac{(\phi_{ns}/\phi_{ng})f_s}{1 + (\phi_{ns}/\phi_{ng})f_s} \tag{21-70}$$

This quantity maximizes at $f_s = \phi_{ns}/\phi_{ng}$.

m_g = gas-particle mass for the primary gas content;

$$\frac{m_g}{m_e} = (\text{molecular weight}) \times 1{,}822 \tag{21-71}$$

here 1,822 is the ratio of the mass of an atom of unit atomic weight (none such exist) to m_e

M_f = Mach number at T_g for the gas at velocity U_g

γ = ratio of specific heats at constant pressure and constant volume, for the gas

β = half the number of degrees of freedom of kinetic energy; $\beta = \frac{3}{2}$ for argon, and $\frac{5}{2}$ for oxygen and nitrogen

Elimination of variables between the equations gives the following first approximation to the volume conductivity, assuming atmospheric-pressure operation:

$$\sigma_n = 6.62 \times 10^{13} \frac{\sqrt{f_s}}{\xi_{n0}(f_s)} V_{Tg}^{1/2} V_{Te}^{1/4} \exp \frac{-V_{Is}}{2V_{Tg}} \tag{21-72}$$

21-36 SYSTEM COMPONENTS

Combining (21-58) and (21-65) gives

$$\begin{bmatrix}\text{Generated}\\\text{field strength}\end{bmatrix} = 4.2 \times 10^5 \frac{\sqrt{\gamma}\, M_f B}{\sqrt{m_g/m_e}} \sqrt{V_{T_g}} \qquad (21\text{-}73)$$

It is reasonably obvious that

$$\begin{bmatrix}\text{Internal power generation}\\\text{per unit volume}\end{bmatrix} = J \begin{bmatrix}\text{generated field}\\\text{strength}\end{bmatrix} \qquad (21\text{-}74a)$$

$$\begin{bmatrix}\text{Internal plasma resistive}\\\text{loss reconverted to heat}\end{bmatrix} = J \begin{bmatrix}\text{field strength used}\\\text{in driving electrons}\end{bmatrix} \qquad (21\text{-}74b)$$

where J is the transverse current density carried by the electrons. Useful power output per unit volume is the difference between these, neglecting sheath-induced boundary-wall losses. Maximum power output occurs when the internal resistive reconversion to heat is half the generated power, making $V_R = V_L$. With J expressed in terms of σ_n from Eq. (21-61), use of (21-72) and (21-73) gives, assuming atmospheric-pressure operation,

$$\frac{W_m}{\text{volume}} = \frac{\sigma_n}{4}\,(\text{generated field})^2$$

$$= 2.92 \times 10^{24} \frac{\gamma M_f^2 B^2}{m_g/m_e} \frac{\sqrt{f_s}}{\xi_{n0}(f_s)}\, V_{T_g}^{3/2} V_{T_e}^{1/4} \exp\frac{-V_{Is}}{2V_{T_g}} \qquad (21\text{-}75)$$

The thermal power brought by the gas into the chamber is the product of the thermal energy $\gamma q_e V_{T_g}$ per entering gas particle by the flow rate $U_g N_g$, giving

$$\begin{bmatrix}\text{Watts per unit area of}\\\text{entering thermal power}\end{bmatrix} = 4.25 \times 10^{10} M_f \frac{\gamma^{3/2}}{\sqrt{m_g/m_e}} \sqrt{V_{T_g}} \qquad (21\text{-}76)$$

Considerable uncertainty exists as to the operating-device magnitudes of ϕ_{ng} and ϕ_{ns}, so also of ξ_{n0}. Clearly performance improves as gas velocity increases, approaching Mach 1; operation above Mach 1 may become feasible.

The following illustrative operating specifications will provide the basis for two contrasting tabulations of first-approximation results based on the simple-model analysis given above:

Mach number $M_f = 0.80$
Magnetic-flux density $B = 1.0$, being 10,000 gauss
Gas temperature $T_g = 2500°K$; $V_{T_g} = 0.216$ ev
Thermal equilibrium; thus $T_e = T_g$; $V_{T_e} = V_{T_g}$

Case I: Primary gas constituent argon, used in a closed-cycle operation;

$$\frac{m_g}{m_e} = 72{,}600,\; \beta = 3/2,\; \gamma = 5/3$$

Seeding: 1 per cent cesium; $f_s = 0.01$, $V_{Is} = 3.87$
ϕ_{ng} (for argon): 0.4×10^{-20} sq m (from Harris[15])
ϕ_{ns} (for cesium) = 300×10^{-20} sq m (from Harris[15]); there is no assurance these are the true values in operating apparatus; from Eq. (21-69)
$\xi_{n0}(f_s) = 0.92 \times 10^6$ (dominated by cesium)

Case II: Primary gas content nitrogen; $m_g/m_e = 51{,}000$, $\beta = 5/2$, $\gamma = 7/5$
Seeding: 1 per cent potassium vapor, $f_s = 0.01$, $V_{Is} = 4.32$
ϕ_{ng} (for nitrogen) is perhaps 6.8×10^{-20} sq m (from Dow[3])
ϕ_{ns} (for potassium) is perhaps 300×10^{-20} as in the argon case
$\xi_{n0}(f_s) = 2.7 \times 10^6$ (dominated by nitrogen)

Using these in the equations above gives as first approximations:

	Case I (argon-cesium)	Case II (nitrogen-potassium)
G_n, m sq/volt-sec	4.8	1.63
σ_n, mhos/m	290	34
U_g, m/sec	750	817
Generated volts per meter	750	817
(W_m/volume), watts/cu cm	40	5.6
Heat-power intake, watts/sq cm	11,370	20,650

Presume uniform performance throughout the chamber in Fig. 21-16, height $b = 20$ cm, width $s = 20$ cm, length $d = 40$ cm. This permits describing performance of the generator as a whole, neglecting boundary-wall (electrode) losses, as follows, being rough first approximations:

	Case I (argon-cesium)	Case II (nitrogen-potassium)
V_g, internal generated volts	150	164
V_L, terminal volts	75	82
Internal resistance, $b/sd\sigma_n$, ohms	0.0086	0.074
Load resistance, ohms	0.0086	0.074
Load current, amp	8,710	1,110
Watts power into the load	650,000	91,000
Heat-power intake, watts	4,550,000	8,270,000
Conversion efficiency, %	14.3	1.1
Net conversion efficiency, % (21-57) assumed to be	100	100

Note that here

$$\begin{bmatrix} \text{Conversion} \\ \text{efficiency} \end{bmatrix} = \frac{[\text{power into the load}]}{\begin{bmatrix} \text{heat accepted} \\ \text{from the source} \end{bmatrix}} \tag{21-77}$$

For Case I, the power drawn from the gas and delivered to the circuit is a large enough fraction of the total thermal input to indicate that the exit temperature must be significantly lower than the entrance temperature. Thus the gas temperature must vary longitudinally, and the internal behavior will be nonuniform. The sharp contrast between the cases has three major causes:

1. ϕ_{ng} for argon is very low in this temperature range, being near a minimum in its Ramsauer variance.
2. The exponential occurrence of the low cesium V_{Is} makes its influence strong.
3. Nitrogen heat intake is large because $\gamma = 5/2$.

Power levels as here described, and much higher, should in time become common practice.

Engineering Design and Operational Problems. High power-generation densities and reasonably satisfactory conversion efficiencies are undoubtedly in prospect from MHD apparatus. However, there are many problems of designing for optimization and long life that will require a great deal of research and development before major economic value can be demonstrated.

The electrodes themselves, and the boundary layers between the plasma stream and the electrode walls, offer severe problems. Substantial heat losses occur to walls as a result of boundary-layer behavior, and the heat so lost does not pass to the heat sink; it has a severely destructive effect on the walls. There are often large internally circulating currents. The electrode walls are always segmented; resistive inserts can be used between segments to discourage the tendency for generation to concentrate near the entry port. The "Hall effect" causing transverse nonuniformities in plasma electron density can be pronounced, and increases heat losses to the walls. With proper connections between segmented electrodes, the Hall effect can be used to generate power.

The factor $\exp(-V_{Is}/2V_{Te})$ has a dominating influence in Eq. (21-75). It causes power generation to fall extremely rapidly with declining electron temperature. For this reason, very substantial research effort is under way toward exploiting the possibility that, with sufficient magnetic flux, the heat input to the electrons, from their being driven violently through the plasma gas, may be so great that the poor heat transfer between electrons and gas particles will result in the electrons having a significantly higher temperature than the gas. Use of (21-75) can show that, if the electron temperature is kept near or above 2500°K, a drop of the gas temperature to 1600 or 1800°K will not cause a serious fall-off in generated power density.

Conceivably high conductivity can be maintained in a relatively cool gas by introducing transient ionization influences, requiring relatively little power, that can produce ionization as fast as normal recombination processes destroy it. Flames contain transient chemical states which introduce electrical conductivity; conceivably gases while actually burning in the plasma chamber might have substantially higher conductivities than the ultimate stable products of combustion. Even with operation at 1500° to 1600°K achieved, MHD generation will still be employing relatively high-temperature heat sources and will therefore be used in systems that give major value to heat rejected at the exit-port sink, except perhaps in space-vehicle applications.

21-7. Thermonuclear-fusion Generation

Introduction; Scientific Basis.[16,19] The potentiality for electric-power generation with controlled nuclear fusion as the primary energy source appears spectacular, with, however, many years of engineering research and development ahead before the promise is realized. The primary energy source will be the very small fraction of deuterium contained in place of hydrogen in ordinary water. Deuterium is "heavy hydrogen," containing one proton and one neutron in the atomic nucleus whereas hydrogen contains only one proton. Processing of ordinary water to obtain heavy water is very inexpensive; thus there exists an unlimited supply of "fuel," available at trivial cost.

The fuel must be raised to a critical ignition temperature before the fusion energy can be released; the "match" must be lit to start the fire. The ignition temperature commonly referred to is 100,000,000°K, kinetic temperature about 10,000 ev, for deuterium containing a very little tritium. The ignition temperature for deuterium alone is perhaps four or more times this.[19] The fusion products for the reactions that can occur consist variously of neutrons, protons, and helium isotopes, plus enormous energy releases appearing as kinetic energy of the fusion products. Thus no harmful radioactive fusion products result directly.

In a 100,000,000°K reacting plasma the average deuteron energy would be about 15,000 ev; electrically speaking this does not represent an extreme energy per particle. The reacting gas at this temperature need not present a survival hazard to its enclosure, because the heat-transfer rate to the enclosure can be kept low. This is partly because the reacting gas will have a low number density, perhaps 10^{20} per cubic meter (10^{14} per cubic centimeter), which is a fairly good vacuum. Beyond this, the high-temperature gas is fully ionized, therefore an electrical conductor, and therefore subject to "confinement" by a strong magnetic field to regions not in contact with its enclosure. Electrical conductors and strong magnetic fields do not rapidly interpenetrate; so a magnetic field can control the location of the conducting gas. The reaction releases neutrons, which are not charged particles, therefore are not magnetically confined and so reach the environmental boundaries. Here they deliver their kinetic energies as heat (millions of electron volts per neutron), thus producing the useful energy output by maintaining the boundary hot—at any desired temperature—during heat-energy withdrawal from the boundary to apparatus that converts the thermal energy to electrical energy.

In all the several major research efforts now in progress toward ultimately "lighting the match," the deuteron gas comprises an electrical-discharge plasma in which the deuteron d-c space-charge forces are neutralized by the presence of electrons at a density equal to that of the deuterons. This need for electrons appears unfortunate, because the electrons undergo acceleration and therefore radiate energy electromagnetically, by both of two distinct mechanisms. One radiative mechanism is that, when an electron makes a "near-miss" passage close to a deuteron, the strong interparticle electrical force gives the electron a very severe acceleration, then deceleration, with resulting short-wave electromagnetic radiation, called "bremsstrahlung radiation." It is an important factor in determining the critical temperature at which the fusion reaction is self-sustaining, by fusion-energy release equaling energy-loss rates. The other mechanism arises because rapid accelerations and decelerations are inherent attributes of the tight helical or trochoidal trajectories that a strong magnetic field compels the electrons to follow in this environment; these accelerations can give rise to major radiative energy in the plasma; if it escapes, there must be compensation by fusion generation. This is variously called cyclotron radiation and "magnetic" radiation;[19] it becomes of primary importance at a somewhat higher temperature than for bremsstrahlung. The reactions of primary importance toward controlled nuclear fusion are[19]

$$_1D^2 + {_1D^2} \rightarrow {_1T^3} + {_1p^1} + 4.03 \text{ mev} \qquad \text{at } \alpha \text{ rate} \qquad (21\text{-}78a)$$
$$_1D^2 + {_1D^2} \rightarrow {_2He^3} + {_0n^1} + 3.25 \text{ mev} \qquad \text{at } \beta \text{ rate} \qquad (21\text{-}78b)$$
$$_1D^2 + {_1T^3} \rightarrow {_2He^4} + {_0n^1} + 17.58 \text{ mev} \qquad \text{at } \gamma \text{ rate} \qquad (21\text{-}78c)$$

The "rate" parameters α, β, γ, are temperature-varying measures of collision cross section for the reaction, averaged over the energy ranges involved in the Maxwellian velocity distribution. γ is larger by perhaps 100 or so than either α or β; note the very great energy release in the deuteron-triton reaction.

Evaluation of these various parameters, together with estimates of bremsstrahlung and of cyclotron-radiation energy-removal rates, and theoretical determination of the approximate energy-removal rates, lead to critical contours shown in Fig. 21-17, for an environment having a magnetic field providing confinement for an adequately dense deuteron plasma; thus[19]

1. The U-shaped contour (drawn presuming that some cyclotron radiation escapes from the plasma) for the critical tritium percentage at which fusion-energy generation rate equals energy-removal rate (at lower temperatures chiefly by bremsstrahlung); for conditions above this contour the reactor generates a net heat-power output; below it the reaction is not self-sustaining.

2. The similar wholly declining critical contour for equal energy loss and generation, presuming cyclotron radiation to be wholly trapped in the plasma, none escaping; the zero-tritium intercept is at 47 kev.

3. The rising contour for the conditions at which the fusion-reaction production of tritium and its fusion-reaction loss are just equal.

Clearly there is a considerable range of small percentages of tritium for which the reaction is expected to become self-sustaining at a kinetic temperature near to 10,000 Kv.

State of the Art. Lighting the thermonuclear match in a controlled environment will require:

1. A magnetic-field configuration and strength adequate to keep the hot, reacting plasma gas confined into a region small enough, to make the density high enough, to produce enough collisions to make the reaction self-sustaining for long enough time, to produce net useful energy output; design plans deal with magnetic-flux densities ranging between 30,000 and 100,000 gauss; deuteron particle densities will be perhaps 10^{14} to 10^{15} per cubic centimeter.

2. Means for causing the initial temperature rise, as for example, by a rapid transient contraction of the magnetic-flux pattern so as to compress the tenuous plasma to the extreme degree necessary to produce the high temperature and density—this is the "scratching of the match."

3. A magnetic-field configuration that produces adequate "stability" of the highly conducting plasma, in the face of tendencies, particularly those known as "kink instability" and "pinch instability," to escape from the desired confinement or from the transient compression.

Various configurations and magnetic transient schedules have been used in efforts to create for at least a few microseconds a self-sustaining fusion-reaction plasma, without as yet success. In various patterns for this: the "Scylla machine" at Los Alamos, the "Mirror machine" at Livermore, Calif., the "C Stellerator" at Princeton, and there are others, plasma kinetic temperatures of perhaps 1,000 ev have been observed, and there is substantial promise that achievement of 3,000 to 4,000 ev will become a familiar pattern before too long. Present research emphasis is on achieving

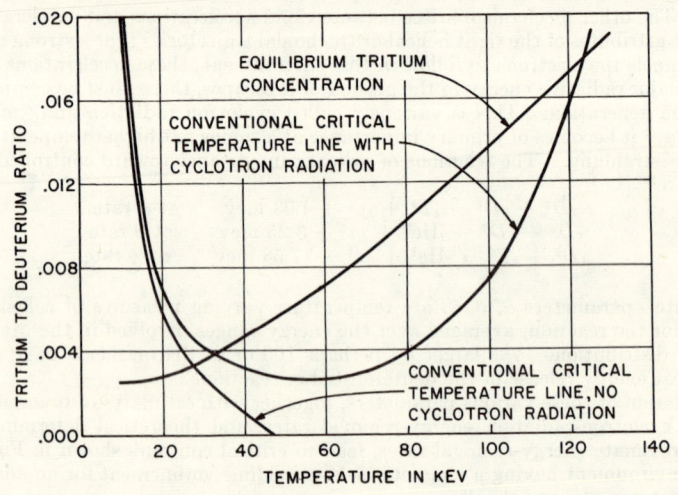

Fig. 21-17. Contours of critical condition for a self-sustaining reaction; equilibrium between using up and creating tritium; both as functions of temperature and tritium content. (*From Ref. 19.*)

mastery of techniques and completeness of understanding of the physics of these high-temperature plasmas, and for this purpose the 3,000-volt kinetic temperature will be adequate.

Obtaining the high magnetic-flux densities necessary for confinement will require cryogenic techniques for obtaining extremely low (in some cases essentially zero) resistance in coils producing the field. Sodium in the neighborhood of 10°K, while not a superconductor, is an extremely good conductor and has sufficient ductility to permit use as a conducting substance encased in a thin stainless-steel cylindrical shell. Niobium-tin and niobium-zirconium have been found to retain superconducting properties, even in the presence of very strong magnetic fields, at temperatures somewhat above the 4°K value for liquid helium. Fabrication is difficult.

The energy output of controlled thermonuclear fusion will appear partly as kinetic energy of neutrons that, being neutral and at extreme energies, leave the plasma and impinge and largely pass into or through enclosing surfaces, and partly as radiative energy (particularly bremsstrahlung) that impinges on the enclosing walls. Both these produce heat in the enclosing walls, *at some specific rate of heat delivery*, not at a specific temperature. By providing means for removal of heat to a thermal-to-electric energy converter of an appropriate type, most of this heat energy can be usefully converted. Clearly the hotter the heat removal, the easier it is to obtain high conversion efficiency in terms of some variant of the theoretical Carnot efficiency $(T_h - T_c)/T_h$. The numbers in Table 21-4 are interesting. Diminishing returns appear for the higher temperatures.

Table 21-4

Heat-removal temp (thermal-converter heat source) T_h		Thermal-converter heat-sink temp T_c		Carnot efficiency $(T_h - T_c)/T_h$, %
°K	°F	°C	°F	
1000	1325	300	32	70
1500	2227	300	32	80
3000	4925	300	32	90

A judgment has been expressed, by some authorities familiar with the heavy-equipment technology of research toward thermonuclear fusion, that once a means for "lighting the match" has been found, the engineering design and construction of the fusion reactor, while complex and challenging, may well be less difficult than for fission reactors, to no small degree because of the minimal radiation hazard.

The problem of reaching the goal of at least briefly self-sustained fusion reaction represents extreme difficulties in research. There is continuing but slow progress toward achieving first a full understanding of the physics of these extremely hot plasmas, which is fully displayed at perhaps 30,000,000°K. Once behavior up to this temperature is fully understood, the extension to ignition should not be difficult. Fulfillment of the research objective of achieving briefly sustained reaction might come within 2 years, or not for 20 years. But it will come.

REFERENCES

Books

1. Fowler, R. H., and E. A. Guggenheim: *Statistical Thermodynamics*, Cambridge University Press, New York, 1961.
2. Smith, R. A.: *Semiconductors*, Cambridge University Press, New York, 1961.
3. Dow, W. G.: *Fundamentals of Engineering Electronics*, 2d ed., John Wiley & Sons, Inc., New York, 1952.
4. Heikes, R. R., and R. W. Ure, Jr.: *Thermoelectricity: Science and Engineering*, Interscience Publishers, New York, 1961.
5. Moore, A. D.: *Fundamentals of Electrical Design*, McGraw-Hill Book Company, New York, 1927 (for heat radiation between facing surfaces).
6. Cowling, T. G.: *Magnetohydrodynamics*, Interscience Publishers, Inc., New York, 1957.

Journal Issues

7. New Energy Sources Issue (includes fuel cells), *Proc. IEEE*, vol. 51, pp. 663–838, May, 1963.

Journal Articles and Reports

8. Conwell, E. M.: Properties of Silicon and Germanium; II, *Proc. IRE*, vol. 46, pp. 1281–1300, June, 1958.
9. Cutler, M.: Thermoelectric Behavior of a p-n Junction, *J. Appl. Phys.*, vol. 32, pp. 222–227, February, 1961.

10. Miller, R. C., R. R. Heikes, and A. E. Fein: Dependence of Thermoelectric Figure of Merit on Energy Bandwidth, *J. Appl. Phys.*, vol. 33, pp. 1928–1932, June, 1962.
11. Blair, J., and D. C. White: Thermoelectric Energy Conversion in Engineering Education, paper presented at the Workshop on Direct Energy Conversion at the June, 1960, Convention of the American Society for Engineering Education.
12. Davison, J. W., and J. Pasternak: Status Report on Thermoelectricity, *NRL Mem. Rept.* 1361, October, 1962.
13. Carabateas, E. N., S. S. Pezaris, and G. N. Hatsopoulos: Interpretation of Experimental Characteristics of Cesium Thermionic Converters, *J. Appl. Phys.*, vol. 32, p. 352, March, 1961.
14. Hernqvist, K. G., M. Kanefsky, and F. H. Norman: Thermionic Energy Converter, *RCA Review*, vol. 19, pp. 244–258, June, 1958.
15. Harris, L. P.: Electrical Conductivity of Cesium-seeded Argon Plasmas Near Thermal Equilibrium, and Electrical Conductivity of Potassium-seeded Argon Plasmas Near Thermal Equilibrium, *Gen. Elec. Research Lab. Rept.* 63-RL-3278G (April, 1963); 63-RL-3334G (May, 1963).
16. Post, Richard F.: Controlled Fusion Research—An Application of the Physics of High Temperature Plasmas, *Proc. IRE*, vol. 45, pp. 134–160, February, 1957.

Papers from the Proceedings of the Several Symposia on Engineering Aspects of Magnetohydrodynamics

17. Brogan, T. R., A. R. Kantrowitz, R. J. Rosa, and Z. T. Stekly (Avco-Everett Research Labs., Everett, Mass., and MIT): Progress in MHD Power Generation, Second Symposium, University of Pennsylvania, Philadelphia, Pa., March, 1961.
18. Harris, L. P., and G. E. Moore (General Electric Research Lab., Schenectady, N.Y.): Some Electrical Measurements on MHD Channels (combustion products), Third Symposium, University of Rochester, Rochester, N.Y., March, 1962.
19. William L. Huss (MIT): Reaction Kinetics of Deuterium-Tritium Mixtures, Third Symposium, University of Rochester, Rochester, N.Y., March, 1962.
20. F. H. Shair (General Electric Space Sciences Lab., King of Prussia, Pa.): Theoretical Performance of MHD Generators Utilizing Magnetically Induced Non-equilibrium Ionization in Pure Alkali Metal Vapors and Seeded Gas Systems, Fourth Symposium, University of California, Berkeley, Calif., April, 1963.

Part IV

SYSTEM THEORY

Chapter 22

INFORMATION THEORY*

FAZLOLLAH M. REZA, *Syracuse University, Syracuse, New York.*

CONTENTS

22-1. Introduction	22-4
22-2. An Outline	22-6
22-3. Average Information per Message of a Discrete Source without Memory	22-7
A Naïve Approach	22-8
A Formal Approach	22-8
22-4. Some Properties of Average Information (Entropy)	22-9
Binary Sources	22-9
22-5. Entropies of Simple Discrete Communication Systems without Memory	22-10
Conditional Entropies	22-10
22-6. A Measure of Mutual Information or Transinformation	22-12
Set-theory Diagrams	22-12
22-7. Capacity of a Discrete Channel	22-12
Binary Symmetric Channel	22-13
Binary Erasure Channel	22-13
Binary Channel	22-13
A Method for Determining the Capacity of an n by n Channel	22-13
Capacity of a Particular Noiseless Channel with Symbol Costs	22-14
22-8. Memoryless Continuous Sources and Channels	22-15
Entropy Maximization	22-16
Gaussian Noisy Channels	22-16
Channels with Additive Noise	22-16
22-9. Entropies and Mutual Information of Systems with Memory	22-17
Mutual Information of Two Stationary Random Processes	22-17
22-10. Entropy of Regular Markov Sources	22-18
Some Definitions	22-19
Some Theorems on Regular Markov Chains	22-19
Entropy of a Regular Markov Chain	22-19
22-11. Mutual Information of Gaussian Random Vectors and Processes	22-20

* This chapter was written during the spring of 1963, when the author was a Visiting Professor at the Laboratoriet for Telegrafi Og Telefoni of the Royal Technical University of Denmark. Acknowledgments and thanks are due to several members of the Laboratoriet for their comments and discussions; and to the Rome Air Development Center which sponsored earlier research work of the author in this area.

The Entropy of a Gaussian n-dimensional Vector.............. 22-20
Mutual Information of Two Gaussian Vectors................. 22-20
Mutual Information of Two Gaussian Processes............... 22-21
22-12. Generalities of Encoding................................... 22-22
22-13. Existence Criterion for Noiseless Encoding................. 22-24
A Noiseless Coding Theorem for Irreducible Codes............ 22-24
McMillan's Theorem for Uniquely Decipherable Codes......... 22-24
A Theorem Describing the Efficiency of Uniquely Decipherable Codes.. 22-24
Shannon-Fano Binary Code................................. 22-24
22-14. Some Methods of Encoding for Noiseless Channels.......... 22-25
Shannon's Irreducible Binary Encoding....................... 22-25
Huffman's Minimum-redundancy Code....................... 22-25
22-15. Hamming Binary Codes.................................... 22-25
t-error Correcting Codes....................................... 22-26
22-16. Principles of Group Codes................................. 22-26

22-1. Introduction

Information theory in the context used here is a part of the statistical theory of communication and is primarily based on Shannon's pioneering work.

The statistical theory of communication has its roots in the mathematical theory of probability and extracts its results, to a large extent, from that source. Methodologically speaking, it appears that development of new techniques in communication theory generally follows or parallels developments in the field of probability theory, in particular stochastic theory (time series).

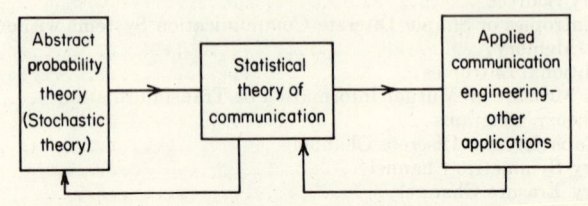

FIG. 22-1. The relative position of statistical theory of communication.

The recognition of the fact that the problem of "transmission of information" is of a statistical nature, in itself, is a significant milestone of communication theory. During the past several decades, scientists facing technological problems of a statistical nature have searched for idealized models and general techniques. These methods have been gradually developed in what today is termed statistical communication theory. The latter field, in turn, draws its resources from pertinent mathematical theories. Of course, it is not always possible to make a clear distinction among these three categories, but the sketch of Fig. 22-1 points to such a possible classification.

Examples of the abstract theories can be found in the work of Doob, Kolmogorov, Wald, Wiener, Kac, Loéve, Wolfowitz, Feller, Fortêt, and many other mathematicians. The works of Blanc-Lapierre, Gabor, Middleton, McMillan, Feinstein, Shannon, Slepian, Rice, and Wiener are examples of contributions which are directed toward the statistical theory of communication. In the third category, one may list hundreds of articles which are primarily of interest to the technologists engaged in the development phase of communication equipments; some of these are covered in Chapter 12.

An important segment of mathematical statistics deals with an investigation of interrelationship between two random variables. When a phenomenon is under study

INFORMATION THEORY

in the field of physical sciences and engineering, one often desires to evaluate the relationship which may exist between certain measurable quantities X and Y. Of course, X and Y may be vectors or matrices comprising a complex set of data. The random variable X may be referred to as the *cause* and the random variable Y as the *effect* (or *input* and *output*). A physical law attempts to describe the nature of the interrelationship between cause and effect.

In the field of statistical studies several such parameters are known. Historically, the coefficient of correlation seems to be one of the first* and by far the best known of these parameters. The correlation coefficient, where applicable, indicates a measure of linear dependence between the cause and the effect. Let E stand for the operation of averaging, the coefficient of correlation between X and Y is defined as

$$\rho(X,Y) = \frac{E\{[X - E(X)][Y - E(Y)]\}}{\sqrt{E[X - E(X)]^2}\sqrt{E[Y - E(Y)]^2}} \qquad (22\text{-}1)$$

Simple properties of the correlation coefficient are described below:

(a) $\qquad \rho(X,Y) = \rho(Y,X)$
(b) $\qquad -1 \leq \rho(X,Y) \leq +1$ $\qquad\qquad\qquad\qquad\qquad$ (22-2)
(c) $\qquad \rho(X,Y) = 0$ if and only if X and Y are uncorrelated

When the coefficient of correlation is 0, the input and the output are uncorrelated. For $\rho = \pm 1$ variables X and Y will be linearly related. Thus in a certain sense ρ gives some "information" about the linear interdependence of X and Y. The use of the correlation coefficient to indicate a measure of interrelationship between two variates has its merits and drawbacks.

The search for suitable measures describing the behavior of probabilistic systems has long been of interest in communication theory. An axiomatic approach to this problem is not contemplated here. However, from an engineering point of view, it appears that a most suitable information measure between a random input and output should subscribe to the following basic properties:

(a) $\qquad I(X;Y) = I(Y;X)$
(b) $\qquad I(X;Y) \geq 0$ $\qquad\qquad\qquad\qquad\qquad$ (22-3)
(c) $\qquad I(X;Y) = 0$

if and only if X and Y are mutually independent—X gives no information about Y and vice versa.

(d) $\qquad I(X;Y) = I(U;V)$

If the random input X is subjected to a suitable transformation $U = h(X)$ and the output Y is subjected to a suitable transformation $V = g(Y)$ then the new pair of input-output should entail the same amount of information as before.†

In this respect, the following measure of mutual information was evolved by C. E. Shannon:

$$I(X;Y) = E\left[\log \frac{f(X,Y)}{f_1(X)f_2(Y)}\right] \qquad (22\text{-}4)$$

where $f(x,y)$ is the joint probability density of (X,Y), $f_1(x)$ and $f_2(y)$ marginal probability densities of X and Y, respectively.

This suggested measure of mutual information between X and Y exists under quite general circumstances (see Ref. 21, pages 289–290). It is to be noted that $|\rho(X,Y)|$ fulfills the requirements a, b, and c (if "independent" is replaced by "uncorrelated") but not necessarily d.

---

* The first use of the term "correlation" in a technical sense is attributed to Sir Francis Galton's paper, Correlations and Their Measurement Chiefly from Anthropometric Data," presented on Dec. 5, 1888.

† The mathematical translation of "suitable transformation" as used here is: Borel measurable functions mapping the real axis in a one-to-one manner into itself.

22-2. An Outline

No exact borders can be drawn for a dynamic field of knowledge, but a tentative sketch of the area may prove profitable.

1. "Information" is conveyed because of "randomness" of the messages transmitted by a "source." A deterministic source conveys no information. Generally some "information measure" can be associated with a statistical source by outlining properties that such a measure should possess and deriving a suitable function.

2. Shannon-Wiener's measure of information associated with a discrete statistical source is referred to as "communication entropy" or entropy of the source. Entropy is a "describing function," a most useful parameter for the overall study of information-processing sources and equipments, much in the same way that the concepts of average "value," "power," and "moment" have found universal usage in problems of physics and engineering. In mathematical terms, Shannon associated a new parameter with a random variable with known probability description.

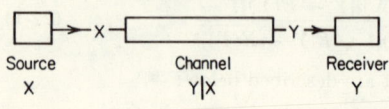

FIG. 22-2. A simple communication system.

3. The application of the concept of entropy to a two-dimensional random variable (X,Y) has special significance. This is due to the fact that such a model describes a communication set-up in its simplest form; see Fig. 22-2.

The bivariate random variable (X,Y) encompasses five probability situations each having a physical interpretation in our described model:

$$P\{X\}, P\{Y\}, P\{Y|X\}, P\{X|Y\}, P\{X,Y\}$$

The corresponding entropies are:

$H(X)$: source entropy, that is, the entropy associated with $P\{X\}$
$H(Y)$: receiver entropy, that is, the entropy associated with $P\{Y\}$
$H(Y|X)$: noise entropy, that is, the entropy associated with $P\{Y|X\}$
$H(X|Y)$: channel entropy, that is, the entropy associated with $P\{X|Y\}$ looking from the receiver's end
$H(X,Y)$: overall system entropy, that is, the entropy associated with $P\{X,Y\}$

These entropies and their mathematical interrelations offer interesting physical interpretations. It can be demonstrated that the quantity

$$I(X;Y) = H(X) - H(X|Y) \tag{22-5}$$

has the special significance of a measure of information transmitted through the system. $I(X;Y)$ is referred to as the transinformation.

4. The concept of transinformation is the central theme of information theory. It can be shown that the transinformation exists under very general conditions, and it preserves its communication value. If (X,Y) is a joint random vector with finite or infinite dimensions, these general concepts will be still applicable. That is, a measure of transinformation can be found for joint random vectors, as well as for large classes of stochastic systems with memory, referred to as ergodic systems.

5. Roughly speaking, the maximum of the transinformation for a given channel may be obtained by choosing the "best" source, that is, the matching source for the specified channel. This maximum value of the transinformation is referred to as the capacity of the channel.

6. Perhaps the most important result of Shannon's theory is the fact that, through a given channel, information can be transmitted at a rate as close to the channel capacity as desired with as low a probability of error as may be specified; it is impossible to surpass this ideal rate and meanwhile remain confined within any preassigned transmission-error probability. Statements of this type have been investigated for different types of sources and channels. These results are generally referred to as

fundamental theorems of information theory. Their exact descriptions require mathematical formalism developed in a later section.

7. The ideal transmission procedure can be accomplished through "coding" the output of the source, possibly into words of another language. For the new source, the probability of messages, and hence the entropy of the new output, may be more appropriately chosen. An existence proof for the possibility of the "ideal transmission of information" is known, but present actual coding procedures are far from being complete.

To sum up, in this chapter, we undertake an introductory study of the simple and powerful concepts of:

The communication entropy
A measure of transinformation
The channel capacity
Elements of coding theory

For the most part, we shall use the notation and sequences of Ref. 21. The reader may find this reference a convenient supplement for filling in the steps which may have been left out in the present condensed form.

It has been pointed out* that "information theory will probably always remain, like thermodynamics, conceptual and guiding rather than definitive . . . the least that one can say for the theory is that it has unified diverse appearing facts into a common conceptual framework."

This serves as a fair warning to the application-oriented reader. Information theory, per se, does not primarily provide new techniques for solving problems. At this stage of development the theory mainly embraces a few new concepts along with a handful of fundamental theorems describing certain ideal rates for the transmission of information. Since the main attributes of this theory are in the providing of new concepts, the domain of the theory is quite broad. It may be used to illuminate the probabilistic behavior of systems without regard to the fact that the input is an optical, electrical, or mechanical quantity. The dominant fact is the probability distribution of the system without regard to the application fields themselves.

There has been some "application" of information theory to a broad spectrum of scientific disciplines ranging from abstract mathematics, physics, optical and electrical communication, biology, and genetics to psychology and linguistics. As an example of the application of information theory to mathematics, attention is drawn to the extensive research activities which have been stimulated by communication-theory considerations. Much of the recent work of Kolmogorov on entropy and ergodicity are in this category. A specific example of such a feedback link maybe found in the work of Yu. V. Linnik, where a novel proof of the central-limit theorem based on information theory is derived.

Examples of the stimulating influence of information theory in other disciplines can be ascertained by the interested reader from the abundant and well-publicized literature that is available.

22-3. Average Information per Message of a Discrete Source without Memory

In problems of communications, we are dealing with sources which transmit signals of a statistical nature. These signals are transmitted at random, but with specified probabilities. That is, while the individual signal probabilities are known, one never knows for sure which signal is going to be transmitted next. The question raised and answered by Wiener-Shannon is the following: Is there a suitable measure of uncertainty or surprise which could be associated with a statistical source of this sort? Such a measure should depend solely on message probabilities and not on the physical nature of the signals, such as their strength or their physical origin.

* B. McMillan and D. Slepian, *Proc. IRE*, 1962, pp. 1151–1157.

A Naïve Approach. For a naïve introduction to the subject, assume that a transmitter X transmits any of n messages $(x_1, x_2, \ldots, x_n)$ with equal probabilities $P\{x_i\} = 1/n$. Also assume that the output of this transmitter will be supplemented by messages $(y_1, y_2, \ldots, y_m)$ which will be transmitted later by a transmitter Y independent of the first set of messages with equal probabilities $P\{Y_i\} = 1/m$.

The desired measure of information, say $f(\ )$, should depend only on message probabilities. Thus, for the first transmitter, the average information per signal should be of the form $f(1/n)$ and for the second transmitter of the form $f(1/m)$.

Next, imagine a composite transmitter which transmits one of the above messages of X along with one of the messages of Y. The average information per new signal is $f(1/mn)$, as each signal now consists of a message of X and one of Y, that is (X, Y).

If the two sets of messages are independent of each other, then it is natural to require that the sum of the average information provided by the first transmitter X and the second transmitter Y would equal the average information given by the messages of the compound (X, Y) transmitter.

$$f\left(\frac{1}{n}\right) + f\left(\frac{1}{m}\right) = f\left(\frac{1}{mn}\right) \tag{22-6}$$

The simplest solution to this functional equation is of logarithmic type:

$$\log \frac{1}{m} + \log \frac{1}{n} = \log \frac{1}{mn}$$
$$-\log m - \log n = -\log mn \tag{22-7}$$

Thus the average information of a binary source with two equiprobable messages is

$$-\log \tfrac{1}{2} = \log 2$$

For most engineering applications, it is customary to consider logarithms to the base 2. The binary unit of information is called a bit, short for binary digit. This unit corresponds to the amount of average information associated with a binary source with two equiprobable signals.

The amount of information per symbol of a source which transmits 2^N equiprobable messages is found to be

$$-\log \frac{1}{2^N} = \log 2^N = N \text{ bits per message} \tag{22-8}$$

A Formal Approach. A more formal derivation of a suitable function for a measure of information requires the assignment of desired properties followed by the necessary mathematical search. The following properties seem quite appropriate for the desired measure.

1. Given a finite complete probability scheme $(p_1, p_2, \ldots, p_n)$, the associated average information $H(p_1, p_2, \ldots, p_n)$ must take on its largest value when all events are equiprobable.

2. For a joint finite complete scheme the associated entropies should satisfy the identity

$$H(X, Y) = H(X) + H(Y|X) \tag{22-9}$$

The average information conveyed by (X, Y) is the sum of the average information given by X and that provided by Y when X is given.

3. Adding an impossible event to a scheme should not change the entropy of the scheme.

$$H(p_1, p_2, \ldots, p_n, 0) = H(p_1, p_2, \ldots, p_n) \tag{22-10}$$

4. The average information is continuous with respect to all its arguments.

INFORMATION THEORY

The only function which satisfies these requirements is

$$H(X) = -\lambda \sum_{k=1}^{n} p_k \log p_k \qquad (22\text{-}11)$$

λ being an arbitrary constant usually taken to be unity. See Refs. 5, 12, 23, and page 124 of Ref. 21.

The average amount of information, or average uncertainty of the system, is also referred to as the entropy of the system. This is in analogy with a similar function which has been used in thermodynamics for several decades. Some authors prefer to use comentropy, a condensed form for communication entropy. For other properties of a measure of information, see Refs. 14, 20, and 23.*

22-4. Some Properties of Average Information (Entropy)

The entropy function has the following properties:

1. The maximum entropy of an independent discrete n-symbol source without memory is $\log n$. This value corresponds to the case when all outcomes are equiprobable

$$H(p_1, p_2, \ldots, p_n) = \log n \qquad (22\text{-}12)$$

2. For independent sources the entropy is an additive function

$$H(X,Y) = H(X) + H(Y) \qquad (22\text{-}13)$$

In other words, independent detailed informations are additive, as suggested by Fig. 22-3.

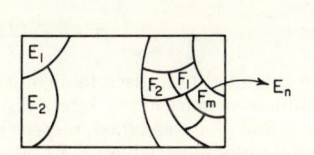

FIG. 22-3. Sample space, events, and sub-events.

FIG. 22-4. Entropy of a binary memoryless source.

If one of the signals, say E_n, from the original set $(E_1, E_2, \ldots, E_n)$ with $P\{E_n\} = p_n$ is subdivided into signals $(F_1, F_2, \ldots, F_m)$ with $P\{F_k\} = q_k$ then

$$H(p_1, p_2, \ldots, p_n) + p_n H\left(\frac{q_1}{p_n}, \frac{q_2}{p_n}, \ldots, \frac{q_n}{p_n}\right)$$
$$= H(p_1, p_2, \ldots, p_{n-1}; q_1, q_2, \ldots, q_m) \qquad (22\text{-}14)$$

3. The entropy is a convex function.

For example, if in the simple binary case, one considers the three transmitters $(p, 1-p)$, $(q, 1-q)$, $\left(\dfrac{p+q}{2}, 1 - \dfrac{p+q}{2}\right)$, then

$$\tfrac{1}{2}[H(p, 1-p) + H(q, 1-q)] \leq H\left(\frac{p+q}{2}, 1 - \frac{p+q}{2}\right) \qquad (22\text{-}15)$$

Binary Sources. The entropy function for the simple case of a binary source is plotted in Fig. 22-4.

* The interested reader is referred to an illuminating article of A. Rényi, *Acta Math. Acad. Sci. Hungaricae*, 1959, pp. 441–451.

For example, consider the following three transmitters which transmit two letters A and B independently; the transmission probabilities for transmitters are

1. $(1/4, 3/4)$
2. $(1/2, 1/2)$
3. $(1/64, 63/64)$

The corresponding entropies are

$H_1 = -\tfrac{1}{4} \log \tfrac{1}{4} - \tfrac{3}{4} \log \tfrac{3}{4} = 0.811$
$H_2 = -\tfrac{1}{2} \log \tfrac{1}{2} - \tfrac{1}{2} \log \tfrac{1}{2} = 1$
$H_3 = -\tfrac{1}{64} \log \tfrac{1}{64} - \tfrac{63}{64} \log \tfrac{63}{64} = 0.116$

The outcomes of the third transmitter, on the average, are easier to predict than those of the first transmitter. The amount of uncertainty per transmitted letter is largest for the second transmitter. As far as the entropy of the transmitted messages is concerned, the second transmitter is the most efficient of the three. That is, the latter transmitter conveys more information per symbol than the other two.

$$H_3 < H_1 < H_2$$

22-5. Entropies of Simple Discrete Communication Systems without Memory

In the previous section, it was pointed out that, to every finite discrete probability scheme, there corresponds an entropy. This entropy, in a way, describes the average uncertainty per event of the independent statistical source which generates the said events.

$$H(X) = -\sum_{k=1}^{n} P\{x_k\} \log P\{x_k\}$$
$$0 \leq H(X) \leq \log n \tag{22-16}$$

In the present section the same concept will be extended to a discrete memoryless two-dimensional scheme, that is, the space of the joint occurrence of two sets of discrete independent events E and F. Random variables X and Y are assigned, respectively, to the sample spaces of E and F; the two-dimensional random variable (X, Y) takes on values in the product space $E \otimes F$ (see Ref. 21, Chapter 3). Thus three entropy functions $H(X)$, $H(Y)$, and $H(X,Y)$ may be immediately envisaged.

$$H(X) = -\sum_{k=1}^{n} P\{X = x_k\} \log P\{X = x_k\} \tag{22-17}$$

$$H(Y) = -\sum_{j=1}^{m} P\{Y = y_j\} \log P\{Y = y_j\} \tag{22-18}$$

$$H(X,Y) = -\sum_{k=1}^{n}\sum_{j=1}^{m} P\{X = x_k, Y = y_j\} \log P\{X = x_k, Y = y_j\} \tag{22-19}$$

Conditional Entropies. In a natural way one may also evaluate conditional entropies associated with conditional probabilities of the product space under consideration; see Fig. 22-5.

When a symbol x_k is transmitted, that is, when an event E_k occurs, the symbol goes through a channel and may be subjected to some sort of noise or error. Therefore, x_k may be received as $y_1, y_2, \ldots,$ or y_m with the following respective probabilities:

$$P\{y_1|x_k\}, P\{y_2|x_k\}, \ldots, P\{y_m|x_k\}$$

INFORMATION THEORY 22-11

These probabilities are specified by the structure of the prevailing noise in the channel. The entropy of this situation is

$$H(Y|x_k) = - \sum_{j=1}^{m} P\{y_j|x_k\} \log P\{y_j|x_k\} \quad (22\text{-}20)$$

When the average of $H(Y|x_k)$ for all transmitted symbols x_k is computed, then one obtains the conditional entropy of the channel, which indicates the entropy of noise in that channel.

$$H(Y|X) = \overline{H(Y|x_k)} = - \sum_{k=1}^{n} P\{x_k\} H(Y|x_k)$$

$$= - \sum_{k=1}^{n} \sum_{j=1}^{m} P\{X = x_k, Y = y_j\} \log P\{Y = y_j | X = x_k\} \quad (22\text{-}21)$$

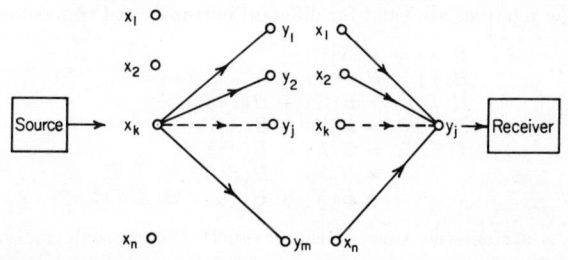

Fig. 22-5. Message transmission and reception in a discrete memoryless channel.

Similarly for the conditional entropy corresponding to $P\{x_k|y_j\}$ (called equivocation entropy of the channel), one obtains

$$H(X|Y) = \overline{H(X|y_j)}$$

$$= - \sum_{k=1}^{n} \sum_{j=1}^{m} P\{X = x_k, Y = y_j\} \log P\{X = x_k | Y = y_j\} \quad (22\text{-}22)$$

To sum up, the following entropies are associated with our described model (see Ref. 21, Chapter 3):

$H(X)$: average information per character at the source, or the entropy of the source.

$H(Y)$: average information per character at the destination, or the entropy of the receiver.

$H(X,Y)$: average information per pair of transmitted and received characters, or the average uncertainty of the communication system as a whole.

$H(Y|X)$: a specific character x_i being transmitted; one of the permissible y_j may be received with a given probability. The entropy associated with this probability scheme when x_i covers sets of all transmitted symbols, that is, $\overline{H(Y|x_i)}$, is the conditional entropy $H(Y|X)$, a measure of information about the receiving port, where it is known that X is transmitted.

$H(X|Y)$: a specific character y_j being received; this may be a result of transmission of one of the x_i with a given probability. The entropy associated with this probability scheme when y_j covers all the received symbols, that is, $\overline{H(X|y_j)}$, is the entropy $H(X|Y)$ or equivocation, a measure of information about the source, where it is known that Y is received.

These five entropies jointly describe all the statistical properties of the communication system. There are some interrelations amongst them, which are discussed in the next section.

22-6. A Measure of Mutual Information or Transinformation

Subsequent to some physical and mathematical considerations, one is led to accept the following expression as a suitable measure of information contained in a discrete random variable Y about another discrete random variable X:

$$I(X;Y) = \overline{\log \frac{P\{x_i,y_j\}}{P\{x_i\}P\{y_j\}}} = \overline{\log \frac{P\{x_i|y_j\}}{P\{x_i\}}} = \overline{\log \frac{P\{y_j|x_i\}}{P\{y_j\}}}$$

$$= \sum_{i=1}^{n} \sum_{j=1}^{m} P\{X = x_i;\, Y = y_j\} \log \frac{P\{X = x_i,\, Y = y_j\}}{P\{X = x_i\}P\{Y = y_j\}} \quad (22\text{-}23)$$

Transinformation, or mutual information, is a nonnegative number which describes the overall flow of information through the system. It is difficult to appreciate the impact of such a statement at once, but there is ample justification for substantiating the choice of this function as a most convenient measure for information flow.

The following relations are valid for different entropies and transinformation:

$$\begin{aligned}
H(X) &\geq H(X|Y) \\
H(Y) &\geq H(Y|X) \\
H(X,Y) &= H(X) + H(Y|X) \\
H(Y,X) &= H(Y) + H(X|Y) \\
I(X;Y) &= H(X) - H(X|Y) \\
&= H(Y) - H(Y|X) \\
&= H(X) + H(Y) - H(X,Y)
\end{aligned} \quad (22\text{-}24)$$

For a lossless system (no noise in the channel), the transinformation equals the entropy of the source.

$$\begin{aligned}
H(Y|X) &= H(X|Y) = 0 \\
I(X;Y) &= H(X) = H(Y) = H(X,Y)
\end{aligned} \quad (22\text{-}25)$$

If and only if the input and output are mutually independent, the transinformation is zero.

$$\begin{aligned}
H(X|Y) &= H(X) \\
I(X;Y) &= 0
\end{aligned} \quad (22\text{-}26)$$

Set-theory Diagrams. A summary of the simplest entropy relations is exhibited in the diagram of Fig. 22-6.

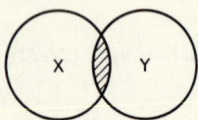

FIG. 22-6. A simple Venn-type diagram for a discrete memoryless communication system.

$(X \cup Y)$	$H(X,Y)$
$(X - Y)$	$H(X|Y)$
$(Y - X)$	$H(Y|X)$
$(X \cap Y)$	$I(X;Y) = H(X) - H(X|Y)$

Such schematic diagrams can be succinctly drawn to exhibit the relations amongst entropies of two, three, or more discrete random variables (Ref. 21, pages 106–108).*

22-7. Capacity of a Discrete Channel

One of the most interesting concepts introduced by Shannon is the concept of channel capacity. A given channel may be driven by many sources which are members of a specified class. Amongst all these sources there will generally be one with

---

* The impact of the use of set theory in this context was pointed out earlier by the author in U.S. Air Force Cambridge Research Center report AFCRC-TN-59-588, September, 1959. Similar results were independently obtained by N. M. Blachman, *Proc. IRE*, vol. 49, pp. 1331–1332, August, 1961. An analytically obliging approach is found in H. K. Ting, *On the Amount of Information; Theory of Probability and Its Applications*, vol. VII, no. 4, 1962, English translation, pp. 439–447.

INFORMATION THEORY

some special symbol distribution which leads to the largest transinformation through the channel. Then we may say that this source and the channel are matched. The largest transinformation through the channel is referred to as the channel capacity. More specifically, the capacity C of a given memoryless discrete channel is defined as the largest value of the transinformation obtained by having the channel driven by all discrete sources without memory.

$$C = \max I(X;Y) = \max [H(X) - H(X|Y)] \qquad (22\text{-}27)$$

Binary Symmetric Channel. The binary symmetric channel is illustrated in Fig. 22-7; a symbol is correctly transmitted with probability p and incorrectly with

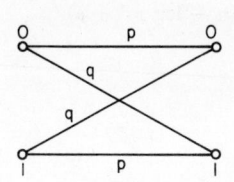

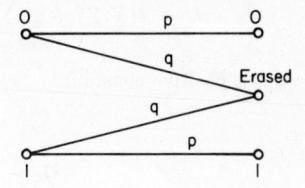

Fig. 22-7. A binary symmetric channel.

Fig. 22-8. A binary erasure channel.

probability $q = 1 - p$. The capacity is obtained when the source has an entropy of 1 bit per symbol.

$$C = 1 + p \log p + q \log q \qquad (22\text{-}28)$$

Binary Erasure Channel. The binary erasure channel is illustrated in Fig. 22-8; a symbol is either correctly received with probability p or not received at all with probability $q = 1 - p$. The capacity is achieved when the source has an entropy of 1 bit per symbol.

Binary Channel. In a binary channel, symbol j is received when symbol i is transmitted with probability p_{ij}. If

$$p_{11} = \alpha \qquad p_{21} = 1 - \beta$$
$$p_{12} = 1 - \alpha \qquad p_{22} = \beta$$

the channel capacity is found to be

$$C = \frac{-\beta H(\alpha) + \alpha H(\beta)}{\beta - \alpha} + \log \left[1 + \exp \frac{H(\alpha) - H(\beta)}{\beta - \alpha}\right] \qquad (22\text{-}29)$$

Fig. 22-9. Message x_k may be received as $y_1, y_2, \ldots, y_n$.

A Method for Determining the Capacity of an n by n Channel*

In the system of Fig. 22-9, let

$$P\{X = x_k\} = p_k$$
$$P\{Y = y_j\} = p_j \qquad k, j = 1, 2, \ldots, n$$

and consider the auxiliary parameters $(Q_1, Q_2, \ldots, Q_n)$ defined by the set of equations

$$-H(Y|x_1) = p_{11}Q_1 + p_{12}Q_2 + \cdots + p_{1n}Q_n$$
$$\cdots\cdots\cdots\cdots\cdots\cdots\cdots\cdots\cdots\cdots \qquad (22\text{-}30)$$
$$-H(Y|x_n) = p_{n1}Q_1 + p_{n2}Q_2 + \cdots + p_{nn}Q_n$$

* This method is originally due to S. Muroga; see Ref. 21, pp. 117–120.

The transinformation is

$$\begin{aligned}I(X;Y) &= H(Y) - H(Y|X) \\ &= -(p_1' \log p_1' + \cdots + p_n' \log p_n') - p_1 H(Y|x_1) - \cdots - p_n H(Y|x_n) \\ &= \sum_{k=1}^{n} p_k'(Q_k - \log p_k')\end{aligned} \qquad (22\text{-}31)$$

where p_k' is the probability of receiving $y_k (k = 1, 2, \ldots, n)$. In order to maximize $I(X;Y)$ we may set forth to maximize the function

$$U = I(X;Y) + \mu \sum_{k=1}^{n} p_k' = \sum_{k=1}^{n} p_k'(Q_k - \log p_k' + \mu) \qquad (22\text{-}32)$$

where μ is a positive constant.

$$\frac{\partial U}{\partial p_k'} = 0$$

gives
$$Q_k - \log p_k' = \log e - \mu \qquad (22\text{-}33)$$

We find that

$$\begin{aligned}C = \max I(X;Y) &= \sum_{k=1}^{n} p_k'(\log e - \mu) \\ &= \log e - \mu \\ &= Q_k - \log p_k'\end{aligned} \qquad (22\text{-}34)$$

$$\sum_{k=1}^{n} p_k' = \sum_{k=1}^{n} \exp(Q_k - e) = 1$$

$$C = \log \left(\sum_{k=1}^{n} 2^{Q_k} \right) \qquad (22\text{-}35)$$

Capacity of a Particular Noiseless Channel with Symbol Costs. In engineering problems the transmission of each symbol may have a time duration or a "cost" associated with it. In such problems, if one assumes that all possible distinct long messages are transmitted with equal probabilities over a noiseless channel, then the channel capacity (in the absence of noise) may be computed as follows:

Consider the alphabet matrix $[a_1, a_2, \ldots, a_n]$ and the corresponding matrix of time duration $[t_1, t_2, \ldots, t_n]$.

Let $N(T)$ be the number of distinct messages of length T. Assuming no extra spaces between successive words, we have

$$N(T) = \sum_{k=1}^{n} N(T - t_k) \qquad (22\text{-}36)$$

The solution to this difference equation is of the type

$$N(T) = \sum_{k=1}^{n} A_k r_k^T \qquad (22\text{-}37)$$

where A_k's are constants depending on the boundary conditions of the problem, and r_k's are the roots of

$$1 - \sum_{k=1}^{n} r^{-t_k} = 0 \qquad (22\text{-}38)$$

INFORMATION THEORY

One can show that no complex root of this equation can have a larger magnitude than its only real positive root r_j.

In his original paper,[24] Shannon dispenses with the possibility of the characteristic equation having complex roots of larger or equal magnitude than the positive real root. Brillouin seems to be the first person who recognized the need for a more accurate proof of the statement. He delves into this matter in an example which is treated in Ref. 2. A general proof of the statement has recently appeared in Reza, Ref. 22. Therefore, it is correct to conclude that

$$\lim_{T \to \infty} N(T) = \lim_{T \to \infty} A_j r_j^T \qquad (22\text{-}39)$$

When all these long messages are transmitted with equal probabilities over a noise-free channel, the capacity (per unit time) becomes

$$C = \lim_{T \to \infty} \frac{\log A_j r_j^T}{T} = \log r_j \qquad (22\text{-}40)$$

22-8. Memoryless Continuous Sources and Channels

Here we consider a source transmitting a signal X with specified probability $f_1(x)$. The values of the signal X are taken from a finite or infinite real interval. As an extension of the definition of entropy in the discrete case, one may define the entropy of such a source as

$$H(X) = E[-\log f_1(X)] = -\int_{-\infty}^{\infty} f_1(x) \log f_1(x) \, dx \qquad (22\text{-}41)$$

When such signals are submitted to a channel, then the following quantities may be defined as an extension of the analogous quantities in the discrete case:

$$H(Y) = E[-\log f_2(Y)] = -\int_{-\infty}^{\infty} \int_{-\infty}^{\infty} f(x,y) \log f_2(y) \, dx \, dy$$

$$H(Y|X) = E[-\log f_x(Y|X)] = -\int_{-\infty}^{\infty} \int_{-\infty}^{\infty} f(x,y) \log \frac{f(x,y)}{f_1(x)} \, dx \, dy$$

$$H(X|Y) = E[-\log f_y(X|Y)] = -\int_{-\infty}^{\infty} \int_{-\infty}^{\infty} f(x,y) \log \frac{f(x,y)}{f_2(x)} \, dx \, dy$$

$$H(X,Y) = E[-\log f(X,Y)] = -\int_{-\infty}^{\infty} \int_{-\infty}^{\infty} f(x,y) \log f(x,y) \, dx \, dy$$

Unlike the discrete case, the entropies associated with continuous schemes no longer have immediate information-theoretic values. The difficulty lies in the fact that these entropies may become zero, infinite, and even negative. Therefore, they cannot always be used as measures—a change of scale also affects their values.*

Despite this basic drawback, the mutual information $I(X;Y)$ remains a true measure of the information conveyed by Y concerning X. This measure is nonnegative,

---

* The drawback that $H(X)$ may become negative can be remedied by adding an additive constant. For example, assume $E[X] = 0$ and $E[X^2]$ finite and let

$$H'(X) = H(X) - \tfrac{1}{2} \log E[X]^2$$

$H'(X)$ represents the entropy measured with respect to a normal distribution, since the entropy of a normal random variable with variance $\sigma^2 = E[X]^2$ is $\sigma \sqrt{2\pi e}$. It is not difficult to show that this quantity is always nonnegative. In fact, Yu. V. Linnik (*Theory of Probability and Its Applications*, vol. IV, no. 3, pp. 288–299, 1959) has shown that

$$-|\log L| - \log \sigma \leq H'(X) \leq \sigma^2 - \log \sigma + c_1$$

where $L = \sup_x P\{x\}$ and c_1 is an absolute constant.

generally finite, and invariant under linear transformations (see Ref. 21, Chapter 8).

$$\begin{aligned}I(X;Y) &= H(X) - H(X|Y) = H(Y) - H(Y|X) \\ &= H(X) + H(Y) - H(X,Y) \\ &= \int_{-\infty}^{\infty} \int_{-\infty}^{\infty} f(x,y) \log \frac{f(x,y)}{f_1(x)f_2(y)} \, dx \, dy \end{aligned} \qquad (22\text{-}42)$$

Entropy Maximization. It is useful to investigate the signal probabilities of a source with the maximum entropy within a class of sources. Few problems of this sort have been studied. The following three cases are well known as they were originally introduced in Shannon's paper.[24]

1. Determine the source with the maximum entropy among all sources without memory-transmitting signals from the interval $[a,b]$.

The source with the largest entropy turns out to be a random variable with uniform distribution in $[a,b]$.

$$f(x) = \frac{1}{b-a} \qquad a \leq x \leq b \qquad (22\text{-}43)$$
$$H(X) = \log_e (b - a)$$

2. Determine the source with the maximum entropy among all sources without memory-transmitting signals from the interval $[0, \infty]$ and having a specified average $\bar{X} = a > 0$.

The maximum entropy corresponds to an exponentially distributed source

$$f(x) = \frac{1}{a} e^{-x/a} \qquad (22\text{-}44)$$
$$H(X) = \log_e ae$$

3. Determine the source with the maximum entropy among all sources without memory-transmitting signal from the interval $[-\infty, +\infty]$, and having

$$\bar{X} = 0 \qquad \overline{X^2} = \sigma^2$$

The maximum entropy is associated with the normal probability-distribution law.

$$f(x) = \frac{1}{\sqrt{2\pi}\,\sigma} e^{-x^2/2\sigma^2} \qquad (22\text{-}45)$$
$$H(X) = \log_e \sqrt{2\pi e}\,\sigma$$

Gaussian Noisy Channels. The input-output performance of the system is specified by a two-dimensional normal law

$$f(x,y) = \frac{1}{2\sigma_x\sigma_y \sqrt{1-\rho^2}} \exp\left[\frac{1}{1-\rho^2}\left(\frac{x^2}{\sigma_x^2} - 2\rho \frac{xy}{\sigma_x\sigma_y} + \frac{y^2}{\sigma_y^2}\right)\right]$$

where ρ is the coefficient of correlation between X and Y, and is assumed to be different from unity: $\rho \neq \pm 1$.

The transinformation turns out to depend only on ρ, a fact which points out another of the many elegant properties of normal distributions.

$$I(X;Y) = -\tfrac{1}{2} \log_e (1 - \rho^2) \qquad |\rho| \neq 1 \qquad (22\text{-}46)$$

When noise is such that the received signal is independent of the transmitted signal, then $\rho = 0$ and $I(X;Y) = 0$.

Channels with Additive Noise. Let X be the input signal, Z the noise, and assume that X and Z are independent and that their sum produces a signal Y which is the output of the channel

$$Y = X + Z$$

INFORMATION THEORY

The transinformation turns out to be

$$I(X;Y) = H(Y) + \int_{-\infty}^{\infty} \phi(z) \log \phi(z) \, dz \tag{22-47}$$

where $\phi(z)$ is the probability distribution of Z.

In the particular case when X and Z are normal with

$$\begin{array}{ll} \bar{X} = 0 & \bar{Z} = 0 \\ \overline{X^2} = S & \overline{Z^2} = N \end{array} \tag{22-48}$$

one arrives at Shannon's well-known formula for the channel capacity,

$$C = \tfrac{1}{2} \log \left(1 + \frac{S}{N}\right) \tag{22-49}$$

This formula may be extended to the analogous case where input and noise each consists of an n-dimensional vector with mutually independent components and equal power in each component (Ref. 21, Chapter 9). For such channels

$$C = \frac{n}{2} \log \left(1 + \frac{S}{N}\right) \tag{22-50}$$

22-9. Entropies and Mutual Information of Systems with Memory

The success of the present development of information theory is due to the fact that the elementary concepts originally introduced are very few, yet their coverage is quite broad. The concepts of communication entropy, mutual information, and channel capacity used in the case of simple memoryless models are almost always valid, and equally useful for the more complex regimes. When the output of a source and received messages are random vector signals, the average mutual information still can be defined as

$$I(X;Y) \doteq E \left[\log \frac{f(X,Y)}{f_1(X)f_2(Y)} \right] \tag{22-51}$$

where X and Y are two random vectors with a known joint probability function $f(x,y)$. Of course, from the mathematical point of view, one is required to examine the finiteness of this quantity. It can be shown that the necessary and sufficient condition for the finiteness of (22-51) is that the joint probability function $f(x,y)$ should exist and be absolutely continuous with respect to $f_1(x)f_2(y)$.

The concept of mutual information can be further applied to encompass the case where X and Y are random processes of a fairly general type. Such a generalization is outlined in this and the two succeeding sections. These generalizations were implied in Shannon's original work, but a detailed examination of the suggested processes became possible only several years later, when the necessary links were provided by other researchers.

B. McMillan, A. N. Kolmogorov, and A. I. Khinchin have provided indispensable links for such generalizations. Subsequently, it appears that the Russian mathematicians I. M. Gel'fand and A. M. Yaglom made the first broad mathematical analysis of the situation and provided more general definitions for the familiar concepts. Section 22-11 gives a summary of their important results, without attempting any proof. We shall use the Gel'fand-Yaglom notations, as much as possible.

Mutual Information of Two Stationary Random Processes. Let X and Y be two stationary random processes depending on the same parameter t (real-time axis). In order to define entropies, one may select sequences of signals at the input and the output, respectively,

$$\ldots, x_0, x_1, x_2, \ldots, x_{n-1}, \ldots$$
$$\ldots, y_0, y_1, y_2, \ldots, y_{n-1}, \ldots$$

Now consider the input and the output of the system as two random vectors and compute their respective entropies.

$$X_n \quad x_0, x_1, \ldots, x_{n-1} \to H(X_n)$$
$$Y_n \quad y_0, y_1, \ldots, y_{n-1} \to H(Y_n) \quad (22\text{-}52)$$

The corresponding channel entropies and the mutual information may be also computed as

$$(x_0|Y_n), (x_1|Y_n), \ldots, (x_{n-1}|Y_n) \to H(X_n|Y_n)$$
$$(y_0|X_n), (y_1|X_n), \ldots, (y_{n-1}|X_n) \to H(Y_n|X_n) \quad (22\text{-}53)$$
$$I(X_n;Y_n) = H(X_n) - H(X_n|Y_n)$$

Subject to some mathematical care, this method of calculation can be presumably extended to the limit when n approaches infinity.[17]

The main assumption is that the source is ergodic; that is, its output sequences are metrically transitive. Essentially, McMillan showed that the following expression is a constant number:

$$\lim_{n \to \infty} -\left[\frac{1}{n} \log P\{X_0, X_1, \ldots, X_{n-1}\}\right] = \lim_{n \to \infty} \frac{1}{n} E[-\log P\{X_0, \ldots, X_{n-1}\}] \quad (22\text{-}54)$$

(the limit being taken in the sense of convergence in the mean).

It is illuminating to introduce the random variable

$$Z_n = -\frac{1}{n} \log P\{X_0, \ldots, X_n\}$$

and rewrite the above result in the form

$$\lim_{n \to \infty} E[|Z_n - E(Z_n)|] = 0 \quad (22\text{-}55)$$

The content of this equation is sometimes referred to as the *asymptotic equipartition property*.[17,12,5] Equation (22-55) shows that Z_n gives a reasonable approximation of its theoretical average value $E[Z]$. The mathematically inclined reader may find interesting discussions on McMillan's theorem and its relation to ergodic theorems in L. Breiman, L. Carleson, K. L. Chung[15], and A. J. Thomasian.[30]

The entropies of the stochastic source and the channel can be defined by finding the limits shown below:

$$H(X) = \lim \frac{1}{n} H(X_n) \quad n \to \infty$$

$$H(Y) = \lim \frac{1}{n} H(Y_n)$$

$$H(X|Y) = \lim \frac{1}{n} H(X_n|Y_n) \quad (22\text{-}56)$$

$$H(Y|X) = \lim \frac{1}{n} H(Y_n|X_n)$$

$$H(X,Y) = \lim \frac{1}{n} H(X_n,Y_n)$$

$$I(X;Y) = H(X) - H(X|Y) = H(Y) - H(Y|X)$$

22-10. Entropy of Regular Markov Sources

This section is devoted to an investigation of the entropy of a very common sort of source with memory, namely, regular sources of the Markov type. Some elementary background of the Markov chain is sketched below and supplemented in Chapter 28.

INFORMATION THEORY

Some Definitions. Finite Markov chain: a stochastic process which moves through a finite number of states $(S_1, S_2, \ldots, S_N)$. The probability of entering any state depends only on the *last state* occupied. The Markov chain is usually specified through a transition probability matrix $[P]$.

Regular chain: A finite Markov chain with a transition probability matrix $[P]$ such that for some positive integer n, $[P]^n$ has no zero entries.

Ergodic chain: A Markov chain for which it is possible to go from every state to every other state in a finite number of steps.

For example, $[P_1]$ is regular; $[P_2]$ is nonregular and nonergodic.

$$[P_2] = \begin{bmatrix} 1 & 0 \\ \tfrac{1}{2} & \tfrac{2}{3} \end{bmatrix} \quad [P_1] = \begin{bmatrix} \tfrac{1}{3} & \tfrac{2}{3} & 0 \\ 0 & 0 & 1 \\ 1 & 0 & 0 \end{bmatrix}$$

Some Theorems on Regular Markov Chains. Theorem 1. If $[P]$ is the transition probability matrix of a regular Markov chain, then $[P]^n$ approaches a probability matrix $[A]$. Each row of $[A]$ is identical with every other, and consists of the probability vector $[T] = (t_1, t_2, \ldots, t_n)$.

Theorem 2. For any initial probability vector $[\pi]$, $[\pi][P]^n$ approaches the vector $[T]$ as n tends to infinity.

Theorem 3. $[T]$ is the unique probability vector $[T][P] = [T]$. (This is due to the fact that if $[T']$ is a suitable probability vector, i.e., $[T'][P] = T$ then $[T'][P]^n$ must approach T. Thus $[T'] = [T]$).

Theorem 4. $[T][P] = [P][T] = [T]$.

Theorem 5. For a regular chain, there is a limiting probability t_j, of being in state s_j, *independent* of the starting state; t_j is also the fraction of times that the process can be expected in state s_j.

Entropy of a Regular Markov Chain. The entropy for going one step ahead from state i is

$$H_i^{(1)} = -\sum_{j=1}^{n} P_{ij} \log P_{ij} \tag{22-57}$$

The average uncertainty for moving one step ahead, given that the initial probability of the ith state is P_i, is

$$H(X) = H^{(1)} = \overline{H_i^{(1)}} = \sum_{i=1}^{n} P_i H_i^{(r)} \tag{22-58}$$

Based on the above theorems, one can show that, for a regular Markov source with initial probability matrix t, the entropy of the system for going $\alpha + \beta$ steps ahead (α, β positive integers) equals the sum of the entropies for going α steps and β steps ahead, respectively.

Theorem. If the initial probability row matrix is the fixed probability vector t of a regular Markov chain, then

$$\begin{aligned} H^{(r+1)} &= H^{(1)} + H^{(r)} = (r+1)H^{(1)} \\ H^{(\alpha+\beta)} &= H^{(\alpha)} + H^{(\beta)} = (\alpha + \beta)H^{(1)} \end{aligned} \tag{22-59}$$

The entropy of a statistical source of this sort can be defined as

$$\lim_{n \to \infty} \frac{H^{(n)}}{n} = H^{(1)} = H(X)$$

As an extension of this idea, the entropy of a regular Markov chain with any arbitrary initial probability matrix can be defined as

$$H(X) = \lim_{n \to \infty} \frac{H^{(n)}}{n}$$

It is not difficult to see that, for regular Markov chains, the imposing effect of the initial-state probability selection soon disappears. Such systems will reach some kind of steady state no matter what the initial probabilities were.

The relative simplicity of computation of the entropy of regular Markov chains has prompted wide applications in the field of linguistics.

22-11. Mutual Information of Gaussian Random Vectors and Processes

The Entropy of a Gaussian n-dimensional Vector. Let the probability distribution function be

$$f(x_1, \ldots, x_n) = \frac{(\det A)^{1/2}}{(2\pi)^{n/2}} \exp -\frac{1}{2}\left[\sum_{i=1}^{n}\sum_{j=1}^{n} a_{ij}(x_i - m_i)(x_j - m_j)\right]$$

where A is the positive definite matrix associated with the quadratic form

$$\sum_{i=1}^{n}\sum_{j=1}^{n} a_{ij} z_i z_j$$

The covariance $E[(x_i - m_i)(x_j - m_j)]$ turns out to be equal to the ijth term of the matrix A^{-1}. The entropy of such a source is found to be

$$H(x) = \log (2\pi e)^{n/2} (\det A)^{-1/2} \tag{22-60}$$

For the simple source with probability density function

$$f(x_1, x_2) = \frac{\sqrt{\alpha\gamma - \beta^2}}{2\pi} \exp[-\tfrac{1}{2}(\alpha x_1^2 + 2\beta x_1 x_2 + \gamma x_2^2)]$$

we have
$$H(x) = \log \frac{2\pi e}{\sqrt{\alpha\gamma - \beta^2}} \tag{22-61}$$

We note in passing that for an n-dimensional source, defined in $(-\infty, +\infty)$, with specified second-order covariance matrix (power), the Gaussian distribution presents the source with the largest entropy.

Mutual Information of Two Gaussian Vectors. Consider a statistical system (X,Y) where the input X and the output Y are jointly normally distributed random vectors

$$X = [X_1, X_2, \ldots, X_n] \qquad Y = [Y_1, Y_2, \ldots, Y_m]$$
$$\overline{X_k} = 0 \qquad k = 1, 2, \ldots, n$$
$$\overline{Y_k} = 0 \qquad k = 1, 2, \ldots, m$$

Let A, B, and D be, respectively, the moment matrices of X, Y, and (X,Y); that is,

$$a_{ij} = [\overline{X_i X_j}] \qquad b_{ij} = [\overline{Y_i Y_j}] \qquad d_{ij} = [\overline{X_i Y_j}]$$

Then the mutual information of the system is found to be

$$I(X;Y) = \frac{1}{2}\frac{\det A \det B}{\det C} \tag{22-62}$$

where
$$C = \begin{bmatrix} A & D \\ D' & B \end{bmatrix}$$

D' is the transpose of D, and we assume that the probability distributions are nonsingular, hence A, B, and C are nonsingular.

This formula is quite interesting and can be also written in an equivalent form as a generalization of Eq. (22-46)

$$I(X;Y) = -\tfrac{1}{2} \log (1 - \rho_1^2)(1 - \rho_2^2)(1 - \rho_3^2) \cdots (1 - \rho_l)^2 \tag{22-63}$$

where ρ_j is the correlation between X_j' and Y_j' and $l = \min(n,m)$; X_j' and Y_j' are obtained by applying some suitable transformation to X and Y; see Ref. 6.

There is an interesting geometrical interpretation of (22-63) by Gel'fand and Yaglom[6] which is based on earlier classical works of Kolmogorov. In case of the bivariate normal random variable (X,Y), let $\rho(X,Y) = \cos \alpha$, and think of α as the angle between vectors X and Y. Then

$$I(X;Y) = -\tfrac{1}{2} \log (1 - \cos^2 \alpha) = -\log |\sin \alpha| \qquad (22\text{-}64)$$

In the general case, when X and Y are vectors in n- and m-dimensional spaces, in analogy with (22-64), one can write

$$I(X;Y) = -\log |\sin \alpha_1 \sin \alpha_2 \cdots \sin \alpha_l| \qquad (22\text{-}65)$$

These angles are referred to as stationary angles between linear spaces associated with vectors X and Y. The second-order moments, say $\overline{X_i X_j}$, are taken as the norm in these subspaces.

The third equivalent formula for (22-63) is also given by Gel'fand and Yaglom. Let E denote the identity matrix and

$$C_1 = \begin{bmatrix} A & 0 \\ 0 & B \end{bmatrix} \qquad C_1^{-1} = \begin{bmatrix} A^{-1} & 0 \\ 0 & B^{-1} \end{bmatrix}$$

Then
$$\frac{\det A \det B}{\det C} = \det [C \cdot C_1^{-1}]^{-1} = \begin{vmatrix} E & DB^{-1} \\ D'A^{-1} & E \end{vmatrix}^{-1} \qquad (22\text{-}66)$$

and

$$I(X;Y) = -\tfrac{1}{2} \log \det (E - DB^{-1}D'A^{-1}) = -\tfrac{1}{2} \log (E - D'A^{-1}DB^{-1}) \qquad (22\text{-}67)$$

Mutual Information of Two Gaussian Processes. A random process $X(t)$ is called a normal process if for all sets of $(t_1, t_2, \ldots, t_n) \epsilon T$ the joint distribution of $(X(t_1), X(t_2), \ldots, X(t_n))$ is multinormal. Two normal random processes $X(t)$ and $Y(s)$ are jointly normal, if all joint distributions for $X(t)$ and $Y(s)$ are also normal. In the following, we assume

$$\overline{X(t)} = \overline{Y(t)} = 0$$

$$\rho_{xx}(t,t_1) \qquad \rho_{yy}(s,s_1) \qquad \text{and} \qquad \rho_{xy}(t,s) \text{ known}$$

One can proceed with computing the mutual information $I(X_n; Y_n)$ between two random vectors X_n and Y_n according to methods described before.

$$\begin{array}{l} X_n: X(t_1), X(t_2), \ldots, X(t_n) \\ Y_m: Y(s_1), Y(s_2), \ldots, Y(s_m) \end{array} \qquad (22\text{-}68)$$

$$I(X_n; Y_m) = -\tfrac{1}{2} \log \det (E - DB^{-1}D'A^{-1}) = -\tfrac{1}{2} \log (E - \Delta_{n,m})$$

where $\Delta_{n,m} = DB^{-1}D'A^{-1}$

Note that the multinormal joint distributions can now be calculated.

$$\begin{bmatrix} \rho_{xx}(t_i,t_j) & \rho_{xy}(t_j,s_n) \\ \rho_{yx}(s_m,t_j) & \rho_{yy}(s_m,s_n) \end{bmatrix} \qquad (22\text{-}69)$$

$$\rho_{xy}(s_m,t_j) = \rho_{yx}(t_j,s_m)$$

Next step is to find the limit of the mutual information between jointly normal vectors X_n and Y_m when $n \to \infty$ and $m \to \infty$. Of course, the reader appreciates that such a step would require a considerable amount of mathematical care concerning the existence of such limits. But we shall not quibble about this. To simplify the matter, in the following, we may let $m = n = l$. The basic idea behind this stems from considerations similar to those for obtaining Eq. (22-63).

$$I(X(t); Y(s)) = \lim_{l \to \infty} I(X_l; Y_l) \qquad (22\text{-}70)$$

It is natural to bring in λ_k's which are the proper values of $(E - \Delta)$ when $n \to \infty$

$$\det (E - \Delta_l) = \prod_{k=1}^{\infty} (1 - \lambda_k) \tag{22-71}$$

This infinite product will converge under very general conditions outlined in Ref. 6. Thus one finally obtains

$$I(X(t); Y(s)) = -\tfrac{1}{2} \log \prod_{k=1}^{\infty} (1 - \lambda_k) \tag{22-72}$$

This result is important because it not only shows that the mutual information of two jointly normal processes defined over continuous spaces is finite but also shows how the transinformation can be computed. More details and examples are given in Ref. 6.

22-12. Generalities of Encoding

In a quite general sense, the word encoding implies some transformation of the input signal prior to its entry into what is usually referred to as the communication channel. Sometimes the encoding is a matter of necessity, such as the encoding of an English text in binary numbers when the information-processing equipments are of binary nature. In general, the encoding may be a matter of choice—the purpose of encoding is to improve the "efficiency" of the transmission of information in some sense. For instance, as we have already seen, for a given channel there is generally a matching source which produces the highest rate of transinformation. Therefore, one may desire to encode the output of a given source in order to achieve a more desirable signal distribution of the channel input. In brief, we may say that the encoder transforms the output of a source into new signals which are most acceptable to the channel. The decoder transforms the output signals of the channel into a form readily acceptable to the receiver.

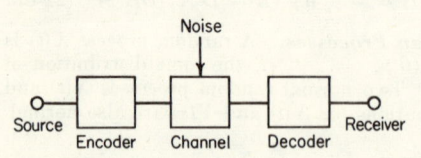

Fig. 22-10. A communication system with encoder-decoder.

In the following, we confine ourselves to a simplified encoding-decoding model as indicated in Fig. 22-10. We consider only discrete sources having a finite alphabet and no memory. The following terms are used:

Letter, symbol, or character: any individual member of a finite set called the alphabet.

Message or word: a finite sequence of letters of the alphabet.

Length of a word: the numbers of letters in a word.

Language: the set of all possible words from a particular alphabet.

Encoding or enciphering: a procedure for associating words from a language with given words of another language in a one-to-one manner.

Decoding or deciphering: the inverse operation of assigning words of the second language to given words of the first.

Uniquely decipherable encoding or decoding: the operation in which the correspondence of all possible sequences of words between the two languages without space marks between the words is one-to-one.

Let an alphabet set be denoted by

$$[A] = [a_1, a_2, \ldots, a_D]$$

Sequences such as $a_1 a_1 a_2$, $a_D a_1 a_2$, and $a_2 a_2 a_2 a_2$ will be referred to as words of this language. The lengths of these words are three, three, and four symbols, respectively.

Similarly the set of letters [0,1] constitutes what is commonly known as the binary alphabet; 001 is a word in the binary language.

By speaking of more efficient encoding, we agree to refer to encoding procedures that improve certain "cost functions." Perhaps the simplest cost function is obtained when we assign a constant cost figure t_i to each message m_i; t_i can be the duration or any other cost factor. Then the average cost per message becomes

$$R_t = \sum_{i=1}^{N} P\{m_i\} t_i$$

Obviously, the most efficient transmission is the one that minimizes the average cost R_t. In this chapter we confine ourselves to the simplest case when all symbols have identical cost. The average cost per message becomes proportional to the average of n_i, the number of symbols per message, or the average length of messages $\bar{L}$:

$$R_t = \bar{L} = \sum_{i=1}^{N} P\{m_i\} n_i \qquad t_i = n_i \qquad i = 1, 2, \ldots, N$$

An increase in transmission efficiency may be achieved by properly encoding messages, that is, by assigning new sequences of symbols to each message m_i such that the statistical distribution of the new symbols reduces the average word length $\bar{L}$.

The efficiency of the encoding procedure can be defined if, and only if, we know the lowest possible bound of $\bar{L}$. Of course, if such a lower bound does not exist, the term efficiency will be meaningless. Thus an important question arises here. For a given set of messages and a given alphabet, what is the lowest possible $\bar{L}$ that can be obtained?

The principal problem is to arrive at efficient overall encoding-decoding procedures. This turns out to be a difficult problem even when there is no noise in the channel. An important result for the noiseless case is the fact that the efficiency of the encoding-decoding procedure is

$$\eta = \frac{H(X)}{\bar{L} \log D} \qquad (22\text{-}73)$$

where $H(X)$ is the entropy of the original message ensemble, D the number of symbols in the alphabet, and $\bar{L}$ the average length of the encoded messages. This definition for efficiency implies that

$$\bar{L} \geq \frac{H(X)}{\log D} \qquad (22\text{-}74)$$

This is rather a curious property. It can be shown that if this inequality is violated then the encoding-decoding will be confusing. That is, the virtue of a one-to-one correspondence between the input and the output will be lost. The next question is to see how we can achieve, or approach, this ideal encoding-decoding scheme. An answer to this question is given in Sec. 22-14.

As an example, suppose that the source has to transmit four messages (m_1, m_2, m_3, m_4), with respective probabilities $(½, ¼, ⅛, ⅛)$, over a noiseless binary channel. The following encoding-decoding scheme is suggested almost by common sense:

$$\begin{aligned} m_1 &\to 0 \\ m_2 &\to 1 \quad 0 \\ m_3 &\to 1 \quad 1 \quad 0 \\ m_4 &\to 1 \quad 1 \quad 1 \end{aligned}$$

For such case we find

$$\bar{L} = 1¾ \qquad H(X) = 1¾$$
$$\eta = 100 \text{ per cent, redundancy} = 1 - \eta = 0$$

Now if a sequence of original messages are replaced by their encoded correspondents, the latter sequence can be uniquely decoded in the binary language. Any attempt to further shorten the average length will violate the aforementioned property, which is usually referred to as unique decipherability or separability.

There is a subclass of uniquely decipherable codes which is referred to as irreducible or prefix codes. The defining property of this class of codes can be exhibited by a simple example. In binary codes, if the sequence 10110 is assigned to a message m_k, then none of its prefixes, that is, 1011, 101, 10, and 1 should be assigned to the encoding of the original messages. If the property holds for all encoded messages, then the code possesses the prefix property.

When noise is present in the channel, the principal aim of encoding is to combat noise. Here the guiding thought is to transmit relatively dissimilar messages such that, when they are subjected to the channel noise, they still can remain reasonably dissimilar. Of course, if we wish to transmit two messages m_1 and m_2 over a binary noisy channel, we may assign a long sequence of zeros to m_1 and an equally long sequence of ones to m_2. Thus, even if a good percentage of zeros (ones) are erroneously received as ones (zeros), there is still room for a reasonably good detection as long as the noise interference is less than 50 per cent. Of course, the overall efficiency of the system may not be high, because the longer sequences will result in higher costs. Nonetheless, this is the type of problem which is being investigated in devising codes for noisy channels.

A simple class of codes which correct one or more single independent errors has been extensively studied in the literature. An indication of codes of this type will be given in Sec. 22-14. For a comprehensive study of coding theory see, for instance, Ref. 19.

22-13. Existence Criterion for Noiseless Encoding

In this section we discuss the conditions required for the existence of uniquely decipherable codes.

A Noiseless Coding Theorem for Irreducible Codes. The necessary and sufficient condition for existence of an irreducible noiseless encoding procedure with specified word length $(n_1, n_2, \ldots, n_N)$ is that the set of positive integers $(n_1, n_2, \ldots, n_N)$ satisfies

$$\sum_{i=1}^{N} D^{-n_i} \leq 1 \qquad (22\text{-}75)$$

McMillan's Theorem* for Uniquely Decipherable Codes. Let $(m_1, m_2, \ldots, m_N)$ be a sequence of messages encoded in uniquely decipherable words of respective symbol length $(n_1, n_2, \ldots, n_N)$ taken from a finite alphabet $(a_1, a_2, \ldots, a_D)$. Then

$$\sum_{i=1}^{N} D^{-n_i} \leq 1 \qquad (22\text{-}76)$$

A Theorem Describing the Efficiency of Uniquely Decipherable Codes. Let $[X]$ be a discrete message source, without memory, and x_i any message of this source with probability of transmission $P\{x_i\}$. If the $[X]$ ensemble is encoded in a sequence of uniquely decipherable characters taken from the alphabet $(a_1, a_2, \ldots, a_D)$ then

$$\bar{L} = \sum_{i=1}^{N} P\{x_i\} n_i \geq \frac{H(X)}{\log D} \qquad (22\text{-}77)$$

Shannon-Fano Binary Code. Divide the message ensemble (s) into two almost equiprobable subsets (s_0, s_1). Assign a zero to all members of s_0 and a one to all members of s_1. Next divide s_0 into two subsets (s_{00}, s_{01}) assigning 00 to all members of s_{00} and 01 to all members of s_{01}. The method is continued until all subsets contain only one message. The efficiency of the method depends on the capability of appro-

---

* Also see Ref. 18.

priately dividing each set

$$\begin{array}{llllllll}
½ & m_1 & 0 \to 0 \\
¼ & m_2 & 1 & 0 \to 1 & 0 \\
⅛ & m_3 & 1 & 1 \to 1 & 1 & 0 \\
⅛ & m_4 & 1 & 1 & 1 \to 1 & 1 & 1
\end{array}$$

22-14. Some Methods of Encoding for Noiseless Channels

Shannon's Irreducible Binary Encoding. Following this method one assigns a length n_k to the message x_k with the probability $P\{x_k\}$ such that

$$-\frac{\log P\{x_k\}}{\log D} \leq n_k = \text{integer} < -\frac{\log P\{x_k\}}{\log D} + 1 \qquad k = 1, 2, \ldots, N \qquad D \geq 2 \tag{22-78}$$

Clearly, if such a code is devised, then its average length will satisfy the following inequality

$$\frac{H(X)}{\log D} \leq \bar{L} < \frac{H(X)}{\log D} + 1 \tag{22-79}$$

The efficiency of the procedure will also be bounded by a similar inequality. Shannon's procedure for devising such a binary code is described in Ref. 21.

Huffman's Minimum-redundancy Code. D. Huffman has suggested a method for constructing prefix codes with minimum redundancy. Huffman's code is an optimum code; that is, no other code with lower average length can be found.

For binary channels, Huffman's method may be briefly described as follows:

1. List messages in order of nonincreasing probabilities:

$$[x_1, x_2, \ldots, x_{N-1}, x_N]$$

2. Assign a 0 to x_N and a one to x_{N-1}.
3. Combine x_N and x_{N-1} to produce a message x'_{N-1}. Now apply steps 1 and 2 to the message ensemble:

$$(x_1, x_2, \ldots, x'_{N-1})$$

[That is, rearrange $(x_1, x_2, \ldots, x'_{N-1})$ in order of nondecreasing probabilities and proceed. This is called an auxiliary message ensemble.]

4. Continue until all messages are combined. Now retrack each message x_k in all auxiliary ensembles in which it appeared and write down all the encountered zeros and ones in their proper orders. For an example see Ref. 21, Chapter 4.

22-15. Hamming Binary Codes

All transmitted words are of length n. Out of n binary digits, m places are assigned to information digits and $k = n - m$ to check digits. That is, the values of the last k digits depend on the values of the first $n - k$ (information) digits. The distance of two words is defined as the number of positions in which the two words differ. More specifically, the distance of $U = (\alpha_1, \alpha_2, \ldots, \alpha_n)$ and $V = (\beta_1, \beta_2, \ldots, \beta_n)$ is defined as

$$D(U, V) = \sum_{k=1}^{n} (\alpha_k \oplus \beta_k)$$

The notation $\oplus$ implies

$$1 \oplus 0 = 0 \oplus 1 = 1$$
$$0 \oplus 0 = 1 \oplus 1 = 0$$

For example, the distance between 001011 and 101001 is 2 units.

Assuming that the channel is not too noisy and at the most one single error may occur in a block of n digits, then in order to be able to correct this possible error, the number of parity checks k must satisfy

$$2^k \geq n + 1$$

For example, if we wish to transmit 8 words, then we may select $m = 3$ and search for the smallest k satisfying

$$2^k \geq k + 3 + 1$$

The desired number of minimum digits is $k = 3$.

The following method for locating a single error, assuming that no more than one single error has occurred, is due to R. W. Hamming.

Step 1. Number the messages to be transmitted as $0, 1, 2, \ldots, N$.

Step 2. Write the message $M(M = 0, 1, \ldots, N)$ in its binary form. Add enough zeros to the left of M so that all transmitted messages have an equal number of information digits n.

Step 3. Select the smallest k satisfying $2^k \geq k + n + 1$.

Step 4. Number the positions in each block from left to right ($x_1, x_2, \ldots$, etc.).

Step 5. Assign positions ($x_1, x_2, x_4, x_8, \ldots$) to parity checks, and the rest to information digits.

Step 6. Include only those position numbers containing a 1 in their ith digits in the ith equation. For example, write

$$x_1 + x_3 + x_5 + x_7 + x_9 + x_{11} + x_{13} + \cdots = 0, s_1$$
$$x_2 + x_3 + x_6 + x_7 + x_{10} + x_{11} + x_{14} + \cdots = 0, s_2$$
$$x_4 + x_5 + x_6 + x_{12} + x_{13} + x_{14} + \cdots = 0, s_3$$

These equations allow the determination of parity checks for transmission.

When a word is received, apply the set of equations in step 6. If equation s_i is valid, then write $s_i = 0$; otherwise $s_i = 1$. The location of the erroneous digit is given by the binary number $l = \cdots s_2 s_1$. An example is given in Ref. 21, Chapter 4.

***t*-error Correcting Codes.** Hamming's single-error correcting codes have been generalized to give code books which can correct t or fewer independent single errors. The mutual distances between all transmitted words must be at least $2t + 1$. When a word is received, it will be detected as the closest word in the transmitter's dictionary. It is for this reason that they are also referred to as Hamming's distance codes. There are at least three types of important questions in the area: (1) What is the dictionary containing the largest number of words for a given distance? (2) How can a method be devised for finding a suitable set of such words? (3) How can these codes be implemented? For a discussion on these and related problems, see, for instance, Ref. 19.

22-16. Principles of Group Codes

The first principal problem in devising distance codes is to find a method for selecting words with a certain minimum mutual distance. The idea of employing some of the suitable properties of finite groups for this selection is due to D. Slepian.[28]

Imagine the simple case of transmission of four messages over a binary symmetric channel. We may assign the following codes to the information digits:

$$m_1 \to 00 \quad m_3 \to 10$$
$$m_2 \to 01 \quad m_4 \to 11$$

In order to devise a single-error correcting group code, we find it necessary to add three parity-check digits. The selection of parity-check equations is based on certain rules which are described in the referred sources. Having selected a suitable parity-check matrix, one writes

$$\begin{bmatrix} 1 & 0 \\ 0 & 1 \\ 1 & 1 \end{bmatrix} \begin{bmatrix} 0 & 0 & 1 & 1 \\ 0 & 1 & 0 & 1 \end{bmatrix} = \begin{bmatrix} 0 & 0 & 1 & 1 \\ 0 & 1 & 0 & 1 \\ 0 & 1 & 1 & 0 \end{bmatrix}$$

The transmitter now has the following four encoded words for transmission:

$$u_1 \to 0\ 0\ 0\ 0\ 0 \qquad u_3 \to 1\ 0\ 1\ 0\ 1$$
$$u_2 \to 0\ 1\ 0\ 1\ 1 \qquad u_4 \to 1\ 1\ 1\ 1\ 0$$

The receiver may receive any one of 2^5 possible binary words of length 5. The code book is arranged in the manner described below. We expand the group of 2^5 binary words according to the elements (u_1, u_2, u_3, u_4), which form a subgroup.

This is the first row of our table. For the leading term of the second row we select a word with the lowest possible weight, say 00001. Elements of the second row are obtained as

$$u_k \oplus 0001 \qquad k = 1, 2, 3, 4$$

Similarly for the third row we select an element with the lowest possible weight which did not appear in the first two rows, say 00010. The elements of the third row are

$$u_k + 0001 + 00010 \qquad k = 1, 2, 3, 4$$

In this manner one obtains the table:

u_1	u_2	u_3	u_4
0 0 0 0 0	0 1 0 1 1	1 0 1 0 1	1 1 1 1 0
0 0 0 0 1	0 1 0 1 0	1 0 1 0 0	1 1 1 1 1
0 0 0 1 0	0 1 0 0 1	1 0 1 1 1	1 1 1 0 0
0 0 1 0 0	0 1 1 1 1	1 0 0 0 1	1 1 0 1 0
0 1 0 0 0	0 0 0 1 1	1 1 1 0 1	1 0 1 1 0
1 0 0 0 0	1 1 0 1 1	0 0 1 0 1	0 1 1 1 0
1 0 0 1 0	1 1 0 0 1	0 0 1 1 1	0 1 1 0 0
1 1 0 0 0	1 0 0 1 1	0 1 1 0 1	0 0 1 1 0

This expansion has the property that for any received word v in column k there does not exist any transmitted u_i closer to v than u_k. Therefore, when a word in the kth column is received, it may be detected as u_k. The detection follows the minimum-distance rule.

In the present scheme, if only a single error has occurred, then the error will certainly be detected. The scheme also will detect 2 out of 10 possible double-error patterns.

R. C. Bose and D. K. R. Chaudhuri have extended the described procedure to t-error correcting group codes.[19]

The field of coding theory, although relatively new, has been enriched by extensive engineering contributions of recent years. The golden promise of an error-free transmission of information has spurred the ingenuity of communication scientists. Yet what has been achieved is only a fragment of what the fundamental theorems of information theory foretold.

REFERENCES*

1. Blanc-Lapierre, André, and Robert Fortêt: *Théorie des fonctions aléatoires*, Masson et Cie, Paris, 1953.
2. Brillouin, L.: *Science and Information Theory*, Academic Press Inc., New York, 1956.
3. Elias, P. J.: Information Theory, Chap. 16 in *Handbook of Automation, Computation and Control*, vol. 1, pp. 16-01 to 16-48, John Wiley & Sons, Inc., New York, 1958.
4. Fano, R. M.: *Statistical Theory of Communication*, John Wiley & Sons, Inc., New York, 1961.
5. Feinstein, A.: *Foundations of Information Theory*, McGraw-Hill Book Company, New York, 1958.
6. Gel'fand, I. M., and A. M. Yaglom: Calculation of the Amount of Information about a Random Function Contained in Another Such Function, *Usp. Mat. Nauk. S.S.S.R.*, new series, vol. 12, 1957, English translation in *Trans. Am. Math. Soc.*, ser. 2, vol. 12, pp. 199–246, 1959.
7. Goldman, S.: *Information Theory*, Prentice-Hall, Inc., Englewood Cliffs, N.J., 1953.

* For more extensive bibliographies, see Refs. 2, 14, 21, and 31.

8. Hamming, R. W.: Error Detecting and Error Correcting Codes, *Bell System Tech. J.* vol. 29, pp. 147–150, 1950; *Bell Lab. Record*, vol. 28, pp. 193–198, 1950.
9. Harman, W. A., *Principles of the Statistical Theory of Communication*, McGraw-Hill Book Company, New York, 1963.
10. Huffman, D. A.: A Method for the Construction of Minimum Redundancy Codes, *Proc. IRE*, September, 1952, pp. 1098–1101.
11. Yaglom, A. M., and I. A. Yaglom: *Probabilité et information*, Dunod, Paris, 1959.
12. Khinchin, A. I.: *Mathematical Foundations of Information Theory*, Dover Publications, Inc., New York, 1957.
13. Kolmogorov, A. N.: On the Shannon Theory of Information Transmission in the Case of Continuous Signals, *IRE Trans. Inform. Theory*, vol. IT-2, pp. 102–108, December, 1956.
14. Kullback, S.: *Information Theory and Statistics*, John Wiley & Sons, Inc., New York, 1959.
15. Machol, R. E., and P. Gray (eds.): *Recent Developments in Information and Decision Processes*, The Macmillan Company, New York, 1962.
16. McMillan, B.: Two Inequalities Implied by Unique Decipherability, *IRE Trans. Inform. Theory*, vol. IT-2, pp. 115–116, December, 1956.
17. McMillan, B.: The Basic Theorems of Information Theory, *Ann. Math. Statist.*, vol. 24, p. 196, 1952.
18. Mandelbrot, B.: On Recurrent Noise Limiting Coding, P. I. B. Symposium of Information Network, 1954, pp. 205–221.
19. Peterson, W. W.: *Error Correcting Codes*, John Wiley & Sons, Inc., New York, 1961.
20. Rényi, A.: Wahrscheinlichkeitzrechnung mit einem Anhang über Informationstheorie, VEB Deutscher Verlag der Wissenschaften, Berlin, 1962.
21. Reza, F. M.: *An Introduction to Information Theory*, McGraw-Hill Book Company, New York, 1961.
22. Reza, F. M.: A Note on the Capacity of Discrete Noiseless Channels, *Z. angew. Math. Phys.*, vol. 14, Fasc. 1 (1963), ZAMP, pp. 175–178.
23. Schutzenberger, M. P.: Contributions aux applications statistiques de la théorie de l'information, Inst. Stat. Univ. Paris. (A), 1954, pp. 1–115.
24. Shannon, C. E.: A Mathematical Theory of Communication, *Bell System Tech. J.*, vol. 27, pp. 379–423, 623–656, 1948.
25. Shannon, C. E.: Communication in the Presence of Noise, *Proc. IRE*, vol. 37, pp. 10–21, 1949.
26. Shannon, C. E.: Probability of Error for Optimal Codes in a Gaussian Channel, *Bell System Tech. J.*, vol. 38, no. 3, pp. 611–656, 1959.
27. Shannon, C. E.: Certain Results in Coding Theory for Noisy Channels, *Inform. Control*, vol. 1, no. 1, pp. 6–25., September, 1957.
28. Slepian, D.: A Class of Binary Signaling Alphabets, *Bell System Tech. J.*, vol. 35, pp. 203–234, January, 1956.
29. Slepian, D.: A Note on Two Binary Signaling Alphabets, *IRE Trans. Inform. Theory*, vol. IT-2, pp. 84–86, June, 1956.
30. Thomasian, A. J.: Elementary Proof of the AEP of Information Theory, *Ann. Math. Statist.*, vol. 31, no. 2, pp. 452–456, 1960.
31. Wolfowitz, J.: *Coding Theorems of Information Theory*, Springer-Verlag, OHG, Berlin, 1961.

Chapter 23

GAME THEORY

ROBERT M. THRALL, *Professor of Mathematics and Operations Analysis, University of Michigan.*

CONTENTS

23-1.	Introduction	23-1
23-2.	Definition of a Game	23-1
23-3.	Two-person Zero-sum Games	23-4
23-4.	Minimax Theorem for Matrix Games	23-9
23-5.	Iterative Approximation Solution for Games	23-12
23-6.	n-person Games	23-13
23-7.	Some Examples of Two-person Non-zero-sum Games	23-16
23-8.	Games against Nature	23-17

23-1. Introduction

The theory of games arises from an attempt to provide a mathematical model of conflict situations. Although forerunners of the theory date back many years, the first attempt to build a mathematical theory was made in 1921 by Emile Borel. The theory was first firmly established by John von Neumann in 1928 with his proof of the minimax theorem, but it was not until the appearance of Ref. 10 in 1944 that the topic received general notice. Since then many facets of game theory have received attention from mathematicians who have been challenged by its many fascinating problems and from others who have seen in this theory an important model for understanding and resolution of economic, military, and political conflicts.

Early attention was focused on ramifications of the two-person zero-sum game, finite or infinite; more recently primary emphasis has shifted to general games where there are still many important unsolved problems and opportunities for useful refinements of the basic definitions.

The reader is referred to the references for a selected list of books on game theory and to these books, in turn, for more complete bibliographies.

23-2. Definition of a Game

A game in *extensive* form is described by a set of players and a set of rules which specify (1) what choices of action are open to each player under all possible circumstances and (2) the payoff to each player at the end of any "play." Games are classified in terms of (1) the number of players, (2) the number of moves, (3) the nature of the payoff, and (4) other characteristics of the rules. A *finite* game is one in which the total number of alternatives (over all moves) is finite. Until further notice, we shall use "game" to mean "finite game."

Before giving a formal definition of a game we develop some auxiliary concepts. Let $A = \{a_1, \ldots, a_p\}$ be a finite set of elements called *nodes* and let R be a symmetric relation in the cartesian product $A \times A$. We call the elements of R *arcs*. The two sets A and R define a graph H. Intuitively, a graph is a set of nodes with certain pairs joined by arcs. A *path* in a graph is a set of nodes $b_1, \ldots, b_s$ such that (b_i, b_{i+1}) is an arc in H for $i = 1, \ldots, s - 1$. A graph is said to be *connected* if for each pair of nodes a, b in A there is a path beginning with a and ending with b. A path is said to be a *loop* if (1) $s > 3$, (2) $b_1 = b_s$, and (3) $b_i = b_j$ holds only for the initial and final nodes. A *tree* is a connected graph having no loops.

A tree is said to be *rooted* if it has one distinguished node. The analogy with botanical trees is obvious if we identify the distinguished node with the foot of the trunk and disregard the root structure. Several trees are illustrated in Fig. 23-1.

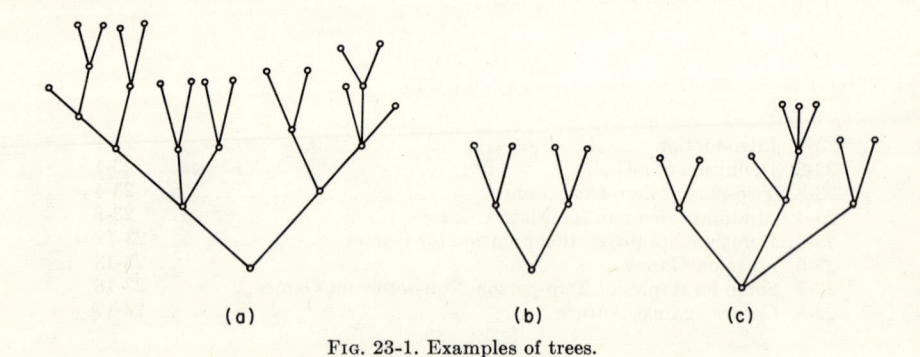

FIG. 23-1. Examples of trees.

Tree a has 27 nodes and 26 arcs; tree b has 7 nodes and 6 arcs; tree c has 13 nodes and 12 arcs. In a tree the number of arcs is always one less than the number of nodes.

A node in a tree is said to be *terminal* if it lies on only one arc. Tree a has 15 terminal nodes.

A game is described in terms of a tree in which each nonterminal node represents a move by some player and a move consists of selecting one of the upward-going arcs from the node. Play begins at the distinguished node and continues until a terminal node is reached. If player i is to move at a certain stage the corresponding node is labeled with the symbol i. A path starting at the distinguished node and ending at a terminal node is called a *play* of the game. Note that there is a one-to-one correspondence between plays and terminal nodes. The tree for chess is extremely complex; tree b belongs to the game of matching pennies; the labels and meanings of the arcs are given in Fig. 23-2.

Suppose that player 1 wishes to match player 2. He then wins in the plays whose arcs are labeled H, H or T, T whereas player 2 wins in the remaining cases H, T and T, H. The components of the vectors at the top of each play indicate the payoffs to the player (ordered by their labels).

Note that if player 2 knew just at which node he was making his choice he could always win. The concept of information sets provides for this. A player is told at any time he has the move merely that the game has reached some node in a given set of nodes called an *information set*. Of course, all the nodes in an information set must have the same label and the same number of upward arcs, and corresponding upward arcs must be identified. In the example, labels H and T provide this identification.

It is customary to require that no information set contain more than one node of any play.

If the game involves chance moves (e.g., dealing the hands in bridge), then the corresponding node is labeled with a 0 and a probability distribution over the upward arcs must be given. Each information set belonging to player 0 consists of exactly one node.

Each play has associated with it an *outcome* which is a vector indicating the return (or payoff) to each player accruing from that play. It is customary to label each terminal node with the associated outcome.

If for every terminal node the sum of the payoffs is zero, the game is said to be *zero-sum*. For example, a parlor game in which all the payoffs are between the players is zero-sum; whereas a labor-management game with the possibility of strikes and lockouts is not zero-sum. Thus a zero-sum game bears a certain resemblance to a conservative system in mechanics.

A *pure strategy* for a player is a selection of one alternative (set of corresponding upward arcs) from each of his information sets. A *mixed strategy* is a probability distribution over the set of all pure strategies.

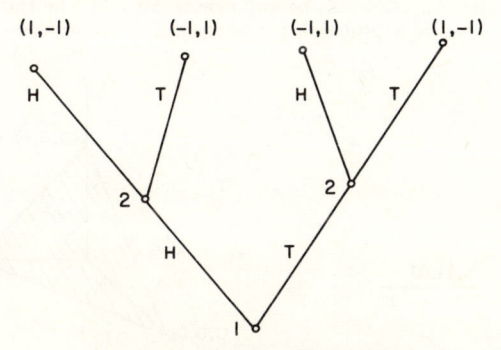

Fig. 23-2. Matching-pennies tree.

In a (finite) game the number of pure strategies for each player is finite, although even for familiar parlor games such as chess, poker, and bridge this number is so large as to be beyond the handling capacity of even our largest computing machines. For example, a pure strategy in chess would consist of selecting a move for each (legal) position of pieces on the chessboard (for complete accuracy one should include as part of a position the sequence of past moves which led to it so as to take account of special rules concerning draws through repetition of positions). Details of each pure strategy could theoretically be recorded in a (large) book and a play of chess would consist of each player's selecting a pure strategy and letting a referee determine the outcome by reference to the book.

In this sense every finite game can be reduced to a single decision for each player; if player i, for $i > 0$, has N_i pure strategies, his decision consists of selecting an integer x_i from the set $I(N_i) = \{1, \ldots, N_i\}$. A referee then carries out the chance moves of player 0 and announces the outcome vector.

A game in *normalized* or normal form is thus characterized by

1. A set of integers $N_0, N_1, \ldots, N_n$
2. A function F on the product set $I(N_0) \times I(N_1) \times \cdots \times I(N_n)$ to euclidean n space
3. A probability distribution p_0 over the set $\{1, \ldots, N_0\}$

Thus $F(x_0, x_1, \ldots, x_n)$ is the outcome vector corresponding to the selections of strategy x_i by player i ($i = 0, \ldots, n$). We denote by $F_i(x_0, x_1, \ldots, x_n)$ the ith component of F. Then the expected return to player i for given choices $x_1, \ldots, x_n$ is

$$G_i(x_1, \ldots, x_n) = \sum_{x_0=1}^{N_0} p_0(x_0) F_i(x_0, x_1, \ldots, x_n) \quad (23\text{-}1)$$

where $p_0(x_0)$ is the probability that player 0 chooses alternative x_0.

In applications of game theory the components of the outcome vector represent utilities to the individual players. Since most games involve chance moves and since a player has no control over them, it is a consequence of (23-1) that these utilities should conform to certain assumptions concerning probability mixtures.

Consider two events A and B and let their utilities to an individual be, respectively, u and v. We construct a new event C called a *probability mixture* of A and B which consists of event A with probability p and event B with probability $1 - p$. Let w be the utility of C. If

$$w = pu + (1 - p)v$$

we feel that the utility is linearly adapted to chance events.

More formally, let $A_1, \ldots, A_r$ be any events, let $u(A_i)$ be the utility of A_i, and let $\Pi = (p_1, \ldots, p_r)$ be a probability vector, i.e., a vector whose components are

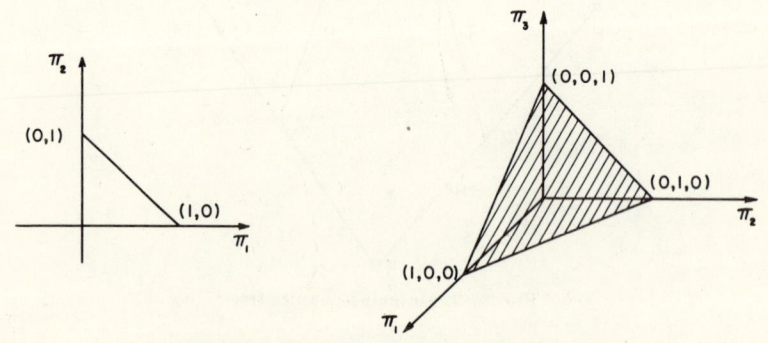

Fig. 23-3. Probability simplices for dimensions 2 and 3.

nonnegative real numbers which have sum one. We introduce a new event A called the Π *mixture* of $A_1, \ldots, A_r$ and designated by $A = [A_1, \ldots, A_r, \Pi]$. We say that u is a *linear utility function* if

$$u(A) = u([A_1, \ldots, A_r, \Pi]) = p_1 u(A_1) + \cdots + p_r u(A_r) \quad (23\text{-}2)$$

for all finite collections $A_1, \ldots, A_r$ and probability vectors Π. Chapter 2 of Ref. 8 discusses axiomatic characterizations of linear utility functions. In all that follows we shall assume that the numbers in each outcome vector arise from linear utility functions. Thus formulas such as (23-1) are meaningful, and $G_i(x_1, \ldots, x_n)$ represents the utility to player i of the simultaneous strategy selections $x_1, \ldots, x_n$, and we need no longer consider player 0.

For a game in normalized form a mixed strategy for player i is a probability vector Π_i with N_i components $\Pi_i(1), \ldots, \Pi_i(N_i)$. The unit vectors correspond to pure strategies, and the set of all mixed strategies can be regarded as an $(N_i - 1)$-dimensional simplex.

The cases for dimensions 2 and 3 are illustrated in Fig. 23-3.

23-3. Two-person Zero-sum Games

Whereas the analysis of general games is still far from complete, for the case of two-person zero-sum games the situation is much more satisfactory. There are (1) an acceptable concept of solution, (2) an existence theorem guaranteeing a solution, (3) a classification of solutions, (4) several reasonably practical methods for determining solutions, and (5) a collection of applications of the theory. Since the computation of solutions is limited by the capacity of computing machines, there is still some interest in studying these games in extensive form in the hope of reducing a larger class to manageable size. However, in the present section we consider only games in normal form.

A two-person game in normal form is characterized by two p by n matrices A and B where A gives the payoffs for player 1, B gives them for player 2, and $N_1 = p$, $N_2 = n$. If the game is also zero-sum then $B = -A$ and we thus speak of a *matrix game* defined by a p by n matrix $A = \|a_{ij}\|$ in which a_{ij} represents what player 1 is paid by player 2 when they choose respective strategies i and j. A negative entry indicates a positive payment from player 1 to player 2.

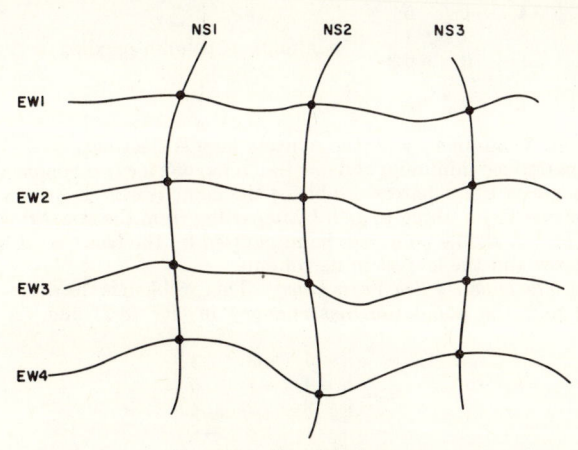

Fig. 23-4. The debaters.

We consider a few examples.

Example 1. The Debaters. Two candidates for the legislature have agreed to hold a public debate somewhere in their district and are negotiating over the location for it. Each wishes to have as friendly an audience as possible, and the 12 possible locations vary considerably in the balance between Republicans and Democrats. Each of the 12 possible sites lies on the intersection of an east-west and a north-south highway, as pictured in Fig. 23-4.

The excess (measured in thousands) of Democrats over Republicans at the EWi and NSj intersection is a_{ij} where

$$A = \begin{Vmatrix} 7 & -3 & 3 \\ -4 & -1 & 5 \\ 3 & 2 & 4 \\ 1 & 0 & -6 \end{Vmatrix}$$

An elder statesman in the district suggests that they settle the question by turning it into a game in which the Democrat chooses an EW route and the Republican a NS route, and then holding the debate at the intersection of these routes. The candidates agree to this.

The Democrat now studies what is the worst that can happen to him if he chooses any route. Thus for $i = 1$, he would like $j = 1$ and $j = 3$, which would give him, respectively, advantages of 7 and 3; however, in the worst case $j = 2$ would lose 3.

Thus he writes -3 at the end of the first row and, similarly, copies the smallest (minimum) entry of each row at its end.

$$\begin{Vmatrix} 7 & -3 & 3 \\ -4 & -1 & 5 \\ 3 & 2 & 4 \\ 1 & 0 & -6 \end{Vmatrix} \begin{matrix} -3 \\ -4 \\ 2 \\ -6 \end{matrix} \qquad i = 3 \quad \text{maximum of row minima} = 2$$

A conservative plan (strategy) would thus be to choose $i = 3$, which guarantees him an advantage of at least 2. This is called a *maximin strategy*.

Similarly, the Republican looks for the largest entry in each column (largest rather than smallest because his interests are exactly opposite those of the Democrat) and chooses the column with the smallest maximum; i.e., he follows a *minimax strategy*.

$$\begin{Vmatrix} 7 & -3 & 3 \\ -4 & -1 & 5 \\ 3 & 2 & 4 \\ 1 & 0 & -6 \end{Vmatrix}$$
$$ 7 2 5 \qquad j = 2 \quad \text{minimum of column maxima} = 2$$

Since maximin = minimax = 2, the solution here is clear-cut.

When the guaranteed minimum and maximum payoffs for two opponents are exactly equal (as they are in the debaters' problem) the game is said to have a *saddle point*; clearly, each player has nothing to gain by departing from the strategy which leads to the saddle point. A saddle point can be recognized by the fact that it is the smallest number in its row and the largest in its column.

Example 2. *The Debaters Ten Years Later.* This problem is the same as in Example 1 except that now the population has changed in city (4,2) and the matrix A is replaced by

$$B = \begin{Vmatrix} 7 & -3 & 3 \\ -4 & -1 & 5 \\ 3 & 2 & 4 \\ 1 & 6 & -6 \end{Vmatrix}$$

Following the previous process, we get

$$\text{Row minima}$$
$$\begin{Vmatrix} 7 & -3 & 3 \\ -4 & -1 & 5 \\ 3 & 2 & 4 \\ 1 & 6 & -6 \end{Vmatrix} \begin{matrix} -3 \\ -4 \\ 2 \\ -6 \end{matrix} \text{ maximin}$$

Column maxima $ 7 6 5$

 minimax

We no longer have minimax = maximin, and there is no longer a clear-cut solution. To handle such problems, we employ a *mixed strategy*, as illustrated by the following example.

Example 3. *Matching Pennies.* This game is familiar to everyone. We have two players, Red and Blue, and assume that Red is trying to match Blue. Each has two strategies: Heads and Tails. The game can be shown schematically as follows:

		Blue H	Blue T
Red	H	1	-1
	T	-1	1

There is no saddle point here, and if either lets the other know his choice in advance, he is certain to lose. However, suppose that Red determines his choice by flipping an honest coin. Then the expected return to Blue if he chooses Heads is

$$\tfrac{1}{2} \times 1 + \tfrac{1}{2}(-1) = 0$$

and if he chooses Tails is

$$\tfrac{1}{2}(-1) + \tfrac{1}{2}(1) = 0$$

and thus by adopting a mixed strategy (i.e., a probability distribution over the two pure strategies) Red can protect himself against loss, at least at the level of mathematical expectation.

A similar mixed strategy is available to Blue.

Example 4. Scissors, Paper, Stone. Another classical game frequently played by children is scissors, paper, stone. Two players, Red and Blue, simultaneously name one of the three objects. If both name the same object, the game is a draw. Otherwise, the winner is determined by the jingle

Scissors cut Paper, Stone breaks Scissors, and Paper covers Stone.

The payoff matrix is accordingly

	Blue Scissors	Blue Paper	Blue Stone	Row minima
Scissors	0	1	−1	−1
Paper	−1	0	1	−1
Stone	1	−1	0	−1
Column maxima	1	1	1	

One sees easily that a mixed strategy which chooses each pure strategy with probability $\tfrac{1}{3}$ protects each player against expected loss, and it can be shown that no other mixed strategy has this property.

We now return to the general case. It is clear that, once the payoff matrix is written down, the context of the problem is no longer relevant to the process of solution. As we have seen, a two-person zero-sum game in normalized form is given by a matrix A. The entry a_{ij} is the payment made by Blue (the column player) to Red (the row player) if Red chooses row i and Blue chooses column j.

For a 5 by 4 matrix we have

		Blue 1	2	3	4	Minimum value in ith row
Red	1	a_{11}	a_{12}	a_{13}	a_{14}	a_1
	2	a_{21}	a_{22}	a_{23}	a_{24}	a_2
	3	a_{31}	a_{32}	a_{33}	a_{34}	a_3
	4	a_{41}	a_{42}	a_{43}	a_{44}	a_4
	5	a_{51}	a_{52}	a_{53}	a_{54}	a_5
Maximum value in jth column		b_1	b_2	b_3	b_4	

23–8 SYSTEM THEORY

Let
$$a_i = \min_j a_{ij} = \text{minimum value in the } i\text{th row}$$
$$a = \max_i a_i = \text{maximum of all the } a_i$$
$$b_j = \max_i a_{ij} = \text{maximum value in the } j\text{th column}$$
$$b = \min_j b_j = \text{minimum of all the } b_j$$

Then
$$a = \max_i \min_j a_{ij}$$
$$b = \min_j \max_i a_{ij}$$

Theorem. $\min_j \max_i a_{ij} \geq \max_i \min_j a_{ij}$.

Proof. We must show $a \leq b$. We have $a = \max_i a_i$, say $a = a_p$, and
$$a = a_p = \min_j a_{pj}$$
say $a = a_p = a_{pq}$. Similarly let $b = b_h = a_{hk}$. Then we have
$$a = a_{pq} \leq a_{pk} \leq a_{hk} = b$$

that is, maximin $\leq$ minimax. This general statement is a universal theorem and is true for all functions of two variables when both sides are defined. If maximin = minimax, we say that the *minimax theorem holds* for pure strategies. The debaters' problem is an example, as are all matrices having saddle points.

Earlier, we saw how to represent chess as a (very large) matrix game. It can be proved that this game has a saddle point in pure strategies, although it is not known whether the result is a win, a loss, or a draw for the first player.

A game in extensive form is said to have *perfect information* if each information set contains exactly one node. It can be proved (Ref. 8, pp. 126ff.) that if a two-person zero-sum game in extensive form has perfect information then the same game in normal form has a saddle point in pure strategies. Ticktacktoe, checkers, and chess are games with perfect information. Games involving chance moves or in which any player has information about past moves which is not shared by others are not games with perfect information.

Consider a game with matrix $\begin{Vmatrix} a & b \\ c & d \end{Vmatrix}$ and having no saddle point, where the elements are arranged in such a way that $a \leq b$, $a \leq c$, $a \leq d$. If $b < d$, then the game has a saddle point; therefore, $b \geq d$; if $c < d$, then c is a saddle point; therefore, $c \geq d$. Consider the following example:

		Blue		
		1	2	
Red	1	3	6	3
	2	5	4	4
		5	6	

Then maximin = 4 and minimax = 5; hence there is no saddle point, and we try a mixed strategy. We would like to decide on some ratio for playing the two strategies. The determination of the strategy to be played would then be made in a random fashion with some preassigned probability.

If Red chooses strategy 2 with probability p and strategy 1 with probability $1 - p$; then when Blue chooses strategy 1, Red's expectation is

$$y_1 = 3(1 - p) + 5p$$

When Blue chooses strategy 2, Red's expectation is

$$y_2 = 6(1-p) + 4p$$

We want to choose p so that y_1 and y_2 are both as big as possible. This will maximize the total expectation. We want to find

$$\max_p \min_i y_i(p)$$

Graphically, we find that this point is at the intersection of the two lines and occurs when (see Fig. 23-5)

$$p = \tfrac{3}{4} \quad \text{and} \quad 1 - p = \tfrac{1}{4} \quad \text{and} \quad y_1 = y_2 = 4\tfrac{1}{2}$$

By using this mixed strategy, the first player guarantees himself an expectation of at least $4\tfrac{1}{2}$. Similarly, there is a mixed strategy for the second player which guarantees that his losses will not be more than $4\tfrac{1}{2}$. Using pure strategies, all Red can guarantee is that he will net a minimum gain of 4, and all Blue can guarantee is that he will suffer a maximum loss of 5.

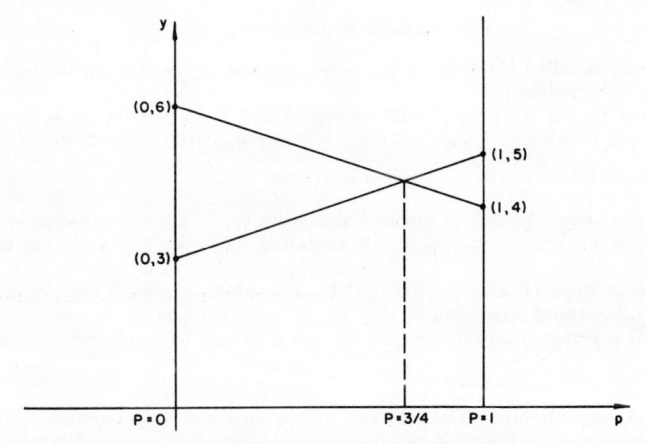

Fig. 23-5. Saddle point with mixed strategy.

$4\tfrac{1}{2}$ is the *value of the game;* that is, using a mixed strategy, the first player can guarantee a gain of at least $4\tfrac{1}{2}$, and the second player can guarantee a loss not exceeding $4\tfrac{1}{2}$. The ratio of the number of times strategy 1 is played to the number of times strategy 2 is played is 1:3. 1 and 3 are called *oddments*.

In a 2 × 2 case without a saddle point, the oddments for a good mixed strategy for Red are found by subtracting the first column from the second, interchanging the order of the pair of numbers, and using their absolute values: $(|d - c|, |b - a|) = (1,3)$ are the oddments. For Blue, subtract the first row from the second, interchange the entries, and use their absolute values.

23-4. Minimax Theorem for Matrix Games

Consider, again, a general matrix game Γ defined by a p by n matrix $A = \|a_{ij}\|$. A mixed strategy for the row player (player 1) is a probability vector $P = (p_1, \ldots, p_p)$; similarly, a mixed strategy for the column player (player 2) is a probability vector $Q = (q_1, \ldots, q_n)$. The goal of the row player is to select a mixed strategy P which maximizes the smallest component of $A^T P$, where A^T is the transpose of A. Let E_t denote the column vector with t components each equal to one. Then the problem for the row player is to select P subject to the conditions

$$\begin{aligned} A^\mathsf{T} P &\geq v_0 E_n \\ E_p^\mathsf{T} P &= 1 \\ P &\geq 0 \end{aligned} \tag{23-3}$$

which maximizes v_0.

Dually, the column player wishes to select Q subject to

$$\begin{aligned} AQ &\leq v^0 E_p \\ E_n^\mathsf{T} Q &= 1 \\ Q &\geq 0 \end{aligned} \tag{23-4}$$

which minimizes v^0.

Clearly, for any strategies P and Q

$$v_0 = v_0(E_n^\mathsf{T} Q) \leq (P^\mathsf{T} A)Q \leq P^\mathsf{T}(v^0 E_p) \leq v^0 \tag{23-5}$$

The minimax theorem (for mixed strategies) states that there exist strategies P_0 and Q_0 and a number $v = v(\Gamma)$ for which

$$A^\mathsf{T} P \geq v E_n \quad \text{and} \quad AQ \leq v E_p$$

This together with (23-5) implies that

$$\max v_0 = \min v^0 = v \tag{23-6}$$

This number v is called the *value* of the game and the corresponding strategies P_0 and Q_0 are said to be *optimal*.

We observe that if Γ is a game with matrix A and Γ' has matrix $A' = A + c E_p E_n^\mathsf{T}$, then given any strategies P and Q we have $PA'Q = PAQ + c$. It follows that

$$v(\Gamma') = v(\Gamma) + c \tag{23-7}$$

and, moreover, that the sets of optimal strategies for Γ' are the same as those for Γ. Thus there is no loss of generality in assuming each $a_{ij} > 0$ and, therefore, also $v(\Gamma) > 0$.

Shapley and Snow (Ref. 6, pp. 27–35) have provided a theory for determining v and the sets of all optimal strategies P_0 and Q_0.

Let p and n be two positive integers and let S denote two ordered subsets

$$i_1, \ldots, i_r \qquad j_1, \ldots, j_r$$

of $I(p)$, $I(n)$, respectively. Then for any p by n matrix A we denote by A_S the square submatrix of degree r whose h, k entry is $a_{i_h j_k}$ ($h, k = 1, \ldots, r$). With any ordered pair (X, Y) of vectors of degree r we associate the ordered pair $(\dot{X}, \dot{Y})$ of degrees p, n, respectively defined by

$$\begin{aligned} \dot{x}_{i_h} &= x_h \qquad \dot{y}_{j_k} = y_k \\ \dot{x}_i &= \dot{y}_j = 0 \text{ otherwise} \end{aligned} \tag{23-8}$$

Let B be any nonsingular matrix of degree r. Then if $E_r^\mathsf{T} B^{-1} E_r \neq 0$ we set

$$\begin{aligned} v_B &= \frac{1}{E_r^\mathsf{T} B^{-1} E_r} \\ P_B &= v_B (B^{-1\mathsf{T}} E_r) \\ Q_B &= v_B (B^{-1} E_r) \end{aligned} \tag{23-9}$$

The Shapley-Snow theorem states that given any matrix game Γ with p by n matrix A there exists S such that if $B = A_S$ then

$$v = v(\Gamma) = v_B \tag{23-10}$$

and $\dot{P}_B$ and $\dot{Q}_B$ are optimal strategies for players 1 and 2, respectively.

The matrix B is called a *kernel* of the game Γ and the corresponding strategies $\dot{P}_B$, $\dot{Q}_B$ are called *kernel strategies*. The set of all optimal strategies for either player is the convex hull of his kernel strategies.

GAME THEORY

Given any game Γ with matrix $A > 0$, i.e., all $a_{ij} > 0$, clearly $v(\Gamma) > 0$. Then to determine whether a given set S determines a kernel B one first calculates $E_r^T B^{-1} E_r$ and if this is not zero one then checks if both $P_B \geqq 0$ and $Q_B \geqq 0$. If so, then finally, one checks if $A^T P_B \geqq v_B E_n$ and $A \dot{Q}_B \leqq v_B E_p$. If these inequalities hold then B is a kernel. By choosing in turn all possible S for $r = 1, \ldots, \min\{p,n\}$ and checking each $B = A_S$ one can, in theory, find all kernels for Γ. For actual computation this method is impractical because of the large number of cases to be considered.

A more practical procedure is to reduce the game to an equivalent pair of dual linear-programming problems (Chapter 26) and then apply the simplex method (Chapter 25). Again assuming $A > 0$, and therefore making it reasonable to require v_0 and v^0 to be positive, we substitute $P = v_0 X$, $Q = v^0 Y$ in (23-3) and (23-4) and obtain the dual systems

$$A^T X \geqq E_n$$
$$X \geqq 0 \tag{23-11}$$
$$\min \frac{1}{v_0} = E_p^T X$$

and
$$A Y \leqq E_p$$
$$Y \geqq 0 \tag{23-12}$$
$$\max \frac{1}{v^0} = E_n^T Y$$

Clearly, both problems are feasible; hence by the duality theorem for linear programming (Sec. 26-3)

$$\max \frac{1}{v^0} = \min \frac{1}{v_0}$$

(This gives a proof of the minimax theorem.) The simplex method applied to (23-12) requires only the addition of slack variables and hence is, in general, somewhat simpler than when applied to (23-11) where artificial variables are required. Of course, an optimal X can be obtained from the final tableau for (23-12).

A converse procedure for passing from a linear-programming problem to an equivalent matrix game is also available and provides a proof of the fundamental duality theorem for linear programming based on the minimax theorem (Ref. 8, A5.3, pp. 419–423).

To illustrate the use of the simplex method in solving games consider the game with matrix

$$A = \begin{Vmatrix} 2 & 1 & 6 \\ 5 & 8 & 3 \end{Vmatrix}$$

For this matrix we consider (23-12), introduce slack variables y_4, y_5, and proceed to the simplex solution.

$1 = y_1$	y_2	y_3	y_4	y_5
1	2	1	6	1	0
1	5	8	3	0	1
0	−1	−1	−1	0	0
3/5	0	−11/5	24/5	1	−2/5
1/5	1	8/5	3/5	0	1/5
1/5	0	3/5	−2/5	0	1/5
1/8	0	−11/24	1	5/24	−1/12
1/8	1	15/8	0	−1/8	1/4
1/4	0	5/12	0	1/12	1/6

23-12 SYSTEM THEORY

From the final tableau we find the optimal solution to be $y_1 = \frac{1}{8}$, $y_2 = 0$, $y_3 = \frac{1}{8}$, $1/v = \frac{1}{4}$; and from the columns of the slack variables we obtain the optimal vector $x_1 = \frac{1}{12}$, $x_2 = \frac{1}{6}$ for the dual problem (23-11). Thus the game has value 4; the optimal strategy for the row player is $p_1 = \frac{1}{3}$, $p_2 = \frac{2}{3}$; and the optimal strategy for the column player is $q_1 = \frac{1}{2}$, $q_2 = 0$, $q_3 = \frac{1}{2}$.

23-5. Iterative Approximation Solution for Games

A method for solving two-person zero-sum games which is often useful is the iterative or successive-approximation method. This is sometimes called the method of fictitious play. Consider the following example where the game matrix A has 3 rows and 4 columns:

$$\begin{bmatrix} 2 & 3 & 0 & 5 \\ 3 & 1 & 4 & 2 \\ 4 & 5 & 3 & 2 \end{bmatrix}_*\quad \begin{array}{cccc} 5^* & 5 & 5 & 8 & 13 \\ 2 & 6^* & 10^* & 11 & 13 \\ 2 & 5 & 8 & 13^* & 15^* \\ 4 & 5 & 3 & 2^* \\ 6 & 8 & 3^* & 7 \\ 9 & 9 & 7^* & 9 \\ 12 & 10^* & 11 & 11 \\ 16 & 15 & 14 & 13^* \end{array}$$

First find the maximin for pure strategies and place a star in a new column just opposite the maximin row (here row 3) and copy this row directly underneath the game matrix. Since the player who controls the columns wants to minimize the payoff, we next mark with a star the *smallest* number in this row (here the 2 in column 4) and copy the corresponding column just beyond the initial star. Now place a star on the largest number in this column and enter the sum of corresponding row of the game matrix (here row 1) and the current bottom row in the tableau just below that row. Once again star the smallest element (here the 3 in the third column) and add the corresponding column to the current last column and enter the result to the right of this column. We can continue this process indefinitely, alternating between row and column expansions. In case of ties at any stage place the star on the tied line (row or column) which has been inactive longest; if any tied lines have never been active choose the inactive one with the smallest index.

The stars play two roles in this process: (1) they serve in obtaining bounds for the value of the game, and (2) they serve as counters from which approximate optimal strategies can be obtained as well as providing a numerical check on the numerical operations involved in the process.

For each positive integer t let v_t denote $1/t$ times the starred (smallest) element in the tth added row and let w_s denote $1/s$ times the starred (largest) element in the sth added column. Then

$$v_t \leq v \leq w_s \qquad s, t = 1, 2, \ldots . \tag{23-13}$$

Hence, after z complete steps we have

$$\max_{t \leq z} v_t \leq v \leq \min_{s \leq z} w_s \tag{23-14}$$

as bounds for v. The process terminates when we are willing to accept these bounds as estimates for v, i.e., when the difference $\min w_s - \max v_t$ is considered small enough to be neglected.

Thus in our illustrative example

$$\begin{array}{llllll} & v_1 = 2 & v_2 = 1\frac{1}{2} & v_3 = 2\frac{1}{3} & v_4 = 2\frac{1}{2} & v_5 = 2\frac{3}{5} \\ \text{and} & w_1 = 5 & w_2 = 3 & w_3 = 3\frac{1}{3} & w_4 = 3\frac{1}{4} & w_5 = 3 \end{array}$$

Thus our best bounds for $z = 5$ steps are

$$2\frac{3}{5} = v_5 \leq v \leq w_2 = w_5 = 3 \tag{23-15}$$

GAME THEORY

Suppose that the game matrix A has n rows and m columns. Let $a_i(t)$ be the number among the first t column stars which lie in row i ($i = 1, \ldots, n$). Here we count the unattached star and those in columns $1, 2, \ldots, t-1$ but not the star in column t. Then the vector P_t having $p_{ti} = (1/t)a_i(t)$ is a mixed strategy for the row player which guarantees an expected value at least v_t for him; indeed if R_t is the tth added row then

$$P_t^\mathsf{T} A = \frac{1}{t} R_t \tag{23-16}$$

Similarly let $b_j(s)$ denote the number among the first s row stars which lie in column j ($j = 1, \ldots, m$). Then the vector Q_s having $q_{sj} = (1/s)b_j(s)$ is a mixed strategy for the column player which guarantees an expected value at most $v_s(B)$ for him, i.e.,

$$A Q_s = \frac{1}{s} C_s \tag{23-17}$$

where C_s is the sth added column. The results (23-16) and (23-17) give checks on numerical accuracy at any stage. Thus in our example

$$P_5 = \tfrac{1}{5} \|0\ 1\ 2\ 2\|^\mathsf{T} \quad \text{and} \quad Q_5 = \tfrac{1}{5} \|1\ 2\ 2\|^\mathsf{T} \tag{23-18}$$

are the best approximate strategies available after $z = 5$ steps; they correspond to (23-15).

It is convenient to add an initial row and column to record the v_t and w_s. Thus after 10 steps we have

$$
\begin{array}{c}
\\
\\
v_t\\
2\\
1\tfrac{1}{2}\\
2\tfrac{1}{3}\\
2\tfrac{1}{2}\\
2\tfrac{3}{5}\\
2\tfrac{1}{2}\\
2\tfrac{3}{7}\\
2\tfrac{1}{2}\\
2\tfrac{2}{3}\\
2\tfrac{3}{5}
\end{array}
\begin{array}{c}
w_s\\
\left[\begin{array}{cccc}
2 & 3 & 0 & 5\\
3 & 1 & 4 & 2\\
4 & 5 & 3 & 2
\end{array}\right]\\
4 & 5 & 3 & 2*\\
6 & 8 & 3* & 7\\
9 & 9 & 7* & 9\\
12 & 10* & 11 & 11\\
16 & 19 & 14 & 13*\\
20 & 20 & 17 & 15*\\
22 & 23 & 17* & 20\\
26 & 28 & 20* & 22\\
29 & 29 & 24 & 24*\\
33 & 34 & 27 & 26*
\end{array}
\begin{array}{c}
\\
\\
*\\
\end{array}
\begin{array}{cccccccccc}
 & 5 & 3 & 3\tfrac{1}{3} & 3\tfrac{1}{4} & 3 & 3 & 2\tfrac{6}{7} & 2\tfrac{7}{8} & 2\tfrac{7}{9} & 2\tfrac{8}{10}\\
 & 5* & 5 & 5 & 5 & 13 & 18* & 18 & 18 & 23 & 28*\\
 & 2 & 6* & 10 & 11 & 13 & 15 & 19 & 23* & 25 & 27\\
 & 2 & 5 & 8 & 13* & 15* & 17 & 20* & 23 & 25* & 27
\end{array}
$$

Note the use of the tie-breaking rule in assigning stars in added columns 8, 9 and row 10. The best bounds are now given by

$$2\tfrac{2}{3} = v_9 \leq v \leq w_9 = 2\tfrac{7}{9} \tag{23-19}$$

The actual optimal value and strategies are

$$v = 2\tfrac{21}{29} \quad P = \tfrac{1}{29}\|7\ 13\ 9\|^\mathsf{T} \quad Q = \tfrac{1}{29}\|0\ 3\ 12\ 14\|^\mathsf{T} \tag{23-20}$$

23-6. n-person Games

Whereas the theory of two-person zero-sum games is both reasonably complete and fairly well accepted, such is not the case for general games. The classical theory begins with the game in the normalized form developed in Sec. 23-2 and passes to what is called *characteristic function form;* it is this formulation which is central in much of the treatment of games with three or more players. The case of two-person non-zero-sum games requires and has had special treatment (see the following section and Chapters 5 and 6 of Ref. 8).

Denote by $N = \{1, \ldots, n\}$ the set of players in a game, and let M be any subset of N. Let $v(M)$ denote the maximum expected total payoff that the members of M can guarantee themselves. The function v thus defined maps the set 2^N of all subsets of N into the real numbers and is called the *characteristic function* of the game.

The function v has two properties

$$v(\phi) = 0 \quad \text{where } \phi \text{ is the empty set} \tag{23-21}$$

and $$v(M_1 \cup M_2) \geq v(M_1) + v(M_2) \quad \text{if } M_1 \cap M_2 = \phi \tag{23-22}$$

This second property is called *superadditivity*. These two properties provide a direct definition of a characteristic function, since given any function v which satisfies (23-21) and (23-22) there is a game (in normalized form) which has v as its characteristic function.

A vector $a = (a_1, \ldots, a_n)$ is said to be an *imputation* for a game with characteristic function v if

$$a_i \geq v(\{i\}) \quad (i = 1, \ldots, n) \tag{23-23}$$

and $$\Sigma a_i = v(N) \tag{23-24}$$

The imputations are regarded as possible settlements at the end of a game. Equation (23-23) is referred to as the condition of *individual rationality* since it limits imputations to those vectors which give each player at least as much as he could assure himself if all the remaining players were allied against him. Equation (23-24), called *Pareto* (or group) *rationality*, is based on the argument that if b is any vector with $\Sigma b_i = b_0 < v(N)$ then the vector a having $a_i = b_i + [v(N) - b_0]/n$ would be preferred to b by every member of N, and therefore that b should not be accepted as a final outcome. This argument is certainly not universally applicable for a descriptive theory since it rules out strikes and wars; however, it does lead to the interesting theory developed by von Neumann and Morgenstern.

The brief space alloted to game theory in this volume does not permit much discussion of the many modifications of the classical theory that have been proposed. The current trend is toward the investigation of several different models without insisting on universal validity for any one of them. References 8 and 11 will provide the interested reader an excellent introduction to modern n-person theory, although there have been extensive and important developments since these references appeared, some of which are not yet published. We continue with a development of the classical theory since, despite its flaws, it provides the best introduction to the general theory.

Let a and b be two imputations and let M be a subset of N. We say that a *dominates* b *relative* to M, written

$$a \underset{M}{>} b \tag{23-25}$$

if $$a_i > b_i \quad \text{for all } i \in M \tag{23-26}$$

and $$\sum_{i \in M} a_i \leq v(M) \tag{23-27}$$

The first of these conditions is called *M-preferability;* note that the inequalities are strict; the use of $\geq$ leads to difficulties. The second condition is called *M-effectiveness* since it does not ask more for members of M than they can collectively assure themselves.

More generally we say that a *dominates* b, written

$$a > b \tag{23-28}$$

if there exists a subset M relative to which a dominates b. Relative domination is asymmetric and transitive; domination is neither.

A set S of imputations is said to be a *solution* or a *stable set*, if

$$a > b \text{ does not hold for any } a, b \text{ in } S \tag{23-29}$$

and $$\text{For each } b \text{ not in } S \text{ there exists an } a \text{ in } S \text{ with } a > b \tag{23-30}$$

These conditions can be stated more concisely in terms of the following concept: Let S be a set of imputations; by the *dominion of S* (written dom S) we mean the set of all imputations dominated by members of S; i.e., b is in dom S if there exists an a in S with $a > b$. Now conditions (23-29) and (23-30) become, respectively,

$$S \cap \text{dom } S = \phi \qquad (23\text{-}29a)$$
and
$$S \cup \text{dom } S = \{\text{all imputations}\} \qquad (23\text{-}30a)$$

The theory of solutions has the following status: There is no general existence theorem, although there are many special games for which solutions are known and there is an existence theorem for $n \leq 3$. The situation is that there are both too few and too many solutions known, in the sense that although we do not have enough solutions to provide a general existence theorem we do have such a broad variety of solutions for many games that the concept of solution loses much of its sharpness. We also face the problem that a solution which contains an infinite number of imputations gives little guidance.

In an attempt to avoid this multiplicity of solutions and imputations, Shapley (Ref. 7, pp. 307–317; Ref. 8, pp. 245–252) has proposed a single vector

$$u = (u_1, \ldots, u_n)$$

now called the *Shapley value*, as a solution. The allocation to player i is determined by his average contribution to all coalitions of which he is a member. The formula is

$$u_i = \sum_{S \subset N} \gamma_n(s)[v(S) - v(S - \{i\})] \qquad (23\text{-}31)$$

where s is the number of members in S and

$$\gamma_n(s) = \frac{(s-1)!(n-s)!}{n!} \qquad (23\text{-}32)$$

is a weighting function. Shapley gives an axiomatic characterization of his value and cites a number of examples where it approximates the payoffs actually observed in society. Thus the Shapley value has both a normative justification in terms of providing for a "fair division of the spoils" and a descriptive verification.

Another approach[13] replaces the characteristic function v by a partition function which for each partition of N assigns a real number to each coalition of the partition. Thus the payoff assigned to a given subset M of N depends on the manner in which the remaining players are grouped.

Let $P = \{P_1, \ldots, P_r\}$ be a partition of N into coalitions (cosets) $P_1, \ldots, P_r$. The set of all partitions of N is denoted by

$$\Pi = \{P\}$$

Then for each partition P assume there is an *outcome function* F_P which assigns a real number $F_P(P_i)$ to the coalition P_i when the partition P forms. The function F which maps P in Π into the function F_P is called the *payoff function* or *partition function* for the game. The ordered pair $\Gamma = (N,F)$ is called an *n-person game in partition-function form*.

For each nonempty subset M of N define the *value* of M as

$$v(M) = \min F_P(M) \qquad (23\text{-}33)$$

where the minimum is taken over all partitions having M as a coset, and define $v(\phi) = 0$. This function v is not necessarily superadditive; indeed, there exists a game Γ having any preassigned set of numbers $v(M)$ as its set of values.

However, if in (23-33) we took the minimum over all partitions P that have M as union of cosets we would obtain a superadditive function; this would correspond to permitting secret coalitions, whereas the v given by (23-33) is based on a prohibition of secret agreements between the players.

23-16   SYSTEM THEORY

A vector $a = (a_1, \ldots, a_n)$ is called an imputation if (23-23) holds and if

$$\sum_{i=1}^{n} a_i = \sum_{j=1}^{r} F_P(P_j) \qquad \text{for some } P \text{ in } \Pi \tag{23-24a}$$

Replacing the equality in (23-24a) by $\leq$ would allow for a disposal of wealth.

The definition of relative domination now requires, in addition to M-preferability (23-26) and M-effectiveness (23-27), also the following condition, called M-*realizability*:

$$(23\text{-}24a) \text{ holds for a } P \text{ which has } M \text{ as a coset} \tag{23-34}$$

The concepts of domination and solution are introduced exactly as in the classical case by (23-28) and by (23-29) and (23-30).

A major feature of games in partition-function form is that the absence of superadditivity makes the theory applicable to cases where certain coalitions would reduce rather than enhance the total strength. For example, a coalition of a radical labor leader and a reactionary politician might reduce the effectiveness of each.

This new theory is more general than the classical one, and it is hoped that, as is frequently the case, study of the more general theory may shed light on the special case. There are some indications that this may be the situation here, but the theory is still too new for any firm conclusions.

23-7. Some Examples of Two-person Non-zero-sum Games

In this section we give three games; each of these illustrates one of the difficulties still remaining in both two-person and n-person games.

Example 1. *The Doghouse Problem.* Mr. Red has a warehouse filled with nails; Mr. Blue has a warehouse filled with lumber. There is a demand for doghouses. There are two pure strategies for Red and two for Blue. If they cooperate, they can build doghouses and gain a profit, x for Red, y for Blue. If either refuses to cooperate, both Red and Blue have to pay storage charges, 1 for the nails and 2 for the lumber, respectively.

We label the two strategies, C for cooperate and N for not cooperate, and write

		Blue	
		C	N
Red	C	(x,y)	$(-1,-2)$
	N	$(-1,-2)$	$(-1,-2)$

where the first number in each pair represents the payoff to Red, the second the payoff to Blue, for their various choices of strategies. If both cooperate, a profit of, say, 100 can be made; that is, $x + y = 100$. There are no generally accepted procedures to deal with the division of the proceeds (i.e., what x and y should be). The classical theory merely requires $x \geq -1$ and $y \geq -2$.

Example 2. *The Prisoners' Dilemma.* Two alleged burglars, Red and Blue, are apprehended; however, there is no real evidence as to their guilt. The detectives separate the men and try to get each to confess by offering him a reward of $1,000 and promising to fine his associate $2,000. However, if both confess, each will be fined $1,000.

	Blue	
	Not confess	Confess
Red — Not confess	(0,0)	(−2,000, 1,000)
Red — Confess	(1,000, −2,000)	(−1,000, −1,000)

Note that no matter what Blue does, Red is a thousand dollars better off if he confesses than if he doesn't; i.e., strategy 2 *dominates* strategy 1 for Red by being uniformly better in every eventuality. The same is true for Blue. Thus, if each player individually maximizes his own return, they end up losing $1,000 each; whereas if neither confesses, both are better off. The dilemma posed by this example has not yet been solved in terms of game theory.

Example 3. *The Loving Couple*. Mr. and Mrs. Brown are newly married and discover that their evaluations of various evening entertainments are not the same. The husband (H) likes hockey but does not care much for going to the movies; whereas the wife (W) adores movies and actively dislikes hockey (her feet get cold, and she can't understand it anyhow).

Being a loving couple, they make a compact to stay home and hold hands in case they don't agree on outside entertainment; this is agreeable to both of them. In the language of game theory, we say that each has two strategies: to vote for hockey or movie. Thus they go out only if they vote for the same activity. We assume that their utilities for these activities are given by the following table:

	Wife	
	Hockey	Movie
Husband — Hockey	(3, −1)	(1, 1)
Husband — Movie	(1, 1)	(0, 4)

where the first component in each square is the husband's appraisal and the second is the wife's. One method of resolving this game would be the use of a *correlated mixed strategy*. Both agree to vote according to the outcome of some random device, perhaps the toss of a coin with heads meaning hockey and tails movie. The expectations would then be $\frac{1}{2}(3, -1) + \frac{1}{2}(0, 4) = (\frac{3}{2}, \frac{3}{2})$ which beats staying home for both.

These examples illustrate only a few of the difficulties in the way of a generally satisfactory theory of n-person non-zero-sum games. In the first, we encounter the bargaining problem; in the second, the fallacy of local maximization; and in the third, the difficulty of nontransferable utility (i.e., there is no way that the husband can give his wife half of his enjoyment of a hockey game).

23-8. Games against Nature

Consider a situation in which a decision maker must select one of p alternatives $R_1, \ldots, R_p$. We suppose that there are n states of nature $S_1, \ldots, S_n$ and consider the matrix $A = \|a_{ij}\|$ in which a_{ij} is the valuation placed by the decision maker on having selected alternative R_i in case nature turns out to be in state S_j. We assume that the decision maker has no information regarding which state of nature will obtain.

23-18　SYSTEM THEORY

The decision maker could take the point of view that A is the matrix of a two-person zero-sum game in which he is the row player and nature is the column player. Then he can employ a maximin strategy to obtain a security level. The maximin can be calculated either for pure strategies or for mixed strategies in accordance with which model better fits his environment.

We consider an example. Two youths plan to set up as vendors at the baseball stadium of their high school. They can sell ginger ale or hamburgers, but to avoid a city tax, vendors must limit their daily purchases to $10. Ginger ale costs 5 cents and sells at 10 cents per cup. Hamburgers cost 10 cents and sell at 25 cents each. On a warm day they can sell unlimited ginger ale and 40 hamburgers; on a cool day they can sell unlimited hamburgers and 100 cups of ginger ale—anything not sold is a total loss. The situation is summarized in Fig. 23-6. The entries are profits in dollars. The maximin strategy is program 1 with probability $3/5$, program 2 with probability $2/5$, and the expected profit then is $6.

	States of nature	
	Warm day	Cool day
Program 1, 200 cups of ginger ale.........	10	0
Program 2, 100 hamburgers.............	0	15

Fig. 23-6. The young vendors.

If physical mixtures are permitted, then

$$5x + 10y = 1{,}000$$

where x = number of cups of ginger ale and y = number of hamburgers. The profit on a warm day is then

$$\frac{x}{10} + \frac{1}{4}\min(40,y) - 10 = 10 - \frac{y}{5} + \frac{1}{4}\min(40,y)$$

and on a cool day is

$$\tfrac{1}{10}\min(x,100) + \frac{y}{4} - 10 = \tfrac{1}{10}\min(200 - 2y,100) + \frac{y}{4} - 10$$

The situation for physical mixtures is illustrated in Fig. 23-7. The mixed program $(111\tfrac{1}{9}, 44\tfrac{4}{9})$ gives a profit of $11\tfrac{1}{9}$ dollars regardless of the weather. [The nearest integer values are $(112,44)$ and $(110,45)$, both with guaranteed profit of 11 dollars.] Note that the physical mixture represents a considerable improvement over the minimax mixture of pure strategies.

If the probability p of a warm day is known, the expected return can be somewhat improved, as shown in Fig. 23-8.

The maximin strategy is not the only possibility for use in games against nature. Indeed, it may represent much too pessimistic an approach. Chapter 13 of Ref. 8 summarizes and discusses a number of alternate decision criteria, among which we consider here only a subset treated by John Milnor (Ref. 8, pp. 297–298).

Laplace. If the probabilities of the different possible states of nature are unknown, we should assume that they are all equal. Thus if the player chooses the ith row, his expectation is given by the average $(a_{i1} + \cdots + a_{in})/n$, and he should choose a row for which this average is maximized.

GAME THEORY

Wald (Maximin Principle). If the player chooses the ith row, then his payoff will certainly be at least min a_{ij}. The safest possible course of action is therefore to choose a row for which min a_{ij} is maximized. This corresponds to the pessimistic hypothesis of expecting the worst.

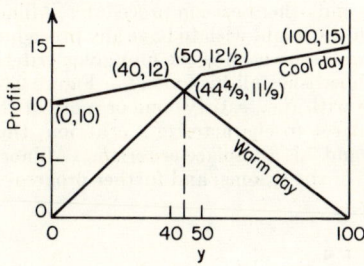

Fig. 23-7. The young vendors with physical mixtures. Numbers in parentheses are (y,profit).

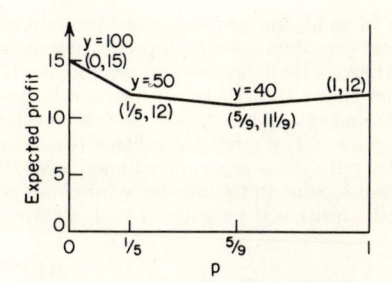

Fig. 23-8. Maximum expected return with known probability. Numbers in parentheses are (p,expected profit).

If mixed strategies for the player are also allowed, then this criterion should be formulated as follows: Choose a probability mixture $(\xi_1, \ldots, \xi_m)$ of the rows so that the quantity $\min_j (\xi_1 a_{1j} + \cdots + \xi_m a_{mj})$ is maximized. In other words, play as if nature were the opposing player in a zero-sum game, i.e., as if nature were malicious.

Hurwicz. Select a constant $0 \leq \alpha \leq 1$ which measures the player's optimism. For each row (or probability mixture of rows), let a denote the smallest component and A the largest. Choose a row (or probability mixture of rows) for which $\alpha A + (1 - \alpha)a$ is maximized. For $\alpha = 0$ this reduces to the Wald criterion.

		Laplace	Wald	Hurwicz	Savage
1.	Ordering	⊗	⊗	⊗	⊗
2.	Symmetry	⊗	⊗	⊗	⊗
3.	Str. domination	⊗	⊗	⊗	⊗
4.	Continuity	×	⊗	⊗	⊗
5.	Linearity	×	×	⊗	×
6.	Row adjunction	⊗	⊗	⊗	
7.	Col. linearity	⊗			⊗
8.	Col. duplication		⊗	⊗	⊗
9.	Convexity	×	⊗		⊗
10.	Special row adj.	×	×	×	⊗

× = Compatibility

Fig. 23-9. Axioms for criteria.

Savage (Minimax Regret). Define the (negative) regret matrix (r_{ij}) by $r_{ij} = a_{ij} - \max_k a_{kj}$. Thus r_{ij} measures the difference between the payoff which actually is obtained and the payoff which could have been obtained if the true state of nature had been known. Now apply the Wald criterion to the matrix (r_{ij}). That is, choose a row (or mixture of rows) for which min r_{ij} [or min $(\xi_1 r_{1j} + \cdots + \xi_m r_{mj})$] is maximized.

These four criteria are certainly different. This is illustrated by the following example, where the preferred row under each criterion is indicated:

2	2	0	1	Laplace
1	1	1	1	Wald
0	4	0	0	Hurwicz (for $\alpha > 1/4$)
1	3	0	0	Savage

In an attempt to throw further light on these and other decision procedures, Milnor next considers certain properties or axioms which one might wish to have any procedure satisfy. He determines those axioms which characterize each of the four given criteria and shows that there can be no criterion which satisfies all his axioms. Figure 23-9 shows that each criterion fails to be compatible with (i.e., satisfy) one or more of the axioms. The circled X's constitute a sufficient set to characterize a criterion; thus any criterion compatible with axioms 1, 2, 3, 6, and 7 is a Laplace criterion. Milnor's results point to the necessity for closer scrutiny of the axioms, and further progress is still awaiting the results of such scrutiny.

REFERENCES

1. Blackwell, David, and M. A. Girshick: *Theory of Games and Statistical Decisions*, John Wiley & Sons, Inc., New York, 1954.
2. Burger, Ewald: *Introduction to the Theory of Games*, Prentice-Hall, Inc., Englewood Cliffs, N.J., 1963.
3. Dresher, Melvin: *Games of Strategy: Theory and Applications*, Prentice-Hall, Inc., Englewood Cliffs, N.J., 1961.
4. Dresher, M., A. W. Tucker, and P. Wolfe: *Contributions to the Theory of Games*, vol. III, Ann. Math. Studies, 39, Princeton University Press, Princeton, N.J., 1957.
5. Karlin, Samuel: *Mathematical Methods and Theory in Games, Programming, and Economics*, vols. I and II, Addison-Wesley Publishing Company, Inc., Reading, Mass., 1959.
6. Kuhn, Harold W., and Albert W. Tucker: *Contributions to the Theory of Games*, vol. I, Ann. Math. Studies, 24, Princeton University Press, Princeton, N.J., 1950.
7. Kuhn, Harold W., and Albert W. Tucker: *Contributions to the Theory of Games*, vol. II, Ann. Math. Studies, 28, Princeton University Press, Princeton, N.J., 1953.
8. Luce, R. Duncan, and Howard Raiffa: *Games and Decisions: Introduction and Critical Survey*, John Wiley & Sons, Inc., New York, 1957.
9. McKinsey, J. C. C.: *Introduction to the Theory of Games*, McGraw-Hill Book Company, New York, 1952.
10. von Neumann, John, and Oskar Morgenstern: *Theory of Games and Economic Behavior*, Princeton University Press, Princeton, N.J., 1944 (2d ed., 1947).
11. Tucker, Albert W., and R. Duncan Luce: *Contributions to the Theory of Games*, vol. IV, Ann. Math. Studies, 40, Princeton University Press, Princeton, N.J., 1959.
12. Williams, John D.: *The Compleat Strategyst: Being a Primer on the Theory of Games of Strategy*, McGraw-Hill Book Company, New York, 1954.
13. Thrall, Robert M., and William F. Lucas: *n*-person Games in Partition Function Form, *Naval Research Logistics Quarterly*, vol. 10, no. 4, pp. 281–298, December, 1963.

Chapter 24

DECISION THEORY

GEORGE R. COOPER, *School of Electrical Engineering, Purdue University, Lafayette, Indiana.*

CONTENTS

24-1. The Decision Problem in System Theory	24-4
Types of Decision Problems in System Theory	24-4
Relation of Decision Theory to System Design	24-5
24-2. Basic Concepts of Decision Theory	24-5
Classification of Decision Processes	24-6
Types of Hypotheses	24-6
Types of Estimation	24-7
Nature of the Observed Data	24-7
Types of Decision Rules	24-8
24-3. System Evaluation and Comparison	24-9
Generalized Evaluation Functions	24-9
Risk Formulation	24-10
Information-loss Formulation	24-10
System Comparisons	24-11
24-4. System Optimization	24-11
Systems Which Minimize Risk	24-11
Systems Which Minimize Information Loss	24-12
24-5. Sequential Decision Processes	24-13
General Theory	24-13
Truncated Sequential Tests	24-13
Comparison of Sequential and Fixed-sample Tests	24-14
24-6. Application of Decision Theory to Signal Detection	24-14
General Theory of Binary Detection	24-14
Neyman-Pearson Test	24-16
Ideal Observer	24-16
Threshold Detection in Gaussian Noise	24-16
Coherent Binary Detection	24-16
Incoherent Binary Detection	24-19
Detection of Stochastic Signals	24-20
Evaluation and Comparison of Binary Detection Systems	24-21
Coherent Detection	24-22
Incoherent Detection	24-22
Stochastic Signal Detection	24-23
Multiple Alternative Detection	24-23
Binary Sequential Detection	24-25

24-7. Application of Decision Theory to Signal Extraction.......... 24-25
   Classical Estimation Theory........................... 24-26
   Decision-theory Formulation.......................... 24-27
   Coherent Estimation of Signal Amplitude............... 24-28
   Waveform Estimation in Gaussian Noise................ 24-29
24-8. Further Applications of Decision Theory.................... 24-29
   Adaptive Systems..................................... 24-29
   Learning Systems..................................... 24-30
   Recognition Systems.................................. 24-31

List of Symbols

a_0	= rms signal-to-noise ratio
$\hat{a}_0$	= an estimate of a_0
A	= upper threshold for binary sequential detection
ASN	= average sample number of a sequential test
B	= lower threshold for binary sequential detection
B_c	= bias term, coherent detection
B_i	= bias term, incoherent detection
B_S	= bias term, stochastic signal detection
B_T	= bias term, continuous case
c	= cost of each sample in a sequential test
C_d	= cost of correct decision of signal present (detection)
C_f	= cost of incorrect decision of signal present (false alarm)
C_m	= cost of incorrect decision of no signal (miss)
C_q	= cost of correct decision of no signal (quiet)
C_0	= component of constant cost function
$C(\mathbf{S},\mathbf{Y})$	= cost function for signal detection
$C(\boldsymbol{\theta},\hat{\boldsymbol{\theta}})$	= cost function for parameter estimation
$C_i(\mathbf{S})$	= cost of rejecting the ith hypothesis given that it is true
$C_j{}^k$	= cost of accepting jth signal when kth signal is present
D	= generalized detection parameter
$D_j{}^k$	= set of cost differences
$E[\ ]$	= mathematical expectation
$E_{\bar{S}}$	= average signal energy
$E_s[\ ]$	= mathematical expectation with respect to the signal ensemble
$f(\mathbf{N})$	= probability density function of noise vector
$F(\mathbf{S},\mathbf{Y})$	= generalized loss function
$g(\mathbf{Y}\|\mathbf{V})$	= decision rule
g^*	= Bayes decision rule
$g_M{}^*$	= minimax decision rule
G	= linear filter for stochastic signal
h	= an exponent in sequential detection
$h(\mathbf{S},g)$	= conditional information loss
$h_M(T - t)$	= matched filter for coherent detection
$\mathbf{H}$	= linear filter for incoherent detection
H^*	= Bayes equivocation
$H(w,g)$	= average information loss
$\mathbf{I}$	= identity matrix
K	= threshold for binary detection
$L(\mathbf{V})$	= generalized likelihood ratio
$L(\mathbf{V}_m)$	= generalized likelihood ratio after the mth sample of a sequential test
$L_k(\mathbf{V})$	= generalized likelihood ratio for kth signal
M	= number of decisions to be made
n	= number of samples taken in a time T
$n(t)$	= a noise waveform
n_k	= kth sample of $n(t)$
$n(\mathbf{S})$	= number of samples taken in a sequential test
N	= maximum length of a truncated sequential test
$\mathbf{N}$	= $(n \times 1)$ column vector representing n noise samples

DECISION THEORY

p	= probability that any signal is present
p_k	= probability that kth signal is present
$p(\mathbf{V}\|\theta)$	= conditional probability density of observed vector $\mathbf{V}$ given a parameter θ
$p(\mathbf{V}\|\mathbf{S})$	= conditional probability density of observed vector $\mathbf{V}$ given signal vector $\mathbf{S}$
$p_k(\mathbf{V})$	= conditional probability density of $\mathbf{V}$ given kth signal
$p_1(\mathbf{V})$	= conditional probability density of $\mathbf{V}$ given any signal
$p_2(\mathbf{S}\|\mathbf{Y})$	= conditional probability density of the signal given the decision
$p_3(\mathbf{Y}\|\mathbf{S})$	= conditional probability density of the decision given the signal
$P(Y_0)$	= probability of accepting Y_0
$P_i(\mathbf{S})$	= conditional probability of rejecting the ith hypothesis given that signal is present
$P_j{}^k$	= probability of accepting jth signal when kth is present
P_e	= total probability of error, binary case
P_f	= conditional probability of false alarm
P_m	= conditional probability of a miss
q	= probability of no signal
Q_G	= quadratic form $\{ = \operatorname{tr}[(\mathbf{\Lambda}_s\mathbf{\Lambda}^{-1})^2]\}$
Q_H	= quadratic form $(= \overline{\bar{\mathbf{s}}\mathbf{H}\mathbf{s}})$
Q_s	= quadratic form $(= \bar{\mathbf{s}}\mathbf{\Lambda}^{-1}\mathbf{s})$
Q_v	= quadratic form $(= \tilde{\mathbf{V}}\mathbf{\Lambda}^{-1}\mathbf{s})$
$r(\mathbf{S},g)$	= conditional risk
$R, R(w,g)$	= average risk
R_0	= average risk for zero error probabilities
$R_{\max}$	= maximum average risk
R^*	= Bayes risk
$R_M{}^*$	= minimax average risk
$\mathbf{s}$	= normalized signal vector $\left(= \dfrac{1}{a_0}\mathbf{S}\right)$
$s(t)$	= a signal waveform
s_k	= kth sample of $s(t)$
$\mathbf{S}$	= $(n \times 1)$ column vector representing n signal samples
$\bar{\mathbf{S}}_0$	= weighted-average signal $(= \bar{a}_0\mathbf{\Lambda}^{-1}\hat{\mathbf{s}})$
t_0	= time at which observation starts
tr[ ]	= trace of a matrix, i.e., sum of elements on the principal diagonal
T	= duration of observation period
$v(t)$	= an observed waveform
v_k	= kth sample of $v(t)$
$\mathbf{V}$	= $(n \times 1)$ column vector representing n observation samples
$\mathbf{V}_F$	= $(n \times 1)$ column vector representing observed data after filtering (incoherent signal)
$\mathbf{V}_G$	= $(n \times 1)$ column vector representing observed data after filtering (stochastic signal)
$\mathcal{V}_0$	= region in observation space leading to a decision of no signal
$\mathcal{V}_1$	= region in observation space leading to a decision of signal present
$w(a_0)$	= probability density of signal amplitude
$w(\mathbf{S})$	= probability density function of signal vector
$w(\theta)$	= probability density of signal parameters
$w_0(\mathbf{S})$	= least-favorable-signal probability density
$w_1(\mathbf{S})$	= conditional probability density of signal given that it is present
$w_k(\mathbf{S})$	= conditional probability density of kth signal given that it is present
Y_0	= decision of no signal present, binary case
Y_1	= decision of signal present, binary case
$\mathbf{Y}$	= $(M \times 1)$ column vector representing M decisions
$Y_{a_0}(\mathbf{V})$	= operation performed on $\mathbf{V}$ to estimate a_0
$Y_s(\mathbf{V})$	= operation performed on $\mathbf{V}$ to estimate $\mathbf{S}$
$Y_\theta(\mathbf{V})$	= operation performed on $\mathbf{V}$ to estimate θ
$z(s)$	= logarithm of a probability ratio in sequential detection

Greek Symbols:

$\delta(\ )$	= Dirac delta function
η	= ratio of Bayes risk to maximum risk
θ	= a signal parameter
$\hat{\theta}$	= an estimate of a signal parameter

$\boldsymbol{\theta}$	$= (M \times 1)$ column vector representing M signal parameters
$\Theta(z)$	= error function
λ_{ij}	= element of the covariance matrix
$\boldsymbol{\Lambda}$	= covariance matrix of noise vector
$\boldsymbol{\Lambda}_e$	= covariance matrix of estimation error vector
$\boldsymbol{\Lambda}_s$	= covariance matrix of signal vector
μ	= a priori likelihood ratio ($= p/q$)
$\varphi_n(t, \ )$	= autocorrelation function of noise
σ	= standard deviation of a Gaussian random variable

24-1. The Decision Problem in System Theory

Statistical decision theory is essentially a generalization and unification of classical statistical theory.[13] As such it contains most of the classical concepts as special cases and hence has a range of application which is even broader than the classical methods. Because it is not possible to mention more than a few applications in a single chapter, the present discussion is limited to those aspects of the subject which pertain most closely to the theory of engineering systems. For more extensive discussions of these and other topics, the interested reader is directed to the list of references at the end of this chapter.

The importance of statistical decision theory to system theory lies in its ability to provide a common framework for three significant aspects of systems: (1) a method for representing optimum system structure, (2) an evaluation of the performance of this optimum system, and (3) a quantitative comparison of actual systems with the theoretical optimum. Statistical decision theory is able to bridge the gap between theory and practice in a way which was not possible, or at least seldom attempted, by classical statistical methods.

Types of Decision Problems in System Theory. From a functional standpoint, there are two major classes of systems in which decision problems arise quite naturally. In one class, the system is required to make a choice among several alternatives, and it must perform this function on the basis of unreliable information. Examples of such systems range from the elementary telegraph system to extremely complex missile-attack-warning systems. However, regardless of the application of the system, or the number of alternatives permitted it, the distinguishing characteristic of such systems is that they must produce an unequivocal answer which is either exactly right or completely wrong.

In the second class, the system is required to make a guess concerning the value of some physical quantity or a number of quantities, and again, it must base this guess on unreliable information. Examples of such systems range from the simple thermometer to an elaborate system for measuring the temperature on the surface of Venus; from standard AM broadcasting to precision-radar tracking systems. The distinguishing characteristic of these systems is that their outputs may assume a continuous range of values which is seldom, if ever, exactly right but which, hopefully, is not far wrong.

It is possible, of course, for a system to perform both functions mentioned above. An outstanding example of this is the usual radar system which must first make a decision regarding the presence of possible targets and must then estimate the range and velocity of those targets accepted as being present. The operations which it must perform on the observed data in order to accomplish these functions are quite different in general. Hence it may be convenient to consider that the complete system is composed of two subsystems, each performing one of the two functions. From this standpoint, therefore, it is sufficient to limit the discussion to the two distinct classes of systems and not attempt to treat composite systems as a separate class.

It will be noted from the above discussion of the decision function that almost any engineering system includes one, or both, of the above operations in accomplishing its mission. In general, however, these decision functions represent only a part of the overall system operation. In particular, the system is usually required to perform some action in response to the decision it produces, and the consequences of this action depend upon the correctness of decision. Hence it is usually possible to assign some sort of "cost" to every combination of decision and physical situation leading to that decision. This cost then leads to an evaluation of system performance which is closely related to the operational requirements of the system rather than being based on an abstract mathematical criterion. It is this aspect of decision theory which makes it a much more practical tool for system designers than the classical statistical methods.

Relation of Decision Theory to System Design. If decision theory is to be of value in the design of engineering systems, it is necessary to relate the mathematical aspects of the theory to the practical requirements of system design. It will be found in the following sections that there are several ways in which this is done.

In the first place, the quite general way in which criteria of optimality can be selected makes it possible to tailor these criteria to fit the operational requirements of the system. Thus one can weight the costs associated with erroneous decisions in accordance with the seriousness of these errors. Although the engineering judgment required to do this is in large measure subjective, the very fact that the mathematical formulation can include such judgments should properly be interpreted as an important practical advantage.

Secondly, the mathematical analysis can usually be carried to the point where it specifies the structure of a system which is optimum with respect to the selected criteria. That is, the mathematical specification of the optimum system can be interpreted in terms of the physical operations that must be performed upon the observed data. In some cases, these operations cannot be achieved with an acceptable amount of hardware even though they are physically realizable, but even in these cases the structure of the optimum system serves a useful purpose in that it establishes the ideal which a practical system should approximate.

Another aspect of the decision-theoretic formulation which is closely related to the above is the possibility of establishing a quantitative evaluation of the performance of the optimum system. This feature is of obvious importance when the optimum system can be realized practically, but it may also be quite important even in those cases in which the optimum system cannot be built. In such situations the evaluation of the optimum system establishes a bound on the performance of any system and, as such, guards against the possibility of wasting design effort in trying to achieve the impossible.

Finally, the decision-theory approach is sufficiently general so that all suboptimum systems can be evaluated on the basis of the same criteria as would be used in determining the structure of optimum systems. Hence the performance of practical systems can be compared with that of optimum systems in order to assess the desirability of making improvements. This feature makes it possible to incorporate economic considerations into the design of engineering systems by relating the performance improvement obtained with a given expenditure to the ultimate improvement obtainable with unlimited funds.

It is of interest to note that the features of generality and unification cited above are usually associated with more abstract and less practical mathematical theories. Yet it is these very features which are largely responsible for the fact that decision theory has greater practical application to system design than do the less general concepts of classical statistics.

24-2. Basic Concepts of Decision Theory

Since decision theory is based upon statistical concepts, its application to system problems requires that these problems be phrased in statistical terms. Thus a fundamental assumption of this approach is that all the system variables are random in nature and can be described as sample functions of suitably defined random processes. Fortunately, it is usually possible to do this, even when these quantities are deterministic; so that the use of statistical concepts becomes quite natural.

The purpose of the present section is to describe some of these statistical concepts and to relate them to the problems arising in system engineering. In addition, this material will serve to indicate more clearly how the physical quantities in a system can have random properties which either arise naturally or are artificially bestowed.

Classification of Decision Processes. There are several ways in which decision processes might be classified, but the present discussion will be limited to only three. In general, the appropriate characteristic in each category must be specified in order to form a complete classification for any given process.

The first category pertains to the nature of the desired decision, and as noted previously, there are three situations which arise in practical system work:
1. A single selection is to be made from a *finite* number of alternatives.
2. A single value is to be chosen from a *continuous* range of possible values.
3. Both the above operations are to be performed.

Item 1 is usually referred to as *signal detection* in the engineering literature and as *hypothesis testing* in the mathematical literature. The term signal detection originated in connection with communication systems but is also appropriate in most other situations in which this type of decision arises. For example, in a system for automatically inspecting a manufactured product, the presence of an imperfection can be considered as a signal to be detected.

Item 2 above is often referred to in engineering terms as *signal extraction* and in mathematical terms as *parameter estimation*, although the latter terminology is also becoming quite common in connection with communication systems. The term signal extraction is appropriate in many other situations, also; for example, in seismographic systems the ground motion represents a signal whose waveform is to be extracted.

Item 3 includes those situations, very common in practice, in which both signal detection and signal extraction must be performed.

Signal detection is discussed quantitatively elsewhere in this handbook in the context of radar (Chapter 14), infrared (Chapter 15), and optical signals (Chapter 16); signal extraction is considered in Chapters 12 (communications), 22 (information theory), 31 (human engineering), and 37 (telemetry).

The second category into which decision processes must be classified pertains to the nature of the data which are observed. These data may consist of samples taken at discrete time instants or may be the result of a continuous observation over a specified time interval. Hence it is appropriate to use the terms *discrete data* and *continuous data* to describe these two cases.

The third category pertains to the length of the observation interval. This length may be fixed before the observation commences or it may be allowed to vary depending upon the outcome of the observation. In the first case a *fixed-interval* decision process results while the latter leads to a *sequential* decision process; for the latter, a special technique, dynamic programming (Chapter 27), has been developed.

The above methods of classification are discussed in greater detail in subsequent sections. A convenient summary of these methods is presented in Table 24-1.

Table 24-1. Classification of Decision Processes

Type of decision	Type of data	Type of observation
1. Detection (hypothesis testing) 2. Extraction (parameter estimation) 3. Combined	1. Discrete 2. Continuous	1. Fixed interval 2. Sequential

Types of Hypotheses. In considering the category of signal detection or hypothesis testing in more detail it is convenient to make a further classification based on the number of hypotheses (Ref. 7, Chap. 18). If a decision is to be made between only two hypotheses, then a *binary detection system* is being considered. A *multiple alternative detection* system results when the decision must select one hypothesis from a group of more than two.

Some authors prefer to restrict the term "detection" to what is here called "binary detection," and use "recognition" or "identification" for what is here called "multiple alternative detection."

A *simple hypothesis* is one which asserts the presence of a single nonrandom signal, while a *class hypothesis* asserts the presence of an unspecified member of some specified class of signals. Either type of hypothesis may arise with either binary detection or multiple alternative detection. A hypothesis which asserts the absence of any signal (noise only present) is frequently referred to as a *null hypothesis*, and this may also arise in either binary detection or multiple alternative detection.

The primary function of a detection system is to select *one* of the available hypotheses as being the one which represents the true state of nature. This decision is to be based on a specified sample of observed data which is processed in a suitable fashion to achieve the decision. Hence the principal problem in the design of decision systems is that of specifying the method in which observed data are processed to yield a decision among the various hypotheses.

Types of Estimation. Two types of estimation have been widely discussed in the literature (Ref. 7, Chap. 21). *Point estimation* assigns a definite value to a signal or some parameter of a signal while *interval estimation* asserts that such a value lies within a certain interval with a given probability. Contrary to the situation which usually prevails in statistical testing, point estimation is employed almost exclusively in signal-extraction systems. The reason for this is that the extraction system must ordinarily produce some physical quantity as its output, and this quantity must necessarily be assigned a definite value. Hence the entire discussion of signal extraction, or parameter estimation, will be limited exclusively to the case of point estimation.

When considering point estimation, the *dimensionality* refers to the number of parameters being estimated. Typical examples of one-dimensional estimates might be such things as signal amplitude, time of arrival, instantaneous frequency, length, temperature, or velocity. When two or more such quantities are to be estimated simultaneously (such as distance, velocity, and acceleration in the radar case), a multidimensional estimate results.

Another important example of multidimensional estimation arises when a signal waveform is to be extracted. This may be done either on a continuous time basis or at specified discrete instants of time. The many different situations which can arise in this type of estimation are discussed in Sec. 24-7, after suitable methods of representing signals have been considered.

Nature of the Observed Data. It is assumed throughout the present discussion that the observed data consist of a random disturbance (noise) to which a signal may be *added*. The additive assumption is typical of most physical situations, but there are some in which the signal and noise are combined by means of some other operation such as multiplication. However, the theoretical analysis of these other situations is extremely difficult in general and the available results are quite fragmentary.

A further assumption which will be imposed is that the signal and noise are functions of time. Again this is true of many physical situations, but there are some in which the independent variable may be some quantity such as distance, angle, or an abstract index. Fortunately, most of these cases can be handled by analogy to similar time-dependent situations with a trivial change in notation, so that the assumption does not limit the range of applicability of the present theory in any significant way.

In all cases discussed here it will be specified that the data are observed over a finite time interval T. The observation may be continuous throughout this interval or may be made at specified time instants within the interval. The first situation is designated as *continuous sampling* while the second is *discrete sampling*. From a mathematical standpoint it is more convenient to deal with the discrete-sampling case first and then treat the continuous case as the limit which is approached when the interval between samples becomes arbitrarily small. Fortunately, this limit exists in almost all cases which are representative of physical situations.

The signal itself may be described in general terms involving both random and deterministic parameters. One method of classifying signals which is commonly employed pertains to the receiver's state of knowledge concerning the time at which the signal occurs. If this time is completely known, the signal is said to be *coherent*; if it is not known, the signal is *incoherent*. Note, however, that even a coherent signal may have random parameters, such as amplitude.

A convenient and compact way of representing a discretely sampled signal is by means of a vector whose components are the sample values (Ref. 14, Chap. 2). In order to make this concept more precise, consider the continuous signal $s(t)$ shown in Fig. 24-1. Within the observation interval T there are n sample points so that the discretely sampled signal can be specified by a set of numbers $s_1, s_2, \ldots, s_n$ where $s_k = s(t_k)$. If the possible signal values at the sample points are taken as the coordinates of an n-dimensional vector space, then every signal can be interpreted as a vector (i.e., a point) in this signal space and can be represented by $\mathbf{S} = (s_1, s_2, \ldots, s_n)$. The case of no signal is simply that in which all the components of $\mathbf{S}$ are zero.

The noise $n(t)$ which is added to the signal can also be represented as a vector in the same way. It is convenient to consider the noise samples at exactly the same times as the signal, and this assumption does not restrict the generality in any way. Hence the noise will be represented by the vector $\mathbf{N} = (n_1, n_2, \ldots, n_n)$ in a noise space of the same dimensionality as the signal space. As a consequence, the observed data $v(t)$ may also be represented at the same times by the vector $\mathbf{V} = (v_1, v_2, \ldots, v_n)$, where $\mathbf{V} = \mathbf{S} + \mathbf{N}$.

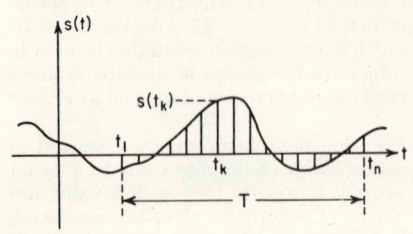

Fig. 24-1. A signal waveform showing the observation interval and the discrete samples.

Since the signal has random parameters, and the noise is random, the occurrence of particular component values in $\mathbf{S}$, $\mathbf{N}$, and $\mathbf{V}$ is governed by a set of multidimensional probability density functions. These density functions will be designated as $w(\mathbf{S})$, $f(\mathbf{N})$, and $p(\mathbf{V}|\mathbf{S})$, where the last is the conditional probability density of the observed data given the signal. Usually $f(\mathbf{N})$ is assumed to be known, and this is sufficient to determine $p(\mathbf{V}|\mathbf{S})$ because of the additive nature of the signal and noise. The signal probability density function $w(\mathbf{S})$ may not be known a priori, and this case raises some difficult questions which are best considered in Sec. 24-4 under Systems Which Minimize Risk.

The observation interval T will be assumed to be fixed throughout most of this chapter. A brief discussion of sequential decision processes, in which T is variable, however, is presented in Sec. 24-5. Likewise, most of the discussion will assume discrete sampling for which n, the number of samples in T, is finite.

Types of Decision Rules. The primary purpose of any decision system is to operate upon the observed data $\mathbf{V}$ and produce a decision of some sort. The mathematical representation of the physical system which accomplishes this result is the *decision rule* (Ref. 7, Chap. 18). It is convenient to represent the decision as a vector in a decision space. The dimensionality of this vector $\mathbf{Y}$ depends upon the type of decision to be made and ranges from one in the case of signal detection to n in case the signal waveform is to be estimated. In the case of parameter estimation the dimensionality of $\mathbf{Y}$ is equal to the number of parameters being estimated. The number of points and, hence, the number of possible vectors in the decision space also depends upon the nature of the decision. For example, in the case of signal detection with M alternatives (i.e., M hypotheses to be tested), the decision space contains only M points. In the case of parameter estimation on M parameters, the decision space is M-dimensional and continuous (i.e., contains an infinite number of points).

In view of the statistical nature of the observed data, it is convenient to represent the decision rule as a conditional probability $g(\mathbf{Y}|\mathbf{V})$. It is usually assumed that this probability is discrete and assumes values of 0 or 1. That is, for each observed $\mathbf{V}$ there is one and only one $\mathbf{Y}$ possible. Such a decision rule is said to be *nonrandomized*. The more general case of a *randomized* decision rule in which $g(\mathbf{Y}|\mathbf{V})$ is a continuous density function is not excluded in the present analysis, but the use of such a function usually offers no advantage.[8]

An important feature of the decision rules formulated here is that the decision $\mathbf{Y}$

depends only upon the observation **V** and not upon the particular signal **S**. That is,
$$g(\mathbf{Y}|\mathbf{V}) = g(\mathbf{Y}|\mathbf{V},\mathbf{S})$$
Of course, the form of the optimum decision rule depends upon the statistics of the signal even though **Y** is algebraically independent of **S**.

A nonrandomized decision rule can be interpreted as a transformation which maps each point in the observation space into a point in the decision space. Many **V**'s may map into each **Y**. For example, in the case of binary detection the observation space is divided into two regions, and all the points in each region map into a corresponding single point in the decision space.

It should be noted that no restrictions have been imposed on the decision rule so far. In order to provide a means for selecting suitable decision rules it is necessary to establish some criterion of excellence. One way of doing this, which will be followed here, is to assign a loss function $F(\mathbf{S},\mathbf{Y})$ to each combination of signal **S** and decision **Y**. This loss function will be small (or zero) for correct decision and, in the case of incorrect decision, will have a value which increases with the seriousness of the error. The mathematical expectation of the loss function will be a measure of the quality of the decision rule, and an optimum decision rule is that one which minimizes this expectation, as discussed in the following section.

It is now possible to state in a precise fashion the nature of the mathematical problem involved in applying statistical decision theory to system evaluation, optimization, and comparison. This may be done in three parts:

1. *System Evaluation.* Given the observed data **V** having the conditional probability density function $p(\mathbf{V}|\mathbf{S})$, the a priori signal probability density function $w(\mathbf{S})$, and the decision rule $g(\mathbf{Y}|\mathbf{V})$, find the mathematical expectation of an assigned loss function $F(\mathbf{S},\mathbf{Y})$.

2. *System Optimization.* Given the observed data **V** having the conditional probability density function $p(\mathbf{V}|\mathbf{S})$, and the a priori signal probability density function $w(\mathbf{S})$, find the decision rule $g(\mathbf{Y}|\mathbf{V})$ which *minimizes* the mathematical expectation of an assigned loss function $F(\mathbf{S},\mathbf{Y})$.

3. *System Comparison.* For an actual system with a specified decision rule, compare the expected value of $F(\mathbf{S},\mathbf{Y})$ as found in item 1, or some function of this quantity, with its minimum possible value for an optimum decision rule as found in item 2, or with the similar quantity for another actual system.

24-3. System Evaluation and Comparison

The most important concept in formulating a general approach to system evaluation is the establishment of a loss function. This loss function may or may not depend upon system operation, and both situations are considered. In the presence of random inputs the loss function is also random, so that system evaluation is carried out in terms of the expected value of the loss function. This mathematical expectation may be obtained either with or without prior knowledge of the signal statistics, and again both cases are considered.

For purposes of system comparison, the probabilities of error often convey a greater physical significance. The relationships among these probabilities and the loss functions are discussed briefly.

Generalized Evaluation Functions. It is convenient to present a general formulation of the system-evaluation problem first and then consider some more specific aspects of it. The first step is the selection of an appropriate function which assigns a weight to every combination of system input and subsequent decision. The nature of this *loss function* $F(\mathbf{S},\mathbf{Y})$ depends upon the nature of the input signal, the type of decision desired, and the operational requirements imposed by the system designer. To a certain degree, its selection is subjective and arbitrary, and this feature is a source of much concern to those first attempting to use statistical decision-theory concepts in system design. However, it is this very feature which, when skillfully exploited, provides the flexibility needed to match system design to operational requirements.

In general, the loss function has a value which increases as the actual decision departs farther, in some sense, from the desired decision for a given signal. There are an infinite number of ways in which this value might be assigned; some examples of reasonable assignments are presented in Secs. 24-6 and 24-7 for specific applications.

Because of the random disturbances added to the signal before observation, the resulting decision for any signal is also random. Hence the value of the loss function which is assigned to the result of any given observation is a random variable and, as such, is not particularly significant of average system performance. It is reasonable, therefore, to base system evaluation on the average, or mathematical expectation, of the loss function. This choice is also arbitrary, and it is certainly possible that other criteria for evaluating loss functions might be considered.

Two different forms of the average loss function may be considered, depending upon whether or not the a priori statistics of the signal are known. If this information is not available the conditional average loss may be represented by (Ref. 7, Chap. 18)

$$E[F(\mathbf{S},\mathbf{Y})|\mathbf{S}] = \int d\mathbf{V} \int p(\mathbf{V}|\mathbf{S})g(\mathbf{Y}|\mathbf{V})F(\mathbf{S},\mathbf{Y}) \, d\mathbf{Y} \tag{24-1}$$

in which the integrals are multidimensional and over the entire spaces occupied by $\mathbf{V}$ and $\mathbf{Y}$. When the a priori probability of the signal is known, the average loss becomes

$$E[F(\mathbf{S},\mathbf{Y})] = \int d\mathbf{S} \int d\mathbf{V} \int w(\mathbf{S})p(\mathbf{V}|\mathbf{S})g(\mathbf{Y}|\mathbf{V})F(\mathbf{S},\mathbf{Y}) \, d\mathbf{Y} \tag{24-2}$$

In the above generalized formulation it has not been specified whether or not the loss function $F(\mathbf{S},\mathbf{Y})$ depends upon the decision rule $g(\mathbf{Y}|\mathbf{V})$, and it is possible to consider loss functions of both types. If the loss function is independent of the decision rule, it is frequently referred to as a *cost function*, and the averages value becomes the *risk*. When the loss function is made a measure of *information loss*, however, it does depend upon the decision rule, and its average value is the usual *equivocation* of information theory [Eq. (22-22)]. Both these more specialized formulations are considered in the following paragraphs.

Risk Formulation. In order to identify the risk formulation, the loss function will be designated as $C(\mathbf{S},\mathbf{Y})$ and will be referred to as a *cost function*. Since the distinguishing characteristic of this function is that it does not depend upon the decision rule, it must be selected before system evaluation can proceed. By direct analogy to the general formulation above, it is possible to define a *conditional risk* $r(\mathbf{S},g)$ and an *average risk* $R(w,g)$ as

$$r(\mathbf{S},g) = \int d\mathbf{V} \int p(\mathbf{V}|\mathbf{S})g(\mathbf{Y}|\mathbf{V})C(\mathbf{S},\mathbf{Y}) \, d\mathbf{Y} \tag{24-3}$$

and
$$R(w,g) = \int d\mathbf{S} \int d\mathbf{V} \int w(\mathbf{S})p(\mathbf{V}|\mathbf{S})g(\mathbf{Y}|\mathbf{V})C(\mathbf{S},\mathbf{Y}) \, d\mathbf{Y} \tag{24-4}$$
$$= \int r(\mathbf{S},g)w(\mathbf{S}) \, d\mathbf{S} \tag{24-5}$$

The notation used above implies that the conditional risk depends upon the given signal and the decision rule, while the average risk depends upon the signal probability density function and the decision rule.

The risk formulation is undoubtedly the most satisfactory one from the standpoint of being able to carry out the desired evaluation under the most general conditions. In addition, it is this formulation which enables the system designer to utilize the arbitrariness in assigning cost functions to achieve desired characteristics of system operation. Hence it is this formulation which is assumed throughout the remaining discussion except where specifically noted otherwise.

Information-loss Formulation. When the information-loss formulation is employed, the loss function is defined by (Ref. 7, Chap. 22)

$$F(\mathbf{Y},\mathbf{S}) = -\log p_2(\mathbf{S}|\mathbf{Y}) \tag{24-6}$$

where $p_2(\mathbf{S}|\mathbf{Y})$ is the conditional probability density of the signal given the decision. This loss function is frequently referred to as the "uncertainty" or "surprisal" about the signal $\mathbf{S}$ when the decision $\mathbf{Y}$ is known. This type of loss function is motivated by the concepts of information theory and is attractive because it is specific and employs an intuitively appealing criterion. However, it is clearly *not* independent of

DECISION THEORY 24-11

the decision rule since the a posteriori probability $p_2(\mathbf{S}|\mathbf{Y})$ depends upon how the decision is made.

When the information loss is averaged, it becomes the equivocation. By analogy to the general formulation, the *conditional information loss* is given by

$$h(\mathbf{S},g) = -\int d\mathbf{V} \int p(\mathbf{V}|\mathbf{S})g(\mathbf{Y}|\mathbf{V}) \log p_2(\mathbf{S}|\mathbf{Y}) \, d\mathbf{Y} \tag{24-7}$$

and the *average information loss* by

$$\begin{aligned} H(w,g) &= -\int d\mathbf{S} \int d\mathbf{V} \int w(\mathbf{S})p(\mathbf{V}|\mathbf{S})g(\mathbf{Y}|\mathbf{V}) \log p_2(\mathbf{S}|\mathbf{Y}) \, d\mathbf{Y} \\ &= \int h(\mathbf{S},g)w(\mathbf{S}) \, d\mathbf{S} \end{aligned} \tag{24-8}$$

Although the information-loss criterion is quite attractive, it is very difficult to deal with in general cases. In fact, complete results are available for only a few specific cases, and for these cases the resluts are essentially equivalent to the risk formulation.

System Comparisons. The above formulations can be used in an obvious manner to compare the operation of two or more systems. For example, if the signal is specified, then two systems having decision rules of g_1 and g_2 can be compared by noting the relative magnitudes of $h(\mathbf{S},g_1)$ and $h(\mathbf{S},g_2)$ or $r(\mathbf{S},g_1)$ and $r(\mathbf{S},g_2)$. The system with the smaller values is the better one. In a similar manner, when the signal probability density function is known, the comparison may be made on the basis of $H(w,g_1)$ and $H(w,g_2)$ or $R(w,g_1)$ and $R(w,g_2)$.

It is often more revealing to compare systems on the basis of the probabilities of error associated with various decisions. For example, the conditional probability density that the system makes a decision $\mathbf{Y}$, when a given signal $\mathbf{S}$ is present, is given by

$$p_3(\mathbf{Y}|\mathbf{S}) = \int p(\mathbf{V}|\mathbf{S})g(\mathbf{Y}|\mathbf{V}) \, d\mathbf{V} \tag{24-9}$$

Integrating this density function over that portion of the decision space which should *not* be associated with the given signal $\mathbf{S}$ leads immediately to the total probability of error for this signal. The above conditional probability can also be related to the conditional risk by

$$r(\mathbf{S},g) = \int p_3(\mathbf{Y}|\mathbf{S})C(\mathbf{S}|\mathbf{Y}) \, d\mathbf{Y} \tag{24-10}$$

24-4. System Optimization

The evaluation functions discussed in the previous section may also be used to provide a quite general approach to the design of optimum systems. The basis for this approach is the assumption that minimizing the expected value of the loss function provides a suitable criterion of optimality. Because of the generality with which loss functions can be formulated, it is possible to make such a criterion fit a wide range of physical situations.

The actual system optimization is carried out by finding the decision rule which minimizes the expected value of the loss function for any given signal or class of signals. In the case of the risk formulation, this operation can usually be carried out in a straightforward manner. The situation is much more complicated, however, when the information-loss criterion is employed, because of the interdependence of the loss function and the decision rule.

Systems Which Minimize Risk. A fundamental problem which arises in connection with system optimization pertains to the available knowledge of the a priori signal statistics. The two extreme cases of complete knowledge and no knowledge will be considered in detail. There are, of course, a multitude of situations represented by an intermediate state of knowledge.

A system which minimizes average risk for a given a priori signal distribution and an assigned cost function is called a *Bayes system* (see Bayes's theorem, Sec. 38-3). The signal probability density function $w(\mathbf{S})$ is assumed to be completely known. The optimum decision rule g^* leads to a *Bayes risk* R^* such that

$$R^* = \min_g R(w,g) = R(w,g^*) \tag{24-11}$$

Examples indicating how this optimum decision rule can be determined in specific cases are presented in Sec. 24-6. These examples also illustrate how the physical structure of the optimum system is determined by the decision rule. It is sufficient here to note that for any a priori distribution $w(\mathbf{S})$ there exists a Bayes decision rule g^*, and that for any cost function there is some *least favorable* distribution $w_0(\mathbf{S})$ such that

$$\min_g R(w_0,g) = \max_w \min_g R(w,g) \tag{24-12}$$

This latter result is significant in estimating the importance of accurate a priori information.

When the signal probability density function is unknown, it is possible to establish a criterion of optimality in connection with the conditional risk $r(\mathbf{S},g)$. For every possible signal there is some decision rule which minimizes the conditional risk. There is some particular signal for which this minimum conditional risk assumes its *largest* value. The corresponding decision rule for this case is the *minimax decision rule* g_M^*, and the resulting system is a minimax system. This decision rule is such that

$$\max_{\mathbf{S}} r(\mathbf{S},g_M^*) = \max_{\mathbf{S}} \min_g r(\mathbf{S},g) \tag{24-13}$$

There will also be an average risk associated with minimax systems whenever the a priori signal statistics exist, even if they are unknown to the receiver. This average risk is simply

$$R_M^*(w_0,g_M^*) = \max_w \min_g R(w,g) = \min_g \max_w R(w,g) \tag{24-14}$$

Comparing this with the Bayes risk as given by (24-11) shows that the *minimax average risk* is the largest possible Bayes risk. Hence the minimax decision rule assumes the least favorable signal distribution $w_0(\mathbf{S})$ since there is no better a priori information available.

The above discussion emphasizes the importance of using whatever a priori information about the signal is available since this information will always reduce the average risk. Obtaining this information in practical situations is one of the major problems in the application of statistical decision theory. In some cases it may even be desirable to *assume* the a priori signal probabilities in a manner which is consistent with the physical limitations of the system. In other cases, no assumption is more reasonable than any other, and one is forced to use the minimax procedure.

Systems Which Minimize Information Loss. When the information-loss formulation is used, the criterion of optimality is the minimization of this quantity, and again the problem of a priori signal statistics arises. When these statistics are known, the Bayes decision rule g^* yields the *Bayes equivocation*

$$H^* = \min_g H(w,g) = H(w,g^*) \tag{24-15}$$

When they are not known, the minimax decision rule minimizes the maximum conditional loss such that

$$\max_{\mathbf{S}} h(\mathbf{S},g_M^*) = \max_{\mathbf{S}} \min_g h(\mathbf{S},g) \tag{24-16}$$

Whether or not the minimax average information loss can be related to the maximum Bayes equivocation, in a manner analogous to that indicated in (24-14) for the risk formulation, has not yet been established in the literature.

It also appears that there are no general procedures for carrying out the required minimization for the information-loss criterion. There are some special cases, however, for which it can be shown that the information-loss criterion is equivalent to a risk criterion with a particular assignment of the cost function. Hence the optimum system structure can be determined in these cases. These results also demonstrate

DECISION THEORY 24-13

that it is possible for a system to be optimum with respect to both criteria simultaneously, although this will not be the case in general.

24-5. Sequential Decision Processes

In the discussion so far it has been assumed that the observation interval T is fixed. This assumption simplifies the analysis, but in a practical situation it may lead to observation intervals that are longer than necessary, because the use of a fixed interval does not take full advantage of particularly favorable pieces of data that may occur during the interval. A sequential decision process consists of a sequence of decisions, one of which, at any time, may be the decision to continue observing, while any other decision terminates the test. Thus for particularly favorable samples of observed data the observation interval terminates early, while for unfavorable samples it will be longer than usual. Nevertheless, it usually turns out that the average length of observation interval required for a given assurance of correct decision is significantly shorter in a sequential decision process than in a fixed-interval process.

General Theory. It is convenient in discussing the general theory of sequential decision processes to assume discrete sampling of the observed data. After each sample the system attempts to make a decision accepting one of the M possible hypotheses. However, if the data are not sufficiently convincing to make this decision, another sample will be taken and the process continued. The test will terminate whenever one of the M hypotheses is actually accepted. The number of samples required to reach this decision is a random number n, and one of the objectives of an optimum sequential decision process is to reduce the average value of this number. Hence, if a risk formulation is used, it is necessary to include a cost which depends upon the value of n.

In general, the mathematical expectation of n will depend upon the particular signal actually present. This dependence can be indicated by designating the number of samples required to terminate the test, given that signal $\mathbf{S}$ is present, as $n(\mathbf{S})$. The average risk R can now be written as[2]

$$R = \sum_{i=1}^{M} \int w(\mathbf{S}) P_i(\mathbf{S}) C_i(\mathbf{S}) \, d\mathbf{S} + c \int w(\mathbf{S}) E[n(\mathbf{S})] \, d\mathbf{S} \qquad (24\text{-}17)$$

where $P_i(\mathbf{S})$ = probability of *rejecting* hypothesis H_i given that signal $\mathbf{S}$ is present
$C_i(\mathbf{S})$ = cost of rejecting H_i given it is true
c = cost of each sample
$E[n(\mathbf{S})]$ = expected value of the number of samples with respect to the ensemble of possible tests

In this expression the costs of correct decision have been taken to be zero.

An optimum decision process is taken to be one which minimizes the average risk R. The general comments of the preceding section pertaining to such minima also apply here. However, it is very difficult to determine the decision rules which lead to minimum average risk in the general case, and explicit results are available only for the binary detection problem in which there are only two possible decisions. In particular, it has been shown that the likelihood-ratio tests, to be discussed in the following section, can be made optimum for the case of binary sequential detectors.[11] Since it is more convenient to discuss the optimum nature of such tests within the framework of specific applications, no further discussion will be given here.

An important characteristic of optimum sequential tests is the *average sample number* (ASN), which is just $E[n(\mathbf{S})]$ in the above expression for average risk. In the binary case, the ASN can be explicitly related to the decision levels associated with the likelihood ratio and thus be used as a basis for comparing decision systems.

Truncated Sequential Tests. A question of fundamental importance in connection with sequential decision processes is whether the test will terminate. It can be shown that such tests will terminate with probability one when the observed samples are independent, and also for a large class of situations when the samples are not independent.[10] However, even with complete assurance that the test will eventually terminate, the standard sequential procedure may be unsatisfactory. This is because any individual test may last longer than can be tolerated, or because the ASN becomes too large if the actual signal level falls below that for which the test was designed. Hence it often becomes necessary to terminate the test at some arbitrary stage and make the best decision possible with the available data. Such tests are designated as *truncated sequential tests*.

The procedure followed in truncated sequential tests is to carry out the normal operation until either a decision is made or stage N of the test is reached. Hence this procedure is an attempt to reconcile the good features of both purely sequential tests and fixed-sample tests. Since the number of samples can never exceed N, and since the test is entirely standard when $n \leq N$, it follows that the ASN for a truncated test is always less than the ASN for a purely sequential test. However, the error probabilities, and hence the average risk, of truncated tests will be somewhat higher.

Comparison of Sequential and Fixed-sample Tests. In the case of binary sequential detection, it can be shown that optimum sequential methods lead to an ASN which is only about 20 to 30 per cent of the required sample size in a fixed-sample test for the same probabilities of error.[2] This represents an appreciable saving in observation time. The price that has to be paid for shortening the observation time is an increase in the variance of the sample size n. It can be shown that as the ASN is made smaller the variance of n increases, and that if the variance of n is made small the ASN approaches the required fixed-sample size.[10] In a typical case the dispersion in n may be about 50 per cent of the ASN when the ASN is 30 per cent of the fixed-sample size.

24-6. Application of Decision Theory to Signal Detection

The central problem of signal detection is that of specifying the operation to be performed upon the observed data in order to decide which one of a finite number of possible signals is most likely present. The simplest situation is that of binary detection, and this is the one which is used here for purposes of illustrating the general theory outlined in the previous sections. However, a brief discussion of multiple alternative detection is also included.

The binary-detection problem arises in a great many practical engineering systems. These include, to name just a few, systems involving radar, sonar, teletype transmission, pulse-code modulation, infrared detection, radio astronomy, and product inspection. Although the physical quantities of interest may be quite different in the various applications, the mathematical models of the systems are essentially identical.

General Theory of Binary Detection. Since the risk formulation is the only one which can be carried through in the majority of cases, it is desirable to consider this situation first. In this formulation, the average risk was defined to be

$$R(w,g) = \int d\mathbf{S} \int d\mathbf{V} \int w(\mathbf{S}) p(\mathbf{V}|\mathbf{S}) g(\mathbf{Y}|\mathbf{V}) C(\mathbf{S},\mathbf{Y}) \, d\mathbf{Y} \qquad (24\text{-}4)$$

where $\mathbf{S}$ and $\mathbf{V}$ are assumed to be n-dimensional vectors for which the appropriate probability density functions are known. Since binary detection is being considered here, $\mathbf{Y}$ is one-dimensional and has only two possible values, Y_0 being the decision that no signal is present and Y_1 the decision that some unspecified signal is present. Hence the decision divides the observation space into two regions v_0 and v_1 and uniquely assigns one decision to each region. This may be expressed mathematically in terms of the Dirac delta-function notation as

$$\begin{aligned} g(\mathbf{Y}|\mathbf{V}) &= \delta(\mathbf{Y} - Y_0) & \mathbf{V} \in v_0 \\ &= \delta(\mathbf{Y} - Y_1) & \mathbf{V} \in v_1 \end{aligned} \qquad (24\text{-}18)$$

The a priori signal statistics contain both a discrete component and a continuous component, since the case of no signal is assumed to occur with finite probability. Hence it is convenient to represent the signal probability density function as

$$w(\mathbf{S}) = q\delta(\mathbf{S} - 0) + p w_1(\mathbf{S}) \qquad (24\text{-}19)$$

where q is the discrete probability of no signal, p is the discrete probability of signal present, and $w_1(\mathbf{S})$ is the conditional probability density function of the signal given that signal is present. If the alternative is not the null hypothesis, but another class of signals (such as might arise in binary phase-reversal modulation, for example), the δ function in (24-19) also becomes a continuous density function.

The cost function may now be assigned four discrete values corresponding to the four possible combinations of signal presence and final decision. The four values will be designated as

$C(\mathbf{S} = 0, Y_0) = C_q$, correct decision of no signal present (quiet)
$C(\mathbf{S} = 0, Y_1) = C_f$, incorrect decision of signal present (false alarm)
$C(\mathbf{S} \neq 0, Y_0) = C_m$, incorrect decision of no signal present (miss)
$C(\mathbf{S} \neq 0, Y_1) = C_d$, correct decision of signal present (detection)

The costs associated with correct decisions C_q and C_d will generally be assigned small values, possibly zero, but, for convenience, never negative. The costs associated with incorrect decisions C_f and C_m will be assigned large positive values.

Using the above definition in (24-4) leads immediately to an average risk which may be written as (Ref. 7, Chap. 19)

$$R(w,g) = qC_f + pC_d + \int_{\mathcal{V}_0} [p(C_m - C_d)p_1(\mathbf{V}) - q(C_f - C_q)p(\mathbf{V}|0)]\, d\mathbf{V} \quad (24\text{-}20)$$

where the integration is over that portion of the observation space which leads to a decision of no signal, and

$$p_1(\mathbf{V}) \equiv \int p(\mathbf{V}|\mathbf{S})w_1(\mathbf{S})\, d\mathbf{S}$$

It is also possible to express the average risk in terms of the conditional error probabilities. There are two such probabilities in the binary case. These are the *probability of false alarm* given by

$$P_f = \int_{\mathcal{V}_1} p(\mathbf{V}|0)\, d\mathbf{V} = 1 - \int_{\mathcal{V}_0} p(\mathbf{V}|0)\, d\mathbf{V} \quad (24\text{-}21)$$

and the *probability of a miss* given by

$$P_m = \int_{\mathcal{V}_0} p_1(\mathbf{V})\, d\mathbf{V} \quad (24\text{-}22)$$

In terms of these quantities, the average risk becomes

$$R(w,g) = qC_q + pC_d + pP_m(C_m - C_d) + qP_f(C_f - C_q) \quad (24\text{-}23)$$

When the average risk is expressed in the form given by (24-20), the problem of finding the optimum detection system becomes one of finding the boundary of the region $\mathcal{V}_0$ which will minimize the integral. This condition obviously occurs if $\mathcal{V}_0$ contains all points for which the integrand is negative and none for which it is positive. A convenient way of expressing this result is in terms of a *generalized likelihood ratio* defined by

$$L(\mathbf{V}) \equiv \frac{pp_1(\mathbf{V})}{qp(\mathbf{V}|0)} \quad (24\text{-}24)$$

and a *threshold* of

$$K \equiv \frac{C_f - C_q}{C_m - C_d} > 0 \quad (24\text{-}25)$$

The Bayes decision rule may then be stated as

Decide Y_1 when $L(\mathbf{V}) \geq K$
Decide Y_0 when $L(\mathbf{V}) < K$

and the boundary between regions v_0 and v_1 in the observation space is defined by

$$L(\mathbf{V}) = K$$

The corresponding Bayes risk becomes

$$R^* = R_0 + p(C_m - C_d)\left(\frac{K}{\mu} P_f + P_m\right) \qquad (24\text{-}26)$$

where $\qquad R_0 \equiv qC_q + pC_d$

is the average risk for zero error probabilities (noiseless case) and

$$\mu \equiv \frac{p}{q}$$

is the a priori likelihood ratio.

The operations which must be performed by the receiver in accomplishing optimum signal detection may now be stated concisely. The receiver observes the data $v(t)$ for a time interval T and samples them n times to obtain the coordinates of the vector **V**. The generalized likelihood ratio $L(\mathbf{V})$ is then computed from **V** and compared with the threshold K, which is determined by the assigned cost values. If the likelihood ratio is smaller than the threshold, the decision of "no signal" is made; otherwise, the decision is made that some unspecified signal is present. In order for the receiver to compute the likelihood ratio, it must have available the a priori probabilities p, q, and $w_1(\mathbf{S})$ and the conditional probability $p(\mathbf{V}|\mathbf{S})$. The actual structure of the receiver depends upon the nature of these probability functions; some specific examples are considered in the following sections.

Before leaving the general theory of binary detectors, it is desirable to compare the results quoted above with some of the more classical formulations of the same problem which, as noted previously, are simply special cases of statistical decision theory. Two such concepts will be noted here.[9]

Neyman-Pearson Test. A detector which is optimum with respect to the Neyman-Pearson criterion is one which minimizes the conditional probability of a miss subject to a fixed conditional probability of false alarm. It is clear from (24-23) that this is equivalent to minimizing the average risk with a threshold chosen so that the region v_0 specified by this threshold leads to the desired probability of false alarm as given by (24-21). Hence the Neyman-Pearson detector is simply a Bayes detector with a particular set of cost values determined by the probability of false alarm.

Ideal Observer. A detector which is optimum with respect to the ideal-observer criterion is one which minimizes the total probability of error as defined by

$$P_e \equiv qP_f + pP_m \qquad (24\text{-}27)$$

It has been shown that this criterion also leads to a Bayes detector in which the threshold is unity.[9] This condition is easily achieved by selecting cost values such that $C_f - C_q = C_m - C_d$.

Threshold Detection in Gaussian Noise. In order to consider some specific examples of optimum binary-detection systems it is necessary to specify the probability density functions needed to compute the generalized likelihood ratio. For this purpose, it will be assumed that the noise which is superimposed upon the signal is Gaussian and has zero mean. The Gaussian assumption is realistic in most physical situations and is mathematically convenient. In fact, it is worth noting that complete analytical results for optimum detectors have been obtained *only* for the Gaussian process or for processes derived from the Gaussian (such as the Rayleigh process). This limitation is not peculiar to the decision-theoretic formulation but is a common one in classical statistics as well.

In writing the joint probability density function for the noise, it is convenient to set its rms value equal to unity. This can be done without loss of generality, since it is the signal-to-noise ratio which is important in determining detectability and not the

DECISION THEORY

absolute values. Thus let the noise probability density function be[3]

$$f(\mathbf{N}) = (2\pi)^{-n/2}|\mathbf{\Lambda}|^{-1/2} \exp(-\tfrac{1}{2}\tilde{\mathbf{N}}\mathbf{\Lambda}^{-1}\mathbf{N}) \qquad (24\text{-}28)$$

where $\mathbf{N} = (n_1, n_2, \ldots, n_n)$ is the noise vector, $\tilde{\mathbf{N}}$ is its transpose, $\mathbf{\Lambda}$ is the covariance matrix whose elements λ_{ij} are given by

$$\lambda_{ij} \equiv E[n_i n_j] = \lambda_{ji} \qquad i,j = 1, 2, \ldots, n$$

$\mathbf{\Lambda}^{-1}$ is its inverse, and $|\mathbf{\Lambda}|$ is the determinant of $\mathbf{\Lambda}$. Since the rms value of the noise is unity, it follows that over the sampling interval $(t_0, t_0 + T)$

$$\frac{1}{T}\int_{t_0}^{t_0+T} n^2(t)\, dt = 1 \qquad (24\text{-}29)$$

If the noise is stationary, it also follows that $\lambda_{ii} = 1$, $(i = 1, 2, \ldots, n)$, but the general formulation does not require stationarity. It will also be noted that there is no assumption that noise samples are statistically independent.

Since signal and noise have been assumed additive, it follows that

$$\mathbf{N} = \mathbf{V} - \mathbf{S}$$

and hence the conditional probability of the observed data given the signal is

$$p(\mathbf{V}|\mathbf{S}) = f(\mathbf{V} - \mathbf{S}) \qquad (24\text{-}30)$$

and, in addition,

$$p_1(\mathbf{V}) = \int f(\mathbf{V} - \mathbf{S}) w_1(\mathbf{S})\, d\mathbf{S} = E_s[f(\mathbf{V} - \mathbf{S})] \qquad (24\text{-}31)$$

where $E_s[\ ]$ implies an expectation with respect to the ensemble of possible signals.

In writing an explicit form for the likelihood ratio $L(\mathbf{V})$ it is convenient to consider its logarithm, because of the exponential form of the density functions, and to normalize the signal amplitude with respect to the noise. Hence one can write

$$\log L(\mathbf{V}) = \log \mu + \log E_s[\exp(a_0 \tilde{\mathbf{V}} \mathbf{\Lambda}^{-1}\mathbf{s} - \tfrac{1}{2}a_0^2 \tilde{\mathbf{s}}\mathbf{\Lambda}^{-1}\mathbf{s})] \qquad (24\text{-}32)$$

where
$$a_0 \equiv \frac{\text{rms value of signal over the sampling interval}}{\text{rms value of noise}}$$

$$\mathbf{S} \equiv a_0 \mathbf{s}$$

It will be noted that a_0 is the input signal-to-noise ratio and may be a random variable. In addition,

$$\frac{1}{T}\int_{t_0}^{t_0+T} s^2(t)\, dt = a_0^2 \qquad (24\text{-}33)$$

since the noise rms value is unity.

The situation which is of greatest interest in the design of optimum detectors is that in which the signal is small compared with the noise, since this is the condition for which optimality yields the biggest dividends. Such situations are designated as *threshold detection* and are characterized by a_0 values less than unity. In this case, the exponential in (24-32) may be expanded as a power series in a_0, the mathematical expectation over the signal ensemble taken, and the logarithm again expanded as a power series. The result is

$$\log L(\mathbf{V}) = \log \mu + \bar{a}_0 \bar{Q}_v + \tfrac{1}{2}\overline{a_0^2}\left(\overline{Q_v^2} - \frac{\bar{a}_0^2}{\overline{a_0^2}}\bar{Q}_v^2 - \bar{Q}_s\right) + O(\overline{a_0^3}) \qquad (24\text{-}34)$$

where $Q_v = \tilde{\mathbf{V}}\mathbf{\Lambda}^{-1}\mathbf{s}$
$Q_s = \tilde{\mathbf{s}}\mathbf{\Lambda}^{-1}\mathbf{s}$

and the bars over the quantities indicate the mathematical expectation over the signal ensemble. The assumption has been made here that the signal amplitude is statistically independent of the other random parameters of the signal. In general, the terms $0(\overline{a_0^3})$ can be neglected in determining the structure of the optimum detector, although they may be significant in evaluating its performance.

In order to determine the structure of optimum binary detectors more specifically, it is now necessary to consider the nature of the signal in greater detail. When this is done, it is found that the detector structure depends significantly upon whether the signal is *coherent* or *incoherent*. In particular, it can be shown that coherent signals lead to *linear* systems while incoherent signals lead to *quadratic* systems.[8] These two situations are discussed under the following two headings.

Coherent Binary Detection. In the case of coherent detection, the time at which the signal occurs is known to the observer. Thus, if the signal is described by $s(t - t_0)$, the time t_0 is not a random variable and hence

$$\bar{Q}_v = \tilde{\mathbf{V}}\boldsymbol{\Lambda}^{-1}\bar{\mathbf{s}} \qquad (24\text{-}35)$$

where $\bar{\mathbf{s}}$ is the mathematical expectation of the signal with respect to parameters other than t_0. In addition, it is convenient to evaluate those terms which are second-order in $\mathbf{V}$ under the assumption of noise only. Under these conditions, (24-34) may be written as

$$\log L(\mathbf{V}) \simeq B_c + \bar{a}_0 \tilde{\mathbf{V}}\boldsymbol{\Lambda}^{-1}\bar{\mathbf{s}} \qquad (24\text{-}36)$$

where $B_c \equiv \log \mu - \tfrac{1}{2}\overline{a_0^2}\tilde{\mathbf{s}}\boldsymbol{\Lambda}^{-1}\bar{\mathbf{s}}$ is a *bias* term.

In order to interpret the operations specified by (24-36), it is convenient to define a *weighted-average signal* as

$$\bar{\mathbf{S}}_0 \equiv \bar{a}_0 \boldsymbol{\Lambda}^{-1}\bar{\mathbf{s}} \qquad (24\text{-}37)$$

and note that this signal is known to the observer. The second term of (24-36) may now be written as

$$\bar{a}_0 \tilde{\mathbf{V}}\boldsymbol{\Lambda}^{-1}\bar{\mathbf{s}} = \tilde{\mathbf{V}}\bar{\mathbf{S}}_0 = \sum_{j=1}^{n} v(t_j)\overline{s_0(t_j)} \qquad (24\text{-}38)$$

where $v(t_j)$, $(j = 1, \ldots, n)$ are the samples of observed data at the sampling times t_j, and $\overline{s_0(t_j)}$, $(j = 1, \ldots, n)$ are corresponding samples of the weighted-average signal. The operation defined by (24-38) is simply a discrete cross correlation of the observed data and the weighted-average signal, and this is a *linear* operation on the data.

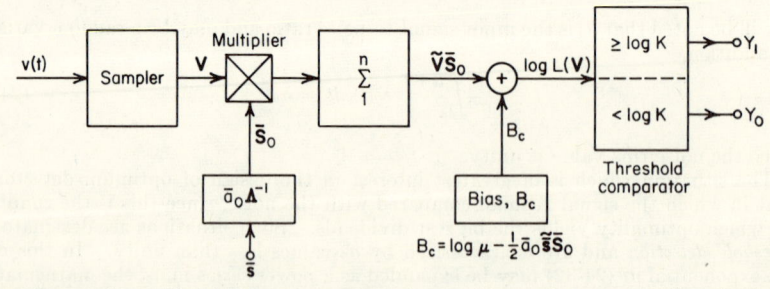

FIG. 24-2. Optimum threshold structure for coherent binary detection in Gaussian noise with discrete sampling.

A block diagram showing one possible structure of the optimum detector for coherent binary signals is shown in Fig. 24-2. This structure is not unique, however, as there are several other possible interpretations of (24-36). One of these is precisely the matched linear filter of conventional communication theory (see Sec. 14-4, and Ref. 7, Chap. 20). It will be noted that the actual decision operation is performed by a threshold device which compares the computed value of $\log L(\mathbf{V})$ with $\log K$ and indicates an output of either Y_0 (no signal) or Y_1 (signal plus noise).

DECISION THEORY

Although all the discussion so far has assumed discrete sampling, it is also possible to formulate the problem on the basis of continuous sampling. The procedure for doing this consists of allowing the number of discrete samples in a fixed time interval $(0,T)$ to approach infinity; when this is done the summations approach integrals in the limit. The system structure is then revealed by a result which is analogous to (24-36) (Ref. 7, Chap. 20).

$$\log L[v(t)] \simeq B_T + \int_0^T v(t)\bar{S}_0(t)\, dt \qquad (24\text{-}39)$$

where $B_T \equiv \log \mu - \frac{1}{2}\bar{a}_0 \int_0^T \bar{s}(t - t_0)\bar{S}_0(t)\, dt$ is the bias, and $\bar{S}_0(t)$ is the mathematical expectation (over the signal ensemble) of a weighted signal defined by the integral equation

$$\int_0^T \phi_n(t,\lambda) S_0(\lambda)\, d\lambda = a_0 s(t - t_0)$$

in which $\phi_n(t,\lambda) = E[n(t)n(\lambda)]$ is the autocorrelation function of the noise.

It is clear from (24-39) that the optimum detector operation again consists of cross-correlating the observed data with a weighted-average signal which is known to the observer. A possible structure for this operation is similar to that shown in Fig. 24-2,

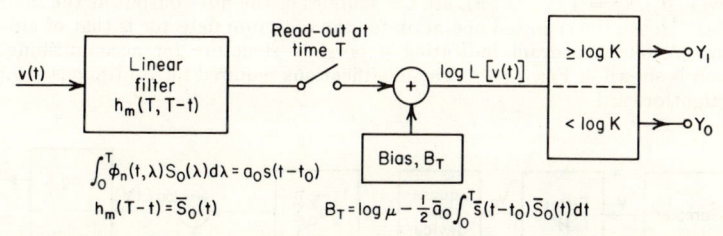

FIG. 24-3. Optimum threshold structure for coherent binary detection in Gaussian noise with continuous sampling.

with the sampler removed and the summation replaced by a true finite-time integrator. An alternative structure which is useful when the noise is stationary is shown in Fig. 24-3. This is simply an invariant matched linear filter with a response at time T to an impulse at time t of

$$h_m(T - t) = \bar{S}_0(t) \qquad 0 \leq t \leq T \qquad (24\text{-}40)$$

The output of this filter at time T is combined with the bias B_T and compared with the threshold.

A significant point in connection with both the discrete and continuous sampling systems is that the operation performed on the observed data is linear. This is a direct consequence of the coherent nature of the signal and is the distinguishing characteristic of optimum systems for this type of signal.

System evaluation in connection with coherent detection is discussed below under Evaluation and Comparison of Binary Detection Systems.

Incoherent Binary Detection. In most practical applications, the exact time at which the signal occurs is not known to the observer. Thus, if the signal is described by $s(t - t_0)$, the time t_0 is a random variable to which a probability density function can be assigned. This density function may or may not be known, but it is frequently possible to assume that t_0 is uniformly distributed over some long interval so that

$$E_{t_0}[\mathbf{s}] = 0$$

and hence $\bar{Q}_v$ vanishes. Under these conditions, (24-36) may be written in the discrete sampling case as (Ref. 7, Chap. 20)

$$\log L(\mathbf{V}) \simeq B_i + \tfrac{1}{2}\overline{a_0{}^2}\tilde{\mathbf{V}}\mathbf{\Lambda}^{-1}\overline{\mathbf{s}\tilde{\mathbf{s}}}\mathbf{\Lambda}^{-1}\mathbf{V} \tag{24-41}$$

where the bias B_i is given by

$$B_i \equiv \log \mu - \tfrac{1}{2}\overline{a_0{}^2}\overline{\tilde{\mathbf{s}}\mathbf{\Lambda}^{-1}\mathbf{s}}$$

and the second term on the right side of (24-41) is simply $\overline{Q_v{}^2}$.

In order to interpret the operation defined by (24-41) it is convenient to define a linear operator (filter) $\mathbf{H}$ such that

$$\tilde{\mathbf{H}}\mathbf{H} = \mathbf{\Lambda}^{-1}\overline{\mathbf{s}\tilde{\mathbf{s}}}\mathbf{\Lambda}^{-1} \tag{24-42}$$

then
$$\mathbf{V}_F = \mathbf{H}\mathbf{V} \tag{24-43}$$

is the result of filtering the observed data and (24-41) becomes

$$\log L(\mathbf{V}) \simeq B_i + \tfrac{1}{2}\overline{a_0{}^2}\tilde{\mathbf{V}}_F\mathbf{V}_F \tag{24-44}$$

where
$$\tilde{\mathbf{V}}_F\mathbf{V}_F = \sum_{j=1}^{n}[v_F(t_j)]^2$$

in which $v_F(t_j)$, $(j = 1, \ldots, n)$, are the samples of the filter output at the sampling instants. Hence the required operation for the optimum detector is that of autocorrelation. A block diagram indicating a possible structure for accomplishing this operation is shown in Fig. 24-4. The modifications required for continuous sampling are straightforward.

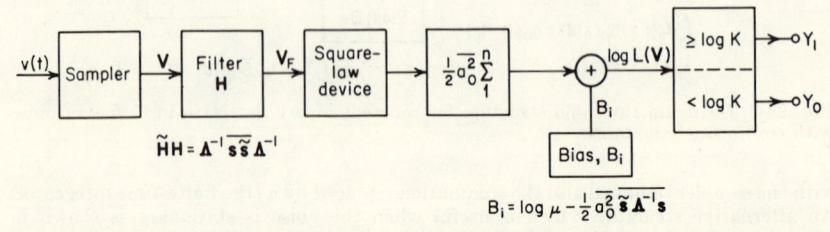

Fig. 24-4. Optimum threshold structure for incoherent binary detection in Gaussian noise with discrete sampling.

The most significant point in connection with the optimum detector for the incoherent case is that it is nonlinear. The square-law characteristic is required in order to correlate the observed data with themselves since a coherent noise-free reference signal is not available. This requirement has important implications in the evaluation of system performance, as will be seen below.

Detection of Stochastic Signals. In the cases of both coherent and incoherent detection, the general form of the signal has been known even though it contained random parameters. Thus it could be considered deterministic in the sense that all future values could be predicted from the complete past history. Situations arise in practice, however, in which the signals are random functions of time and the future values cannot be predicted from the past history. Examples of this sort occur in scatter-path communication systems, antijam communication systems, seismographic detection systems, radio spectroscopy, and radio astronomy, to name just a few. Although the statistical structure of such signals may not be known exactly, it is often reasonable to assume that they come from Gaussian processes or processes derived from the Gaussian. This assumption is the only one for which specific results are available, and it will be followed here.

DECISION THEORY

Since the signal is now random, the previous expression for the likelihood ratio (24-32) no longer applies. If the signal is assumed to be Gaussian with zero mean and have a covariance matrix of $a_0^2 \mathbf{\Lambda}_s$, the likelihood ratio may be written as (Ref. 7, Chap. 20)

$$\log L(\mathbf{V}) = \log \mu - \tfrac{1}{2} \log |\mathbf{I} + a_0^2 \mathbf{\Lambda}_s \mathbf{\Lambda}^{-1}| + \tfrac{1}{2} \tilde{\mathbf{V}}[\mathbf{\Lambda}^{-1} - (\mathbf{\Lambda} + a_0^2 \mathbf{\Lambda}_s)^{-1}]\mathbf{V} \quad (24\text{-}45)$$

where $\mathbf{I}$ is the identity matrix. The first two terms on the right side of (24-45) may be interpreted as a bias, and the last term may be written in terms of a linear filter defined by

$$\tilde{\mathbf{G}} \mathbf{G} \equiv \mathbf{\Lambda}^{-1} - (\mathbf{\Lambda} + a_0^2 \mathbf{\Lambda}_s)^{-1}$$

and letting

$$\mathbf{V}_G = \mathbf{G}\mathbf{V}$$

Hence the equation revealing the structure becomes

$$\log L(\mathbf{V}) = B_s + \tfrac{1}{2} \tilde{\mathbf{V}}_G \mathbf{V}_G \quad (24\text{-}46)$$

where
$$B_s \equiv \log \mu - \tfrac{1}{2} \log |\mathbf{I} + a_0^2 \mathbf{\Lambda}_s \mathbf{\Lambda}^{-1}|$$

and
$$\tilde{\mathbf{V}}_G \mathbf{V}_G = \sum_{j=1}^{n} [v_G(t_j)]^2$$

The block diagram showing this structure is given in Fig. 24-5.

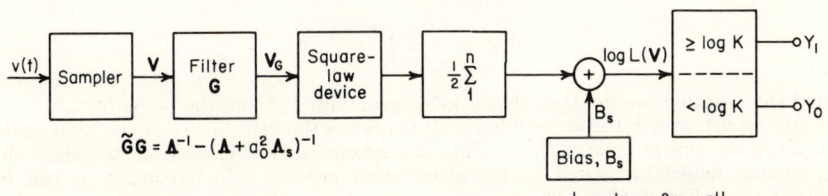

FIG. 24-5. Optimum threshold structure for detection of Gaussian stochastic signals in Gaussian noise with discrete sampling.

The optimum detector for stochastic signals is seen to bear a strong resemblance to that for incoherent detection of deterministic signals. This result is to be expected since stochastic signals are necessarily incoherent. It should also be noted that the optimum receiver for this case is essentially an energy detector. Finally, it should be pointed out that the structure of the optimum detector does *not* depend upon the assumption that the signal-to-noise ratio is small, since (24-45) is exact for all values of a_0. This is in contrast to the two previous results for which this assumption was required.

Evaluation and Comparison of Binary Detection Systems. As noted previously, the two principal ways in which detection systems can be compared are in terms of the average risk or in terms of the error probabilities. Since the average risk per se is not a particularly meaningful number, it is more significant to express the same information in terms of the *minimum detectable signal*. This quantity is defined to be the input signal-to-noise ratio $(a_0)_{\min}$ for which the average risk is some specified fraction of the maximum average risk. Obviously, the system having the smallest value of $(a_0)_{\min}$ is considered to be the best. It is often desirable to relate $(a_0)_{\min}$ to the observation interval T or, equivalently, to the number of observed samples n.

In the case of binary systems the maximum average risk depends only upon the a priori probabilities of signal occurrence and the costs assigned to incorrect decisions. Thus,

$$R_{\max} = pC_m + qC_f \quad (24\text{-}47)$$

and $(a_0)_{\min}$ is that signal-to-noise ratio for which the optimum detector has a Bayes risk of

$$R^* = \eta R_{\max} \quad 0 < \eta < 1 \quad (24\text{-}48)$$

where η is any specified fraction. It is apparent that meaningful comparisons among systems can be made only when the systems in question have the same a priori probabilities and costs and for the same value of η.

The two error probabilities for binary systems are the conditional probability of false alarm P_f and the conditional probability of miss P_m. These quantities were defined explicitly in (24-21) and (24-22), and they also depend upon the signal-to-noise ratio and the observation time. If these probabilities for the optimum detection system are designated as $P_f{}^*$ and $P_m{}^*$, the corresponding average risk is

$$R^* = R_0 + p(C_m - C_d)\left(\frac{K}{\mu}P_f{}^* + P_m{}^*\right) \qquad (24\text{-}49)$$

where
$$R_0 \equiv qC_q + pC_d$$

Coherent Detection. When coherent detection in Gaussian noise is considered for the optimum threshold system, the error probabilities can be obtained from (24-21) and (24-22) by putting in the appropriate density functions as obtained from (24-30) and (24-31). The results are (Ref. 7, Chap. 20)

$$\left\{\begin{matrix}P_f{}^*\\P_m{}^*\end{matrix}\right\} \simeq \frac{1}{2}\left\{1 - \Theta\left[\frac{\bar{a}_0 Q_{\bar{s}}^{1/2}}{2\sqrt{2}} \pm \frac{\log(K/\mu)}{\sqrt{2}\,\bar{a}_0 Q_{\bar{s}}^{1/2}}\right]\right\} \qquad (24\text{-}50)$$

where $Q_{\bar{s}} = \bar{\mathbf{s}}\mathbf{\Lambda}^{-1}\bar{\mathbf{s}}$ and the error function $\Theta(z)$ is defined by

$$\Theta(z) = \frac{2}{\sqrt{\pi}}\int_0^z e^{-u^2}\,du$$

and the notation implies that the $+$ in $\pm$ goes with $P_f{}^*$ and the $-$ with $P_m{}^*$.

An important point in connection with (24-50) is the dependence of the error probabilities on the product $\bar{a}_0 Q_{\bar{s}}^{1/2}$. This dependence makes it possible to relate the minimum detectable signal to the observation interval. In particular, it can be shown that (Ref. 7, Chap. 20)

$$\overline{(a_0{}^2)}_{\min} \sim \frac{1}{n} \qquad \text{discrete sampling}$$
$$\sim \frac{1}{T} \qquad \text{continuous sampling} \qquad (24\text{-}51)$$

where n is the number of samples in the observation interval T. For the special case in which the noise background is white, the square of this product term becomes

$$\bar{a}_0{}^2 Q_{\bar{s}} = \frac{2E_{\bar{s}}}{N_0} \qquad (24\text{-}52)$$

where $E_{\bar{s}}$ is the average signal energy in the observation interval and N_0 is the power spectral density of the white noise. This ratio is one which arises very commonly in connection with system evaluation.

Incoherent Detection. In the case of incoherent detection in Gaussian noise, the error probabilities for the optimum threshold system are given by (Ref. 7, Chap. 20)

$$\left\{\begin{matrix}P_f{}^*\\P_m{}^*\end{matrix}\right\} \simeq \frac{1}{2}\left\{1 - \Theta\left[\frac{\overline{a_0{}^2}Q_H^{1/2}}{4} \pm \frac{\log(K/\mu)}{\overline{a_0{}^2}Q_H^{1/2}}\right]\right\} \qquad (24\text{-}53)$$

where $Q_H \equiv \overline{\bar{\mathbf{s}}\mathbf{H}\bar{\mathbf{s}}}$ and $\mathbf{H}$ is the linear operator defined in (24-42).

The minimum detectable signal can be related to the observation interval by

$$\overline{(a_0{}^2)}_{\min} \sim \frac{1}{\sqrt{n}} \qquad \text{discrete sampling}$$
$$\sim \frac{1}{\sqrt{T}} \qquad \text{continuous sampling} \qquad (24\text{-}54)$$

This behavior is characteristic of most binary systems for incoherent detection and is a direct consequence of the square-law operation required in such systems.

Stochastic Signal Detection. When the signals to be detected are random in nature, the error probabilities may still be obtained in the threshold case and expressed as (Ref. 7, Chap. 20)

$$\begin{Bmatrix} P_f^* \\ P_m^* \end{Bmatrix} \simeq \frac{1}{2} \left\{ 1 - \Theta \left[\frac{a_0^2 Q_G^{1/2}}{4} \pm \frac{\log (K/\mu)}{a_0^2 Q_G^{1/2}} \right] \right\} \qquad (24\text{-}55)$$

where $Q_G \equiv \text{tr}[(\mathbf{\Lambda}_s \mathbf{\Lambda}^{-1})^2]$, in which $\text{tr}[\ ]$ implies the trace of the matrix in the argument, i.e., the sum of the elements on the principal diagonal. As in the case of incoherent detection, the minimum detectable signal is given by (24-54).

A comparison of Eqs. (24-50), (24-53), and (24-55) reveals that they all have the general form

$$\begin{Bmatrix} P_f^* \\ P_m^* \end{Bmatrix} \simeq \frac{1}{2} \left\{ 1 - \Theta \left[\frac{D}{4} \pm \frac{\log (K/\mu)}{D} \right] \right\} \qquad (24\text{-}56)$$

where the parameter D is defined for the various cases in the following manner:

1. Coherent detection; $D = \sqrt{2}\, \bar{a}_0 Q_s^{1/2}$
2. Incoherent detection; $D = \overline{a_0^2} Q_H^{1/2}$
3. Stochastic signal detection; $D = a_0^2 Q_G^{1/2}$

Hence the error probabilities for all three cases can be represented graphically by the same family of universal curves, as shown in Fig. 24-6.

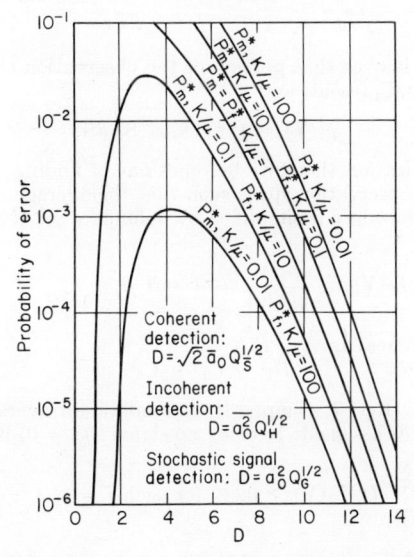

Fig. 24-6. Error probabilities for optimum binary detection in Gaussian noise.

Multiple Alternative Detection. It is of interest to note briefly how the concepts associated with binary detection can be extended to multiple alternative detection. The basic assumption made here is that only one of a finite number of possible signals can appear during any one observation period, and a decision is to be made as to which one is most likely to be present. Hence, if there are M possible signals, plus the null signal, the decision rule must divide the observation space into $M + 1$ regions $v_0, v_1, \ldots, v_M$. By analogy to (24-18), this generalized decision rule may be expressed as

$$g(\mathbf{Y}|\mathbf{V}) = \delta(\mathbf{Y} - Y_i) \qquad \mathbf{V} \in v_i \qquad i = 0, 1, \ldots, M \qquad (24\text{-}57)$$

In a manner similar to (24-19), the a priori signal statistics can be generalized to

$$w(\mathbf{S}) = q\delta(\mathbf{S} - 0) + \sum_{k=1}^{M} p_k w_k(\mathbf{S}) \tag{24-58}$$

where
$$\sum_{k=1}^{M} p_k = 1 - q$$

and $w_k(\mathbf{S})$ is the conditional probability density function of the kth signal given that this signal is present while p_k is the discrete probability that the kth signal is present.

The cost function is again assigned discrete values, but in this case there are $(M + 1)^2$ such values required. These cost values may be designated as $C_j{}^k$, which is the cost of deciding signal j is present when signal k actually occurs. Since the cases $j \neq k$ represent the cost of errors, it is necessary that

$$C_j{}^k \Big|_{j \neq k} > C_j{}^j$$

In this notation, $k, j = 0$ corresponds to the case of no signal.

On the basis of the above definitions, the average risk may be written as (Ref. 7, Chap. 23)

$$R(w,g) = \sum_{j=0}^{M} \int_{\mathbb{U}_j} [qC_j{}^0 p(\mathbf{V}|0) + \sum_{k=1}^{M} p_k C_j{}^k p_k(\mathbf{V})] \, d\mathbf{V} \tag{24-59}$$

where the integration is over that portion of the observation space which leads to a decision of the jth signal present, and

$$p_k(\mathbf{V}) = \int p(\mathbf{V}|\mathbf{S}) w_k(\mathbf{S}) \, d\mathbf{S}$$

The problem of optimization therefore becomes one of finding the boundaries of the various regions in the observation space such that the average risk is minimized.

The minimization procedure is simplified by defining a set of generalized likelihood ratios as

$$L_k(\mathbf{V}) \equiv \frac{p_k p_k(\mathbf{V})}{q p(\mathbf{V}|0)} \qquad k = 1, 2, \ldots, M \tag{24-60}$$

and a set of cost differences as

$$D_j{}^k = C_j{}^k - C_0{}^k \tag{24-61}$$

It can then be shown (Ref. 7, Chap. 23) that minimum average risk occurs if the following decision rule is assigned: Decide "no signal" $(j = 0)$ if

$$D_j{}^0 + \sum_{k=1}^{M} D_j{}^k L_k(\mathbf{V}) \geq 0 \qquad \text{for each } j = 1, 2, \ldots, M$$

Decide signal j present if

$$D_j{}^0 + \sum_{k=1}^{M} D_j{}^k L_k(\mathbf{V}) \leq 0$$

and

$$D_j{}^0 + \sum_{k=1}^{M} D_j{}^k L_k(\mathbf{V}) \leq D_i{}^0 + \sum_{k=1}^{M} D_i{}^k L_k(\mathbf{V})$$

for all $i \neq j$ $(i = 1, 2, \ldots, M)$

DECISION THEORY

System evaluation can again be carried out by determining a set of conditional error probabilities. These may be defined by

$$P_j{}^k = \int_{\mathcal{V}_j} p_k(\mathbf{V}) \, d\mathbf{V} \qquad j \neq k \qquad \begin{matrix} j = 0, 1, \ldots, M \\ k = 1, 2, \ldots, M \end{matrix} \qquad (24\text{-}62)$$

which is the probability of deciding j when k is present, and by

$$P_k{}^0 = \int_{\mathcal{V}_k} p(\mathbf{V}|0) \, d\mathbf{V} \qquad k = 1, 2, \ldots, M \qquad (24\text{-}63)$$

which is the probability of deciding k when no signal is present.

For the case of Gaussian noise it is possible to determine the structure of optimum detectors for specific situations by following the same general procedures as were employed in the binary case (Ref. 7, Chap. 23). The calculations are much more involved, however, unless very special cost assignments are made.

Binary Sequential Detection. It was noted in Sec. 24-5 that the likelihood-ratio test could also be used in connection with sequential tests and would lead to optimum tests in the binary case. The ratio $L(\mathbf{V}_m)$ is computed after the mth sample is taken, compared with *two* threshold values, and the following decision made:

Decide Y_1 when $L(\mathbf{V}_m) \geq A$
Decide Y_0 when $L(\mathbf{V}_m) \leq B$
Sample again when $B < L(\mathbf{V}_m) < A$

This process is continued until a decision of either Y_1 or Y_0 is made; the number of samples required to reach this decision is designated n.

An important parameter of binary sequential tests is the *operating characteristic function* (OCF). This is defined to be the probability of accepting Y_0 and will be designated as $P(Y_0)$. It can be related to the threshold values by[2]

$$P(Y_0) = \frac{A^h - 1}{A^h - B^h} \qquad (24\text{-}64)$$

where h satisfies the relation

$$\int \left[\frac{p(\mathbf{V}_m|\mathbf{S}_1)}{p(\mathbf{V}_m|0)}\right]^h p(\mathbf{V}_m|\mathbf{S}) \, d\mathbf{V}_m = 1 \qquad (24\text{-}65)$$

The average length of the sequential test (ASN) can in turn be obtained from the OCF by

$$\text{ASN} = \frac{P(Y_0) \log B + [1 - P(Y_0)] \log A}{\overline{z(s)}} \qquad (24\text{-}66)$$

where

$$\overline{z(s)} = E\left[\log \frac{p(v_i|\mathbf{S}_1)}{p(v_i|0)}\right] \qquad v_i = \text{any sample}$$

This average sample number can be compared with the length of fixed-sample tests having the same error probabilities. For example, it can be shown that, for coherent binary detection in which $P_f = P_m = 10^{-5}$, the ASN is only 30 per cent of the sample length required for a fixed-sample detector having the same error probabilities. As noted in Sec. 24-5, however, the penalty for this reduction in sample length is an increase in the variance of n. For the example quoted the dispersion in n is about 50 per cent of the ASN.

24-7. Application of Decision Theory to Signal Extraction

The central problem of signal extraction is that of specifying the operation to be performed upon the observed data in order to estimate one or more parameters of the signal. In case the signal waveform is to be extracted, the parameters to be estimated are simply the amplitudes of the signal at the discrete sampling times. In other situations the parameter of interest may be the frequency or phase of a sinusoidal waveform, the arrival time of a pulse, or the average power of a stochastic signal.

The problem of signal extraction arises in many practical engineering systems. For example, all communication systems employing analog modulation methods require this operation. Many observation systems such as radar, sonar, and radio astronomy are required to produce an estimation of one or more signal parameters in addition to detecting the presence of a signal.

Classical Estimation Theory. The mathematical literature dealing with the theory of estimation is quite extensive; only a few major results are noted here.

In general the estimation is to be based on an observed sample $\mathbf{V}$ and it is assumed that the observer has available the conditional probability density function $p(\mathbf{V}|\mathbf{S})$. The observer may also have available the a priori signal probability density function $w(\mathbf{S})$, in which case an *unconditional estimate* can be made. If the a priori probability is not available, then a *conditional estimate* is made.

In terms of the previous notation, the estimate is designated as $\mathbf{Y}$ and the estimator (that is, the operation to be performed on $\mathbf{V}$) is $g(\mathbf{Y}|\mathbf{V})$. The dimensionality of $\mathbf{Y}$ is the same as that of $\mathbf{V}$ in case signal waveform is to be extracted, and is equal to the number of parameters being estimated otherwise. The estimate $\mathbf{Y}$ is, of course, a random variable (or a set of random variables when $\mathbf{Y}$ is multidimensional) so that the goodness of the estimator must be judged on the statistical properties of $\mathbf{Y}$. Of particular interest are the mean and variance of the estimates.

A *conditional unbiased estimator* is one for which the expected value of the estimate is equal to the true value of the parameter. Similarly, an *unconditional unbiased estimator* is one for which the expected value of the estimate is equal to the mean value of the parameter. A *minimum-variance estimator* is one for which the mean-square deviation of the estimate from its expected value is a minimum over some class of estimators. A condition of optimality is that an estimator be minimum-variance and unbiased. A lower bound on the variance of such conditional estimators is given by the Cramér-Rao inequality which, for the case of a single parameter θ, has the form[3]

$$E[(\hat{\theta} - \theta)^2] \geq \frac{1}{\int \left[\dfrac{\partial \log p(\mathbf{V}|\theta)}{\partial \theta}\right]^2 p(\mathbf{V}|\theta) \, d\mathbf{V}} \qquad (24\text{-}67)$$

in which $\hat{\theta}$ is the estimate, θ is the true value, and $p(\mathbf{V}|\theta)$ is simply $p[\mathbf{V}|\mathbf{S}(\theta)]$. In the case of unconditional estimations, this bound becomes

$$E[(\hat{\theta} - \bar{\theta})^2] \geq \frac{1}{\int d\theta \int \left[\dfrac{\partial \log [w(\theta)p(\mathbf{V}|\theta)]}{\partial \theta}\right]^2 w(\theta) p(\mathbf{V}|\theta) \, d\mathbf{V}} \qquad (24\text{-}68)$$

where $w(\theta)$ is the a priori probability of θ and $\bar{\theta}$ is the expected value of θ.

A method of great importance in classical estimation theory is that of maximum likelihood.[3] The reason for this importance is that such estimators are minimum-variance and unbiased under rather general conditions. Conditional maximum-likelihood estimates are those which satisfy the equation

$$\left. \frac{\partial \log p(\mathbf{V}|\theta)}{\partial \theta} \right|_{\theta = \hat{\theta}} = 0 \qquad (24\text{-}69)$$

while unconditional maximum-likelihood estimates satisfy the equation

$$\left. \frac{\partial \log [w(\theta)p(\mathbf{V}|\theta)]}{\partial \theta} \right|_{\theta = \hat{\theta}} = 0 \qquad (24\text{-}70)$$

It can be shown (Ref. 7, Chap. 21) that as the observation time becomes long these estimates are normally distributed with variances given by (24-67) and (24-68) written as equalities.

DECISION THEORY

The results given above can be extended to the case of two or more parameters without difficulty. However, they become much more involved, and the variances must be expressed in terms of a covariance matrix.

Maximum-likelihood estimators are most useful when the parameters to be estimated appear linearly in the signal waveform (e.g., signal amplitude). In most other situations the likelihood equations (24-69) and (24-70) are transcendental, and explicit results are not possible in general except for certain limiting cases such as weak-signal, large-sample situations.

Decision-theory Formulation. The decision-theoretic formulation of signal extraction is again based on the average risk. In order to include the case of more than one parameter, it is convenient to consider the parameter vector $\theta = (\theta_1, \theta_2, \ldots, \theta_M)$ having an a priori probability density function $w(\theta)$. In the case of signal-waveform estimation the parameters are just the signal sample amplitudes so that $\theta = \mathbf{S}$ and $M = n$, the number of samples. By analogy to (24-4) the average risk can be written as

$$R(w,g) = \int d\theta \int d\mathbf{V} \int w(\theta) p(\mathbf{V}|\theta) g(\mathbf{Y}|\mathbf{V}) C(\theta,\mathbf{Y}) \, d\mathbf{Y} \tag{24-71}$$

When a nonrandomized decision rule is employed, it uniquely assigns one point in the decision space to each point in the observation space. Thus it may be represented by

$$g(\mathbf{Y}|\mathbf{V}) = \delta[\mathbf{Y} - Y_\theta(\mathbf{V})] \tag{24-72}$$

where $Y_\theta(\mathbf{V})$ is the functional operation performed on the data $\mathbf{V}$ to obtain the parameter estimate $\hat{\theta}$. Hence the average risk becomes

$$R(w,g) = \int d\theta \int w(\theta) p(\mathbf{V}|\theta) C(\theta,\hat{\theta}) \, d\mathbf{V} \tag{24-73}$$

The cost function $C(\theta,\hat{\theta})$ may be selected arbitrarily, but it is reasonable to select a function such that the cost of an error increases with the magnitude of the error. The quadratic cost function satisfies this condition and has the added feature of being convenient to handle mathematically. Thus, let

$$C(\theta,\hat{\theta}) = C_0(\tilde{\theta} - \tilde{\hat{\theta}})(\theta - \hat{\theta}) = C_0 \sum_{j=1}^{M} (\theta_j - \hat{\theta}_j)^2 \tag{24-74}$$

and the average risk becomes

$$R(w,g) = C_0 \int d\theta \int (\tilde{\theta} - \tilde{\hat{\theta}})(\theta - \hat{\theta}) w(\theta) p(\mathbf{V}|\theta) \, d\mathbf{V} \tag{24-75}$$

The criterion of optimality is again chosen to be that of minimum average risk, and the estimator which accomplishes this can be obtained by differentiating (24-75) with respect to $\hat{\theta}$ and equating the derivative to zero. The result is (Ref. 7, Chap. 21)

$$\hat{\theta} = Y_\theta(\mathbf{V}) = \frac{\int \theta w(\theta) p(\mathbf{V}|\theta) \, d\theta}{\int w(\theta) p(\mathbf{V}|\theta) \, d\theta} \tag{24-76}$$

which is just the conditional expectation of θ given $\mathbf{V}$. This is the *Bayes estimator* for a quadratic cost function. It can be shown that such an estimator has the smallest average variance among all unbiased estimators. Hence the variance of this estimate is given by (24-68) with an equality sign.

Before considering the above estimation procedure in more detail, it is worth commenting on other possible cost functions. One of considerable importance is the *constant cost function* defined by

$$C(\theta,\hat{\theta}) = C_0 \sum_{j=1}^{M} A_j - \delta(\theta_j - \hat{\theta}_j) \tag{24-77}$$

Its importance lies in the fact that it leads to unconditional maximum-likelihood estimates. Hence the maximum-likelihood method of classical estimation theory is seen to be a special case of the more general decision-theory approach.

The feature of mathematical convenience was listed above as one of the reasons for using the quadratic cost function. However, it can be shown that many other cost functions are also optimum with respect to the Bayes estimator for a quadratic cost function. The conditions that must be satisfied for this conclusion to hold, are (Ref. 7, Chap. 21):

1. Each component of the cost function must be symmetrical about the point of zero error.
2. Each component of the cost function must be a nondecreasing function of the absolute error.
3. The conditional density function $p(\theta|\mathbf{V})$ must be unimodal and symmetric about its mode.

These conditions are satisfied by almost all reasonable cost functions combined with a symmetrical a priori probability and Gaussian-noise statistics. Hence the use of a quadratic cost function is far less restrictive than might otherwise appear.

Coherent Estimation of Signal Amplitude. As an illustration of the decision-theoretic approach to parameter estimation, two simple examples will be presented. The first of these is the problem of estimating the amplitude factor a_0 of a deterministic signal. It is assumed that the amplitude of the observed signal is one realization of an ensemble of possible amplitudes to which a known Gaussian probability density function is assigned. Thus the a priori probability of the parameter $\theta = a_0$ is

$$w(a_0) = \frac{1}{\sqrt{2\pi\sigma^2}} e^{\frac{-(a_0 - \bar{a}_0)^2}{2\sigma^2}} \tag{24-78}$$

where $\bar{a}_0$ is the mean value of the ensemble and σ^2 is its variance. As described above under Threshold Detection in Gaussian Noise, the signal is related to a normalized signal by

$$\mathbf{S} = a_0 \mathbf{s}$$

and the observed data are

$$\mathbf{V} = \mathbf{S} + \mathbf{N}$$

where $\mathbf{N}$ is normalized Gaussian noise with a covariance matrix $\mathbf{\Lambda}$.

Under the above assumptions, the Bayes estimator for a quadratic cost function, as given by (24-76), becomes (Ref. 7, Chap. 21)

$$Y_{a0}(\mathbf{V}) = \frac{Q_v + \bar{a}_0/\sigma^2}{Q_s + 1/\sigma^2} = \hat{a}_0 \tag{24-79}$$

where
$$Q_v = \tilde{\mathbf{V}}\mathbf{\Lambda}^{-1}\mathbf{s}$$
$$Q_s = \tilde{\mathbf{s}}\mathbf{\Lambda}^{-1}\mathbf{s}$$

just as they were in (24-34). It will be noted from the form of Q_v that this estimator is a linear function of the received data $\mathbf{V}$.

The structure of the Bayes estimator is very similar to that required for optimum coherent binary detection as shown in Fig. 24-2. Using the same interpretation of Q_v and Q_s as discussed under Coherent Binary Detection, it is straightforward to show that the Bayes estimator for this case takes on the form shown in Fig. 24-7 (Ref. 7, Chap. 21).

The average risk of the Bayes estimator can be obtained from (24-75) and becomes

$$R^* = \frac{C_0 \sigma^2}{\sigma^2 Q_s + 1} \tag{24-80}$$

This is seen to vanish as $\sigma^2 \to 0$. Perhaps of greater interest, however, are the mean and variance of the estimate. These are easily shown to be (Ref. 7, Chap. 21)

$$E[a_0] = \bar{a}_0 \tag{24-81}$$

and
$$E[(\hat{a}_0 - \bar{a}_0)^2] = \frac{\sigma^4 Q_s}{\sigma^2 Q_s + 1} \tag{24-82}$$

It may be observed that the Bayes estimator is unbiased in this case.

Waveform Estimation in Gaussian Noise. The problem of estimating the waveform of a stochastic signal in the presence of Gaussian noise will be used as a final illustrative example. In this case the parameters being estimated are the sample values of the signal so that $\theta = \mathbf{S}$. In addition, it will be assumed that the signal is also Gaussian with a covariance matrix of $\mathbf{\Lambda}_s$ and zero mean. The noise covariance

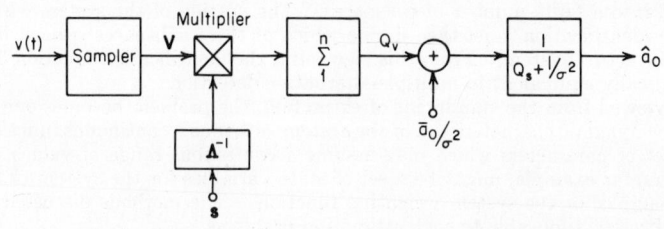

FIG. 24-7. Structure of the Bayes estimator for coherent estimation of signal amplitude in Gaussian noise with discrete sampling.

matrix is again just $\mathbf{\Lambda}$. The Bayes estimator for a quadratic cost function, as given by (24-76), becomes

$$Y_S(\mathbf{V}) = \frac{\int \mathbf{S} \exp\left[-\tfrac{1}{2}\tilde{\mathbf{S}}(\mathbf{\Lambda}_s^{-1} + \mathbf{\Lambda}^{-1})\mathbf{S} + \tilde{\mathbf{V}}\mathbf{\Lambda}^{-1}\mathbf{S}\right] d\mathbf{S}}{\int \exp\left[-\tfrac{1}{2}\tilde{\mathbf{S}}(\mathbf{\Lambda}_s^{-1} + \mathbf{\Lambda}^{-1})\mathbf{S} + \tilde{\mathbf{V}}\mathbf{\Lambda}^{-1}\mathbf{S}\right] d\mathbf{S}} \qquad (24\text{-}83)$$

Evaluation of (24-83) leads immediately to

$$Y_S(\mathbf{V}) = (\mathbf{\Lambda}_s^{-1} + \mathbf{\Lambda}^{-1})^{-1}\mathbf{\Lambda}^{-1}\mathbf{V} \qquad (24\text{-}84)$$

It can also be shown that this estimator is an unconditional maximum-likelihood estimator.

The estimation error is the difference between the true signal and the estimate and is also Gaussian with zero mean. It has a covariance matrix of

$$\mathbf{\Lambda}_e = E[(\hat{\mathbf{S}} - \mathbf{S})(\tilde{\hat{\mathbf{S}}} - \tilde{\mathbf{S}})] = \mathbf{\Lambda}_s[\mathbf{I} - (\mathbf{\Lambda}_s^{-1} + \mathbf{\Lambda}^{-1})^{-1}\mathbf{\Lambda}^{-1}] \qquad (24\text{-}85)$$

The average risk associated with this estimate can be evaluated from (24-75) as

$$R^* = C_0 \text{tr}(\mathbf{\Lambda}_e) \qquad (24\text{-}86)$$

It will be noted that the operation defined by (24-84) is linear. In the case of continuous sampling, this operation can be performed by a time-varying linear realizable filter. For the special case in which the signal and noise are stationary, this linear filter is invariant and closely related to the classical Wiener filter.

24-8. Further Applications of Decision Theory

In the preceding sections, the applications of decision theory which have been discussed have all been associated with communication systems or systems which perform analogous functions. The purpose of this section is to mention briefly some other types of systems in which these concepts are useful for purposes of evaluating their performance or optimizing system design, and to indicate how decision-theory concepts can be applied. In most cases, significant general results are not available, since application of the theory has not yet progressed that far. In this sense, the present section is somewhat speculative and indicates what appear to be profitable applications of decision theory.

Adaptive Systems. So far as the present discussion is concerned, an adaptive system may be defined as one which is capable of modifying its operating characteristics in response to changes in environment or input signal in such a way as to improve some performance specification. A more complete definition and discussion of the adaptive concept will be found in Chapter 30. This concept has found extensive application in both control systems and communication systems.

At least two of the functions which any adaptive system must perform can be discussed in terms of decision theory. One of these functions is that of identifying the dynamic characteristics of the system by making measurements in the presence of normal operational inputs and random disturbances. This function can be treated either as a detection problem or as an extraction problem. When viewed from the standpoint of detection, the function space representing all possible system functions is divided into a finite number of subspaces. The portion of the system which performs the identification must then decide which of these subspaces is most likely to contain the actual system. From this viewpoint, the identification function becomes mathematically equivalent to multiple-alternative detection.

When viewed from the standpoint of extraction, the problem becomes one of estimating the dynamic characteristics of the system, either as a continuous function or as a finite set of parameters which may assume a continuous range of values. These parameters, for example, might be a set of state variables for the system or a set of discrete samples of the system weighting function. The methods discussed in Sec. 24-7 are directly applicable to such estimation problems.

The second function which is of interest here has been referred to in the literature as the "decision function." This portion of the adaptive system must operate upon the result of the identification function in order to determine how the overall system should be modified. In order to make this determination, it is necessary to compare the observed system with some preassigned standard. In almost all adaptive systems which have been considered to date, this comparison is one-dimensional; that is, it is based on a single number.[12] Such a simple approach destroys much of the information contained in the identification procedure and is hardly adequate for complex systems. A systematic method for extending this operation to multidimensional comparisons is provided by decision-theory concepts. For example, the desired system characteristic can be associated with the quantity S, previously identified as the signal, and the observed system characteristic can be associated with V. A decision rule $g(Y|V)$ relates a set of possible system modifications Y to each V, and a cost function $C(S,Y)$ evaluates the desirability of these modifications. With this formulation of the problem, it becomes possible to optimize the "decision function" of the adaptive system in a straightforward manner. Such an approach does not appear to have been used thus far but seems to provide a much-needed extension to existing procedures.

Since the need for adaptive systems arises only when the basic system changes its characteristics with time, it is important that the identification and decision functions noted above be carried out in the minimum possible time. It seems likely, therefore, that the sequential methods discussed in Sec. 24-5 should find important applications here. However, even in the case of a fixed observation time, the ability of the decision-theoretic approach to deal with nonstationary processes and time-varying systems should provide a distinct advantage.

Learning Systems. For purposes of the present discussion, a learning system may be described as one that is capable of improving its operating characteristics by remembering and evaluating its response to previous inputs. Thus a learning system is adaptive, but the converse is not necessarily true (Sec. 30-6). In addition to the applications noted above in connection with adaptive systems, the introduction of the learning function adds two more possible applications of decision theory.

It has been noted previously that lack of complete knowledge concerning the a priori statistics of the signal is a major limitation in many practical applications of decision theory. In the case of an adaptive control system, the pertinent a priori statistics relate to the way in which the basic system changes its performance characteristics, while in a communication system these statistics may describe the manner in which the transmission channel changes. Any process by which improved a priori statistics are obtained and used can be interpreted as a learning operation.

This type of learning can be envisioned as being accomplished in a subsystem which may employ decision-theoretic concepts to make an estimate of the desired a priori statistics. This may be done on a detection basis by dividing the space of possible statistics into a finite number of regions, or on an extraction basis by estimating a finite number of parameters which describe the a priori statistics.

The second way in which learning can be incorporated in an adaptive system is to provide a means for changing the decision rules by which the system decides upon proper modifications. These rules will, of course, change as a result of learning more about the a priori statistics, but further changes can be made by evaluating the effectiveness of previous operation. This is a particularly significant form of learning because it provides a means for optimizing the decision rules in situations where analytical determination of these rules is too difficult or requires excessive computation. Thus it is not necessary to construct a system with the optimum decision rules built in, but instead it allows the system to hunt for its own rules.

Recognition Systems. As used here, the term recognition system implies one which examines a given temporal or spatial input and classifies it into one of a finite number of categories (see Sec. 30-7). Hence the decision operation is required in systems of this type. There are some important differences, however, between this type of operation and that for which the basic theory was outlined.

Consider, for example, the problem of recognizing two-dimensional patterns. It is usually required that such patterns be properly classified on the basis of their shape but not dependent on such things as size or orientation. Thus the classes of input signals which should lead to the same final decision may bear little relationship to one another. This feature leads to extremely complex decision rules and has led some workers to make classifications based on a few salient features of the pattern only.[1] The penalty that must be paid for this type of simplification is an increased probability of making an erroneous decision.

The difficulty, of course, is a conceptual one rather than a fundamental one. The fact that the human observer can make such decisions with a very small probability of error demonstrates that sufficient information is available. Continuing this analogy further implies that a decision system which exhibits comparable capability must necessarily include a learning function which continually refines the decision rules. Experimental systems of this sort have been demonstrated,[6] and a mathematical proof that the learning operation converges is available.[5] However, the problem of establishing conditions for optimum learning systems is largely untouched and represents a fruitful avenue for future work.

REFERENCES

1. Braverman, D.: A Decision Theoretic Approach to Machine Learning and Pattern Recognition, *WESCON Conv. Record*, 38/3, 1961.
2. Bussgang, J., and D. Middleton: Optimum Sequential Detection of Signals in Noise, *IRE Trans. Inform. Theory*, IT-1, p. 5, 1955.
3. Cramér, H.: *Mathematical Methods of Statistics*, Princeton University Press, Princeton, N.J., 1946.
4. Fano, R. M.: *Transmission of Information*, John Wiley & Sons, Inc., New York, 1961.
5. Joseph, R. D.: Contributions to Perceptron Theory, Ph.D. Thesis, Cornell University, 1961.
6. Minsky, M.: Steps toward Artificial Intelligence, *Proc. IRE*, vol. 49, no. 1, p. 8, January, 1961.
7. Middleton, D.: *An Introduction to Statistical Communication Theory*, McGraw-Hill Book Company, New York, 1960.
8. Middleton, D., and D. Van Meter: Detection and Extraction of Signals in Noise from the Point of View of Statistical Decision Theory, II, *J. Soc. Ind. Appl. Math.*, vol. 4, no. 2, June, 1956.
9. Peterson, W. W., T. G. Birdsall, and W. C. Fox: The Theory of Signal Detectability, *IRE Trans. Inform. Theory*, PGIT-4, p. 171, 1954.
10. Wald, A.: *Sequential Analysis*, John Wiley & Sons, Inc., New York, 1947.
11. Wald, A., and J. Wolfowitz: Optimum Character of the Sequential Probability Ratio Test, *Ann. Math. Statist.*, vol. 19, p. 326, 1948.
12. Weaver, C. S.: Adaptive Communication Filtering, *IRE Trans. Inform. Theory*, IT-8, no. 5, September, 1962.
13. Weiss, L.: *Statistical Decision Theory*, McGraw-Hill Book Company, New York, 1961.
14. Woodward, P. M.: *Probability and Information Theory*, McGraw-Hill Book Company, New York, 1953.

Chapter 25

THE SIMPLEX METHOD†

GEORGE B. DANTZIG, *Chairman, Operations Research Center, University of California, Berkeley, California.*

CONTENTS

25-1. Origin and General Description........................... 25-1
25-2. Test for Optimal Feasible Solution......................... 25-2
25-3. Improving a Nonoptimal Basic Feasible Solution........... 25-4
25-4. Iterative Procedure....................................... 25-6
25-5. Phase I: Finding an Initial Basic Feasible Solution........... 25-7
25-6. Illustration... 25-8

25-1. Origin and General Description

After World War II, the United States Air Force Headquarters became interested in scientific programming techniques in order to coordinate more efficiently the energies of whole nations in the event of a total war. Intensive work began in June, 1947, in a group called Project SCOOP (Scientific Computation of Optimum Programs). Early participants included Marshall Wood and the author and soon thereafter John Norton and Murray Geisler. The Air Force research team approach to the Air Force planning problems was greatly influenced by the interindustry model of the United States economy developed by Prof. Wassily Leontief and by W. D. Evans, J. Cornfield, and M. Hoffenberg of the Bureau of Labor Statistics. However, the dynamic character of Air Force planning required a more general formulation and resulted in the *linear-programming model*. Since 1950, linear programming has found extensive application in industry (see Chapter 26).

This mathematical model consists typically of a number of linear equations each of which represents a material balance or accounting of some "item" (such as a pilot or some commodity). The unknowns are the quantities of various possible "activities" or functions to be performed. The set of activity levels is called the *program*. As a rule the values are restricted to be positive or zero. Any set of values that (1) satisfy the equations, and (2) are nonnegative is called a *feasible program*. Because the number of unknowns typically exceeds the number of equations, there is usually more than one admissible solution to the system, and each one corresponds to a different possible

† For a more extensive treatment of the material found here, see *Linear Programming and Extensions* (Princeton University Press, Princeton, N.J., 1963) by the author. See also the following chapter in this handbook.

feasible program. In order to select from among the alternative possibilities, a linear (objective) form or "cost form" is used to measure the "cost" of a particular program. The linear-programming problem mathematically thus becomes one of finding a solution to the system of linear constraints which minimizes the total cost. The simplex method for solving such systems was developed by the end of the summer of 1947 with helpful suggestions from T. C. Koopmans and Leonid Hurwicz of the Cowles Foundation.

The *central mathematical problem*, stated a little more precisely, is to find the values of $x_1, x_2, \ldots, x_n$ and minimum z satisfying the simultaneous system of equations and inequalities ($x_j \geq 0$).

Standard Form of the Problem

$$\begin{aligned}
a_{11}x_1 + a_{12}x_2 + \cdots + a_{1n}x_n &= b_1 \quad (x_j \geq 0) \\
a_{21}x_1 + a_{22}x_2 + \cdots + a_{2n}x_n &= b_2 \\
&\vdots \\
a_{m1}x_1 + a_{m2}x_2 + \cdots + a_{mn}x_n &= b_m \\
c_1 x_1 + c_2 x_2 + \cdots + c_n x_n &= z \text{ (min)}
\end{aligned} \quad (25\text{-}1)$$

The simplex method makes use of two important characteristics of solutions:

1. *If a feasible solution exists (not necessarily minimal) then one exists with at least $n - m$ variables equal to zero.*
2. *If, in addition, a lower bound for the objective form z exists, then a minimal feasible solution exists with at least $n - m$ variables equal to zero.*

Suppose a *feasible solution* is at hand with $n - m$ variables equal to zero, that is, say one that satisfies (25-1) with z not necessarily minimum. The chief feature of the simplex method is that it calls for an elimination of each of the remaining variables from all but one of the equations by choosing a *pivot* term in a manner quite analogous to ordinary elimination for solving m equations in m unknowns. If the elimination is possible, these remaining variables are called "basic." The elimination permits one to examine the special solution and to apply a simple test to decide whether or not it is optimal. If it fails the test, a corrective procedure is applied which substitutes a certain one of the nonbasic variables for one of the basic variables. This can be done with relatively little computational effort, and the test is then applied again. The procedure is iterated a finite number of times until a terminal situation is obtained.

The net effect of an elimination with respect to a set of basic variables is to replace the original system by a second *equivalent* system of equations called the *canonical system*. We shall also refer to this process as reduction to *canonical form*. If $x_1, x_2, \ldots, x_m$ say, are the variables selected for elimination, then the canonical system takes the form

Canonical Form Relative to $x_1, x_2, \ldots, x_m$

$$\begin{aligned}
x_1 \quad\quad\quad + \bar{a}_{1m+1}x_{m+1} + \cdots + \bar{a}_{1n}x_n &= \bar{b}_1 \quad (x_j \geq 0) \\
x_2 \quad\quad + \bar{a}_{2m+1}x_{m+1} + \cdots + \bar{a}_{2n}x_n &= \bar{b}_2 \\
&\vdots \\
x_m + \bar{a}_{mm+1}x_{m+1} + \cdots + \bar{a}_{mn}x_n &= \bar{b}_m \\
(-z) + \bar{c}_{m+1}x_{m+1} + \cdots + \bar{c}_n x_n &= -\bar{z}_0
\end{aligned} \quad (25\text{-}2)$$

where $\bar{a}_{ij}$, $\bar{b}_i$ and $\bar{z}_0$ are constants.

It is important to note that this system is *equivalent* to the original system of Eqs. (25-1) and may be used in its place for any further discussions. In this form it is easy to obtain a particular solution to the equations and to test whether or not it is feasible and optimal.

25-2. Test for Optimal Feasible Solution

Theorem 1. If in the canonical form the values of the constant terms $\bar{b}_i$ and the coefficients $\bar{c}_j$ in the cost equation are nonnegative, the basic solution obtained by setting nonbasic

THE SIMPLEX METHOD

variables to zero and solving for values of basic variables; namely,

$$\begin{aligned} x_j &= 0 & \text{for } j &= m+1, \ldots, n \\ x_i &= \bar{b}_i & \text{for } i &= 1, 2, \ldots, m \\ z &= \bar{z}_0 \end{aligned} \qquad (25\text{-}3)$$

is optimal.

Proof. Since x_j must be nonnegative and $\bar{c}_j \geq 0$ *whatever be the feasible solution* it follows from the last equation of (25-2) that the values of z for the class of feasible solutions must always exceed $\bar{z}_0$ by the sum of the nonnegative terms on the left. Hence any particular feasible solution with $z = \bar{z}_0$ such as (25-3) would be a minimizing solution.

For example, consider the problem of finding min z and nonnegative x_1, x_2, x_3, x_4 satisfying

$$\begin{aligned} \mathbf{x_1} - 2x_2 + 2x_3 - 4x_4 &= 1 & (x_j \geq 0) \\ -x_1 + 3x_2 - 4x_3 + 5x_4 &= 2 \\ x_1 + 2x_2 + 3x_3 + 4x_4 &= z \text{ (min)} \end{aligned} \qquad (25\text{-}4)$$

Suppose we would like to check if the particular solution obtained by setting

$$x_3 = x_4 = 0$$

and solving for x_1, x_2, z is feasible and optimal. If we pivot on x_1 (boldface term) by dividing through (if necessary) by its coefficient so that it becomes unity and then eliminating x_1 from the other equations we obtain

$$\begin{aligned} x_1 - 2x_2 + 2x_3 - 4x_4 &= 1 \\ + \mathbf{x_2} - 2x_3 + x_4 &= 3 \\ 4x_2 + x_3 + 8x_4 &= z - 1 \end{aligned} \qquad (25\text{-}5)$$

If next we pivot on the x_2 term in the second equation, eliminating x_2 from all the other equations including the first, we obtain the equivalent canonical form with respect to x_1 and x_2:

$$\begin{aligned} x_1 \phantom{{}+x_2} - 2x_3 - 2x_4 &= 7 \\ x_2 - 2x_3 + x_4 &= 3 \\ 9x_3 + 4x_4 &= z - 13 \end{aligned} \qquad (25\text{-}6)$$

Note that (25-6) fulfills the conditions of Theorem 1; hence an optimal solution is obtained by setting nonbasic variables $x_3 = x_4 = 0$ and solving for x_1, x_2, z:

$$x_1 = 7 \qquad x_2 = 3 \qquad x_3 = x_4 = 0 \qquad z = 13$$

This solution is also *the unique optimal solution,* as will become clear by applying the following:

Theorem 2. If in the canonical system the basic solution is feasible and the coefficients $\bar{c}_j$ in the cost form are all positive for nonbasic variables, the solution is the unique feasible optimal solution.

Corollary. If in the canonical system the basic solution is feasible and the coefficients $\bar{c}_j \geq 0$ for all j, then no other feasible solution can also be optimal unless the values of $x_j = 0$ whenever $\bar{c}_j > 0$.

Proof. Consider another *feasible solution* and suppose the values of some $x_j > 0$ for a $\bar{c}_j > 0$, then the value of z must be *greater* than $\bar{z}_0$ since the term $\bar{c}_j x_j > 0$ in the last equation of (25-2). Hence any feasible solution which is also minimal must have $x_j = 0$ for all variables corresponding to $\bar{c}_j > 0$, establishing the corollary. In the event *all* $\bar{c}_j > 0$ corresponding to nonbasic x_j (hypothesis of Theorem 2), this means that all nonbasic $x_j = 0$; but this leaves only the values $x_i = \bar{b}_i$, $z = \bar{z}_0$ for basic variables upon substitution.

25-3. Improving a Nonoptimal Basic Feasible Solution

The standard simplex method works only with basic solutions which are feasible (i.e., $\bar{b}_i \geq 0$).† To understand the underlying principle let us suppose that one or more of the coefficients $\bar{c}_j$ for nonbasic variables are negative in the canonical form. In this case, the basic feasible solution fails the test for optimality.

If all $\bar{b}_i > 0$ (note zero is excluded) we can immediately improve the basic feasible solution by increasing the value of one of the corresponding nonbasic variables whose $\bar{c}_j < 0$, *keeping the other nonbasic variables at value zero.* As a good empirical rule the variable x_s is chosen for increase such that

$$\bar{c}_s = \min \bar{c}_j < 0$$

For example, suppose the canonical form is

$$\begin{aligned} x_1 \quad\quad - 2x_3 - 2x_4 &= 7 \\ x_2 - 2x_3 + \mathbf{x}_4 &= 3 \\ + 9x_3 - 4x_4 &= z - 13 \end{aligned} \quad (25\text{-}7)$$

which is the same as our earlier example except that the sign of the coefficient $\bar{c}_4$ has been reversed. The basic solution formed by setting nonbasic variables $x_3 = x_4 = 0$ and solving for x_1, x_2, and z yields $x_1 = 7$, $x_2 = 3$, $z = 13$, which is feasible.

Since $\bar{c}_4 = -4 = \min \bar{c}_j$, we now choose to increase x_4, maintaining all other nonbasic variables (in this case x_3) at zero values. The values of the basic variables and z will now depend on the value chosen for x_4. For example, the value of z is given by

$$z = 13 - 4x_4 \quad (25\text{-}8)$$

so that the larger the value x_4 the lower will be the value of z. On the other hand, the values of the basic variables become

$$\begin{aligned} x_1 &= 7 - (-2x_4) \\ x_2 &= 3 - x_4 \end{aligned} \quad (25\text{-}9)$$

so that the largest value that x_4 can take, without destroying the feasibility of the solution, is $x_4 = 3$. Above this value, x_2 would be negative. By setting $x_4 = 3$ the value of z is lowered by $4x_4$ or 12 units and the new solution becomes

$$x_1 = 13 \quad x_4 = 3 \quad x_2 = x_3 = 0 \quad z = 1$$

Notice that x_2 has vanished at the critical value $x_4 = 3$.

Accordingly *in the next iteration* we shall *interchange* the role of x_2 and x_4 as basic and nonbasic variables by pivoting on the boldface term x_4 in (25-7) to eliminate x_4 from the other equations. The pivot term is always in the column of the variable to be introduced into the basic set and the row of the basic variable to be dropped (in this case x_2). This yields

$$\begin{aligned} x_1 + 2x_2 - 6x_3 \quad\quad &= 13 \\ + \; x_2 - 2x_3 + x_4 &= 3 \\ 4x_2 + \; x_3 \quad\quad &= z - 1 \end{aligned} \quad (25\text{-}10)$$

Notice that this is now in canonical form relative to x_1 and x_4 except that we have not bothered to rearrange terms.

The new basic solution is obtained by setting the new nonbasic variables $x_2 = x_3 = 0$ and solving for the remaining variables: $x_1 = 13$, $x_4 = 3$, $z = 1$. Applying our test for optimality we see that the solution is *optimal* because all $\bar{b}_i \geq 0$ and $\bar{c}_j \geq 0$. Moreover it is *unique* since all $\bar{c}_j > 0$ (not zero) for nonbasic variables.

† Important variants of the simplex method exist—called composite routines—when the basic solution is not feasible. When all $c_j \geq 0$ the dual-simplex method of Lemke can be used.

THE SIMPLEX METHOD

As we have seen in the numerical example, the canonical form provides in general an immediate criterion for testing optimality of a basic feasible solution. Furthermore, if the solution is not optimal, it leads to another solution if all $\bar{b}_i > 0$, which reduces the value of the cost or objective function.

Let us now formalize in general this procedure of improving a nonoptimal basic feasible solution. If at least one *relative cost factor* $\bar{c}_j$ is negative in the canonical form (25-2), it is possible to construct a new basic feasible solution with a total cost lower than $z = \bar{z}_0$. Indeed, a lower-cost solution can be obtained by increasing the value of one of the nonbasic variables x_s and adjusting the values of the basic variables accordingly, where x_s is *any* variable whose relative cost factor $\bar{c}_s$ is negative. In particular the index s can be chosen such that

$$\bar{c}_s = \min \bar{c}_j < 0 \tag{25-11}$$

The latter is the rule for choice of s followed in practical computational work because it is convenient and because it has been found that it usually leads to fewer iterations of the algorithm than just choosing *any* $\bar{c}_s < 0$.

Using the canonical form (25-2) we construct a solution in which x_s takes on some positive value, the values of all other nonbasic variables are still zero, and the values of the basic variables including z are adjusted to take care of the increase in x_s:

$$\begin{aligned} x_1 &= \bar{b}_1 - \bar{a}_{1s}x_s \\ x_2 &= \bar{b}_2 - \bar{a}_{2s}x_s \\ &\cdots\cdots\cdots\cdots \\ x_m &= \bar{b}_m - \bar{a}_{ms}x_s \end{aligned} \tag{25-12}$$

$$z = \bar{z}_0 + \bar{c}_s x_s \qquad (\bar{c}_s < 0) \tag{25-13}$$

Since $\bar{c}_s$ has been chosen negative, it is clear that the value of x_s should be made as large as possible in order to make the value of z as small as possible. The only thing that prevents setting x_s infinitely large is the possibility that the value of one of the basic variables in (25-12) will become negative. However, if all $\bar{a}_{is} \leq 0$, then x_s can be made arbitrarily large, establishing

Theorem 3. *If in the canonical system, for some s, all coefficients $\bar{a}_{is}$ of x_s are negative (or zero), including its relative cost factor $\bar{c}_s$, then a class of feasible solutions can be constructed whose z values have no lower bound.*

On the other hand, if at least one $\bar{a}_{is}$ is positive, it will not be possible to increase the value of x_s indefinitely, because beyond the value $x_s = \bar{b}_i/\bar{a}_{is}$ the value of x_i will be negative. If $\bar{a}_{is}$ is positive for several i, then the smallest of such ratios, whose subscript will be denoted by $i = r$, will determine the largest value of x_s possible such that all values of x_i in (25-12) remain nonnegative. Let $x_s^* = \max x_s$ possible; then

$$x_s^* = \frac{\bar{b}_r}{\bar{a}_{rs}} = \min_{\bar{a}_{is} > 0} \frac{\bar{b}_i}{\bar{a}_{is}} \geq 0 \qquad (\bar{a}_{rs} > 0) \tag{25-14}$$

where only those i and $i = r$ are considered for which $\bar{a}_{is} > 0$, and where r, in case of a tie, may be chosen at random from among those tied. For example, if $\bar{a}_{1s} > 0$ and $\bar{a}_{2s} > 0$ but $\bar{b}_1 = \bar{b}_2 = 0$, then *flip a coin* to decide† whether $r = 1$ or $r = 2$.

Definition. *The basic solution is degenerate if the values of one or more of the basic variables is zero.*

In this case it is clear by (25-14) that if for some $\bar{a}_{is} > 0$, it happens that the corresponding value $\bar{b}_i$ of the basic variable is zero, then no increase in x_s is possible that will maintain the values of the basic variables nonnegative and therefore there will be no decrease in z. However, if the basic solution is *nondegenerate* we have the following result:

---

† The choice of r in case of tie has been subject to much investigation because of the theoretical possibility that a poor choice could lead to a repetition of the same basic solution after a finite number of iterations. Orchard-Hays in his electronic computer code chooses r such that max $\bar{a}_{rs} = \bar{a}_{is}$ among those tying i and reports excellent results in practice. While this rule does not have a firm theoretical foundation, it is probably a better working rule than the random rule for most practical cases.

Theorem 4. If in the canonical system for some s the relative cost factor $\bar{c}_s$ is negative and at least one other coefficient $\bar{a}_{is}$ is positive, then from a nondegenerate basic feasible solution a new basic feasible solution can be constructed with lower total cost z.

Specifically we shall show that by replacing x_r by x_s in the set of basic variables $x_1, x_2, \ldots, x_m$, the new set is basic and the basic solution is feasible. We shall show feasibility first. Indeed, from the assumption that all $\bar{b}_i > 0$, it follows that the value of $x_s = x_s{}^*$ determined by (25-14) is positive (not zero). Substituting this value of $x_s{}^*$ into (25-12), (25-13) gives a new feasible solution

$$x_i = \bar{b}_i - \bar{a}_{is}x_s{}^* \geq 0 \qquad (i = 1, 2, \ldots, m)$$
$$x_s = \frac{\bar{b}_r}{\bar{a}_{rs}} \qquad \left(x_s{}^* = \frac{\bar{b}_r}{\bar{a}_{rs}} > 0\right) \qquad (25\text{-}15)$$
$$x_j = 0 \qquad (j = m+1, \ldots, n \text{ except } j \neq s)$$

with total cost

$$z = \bar{z}_0 + \bar{c}_s x_s{}^* < \bar{z}_0 \qquad (\bar{c}_s < 0) \qquad (25\text{-}16)$$

There remains only to show that the new feasible solution is *basic*. It is clear from the definition of the index $i = r$ that

$$x_r = \bar{b}_r - \bar{a}_{rs}x_s{}^* = 0 \qquad (25\text{-}17)$$

and that we are trying to show that $x_1, x_2, \ldots, x_m$ (excluding x_r) and x_s constitute a new basic set of m variables. To prove this we simply observe that, since $\bar{a}_{rs} < 0$, we may use the rth equation of (25-2) and $\bar{a}_{rs}$ as "pivot element" to eliminate the variable x_s from the other equations and the minimizing form. It is clear that only one elimination is needed to reduce the system to canonical form relative to the new set of variables.

This ability to obtain a new canonical form from the previous one by means of a single additional elimination constitutes the key to the computational efficiency of the simplex method.

25-4. Iterative Procedure

The new basic feasible solution can again be tested for optimality by seeing if all $\bar{c}_j \geq 0$ (Theorem 1). If it is not optimal then one may choose by criterion (25-11) a new variable x_s to increase and proceed to construct either (1) a class of solutions in which there is no finite lower bound for z (if all $\bar{a}_{is} \leq 0$) or (2) a new basic feasible solution in which the cost z is lower than the previous one (provided the values of the basic variables for the latter are strictly positive, for otherwise the new value of z may be equal to the previous value).

The simplex algorithm consists of repeating this cycle again and again, terminating only when there has been constructed (1) a class of feasible solutions in which $z \to -\infty$ or (2) an optimal basic feasible solution.

Assuming nondegeneracy at each cycle, *the entire process will terminate in a finite number of cycles*. The reason for this is that there is only a finite number of ways to choose a set of m basic variables out of n variables. If the algorithm were to continue indefinitely, it could only do so by repeating the same basic set of variables—hence the same cost z. The latter is prevented from happening because on each cycle the value of z decreases a nonzero amount.

However, when degenerate solutions occur one can no longer argue that the procedure will terminate in a finite number of iterations because with no change in the value of z it is conceivable that the same basic set of variables may recur. Hence if one were to continue with the same selection of s and r for each iteration as before, the same basic set would recur after, say, k iterations, and again after $2k$ iterations, etc., indefinitely. There is therefore the possibility of *circling* in the simplex algorithm.†
In fact, examples have been constructed by A. Hoffman and E. M. L. Beale.

† We prefer the term "circling" rather than "cycling" commonly found in the literature, to avoid confusion with the basic cycle of the simplex algorithm.

We have therefore shown the convergence of the simplex method to an optimal solution in a finite number of cycles only for the case of nondegenerate basic solutions. There are simple ways to change the constant terms slightly so as to assure nondegeneracy and to prove that the procedure given there is valid even under degeneracy.

25-5. Phase I: Finding an Initial Basic Feasible Solution

Up to the present we have been assuming that it is possible to specify a basic set of variables which could be used to perform the necessary initial eliminations and reduce the problem to canonical form. It has also been assumed that the associated basic solution is feasible.

It is true that many problems encountered in practice often have a starting solution readily at hand. For example one can immediately construct a great variety of starting basic solutions for the important class called "transportation" problems. Economic models often contain storage and slack activities which permit an obvious starting solution in which nothing but these activities takes place. Such a solution may be a long way from the optimum solution but at least it is an easy start. Usually little or no effort is required in these cases to reduce the problem to canonical form.

However, many problems encountered in practice do not provide any obvious starting basic set of variables. This is true whenever the model does not have slack variables for some equations, or whenever the slack variables have negative coefficients.

Furthermore, whether or not a starting solution is readily available, little or nothing may be known (mathematically speaking) about the problem given to a computer for solution. It may have

1. *Redundancies*—for example, the equation balancing money flow may have been obtained from the equations balancing material flows by multiplying price by quantity and summing. The classical transportation problem contains a redundant equation.

2. *Inconsistencies*—due to outright clerical errors or to use of inconsistent data such as impossible requirements relative to the available resources. For example, one may pose a problem in which resources are known to be in short supply, and the main question asked is really whether a feasible solution exists at all!

It is clear that a general mathematical technique must be developed to solve linear-programming problems free of any prior knowledge or assumptions about the system being solved. In fact, if there are inconsistencies or redundancies, these are important facts to be discovered.

A simple device will now be given which uses the simplex algorithm itself to provide (if it exists) a starting basic feasible solution. This part of the process is referred to as *Phase* I. The second part of the process, obtaining an optimal basic feasible solution, is then referred to as *Phase* II. The device has several important features that should be noted:

1. No assumptions are made regarding the original system; it may be redundant, inconsistent, or not solvable in nonnegative numbers.

2. No eliminations are required to obtain an initial solution in canonical form for Phase I.

3. End product of Phase I is a basic feasible solution (if it exists) in canonical form ready to initiate Phase II. The procedure for Phase I is this:

Step 1. *Arrange the original system of equations so that all constant terms b_i are positive or zero by changing, where necessary, the signs on both sides of any of the equations.*

Step 2. *Augment the system to include a basic set of "artificial" variables $x_{n+1} \geq 0$, $x_{n+2} \geq 0, \ldots, x_{n+m} \geq 0$ so that it becomes*

$$\begin{aligned}
a_{11}x_1 + a_{12}x_2 + \cdots + a_{1n}x_n + x_{n+1} &= b_1 \\
a_{21}x_1 + a_{22}x_2 + \cdots + a_{2n}x_n \phantom{{}+{}} + x_{n+2} &= b_2 \\
\vdots & \\
a_{m1}x_1 + a_{m2}x_2 + \cdots + a_{mn}x_n + x_{n+m} &= b_m
\end{aligned} \qquad (25\text{-}18)$$

where all $\bar{b}_i \geq 0$ and for $j = 1, 2, \ldots, n; n+1, \ldots, n+m$.

$$x_j \geq 0 \qquad (25\text{-}19)$$

Step 3. Using the simplex algorithm, find a solution to (25-18) and (25-19) which minimizes the sum of the artificial variables, denoted by w:

$$x_{n+1} + x_{n+2} + \cdots + x_{n+m} = w \qquad (25\text{-}20)$$

We call this the *infeasibility form*. The coefficients of x_j analogous to $\bar{c}_j$ for the cost form are referred to as $\bar{d}_j$.

Step 4. If min $w > 0$, then no feasible solution exists and the procedure is terminated. On the other hand, if min $w = 0$, initiate Phase II for the simplex algorithm by:

  a. *Dropping from further consideration all nonbasic variables x_j whose corresponding coefficients $\bar{d}_j$ are positive (not zero) in the final modified w equation*
  b. *Replacing the linear form w (as modified by various eliminations) by the linear form z, after first eliminating from z all basic variables*

In practical computational work the elimination of the basic variables from the z form is usually done on each cycle of Phase I. If this is the case, then the modified z form may be used immediately to initiate Phase II.

25-6. Illustration

Solve the system $x_1 + 2x_2 \geq 20$, $x_1 - 3x_2 \leq 8$, $x_1 + x_2 = z$ for nonnegative x_1 and x_2 and minimum z. Introduce an "excess" variable x_4 and a "slack" variable x_3 into the linear inequalities, converting them into a system of equations in nonnegative variables in standard form

$$\begin{aligned} x_1 + 2x_2 - x_3 &= 20 \qquad (x_j \geq 0) \\ x_1 - 3x_2 + x_4 &= 8 \\ x_1 + x_2 &= z \ (\min) \end{aligned}$$

Before initiating Phase I, check to see that the constant term of each equation is nonnegative—if not change signs of all terms. Since the constant terms above are nonnegative, we augment the system with nonnegative variables x_5 and x_6 and an infeasibility form $x_5 + x_6$ equal to w, obtaining:

Admissible variables	Artificial variables	
$x_1 + 2x_2 - x_3$	$+ x_5$	$= 20$
$x_1 - 3x_2 + x_4$	$+ x_6$	$= 8$
$x_1 + x_2$		$= z$
	$x_5 + x_6$	$= w$

This is reduced to canonical form by subtracting the sum of the first two equations from the last:

Cycle 0—Phase I—Equation Form

Nonbasic variables	Basic variables	Constants	
$x_1 + 2x_2 - x_3$	$+ x_5$	$=$	20
$\mathbf{x_1} - 3x_2 + x_4$	$+ x_6$	$=$	8
$x_1 + x_2$	$(-z)$	$=$	0
$-2x_1 + x_2 + x_3 - x_4$	$(-w)$	$=$	-28
$*$	$\bullet \bullet \bullet \bullet$		

We use dots $\bullet$ to indicate the position of basic variables and the symbol $*$ for the variable entering the basic set on the next iteration. When the above is written in "detached coefficient form" it is called the simplex tableau:

THE SIMPLEX METHOD

Cycle 0—Phase I—Tableau Form

Basic variables	Nonbasic variables x_1 x_2 x_3 x_4	Basic variables x_5 x_6 $-z$ $-w$	Constants
x_5	1 2 −1	1	20
x_6	1 −3 1	1	8
$-z$	1 1	1	0
$-w$	−2 +1 +1 −1	1	28
	• • • •		

Since $w > 0$ our objective in Phase I is to reduce w to a minimum (zero, if a feasible solution exists). The coefficient $\bar{d}_1 = -2$ in the w form while all other $\bar{d}_j$ are larger; hence our next step is to increase x_1. The explicit rule for deciding which variable to drop is given by (25-14). Indeed the maximum increase is $x_1 = 8$ beyond which the value of x_6 is negative; therefore, in the next iteration x_1 replaces x_6 as basic variable. Hence the new *pivot* appears in column 1 and row 2 where x_6 appears. Using the pivot, x_1 is eliminated from the other equations, resulting in a canonical form relative to the new basic variables x_1 and x_5 as given for cycle 1.

Cycle 1—Phase I—Equation Form

$$\begin{aligned}
+\,5x_2 - x_3 - x_4 \;&\;+ x_5 - x_6 \;&\;&= 12 \\
x_1 - 3x_2 + x_4 \;& + x_6 \;&\;&= 8 \\
\hline
+\,4x_2 - x_4 \;& - x_6 \;& -z \;&= -8 \\
-\,5x_2 + x_3 + x_4 \;& + 2x_6 \;& -w \;&= -12
\end{aligned}$$

Cycle 1—Phase I—Tableau Form

Basic variables	x_1 x_2 x_3 x_4	x_5 x_6 $-z$ $-w$	Constants
x_5	5 −1 −1	1 −1	12
x_1	1 −3 1	1	8
$-z$	4 −1	−1 1	−8
$-w$	−5 1 1	2 1	−12
	*		

The subsequent iterations are shown in detached coefficient or tableau form.

Cycle 2—Phase II—Tableau Form

Basic variables	x_1 x_2 x_3 x_4	x_5 x_6 $-z$ $-w$	Constants	
x_2	1 −0.2 −0.2	0.2 −0.2	2.4	
x_1	1 −0.6 +0.4	0.6 +0.4	15.2	
$-z$	+0.8 −0.2	−0.8 −0.2 1	−17.6	←drop
$-w$		1 1 1	0	
	• • *	←drop→		

Cycle 3—Phase II—Optimal

	x_1 x_2 x_3 x_4	x_5 x_6 $-z$ $-w$	Constants	
x_2	0.5 1 −0.5	0.5	10	
x_4	2.5 −1.5 1	1.5 1	38	
$-z$	0.5 0.5	−0.5 1	−10	
$-w$		1 1 1	0	←drop
	• •	←drop→		

Optimal solution: $x_2 = 10$, $x_4 = 38$, $z = 10$, $w = 0$; *all other* $x_j = 0$.

Short Cuts

1. The columns relating to $(-z)$ and $(-w)$ remain the same from cycle to cycle. Accordingly these columns are often dropped and assumed implicitly or the variables z and w are moved to the constant-term side.

2. As soon as an artificial variable is dropped from the basis, it is often dropped from further consideration.

3. If artificial variables are still contained in the basis at the end of Phase I, their values must be made to stay zero during Phase II. This can be done by Step 4a or by restricting the value of w to the range $(-w) > 0$ just like any other variable $x_j \geq 0$.

4. If *all* artificial variables are dropped from further consideration, or if Step 4a is used, then w and the w form can also be dropped.

Chapter 26

ELEMENTS OF A STRATEGY FOR MAKING MODELS IN LINEAR PROGRAMMING

A. CHARNES, *Northwestern University.*

W. W. COOPER, *Carnegie Institute of Technology.*

CONTENTS

26-1. Introduction	26-1
Model Equivalences	26-2
Extremization	26-2
Algorithmic Completion	26-2
26-2. Some Illustrations	26-3
Solving Systems of Simultaneous Equations	26-3
Characterizing the Rank of a Matrix	26-5
26-3. Duality in Linear Programming	26-10
26-4. Constrained Regressions	26-15
26-5. Some Transformations and Reductions	26-19
A Single Objective Design Problem	26-19
Compound Objectives	26-21
Transformations and Reductions for Designs under Inequality Constraints—General Case	26-22
26-6. Algorithmic Alterations	26-24
26-7. Summary and Conclusion	26-28

26-1. Introduction

In the United States—as in Russia and elsewhere—progress in linear programming has been accompanied by a wealth of management applications.† These have guided and stimulated research in linear-programming theory and methods and have led to progress in other areas as well. Only a bare decade ago, for instance, almost nothing was known about the nature of the models that could be used in management. Today,

---

† See, for example, the opening remarks in Chapter 25, The Simplex Method, by G. B. Dantzig, and see Ref. 12 for the first published report of an industrial application in the United States. The earlier work in Russia, by L. V. Kantorovich[29,30] is also developed in a context of management applications. See Refs. 10 and 33 for further discussion.

by contrast, a variety of standard (special) model types,† associated algorithms (solution methods), and theorems are available for a range of managerial applications. Something may also now be said about strategies for synthesizing models in applied work, and doubtless still further progress will be made along these lines (as well as in model typology) as the applications in management, social science, etc., are extended along with recent work which is opening a way toward applications in engineering analysis and designs, the natural sciences, etc.‡

In order to obtain access to some of these applications—e.g., in engineering designs—three ingredients of a modeling strategy will be singled out for special examination in this chapter: (1) model equivalences, (2) optimization (extremal) principles, and (3) algorithmic completion of a model. The examination will proceed by means of examples in such areas as (a) solving simultaneous equations, (b) fitting curves, and (c) treating certain problems which may be encountered in engineering designs.

This mode of development should provide insight into ways by which the ideas and methods of linear programming can be extended. Notice, for instance, that this approach to model synthesis is intended to carry with it a distinction between a problem and its model. Thus it may be possible to utilize model equivalences in order to secure a wanted subset of solutions to a nonlinear *problem* by means of a linear *model* if only matters can be arranged so that the solution set of the model intersects the solution set of the problem at the points which are of interest.§

Model Equivalences. More generally, a model equivalence may be said to have been achieved whenever subsets in the solutions for one model (or series of models) can be used to generate the characteristics which are of interest in another model.

Extremization. Similar remarks apply for the use that may be made of the extremal principles involved in maximization, minimization, etc. Here it may be recalled that the focus is on optimization under inequality constraints. The two—i.e., optimization and constraints—may thus be joined in a variety of ways in order to deal with problems in which optimization per se is not an issue.‖ Moreover, by means of constraint adjunctions, successive optimizations, parameterizations, etc., it is possible to secure still further results. Thus the mere solution of an initially stated optimization need not be the terminal point of an analysis. Still further results may be obtained by introducing new constraints or undertaking parameter-variation analysis and new optimizations, etc.

Algorithmic Completion. The fact that any algorithm is also a mathematical construct—hence has special mathematical properties—may also be used to advantage, as in the case of algorithmic completion of a model. The simplex method, for instance, has the property of utilizing only basic solutions. Such basic solutions are associated with linearly independent sets of vectors, which span a space of dimension m when this is the number of rows in the corresponding tableau. Furthermore, it is possible here, too, to adjoin new rows or columns to any tableau in order to obtain still further characterizations in higher dimensions when these are wanted.¶ Hence it is possible, on occasion, to utilize these "basis" properties of the simplex method and to construct models for its application in such a way that the solutions which are obtained can be associated with other characterizations that are wanted.

† For example, the so-called "critical path-PERT" scheduling methods along with "transportation-distribution" models are perhaps the best-known examples of such special linear-programming types. See, for example, Chapters 14 and 17 in Ref. 7 as well as Ref. 8 and Sec. 36-4 of this handbook.
‡ See, for example, Refs. 36 and 41 for recent examples.
§ This idea has, of course, been used for a considerable period of time—as witness the use of linear methods (including its optimization conditions and algorithms) to obtain solutions for nonlinear problems by studying possible intersections with related tangent planes, etc.
‖ For example, as in the traffic-simulation example reported in Chapters 17 and 20 of Ref. 7 where "equilibrium" traffic patterns were determined by means of suitably arranged constrained optimization models.
¶ The Phase I–Phase II computing procedure discussed in Sec. 25-5 provides one example of such a tableau expansion which is there designed to handle the possibility of nonsolvability as well as to secure an easy way of starting the simplex iterations.

26-2. Some Illustrations

Solving Systems of Simultaneous Equations. Some immediate illustrations of these ideas can be obtained by reference to issues that are confronted when attempting to solve systems of simultaneous equations. Thus, suppose we want to find values for n variables x_j, $j = 1, \ldots, n$, which simultaneously satisfy

$$\sum_{j=1}^{n} a_{ij} x_j = b_i \qquad i = 1, \ldots, m \tag{26-1}$$

where the a_{ij} and b_i are known constants.[†]

No issue of optimization is evident here. Yet the following optimization model is easily seen to be equivalent to (26-1):

$$\min z \equiv \sum_{i=1}^{m} y_i$$

subject to

$$\sum_{j=1}^{n} a_{ij} x_j + y_i = b_i \qquad y_i \geq 0 \qquad i = 1, \ldots, m \tag{26-2}$$

Since each y_i (but not x_j) is to be nonnegative,[‡] it follows that $z \equiv \Sigma y_i \geq 0$. If the minimization yields $z^* = 0$ then the associated x_j^* obtained from (26-2) will also satisfy (26-1). If, on the other hand, $z^* > 0$ is obtained then there are no x_j values which can simultaneously satisfy all the relations in (26-1); i.e., Eqs. (26-1) are inconsistent. If desired, the x_j^* values secured via $z^* > 0$ in (26-2) may then be interpreted as representing values which are "as close as it is possible to come" for a solution to (26-1). That is, we may then interpret (26-2) as an optimizing equivalent wherein we are to come "as close as possible" to a solution for (26-1) with $z^* \geq 0$ providing a measure of the resulting deviation, if any, and each $y_i^* \geq 0$ providing still further detail on specific constraints where discrepancies occur and the amount, or "measure," of each such discrepancy.

It is possible to refine these last ideas even further[§] by considering the case in which a $z^* > 0$ occurs. Of course, this means that no solutions are available which satisfy (26-1). But then it may be the case that some of the constraints in (26-1) are to be accorded a preferred status.

To see what is involved here, notice that each y_i is accorded the same weight, unity, via its coefficient in the functional of (26-2). But this equal weighting is readily remedied by replacing the functional $z = \Sigma y_i$ with a new $\hat{z} = \Sigma c_i y_i$ where the constants c_i are any suitable set of nonnegative weights.[||] If "absolute" or "preemptive," preferences are at issue, however, then recourse may be had to other optimizations. For instance, a set of "non-Archimedean" weights M_i may be used in place of ordinary real numbers for the c_i. These "preemptive" weightings are represented by $M_i \gg M_{i+s}$, $s \geq 1$, in order to mean that there is no positive number k with the

---

[†] See the first m expressions in Eqs. (25-1), but observe that the variables are now no longer restricted to be nonnegative. We are assuming all problems to be "regular" or "regularized," if required. See Chapter 12 of Ref. 7 for further discussion.

[‡] These y_i serve the same purpose as the artificial variables discussed in the final examples of Chapter 25. The symbol * will generally be used to distinguish optimal values.

[§] See also Ref. 21 for use of linear-programming methods to scale or condition matrices preparatory to solving equation systems.

[||] These terms would appear as constants in the final row of Table 25-1. The terms "functional," "linear form," and "function" are used interchangeably throughout this chapter. They have the same meaning as the terms "cost form," "linear form," etc., used in Chapter 25.

property $kM_{i+s} \geq M_i$.† Thus, if $M_1 \gg M_2 \gg \cdots \gg M_i \gg \cdots \gg M_m$ then the association in $\sum_{i=1}^{m} M_i y_i$ means y_1 is to be made as small as possible; also, without increasing y_1, the objective is to make y_2 as small as possible; and so on. In any case, however, the attainment of a solution with $z^* \geq 0$ may first be secured. Then, via the transformations indicated by Dantzig, the computations may be continued (without backtracking) after a $z^* > 0$ is secured, say, followed by an introduction of a suitable mixture of c_j and M_i weights.

Up to this point only issues of optimization and model equivalence have been examined in (26-1) and (26-2). Nothing has been said about the properties of any particular algorithm. Suppose, however, that the simplex method is now prescribed for (26-2). If only the standard simplex method‡ is to be used, then some allowance must be made for the fact that the x_j variables in (26-2) are not constrained in sign. One way to do this is to represent each unconstrained variable as a difference of two nonnegative variables—viz.,

$$x_j = x_j^+ - x_j^- \qquad x_j^+, x_j^- \geq 0 \qquad j = 1, \ldots, n \qquad (26\text{-}3)$$

and then utilize the latter variables in place of the former when solving (26-2).

In order to underscore that we are not concerned only with nonnegative values of x_j which satisfy (26-1), we now alter the problem to one wherein we are to find all basic solutions that satisfy (26-1). For convenience in examining this problem in a way that utilizes (26-3) we first rewrite (26-1) in vector notation as

$$\sum_{j=1}^{n} P_j x_j = P_0 \qquad (26\text{-}4)$$

Here P_j is the m-rowed column vector consisting of the elements drawn from the jth column of the $m \times n$ matrix

$$A = \begin{bmatrix} a_{11} & \cdots & a_{1j} & \cdots & a_{1n} \\ \vdots & & \vdots & & \vdots \\ a_{i1} & \cdots & a_{ij} & \cdots & a_{in} \\ \vdots & & \vdots & & \vdots \\ a_{m1} & \cdots & a_{mj} & \cdots & a_{mn} \end{bmatrix} \qquad (26\text{-}5)$$

which are the coefficients in (26-1). P_0, called a "stipulations vector," is also m-rowed with the constants b_i, $i = 1, \ldots, m$, for its elements.

† A more precise description is given in Chapter 19 of Ref. 7. The Phase I routine discussed by Dantzig may be regarded as one special kind of preemptive in that all the artificial variables—viz., his $x_{n+1}, \ldots, x_{n+m}$ which are the same as our y_i, $i = 1, \ldots, m$ —are driven to zero before any attention is paid to the remaining variables. This special case corresponds to $M_i = M_j = M$ so that the functional is

$$\min z = \sum_{j=1}^{n} c_j x_j + M \sum_{i=1}^{m} y_i$$

where M is a "dominatingly large" penalty factor relative to any possible combination of the real numbers c_j.

‡ See the opening sentence of Sec. 25-3, where it is observed that the standard simplex method works only with nonnegative-valued variables.

MAKING MODELS IN LINEAR PROGRAMMING   26-5

In this same notation, (26-2) is represented by

$$\min z \equiv \sum_{i=1}^{m} y_i$$

subject to

$$\sum_{j=1}^{n} P_j x_j + \sum_{i=1}^{m} Q_i y_i = P_0 \qquad y_i \geq 0 \qquad i = 1, \ldots, m \qquad (26\text{-}6)$$

Here Q_i is the ith unit vector; i.e., Q_i is an m-rowed vector with unity in its ith row and zero everywhere else.

Utilizing (26-3) we obtain

$$\min z \equiv \sum_{i=1}^{m} y_i$$

subject to

$$\sum_{j=1}^{n} P_j x_j^+ - \sum_{j=1}^{n} P_j x_j^- + \sum_{i=1}^{m} Q_i y_i = P_0 \qquad x_j^+, x_j^-, y_i \geq 0 \qquad (26\text{-}7)$$

for use in place of (26-6). We are then in position to observe that x_j^+ and x_j^-, for each j, have P_j and $-P_j$ as corresponding vectors. These are simply the negatives of each other and hence linearly dependent. But the simplex method proceeds by means of basic solutions only, and such solutions are associated only with linearly independent sets of vectors. This means that no basic solution can have both x_j^+ and x_j^- positive. Hence when the simplex method is specified for use in (26-7)—a simple illustration of "algorithmic completion"—we shall always have

$$x_j^+ x_j^- = 0 \quad \text{and} \quad x_j = 0 \quad \text{if and only if } x_j^+ = x_j^- = 0 \qquad (26\text{-}8)$$

Notice now that specification of the simplex method resolves any possible ambiguity in the relations (26-3) and hence, also, any similar ambiguity in the solutions of (26-7) and (26-6). Since (26-6) and (26-4) are only different ways of representing (26-2) and (26-1), it is evident that all of our previous discussion applies. Furthermore if we can secure $z^* = 0$ in the minimization† then we can also obtain *all* basic solutions, as wanted, by utilizing (26-3) to produce the corresponding unconstrained (in sign) values of x. Indeed, (26-1) will have more than one (unique) solution if and only if the optimum tableau shows that at least one alternate optimum is present. An alternate optimum is one with a different set of basic variables which also has the same min $z = z^*$ value. These alternate optima may be generated, as required, by means of the standard simplex algorithm.‡ In fact all of them may be obtained in this fashion, and thus the simplex method applied to (26-7) permits us to obtain all the basic solutions to (26-1). Finally, if further solutions (basic or not) are wanted, these may also be obtained by taking "convex combinations"§ of the x_j^{*+} and x_j^{*-} which are in the basic optimal sets of (26-7).

Characterizing the Rank of a Matrix. Up to this point we have focused on corresponding solution sets in order to demonstrate some parts of a model equivalence for linear-programming analyses. We now want to broaden this by showing how properties *associated* with a solution can be used for other purposes by means of a specified algorithm applied to a suitably arranged model. In particular we want an illustration that utilizes the simplex method for algorithmic completion of a model, and for this we shall turn to a model that utilizes the basis properties of this method to characterize the rank of the matrix A in (26-5).

---

† As observed in the preceding discussion, $z^* = 0$ is a necessary condition for the existence of *any* solutions to (26-1).
‡ See Chapter 12 of Ref. 7 for a systematic means of obtaining these alternate optima by so-called "Tarry traversals" in an ordered array.
§ That is, linear combinations with nonnegative coefficients that sum to unity.

It is now convenient to utilize notation which differs from that of Chapter 25, and also to introduce a suitable set of names for the symbols we shall be using. We refer to the P_j formed from A as "structural vectors" (and their x_j variables as "structural variables") in order to distinguish them from the "artificial" or "slack" vectors Q_i (with slack or artificial variables y_i). At a stage like that of Eq. (25-2), we replace the symbols $\bar{a}_{ij}$ by x_{ij} or y_{ij} according to whether a structural or artificial vector is involved.

Our notation and the "names" we are using may be related to that of (25-2) by means of the "dictionary" of Table 26-1. Here, as before, we retain the symbol z,

Table 26-1. Notation Dictionary

Notation of Chapter 25	Notation used in this chapter	
	Structure	Slack or artificial
$\bar{a}_{ij}$	x_{ij}	y_{ij}
$\bar{b}_r$	x_r	y_r
$\bar{c}_k$	$c_k - z_k$	$M_k - \omega_k$†

† For artificial vectors only. Slack will have only the term ω_k and no M_k.

a so-called "figure of merit," which is obtained by assigning "criterion elements" c_j, in original data, to the program values x_j. Finally we designate as the "objective" any statement such as "min z" or "max z" so that, in general, our "objective" is to "optimize a specified figure of merit."‡

Returning now to the issue of characterizing A by reference to its rank, we note that there is no immediate issue of a solution in the sense of a set of program values for the variables. On the other hand, the rank of A is equal to the maximum number of linearly independent vectors which the structure of A can be made to yield. This, in turn, suggests the possibility of stating some kind of optimization model with solutions that will have the property of being associated with a maximum number of linearly independent structural vectors P_j. A method of solution which utilizes only linearly independent vectors may then also be joined to such a model in order to complete it for the purposes at hand.

Each basic solution of the simplex method is uniquely associated with a set of linearly independent vectors. Indeed if the simplex method is applied to a matrix A which is $m \times n$ then the basis vectors will be m in number. If A is of rank $r < m \leq n$, the simplex method can utilize at most r vectors from A, in which event $m - r$ vectors Q_i—slack or artificial—may be used to complete the basis. Hence if the simplex method is to be used our problem may equally well be one in which this algorithm may be used to minimize the number of slack or artificial vectors needed to complete a basis.

To erect such a model, where no issue of solvability is present, we have recourse to one version of the perturbation theory that was originally developed to resolve the issue of degeneracy for the simplex method.§ For this purpose we utilize the same constraints as in (26-7) except that we replace the P_0 vector by a new stipulations

‡ Or possibly a vector of such figures of merit. See, for example, Ref. 6 for a discussion of such vector optimizations—as distinguished from scalar optimizations—in the context of economics and city-street-planning models.

§ See Ref. 5. Actually the motivation for this research was more general in that it was also desired to render the simplex method entirely unexceptional for proving mathematical theorems—and obtaining other kinds of precise characterizations—and thereby elevating the simplex method to something more than a tool for numerical calculations. (See Chapter 12 of Ref. 7 for further discussion.)

vector $\hat{P}_0$—the null or zero vector—and then "perturb" the latter to a value $\hat{P}_0(\epsilon)$. We also alter the functional of (26-7) by introducing the preemptive priority coefficients $M_1 \gg M_2$, etc., as previously discussed, in order to obtain

$$\min \sum_{i=1}^{m} M_i y_i$$

subject to

$$\sum_{i=1}^{m} Q_i y_i + \sum_{j=1}^{n} P_j x_j^+ - \sum_{j=1}^{n} P_j x_j^- = \hat{P}_0(\epsilon)$$

$$\equiv \sum_{i=1}^{m} Q_i \epsilon^i + \sum_{j=1}^{n} P_j \epsilon^{m+j} - \sum_{j=1}^{n} P_j \epsilon^{m+n+j} \qquad y_i, x_j^+, x_j^- \geq 0 \quad (26\text{-}9)$$

If we choose any $\epsilon > 0$ it is then evident that the problem (26-9) always has at least one solution. For instance, $y_i = \epsilon^i > 0$, $x_j^+ = \epsilon^{m+j} > 0$, and $x_j^- = \epsilon^{m+n+j} > 0$ is a solution, although not necessarily basic, in positive values. Alternatively we can commence with artificial vectors only and be assured of a basic solution by setting

$$y_i = \epsilon^i + \sum_{j=1}^{n} a_{ij} \epsilon^{m+j} - \sum_{j=1}^{n} a_{ij} \epsilon^{m+n+j} \qquad (26\text{-}10)$$

for each $i = 1, 2, \ldots, m$. Moreover we are assured that y_i values will all be positive if we choose $\epsilon > 0$ to be "preemptively small." By this we mean that $\epsilon^r \gg \epsilon^s$ whenever $r < s$ and furthermore there is no positive number k with the property $k\epsilon^s \geq \epsilon^r$. In particular, since ϵ^i dominates all succeeding terms in (26-10), we see that each of these y_i will be positive under the indicated definition of $\epsilon > 0$. Furthermore, one of these terms will be smallest, as is the case here for $i = m$.

This ϵ^i ordering is completely equivalent to a lexicographic ordering.† No trouble will be encountered in the removal-replacement procedure that is employed by the simplex algorithm when generating a successor tableau. The powers of $\epsilon > 0$ will continue to be decisive in succeeding tableaus, and this same situation will continue to obtain; degeneracy will never be encountered, since the basis properties of the simplex algorithm will themselves ensure that each variable in a basic solution will be positive and differ in at least one place from every other variable in its ϵ coefficient.‡

We now observe that if a start is made only with artificial vectors we must have $z_k - c_k \neq 0$, for some k, and hence either $z_k - c_k > 0$ under P_k or under $-P_k$, unless A is null.§ This means that such a start cannot be optimal unless r, the rank of A, is zero—in which event no structural vector from A can be used as part of any basis and the start itself is optimal and exhibits $r = 0$ for A.

Bearing these ideas in mind we now refer to Table 26-2, which represents a tableau at some stage subsequent to a start with artificial vectors. We shall suppose that A is any matrix with rank $r > 0$ and that at the stage where we are considering the tableau some of the artificial vectors may already have been removed from the basis.

Here we have depicted one of the artificial vectors Q_t which remains in the basis with its associated positive basis variable y_t alongside it. The y_{ti} and x_{tj} that also appear in this same row have the meanings which were assigned to them in Table 26-1.

† It should perhaps be noted that the relative magnitudes of the M_i and ϵ^j are always taken to be such that the smallest M_i is greater than any real multiple of the reciprocal of the largest power of ϵ.

‡ Notice, for instance, the diagonal of unit coefficients associated with $x_1, \ldots, x_m$ in Eq. (25-2). These variables will therefore have the m unit vectors associated with their solution values in the tableau at this stage and hence only *one* of the $\epsilon^k > 0$ will have a nonzero coefficient in each such column.

§ Hence has all $a_{ij} = 0$ for its entries. Since the objective for (26-9) involves a minimization it is convenient to switch from $c_k - z_k$, as in Table 26-1, to $z_k - c_k$, and hence an optimum is now achieved when $z_k - c_k \leq 0$, all k, applies.

Table 26-2

	$\hat{P}_0(\epsilon)$		M_1 Q_1	...	M_i Q_i	...	M_m Q_m		c_1 P_1	...	c_j P_j	...	c_n P_n
M_t, Q_t	y_t		y_{t1}	...	y_{ti}	...	y_{tm}		$-x_{t1}$	...	$-x_{tj}$	...	$-x_{tn}$
$0, P_s$	$\hat{x}_s^\pm$		$\hat{x}_{s1}$	...	$\hat{x}_{si}$	...	$\hat{x}_{sm}$		$-\hat{x}_{s1}$	...	$-\hat{x}_{sj}$	...	$-\hat{x}_{sm}$
	$z(\epsilon)$		ω_1 $-M_1$	...	ω_i $-M_i$	...	ω_m $-M_m$		$-z_1$ $+c_1$	...	$-z_j$ $+c_j$	...	$-z_n$ $+c_n$

NOTE: 1. In the example of this section all $c_j \equiv 0$.
2. When Q_i is a slack vector then no M_i is assigned to it.

26-8

MAKING MODELS IN LINEAR PROGRAMMING

The structural vector P_s generically depicts the situation which applies when some of the vectors P_j from A are also in the basis. Since P_s may be associated with either an x_j^+ or an x_j^- we have written $x_j^\pm$, to cover either $x_j^+ > 0$ or $x_j^- > 0$. In fact, for each y_t and $x_s^\pm$ we have

$$y_t = \sum_{i=1}^{m} y_{ti}\epsilon^i + \sum_{j=1}^{n} x_{tj}\epsilon^{m+j} - \sum_{j=1}^{n} x_{tj}\epsilon^{m+n+j} > 0 \quad \text{all } t \, \epsilon \, I_q$$

$$x_s^\pm = \sum_{i=1}^{m} \hat{y}_{si}\epsilon^i + \sum_{j=1}^{n} \hat{x}_{sj}\epsilon^{m+j} - \sum_{j=1}^{n} \hat{x}_{sj}\epsilon^{m+n+j} > 0 \quad \text{all } s \, \epsilon \, I_p \tag{26-11}$$

where I_q is the set of indices for the artificial variable in this basis and I_p is the set of indices for the structural variables. [The caret or "hat" over the latter expressions in (26-11) distinguishes them from the corresponding y_{ti} and x_{tj} for the artificial variable y_t.]

In this notation we can also represent the solution at this stage as

$$P_0(\epsilon) = \sum_{t \epsilon I_q} Q_t y_t + \sum_{s \epsilon I_p^+} P_s x_s^+ - \sum_{s \epsilon I_p^-} P_s x_s^- \tag{26-12}$$

where, for clarity, we have further decomposed I_p into an I_p^+ for the set where $x_s^+ > 0$ applies, and into I_p^- where $x_s^- > 0$ applies. Of course, the summations are to be taken over all the indicated indices.

The functional value of $z = z(\epsilon)$ at this stage is obtained by simply positioning the criterion elements written in the left-hand stub of Table 26-2 alongside their respective basis variables

$$z(\epsilon) = \sum_{t \epsilon I_q} M_t y_t > 0 \tag{26-13}$$

We have here omitted the criterion coefficients for the $x_s^\pm$ since these are all zero. Hence, as long as any Q_t are in the basis we shall have $z(\epsilon) > 0$, as indicated in (26-13), and in any event, we shall have $z(\epsilon) \geq 0$ with $z(\epsilon) = 0$ attainable when A is of rank m.

We next write

$$Q_i = \sum_{t \epsilon I_q} Q_t y_{ti} + \sum_{s \epsilon I_p} P_s \hat{y}_{si} \quad i = 1, \ldots, m$$

$$\pm P_j = \pm \left(\sum_{t \epsilon I_q} Q_t x_{tj} + \sum_{s \epsilon I_p} P_s \hat{x}_{sj} \right) \quad j = 1, \ldots, n \tag{26-14}$$

These expressions, which are formed, column by column, from the data of Table 26-2, relate all vectors in (26-9) to the subset Q_t, $t \, \epsilon \, I_q$; P_s, $s \, \epsilon \, I_p$; which are positioned in the stub. Because this subset is a basis it is possible to express every m vector (or point in m space) in terms of this set, and moreover, these expressions are unique.

To complete our characterizations for this tableau we then introduce the definitions

$$\omega_i \equiv \sum_{t \epsilon I_q} M_t y_{ti} + \sum_{s \epsilon I_p} c_s \hat{y}_{si} \equiv \sum_{t \epsilon I_q} M_t y_{ti}$$

$$i = 1, \ldots, m, \text{ as exhibited under the } Q_i \tag{26-15}$$

$$\pm z_j \equiv \pm \left(\sum_{t \epsilon I_q} M_t x_{tj} x_{tj} + \sum_{s \epsilon I_p} c_s \hat{x}_{sj} \right) \equiv \pm \sum_{t \epsilon I_q} M_t x_{jt}$$

$$j = 1, \ldots, n, \text{ as exhibited under the } \pm P_j$$

Here all c_j and thus all c_s are zero. Hence the criteria of optimality for the corresponding program may be established according to whether $\omega_i > M_i$, $\pm z_j > 0$, or their opposites apply. (See Table 26-1, above, as it relates to the preceding chapter.)

If A is of rank $r < m$, we shall only need r of the P_j vectors to express every vector in A in the manner of (26-14). We want to show, however, that an optimal tableau will have this characteristic. Hence suppose that the tableau of Table 26-2 has only $k < r$ structural vectors in the basis. But then,

Theorem 1. If A is of rank r then any tableau which contains fewer than r structural vectors in the basis will have $x_{jt} \neq 0$ for some j and t.

Proof.† Refer to the second set of expressions in (26-14) and assume that there are only $k < r$ vectors P_s in the set $s \, \varepsilon \, I_p$.

If, in contradiction to the theorem, we had $x_{tj} = 0$ all j and t then these expressions would become

$$\pm P_j = \pm \sum_{s \varepsilon I_p} P_s \hat{x}_{sj} \qquad j = 1, \ldots, n$$

and we would have expressed *every* vector in A in terms of a subset of only $k < r$ vectors. This would imply rank $A \leq k < r$, contradicting the hypothesized rank of $A = r$. Hence to avoid this contradiction we must assume that the hypothesis of the theorem is true when A is of rank $r > k$. Q.E.D.

But now we observe that if A is of rank r we cannot have more than r of its P_j vectors in any basis. Hence we also have

Theorem 2. If A is of rank r then a necessary condition for optimality of Table 26-2 under the simplex algorithm is that there be r vectors in the set $\{P_s/s \, \varepsilon \, I_p\}$.

Proof. By virtue of the hypothesized rank we cannot have more than r such vectors. By Theorem 1 we shall have $x_{jt} \neq 0$ for some j and t when there are fewer than r structural vectors in the basis. But now consider any such j. Its nonzero x_{jt} are each associated with a preemptive coefficient M_t in the stub. See the expressions for Q_i in (26-14) and for ω_i in (26-15). Among its $x_{jt} \neq 0$ there must be a first one. If $x_{jt} > 0$ then, preemptively, $z_j > 0$. On the other hand, if this first $x_{jt} < 0$ then we shall have, preemptively, $-z_j > 0$. But an optimal simplex tableau for (26-9) requires $z_j, -z_j \leq 0$; hence $z_j = 0$, all j. Q.E.D.

Corollary 1. An optimal simplex tableau for (26-8) will have $x_{tj} = 0$ all j and t.

But now, combining the above results, we have

Corollary 2. If A is of rank r then an optimum simplex tableau for (26-8) will necessarily have exactly r structural vectors in its associated basis along with $m - r$ artificial vectors.

We are assuming (a) that (26-9) has a finite optimum and (b) that the simplex algorithm will achieve an optimum tableau for this problem in a finite number of iterations. Assumption (a) may be dropped, as will be shown via the dual theorem of linear programming in the next section. If (a) is true, however, then (b) follows readily from the perturbation we are employing in connection with the simplex method. The latter, as we have observed, works only with basic solutions, and only a finite number of bases can be formed from the vectors in (26-9). The perturbation, on the other hand, ensures that each iteration will produce a $z(\epsilon)$ which is strictly smaller than its predecessors. Since in (26-9) we are concerned only with an optimal basic solution‡—more precisely, with its associated set of basis vectors—we are assured that a uniquely smallest $z(\epsilon)$ will be achieved in a finite number of iterations.

26-3. Duality in Linear Programming

Publication of the simplex method in Ref. 32 may be regarded as starting the very active progress that has since been made in linear programming. Up to that time there had not been available any really efficient and generally satisfactory method of

---

† A somewhat more elaborate proof can be erected by means of simplex computation theory only. We use the above, however, for the sake of brevity. The same is also true in the subsequent proofs; so that one can build up all these ideas if one starts only with the definition of a basis as is done on p. 124 in Ref. 7 rather than the more usual start to introduce the ideas of a basis via the concept of linear independence.

‡ See Sec. 25-1 for a discussion that relates optimal basic solutions to the solutions of linear-programming problems in general.

computation, and hence, as far as applied work was concerned, there was little incentive for directing developments toward linear-programming models.† The important by-products of insight into management planning (and other) problems via their systematic modeling by linear programming, which has since been supplied, could hardly have been forthcoming without this important innovation by G. B. Dantzig.

The power of linear programming for theoretical analysis, as well as immediate application, would be considerably weaker without the duality relations that were discovered and published by Profs. Gale, Kuhn, and Tucker, also in Ref. 32. The resulting theory of duality provided direct contact between linear programming and other important disciplines, such as the theory of games (Chapter 23) and statistical decision theory (Chapter 24). See also the discussion by Gale, Kuhn, and Tucker in Ref. 32. It also opened a way for dealing with many kinds of nonlinear problems.

To bring selected aspects of this duality theory into prominence we first utilize matrix notation and write

$$\max z \equiv c'x$$

subject to
$$Ax \leq P_0$$
$$x \geq 0 \tag{26-16}$$

where A is, as in (26-5), an arbitrary $m \times n$ matrix which "structures" the problem—i.e., the vectors P_j in A are structural vectors—and x is the associated $n \times 1$ vector of structural variables. The vector c is a corresponding $n \times 1$ (column) vector of criterion elements which is here written in transpose (row) form, as indicated by the prime ($'$), so that the resulting scalar product ($c'x$) generates a figure of merit symbolized by the scalar quantity z. Finally, P_0 is an $m \times 1$ column vector of stipulations with components b_i, $i = 1, \ldots, m$.

The problem (26-16) is called the "direct" or "primal" problem of linear programming in order to distinguish it from a related "dual" problem which can be obtained by transposing the data and reorienting the objective as in the following representation:

$$\min g \equiv w'P_0$$

subject to
$$w'A \geq c'$$
$$w' \geq 0 \tag{26-17}$$

Of course, g, the figure of merit for the dual, is the scalar quantity obtained from $w'P_0 \equiv \sum_{i=1}^{m} w_i b_i$. All data in (26-16) are represented in (26-17) as well. We now emphasize that each dual pair‡ of linear-programming problems is stated in entirely different variables. A good deal of the sharpness of linear programming may be attributed to this separation of problem variables which are applied to the same data.

In (26-17) we have used the symbol w for the $1 \times m$ vector of dual variables ($w_1, \ldots, w_m$) since they are often used to evaluate the "worth" or desirability of altering elements of the stipulations vector P_0 which appears in the constraints of (26-16) and the functional in (26-17). Notice that the vector c also enters only in the functional of one problem and the constraints of the other.

In applied work this interchange between criterion and stipulation elements can be an important convenience as when, for instance, it is desired to study the sensitivity of an optimal program relative to possible changes in stipulations or figure-of-merit data. Thus, for instance, the following Theorems 3 and 4—from duality theory—make it possible to study z^* variations in just this manner by altering P_0, while holding c and w^* constant, without having to recompute the x^* values for each such alteration:

† See Ref. 10 for some of the "shortcomings" of earlier Russian work by L. V. Kantorovich in this and other respects, as reported in Refs. 29 and 30, and see also the further comments by T. C. Koopmans in Ref. 33.

‡ Either problem may be regarded as primal or dual as desired. The important point is that every linear-programming problem, irrespective of its objective (maximization or minimization) always has a dual with the opposite objective (minimization or maximization).

Theorem 3. *If both the problems (26-16) and (26-17) have solutions then every one of these solutions must satisfy*

$$c'x \leq w'P_0$$

Theorem 4. *On the hypothesis of Theorem 3 the optimal solutions of (26-16) and (26-17) satisfy*

$$z^* \equiv c'x^* = w^{*\prime}P_0 \equiv g^*$$

Even though the simplex method was developed prior to any knowledge of these theorems it has, fortunately, the property that it will solve both members of a pair of dual problems whenever it solves either of them. Furthermore, by means of further refinements in the simplex method that were subsequently forthcoming,† it is possible to prove these theorems, even without the indicated qualifications, in such a way that the simplex method may be used to characterize the status of any problem (solvable or not) by reference to details that are automatically available from the resulting tableaus. This, too, can be advantageous in applied work where the solution status of a problem and its possible further alterations are not always immediately clear.

We propose now to develop a proof of Theorems 3 and 4 in a way that will provide direct access to simplex-tableau details. First we prove Theorem 3 in a relatively straightforward fashion. We have assumed that solutions to (26-16) exist. Hence we can write

$$Ax \leq P_0 \qquad (26\text{-}18)$$

for some x. But then any nonnegative vector $w' \geq 0$ may be applied to give

$$w'Ax \leq w'P_0 \qquad (26\text{-}19)$$

Similarly, if for some w we can write $w'A \geq c'$ then for any $x \geq 0$ we have

$$w'Ax \geq c'x \qquad (26\text{-}20)$$

But now if we have used w', $x \geq 0$ which satisfy (26-16) and (26-17) then we can combine (26-19) and (26-20) to obtain

$$c'x \leq w'Ax \leq w'P_0 \qquad (26\text{-}21)$$

We immediately have the consequence noted in Theorem 2.

$$c'w \leq w_i P_0 \qquad (26\text{-}22)$$

for any x and w satisfying the constraints of (26-16) and (26-17).

We have written (26-21) to indicate that these duality relations form part of a more general saddle-point theory, as is true of the theory of games (Chapter 23); and indeed, by choosing suitable functionals and specializing the matrix A in certain ways we can obtain the so-called minimax theorem of the theory of games. That is, if this route is followed, the optimal vectors w^* and x^* will in fact be the optimal strategies for the opposing players in a two-person zero-sum game.‡

It is not proposed to follow this route here. Instead, we prove Theorem 4 in a way that provides direct access to details of the simplex tableaus.§

Recall, from the previous section, that we have defined n scalars

$$z_j \equiv \sum_{i \epsilon I} c_i x_{ij} \qquad j = 1, \ldots, n \qquad (26\text{-}23)$$

---

† For example, by the regularization procedures set forth in Ref. 5. See also Ref. 7 for further discussion.
‡ See the discussion in Sec. 23-4 where, as Thrall notes, the simplex method provides an efficient method for obtaining the solutions to a game.
§ This proof is adapted from the one given on pp. 559–560 in Ref. 7.

MAKING MODELS IN LINEAR PROGRAMMING

which are used to test optimality, as in Table 26-2, according to whether

$$z_j - c_j \geq 0 \tag{26-24}$$

is satisfied in all the n structural columns.†

Here, of course, we do not need any further distinction, and so "$i \, \varepsilon \, I$" represents the set of all indices in the basis for the tableau at an arbitrary stage in the iterations. Since $\{P_i | i \, \varepsilon \, I\}$ are a basis we also have, uniquely,

$$P_j = \sum_{i \varepsilon I} P_i x_{ij} \qquad j = 1, \ldots, n \tag{26-25}$$

On the other hand, the unit vectors are also a basis, and so we can write

$$P_j = \sum_{s=1}^{m} Q_s a_{sj} \qquad j = 1, \ldots, n \tag{26-26}$$

wherein a_{sj} is the sth component, in original data, for the vector P_j. Among the expressions (26-26) we distinguish those for the basis set as

$$P_i = \sum_{s=1}^{m} Q_s a_{si} \qquad i \, \varepsilon \, I \tag{26-27}$$

Hence, also, by direct substitution in (26-25),

$$P_j = \sum_{i \varepsilon I} P_i x_{ij} = \sum_{i \varepsilon I} \left(\sum_{s=1}^{m} Q_s a_{si} \right) x_{ij}$$

$$= \sum_{s=1}^{m} Q_s \sum_{i \varepsilon I} a_{si} x_{ij} \tag{26-28}$$

for each of the $j = 1, \ldots, n$ vectors P_j. Now all these expressions are unique, and hence a direct comparison with (26-26) produces

$$\sum_{i \varepsilon I} a_{si} x_{ij} = a_{sj} \tag{26-29}$$

for $s = 1, \ldots, m$, in each j.

The simplex method always chooses a vector

$$\omega' = (\omega_1, \ldots, \omega_s, \ldots, \omega_m) \tag{26-30}$$

which satisfies

$$\sum_{s=1}^{m} \omega_s a_{si} = c_i \tag{26-31}$$

for each $i \, \varepsilon \, I$. That is, the ω vector is chosen so that for $j = i \, \varepsilon \, I$ we shall have $z_i - c_i = 0$ under the basis vectors at this tableau stage. At any iteration we also have, by definition,

$$z_j \equiv \sum_{i \varepsilon I} c_i x_{ij} \qquad j = 1, \ldots, n \tag{26-32}$$

with

$$z_j - c_j \geq 0 \qquad \text{all } j \tag{26-33}$$

† We here switch to the optimality criterion for a maximization problem in order to underscore that we are no longer confined to (26-9) when we refer to Table 26-2. We are referring now to any problem of the form (26-16). Since the x_j are now all constrained to be nonnegative we eliminate all the last n columns—for $-P_j$—in Table 26-2. Also, the Q_i now become slack vectors, and hence we eliminate all the M_i as well.

required for establishing optimality. But substitution of (26-31) into (26-32) produces

$$\sum_{i \varepsilon I} \left(\sum_{s=1}^{m} \omega_s a_{si} \right) x_{ij} = \sum_{i \varepsilon I}^{m} \omega_s \sum_{i \varepsilon I} a_{si} x_{ij} = \sum_{s=1}^{m} \omega_s a_{sj} \qquad (26\text{-}34)$$

as follows from (26-29).

Now we observe that

$$\sum_{s=1}^{m} \omega_s a_{sj} \equiv \omega' P_j \qquad (26\text{-}35)$$

Hence (26-33) will be satisfied if and only if

$$\omega' P_j \geq c_j \qquad j = 1, \ldots, n \qquad (26\text{-}36)$$

Furthermore, for the unit vectors, represented as slack, Q_s, we have $c_s \equiv 0$, and thus in this case (26-36) becomes

$$\omega_s \geq 0 \qquad s = 1, \ldots, m \qquad (26\text{-}37)$$

Comparing (26-36) and (26-37) with (26-17) and observing that $A = (P_1, \ldots, P_n)$ we see that we are entitled to add a "tail" to the ω vector and write $\omega' = w'$ at the stage when an optimal tableau is reached. Furthermore each ω_s appears under a Q_s, and so we also see that the indicated w_s values are automatically available under the columns for the slack vector in the z row of the tableau. That is, the simplex method always generates w' values for (26-17) whenever it solves (26-16).†

There remains now only the task of showing that these $\omega_s = w_s$ are optimal so that we can assert, further, that $\omega_s^* = w_s^*$ and hence, as in Theorem 3, $c'x^* = w^{*'}P_0$ where the x^* are to be found under P_0 and the w^* are to be found under Q in the z row of an optimal tableau. For this we first specialize (26-32) by writing‡

$$z_0 = \sum_{i \varepsilon I} c_i x_i \qquad (26\text{-}38)$$

to indicate the z value (figure of merit) which the basic program variables x_i will generate at this stage. But then, via (26-31), which is $\omega' P_i = c_i$, all $i \varepsilon I$, we have, by substitution in (26-38),

$$\sum_{i \varepsilon I} (\omega' P_i) x_i = \omega' \sum_{i \varepsilon I} P_i x_i \qquad (26\text{-}39)$$

But

$$\sum_{i \varepsilon I} P_i x_i = P_0 \qquad (26\text{-}40)$$

and hence if we use

$$g_0 \equiv \omega' P_0 \qquad (26\text{-}41)$$

to indicate the corresponding value of the functional, g in (26-17), we see that (26-38) and (26-39) together give $z_0 = g_0$. This equality via the vector ω' is maintained in all tableaus,§ and an optimal tableau produces $\omega' = w'$. But, as was established for Theorem 3, every solution must satisfy $z \leq g$. Evidently an optimal tableau produces the highest value of $z_0 = z^*$. It also produces the lowest value $g_0 = g^*$ and hence, as Theorem 3 asserts, we shall have an optimum available in the tableau with $c'x^* = w^{*'}P_0$.

† The converse is also true, as can be shown by analogous reasoning, and hence either problem may be chosen, at a user's pleasure, whenever the simplex method is to be used for its solution.

‡ It would be more elegant to write this as $z_0 = \sum_{i \varepsilon I} c_i x_{i0}$, as is usually done in the so-called uniform tableau notation. See, for example, p. 426 in Ref. 7.

§ This may be made the basis of producing new algorithms which are of the so-called subdual variety, as is explained in Chapter 14 of Ref. 7.

The following are now easy consequences of Theorems 3 and 4:

Corollary 3. If solutions exist for both members of a dual pair of linear-programming problems then they will both have finite and equal optimal figures of merit.

Corollary 4. If either member of a dual pair of linear-programming problems has a finite optimum then the other will also have a finite optimum.

As a consequence of the above developments, we can prove that (26-9) has a finite optimum. We have already observed that it has a solution. But now using I for the identity matrix and $\mathfrak{M}$ for the m-rowed vector with M_i, $i = 1, \ldots, m$, for its components we can write the dual to (26-9) as

$$\max w'\hat{P}_0(\epsilon)$$

subject to

$$\begin{aligned} w'I &\leq \mathfrak{M} \\ w'A &\leq 0 \\ -w'A &\leq 0 \end{aligned} \qquad (26\text{-}42)$$

with w otherwise unrestricted.† This may equally‡ well be rendered as

$$\max w'P_0(\epsilon)$$

subject to

$$\begin{aligned} w'I &\leq \mathfrak{M} \\ w'A &= 0 \end{aligned} \qquad (26\text{-}43)$$

But now we observe that (26-43) always has the solution $w' = 0$. Hence, by Corollary 3 we see that (26-9), which is solvable, must also be bounded.§ The theorems and corollaries in our preceding section therefore hold without qualification, and the model (26-9) may always be used to characterize the rank of any matrix A when it is completed by the simplex algorithm.

26-4. Constrained Regressions

The immediately preceding sections have emphasized algorithmic completion by reference to the simplex method. We shall want to return to this topic for some further extensions that can be secured by altering one or more aspects of the simplex

Table 26-3. Regression Data†

y_j	3	3	7	6	12
x_j	1	2	4	5	8

† From B. W. Lindgren and G. W. McElrath, *Introduction to Probability and Statistics*, p. 231, The Macmillan Company, New York, 1959.

algorithm. Before doing this, however, it is useful to turn once more to the properties of an optimization and study ways in which they might also be employed to deal with other kinds of nonlinear problems irrespective of the algorithms that are utilized.

Consider as an example the situation which obtains when a curve is to be fitted to the observations obtained from an experiment. The data of Table 26-3 provide a simple illustration. Here it is assumed that the curve to be fitted is of the form

$$y = a + bx \qquad (26\text{-}44)$$

relative to the data supplied in Table 26-3.

† In duality we have the variables in one problem restricted to be nonnegative only when the corresponding constraints in the other problem are inequalities. See Table 7 on p. 191 of Ref. 7 for a suggested mnemonic and proof.

‡ That is, (26-41) and (26-42) are wholly equivalent in the sense used for (25-1) and (25-2); i.e., *all* w which satisfy (26-42) will satisfy (26-43) and vice versa.

§ Equivalently we may observe that (26-9) is bounded by the nonnegativity conditions on the y_i and then proceed via Corollary 4 to establish that (26-42) and (26-43) also have a finite optimum.

One approach is to choose a "best" curve in accordance with the least-squares principle, viz.,

$$\min_{a,b} \sum_{j=1}^{5} (y_j - a - bx_j)^2 \qquad (26\text{-}45)$$

which means that we are to consider all possible values for a and b—which are the variables here—and choose the ones that minimize this sum.

An equivalent way of choosing the minimizing a and b values is secured by solving for a and b in

$$\sum_{j=1}^{5} y_j = 5a + b \sum_{j=1}^{5} x_j$$
$$\sum_{j=1}^{5} y_j x_j = a \sum_{j=1}^{5} x_j + b \sum_{j=1}^{5} x_j^2 \qquad (26\text{-}46)$$

which are the so-called normal equations obtained by differentiating (26-45) first with respect to a and then with respect to b, setting the results equal to zero, and solving.

The result of solving (26-46), with the data of Table 26-3, is $a = 1$, $b = 1.3$; substi-

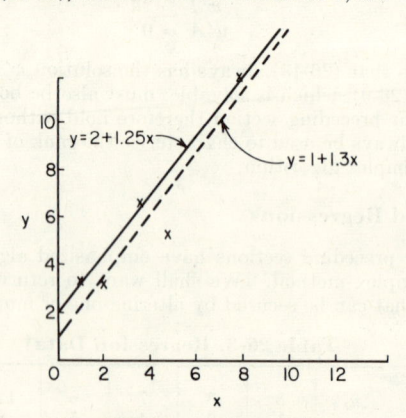

FIG. 26-1. Regression examples.

tuting in (26-44) yields $y = 1 + 1.3x$, as plotted in Fig. 26-1. The resulting expression may be thought of as a "mean regression" by reference to the fact that the minimum of a sum of squares is taken on at the mean. Alternatively, a "median-regression" approach may be developed by minimizing a sum of absolute deviations—viz.,

$$\min_{a,b} \sum_{j=1}^{5} |y_j - a - bx_j| \qquad (26\text{-}47)$$

wherein the vertical strokes mean that the absolute values are to be taken for the corresponding expressions in the sum.

A linear-programming equivalent for (26-47) is

$$\min g \equiv \delta_1^+ + \delta_1^- + \delta_2^+ + \delta_2^- + \delta_3^+ + \delta_3^- + \delta_4^+ + \delta_4^- + \delta_5^+ + \delta_5^-$$

subject to

$$\begin{aligned}
3 &= a + b - \delta_1^+ + \delta_1^- \\
3 &= a + 2b \phantom{{}+ b} - \delta_2^+ + \delta_2^- \\
7 &= a + 4b \phantom{{}+ b - \delta_2^+ + \delta_2^-} - \delta_3^+ + \delta_3^- \\
6 &= a + 5b \phantom{{}+ b - \delta_2^+ + \delta_2^- - \delta_3^+ + \delta_3^-} - \delta_4^+ + \delta_4^- \\
12 &= a + 8b \phantom{{}+ b - \delta_2^+ + \delta_2^- - \delta_3^+ + \delta_3^- - \delta_4^+ + \delta_4^-} - \delta_5^+ + \delta_5^-
\end{aligned} \qquad (26\text{-}48)$$

$$\delta_1^+,\ \delta_1^-,\ \delta_2^+,\ \ldots,\ \delta_5^- \geq 0$$

which has as its solution $a = 2$, $b = 1.25$ with the result $y = 2 + 1.25x$, which is also plotted in Fig. 26-1.

Here we observe that these two "best" (or optimal) fits are different. We do not, however, discuss the relative merits of mean regressions vs. median regressions, or the other alternatives that are available.† Instead, we want to show that it is the optimization in (26-48) that makes it equivalent to (26-47).

Instead of proceeding via (26-47) and (26-48), however, we move to a more general formulation and suppose that, in general, we want to fit an expression of the form

$$s = \sum_{i=1}^{n} a_i y_i \qquad (26\text{-}49)$$

by means of a median regression. We have now effected a complete change in notation so that in (26-49) the y_i, $i = 1, \ldots, n$ are independent variables against which the one dependent variable s is to be regressed. We also want to emphasize that the relation (26-49) need not be linear. In fact, the y_i values may themselves be deducible from vectors of other variables $x = (x_1, \ldots, x_n)$ only via the known relations

$$y_i = f_i(x) \qquad (26\text{-}50)$$

The immediate problem, in any event, is to estimate the a_i values.

For our present discussion we assume that there are observations $y_{ik}, k = 1, \ldots, m$, on each y_i—generated perhaps via (26-50)—and that these are associated with corresponding s_k observations so that, via a suitable pairing, we can arrange expressions like

$$\sum_{i=1}^{n} a_i y_{ik} - s_k = \epsilon_k \qquad k = 1, \ldots, m \qquad (26\text{-}51)$$

from which to form our estimates. Here we admit any choice of a_i values‡ and then measure the "observational fit" by reference to the ϵ_k deviation value which is secured for each of these $k = 1, \ldots, m$ expressions.

Given the observations y_{ik} and s_k, $k = 1, \ldots, m$, we can form our median-regression estimation problem as

$$\min \sum_{k=1}^{m} \left| \sum_{i=1}^{n} a_i y_{ik} - s_k \right| = \min \sum_{k=1}^{m} |\epsilon_k| \qquad (26\text{-}52)$$

which means that we are to choose the a_i, $i = 1, \ldots, n$, values so that the resulting ϵ_k deviations, whether positive or negative, are to be minimized by reference to their absolute values.

A "goal-programming" model§ is said to be involved whenever some of the constraints are incorporated in the functional, and this is the way that we shall develop the model for our regression estimation problem. First, we replace the constraints of (26-51) by

$$\sum_{i=1}^{n} a_i y_{ik} - s_k = \epsilon_k^+ - \epsilon_k^- \qquad \epsilon_k^+, \epsilon_k^- \geq 0 \qquad k = 1, \ldots, m \qquad (26\text{-}53)$$

where the indicated nonnegativity requirements may be used to define $\epsilon_k^+ - \epsilon_k^- = \epsilon_k$ with ϵ_k, as in (26-51), being unconstrained in its sign. Next we use an "artifact" and

---

† See Ref. 1 for simulation studies on this topic.

‡ That is, we propose to restrict ourselves to a case in which all observations are available, etc., and avoid the further (constrained-regression) issue associated with replacing or supplementing some of these observational constraints with other qualitative requirements. See, for example, Ref. 12 or Chapter 10 in Ref. 7.

§ This terminology is adopted from Appendix B of Ref. 7.

replace each term in (26-52) by

$$\left| \sum_{i=1}^{n} a_i y_{ik} - s_k \right| = \epsilon_k^+ + \epsilon_k^- \tag{26-54}$$

Then we orient the objective to minimization, as in (26-52), and combine this with the constraint set of (26-53). This produces the following as a model for our problem:†

$$\min \sum_{k=1}^{m} (\epsilon_k^+ + \epsilon_k^-)$$

subject to

$$\sum_{i=1}^{n} a_i y_{ik} - s_k - \epsilon_k^+ + \epsilon_k^- = 0 \qquad \epsilon_k^+, \epsilon_k^- \geq 0 \qquad k = 1, \ldots, m \tag{26-55}$$

Of course, if the ϵ_k^+ and ϵ_k^- are to be interpreted as deviations for the regression surface relative to the kth observation then $\epsilon_k^+, \epsilon_k^- > 0$ for any k could not hold since this would mean that the resulting surface deviated simultaneously above and below (in nonzero amount) the corresponding observation s_k. As was seen for (26-8), this kind of ambiguity can be eliminated if the simplex algorithm is specified for completing (26-55). That is, we could then be sure that in every iteration

$$\epsilon_k^+ \epsilon_k^- = 0 \tag{26-56}$$

would be satisfied. On the other hand, we are not concerned with every iteration. We only want to be sure that (26-56) holds at an optimum, as we shall show to be the case for (26-55), no matter what algorithm is employed.

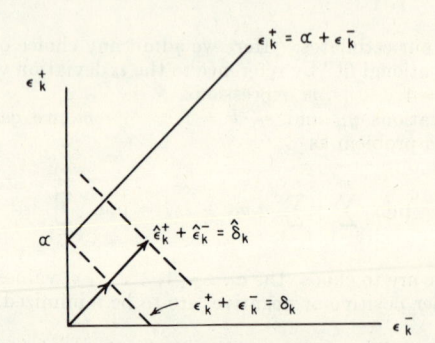

Fig. 26-2. Optimization to satisfy nonlinear conditions.

To prove that this is the case let *any* admissible set of values be specified for the variables a_i. But when this is done we can then consider each of the pairs $\epsilon_k^+, \epsilon_k^- \geq 0$ independently. Thus refer to Fig. 26-2, which reflects the situation for any constraint with $\Sigma a_i y_{ik} - s_k = \alpha = \epsilon_k^+ - \epsilon_k^-$, and so, for any α, the corresponding ϵ_k^+, ϵ_k^- must lie on a 45° line such as the one shown in Fig. 26-2. If now both ϵ_k^+ and ϵ_k^- are chosen to be positive then the orthogonal to the line defines a value like

$$\hat{\delta}_k = \hat{\epsilon}_k^+ + \hat{\epsilon}_k^- \tag{26-57}$$

where $\hat{\delta}_k$ measures its distance from the origin. But utilizing $\hat{\epsilon}_k^+ = \alpha + \epsilon_k^-$, we see

---

† That is, min $g \equiv \sum_{k=1}^{m} (\epsilon_k^+ + \epsilon_k^-)$.

that (26-57) produces $\hat{\delta}_k = \alpha + 2\epsilon_k^-$, and hence we could reduce this value to

$$\delta_k = \epsilon_k^+ + \epsilon_k^- = \alpha < \hat{\delta}_k \tag{26-58}$$

by choosing $\epsilon_k^- = 0$ and $\epsilon_k^+ = \alpha$ when $\alpha > 0$ applies. If $\alpha = 0$ then we have also $\epsilon_k^+ = \epsilon_k^- = 0$ while if $\alpha < 0$ then the alternate substitution would produce

$$\hat{\delta}_k = \hat{\epsilon}_k^+ + \hat{\epsilon}_k^- = 2\hat{\epsilon}_k^+ - \alpha \tag{26-59}$$
and so
$$\hat{\delta}_k > \delta_k = \epsilon_k^- = -\alpha > 0 \tag{26-60}$$

would be the uniquely smallest value attainable by setting $\epsilon_k^+ = 0$.

Of course, for any α we can always choose the ϵ_k^+ and ϵ_k^- for each constraint in the indicated manner. But then each constraint is thereby satisfied in (26-55). Thus solutions always exist for this model. Also $\epsilon_k^+, \epsilon_k^- \geq 0$ all k, and hence the functional is bounded for minimization. Hence we can call on Corollary 4 of the preceding section and assert that we have now proved

Theorem 5. The condition (26-56) *for all k must hold as a necessary condition for an optimum to the model* (26-55) *with* $\epsilon_k^+ = \epsilon_k^- = 0$ *whenever* $\sum_{i=1}^{n} a_i y_{ik} = s_k$ *for any constraint.*

This means that any algorithm may be used for solving (26-55) without any ambiguity for interpreting the resulting regression when an optimum has been achieved.

We should now observe that other approaches are also possible besides those which have been indicated. For instance, a "regression game" might be formed and one might then proceed to choose the a_i in order to minimize the maximum deviation.† Also one might add conditions like $a_i \geq 0$ for some i or, similarly, constrain sums of subsets of the coefficients and so on, which is the reason for calling these "constrained regressions."‡ Indeed one might impose various "symmetry conditions" such as

$$\sum_{k=1}^{m} \epsilon_k^+ = \sum_{k=1}^{m} \epsilon_k^- \tag{26-61}$$

on the deviations themselves. When conditions like (26-61) are imposed, however, the independence assumed for the proof of Theorem 5 no longer holds, and one would then have need for recourse to other devices such as "algorithmic completion" in order to guarantee that the resulting regression had all the properties that are wanted.

26-5. Some Transformations and Reductions

A Single Objective Design Problem. Before turning to some of the ways in which the simplex method can be modified in order to extend its power for algorithmic completion and treatment of nonlinear problems, we want to notice that various transformations can be used to reduce supposedly nonlinear problems to a linear equivalent. A commonly used example is the so-called logarithmic transformation which we now want to employ in association with some constrained optimization models to illustrate some of the indicated equivalences.

We shall shortly be concerned with certain formulations for modeling engineering designs along with transformations which are related to ones we shall first employ in the context of some concrete examples. Hence, in the spirit of our preceding sections, we commence with a very simple example to illustrate some of these ideas. These will be generalized directly thereafter and extended to more complex cases.

The example we shall use is formulated from the following considerations:§ A simple

---

† See Refs. 38, 39, and 1.

‡ It has also now been shown that when the simplex method is suitably modified it can be used to solve problems involved in minimizing a quadratic functional subject to linear inequality constraints. See Ref. 42. Hence a least-squares approach to the topic of constrained regressions is now also possible via this route.

§ The source of this example is Chapter 7 of Ref. 28, which should be referred to for an alternative development.

tensile bar is to be fabricated from a specified material. This bar must be designed so that it will have a fixed length L and transfer a specified constant force P.

If we let x_1 be the cross-sectional area, which is variable for the indicated bar, we might write our design objective as

$$\min C_m \equiv c_1 d x_1 L + c_2 x_1^{1/2} \qquad (26\text{-}62)$$

so that this nonlinear functional defines C_m as our figure of merit. The constant c_1, which is nonnegative, reflects the material cost in dollars per unit weight and c_2 (also a nonnegative constant) represents the production costs per unit. The production costs are thus seen to be directly proportional to the square root of the cross-sectional area. The material costs per unit, are, on the other hand, directly proportional to the weight $dx_1 L$, where L (as already indicated) is a fixed length and d is a weight-density constant for the material that has been specified.

It is now assumed that upper and lower limits for the cross-sectional area are given by constants d_2 and d_3, respectively, so that constraints of the form

$$d_2 \leq x_1 \leq d_3 \qquad (26\text{-}63)$$

are imposed on the variations that are permitted. Certain rigidity conditions must also be met, and these can be conveniently developed by reference to a new variable x_3, which is defined (and limited) by reference to x_1 in the following manner:

$$x_3 = \frac{x_1 E}{L} \geq k_1 \qquad (26\text{-}64)$$

Here E is Young's modulus for the specified material† and the constant k_1 stipulates a minimum permissible value for the bar's axial rigidity. Finally, to complete our constraints we may suppose that yielding cannot be tolerated. Again introducing a new variable we write

$$x_2 \leq d_1 \qquad (26\text{-}65)$$

for the indicated condition and then, from elementary strength of materials, we define x_2 by

$$x_2 \equiv \tau_{\max} = \frac{P}{2x_1}$$

$$\tau_{\max} \leq \frac{S}{2N} = d_1 \qquad (26\text{-}66)$$

where the constant S is the yield strength of the designated material and N is a factor of safety whose value has been presupposed for this design.‡

Collecting the above conditions together we can now write the model for our design problem as

$$\min C_m \equiv c_1 d x_1 L + c_2 x_1^{1/2}$$

subject to

$$\begin{aligned} x_2 - \frac{P}{2x_1} &= 0 \\ x_2 &\leq d_1 \\ x_1 &\geq d_2 \\ x_1 &\leq d_3 \\ x_3 - \frac{E}{L} x_1 &= 0 \\ x_3 &\geq k_1 \\ x_1, x_2, x_3 &\geq 0 \end{aligned} \qquad (26\text{-}67)$$

† The values of E and d_1 are constants fixed by the material. But the duality relations of linear programming permit these to be studied and evaluated parametrically, if desired.

‡ But cf. the preceding footnote for comments that are also pertinent to a treatment of variations in S and N.

MAKING MODELS IN LINEAR PROGRAMMING

Because C_m is monotonic in x_1 we can evidently simplify the above expressions and achieve an equivalent, but much simpler, model for direct minimization of the variable x_1. We want to illustrate another process, however, and so we proceed via a different route and apply logarithmic transformations to the above constraints and obtain

$$
\begin{aligned}
\ln x_1 + \ln x_2 &= \ln \frac{P}{2} \\
-\ln x_1 &\geq -\ln d_3 \\
\ln x_1 &\geq \ln d_2 \\
-\ln x_2 &\geq -\ln d_1 \\
\ln x_1 - \ln x_3 &= -\ln \frac{E}{L} \\
\ln x_3 &\geq \ln k_1 \\
x_1, x_2, x_3 &\geq 0
\end{aligned}
\tag{26-68}
$$

These are evidently standard linear inequality constraints, as can be seen by writing

$$y_j \equiv \ln x_j \tag{26-69}$$

and observing that these y_j values are not constrained to be nonnegative.

The functional in (26-67) does not admit of a similar reduction. On the other hand, the functional is strictly increasing in x_1. Also the logarithmic transformation is strictly monotone. Therefore, the functional is also strictly increasing in $y_1 = \ln x_1$.

This means that we can attain an equivalent objective for (26-67), and in fact, we can write our new problem as

$$\min y_1$$

subject to

$$
\begin{aligned}
y_1 + y_2 &= \ln \frac{P}{2} \\
-y_1 &\geq -\ln d_3 \\
y_1 &\geq \ln d_2 \\
-y_2 &\geq -\ln d_1 \\
y_1 - y_3 &= -\ln \frac{E}{L} \\
y_3 &\geq k_1
\end{aligned}
\tag{26-70}
$$

and be sure that the resulting optimal y^* values will, via (26-69), produce x^* values that are optimal for (26-67).

Compound Objectives. We want now to carry these same kinds of reductions one stage further in order to cast further light on possible model equivalences. For this purpose we refer again to Johnson,[28] Chapter 7, and observe that it may be necessary to consider "subsidiary" as well as "primary" design objectives. Then we propose to utilize these parts of an engineering-design optimization by extending the preceding example in a way that will also effect elision with the topic of minimax† and other kinds of compound objectives in linear programming.

For illustration we suppose that there is some uncertainty with respect to the customer demands that will emerge for the bar that is being designed. In particular, we assume that the bar either will be shipped by rail or will be sold over the counter. It is not known, in advance, which set of these two demands will be greater, and to handle these uncertainties, we can suppose that a "minimax" objective is prescribed for the designer. That is, the designer is to minimize the maximum costs that may be experienced for either of the two types of demand.

To develop the indicated illustration, let it be supposed that items sold over the counter will have a unit cost given by

$$C_m \equiv c_1 x_1 x_4 L + c_2 x_1^{1/2} \tag{26-71}$$

† For a further treatment of minimax objectives, see Chapter 23.

This expression is the same as (26-62) except that the weight-density factor is now represented by x_4, which is here allowed to be variable. These C_m costs are assumed to be applicable irrespective of whether the item is shipped by rail or sold over the counter. In the former case, however, an additional unit shipment cost C_s will be experienced as given by

$$C_s \equiv c_3(x_1 x_4 L)^{3/4} \tag{26-72}$$

In order to secure access to the rates defined by $c_3 \geq 0$, however, the weight-density conditions used in the design must conform to

$$x_4 \geq d \tag{26-73}$$

If we now define

$$\begin{aligned} f_1(x) &\equiv C_m \\ f_2(x) &\equiv C_m + C_s \end{aligned} \tag{26-74}$$

we can then write our compound-objective problem in the form

$$\min_x \max \{f_1(x), f_2(x)\} \tag{26-75}$$

where the choice of x values is restricted by the constraints given in (26-67) and (26-73). Alternatively, we can achieve an equivalent problem by writing

$$\min z$$

subject to

$$\begin{aligned} f_1(x) &\leq z \\ f_2(x) &\leq z \end{aligned}$$

and

$$\begin{aligned} x_2 - \frac{P}{2x_1} &= 0 \\ x_2 &\leq d_1 \\ x_1 &\geq d_2 \\ x_1 &\leq d_3 \\ x_3 - \frac{E}{L} x_1 &= 0 \\ x_3 &\geq k_1 \\ x_4 &\geq d \end{aligned} \tag{26-76}$$

Here the constraints are, as before, drawn from (26-67) and (26-73) while, additionally—via (26-74)—two new constraints are introduced in place of the compound (minimax) objective prescribed in (26-75). To see that this new model will have solutions which achieve the indicated objective it is only necessary to observe that *any* choice of x values will prescribe a lower limit to z by reference to the maximum of the two values f_1 and f_2. Because the objective is to minimize z, however, it is evident that an optimum will be achieved only by effecting an x choice, satisfying the constraints, that will minimize over the set of maximum values for f_1 and f_2 as in (26-75).

In this simple case we can, of course, effect a reduction back to (26-70) as a model equivalent for (26-76). To see that this is so we observe that we can first set $x_4 = d$ without worsening any other choice of x in (26-76). This means, in turn, that we can set $x_4 = d$ in each of the functionals defined by (26-74). Then eliminating the last constraint in (26-76) and observing that C_m and C_s are both increasing functions of x_1 it is evident that we are back again to (26-67) and thus, via the preceding development, we can also use (26-70) and (26-68) as a model equivalent for (26-76).

Transformations and Reductions for Designs under Inequality Constraints—General Case. The preceding examples illustrated certain aspects of linear-programming model equivalences. They are not meant to imply that all problems in engineering design can be reduced to ordinary linear-programming problems.

Some of the ideas in the preceding developments can be generalized, however, and clarified in a way that leads naturally into some of the algorithmic adjustment and completion procedures that we want to examine. Consider, for this purpose, the following generalization of a design objective like, say, (26-62) or (26-75):†

$$\min g \equiv \sum_{i=1}^{m} c_i \prod_{j=1}^{n} t_j{}^{d_{ij}} \equiv c_1(t_1{}^{d_{11}} \cdots t_n{}^{d_{1n}}) + \cdots + c_m(t_1{}^{d_{m1}} \cdots t_n{}^{d_{mn}}) \quad (26\text{-}77)$$

Here, as before, we assume that the c_i are nonnegative constants but the d_{ij}, also constants, may have any sign. R. J. Duffin[19,20] and C. Zener[43,44] have dealt with this class of problems where the t_j choices are restricted only by $t_j \geq 0, j = 1, \ldots, n$. Subsequent work, as in Refs. 11 and 16, has extended this to constraints of the form

$$L_i \leq \prod_{k=1}^{n} t_k{}^{d_{ik}} \leq U_i \quad (26\text{-}78)$$

or

$$\ln L_i \leq \sum_{k=1}^{n} d_{ik} \ln t_k \leq \ln U_i \quad (26\text{-}79)$$

where the L_i and U_i values are known constants.

In effecting reductions for the rather involved functional in (26-77) we first take advantage of the nonnegativity imposed on the t_j choices and utilize the definition of a logarithm to write

$$0 \leq t_j \equiv e^{\ln t_j} \equiv e^{u_j} \quad (26\text{-}80)$$

Evidently $u_j \equiv \ln t_j$ is not restricted to nonnegative ranges.

Via this transformation we can now replace the function of (26-77) and achieve an equivalent model by writing

$$\min \mathcal{C}(u) = \sum_{k=1}^{m} c_k e^{\sum_{j=1}^{n} d_{kj} u_j}$$

subject to

$$\sum_{j=1}^{m} a_{ij} u_j \geq b_i$$

(26-81)

where each b_i now designates a value for the corresponding $\ln L_i$ or $\ln U_i$ in (26-78) or (26-79) and the constraints are oriented, as indicated, by effecting suitable sign adjustments.

The problem achieved in (26-81) is still rather recondite; so we further simplify it by writing

$$\min \mathcal{C}(w) = \sum_{k=1}^{m} c_k e^{w_k}$$

subject to

$$\sum_{j=1}^{n} d_{kj} u_j - w_k = 0$$

$$\sum_{j=1}^{n} a_{ij} u_j \geq b_i$$

(26-82)

To see that this is, in fact, an equivalent for (26-81) we need only note that the constraints indexed by i are the same in both problems whereas the constraints indexed

† This is also a generalization of certain functions which are of interest in economics. See Ref. 2.

by k in (26-82) are definitional in the sense that these w_k values are equal to the sums forming the exponents for the functional in (26-81).

The chain leading from (26-77) to (26-82) now establishes that (26-77) subject to (26-78) is equivalent to (26-82). A fortiori, the functional of (26-77), subject only to nonnegativity on the t_j choices, is equivalent to (26-82) after eliminating all constraints indexed by i. Observe then that the latter separates the model's elements in a way that makes it possible to take advantage of certain properties that are possessed

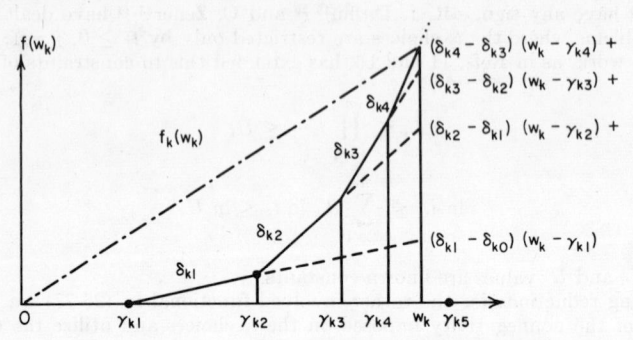

FIG. 26-3. Formulation for piecewise linear function.

by the new functional $\mathcal{C}(w) \equiv \mathcal{C}(w_1, \ldots, w_m)$. First, we observe that $\mathcal{C}$ is convex: its graph has a shape like the ones exhibited for the curve of solid lines in Fig. 26-3. Second, we have achieved a functional which is separable: $\mathcal{C}(w)$ can be written as a sum of m functions each involving only one variable in its argument—viz.,†

$$\mathcal{C}(w_1, \ldots, w_m) = \mathcal{C}_1(w_1) + \cdots + \mathcal{C}_m(w_m) \qquad (26\text{-}83)$$

The attainment of separability means that we can utilize piecewise linear approximations to achieve a solution, if wanted,‡ via the nonlinear power of so-called adjacent-extreme-point methods, of which the simplex method is only one example.§ The fact that a convex function in w has been achieved means, moreover, that we do not have to utilize modifications like the restricted-basis-entry procedures, shortly to be discussed, and hence we may immediately apply, say, the simplex method for the minimizing objective that is here specified.

26-6. Algorithmic Alterations

We now turn to possible extensions in our algorithmic model completions where we again focus on the simplex method and initiate our discussion by the relatively simple example portrayed in Fig. 26-3. Here we wish to minimize a piecewise linear functional obtained by transforming a function of the form $F(y_1, \ldots, y_n)$ into a series of possibly approximating piecewise linear functions, each involving only one variable in its argument.‖ That is, we wish to minimize, say, the resulting function $\sum_{i=1}^{p} \phi_i(w_i)$ subject to a set of linear inequality constraints on the admissible w_i choices.

In one approach to this topic we may proceed as follows:¶ Let $f_1(w_k), \ldots, f_4(w_k)$

---

† In this case the functionals $\mathcal{C}_1, \ldots, \mathcal{C}_m$ have the further important property that they are all exponentials which differ only by their associated c_k constants. See Ref. 11.

‡ Alternatively exact characterizations and duality interpretations can be achieved via the general convex programming theory set forth in Ref. 14.

§ Another equally general method is C. E. Lemke's dual method. See Ref. 35.

‖ Such functions are said to be "separable." See, for example, Ref. 7, pp. 351ff.

¶ See Chapter 10 in Ref. 7 for another development which, more generally, transforms this into a problem in high dimensions.

with slopes $\delta_{k1}, \ldots, \delta_{k4}$, respectively, be expressions for successive segments of one $\phi_i(w_i)$, as represented by the solid lines of Fig. 26-3. Now consider the function $f_k(w_k)$ represented by the sequence of dots and dashes from the origin to $f_4(w_k)$ for some $\gamma_{k4} \leq w_k \leq \gamma_{k5}$. Here we have, then,

$$f_k(w_k) = f_4(w_k) \tag{26-84}$$

and also
$$\delta_{k4} = \frac{f_4(w_k) - f_4(\gamma_{k4})}{w_k - \gamma_{k4}} \tag{26-85}$$

or
$$\begin{aligned} f_4(w_k) &= \delta_{k4}(w_k - \gamma_{k4}) + f_4(\gamma_{k4}) \\ &= \delta_{k4}(w_k - \gamma_{k4}) + f_3(\gamma_{k4}) \end{aligned} \tag{26-86}$$

since $f_4(\gamma_{k4}) = f_3(\gamma_{k4})$. (See the intersection of the graphs of f_3 and f_4 at γ_{k4} in Fig. 26-3.) However,

$$\delta_{k3} = \frac{f_3(\gamma_{k4}) - f_3(\gamma_{k3})}{\gamma_{k4} - \gamma_{k3}} \tag{26-87}$$

and so, clearing fractions, and adding and subtracting $w_k \delta_{k3}$, we obtain

$$f_3(\gamma_{k4}) = -\delta_{k3}(w_k - \gamma_{k4}) + \delta_{k3}(w_k - \gamma_{k3}) + f_3(\gamma_{k3}) \tag{26-88}$$

Direct substitution in (26-86) then produces

$$f_4(w_k) = (\delta_{k4} - \delta_{k3})(w_k - \gamma_{k4}) + \delta_{k3}(w_k - \gamma_{k3}) + f_3(\gamma_{k3}) \tag{26-89}$$

But now we observe that $f_3(\gamma_{k3}) = f_2(\gamma_{k3})$ and

$$\delta_{k2} = \frac{f_2(\gamma_{k3}) - f_2(\gamma_{k2})}{\gamma_{k3} - \gamma_{k2}} \tag{26-90}$$

or
$$f_3(\gamma_{k3}) = f_2(\gamma_{k3}) = -\delta_{k2}(w_k - \gamma_{k3}) + \delta_{k2}(w_k - \gamma_{k2}) + f_2(\gamma_{k2}) \tag{26-91}$$

and, hence,

$$f_4(w_k) = (\delta_{k4} - \delta_{k3})(w_k - \gamma_{k4}) + (\delta_{k3} - \delta_{k2})(w_k - \gamma_{k3}) + \delta_{k2}(w_k - \gamma_{k2}) + f_2(\gamma_{k2}) \tag{26-92}$$

Evidently, then, we can continue, as before, and write

$$f_4(w_k) = \sum_{s=1}^{4} (\delta_{ks} - \delta_{k(s-1)})(w_k - \gamma_{ks}) \tag{26-93}$$

with $\delta_{k0} \equiv 0$ and so, via (26-84), we have

$$f_k(w_k) = \sum_{s=1}^{4} (\delta_{ks} - \delta_{k(s-1)})(w_k - \gamma_{ks}) \tag{26-94}$$

for any $\gamma_{k4} \leq w_k \leq \gamma_{k5}$ in this case.

We now proceed more generally and suppose that we have, at most, l segments to consider for any $\phi_i(w_i)$.† Thus we write our function as

$$\sum_{k=1}^{m} \sum_{s=1}^{l} (\delta_{ks} - \delta_{k(s-1)})(w_{ks}^{+} - w_{ks}^{-}) \tag{26-95}$$

where
$$w_k - w_{ks}^{+} + w_{ks}^{-} = \gamma_{ks} \qquad w_{ks}^{+}, w_{ks}^{-} \geq 0 \qquad s = 1, \ldots, l \tag{26-96}$$

Since, by hypothesis, the simplex algorithm is specified for use, we are also assured that, always,

$$w_{ks}^{+} w_{ks}^{-} = 0 \tag{26-97}$$

---

† In cases where some segments are to be interdicted we assign a coefficient $\delta_{ks} = M$ for the minimizations being considered.

and $w_{ks}{}^+ = w_{ks}{}^- = 0$ if and only if $w_k = \gamma_{ks}$. We also further require

$$w_k - \gamma_{ks} = \begin{matrix} w_k - \gamma_{ks} & \text{if } w_k \geq \gamma_{ks} \\ 0 & \text{if } w_k < \gamma_{ks} \end{matrix} \qquad (26\text{-}98)$$

for each k and s. However, our model has been developed so that, with the specified algorithmic completion, we can satisfy the nonlinear requirements in (26-98) by merely omitting $w_{ks}{}^-$ from the functional. That is, we replace (26-95) by the simpler

$$\sum_{k=1}^{m} \sum_{s=1}^{l} (\delta_{ks} - \delta_{k(s-1)}) w_{ks}{}^+ \qquad (26\text{-}99)$$

and leave (26-96) and (26-97) unaltered. To verify that satisfaction of (26-98) is thereby achieved we observe that if $\gamma_{kt} < w_k < \gamma_{kt+1}$, for any t, we can have only $w_{ks}{}^+ = w_k - \gamma_{ks} > 0$ for $s = 1, \ldots, t$ in the functional while all other $w_{ks}{}^+ = 0$ and $w_{ks}{}^- = \gamma_{ks} - w_k > 0$, $s = t+1, \ldots, l$, appears only in the constraints.

We can also use the above mode of development to secure piecewise linear approximations for solving problems of the kind posed in the preceding section since, as we saw, we were able to effect a transformation into a sum of functions of the form $\mathcal{C}_k(w_k) = c_k e^{w_k}$, and hence thereby achieved the requisite separability. Thus by choosing a suitable approximating grid† we could replace (26-82) by

$$\min \sum_{k=1}^{m} \sum_{s=1}^{l} (\delta_{ks} - \delta_{k(s-1)}) w_{ks}{}^+$$

subject to

$$\sum_{j=1}^{n} d_{kj} u_j - w_k = 0$$
$$\sum_{j=1}^{n} a_{ij} u_j \geq b_i \qquad (26\text{-}100)$$
$$w_k - w_{ks}{}^+ + w_{ks}{}^- = \gamma_{ks}$$
$$w_{ks}{}^+, w_{ks}{}^- \geq 0$$

in which we can observe that the original constraints are further augmented, via (26-96), with new constraints which originate from the functional approximations utilized. But, of course, we must also specify a suitable algorithm such as the simplex method if we are to ensure, always, that (26-100) will be a valid representation for (26-82), to the degree of approximation that is desired, since we are also implicitly requiring the satisfaction of (26-97).

In this case the indicated algorithmic completion may be attained by reference to the simplex method because the minimization is here directed to a piecewise linear functional which is convex and increasing.‡ Of course, this very simple kind of algorithmic completion procedure will not always be obtainable. Simply changing the objective to maximizing $\mathcal{C}(w)$ in (26-82), for example, would (in general) introduce unbounded solution possibilities in the correspondingly altered approximating model of (26-100) even though no such possibilities were present in (26-82).§ These can be handled by suitably altering the admitted ways of forming bases in the simplex algorithm, however, and still further ways of restricting the vectors that may be

---

† Since minimization is the objective one would naturally use a finer grid to the left of the origin and a coarser grid to the right. Alternatively one could proceed via a sequence of models to obtain checks and further approximating refinements as required.

‡ The relevant theorem and proof may be found in the appendix to A. Charnes, W. W. Cooper, and B. Mellon, Programming and Sensitivity Analysis in an Integrated Oil Company, *Econometrica*, vol. 22, no. 2, pp. 193–217, April, 1954.

§ Cf. Chapter 10 in Ref. 7.

admitted as basis entries can serve to expand the nonlinear power of the simplex method.

It is not possible to explore all aspects of this topic, and so we shall proceed as follows:† The elaboration from (26-82) to (26-100) replaced a model of the form

$$\max \sum_{j=1}^{n} f_j(x_j)$$

subject to

$$\sum_{j=1}^{n} P_j x_j = P_0 \qquad x_j \geq 0 \qquad (26\text{-}101)$$

with one that (a) utilized piecewise linear approximations for the $f_j(x_j)$ and (b) expanded the constraints and variables to a new model that can be represented by

$$\max \sum_{j=1}^{n} c_j \lambda_j + \sum_{j=n+1}^{n'} c_j \lambda_j$$

subject to

$$\sum_{j=1}^{n} \binom{P_j}{D_j} \lambda_j + \sum_{j=n+1}^{n'} \binom{0}{G_j} \lambda_j = \binom{P_0}{G_0}$$

$$\lambda_j \geq 0 \qquad j = 1, \ldots, n, n+1, \ldots, n' \qquad (26\text{-}102)$$

In order to clarify what is involved here we can also write this as

$$\max \sum_{j=1}^{n} c_j \lambda_j + \sum_{j=n+1}^{n'} c_j \lambda_j$$

subject to

(i) $$\sum_{j=1}^{n} a_{ij} \lambda_j = b_i$$

(26-103)

(ii) $$\sum_{j=1}^{n} \gamma_{(i+m)j} \lambda_j + \sum_{j=n+1}^{n'} \gamma_{(i+m)j} \lambda_j = \gamma_{i+m}$$

where for (i) we have $i = 1, \ldots, m$ constraints associated with the initial‡ vectors P_j, P_0 while for (ii) we have $i = 1, \ldots, m'$ constraints arising from the approximations.

This model is written so that it admits of treatment for nonlinear constraint approximations§ as well as for nonlinear functionals in a simplex type of solution. Adjustments to the ordinary simplex algorithm will, in general, be required, however, along with alterations in the usual optimality criteria. There are numerous ways of effecting these alterations, but the key features involve so-called "restricted-basis-entry procedures" whose salient properties—under the indicated alterations from (26-101) to (26-102)—may be summarized as follows:

(a) If any $z_k - c_k < 0$ values appear in a simplex tableau for (26-102) then the

---

† The details of the following models are reproduced from Ref. 9, where these restricted-basis-entry procedures (and model alterations) were first presented as part of a systematic exploration of the nonlinear power of adjacent-extreme-point methods in general.

‡ That is, these initial P_j, P_0 vectors also reflect the transformations, constraints, etc., needed to achieve the separability for the $f_j(x_j)$ of (26-101).

§ Cf., for example, the further generalizations to nonlinear (but convex and separable) functions and constraints as given in Ref. 27. The so-called "mesh point" approaches to piecewise linear approximations for nonlinear characteristics of TEL in blending-type models as given in Ref. 31 extend the idea of restricted-basis-entry procedures to deal with piecewise linear approximation of nonlinear constraints and associated algorithmic alterations.

only replacements which are permitted will be either (i) a $\begin{pmatrix} P_k \\ D_k \end{pmatrix}$ for a $\begin{pmatrix} P_r \\ D_r \end{pmatrix}$ or (ii) a $\begin{pmatrix} 0 \\ G_k \end{pmatrix}$ for a $\begin{pmatrix} 0 \\ G_r \end{pmatrix}$.

(b) An optimum† is attained for (26-102) when $z_j - c_j \geq 0$ for all *permissible* replacements.

All restricted-basis-entry procedures conform to the principles underlying these rules, although, of course, numerous added variations and further model alterations are possible and different kinds of algorithmic completion properties can thereby be obtained for use in model-construction strategies. When the indicated restrictions are applied, however, there can be no confusion—as witness, for instance, the fact that (a) a $\begin{pmatrix} P_r \\ D_r \end{pmatrix}$ and $\begin{pmatrix} P_k \\ D_k \end{pmatrix}$ interchange can open an unbounded solution possibility only if this was already present in the original constraints of (26-101), and hence (b) other such possibilities that are encountered when maximizing a convex function, say, can then be ignored as belonging to the class of nonpermitted replacements. Finally, it should be noted that (a) the extreme-point properties of the simplex method continue to be exploited in a way that satisfies conditions like (26-97) and (b) this is done while permitting the appearance of optima that do not necessarily occur on the extreme points associated with the original constraints of (26-101).

26-7. Summary and Conclusion

Comprehensive coverage of all aspects of model-making theory was not an intended objective in this chapter. The intent was rather to focus on certain possibilities which are easily overlooked when "engineering"‡ a linear-programming application. For this purpose, we first stressed the need for distinguishing between a problem and the way it might be modeled. This included, for example, a possible search for intersections between the wanted solutions of a nonlinear problem, say, and a linear model erected to deal with it. It might also have included the use of approximating model types (e.g., transportation model types) for obtaining bounds to possible solutions, advanced starts, and other kinds of useful characterizations.§

Bearing these kinds of considerations in mind we also turned to possible uses of optimization properties as another aspect of a modeling strategy. This kind of development, and the preceding ones, led us rather naturally into a discussion of some of the model equivalents that might be obtained by effecting various transformations. These transformations, together with the properties of a constrained optimization, could also be used to obtain still further extensions—as when we were able, for example, to obtain the replacement of a compound objective with a simpler one to deal with a larger problem in engineering design.

A great deal of linear-programming theory has been "computationally oriented," and it therefore seemed fitting to spend some time—if only briefly—on one (but only one) of the important linear-programming algorithms in order to see how it, too, could be incorporated in a model-construction strategy. Main emphasis was placed on the so-called restricted-basis-entry procedures, along with their related model alterations, with special reference to simplex-method extensions. This evidently does not exhaust the topic which we have referred to as algorithmic completion of a model. Thus we have not dealt with topics such as joining two or more algorithms for use at different stages of analysis, as is done in diophantine linear programming, for instance, where the algorithms may be used, *inter alia*, to generate new constraints in a way that ultimately generates the model equivalents that are wanted even when the exact model is not explicitly known *ab initio*.‖ Alternatively, we have also not considered cases where two or more linear-programming models can be joined and sepa-

---

† This may be only a local star (= local extreme point) optimum, however. Cf. Ref. 9.
‡ This term is also related to the ideas of "modeling for understanding" and "modeling for operational feasibility," as discussed in Sec. 26-1 and Ref. 6.
§ See Chapter 14 in Ref. 7.
‖ See Refs. 22, 23, 24, and 25 as well as 3 and 18.

rately solved, perhaps by a single algorithm, in order to obtain solutions to other kinds of nonlinear problems.† But, as these concluding remarks suggest, progress continues to be made in these and other aspects of linear programming, and the interested reader is best referred to the literature in which this progress is constantly being reported. This includes, as prime sources, the following three journals: *Management Science, Operations Research,* and *Naval Research Logistics Quarterly* which are published by, respectively, The Institute of Management Sciences, Operations Research Society of America, and U.S. Office of Naval Research.

REFERENCES

1. Agin, N.: *A Comparative Study of Regression Models: Least Squares, Least Deviations, Chebyshev and Non-parametric,* Carnegie Institute of Technology, Graduate School of Industrial Administration, Pittsburgh, Pa., May, 1960.
2. Arrow, K., H. B. Chenery, B. Minas, and R. M. Solow: Capital-Labor Substitution and Economic Efficiency, *Rev. Econ. Statist.*, vol. 43, no. 3, pp. 225–250, August, 1961.
3. Beale, E. M. L.: A Method for Solving Linear Programming Problems When Some But Not All of the Variables Must Take Integral Values, Princeton University Statistical Techniques Research Group *Tech. Rept.* 19, July, 1958.
4. Bowman, E. H., and R. B. Fetter (eds.): *Analysis for Production Management,* Richard D. Irwin, Inc., Homewood, Ill., 1957.
5. Charnes, A.: Optimality and Degeneracy in Linear Programming, *Econometrica,* vol. 20, no. 2, pp. 160–170, 1952.
6. Charnes, A., and W. W. Cooper: Constrained Extremization Models and Their Use in Developing System Evaluation Measures, in *General Systems Yearbook,* John Wiley & Sons, Inc., New York, 1964.
7. Charnes, A., and W. W. Cooper: *Management Models and Industrial Applications of Linear Programming,* 2 vols., John Wiley & Sons, Inc., New York, 1961.
8. Charnes, A., and W. W. Cooper: Network Interpretation and a Directed Sub-dual Algorithm for Critical Path Scheduling, *J. Ind. Eng.*, vol. 13, no. 4, pp. 213–218, July–August, 1962.
9. Charnes, A., and W. W. Cooper: Nonlinear Power of Adjacent Extreme Point Methods in Linear Programming, *Econometrica,* vol. 25, pp. 132–153, January, 1956.
10. Charnes, A., and W. W. Cooper: On Some Works of Kantorovich, Koopmans and Others, *Management Sci.*, vol. 8, no. 3, pp. 246–263, April, 1962.
11. Charnes, A., and W. W. Cooper: Optimizing Engineering Designs under Inequality Constraints, *ONR Research Memo* 64, Northwestern University, Evanston, Ill., August, 1962.
12. Charnes, A., W. W. Cooper, and R. Ferguson: Optimal Estimation of Executive Compensation by Linear Programming, *Management Sci.*, vol. 1, no. 2, pp. 138–151, January, 1955.
13. Charnes, A., W. W. Cooper, and A. Henderson: *An Introduction to Linear Programming,* John Wiley & Sons, Inc., New York, 1953.
14. Charnes, A., W. W. Cooper, and K. Kortanek: Duality in Semi-infinite Programs and Some Works of Haar and Carathéodory, *Management Sci.*, vol. 9, no. 2, pp. 209–228, January, 1963.
15. Charnes, A., W. W. Cooper, and B. Mellon: Blending Aviation Gasolines—A Study in Programming Interdependent Activities, *Econometrica,* vol. 20, no. 2, pp. 135–159, April, 1952.
16. Charnes, A., W. W. Cooper, and E. Schatz: Management Mathematics and Engineering Designs, *Proc. Second Conf. Engineering Designs,* University of California, Los Angeles, August, 1962.
17. Churchman, C. W.: Reliability of Models in the Social Sciences, Working Paper no. 7, Center for Research in Management Science, University of California, Berkeley, July 1, 1962.
18. Dantzig, G. B.: On the Significance of Solving Linear Programming Problems with Some Integer Variables, *Econometrica,* vol. 28, no. 1, pp. 30–44, January, 1960.
19. Duffin, R. J.: Cost Minimization Problems Treated by Geometric Means, *Operations Res.*, vol. 10, no. 5, pp. 668–675, September–October, 1962.
20. Duffin, R. J.: Dual Programs and Minimum Cost, *J. Soc. Ind. Appl. Math.*, vol. 10, no. 1, pp. 119–123, March, 1962.

† See, for example, A. Charnes and W. W. Cooper, Programming with Linear Fractional Functionals, *Naval Research Logistics Quart.*, vol. 9, nos. 3, 4, September–December, 1962, as well as Appendixes F and G in Ref. 7.

21. Fulkerson, D. R., and P. Wolfe: An Algorithm for Scaling Matrices, *Rept.* RM-2956, ASTIA no. AD 279143, The Rand Corporation, Santa Monica, February, 1962.
22. Gomory, R. A.: An Algorithm for Integer Solutions to Linear Programs, Princeton–IBM Mathematics Research Project, *Tech. Rept.* 1, November, 1958.
23. Gomory, R. A.: *All-integer Integer Programming Algorithm*, IBM Research Center, New York, January, 1960.
24. Gomory, R. A.: *A Method for the Mixed Integer Problem*, The Rand Corporation, Santa Monica, 1960.
25. Gomory, R. A., and W. J. Baumol: Integer Programming and Pricing, *Econometrica*, July, 1960.
26. Gomory, R. A., and A. Hoffman: Finding Optimum Combinations, *Intern. Sci. Technol.*, pp. 26–33, July, 1962.
27. Hartley, H. O.: Nonlinear Programming by the Simplex Method, *Econometrica*, vol. 29, no. 2, pp. 223–237, April, 1961.
28. Johnson, R. C.: *Optimum Design of Mechanical Elements*, John Wiley & Sons, Inc., New York, 1961.
29. Kantorovich, L. V.: *Mathematical Methods of Organizing and Planning Production*, Leningrad University, 1939 (Russian). Translated into English and republished under the same title, with a foreword by T. C. Koopmans, in *Management Sci.*, vol. 6, no. 4, pp. 363–422, July, 1960.
30. Kantorovich, L. V.: On the Translocation of Masses, *Dokl. Akad. Nauk. SSSR*, vol. 37, nos. 7–8, 1942. Republished under the same title with a foreword by A. Charnes in *Management Sci.*, vol. 5, no. 1, October, 1958.
31. Kawaratani, T. T., R. J. Ullman, and G. B. Dantzig: Computing Tetraethyl-lead Requirements in a Linear Programming Format, *Operations Res.*, vol. 8, no. 1, pp. 24–29, January–February, 1960.
32. Koopmans, T. C. (ed.): *Activity Analysis of Production and Allocation*, Cowles Commission Monograph 13, John Wiley & Sons, Inc., New York, 1951.
33. Koopmans, T. C.: On the Evaluation of Kantorovich's Work of 1939, *Management Sci.*, vol. 8, no. 3, pp. 264–265, April, 1962.
34. Kuhn, H. W., and A. W. Tucker (eds.): *Linear Inequalities and Related Systems*, Annals of Mathematics Studies, vol. 38, Princeton University Press, Princeton, N.J., 1956.
35. Lemke, C. E.: The Dual Method of Solving the Linear Programming Problem, *Naval Res. Logistics Quart.*, vol. 1, no. 1, pp. 36–47, 1954.
36. Mullins, W. W., and R. F. Sekerka: Application of Linear Programming Theory to Crystal Faceting, *J. Phys. Chem. Solids*, vol. 23, pp. 801–803, 1962.
37. Riley, V., and S. Gass: *Linear Programming and Associated Techniques: A Comprehensive Bibliography of Linear, Nonlinear and Dynamic Programming*, The Johns Hopkins Press, Baltimore, 1958.
38. Wagner, H.: Linear Programming Techniques for Regression Analysis, *J. Am. Statist. Assoc.*, vol. 54, pp. 206–212, 1959.
39. Wagner, H.: Non-linear Regression with Minimal Assumptions, *J. Am. Statist. Assoc.*, vol. 57, no. 299, pp. 572–578, September, 1962.
40. Weingartner, H. M.: *Mathematical Programming and the Analysis of Capital Budgeting Problems*, Prentice-Hall, Inc., Englewood Cliffs, N.J., 1963.
41. White, W. C., and M. B. Shapiro: Linear Programming Applied to Ultraviolet Absorption Spectroscopy, *Commun. Assoc. Computing Machinery*, vol. 6, no. 2, pp. 66–67, February, 1963.
42. Wolfe, P.: The Simplex Method for Quadratic Programming, *Econometrica*, vol. 27, no. 3, July, 1959.
43. Zener, C.: A Mathematical Aid in Optimizing Engineering Designs, *Proc. Natl. Acad. Sci. U.S.*, vol. 47, pp. 537–539, 1961.
44. Zener, C.: A Further Mathematical Aid in Optimizing Engineering Designs, Westinghouse Research Laboratories, *Rept.* 914-001-1000-P1, Pittsburgh, Pa., Jan. 13, 1962.

Chapter 27

DYNAMIC PROGRAMMING—THEORY AND APPLICATION

RICHARD BELLMAN, *The RAND Corporation.*

CONTENTS

27-1. Introduction	27-1
27-2. Multistage Processes	27-1
27-3. Functional Equations	27-2
27-4. Multistage Decision Processes	27-3
27-5. Functional-equation Approach	27-3
27-6. Principle of Optimality	27-4
27-7. Control Theory	27-4
27-8. Constraints	27-4
27-9. Feedback Control	27-5
27-10. Optimal Trajectories	27-5
27-11. Approximation in Function Space and Policy Space	27-5
27-12. Allocation Processes	27-5
27-13. Multistage Decision Processes of Continuous Type	27-6
27-14. Multistage Decision Processes of Stochastic Type	27-7
27-15. Applications	27-7
27-16. Adaptive Processes	27-8

27-1. Introduction

The term "dynamic programming" was coined to describe a new mathematical theory created to treat multistage decision processes and those processes which can be interpreted in this fashion. In order to indicate the strong link with classical analysis, we shall begin with a brief discussion of multistage processes and then turn to decision processes.

Our initial treatment will be abstract in order to illuminate the high points without any undue distraction due to detail. Following this, we shall present some detailed applications and numerous references to further applications. Extensive coverage will be found in the books.[1,2,3]

27-2. Multistage Processes

Let p be a point in an abstract space S and T be a transformation which takes S into itself, by which we mean that $T(p)$ belongs to S whenever p does. Applying T

repeatedly, starting with an initial point p, we generate a sequence of points

$$p_1 = T(p),\ p_2 = T(p_1),\ \ldots,\ p_{N+1} = T(p_N),\ \ldots \qquad (27\text{-}1)$$

We call the set of points generated in this manner a *multistage process*, and the point p is called a *state variable*.

With this multistage process, we associate a scalar function $g(p, p_1, \ldots, p_N, \ldots)$. It is clear that in the study of this function we cannot expect to obtain results of much interest unless we impose some structure. Let us then take g to have a simple form. Of frequent occurrence in mathematics itself and in applications, and therefore of primary importance, are functions of the type

$$\begin{aligned} & h(p) + h(p_1) + \cdots + h(p_N) + k(p_N) \\ & h(p, p_1) + h(p_1, p_2) + \cdots \\ & \max_i h(p_i) \end{aligned} \qquad (27\text{-}2)$$

These functions possess certain separable features which make their analytic and computational study feasible.

27-3. Functional Equations

To illustrate the method we shall use to study these functions, let us begin with the first function of (27-2),

$$g(p, p_1, p_2, \ldots, p_N) = h(p) + h(p_1) + \cdots + h(p_N) + k(p_N) \qquad (27\text{-}3)$$

a function associated with a finite multistage process. Since the points p_i are determined by means of the recurrence relation

$$p_{i+1} = T(p_i) \qquad i = 0, 1, 2, \ldots \qquad (p_0 \equiv p) \qquad (27\text{-}4)$$

it follows that g is a function of p and N. Hence, we may write

$$g(p, p_1, \ldots, p_N) = f_N(p) \qquad (27\text{-}5)$$

defined for $N = 0, 1, 2, \ldots$, and all p in S. Observe that we can derive a very simple recurrence relation for this sequence of functions, namely,

$$\begin{aligned} f_N(p) &= g(p) + [g(p_1) + g(p_2) + \cdots + g(p_N) + k(p_N)] \\ &= g(p) + f_{N-1}(p_1) \\ &= g(p) + f_{N-1}(T(p)) \end{aligned} \qquad (27\text{-}6)$$

for $N \geq 1$ with

$$f_0(p) = g(p) + k(p) \qquad (27\text{-}7)$$

Let us apply the same approach to the study of the second function of (27-2),

$$g = h(p, p_1) + h(p_1, p_2) + \cdots + h(p_{N-1}, p_N) \qquad (27\text{-}8)$$

Writing

$$f_N(p) = h(p, p_1) + h(p_1, p_2) + \cdots + h(p_{N-1}, p_N) \qquad (27\text{-}9)$$

we have

$$\begin{aligned} f_N(p) &= h(p, p_1) + [h(p_1, p_2) + \cdots + h(p_{N-1}, p_N)] \\ &= h(p, p_1) + f_{N-1}(p_1) \\ &= h(p, T(p)) + f_{N-1}(T(p)) \end{aligned} \qquad (27\text{-}10)$$

for $N \geq 2$, with $f_1(p) = h(p, T(p))$.

Finally, let us discuss the function $\max_i h(p_i)$. Since this depends only upon p, the initial point, we write

$$\begin{aligned} f(p) &= \max_i h(p_i) \\ &= \max\ [h(p), h(p_1), h(p_2), \ldots] \\ &= \max\ [h(p), \max_{i \geq 1} h(p_i)] \\ &= \max\ [h(p), f(T(p))] \end{aligned} \qquad (27\text{-}11)$$

DYNAMIC PROGRAMMING—THEORY AND APPLICATION

What we have done so far has strong connections with the classical theory of iteration and, more generally, with abstract dynamics; see Ref. 3 for further references and discussion.

27-4. Multistage Decision Processes

Let us now introduce the element of decision making (see Chapter 24). We suppose that at each stage we have a choice of transformations to employ. Thus

$$p_{n+1} = T(p_n, q_{n+1}) \tag{27-12}$$

where q_n is an element of a set R, and $T(p,q)$ is a transformation with the property that $T(p,q)$ belongs to S whenever p belongs to S and q belongs to R.

We consider the process to proceed in the following fashion: Starting in state p, we choose q_1, which is equivalent to a choice of a transformation $T(p,q_1)$. Observe that we equate *decision* to *transformation*. Beginning in the new state p_1, we choose q_2, and so on. The value of q that is chosen at each time or at each *stage*, as we shall say, depends upon the current state of the system.

To determine the choice of the q_i we construct a *criterion function*, a scalar function

$$R = R(p, p_1, \ldots, p_N, \ldots, q_1, q_2, \ldots, q_N, \ldots) \tag{27-13}$$

which is to be maximized. A set of decisions $q_{N+1} = q_{N+1}(p_N)$ is called a *policy*; a set which maximizes R is called an *optimal policy*. In other words, an optimal policy is a function of the current state, in phase space and time, which yields the maximum of the criterion function. Our aim is to characterize optimal policies and to obtain analytic and computational techniques for their determination.

Once again, in order to obtain results of significance, we must introduce some structure into the function R. Let us initially take R to have the form

$$g(p, q_1) + g(p_1, q_2) + \cdots + g(p_{N-1}, q_N) + h(p_N) \tag{27-14}$$

As we shall see, a large number of processes can be treated in this general category.

For reasons which we shall not discuss in detail here (see Refs. 1, 2, and 3), the classical techniques of calculus are not particularly useful in the treatment of these general multistage maximization problems. Let us then proceed to present a different approach.

27-5. Functional-equation Approach

Once again, we make the observation that the maximum value of R depends only upon the initial state and the number of stages of the process. Let us then write

$$f_N(p) = \max_{\{q\}} [g(p,q_1) + g(p_1,q_2) + \cdots + g(p_{N-1}, q_N) + h(p_N)] \tag{27-15}$$

for $N \geq 1$ and p in S. To obtain a recurrence relation for the sequence $\{f_N(p)\}$, we observe that

$$\begin{aligned} f_N(p) &= \max_{q_1} \max_{q_2} \cdots \max_{q_N} [g(p,q_1) + \cdots + g(p_{N-1}, q_N) + h(p_N)] \\ &= \max_{q_1} \{g(p,q_1) + \max_{q_2} \max_{q_3} \cdots \max_{q_N} [g(p_1, q_2) + \cdots + h(p_N)]\} \\ &= \max_{q_1} [g(p,q_1) + f_{N-1}(p_1)] \\ &= \max_{q_1} [g(p,q_1) + f_{N-1}(T(p,q_1))] \end{aligned} \tag{27-16}$$

for $N \geq 2$, with $f_1(p) = \max_{q_1} g(p, q_1)$.

We have thus reduced a multidimensional maximization problem to a sequence of lower-dimensional problems.

27-6. Principle of Optimality

The recurrence relation of (27-16), obtained by means of analytic manipulation, can be derived in a uniform fashion by means of a fundamental, yet simple and intuitive, principle which governs the theory of dynamic programming:
  Principle of Optimality. An optimal policy has the property that, whatever the initial state and initial decision are, the remaining decisions must constitute an optimal policy with regard to the state resulting from the first decision.

To illustrate the use of these ideas, we shall now consider some processes arising in control engineering, trajectory analysis, and mathematical economics.

27-7. Control Theory

Suppose that we have a physical system which is not behaving in a completely desirable fashion. To improve its performance, we exert some additional forces, controlling forces. The basic problem of control (Chapter 29) is that of striking a balance between the cost of control and the cost of deviation of the system from its ideal performance. The translation of this into analytic terms gives rise to a host of fascinating and difficult optimization problems. Let us consider here one of the simplest.

Suppose that the physical system is described at time n, $n = 0, 1, 2, \ldots$, by the scalar quantity u_n, and the equation governing this quantity is

$$u_{n+1} = g(u_n, v_n) \qquad u_0 = c \qquad (27\text{-}17)$$

where v_n is the control variable and c is the initial state of the system. In this case, we ask that the values of v_n be chosen so as to keep the system as closely as possible in a constant state a. Let the cost of deviation at the nth stage be $k(u_n - a)$ and the cost of control be $h(v_n)$.

In accordance with our stated aim, we wish to minimize the expression

$$\sum_{n=0}^{N} [k(u_n - a) + h(v_n)] \qquad (27\text{-}18)$$

Denote the minimum value by $f_N(c)$, for $N = 0, 1, 2, \ldots, -\infty < c < \infty$. Then

$$f_0(c) = \min_v \, [k(c - a) + h(v)] \qquad (27\text{-}19)$$

and

$$f_N(c) = \min_v \, [k(c - a) + h(v) + f_{N-1}(g(c,v))] \qquad (27\text{-}20)$$

For numerical results and further discussion, see Refs. 1 and 2.

27-8. Constraints

In many processes, there is not a free choice of the control variable at each stage. Suppose that there is a constraint of the type $|v_n| \leq k$. In this case, the recurrence relation of (27-20) becomes

$$f_N(c) = \min_{|v| \leq k} \, [k(c - a) + h(v) + f_{N-1}(g(c,v))] \qquad (27\text{-}21)$$

Generally, it is very easy to incorporate constraints on policies into dynamic programming.

The more general constraint

$$k(u_n) \leq v_n \leq K(u_n) \qquad (27\text{-}22)$$

or the constraint $v_n = \pm k$ ("bang-bang") control can easily be taken account of in the functional equation.

These relations are seldom analytically tractable, but they do lead to simple computational algorithms, granted a digital computer (see Ref. 2).

27-9. Feedback Control

The value of v which yields the minimum in (27-20) depends upon both N and c, $v = v_N(c)$, the policy function. In other words, the optimal decision depends upon the time and the state at this time. This is the essence of "feedback control," one of the basic principles of engineering.

We shall discuss control theory again below in connection with multistage decision processes of continuous type.

27-10. Optimal Trajectories

A class of problems which lend themselves particularly well to a dynamic-programming formulation arises in the determination of optimal trajectories. The general question is that of going from a point p in phase space to a point p' in phase space at minimum cost, where the cost is measured in time, fuel, or other economic units.

Let $f(p)$ denote this minimum time. Then the principle of optimality yields the functional equation

$$f(p) = \min_q [t(p,q) + f(q)] \tag{27-23}$$

This equation can be used to study classical trajectory problems,[2] and also problems involved in optimal traversal of networks.[2,4]

The computational solution of (27-23) is not routine if p is of large dimension, and a number of techniques have been developed to treat various cases.[2,5] All in all, the dimensionality barrier is the principal obstacle to the routine application of dynamic programming to the numerical treatment of multistage decision processes.

The method of polynomial approximation seems promising; see Ref. 2.

27-11. Approximation in Function Space and Policy Space

In studying (27-23), we can employ the classical method of successive approximations, i.e., guess $f_0(p)$ and then determine the sequence $\{f(p)\}$ inductively by means of the recurrence relation

$$f_{n+1}(p) = \max_q [t(p,q) + f_n(q)] \tag{27-24}$$

$n = 0, 1, 2, \ldots$ (approximation in function space), or we can guess a policy $q = q_0(p)$ and then determine $f_0(p)$ by means of

$$f_0(p) = t(p,q_0(p)) + f_0(q_0(p)) \tag{27-25}$$

(approximation in policy space). The next policy, $q_1(p)$, is obtained by minimizing the expression

$$t(p,q) + f_0(q) \tag{27-26}$$

and so on. This approach yields monotone convergence of return functions. It has many applications in classical analysis, particularly in connection with quasilinearization; see Refs. 6 and 7.

27-12. Allocation Processes

In operations research and mathematical economics, one frequently encounters the problem of allocating a quantity of resources in an optimal fashion among a number of different activities (Chapter 26). Analytically, the problem reduces to maximizing an expression such as

$$\sum_{i=1}^{N} g_i(x_i, y_i, z_i) \qquad (27\text{-}27)$$

subject to constraints such as

$$\sum_i x_i \le x \qquad x_i \ge 0$$
$$\sum_i y_i \le y \qquad y_i \ge 0 \qquad (27\text{-}28)$$
$$\sum_i z_i \le z \qquad z_i \ge 0$$

Calling the maximum value $f_N(x,y,z)$, we readily obtain the recurrence relation

$$f_N(x,y,z) = \max \, [g_N(x_N, y_N, z_N) + f_{N-1}(x - x_N, y - y_N, z - z_N)] \qquad (27\text{-}29)$$

where the maximum is over the region $0 \le x_N \le x$, $0 \le y_N \le y$, $0 \le z_N \le z$.

For a small number of resources, a direct computational solution is available. For larger numbers we can use Lagrange multipliers[2] and successive approximations.[2] Transportation and scheduling problems of the Hitchcock-Koopmans variety can be handled in the important case of a small number of depots and a large number of demand points; see Ref. 2.

The functional-equation approach can be used to treat the duality theorems (Sec. 26-3) of linear and nonlinear programming; see the Appendix in Ref. 2 by Dreyfus and Freimer.

27-13. Multistage Decision Processes of Continuous Type

If we make the points in time at which decisions are made closer together, say at times $0, \Delta, 2\Delta, \ldots$, let the return per stage have the form $g(p,q)\Delta$, and let $T(p)$ have the form $p + S(p,q)\Delta$, we obtain a functional equation of the form

$$f_{N\Delta + \Delta}(p) = \max_q \, [g(p,q)\Delta + f_N(p + S(p,q)\Delta)] \qquad (27\text{-}30)$$

Passing to the limit as $\Delta \to 0$, we obtain a functional equation of the form

$$\frac{\partial f}{\partial t}(p,t) = \max_q \, [g(p,q) + (\text{grad } f, S(p,q))] \qquad (27\text{-}31)$$

where $f_{N\Delta}(p) \to f(p,t)$.

The most important application of this formalism is to the calculus of variations; see Refs. 1, 2, and 8, where the connection with Hamilton-Jacobi formalism is discussed. References may also be found in Ref. 3 to the connections with Huygens' principle.

The maximum principle of Pontryagin[9] can be derived from (27-31) and conversely. For derivations of both from the classical calculus of variations, see Ref. 10.

A class of problems particularly amenable to explicit analytic treatment by means of dynamic programming is those in which it is required to minimize a quadratic functional subject to linear equations connecting the state and control variables. An example of this is the problem of minimizing

$$\int_0^T [(x,Bx) + (y,Cy)] \, dt \qquad (27\text{-}32)$$

over all vectors $y(t)$ where x and y are related by means of

$$\frac{dx}{dt} = Ax + y \qquad x(0) = c \qquad (27\text{-}33)$$

DYNAMIC PROGRAMMING—THEORY AND APPLICATION

For general results in this direction, see Ref. 11, and for particular applications see Refs. 12 and 2. For an elegant treatment of prediction theory by these techniques, see Ref. 13. Applications have also been made to study of the variation of Green's functions, characteristic values, and characteristic functions, with change in the region; see Ref. 14.

27-14. Multistage Decision Processes of Stochastic Type

Let us now turn our attention to those processes where complexity or uncertainty force us to introduce stochastic elements. Suppose that the relationship now between successive state variables is

$$p_{n+1} = T(p_n, q_{n+1}, r_n) \qquad (27\text{-}34)$$

where the r_n are random variables, and that the criterion function has the form

$$R_N = \sum_{n=0}^{N} g(p_n, q_n, r_n) \qquad (27\text{-}35)$$

For the sake of simplicity, suppose that the r_n are independent random variables, although correlations of various types can also be treated.

Since the criterion function is a random variable, let us agree to maximize an expected value. We say *an* expected value since it is essential that we define what we mean by a policy, and an optimal policy, before we consider taking averages, and maximizing averages.

Keeping the feedback-control concept in mind, we postulate that the process unfolds in the following manner: The initial state p is observed, an initial decision q_1 is made, and a random variable r_1 is drawn from a given distribution function $dG(r)$. The new state, $p_1 = T(p, q_1, r_0)$, is observed, and a new decision q_2 is made. A policy is once again a function $q = q(p, N)$, and an optimal policy is one which maximizes the expected value of R_N.

As before, we write

$$f_N(p) = \max_q \; (\exp_r R_N) \qquad (27\text{-}36)$$

and obtain from the principle of optimality the functional equation

$$f_N(p) = \max_q \; [\int [g(p,q,r) + f_{N-1}(T(p,q,r))] dG(r)] \qquad (27\text{-}37)$$

for $N \geq 2$, with

$$f_1(p) = \max_q \; [\int g(p,q,r) dG(r)] \qquad (27\text{-}38)$$

The point to emphasize is that the formalism of dynamic programming enables us to handle stochastic decision processes in precisely the same way that deterministic processes are treated. Abstractly and conceptually, there is no difference in the formulation.

27-15. Applications

The principal applications to date have been in the area of stochastic control theory, inventory processes, and Markovian decision processes in general.

Some of the applications of dynamic programming to stochastic variational processes are quite straightforward;[2,3] others require more of the classical theory of stochastic processes.

The study of inventory processes has been intensively developed by means of dynamic-programming techniques.[1,2,15]

These processes are particular examples of Markovian decision processes.[1,2,16]

The study of linear systems with quadratic criteria has been carried out to some degree; see the references in Refs. 2 and 3. For computational results, see Ref. 2.

27-16. Adaptive Processes

So far we have discussed processes in which the relevant cause-and-effect relations were known or, if not, replaced by known distributions of outcomes. To treat decision processes outside of this format requires some new concepts and techniques, a blending and extension of classical variational theory and classical probability theory. Since the field of adaptive processes (Chapter 30), processes which involve learning and controlling at the same time, is so new and relatively unformed, and simultaneously complex, we shall refer the reader to Ref. 3 for further details and references on the application of dynamic programming to this field.

One way of avoiding some of the aspects of the unknown is to introduce the theory of games (Chapter 23). For a treatment of multistage games by means of dynamic programming, see Ref. 1.

REFERENCES

1. Bellman, R.: *Dynamic Programming*, Princeton University Press, Princeton, N.J., 1957.
2. Bellman, R., and S. Dreyfus: *Applied Dynamic Programming*, Princeton University Press, Princeton, N.J., 1962.
3. Bellman, R.: *Adaptive Control Processes: A Guided Tour*, Princeton University Press, Princeton, N.J., 1961.
4. Kalaba, R.: Network Optimization, *Proc. Symp. Appl. Math.*, American Mathematics Society, Providence, R.I., 1961.
5. Bellman, R., and R. Kalaba: Reduction of Dimensionality, Dynamic Programming, and Control Processes, *J. Basic Eng.*, March, 1961, pp. 82–84.
6. Kalaba, R., On Nonlinear Differential Equations, the Maximum Operation, and Monotone Convergence, *J. Math. Mech.*, vol. 8, pp. 519–574, 1959.
7. Beckenbach, E. F., and R. Bellman: Inequalities, *Ergeb. Math.*, Springer, Berlin, 1962.
8. Dreyfus, S.: Dynamic Programming and the Calculus of Variations, *J. Math. Anal. Appl.*, vol. 1, pp. 228–239, 1960.
9. Pontryagin, L. S., et al.: *The Mathematical Theory of Control Processes*, Interscience Publishers, Inc., New York, 1962.
10. Berkovitz, L.: Variational Methods in Control and Programming, *J. Math. Anal. Appl.*, vol. 3, pp. 145–169, 1961.
11. Bellman, R., I. Glicksberg, and O. Gross: *Some Aspects of the Mathematical Theory of Control Processes*, The RAND Corporation, R-313, 1958.
12. Bellman, R.: *Introduction to Matrix Analysis*, McGraw-Hill Book Company, New York, 1960.
13. Kalman, R.: *New Methods and Results in Linear Prediction and Filtering Theory*, RIAS, Technical Report 61-1, 1960.
14. Bellman, R., and R. S. Lehman: Functional Equations in the Theory of Dynamic Programming—X: Resolvents, Characteristic Functions and Values, *Duke Math. J.*, vol. 27, pp. 55–70, 1960.
15. Arrow, K., S. Karlin, and H. Scarf: *Studies in the Mathematical Theory of Inventory and Production*, Stanford University Press, Stanford, Calif., 1961.
16. Howard, R.: *Dynamic Programming and Markov Processes*, John Wiley & Sons, Inc., New York, 1960.

Chapter 28

QUEUES AND MARKOV PROCESSES
THE RESPONSE OF OPERATING SYSTEMS TO FLUCTUATING DEMAND AND SUPPLY

PHILIP M. MORSE, *Director, Operations Research Center, Massachusetts Institute of Technology.*

CONTENTS

28-1. Introduction	28-1
28-2. Definitions	28-2
28-3. Discrete Time	28-2
28-4. Simple Markov Processes	28-4
28-5. Stochastic Service Systems	28-8
28-6. The Formation of Queues	28-10
28-7. Continuous Time	28-12
28-8. Arrival and Service Statistics	28-14
28-9. Bottlenecks and Queues	28-16
28-10. Steady-state Operation	28-18
28-11. Effect of Service and Arrival Statistics	28-22
28-12. Queue Discipline	28-25
28-13. Multiple Service Channels	28-27

28-1. Introduction

Systems are designed for use by men. When so used, the resulting activity, of both men and equipment, is called an *operation*. The dynamics of the operational system is only partly prescribed by the purely physical properties of the equipment constituting the system. It is also, and often predominantly, prescribed by the human characteristics of the operators, by the directives laid down by the executive in charge, and by the fluctuating demands made on the operation by external forces, both physical and human.

Many of these internal and external forces exhibit fluctuations so complex that we are forced to classify them as random and to express them in terms of probability distributions. It is thus important to devise a dynamical theory which can predict the behavior of an operation under the influence of such fluctuating forces, the predictions also being given in probabilistic terms. Processes subject to random forces are called *stochastic processes*. In this chapter we shall deal with only a few of the simpler aspects of the theory of stochastic processes; those interested in further details should consult the many books and articles on the subject (see the references for a sample). We assume that the reader is familiar with elementary probability theory, to the extent covered in Chapter 38.

28-2. Definitions

The systems we shall discuss are those where the stochastic effects predominate and where the dynamic characteristics of the operational system can be expressed by a few parameters. The examples given will, of course, be simple ones, as are the corresponding examples given in an introductory text on dynamics. But the basic theory discussed here can be extended to deal with more complex operational systems, just as the principles of dynamics can deal with complex physical systems.

For our present purposes we assume that the operational system can exist in a number of possible *states*. A fleet of trucks will have so many trucks in one terminus, so many in another, so many en route, and so many in the garage for maintenance. A warehouse will contain specified numbers of each item it is supposed to store; a machine shop will have machine A working on job 3, machine B working on job 1, etc. In many cases these states will constitute a denumerable set, and in some cases they may be finite in number. As the system reacts to the external and internal forces acting on it, its state will change with time. Our theory must be able to predict these changes, at least in terms of a probability distribution, $p_i(t)$ being the probability that the system is in state i at time t.

At the instant $t = 0$ we may know that the system is in state j; we represent this fact by setting $p_j(0) = 1$ and $p_i(0) = 0$ for $i \neq j$. At a later time t, because of the variability of the forces acting, we cannot be sure in which state the system will find itself. All we can do is to predict the probability $p_i(t)$ that it will be in state i at time t. The set of probabilities $p_i(t)$ are limited in magnitude by the requirement that they cannot be negative, and their sum must be unity,

$$p_i(t) \geq 0 \qquad \sum_i p_i(t) = 1 \qquad (28\text{-}1)$$

which are the basic requirements of any denumerable probability distribution. Our dynamical theory for stochastic systems must provide equations which will enable us to compute the changes in these probabilities as time advances. Related to the state probabilities are the conditional probabilities of transition $T_{ij}(t_0,t_1)$, that the system be in state j at time t_1 if it is in state i at time t_0.

28-3. Discrete Time

In many cases it is necessary only to know the state of the system at the ends of discrete periods of time, each of length τ (a day or week or hour). Either because the state is known only at the ends of successive periods, or because of administrative reasons, decisions regarding the operation of such systems are only made at the ends of periods. In such cases t can be considered to be a discrete variable, and if we use units of τ, the variable t in the probabilities $p_i(t)$ takes on only integral values ($t = 0$, 1, 2, . . .). The dynamical equations for the system must therefore be concerned with the transition probability $M_{ij} = T_{ij}(t, t+1)$, which is the conditional probability that the system will be in state j in the $(t+1)$st period if it is in the ith state in the tth period. In general M_{ij} depends on t and also depends on the past history of the system, but in a number of cases of interest M_{ij} is more or less independent of time and is roughly dependent only on i and j.

In this limiting situation, when the transition probability M_{ij}, from initial state i to final state j in one period, is independent of t and also of past history, the system is called a *Markovian* system, and its behavior is called a *Markov process*. In this case

QUEUES AND MARKOV PROCESSES

we can show that $T_{ij}(t_0,t_1)$ is a function of i, j and $t_1 - t_0$ only and that T_{ij} is related to the elements M_{ij} by the Chapman-Kolmogorov equations

$$T_{ij}(t_0,t_1) = \sum_k T_{ik}(t_0,t_1-1)M_{kj}$$
$$= \sum_k M_{ik}T_{kj}(t_0+1,t_1) \qquad (28\text{-}2)$$

The simplest mechanical model of a Markov process is the simplified one-dimensional Brownian motion called a *random walk*. In this model the point, representing the system, moves a distance h in each period, but the direction of the motion is chosen at random, being equally likely to be to the left as to the right. In this case the state of the system can be the number i of distances h by which the representative point is displaced to the right of the origin (displacements to the left are negative values of i). The position of the point at time $t+1$ depends only on its position at time t, being equally likely to be either one unit to the right or one unit to the left of its position at t. Thus the transition probability M_{ij} is zero unless $j = i \pm 1$ and $M_{ij} = \frac{1}{2}$ when $j = i \pm 1$. Since M_{ij} is independent of t, the random walk is a Markov process, and by the rules of combination of conditional probabilities, the probability distribution of the point at time $t+1$ is related to that at time t by the equation

$$p_j(t+1) = \sum_i p_i(t)M_{ij} = \frac{1}{2}[p_{j-1}(t) + p_{j+1}(t)]$$

The solution of this set of equations, corresponding to a particular choice of initial values of $p_i(0)$, describes, in terms of probability, the random diffusion of the point along the line. In the limit of infinitesimal periods and displacements the equation goes over into the one-dimensional diffusion equation, and the resulting distribution, if the point is definitely at the origin at $t = 0$, predicts a zero mean value of displacement and a mean-square displacement (displacement variance) proportional to time, typical of Brownian motion.[3] Although the *ratio* between the number of steps to the left and those to the right approaches unity as t increases (as required by the random nature of the motion) the *difference* between the two does not approach zero; in fact the mean square of this difference increases proportional to t. The diffusing point drifts gradually away from the origin, the drift being equally likely in either direction.

Other operational systems exhibit the stochastic properties of the random walk. For example, consider the idealized production line, which can produce either item A (a sedan, for instance) or item B (a convertible, maybe) in one period of time. Also, during each period, an order for the delivery of one item is received. In the long run the number of orders for A will equal the number of orders for B, but the sequence is a random one; all we can say is that during any period it is equally likely that the order received is for A or for B. If the order is for A and there are n completed items A on hand, one A will be shipped during the period; if no A's are on hand the item will be back-ordered, to be shipped when an A is produced. We can thus define the *net stock* of A's as the actual number on hand at the end of a period, if any are on hand, or minus the number of back orders for A if none are on hand; similarly for B.

Since, in the long run, equal numbers of A's and B's are ordered, the manager of production might decide to alternate his production, making an A during the first period, a B during the second, and so on. The number designating the state of the system can then be chosen to reflect the balance between the stocks of A and B on hand at any time. At time t the state number i is chosen to be equal to (net stock of A) − (net stock of B) if t is an even integer, or (net stock of A) − (net stock of B) − 1 if t is an odd integer. After $2n$ periods, for example, if n A's and m B's are on hand, i will equal $n - m$; if n A's are on hand and m back orders for B are on hand, i will equal $n + m$. Because of the alternation in production between A and B there will be one extra A at the end of odd-numbered periods; thus we subtract 1 to obtain i.

It might seem that the value of i is not sufficient to determine the values of net stock of A or B separately. However, since one item (either A or B) is withdrawn or back-ordered each period, we see that (total stock), which equals (net stock of A) + (net stock of B), equals a constant, independent of t. Thus a knowledge of i serves to determine both net stocks. If the initial value of total stock is zero, for example, and i is positive at some later time, then some A's are on hand and an equal number of back orders for B are on hand; if i is negative, too many B's and not enough A's have been produced.

It should now be apparent that this operation can be described in terms of a random walk; the equation for the probability $p_i(t)$ that the system is in state i at the end of period t is that given previously. If an order for A is received, the value of i reduces by 1; if B is ordered during a period, i increases by 1. Therefore, the expected value of i will stay zero if it is zero at $t = 0$, but i actually will fluctuate over a wider and wider range of values as t increases. In fact the mean-square value of i will increase, proportional to t, without limit.

It often is instructive to simulate a sample run of a model under study, to see whether the general behavior corresponds to that of the system it is supposed to represent. In the present case it is simple to perform the simulation by tossing a coin to determine the sequence of orders; heads for an A, tails for a B. A sample run for 20 periods goes as follows:

Period t......	1	2	3	4	5	6	7	8	9	10	11	12	13	14	15	16	17	18	19	20
Producing......	A	B	A	B	A	B	A	B	A	B	A	B	A	B	A	B	A	B	A	B
Receive order for....		B	A	A	B	B	A	B	A	B	B	B	B	A	B	A	B	B	A	A
Net stock A......	0	1	0	0	0	1	0	1	0	1	1	2	2	2	2	2	2	3	2	2
Net stock B......	0	-1	0	0	0	-1	0	-1	0	-1	-1	-2	-2	-2	-2	-2	-2	-3	-2	-2
State i........	0	1	0	-1	0	1	0	1	0	1	2	3	4	3	4	3	4	5	4	3

The random-walk nature of the fluctuations in the state index i is apparent.

In spite of the fact that the probability of receiving an order for an A is equal to that of receiving an order for a B in any period, the chance that a random unbalance occurs between actually received orders for A and for B is not vanishingly small; in fact the mean square of the unbalance is expected to increase with t. To protect the operation from such random unbalances the manager should not blindly alternate production but should use a production policy more closely coupled to the actual arrival of demands. However, before we demonstrate how a change in the system's internal forces (i.e., production schedule) will change the dynamic behavior of the system, we should discuss the general nature of the dynamical equations for a Markov process.

28-4. Simple Markov Processes

The system we are studying may be in any of a denumerable number of states; the probability that it is in state i at the end of period t (t an integer) is $p_i(t)$, where $\sum_i p_i(t) = 1$. If the system is Markovian, the conditional probability M_{ij}, that the system will be in state j at $t + 1$ if it is in state i at t, depends only on i and j. If this is so, then the "equation of motion" of the system, relating $p_j(t + 1)$ with the $p_i(t)$'s, is

$$p_j(t + 1) = \sum_i p_i(t) M_{ij} \tag{28-3}$$

Since M_{ij}, as a function of j, is a probability distribution, it must be limited by the same equations as (28-1),

$$M_{ij} \geq 0 \qquad \sum_j M_{ij} = 1 \tag{28-4}$$

QUEUES AND MARKOV PROCESSES   28–5

If the system is in state i at t, it must be in one or the other of the states j at $t + 1$; so the sum over j of M_{ij} must be 1.

Probabilities M_{ij} are elements of a *matrix*

$$\mathfrak{M} = \begin{pmatrix} M_{11} & M_{12} & M_{13} & — & — \\ M_{21} & M_{22} & M_{23} & — & — \\ M_{31} & M_{32} & M_{33} & — & — \\ — & — & — & — & — \end{pmatrix} \qquad (\mathfrak{M})_{ij} = M_{ij} \qquad (28\text{-}5)$$

with the usual properties of matrices,

$$(\mathfrak{M} + \mathfrak{N})_{ij} = M_{ij} + N_{ij} \qquad (\mathfrak{M} \cdot \mathfrak{N})_{ij} = \sum_k M_{ik} N_{kj} \qquad (28\text{-}6)$$

A matrix with elements subject to the limitations of (28-4) (no negative elements, each row adding up to unity) is called a *stochastic* matrix. We note that, if $\mathfrak{M}$ is stochastic $\mathfrak{M}^n$ is also stochastic, and if $p_j(t)$ satisfies (28-1) then $p_i(t + 1)$ also satisfies (28-1).

Equation (28-3) can be extended to obtain $p_j(t)$ in terms of the initial probability distribution $p_i(0)$, for

$$p_j(t) = \sum_k p_k(t-1) M_{kj} = \sum_{lk} p_l(t-2) M_{lk} M_{kj}$$
$$= \sum_l p_l(t-2) (\mathfrak{M}^2)_{lj} = \cdots = \sum_i p_i(0) (\mathfrak{M}^t)_{ij} \qquad (28\text{-}7)$$

that is, $T_{ij}(0,t) = (\mathfrak{M}^t)_{ij} = T_{ij}(t_0, t_0 + t)$, from (28-2). Therefore, if we can compute the tth power (t is an integer) of matrix $\mathfrak{M}$ we can compute directly the probability distribution for the system at time t in terms of its initial distribution. In particular, the conditional probability that the system is in state j at time t if it initially were in state i is $T_{ij}(0,t) = (\mathfrak{M}^t)_{ij}$.

We have not the space to discuss the techniques for computing the transient behavior of a Markov process at the beginning of its motion (see Refs. 2 and 5 for details). In many cases (the random walk discussed earlier is an example) the system does not settle down to steady-state motion but continues to diverge from its initial state. But in a large number of operations the rules of operation (the internal forces) are specifically chosen so the motion does settle down. In these cases matrix $\mathfrak{M}^t$ turns out to be the sum of a steady-state matrix, independent of t, and a transient matrix, which approaches zero as t increases.

In these cases we can solve directly for the steady-state probabilities P_i, the limiting values of $p_i(t)$ as $t \to \infty$. When a steady state occurs, the P_i's are independent of t and also of the initial state of the system; they represent the state distribution when the system has forgotten how it started. The equations for the P's are obtained from (28-3) by setting $p_i(t + 1) \to p_i(t) \to P_i$,

$$P_j = \sum_i P_i M_{ij} \qquad \sum_i P_i = 1 \qquad (28\text{-}8)$$

which is a set of simultaneous equations, soluble for the P's as long as the M_{ij}'s satisfy (28-4). Note that the right-hand side of the equation for P_j involves all the elements in the jth *column* of matrix $\mathfrak{M}$.

In each case we must test to see whether a steady-state solution is possible. For example, for the random walk of Sec. 28-3 the steady-state solution is for all the p's to be equal, for the moving point to be equally likely to be anywhere. Since the point can move along a line of infinite extent, each steady state p_i would have to be infinitesimally small, and it would take an infinite time for such a state to be reached; so such a solution is of little practical interest.

For systems designed to occupy only a finite number of states, the transient motion usually becomes negligible in a finite time. As an example, we return to the production facility of Sec. 28-3, which turns out item A or item B in one period. Suppose again that one order, for either A or B, comes in each period, with no regular pattern in the sequence except, this time, a fraction α of the time A is ordered and $1 - \alpha$ of the periods the order is for a B.

To make the system responsive to these random fluctuations of orders for A or B, we shall arrange that the line keeps M items ahead of the orders. Usually this "buffer" inventory has some of each item, so that orders can always be filled; if inventory of one item happens to go to zero, production switches to making that item; so the system is always ahead of the demand. For example, the operating rule could be that, once started making A's, we continue to make A's until the buffer inventory no longer has a B, when we change over to making B's until the last A is sold, then we change back to making A's.

The $2M$ possible states for such a system can be ordered and designated as follows:

State index i	Inventory during period Of A	Inventory during period Of B	Item made during period
0	0	M	A
1	1	$M - 1$	A
......	......	......	...
$M - 2$	$M - 2$	2	A
$M - 1$	$M - 1$	1	A
M	M	0	B
$M + 1$	$M - 1$	1	B
......	......	......	...
$2M - 2$	2	$M - 2$	B
$2M - 1$	1	$M - 1$	B

To determine the transition probabilities, we consider what can happen to the system if it is in state $i < M - 1$ at the beginning of a period, with i items A and $M - i$ items B on hand and an item A being made. During the period an order arrives; α of the time it is for an A, in which case an A is shipped, but by the end of the period a new A is finished; so at the beginning of the next period the system's state is again i. However, in $1 - \alpha$ of the periods the order is for a B, in which case the end of the period will see one more A and one less B in stock; the next state is $i + 1$. If the initial state is between M and $2M - 1$, with a B being made and $2M - i$ items A, $i - M$ items B in stock, then an order for B will not change the state, and an order for an A will increase i by 1.

If the initial state is $i = M - 1$ and an order for A arrives, the state is unchanged and another A will be made next period. But if the order is for the only B in stock then production is changed over and during the next several periods B's are made, until state $2M - 1$ is reached. When in this state, if an order for A is received, production is changed over to making A's again and the system is in state 0 once more.

It is not difficult to show that the elements of the transition matrix for this system are

$$M_{ij} = \begin{cases} \alpha(j = i), = 1 - \alpha(j = i + 1), = 0(j \neq i \text{ or } i + 1)(0 \leq i < M) \\ 1 - \alpha(j = i), = \alpha(j = i + 1), = 0(j \neq i \text{ or } i + 1)(M \leq i < 2M - 1) \\ \alpha(j = 0), = 1 - \alpha(j = i), = 0(j \neq 0 \text{ or } i)(i = 2M - 1) \end{cases}$$

Equations (28-8) for the steady-state behavior of this system then have the form

$$P_0 = P_0 M_{0,0} + P_{2M-1} M_{2M-1,0} = \alpha P_0 + \alpha P_{2M-1}$$
$$P_M = (1 - \alpha) P_M + (1 - \alpha) P_{M-1}$$
$$P_j = P_j M_{j,j} + P_{j-1} M_{j-1,j} = \begin{cases} \alpha P_j + (1 - \alpha) P_{j-1} & (0 < j < M) \\ (1 - \alpha) P_j + \alpha P_{j-1} & (M < j < 2M) \end{cases}$$

which has the solution

$$P_0 = P_1 = \cdots = P_{M-1} = \frac{\alpha}{M}$$
$$P_M = P_{M+1} = \cdots = P_{2M-1} = \frac{1 - \alpha}{M} \tag{28-9}$$

In this case the rules of operation have been adjusted so that, in the long run, in α of the periods an A is made and in $(1 - \alpha)$ of the periods a B is made, as is required by the average ratio between orders. But, in contrast to the case considered earlier, with fixed schedule alternating A's and B's, the present case responds to the fluctuations in demand, the change in production occurring only after M items B have been withdrawn, no matter how many A's were withdrawn meanwhile (and likewise for the change from B to A).

In fact the change from producing A's to producing B's can occur only when the system is in state $M - 1$ and then only when the next order is for a B; so the fraction of periods during which a changeover from making A to making B occurs is $(1 - \alpha) P_{M-1} = \alpha(1 - \alpha)/M$. The fraction of periods during which a change from making B to making A occurs is also $\alpha(1 - \alpha)/M$; so that the chance of a change in production occurring in a given period is $2\alpha(1 - \alpha)/M$. If the additional set-up cost per changeover is C_c then the expected changeover cost per period is $[2\alpha(1 - \alpha)/M]C_c$, which diminishes as M is increased. The mean time between changeovers is $M/2\alpha(1 - \alpha)$ periods, which is never less than $2M$.

Thus we can reduce changeover costs at the expense of having a buffer inventory M of finished items. On the average there will be $\frac{1}{2}M$ item A's and $\frac{1}{2}M$ item B's in stock. If the cost per period of keeping one item (either A or B) is C_i, then the mean inventory cost per period is MC_i and that part of the production cost per period which depends on M is $C_p(M) = MC_i + [2\alpha(1 - \alpha)/M]C_c$. This is minimal when M is the nearest in value to

$$M_{\min} = \frac{1}{2} + \sqrt{\frac{1}{4} + 2\alpha(1 - \alpha) \frac{C_c}{C_i}} \tag{28-10}$$

for which $C_p(M_{\min} + 1) - C_p(M_{\min}) = 0$. Thus a solution for the steady-state behavior of the system enables us to find the most effective operating policy in the face of fluctuating external forces.

As a demonstration of the nature of the Markov model we can run a simulation, using a table of random integers to simulate the fluctuations in arriving orders. For $\alpha = 0.3$, for example, if the next digit is 0, 1, or 2 we take the next order to be for an A; if it is 3, 4, . . . , 9 the next order is for a B. A sample sequence, for $M = 2$, for 20 periods, is

Period t	1	2	3	4	5	6	7	8	9	10	11	12	13	14	15	16	17	18	19	20
A on hand	0	1	2	2	1	1	1	0	1	1	2	2	2	2	1	0	1	1	2	2
B on hand	2	1	0	0	1	1	1	2	1	1	0	0	0	0	1	2	1	1	0	0
Producing	A	A	B	B	B	B	B	A	A	A	B	B	B	B	B	A	A	A	B	B
State i	0	1	2	2	3	3	3	0	1	1	2	2	2	2	3	0	1	1	2	2
Order for received	B	B	B	A	B	B	A	B	A	B	B	B	B	A	A	B	A	B	B	

In this sample run, five changeovers of production occurred, or 0.25 changes per period, as compared with the expected value of $2\alpha(1 - \alpha)/M = 0.21$. We note how production adjusts itself to fluctuations in demand. The system performs a sort of

28–8   SYSTEM THEORY

circular random walk, either standing still or advancing a step each period, going around the circle from 0 to 3 and then on to zero for another cycle.

28-5. Stochastic Service Systems

Most operations contain "bottlenecks" of one sort or another, elements of the operation wherein the flow of things or persons or messages is subjected to delays of more or less random duration. Aircraft, arriving at a landing field, may be delayed in landing by weather or by the presence of another plane in the process of landing. Messages may momentarily arrive at a message center faster than they can be sent along. A machine, in need of repair, may have to remain idle until the repair crew finishes another job. A customer desiring to purchase some item may have to wait because the store is temporarily out of stock. An understanding of the dynamics of these various "bottleneck" situations is essential for an understanding of the whole operation; alleviation of the worst bottleneck often has a marked effect on the behavior of the whole system.

In each of the situations mentioned a sequence of *units* (aircraft, messages, machines, or customers) arrives at a *service facility* in a more or less random sequence in time; their further progress may be delayed because of delay in service or because other units are ahead of them or both. Such systems, or parts of systems, are called *stochastic service systems*. For obvious reasons the theory of their behavior is often called *queuing theory* (though in many cases no queues are formed). It is a special, and rather elementary, part of the theory of stochastic processes. We shall discuss in this section a few typical examples and illustrate some of the techniques of solution of the dynamical equations.

In a few cases the discrete-time model of the Markov process can be used. Consider a simplified service operation of the following sort: A shop (or a particular machine in a shop) can carry out either task A or task B. Task A takes one period to complete; task B takes two periods. During each period, while the shop is working on the immediate task, orders for the succeeding period are received; the mean rate of arrival per period of orders for task A is α, the mean rate of orders for B is β, and their distribution in time is random (Poisson distribution; see Sec. 28-8). Therefore, the probability that at least one order for A arrives in a period is $a = 1 - e^{-\alpha}$, that at least one order for B arrives is $b = 1 - e^{-\beta}$. The probability that at least one order for A arrives in a period, but no order for B, is $a(1-b)$, and the chance that at least one order, for either A or B, arrives is $a + b - ab$. Also the probability that an order for A arrives and is the first order which has arrived during the period is $[\alpha/(\alpha + \beta)](a + b - ab)$, etc.

Since the shop (or the machine) cannot execute more than one order per period, the manager must decide what to do about the extra orders which may arrive. One possibility is to reject them, in effect allowing them to be executed by some other shop, so that no orders are held over from period to period. Furthermore, if both kinds of orders arrive in a period, he must decide which order to execute during the next period. For example, he may choose to take the order for job B (the two-period job) whenever he has a choice, and to execute an A only if none for B arrives during that period (and, of course, if one for A does arrive!).

In this case there are four possible states for the shop during a period: 0, the shop is idle; 1, it is executing A; 2, it is executing the first period of job B; 3, it is executing the second period of B. The transition matrix is

$$\mathfrak{M} = \begin{pmatrix} (1-a)(1-b) & a(1-b) & b & 0 \\ (1-a)(1-b) & a(1-b) & b & 0 \\ 0 & 0 & 0 & 1 \\ (1-a)(1-b) & a(1-b) & b & 0 \end{pmatrix} \quad (B \text{ preferred}) \qquad (28\text{-}11)$$

For example, if the system is in state 0, 1, or 3, the chance that it will be idle next period is $M_{00} = M_{10} = M_{30} = (1-a)(1-b)$, the chance that no orders arrive during the period. On the other hand, if the system is in state 2 it will certainly go on to finish task B (i.e., to state 3).

QUEUES AND MARKOV PROCESSES   28–9

According to Eq. (28-7), the transient behavior of the system is obtained from the tth power of matrix M, which can be shown to be (for $t = 1, 2, 3, \ldots$)

$$\mathfrak{M}^t = \frac{1}{1+b} \begin{pmatrix} (1-a)(1-b) & a(1-b) & b & b \\ (1-a)(1-b) & a(1-b) & b & b \\ (1-a)(1-b) & a(1-b) & b & b \\ (1-a)(1-b) & a(1-b) & b & b \end{pmatrix}$$

$$- \frac{(-b)^t}{b(1+b)} \begin{pmatrix} (1-a)b(1-b) & ab(1-b) & b^2 & -b \\ (1-a)b(1-b) & ab(1-b) & b^2 & -b \\ -(1-a)(1-b) & -a(1-b) & -b & 1 \\ (1-a)b(1-b) & ab(1-b) & b^2 & -b \end{pmatrix} \quad (28\text{-}12)$$

In other words, if the shop is in state 0 at $t = 0$, the probability that it is idle during the tth period is

$$(\mathfrak{M}^t)_{00} = \frac{(1-a)(1-b)}{1+b} [1 - (-b)^t] = (1-a)(1-b) \ (t = 1),$$

$$= (1-a)(1-b)^2 \ (t = 2), \ = (1-a)(1-b)(1-b+b^2) \ (t = 3)$$

and so on. The chance that the shop is busy on the second period of job B, if initially it is in state 0, is

$$(\mathfrak{M}^t)_{03} = \frac{b + (-b)^t}{1+b} = 0 \ (t = 1), \ = b \ (t = 2), \ = b(1-b) \ (t = 3), \text{ etc.}$$

These probabilities oscillate somewhat at first, as t increases, because of the possible two-period cycle of B tasks. But, as time goes on, these periods lose any phase relation with initial conditions and the system settles down to the steady-state distribution of probabilities $P_j = \lim [(\mathfrak{M}^t)_{ij}]$ for $t \to \infty$, where

$$P_0 = \frac{(1-a)(1-b)}{1+b} \qquad P_1 = \frac{a(1-b)}{1+b} \qquad P_2 = P_3 = \frac{b}{1+b} \quad (28\text{-}13)$$

which are the solutions of Eq. (28-8) for this case. In the long run the fraction of periods the shop is idle is $[(1-a)(1-b)/(1+b)]$, for example. We note that the rate at which the system settles down to its steady state depends on the value of b, the chance that an order for job B arrives. If b is nearly unity it will take many periods to reach steady state; if b is small the system will settle down in a few periods. A number of A's must intersperse between the B jobs before regularity is lost; if b is near 1 this will take several periods.

On the other hand, the manager may decide to take on task A in preference to task B, even to the extent of postponing the second period of B, if an order for A arrives while the first period of B is under way. In this case there will be five possible states of the system: 0, idle; 1, task A (no B half done); 2, first period of B; 3, second period of task B; 4, task A (a half-done B waiting). The matrix is

$$\mathfrak{M} = \begin{pmatrix} (1-a)(1-b) & a & b(1-a) & 0 & 0 \\ (1-a)(1-b) & a & b(1-a) & 0 & 0 \\ 0 & 0 & 0 & 1-a & a \\ (1-a)(1-b) & a & b(1-a) & 0 & 0 \\ 0 & 0 & 0 & 1-a & a \end{pmatrix} \quad (A \text{ preemptive})$$

with steady-state solution

$$P_1 = \frac{(1-a)(1-b)}{1+b} \qquad P_1 = \frac{a}{1+b} \qquad P_2 = P_3 = \frac{b(1-a)}{1+b} \qquad P_4 = \frac{ab}{1+b} \quad (28\text{-}14)$$

A decision as to which policy to adopt might be based on relative profits. Designate the profit per task A as C_a and that per task B as $2C_b$ (i.e., the profit per period for B is C_b); then the former policy (priority to B) is to be preferred over the latter (preemptive priority to A) if $a(1 - b)C_a + 2bC_b > a(1 + b)C_a + 2b(1 - a)C_b$, i.e., if $C_b > C_a$, a not surprising result. On the other hand, if a penalty C_p must be paid each week that a half-finished B is waiting, then priority to B is to be preferred to preemptive priority to A if $C_b > C_a - \frac{1}{2}C_p$. Other production policies can be devised; they also can be tested out by solving the corresponding equations of motion.

28-6. The Formation of Queues

Alternatively, the shop manager may be in a position to hold all orders for jobs, to be executed as soon as the shop can get to them. In this case, of course, the mean number of orders for A, plus twice the mean number of orders for B, arriving each period, $\alpha + 2\beta = -\ln[(1 - a)(1 - b)^2]$, must be less than unity; otherwise the back orders would continually pile up and no steady-state situation could be reached. Even though $(\alpha + 2\beta) < 1$ there will be periods when several orders arrive, and back orders will begin to pile up; these will be worked off during the randomly occurring periods when no orders turn up.

There are an infinite number of possible states for this system, since the chance that j back orders have accumulated, though diminishing as j increases (when $\alpha + 2\beta < 1$) is not zero for any finite j. They can most conveniently be denumerated in terms of two indices

State 0 = shop idle for the period
State $j1$ = busy on A, $j - 1$ back orders on hand
State $j2$ = busy on B, first part, $j - 1$ back orders on hand
State $j3$ = busy on B, second part, $j - 1$ back orders on hand

where j is any integer larger than zero. If the shop is to carry out each back order on a first-come-first-served basis, it need not know the particular distribution of the orders for A or for B in the $j - 1$ back orders; only the total number $j - 1$ of back orders and the nature of the next order to be executed are important.

Since all orders are kept, we must know the probability that n orders will arrive in one period, not just the chance that 1 or more arrive. If the mean rate of arrival of orders for A is α and that for orders for B is β, and if both are distributed at random in time (Poisson distribution, see Sec. 28-8), then the chance that n orders (for either A or B) arrive in one period is

$$U_n = (\lambda^n/n!)e^{-\lambda} \qquad \lambda = \alpha + \beta \qquad (28\text{-}15)$$

and the chance that the next order turned up is for A, rather than for B, is $\gamma = \alpha/\lambda$.

Rather than write out the transition matrix $\mathfrak{M}$ for this system, we shall write out Eqs. (28-8) for steady-state behavior:

$$\begin{aligned}
P_0 &= P_0 U_0 + (P_{11} + P_{13})U_0 \\
P_{j1} &= \gamma P_0 U_j + \gamma \sum_{i=1}^{j+1} (P_{i1} + P_{i3})U_{j+1-i} \\
P_{j2} &= (1 - \gamma)P_0 U_j + (1 - \gamma) \sum_{i=1}^{j+1} (P_{i1} + P_{i3})U_{j+1-i} \\
P_{j3} &= \sum_{i=1}^{j} P_{i2} U_{j-i}
\end{aligned} \qquad (28\text{-}16)$$

For example, to be in state zero, the system could have been idle the previous period and no orders arrived, or it could be working on task A or finishing a task B and no order arrived (probability U_0). Or, to be working on task A, with $j - 1$ back orders, the shop could have been idle the previous period and j orders arrived, or it could have been working on A or B, with $i - 1$ back orders and $j + 1 - i$ orders arrived during the period, *if* the next order to be carried out had happened to be for job A

(probability γ). To solve these equations we use the method of generating functions, defining the following functions of the auxiliary variable z ($|z| < 1$):

$$F_s(z) = \sum_{n=1}^{\infty} z^n P_{ns} \quad (s = 1, 2, 3) \qquad P_0 + F_1(1) + F_2(1) + F_3(1) = 1$$

$$K(z) = \sum_{n=0}^{\infty} z^n U_n = e^{\lambda(z-1)} \qquad K(1) = 1$$

From these generating functions can be obtained the behavior of the system. For example, $P_{ns} = (1/n!)(d^n F_s/dz^n)_{z=0}$, and the probability that the shop is busy on task A is $F_1(1)$. Also the mean number of back orders on hand at any period is

$$L = \sum_{j=1}^{\infty} (j-1)(P_{j1} + P_{j2} + P_{j3}) = \sum_{s=1}^{3} F_s'(1) - (1 - P_0)$$

where $F_s'(z) = (dF_s/dz)$.

Multiplying the second equation of (28-16) by z^j and summing over j from 1 to ∞, and adding γ times the first equation, gives

$$\gamma P_0 + F_1 = \gamma P_0 K + \frac{\gamma}{z}(F_1 + F_3)K$$

whereas multiplying the first equation by $(1 - \gamma)$ and adding z^j times the third equation, summed over j, produces

$$(1 - \gamma)P_0 + F_2 = (1 - \gamma)P_0 K + \frac{1-\gamma}{z}(F_1 + F_3)K$$

Finally, multiplication of the fourth equation by z^j and summing produces $F_3 = KF_2$. Combining these equations and solving for the F's, we obtain

$$F_1 = \gamma F(z) \qquad F_2 = (1-\gamma)F(z) \qquad F_3 = (1-\gamma)K(z)F(z)$$

where
$$F(z) = \frac{z(1-K)P_0}{\gamma K + (1-\gamma)K^2 - z} \to \frac{\lambda P_0}{1 - (2-\gamma)\lambda} \quad (z \to 1)$$

also
$$F'(1) = \frac{\lambda - (3/2 - \gamma)\gamma^2 + (1-\gamma)\lambda^3}{1 - (2-\gamma)\lambda}$$

The value of P_0 can be obtained from the requirement that the sum of all the probabilities, $P_0 + F_1 + F_2 + F_3$ at $z = 1$, be unity. We find that

$$P_0 = 1 - (2-\gamma)\lambda = 1 - \alpha - 2\beta$$

which is the fraction of the periods the shop is idle. As α is the mean number of orders for single-period task A and β is the mean number of orders for double-period task B which arrive per period, this formula is not unexpected. The solutions are valid only when $\alpha + 2\beta < 1$; otherwise the series $F(1)$ diverges and there is no steady-state solution.

An operational parameter of interest is the mean number of back orders on hand in any period,

$$L = F'(1) + (1-\gamma)[K'(1)F(1) + K(1)F'(1)] - (2-\gamma)\lambda$$
$$= \frac{(2 - 3/2\gamma)\lambda^2}{1 - (2-\gamma)\lambda} = \frac{(\alpha + \beta)(1/2\alpha + 2\beta)}{1 - \alpha - 2\beta} \qquad (28\text{-}17)$$

Another important parameter is the mean number of periods W an order has to wait before it is executed. If an order arrives when the system is in state $j1$, it will have to wait a period for each A order ahead of it, $\gamma\lambda(j-1) = \alpha(j-1)$, and 2 periods for each B order ahead of it, $\beta(j-1)$. This is also the case if the order arrives when the system is in state $j3$; but if the system is in state $j2$ the delay will be an additional period, to finish the B order just started. Thus

$$W = \sum_{j=1}^{\infty} [(2-\gamma)(j-1)(P_{j1} + P_{j2} + P_{j3}) + P_{j2}]$$

$$= \frac{(1-\gamma)\lambda + \tfrac{1}{2}(2-\gamma)\lambda^2}{1-(2-\gamma)\lambda} = \frac{\beta + \tfrac{1}{2}(\alpha + 2\beta)(\alpha + \beta)}{1 - \alpha - 2\beta} \tag{28-18}$$

We note that the mean number of back orders L and the expected delay in executing an order are both inversely proportional to P_0, the fraction of periods the shop is idle. Put another way, the only shop which can give rapid service is one which is idle an appreciable fraction of periods. This reciprocal relation between length of wait and fraction of free time is a general property of all stochastic service systems, of all "bottlenecks," incipient or actual. Queue length and waiting time can sometimes be reduced by rearranging the service operation, making it less subject to random fluctuations, or carrying out the service in parallel rather than in series, or the like. But usually a large reduction in queue length or mean wait can be achieved *only* by reducing the arrival rate or by speeding up the service rate, in either case by effectively *increasing the idle time.*

A standard example of the fallacy of "suboptimization" (see Sec. 1-4) is the story of the "efficiency expert," who creates a whole new set of bottlenecks by going around eliminating idle time in a number of strategically placed portions of an operation, thus drastically reducing the efficiency of the operation as a whole by improving the efficiency of some of its parts.

28-7. Continuous Time

The discrete-time examples discussed so far have been useful as simple examples, and correspond to some operations encountered in practice. In many operational systems, however, changes in the state of the system may occur at any time, not just at the end of discrete periods. The effects of external forces are noted, and decisions are made regarding the system's resultant actions, whenever the forces act, not at the ends of predetermined periods. In order to use the Markov model for such systems we have to subdivide time into infinitesimal periods dt in length, and the matrix elements M_{ij} must equal the probability of transition from state i to state j in time dt. If the occurrences causing this change are randomly distributed in time, these probabilities should be proportional to dt (except for the transition $i \to i$). Therefore, to satisfy (28-4),

$$M_{ij} \to \begin{cases} R_{ij}\, dt & (j \neq i) \\ 1 - \sum_{k \neq i} R_{ik}\, dt & (j = i) \end{cases} \tag{28-19}$$

where R_{ij} is the mean *rate of transition* of the system from state i to state j. For the system to be Markovian, the R's should be independent of t.

Therefore, the equation of motion corresponding to (28-3) for the continuous-time Markov process is

$$p_j(t + dt) = \left[1 - \sum_{i \neq j} R_{ji}\, dt\right] p_j(t) + \sum_{i \neq j} p_i(t) R_{ij}\, dt$$

or

$$\frac{d}{dt} p_j(t) = \sum_i p_i(t) R_{ij} \qquad R_{jj} = -\sum_{i \neq j} R_{ji} \tag{28-20}$$

which is a set of differential equations governing the changes in the probability distribution $p_i(t)$ as a function of time. The corresponding rates of change are the elements of a *differential matrix*, having the general properties

$$R_{ij} \geq 0 \quad (j \neq i) \qquad R_{ii} = - \sum_{j \neq i} R_{ij} \qquad (28\text{-}21)$$

For the matrix elements to be independent of time, the events which cause the transitions must be distributed in time with the random incidence but uniform mean density which is characteristic of the Poisson distribution (see Sec. 28-8).

The formal solution of (28-20) is

$$p_j(t) = \sum_i p_i(0)(e^{\mathcal{R}t})_{ij} \qquad (28\text{-}22)$$

where $e^{\mathcal{R}t}$ is a stochastic matrix (Sec. 28-4) with the following series expansion:

$$e^{\mathcal{R}t} = \mathcal{I} + \mathcal{R}t + \tfrac{1}{2}\mathcal{R}^2 t^2 + \tfrac{1}{6}\mathcal{R}^3 t^3 + \cdots$$

where $\mathcal{I}$ is the identity matrix, and where $\mathcal{R}^2$, $\mathcal{R}^3$, etc., are matrices formed from $\mathcal{R}$ by repeated multiplication. Matrix calculus shows that this series has transient terms but that it usually settles down to a steady-state matrix, independent of t, in which every row is identical.

As a simple example we consider the elementary transport system consisting of a single taxicab, which can wait at station A or at station B for a fare. Prospective passengers arrive at either station at random in time; they wish to go either to the other station (a short distance away) or else to some place in the city or its suburbs (call it "elsewhere" for short), in which case we shall assume that the probability of the taxi's round trip being longer than t is $e^{-\mu t}$, μ being the "rate of completion" of the service operation of delivery of the passenger. When the taxi is on station at either station A or B, the mean rate at which it picks up passengers bound for the other station is α and the mean rate at which it picks up someone to take elsewhere is λ. The differential matrix for this system is therefore

$$\mathcal{R} = \begin{pmatrix} -\alpha - \lambda & \alpha & \lambda \\ \alpha & -\alpha - \lambda & \lambda \\ \mu/2 & \mu/2 & -\mu \end{pmatrix}$$

with the taxi in state 1 being stationed at A, in state 2 being at B, and in state 3 being en route elsewhere. It has been assumed that the taxi returns from such trips to either station at random, with equal frequency.

The solution of (28-20) for this case is contained in the stochastic matrix

$$e^{\mathcal{R}t} = \frac{1}{\mu + \lambda}\begin{pmatrix} \mu/2 & \mu/2 & \lambda \\ \mu/2 & \mu/2 & \lambda \\ \mu/2 & \mu/2 & \lambda \end{pmatrix} + \frac{e^{-(2\alpha+\lambda)t}}{2}\begin{pmatrix} 1 & -1 & 0 \\ -1 & 1 & 0 \\ 0 & 0 & 0 \end{pmatrix}$$

$$+ \frac{e^{-(\mu+\lambda)t}}{\mu + \lambda}\begin{pmatrix} \lambda/2 & \lambda/2 & -\lambda \\ \lambda/2 & \lambda/2 & -\lambda \\ -\mu/2 & -\mu/2 & \mu \end{pmatrix}$$

The element $(e^{\mathcal{R}t})_{ij}$ is the probability that, if the system is in state i initially, it will be in state j at time t. In the long run, the chance that the taxi is at station A is $\mu/2(\mu + \lambda)$, the chance that it is at B is the same, and the chance that it is taking a passenger elsewhere is $\lambda/(\mu + \lambda)$. These probabilities are independent of α, the rate of pickup of interstation passengers, because we have assumed that the transit time τ between stations A and B is short compared with $1/\alpha$ or $1/\mu$. If this is not true then the fractions of time the taxi is on station or taking a fare elsewhere are $\mu/2(\mu + \lambda + \alpha\mu\tau)$ and $\lambda/(\mu + \lambda + \alpha\mu\tau)$, respectively, and the fraction of time it spends in going between stations is then $\alpha\mu\tau/(\mu + \lambda + \alpha\mu\tau)$ in the steady state.

28-8. Arrival and Service Statistics

The earlier discussion has indicated that the dynamic behavior of a stochastic service system is determined by the interplay between the arrivals of units to be serviced and the completions of "service" of units. Both arrivals and service are usually subject to random variations, and the reaction between these two stochastic flows produces the fluctuating queues and delays. Therefore, in the study of such operations, it is necessary to determine the statistics of these two processes, arrival and service.

For continuous time, the statistics of arrivals to a given service system can be described either by the *arrival distribution* $A_0(t)$, the probability that the time interval between two successive arrivals is greater than t, or else by the *postarrival density* $a(t)$, such that $a(t)\,dt$ is the probability that an arrival occurs between t and $t + dt$ after the previous arrival. It is not difficult to show that $a(t) = -[dA_0(t)/dt]$ and that

$$A_0(t) = \int_t^\infty a(x)\,dx \qquad A_n(t) = \int_0^t a(x) A_{n-1}(t - x)\,dx \qquad (28\text{-}23)$$

where $A_n(t)$ is the probability that precisely n arrivals ($n > 0$) occur in an interval t after some initial arrival (the A's are thus called *postarrival distributions*). The function $A_0(t)$ can be measured, for a particular operation encountered in practice, by timing a reasonably long sequence of arrivals. The fraction of measured interarrival intervals which are longer than t would approach $A_0(t)$ as more and more arrivals are timed.

It is also not difficult to show that, in general, the probability $U_n(t)$, that precisely n arrivals occur in an interval t *chosen at random in time*, differs from $A_n(t)$ and is related to the A's by the equations

$$U_0(t) = \lambda \int_t^\infty A_0(x)\,dx \qquad U_n(t) = \lambda \int_0^t A_0(x) A_{n-1}(t - x)\,dx \qquad (28\text{-}24)$$

where λ, the mean rate of arrival, is the reciprocal of the mean time between arrivals

$$T_a = \frac{1}{\lambda} = \int_0^\infty t a(t)\,dt = \int_0^\infty A_0(t)\,dt \qquad (28\text{-}25)$$

The functions U_n are called *random-interval distributions*. They may be either separately measured or else computed from the measured $A_0(t)$. [Note that they have been used in connection with Eqs. (28-15) and (28-16).]

Arrivals can be said to be completely random in time *only if* $A_n(t) = U_n(t)$. Comparison of (28-23) and (28-24) shows that this can be true only if

$$\lambda A_0(t) = a(t) = -\frac{dA_0(t)}{dt}$$

i.e., only if $A_0 = e^{-\lambda t}$, in which case

$$A_n(t) = U_n(t) = \frac{(\lambda t)^n}{n!} e^{-\lambda t} \qquad (28\text{-}26)$$

which is the Poisson distribution, as mentioned earlier. For this reason, completely random arrivals are usually called *Poisson arrivals*.

The degree of variability of the interarrival time is measured by its *variance*

$$\sigma_a^2 = \int_0^\infty (t - T_a)^2 a(t)\,dt = \int_0^\infty (2t - T_a) A_0(t)\,dt \qquad (28\text{-}27)$$

The square root of the variance σ_a is called the standard deviation of the interarrival time. For Poisson arrivals, since this variance turns out to be $(1/\lambda)^2$, the standard deviation of the interarrival interval is equal to the interval T_a itself. If arrivals are more regularly spaced, σ_a is smaller than T_a. It is also possible to have σ_a larger than T_a.

Service statistics can be described in much the same way as those for arrivals. We can imagine an infinite supply of units to be served, so the service facility can start work on the next unit as soon as it has finished with the previous one, in which case the startings of service will be distributed in time analogously with arrivals. There will be a probability density $s(t)$, such that $s(t)\,dt$ is the probability that a particular service is completed between times t and $t + dt$ after its start. And there will be a probability $S_0(t)$, a mean duration T_s of service, a mean service rate μ, and a variance σ_s^2 of service time,

$$S_0(t) = \int_t^\infty s(x)\,dx \qquad s(t) = -\frac{dS_0(t)}{dt}$$

$$T_s = \frac{1}{\mu} = \int_0^\infty S_0(x)\,dx \qquad \sigma_s^2 = \int_0^\infty (t - T_s)^2 s(t)\,dt$$

(28-28)

The function $S_0(t)$, the chance that the service operation will take longer than t to complete, is usually the easiest statistic to determine experimentally.

A service operation analogous to Poisson arrivals in regard to service-time distribution is *exponential service*, where $S_0(t) = e^{-\mu t}$. Its variance of service time is $(1/\mu)^2$. Some service operations consist (or behave as though they consist) of a combination of elementary service phases, each of which is exponential. For example, there may be k identical service phases, each of which has a mean duration $T_s/k = 1/\mu k$. The probability density for completion of this sequence is proportional to the Poisson probability of $k - 1$ arrivals, given by (28-26) with $n = k - 1$.

$s(t) = \mu k[(\mu kt)^{k-1}/(k-1)!]e^{-\mu kt}$, so that

$$S_0(t) = \sum_{n=0}^{k-1} \frac{(\mu kt)^n}{n!} e^{-\mu kt} \equiv E_{k-1}(\mu kt)$$

(28-29)

(see Sec. 28-10 for further discussion of the function E_m). The case $k = 1$ is the exponential service. As k increases the regularity of the service times increases as shown in Fig. 28-1 (the variance of these service times is $1/k\mu^2$) until, as $k \to \infty$, the service becomes completely regular, each service being completed in exactly $1/\mu$ units of time. Function $E_k(x)$ is called the kth *Erlang function*, and systems with such service-time distributions are called *Erlang* service facilities.

Other service operations, having variance of service times greater than $1/\mu^2$ (exponential service) can be simulated by having several alternate service operations, each exponential but with a different rate and with a random assignment of the incoming unit to one or the other alternate service operation. For example, if the mean rate of one of two alternate exponential services is β and the mean fraction of units assigned to this service is α, the mean rate of the other is γ and the fraction assigned to it is $1 - \alpha$, then the probability that the assigned operation is not completed in time t is $\alpha e^{-\beta t} + (1-\alpha)e^{-\gamma t}$. Adjusting the value of α so that the mean time for the combined operation is $1/\mu$, we get

$$s(t) = \frac{1}{\mu(\beta - \gamma)}[\beta^2(\mu - \gamma)e^{-\beta t} + \gamma^2(\beta - \mu)e^{-\gamma t}]$$

$$S_0(t) = \frac{1}{\mu(\beta - \gamma)}[\beta(\mu - \gamma)e^{-\beta t} + \gamma(\beta - \mu)e^{-\gamma t}]$$

(28-30)

where we lose no generality in assuming that $\beta > \mu > \gamma$. Service facilities with these statistics are called *hyperexponential* channels. Their variance in service times is $(1/\mu^2\beta\gamma)[\beta\gamma + 2(\beta - \mu)(\mu - \gamma)]$, which is greater than the value $1/\mu^2$, for exponential service, as long as $\beta > \mu > \gamma$. One such curve is shown in Fig. 28-1.

Non-Poisson arrivals may be simulated in a similar manner, by imagining an infinite pool of units, which must pass through a system of "Poisson gates" before being freed to "arrive" at the service facility. Poisson arrivals would be obtained by using a single gate of mean rate λ, which would open to allow a single unit to escape at random times, corresponding to the distribution of (28-26). Erlang arrivals would be simulated by a sequence of k gates, each of rate $k\lambda$, all of which would have to pass a unit before that unit would be free to "arrive" and before a next unit could enter the sequence. Hyperexponential arrivals could be simulated by having two alternate gates, with random choice as to which gate would allow the proximate unit to escape, the other gate being barred from use by another unit until the proximate unit has escaped.

The reason, of course, that it is useful to seek ways of simulating general arrival or service distributions by combinations of exponential elements lies in our desire to

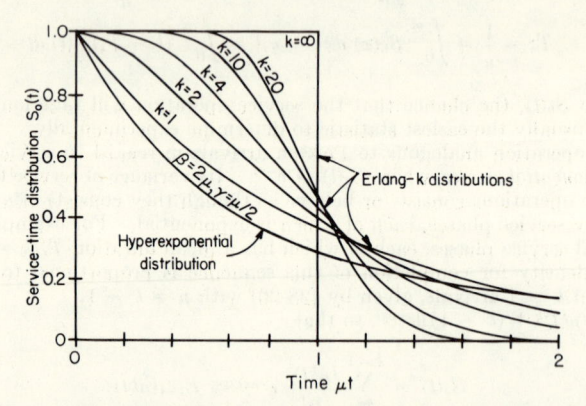

FIG. 28-1. Typical service-time distributions, $S_0(t)$ being the probability that service takes a time greater than t. Curve $k = 1$ is the exponential distribution, $k > 1$ the Erlang distributions $E_{k-1}(k\mu t)$ of Eq. (28-29), and the lowest curve is the hyperexponential distribution of Eq. (28-30).

use Eqs. (28-20), to reduce the system to an equivalent Markovian system. Equations (28-20) can be used only when all transitions from state to state, represented by matrix $\mathcal{R}$, are distributed at random in time, according to a set of probabilities like Eqs. (28-26).

28-9. Bottlenecks and Queues

It should be clear by now that the dynamic characteristics of a stochastic service system are determined by the interaction between the statistics of the arriving units and the statistics of the service facility, modified by the rules governing the behavior of the arriving units before they are serviced (i.e., the *queue discipline*). In fact, enumeration of the distribution functions $A_0(t)$ and $S_0(t)$, or the densities $a(t)$ and $s(t)$, plus specification of queue discipline, plus an indication of whether the arriving units are unlimited in number or not and whether the service facility can accommodate several units in parallel (multiple service channels) or not, must all be given or assumed before the dynamic characteristics of the system can be worked out.

In some cases, if the arriving unit cannot be serviced immediately, it departs, seeking service elsewhere or postponing service. Here the unit acts like a free agent, avoiding any wait; such systems might be said to have *free arrivals*. Alternatively the arriving units may all be *bound* to the particular service facility; they may have to wait until the facility is free to service them. These bound units, waiting for service, of course constitute the *queue*. Systems with bound arrivals are responsible for most of the bottlenecks impeding the flow of people, goods, and information. Part of the interest in the theory of queues lies in a desire to reduce their length.

Of course, there are many examples of arrival behavior intermediate between the free and the bound cases. For example, the length of the queue may be restricted, so that units arriving when the queue length is less than some number N will enter the queue and stay till serviced but units arriving when the queue is of length N do not stay. In other cases units may exhibit impatience, leaving the queue (reneging) if the delay is too great. Individual units may differ in their behavior; so the effect may have to be expressed in terms of probabilities.

The simplest stochastic service system is a single, exponential service facility of mean service rate μ, with free Poisson arrivals having a mean arrival rate λ. In this case the probability that an arrival occurs in any time interval dt is $\lambda\, dt$, and when the service facility is operating, the chance that service is completed in an interval dt is $\mu\, dt$. There are two states of the system, state zero when the system is idle, waiting for the next arrival, and state one when the facility is busy and any other arriving unit goes elsewhere. The equations of motion, (28-20), are

$$\frac{d}{dt} p_0(t) = -\lambda p_0 + \mu p_1$$
$$\frac{d}{dt} p_1(t) = \lambda p_0 - \mu p_1$$

$$\mathcal{R} = \begin{pmatrix} -\lambda & \lambda \\ \mu & -\mu \end{pmatrix}$$

Straightforward solution of these two equations gives

$$p_0(t) = \frac{p_0(0)}{\mu + \lambda} [\mu + \lambda e^{-(\lambda+\mu)t}] + \frac{p_1(0)}{\mu + \lambda} [\mu - \mu e^{-(\lambda+\mu)t}]$$
$$p_1(t) = \frac{p_0(0)}{\mu + \lambda} [\lambda - \lambda e^{-(\lambda+\mu)t}] + \frac{p_1(0)}{\mu + \lambda} [\lambda + \mu e^{-(\lambda+\mu)t}]$$

which can be written in the matrix form

$$p_j(t) = \sum_i p_i(0) (e^{\mathcal{R}t})_{ij}$$

$$e^{\mathcal{R}t} = \frac{1}{\mu + \lambda} \begin{pmatrix} \mu & \lambda \\ \mu & \lambda \end{pmatrix} + \frac{e^{-(\lambda+\mu)t}}{\lambda + \mu} \begin{pmatrix} \lambda & -\lambda \\ -\mu & \mu \end{pmatrix}$$

where $(e^{\mathcal{R}t})_{ij}$ is the conditional probability that, if the system is in state i at $t = 0$, it will be in state j at time t.

The transient term in this probability dies out according to the exponential $e^{-(\lambda+\mu)t}$. After a few arrivals the system has effectively forgotten how it started. The steady-state probability that the facility is busy is $p_1(\infty) = \lambda/(\lambda + \mu)$, and the corresponding probability that it is idle is $p_0(\infty) = \mu/(\lambda + \mu)$. Since arrivals come at random, the fraction of the arriving units which turn away (are *lost* to the system) is equal to $\lambda/(\lambda + \mu)$, the fraction of the time the system is busy. As with the system of Eq. (28-18), the larger the ratio of arrival to service rate $\rho = \lambda/\mu$, the smaller is the fraction of arriving units which are actually serviced. In this case of free units, a steady state is possible even when $\rho > 1$, since excess units leave rather than form a queue.

One further example, to show how an intermediate case behaves, is that of a single, exponential service channel, of mean rate μ, with Poisson arrivals of rate λ, having room for a queue of length 2; if a unit finds less than 2 in the queue when it arrives, it stays until serviced, but if 2 are in the queue (3 are in the system) it does not stay. We denote the states by n, the number in the system ($n = 0, 1, 2, 3$), and find that the transition-rate matrix is

$$\mathcal{R} = \begin{pmatrix} -\lambda & \lambda & 0 & 0 \\ \mu & -\lambda - \mu & \lambda & 0 \\ 0 & \mu & -\lambda - \mu & \lambda \\ 0 & 0 & \mu & -\mu \end{pmatrix}$$

Solution of the equations of motion (28-20) allows one to obtain the transition matrix $e^{\Re t}$, the (i, j)th element of which is the probability that the system will be in state j at time t if it had been in state i at time 0. This matrix is

$$e^{\Re t} = \frac{\mu - \lambda}{\mu^4 - \lambda^4} \begin{pmatrix} \mu^3 & \mu^2\lambda & \mu\lambda^2 & \lambda^3 \\ \mu^3 & \mu^2\lambda & \mu\lambda^2 & \lambda^3 \\ \mu^3 & \mu^2\lambda & \mu\lambda^2 & \lambda^3 \\ \mu^3 & \mu^2\lambda & \mu\lambda^2 & \lambda^3 \end{pmatrix} + \frac{e^{-(\lambda+\mu)t}}{2\lambda\mu(\lambda+\mu)} \begin{pmatrix} \lambda^2\mu & -\lambda^2\mu & -\lambda^3 & \lambda^3 \\ -\lambda\mu^2 & \lambda\mu^2 & \lambda^2\mu & -\lambda^2\mu \\ -\lambda\mu^2 & \lambda\mu^2 & \lambda^2\mu & -\lambda^2\mu \\ \mu^3 & -\mu^3 & -\lambda\mu^2 & \lambda\mu^2 \end{pmatrix}$$

$$+ \sum_{n=1}^{2} \frac{e^{-\omega_n t}}{4\lambda\mu\omega_n} \begin{pmatrix} \lambda^2\mu & \lambda^2(\lambda - \omega_n) & \lambda^2(\omega_n - \mu) & -\lambda^3 \\ \lambda\mu(\lambda - \omega_n) & \lambda\mu(2\omega_n - \mu) & -\lambda(\omega_n - \mu)(\omega_n - \lambda) & \lambda^2(\lambda - \omega_n) \\ \mu^2(\omega_n - \mu) & -\mu(\omega_n - \mu)(\omega_n - \lambda) & \lambda\mu(2\omega_n - \lambda) & \lambda\mu(\mu - \omega_n) \\ -\mu^3 & \mu^2(\lambda - \omega_n) & \mu^2(\mu - \omega_n) & \lambda\mu^2 \end{pmatrix}$$

where $\omega_1 = \lambda + \mu - \sqrt{2\lambda\mu}$ and $\omega_2 = \lambda + \mu + \sqrt{2\lambda\mu}$. The last three terms are transients, the slowest of which dies out with a factor $e^{-\omega_1 t}$. Here the fraction of arriving units which are lost to the system (in the steady state) is equal to the probability $\lambda^3/(\lambda^3 + \lambda^2\mu + \lambda\mu^2 + \mu^3)$ that the system is full $(n = 3)$. This approaches

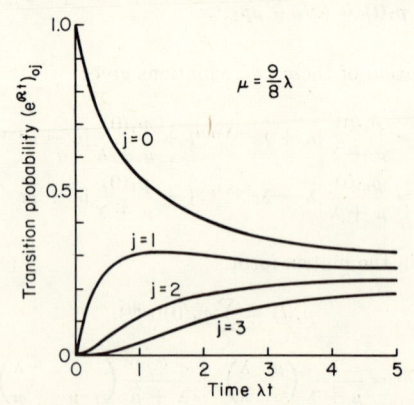

FIG. 28-2. Probability of transition $(e^{\Re t})_{0j}$ from initially idle state 0 to state with j units in system at time t, for one-channel system with maximum queue length $N = 2$, as discussed in text.

unity as λ is made larger than μ, drops rapidly to zero as λ is made smaller than μ. It is thus possible to hold and service a larger percentage of the arriving units, for the same value of $\rho < 1$, than could be achieved with the zero-queue system, at the expense of delays in queue of the bound units, of course (see Fig. 28-2). The mean number of units in the system is $(\rho + 2\rho^2 + 3\rho^3)/(1 + \rho + \rho^2 + \rho^3)$, which is roughly equal to $\rho = \lambda/\mu$ when $\rho < 1$ and approaches 3 when ρ is much larger than 1. Analysis of systems with maximum queue length N greater than 2 shows that, as N is increased, it takes longer and longer for the transient term to die out (if $\mu > \lambda$), until finally, for bound units $(N \to \infty)$, there is no steady state for $\rho > 1$.

28-10. Steady-state Operation

Whenever the arrival and service statistics can be simulated by a combination of exponential elements, then the behavior of the system can be described as a continuous-time Markov process, with equation of motion of the type of (28-20). In general the state probabilities $p_j(t)$ will have transient terms, as illustrated in the preceding examples, but in many cases these will die out exponentially and the p's will take on their steady-state values. These values $P_i = p_i(\infty)$ usually suffice to determine most of the properties of the system which have practical interest. The equations for their determination are

$$P_j \sum_{i \neq j} R_{ji} = \sum_{i \neq j} P_i R_{ij} \qquad \sum_j P_j = 1 \qquad (28\text{-}31)$$

which are obtained from (28-20) by requiring that the P_j's be independent of time. The element R_{ij} is the rate of transition from state i to state j. Equations (28-31) thus state that, for steady state, the rate of transition out of state j to any other state i $\left(P_i \sum_{i \neq j} R_{ji}\right)$ must be equal to the rate of transition into state j, from all other states i $\left(\sum_{i \neq j} P_i R_{ij}\right)$.

The particular choice of index numbers to assign to the various states has some effect on the form of matrix $\mathcal{R}$ and on the ease of solution, though of course it has no effect on the behavior of the system. In many cases, where queues are likely to form, several indices may be advisable (k, m, n, say) with index n equal to the number of units in the system (or in the queue). In these cases it may be desirable to express the solution in terms of generating functions (see Sec. 28-6)

$$F_{km}(z) = \sum_n z^n P_{kmn} \qquad \sum_{kmn} P_{kmn} = \sum_{km} F_{km}(1) = 1$$

In this case the expected number L in the system (or queue), and its variance are

$$L = \sum_{kmn} n P_{kmn} = \sum_{km} F'_{km}(1) \qquad F'_{km}(z) = \frac{dF_{km}(z)}{dz}$$
$$(\Delta L)^2 = \sum_{kmn} (n - L)^2 P_{kmn} = \sum_{km} F''_{km}(1) + L - L^2 \qquad (28\text{-}32)$$

A few examples will illustrate these general comments.

The first example is that of a parking lot, with places for M automobiles. It is determined that the cars arrive with a Poisson distribution in time, with mean arrival rate λ, and that the length of stay of a car is exponentially distributed, departure rate per car being μ. The state of the system at time t can be given by j, the number of cars in the lot at time t. For this system the transition rate from state $j - 1$ to state j, because of the arrival of a car, is λ. The rate of transition from state $j + 1$ to state j, because of the departure of a car, is $(j + 1)\mu$, since $(j + 1)$ cars are in the lot in state $(j + 1)$ and any one of them may be the one which leaves. The chance of two cars arriving, or of two cars leaving, in the same element dt of time is proportional to $(dt)^2$ (and even less for 3 or more); thus the transition rates R_{ij} are zero for $i > j + 1$ or $i < j - 1$. Since arrivals must be turned away when the lot is full, $R_{M,M+1}$ must also be zero. Therefore, matrix $\mathcal{R}$ is

$$\mathcal{R} = \begin{pmatrix} -\lambda & \lambda & 0 & 0 & - & - & 0 & 0 \\ \mu & -\lambda - \mu & \lambda & 0 & - & - & 0 & 0 \\ 0 & 2\mu & -\lambda - 2\mu & \lambda & - & - & 0 & 0 \\ \hline 0 & 0 & 0 & 0 & - & - & M\mu & -M\mu \end{pmatrix}$$

and the equations for the steady-state probabilities are

$$\lambda P_0 = \mu P_1 \qquad M\mu P_M = \lambda P_{M-1}$$
$$(\lambda + j\mu)P_j = \lambda P_{j-1} + (j + 1)\mu P_{j+1} \qquad (0 < j < M) \qquad (28\text{-}33)$$

These equations can be solved sequentially, to obtain

$$P_n = \frac{(\lambda/\mu)^n}{n!} \frac{e^{-\lambda/\mu}}{E_M(\lambda/\mu)} \qquad E_k(x) = \sum_{m=0}^{k} \frac{x^m}{m!} e^{-x} \qquad (28\text{-}34)$$

where $E_k(x)$ is the Erlang function, defined in (28-29) and plotted in Fig. 28-1, which is nearly unity as long as x is smaller than k and is quite small when x is larger than k. Put another way, the probability P_M, that the lot is full and cars have to be turned away, is large only when λ/μ is larger than M, i.e., when the mean rate λ of arrival is larger than the mean rate $M\mu$ of departure when the lot is full. The average number of cars in the lot is

$$L = \sum_{n=0}^{\infty} nP_n = \frac{\lambda E_{M-1}(\lambda/\mu)}{\mu E_M(\lambda/\mu)}$$

which is approximately equal to λ/μ when λ is less than $M\mu$ but approaches M asymptotically as λ gets larger than $M\mu$, as shown in Fig. 28-3.

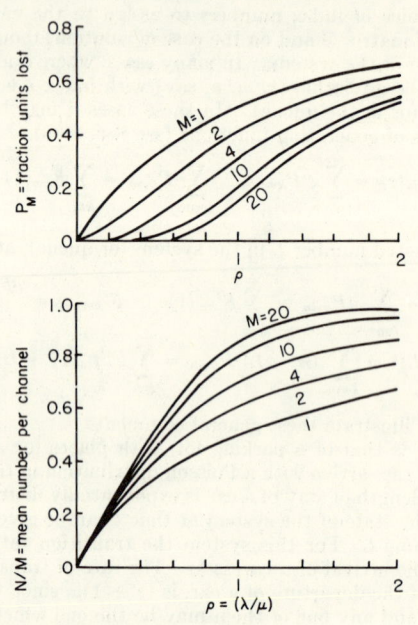

Fig. 28-3. Steady-state probability P_M that all M channels are busy and $1/M$ times the mean number N of units in the system, for the system of Eqs. (28-47).

The prototype of the service system with bound arriving units is the waiting line which forms in front of a single ticket window or the line of cars which waits in front of a single toll booth. If arrivals are Poisson and the service times are distributed exponentially, with mean rate μ, then the equations for the steady-state probabilities are

$$\lambda P_0 = \mu P_1 \qquad (\lambda + \mu)P_j = \mu P_{j+1} + \lambda P_{j-1} \qquad (j > 0) \qquad (28\text{-}35)$$

where state j represents the state with j units in the system, $j - 1$ waiting in queue, and 1 being served (in state 0 no unit is present and the service unit is idle). The generating function for this system is $F(z) = \Sigma z^n P_n$. The equation for F can be obtained by multiplying the jth equation of (28-35) by z^j and summing:

$$(\lambda + \mu)F - \mu P_0 = (\mu/z)F - (\mu/z)P_0 + \lambda zF$$

or
$$F(z) = P_0/(1 - \rho z), \text{ where } \rho = \lambda/\mu$$

Since $F(1) = \Sigma P_n$ must equal 1, the probability P_0 that the facility is idle must equal $1 - \rho$. Thus $1 - P_0 = \rho = \lambda/\mu$ is the fraction of time that the facility is busy; ρ is often called the *utilization factor* of the system. The solution, of course, is valid only when $\rho < 1$.

Expanding $F(z)$ and equating the coefficient of z^n to P_n we obtain

Probability that exactly n units are in system $= P_n = (1 - \rho)\rho^n$

Probability that n or more units are in system $= \sum_{m=n}^{\infty} P_m = \rho^n$ (28-36)

The mean number L in the system and the mean number L_q in the queue can be obtained by differentiating $F(z)$,

$$L = \sum_{n=1}^{\infty} nP_n = F'(1) = \frac{\rho}{1-\rho} = \frac{\lambda}{\mu - \lambda}$$

$$L_q = \sum_{n=1}^{\infty} (n-1)P_n = F'(1) - (1 - P_0) = \frac{\rho^2}{1-\rho}$$
(28-37)

These results should be compared with Eq. (28-17) for a discrete-time system with a queue of back orders. The amount of variability in queue length $(\sigma_n)^2$ is equal to $\rho/(1-\rho)^2$, which is larger than L^2 when $\rho < 1$.

The probability $G(t)$ that an arriving unit will wait a time longer than t before it is through the system, with service completed, depends on the "queue discipline." When service is in order of arrival, then if, at the unit's arrival, there were j units in the system, the probability that the arriving unit will not yet be through the system a time t later equals the probability that j or fewer service operations will be completed in time t. Since each service operation is exponentially distributed, this is the sum of the Poisson elements $(\mu t)^n e^{-\mu t}/n!$ for n from 0 to j. Therefore, the probability of wait longer than t is

$$G(t) = \sum_{j=0}^{\infty} P_j \sum_{n=0}^{j} \frac{(\mu t)^n}{n!} e^{-\mu t} = \sum_{n=0}^{\infty} \frac{(\mu t)^n}{n!} e^{-\mu t} \sum_{j=n}^{\infty} P_j$$

$$= \sum_{n=0}^{\infty} \frac{(\lambda t)^n}{n!} e^{-\mu t} = e^{(\lambda - \mu)t} \qquad (\lambda < \mu)$$
(28-38)

The mean wait in the system of a unit is thus

$$W = \int_0^{\infty} G(t)\, dt = \frac{1}{\mu - \lambda} = \frac{L}{\lambda}$$
(28-39)

The relation $W = L/\lambda$ is a general one, holding for quite general arrival and service distributions, as long as all arriving units stay in the queue. All it says is that the mean wait W, divided by the mean time between arrivals $1/\lambda$, is equal to the mean number L in the system, in the steady-state condition.

The probability $G_q(t)$ that the unit stays longer than time t *in the queue*, before entering service, and the mean wait in queue W_q are obtained in a similar manner,

$$G_q(t) = \sum_{j=0}^{\infty} P_j \sum_{n=0}^{j-1} \frac{(\mu t)^n}{n!} e^{-\mu t} = \frac{\lambda}{\mu} e^{(\lambda - \mu)t}$$

$$W_q = \frac{\lambda}{\mu(\mu - \lambda)} = \frac{L_q}{\lambda}$$
(28-40)

Here again, we see that length of queue and mean length of time spent in queue are inversely proportional to the fraction of time $1 - \rho = P_0$ the system is idle. Unless this fraction is larger than about $\frac{1}{4}$, both W and L are uncomfortably large.

Another fruitful way of considering the behavior of the stochastic service system is in terms of its idle and busy periods. The system is idle a fraction $1 - \rho = 1 - (\lambda/\mu)$ of the time. An *idle period* starts when the last of a batch of units has its service completed and no additional unit has yet arrived; the period stops when the next unit arrives. Thus the distribution of duration of idle time is related to the interarrival distribution $A_0(t)$. For Poisson arrivals, the probability that the idle period lasts longer than t is $e^{-\lambda t}$, the mean length of an idle period is $1/\lambda$, and the variance of this length is $(1/\lambda)^2$, just as with the times between arrivals.

During a *busy period* the service facility is continuously busy, turning out serviced units at a rate μ. Sometimes only one unit is serviced before another idle period occurs; sometimes many units must be serviced before the facility gets a rest. The distribution of duration of the busy period depends on both arrival and service distributions. It is not a simple exponential, even for Poisson arrivals and exponential service.

The mean duration of a busy period T_b can be determined from the general requirement that the fraction of time the facility is busy $\rho = \lambda/\mu$ must equal T_b divided by T_b plus the mean length $1/\lambda$ of an idle period,

$$\frac{\lambda}{\mu} = \frac{\lambda T_b}{1 + \lambda T_b} \quad \text{or} \quad T_b = \frac{1}{\mu - \lambda}$$

The variance of this time, of course, depends on the arrival and service statistics. For Poisson arrivals and exponential service (see Ref. 7, for example) it turns out to be

$$(\Delta T_b)^2 = \frac{\lambda + \mu}{(\mu - \lambda)^3} \tag{28-41}$$

This is not equal to T_b^2, as it would be if the busy-period distribution were exponential; it is larger than T_b^2 for $0 < \lambda < \mu$ and becomes very much larger when the utilization factor ρ is near unity.

28-11. Effect of Service and Arrival Statistics

The simple system we just analyzed is that of bound Poisson arrivals, which stay in order of arrival until serviced, by a single service facility with an exponential distribution of service time. With this system we found that the way to decrease the mean waiting time or the mean length of queue is to decrease the ratio ρ between the mean arrival rate and the mean service rate and thus to increase the mean fraction of time $1 - \rho$ during which the service facility is idle. Because of the variability of the arrival and service times, service must be at times idle in order that it can take care of random peaks of arrivals or delays in service, without occasioning long waits in queue. We now ask whether any other action might be taken, aside from change in ρ, to decrease queue length or mean delay.

For example, we can ask what effect a change in the statistics of the service operation would have; whether making the service operation more regular (decreasing its variance) without changing its mean rate would decrease mean queue length. This effect may be computed by simulating the nonexponential service in terms of a combination of exponential phases. For example, a service with Erlang-k service-time distribution $E_{k-1}(k\mu t)$ corresponds to a sequence of k service phases, each of rate $k\mu$, which must be completed sequentially before the unit is discharged and the next unit can enter service [Eq. (28-29)]. In this case, the state of the system may be designated by n, the number of units in the system (in queue or in service) and s, the phase the unit in service happens to be in, the unit having started with phase k and proceeding stepwise to phase 1, after which it is discharged (in state 0 the service is idle and s is meaningless).

The equations for the steady-state behavior of this system are

$$\lambda P_0 = k\mu P_{1,1} \qquad (\lambda + k\mu)P_{1,s} = k\mu P_{1,s+1} \qquad (s < k)$$
$$(\lambda + k\mu)P_{1,k} = k\mu P_{2,1} + \lambda P_0$$
$$(\lambda + k\mu)P_{n,s} = k\mu P_{n,s+1} + \lambda P_{n-1,s} \qquad (s < k) \qquad (28\text{-}42)$$
$$(\lambda + k\mu)P_{n,k} = k\mu P_{n+1,1} + \lambda P_{n-1,k} \qquad (n > 1)$$

These equations may be solved by the use of generating functions (see Ref. 6) to obtain the mean length of queue, mean waiting time, and other steady-state properties. But these same properties can be worked out, for bound Poisson arrivals and for a single service channel with quite general service-time density $s(t)$, by considering the changes in the system as service is completed on each successive unit.

The time interval between the beginning and end of a single service is called a *service period*, to be distinguished from a busy period, as discussed earlier. The probability π_n of n arrivals during a service period is equal to the probability $[(\lambda t)^n/n!]e^{-\lambda t}$ that n units arrive in time t, multiplied by the density $s(t)$ that the service period is of duration t, integrated over t. Therefore, the generating function for the π's is

$$B(z) = \sum_{n=0}^{\infty} z^n \pi_n = \int_0^{\infty} \sum_n \frac{(\lambda z t)^n}{n!} e^{-\lambda t} s(t) \, dt = \int_0^{\infty} e^{(z-1)\lambda t} s(t) \, dt$$
$$= \Psi_s(\lambda - \lambda z) \quad \text{where} \quad \Psi_s(x) = \int_0^{\infty} e^{-xt} s(t) \, dt$$

is the Laplace transform of the service-completion probability density $s(t)$. We note that $\Psi_s(0) = 1$, $\Psi_s'(0) = -T_s = -(1/\mu)$, and that $\Psi_s''(0) = T_s^2 + \sigma_s^2$. Therefore, $B(1) = 1$ as it should, the mean number $\langle \alpha \rangle$ and the mean square of the number of arrivals per service period are

$$\langle \alpha \rangle = \sum_{n=1}^{\infty} n\pi_n = B'(1) = -\lambda \Psi_s'(0) = \lambda T_s = \rho$$
$$\langle \alpha^2 \rangle = \sum_{n=1}^{\infty} n^2 \pi_n = \sum_n n(n-1)\pi_n + \rho = B''(1) + \rho \qquad (28\text{-}43)$$
$$= \lambda^2(T_s^2 + \sigma_s^2) + \rho = \rho(1 + \rho) + \lambda^2 \sigma_s^2$$

where $\rho = \lambda/\mu$ is the utilization factor.

Next consider the changes in the number l_k of units in the system just *after* the kth service period is completed. If the unit then in service leaves $l_k > 0$ units behind it, then l_{k+1} will equal $l_k - 1$ plus the number α_{k+1} of units which happen to arrive in the $(k + 1)$st service period. On the other hand, if $l_k = 0$ the system is idle after that unit leaves; the $(k + 1)$st service period starts only after the next unit arrives and ends only after that unit leaves. When this unit leaves, it leaves behind it only the number α_{k+1} of units which arrived while it was being serviced (not counting itself). Therefore, we have

$$l_{k+1} = l_k - U(l_k) + \alpha_{k+1} \quad \text{where} \quad U(n) = \begin{cases} 0 & (n = 0) \\ 1 & (n > 0) \end{cases}$$

Each of the quantities l_k, α_k in this equation is an independent random variable; in steady state the average value of l_k must equal that of l_{k+1}, and similarly for the α's. Averaging the equation results in $\langle U \rangle = \langle \alpha \rangle = \rho$ [from (28-43)]. But, from the definition of U, we see that $\langle U(l) \rangle$ is the fraction of time the service facility is busy. Therefore, even for nonexponential service, $\rho = \lambda/\mu$ is appropriately called the utilization factor.

Now square the equation relating l_{k+1} and l_k and average this over time for the steady-state case. We obtain

$$\langle l^2 \rangle = \langle l^2 \rangle + \langle U^2 \rangle + \langle \alpha^2 \rangle - 2\langle lU \rangle - 2\langle \alpha U \rangle + 2\langle l\alpha \rangle$$

However, $U^2 = U$ and $lU = l$, and since α and l are independent random variables, $\langle l\alpha \rangle = \langle l \rangle \langle \alpha \rangle$ and $\langle \alpha U \rangle = \rho \langle \alpha \rangle = \rho^2$; so we finally obtain an equation for the mean number of units in the system, the mean number L_q in the queue, and the mean delays,

$$\langle l \rangle = L = \rho + L_q = \frac{\langle U \rangle - 2\langle \alpha U \rangle + \langle \alpha^2 \rangle}{2 - 2\langle \alpha \rangle} = \rho + \frac{\lambda^2(T_s{}^2 + \sigma_s{}^2)}{2(1 - \rho)}$$

where we have used (28-43) to evaluate the averages. Thus

$$L_q = \frac{\lambda^2/2}{1 - \rho}(T_s{}^2 + \sigma_s{}^2) \qquad W_q = \frac{L_q}{\lambda} \qquad L = \lambda W = L_q + \rho \qquad (28\text{-}44)$$

This is known as the Pollaczek-Khinchin formula. When the service-time distribution is exponential, $T_s = 1/\mu = \sigma_s$, and these equations become identical with those of Sec. 28-10. If the distribution is Erlang, with $T_s = 1/\mu$ and $\sigma_s{}^2 = 1/k\mu^2$, then $L_q = \lambda^2(k+1)/2k\mu(\mu - \lambda)$, which is the formula obtained by solving (28-42). If the service operation is completely regular, with $T_s = 1/\mu$ and $\sigma_s = 0$, then $L_q = \lambda^2/2\mu(\mu - \lambda)$, which is just half that for exponential service.

Thus mean waits and mean numbers in queue can be increased or decreased by increasing or decreasing the variance of service times. However, as long as arrivals are Poisson, the reduction in L_q or W_q is at most a factor of 2, even if σ_s is reduced to zero. Removing variability of service times reduces mean delays but cannot do better than halve them; and usually it is more difficult to reduce σ_s to zero than it is to increase μ sufficiently to double $\mu - \lambda$, which also halves L_q. In other words, reducing variability of service is usually the hard way to reduce L_q and W_q.

A greater effect on L_q can be obtained by reducing the variability of arrivals, if this is possible. If arrivals can be made more regular than Poisson, L_q is also reduced. Calculation of the case for regular arrivals ($\sigma_a = 0$) and exponential service ($\sigma_s = 1/\mu$) yields an L_q which is less than that for Poisson arrivals—exponential service by a factor which is about 4 when $\rho = 0.5$ and is about 3 for $\rho = 0.8$. Increasing variability of arrival increases L_q.

A case of interest, with arrivals of greater interarrival variance than Poisson is that of variable batch arrivals and exponential service. Suppose the arrival of the batches of units has a Poisson distribution in time with mean rate γ, and the number of units per batch has a geometric distribution with probability $(1 - \alpha)\alpha^{n-1}$ that a given batch consists of n units ($\alpha < 1$). The mean number of units per batch is $1/(1 - \alpha)$, and the mean rate of arrival of units (as contrasted with the rate γ of arrival of batches) is $\lambda = \gamma/(1 - \alpha)$. Comparison with Eqs. (28-30) indicates that the arrival distribution for this case is the limiting form of the hyperexponential distribution

$$A_0(t) = \frac{1}{\lambda(\beta - \lambda + \alpha\lambda)}[\alpha\beta\lambda e^{-\beta t} + (1 - \alpha)\lambda(\beta - \lambda)e^{-(1-\alpha)\lambda t}]$$

when $\beta \to \infty$. In other words, the arrivals can be simulated by having alternate "arrival gates" (Sec. 28-8), the one chosen with frequency α releasing a unit instantaneously, whereas the unit which chooses the other gate must wait a time $1/\gamma$, on the average, before being allowed to "arrive." The variance $\sigma_a{}^2$ of interarrival times for such batch arrivals turns out to be $1/(1 - \alpha)\lambda^2$, as compared with the $1/\lambda^2$ of Poisson arrivals. This distribution is sometimes called the stuttering Poisson distribution.

The states for this system are again labeled by two indices, s indicating the "arrival gate" through which the next unit will pass ($s = 1$ for the high-speed gate, $s = 2$ for the slower one) and n indicating the number of units in the system (but not including the one behind the "arrival gate"). The equations for steady-state probabilities are

$$\beta P_{1,0} = \mu P_{1,1} \qquad \gamma P_{2,0} = \mu P_{2,1} \qquad \gamma = (1 - \alpha)\lambda$$
$$(\mu + \beta)P_{1,n} = \alpha(\beta P_{1,n-1} + \gamma P_{2,n-1}) + \mu P_{1,n+1}$$
$$(\mu + \gamma)P_{2,n} = (1 - \alpha)(\beta P_{1,n-1} + \gamma P_{2,n-1}) + \mu P_{2,n+1}$$

The solution can most easily be expressed (see Ref. 6) in terms of powers of the quantity

$$v = \frac{1}{2\mu}[(\mu + \beta + \gamma) - \sqrt{(\mu + \beta - \gamma)^2 - 4\alpha\mu(\beta - \gamma)^2}]$$

Substitution in the equations for the P's shows that

$$P_{sn} = B_s v^{n-1} \qquad (n > 0) \qquad P_{10} = \frac{\mu}{\beta} B_1 \qquad P_{20} = \frac{\mu}{\gamma} B_2$$

The B's are computed from the equation

$$\alpha\gamma\mu B_1 = [\alpha\gamma v - (1 - \alpha)\gamma\mu - \beta\gamma]B_2$$

and the requirement that the sum of all the P's be unity.

In the desired limit $\beta \to \infty$, we find that $v \to \alpha + (1 - \alpha)\rho$ where, as before, $\rho = \lambda/\mu$ is the ratio between arrival rate (of units, not batches) and service rate. Also $B_1 \to 0$, $B_2 \to (1 - \alpha)\rho(1 - \rho)$, $P_{20} = 1 - \rho$, and the mean lengths and mean waits become

$$L_q = \frac{\rho\alpha + \rho^2(1 - \alpha)}{(1 - \alpha)(1 - \rho)} = \frac{\lambda^2}{\mu(\mu - \lambda)} + \frac{\lambda\alpha}{(1 - \alpha)(\mu - \lambda)}. \qquad (28\text{-}45)$$
$$W_q = \frac{L_q}{\lambda} \qquad L = \lambda W = L_q + \rho$$

Therefore, the fraction of times the service facility is idle is $P_{20} = 1 - \rho$ as before. The mean queue length is longer than that for individual Poisson unit arrivals (Sec. 28-10) by an amount proportional to $\alpha/(1 - \alpha)$. Arrival in batches does increase mean wait; the larger the mean batch the greater the increase.

28-12. Queue Discipline

So far we have assumed that the arriving, bound units stay in order in the queue and that service takes place in order of arrival. This type of queue discipline is called "first-come-first-served" or "strict" service acceptance. Other orders of acceptance are encountered in practice. For example, the service facility may pick a unit at random, from those waiting, whenever it has completed service on the preceding unit (this is called "random acceptance"), or it can take the *last* unit which has arrived (called "last-come-first-served" or "reverse" acceptance).

A little thought will persuade one that a change in the order of acceptance within the queue has no effect on the steady-state behavior of the system as a whole, on the mean number in queue or on the mean waiting time, for example. It will, of course, have an effect on the distribution $G(t)$ of time a unit spends in the system and thus on the variance of this time. Equation (28-38) showed that, for Poisson arrivals and exponential service, if first-come-first-served acceptance is practiced, this distribution is exponential, $G(t) = e^{-(\mu-\lambda)t}$, with mean wait $W = 1/(\mu - \lambda)$ and variance equal to $1/(\mu - \lambda)^2$. For last-come-first-served acceptance the distribution of waiting times is not exponential (see Ref. 8). The mean wait is still $1/(\mu - \lambda)$, as before, but the variance of the wait is larger than W^2. For Poisson arrivals and exponential service this distribution turns out to be the same as the distribution of lengths of busy periods, and the variance in waiting times is equal to the value $\lambda/(\mu - \lambda)^3$, given in Eq. (28-41).

When the utilization factor ρ is nearly unity an arriving unit may get serviced right away, or else it may get to the bottom of the pile and have to wait until all the later arrivals are serviced, which accounts for the large variance. The case of random acceptance is intermediate between first-come-first-served and last-come-first-served acceptance. The mean wait is $1/(\mu - \lambda)$ as always, but the variance of the wait lies between $1/(\mu - \lambda)^2$ and $\lambda/(\mu - \lambda)^3$.

In some operations it may be desirable to differentiate between arriving bound units, to give one or more classes of arrivals the preference (or *priority*) of acceptance. Suppose we are able to differentiate between k different arrival types, and we find that the ith class has Poisson arrivals with mean arrival rate λ_i. Our differentiation may or may not be according to estimated service time; in any case our measurements show that the mean service time for class i is $\tau_i = 1/\mu_i$, with a variance σ_i^2. Suppose also that all arriving units are bound and that we assign priorities in order of the class index i, so that all the units in class 1 must be serviced before a unit of class 2 is serviced, and so on up to class k. If a lower-priority unit is in service when a higher-priority unit arrives we do not interrupt service but allow it to be completed before the higher-priority unit is taken on (provided, of course, that a still higher priority unit has not meanwhile arrived). This type of priority is called nonpreemptive priority; we shall not discuss the preemptive case, where service on a low-priority unit is interrupted whenever a higher-priority unit arrives.

The distributions of waiting times for the various classes of unit are fairly complicated functions (see Ref. 8), but the mean waiting times and the mean numbers of various types of units in the system may be worked out without too much trouble. The mean delay in queue W_{qi} and the mean number in queue L_{qi} of the ith class are

$$W_{qi} = \frac{T}{(1 - \rho_{i-1})(1 - \rho_i)} \qquad L_{qi} = \lambda_i W_{qi} \qquad (28\text{-}46)$$

where
$$\rho_i = \sum_{n=1}^{i} \lambda_n \tau_n = \sum_{n=1}^{i} (\lambda_n/\mu_n) \qquad (0 < i \leq k) \qquad \rho_0 = 0$$

$$T = \frac{1}{2} \sum_{n=1}^{k} \lambda_n (\tau_n^2 + \sigma_n^2)$$

Of course, ρ_k must be less than unity for a steady state to be possible. The quantity ρ_i which determines the size of W_{qi} and L_{qi} is thus the *sum* of the utilization factors λ_n/μ_n of all classes with equal or higher priority than the ith. As usual with a single-server system, the mean wait in the system and the mean number in the system, for the ith class, are $W_i = W_{qi} + \tau_i$ and $L_i = \lambda_i W_i$.

The mean time in queue for a single-priority system ($k = 1$) is thus $[\lambda_1(\tau_1^2 + \sigma_1^2)/2(1 - \rho_1)]$, which is the Pollaczek-Khinchin formula (28-44). When other classes are present ($k > 1$) then even the highest-priority units are delayed by the presence of the others, for T is a sum over all classes. This is the case for nonpreemptive priorities; it would not be true for preemptive priorities, for then class 1 units would never be delayed, when they arrive, by the occasional presence of a lower-class unit already in service. The delay W_{q1} (nonpreemptive) is, however, smaller than would be the case if service acceptance were first-come-first-served for all classes alike. In other words, if we wish to reduce delays for a particular class of units, we can assign them a priority of acceptance into service. The reduction of delay, for these units, is appreciable if they constitute only a small fraction of the total arriving units and/or if their mean service time τ_i is smaller than the mean service times for other classes.

On the other hand, if our aim is to reduce the mean wait of all units, not just that of the priority units, then we must assign priority to those units having the shortest mean service time $\tau = 1/\mu$. In other words, assignment of priorities, so that the sequence τ_i is a monotonically increasing sequence with increasing i, yields the smallest value of $L = \Sigma L_i$. The gain is not large, however, unless there is a wide range of values among the τ_i's (see Ref. 6).

28-13. Multiple Service Channels

Another modification of the internal structure of the service system, possible in some cases, is the multiplication of service channels so that several units can be serviced in parallel. Suppose it is possible to divide a service facility, originally exponential, with mean service rate μ, into M equal facilities, each exponential with mean service rate μ/M. The advantage which may thus be gained will of course depend on the behavior of the arriving units, which we assume are Poisson-distributed, with mean rate λ.

For example, for free units, which stay in the system only if there is an idle channel available on arrival, there are now $M + 1$ states, labeled by the number n of busy channels ($0 \leq n \leq M$). This is the case of the parking lot of Sec. 28-10, with the μ of equations such as (28-34) changed to μ/M. Thus the probability (steady-state) P_n that n of the service channels are busy and the mean number N of busy channels are

$$P_n = \frac{(M\rho)^n}{n!} \frac{e^{-M\rho}}{E_M(M\rho)} \qquad N = \frac{M\rho E_{M-1}(M\rho)}{E_M(M\rho)} \qquad (28\text{-}47)$$

where $\rho = \lambda/\mu$ is the ratio of the total arrival rate to the total service rate of all M channels. The fraction of units lost to the system is equal to P_M, the probability that all M channels are busy. Figure 28-3 shows that this is small when $\rho < 1$ and approaches unity as ρ is made larger than 1. The larger M is, the larger ρ must be before P_M becomes appreciably larger than zero.

Thus by subdividing the service facility (if this is possible without loss of efficiency, i.e., if the rate of each of the M channels is μ/M) then an increase of M will decrease the fraction of units lost. With M channels, the chance that at least one channel will be idle is larger than the chance that a single fast channel (of rate μ) will be idle, and so more free units are "trapped" into staying, though they have to stay proportionally longer to be serviced than with the single-channel system. The mean number N in the system, for large values of M, is practically equal to $M\rho = M(\lambda/\mu)$ when $\rho < 1$ and is practically equal to M when $\rho > 1$, as is shown in Fig. 28-3.

If the arriving units are bound, having to stay in the system until serviced, the resulting steady-state behavior of the multichannel system depends on the queue discipline. If the queue forms in front of each service channel, each arriving unit choosing a channel at random, and sticking to it, then the system consists of M independent, single-channel systems of the sort analyzed in Sec. 28-10, each exponential service channel of rate μ/M being fed by Poisson arrivals of rate λ/M, so that the utilization factor for each channel is $\lambda/\mu = \rho$, the utilization factor for the whole system. The mean number in each queue is $L_q = \rho^2/(1 - \rho)$, but since there are M queues, the total number waiting for service is M times greater than for the single-channel system of equal total service rate. Likewise the mean wait in queue of a unit is $(M/\lambda)L_q$, which is M times longer than the wait in front of a single channel of M times the service rate.

Thus when individual queues are randomly formed in front of each channel, the subdivision of a service facility into M parallel channels has no advantage whatsoever. When units choose a channel at random on arrival and are not allowed to change channels later the channels are very inefficiently used; quite often one or more channels are idle while units are in queue in front of adjacent channels. Almost any other queue discipline would utilize the M channels more efficiently. For example, the arriving unit could choose the shortest queue or, if two or more queues were minimal, could choose at random between them.

The most efficient utilization is to have a single queue, which feeds all M channels, replenishing each channel as soon as it becomes free, until the queue is exhausted. The equations for the steady-state values of P_n, the probability that n units are in the system, either in queue or in service, for this system are

$$\lambda P_0 = \mu P_1 \qquad \left(\lambda + \frac{n\mu}{M}\right) P_n = \lambda P_{n-1} + (n+1)\frac{\mu}{M} P_{n+1} \qquad (0 < n < M)$$

$$(\lambda + \mu) P_n = \lambda P_{n-1} + \mu P_{n+1} \qquad (M \leq n)$$

If $n < M$ all n units are in service and there is no queue; if $n > M$ all the M channels are busy and $n - M$ units are in the consolidated queue.

The solution of this set of equations (see Ref. 6) and the resulting expressions for the total number L in the system, the number L_q in the queue, and the corresponding waiting times are

$$P_n = \begin{cases} P_0 \dfrac{(M\rho)^n}{n!} & (0 \leq n \leq M) \\ P_0 \rho_n \dfrac{M^M}{M!} & (M \leq n) \end{cases} \qquad P_0 = \frac{(1-\rho)e^{-M\rho}}{D_{M-1}(M\rho)} \qquad (28\text{-}48)$$

$$L_q = \frac{\rho^{M+1} M^M e^{-M\rho}}{(1-\rho)M! D_{M-1}(M\rho)} \qquad L = L_q + M\rho \qquad L = \lambda W \qquad \text{etc.}$$

where

$$D_m(x) = \frac{1}{m+1} \sum_{n=0}^{m} E_n(x) = \sum_{n=0}^{m} \left(1 - \frac{n}{m+1}\right) \frac{x^n}{n!} e^{-x}$$

is unity at $x = 0$, is approximately $1 - x$ for $x < 1$ and approximately 0 for $x > 1$, the approximations holding closer and closer to $x = 1$ the larger m is. Function $E_n(x)$ is defined in Eq. (28-34).

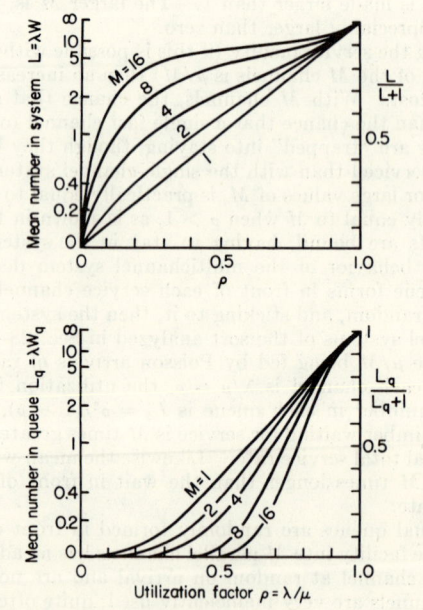

FIG. 28-4. Mean number L in system and mean number L_q in queue (and mean waiting times W and W_q) for the system of Eqs. (28-48), plotted as functions of ρ, the ratio between arrival rate λ and total service rate μ, for different values of M, the number of service channels.

Values of L and L_q are plotted in Fig. 28-4 for different values of M, the number of service channels, as functions of ρ, the ratio of mean arrival rate λ to mean total service rate μ. We see that, if we wish to keep small the mean stay W of a unit in the system (i.e., in queue *and* in service) we should use a single fast service channel of rate μ (if this is possible) rather than M channels each of rate μ/M. On the other hand, if we wish to reduce the mean delay *in queue* W_q the M slower channels are to be preferred to the single fast channel. Fragmenting the single channel into M channels, each $1/M$ times as fast, reduces the queue length and delay in queue at the expense of increased time in service. Whether this is desirable or not depends on what operational characteristic of the system is important.

The effects, on the behavior of a multichannel system, of modification of arrival or of service statistics, or of changes in queue discipline, are all problems which have been studied and which are reported in the references given at the end of this chapter. There is no space to discuss them here. The theory of stochastic service systems (with or without queues) now covers a wide variety of operational situations, of ever-increasing importance in our way of life. Its ramifications deserve to become as thoroughly studied, and as well known to students, as are those of celestial mechanics, for example.

List of Symbols*

$a(t)$ = arrival density (28-23)
$A_n(t)$ = probability n arrivals in postarrival time t (28-23)
$B(z) = \Sigma z^n \pi_n$, generating function for the π_n's (28-43)
C = operating cost per period
$D_m(x)$ = see Eq. (28-48)
$E_k(x)$ = Erlang function (28-29)
$F(z) = \Sigma z^n P_n$, generating function for the P_n's (28-32)
$G(t)$ = probability unit waits in system longer than t (28-38)
i, j = state indices
L = mean number in system (28-32)
L_q = mean number in queue (28-37)
m, n = number in system at time t
M = number of service channels
M_{ij} = element of Markov matrix $\mathfrak{M}$ [(28-4), (28-5)]
$p_i(t)$ = probability system is in state i at time t (28-1)
P_i = steady-state probability of being in state i (28-8)
R_{ij} = element of transition-rate matrix $\mathfrak{R}$
$s(t)$ = service density (28-28)
$S_n(t)$ = probability n services completed in time t (28-28)
t = time
T_a, T_s = mean interarrival or service time [(28-25), (28-28)]
$T_{ij}(t_0, t_1)$ = probability transition state i at t_0 to state j at t_1
$U_n(t)$ = probability n arrivals in random time interval t (28-24)
$W = L/\lambda$, mean wait in system
$W_q = L_q/\lambda$, mean wait in queue
α, β, γ = rates of transition
λ = mean arrival rate [(28-24), (28-25)]
μ = mean service rate (28-28)
π_n = probability n arrivals in a service period (28-43)
$\rho = \lambda/\mu$, utilization factor (28-36)
σ = standard deviation
σ^2 = variance of times (28-27)
$\Psi_s(x)$ = Laplace transform of service distribution (28-43)

* Numbers in parentheses are equation numbers where symbol is first defined or used.

REFERENCES

The general theory of stochastic processes has a very large literature. A few recent books which will serve the reader to gain a general concept of the field are:

1. Feller, W.: *Introduction to Probability Theory*, 2d ed., John Wiley & Sons, Inc., New York, 1957.
2. Barucha-Reid, A. T.: *Markov Processes and Their Applications*, McGraw-Hill Book Company, New York, 1960.
3. *Selected Papers on Noise and Stochastic Processes*, Dover Publications, Inc., New York, 1954.
4. Bartlett, M. S.: *Stochastic Processes*, Cambridge University Press, New York, 1955.
5. Bellman, R.: *Introduction to Matrix Analysis*, especially Chaps. 14 and 15, McGraw-Hill Book Company, New York, 1960.

Several books have recently appeared which deal with stochastic service processes. They are, in order of increasing mathematical generality and difficulty:

6. Morse, P. M.: *Queues, Inventories and Maintenance*, John Wiley & Sons, Inc., New York, 1958.
7. Saaty, T. L.: *Elements of Queueing Theory*, McGraw-Hill Book Company, New York, 1961.
8. Riordan, J.: *Stochastic Service Systems*, John Wiley & Sons, Inc., New York, 1962.
9. Takacs, L.: *Introduction to the Theory of Queues*, Oxford University Press, Fair Lawn, N.J., 1962.

The books of Riordan and Saaty have good bibliographies of articles on the subject.

Chapter 29

FEEDBACK THEORY

WILLIAM A. LYNCH, *Head, Mechanical Engineering Department, Polytechnic Institute of Brooklyn.*

JOHN G. TRUXAL, *Dean of Engineering, Polytechnic Institute of Brooklyn.*

LUDWIG BRAUN, *Associate Professor of Electrical Engineering, Polytechnic Institute of Brooklyn.*

CONTENTS

29-1. Introduction	29-2
29-2. Definition of Feedback Systems	29-2
29-3. What Does Feedback Accomplish?	29-3
29-4. Models of Systems	29-5
Mathematical Models	29-5
Circuit Models	29-6
Flow Diagrams	29-6
Block Diagrams	29-7
Signal-flow Graphs	29-7
29-5. Four Definitions	29-9
System Transfer Function or Transmittance	29-9
Loop Transmittance	29-9
Return Difference	29-10
Sensitivity	29-10
29-6. Evaluation of the Transfer Function of a Flow Diagram	29-11
Step-by-step Reduction	29-12
Mason's Reduction Formula	29-15
The Evaluation of the Graph Determinant in Compact Form	29-16
29-7. Sensitivity	29-19
Example of Sensitivity Control with Feedback	29-20
Design Control of Sensitivity	29-23
Topological Interpretation of Sensitivity	29-24
Example of Use of Topological Formulation	29-25
Comprehensive Calculation Techniques	29-29
Concluding Remarks	29-32
29-8. The Significance of Sensitivity	29-32
"Parameter Margins"	29-33
Relation between S_x^T and Parameter Margin for x if $F_x' = 1$	29-34
A Nontrivial Example	29-35
Relation between S_x^T and Parameter Margins if $F_x' = \infty$	29-39

Relation between $S_x{}^T$ and Parameter Margin in General........... 29-40
Concluding Comments... 29-42
29-9. Multiloop Systems.. 29-42
Positive and Negative Feedback............................... 29-44
Feed-forward Systems... 29-46
Example of Double Control of Sensitivity...................... 29-47
Example of Stabilization of Nonlinear Oscillation............. 29-48
Data-recovery System... 29-49
29-10. Concluding Comments.................................... 29-50

29-1. Introduction

In the analysis of physical systems, we persistently encounter examples which are conveniently described by considering them as *feedback systems*—systems which, broadly speaking, are self-supervisory (compare the following chapter on Adaptive and Learning Control Systems). In these systems, which are capable of performing their appointed tasks despite adverse influences, the output signal is compared with the desired value of the output in order to generate an "error" signal which can be used to change the output (usually in such a direction as to decrease the error toward zero). In this chapter, we attempt to summarize certain basic properties of feedback systems, and methods of analysis and design which are significant for engineers working with feedback. In short, in the following pages, we present the fundamental elements of *feedback theory*.

29-2. Definition of Feedback Systems

According to the definition given in the above paragraph, a feedback system includes at least the elements depicted in Fig. 29-1, which also serves to introduce our notation. The output signal c is measured by the sensor to yield the feedback signal b. The comparator generates an error e which depends on a comparison of the input or desired output r and the measured output b; in the simplest case, e is simply given by the difference

$$e = r - b \qquad (29\text{-}1)$$

as depicted by the $+$ and $-$ signs in the figure. More generally, the comparator might measure any function of r and b, e.g., $(r - b)^3$, but in this chapter we consider only comparators described by Eq. (29-1). Finally, the error signal e drives (or modifies) the output c through the *process*.

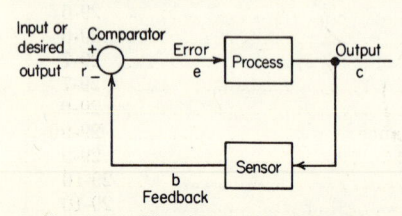

Fig. 29-1. Simplest feedback system.

Illustrations of the basic system of Fig. 29-1 are plentiful throughout our daily life. In a broad class of problems, the human being is an important element of the feedback system. For example, if a man reaches forward to pick up a pencil lying on a desk, a feedback system drives the grasping hand. In this case, the desired output (the desired position of the hand) is the pencil position; the actual output is the instantaneous position of the hand. The sensor is now the eye visualizing the present position of the hand; this position, b, is compared (in the brain) with the pencil position (also visualized by the eye), and an error signal actuates the muscles which move the hand (Fig. 29-2).

The *feedback* aspect of the operation arises because of the continual measurement of the hand position and comparison with the measured pencil location; actually the eye functions to measure directly the error, which the brain then converts to the appropriate drive signal for the muscles. The improvement in performance which results from the feedback can be demonstrated if the same task is attempted with the eyes closed after only an initial measurement of the pencil position relative to the initial hand position.

In spite of the obvious merits of such feedback structures (and a primary objective of this chapter is to investigate analytically the quantitative reasons for such merits), it was not until very recently that engineers recognized the importance of feedback as a design tool. Feedback theory, as we know it today, was born in the 1920s with the work of the Bell Telephone Laboratories, and received its greatest impetus during World War II with the extensive work on automatic control systems for such military purposes as fire control (the positioning of gun turrets), radar-antenna control (the

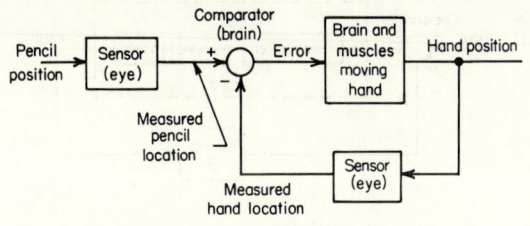

FIG. 29-2. A human feedback system.

positioning of the antenna), and aircraft control (for stabilization as well as navigation). During the last decade, major advances in feedback theory have been associated with missile guidance and control, and with industrial automation (for example, the automatic production line, or the automatic adjustment of steam-boiler operating conditions to maximize the economy of operation).

29-3. What Does Feedback Accomplish?

Before we proceed further with the detailed analysis of various classes and types of feedback systems, it is useful to attempt to describe the properties of feedback systems in quantitative terms. In other words, we should pose the question: Quantitatively, what does feedback accomplish in such systems? In what way can we measure the effectiveness of the feedback?

In the vast majority of feedback systems which we encounter in engineering, the feedback is inserted in order to control the degree to which the overall transmission characteristics of the system depend upon the variation in the values of one or more of the system parameters. The early work on feedback theory by the engineers at Bell Telephone Laboratories was largely motivated by the advent of transcontinental telephone transmission systems, in which a large number of repeater amplifiers were used (Sec. 13-5). In order to maintain the overall characteristics of the system constant as the gains of the individual amplifiers varied because of tube aging, temperature and voltage variations, and the like, the design engineers turned to the utilization of large amounts of feedback around each of the amplifiers. The reliability requirements are illustrated by the present-day demand that amplifiers in the transatlantic telephone cable should operate for 20 years without maintenance (because of the expense and difficulty of pulling up the cable from the bottom of the ocean in order to make repairs and adjustments).

Thus the critical characteristic of feedback systems is most often the tendency of the system to operate satisfactorily even when the specific parameters of the various components are changing radically. In order to place this discussion on a more quantitative basis, it is helpful to consider the characteristics of the two systems of Fig. 29-3—the first involving no feedback, and the second utilizing feedback embodied in the sensor and comparator elements. (We conventionally term the former system *open-loop,* and the latter *closed-loop* to emphasize that there is a closed signal-transmission loop around which signals may flow.)

29-4 SYSTEM THEORY

In order to consider a specific example, let us specify that the gain (i.e., the ratio of the output to the input signals) of each of the two systems should be 10. In the open-loop system, this specification may be realized by a single amplifier with a gain of 10; however, when the amplifier gain changes by 2 per cent, the overall transmission of the system obviously also changes by the same 2 per cent.

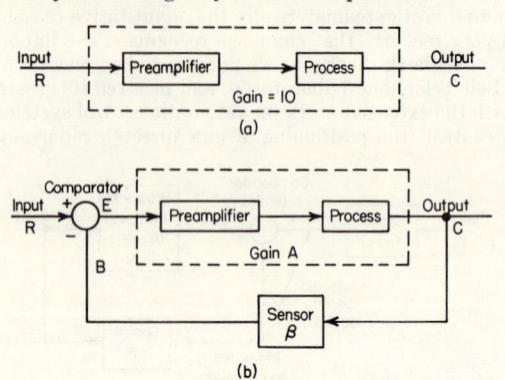

FIG. 29-3. Evaluation of feedback. (a) Open-loop system. (b) Feedback system.

The closed-loop system is quite different in its behavior. In this case, the overall transmission can be found by solving simultaneously the three equations which describe the performance of each of the elements of the system:

$$E = R - B \qquad (29\text{-}2)$$
$$B = \beta C \qquad (29\text{-}3)$$
$$C = AE \qquad (29\text{-}4)$$

When the variables E and B are eliminated from these three equations (we shall see in Sec. 29-6 how to effect this elimination in a straightforward manner), the overall transmission is

$$\frac{C}{R} = \frac{A}{1 + \beta A} \qquad (29\text{-}5)$$

If β is selected (arbitrarily for the moment) as 99/1,000, the realization of an overall gain of 10 requires that the quantity A be 1,000; hence we have the feedback configuration shown in Fig. 29-4.

In order to realize our overall gain of 10, we now require an amplifier with a gain of 1,000 (or three tandem amplifiers, each with a gain of 10). The introduction of feedback has obviously complicated our system design—not only is the number of required amplifiers raised from one to three, but we now need the sensor or β circuit, and the comparator. What have we gained from this extra complexity?

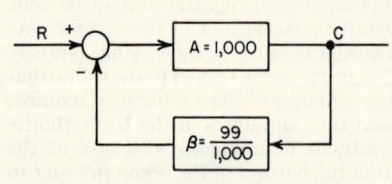

FIG. 29-4. System for $C/R = 10$.

The advantage of feedback is demonstrated if we consider now the effects of variation in the gain of the amplifiers. If the total amplifier gain of 1,000 changes by 2 per cent (e.g., if A drops to 980), the overall transmission of our complete system decreases to the value

$$T = \frac{980}{1 + (99/1{,}000)980} = 9.998$$

As a result of the feedback, the overall gain drops only 0.02 per cent, even though the gain of the amplifier falls 2 per cent—we have achieved an improvement in performance by a factor of 100, in terms of making the system insensitive to the variations of system parameters.

It is natural to object that insertion of the feedback actually requires more amplifiers and that hence the above improvement is deceptive; however, even if all three amplifiers change in the same direction by the full 2 per cent in our example, the total amplifier gain changes by approximately 6 per cent, and the gain of our overall system changes by only 0.06 per cent—so that we still achieve a marked improvement in comparison with the system with no feedback.

29-4. Models of Systems

From these introductory ideas, we begin to see the potentialities of feedback configurations in the creation of systems which are relatively insensitive to the characteristics of the devices of which they are constituted. We are now ready to deal with feedback on a quantitative or analytical basis with a view to establishing a theory for feedback systems that will be capable of accounting for all the various aspects of system performance that are influenced by the presence of a feedback configuration. As an essential first step, we turn to the important question of how to describe a system analytically and to the corollary problem of how to develop the feedback point of view advantageously.

It is very easy to lose sight of the central fact of analysis—that we analyze an idealization or *model* of a system and not the system itself. Before we can embark upon any phase of analysis, we must somehow contrive to effect a transformation from the physical system to a corresponding (but not equivalent) description that is phrased mathematically and therefore is amenable to the methods of mathematical analysis. In addition to accepting the fact that our analytical representation of the system is a frank idealization involving various degrees of judicious approximation, we must accept the associated fact that the signals with which we drive or excite the analytical model must also be idealizations. Thus the whole process of analysis is completely removed from the realm of reality, and at the conclusion of our analysis, we have the engineering task of using our results as a means for understanding the behavior of the physical system.

The very impressive achievements of present-day engineering technology stem in no small measure from three developments of analysis in which the model plays a dominant part:

1. The recognition of a clear distinction between physical systems and signals, and their corresponding models
2. The development of a wide variety of model concepts and associated symbolisms
3. The development of the several bodies of theory which guide our theoretical utilization of the various models

In this section, in which we consider several different model concepts, we shall attempt to show that each has a sphere of applicability in which it may excel in a specific problem. We hope also to underline the suggestion that no single model can satisfy all our analytical needs and that our best preparation for analysis and design is therefore achieved when we have a working facility with all relevant models. The models that we shall consider can be placed in three categories: mathematical models, circuit models, and flow diagrams.

Mathematical Models. Any set of mathematical relations that describe our analytical evaluation of a physical system constitutes a mathematical model. That such a model represents an interpretative concept is seen when, for example, we write the relation

$$e = ri \tag{29-6}$$

as a model of a physical resistor. In so doing, we have already decided that for our purposes we may neglect any residual effects which are capacitive or inductive. We are therefore idealizing the resistor in terms of a purely dissipative device. If we view r as invariant with the current, we are assuming a linear, time-invariant system.

Thus, in writing this very familiar expression, we make (although usually without conscious thought) many of the idealizations that are the ingredients of every model formulation. Because of the nearly ideal nature of electric-circuit elements, we are actually on much firmer ground with the resistor model of Eq. (29-6) than we are with the corresponding damper model of translational mechanical systems:

$$v = \frac{1}{B} f \qquad (29\text{-}7)$$

where v and f are velocity and force and B is the damping constant (which usually is a strongly nonlinear function of f).

Circuit Models. The familiar electric-circuit model is a highly developed pictorial model which actually symbolizes a number of mathematical models. Its symbolism is such that each of the ideal elements of which it is constituted represents a physical law [such as that of Eq. (29-6)] governing the terminal behavior in terms of voltage and current, and implying all the previously mentioned aspects of idealization. It also symbolizes the energy-storage and dissipative properties of its elements. Therefore, a configuration of ideal elements and ideal sources characterizing a model of an electric circuit represents sets of interrelations that may be transcribed in terms of voltage or current equilibrium (i.e., loop or node equations) or, alternatively, in terms of energy or power equilibrium. Furthermore, we may transcribe from the circuit model a mathematical model which is descriptive of the transient behavior or of the steady-state behavior as desired, and all from the same circuit model.

Although the circuit model frequently serves as the basis for writing a set of equations describing an appropriate aspect of circuit behavior, it is sometimes more advantageous to transform the model itself into equivalent forms having a simpler topology. Used in this way, the model retains its integrity and the reductions or simplification are accomplished without recourse to a mathematical version. Thus it is useful to think in terms of an "algebra" of circuit models which consists in successive applications of network theorems (such as Thévenin's and Norton's theorems or the substitution theorem) and equivalent-circuit transformations, such as the ubiquitous tee-to-pi transformation. The ultimate exploitation of this kind of transformation brings the given circuit model to a state of simplification whereby the desired solution can be transcribed from the model by inspection.

Flow Diagrams. Quite apart from the familiar circuit models and their mathematical counterparts are the models based upon signal-flow concepts. These models, which derive their chief analytical value from vivid pictorial properties, focus attention on a causal viewpoint expressing the analyst's visualization of various cause-and-effect relations with which the system may be described. The interjection of the cause-and-effect point of view forces the analyst to make certain decisions that he might not have to make in using a more conventional symbolism. For example, a cause-and-effect interpretation of Eq. (29-6) could take the form of Fig. 29-5. Here the current i is arbitrarily taken to be the cause, and the terminal voltage e the effect. The parameter r takes on the role of an operator, and there is a clear progression of events from cause to effect, with the signal i being operated upon by the parameter r (here merely multiplication by r) with the emergent effect the terminal voltage e. If we wish to portray the related expression

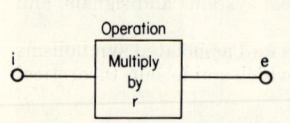

$$i = \frac{e}{r} \qquad (29\text{-}8)$$

Fig. 29-5. A cause-and-effect interpretation of Eq. (29-6).

a new flow diagram is required, because now the roles of cause and effect have become reversed. When formulating a circuit model of the resistor, we merely draw the familiar zigzag symbol for the resistance element without taking heed of causal implications.

Any rebellion we may experience in having to cope with this seeming artificiality is more than offset when the benefits of causal representation are properly assessed. It is possible to visualize almost any situation in terms of cause-and-effect relations because this is a basic pattern of human experience. In many problems, such a viewpoint produces no particular advantage over other representations; but in feedback systems, this way of portraying a system has tremendous advantages over mathematical and circuit-model representations. The unifying motif of the systems with which this chapter is concerned is feedback. In a cause-and-effect portrayal, feedback is shown unmistakably as a closed causal chain which appears diagrammatically as a closed signal path or *feedback loop*.

The basic idea of showing cause-and-effect relations in terms of a graphical portrayal has provided the inspiration for several different, though related, flow diagrams. The best known of these is the *block diagram* used in the illustrations of Secs. 29-2 and 29-3. A more abstract interpretation of the flow diagram is the *signal-flow graph*, which utilizes two very simple graphical elements, instead of lines and blocks and summing elements, to portray a set of simultaneous algebraic equations in a preconceived pattern of cause and effect. The third widely used flow diagram is the *analog simulation model*, which in effect literally symbolizes a set of integrodifferential equations in terms suitable for programming a physical analog computer.

Block Diagrams. We have already had some experience with block diagrams in Secs. 29-2 and 29-3, and the illustrations of these sections are sufficient to show the symbolism. There are actually four graphical elements used: blocks, directed lines, summation devices (also referred to as comparators or differencing devices), and finally junctions or pick-off points. The block diagram has a great deal of appeal as a means for boxing off the principal component parts of a system and indicating the way in which these main subdivisions are interrelated. As we have already seen, this kind of model lends itself to both qualitative and quantitative thinking. In Fig. 29-2, illustrating a human feedback system, we were able to gain some insight into the human system and how it possibly functions without being able to describe the actions of the individual blocks quantitatively.

When each of the component parts of a system is understood well enough to permit a model to be formulated, each block of a system representation can be labeled in terms of a pertinent mathematical model (a set of differential equations or, in a linear case, a transfer function). The block diagram is now a quantitative portrayal which still retains the features of segregating the noninteracting subdivisions of the system as well as retaining a strong visual appeal in terms of signal-flow paths and closed feedback loops.

Signal-flow Graphs. A signal-flow graph might be thought of as a highly stylized interpretation of a block diagram. Only two graphical elements are used: small circles, called *nodes*, and directed lines, called *branches*, which interconnect the nodes. The branches replace the combination of directed lines and blocks of the block diagram, and the nodes replace both the summing devices and pick-off points. A block diagram is easily translated into an equivalent signal-flow graph by transferring block labels to corresponding branches, replacing both the comparators and pick-off points by nodes, and finally introducing minus signs in appropriate branches when signals are to be subtracted. In Fig. 29-6, the details of two such translations are shown. It is noteworthy that the minor loops (involving G_1H_1 and G_2H_2) are noninteracting or nontouching in the models of parts a and b of the figure, whereas they are touching in the second example (parts c and d).

Although signal-flow graphs serve a useful purpose as alternative ways of displaying the information of a block diagram, this is not their primary role. They were first devised as an aid to analyzing electronic circuits employing feedback and were formulated directly from a mathematical model, or directly from a circuit model. In either case, they serve the very valuable purpose of placing the visualized feedback clearly in evidence. When used in this way, a signal-flow graph may be regarded as a graphical portrayal of a set of simultaneous algebraic equations which are ordered in a specific cause-and-effect pattern.

The nodes register the values of the variables of the equations, and the branches represent relations between pairs of variables. The arrows on the branches point away from cause toward effect, and signals always flow in the arrow direction. As an example, the equation

$$ax_1 + bx_2 + cx_3 = x_4 \tag{29-9}$$

is shown graphically in Fig. 29-7. The ordering of the equation suggests x_4 as an effect arising from three independent causes. Each of the terms on the left-hand side of (29-9) constitutes a *branch signal*, and the three branch signals incident upon node 4 are summed to produce the *node signal* x_4. This example illustrates the fact that each node symbol performs two separate tasks: it sums all incoming branch signals, and it serves as a source of its registered node signal for all outgoing branches.

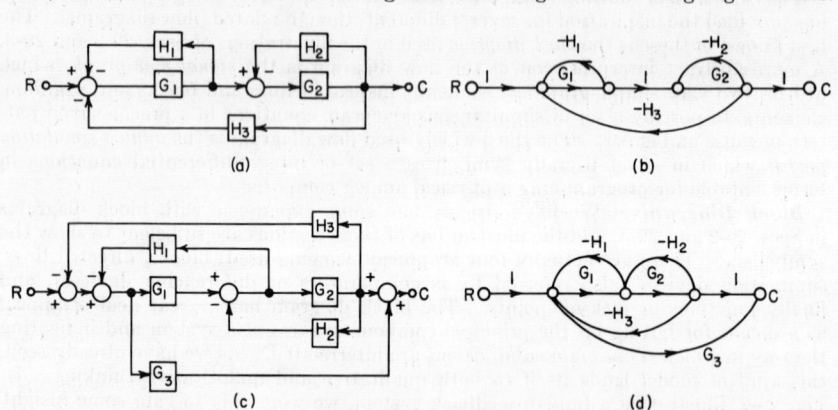

FIG. 29-6. Two examples of conversions from block-diagram to signal-flow-graph symbolism.

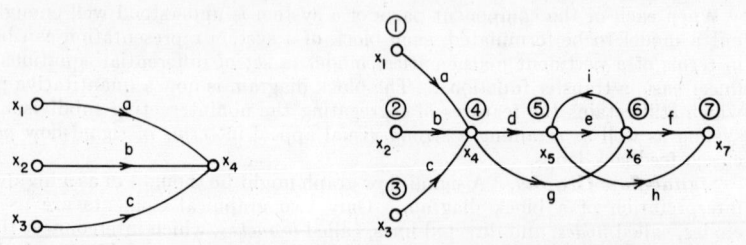

FIG. 29-7. A graph of Eq. (29-9).

FIG. 29-8. A flow graph representing a set of simultaneous algebraic equations.

In general, each node performs these tasks simultaneously, Fig. 29-7 being a rather special case. The graph of Fig. 29-8 shows node 4 as registering the node signal x_4, and in turn transmitting this signal along the outgoing branches d and g. At node 5, a second dependent node signal x_5 represents a second equation

$$dx_4 + ix_6 = x_5 \tag{29-10}$$

Similarly at nodes 6 and 7, the node signals represent, respectively, the equations

$$gx_4 + ex_5 = x_6 \tag{29-11}$$
$$fx_6 + hx_5 = x_7 \tag{29-12}$$

Thus the signal flow graph of Fig. 29-8 is a graphical portrait of Eqs. (29-9) through (29-12), showing the particular indicated ordering. It is noteworthy that there is a feedback loop in the graph formed of branches e and i, which indicates that x_5 is a function of x_4 and also of x_6, which in turn is a function of x_5. We may therefore designate any system that this model represents as being a feedback system, although such a designation may or may not prove to be a useful viewpoint.

Some concluding comments regarding the nodes may be useful. The nodes numbered 4, 5, 6, and 7 in Fig. 29-8 are *dependent nodes* because each has at least one incoming branch symbolizing a dependency upon some other node. Nodes 1, 2, and 3 are independent or *source nodes* since clearly their node signals do not depend upon the signals at any of the other nodes. These source variables are therefore independently specified. Nodes 4, 5, and 6 serve simultaneously as summing points and as signal sources for outgoing branches. Node 7, which is called a *sink node*, is unique in that it has only incoming branches.

29-5. Four Definitions

Having dealt with the question of how to describe a system analytically, we turn next to the definition of four terms—four functions actually—which play a dominant role in feedback theory, and then in Secs. 29-6 and 29-7 we develop various techniques which can be used in evaluating these functions.

System Transfer Function or Transmittance. Since in system analysis we are ordinarily interested in system dynamics, we describe systems in terms of the transmittance $T(s)$, which is the overall transfer function relating system output to

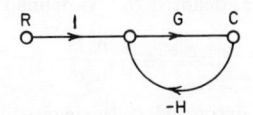

FIG. 29-9. Basic single-loop system.

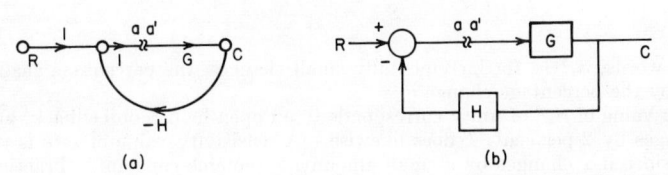

FIG. 29-10. Loop broken at a-a'. (*a*) Flow graph. (*b*) Equivalent block diagram.

system input. Analytically, we express $T(s)$ as the ratio of $C(s)$, the Laplace transform of system output $c(t)$ to $R(s)$, the Laplace transform of system input $r(t)$. The variable s is the complex frequency variable.

For the single-loop system of Fig. 29-9 we have seen* in Sec. 29-3 that the transmittance is simply

$$T(s) = \frac{G(s)}{1 + G(s)H(s)} \tag{29-13}$$

In the next section we demonstrate the general formula for the transmittance of any complex feedback configuration.

Loop Transmittance. The loop transmittance $L(s)$ is the transfer function all the way around the loop (with more than one loop, there are, of course, several loop transmittances). For example, in the system of Fig. 29-9, if we break the loop at a-a', shown in Fig. 29-10, $L(s)$ is the transmittance from a' back to a (with R assumed to be equal to zero); i.e.,

$$L(s) = -G(s)H(s) \tag{29-14}$$

(The minus sign arises when we pass through the comparator in the block diagram.

---

* In Eq. (29-5), the expression derived was $A/(1 + \beta A)$, where A was the forward amplifier transfer function, and $-\beta$ the feedback transfer function. In that example, both A and β were constants; more generally, both are transfer functions.

In the flow graph, the minus sign is explicitly associated with one of the branches of the loop.)

Return Difference. Return difference $F(s)$ can be defined conveniently with the aid of Fig. 29-10. If a' is a unit signal, then a is a response signal having the value $L(s)$ after having traversed the loop. The difference between a' and a is the return difference, and when there is a single loop, we may write in general

$$F(s) = 1 - L(s) \tag{29-15}$$

For the particular example of Fig. 29-10,

$$F(s) = 1 - (-GH) = 1 + GH \tag{29-16}$$

When there is more than one loop, there are several return differences (since there are then several different loop transmittances). It is necessary in such cases to identify in some way which of the several functions is being considered. This is usually accomplished by using a subscript on both L and F to identify the reference element to which the functions refer. (This point is discussed further in Sec. 29-6.)

Sensitivity. The sensitivity measures the effectiveness of the feedback in reducing the effects of variation of a specific parameter, which we shall denote by x (x may be an amplifier gain, a time constant, a moment of inertia, a plate-supply voltage, or any other parameter affecting system performance). The sensitivity of the transmission T with respect to the parameter x, denoted $S_x{}^T$, is defined as

$$S_x{}^T = \frac{d \ln T}{d \ln x} \tag{29-17}$$

Alternatively, from the known derivative of the logarithm of a function, we can write

$$S_x{}^T = \frac{dT/T}{dx/x} \tag{29-18}$$

In other words, $S_x{}^T$ is, for incrementally small changes, the percentage change in T divided by the percentage change in x.

Thus a value of $S_x{}^T$ of unity corresponds to an open-loop (nonfeedback) system—if x changes by 2 per cent, T does likewise. A sensitivity value of zero is the ideal system, since if x changes by a small amount, T remains constant. Practical feedback systems which we consider in the following sections frequently realize values of sensitivity as small as 0.001 for all signals of interest. In such cases, a 2 per cent change in x results in a change of 0.002 per cent in T.

In our single-loop feedback system of Fig. 29-9, we may calculate the sensitivity of T with respect to G by straightforward differentiation of the expression for T in Eq. (29-13):

$$\begin{aligned} S_G{}^T &= \frac{G}{T} \frac{dT}{dG} \\ &= \frac{G}{G/(1+GH)} \left[\frac{1}{1+GH} - \frac{GH}{(1+GH)^2} \right] \\ &= \frac{1}{1+GH} \end{aligned} \tag{29-19}$$

In other words, the sensitivity in this case is simply related to the overall transmission. As a numerical example, we can return to our feedback amplifier of Sec. 29-3, in which we had

$$G = 1{,}000 \quad \text{and} \quad H = 99/1{,}000$$

or

$$S_G{}^T = \frac{1}{100}$$

Thus (as we saw previously) a change in G of 2 per cent results in a change of T of only 0.02 per cent.

The four terms which we have defined in this section describe fundamental concepts of the feedback engineer. The basic problem in feedback-system design is the simultaneous satisfaction of the given specifications on overall transmittance T and sensitivity S: we are essentially trying to realize suitable dynamic transmission while controlling the effects of parameter variations. Thus T and S are fundamental measures of significant system properties. In the calculation of both T and S, the loop gain L plays a central role; furthermore, as we shall see in Sec. 29-8, a rather general problem of design arises from the tendency of feedback systems to oscillate, and the stability is determined by the loop transfer functions.

29-6. Evaluation of the Transfer Function of a Flow Diagram

Our objective in this section is to develop a procedure for evaluating by inspection a desired transfer function describing the ratio of any designated dependent node signal to any specified source. There are often several different transfer functions that are of possible interest in a given system, and the inspection technique to be described therefore offers a significant advantage. For the inspection method to be applicable, it is necessary to represent the excitation in any graph transmittance evaluation as a pure source node.

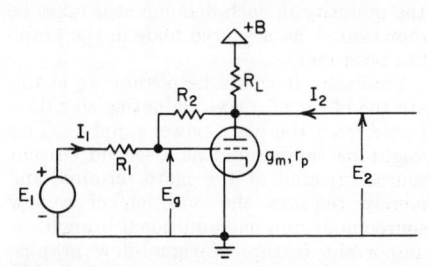

FIG. 29-11. A triode amplifier with plate-to-grid feedback.

FIG. 29-12. Linear incremental model of the triode amplifier.

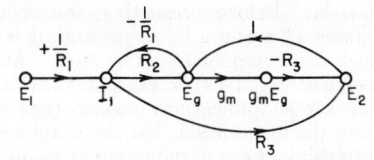

FIG. 29-13. Flow-graph interpretation of the model of Fig. 29-12.

As an example, let us consider the vacuum-tube amplifier of Fig. 29-11. Our primary interest might lie in the voltage gain E_2/E_1, and for this purpose a flow graph can be formulated with E_1 as the source and E_2 as the dependent node of interest. As an intermediate step, it is usually helpful to draw a linear incremental model as in Fig. 29-12.

A flow graph may be based on any one of several possible sets of equations which may be transcribed from the circuit. Writing node equations, for example,

$$(E_g - E_1)\frac{1}{R_1} + (E_g - E_2)\frac{1}{R_2} = 0 \tag{29-20}$$

$$g_m E_g + \frac{E_2}{R_3} - I_1 = 0 \tag{29-21}$$

where

$$I_1 = (E_g - E_2)\frac{1}{R_2} = (E_1 - E_g)\frac{1}{R_1}$$

and

$$g_m E_g = g_m(E_g)$$

The resulting graph appears as in Fig. 29-13, and from this we may evaluate the voltage gain E_2/E_1 or the input-terminal admittance I_1/E_1. Should we wish to determine the output-terminal impedance, we must modify the graph—though only in detail. For this last determination, we require a source node symbolizing the injection of the current I_2 at the output terminals. The presence of this new source simply adds one new term to (29-21), which now reads

$$g_m E_g + \frac{E_2}{R_3} - I_1 - I_2 = 0$$

from which
$$E_2 = (I_1 + I_2 - g_m E_g) R_3 \tag{29-22}$$

This shows E_2 to be the linear superposition of three branch signals. All but one of the branches already exist; therefore, only a branch R_3 from I_2 needs to be added. The modified graph is shown in Fig. 29-14. With this graph available, we may evaluate any of the following transmittances:*

$$\frac{E_2}{E_1}, \frac{E_2}{I_2}, \frac{I_1}{I_2}, \frac{I_1}{E_1} \tag{29-23}$$

and we observe that the requirement that the quantity in each denominator must be represented as a source node in the graph has been met.

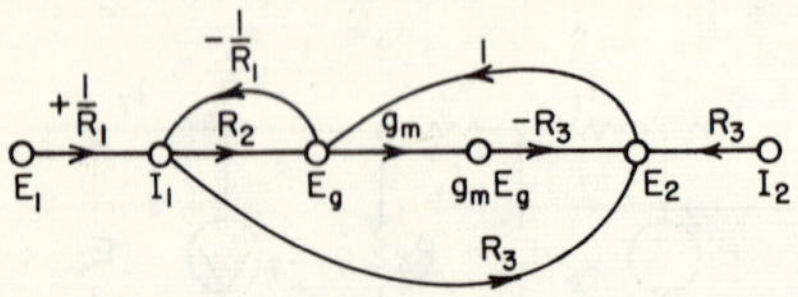

Similarly, it would be possible to evaluate the effect of a contaminating signal injected from the plate power supply. This might be represented as a second current source injected at the plate terminal and merely requires the addition of a new source node and one additional branch.

Fig. 29-14. Inclusion of I_2 as a source.

This discussion has revealed a rather important feature of signal-flow graphs; additional information about the system may be obtained by making minor additions to an existing graph, without the necessity of repeating an analysis from the beginning.

Step-by-step Reduction. Before presenting the reduction formula for the evaluation of a transmittance of a graph by inspection, it is useful to consider some reduction techniques which can be applied step by step. Algebraic reductions (such as the elimination of dependent variables, for example) have their flow-graph counterparts, and several elementary simplifications suggest themselves as a result of the properties of the nodes and the assumption that the graphs are linear.

In general, it is useful to think of graph reduction or simplification as involving the absorption of nodes or the elimination of loops or both. We shall first consider the reduction of single loops and then present a comprehensive step-by-step procedure for the absorption of any desired group of dependent nodes, with the end result, if desired, the reduction of the graph to a single branch. The transmittance of this final branch is, of course, the graph transmittance or transfer function being sought.

An elementary single-loop graph may be reduced to a single branch by dividing the forward path transmittance by one minus the loop transmittance. This simple rule, which is illustrated in Fig. 29-15, may easily be derived by reconstructing the set of equations which the graph represents and solving these for T. Thus, for $x_1 = 1$,

$$T_{12} = \frac{x_2}{1} = P \frac{1}{1 - L} \tag{29-24}$$

This simple rule for a single loop is applied to any paths which share nodes in common (i.e., touch) the loop. Thus, in Fig. 29-16, there are two paths from node 1 to node 2 and a third path from node 1' to node 2. The two graph transmittances are

---

* It may strike the reader as unorthodox to regard quantities such as I_1/E_1 and E_2/I_2 as *transfer* quantities. However, in the cause-and-effect symbolism of a flow graph they become transmittances and structurally are not distinguished from E_2/E_1, for example.

FEEDBACK THEORY

therefore

$$T_{12} = \frac{abc}{1-bf} + d = \frac{P_1}{1-L} + P_2 \qquad (29\text{-}25)$$

and

$$T_{1'2} = \frac{ebc}{1-bf} = \frac{P_3}{1-L} \qquad (29\text{-}26)$$

The path* d does not touch the loop and is therefore not modified by it.

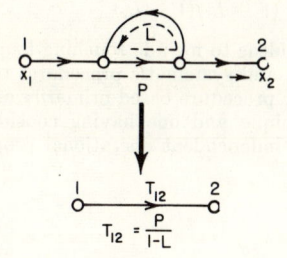

FIG. 29-15. Reduction of a single loop.

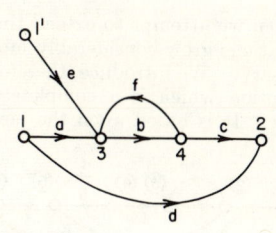

FIG. 29-16. Illustrating touching and nontouching paths.

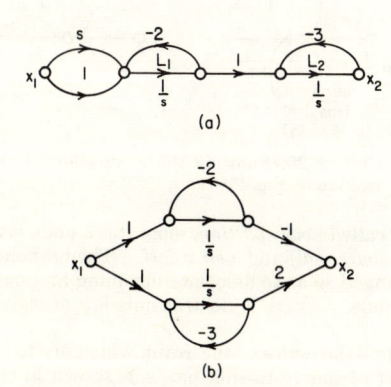

FIG. 29-17. Two graphs of the transfer function of Eq. (29-27).

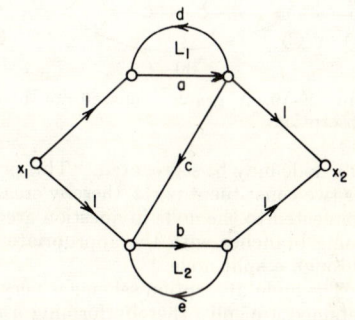

FIG. 29-18. Another example of nontouching loops.

With the aid of the simple rule, any flow graph involving nontouching loops can be reduced to a cascade (nonfeedback) graph and ultimately to a single branch. As an example, the transfer function

$$T = \frac{s+1}{s^2 + 5s + 6} = \frac{s+1}{(s+2)(s+3)} \qquad (29\text{-}27)$$

can be represented by either of the two graphs of Fig. 29-17. In part a each loop is reduced using the simple loop rule with the result

$$T = \frac{x_2}{x_1} = \frac{P}{(1-L_1)(1-L_2)} = \frac{(s+1)(1/s^2)}{(1+2/s)(1+3/s)} \qquad (29\text{-}28)$$

In part b, which is a representation of the partial-fraction expansion of Eq. (29-27), the graph transmittance is found to consist of the sum of two weighted path transmittances:

$$T = \frac{P_1}{1-L_1} + \frac{P_2}{1-L_2} = \frac{-1/s}{1+2/s} + \frac{2/s}{1+3/s} \qquad (29\text{-}29)$$

* We shall usually not distinguish between the terms *path* and *path transmittance* or *loop* and *loop transmittance*. Thus, if we speak of the sum of two paths, we mean the sum of their transmittances.

The graph of Fig. 29-18 is used to illustrate the interrelations between the paths and loops. There are three source-to-sink paths, and the graph transmittance is the sum of these paths with a proper weighting of each to account for touching loops. The path 1-a-1, which we shall designate P_1, touches only L_1; the second path, P_2, which is 1-b-1, touches only L_2; but the third path, P_3, or 1-a-c-b-1, touches both loops. The graph transmittance is written

$$T = \frac{x_2}{x_1} = \frac{P_1}{1 - L_1} + \frac{P_2}{1 - L_2} + \frac{P_3}{(1 - L_1)(1 - L_2)} \qquad (29\text{-}30)$$

When we attempt to extend these elementary ideas to include touching loops, the evaluations grow considerably more complicated, and systematic procedures become necessary. We introduce, therefore, a systematic procedure based primarily on node absorption which is a completely general technique and one having considerable utility. It is based upon the fact that the two independent operational properties

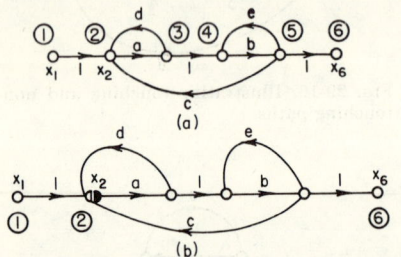

FIG. 29-19. A graph prepared for node absorption.

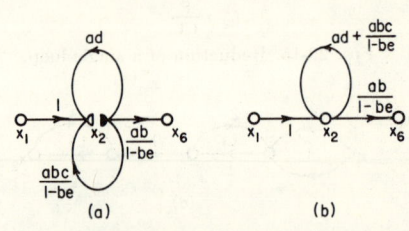

FIG. 29-20. Showing the evaluation of the residue of Fig. 29-19.

of a node may be segregated. This process, called *node splitting*, separates a node into its two constituent parts, thereby creating a *source half* and a *sink half*. The branches connected to the node in question are rearranged so as to associate incoming and outgoing branches with the appropriate half node. There is no transmission of signal through a split node.

The node-absorption scheme is very simple—the nodes of the graph which are to be retained are split, thereby forming a new set of source-to-sink paths as shown in the example of Fig. 29-19. Here the source half is shown in black and the sink in white. In the example, the arbitrary decision has been taken to absorb all nodes with the exception of the system source and sink (nodes 1 and 6) and the intermediate node, 2. Signal transmission between nodes 1 and 6 has been interrupted and the remaining signal paths are between nodes 1 and 2, nodes 2 and 6, and between the halves of node 2. A simplified graph or *residue* may now be constructed in which the branches are the source-to-sink paths of Fig. 29-19b. The branch evaluations of the residue are shown in Fig. 29-20a and the reconstituted graph in Fig. 29-20b (the reconstitution here consisting of the reconnection of the two half nodes).

The graph transmittance may now be evaluated by using the simple loop rule; here

$$T_{16} = \frac{x_6}{x_1} = \frac{ab/(1 - be)}{1 - ad - [abc/(1 - be)]} = \frac{ab}{1 - be - ad - abc + adbe} \qquad (29\text{-}31)$$

It is noteworthy that the simple loop rule had to be employed in the evaluation of the residue branches, since two of the source-to-sink paths have a touching loop eb which has remained intact in the node splitting.

Sometimes it is desired to absorb as many nodes as possible, yet to preserve explicitly the integrity of the loop structure. To accomplish this we must open every loop in the graph by splitting an appropriate set of nodes. For the example of Fig. 29-19 it is necessary to split at least two intermediate nodes in order to open the three loops. Several choices exist, and choosing nodes 3 and 5, the residue takes the form shown in

Fig. 29-21. This kind of graph in which the essential loop structure is preserved and described with a minimum number of nodes is called an *index* residue. The index, which is the number of nodes additional to the source and sink which is required, characterizes the essential complexity of the system and is a useful means for classifying systems.

Mason's Reduction Formula. With the aid of the foregoing examples, it is possible to visualize a specified graph transmittance in terms of two basic topological elements—the open paths and the closed paths. As we have already seen, the transmittance of a graph is the sum of all different paths leading from a specified source to some chosen dependent node, where each path transmittance becomes weighted by factors arising from the configuration of loops. When there are no loops, the weighting factor for each path is unity.

Based upon this view of graph topology, Mason's reduction formula for graph transmittance† takes the form of the summation:

$$T_{ij} = P_1 \frac{\Delta_1}{\Delta} + P_2 \frac{\Delta_2}{\Delta} + \cdots + P_n \frac{\Delta_n}{\Delta}$$

$$= \sum_k \frac{P_k \Delta_k}{\Delta} \qquad (29\text{-}32)$$

the summation being taken over all k different open paths from source i to dependent node j.

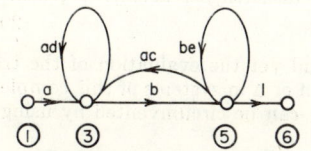

Fig. 29-21. Index residue of Fig. 29-19a which retains the essential loop structure.

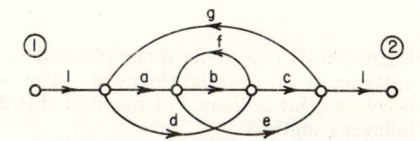

Fig. 29-22. Example for evaluation.

The common denominator Δ, called the *graph determinant*, is a function of only the loops and the way in which they interact with one another. Formally Δ is described by the function

$$\Delta = [(1 - L_1)(1 - L_2) \cdots (1 - L_m)]^* \qquad (29\text{-}33)$$

where * indicates that, of the 2^m possible terms in the expansion of Δ, only those terms are to be retained which *do not contain loops which touch in the graph*. To follow the indicated procedure of (29-33) would usually involve the generation of many terms which would eventually be discarded; therefore, this approach is seldom used. It is easy to write the expansion in the following form, however:

$$\begin{aligned}\Delta = 1 &- (\Sigma \text{ all different loops}) \\ &+ (\Sigma \text{ all different pairs of nontouching loops}) \\ &- (\Sigma \text{ all different triplets of nontouching loops}) \\ &+ \cdots \end{aligned} \qquad (29\text{-}34)$$

The *path factor* Δ_k is a function of only the loops which do not touch the kth path and is the determinant of the remaining graph after the kth path (and its attached loops) has been deleted from the graph.

Before trying some examples we should observe that an open path is a succession of branches which never includes a loop; therefore, no node is ever encountered more than once. Similarly loops are simple closed paths in which no node is encountered more than once per cycle of traversal. With these agreements in mind, let us find the graph transmittance of the example shown in Fig. 29-22.

† S. J. Mason, Feedback Theory—Further Properties of Signal Flow Graphs, *Proc. IRE*, vol. 44, no. 7, pp. 920–927, July, 1956.

There are four open paths:

$$abc \quad dc \quad ae \quad dfe$$

and five loops:

$$bf \quad abcg \quad dfeg \quad dcg \quad aeg$$

In this example, all the loops touch each other and all paths touch all loops; therefore, we write

$$T_{12} = \frac{abc + dc + ae + dfe}{1 - bf - abcg - dfeg - dcg - aeg} \qquad (29\text{-}35)$$

As a second example we consider the graph of Fig. 29-23. To evaluate the transmittance T_{12}, we observe that there are two paths:

$$ab \quad gfab$$

and five loops:

$$ac \quad gi \quad abd \quad ghj \quad gfae$$

The path $gfab$ touches all loops, but ab fails to touch gi and ghj. Furthermore, not all the loops are touching, and we can visualize four different *pairs* of nontouching loops but no higher-order combinations. The graph transmittance may be written on the basis of these observations as follows:

$$T_{12} = \frac{ab(1 - gi - ghj) + gfab}{1 - ac - gi - abd - ghj - gfae + acgi + (abd)(ghj) + ac(ghj) + gi(abd)} \qquad (29\text{-}36)$$

Figure 29-23 represents a complicated system, and yet the evaluation of the transmittance is quite straightforward. The evaluation of Δ in systems of this complexity becomes a bit tedious, but much of this difficulty can be circumvented by using the following method.

The Evaluation of the Graph Determinant in Compact Form. The evaluation of Δ in expanded form [Eq. (29-34)] sometimes becomes tedious and involved. This is usually the case when a graph has several nontouching loops together with coupled loops so that the analyst is forced to concentrate, in order not to overlook any of the nontouching-loop sets. Recognizing this difficulty, Mason* proposed a way of writing the graph determinant in factored form.

FIG. 29-23. Second example for evaluation.

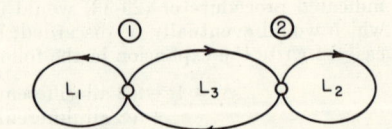

FIG. 29-24. Loop configuration.

Although we shall not attempt a formal development and proof, we shall offer an example from which the general procedure may be conjectured. Let us consider the familiar three-loop complex shown in Fig. 29-24.

Using the expansion technique of Eq. (29-34) we write

$$\Delta = 1 - L_1 - L_2 - L_3 + L_1 L_2 \qquad (29\text{-}37)$$

which could be rearranged to read

$$\Delta = (1 - L_1)(1 - L_2) - L_3 \qquad (29\text{-}38)$$

* Op. cit.

FEEDBACK THEORY

This suggests that the graph might be partitioned by opening the central loop L_3. The determinant of the remaining loop structure is then simply the product of the separate determinants of the disparate loops L_1 and L_2. Thus (29-38) might be written

$$\Delta = \Delta_1 \Delta_2 - L_3 \tag{29-39}$$

The idea suggested can be developed formally by inserting an interior node in L_3, splitting this node, and then evaluating the resulting source-to-sink path, as in Fig. 29-25.

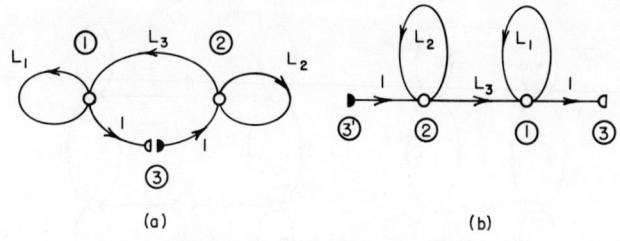

FIG. 29-25. Loop transmittance of node 3.

The transmittance $T_{3'3}$, which is the loop transmittance of node 3, is

$$T_{3'3} = \frac{L_3}{(1 - L_1)(1 - L_2)} \tag{29-40}$$

The quantity $1 - T_{3'3}$, which is the return difference of node 3, is found as

$$1 - T_{3'3} = \frac{(1 - L_1)(1 - L_2) - L_3}{(1 - L_1)(1 - L_2)} \tag{29-41}$$

The numerator is Δ for the system with node 3 closed, and the denominator is the determinant of the system *after* node 3 has been opened (or, in effect, removed). Designating the latter determinant by Δ_0, we note that

$$F_{3'3} = 1 - T_{3'3} = \frac{\Delta}{\Delta_0} = 1 - \frac{\sum_k L_k \Delta_k}{\Delta_0} \tag{29-42}$$

where L_k is the kth opened loop resulting from splitting the jth node and Δ_k is the loop (path) factor. Solving for Δ we obtain

$$\Delta = \Delta_0 - \sum_k L_k \Delta_k \tag{29-43}$$

Before commenting on the significance of this relation, let us apply it to node 1 of Fig. 29-24. Splitting node 1 and opening out the resulting source-to-sink paths L_k we obtain Fig. 29-26. In this case, $\Delta_0 = 1 - L_2$ and the path summation, when subtracted from Δ_0, yields

$$\Delta = (1 - L_2) - L_1(1 - L_2) - L_3 \tag{29-44}$$

which again is the known value of Δ.

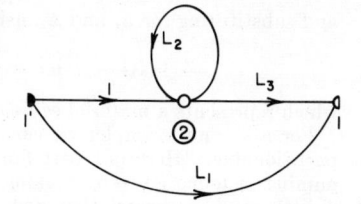

FIG. 29-26. The result of splitting node 1.

It turns out that Eq. (29-43) is general and may be applied with respect to any chosen node of a linear graph, whether that node be one in the given graph, or one inserted specifically for the evaluation of Δ.

29-18 SYSTEM THEORY

Let us try the idea on two examples which would ordinarily require rather complicated expansions in the determinant evaluation. In the first, shown in Fig. 29-27a, there are seven loops, eleven separate nontouching pairs, six nontouching triplets, and one nontouching quadruplet. Thus the determinant is the algebraic sum of unity and 25 nontouching loop products. Here, aside from the tedium of writing this number of terms, there is danger of omitting or misinterpreting a term.

This example lends itself beautifully to the technique just described. The insertion of an interior node as shown in Fig. 29-27b permits the graph to be partitioned into two separate loop structures of the form shown in Fig. 29-24.

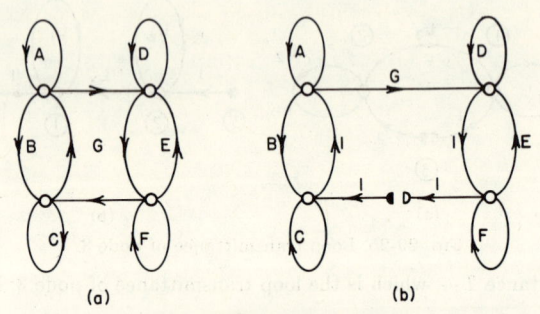

Fig. 29-27. First example.

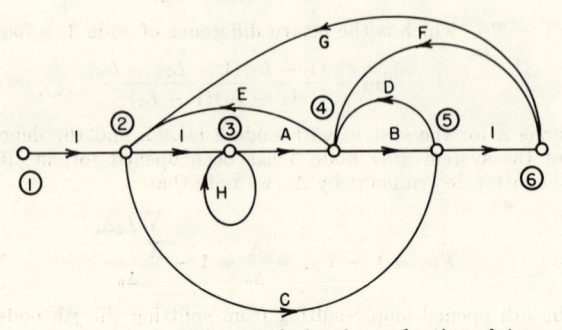

Fig. 29-28. Second example for the evaluation of Δ.

Designating Δ_1 and Δ_2 as the determinants of the partitioned subgraphs, we employ (29-43) to write by inspection

$$\Delta = \Delta_1 \Delta_2 - G$$

and substituting for Δ_1 and Δ_2, using (29-38),

$$\Delta = [(1 - A)(1 - C) - B][(1 - D)(1 - F) - E] - G \qquad (29\text{-}45)$$

which represents a material saving in effort over the more conventional expansion.

For a second example, we consider the graph of Fig. 29-28, which is clearly not partitionable. Here our best hope is to split a node which will open the greatest number of loops (thereby making the evaluations of Δ_0 and Δ_k as easy as possible). A little study indicates that node 4 is a good choice, since splitting it leaves intact only two loops, as shown in Fig. 29-29.

Since the two remaining loops of Fig. 29-29 are nontouching,

$$\Delta_0 = (1 - H)(1 - CG) \qquad (29\text{-}46)$$

FEEDBACK THEORY

The forward paths L_k and their factors yield

$$\sum_k L_k \Delta_k = BGA + EA + (1 - H)(BF + BD + ECD + ECF) \tag{29-47}$$

and the determinant of the system is

$$\Delta = (1 - H)(1 - CG - BF - BD - ECD - ECF) - BGA - EA \tag{29-48}$$

In summary, the rules for the evaluation of the determinant of a system of coupled loops may be stated as follows:

1. Choose a single node j to be split (mentally) which will either partition the loop graph advantageously or open the majority of the loops. (The node may be one of

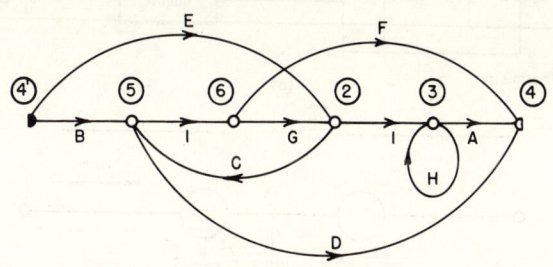

FIG. 29-29. Graph with node 4 split.

the original nodes of the given graph, or a node inserted specifically for purposes of facilitating the evaluation.)

2. Evaluate Δ_0 as the determinant of all loops which *do not* contain node j.
3. Sum all different products of loops and loop factors which *do* contain node j.
4. Substitute values in

$$\Delta = \Delta_0 - \sum_k L_k \Delta_k$$

It is unnecessary to split the chosen node or to redraw the resulting graph, since the steps outlined above serve as a guide to the meanings of the various entries in the formulation of Δ; in some cases, however, it is easier to evaluate Δ after the graph is redrawn.

It should be noted finally that some systems contain loop subgraphs that are not coupled. When all branches that are not components of loops are deleted from the graph, there may remain several partitioned loop subgraphs. The foregoing procedure applies then to the evaluation of Δ for each of the several subgraphs. The system determinant is then the product of the several subgraph determinants.

29-7. Sensitivity

As we have emphasized in the preceding sections, feedback is used primarily for the control of sensitivity—i.e., in order that the system may possess the self-calibration property of automatically compensating for parameter variations. Evaluation of a feedback system must then rely on a quantitative measure of this sensitivity; in Sec. 29-5 we defined the sensitivity of the overall transmittance T to changes in a parameter x as

$$S_x^T = \frac{d \ln T}{d \ln x} = \frac{dT/T}{dx/x} \tag{29-49}$$

In this section, we wish to consider first the sensitivity-control features of one typical feedback system (as an example); then we consider the techniques available for the calculation of sensitivity.

29-20 SYSTEM THEORY

Example of Sensitivity Control with Feedback. Examples of the importance of feedback in the control of sensitivity are available throughout the field of electronic instrumentation. As an illustration, we consider in this section an instrument for the measurement of temperature, involving the generation of an electric voltage proportional to the instantaneous temperature within an enclosure.

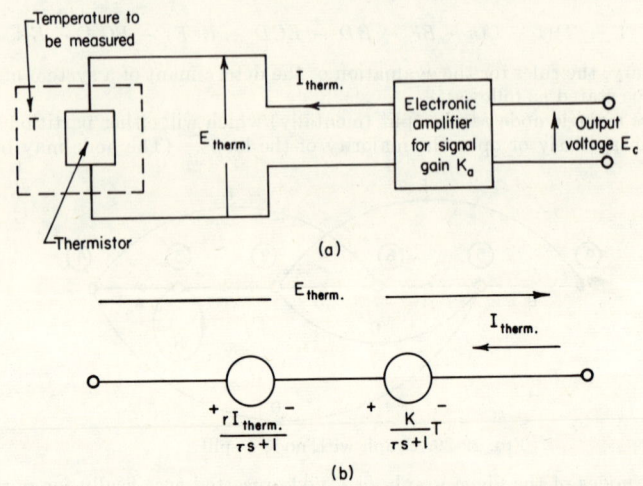

FIG. 29-30. Temperature instrument. (*a*) Measuring system. (*b*) Model for the thermistor (*T* incremental temperature; *r*, *K* positive constants).

The basic, nonfeedback system is depicted in Fig. 29-30. The temperature-sensitive element is a thermistor, with the model, shown in part *b* of the figure, including a voltage source proportional to the change in temperature from a quiescent or normal operating value. In series with this voltage source there is a second voltage source dependent on the thermistor current, since an increase of the current causes an increase in temperature as a result of the power dissipated in this resistive element. The voltage source is related to the temperature through the transfer function $-K/(\tau s + 1)$, where τ is the thermal time constant. In other words, when the ambient temperature changes abruptly in the oven or container within which the thermistor is located, the terminal voltage approaches exponentially the new value as depicted in Fig. 29-31.

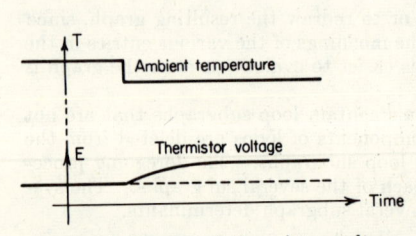

FIG. 29-31. Change of thermistor voltage with temperature.

If the amplifier of Fig. 29-30*a* draws no input current, the input voltage equals the voltage of the source in the thermistor model; consequently the output E_0 is related to the incremental temperature by the overall transfer function

$$\frac{E_0}{T} = -\frac{K_a K}{\tau s + 1} \qquad (29\text{-}50)$$

Since the thermal time constant is typically of the order of magnitude of 2 sec (or greater), the full reading of the output voltage is not approached until about 8 sec (four time constants) after a change in temperature. In an alternative viewpoint, we can say that the system does not respond to temperature variations at frequencies greater than ½ radian/sec (the reciprocal of the time constant), or about 1/12 cycle/sec.

FEEDBACK THEORY 29-21

In order to speed up the response of our instrument, we can pass the output voltage E_0 of Fig. 29-30 through an electrical network, as depicted in Fig. 29-32. If this compensation network possesses the transfer function $(\tau s + 1)/(\tau_c s + 1)$, the overall instrument transfer function is

$$\frac{E_{\text{out}}}{T} = -\frac{KK_a}{\tau s + 1}\frac{\tau s + 1}{\tau_c s + 1} = -\frac{KK_a}{\tau_c s + 1} \tag{29-51}$$

In other words, the thermal time constant τ is replaced by the electrical time constant τ_c. If τ_c is made 1/100th of τ, we succeed in speeding up the instrument response by the factor of 100; in other words, the instrument responds to signals at frequencies 100 times greater.

We seldom achieve such significant results without an associated penalty; in the present case, the penalty is apparent if we consider the realization of the compensation network required in Fig. 29-32. Search for a suitable network yields the system drawn in Fig. 29-33; in order to effect a 100:1 decrease in the instrument time

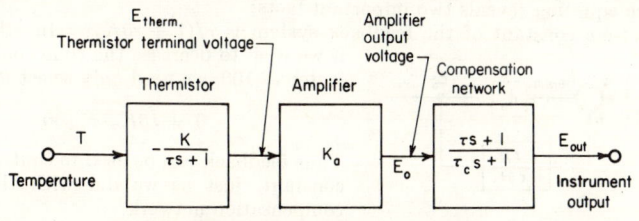

FIG. 29-32. Improved instrument.

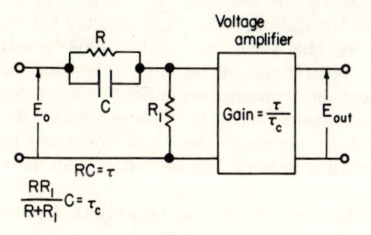

FIG. 29-33. $\dfrac{E_{\text{out}}}{E_0} = \dfrac{\tau s + 1}{\tau_c s + 1}$.

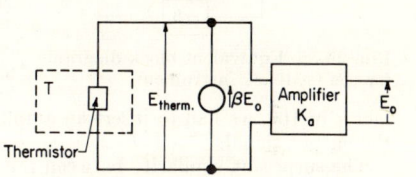

FIG. 29-34. Instrument with feedback.

constant, we need to include an amplifier of gain 100. The greater the improvement demanded in dynamic response, the larger the gain required.

The instrument of Fig. 29-32 is quite satisfactory for many purposes, but any changes in the gain of either amplifier result in corresponding (and equal) changes in the overall instrument transfer function. Hence, as amplifiers age or environmental conditions vary, the instrument requires continual recalibration. If feedback is used, not only can the speed of response be controlled—but also the recalibration can be made automatic and self-contained.

An appropriate feedback configuration is shown in Fig. 29-34, where a current proportional to the output voltage is passed through the thermistor. Under these conditions, the terminal voltage of the thermistor is determined from Fig. 29-30b as

$$E_{\text{therm}} = \frac{-K}{\tau s + 1}T - \frac{r}{\tau s + 1}\beta E_0$$

and the system is described by the block diagrams of Fig. 29-35 [the second diagram is derived from the first if we recognize that addition after multiplication of each variable by $1/(\tau s + 1)$ is equivalent to addition followed by multiplication of the sum].

The dynamic characteristics of the feedback instrument are given by the overall transmittance, which can be written directly from the relations of Sec. 29-6:

$$\frac{E_0}{T} = (-K) \frac{K_a/(\tau s + 1)}{1 + [r\beta K_a/(\tau s + 1)]} \qquad (29\text{-}52)$$

or, if we clear of fractions,

$$\frac{E_0}{T} = \frac{-KK_a}{\tau s + 1 + r\beta K_a} \qquad (29\text{-}53)$$

Equation (29-53) can be placed in somewhat more convenient form if we divide numerator and denominator by $1 + r\beta K_a$

$$\frac{E_0}{T} = \frac{-KK_a/(1 + r\beta K_a)}{[\tau/(1 + r\beta K_a) s + 1]} \qquad (29\text{-}54)$$

The above equation reveals two important facts:

1. The time constant of the feedback system is $\tau/(1 + r\beta K_a)$. In other words, if we wish to decrease the time constant by a factor of 100, we need only select β such that

$$1 + r\beta K_a = 100 \qquad (29\text{-}55)$$

Thus feedback can be used to control the time constant, just as we did with the tandem compensation network.

2. As a result of the feedback, the gain is reduced by division by $1 + r\beta K_a$; in other words, the gain is reduced in exactly the same ratio as the time constant. This result is rather startling, since it is identical with our earlier conclusions about the tandem compensation scheme. There, too, if we wished to decrease the time constant by a factor of 100, we had to insert an amplifier with a gain of 100 to restore the level of the output signal.

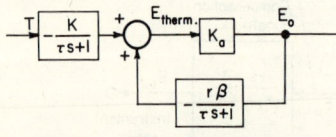

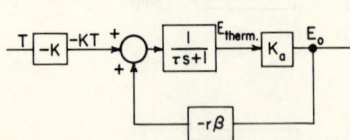

FIG. 29-35. Equivalent block diagrams for the feedback instrument.

The surprising similarity between the two instruments ends at this point, however. In the tandem scheme, a 1 per cent change in amplifier gain results in 1 per cent change in output; i.e.,

$$S_{K_a}{}^{E_0/T} = 1$$

In the feedback configuration of Fig. 29-34, however,

$$S_{K_a}{}^{E_0/T} = \frac{1}{1 + [r\beta K_a/(\tau s + 1)]} = \frac{\tau s + 1}{\tau s + 1 + r\beta K_a} \qquad (29\text{-}56)$$

At zero frequency (i.e., in the steady state),

$$S_{K_a}{}^{E_0/T} = \frac{1}{1 + r\beta K_a} \quad (= \tfrac{1}{100} \text{ for our numerical example}) \qquad (29\text{-}57)$$

The sensitivity of the feedback system is reduced by the *same* factor as the time constant and gain! The feedback instrument contains automatic self-calibration; i.e., the feedback instrument is able automatically to compensate for changes in amplifier gain.

Thus the feedback system possesses all the engineering-design flexibility of the tandem compensation scheme and, in addition, the possibility of design control over the sensitivity of the instrument to changes in system parameters. Because of this added, significant advantage, feedback systems are used extensively in instrumentation. The only penalty we must pay for this sensitivity control is the extra equipment (often trivial) to realize the feedback path—in Fig. 29-34, the current generator with the value of current proportional to the output voltage.

Design Control of Sensitivity. The two control systems of Fig. 29-36 illustrate two different design interpretations for achieving a specified overall transfer function with a particular *process* given. In case *a* the desired transfer function is achieved by utilizing a *controller* placed in tandem with the process; in *b* the controller function is supplied by a local feedback path encircling the plant and the tandem element is merely an amplifier. Our interest is centered in the comparison of the two systems from the point of view of how the overall transfer function is affected by variations in the plant parameters K and a over which we presumably have no direct design control. In other words, we want to investigate the sensitivity of $T(s)$ to both K and a for the two cases.

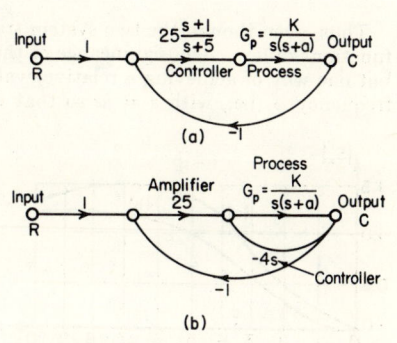

Fig. 29-36. Control systems with normal process-parameter values $K = 1, a = 1$. (a) Tandem controller. (b) Minor-loop controller.

For case *a*, the overall transmittance of the flow graph is

$$T(s) = \frac{C}{R} = \frac{25K(s+1)}{s(s+5)(s+a) + 25K(s+1)} \qquad (29\text{-}58)$$

If this is placed in the form

$$T(s) = \frac{1}{1 + \dfrac{s(s+5)(s+a)}{25K(s+1)}} \qquad (29\text{-}59)$$

the sensitivity $S_K{}^T$ can be evaluated by direct differentiation to produce

$$S_K{}^T = \frac{s(s+5)(s+a)}{25K(s+1) + s(s+a)(s+5)} \qquad (29\text{-}60)$$

and when $a = 1$ and $K = 1$, this reduces to

$$S_K{}^T = \frac{s(s+5)}{s^2 + 5s + 25}$$

Similarly, the sensitivity $S_K{}^T$ for case *b* is carried out by operating with the transmittance

$$T(s) = \frac{25K/s(s+a)}{1 + [4Ks/s(s+a)] + [25K/s(s+a)]}$$

$$= \frac{25K}{s(s+a) + K(25 + 4s)} \qquad (29\text{-}61)$$

Dividing through by K, and then computing the sensitivity $S_K{}^T$, we obtain

$$S_K{}^T = \frac{s(s+a)/K}{25 + 4s + s(s+a)/K} \qquad (29\text{-}62)$$

and again inserting the normal values of the process parameters, we find

$$S_K{}^T = \frac{s(s+1)}{s^2 + 5s + 25}$$

Thus, even though the two system transmittances are identical, the two sensitivity functions differ. The significance of the difference is somewhat difficult to interpret, but one way of achieving a relative evaluation might be to plot $|S_K{}^T|$ vs. the angular frequency ω (i.e., with $s = j\omega$ so that we are focusing attention on the system performance under sinusoidal excitation). The results shown in Fig. 29-37 demonstrate the superiority of case b—which utilizes local feedback around the process—when the angular frequencies are low. Thus, through the definition of the sensitivity function, we have expressed in quantitative terms the degree to which the feedback accomplishes its appointed task and we have a firm basis for the comparison of alternate designs. (It is not proper to infer, from this example, that minor-loop feedback always improves the system sensitivity. To the authors' knowledge, essentially no attempt has been made to study the relationship between sensitivity and configuration. Therefore, this relationship is not clearly understood.)

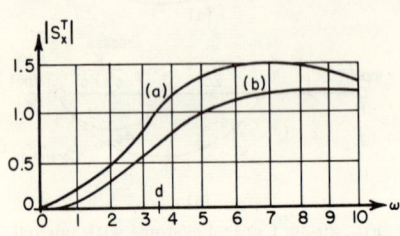

FIG. 29-37. Sensitivities of the systems of Fig. 29-36.

Evaluation of the competitive designs of Fig. 29-36 might also require determination of the sensitivities with respect to parameter a. By direct differentiation, we learn that for case a

$$S_a{}^T = \frac{-s(s+5)}{(s+1)(s^2 + 5s + 25)}$$

and for case b

$$S_a{}^T = \frac{-s}{s^2 + 5s + 25}$$

and plotting graphs of $|S_a{}^T|$ vs. ω would again reveal the superiority of the minor-loop compensation (case b) design at low frequencies (for $\omega < 1$, case a is five times as sensitive to changes in the parameter a as case b).

In this section our objective is not, however, to consider a specific example in great detail but rather to indicate the general way in which the feedback engineer attempts to make a quantitative assessment of the role of feedback in a system design—an assessment which is based on the sensitivity function.

Although the value of the sensitivity function is readily demonstrated, its computation has been shown to be tedious even in the simple examples that we have chosen. If we are to make any real use of this concept, it is obvious that attention must be given to the problem of making sensitivity computations much simpler. Fortunately, several solutions to the computation problem have been discovered, and because there is a repertoire of these, the remainder of this section is devoted largely to developing and illustrating the several different useful techniques and viewpoints.

Topological Interpretation of Sensitivity. The development of Mason's reduction theorem has been such a decided boon to the evaluation of graph transmittances that it was natural to seek a similar evaluation of sensitivity. This is readily accomplished by carrying out on Mason's formula for transmittance the steps of the formal derivative definition of sensitivity. Thus

$$S_x{}^T = \frac{x}{T} \frac{\partial}{\partial x} \frac{\sum_k P_k \Delta_k}{\Delta} \qquad (29\text{-}63)$$

FEEDBACK THEORY

and for convenience, we introduce the symbol

$$\Sigma \equiv \sum_k P_k \Delta_k \tag{29-64}$$

Then
$$S_x^T = \frac{\Delta(\partial \Sigma/\partial x) - \Sigma(\partial \Delta/\partial x)}{\Delta^2} \frac{x\Delta}{\Sigma}$$

$$= \frac{x(\partial \Sigma/\partial x)}{\Sigma} - \frac{x(\partial \Delta/\partial x)}{\Delta} \tag{29-65}$$

We next take note of a significant property of the functions Δ and Σ; each is a linear function of any path or loop transmittance, and hence of any branch thereof.* In other words, in the expansions of Δ and Σ, no path or loop appears more than once in any product term; this is a consequence of the restriction of such terms to products of nontouching loops and paths. This makes it possible to conclude that the functions

$$x \frac{\partial \Sigma}{\partial x} \quad \text{and} \quad x \frac{\partial \Delta}{\partial x}$$

contain only those product terms in which x was originally a factor. Therefore, we might write

$$x \frac{\partial \Delta}{\partial x} = \Delta - \Delta_0 \tag{29-66}$$

and
$$x \frac{\partial \Sigma}{\partial x} = \Sigma - \Sigma_0 \tag{29-67}$$

where Δ_0 and Σ_0 signify the expansions of Δ and Σ evaluated *with the branch x removed from the graph.*

Returning to Eq. (29-65) and substituting (29-66) and (29-67), we arrive at the following result:

$$S_x^T = \frac{\Delta_0}{\Delta} - \frac{\Sigma_0}{\Sigma} \tag{29-68}\dagger$$

Example of Use of Topological Formulation. The middle-frequency behavior of a common-emitter transistor amplifier with collector-to-base feedback is described by the flow graph of Fig. 29-38. Using (29-68), the sensitivity of the current gain I_2/I_1 can be determined readily with respect to any of the branches, and to illustrate, we evaluate the sensitivity S_c^T. We observe first that there are two source-to-sink paths $abcd$ and aed each of which carries a unit path factor. Only the former is interrupted when the branch c is removed; therefore,

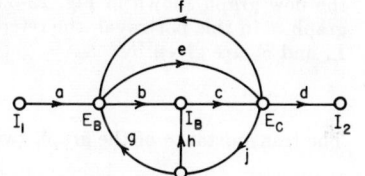

Fig. 29-38. Example for sensitivity evaluation.

$$\frac{\Sigma_0}{\Sigma} = \frac{aed}{abcd + aed}$$

We note further that there are five distinct loops (all touching one another): the principal loop, bcf, involving the collector-to-base feedback; and, additionally, the

---

* We are assuming here that the parameter x appears in the flow diagram only once, and then as a multiplier of a branch. If x is a circuit parameter (R, L, C, g_m, μ, a, or β, for example) we can always draw the flow diagram to satisfy this assumption, since the overall transmission T is a bilinear function of x. In the exceptional cases when the assumption cannot be satisfied (e.g., when x is the ambient temperature, and hence affects several parameters simultaneously), we must return to the basic definition for the calculation of S.

† As noted in the previous footnote, Eq. (29-68) is valid only if the signal-flow diagram is drawn in such a way that x appears only as the multiplier of a single branch.

loops $bcjg$, cjh, fe, and ejg. Removing the branch c opens three of these, and we have

$$\frac{\Delta_0}{\Delta} = \frac{1 - fe - ejg}{1 - bcf - bcjg - cjh - fe - ejg}$$

The complete sensitivity function can now be written as

$$S_c^T = \frac{1 - e(f + jg)}{1 - c(bf + bjg + jh) - e(f + jg)} - \frac{aed}{abcd + aed}$$

A numerical evaluation would ordinarily eliminate a number of insignificant factors and further simplify the function.

The topological formulation of (29-68) is amenable to further interpretation which sheds light on the significance of the sensitivity function. It will be recalled that the term Δ_0/Δ is the reciprocal of the return difference referred to a specified element. Thus

$$F_x = \frac{\Delta}{\Delta_0} = 1 - L_x$$

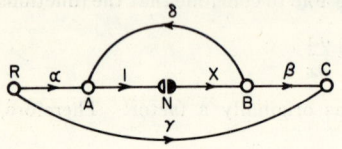

Fig. 29-39. Canonical sensitivity graph.

is the return difference referred to the element (branch) x, or to be consistent with our previous reference to return difference (see Sec. 29-6), F_x is the return difference referred to an interior node introduced into branch x and L_x is the loop transmittance of the node. If we factor the term $1/F_x$ out of (29-68) we may write the sensitivity in the form

$$S_x^T = \frac{1}{F_x}\left(1 - \frac{T_0}{T}\right) \tag{29-69}$$

where T is the graph transmittance and $T_0 = T_{(x=0)}$. The form of (29-69) suggests the flow graph shown in Fig. 29-39, which is referred to as the canonical sensitivity graph. In this portrayal, the reference element x is shown in an isolated branch and L_x and F_x are given by

$$L_x = x\delta$$
$$F_x = 1 - x\delta$$

The transmittance of the graph (with node N closed) is

$$T = \frac{\gamma(1 - x\delta) + \alpha\beta x}{1 - x\delta}$$

and the leakage transmittance (with node N open) is

$$T_0 = \gamma$$

The canonical graph of Fig. 29-39 is not important because we wish to convert every flow graph to this form; rather, we may, in general, visualize any system in this form, with the focus of attention on the reference element. Thus we gain some further insight into the meanings of sensitivity.

To demonstrate the generality of Fig. 29-39, we write equations describing the three node signals A, B, and C:

$$A = \alpha R + \delta B$$
$$B = xA \quad \text{(with } N \text{ closed)}$$
$$C = \gamma R + \beta B$$

From these relations we can express the branch transmittances (which in general are rational functions of the complex frequency) in the following forms:

$$\alpha = \frac{A}{R}\bigg|_{B=0} \tag{29-70}$$

FEEDBACK THEORY 29-27

(The requirement that $B = 0$ can be met by opening node N so that no transmission can occur around the loop. Basically what is required in the evaluation of α is that no signal shall enter node A except that arriving via branch α. The way in which this requirement is satisfied is of secondary importance.)

$$\beta = \frac{C}{B}\bigg|_{R=0} \tag{29-71}$$

(Here the implication is that unit signal is injected at node B and that no signal arrives at node C except that delivered by branch β.)

$$\delta = \frac{A}{B}\bigg|_{R=0} \tag{29-72}$$

(Again the source R is made zero and unit signal is imagined to be injected at node B.)

$$\gamma = \frac{C}{R}\bigg|_{B=0} \tag{29-73}$$

In each of the evaluations in (29-70) to (29-73) it is helpful to think of a node signal (such as the signal at A, B, or C) as representing a dependent variable resulting

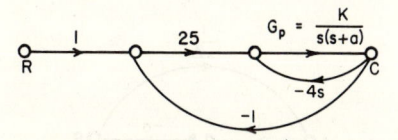

Fig. 29-40. System of Fig. 29-36b repeated.   Fig. 29-41. Isolation of branch a.

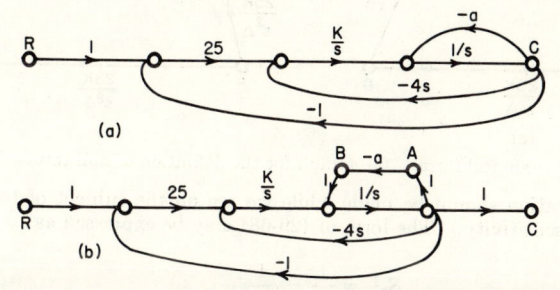

Fig. 29-42. (a) The modified graph. (b) Further isolation of element.

from the superposition of several independent variables acting through appropriate branches. The graph as a whole portrays a *set* of such equations, but in the case of each individual node equation there is but one dependent variable—the node signal.

Furthermore, when we actually evaluate the four branch transmittances of Fig. 29-39 from the original flow diagram, it is usually helpful to utilize the above equations with, in each case, x set equal to zero. Although it is not necessary to make $x = 0$, visualization of δ as A/B with $R = 0$ is ordinarily simpler if $x = 0$.

In order to illustrate the procedure of reduction to the canonical form we take the example of Fig. 29-36b which is reproduced as Fig. 29-40. Let us assume that we wish to choose a as the reference element. The first step is to modify the graph so that the element a appears alone in a branch. This is accomplished as shown in Fig. 29-41. The modified graph is shown in Fig. 29-42a, but there is one more step required inasmuch as we do not wish any other branches or paths to appear in parallel with branch a.

The final isolation is accomplished in Fig. 29-42b by arranging for branch a to occupy the central position in a tandem arrangement of three branches two of which are unity-transmission branches (actually only one isolator is required here, but if two are always used, all parallel-path difficulties are automatically avoided). With this form, the only branch leaving A is $-a$, and the only branch entering B is $-a$.

The evaluation of the canonical branches is now routine:

$$\alpha = \frac{A}{R}\bigg|_{B=0} = \frac{25K/s^2}{\Delta}$$

$$\beta = \frac{C}{B}\bigg|_{R=0} = \frac{1/s}{\Delta}$$

$$\delta = \frac{A}{B}\bigg|_{R=0} = \frac{1/s}{\Delta} \quad (29\text{-}74)$$

$$\gamma = \frac{C}{R}\bigg|_{B=0} = \frac{25K/s^2}{\Delta}$$

where

$$\Delta = 1 + \frac{4K}{s} + \frac{25K}{s^2}$$

is the determinant of the flow graph of Fig. 29-42 with $a = 0$. The canonical graph appears finally as shown in Fig. 29-43a.

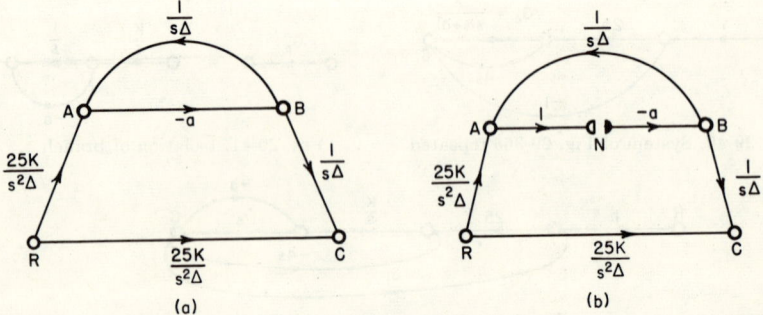

FIG. 29-43. (a) The canonical form. (b) Graph for the definition of null return difference.

One final observation should be made while we are on the subject of topological interpretations of sensitivity. The form of (29-68) may be expressed as

$$S_x^T = \frac{1}{F_x} - \frac{1}{F_x'} \quad (29\text{-}75)$$

where $F_x' = \Sigma/\Sigma_0$ is called the *null return difference* referred to the element x. Null return difference is most conveniently evaluated in terms of the topological definition Σ/Σ_0 but can be computed independently. The normal return difference is defined with reference to a node (whether one of the given nodes of a graph or one inserted interiorly in a branch, as in the case of node N) as

$$F_N = 1 - L_N$$

where N is a node which has been split and where L_N is the source-to-sink transmittance from the source half back to the sink half with all other excitation signals removed ($R = 0$ in this case). Here, since the node N has been inserted into branch $-a$, $F_N = F_{-a}$.

The concept of null return difference is somewhat more complicated, inasmuch as it involves not just the feedback loop, but the entire graph. It is again referred to a reference node and has the form

$$F_N' = 1 - L_N'$$

where L_N' is the signal at the sink half of the split node per unit signal at the source half with the system source R adjusted to make the system sink C exactly equal to zero. Evidently null return difference depends on the existence of a leakage path (as also shown by the definition Σ/Σ_0) in order that a cancellation of signals through the leakage path and through the path from N to C may take place.

To illustrate the direct evaluation of null return difference referred to element $-a$ of Fig. 29-43a, we use the graph with node N split (Fig. 29-43b). To determine the null-signal value of R we note that the signal at node A is

$$(-a)\frac{1}{s\Delta} + R\frac{25K}{s^2\Delta} = 0$$

and hence

$$R = \frac{as}{25K}$$

Then

$$F_{-a}' = 1 - \left[(-a)\frac{1}{s\Delta} + \frac{as}{25K}\frac{25K}{s^2\Delta}\right] = 1$$

which can be checked by evaluating Σ_0/Σ in the canonical graph or in Fig. 29-42.

The importance of the concept of null return difference (and the recognition that Σ/Σ_0 equals F') derives from the occasional possibility of calculating F' by inspection

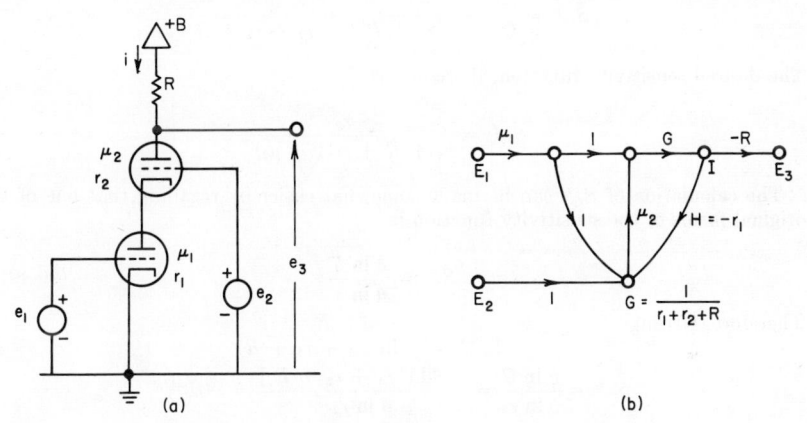

FIG. 29-44. (a) Cascade amplifier. (b) Signal-flow graph.

of the circuit diagram, without formulation of the flow diagram. For example, if the system is such that adjustment of the input R to yield zero output results in zero returned signal for the reference node or element under consideration, we can state at once that F' is unity.

Comprehensive Calculation Techniques. The topological formulation of the sensitivity function (29-68) is convenient and efficient when the reference element is isolated in a single branch, and we have demonstrated the fact that this is a condition which can be brought about by transformation of the given graph; however, it is decidedly inconvenient to have to modify a given graph in order to be able to evaluate a sensitivity function. We are strongly interested, therefore, in exploring computational techniques which avoid the necessity for graph transformations.

Although the system of Fig. 29-40 illustrates the need for a graph transformation in order to employ the topological formulation of sensitivity, we turn to a new example furnished by the cascade amplifier of Fig. 29-44a, a flow graph for which is given in part b of the figure.

The calculation of the sensitivity of $T_{13} = E_3/E_1$ to variation in r_2 presents a difficulty since r_2 is a term in the function G. We cannot employ (29-68) directly unless we first modify the graph as we did in Fig. 29-42a, a step we now seek to avoid. We can, however, utilize a property of the partial derivative to write

$$S_{r_2}{}^T = \frac{r_2}{T}\frac{\partial T}{\partial r_2} = \frac{r_2}{G}\frac{G}{T}\frac{\partial T}{\partial G}\frac{\partial G}{\partial r_2}$$

which, when rearranged becomes

$$S_{r_2}{}^T = \frac{\partial T}{\partial G}\frac{G}{T}\frac{\partial G}{\partial r_2}\frac{r_2}{G}$$

and this is recognized to be

$$S_{r_2}{}^T = S_G{}^T S_{r_2}{}^G \tag{29-76}$$

The first factor on the right is very easily evaluated using the topological formulation of (29-68) and since removal of the branch G opens both paths as well as the single loop.

$$S_G{}^T = \frac{1}{\Delta} = \frac{1}{1 + \mu_2 G r_1} \tag{29-77}$$

The second factor on the right side of (29-76) can be found easily by straightforward differentiation to yield

$$S_{r_2}{}^G = \frac{r_2}{G}\frac{-1}{(r_1 + r_2 + R)^2} = \frac{-r_2}{r_1 + r_2 + R}$$

The desired sensitivity function, therefore, is

$$S_{r_2}{}^T = \frac{-r_2}{r_2 + R + r_1(1 + \mu_2)}$$

The calculation of $S_{r_2}{}^G$ can be made somewhat easier by recalling that one of the original forms of the sensitivity function is

$$S_x{}^T = \frac{\partial \ln T}{\partial \ln x} \tag{29-49}$$

Therefore, writing

$$\ln G = -\ln(r_1 + r_2 + R)$$

$$S_{r_2}{}^G = \frac{\partial \ln G}{\partial \ln r_2} = \frac{-\partial[\ln(r_1 + r_2 + R)]}{\partial \ln r_2} = -S_{r_2}{}^{1/G}$$

Utilizing this convenient relation, we invert G, and then, since the differentiation is now trivially simple, we write by inspection

$$S_{r_2}{}^G = -S_{r_2}{}^{1/G} = \frac{r_2}{r_1 + r_2 + R}(-1)$$

Incidentally, we should also recognize at this point another convenient relation:

$$S_x{}^T = -S_{1/x}{}^T \tag{29-78}$$

The circuit of Fig. 29-44 affords a second example of importance. When we attempt to evaluate the sensitivity of T_{13} with respect to r_1 we encounter a new difficulty in that r_1 appears in two different branches—G and H. Again we are anxious to avoid modification of the flow graph; therefore, we consider the effect upon T_{13} of the variation of r_1 first in branch G and then separately in branch H and then we superpose the results. Phrased in terms of partial derivatives, what we propose is to write

$$S_{r_1}{}^T = \frac{r_1}{T}\left(\frac{\partial T}{\partial G}\frac{\partial G}{\partial r_1} + \frac{\partial T}{\partial H}\frac{\partial H}{\partial r_1}\right)$$

which may be rearranged as

$$S_{r_1}{}^T = \frac{\partial T}{\partial G}\frac{G}{T}\frac{\partial G}{\partial r_1}\frac{r_1}{G} + \frac{\partial T}{\partial H}\frac{H}{T}\frac{\partial H}{\partial r_1}\frac{r_1}{H}$$

But this is now recognized to be

$$S_{r_1}{}^T = S_G{}^T S_{r_1}{}^G + S_H{}^T S_{r_1}{}^H \qquad (29\text{-}79)$$

The first pair of factors on the right of (29-79) yields

$$S_G{}^T S_{r_1}{}^G = \frac{1}{\Delta}\frac{-r_1}{r_1 + r_2 + R}$$

From (29-77) we substitute the value of $1/\Delta$ to obtain

$$S_G{}^T S_{r_1}{}^G = \frac{-r_1}{r_2 + R + r_1(1 + \mu_2)} \qquad (29\text{-}80)$$

The second pair of factors yields

$$S_H{}^T S_{r_1}{}^H = \left(\frac{1}{\Delta} - 1\right)(1) = \frac{-\mu_2 r_1}{r_2 + R + r_1(1 + \mu_2)} \qquad (29\text{-}81)$$

and combining (29-80) and (29-81) we obtain

$$S_{r_1}{}^T = \frac{-r_1(1 + \mu_2)}{r_2 + R + r_1(1 + \mu_2)} \qquad (29\text{-}82)$$

From this example we observe that, when an element appears in several branches of a flow graph, the sensitivity of a specified transmittance with respect to the element is a sum of sensitivities, with each calculated as though the element varied in

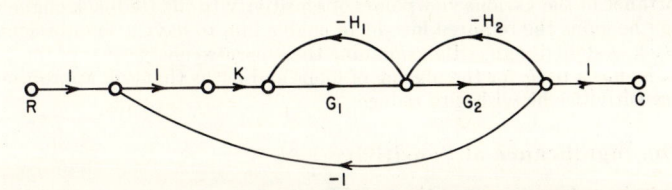

FIG. 29-45. Flow graph of a position-control system.

that branch and remained constant in all others. To express this situation somewhat generally, if the element x is embedded functionally in several branches G_1, $G_2 \ldots H_1, H_2 \ldots$

$$S_x{}^T = S_{G_1}{}^T S_x{}^{G_1} + S_{G_2}{}^T S_x{}^{G_2} + \cdots + S_{H_1}{}^T S_x{}^{H_1} + \cdots \qquad (29\text{-}83)$$

A somewhat different situation is encountered in the flow graph of Fig. 29-45 representing a position-control system. Here it is specified that

$$G_2 = \frac{K_2}{s(s + c)}\frac{1 + \zeta\omega_n s}{s^2 + 2\zeta\omega_n s + \omega_n{}^2}$$

and we wish to investigate the sensitivity of C/R to variations in the damping constant ζ. We recognize the possibility of expressing the desired sensitivity in the form

$$S_\zeta{}^T = S_{G_2}{}^T S_\zeta{}^{G_2} \qquad (29\text{-}84)$$

but the evaluation of $S_\zeta^{G_2}$ appears to be somewhat messy. In accordance with our earlier handling of a rational function, we may write

$$S_x^{F(s)} = \frac{\partial \ln F(s)}{\partial \ln x} = \frac{\partial \ln N(s)}{\partial \ln x} - \frac{\partial \ln D(s)}{\partial \ln x}$$
$$= S_x^{N(s)} - S_x^{D(s)}$$

where $N(s)$ and $D(s)$ are, respectively, the numerator and denominator of $F(s)$. This makes the evaluation one that can be written by inspection:

$$S_\zeta^{G_2} = \left[\frac{\zeta}{1 + \zeta\omega_n s}(s\omega_n) - \frac{\zeta(2\omega_n s)}{s^2 + 2\zeta\omega_n s + \omega_n^2}\right]$$

The sensitivity $S_{G_2}^T$ is most easily obtained using the topological formulation of Eq. (29-68):

$$S_{G_2}^T = \frac{\Delta_0}{\Delta}$$
$$= \frac{1 + G_1 H_1}{1 + G_1 H_1 + G_2 H_2 + K G_1 G_2}$$

Finally, substituting in (29-84) we obtain

$$S_\zeta^T = \frac{1 + G_1 H_1}{1 + G_1 H_1 + G_2 H_2 + K G_1 G_2}\left(\frac{\zeta\omega_n s}{1 + \zeta\omega_n s} - \frac{2\zeta\omega_n s}{s^2 + 2\zeta\omega_n s + \omega_n^2}\right) \quad (29\text{-}85)$$

Concluding Remarks. In our discussion of sensitivity we have tried to underscore the importance of the concept of sensitivity as a basis for making a quantitative comparison of competitive feedback-system designs. We have then presented a repertoire of computational techniques so that the concept need never be lost in the complexities of its evaluation. Finally, although we have shown this only by implication, the importance of the various viewpoints of sensitivity to the feedback engineer is that from them he gains the required insight to enable him to have a sound scientific basis for feedback-system design. By examining the separate constituents of the sensitivity functions he has a basis for the placing of loops and paths that will appropriately control the sensitivities he wishes to reduce.

29-8. The Significance of Sensitivity

The sensitivity functions for a linear feedback system or active network provide the fundamental evaluation of the effectiveness of the feedback. Specifically, when we measure the sensitivity of an overall transmission T to a parameter x, we determine the dependence of T on x in very general terms. In terms of the formal definition, S_x^T measures only the percentage change in T resulting from a given (small) percentage change in x; actually, however, we are measuring indirectly the relation between T and x, the allowable tolerances in design and manufacture, the effects of small nonlinearities in the element x, and the effects of noise introduced at the point in the system where x appears.

In order to utilize the sensitivity concept in system evaluation and design, we need to develop correlations between the numerical value of the sensitivity and the more familiar criteria of system performance. For example, if the S_x^T is evaluated as -0.4, what is the significance in terms of system characteristics? From the definition of sensitivity, we can state at once that a 1 per cent increase in x results in a 0.4 per cent decrease in T. As we indicate below, however, there are several other important deductions possible from this value of -0.4.

The problem of interpreting S_x^T is emphasized if we recognize that, in almost all cases of significance, the sensitivity is a function of the complex frequency s. For example, the simple system of Fig. 29-46 is characterized by

$$S_K{}^T = \frac{s(s+1)}{s^2 + s + K} \tag{29-86}$$

At zero frequency, the sensitivity is zero (the steady-state value of the step-function response is independent of K as long as K is greater than zero); when $s = -1$, the sensitivity is likewise zero. At any other value of $s = j\omega$, S involves both a magnitude and a phase angle, or both real and imaginary parts. Our basic question is: what is the significance of the complex-frequency dependence of S?

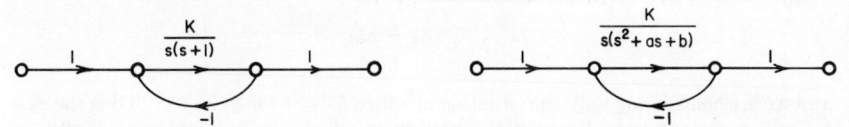

FIG. 29-46. Simple feedback system.   FIG. 29-47. Third-order feedback system.

"*Parameter Margins.*" In system design, x is allowed to assume its normal value x_n. One or more parameters are then adjusted to yield a feedback system with satisfactory overall characteristics. After the design is completed, we determine $S_x{}^T(s)$. In this section, we demonstrate how $S_x{}^T$ can be used to determine simply the minimum change in x which may result in system instability.

The above concept is illustrated by the simple system of Fig. 29-47. If the normal values of a and b are each 2, the system design involves the choice of K to yield satisfactory overall system performance. By use of root-locus techniques or Bode diagrams, we might select the value of K as 2, with result

$$T = \frac{2}{s^3 + 2s^2 + 2s + 2}$$
$$\phantom{T = \frac{2}{s^3 + }} \uparrow \uparrow \uparrow$$
$$\phantom{T = \frac{2}{s^3 + }} a b K$$

With the system design completed, evaluation of the effectiveness of the feedback must include consideration of the effects of variation in a and b (perhaps due to manufacturing tolerances, aging, environmental changes, etc.).

In this trivial example, such an investigation can be initiated very simply. The Routh test, for example, indicates that the system becomes unstable if a is reduced from the normal value of 2 to a value of unity (the corresponding frequency of oscillation is $\sqrt{2}$ radian/sec). In order to describe this change quantitatively, we can state that the "parameter margin" for a is -6 db*—i.e., a 6-db decrease in a causes instability. The use of the term "parameter margin" here is exactly analogous to the familiar term gain margin: in our example, an increase of K from the value 2 to 4 (i.e., a 6-db increase) causes instability; hence we say the system possesses a *gain margin* of 6 db.

Thus *the parameter margin for x is the required change in x to cause system instability.* Conventionally, we measure the parameter margin in decibels.

In the trivial example of Fig. 29-47, the parameter margins for both a and b are easily calculated. In more significant feedback systems, however, the Routh test is of little help, and evaluation demands extensive plotting of Bode diagrams or root loci as well as the factoring of high-degree polynomials (or tedious calculation). The sensitivity-function concept provides a simple, convenient basis for the evaluation of the parameter margins. In order to develop this approach, we consider first the special case with the null return difference equal to unity.

* In this chapter, the value in decibels of a ratio a/b is given by

$$\left(\frac{a}{b}\right)_{db} = 20 \log_{10}\left(\frac{a}{b}\right)$$

Relation between S_x^T and Parameter Margin for x if $F_x' = 1$. If the null return difference is unity, the overall transmission and sensitivity are given by expressions of the form

$$T = \frac{\Sigma_0}{\Delta_0 + x\Delta_1} \qquad S_x^T = \frac{\Delta_0}{\Delta_0 + x\Delta_1} - 1 \qquad (29\text{-}87)$$

where the determinant Δ of the flow graph is written as a linear function of x:

$$\Delta = \Delta_0 + x\Delta_1 \qquad (29\text{-}88)$$

The numerator of T is simply Σ_0 (rather than the general $\Sigma_0 + x\Sigma_1$) since

$$F' = \frac{\Sigma_0 + x\Sigma_1}{\Sigma_0}$$

and we are considering only the situation in which $F' = 1$ (or $\Sigma_1 = 0$). Thus the case $F_x' = 1$ corresponds to the special situation in which the numerator of T is independent of the parameter.

As x is varied, the transmission T of (29-87) changes, with the poles described by the relation

$$\Delta_0 + x\Delta_1 = 0 \quad \text{or} \quad x = -\frac{\Delta_0}{\Delta_1} \qquad (29\text{-}89)$$

Since the poles of T are continuous functions of the parameter x, the transition from stability to instability occurs when (29-89) yields poles on the $j\omega$ axis at ω_1, or when

$$x_1 = -\frac{\Delta_0(j\omega_1)}{\Delta_1(j\omega_1)}$$

where x_1 is the value of x resulting in oscillation at the frequency ω_1. Thus instability occurs when

$$\Delta_1 = -\frac{\Delta_0}{x_1}$$

or, if we substitute in the S_x^T expression of (29-87), when

$$S_x^T = \frac{x/x_1}{1 - x/x_1} = \frac{x}{x_1 - x} = \frac{1}{(x_1 - x)/x} \qquad (29\text{-}90)$$

The denominator of the last term of (29-90) is simply the fractional change in x required to cause instability.

From the above discussion, we can state the following theorem:

If $F_x' = 1$, the necessary and sufficient condition for a change in x to cause instability at a frequency ω_1 radians/sec is that $S_x^T(j\omega_1)$ should be real. The required fractional change in x is

$$\frac{x_1 - x_n}{x_n} = \frac{1}{S_x^T(j\omega_1)} \qquad (29\text{-}91)$$

where x_n is the normal design value of x and x_1 is the value which results in instability.

The significance of the theorem can be illustrated by the simple example depicted in Fig. 29-48, where a_n, the normal value of a, is 2. In this case, the sensitivity of T to a is calculated directly by the techniques of the last section:

$$S_a^T = S_G^T S_a^G$$
$$S_a^T = -\frac{as^2}{2}\frac{G}{1+G} \qquad (29\text{-}92)$$

Thus the condition for oscillation in this case is that $G/(1+G)$ should be real (since s^2 is real when $s = j\omega$).

FEEDBACK THEORY

In the specific system of Fig. 29-48,

$$\frac{G}{1+G} = \frac{2}{s^3 + as^2 + 2s + 2}$$

If this function is to be real, the denominator polynomial must be real:

$$s^3 + 2s = 0 \qquad \begin{aligned} \omega_1 &= 0 \\ \omega_2 &= \sqrt{2} \end{aligned}$$

At these two frequencies at which S_a^T is real, the sensitivity assumes the values

$$\omega_1 = 0 \qquad S_a^T(j\omega_1) = 0$$
$$\omega_2 = \sqrt{2} \qquad S_a^T(j\omega_2) = \frac{a}{1-a} = -2$$

The fractional changes in a required to cause instability are the reciprocals of the sensitivity values, or

$$\omega_1 = 0 \qquad \frac{a_1 - 2}{2} = \infty \qquad a_1 = \infty$$
$$\omega_2 = \sqrt{2} \qquad \frac{a_2 - 2}{2} = -\tfrac{1}{2}(-6 \text{ db}) \qquad a_2 = 1$$

Thus the only solution of interest (that for ω_2 and a_2) indicates instability at a frequency $\sqrt{2}$ radian/sec results if a is changed by -6 db from the normal value of 2. The above result is, of course, no different from that obtained previously from the

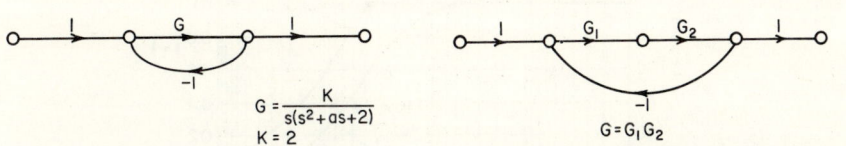

FIG. 29-48. Example for theorem.

FIG. 29-49. Complex system.

Routh test; indeed, in this trivial example, the Routh test is considerably simpler than the analysis via the sensitivity function. The important feature of the method, however, lies in the form of Eq. (29-92). The parameter margin for a is determined entirely by the value of S_a^T when this function is real.

The parameter margin for a in the much more complicated (and more interesting) system of Fig. 29-49, in which the $G_1(s)$ block has been added to the earlier system of Fig. 29-48, is determined from an equation of almost the same form. In this case

$$S_a^T = -\frac{as^2}{2} \frac{G_2}{1+G} \tag{29-93}$$

For this fifth-order system, the Routh test immediately embroils the engineer in depressingly complicated calculations culminating in the determination of the real zeros of a quartic polynomial. Once the gain and phase plots of $G_2/(1+G)$ are available [actually we need the plots only in the vicinity of the frequencies at which $G_2/(1+G)$ is real], determination of the parameter margin for a is trivial. Furthermore, the plots of $1/(1+G)$ and G_2 are ordinarily available from the frequency-domain design of the system, so that the plots for $G_2/(1+G)$ can be drawn without a significant amount of extra work.

Thus, if the system is initially designed on the basis of frequency analysis, the determination of the parameter margins is simple and straightforward if we exploit the sensitivity concept.

A Nontrivial Example. As indicated in the preceding paragraphs, the method described above is primarily useful in the study of reasonably complex systems. Once we plot the pertinent transfer functions for determination of the sensitivity (plots which ordinarily are developed in the course of normal system design), the evaluation of the various parameter margins is straightforward.

In order to illustrate the potency of the method, we consider the system in Fig. 29-50.* Conventional frequency-domain design in terms of the Bode plots and the closed-loop characteristics determined from the Nichols chart results in the selection of the parameter values shown in Fig. 29-50. The resulting closed-loop gain and phase characteristics for $T(s) = (C/R)(s)$ are shown in Fig. 29-51.

In order to evaluate the final system which has been designed, we next need to determine the effects of nonlinearity, of environmental changes, and of aging and manufacturing tolerances. From the expected variations and our understanding of the physical principles underlying the operation of the various elements of the system, we might decide that primary interest lies in the effects resulting from variation of the three parameters τ_1, τ_2, and K. A simple measure of the dependence of system performance on each parameter is the parameter margin or the fractional parameter change required to cause instability.

Normal Values
$\tau_1 = \frac{1}{4}$
$\tau_2 = \frac{1}{16}$
$K = \frac{1}{6}$

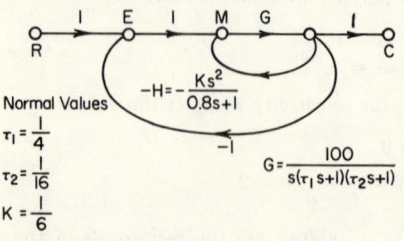

$-H = -\frac{Ks^2}{0.8s+1}$

$G = \frac{100}{s(\tau_1 s+1)(\tau_2 s+1)}$

FIG. 29-50. Control system.

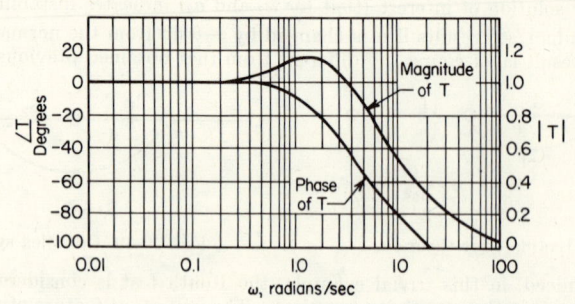

FIG. 29-51. Closed-loop gain and phase for system of Fig. 29-50.

The first step in the evaluation of the three parameter margins is the determination of the corresponding three sensitivity functions. On the basis of the methods described above, we can write

$$S_{\tau_1}^T = S_{\tau_1}^G S_G^T$$
$$= -S_{\tau_1}^{\tau_1 s+1} S_G^T$$
$$= -\frac{\tau_1 s}{\tau_1 s + 1} \frac{1}{1 + G + GH}$$
$$= -\frac{\tau_1 s^2 (\tau_2 s + 1)}{100} \frac{G}{1 + G + GH} = -\frac{\tau_1 s^2 (\tau_2 s + 1)}{100} T$$

At the normal operating values ($\tau_1 = \frac{1}{4}$, $\tau_2 = \frac{1}{16}$),

$$S_{\tau_1}^T = -\frac{s^2(s + 16)}{6{,}400} T$$

* The system and closed-loop characteristics are taken from the book by H. Chestnut and R. Mayer, *Servomechanism and Regulating System Design*, 2d ed., p. 373, John Wiley & Sons, Inc., New York, 1958.

Correspondingly,

$$S_{\tau_2}{}^T = -\frac{s^2(s+4)}{6{,}400}T$$

The sensitivity of T to K is

$$\begin{aligned}S_K{}^T &= S_K{}^H S_H{}^T \\ &= S_H{}^T \\ &= \frac{\Delta(H=0)}{\Delta} - 1\end{aligned} \qquad (29\text{-}94)$$

since the null return difference with reference to H is unity by inspection of the flow diagram of Fig. 29-50. Thus, with Δ simply $1 + G + GH$, we have

$$S_K{}^T = \frac{-GH}{1+G+GH} = -HT$$

or, at the normal value of K of $\frac{1}{6}$,

$$S_K{}^T = \frac{1}{4.8}\frac{s^2}{s+1.25}T$$

In order to determine the three parameter margins, we first note that each null return difference is unity. According to the theorem, then, we need only find the value of the sensitivity when S is real. Thus, in the three cases, we seek real values of

$$S_{\tau_1}{}^T = -\frac{s^2}{6{,}400}(s+16)T \qquad S_{\tau_2}{}^T = -\frac{s^2}{6{,}400}(s+4)T \qquad S_K{}^T = \frac{s^2}{4.8}\frac{1}{s+1.25}T$$
(29-95)

when $s = j\omega$. Rather than considering the entire functions, we can seek real values of

For τ_1: $(s+16)T$ \qquad For τ_2: $(s+4)T$ \qquad For K: $\dfrac{1}{s+1.25}T$

The simplest procedure is to sketch, directly on the plot of Fig. 29-51, the phase angles of $(s+16)T$, $(s+4)T$, and $T/(s+1.25)$ in the vicinity where these phase

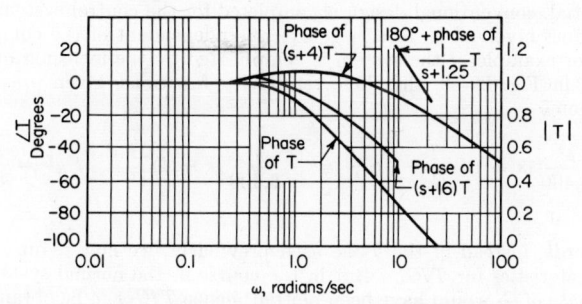

Fig. 29-52. Phase characteristics for calculation of parameter margin in system of Fig. 29-50.

angles are 0 or 180°. (We note that a high degree of accuracy is not necessary since the magnitude of T is not changing rapidly.) The three sketches, shown in Fig. 29-52, indicate that for each parameter there is only one frequency at which the relevant function is real. The frequencies and the corresponding values of $|T|$ are

For τ_1: \qquad $\omega_1 = 0.7$ radian/sec \qquad $|T(j\omega_1)| = 1.07$
For τ_2: \qquad $\omega_2 = 4.3$ radians/sec \qquad $|T(j\omega_2)| = 0.93$
For K: \qquad $\omega_3 = 18$ radians/sec \qquad $|T(j\omega_3)| = 0.33$

We are now ready to calculate the three values of sensitivity from the expressions of (29-95). Substitution of the above values of ω and $|T|$ gives

For τ_1:
$$S_{\tau_1}{}^T(j\omega_1) = \frac{0.5}{6{,}400} 16(1.07)$$
$$= \frac{1}{748}$$

For τ_2:
$$S_{\tau_2}{}^T(j\omega_2) = \frac{(4.3)^2|4 + j4.3|}{6{,}400} 0.93$$
$$= \frac{1}{63}$$
(29-96)

For K:
$$S_K{}^T(j\omega_3) = \frac{-18}{4.8} 0.33$$
$$= -\frac{1}{0.8}$$

Therefore, we can move the system to the border line of instability with any one of the following fractional changes in a parameter:

$$\frac{\tau_{1\text{osc}} - \tau_1}{\tau_1} = 748 \qquad \text{increase of 57.5 db}$$

$$\frac{\tau_{2\text{osc}} - \tau_2}{\tau_2} = 63 \qquad \text{increase of 36 db} \qquad (29\text{-}97)$$

$$\frac{K_{\text{osc}} - K}{K} = -0.8 \qquad \text{decrease of 14 db}$$

If we wish, we may solve each of Eqs. (29-97) for the critical values of the three parameters—i.e., for the values, any one of which causes instability if the other two parameters assume normal values:

$$\tau_{1\text{osc}} = 187 \qquad \tau_{2\text{osc}} = 4 \qquad K_{\text{osc}} = \tfrac{1}{30} \qquad (29\text{-}98)$$

Several aspects of the above example are noteworthy in conclusion:

1. The results, Eq. (29-97) or (29-98), are obtained with surprisingly little work once the initial, conventional design is completed for the control system.

2. The effort involved is, to a large extent, independent of the complexity of the system. For example, if the system is complicated by the inclusion of the G_1 block as depicted in Fig. 29-53, the three sensitivity functions, given previously in Eq. (29-95), become

$$S_{\tau_1}{}^T = -\frac{s^2}{6{,}400}(s + 16)\frac{T}{G_1} \qquad S_{\tau_2}{}^T = -\frac{s^2}{6{,}400}(s + 4)\frac{T}{G_1} \qquad S_K{}^T = \frac{s^2}{4.8}\frac{1}{s + 1.25}\frac{T}{G_1}$$
(29-99)

In other words, instead of the T we used previously, we now require the gain and phase characteristics for T/G_1. But in the course of the normal system design, the gain and phase of G_1 would have been plotted; hence T/G_1 can be obtained simply by combining the T and the G_1 characteristics. Once T/G_1 is evaluated, the entire analysis for the determination of the parameter margins proceeds exactly as in the above example. The complexity of G_1 is irrelevant!

3. The accuracy required in the evaluation of the frequency at which the sensitivity is real depends on the value of S and on the behavior of T in the neighborhood of the frequency in question. For example, in the above example with the three sensitivities

$$S_{\tau_1}{}^T = \frac{-s^2(s + 16)}{6{,}400}T \qquad S_{\tau_2}{}^T = \frac{-s^2(s + 4)}{6{,}400}T \qquad S_K{}^T = \frac{s^2}{4.8(s + 1.25)}T \qquad (29\text{-}100)$$

the first two sensitivities are very small (i.e., very large fractional changes in τ_1 and τ_2 are required to cause instability). Under these circumstances, we would ordinarily be relatively unconcerned about accuracy. Furthermore, S is real in the frequency range in which T is changing rather slowly, so that the principal effect of inaccurate frequency determination is evidenced in error in the s factors $s^2(s+4)$. In the case of the K variation, however, the sensitivity is close to unity; in this case, reasonable accuracy is required. For example, if $S_K{}^T$ is measured as $-1/0.9$ rather than $-1/0.8$, the K for instability is $\frac{1}{60}$, or half of the value actually obtained. This demand for accuracy is largely offset, however, by the fact that the large sensitivity magnitude is associated with a frequency range well outside the bandwidth of the system; hence the phase of T is changing rapidly in the frequency range of interest, and an accurate determination of the frequency at which $S_K{}^T$ is real is possible. Thus, when the feedback is effective in reducing sensitivity, poor accuracy is permissible; when the feedback is ineffective and accuracy is required, we can evaluate with reasonable precision.

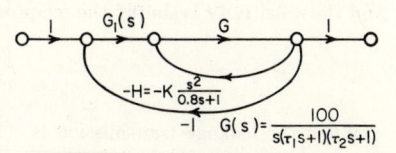

FIG. 29-53. More complex system.

4. Finally, we should emphasize once more the condition originally imposed; the null return difference must be unity if the theorem is to be applicable (we next consider the case when $F_x{}'$ is not unity). Unfortunately, there does not seem to be any simple technique for determination from the signal-flow graph of the conditions under which the null return difference is unity. From the basic definition, we know that $F_x{}'$ is unity if the input of the x branch is proportional to the system output when the flow diagram is drawn in such a way that x appears only once and as the multiplier of a single branch.*

More generally, from the formula of Sec. 29-7, we know that

$$F_x{}' = \frac{\Sigma}{\Sigma_0} = \frac{\Sigma P_i \Delta_i}{\Sigma P_i \Delta_i |_{x=0}} \qquad (29\text{-}101)$$

Hence $F_x{}'$ is unity when x does not appear in any direct path (any P_i above) or in any loop which does not touch all the direct paths.

Relation between $S_x{}^T$ and Parameter Margin if $F_x{}' = \infty$. The entire discussion above is restricted to the special case when $F_x{}'$ equals unity—in other words, when

$$\Sigma = \Sigma_0$$

where Σ represents $\Sigma P_i \Delta_i$ in Mason's formula and Σ_0 is the value of Σ when $x = 0$. The general form of the overall transmission T for a feedback system is

$$T = \frac{\Sigma_0 + x\Sigma_1}{\Delta_0 + x\Delta_1} = \frac{\Sigma}{\Delta} \qquad (29\text{-}102)$$

where Σ_0, Σ_1, Δ_0, and Δ_1 are independent of x.† The special case we have been considering ($F_x{}' = 1$) corresponds to $\Sigma_1 = 0$, or

$$T = \frac{\Sigma_0}{\Delta_0 + x\Delta_1}$$

---

* The null return difference $F_x{}'$ is unity minus the returned signal when the x branch is broken at the input, a unit signal is transmitted through x, and the system input is adjusted to yield zero output. The "returned signal" is the signal at the receiving side of the break.

† Equation (29-102) is valid only when T is a bilinear function of x, but this constraint does not seriously limit the applicability of the entire discussion here.

An equally important special case occurs when $\Sigma_0 = 0$, or

$$T = \frac{x\Sigma_1}{\Delta_0 + x\Delta_1}$$

Under these circumstances, there is no leakage transmission: when x is set equal to zero, the overall transmission becomes zero. The null return difference F_x' is infinite, and the sensitivity is simply the reciprocal of the return difference with reference to x:

$$S_x^T = \frac{1}{F_x}$$

When the leakage transmission is zero, the sensitivity is again directly related to the corresponding parameter margin. The theorem is derived in exactly the same manner as in the previous case. Instability again depends upon the change in x moving the zeros of the denominator of T

$$T = \frac{x\Sigma_1}{\Delta_0 + x\Delta_1}$$

onto the imaginary axis. For this value of x and the particular frequency ω_1

$$\Delta_0(j\omega_1) + x\Delta_1(j\omega_1) = 0 \tag{29-103}$$

The corresponding value of sensitivity is, since the leakage is zero,

$$S_x^T = \frac{\Delta_0(j\omega_1)}{\Delta_0(j\omega_1) + x_n\Delta_1(j\omega_1)} \tag{29-104}$$

where x_n is the normal value of x. Substitution of (29-103) in (29-104) gives

$$S_x^T = \frac{-x\Delta_1(j\omega_1)}{-x\Delta_1(j\omega_1) + x_n\Delta_1(j\omega_1)} \tag{29-105}$$

If $\Delta_1(j\omega_1) \neq 0$, the sensitivity is real and has the form

$$S_x^T = \frac{x}{x - x_n} \tag{29-106}$$

where x is the value which causes instability. From Eqs. (29-103) and (29-106), we can state the following theorem:

If $F_x' = \infty$, the necessary and sufficient condition for a change in x to cause instability at a frequency ω_1 radians/sec is that $S_x^T(j\omega_1)$ should be real, and the required fractional change in x is

$$\frac{x_1 - x_n}{x_n} = \frac{1}{S_x^T(j\omega_1) - 1} \tag{29-107}$$

where x_n is the normal design value of x and x_1 is the value which results in instability.

Thus the only difference between this case ($F_x' = \infty$) and the previous special case ($F_x' = 1$) resides in the form of the equation for the fractional change in x. In the $F_x' = 1$ case, the fractional change is $1/S_x^T(j\omega_1)$, in contrast to Eq. (29-107). Thus, from a design viewpoint, when we wish a large required fractional change in x, we seek near-zero values of S when $F_x' = 1$, but near-unity values of S when $F_x' = \infty$.

The analysis when $F_x' = \infty$ follows exactly the steps described in the examples illustrating the theorem when $F_x' = 1$.

Relation between S_x^T and Parameter Margin in General. We have now developed two theorems covering the special cases of unity and infinite null return

FEEDBACK THEORY

differences. In the general situation, F_x' assumes neither value. In general, however, T can be written as

$$T = \frac{\Sigma_0 + x\Sigma_1}{\Delta_0 + x\Delta_1} = \frac{\Sigma_0}{\Delta_0 + x\Delta_1} + \frac{x\Sigma_1}{\Delta_0 + x\Delta_1} \qquad (29\text{-}108)$$

In other words, T can always be considered as the sum of two transmissions: T_1 in which $F_x' = 1$; and T_2 in which $F_x' = \infty$. Instability which occurs because, as x changes, $\Delta_0 + x\Delta_1$ possesses a zero on the $j\omega$ axis, results in both T_1 and T_2 exhibiting this oscillatory behavior. Consequently, we can investigate either T_1 or T_2 in order to determine the parameter margin.

Alternatively, if we recognize that it is only the character of $\Delta_0 + x\Delta_1 = \Delta$ which we seek to study, we can study any system which possesses the same determinant as the original system. The technique is illustrated by the example depicted in Fig. 29-54, in which the parameter of interest is a. In order to simplify the determination of an appropriate system for evaluation of the parameter margin, we redraw the flow diagram as shown in Fig. 29-55.

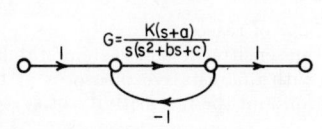

Fig. 29-54. Simple system with $F_x' \neq 1$, $F_x' \neq \infty$.

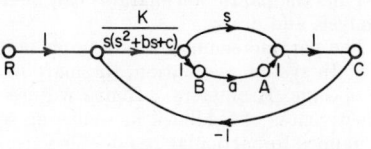

Fig. 29-55. Modified form of Fig. 29-54 with a appearing alone as multiplier of one branch.

If we now consider point A as the output of the system, we are dealing with the special case of no leakage ($F_a' = \infty$), so that the second theorem above is directly applicable. Alternatively, we can consider B as the output, with the result that $F_a' = 1$ and the Δ is again unchanged.

To complete this example, we choose the latter alternative (B as the output). The resulting sensitivity of the transmission B/R to changes in a is determined by the same determinant as the Δ characterizing the original system:

$$S_a^{B/R} = \frac{\Delta_0}{\Delta} - 1$$

$$= \frac{1 + \dfrac{Ks}{s(s^2 + bs + c)}}{1 + \dfrac{K(s + a)}{s(s^2 + bs + c)}} - 1$$

$$S_a^{B/R} = \frac{-aK}{s^3 + bs^2 + (c + K)s + aK} \qquad (29\text{-}109)$$

$$S_a^{B/R} = \frac{-a}{s + a} \frac{G}{1 + G} \qquad (29\text{-}110)$$

Equation (29-110) is directly useful in the general case in which $G/(1 + G)$ is already plotted; in this simple example, we can proceed analytically from (29-109).

The sensitivity is real at the frequency

$$-\omega^3 + (c + K)\omega = 0 \qquad \text{or} \qquad \omega_1 = \sqrt{c + K}$$

The corresponding value of sensitivity is

$$S_a^{B/R} = \frac{-aK}{aK - b(c + K)}$$

Hence the value of a causing instability (i.e., a_1) is given by

$$\frac{a_1 - a}{a} = \frac{aK - b(c + K)}{-aK}$$

or
$$a_1 = b\frac{c + K}{K}$$

Concluding Comments. In these paragraphs we have outlined a straightforward analytic and graphical technique for the determination of the parameter margins. The simplicity of the method derives from:
1. The ease with which the various sensitivity functions can be evaluated.
2. The fact that the various sensitivities are closely related to the functions which are plotted in normal closed-loop system design.
3. The simplicity of the two theorems—in particular the fact that we need determine the value of the sensitivity only when this S is real. Even in the most general cases we are required to evaluate S only in a small portion of the frequency band.

Thus the parameter margins can be determined in the course of the normal system analysis and design.

The parameter margins are important for a variety of reasons:
1. In system evaluation, the margins measure the sensitiveness of system stability to specific parameters. Hence we are provided with quantitative measures of the effectiveness of feedback in achieving self-calibration and the insensitivity of system dynamics to particular parameter variations.
2. The simplicity of the expressions for the parameter margins allows the guiding of feedback-system design toward the realization of specified minimum values for the margin of those parameters which vary significantly, because of either manufacturing tolerances or environmental changes.
3. The simplicity of the analysis also permits evaluation of the relative values of various parameter margins. In most cases, the different sensitivity functions possess common factors as well as the distinctive factors, so that the designer can see at once the possibility of controlling either simultaneously or independently the significant parameter margins. As a simple example, the difference in the two theorems pinpoints the difficulty of achieving simultaneously large parameter margins for two elements, one of which has $F_x' = \infty$, the other $F_x' = 1$. In this case, we must choose a configuration resulting in significantly different forms for the two sensitivity functions.
4. The theory presented here provides a possible approach to the identification problem. Experimental measurement of the sensitivity and the transmission, for example, permits evaluation of specific coefficients or parameters in the internal transfer function, even though we may not be free to vary a parameter to the point at which the system becomes oscillatory.
5. Finally, the theory is directly applicable to the identification and evaluation of multiple nonlinearities within the system. When the nonlinearity can be represented by a parameter variation with signal level (e.g., by a describing function), the theory presented above yields quantitative measures of the tendency of each nonlinearity to cause instability, the relative significance of the different nonlinearities, or the magnitude of the nonlinearity from the measured frequency and amplitude of oscillation.

29-9. Multiloop Systems

Throughout the preceding sections our principal interest has focused on feedback systems which are *single-loop*: there is only one feedback loop or closed loop around which signals may circulate. In Sec. 29-6 we considered multiloop systems briefly as we presented the general techniques for the calculation of transmission; in this section we wish to look further at the properties and characteristics of such multiloop systems.

As we noted in Sec. 29-3, the primary motivation for the use of feedback arises, if we exclude the special case of oscillators, from the possibility of controlling the sensitivity of the overall transmission to changes in various system parameters. Alternatively, we can consider this advantage of feedback as the possibility of controlling two different transmissions independently. The equivalence of the two viewpoints is illustrated by Fig. 29-56. Here we have a process (or known system); at the input of this process, there enters a corruption or noise signal U which affects the output; we desire an overall system in which a specified relation between C and R is realized, and at the same time the effect of U on the output C is controlled.

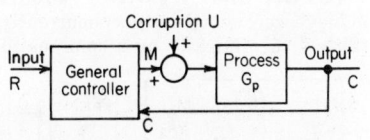

FIG. 29-56. System with two inputs.

Independent control of the two transmissions C/R and C/U is possible if we insert a general, linear controller as shown (the inputs to the controller are R and C, the output M). For this system, we can write the relation

$$C = G_p(M + U) \tag{29-111}$$

If the system is linear, M is some linear function of R and C; in other words,

$$M = AR - BC \tag{29-112}$$

where A and B are two transfer functions which depend upon the particular form of the controller. Substitution of (29-112) into (29-111) and solution for C yields

$$C = \frac{AG_p}{1 + BG_p} R + \frac{G_p}{1 + BG_p} U \tag{29-113}$$

Equation (29-113) demonstrates the equivalence of control over sensitivity and control over the effects of corrupting signals, since the feedback results in reduction of the effect of U by division of C/U by the factor $1 + BG_p$, and the sensitivity of C/R with respect to G_p is likewise

$$S_{G_p}^{C/R} = \frac{1}{1 + BG_p} \tag{29-114}$$

In both cases, the return difference $1 + BG_p$—i.e., unity minus the loop gain—measures the effectiveness of feedback.

Equation (29-113) indicates at once the possibility of independent control over the two transmissions C/R and C/U. (The transmission C/R is obtained with $U = 0$, while C/U is obtained with $R = 0$.) If we specify both functions, the controller function A is given simply by

$$A = \frac{C/R}{C/U} \tag{29-115}$$

and B can be determined from either C/R or C/U. For example,

$$B = \frac{1}{C/U} - \frac{1}{G_p} \tag{29-116}$$

The concept of such independent control is intriguing, but the design equations (29-115) and (29-116) are somewhat misleading in their simplicity. For example, if we wish to have our system reject completely the effects of U (or have zero sensitivity to changes in G_p), we must make C/U identically zero—a value which, with G_p given, can be achieved only if B is infinite. But, with B infinite, C/R is nonzero only if A likewise is infinite so that C/R reduces to

$$\frac{C}{R} \to \frac{A}{B}$$

The possibility of realizing such an ideal situation in actual practice demands a more careful investigation of the available techniques for realizing the controller of Fig. 29-56.

Positive and Negative Feedback. The realization of a suitable controller to achieve zero sensitivity concurrently with a specified transmission can be accomplished theoretically by a combination of at least two feedback paths with different signs on the loop gains, as depicted in one simple case in Fig. 29-57. Here the controller consists of the three separate transfer functions G_1, G_2, and G_3; the overall system contains two loops, with the loop gains $-G_2 G_p G_3$ and $+G_2 G_1 G_3$ (one with a positive sign, the other with a negative sign, or the former negative feedback and the latter positive feedback in the vernacular*).

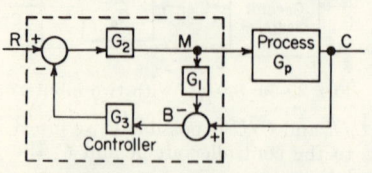

Fig. 29-57. A two-loop system.

The characteristics of the system are clear if we write the expressions for overall transmission and sensitivity with respect to G_p:

$$\frac{C}{R} = \frac{G_2 G_p}{1 + G_2 G_3 (G_p - G_1)} \tag{29-117}$$

$$S_{G_p}^{C/R} = \frac{1}{1 + [G_2 G_3 G_p/(1 - G_1 G_2 G_3)]} = \frac{1 - G_1 G_2 G_3}{1 + G_2 G_3 (G_p - G_1)} \tag{29-118}$$

Thus, if we make

$$G_1 G_2 G_3 = 1$$

we can realize a sensitivity of zero with an overall transmission

$$\frac{C}{R} = \frac{1}{G_3}$$

In other words, this configuration permits the realization of *zero sensitivity* simultaneously with the achievement of any desired overall transmission, subject only to the requirement that G_3, G_2, and G_1 be realizable transfer functions.

The situation is even more exciting than we have suggested so far. If we consider the denominator of C/R (or of S), which determines the stability problem, we find that the choice of G_1 as

$$G_1 = G_p$$

results in a remarkable simplification: the denominator is now just unity; so there is no possibility of poles lying in the right half plane. In other words, stability is assured if G_2 and G_p are themselves stable functions. Hence, if G_3, G_1, and G_2 can be chosen to satisfy the relations

$$G_3 = \frac{1}{C/R} \qquad G_1 = G_p \qquad G_2 = \frac{1}{G_1 G_3}$$

we have designed a system which simultaneously has

An arbitrary overall transmission
Zero sensitivity to changes in the process
No stability problem

---

* The electronics circuit designer often refers to such systems as having combined positive and negative feedback, with positive feedback referring to a loop gain which has a positive sign multiplying various transfer functions. Since the phase shift associated with the usual transfer function varies with frequency (and may pass through 180°), it is difficult to make a general definition of positive and negative feedback. In most cases, however, the implication of the terms is clear.

Before we consider the pitfalls associated with this magical configuration of Fig. 29-57, which simultaneously satisfies all conditions one might want in control-system design, let us consider what we have done in terms of the qualitative operation of the system. By selecting $G_1 = G_p$, we have made the signal B zero under normal operation (since B is $G_p M - G_1 M$); thus there is *no feedback at all* unless G_p varies from the value for which G_1 was designed. In other words, with the normal value of G_p, the system is open-loop and not a feedback structure. With regard to variations of G_p, however, the feedback does exist; indeed we have adjusted $G_1 G_2 G_3$ to unity, so that the feedback around G_p is infinite (the loop $G_1 G_2 G_3$ is on the verge of instability); by this technique we have achieved the zero sensitivity. Insofar as overall transmission and sensitivity are concerned, the system we have designed is equivalent to that of Fig. 29-58; we have merely discovered a "simple" way to realize the infinite gain through the medium of positive feedback.

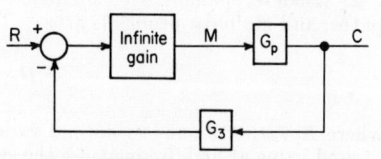

Fig. 29-58. System equivalent to our two-loop configuration.

When we turn to the actual problem of constructing the system of Fig. 29-57, however, we find that this "simple" solution contains a few difficulties. In order to understand the practical limitations, we should consider typical transfer functions for the various blocks. For example, let us assume the process is second-order, with

$$G_p = \frac{2}{s(s+5)}$$

If we simply closed a single loop around such a process and adjusted the gain for a closed-loop system with a relative damping ratio of 0.5, we could obtain an undamped natural resonant frequency of 5; hence let us select, as the desired overall transfer function, the same

$$\frac{C}{R} = \frac{25}{s^2 + 5s + 25}$$

What are the required three transfer functions G_1, G_2, and G_3?

G_1 must be a model of the process, or

$$G_1 = \frac{2}{s(s+5)} \tag{29-119}$$

the feedback transfer function G_3 is to be the reciprocal of C/R:

$$G_3 = \frac{1}{25} s^2 + \frac{1}{5} s + 1 \tag{29-120}$$

and G_2 is determined from the requirement that $G_1 G_2 G_3$ be unity, or

$$G_2 = \frac{25}{2} \frac{s(s+5)}{s^2 + 5s + 25} \tag{29-121}$$

Two difficulties arise in the practical implementation of our system:

1. The requirement that G_3 include a second-derivative term causes trouble if the signal C contains any high-frequency noise (since differentiation tends to accentuate high-frequency components). A practical compromise ordinarily involves realizing, not the G_3 of (29-120) but a function which is approximately equal to this G_3 over the frequency band of interest, e.g.,

$$G_{31} = \frac{\frac{1}{25}s^2 + \frac{1}{5}s + 1}{\left(\frac{1}{25}s + 1\right)\left(\frac{1}{40}s + 1\right)}$$

Even this presents problems since at high frequencies the transfer function is 40, meaning that an amplifier gain of 40 is required; if the amplifier gain falls off as s increases, G_3 may have to include an additional denominator factor. In other words, we encounter very major, practical design difficulties in constructing a minor loop which possesses a loop gain sufficiently close to unity over the band of frequencies of interest.

2. When G_p changes, even slightly, we no longer realize the cancellation of feedback paths, and stability problems arise. For example, if

$$G_p = \frac{K}{s(s+5)}$$

where K varies about the normal value of 2, and if G_3 is not precisely equal to the desired value at high frequencies, the system may become unstable or at least undesirably underdamped. This difficulty is particularly poignant when we realize that in practice we never know G_p with great precision; indeed a primary purpose of using feedback is to permit satisfactory system design with G_p only crudely known. Particularly troublesome are the nonlinearities which may be present in the process, especially if we are not aware of the specific nature of the nonlinearities at the outset of the design.

In view of these difficulties, the usefulness of our two-loop configuration of Fig. 29-57 depends markedly upon the specific transfer functions which are used; in the design of complex systems, we frequently require extensive analog-computer studies before we can evaluate the complete performance characteristics of the system. Our purpose here, however, is not to present a design theory but rather to illustrate the flexibility achievable with multiloop-feedback systems.

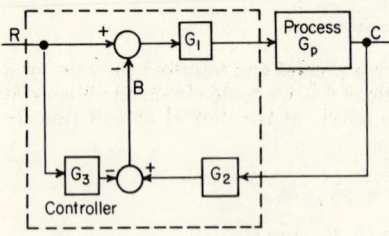

Fig. 29-59. Ham-Lang system.

Feed-forward Systems. A somewhat different type of feedback configuration is illustrated in Fig. 29-59; although this system is single-loop, it possesses many of the properties associated with our two-loop system of Fig. 29-57 and hence is conveniently discussed briefly at this point. The system differs from a conventional single-loop system as a result of the feed-forward path from R through G_3; the signal R affects B not only through the $G_1 G_p G_2$ path but also via $-G_3$.*

The characteristics of the system are apparent from the expressions for the overall transmission and the sensitivity of C/R with respect to G_p:

$$\frac{C}{R} = \frac{(1+G_3)G_1 G_p}{1+G_1 G_p G_2} \qquad (29\text{-}122)$$

$$S_{G_p}{}^{C/R} = \frac{1}{1+G_1 G_p G_2} \qquad (29\text{-}123)$$

If we select G_3 to satisfy the relation

$$G_3 = G_1 G_p G_2$$

The overall transmission becomes

$$\frac{C}{R} = G_1 G_p$$

and again we essentially possess an open-loop system. The two signals entering the lower comparator of Fig. 29-59 are equal and cancel; therefore, $B = 0$, and there is

---

* The system is termed "feed-forward" because R enters the main loop not only at the comparator but also forward in the system after $G_1 G_p G_2$. We might equivalently consider the system in which R was passed through a block G_4 and then added to the output of G_1.

FEEDBACK THEORY 29-47

no feedback, no chance for instability. The sensitivity, however, is exactly that of a conventional single-loop system (with the G_3 path omitted). Thus we obtain the advantages of feedback without the associated difficulties of stabilization (again, of course, only as long as we can satisfy the cancellation condition implied by the above formula for G_3).

Figure 29-59 is merely one of several equivalent configurations in which the feed-forward technique is utilized to simplify the problem of designing a feedback configuration with, simultaneously, a high degree of relative stability and a low sensitivity. Figure 29-60 shows two equivalent configurations; additional schemes can be derived from the basic idea that, having chosen a configuration to realize the sensitivity specifications, we utilize an insertion of a function of the input signal into the loop in such a way as to simplify the stability problem.

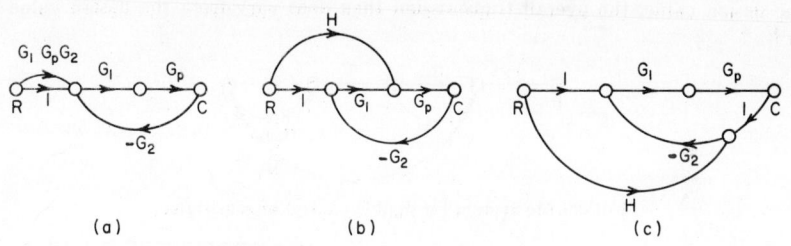

Fig. 29-60. Simple feed-forward schemes:
(a) Ham-Lang system of Fig. 29-59.

$$S_{G_p}{}^{C/R} = \frac{1}{1 + G_1 G_p G_2} \qquad \frac{C}{R} = G_1 G_p$$

(b) Feed-forward to process input.

$$S_{G_p}{}^{C/R} = \frac{1}{1 + G_1 G_p G_2} \qquad \frac{C}{R} = \frac{(H + G_1) G_p}{1 + G_1 G_2 G_p}$$

If $H = G_1{}^2 G_2 G_p$, same as (a)

If $H = \dfrac{1}{G_2 G_p}$, $\dfrac{C}{R} = \dfrac{1}{G_2}$

(c) Feed-forward into feedback path.

$$S_{G_p}{}^{C/R} = \frac{1}{1 + G_1 G_p G_2} \qquad \frac{C}{R} = \frac{G_1 G_p (1 + H G_2)}{1 + G_1 G_p G_2}$$

If $H = G_1 G_p$, same as (a).

The above examples illustrate the general technique of using parallel transmission paths or multiloop structures in order to achieve in design at least some degree of independence in the control of the two functions: the overall transmission and the sensitivity (or a subsidiary transmission). In more complicated situations (in which there may be many loops inherent in the structure before the feedback engineer initiates his phase of the design, or in which there may be many inputs and outputs), the determination of the characteristics of the feedback system or the design selection of various loops and transfer functions often taxes the ingenuity and theoretical background of the engineer; such problems constitute the core of linear feedback theory.

In this chapter, however, our interest is merely in indicating the scope of the capabilities of feedback systems, and we leave to advanced texts the details of the general theory. Thus we conclude this section with an extremely brief description of three multiloop systems which are of considerable practical importance and which are related to the two systems we have discussed in more detail above.

Example of Double Control of Sensitivity. The two-loop zero-sensitivity system of Fig. 29-57 can be extended when two active elements G_1 and G_2 are present, as in Fig. 29-61. In this configuration, if H_1 and H_2 are selected to satisfy the criteria

$$H_1 = \frac{1}{G_1}$$
$$H_2 = \frac{1}{G_2}$$
(29-124)

over the frequency band of interest, the overall transmission and sensitivities are

$$\frac{C}{R} = \frac{1}{H_3} \qquad S_{G_1}{}^{C/R} = 0 \qquad S_{G_2}{}^{C/R} = 0 \qquad (29\text{-}125)$$

If either G_1 or G_2 changes, the overall transmission is not affected; if both change from the design value, the overall transmission then does vary from the design value of $1/H_3$.*

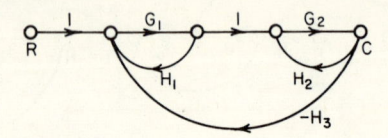

FIG. 29-61. System for double control of sensitivity.

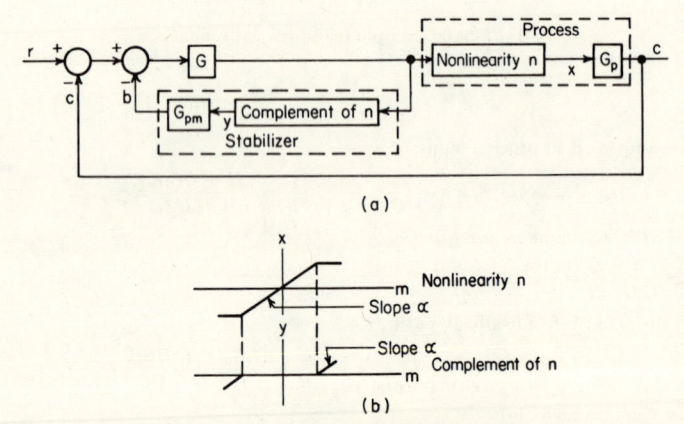

FIG. 29-62. Nonlinear feedback system.

Example of Stabilization of Nonlinear Oscillation. The feedback control system of Fig. 29-62 (omitting the stabilization section) includes a nonlinear process consisting of saturation followed by a linear section. When the controller G is designed to give appropriate small-signal performance and the loop is closed, the system is found to be unstable for large signals (when the nonlinearity affects the operation). Thus the instability can be directly associated with the presence of the nonlinearity.

Under such circumstances, the system is stabilized by the insertion of the nonlinear stabilizer network consisting of two parts:

1. A nonlinear element which *complements* the process nonlinearity such that the parallel combination of n and its complement represents a linear component.

* This configuration, when utilized for a transistor amplifier, for example, usually seems to result in a conditionally stable system: as one transistor gain falls, the overall system eventually becomes unstable for slight decreases in the gain of the other transistor. Such a characteristic results in the necessity for considerable care in the choice of actual transfer functions in order to realize satisfactory operation over wide operating ranges.

FEEDBACK THEORY   29-49

2. A linear block with a transfer function G_{pm} which equals, to the best of our knowledge, the linear G_p. If G_{pm} accurately models G_p and the two nonlinearities are complementary, the total feedback signal $b + c$ is exactly that which would be present in the *linear* feedback system depicted in Fig. 29-63. Thus, even though the nonlinearity n affects the output c, n has no effect on the stability of the closed-loop system. The stabilizer loop removes the instability resulting from the nonlinearity.

The basic nonlinear-compensation technique illustrated in Fig. 29-62 can be extended to the stabilization of oscillating systems involving a wide variety of nonlinearities (backlash, hysteresis, static friction, etc.). Actually, it can be demonstrated that the model G_{pm} need be only a very crude approximation of G_p and the nonlinear compensator may deviate significantly from the true complement of n. Stabilization of this type has been used extensively in operating control systems (e.g., in removing the low-amplitude oscillations often associated with nonlinearities in the valve of a hydraulic power amplifier).

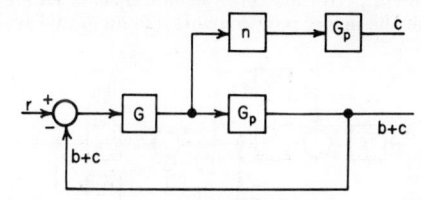

Fig. 29-63. Equivalent of Fig. 29-62 with perfect model.

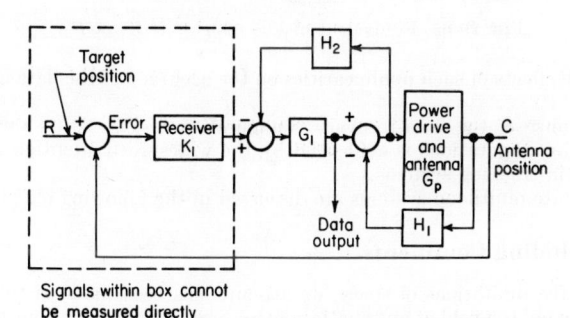

Fig. 29-64. Data-recovery system.

Data-recovery System. In a tracking radar, the radar receiver provides an output signal measuring the error between target and antenna positions (in azimuth, for example; see Sec. 14-3). For the computer used to calculate future target position, we require data on the present target location. Thus the tracking system works under two requirements:

1. The antenna must follow the target with sufficient accuracy to ensure that the target never leaves the antenna beam.
2. The system must provide a separate data output with as accurate target-position data as possible.

The system of Fig. 29-64 is widely used in this application.

The basic control system consists of all the elements except H_2: G_p is the power amplifier and antenna system, G_1 a preamplifier controller, and H_1 a minor feedback path to linearize and improve the performance of the power elements. If the data-recovery block H_2 is inserted and we choose H_2 so that

$$H_1 H_2 = K_1 \tag{29-126}$$

the feedbacks of C to the input of G_1 along the two possible paths ($-K_1$ through the main feedback path, the input comparator, and K_1; and $+H_1H_2$ through H_1, the comparator H_2, and the second comparator) exactly cancel. Consequently, the only two closed loops in the system are those with the loop gains $-G_1H_2$ and $-G_pH_1$, and the system is equivalent to the configuration of Fig. 29-65. In other words, we have inserted H_2 in order that the "negative" feedback of C to the input of G_1 (feedback arising because of the nature of the radar operation) may be canceled by the "positive" feedback through H_1 and H_2.

The data on target position R can be obtained at the output of G_1 (in either Fig. 29-65 or 29-64); since the preamplifier G_1 needs to provide almost no power, the G_1H_2 feedback system can be a wideband and high-accuracy system, so that the output of G_1 is an accurate reproduction of R.

There are two practical difficulties with such a system, both of which must be investigated in terms of the specific transfer functions when we are designing an actual system. First, Eq. (29-126) requires that H_2 track K_1: as K_1 varies (because of signal fading, for example), (29-126) is violated, and we must evaluate the effects of this deviation on system performance. Second, G_p is often highly nonlinear (as we might expect because of the power requirements); in an actual system design, we must

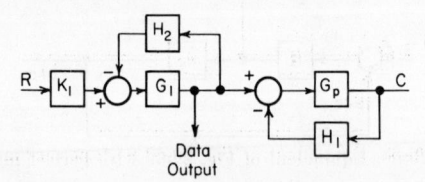

FIG. 29-65. Equivalent of Fig. 29-64 if $H_1H_2 = K_1$.

investigate the effects of such nonlinearities on the accuracy of the data at the output of G_1.

The importance of the data-recovery configuration of Fig. 29-64 and the several closely related configurations is evidenced by the widespread adoption of such feedback schemes in tracking systems.

More elaborate multiloop systems are discussed in the following chapter.

29-10. Concluding Comments

Because of the limitations of space, no attempt has been made in this chapter to cover exhaustively the field of automatic control. Such important topics as stability, control-system design, and performance measures have not been presented here. The authors have chosen to concentrate on modeling and sensitivity—two important aspects of automatic control which are treated inadequately in the literature.

The annotated bibliography below is intended to furnish the reader with a brief list of references which provide coverage of those aspects of automatic-control theory which have not been included here.

REFERENCES

1. Bode, H.: *Network Analysis and Feedback Amplifier Design*, D. Van Nostrand Company, Inc., Princeton, N.J. 1945. Classical book on feedback theory.
2. Chestnut, H., and R. Mayer: *Servomechanisms and Regulating System Design*, 2 vols., John Wiley & Sons, Inc., New York, 1955. Good practice-oriented book with major emphasis on Nyquist- and Bode-plot approach to design.
3. Truxal, J. G.: *Automatic Feedback Control System Synthesis*, McGraw-Hill Book Company, New York, 1955. Theoretical treatment of synthesis and analysis of linear, nonlinear, and sampled-data systems, and systems with random inputs.
4. Gille, J. C., M. J. Pélegrin, and P. Decaulne: *Feedback Control Systems*, McGraw-Hill Book Company, New York, 1959. Good practice-oriented book, briefly covering most of the field of control. Many examples of practical systems (mostly aircraft) used for illustration.

5. Graham, D., and D. McRuer: *Analysis of Nonlinear Control Systems*, John Wiley & Sons, Inc., New York, 1961. Good treatment of nonlinear control systems from both theoretical and practical viewpoints.
6. Seifert, W. W., and C. W. Steeg: *Control Systems Engineering*, McGraw-Hill Book Company, New York, 1960. Good comprehensive treatment of wide range of topics related to control (e.g., numerical analysis and solution of differential equations) as well as feedback theory.
7. Ashby, W. Ross: *Design for a Brain*, 2d ed., John Wiley & Sons, Inc., New York, 1960. Not strictly control, but a fascinating book on cybernetics (which is related to control).
8. Truxal, J. G. (ed.): *Control Engineers' Handbook*, McGraw-Hill Book Company, New York, 1958. Extensive collection of theoretical and practical control information. Brief discussions of some aspects of feedback theory, and of components—such as a-c and d-c servomotors, hydraulic and pneumatic components, and gears.
9. Smith, O. J. M.: *Feedback Control Systems*, McGraw-Hill Book Company, New York, 1958. Difficult to read, because of the author's style of writing, and unconventional nomenclature; however, it contains discussions of subjects seldom or never covered in other books, such as transport lags, lightly damped systems, and a-c systems.
10. Ragazzini, J. R., and G. F. Franklin: *Sampled-data Control Systems*, McGraw-Hill Book Company, New York, 1958. Good treatment of sampled-data systems.
11. Tou, J. T.: *Digital and Sampled-data Control Systems*, McGraw-Hill Book Company, New York, 1959. Good treatment. More extensive coverage than Ragazzini and Franklin.
12. Savant, C. J., Jr.: *Basic Feedback Control System Design*, McGraw-Hill Book Company, New York, 1958. Good elementary treatment.
13. Mishkin, E., and L. Braun, Jr.: *Adaptive Control Systems*, McGraw-Hill Book Company, New York, 1961. Discussion of adaptive control systems and of some unusual topics in control; e.g., process identification, intentionally designed nonlinear control systems. Also presents some of the tools of system engineering, e.g., digital techniques, decision theory, game theory, and dynamic programming.
14. Cunningham, W. J.: *Introduction to Nonlinear Analysis*, McGraw-Hill Book Company, New York, 1958. Good treatment of wide range of nonlinear techniques including numerical and graphical methods of solution.
15. Gibson, J. E.: *Nonlinear Automatic Control*, McGraw-Hill Book Company, New York, 1963. Extensive coverage of nonlinear control, including adaptive and optimizing systems.
16. Gibson, J. E., and F. B. Tuteur: *Control System Components*, McGraw-Hill Book Company, New York, 1958. Discusses theory of operation of useful models for electrical, electronic, mechanical, hydraulic, and pneumatic control-system components.
17. Laning, J. H., Jr., and R. H. Battin: *Random Processes in Automatic Control*, McGraw-Hill Book Company, New York, 1956. Comprehensive treatment of aspects of statistics of interest to control engineer including stationary and nonstationary random processes, Wiener-filter theory, and operations with finite data.
18. Hahn, H.: *Theory and Application of Liapunov's Direct Method*, Prentice-Hall, Inc., Englewood Cliffs, N.J., 1963. English translation of classical text which makes a comprehensive presentation of this aspect of stability theory. Includes an extensive bibliography containing publications as recent as 1962.

Chapter 30

ADAPTIVE AND LEARNING CONTROL SYSTEMS

JOHN E. GIBSON, *Director, Control and Information Systems Laboratory, and Professor of Electrical Engineering, Purdue University.*

CONTENTS

30-1. Introduction to Adaptive Control	30-1
30-2. The Identification Problem	30-2
30-3. The Decision Problem	30-5
30-4. The Modification Problem	30-8
30-5. Example Systems	30-10
30-6. Adaptive and Learning Systems Compared	30-10
30-7. Pattern Recognition	30-11
30-8. Simple Reinforcement Learning Model and Learning Curve	30-13
30-9. Some Proposed Learning Automata	30-14
30-10. Learning Control Systems	30-17

30-1. Introduction to Adaptive Control

In a broad sense, any system in which a parameter is adjusted to counteract a degradation in performance brought about by a change in the system's environment could be called adaptive. In this sense, there are hundreds of adaptive control systems in present-day operation. Almost every jet-aircraft fuel control is temperature-compensated, for example. For that matter every automobile engine with an automatic choke must be considered adaptive.

Or suppose that by clever design the operating range of a system can be broadened to accommodate the expected variation of the environment without any "air data" or parameter compensation. Could this be called adaptive? If so, there are millions of adaptive systems in successful operation today.

If these systems are called "adaptive" then an entirely new term is needed to characterize the systems to be discussed in this chapter. We prefer to acknowledge that the above systems can be excellent design solutions to a given problem but to reserve the word "adaptive" for a class of systems which remains to be described.

An adaptive control differs philosophically from a conventional closed-loop control as the closed-loop system differs from open-loop control (Sec. 29-2). If closed-loop control is merely a "concept" then so is adaptive control. If a conventional feedback system has objective reality, then so does adaptive control. Occasionally we have difficulty in deciding whether "feedback" is only a concept. Borderline cases such as the unbypassed resistor of a vacuum-tube amplifier exist. Other examples which few engineers will accept, such as the conventional electric-power transformer, have been proposed for feedback analysis. The adaptive concept is as controversial and nonobjective as this but no more so.

The control engineer is often faced with the following problem: Given a portion of the overall system (often called the "plant") which may not be changed by a designer, and given certain performance specifications along with other constraints (e.g., maximum forces or accelerations), find a satisfactory controller to actuate the plant. If the input of the resulting controller is sensitive only to the system input, the resulting system is open-loop. This is a perfectly satisfactory approach, provided the system and its environment and the expected inputs can be completely defined beforehand. If, however, these elements are not completely known (for example, the class of expected inputs or load disturbances may be known only statistically), the controller in addition should be a function of the plant output; the system is then called closed-loop. In order to design the closed-loop controller successfully, the engineer must be given the plant characteristics and how they vary under environmental changes. Conventional control-system design will fail without these.

A number of imaginative solutions have been proposed for the problem of incompletely defined system dynamics. The solutions suggested in opening this section compensated for anticipated changes from an initially well-defined plant. Even this approach is difficult or impossible, however, if the environment cannot be sensed directly but only insofar as it is reflected in a degradation in plant performance, or if little or nothing is initially known about the plant or the environment.

In such cases it may be advisable to so organize the overall system that it performs certain of the functions conventionally exercised by the designer himself during the design phase. That is, the controller may be called on to compute or *identify* the characteristics of the plant while the system is in normal operation. The controller must then make a *decision* concerning the way in which the available parameters of the system should be adjusted so as to improve the operation with respect to a defined index of performance. Finally certain signals or parameters must undergo *modification* to accomplish this result. These three functions of identification, decision, and modification, when accomplished in a closed-loop fashion by the system, constitute the essence of adaptive control.

Naturally in an area as topical as adaptive control, it is difficult, if not impossible, to be definitive. Recent research is referenced throughout the chapter. In addition, there have appeared several review-type articles that attempt to establish the state of the art. See, for example, Truxal.[1]

The National Electronics Conference in 1959 held a session on adaptive control in which four more or less basic papers were presented,[2] and in February of 1959 a conference on adaptive control systems was held at Wright Field by the Air Force at which many of the Air Force contractors who had been working on adaptive systems reported on their progress. The proceedings of this conference are available.[3] Mishkin and Braun have edited a book[4] which, while only in part concerned with adaptive control, will provide interesting background reading. This is also true of the book by Bellman.[43] Both Refs. 3 and 4 discuss a number of more or less practical systems that have been proposed as adaptive. Probably the most complete bibliography covering early work on adaptive control has been prepared by Stromer.[5]

30-2. The Identification Problem

From a philosophical point of view the most distinctive aspect of an adaptive control system is the identification feature. Considerable research effort has been expended in this area, and certain theoretical results are available. Much more, however, remains undone. The most severe restriction placed on present identification schemes is that the plant is assumed linear and only slowly time-variant. Schemes may be classified as to whether or not a priori knowledge of the plant is needed for successful operation. Of course, it is assumed throughout that the method must work in real time and while the plant is subject to normal control inputs; thus direct application of step inputs or impulses and the like cannot be tolerated. An identification scheme is called complete if the entire weighting function of the (assumed) linear plant is the object, and partial if only certain lesser characteristics are sought.

In a theoretical sense the identification problem has meaning only if the system is subjected to extraneous disturbances during the identification procedure. Otherwise the problem is trivial. A number of authors have discussed the general problem of obtaining a functional representation of a four-terminal network given only certain input and output data on the network. For the problem with the additional restrictions implied by adaptive control three approaches merit detailed consideration: cross-correlation identification, which was originally suggested by Lee[6] and reduced to practice by Anderson, Buland, and Cooper;[7] Levin's sampling method;[8] and Turin's matched-filter technique.[9] For a more detailed discussion of these and other less general techniques see Chapter 11 of the recent text by the present author.[10]

The basic linear identification problem is illustrated in Fig. 30-1, along with the notation used for the input test signal, the output signal, external noise, and the observed signal. In order to measure the high-frequency characteristics of $g(\lambda)$, it is necessary that $x(t)$ be a wideband signal. An exact representation of $g(\lambda)$ would require an infinite-bandwidth signal. although practically speaking this is neither possible nor necessary. Since the main interest of this discussion is the effect of external noise errors, any errors due to the practical limitation of the test signal will be assumed to be small compared with the effect of $n(t)$.

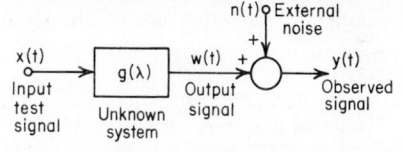

FIG. 30-1. Basic identification problem.

The output signal $w(t)$ is assumed to be unmeasurable; thus the identification must be based upon measurements of the observed signal $y(t)$. A stationary random process with zero mean is assumed for the external noise. The unknown impulse response is assumed to be essentially constant over the identification period, and $g(\lambda)$ is assumed to approach zero for large values of λ.

The signal-to-noise ratio at the output of $g(\lambda)$ is a useful measure of the effects of $x(t)$ upon normal system operation. This is defined as

$$\gamma = \frac{\overline{w^2}}{\overline{n_{\text{eff}}^2}} \tag{30-1}$$

where $\overline{w^2}$ is the mean-square value of the response of the system to $x(t)$; $\overline{n_{\text{eff}}^2}$, the effective mean-square value of the external noise, is in turn defined as

$$\overline{n_{\text{eff}}^2} = \frac{1}{2\pi} \int_{-2\pi W_x}^{+2\pi W_x} \Phi_n(\omega) \, d\omega \tag{30-2}$$

where W_x is the equivalent noise bandwidth of $x(t)$, and $\Phi_n(\omega)$ is the Fourier transform of the noise autocorrelation function. For $\gamma \leq 1$, an operator would probably fail to notice any change in the normal operation of the system due to the test signal, because of the ever-present background noise $n(t)$.

Equation (30-1) may also be expressed in terms of $\overline{x^2}$ and the low-frequency power gain K_G and equivalent noise bandwidth W_G of the system.

$$\gamma = \frac{\overline{x^2} W_G K_G}{\overline{n_{\text{eff}}^2} W_x} \tag{30-3}$$

A block diagram of cross-correlation identification is shown in Fig. 30-2. The critical nature of the identification problem is due to the lower limit on *identification time*. This is defined as the time required to evaluate a point on the weighting function of the plant to within a given variance σ^2 for a given signal-to-noise ratio and for a given form of test signal.

Suppose $x(t)$ is a sample of an ergodic random process. If $n(t)$ and $x(t)$ are uncorrelated, the expected value of the signal at the output of the multiplier is

$$z(t) = \phi_{xy}(\tau_i) = \int_0^\infty \phi_x(\tau_i - \lambda) g(\lambda) \, d\lambda \tag{30-4}$$

When $x(t)$ is wideband compared with $g(\lambda)$, the terms in this convolution integral appear as shown in Fig. 30-2b, and

$$\phi_{xy}(\tau_i) \approx \Phi_x g(\tau_i) \tag{30-5}$$

where Φ_x is the area under the $\phi_x(\tau)$ function. The approximation used to obtain (30-5) is equivalent to assuming

$$\phi_x(\tau) = \Phi_x \delta(\tau) \tag{30-6}$$

where δ is the Dirac delta function. Complete identification is achieved by using a number of correlation channels in parallel.

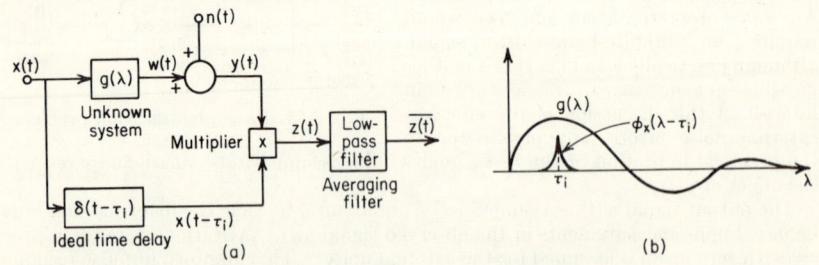

Fig. 30-2. (a) Cross-correlation identification. (b) Convolution of $\phi_x(\tau)$ and $g(\lambda)$.

The noise components of $z(t)$ can be studied by obtaining the autocorrelation function of $z(t)$. Examination of this autocorrelation function reveals that, while certain error terms arise from the assumed external noise and cannot be eliminated, other error terms are a result of the random properties of the test signal alone. If instead of using a continuous noise sample, we generate $x(t)$ by taking a noise sample of duration T_x and repeating it periodically, the autocorrelation function $\phi_x(\tau)$ as well as the noise components in $z(t)$ due to $x(t)$ will be periodic with period T_x. Hence the average of these noise terms over one period will be equal to the average over all time. If the test signal has a zero mean value, the average will also be zero. In order to preserve the quality of the signal component of $z(t)$, the periodic noise sample must be chosen so that each period of $\phi_x(\tau)$ is a narrow pulse.

The optimum averaging filter for a periodic test signal is an ideal finite-memory integrator with memory time nT_x, where n is a positive integer. By using a periodic test signal and an ideal finite-memory integrator, one may eliminate all noise terms due to the test signal. Thus when $x(t)$ is periodic,

$$\phi_z(\tau) = \phi_x(\tau)\phi_n(\tau) \tag{30-7}$$

Since the averaging filter will have a narrow bandwidth compared with $z(t)$, the variance of the impulse-response estimate is very nearly

$$\sigma^2 = \frac{1}{\Phi_x^2 T_I} \int_{-\infty}^{+\infty} \phi_x(\tau)\phi_n(\tau) \, d\tau \tag{30-8}$$

where the identification time T_I is the equivalent smoothing time of the averaging filter. The integral in (30-8) reduces to $\overline{x^2}\Phi_n(0)$ if the external noise is white, or to

$n^2\Phi_x$ if W_x is much greater than the bandwidth of $n(t)$. For either of these conditions the identification time can be obtained in a convenient form by introducing the signal-to-noise-ratio constraint and the fact that $\Phi_x = \overline{x^2}/2W_x$ into (30-8). Thus

$$T_I = \frac{2W_G K_G}{\gamma \sigma^2} \tag{30-9}$$

It can be shown[11] that the identification time using a Gaussian test signal is bounded by

$$T_I \leq \frac{2W_G K_G}{\gamma \sigma^2}(1 + 2\gamma) \tag{30-10}$$

The advantage of periodic test signals over arbitrary random test signals is apparent.

Since Lindenlaub has shown that this and the other two basic methods of identification all require the theoretical minimum identification time, the above results are quite general.

In Table 30-1 are shown sample calculations for the identification times of two systems and for two signal-to-noise ratios. The allowed standard deviation σ in the impulse response is 1 per cent. This is perhaps an overprecise determination and in part accounts for the very large values in the table. T_I varies inversely as the square of the standard deviation, and so if we double the value of σ (to 2 per cent), the T_I will be reduced by a factor of 4. Since the impulse response should vary only slowly compared with the identification time, it is apparent that complete non-a-priori identification is not suitable for plants which are rapidly varying.

Table 30-1. Complete, Linear, Realistic, Non-a-priori Identification Times for Certain Specific Systems and a Standard Deviation σ of 1 Per Cent of Maximum Value of the Impulse Response

Input test signal	γ	$G(s) = \dfrac{1}{s^2 + s + 1}$	$G(s) = \dfrac{1}{s + 1}$
Gaussian white noise	10	35,100 sec = 9 hr 45 min	10,500 sec = 2 hr 55 min
	0.1	200,000 sec = 55 hr 33 min	60,000 sec = 16 hr 40 min
Discrete internal binary noise	10	1,672 sec = 29 min	500 sec = 8 min 20 sec
	0.1	166,700 sec = 46 hr 20 min	50,000 sec = 13 hr 54 min

Identification with a priori knowledge and the identification of nonlinear plants are active research areas at present.

30-3. The Decision Problem

The second of the three elements basic to adaptive control is the decision function. The decision problem deals with the development and specification of analytical methods by which system performance can be evaluated and from which a strategy for achieving optimization can be evolved (Sec. 24-4). The most common method of system evaluation is the use of an IP (index of performance), a functional relationship involving system characteristics in such a manner that optimum performance may be determined from it. The strategy of optimization is perhaps best discussed in terms of an abstract space called the adaptive space whose basis is the parameters available for adjustment and in which contours of constant IP value exist.

An IP can be considered to be any criterion by which system quality is measured. It is convenient to divide presently used criteria into the following three categories: first, conventional system specifications which involve specific measurements such as phase margin or rise time; second, general criteria which attempt to sum up the overall performance (these are quite often expressed as some integral function of error); and third, compound criteria based on some functional relationship of system error and other requirements such as the cost of control. Schultz and Rideout have presented what is probably the most complete discussion of general integral criteria.[12] See Gibson et al.[13] for a discussion of specific specifications.

The decision problem involves the design of a logic by which the system evaluates its present condition with respect to the optimum (by the use of the information gained from identification) and arrives at a plan of action for driving toward the optimum. A convenient way of visualizing this procedure is in a coordinate or parameter space in which contours of constant IP are drawn. The problem is to find the most efficient way of climbing (or descending) the hill to find the maximum (or minimum). This is called a minimax problem. Figure 30-3 shows a hypothetical diagram of this type for an internal-combustion engine under fixed load. The optimum point varies, of course, for various load and environmental conditions. The system must proceed to the optimum (and follow it if it moves) in an efficient manner. Perhaps this procedure will make use of the mathematics of steep descent (Sec. 40-7), but most engineering systems studied up to now have used one of the two general

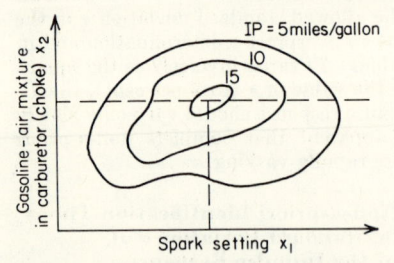

FIG. 30-3. Hypothetical plot of constant IP contours for an internal-combustion engine.

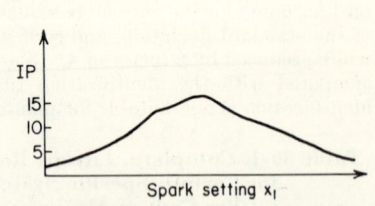

FIG. 30-4. Gas-air ratio held constant, thus reducing the problem of optimization in a single dimension.

schemes discussed below. In order to study various decision schemes, it would be well to develop a mathematical model of the contour plot. Feldbaum[14] and Stankovskii[15] have used a model in which the contours of constant IP are nested ellipses:

$$\text{IP} = c_1 x_1^2 + c_2 x_2^2 \tag{30-11}$$

A more general model, and one that permits a discussion of the rather important phenomenon of skewing of the contours with respect to the axes, is

$$\text{IP} = c_0[x_1^2 + \alpha x_2^2 + \beta(x_2 - m x_1)^2] \tag{30-12}$$

The decision logic must operate on the basis, initially at least, of only the local gradient of the IP contours at the initial condition. Contour shapes must be assumed unknown, since if this is not so the problem is trivial. Two general procedures are available for determining the local gradient; both are in common use by a number of workers. First there is the continuous (usually sinusoidal) perturbation, discussed, for example, by Draper and Li[16] and Vasu;[17] and second, there is the incremental perturbation discussed, for example, by Feldbaum[18] and Gibson.[19] Either of these methods might be superior to the other under certain practical circumstances, but it would be interesting to compare the two approaches for equivalent systems and equal hunting losses on a theoretical basis.

It is common practice to optimize the available parameters of the system sequentially rather than in parallel. No doubt this reduces the speed of adaptation, but under the present state of the art there is essentially a one-to-one correspondence between the complexity of the physical equipment and the numbers of parameters to be simultaneously adjusted. Single-dimensional adjustment presents the simplest analysis problem and provides a certain insight into the multidimensional case. In

Fig. 30-3, for example, let it be assumed that only the x_1 parameter is to be varied. Figure 30-4 represents this situation. Whenever a physical minimax is available, it is not necessary to employ a contrived IP. Unfortunately, however, it is only in the regulator problem that this situation occurs.

Note that, in order to obtain the slope, there must be a variation of the independent variable. Theoretically, the small variations that will occur in normal operation may be employed to generate this slope, and at least one practical system has been operated on this basis.[20] Usually, however, a test signal or perturbation is introduced into the loop from an external generator to ensure a proper slope indication. In effect, we desire to introduce a controlled hunting or dither with which to perform the optimization.

Let us assume that it is desired to operate a jet-aircraft engine at the maximum value of the ratio of combustion-chamber pressure to external atmospheric pressure in order to optimize fuel consumption.[17] This ratio depends on a number of environmental factors, and so it may be undesirable to preprogram the jet-fuel control system to utilize air data to accomplish this optimization. Let us assume that near the

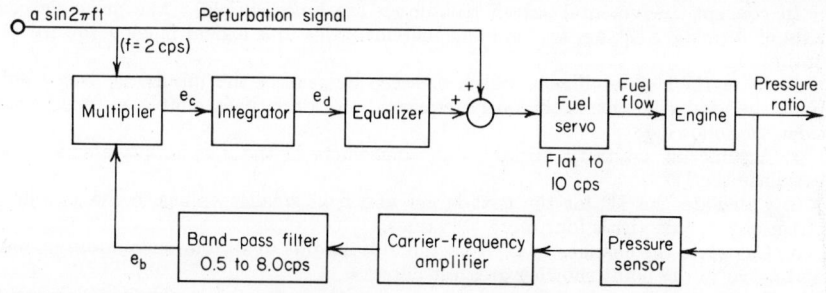

Fig. 30-5. Sine-wave perturbation of a jet-fuel control system. (*From Vasu.*[17])

optimum a parabolic relationship holds between fuel flow and pressure ratio in the system under consideration, as in Fig. 30-4.

$$p = -Kf^2 \qquad (30\text{-}13)$$

Use a sinusoidal perturbation of the fuel flow as a test signal,

$$f = f_{av} + a \sin \omega t \qquad (30\text{-}14)$$

Substituting (30-14) into (30-13), and using the identity $\frac{1}{2}(1 - \cos 2x) \equiv \sin^2 x$, yields

$$p = -K\left(f_{av}^2 + \frac{a^2}{2}\right) - 2Kf_{av}a \sin \omega t + \frac{Ka^2}{2}\cos 2\omega t \qquad (30\text{-}15)$$

Remember that f_{av} need not be constant; in fact, the whole point of the system is that it is slowly varying. As shown in Fig. 30-5, the fundamental components of the pressure wave may be recovered by a band-pass filter.

$$e_c = -2Kf_{av}a^2 \sin^2 \omega t = -2Kf_{av}a^2 \frac{1 - \cos 2\omega t}{2} \qquad (30\text{-}16)$$

The low-pass filter is designed to remove the double-frequency terms; thus

$$e_d = a^2 K f_{av} \qquad (30\text{-}17)$$

After being passed through an equalization network, if required, this signal will be used to drive the average value closer to the value which yields a maximum on the pressure-ratio curve.

On first examination, Fig. 30-5 seems rather complicated; further analysis does little to change this initial impression. In practice, the band-pass filter can be incorporated into the carrier amplifier, and the low-pass filter may be a simple single-stage RC network. The multiplier, however, represents a certain problem, as does the external generator. In this system, as in the previous one, calculation of the hunting loss is not difficult, provided a parabolic relationship is assumed.

A continuous perturbation method that seems to offer several advantages over the various deterministic signal schemes is perturbation with a random noise signal. The approach has been considered by a number of workers; see, for example, O. J. M. Smith (discussion of Ref. 19). Parallel search in a number of dimensions can take place simultaneously, provided the random perturbations are uncorrelated with each other. Probably the major advantage of white Gaussian noise, say, as a perturbation signal is that there is a finite probability that the system will detect the comparative inferiority of a local minimax at which it is operating, and will then move to a superior minimax nearby. As one might suspect, the waiting time for such an event is inversely related to the intensity of the noise, while the hunting loss is directly related to it. Thus the usual engineering compromise must be made.

In concept, incremental search techniques are quite simple. The principle consists of forming $\Delta IP/\Delta x_i$ and moving incrementally in a logical fashion toward the optimum.

Some method of intelligent search or steep descent to the minimum, based only upon the local gradient, is obviously required. As a relatively simple approach, consider the following:

1. Assume an arbitrary initial point, somewhere in the $x_1 x_2$ adaptive space, and calculate the IP.
2. Calculate the IP for the next larger and next smaller values in the x_1 and x_2 directions. This yields four more IP values.
3. Calculate the changes in IP from the initial point to all four surrounding points and move to the point showing greatest decrease.
4. From the new center point, again calculate the surrounding points. This yields three new data points. Continue to move in the direction of steepest descent.

This procedure can be elaborated on for any number of coordinates. Also, it does not depend on initial knowledge of the location of the minimum. Finally, it will continue to track the minimum, even if the minimum changes its location because of parameter variation. This last point is the essence of adaptive control.

At least three improvements on this simplified method of steep descent suggest themselves:

1. Let the size of the increment be a function of the rate of decrease. When the rate of decrease is large, perhaps the step increments should be large; then as the optimum is approached, the steps could be decreased in size. This would combine economical search far from optimum with accurate adjustment near the optimum. There is a danger, however, in this procedure. If the adaptive control approaches a local minimum and the increment size is decreased, the system will remain at the local minimum.
2. Allow diagonal motion through the adaptive space, i.e., parallel operation. This would more closely approximate a true steep-descent procedure.
3. Incorporate some memory in the descent procedure so as to allow a choice of the most likely direction of descent from any initial point. Calculate and move in only this direction so long as it does yield a descent. Make a dimensional search only when it does not yield a descent. A simplification of this procedure has been used in several studies of adaptive systems.[19,21]

30-4. The Modification Problem

The third and final step in the adaptive process is the actual adjustment of the system so as to optimize its performance. The general process might be called control-signal modification, since the object is to modify the error or control signal to achieve optimum performance. There are two ways to accomplish this. The first approach is called parameter adjustment. This is a direct extension of conventional procedures in which a fixed compensator is designed to optimize some aspect of system performance for fixed conditions. In the adaptive system, the parameters of the compensator are adjusted by the system itself so as to maintain optimum performance under varying conditions. This technique has been used by a number of workers.[21] A more general and powerful approach is control-signal synthesis. Here it is realized that the adaptive control is making adjustments to provide a signal at m so as to drive the plant in an optimum manner, and that adjusting the parameters of a series equalizer is an artificial limitation due only to tradition. Meditch[22] has used the latter approach in the design of adaptive controls. When this view is taken, it becomes convenient to work on an interval or sampled basis, and a digital computer applies very naturally.

Since parameter adjustment employs conventional feedback-control concepts and has been mentioned above, an extended discussion is probably not warranted here. The basic difference between the two points of view is brought out in Fig. 30-6. In parameter adjustment, the basic configuration of the conventional control system is maintained intact, and the means of automatically adjusting the parameters toward the optimum is superimposed. In control-signal modification the conventional configuration is entirely abandoned, and the whole control function is assumed by the

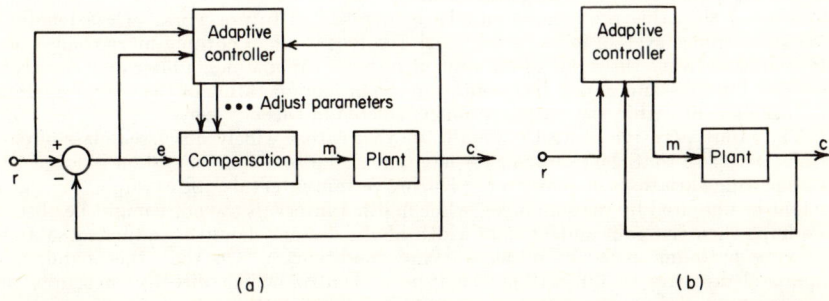

FIG. 30-6. Control-signal modification. (*a*) Parameter adjustment. (*b*) Control-signal synthesis.

adaptive computer. If the physical systems follow this philosophical classification, parameter adjustment is undoubtedly superior from reliability and fail-safe points of view. In a theoretical treatment, however, control-signal synthesis is preferred for its more general conception of the problem and for its possible stimulation of future thinking. It is apparent that parameter adjustment may be applied directly in all the systems discussed in the chapter thus far.

A number of workers have suggested approaches that can be classified under control-signal synthesis. Braun[23] makes use of the Maclaurin-series expansions of the plant impulse response, the plant forcing function, and the response of this forcing function. A new signal is synthesized which, when added to the original forcing function, will force the plant output to zero error. The corrective signal is in the form of a finite sum of singularity functions, including impulses, which would prevent practical application of the technique. The correction is on an interval basis. Merriam[24] uses dynamic programming (Chapter 27) to obtain the time-varying gains required in a feedback configuration for an optimum adaptive system. Bellman's compound quadratic IP is employed. Unfortunately, to obtain the values for the time-varying gains, it is necessary to solve a set of simultaneous nonlinear differential equations. Meditch[22] constructs an optimum plant input on an interval basis by summing an orthonormal polynomial. The particular polynomial chosen was the Legendre, for ease in computation and physical realizability of the system. The terms of the polynomial are found for a given interval by the use of Wiener prediction from the preceding interval in such a fashion as to minimize the predicted value of a compound quadratic IP.

30-5. Example Systems

A number of forms have been proposed as adaptive controls. Among these are model-reference systems and input adaptive systems. The philosophy of each is simple.

The model-reference concept presumes that, rather than embodying the ideal performance in a mathematical index of performance, a model plant or system is defined whose response is agreed to be ideal. It might be argued that this is an extension of conventional practice, since most of the conventional criteria (such as 0.707 damping) are picked because of their known effect in an ideal second-order system. In the model-reference system, however, it is proposed to construct an analog or digital model of the ideal system and to subject this model and the actual system to the same stimulus. The difference between the model response and the actual plant or system response can then be used to actuate a steep-descent procedure which adjusts the values of the available parameters of the actual system in such a way as to minimize this difference.

While this approach appears simple and conceptually appealing, analysis is incomplete at this time; hence a definitive position is premature. In one simple first-order model-reference system, Margolis[25] pointed out rather severe stability limitations, indicating that this problem cannot be overlooked in future work. Occasionally a model-reference system is proposed which resolves itself to nothing more than a prefilter in series with the input to the control system. While such a filter may represent excellent design, and in fact be essential to the proper operation of the overall system, it does not fall within the classification considered in this chapter.

The input adaptive control system is another rather widely discussed class of control. Suppose that the character of the input signal to a control system is subject to rather wide variations in character. During certain intervals, for example, the signal might be obscured by random noise, while in other intervals the noise might be absent. Perhaps then the gain and/or bandwidth of the system should be adjusted so as to provide optimum response under varying conditions.[26] On the other hand, this approach appears to fall within the type of control which directly measures the environment and adjusts a system parameter appropriately.

30-6. Adaptive and Learning Systems Compared

It may be said that an adaptive control exhibits a crude form of learning in its self-adjustment toward an established goal; see Sec. 24-8. However, we prefer to define learning somewhat more formally as follows.

A learning system is a higher form of control, as is an adaptive system. The learning system differs from the adaptive system, however, not only in function, but also in area of application. The adaptive system is designed to modify itself in the face of a new environment so as to optimize its performance. A learning system, on the other hand, is designed to recognize familiar features and patterns in a situation and then, from its past experience or learned behavior, to react in an optimum manner. In the former, emphasis is placed on reacting to a new situation, while in the latter, the emphasis is on remembering and recognizing old situations.

A *gedanken* experiment might be proposed to determine if a given system is a learning system. Under a given set of environmental conditions, the system parameters are given an initial offset from optimum, and the system is allowed to operate. If it adjusts its parameters so as to optimize its performance in accordance with a given IP, as previously discussed, it is adaptive. As yet it is impossible to know if it also includes a learning feature. Now return the system to the initial parameter settings and allow it to proceed. If the same gradual process of adaptation takes place, the system is not a learning system. If, on the other hand, it recognizes familiar patterns and utilizes this information to move more closely (or more rapidly) to the optimum, it is a learning system.

Not only will a learning control system require more extensive logic than an adaptive control system, but inevitably it will be limited by what must be a limited memory capacity. Proper use of logic and memory capacity forms only one aspect of a generally undeveloped research area. Because the learning system bypasses the complete identification procedure necessary in adaptive control, or in other words because for each run it has more a priori information at its command, it can in theory react more rapidly. On the other hand, for the same amount of time, the learning system should approach more closely to the optimum by properly utilizing its past experience.

The terms "adaptive" and "learning" have been used by various investigators to describe the behavior of certain inanimate objects presumably by analogy to the actions of various forms of living organisms. Perhaps the analogy need be only loose or even nonexistent. However, if the physical scientist is so undisciplined as to abuse this analogy he runs the risk of ignoring the fruits of about three-quarters of a century's research on the learning problem. It would seem that proper use of this material will aid research in cybernetics, and then in turn that this research should illuminate the learning problem in psychology and the behavioral sciences in general.

The difficulties of an engineer using the psychology of learning must be acknowledged, however. As Woodworth and Marquis pointed out some years ago in their well-known basic text,[27] "No general theory of learning commands the universal approval of psychologists at the present time. Approaching the problem from different sides and by the aid of different experiments they are inclined to emphasize different factors." More recently, George (Ref. 28, p. 181) states, "Now it can be seen that there are large possibilities for confusion in our basic understanding of 'learning' and more careful definition is necessary. If one turns to some modern definitions of 'learning' it may be seen that greater clarity has been only partially effected." In the 1940s psychologists began to emphasize the mathematical theories of learning, and more recently certain of the concepts of cybernetics have begun to be discussed.

In the last 15 years a growing interest on the part of psychologists in the concepts of feedback control (Ref. 28, Ref. 29, Chaps. 6 and 7) and mathematical learning theory[30] is evident (see Chapter 31). The cross-fertilization between psychology and automata theory is now beginning to yield a few devices of real interest to both fields, and some of these will be discussed below.

The concepts of learning which have been developed by psychologists and which appear to be of immediate use in constructing superior control systems seem to be limited to perhaps the following three: pattern recognition, the simple reinforcement model, and the idea of a learning curve.

30-7. Pattern Recognition

In order to deal with an input the system has to recognize what kind of thing it is; see Sec. 24-8. By recognition is meant the extraction of significant features from a background of irrelevant detail. The need to choose among several possible actions compels us to equip the system with recognition tools.

Pattern recognition has been defined as a mapping of a large, detailed complex description of an event into a simple description. In other words, the essential characteristics are extracted from a mass of data which will not be needed to recognize the event when it again occurs. Among the many approaches to recognition, the simplest method is *template matching*, by which the objects are matched against standards or prototypes. Both element matching and feature matching can be used. A more powerful approach is the *property-listing* method. This can be carried out by searching specific areas of the signal field, such as edge tracing, double-dot writing, or by extracting features by spatial transformations, such as edging, averaging, and vertex hunting.[31] The details of such methods are given in the reference, but for our purposes it will suffice to say that the input object is subjected to a sequence of tests, each of which detects certain features which are invarient under certain types of distortion.

When recognizing a new object one will first try answers which apply to similar patterns; that is, the system should benefit from its past experience. Since we cannot expect new objects to be precisely the same as old ones, the learning process must involve generalization on past experience. Selfridge has brought out the relationship of pattern recognition to learning. He defines pattern recognition as the classification of a data set into learned categories, and learning as the acquisition of feasible operational definitions of the categories.[36]

It has been recognized in recent years that the problem of pattern recognition can be viewed as a test of statistical hypotheses. A concise explanation of this theoretical approach to the recognition problem may be obtained by considering the statistical decision process (Chapter 24) as a game against nature (Sec. 23-8). A nonnegative weighting function measures the loss incurred by the decision maker as a consequence of each possible decision. The aim of the decision maker is to minimize the expectation of the loss in recognition.

A neural-net model is taken as an example for illustration. For every input x, a sequence of measurements $x_1, x_2, \ldots, x_n$ is performed, each of which has two possible outcomes, 0 or 1, thus indicating which class of input x actually belongs to. Assume that all input measurements are statistically independent and that we know (1) the a priori probabilities P_j that an input x is in class ω_j, and (2) the conditional

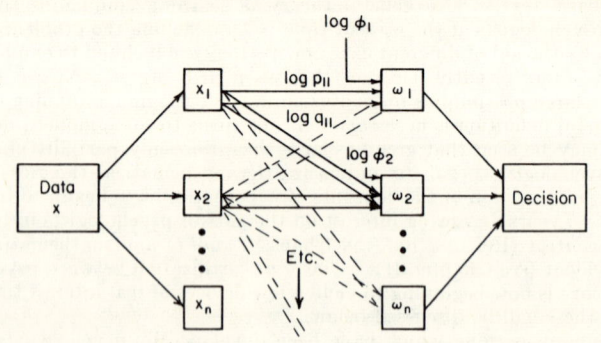

Fig. 30-7. A simple pattern-recognition device similar to Pandemonium (compare Fig. 30-11).

probabilities p_{ij} that if $x \in \omega_j$ then $x_i = 1$. For a given input, v is defined as a set of i's for which $x_i = 1$, and $\bar{v}$ as its complement. The proper choice of the element ω_j within which x_j belongs must now be made. The best choice is that choice which has the least chance of being wrong; i.e., ω_j is to be determined such that $p(\omega_j|v)$, the probability of ω_j for a given v, is the largest. By Bayes's rule (Sec. 38-3), we have

$$p(\omega_j v) = \frac{p(v|\omega_j)P_j}{P(v)} = \frac{p(\omega_j)}{P(v)} p(v|\omega_j) \qquad (30\text{-}18)$$

Since $P(v)$ is not a function of j, the comparison depends upon the computation of $p(v|\omega_j)$. In view of the assumption that all input measurements are independent, we have

$$P(v)p(\omega_j|v) = P_j \sum_v p_{ij}(1 - p_{ij}) \qquad (30\text{-}19)$$

Let

$$W_j = \log P(v)p(\omega_j|v) = \log P_j + \sum_v \log p_{ij} + \sum_v \log q_{ij} \qquad (30\text{-}20)$$

where $q_{ij} = 1 - p_{ij}$. W_j may be regarded as the "weight of the evidence" that $x \in \omega_j$. Thus we want to find the maximum W_j ($j = 1, 2, \ldots, m$). These "maxi-

ADAPTIVE AND LEARNING CONTROL SYSTEMS 30-13

mum-likelihood" decisions can be made by a simple neuron network model as shown in Fig. 30-7.

A rigidly programmed pattern-recognition device, no matter how elaborate, is at best of only marginal interest to the control engineer. An acceptable device must include the ability to generate invariant properties internally or at least to generalize on its stored patterns so as to operate properly upon patterns similar to, but not exactly the same as, those it has worked with previously. Causing the machine to generate its own list of invariant properties is the subject of intense research activity at the present time. Generalization of stored properties is widely accepted as a test of the "I.Q." of the machine. We shall discuss a number of proposed devices below.

30-8. Simple Reinforcement Learning Model and Learning Curve

Bush (Ref. 32, p. 126) points out that the learning curve has been in use for over 50 years. Early attempts were made to derive formally from rational considerations the "correct" analytic learning function, usually the exponential, and to match this to experimental data. The meaning of the two variables that were to be functionally related is usually evident; for example, the time required for a rat to run a maze for a reward is related to the number of trials. The *gendanken* experiment proposed in

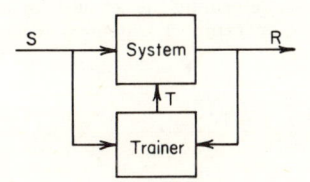

FIG. 30-8. A simple reinforcement learning model.

Sec. 30-6 which distinguishes a learning control system from an adaptive system employs the learning-curve concept.

A reinforcement learning process is one in which certain aspects of the behavior of a system are caused to become more (or less) prominent as time unfolds, as a consequence of the application of a "reinforcement operator" T. This operator is required to affect only those aspects of behavior for which instances have actually occurred recently. The analogy is with "reward" or "extinction" in animal behavior. A simple reinforcement learning model is shown in Fig. 30-8. Let us consider a very simple example. Suppose that on each presentation of an input stimulus S, an animal or a system has to make a choice (e.g., to turn left or right) and that its probability of turning right, at the nth trial, is $p(n)$. Suppose that the Trainer wants it to turn right. Whenever it does this the Trainer might "reward" it by applying the reinforcement operator T_+

$$p(n + 1) = T_+[p(n)] = \alpha p(n) + (1 - \alpha) \qquad 0 < \alpha < 1 \qquad (30\text{-}21)$$

which moves $p(n)$ a fraction $(1 - \alpha)$ of the way toward 1. If the Trainer dislikes what is done, T_- is applied,

$$p(n + 1) = T_-[p(n)] = \alpha p(n) \qquad (30\text{-}22)$$

moving $p(n)$ the same fraction of the way toward 0.

The mathematical model of such "linear" learning operators, generalized to several stimuli and responses, has been considered by Fu.[42] In general, r mutually exclusive and exhaustive classes of responses are considered. Each response class is represented by an alternative A_j, $j = 1, 2, \ldots, r$. A set of probabilities p_j are chosen as an index of behavior; p_j is the probability that the jth class of response will occur on a particular trial. Let the set of r probabilities be represented by a column vector $\mathbf{p}$:

$$\mathbf{p} = \begin{bmatrix} p_1 \\ p_2 \\ \cdot \\ \cdot \\ \cdot \\ p_r \end{bmatrix} \tag{30-23}$$

Then
$$\sum_{j=1}^{r} p_j = 1 \quad \text{and} \quad 0 \leq p_j \leq 1 \tag{30-24}$$

Consider the behavior change being the change of the set of probabilities $\mathbf{p}$. In order to specify what factors in the learning process change $\mathbf{p}$, begin with a general view: whenever certain events $E_1, E_2, \ldots, E_m$ occur, the probabilities are altered in a determined way. The process is assumed Markovian (Sec 28-4). Every time a stimulus occurs it has an outcome, and a particular outcome following a given stimulus changes the set of r probabilities in a unique way which is independent of earlier events in the process. In other words, given a set of r probabilities on trial n, the new set on trial $(n + 1)$ is completely determined by the outcome occurring on trial n. Earlier events will, of course, determine the set on trial n.

A reinforcement "learning operator" $\mathbf{T}$ is defined as an $r \times r$ matrix

$$\mathbf{T} = \|u_{ij}\| \tag{30-25}$$

where
$$\sum_{i=1}^{r} u_{ij} = 1 \quad (j = 1, 2, 3, \ldots, r) \tag{30-26}$$

When the matrix operator $\mathbf{T}$ is applied to $\mathbf{p}$ on trial n, $\mathbf{p}(n)$, we obtain a new column vector $\mathbf{p}(n + 1)$ which represents the new set of r probabilities on trial $(n + 1)$.

$$\mathbf{p}(n + 1) = \mathbf{T}\mathbf{p}(n) \tag{30-27}$$

From this it is not difficult to obtain a learning curve for the system.

30-9. Some Proposed Learning Automata

Because of the central position occupied by pattern recognition in the learning problem, it is not surprising to find that most automata designs are presently concerned with this aspect of the problem. In Rosenblatt's Perceptron[33,34,35] three functional classes of elements are identified: sensory units (S units), association units (A units), and response units (R units); see Fig. 30-9. Generally speaking, all these units perform threshold logic; that is, they sum positive and negative inputs and generate an output only if the sum is greater or less than a given threshold or bias level.

As an example, suppose the Perceptron is to recognize visual patterns and that the S units are photocells arranged in a bank or matrix. Each S unit is connected to several randomly chosen A units. These connections are not subsequently changed. The A units are then connected to a single R unit for each pattern to be recognized. The strength or gain of each of the connections of these A units to the R unit must be adjusted by a relatively long training sequence which acts to reinforce connections which properly respond to test stimuli and to inhibit the remaining connections.

A simple training scheme for a Perceptron is shown in Fig. 30-10. The reinforcement control system (r.c.s.) which receives information both from the outputs of the Perceptron and from the environment conducts the learning process by applying a proper learning operator T to the Perceptron. The learning process is usually operated by modifying the weights of connections from A units to R unit. The interesting features of the Perceptron are the initial random connection of the units and then the modification of their thresholds by a training sequence. Both these concepts seem similar to biological organization.

ADAPTIVE AND LEARNING CONTROL SYSTEMS

Selfridge's Pandemonium,[36] diagramed in Fig. 30-11, is also designed for pattern recognition. Here, however, the connections between layers are not random, and elementary recognition is accomplished at a lower level. In the second layer, comparable with the A unit in the Perceptron, each unit is designed to recognize some invariant

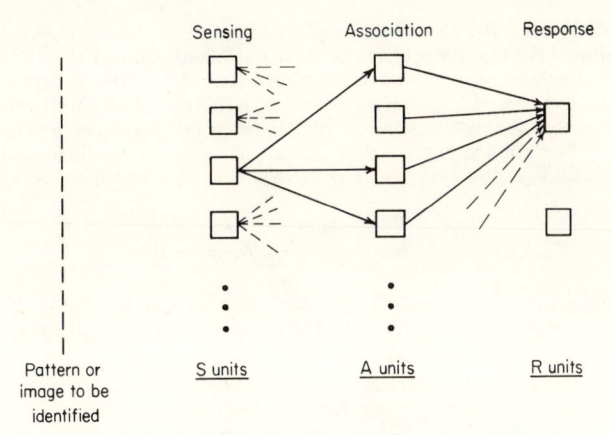

Fig. 30-9. A simple block diagram of a three-layer Perceptron.

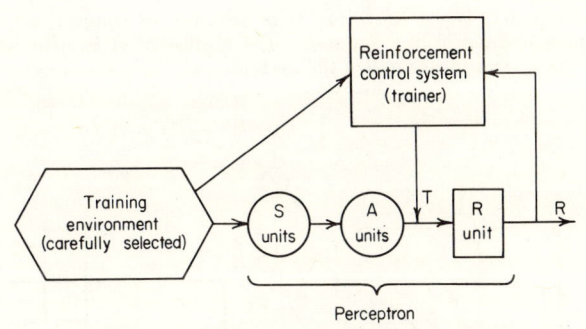

Fig. 30-10. A training scheme for a Perceptron.

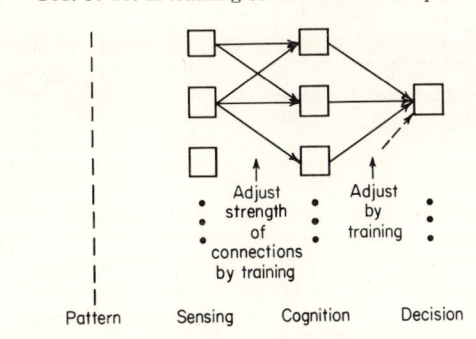

Fig. 30-11. A simplified block diagram of Pandemonium.

feature of a given pattern and to generate an output proportional to the closeness of the match between it and a previously studied training pattern. The single third-layer element then selects the second-layer element with the highest output and indicates that output as its decision. The initial multiple connections between the sensing elements and the second-level elements will be modified in the training period, as will the gains between the second-level units and the decision element. In contrast to the Perceptron, the output of the first- and second-level units will be monotonically increasing functions of the probability of assurance in the identification of the observed image.

The example in Sec. 30-7 should serve as an adequate explanation of the operation of Pandemonium. In the above discussions it has been assumed that P_j and p_{ij} are known for the weighted connections in the network model. Now consider putting to the connections the task of learning a good set of weights. The simple reinforcement learning concept that the path which contributes to "good" responses will be reinforced will be used. The connection weights w_{ij} are thus to be modified. If $x \in w_k$, no change is made on w_{ij}, provided that $j \neq k$. On the other hand, if $j = k$, we change w_{ij} to

$$\begin{aligned} w_{ij}' &= K(w_{ij} + 1) && \text{for } x_i = 1 \\ w_{ij}' &= K w_{ij} && \text{for } x_i = 0 \end{aligned} \quad (30\text{-}28)$$

This is essentially the "reinforcement operator" discussed in Sec. 30-8. The value of w_{ij} is then a simple Markov process with the expected values given by

$$\bar{w}_{ij} = p_{ij} K(\bar{w}_{ij} + 1) + q_{ij} K \bar{w}_{ij}$$

or
$$\bar{w}_{ij} = p_{ij} \frac{K}{1 - K} \quad (30\text{-}29)$$

which is proportional to p_{ij} if the same K is used for all connections. The probabilities P_j can be learned in like manner. The coefficient K is a time-decay factor which helps to stabilize and normalize the system.

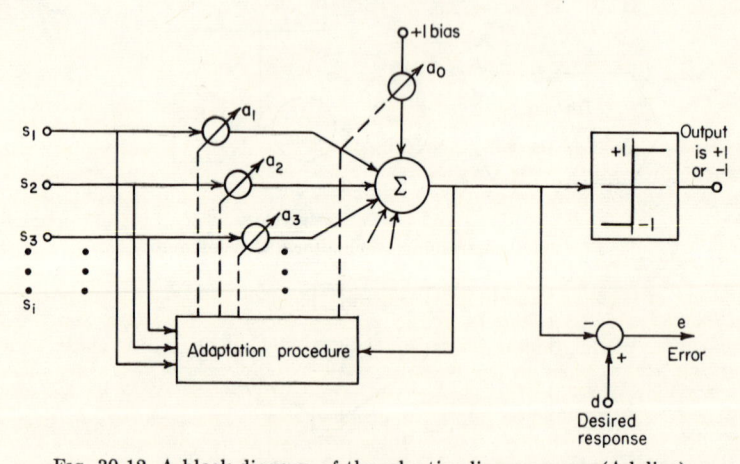

FIG. 30-12. A block diagram of the adaptive linear neuron (Adaline).

Widrow has attempted to avoid an important difficulty in earlier proposed systems by concentrating on a single operational threshold logic unit called Adaline[37] and explicitly accounting for the critical step of adjusting the strength of connections between elements. The so-called ADAptive LInear NEuron (Adaline) is diagramed in Fig. 30-12.

The binary input signals s_i on the individual lines have values of $+1$ or -1. Within the neuron (so-called), a linear combination of the input signals is formed. The weights are the gains $a_1, a_2, \ldots, a_n$, which could have both positive and negative values. The output signal is $+1$ if this weighted sum is greater than a certain

ADAPTIVE AND LEARNING CONTROL SYSTEMS   30–17

threshold and -1 otherwise. The threshold level is determined by the setting of a_0, whose input is permanently connected to a $+1$ source.

$$\text{Output} = +1 \quad \text{if} \quad a_0 + \sum_{i=1}^{n} a_i s_i > 0$$
$$\text{Output} = -1 \quad \text{otherwise}$$

The input-output relationship is determined by choice of the gains $a_0, a_1, \ldots, a_n$. These gains are set during the training procedure.

During a training phase, patterns of ± 1's are fed to the Adaline. With each pattern the parameters are adjusted slightly so as to cause the output to approach its desired value. One might say that the device "learns a little" from each pattern. The machine's total experience is stored in the values of the weights $a_0, a_1, \ldots, a_n$. The machine can be trained on undistorted noise-free patterns by repeating them over and over until the iterative search process converges, or it can be trained on a sequence of noisy patterns on a one-pass basis such that the iterative process converges statistically. After training, the machine can be used to classify the original patterns and noisy or distorted versions of these patterns.

The objective of adjusting the gains $a_0, a_1, \ldots, a_n$ is the following: Given a collection of input patterns and the associated desired outputs, find the best set of gains $a_0, a_1, \ldots, a_n$ to minimize the mean square of the neuron error $\overline{\epsilon_n^2}$. Individual neuron errors could have only the values of $+2, 0,$ and -2 with a two-level quantizer. Minimization of $\overline{\epsilon_n^2}$ is therefore equivalent to minimizing the average number of neuron errors. The error signal ϵ actually measured is the difference between the desired output and the sum before quantization (Fig. 30-12). Widrow has shown that $\overline{\epsilon_n^2}$ is a monotonic function of $\overline{\epsilon^2}$, and that minimization of $\overline{\epsilon^2}$ is equivalent to minimization of $\overline{\epsilon_n^2}$. From Fig. 30-12 the error ϵ is

$$\epsilon = d - a_0 - \sum_{i=1}^{n} a_i s_i \tag{30-30}$$

The problem of adjusting the a_i's to minimize $\overline{\epsilon^2}$ is equivalent to searching a parabolic stochastic hypersurface for a minimum. Like the other devices mentioned, Adaline requires a long training sequence, and like them it is still under development.

30-10. Learning Control Systems

The use of learning in automatic control systems differs in one fundamental way from the learning expected from the various automata discussed above. In the previous situation a training phase is required before the device is made to operate. "Learning" takes place in the training phase but not in the operating phase. In fact, if a pattern-recognition device were to change significantly its response to a familiar pattern it would be said to be degenerating. This is not at all the situation in learning control systems. They will be expected to accomplish learning "on line" and while responding to normal command inputs. This requirement appears unreasonably difficult in view of the thousands of cycles required in the training period of the devices discussed previously. But a second difference in the learning problem as applied to automatic control comes to our rescue. Whereas, previously, corrective reinforcement and inhibition were possible only in the training phase, in the control problem a continuous check during operation may be made by calculation of the index of performance. The analogy to open-loop control vs. closed-loop control is obvious. In lieu of an off-line training phase, a learning control system will employ the adaptive mode of operation to obtain the characteristic patterns required.

No learning control systems are known to be in field operation at the present time. Krug and Letskii have suggested the application of a learning technique for the optimum control of slow but complex dynamic processes, such as chemical processes.[38] They suggest that the optimum control might be found by a systematic evaluation of input and output data and an index of performance. The system as outlined is very general, and a probabilistic approach is proposed for organizing the memory for optimum search procedure. A human operator is needed to make the final decision about the satisfactoriness of the system performance.

Several learning control systems are currently under study in the Control and Information Systems Laboratory of Purdue University. A learning process has been applied to the control of a unit-gain amplifier in the presence of an additive random disturbance,[39] for example. Results show that the learning system yields better performance in an integrated-absolute-error sense than the corresponding adaptive or regulating system. The use of signal modification rather than system-parameter variation appears to have some advantages with respect to learning time and integrated absolute error. Efficient use of finite computer memory appears to be one of the more important questions it is necessary to take into consideration. The application of pattern recognition to learning in control systems is under investigation. A technique which makes use of the partitioning properties of a hyperplane has been studied for the classification of the system input measurement.[40] Likewise a scheme for partitioning the state space of a control system into regions has been developed. The controlled plant may be nonlinear or time-varying. Wonham and Fu have proposed that an adaptive system with prediction of the environment will learn if the predictions improve with time and the prediction data are used to make the adaptation more efficient.[41] The design of the parameter estimator and predictor in learning control systems has been studied in terms of statistical estimation theory.[42] A learning process has been applied to the parameter-estimation process and the decision part of the system, respectively. The concept of using learning curves as a performance measure for a learning control system has been suggested. A learning curve could be defined in this connection as a plot of index of performance or adaptation time of the system vs. the number of trials for the same environmental situation. The system will be said to learn if the curve monotonically approaches the optimum value of the quantity. The learning time may be expressed as the number of trials required to bring the quantity to its optimum value (or a certain percentage of its optimum value) multiplied by the time required for each trial. The time required for each trial usually includes the time required for recognition or identification of the plant characteristics and for decision computation.

All these preliminary investigations are steps toward the realization of a complete learning control system; nevertheless it should be obvious that the complete system is some distance away. If experience in the speed at which new ideas are absorbed in our technology is any guide, however, learning systems will begin to be operational very soon.

REFERENCES

1. Truxal, J. G.: Computers in Automatic Control Systems, *Proc. IRE*, vol. 49, pp. 305–312, 1961.
2. *Proc. Natl. Electron. Conf.*, vol. 15, pp. 1–45, 1959.
3. Gregory, P. C. (ed.): *Proc. Symposium on Self Adaptive Automatic Flight Control Systems*, TR-59-49, ARDC, WADC, Dayton, Ohio, 1959.
4. Mishkin, E., and L. Braun: *Adaptive Control Systems*, McGraw-Hill Book Company, New York, 1961.
5. Stromer, P. R.: Adaptive or Self-optimizing Control Systems—A Bibliography, *IRE Trans. Auto. Control*, vol. AC-4, pp. 65–68, 1959.
6. Lee, Y. W.: Applications of Statistical Methods to Communications Problems, *MIT Res. Lab. Electron. Tech. Rept.* 181, September, 1951.
7. Anderson, G. W., R. N. Buland, and G. R. Cooper: Use of Crosscorrelation in an Adaptive Control System, *Proc. Natl. Electron. Conf.*, vol. 15, pp. 34–45, 1959.
8. Levin, M. J.: Optimum Estimation of Impulse Response in the Presence of Noise, *IRE Trans. Circuit Theory*, vol. CT-7, no. 1, March, 1960.
9. Turin, G. L.: On the Estimation in the Presence of Noise of the Impulse Response of a Random Linear Filter, *IRE Trans. Inform. Theory*, vol. IT-3, no. 1, March, 1957.

10. Gibson, J. E.: *Nonlinear Automatic Control*, McGraw-Hill Book Company, New York, 1963.
11. Lindenlaub, J. C.: Limits on the Identification Time for Linear Systems, Ph.D. Thesis, Purdue University, Lafayette, Ind., June, 1961.
12. Schultz, W. C., and V. C. Rideout: Control System Performance Measures: Past, Present, and Future, *IRE Trans. Auto. Control*, vol. AC-6, pp. 22–35, 1961.
13. Gibson, J. E., Z. V. Rekasius, E. S. McVey, R. Sridhar, and C. D. Leedham: A Set of Standard Specifications for Linear Automatic Control Systems, *Trans. AIEE*, vol. 80, part II, pp. 65–77, 1961.
14. Feldbaum, A. A.: Automatic Optimalizer, *Automation and Remote Control*, vol. 19, no. 8, pp. 718–723, August, 1958.
15. Stankovskii, R. I.: Twin Channel Automatic Optimalizer, *Automation and Remote Control*, vol. 19, no. 8, pp. 729–740, August, 1958; A Multichannel Automatic Optimalizer for Solving Variational Problems, *Automation and Remote Control*, vol. 20, no. 11, pp. 1435–1445, November, 1959.
16. Draper, C. S., and Y. T. Li: *Principles of Optimalizing Control Systems and an Application to the Internal Combustion Engine*, American Society of Mechanical Engineers, New York, 1951.
17. Vasu, G.: Optimalizing Controls Applied to the Control of Engine Pressure, *Trans. ASME*, vol. 79, pp. 481–488, 1957.
18. Feldbaum, A. A.: *Computational Methods in Automatic Control*, Gostekhizdat, Moscow, 1959.
19. Gibson, J. E.: Self-optimizing or Adaptive Control Systems, *Automatic and Remote Control, Proc. First IFAL Congress*, vol. 2, pp. 586–595, Butterworths, London, 1961.
20. Whitaker, H. P., J. Yamron, and A. Kezer: Design of Model-reference Adaptive Control Systems for Aircraft, *MIT, Instrumentation Lab. Rept.* R-164, September, 1958.
21. Gibson, J. E., and E. S. McVey: Multidimensional Adaptive Control, *Proc. Natl. Electron. Conf.*, vol. 15, pp. 12–26, 1959.
22. Meditch, J. S., and J. E. Gibson: On the Real-time Control of Time-varying Linear Systems, *IRE Trans. Auto. Control*, vol. AC-7, no. 4, pp. 3–10, 1962.
23. Braun, L., Jr.: On Adaptive Control Systems, *IRE Trans. Auto. Control*, vol. AC-4, no. 2, pp. 30–42, 1959.
24. Merriam, C. W., III: Use of a Mathematical Error Criterion in the Design of Adaptive Control Systems, *Trans. AIEE*, vol. 78, part II, pp. 506–512, 1959.
25. Margolis, M., and C. J. Leondes: A Parameter Tracking Servo for Adaptive Control Systems, *IRE Trans. Auto. Control*, vol. AC-4, no. 2, November, 1959.
26. Drenick, R. F., and R. A. Shahbender: Adaptive Servomechanisms, *Trans. AIEE*, vol. 76, part II, pp. 286–291, 1957.
27. Woodworth, R. S., and D. G. Marquis: *Psychology*, 5th ed., p. 534, Henry Holt and Company, Inc., New York, 1947.
28. George, F. H.: *The Brain as a Computer*, Addison-Wesley Publishing Company, Inc., Reading, Mass., 1962.
29. Mowrer, O. H.: *Learning Theory and the Symbolic Processes*, John Wiley & Sons, Inc., New York, 1960.
30. Bush, R. R., and F. Mosteller: *Stochastic Models for Learning*, John Wiley & Sons, Inc., New York, 1955.
31. Minsky, M.: Steps toward Artificial Intelligence, *Proc. IRE*, vol. 49, pp. 8–30, January, 1961.
32. Bush, R. R.: A Survey of Mathematical Learning Theory, part II in R. D. Luce (ed.), *Developments in Mathematical Psychology*, The Free Press of Glencoe, New York, 1960.
33. Rosenblatt, F.: Two Theorems of Statistical Separability in the Perceptron, in *Mechanization of Thought Processes*, vol. I, pp. 421–472, Her Majesty's Stationery Office, London, 1959.
34. Rosenblatt, F.: A Comparison of Several Perceptron Models, in M. C. Yovitts, G. T. Jacobi, and G. D. Goldstein (eds.), *1962 Self-organizing Systems*, pp. 463–484, Spartan Books, Washington, D.C., 1962.
35. Rosenblatt, F.: *Principles of Neurodynamics—Perceptrons and the Theory of Brain Mechanisms*, Spartan Books, Washington, D.C., 1962.
36. Selfridge, O. G.: Pandemonium: A Paradigm for Learning, in *Mechanization of Thought Processes*, vol. I, pp. 513–530, Her Majesty's Stationery Office, London, 1959.
37. Widrow, B.: Generalization and Information Storage in Networks of Adaline Neurons, in M. C. Yovitts, G. T. Jacobi, and G. D. Goldstein (eds.), *1962 Self-organizing Systems*, pp. 435–461, Spartan Books, Washington, D.C., 1962.
38. Krug, G. K., and E. K. Letskii: A Learning Automaton of the Tabular Type, *Automation and Remote Control*, vol. 22, no. 10, March, 1962.

39. Hill, J. D., and G. J. McMurtry: A Learning System Proposal with Examples, School of Electrical Engineering, Purdue University, CISL Memo 62-12, August, 1962.
40. Luisi, J. A.: Pattern Recognition Applied to Learning in Control Systems, School of Electrical Engineering, Purdue University, CISL Memo 62-15, August, 1962.
41. Fu, K. S., and W. M. Wonham: Proposal for an Adaptive Control System with Learning, School of Electrical Engineering, Purdue University, CISL Memo 62-5, May, 1962.
42. Fu, K. S.: Learning Control Systems, School of Electrical Engineering, Purdue University, CISL Memo 62-14, August, 1962.
43. Bellman, R.: *Adaptive Control Processes: A Guided Tour*, Princeton University Press, Princeton, N.J., 1961.

Part V

SYSTEM TECHNIQUES

Chapter 31*

HUMAN INFORMATION-PROCESSING CONCEPTS FOR SYSTEM ENGINEERS

RICHARD W. PEW, *Human Performance Center, Department of Psychology, University of Michigan.*

CONTENTS

31-1. Introduction	31-3
31-2. Read-in Subsystem	31-4
Psychophysical Principles	31-4
Detection Sensitivity	31-4
Differential Sensitivity	31-5
Scales of Perceived Magnitude	31-5
Coding for Visual Read-in	31-5
Coding for Auditory Read-in	31-7
Other Sense Modalities	31-10
31-3. Storage Subsystem	31-10
31-4. Information-processing and Decision-making Subsystem	31-11
Variables Affecting Rate of Information Processing	31-11
Decision Making	31-13
31-5. Read-out Subsystem	31-14
31-6. Integrated System Operation	31-15
Man as a Continuous Controller	31-15
Man as a Single-channel, Limited-capacity, Information-processing System	31-17

31-1. Introduction

Even in the increasingly automated systems of today and tomorrow men will continue to play an active, important role. But as the system requirements imposed on design engineers are becoming more demanding, it is becoming increasingly important to consider alternative allocations of functions to men and machines early in the design

---

* I wish to thank Paul M. Fitts for a critical reading of the manuscript and for many helpful suggestions. This chapter was supported in part by the Advanced Research Projects Agency, under a contract monitored by the Behavioral Sciences Division, Air Force Office of Scientific Research, under Contract Number AF 49(638)-1235.

process so as to maximize the contributions of each to overall system performance. In order to take advantage of man's potential contributions as a system component, the engineer needs to have an integrated picture of the capabilities with which men are endowed and of the design limits imposed by them. This chapter will provide little detailed design data—whole textbooks and handbooks of just such data have been compiled.[1-10] Rather, the approach will be to present a point of view or design philosophy for including human operators in systems and to point out a few typical applications.

Recent developments in the study of human performance and the accumulating experience of human-factors designers suggests that an important unifying concept for design of man-machine systems lies in the view of man as a single-channel, limited-transmission-capacity, information-processing system. This view regards man as an information channel in the cybernetic sense rather than as an energy converter or power supply. Man is certainly capable of power generation, and the limits of this ability are of interest, but in modern systems most use is made of man's information-handling and -controlling abilities. It is dangerous to take this picture of a single-channel system too literally. It is too easy to find exceptions. Nevertheless, if the reader obtains a perspective of the sense in which it is valid, he will be well on the way to effective utilization of men in systems.

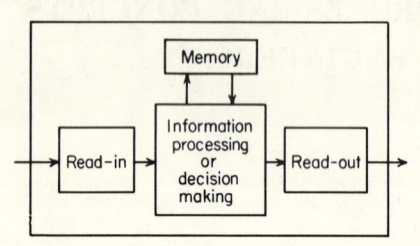

Fig. 31-1. Block diagram showing the four subsystems of the human information-processing system.

Primarily for purposes of exposition, the human information-processing system has been broken down into the four subsystems shown in Fig. 31-1: read-in, memory or storage, information processing or decision making, and read-out. Practically speaking, it is virtually impossible, even with physiological techniques, to examine the operation of any subsystem in isolation, and very difficult even to perform experiments which lead to valid inferences about independent subsystems. This breakdown is mainly a device for bringing some of the important functional characteristics of human operators into focus. Each of these subsystems will be taken up in turn, and then man as an element in a larger system context will be considered.

31-2. Read-in Subsystem

Psychophysical Principles. Before turning to the discussion of specific read-in modes, it is necessary to understand some fundamental definitions and distinctions which traditionally have been identified with the topic of *psychophysics*. Physics enters the title because the energy impinging on the input system is capable of being measured and described in physical terms. The prefix *psycho-* implies that the physical energy is in some way converted into a subjective quantity which it is not possible to measure physically. An observer reports the presence or absence of a sensation, or he categorizes or estimates the magnitude of the subjective effect produced by a physical stimulus. It is very important to distinguish between the public, physically measurable, characteristic of a sensory input and its private representation as a subjective experience. For example, in the case of sound, the physical dimensions of intensity and frequency correlate with the subjective attributes of loudness and pitch, respectively.[11] In the case of visual information, luminance, dominant wavelength, and purity are the physical correlates of brightness, hue, and saturation;[12] see Sec. 16-3. For those purposes in which the observers' reports of the effects of stimulation are the important design variables, the methods of psychophysics make it possible to describe these effects in terms of the physical characteristics of the input signal.

Detection Sensitivity. Perhaps the most basic of sensory tasks is the detection of the presence or absence of signals in noise. This process can represent the task of the search radar or sonar operator, or it can describe the task of the astronomer looking for a sixth-magnitude star. Traditionally an observer's ability to detect the presence of small amounts of energy in quiet, or stronger signals in appreciable noise, has been described by specifying his *threshold;* by one definition, this is the level of physical energy at which he will correctly identify the presence of the signal 50 per cent of the time. More recent research suggests that this concept of the threshold has limitations.[13] Observers can make two kinds of errors: a false alarm, the report of the presence of a signal when it was, in fact, absent; and a miss, the failure to detect a signal that was actually present (Sec. 24-6). This threshold concept confounds detectability with the observer's expectations about when signals will occur and with his assessment of the relative importance of the two kinds of errors. It is clear, for example, that if the preservation of a 1,000-megawatt turbogenerator system hinges on early detection of a malfunction signal, then an observer's assessed importance of correct detections will lead to a higher incidence of false alarms but would not necessarily change his basic capacity to detect the malfunction warning. Such an importance rating would, however, show a shift in the signal level at which 50 per cent correct detections were obtained. By specifying an observer's sensitivity in terms of the effective signal strength required by an "ideal" observer (Sec. 24-6) to achieve equivalent performance using statistical decision theory, rather than in terms of an all-or-none threshold, this problem is avoided and a pure measure of sensitivity to physical energy is obtained.[14] The detection-theory model on which this measurement is based calls the detection-effectiveness parameter d', and that parameter is now frequently employed to describe human observers' detection sensitivity.[15]

Differential Sensitivity. Perhaps the second most basic characteristic of subjective experience which is of interest for design work is the size of the just detectable *change* of some physical attribute of a signal or stimulus. Traditionally the size of this step has been named the JND (just noticeable difference), the *difference limen*, or the *differential threshold*. It has been found that the size of the JND is strongly a function of the base-level magnitude along the stimulus dimension from which deviations are to be detected. A rule of thumb for predicting the size of the JND is given by *Weber's law*, which states that $\Delta I/I = $ constant, where I is the reference-level magnitude along the stimulus dimension and ΔI is the size of the just noticeable change. Weber's law implies that, as the reference magnitude I increases, ΔI must be increased in a constant ratio to I in order for the change in stimulation to be maintained at the just detectable level. This relation has been shown to be precisely correct only over a rather narrow range on a few stimulus dimensions, but this does not negate its practical usefulness as an approximation. The theory of signal detectability also has been extended to include measurement of sensitivity to differences between two signals. While rather limited experimental data are available in this domain, the same concepts are applicable, and it remains important to distinguish changes in differential sensitivity dependent on the signal characteristics from the manifestations of the measurement context.

Scales of Perceived Magnitude. Frequently the engineer has a requirement for a "meter" which might read directly in units of intensity *as perceived by a human observer*. Research on the subject of psychological scaling has provided the scales on which such a meter could be based in cases of certain frequently used dimensions, such as the sone scale of perceived sound intensity given in Fig. 31-2.[11] This scale has been proposed as an international standard. The reference value for one sone corresponds to the loudness of a 1,000-cps tone at 40 db re 0.0002 dyne/sq cm. In other cases it is the *methods* of psychophysical scaling that have proved most valuable. They have been used for obtaining a variety of kinds of scales for application to problems ranging from the determination of the physical correlates of ride comfort in an automobile to the development of isopreference contours in the speech-level vs. noise-level plane for the evaluation of telephone communication systems.[16]

Coding for Visual Read-in. The ability to encode information in visual displays is dependent on basic visual capacities. Just as photographic film is sensitive only to certain wavelengths of light, so is the eye restricted in its sensitivity. There are two basic receptor systems in the human retina, the *photopic* or cone system and the *scotopic* or rod system. The sensitivities to light energy as a function of wavelength for the two systems are shown in Fig. 31-3.[18] The scotopic system is the more sensitive of the two, and its sensitivity is shifted slightly toward the blue end of the spectrum. The photopic system is dominant under daylight-illumination conditions, while only the scotopic system is effective in the dark-adapted eye. The photopic sensitivity curve has been adopted as the standard visibility or luminosity function specifying human sensitivity to light as a function of wavelength (see Sec. 16-3 and Fig. 16-3).

It is a more complicated process to specify colors and to tie the hue and saturation specifications to their physical correlates. The Munsell system of color chips which may be matched to any given color is typical of one technique for describing color

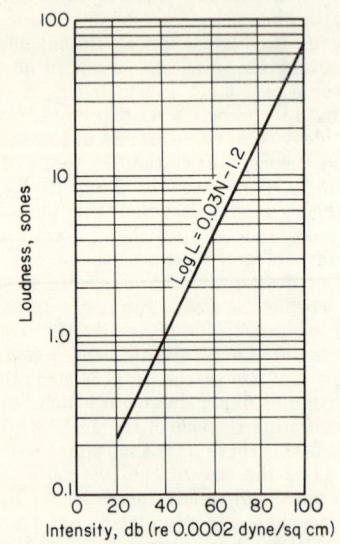

Fig. 31-2. Sone scale of perceived sound intensity (loudness) as a function of the physical intensity of the stimulating tone. These data are for a 1,000-cps tone. The equation of this function is given in the figure. (*After Morgan et al.*[7])

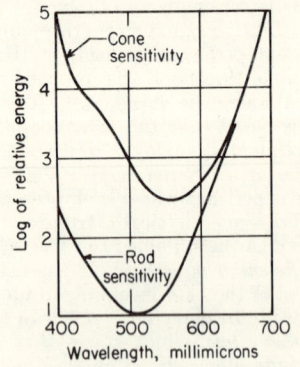

Fig. 31-3. Relative visual sensitivity of the rod and cone receptors as a function of wavelength. These curves show the relative energy required to produce a threshold response. In absolute terms, a minimum energy of 5 to 10 photons is sufficient to produce a response of seeing with the rod system at 510 mμ. (*After Bartley.*[18])

characteristics.[17] The method of matching a given color to a mixture of appropriate amounts of a set of standardized primary colors, such as the red, green, and blue of the CIE tristimulus specification, is another.[12] The associated CIE chromaticity diagram makes it possible to specify the physical correlates of hue and saturation with this method.[12]

It goes without saying that the human visual system is sensitive to differences in spatial location and to spatial and temporal patterning. Ability to distinguish temporal patterns is limited by the *critical flicker fusion frequency* (cff), which specifies the rate at which temporal changes in intensity are no longer detectable. The cff depends on a number of parameters of the stimulus such as size, differential intensity, and on-time fraction, as well as on the part of the retina being stimulated and the state of dark adaptation.[18] Spatial pattern discrimination abilities also depend strongly on the viewing conditions under which the judgments are made.

Spatial discrimination is at the heart of most coding systems utilizing the visual read-in mode. One can evaluate the quality of visual coding at several levels. At the most primitive level are questions of *visibility*. At this level one asks if the coding dimensions, be they letters, numerals, or shapes, are discriminable given unlimited viewing time. However, if one is interested in the efficiency of coding, a time constraint must be imposed and further coding variables come into play. The term *legibility* has been used to describe studies which have examined the effects of coding variables when both speed and accuracy of discrimination are important.[19] At the next higher level it is necessary to inquire beyond simple discriminability of the coding elements to ask about the interpretability of the code. With what speed and accuracy can the observer extract information from the visual display? It is generally true that visibility and legibility enhance interpretability and that designers should consider each level in turn, but these are not the only considerations. It is difficult to make a bad coding system good simply by enhancing visibility and legibility factors.

If optimum interpretability of quantitative readings were all that was required of good visual displays, then most would employ in-line alphanumeric character presentation for digital read-out. However, most visual displays serve a multitude of purposes. Early in a mission they may be required to provide detailed quantitative readings, while later they may be called upon to indicate trends as displays for a controlling task. At one stage of an operation they are check-read for go–no-go information while sometime later it is necessary to set in new critical values defining the "go" region. From the design standpoint these activities impose conflicting display requirements. Table 31-1 indicates in a qualitative way the applicability of each of three kinds of visual displays pictured in Fig. 31-4 to the four kinds of tasks just enumerated. It can be seen that no one display is optimum for all tasks, and hence the relative importance of these kinds of operations in a particular application should influence the choice of instrument design.[7] In the design of individual visual displays, whether they are pictorial or alphanumeric, the most general design principle suggests that better interpretability will result if they require as little recoding or as few translational steps by the operator as possible. This accounts for the trend toward pictorial situation displays, but it must be remembered that, for many situations calling for quantitative information, a numerical read-out remains the most directly meaningful mode.

The system approach to display design must go beyond the individual display with the operator in the display-control loop, and consider that inevitably the operator multiplexes his read-in operation, utilizing several displays in combination. Research has made it clear that it takes time for people to shift their attention from one display code to another, and this time must be added to the time required for a shift in eye fixation. One of the purposes of integrated display-system design—the so-called whole-panel approach—is to minimize the time it takes to read in information from multiple displays. At the same time that interpretability is enhanced by reducing the recoding requirement on individual displays, it is equally important to design display systems so as to ensure compatability among the different instruments, thus avoiding severe shifts in orientation or "set." Data obtained from eye-movement records concerning the number and duration of eye fixations on each display, and the probability of fixation shift from each display to the others, provide one kind of measure of success in achieving display-system compatibility.[20]

Coding for Auditory Read-in. As with any other kind of wave motion, the frequency-intensity-time pattern uniquely specifies the physical characteristics of a sound.[11] It is not surprising, then, that the ear is a remarkably high-fidelity encoder of frequency-intensity-time patterns. Figure 31-5 shows typical auditory-sensitivity curves as a function of frequency.[21] These curves indicate the hearing abilities of an adult population on a percentile basis. The 1 per cent curve implies that, of the sample tested, 1 per cent could detect tones of the given frequency at the given level or better. The remaining 99 per cent were less sensitive.

These curves define detection sensitivity. Above this level the auditory system responds to a remarkably wide intensity range, approximately 14 orders of magnitude from a sound pressure just greater than the thermal noise associated with molecular motion (10^{-16} watt/sq cm) to the intensity of the sound field adjacent to a turbojet aircraft (10^{-2} watt/sq cm). Within this range the perceived loudness of a pure tone

31-8 SYSTEM TECHNIQUES

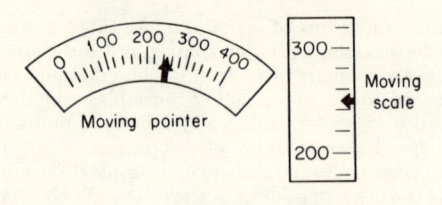

Fig. 31-4. Illustration of three typical kinds of visual displays which are evaluated in Table 31-1.

Table 31-1. Qualitative Evaluation of Three Types of Displays Shown in Fig. 31-4 for Four Kinds of Applications*

For ...	Moving pointer is ...	Moving scale is ...	Counter is ...
Quantitative reading	Fair	Fair	Good (requires minimum reading time with minimum reading error)
Qualitative and check reading	Good (location of pointer and change in position are easily detected)	Poor (difficult to judge direction and magnitude of pointer deviation)	Poor (position changes not easily detected)
Setting	Good (has simple and direct relation between pointer motion and motion of setting knob, and pointer-position change aids monitoring)	Fair (has somewhat ambiguous relation between pointer motion and motion of setting knob)	Good (most accurate method of monitoring numerical settings, but relation between pointer motion and motion of setting knob is less direct)
Tracking	Good (pointer position is readily monitored and controlled, provides simple relationship to manual-control motion, and provides some information about rate)	Fair (not readily monitored and has somewhat ambiguous relationship to manual-control motion)	Poor (not readily monitored, and has ambiguous relationship to manual-control motion)
General	Good (but requires greatest exposed and illuminated area on panel, and scale length is limited)	Fair (offers saving in panel space because only small section of scale need be exposed and illuminated, and long scale is possible)	Fair (most economical in use of space and illuminated area, scale length limited only by number of counter drums, but is difficult to illuminate properly)

* Reproduced from Ref. 7, p. 95. Copyright 1963 by McGraw-Hill, Inc. Used by permission.

HUMAN INFORMATION-PROCESSING CONCEPTS

is specified by the sone scale illustrated in Fig. 31-2. To assess the effective loudness of complex sounds is a more difficult task.

A typical approach to defining the loudness of complex sounds or noises involves empirically determining the width of a series of contiguous frequency bands which span the audible spectrum and which contribute equally to overall loudness, and then evaluating the intensity level of the unknown sound within each of these bands. The sum of the loudness contribution of each band provides an index of overall perceived loudness.[22] The bands determined in this way are closely related to the maximum width of a band of noise which is effective in masking an observer's detection of a pure tone centered in the middle of the noise band, called the *critical bandwidth*. It is as if the auditory system were able to set a filter of the critical bandwidth about the tone and reject the noise outside this band. The width of this theoretical filter is not

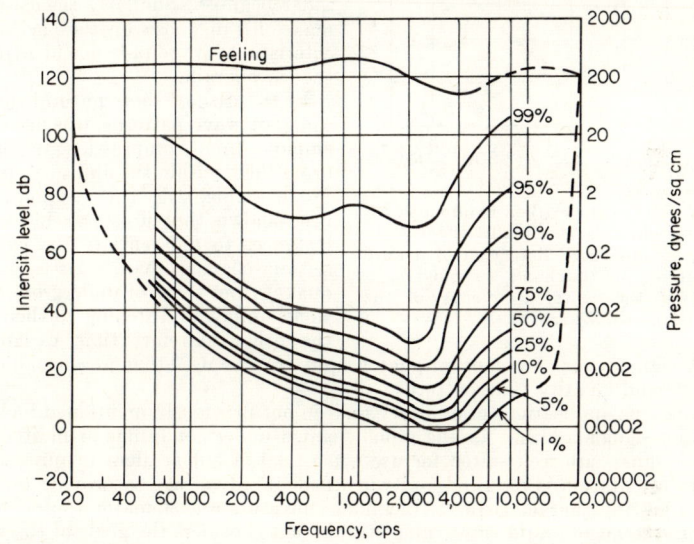

FIG. 31-5. Auditory-detection sensitivity for pure tones as a function of frequency of the tone. Just detectable intensity is given in decibels re 0.0002 dyne/sq cm on the left-hand ordinate and in absolute pressure units on the right-hand ordinate. The parameter of the curves is the percentage of the measured population having a threshold as low or lower than the given curve. (*Reproduced from Pierce and David,*[21] *p. 105, copyright © 1958 by John R. Pierce and Edward E. David, Jr. Reprinted by permission of Doubleday & Company, Inc.*)

constant over the frequency spectrum. Figure 31-6 shows an estimate of critical bandwidth as a function of its center frequency.[23] These same bands are useful also in defining regions along the frequency spectrum contributing equally to the intelligibility of speech.[13] The concept of the *critical band* provides a valuable way to conceptualize one aspect of the operation of the auditory read-in mode which is useful for thinking about problems involving the effects of summing energy distributed along the auditory spectrum. For example, it suggests that in a communication system little gain in speech intelligibility will be achieved by filtering out noise which is separated from the frequency region containing significant speech power by more than one critical bandwidth.

While the intensity and frequency dimensions are prerequisite, it is the integration of these into the time dimension, the ability to detect and identify coherent temporal patterning, which gives the greatest importance to the auditory mode. Speech communication is by far the most important instance in which man exploits this ability. There is a large literature on the topic of speech intelligibility under various conditions of noise, distortion, bandwidth compression, etc., as well as reference material dealing with methods of measuring and predicting speech intelligibility under these conditions, which can only be referenced here.[7,24] Among other applications, temporal patterning has long been the important dimension for the sonar listener and, more recently, for the radar operator as well. Just as the trained sonar operator can identify a variety of underwater objects by their characteristic sound patterns, so can the radar operator, by listening to a signal derived from the envelope of a Doppler or pulse-Doppler tracking-radar return, distinguish target signatures which provide information about the characteristics of the target under surveillance in the radar beam.[25] Similarly the cardiologist makes his diagnosis on the basis of characteristic heart sounds not identifiable in the cardiogram.

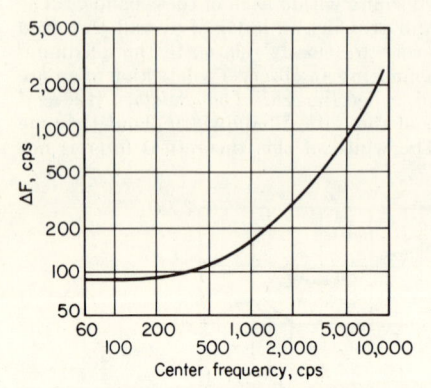

FIG. 31-6. Size of the critical bandwidth ΔF of a noise which is maximally effective in masking an observer's detection of a pure tone centered in the band, graphed as a function of the center frequency of the band. (*After Greenwood.*[23])

In the case of some natural sources of coherent wave patterns, it is necessary to employ time compression in order to bring them within the auditory spectrum. For example, with this technique it has been shown that it is possible in many instances to differentiate the pattern of seismic signals characteristic of earthquakes from those of underground explosions simply by listening to them.[26] At the other extreme, time expansion of auditory signals, like slow-motion photography, can be used to increase man's ability to analyze and identify time patterns.

Two obvious applications of auditory read-in not previously mentioned include use as warning signals and use as the input channel under conditions of limited vision. Auditory signals are well suited for use when read-in independent of physical body and head orientation is required and when it is an advantage to provide immediate attention-getting characteristics. Licklider[27] presents a systematic approach to the design of systems of auditory warning signals that considers the goals of the warning system as a whole. The appropriateness of speech and other auditory displays for use when vision is restricted or impaired by physical constraints is self-evident.

Other Sense Modalities. While hearing and vision are used predominately for information read-in, the skin senses of touch, pain, temperature, or electrical stimulation, and the chemical senses of taste or smell also can serve as information-input transducers. Geldard[28] has done extensive research on the vibration sensitivity of the skin and has designed a simple coding system utilizing vibration location, amplitude, and duration as dimensions for coding information. His system has an alphabet of 45 discriminable symbols. One potential application is for communication with the deaf and blind. The sensory capacities of the other modalities have been studied, but their application as information-input channels has thus far been limited.

31-3. Storage Subsystem

Memory has, of course, a major role in making it possible for man to integrate the effects of previous experience into ongoing activities. Without memory there could be no learning or adaptation to the environment. However, the aspect of memory most relevant to this discussion is more analogous to the computer buffer-storage capacity. In the process of its acquisition, information is in some sense encoded for longer-term storage or for further processing. Most practical tasks require a short-term memory in some form. The air-traffic controller should have in store the status of a number of aircraft, current weather information, and a plan for preserving safe separation.

The ability to utilize trend information and predict future states of a system on the basis of events in the recent past depends on available storage capacity.

Recent research in the area of short-term memory points to the importance of the structural coherence of the units to be remembered in determining storage capacity for these units.[29] In contrast with other human information-handling abilities in which it is the amount of information in the technical sense (see Sec. 22-3) which governs limiting capacity, in the case of memory it is the number of integrated units to be remembered that is particularly important. With verbal materials, three monosyllabic words are about as easily recalled as three randomly selected consonants. However, for engineers the consonant trigram, CGS, would act more like a single unit than three units. These observations have led to the concept of the *chunk* of information, the name given to the element which is encoded in memory as a single unit. While certain kinds of units clearly constitute chunks, no defining operation has yet been brought forward to make the meaning precise. Nevertheless, using monosyllabic words which may be considered an approximation to chunks, the curves of Fig. 31-7 have been obtained. They describe the effective short-term memory capacity for two to six words as a function of the time in storage. The curves for two, three, and five words are based on experimental data obtained by Noyd;[30] the others are theoretical extrapolations. These curves represent the task in which the observer is presented with the to-be-remembered units once and is then immediately engaged in an irrelevant task designed to prevent further overt or covert study of the test item until the time when he is required to recall it. If controlled repetition of the items is allowed before the irrelevant task is introduced, the slope of these curves is systematically shifted in the direction of slower forgetting of the test items.[29] Finally, the short-term memory function is also influenced by the interference both from the immediately preceding mental activities, called *proactive interference*, and from the activity interpolated between the presentation and recall of a given unit, called *retroactive interference*. The curves of Fig. 31-7 were obtained when the level of proactive interference was high and in a steady state;

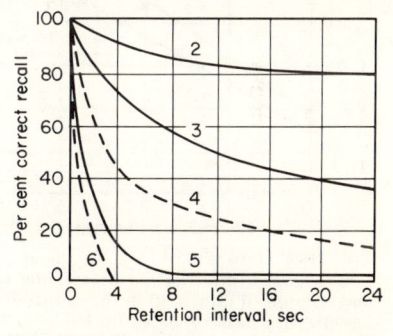

FIG. 31-7. Short-term memory capacity (in per cent of items recalled completely correct) for monosyllabic nouns as a function of the length of the interval between presentation and recall. The parameter of the curves is the number of words held in store on each trial. The curves for two, three, and five words are from the data of Noyd.[30] The dashed curves for four and six words are extrapolations from these data.

that is, the measurements were made in the context of many prior trials of presentation and recall of other units. The interpolated activity, in this case the irrelevant intervening task, provided a constant level of retroactive interference throughout the period between presentation and recall.

Since by these methods it now appears possible to define the basic limitations of short-term memory capacity, it is important to consider human operators' tasks in light of these limitations.

31-4. Information-processing and Decision-making Subsystem

Man's abilities for information processing, the qualitative and quantitative manipulation and combination of input and stored data, form the primary basis for his desirability as a system component. While there is much in the processes of decision making, thinking, and creativity which remains undescribed in terms useful for engineering analysis, advances are being made rapidly.

Variables Affecting Rate of Information Processing. It is incumbent upon the design engineer to make efficient use of his resources. When these resources include men, the answers to such questions as "How much can a man do?" and "How many men will be required for this set of tasks?" depend intimately on knowledge of human information-handling rates. The basic framework for discussing information rate is founded in the concepts describing the human operator as an information channel in the Shannon sense (see Sec. 12-2 and Chapter 22). This framework suggests that the operator transmits information at a rate that is dependent on the particular task; when more information (measured in bits) is transmitted in a given task, a longer time is required.

Hick's[31] early data, as shown in Fig. 31-8, are typical of the linear relationship that has been obtained empirically in reaction-time tasks in which the operator is required to respond appropriately and as rapidly as possible to a particular input signal drawn at random from the set of possible signals. The information transmitted was varied by changing the size of the "alphabet" of alternative signals. This linear relationship has been shown to hold also when subjects speed up, but make errors.[31] The occurrence of errors reduces the amount of information transmitted by an amount which just compensates for the increase in speed. Similarly when the amount of information transmitted is reduced by introducing redundancy in the stimulus sequence, either by making some alternatives more likely than others or by imposing sequential dependencies on the series, the time to process the information often remains a linear function of the average information transmitted.[32] The simple equation $P_t = a + bH_t$ where P_t represents the processing time, H_t the amount of information processed, and a and b are constants dependent on the task context, describes the results remarkably well. This relation has been called *Hick's law*.

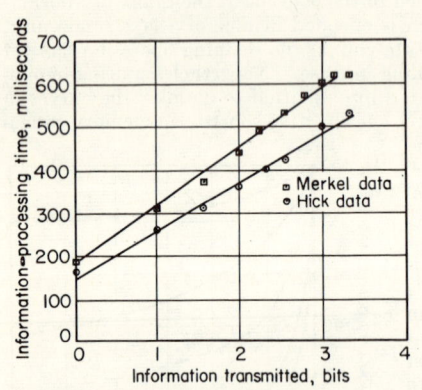

FIG. 31-8. Data illustrating the linear relation between information-processing time and amount of information transmitted in a key-pressing task reported by Hick.[31] The two curves represent two sets of data, one obtained by Merkel in 1885 and the other by Hick himself. The slightly different slopes and intercepts of the two curves reflect the effects of different stimulus and response codes used in the two experiments.

These findings are only the beginning of the story, however. Several other independent variables have been shown to affect rate of transmission in information-conserving tasks in which the operator's purpose is to make a one-to-one correspondence between the input and output information. These include the following:

1. If the input signal-code symbols are less than perfectly discriminable, the rate of processing will be decreased.[33]

2. The *compatibility*, or congruence, of the relationship between the input code and the output code is an important determinant of processing rate.[34,35] For example, if the operator simply responds by touching the appropriate input light, the two codes are highly compatible, and the value of b in the equation defining Hick's law will be small. If, on the other hand, the input information is expressed numerically and the output information is represented by spatially distributed keys to be pressed, the value of b will be larger.

3. The effect of practice at a task will be to decrease the value of b as well as the value of a. It is likely that a greater decrease in b will result if the task was initially an incompatible one than if it was a compatible one.[35]

4. There are a variety of other factors having to do with the operator's orientation to the task, such as his expectation or set concerning which input signals will occur,[36] when they will occur,[37] and the relative importance of speed and accuracy of performance, which also affect his rate of information processing.[38]

The data thus far considered apply mainly to information-conservation tasks, but many of the "interesting" operations men perform on input and stored information involve reduction of information in one way or another. The system monitor acts as an information-scanning gate or filter, always alert to those data which imply a deviation from usual system operation. Certain intellectual operations involve information compression. The operator combines information from a variety of sources and reflects it in a single output operation or decision. Compression for the human operator is equivalent to data reduction for a computer. The importance of the concept of information compression is that it shows promise as a way to quantify the amount of thinking that a task requires. For some simple cases, like accumulating the sum of pairs of two-digit numbers and classifying two-digit numbers into two categories on the basis of high-odd or low-even vs. high-even or low-odd, Posner[39] has found that the amount of information compression required, i.e., the difference between the input information and the appropriate response information (in bits), bears a linear relation to the time required to accomplish the compression task: the more compression, the slower the processing. We are a long way from quantifying thinking in all its intimate details, but information reduction—filtering and compressing—provides useful concepts for considering the rate of information processing in heretofore unquantified domains.

Decision Making. More and more frequently the tasks assigned to men in systems involve information processing in situations in which speed of processing is not the most important dimension, but the quality of the decision is. In automated systems the emphasis is on the monitoring function and, in economic terms, large values and costs are associated with decisions about whether shutdown or corrective action is required. The models based on game theory (Sec. 23-8) and statistical decision theory (Chapter 24) form bases for systematic formulation of rational decisions under uncertainty which have been applied in human decision-making research.[40]

In a game against nature the possible states of nature (indicated by index i; $i = 1, 2, \ldots, n$) and the available courses of action (indicated by index j; $j = 1, 2, \ldots, m$) form a two-dimensional decision matrix defining the outcome structure of the game. To each cell of this matrix is assigned a value V_{ij}, the amount to be gained or lost, associated with each outcome. A further relevant variable is the probability P_i associated with each of the alternative states of nature. This probability reflects the likelihood that any particular state is, in fact, the true state at the time the decision is to be made.

It is possible to propose many different criteria of good decisions based on the information given.[40] Not all will lead to the same decision. Maximizing expected value (EV) is one criterion which is appropriate to many situations. Under this criterion the operator selects the action (j) so as to maximize over m the function $\text{EV} = \sum_{i=1}^{n} V_{ij}P_i$, thus taking into account both the probability of each possible state and the values associated with all possible outcomes.

We may ask two direct questions about this approach to rational decisions under uncertainty: (1) Is it a good guide for men to follow in making decisions? and (2) Is it descriptive of what men actually do? The answer to the first is a qualified "yes" and to the second an unqualified "no." The logic of maximizing expected value is sound. The problem is that rarely in the real world are objective data about P_i or V_{ij} available. Since each decision situation is unique and we see only a single outcome revealed, it is often difficult to obtain the classical kind of relative-frequency assessment of P_i. Similarly, only in certain limited circumstances can a well-defined value be placed on an outcome. As for what men actually do, research has shown large deviations from expected value predictions. Men occasionally fail to satisfy even the simplest axioms of rational behavior.[40]

A popular way to circumvent the limitations of expected-value maximization as a description of actual decision-making behavior involves two modifications. The first is to replace the objective-probability term with a *subjective* or perceived *probability* in the same sense that the perceived attributes of sensory stimuli were discussed in Sec. 31-2. The second is to replace actual value with the subjective, perceived equiva-

31-14   SYSTEM TECHNIQUES

lent of value, *utility*. Subjective probability and utility usually are assumed to bear a monotonic relation to their respective objective quantities. The resultant theory is called subjectively expected utility maximization (SEU), and it has been shown to be capable of describing decision-making behavior over a wide range of conditions.[40] Its limitation lies in the looseness of the description inherent in a model with two parameters which are only indirectly measurable and difficult to infer independently.

The SEU model has been applied mainly to *static* decision-making tasks, the cases in which all the evidence that is to be available arrives at once and a single decision is required. In contrast with this type of task is the more general case of *dynamic* decision making in which the operator processes information as it arrives and makes a sequence of one or more decisions which may or may not be interdependent. While this description is more typical of the real world, it is also more difficult to analyze. The dynamic-programming methods of Bellman (see Chapter 25) are relevant here, as is Bayesian statistical inference.[41]

One fruitful result of recent research in the area of decision making has been to question seriously the desirability of having human operators assimilate all the relevant data as well as serve in the traditional role as the final link, the action selector, in decision-making systems. Now that mathematical models of rational decision processes are available, it is possible to consider giving human operators a more direct evaluative role in such systems. For example, if an operator receives the datum that defensive radar has identified an enemy aircraft vectored toward North America, and from experience can assess the likelihood ratio based on this datum between two hypotheses—enemy attack and friendly visit by enemy—then a computer can utilize this information in combination with similar evaluative reports from other specialists to arrive at the most reasonable conclusion about the true state of the world. Research on this kind of approach is just beginning.[41] Those pursuing it believe that experts making even relatively imprecise estimates of the relative impact of data for each of the alternative hypotheses will, in conjunction with a computer applying optimum processing, arrive at better decisions faster.

At the present time the data on human decision-making capabilities are provocative. One of the important challenges remaining is to interpret and apply them to management decisions in business and government contexts.

31-5. Read-out Subsystem

The read-out subsystem is perhaps the most difficult to discuss in isolation. Any human output depends on the operation of some part of man's musculature. The muscle system involves feedback loops at several levels, and all but the lowest level interact with the central processing system. Thus responding, whether it involves the speech musculature or an arm or foot output, is intimately tied to the operation of the system as a whole.

One property of the read-out subsystem which is of interest is the maximum response rate when the system is relatively free of central-processing operations. The maximum rate of tapping with the fingers can be readily demonstrated to be about 8 to 10 taps per second. Similarly the maximum rate of repeating well-familiarized monosyllables is about 8 per second. A singer's vibratto has a frequency of 5 to 7 cps. The predominate periodicity of muscle tremor is in the range of 10 cps. These and other data suggest that there is a maximum rate of about 8 to 10 per second at which systematic changes in muscle activity can be organized.

When an accuracy constraint is imposed on movements, these rapid rates cannot be maintained. As the accuracy tolerance of a movement is made more strict relative to its amplitude, it takes longer and longer to make the movement. The implication is that more information processing is required to carry out accurate movements. Fitts[42,43] has shown that the concept of channel capacity applies here. To identify the trade-off between the amplitude, speed, and accuracy of a movement, Fitts has defined an index of difficulty: $I_D = -\log_2(W/2A)$, where W is the width of the target zone constraining the accuracy and A is the amplitude of the movement. For a given I_D, movement time will be constant over a range of values of W and A.

HUMAN INFORMATION-PROCESSING CONCEPTS   31-15

Further, movement time T_M has been shown to bear a very precise linear relation to I_D. Product-moment correlations between T_M and I_D are in the range of 0.98; the larger the value of I_D, the longer the movement time. This unambiguous relation between speed and I_D has been interpreted as implying a fixed informational capacity of the central processes for controlling skilled movements. Within this boundary operators can trade speed for accuracy.

It has long been supposed that some "voluntary" movements do not depend heavily on central information-processing control, particularly after they are well practiced. Today we say that such movements can be preprogrammed and carried out as a subroutine without reference back to the executive program until they are completed. It is clear that such activities as walking do not require much executive monitoring until a sidewalk irregularity calls our attention to the requirement for compensation. In industrial assembly operations, one index of improved assembly cycle time is the systematic decrease in the number of eye fixations required per cycle. This reflects the development of preprogrammed elements in the cycle which rely on more direct kinesthetic feedback paths rather than on the slower-acting visual monitoring loop.

The conclusion which emerges from this discussion is that the organization and control of skilled movements develop hierarchically. At the highest level, the executive information-processing and decision-making subsystem is involved in closed-loop control of unpredictable or unpracticed movements no matter what part of the musculature is involved. However, as the movement is learned or made more predictable, the executive levels are freed from the need to control the movement continuously, and larger and larger sequences are carried out as subroutines with less frequent executive monitoring.[44] This results in faster movements which are ultimately limited by the fundamental rate limits on organized muscle activity and not by central information-processing-rate limits.

31-6. Integrated System Operation

Man as a Continuous Controller. Manual-control systems such as those typified by aircraft control or automotive steering continue to be important areas of application of human operators. While there has been a great deal of research on the effects of the dynamic characteristics of the physical components of the man-machine system on overall closed-loop performance,[7] useful results for system designers also emerge from the research directed toward developing representations for the human operator in the same terms used to describe other system elements, namely, the models of automatic-control-system theory (see Chapter 29).

The block diagram of Fig. 31-9 shows the components of a man-machine control system. The human-operator system function shown there is the best quasilinear representation of man's dynamic characteristics that is available from empirical measurement to date.[45] The term quasilinear implies an averaged linear description, smoothed over short periods of time, usually 2 to 5 min. This system function has been derived in the context of a compensatory display which provides the operator with error information only and with no knowledge of system output performance independent of error. It applies only to input signals which have the properties of band-limited random noise, or are random-appearing; periodic signals often allow the operator to anticipate exactly the appropriate response, thus grossly changing his apparent dynamic characteristics. In spite of these restrictions in generality which were required to provide a tractable measurement problem, the system function thus derived allows major advances to be made in manual-control-system design. It makes it possible to include human-operator dynamics in preliminary design considerations where vehicle configurations are determined. McRuer, Krendel, Ashkenas, and their colleagues are largely responsible for developing the interpretations of the research findings which make such applications possible.[46,47]

The data which are summarized in the human-operator block of Fig. 31-9 suggest that man performs as an adaptive, optimizing controller. The terms indicated by the right-hand inner block represent the fixed liabilities with which man is endowed.

He has a pure transmission lag, given by $e^{-\tau s}$ in which τ is taken to be 0.14 sec. The denominator term $1/(T_N s + 1)$ represents a neuromuscular lag. The value of T_N is roughly 0.10 sec. The left-hand block indicates man's adaptive assets for adjusting to the demands of the system he is controlling. The lead time constant T_L has been observed to be adjustable over the range 0 to 5 sec, but generation of values larger than 1 sec appears to require more "effort" on the part of the operator. This lead equalization is utilized by the operator to compensate for his built-in time delay and neuromuscular lag and to improve the system high-frequency performance. The lag time constant T_I can be adjusted over a wider range. Values at least as high as 10 to 15 sec are obtainable. Introduction and adjustment of this equalization lag serve to improve low-frequency characteristics. The operator can adjust his gain K over a wide range, but the limiting values have not been determined.[46]

Given these adaptive capabilities, McRuer and Ashkenas suggest that the operator attempts to minimize the rms error generated and maintain a phase margin, the value of phase shift at the frequency corresponding to 0 db closed-loop gain, at about 40°.[46] So long as the combination of input spectrum bandwidth and plant dynamics permits these optimization goals to be achieved without exceeding the equalization capability which the operator has available, he will do a remarkably good job of optimizing overall system performance in terms of the rms error criterion.

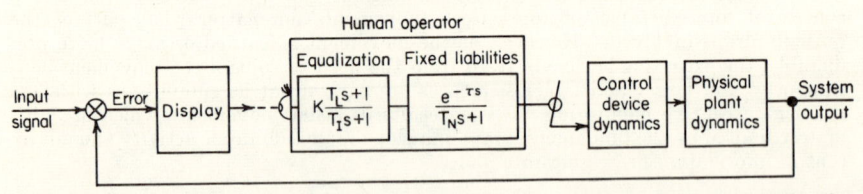

FIG. 31-9. Block diagram of man-machine control system showing the system function for the best available quasilinear representation of human-operator dynamic characteristics in a compensatory tracking system.

The engineer should not consider that his design task is hopeless simply because the unalterable physical plant to be controlled does not lead to a stable system with good performance characteristics when the loop is closed around the man. Among the many ways to reduce the equalization demands on the operator, three will be suggested which have been shown to be applicable to manual-control systems.

1. The operator's ability to introduce lead equalization improves greatly if derivative feedback of the system output to the display, called *display quickening*, is provided.[7]

2. When it is possible to construct a fast-time simulation model of the physical plant, this model can be used to generate in real time a prediction of the future response of the system given its present initial conditions and the plant dynamics. Presentation of this anticipatory information, called a *predictor display*, permits the operator to evaluate the consequences of his control actions before divergent conditions actually develop in the real plant.[48]

3. When it is possible to allow the operator a preview of the input signal, much like the forward view from an automobile of the road ahead, he is better able to compensate for his own transmission delay.[49]

It is becoming more common to inquire whether a nonlinear or a sampled-data model would not be a more accurate and hence more useful model of operator performance than the type of quasilinear model considered here. Such models have been developed, and under similar restricted conditions have been shown to provide a good representation of operator performance. With the possible exception of Bekey's sampled-data model (see Sec. 32-4) they are usually elaborated only in terms of analog-computer mechanizations and are not so easily applicable to the system-synthesis domain as the quasilinear models. When the input-signal bandwidth is less than 1 cps, as is frequently true in practical applications, the quasilinear model provides a surprisingly good description of short-term average performance in single-loop compensatory systems with random-appearing input signals.

Man as a Single-channel, Limited-capacity, Information-processing System. This chapter has been organized around the central concept of the human operator as a single-channel system. While man has multiple read-in and read-out modes, the information-processing subsystem is best thought of as having a limited-capacity single channel. The limited-capacity interpretation of the trade-off between speed and accuracy has been discussed with reference to both information-transmission tasks and skilled movement. The effects described in the simplest terms as "focusing of attention" imply that the processing channel can be programmed to attend to sensory events, to central-processing operations, or to feedback evaluation, but in so doing other environmental changes will be neglected. With respect to skilled movements it was pointed out that highly practiced muscular control can be carried on with a minimum of "executive routine" supervision, thus releasing the processing channel for other tasks.

In continuous manual-control systems, the extent to which equalization is required on the part of the operator, particularly lead equalization, determines in some sense the information-processing "demands" of the task and hence the extent to which other activities can be carried on concurrently. Thus it behooves the system designer to pay careful attention to the equalization requirements of control tasks in situations where overall system performance places other task demands on the operator.

The single-channel, limited-capacity concept carries with it the logical implication that the performance of more than one task at a time will involve a multiplexing operation, whenever higher-level control or monitoring are required, since parallel processing capability is not available. The practical importance of this mode for display design was touched upon with reference to integrated instrument-panel development. The same philosophy can be extended to integrated system operations. Men multiplex most efficiently when the tasks assigned require minimum transition time. The total time spent in transition depends on the relative distribution of load among tasks, the required "dwell time" on any particular task, and the nature and compatibilities of the various sets and expectancies required for performance of the set of tasks to be multiplexed. Creative system design at this level embodies all the principles set forth in the preceding sections of this chapter.

REFERENCES

Handbooks and General Texts

1. Bennett, E., J. Degan, and J. Spiegel (eds.): *Human Factors in Technology*, McGraw-Hill Book Company, New York, 1963.
2. Chapanis, A., W. R. Garner, and C. T. Morgan: *Applied Experimental Psychology*, John Wiley & Sons, Inc., New York, 1949.
3. Fogel, L. J.: *Biotechnology: Concepts and Applications*, Prentice-Hall, Inc., Englewood Cliffs, N.J., 1963.
4. Gagne, R. M.: *Psychological Principles in System Development*, Holt, Rinehart and Winston, Inc., New York, 1961.
5. McCormick, E. J.: *Human Factors Engineering*, McGraw-Hill Book Company, New York, 1964.
6. McFarland, R. A.: *Human Factors in Air Transport Design*, McGraw-Hill Book Company, New York, 1946.
7. Morgan, C. T., J. S. Cook, A. Chapanis, and M. W. Lund (eds.): *Human Engineering Guide to Equipment Design*, McGraw-Hill Book Company, New York, 1963.
8. Stevens, S. S. (ed.): *Handbook of Experimental Psychology*, John Wiley & Sons, Inc., New York, 1952.
9. Woodson, W. E.: *Human Engineering Guide for Equipment Designers*, University of California Press, Berkeley, Calif., 1954.
10. *Handbook of Human Engineering Data for Design Engineers*, Tufts College, Medford, Mass., 1949.

Specific References in Text

11. Licklider, J. C. R.: Basic Correlates of the Auditory Stimulus, chap. 25 in S. S. Stevens (ed.), *Handbook of Experimental Psychology*, John Wiley & Sons, Inc., New York, 1952.

12. Judd, D. B.: Basic Correlates of the Visual Stimulus, chap. 22 in S. S. Stevens (ed.), *Handbook of Experimental Psychology*, John Wiley & Sons, Inc., New York, 1952.
13. Licklider, J. C. R.: Three Auditory Theories, in S. Koch (ed.), *Psychology: A Study of a Science*, vol. 1, pp. 42–139, McGraw-Hill Book Company, New York, 1958.
14. Egan, J. P., and F. R. Clarke: Psychophysics and Signal Detection, chap. 4 in J. Sidowski (ed.), *Experimental Methods and Instrumentation in Psychology*, McGraw-Hill Book Company, New York, 1964.
15. Swets, J. W., W. P. Tanner, and T. G. Birdsall: Decision Processes in Perception, *Psychol. Rev.*, vol. 68, pp. 301–340, 1961.
16. Munson, W. A., and J. E. Karlin: Isopreference Method for Evaluating Speech Transmission Circuits, *J. Acoust. Soc. Am.*, vol. 34, pp. 762–774, 1962.
17. Munsell, A. H.: *A Color Notation*, Munsell Color Co., Baltimore, Md., 1936.
18. Bartley, S. H.: The Psychophysiology of Vision, chap. 24 in S. S. Stevens (ed.), *Handbook of Experimental Psychology*, John Wiley & Sons, Inc., New York, 1952.
19. Fitts, P. M.: Engineering Psychology and Equipment Design, chap. 35 in S. S. Stevens (ed.), *Handbook of Experimental Psychology*, John Wiley & Sons, Inc., New York, 1952.
20. Fitts, P. M., R. E. Jones, and J. L. Milton: Eye Movements of Aircraft Pilots during Instrument-landing Approaches, *Aeron. Eng. Rev.*, vol. 9, pp. 1–16, 1950.
21. Pierce, J. R., and E. E. David: *Man's World of Sound*, Doubleday & Company, Inc., Garden City, N.Y., 1958.
22. Stevens, S. S.: Procedure for Calculating Loudness, Mark VI, *J. Acoust. Soc. Am.*, vol. 33, pp. 1577–1585, 1961.
23. Greenwood, D. D.: Critical Bandwidth and the Frequency Coordinates of the Basilar Membrane, *J. Acoust. Soc. Am.*, vol. 33, pp. 1344–1356, 1961.
24. Licklider, J. C. R., and G. A. Miller: The Perception of Speech, chap. 26 in S. S. Stevens (ed.), *Handbook of Experimental Psychology*, John Wiley & Sons, Inc., New York, 1952.
25. Pew, R. W., and J. I. Elkind: Radar Target Identification by Aural Display, *IRE Intern. Conv. Record*, part 5, vol. 9, pp. 294–301, 1961.
26. Spieth, S. D.: Seismometer Sounds, *J. Acoust. Soc. Am.*, vol. 33, pp. 909–916, 1961.
27. Licklider, J. C. R.: Audio Warning Signals for Air Force Weapon Systems, *WADD Tech. Rept.* 60-814, Wright Patterson Air Force Base, Ohio, March, 1961.
28. Geldard, F. A.: Adventures in Tactile Literacy, *Am. Psychol.*, vol. 12, pp. 115–124, 1957.
29. Melton, A. W.: Implications of Short-term Memory for a General Theory of Memory, *J. Verb. Learn. Verb. Behav.*, vol. 2, pp. 1–21, 1963.
30. Noyd, D. E.: Proactive and Intra-stimulus Interference in Short Term Memory, Unpublished Ph.D. Thesis, University of Michigan, Ann Arbor, Mich., 1965.
31. Hick, W. E.: On the Rate of Gain of Information, *Quart. J. Exptl. Psychol.*, vol. 4, pp. 11–26, 1952.
32. Fitts, P. M., J. R. Peterson, and G. Wolpe: Cognitive Aspects of Information Processing. II, Adjustments to Stimulus Redundancy, *J. Exptl. Psychol.*, vol. 65, pp. 507–514, 1963.
33. Welford, A. T.: The Measurement of Sensory-motor Performance: Survey and Reappraisal of Twelve Years' Progress, *Ergonomics*, vol. 3, pp. 189–229, 1960.
34. Fitts, P. M., and C. M. Seeger: S-R Compatibility: Spatial Characteristics of Stimulus and Response Codes, *J. Exptl. Psychol.*, vol. 46, pp. 199–210, 1953.
35. Deininger, R. L., and P. M. Fitts: Stimulus-response Compatibility, Information Theory, and Perceptual-motor Performance, in H. Quastler (ed.), *Information Theory in Psychology*, pp. 316–349, The Free Press, Glencoe, Ill., 1955.
36. Fitts, P. M., and G. Switzer: Cognitive Aspects of Information Processing: I. The Familiarity of S-R Sets and Subsets, *J. Exptl. Psychol.*, vol. 63, pp. 321–329, 1962.
37. Klemmer, E. T.: Simple Reaction Time as a Function of Time Uncertainty, *J. Exptl. Psychol.*, vol. 54, pp. 195–200, 1957.
38. Fitts, P. M.: Cognitive Factors in Information Processing, *Proc. 14th Intern. Congr. Psychol.*, Washington, D.C., August, 1963 (abstract).
39. Posner, M. I.: Information Reduction in the Analysis of Sequential Tasks, *Psychol. Rev.* (in press).
40. Edwards, W.: The Theory of Decision Making, *Psychol. Bull.*, vol. 51, pp. 380–417, 1954.
41. Edwards, W.: Dynamic Decision Theory and Probabilistic Information Processing, *Human Factors*, vol. 72, pp. 59–73, 1962.
42. Fitts, P. M.: Information Capacity of the Human Motor System in Controlling the Amplitude of Movement, *J. Exptl. Psychol.*, vol. 47, pp. 381–391, 1954.
43. Fitts, P. M., and J. R. Peterson: The Information Capacity of Discrete Motor Responses, *J. Exptl. Psychol.*, vol. 67, pp. 103–112, 1964.

44. Pew, R. W.: Temporal Organization in Skilled Performance, *Office of Research Administration Tech. Rept.* 2814-11-T, University of Michigan, Ann Arbor, Mich., May, 1963.
45. McRuer, D. T., and E. S. Krendel: Dynamic Response of Human Operators, *WADC Tech. Rept.* 56-524, Wright Patterson Air Force Base, Ohio, October, 1957.
46. McRuer, D. T., and I. L. Ashkenas: Design Implications of the Human Transfer Function, *Aerospace Eng.*, vol. 21, pp. 76–77, 144–147, 1962.
47. McRuer, D. T., and E. S. Krendel: The Man-machine System Concept, *Proc. IRE*, vol. 50, pp. 1117–1123, 1962.
48. Kelley, C. R.: Predictor Instruments Look into the Future, *Control Eng.*, March, 1962, pp. 86–90.
49. Crossman, E. R. F. W.: The Information-capacity of the Human Motor-system in Pursuit Tracking, *Quart. J. Exptl. Psychol.*, vol. 12, pp. 1–16, 1960.

Chapter 32

SIMULATION

G. A. BEKEY, *Electrical Engineering Department, University of Southern California.*

D. L. GERLOUGH, *Planning Research Corporation.*

CONTENTS

32-1. Introduction: The Nature of Simulation	32-2
The Development of Models	32-2
Stochastic and Deterministic Models	32-2
The Uses of Simulation	32-3
Simulation Equipment	32-3
32-2. Analog Simulation	32-4
Analogs and Analogies	32-4
Physical Simulation	32-4
Scale Models	32-4
Analogous Models	32-5
Partial-system Tests	32-5
Mathematical Analogs	32-7
Comparison of Analog Simulation Techniques	32-9
Special Analog Simulation Techniques	32-10
Simulation of Random Processes	32-10
Simulation of Pure Time Delays	32-10
Simulation of Sampled-data Systems and Digital Programs	32-10
Simulation of a Biological System	32-12
32-3. Digital Simulation	32-15
Nature of Digital Simulation	32-15
Simulation Involving Mathematical Models	32-16
Simulation of Processes Having Only Verbal-logical Models	32-16
Model of Freeway Traffic Flow	32-16
State Representation	32-16
Bit-configuration Representation of Freeway Traffic	32-16
Manipulation with Bit-configuration Representation	32-17
Memorandum Representation of Freeway Traffic	32-18
Simulation of Random Events (Monte Carlo Methods)	32-19
Scanning the Digital Model	32-22
32-4. Simulation of Manned Systems	32-22
Characteristics of Manned Simulation	32-22
Environmental Simulation	32-23

Flight Trainers and Piloted Simulators........................ 32-24
  The Moving-base Simulator................................ 32-26
  The Variable-stability Airplane............................ 32-26
  Increasing Sophistication in Physical Simulation.............. 32-26
Computers Used with Manned Simulators...................... 32-27
Simulation of the Human Operator........................... 32-27
32-5. Hybrid Analog-Digital Simulation........................... 32-28
Analog Techniques in Digital Simulation....................... 32-29
Digital Techniques in Analog Simulation....................... 32-29
Hybrid Systems... 32-29
32-6. Summary... 32-30

32-1. Introduction: The Nature of Simulation

As used in system engineering, the word simulation refers to the construction of a simplified representation of a process or system in order to facilitate its analysis. Such a representation or model may be qualitative or quantitative. In either case, it is characterized by the fact that it does not include all the features and characteristics of the original system or process. Rather the purpose of the simulation is to show the effect of particular factors which are being investigated. For example, an equivalent circuit of a vacuum tube is a model or simulation of the vacuum tube which correctly represents the voltage and current relationships within the tube to some appropriate degree of approximation but gives little or no information on its physical size or the environmental conditions under which it is designed to operate.

The Development of Models. The process of constructing a simulation or representation of a physical system thus involves one or more abstractions from the real world. These abstractions may be of varying degrees of severity. For example, the simplest type of simulation is that which involves the construction of a scale model of the physical system, and this procedure is common in hydraulics and the chemical industry. While there are important problems of scaling in the construction of such a model, in principle a scale model of a harbor, for example, represents a relatively small departure from the real world. On the other hand, the representation of the temperature distribution in a missile nose cone by means of electric circuits consisting of resistors and capacitors represents much more severe abstractions from the real world. First of all, the original system is a distributed-parameter system while its representation consists of lumped-parameter elements. In the second place, electrical elements and electrical variables are used to represent the original thermal system. A further step in abstraction results if, instead of constructing such an electrical analog, the equations representing its behavior are written and solved. A general-purpose computer, analog or digital, programmed for the solution of these equations, is in fact a mathematical simulation of the original physical problem.

All the types of models described above may be referred to as simulation, and will be so considered in this chapter.

Stochastic and Deterministic Models. The construction of a model is based on information obtained from the physical world by observation or measurement. Consequently, measurement errors will result in erroneous models. Furthermore, measurement is often corrupted by noise; it is characterized by the fact that it is never exactly repeatable either because the process itself is subject to random variation in time or because the measurement includes some random variability, or both. Consequently, one of the serious problems in the simulation of a process is the selection of those random elements which one desires to incorporate in the model. Many models are constructed on a purely deterministic basis with the understanding that the results obtained from the models may represent statistical averages of certain variables in the physical systems.

In many problems the stochastic nature of certain variables or the presence of random disturbances represent important aspects of the system. In such cases the simulation will include noise sources, and Monte Carlo techniques may be used to determine satisfactory confidence limits on the system response. Digital simulation of random events by Monte Carlo methods is discussed below in Sec. 32-3.

Statistical variations between successive trials of the same experiment are also unavoidable in systems where human beings perform or operate as parts of the system, e.g., the pilot of a high-performance aircraft or the operator of a radar console. In such cases, human beings must also be included as portions of simulated systems, and reliable results can be obtained only if large numbers of simulated experiments are conducted in order to take into account the statistical variability of human performance. The particular problems concerned with the simulation of manned systems are discussed in Sec. 32-4.

The Uses of Simulation. Simulation in its various forms is of great importance in system engineering. Some of the important applications are listed below:

1. Design evaluation—a simulated system may be used to evaluate the validity of the preliminary design of a portion of a system, such as a controller.

2. Interrelationships among the parts—one of the key uses of simulation is in the evaluation of the effect of various portions or subsystems of a system upon each other and upon the performance of the system as a whole. For example, tests of a simulated aircraft cockpit may be used to evaluate the effect of proposed displays upon the pilot's performance or the effects of the pilot's performance limitations upon overall vehicle stability.

3. Costs—clearly an important part of simulation is in the attempt to reduce overall costs. This may be accomplished by the evaluation of alternative designs by means of simulation.

4. Study of failure sources—one of the uses of simulation is in the attempt to reconstruct a system failure after its occurrence. Without some type of simulation, it is extremely difficult in many cases to determine possible sources of airplane crashes or controller instability.

5. Hypothesis testing—simulation makes possible experiments which in the physical system may be difficult or impossible to test. For example, a computer simulation of a biological system may make it possible to test the effect of a drug which is known to result in the death of the organism. Hypotheses on probable causes of death can be studied on the simulated system. A mathematical model is a formal statement of a hypothesis on the structure and operation of a system. The hypothesis must be tested by comparing results of experiments in the real world and on the model. Such experimentation is conveniently done using simulation.

Many other uses and applications of simulation can be listed. Among the subsidiary advantages of a simulation are determination of problem areas, determination of significant and insignificant variables, and study of the effect of environmental variations upon performance. Suffice it to say that simulation now has an established place within the overall field of system engineering and that the complexity of large industrial and military systems is such that simulation in one form or another is required at one or more phases between the original conceptual design and the final manufacture.

Simulation Equipment. Simulation equipment may be classified in several ways: it may be general-purpose or special-purpose, physical or mathematical, analog or digital. When some physical system is simulated by the construction of another physical system which obeys similar physical laws, the process is termed *physical simulation.* Examples of physical simulators are scale models of hydraulic systems, gimbal-mounted cockpits, resistor-capacitor networks for study of thermal problems, electrolytic tanks for representing magnetic fields, and environmental-control chambers to simulate the temperature and radiation of space. When the simulation is based on the solution of the equations which describe the performance of a system, it may be termed *mathematical simulation.* Thus an analog or a digital computer may be considered a simulator when it is used to solve equations which represent a mathematical model of a physical system. Clearly, such computers may be general-purpose, or they may be constructed for specific purposes. Models of neural processes are good examples of this distinction. The "Homeostat"[1] is a device constructed to simulate an aspect of neural systems which Ashby calls "ultrastability." It is clear that the device, built of magnets, potentiometers, etc., is a special-purpose simulation. Since continuous displacements of the pivoted magnets correspond to the system variables, this device may be called a special-purpose analog physical simulator. On the other hand, the "Perceptron" (see Sec. 30-9 and Ref. 2) is a mathematical statement of a hypothesis on neural behavior which is based on the assumption that an initially randomly connected system can "learn" to recognize patterns. The model was simulated using a general-purpose digital computer.

The above examples should be sufficient to indicate that alternative methods and equipment are often possible for a particular simulation. In practice, the choice of equipment is dictated by considerations of (1) cost, (2) availability of computers, (3) required fidelity of simulation, and (4) information required from the simulator. In many cases technical factors do not favor any one method, and subjective factors may enter into the selection of simulation equipment.

The computers which are a principal part of simulation equipment are discussed in Chapters 8 to 11 of this handbook.

32-2. Analog Simulation

When a simulation is characterized primarily by continuous signals or elements it is referred to as *analog simulation*. When a substantial portion of a simulation involves discrete signals, elements, or processes, it is referred to as *hybrid analog-digital simulation* (see Sec. 32-5). Analog simulation can take both the forms outlined above, i.e., physical and mathematical simulation. In this section we consider the origin of analogs and analog simulators and present a brief discussion of both physical and mathematical analogs.

Analogs and Analogies. An analog or continuous model of a physical system or process can arise in several ways:

1. The variables of interest in a physical system to be simulated may be continuous variables, e.g., the speed and altitude of an aircraft, the intensity of a light source, the motion of a dial, the temperature of a gas.

2. The physical variables may in fact be discrete or quantized or both, but be approximated by continuous variables. For example, the resistance of a wirewound potentiometer is quantized but may be considered approximately continuous. Units of money are discrete, but for the study of economic cycles, money in circulation can be considered continuous. The sampling rate of an essentially digital system may be sufficiently high compared with the frequencies of interest that a continuous approximation may be valid.

It should be apparent from the above that the choice of continuous or discrete models is dependent only in part on the physical system itself. Questions of precision, convenience, and cost may dictate the approximation of a continuous variable by a discrete one, or vice versa. Furthermore, the distinction between discrete and continuous variables is sometimes primarily a question of the point of view.

Physical Simulation. Analog simulation may involve the study of a physical system by means of an analogous system whose behavior closely approximates that of the system under study *for the particular phenomena being investigated*. Three forms of such simulation are used: scale models, analogous models, and partial-system tests.

Scale models are commonly used in hydraulics and in the process industries. In order to facilitate the study of hydraulic phenomena in a harbor or inner basin, the U.S. Army Corps of Engineers sometimes constructs scale models. Important problems in scale models are:

1. Scaling. In a hydraulic model, if the vertical or depth scale is different from the horizontal scales, adjustments may be required in viscosity of the fluid and friction of channels if the overall behavior is to approximate the original system.

2. Instrumentation. The scale model must be instrumented to provide the desired information, and care must be taken that the instruments do not alter the model significantly.

SIMULATION　　　　　　　　　　　　　　　　　32–5

The *a-c network analyzer*, a device used for the study of electrical power transmission and distribution problems, can also be considered a scale model, which includes a change from distributed to lumped parameters for representation of a transmission line.

Analogous models are systems which are easier to instrument than the original and obey similar physical laws. The study of heat flow by an electrical analog or the study of gaseous flow by means of an electrolytic tank or a resistive sheet are examples of analogous models. Figure 32-1 shows a resistive sheet which is used to represent the flow of gas in the exhaust stream of a rocket engine. The boundaries of a cross section of the nozzle are painted on the sheet in silver ink. An electrical potential is applied across the boundaries, and the equipotential lines along the sheet then correspond to streamlines in the flow pattern. An extensive treatment of analogous models for the study of field problems is given by Karplus.[3] The use of electrical

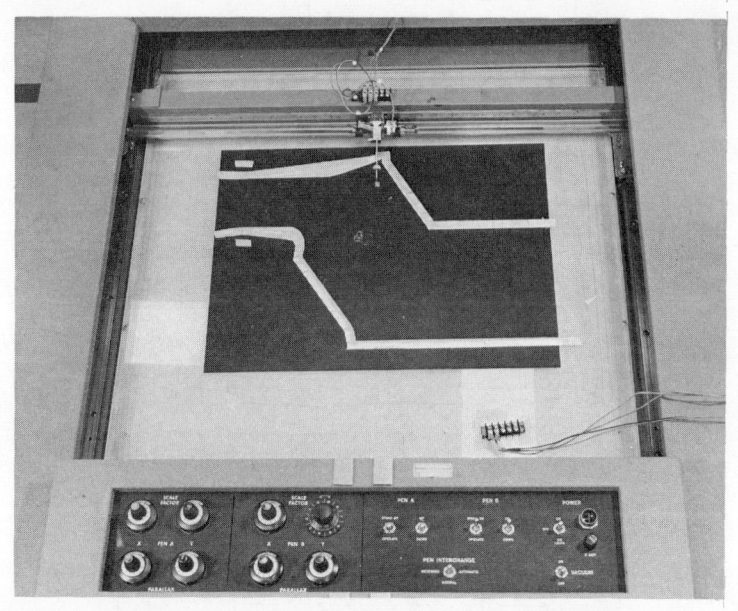

Fig. 32-1. Resistive sheet used to simulate gas flow in a rocket-engine nozzle. (*Courtesy of TRW Space Technology Laboratories.*)

analogs for the study of mechanical, thermal, acoustic, and other problems characterized by distributed parameters is often facilitated if all the continuous variables *except one* are discretized. The resulting lumped-parameter system, which approximates the original system, can be represented by an electrical analog. For example, the continuous temperature distribution in a slab can be approximated by the temperature at a finite set of points on the slab, interconnected by lumped thermal resistance and capacitance. This approximate system can then be considered analogous to an electrical network where the grid points are interconnected by electrical resistors and capacitors, and voltage corresponds to temperature, and current to heat flow; see Fig. 32-4.

Partial-system Tests. This phrase refers to the interconnection of an element of a physical system with a general-purpose computer, which represents a mathematical analog of the rest of the system. This type of simulation is employed primarily in two sets of circumstances:

　1. When a mathematical description of an element is not available or is difficult to obtain (e.g., when a human operator is involved).

2. When the performance of a system element must be evaluated under conditions which simulate actual system performance, but the rest of the system does not yet exist or may be too difficult or expensive to use (for example, testing a satellite-borne radio transmitter under conditions which simulate earth reentry). The simulation of manned systems includes a number of unusual factors and is considered separately in Sec. 32-4.

Partial-system tests of guidance and inertial components are often used in the aerospace industry. The *three-axis flight table* is a servo-controlled, gimbal-mounted device which follows command inputs from a computer (see Fig. 32-2). Components such as sections of missiles, complete flight-control systems, etc., are subjected to the

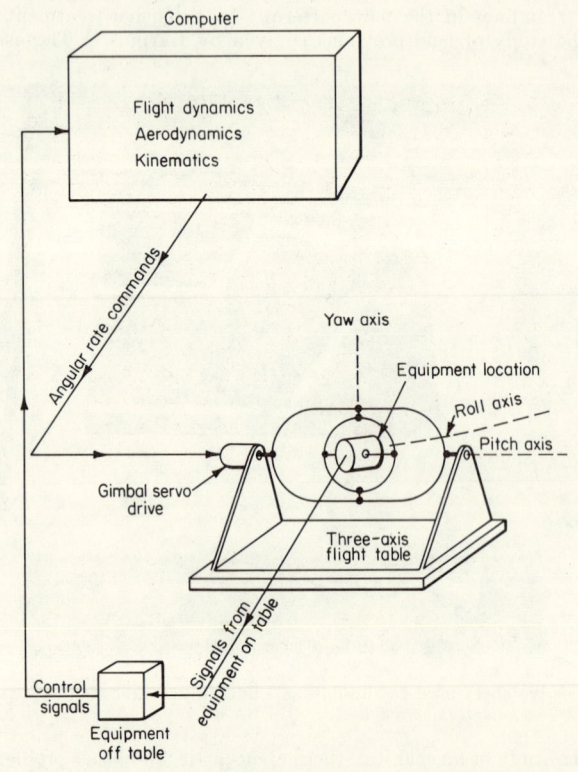

Fig. 32-2. Schematic diagram of three-axis flight table.

torques, angular velocities, and angular accelerations which they would experience in flight. Instruments mounted on the gimbals provide an indication of position and velocity. In some cases, resolvers may be provided to compute Euler-angle transformations. If the component being tested is to be studied in operation, then outputs from the component (e.g., control signals) are fed back to the computer. The system shown in Fig. 32-2 includes a computer for representing both vehicle dynamics and kinematics. A flight-control system, an inertial-guidance system, or a single gyroscope may be mounted on the flight table.

Partial-system tests have been employed recently in the automobile industry in order to measure the effects of simulated road inputs on portions of an automobile.[4]

Problems connected with partial-system tests are the following:

1. The effect of the dynamics of the interconnection between computer and physical component must be understood. For example, in the case of the three-axis flight

table, the acceleration limits of the table itself should be large compared with the desired acceleration of the component to be mounted on it.

2. The interconnection will usually be limited to particular tasks; e.g., a flight table will inherently be limited to carrying packages of a particular volume and weight.

3. The test will inherently be limited to particular aspects of the component's eventual environment. For example, a flight table provides the desired angular motion but does not provide any translation motion. Furthermore, the effects of changes in atmospheric pressure, temperature, or solar radiation may not be included. In each partial-system test, decisions must be made concerning the inclusion of additional effects which would make the simulation more faithful at the expense of increasing its cost.

Mathematical Analogs. Mathematical analogs represent a further level of abstraction than physical analogs. A mathematical model of the system to be studied must be formulated. The resulting equations are solved by means of analog-computer elements which perform specific operations, such as summation, integration, or multiplication. The physical variable in the computer element is generally either an electrical voltage or a shaft rotation. The simulation can be performed using a general-purpose electronic analog computer, or it may be performed using special-purpose elements. To illustrate the difference between physical and mathematical analogs, consider the simple spring-mass system shown in Fig. 32-3a. The mathematical model of the system is given by the equations

$$M_1 \frac{d^2 x_1}{dt^2} = K_1(x_2 - x_1) - B_1 \frac{dx_1}{dt} \qquad (32\text{-}1)$$

$$M_2 \frac{d^2 x_2}{dt^2} = K_1(x_1 - x_2) - K_2 x_2 - B_2 \frac{dx_2}{dt} \qquad (32\text{-}2)$$

where x_1 and x_2 represent the translations of the two masses, the initial velocities of the two masses are zero, and the initial positions are given by x_{10} and x_{20}, respectively. This system can be studied by means of the analogous electrical system of Fig. 32-3b, which is generally easier to instrument than the mechanical system and where variations of parameters are much more convenient. The analog is based on the following analogy:

Velocity ↔ current
Force ↔ voltage
Mass ↔ inductance
Friction factor ↔ resistance
Spring constant ↔ capacitance

This electric circuit is a physical analog.

If now instead of building the equivalent electric circuit, one solves Eqs. (32-1) and (32-2) on a general-purpose analog computer, the result is the mathematical analog of Fig. 32-3c. In some cases (including this example) it is possible to rearrange the analog simulation diagram of Fig. 32-3c such that portions of the computer diagram correspond to portions of the physical system (such as masses, springs, and dashpots). This approach to analog simulation has been called "direct simulation" by Goode and Machol.[5]

Another simple example of the possible approaches to analog simulation is given by the thermal problem of Fig. 32-4a, which represents the temperature distribution along a uniform bar. Starting from a given set of initial conditions, it is desired to find the variation of temperature along the bar as a function of time. This is a distributed-parameter problem and may be described by the equation

$$\frac{\partial^2 T}{\partial x^2} = k \frac{\partial T}{\partial t} \qquad (32\text{-}3)$$

where T is the temperature and t the time. If we focus our attention at the N points indicated in Fig. 32-4a, the continuous variation along the bar is replaced by a set of

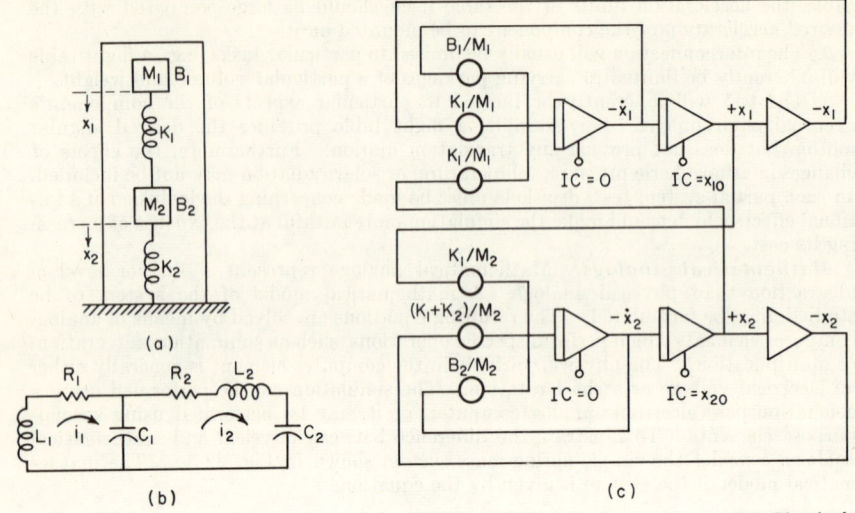

FIG. 32-3. Analog simulation of two-degrees-of-freedom mechanical system. (a) Physical system.

$$M_1 \frac{d^2x_1}{dt^2} + B_1 \frac{dx_1}{dt} + K_1(x_1 - x_2) = 0$$

$$M_2 \frac{d^2x_2}{dt^2} + B_2 \frac{dx_2}{dt} + K_2 x_2 + K_1(x_2 - x_1) = 0$$

$$x_1(0) = x_{10} \qquad x_2(0) = x_{20}$$

(b) Electrical analog.

$$L_1 \frac{d^2q_1}{dt^2} + R_1 \frac{dq_1}{dt} + C_1(q_1 - q_2) = 0$$

$$L_2 \frac{d^2q_2}{dt^2} + R_2 \frac{dq_2}{dt} + C_2 q_2 + C_1(q_2 - q_1) = 0$$

$$q_1(0) = q_{10} \qquad q_2(0) = q_{20}$$

(c) Mathematical analog.

finite differences.[3] The resulting equations are

$$\frac{\partial^2 T}{\partial x^2} \cong \frac{T_{n-1} + T_{n+1} - 2T_n}{(\Delta x)^2} = k \frac{dT_n}{dt} \qquad n = 1, 2, \ldots, N \qquad (32\text{-}4)$$

for the nth node. This equation has the same form as the equation

$$\frac{E_{n-1} + E_{n+1} - 2E_n}{R} = C \frac{dE_n}{dt} \qquad n = 1, 2, \ldots, N \qquad (32\text{-}5)$$

which characterizes the physical analog of Fig. 32-4b, if we choose the correspondences

$$T \leftrightarrow E$$
$$(\Delta x)^2 \leftrightarrow R$$
$$k \leftrightarrow C$$

The equation can also be solved directly using a mathematical analog computer, by N circuits of the form of Fig. 32-4c.

Since mathematical analogs are solved by means of analog elements (Chapter 10) using analog computers (Chapter 11), the problems of instrumentation and scaling are not discussed in this chapter.

Comparison of Analog Simulation Techniques.

Table 32-1 compares the major features of physical and mathematical simulation. It should be noted primarily that the mathematical analog is more versatile, since variation of parameters and alteration of system structure are more convenient. However, a mathematical

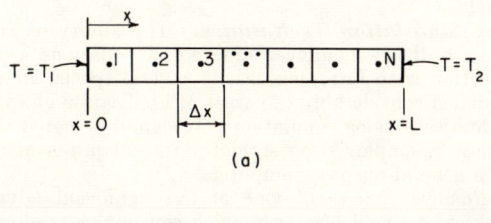

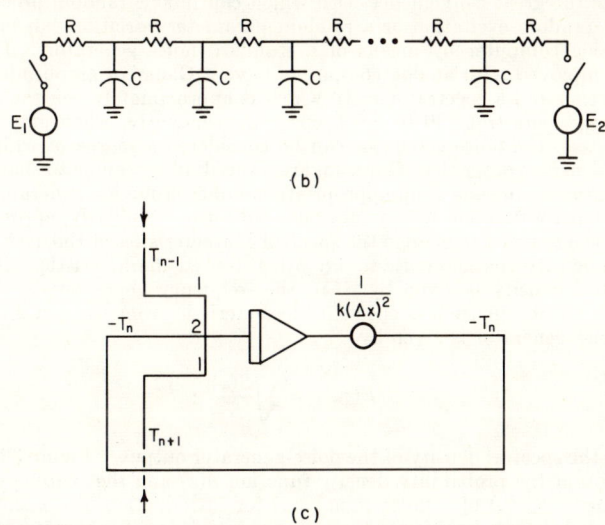

Fig. 32-4. Analog simulation of temperature distribution in a uniform bar. (a) Physical system. (b) Lumped-parameter physical analog. (c) One section of mathematical-analog simulation.

Table 32-1. Comparison of Physical and Mathematical Simulation

Mathematical analog	Physical analog
Easy parameter variations	Parameter variations may be difficult
Mathematical description required of all system elements	Mathematical description not required of all system elements
Time scale can be varied by selection of computer components	Generally designed for a fixed time scale
Well suited to fast-time simulation	Well suited to real-time simulation with human operators
Results affected by selection of model and quality of computer components	Results affected by selection of model and validity of analog
Possibility of false solutions due to the characteristics of the equations themselves	No such possibility

model for each major system element must be formulated. In cases where such models are not well known, or where the performance of particular subsystems is difficult to describe mathematically because of complexity, or where statistical variability of special types is important, it may be necessary to include such elements

32–10 SYSTEM TECHNIQUES

physically in the simulation. Such is the case in the simulation of piloted-vehicle control problems, which are considered separately below in Sec. 32-4.

The difference between analog *computation* and *simulation* is a subtle one, since the process of solving equations may be both computation and simulation. The difference is, in fact, largely a question of point of view when general-purpose analog computers are involved.

Special Analog Simulation Techniques. The ability of analog computation equipment to solve ordinary nonlinear differential equations is well known. The purpose of this section is to introduce briefly several special simulation techniques which serve to broaden considerably the areas of application of analog simulation, in particular, mathematical analog simulation. It should be noted that the techniques to be discussed may be employed on special-purpose equipment or, by appropriate programming, on a general-purpose computer.

Simulation of Random Processes. One of the significant advantages of analog equipment is the great convenience with which continuous random processes can be studied. A random excitation or a random parameter variation can be introduced into an analog computer by means of a standard noise generator. Typical noise generators employed with analog computers have a Gaussian amplitude probability density function and a spectral density which is approximately constant from direct current to anywhere from 30 to 2,000 cps. Consequently, when employed with a low-pass system, the noise generator can be considered a source of white noise. If probability densities other than Gaussian are desired, it is possible to shape the amplitude distribution by means of an appropriate nonlinear function generator, which can change the density function to any other desired form. Similarly, appropriate linear filters may be employed to change the spectral characteristics of the excitation from a flat spectral-density characteristic to any other desired characteristic. If the desired power spectral density is given by $S(\omega)$, the frequency characteristic $F(j\omega)$ of the desired filter to obtain such a spectral characteristic from the output of a white Gaussian noise generator is given by

$$|F(j\omega)| = \sqrt{\frac{S(\omega)}{N_0}} \qquad (32\text{-}6)$$

where N_0 is the spectral density of the noise-generator output. Figure 32-5 illustrates the alteration of the probability density function $p(x)$ and the shaping of the power spectral density $S_{xx}(\omega)$ of a noise signal $x(t)$ for introduction into a simulation. It should be noted that the shaping of the power-spectral-density characteristic by means of a linear filter does not alter the Gaussian probability density function of the signal. Nonlinear filtering is required to obtain a change in the density function.

In recent years, high-speed sampling and repetition circuits have been used with general-purpose analog computers. The availability of high-speed repetitive operation, e.g., 30 to 300 computation cycles per second, has made possible the performing of Monte Carlo studies by means of analog equipment. The use of digital simulation equipment for performing Monte Carlo studies is discussed in Sec. 32-3.

Simulation of Pure Time Delays. Many industrial processes are characterized by the presence of time delays or transportation lags which arise in the movement of material through long distances. Such is the case in the flow of liquids through pipelines. In order to simulate processes which involve pure time delays by means of analog equipment, it is necessary to represent accurately or approximately the effect of the time-delay term. An accurate representation of the time-delay term requires storage of information and can be accomplished by the use of special magnetic-tape recorders with adjustable spacing of record and read heads. The time required for the tape to travel between the two heads is then proportional to the delay term.

An alternate method of representing time delays for simulation purposes is to use frequency-domain approximations, as discussed in Sec. 10-2, Eqs. (10-7) to (10-9).

Simulation of Sampled-data Systems and Digital Programs. An increasing number of automatic control and guidance systems for both airborne and space applications as well as process-control applications are making use of digital controllers. However, an accurate simulation of such a system requires the ability to handle both the non-linearities and high frequencies present in the process and the characteristics of the controller itself. In order to perform such simulations a number of techniques have

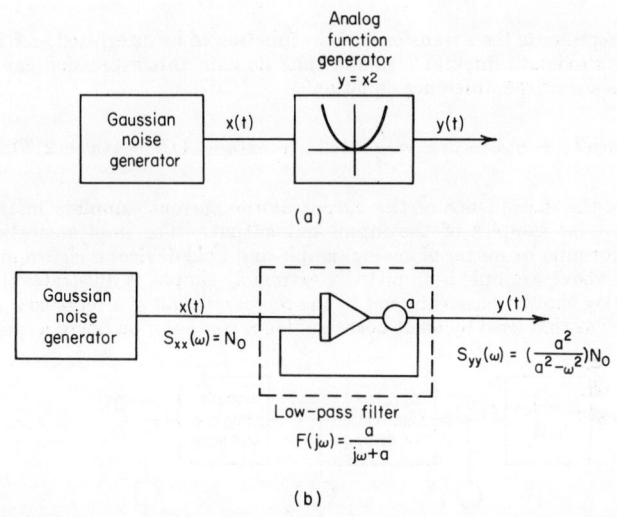

Fig. 32-5. Shaping of noise-generator characteristics. (a) Chi-square probability density obtained from Gaussian probability density. If $x(t)$ is a sample function of a real Gaussian process with zero mean and variance σ_x^2, its probability density function $P(x)$ is

$$P(x) = \frac{1}{\sqrt{2\pi}\,\sigma_x} e^{-x^2/2\sigma_x^2}$$

Then $y(t)$ has a chi-square probability density:

$$P(y) = \frac{1}{\sqrt{2\pi y}\,\sigma_x} e^{-y/2\sigma_x^2}$$

(b) Shaping of power spectral density.

been developed for the simulation of a digital program by means of basically analog elements accompanied by sampling switches. The basic element in such a simulation is the analog sample-and-hold channel, which can be illustrated schematically as in Fig. 32-6. In high-performance systems the switch indicated in this figure is a solid-state switch and the amplifier is capable of sampling in a few microseconds. When the switch closes, the amplifier acts as a tracking device; when the switch opens it holds. The last value sampled equals the charge on the capacitor. The sample-and-hold channel in combination with appropriate timing and switching circuits for energizing and deenergizing the sampling switches has made possible the inclusion of a significant amount of digital logic into analog simulations. While a detailed discussion of the simulation techniques possible with sample-hold devices is beyond the scope of this chapter, its application to the solution of a simple difference equation will be illustrated.

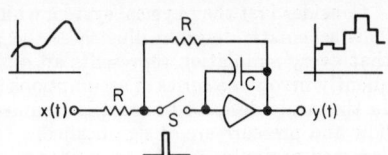

Fig. 32-6. Analog sample-and-hold channel.

Example. Consider the well-known Simpson's one-third rule for integration. This numerical integration formula can be described in z transforms as the relationship

$$\frac{Y(z)}{X(z)} = D(z) = \frac{T}{3} \frac{1 + 4z^{-1} + z^{-2}}{1 - z^{-2}} \qquad (32\text{-}7)$$

where $X(z)$ represents the z transform of the function to be integrated and $Y(z)$ represents the approximate integral. In the time domain this expression can be represented by means of the difference equation

$$y(nT) - y(n-2)T = \frac{T}{3}[x(nT) + 4x(n-1)T + x(n-2)T] \qquad (32\text{-}8)$$

which shows the dependence of the output at the current sampling instant on the present and past samples of the input and output. The implementation of this integration formula by means of analog sample-and-hold devices is shown in Fig. 32-7. While the above example is admittedly extremely simple, it illustrates the applicability of analog simulation techniques to the representation of a more complex digital program such as that used by an airborne guidance computer on board a space vehicle.

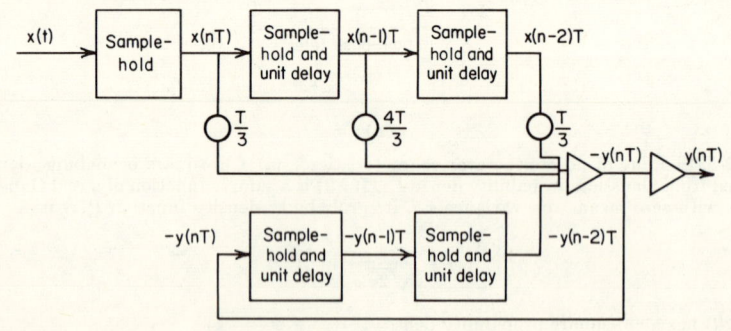

FIG. 32-7. Analog simulation of Simpson's one-third rule for numerical integration.

The ability of analog circuits to simulate digital operations makes possible the investigation of the effects of the digital computer upon the dynamic performance of a system.

Simulation of a Biological System. In order to illustrate the principles involved in the construction of an analog simulation, an example from the study of blood circulation will be presented. The biological example contains all the elements of a typical problem which is to be attacked by simulation techniques and consequently serves as an illustration of the simulation method.

Consider first the physical system which is illustrated in Fig. 32-8 in schematic form. This schematic diagram illustrates the first principle of the simulation method, i.e., that every simulation represents an abstraction from the physical world and consequently involves a series of assumptions regarding the significance of particular effects. In the study described here it is assumed that only physical effects concerning blood flow and pressure are of significance. Nervous and chemical control mechanisms are assumed negligible for the particular purpose of this study. It should again be noted that this does not imply that these effects *are* actually negligible, but that they are so considered for the purpose of this study. In order to study the flow and pressure relationships in the vicinity of the heart the following assumptions were made:

1. The system was lumped into six chambers or vessels which are the right and left ventricles, the right and left atria, the lungs, and the systemic vascular bed.

2. The ventricular volume change during each cardiac cycle was assumed to be sinusoidal.

3. All nervous and chemical control mechanisms were ignored.
4. The heart rate was assumed to be constant.

Clearly, the above assumptions will make it impossible for this particular simulation to provide information on the changes of blood pressure between a location, say, near the shoulder and near a finger or foot. However, blood flow and pressure in the four heart chambers and in the vicinity of the heart can be simulated. Blood entering each

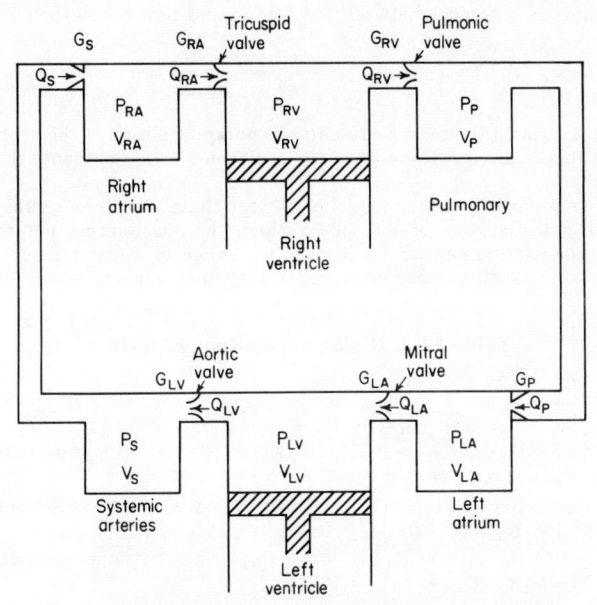

FIG. 32-8. Schematic diagram of adult blood circulation.

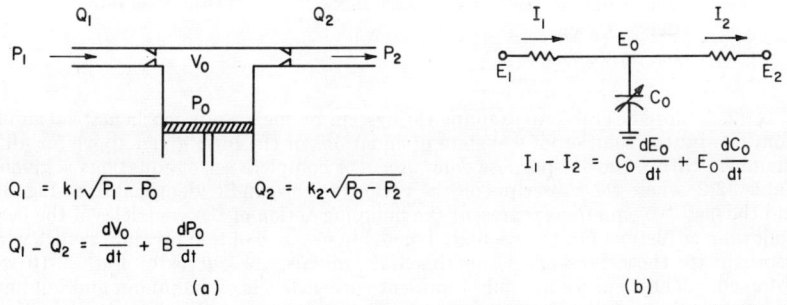

FIG. 32-9. Simplified schematic diagram of one ventricle of the heart. (a) Hydraulic model. (b) Electrical analog.

of the six chambers of Fig. 32-8 is assumed to enter through a valve or constriction. The flow through these valves is assumed to be turbulent, and each chamber is assumed to be elastic. This combination of assumptions can be represented mathematically if one considers a single pumping heart chamber (a ventricle) as indicated in Fig. 32-9a. The flow Q_1 into the chamber is given by

$$Q_1 = G_1 \sqrt{P_1 - P_0} \quad P_1 \geq P_0 \qquad (32\text{-}9)$$
$$= 0 \quad P_1 \leq P_0$$

where the pressures P_1 and P_0 are indicated in Fig. 32-9 and G_1 is the lumped conductance of the valve. The square-root relationship characterizes turbulent flow through the valve. The flow out of the ventricle is given by

$$Q_2 = G_2 \sqrt{P_0 - P_2} \quad P_0 \geq P_2$$
$$= 0 \quad P_0 \leq P_2 \qquad (32\text{-}10)$$

The instantaneous difference between the flow in and flow out is then given by

$$Q_1 - Q_2 = \frac{dV_0}{dt} + \frac{1}{B_0}\frac{dP_0}{dt} \qquad (32\text{-}11)$$

where the first term on the right represents the pumping action of the ventricle and the second term represents its elasticity. The coefficient $1/B_0$ represents the compliance of the chamber.

The above assumed hydraulic model of a heart chamber can be represented by the analogous electrical system of Fig. 32-9b where the resistors are nonlinear and the pumping action is represented by a rate of change of capacitance. Clearly, the instrumentation of such an analogous electrical system is not particularly convenient.

Table 32-2. Biological-system Equations

$\dot{P}_{RA} = B_{RA}(Q_S - Q_{RA})$
$Q_S = G_S \sqrt{P_S - P_{RA}} \quad P_S > P_{RA}$ Otherwise zero
$\dot{P}_{RV} = B_{RV}(Q_{RA} - Q_{RV} - \dot{V}_{RV})$
$Q_{RA} = G_{RA} \sqrt{P_{RA} - P_{RV}} \quad P_{RA} > P_{RV}$ Otherwise zero
$\dot{P}_p = B_p(Q_{RV} - Q_P)$
$Q_{RV} = G_{RV} \sqrt{P_{RV} - P_p} \quad P_{RV} > P_p$ Otherwise zero
$\dot{P}_{LA} = B_{LA}(Q_P - Q_{LA})$
$Q_p = G_P \sqrt{P_P - P_{LA}} \quad P_P > P_{LA}$ Otherwise zero
$\dot{P}_{LV} = B_{LV}(Q_{LA} - Q_{LV} - \dot{V}_{LV})$
$Q_{LA} = G_{LA} \sqrt{P_{LA} - P_{LV}} \quad P_{LA} > P_{LV}$ Otherwise zero
$\dot{P}_S = B_S(Q_{LV} - Q_S)$
$Q_{LV} = G_{LV} \sqrt{P_{LV} - P_S} \quad P_{LV} > P_S$ Otherwise zero
$V_{RV} = C_R \sin \omega t$
$V_{LV} = C_L \sin \omega t$

It is much more desirable to examine the system by means of a mathematical simulation, i.e., by the solution of a system of equations of the form given above for all six chambers with a general-purpose computer. A complete set of equations is given in Table 32-2, where the subscripts on the various terms can be identified from Fig. 32-8 and the last two equations represent the pumping action of the ventricles of the heart. Following adjustment of the simulated model in order to obtain pressure swings which approximate those present in experimental animals, the curves of Fig. 32-10 were obtained. These curves not only represent correct levels of maximum and minimum pressure in the vicinity of the heart but also give a reasonable approximation to the actual waveshapes obtained from the physiological literature. Further details on this simulation can be found in the references.[15,30]

This example serves to illustrate the fact that, while a simulation may not be faithful in every regard, it may be used to study important dynamic relationships in a particular system. In the example cited here, the effect of increased resistance in a portion of the system can be studied directly. In each simulation problem, the most crucial step is the initial one of determining the significant variables to be investigated. Following this investigation it may be necessary to reexamine the initial assumptions and formulate a more complete model which will take into account additional effects in order to render the simulation more faithful.

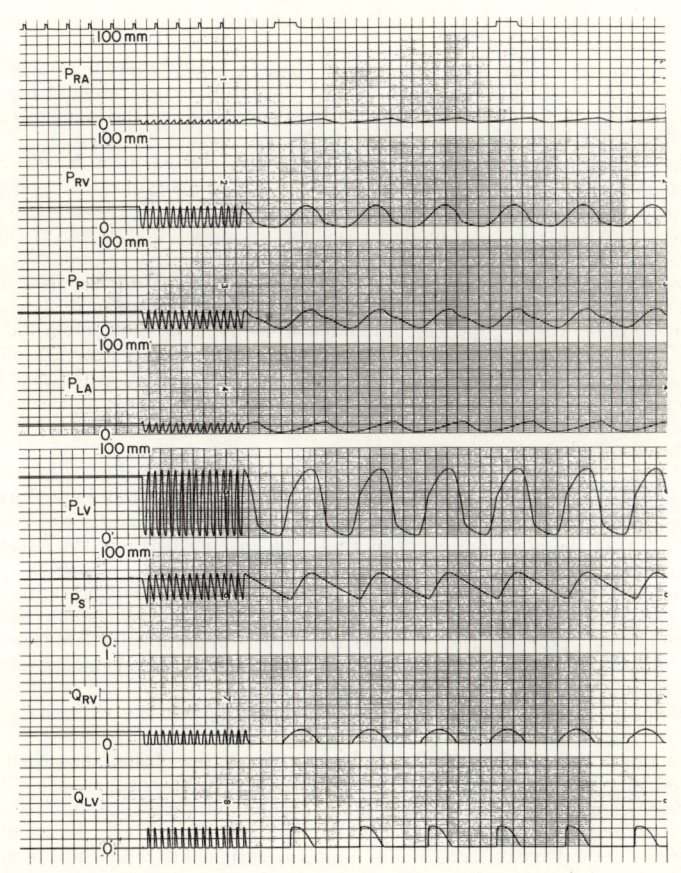

FIG. 32-10. Simulated flow and pressure from circulation model (1-sec time marks appear at the top of curves).

32-3. Digital Simulation

When simulation consists of the manipulation of phenomena which occur with discrete values, it is referred to as *digital simulation*.

Nature of Digital Simulation. While digital simulation can be both physical and mathematical simulation, mathematical simulation is more common. Digital simulation lends itself to the study of systems wherein there are many decision functions to be implemented, and to phenomena which are more easily characterized by a word description than by a set of conventional mathematical equations.

Continuous phenomena may be treated by digital simulation by digitizing or sampling continuous functions.[20] (The design of sampled-data control systems has become an important field of engineering study.)

Digital simulation is usually carried out on a general-purpose digital computer. The process of preparing the simulation model for the digital computer follows many of the general rules which arise in programming any problem for the digital computer.

Whereas in analog simulation all parts of a system are usually simulated in parallel by adding more and more integrators or other pieces of simulation equipment, the digital simulation of the system must proceed serially in time because of the nature of operation of the digital computer. Thus, as the simulated system becomes more complex, the simulation program becomes larger, and the time to perform one simulation cycle becomes longer.

Simulation Involving Mathematical Models. Any system which can be represented by a set of equations, such as with the spring-mass system of Eqs. (32-1) and (32-2), the thermal-diffusion system of (32-4) and (32-5), or the biological systems of Table 32-2, can be readily subjected to digital simulation.

This class of problems is very broad and can include systems from hardware items to management games.[18] Some such simulations may employ linear programming,[19] queuing theory, multiple correlation, transition probabilities, or other mathematical disciplines in the formation of a model. Techniques for solution of sets of equations are treated in Chapter 40 and need not be covered further here.

Simulation of Processes Having Only Verbal-logical Models. In the study of certain types of processes such as the flow of packages by conveyor or flow of automobiles in a traffic stream, the first approach to a model of the system may be in terms of a verbal-logical description not reduced to mathematical statements.[21,22] A rudimentary simulation of automobiles flowing along a freeway is described as an example of a verbal-logical model in the simulation of discrete flow processes.*

Model of Freeway Traffic Flow. The verbal-logical model for a freeway simulation may be stated as follows:

1. Each vehicle proceeds in a particular lane at its desired speed until it encounters another vehicle in the same lane, and then proceeds at the maximum speed available to it (as governed by the slower cars in front). Encountering consists of coming so close that, during the next increment of time, the spacing between the encountered and encountering vehicles would become less than a safe gap.

2. On encountering another vehicle in the same lane the encountering vehicle examines the lane or lanes to its left. If it is safe to do so, the vehicle moves into the lane to the left. If possible, the encountering vehicle maintains its same speed after the lane change. If a change of lane is not safe, the encountering vehicle decreases its speed to that of the encountered vehicle.

3. During each increment of time all vehicles except those in the extreme right-hand lane look for opportunities to move to the right.

4. During each time increment all vehicles traveling at speeds less than their respective desired speeds look for opportunities to increase their speeds.

State Representation. When the flow of any system of discrete objects is simulated, there are two general ways in which the states of the objects may be represented. With the first method one or more binary digits are assigned to represent the presence, position, and perhaps size of the item to be simulated. Certain areas of the computer memory are assigned and organized in such a way as to represent the flow network, and the groups of bits representing the items are caused to flow in the network by suitable manipulations. This technique is primarily useful with binary computers. The second technique consists of representing all the conditions pertaining to a given object by a coded word, the parts of which can be extracted and interpreted by suitable computer routines. This technique is applicable to computers of any number base. These two techniques may be designated bit-configuration representation and memorandum representation, respectively.

Bit-configuration Representation of Freeway Traffic. In the bit-configuration representation of traffic moving on one roadway of a freeway where all vehicles are moving in the same direction and no intersections are present, the roadway may be considered to be made up of several memory cells laid end to end. Forward movement may be accomplished by any mathematical operation or group of operations which causes the binary digits representing a vehicle to move in a forward direction within any given memory cell and, on reaching the boundary of a memory cell, to move across into the next cell, if so required by continuation of the roadway.

In binary arithmetic, multiplying any given value by a power of two shifts the value to the left by the number of places corresponding to the exponent of the power. Thus multiplications may be used to produce the phenomenon of forward movement

---

* The fact that a verbal-logical model is used in this example does not imply that mathematical models are not usable for traffic simulation.

SIMULATION 32–17

if the direction of forward movement is taken to be from the least significant to the more significant ends of the memory cells representing the roadway. Since any given channel may be made up of several memory cells, digits which pass through the more significant (left) end must be added to the less significant (right) end of the memory cell which represents the next section of the channel. Use is made of a multiply operation which results in a full-precision product.

Figure 32-11 shows the representation of each vehicle by a group of bits. Figure 32-12 shows how forward movement of b units is produced by multiplication by 2^b. The partial products resulting from the multiplication of 2^b and the contents of cell 1 go to cells 11 and 21, the more significant partial product going to cell 21 and the less significant going to cell 11. Likewise, the partial products of 2^b and the contents of cell 2 go to cells 12 and 22, etc. Cell 20 contains all zeros. The contents of 20 and 11 are added and placed in cell 1; likewise, the sum of the contents of 21 and 12 is placed

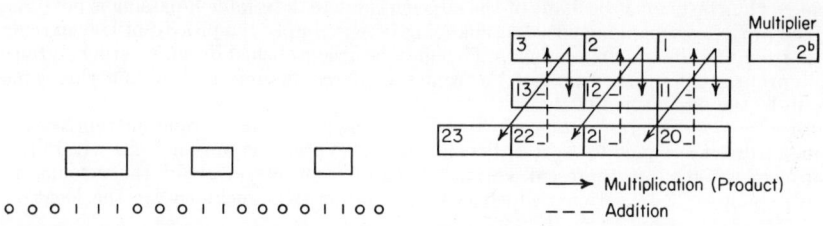

o o o ı ı ı o o o ı ı o o o o ı ı o o

FIG. 32-11. Bit configurations representing vehicles.

FIG. 32-12. Use of multiply operation to move bit configuration forward.

	Number of speed increments				
Number of lanes	3	4	5	10	16
2	9	16	25	100	256
3		19	31	136	361
4			34	164	452
5				185	530

FIG. 32-13. Number of channels for separation of bit configurations having various desired speeds.

in cell 2, etc. Thus, following the procedure shown in Fig. 32-12, it is possible to combine the results of a series of multiplications to give the effect of moving a line of binary 1's as though they were a unit, even though they may be distributed over several memory cells.

Manipulation with Bit-configuration Representation. In the case of bit-configuration representation, special record keeping is required to keep track of vehicles which have desired speeds of some given value but which may be traveling at any of several lesser speeds. This may be accomplished by using many series of memory cells for each traffic lane, each such series being known as a channel. There will be one group of channels for each increment of the distribution of desired speeds. Each of these groups will contain a sufficient number of channels to represent the actual speeds at which vehicles with a given desired speed may travel.

It is possible to compute how many channels will be needed for a specific simulation situation if the number of discrete speeds permitted and the number of lanes are known.[21] From Fig. 32-13 it will be clear that, if each channel is as long as the roadway, and if the roadway is several memory cells long, a computer of large memory capacity will be required to handle situations having many lanes and/or speed increments.

To check whether cars in any channel will come too close to any car in another channel during the next time interval, the cars in the faster channel may be copied into a dummy channel in such a way that they are shifted forward by an amount equal to the distance they would travel during the next time interval plus the minimum safe following distance. Then, using an extract operation, the contents of the slower channel may be compared with the contents of the faster channel (shifted forward) to determine if potential conflicts exist. An alternative method is to establish a mask representing the safe spacing in front of each vehicle present in the system. By an extract operation a check is made to determine whether another vehicle is within the minimum safe area.

When one vehicle gets too close to another one ahead, it must pass or slow down. The ability to pass is contingent on the absence of cars in a critical zone of the adjacent lane. Thus tests must be made of the adjacent lane to determine if passing is possible.

The process of changing speed or changing lane is simply the process of transferring a car from one channel to another. This may be accomplished by subtracting a group of appropriately situated binary 1's from the source channel and adding this same group to the destination channel.

Memorandum Representation of Freeway Traffic. With memorandum representation, a word is assigned to represent each vehicle. Various groups of digits within the word are assigned to represent various items of information about the vehicle, as shown in Fig. 32-14. Items which can be appropriately included are the location

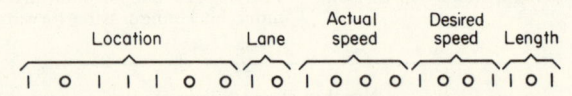

FIG. 32-14. Typical arrangement of information in code word.

(coordinate) of the vehicle along the roadway, the lane in which the vehicle is traveling, the desired speed of the vehicle, the actual speed of the vehicle, the length of the vehicle, the time at which the vehicle entered the roadway section under observation, and the minimum following distance which the vehicle will maintain behind the vehicle ahead. Manipulations necessary to produce movement through the system, to detect conflicts, to accomplish passing, etc., are all simple arithmetic operations. Forward movement for one unit of time, for instance, is accomplished by adding the actual speed to the position of the vehicle on the roadway. Lane changing is accomplished simply by changing the digits which represent the lane. Detection of impending encounters may be accomplished by first adding the actual speed of the front vehicle to the position of the front vehicle and subtracting the length of the front vehicle, the actual speed of the rear vehicle, and following distance maintained by the rear vehicle, as indicated in Fig. 32-15. If the result is negative, it is unsafe for the rear vehicle to continue in its present lane at its present speed. Since, in this form of representation, the time at which each vehicle enters the system may be easily determined, it becomes relatively easy to compute a suitable figure of merit.

Restating the comparison with memorandum representation: Instead of subdividing the flow network in order to keep track of differences in speeds, desired speeds, etc., and to provide for interchange between various lanes for groups of binary digits which are physically representing vehicles, it turns out to be much simpler to keep a memorandum for each item, stating its present location within a coordinate system, its desired speed, its actual speed, and any other characteristics which may be needed. The amount of memory space required for the storage of vehicles may not necessarily be decreased, but there is an economy in complexity of organization, and thus an economy in the storage of instructions. Here a very simple discipline for the treatment of interaction takes place.

A series of studies of freeway problems has demonstrated that memorandum

SIMULATION   32-19

representation is much easier to program than is physical representation. Furthermore, memorandum representation can speed up the computing time by as much as a factor of 20.[23]

Simulation of Random Events (Monte Carlo Methods). One of the most important aspects of digital simulation is the ability to handle stochastic events; i.e., events which occur randomly. Techniques for such problems are often known as

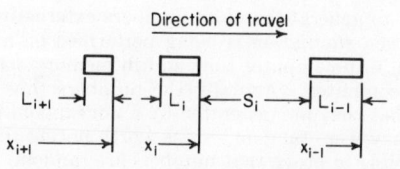

FIG. 32-15. Method of testing space between vehicles.

x_i = distance of ith car from reference (at time t)
L_i = length of ith car
V_i = speed of ith car (at time t)
$S_i \big|_t$ = space ahead of ith car (at time t)
     = $X_{i-1} - L_{i-1} - X_i$
$S_i \big|_{t+\Delta t}$ = estimated space ahead of ith car at time $(t + \Delta t)$
     = $(X_{i-1} + V_{i-1}\Delta t) - L_{i-1} - (X_i + V_i \Delta t)$

"Monte Carlo methods." The paragraphs which follow supplement the discussion in Sec. 40-8.

An approach to the understanding of Monte Carlo techniques for the generation of random events during digital simulation may be based on frequency distributions. Figure 32-16 is a generalized probability distribution function (cumulative probability distribution). Along the horizontal axis are values of the random variable X. Along the vertical axis is a scale representing a probability (fraction) of the total events which will have the value X or less. Thus all events will have a value of X_{max} or less, with probability 1.0. In ordinary usage, this probability distribution function states the likelihood of various results when a particular experiment is conducted a large number of times. For the purposes of Monte Carlo methods, however, the cumulative probability distribution may be regarded as a functional relationship between probabilities and associated values of the variable X. Then, if by some method a probability may be obtained, it is possible to obtain the value of X by entering the distribution using the probability as the Y value and obtaining the corresponding X value as the generated random measurement.

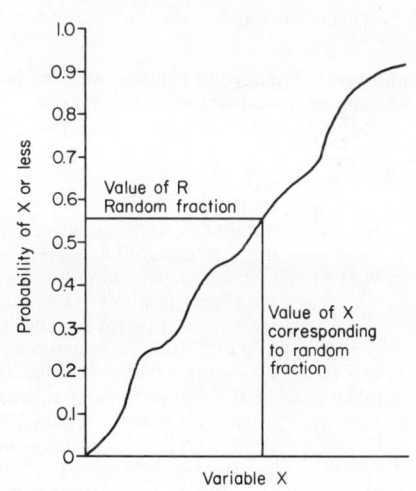

FIG. 32-16. Generalized cumulative-probability distribution.

Another situation calling for the application of Monte Carlo methods is as follows: Consider an event which has a probability of occurrence p. If it is possible to produce a series of uniformly distributed random numbers $R_1, R_2, \ldots, R_n$, $0 \leq R_i \leq 1$, then in all trials for which

$$R_i \leq p$$

the event is assumed to occur. When

$$R_i > p$$

the event is assumed *not* to occur.

For the execution of Monte Carlo methods, probabilities are obtained by producing random numbers. The random (stochastic) variable X is then obtained by treating the random numbers with some procedure which is appropriate to the nature of the distribution function.

While it is possible to generate random numbers externally and supply them to a simulation process, when simulation is being performed on a digital computer, it is usually more economical in computer time and in memory space to generate random numbers within the computer. Actually the numbers thus generated are pseudorandom numbers in that they are generated by a nonrandom process but behave statistically as though they were random. It is worth noting that in testing a series of numbers it is not possible to prove that numbers are random. The results of a negative test should be stated, "There is no evidence that the numbers are not random."

Several techniques have been described in the literature as suitable for generating pseudorandom numbers.[24,25,26] The most common is known as the multiplicative congruential method. A starting number is multiplied by an appropriate multiplier, and the low order or less significant half of the product is taken for the random

Table 32-3. Multipliers for Random-number Generation

Number base and capacity of computer b^n	Random-number multiplier ρ
10^{10}	7^{11}
10^{8}	7^{9} *
2^{35}	5^{13}
2^{36}	5^{13}

* This is the number used in Sec. 40-8.

number. The second random number is formed using the first as a starting number and the same multiplier as before, etc. This technique is usually stated as

$$R_m \equiv \rho R_{m-1} \ (\text{Mod } b^n) \qquad (32\text{-}12)$$

where R_m = mth random number
ρ = multiplier
n = number of digits in a normal word on the particular computer used
b = number base of computer
Mod b^n = instruction to use only the low order or less significant half of the full ($2n$-digit) product (the remainder after dividing the product by b^n the maximum integral number of times)
R_0 = any odd number selected as a starting number

The multiplier ρ may be selected by taking a base which is prime relative to the number base of the computer and raising it to the highest power which can be held by one word of the computer. Table 32-3 gives a few appropriate multipliers.

Random numbers so generated may be interpreted either as random integers with the point at the extreme right or as random fractions with the point at the extreme left. If the random numbers are interpreted as fractions, the result is a rectangular or uniform probability distribution between zero and one. See Table 40-8.

Uniformly distributed pseudorandom numbers can also be generated using a digital shift register with modulo-2 adder feedback. That is, the output of a shift register is summed with the value of an intermediate stage and the resulting sum is used as input to the shift register. It can be shown that with appropriate choices of the feedback point, an n-stage shift register will produce a sequence ($2^n - 1$) bits long before repeating. If such a sequence possesses certain statistical properties (e.g., if the sequence probability density and autocorrelation functions satisfy appropriate conditions) the sequence is termed pseudorandom. The properties of pseudorandom sequences are discussed by Golomb[31] and Hampton.[32] If such a binary sequence has a period sufficiently long compared to a series of computer runs, it may be treated as a random sequence.

Conversion of random fractions (from a uniform distribution) to random deviates of the desired form through the use of the cumulative probability distribution can be accomplished in a variety of ways.[29,27,32] If the cumulative probability distribution is a continuous function which can be represented easily by a manipulated equation, the operation of inserting the random fraction and obtaining the random deviate is purely a matter of calculation. If the cumulative probability distribution is not represented by an equation but by tabular data, the generation of the random deviates may be performed by a form of table-lookup operation, as illustrated in Fig. 32-17. The random fraction is compared with the ordinates of the various points until the first point satisfying the following condition is found:

$$R \leq P_i \quad (32\text{-}13)$$

where R = random fraction
P_i = probability (ordinate) value of the ith point

The value of x for the ith point is taken as the desired.

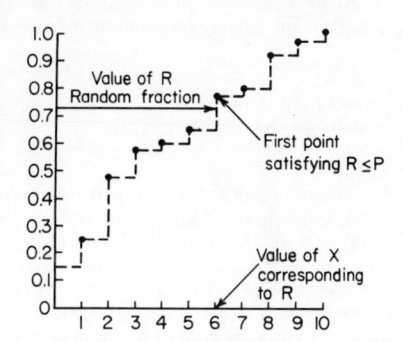

Fig. 32-17. Cumulative probability distribution in tabular form.

If the random deviate x is a continuous distribution, the tabular values may be interpolated along straight-line segments between the points, obtaining a unique value of x for each value of random fraction.

Two examples follow:

Example 1. The Poisson distribution is used for describing various phenomena, such as traffic. The cumulative Poisson distribution is expressed

$$P(x) = \sum_{i=0}^{x} \frac{m^i e^{-m}}{i!} \quad (32\text{-}14)$$

where $P(x)$ = probability of x or fewer occurrences within a sample, and m is the Poisson parameter (mean number of occurrences per sample); e is, of course, the base of natural logarithms. Where the sampling is by time intervals,

$$m = \beta T$$

where β = average number of events per unit time
T = size of time intervals being sampled

The generation of random deviates (arrivals, for instance) obeying the Poisson distribution must be carried out by a trial-and-error process. First a random fraction R is generated. The cumulative Poisson distribution is then formed term by term, using (32-14). At each step the cumulation is compared with R. When the first value of $P(x)$ satisfying the relationship

$$P(x) \geq R \quad (32\text{-}15)$$

is found, the corresponding value of x is taken as the random variate (number of arrivals).

Example 2. Many phenomena characterized by sequences of arrivals, as in traffic situations, may be treated by means of the exponential distribution

$$P(g \geq t) = e^{-t/k} \quad (32\text{-}16)$$

in which g = gap between successive arrivals, in time units
t = time, usually in seconds
k = average time spacing between arrivals (may be thought of as the abscissa of the center of gravity of the area under the exponential curve)
$1/k$ = volume (number of arrivals per unit time, the unit time being the same as that used for t)
$P(g \geq t)$ = probability that $g \geq t$

Equation (32-16) expresses the probability that the spacing between arrivals is equal to or greater than the specified time. The cumulative exponential probability distribution is the complement of this relationship; namely,

$$P(g < t) = 1 - e^{-t/k} \qquad (32\text{-}17)$$

where t is taken as the time spacing between arrivals.

Solving (32-17) for t gives

$$t = -k \ln (1 - P) \qquad (32\text{-}18)$$

By substituting the random fraction R for $(1 - P)$ it is possible to solve for t.

Still another method of conversion from a uniform distribution to another distribution can be accomplished by drawing several random numbers and performing appropriate manipulations upon them. This approach is particularly useful in the case of the normal distribution. Such techniques are treated in Sec. 40-8.

Scanning the Digital Model. Since system simulation must proceed in a serial fashion, the experimenter must give considerable attention to the way in which the system is to be scanned or sampled. The two common approaches to scanning may be termed periodic scanning and event scanning. Periodic scanning consists of observing or updating each portion of the simulated system, advancing the clock by one time unit, and repeating the cycle. This form of scanning is necessary when simulating systems which are in fact sampled data or otherwise digital in nature. Periodic scanning is, furthermore, straightforward to program and to check out.

In systems where important events occur at random or otherwise irregular intervals, many scan cycles may take place with no significant event occurring. In such cases, it is often possible to save substantial computing time by using event scanning. This scanning technique involves determining when the next significant event will occur, extrapolating all system processes to the time of that next significant event, and advancing the clock to that time without any intervening stops. Some simulation programs have been speeded up by factors as large as 10 when changed from periodic scanning to event scanning. The amount of programming to implement event scanning may be very large and time-consuming. If a particular simulation program is to be used only once, it may be advisable to use periodic scanning. If, however, the program is to be run many times so that the expenditure of programming time may be justified in terms of large amounts of computer time to be saved, then event scanning should be used.

In some problems it may turn out to be advantageous to use a combination of event scanning and periodic scanning.[28]

32-4. Simulation of Manned Systems

One of the most important areas of application of the simulation method is in the study of systems in which a human being participates, either as an element of the system such as the pilot of a vehicle or as a passenger whose tolerance to environmental characteristics is limited. In the design of piloted vehicles simulation techniques are so prevalent that in some quarters the word "simulator" is reserved for this type of activity.

Characteristics of Manned Simulation. Simulation involving man includes all the characteristics of unmanned simulation, with the following additional ones which are introduced by the particular characteristics of human performance:

1. Human performance is inherently variable. There is variation of successive trials of the same task by the same operator, and there is a variation in the responses of several operators trying the same task.

2. Human response includes elements which are apparently not determined by the input and can be accounted for only by statistical descriptions. Consequently, the description of systems involving human operators must make use of statistical methods, and the resulting descriptions will be in some sense statistical averages defined over particular populations.

3. The inherent variability of human performance implies that many repetitions of each particular experiment must be tried.

4. Simulation studies involving human operators must be run in real time, whereas studies involving inorganic elements may be run in an accelerated time scale in many cases.

5. The simulation method and the experimental situation must be selected in such a way as to avoid any possible injury to the operators involved.

Simulation of manned systems takes on two primary forms: environmental simulation and man-in-the-loop simulation. Environmental simulation involves creation of an environment which reproduces one or more unusual situations in which human beings may find themselves in a system undergoing design. Man-in-the-loop simulation involves an interaction between man and equipment. Both these types of simulation will be examined briefly in the following paragraphs.

Environmental Simulation. Environmental simulators are needed because human beings are being subjected to environments drastically different from those of ordinary life. For example:

1. Man may be exposed to situations where high temperatures and high levels of pressure are involved, such as in certain types of mining or underground operations.

2. Man may be asked to undergo long periods of weightlessness such as those occurring in interplanetary flight.

3. Man may be asked to operate in atmospheres of different composition from that of his normal habitat.

In order to test the adequacy of the proposed design techniques and to ensure human survival, it is necessary to simulate the particular characteristics of this environment before a completed vehicle is constructed. Once again, a characteristic problem in the design of such simulators is the selection of the particular quantities or variables to be investigated. For example, it may be decided to construct a simulated space cabin for an interplanetary voyage in which human passengers may be subjected to temperatures, radiation levels, and illumination levels similar to those encountered in the actual flight. It may, however, be decided to avoid any attempt to simulate the gravitational environment of free space. On the other hand, other simulations may involve attempts to examine the ability of operators to perform certain tasks under conditions of reduced gravity, and certain kinds of underwater simulations have been proposed for this purpose. Once again, the decision of what is simulated and what is omitted, what is important and what is negligible, rests largely with the designer.

Environmental simulation has included the following major characteristics:

1. Temperature simulation. Variable climate chambers and hangars have been constructed, some with temperatures which range from -300 to $+1000°F$. The dimensions of such a chamber may range from a cell barely adequate to accommodate one man to a chamber of sufficient size to accommodate an entire airplane or space vehicle.

2. Acceleration. The effect of acceleration and deceleration on human operators and passengers (as well as equipment) is usually measured using centrifuges and rocket sleds which are capable of imparting wide ranges of acceleration and deceleration to their payloads. For example, the human centrifuge at Johnsville, Pa., has a cabin located at the end of a 50-ft arm. The centrifugal acceleration to which the operator is exposed may reach 40 to 50 g's. Rocket sleds, such as one located at Edwards Air Force Base, may provide acceleration as high as 50 or 60 g's.

3. Unusual atmospheric conditions. Altitude chambers and environmental chambers have been constructed with a capability of generating ice and snow with atmospheric conditions ranging from sea level to 100,000 ft altitude. Simulated desert sand and dust storms can be generated in certain simulators. Humidity ranges from 0 to 100 per cent, salt spray, tropical rainstorms, and similar unusual conditions have been produced in the laboratory.

4. Vibration. Simulators have been constructed which provide vibration and shock excitation ranging from 5 to 2,000 cycles as well as random vibration sources with various spectral characteristics. Shock in the range of 0 to 100 g's and of various durations has also been simulated.

5. Zero gravity. Conditions of null or zero gravity have been simulated in airplane cabins while the airplane flies a parabolic trajectory during which gravitational and centrifugal accelerations exactly cancel, resulting in periods of weightlessness as long as 15 to 30 sec. Zero g has also been simulated by spinning a man submerged in a fluid.

6. Lack of atmosphere. The lack of atmospheric friction and resistance in space for the performance of particular tasks has been simulated by means of minimum friction and air bearing tables.

7. Complete cabin simulations. A number of tests have been and are being performed in which simulated space cabins including complete closed-cycle ecological systems have been constructed. Human volunteers have stayed under simulated space-cabin conditions for a number of days. In many cases such simulated cabins have included temperature, atmospheric composition, and other aspects of the environment in simulated form.

Other environmental simulators have been constructed for the testing of equipment which does not involve human operators. Such simulators include methods for determining the effect of extreme levels of solar radiation, nuclear-explosion effects, and so forth.

Since no simulator takes into account each and every effect encountered by a human operator or a human passenger in a particular task, the addition or superposition of effects observed in various portions of the simulation must be handled with great care. In many cases a simple linear superposition of effects may not be valid.

Flight Trainers and Piloted Simulators. Where a human pilot performs control or guidance functions in the operation of a system, some form of simulation is essential during the design phase. The simulation may be entirely an analog simulation, since in a control task the operator's input and output are generally continuous, or it may be partially or entirely digital, in which case some form of analog-to-digital and digital-to-analog conversion may be required. In the design of flight-control systems, the simulation generally becomes some form of physical simulation in which there is an interrelationship between a human pilot, an actual or simulated portion of a vehicle-control system (including manual controls, displays, dials, knobs, and so forth), and a general-purpose computer (analog or digital) which provides inputs to the cockpit and operator which represent the variation of environmental characteristics during a particular flight mission. For purposes of illustration, consider the block diagram of a lunar-landing simulator as illustrated in Fig. 32-18. The three essential parts of a manned-flight control-system simulator are seen in this figure: (1) the pilot, (2) the cockpit, and (3) equipment required to provide input signals to the cockpit and process output signals from the operator. Where the pilot responds to simple dial movements, a general-purpose or a special-purpose computer may be adequate to provide the input signals. Where a more realistic simulation of the external environment is required, more elaborate equipment is also necessary. The diagram of Fig. 32-18 shows a television system in which a camera movable in six degrees of freedom approaches a simulated lunar landscape. The resulting television picture is displayed on a monitor located inside the cockpit. Figure 32-19 gives an indication of the degree of fidelity which is possible with this simulation.

The following points can be noted about this simulation: (1) It is a partial simulation. Many effects are not included. In this particular diagram the cockpit is fixed and not subject to either rotational or translational movement. Even if such movement were permitted (as in the so-called moving-base simulator) such movement is always of limited dynamic range and the simulation is only a partial simulation. (2) The simulation is designed for a specific mission. In this case it is clear that the simulation is adequate for only a particular range of altitudes and for a particular lunar landscape without changing either the map or the scaling of the entire system. Comparable limitations are characteristic of all such simulators. (3) The fidelity of simulation is variable and questionable. The choice of fidelity is one of the serious problems in the design of all manned simulators. For a particular task an extremely simple simulated cockpit may be adequate provided that it includes those display and control components essential for a particular mission. For other simulation tasks it may be

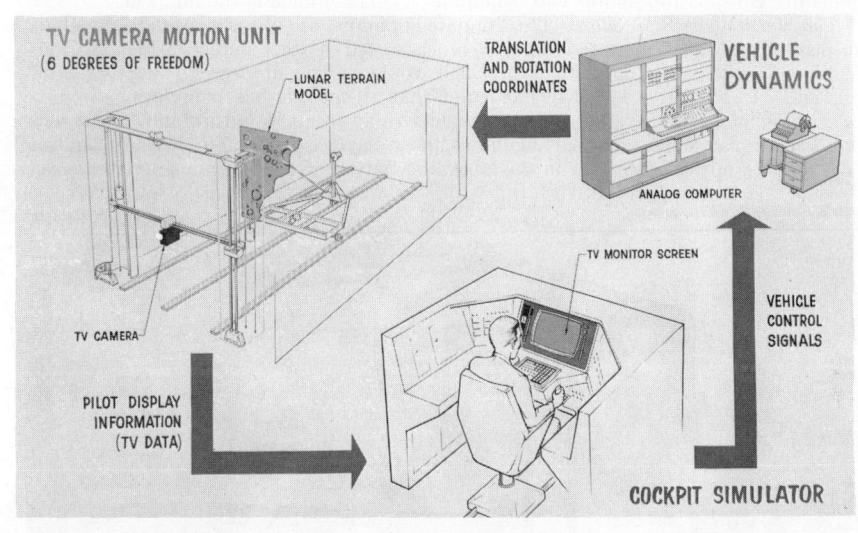

FIG. 32-18. Block diagram of lunar-landing simulator. (*Courtesy of TRW Space Technology Laboratories.*)

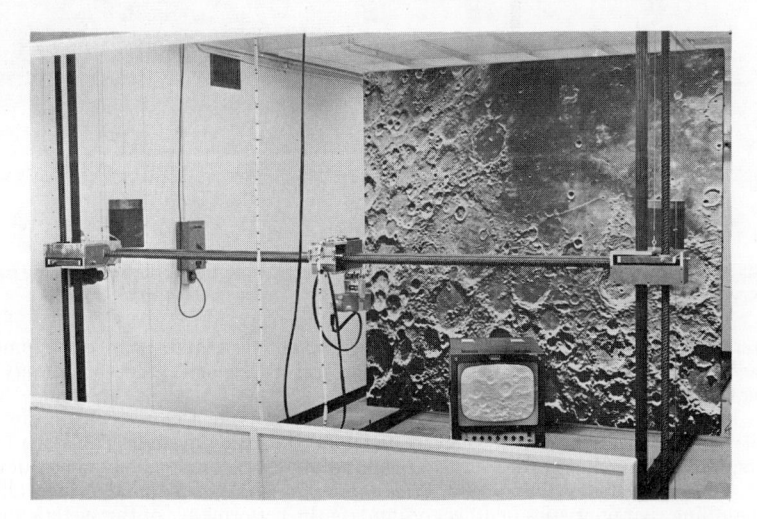

FIG. 32-19. Television picture of simulated lunar surface. (*Courtesy of TRW Space Technology Laboratories.*)

necessary to include an actual production cockpit. In the present example, it can be noted that a television screen is used for an optical input to the operator. If this screen is meant to represent a television or periscope view, its fidelity may be adequate, but it is clearly no substitute for a window. However, the simulation may provide useful results in determining certain ranges of human capability and certain limitations of human performance even if the television picture is used as a simulated window view.

Attempts to overcome various of the limitations of the fixed-base laboratory simulator of the type discussed above have resulted in a variety of more complex and generally considerably more costly simulators. These include the following:

The Moving-base Simulator. The simulated or actual cockpit is mounted on gimbals, suspended on chains, mounted in a sled, or supported in other similar fashion and subjected to movement similar to that which would be encountered during the actual mission. It should be noted in particular that all moving-base simulations involve limitations of dynamic range and consequently may result in faithful movement over only certain particular ranges of angular or linear displacement. Furthermore, motion cues may be misleading, since in the laboratory situation a pilot in a simulated space

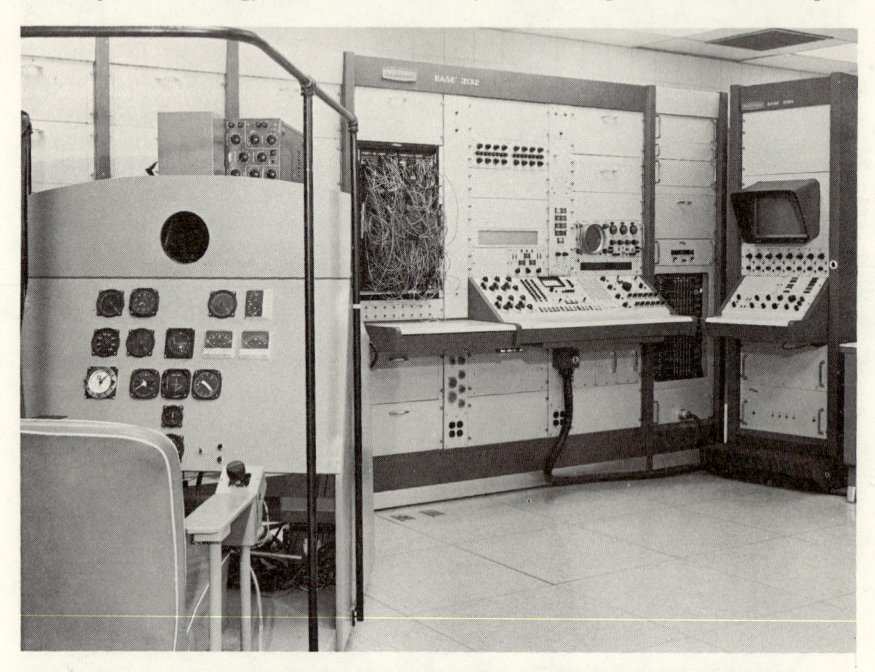

Fig. 32-20. Typical simple cockpit simulator with analog computer. Fingertip controller can be seen on the armrest. (*Courtesy of TRW Space Technology Laboratories.*)

mission will be subjected to the motion in space without the gravitational environment of space, and the effects upon his performance and physiological well-being may be different.

The Variable-stability Airplane. In an attempt to provide a more realistic simulation of flight-control systems of vehicles under investigation, certain airplanes, helicopters, and other vehicles with adjustable handling characteristics have been developed. These vehicles include an airborne computer, analog or digital, which alters their handling characteristics in order to simulate the performance of the system under design. Many such variable-stability aircraft have been built, and they have proved to be an invaluable research and design tool in the aircraft industry. In fact, the simulation of certain phases of reentry from space have been and can be accomplished using the variable-stability aircraft as a simulator.

Increasing Sophistication in Physical Simulation. It is possible to include in the simulation a whole range of equipment from a simple simulator cockpit to a complete mock-up of the actual vehicle. In airplane simulators, for example, it is not uncommon to include not only the cockpit itself, but also the servos, actuators, tail assemblies, hydraulic mechanisms, and similar devices as portions of the simulation in order to ensure that the performance of the pilot will not be distorted by a possibly inaccurate mathematical description included in a computer.

It is clear that simulation in one form or another is essential for the development of manned vehicles since it is important to subject man to simulated conditions before exposing him to actual and possibly hazardous operating conditions. Thus manned simulation has a dual purpose, as a research tool and as a design tool. As a research tool it enables us to determine conditions which will govern the design of future systems by providing envelopes of satisfactory performance. As a design tool it is invaluable for providing the absolutely necessary verification by human subjects of a proposed system configuration. It should be noted, however, that simulation cannot and should not be a substitute for design.

Computers Used with Manned Simulators. It has been noted above that some form of computer is required to generate the input signals to the cockpit and process the pilot's output signals in accordance with a predetermined mission such as a particular flight trajectory, a landing on a carrier deck, or a reentry from space. Historically, analog computers have been used for flight-control simulators for two reasons: (1) bandwidth requirements, since the mission characteristics as well as the input and output signals contained frequencies sufficiently high so that real-time digital computation was impossible and (2) accuracy compatibility, since in many cases the physical characteristics of the airframe and the atmosphere were known only to levels of accuracy compatible with those of analog elements.

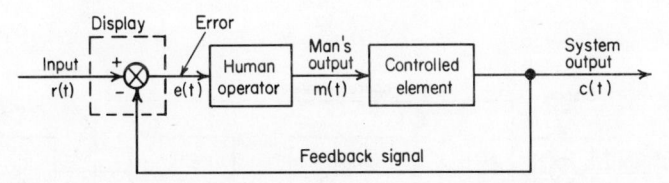

FIG. 32-21. Block diagram of simple tracking task.

Recently, the picture has changed for two reasons: first, the increasing speed of digital computers has made possible the real-time digital simulation of certain portions of aerospace missions; and second, airborne digital computers are being used to an increasing degree to handle the complex levels of data processing and computation which are characteristic of modern high-performance aerospace vehicles. Consequently, it is expected that an increasing use of digital computers in flight simulators will be seen in the future. In many cases, this simulation will take the form of the utilization of hybrid analog-digital equipment (see Secs. 11-14 and 32-5). Figure 32-20 shows an analog computer used with a fixed-base cockpit simulator.

Simulation of the Human Operator. It has been pointed out above that one of the limitations of manned simulation is that it must be performed in real time. Furthermore, the presence of the human operator renders each task different from a succeeding repetition of the same task. In preliminary design of manned systems the simulation could be performed considerably more rapidly and efficiently if the operator himself could be simulated, i.e., if a mathematical model of the operator's performance were available. In recent years a number of such mathematical models have been proposed. These models in general are adequate for representing the input-output characteristics of human operators performing certain particular simple tasks, such as the tracking of low-frequency signals in one dimension. Provided that the input signal in a task illustrated in Fig. 32-21 contains no appreciable energy above approximately $\frac{1}{2}$ cps, and provided further that the input signal is random-appearing, it is possible to represent the input-output characteristics of the operator by means of a quasilinear characteristic. This linear description is a linear differential equation whose coefficients depend on the input bandwidth and on the controlled-element

32-28 SYSTEM TECHNIQUES

dynamics. A number of such characteristics have been obtained experimentally and tabulated. For a considerably more restricted range of tasks, models have been proposed which include discrete as well as continuous operations. For particular simple controlled-element dynamics, such models have shown excellent agreement with experimental data. Nonlinear representations of the operator's performance have also been proposed. Figure 32-22 illustrates the tracking loop with simple mathematical models for the human operator included. Since in general the tasks required of human pilots in actual missions are considerably more complex than simple tracking tasks, it is probably fair to say that the use of human-operator models is limited at the present time to certain preliminary investigations, some of which are described in

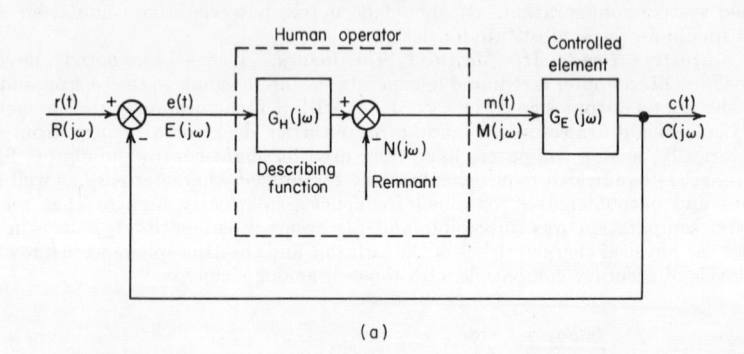

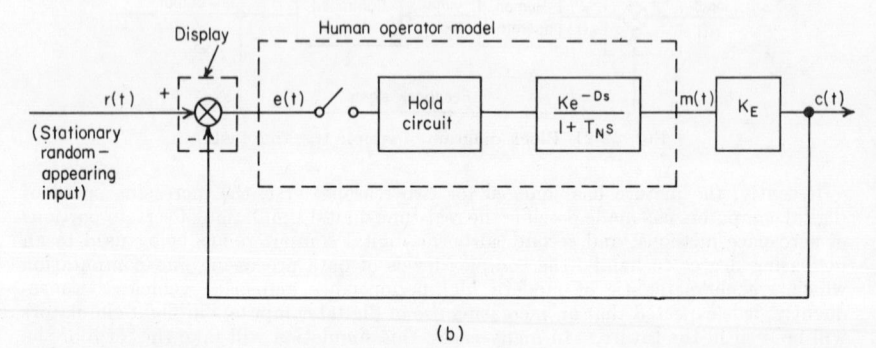

FIG. 32-22. Tracking task with human-operator model. (a) Continuous model. (b) Discrete model.

Sec. 31-6. Detailed design of a vehicle still requires real-time simulation involving actual human operators.

32-5. Hybrid Analog-Digital Simulation

In the above sections analog and digital simulation has been outlined briefly. It has been noted that in many cases the use of analog or digital simulation is determined in part by availability, cost, and similar considerations. In many cases the resulting simulation equipment is neither purely analog nor purely digital. Just as in the field of computation itself there has been a trend toward hybrid equipment (Sec. 11-14), so in the simulation area there has been a tendency toward the utilization of hybrid techniques. Some of the elementary considerations involved in hybrid simulation will be outlined in this section.

SIMULATION   32-29

Analog Techniques in Digital Simulation. Whenever a digital simulation makes use of physical elements which involve continuous inputs or outputs, the simulation may be considered hybrid. For example, if the manned simulation outlined in the preceding section involves continuous commands from the operator, it will be in fact a hybrid simulation since the commands must be digitized before introduction to the computer. Conversely, a digital-computer output must be processed by a digital-to-analog converter in order to result in smooth continuous signals for the display equipment. The simulation of process-control systems which use digital controllers in connection with actual physical equipment may result in hybrid techniques if analog-to-digital or digital-to-analog conversion is required. Clearly, many process variables such as flow, pressure, or temperature are in fact continuous variables. They may, however, be digitized after measurement for use by a digital computer.

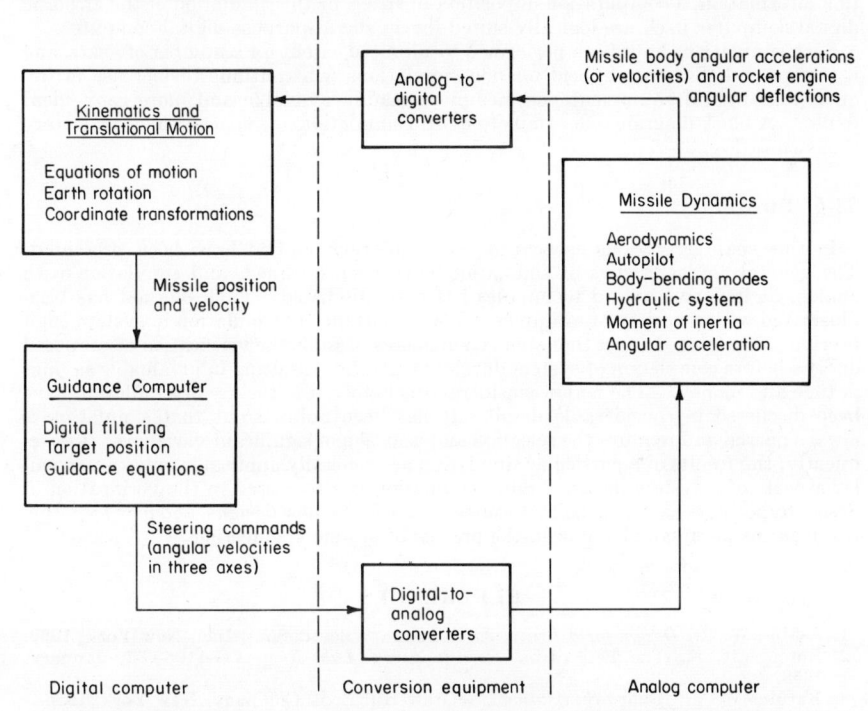

Fig. 32-23. Block diagram of hybrid simulation of a guided-missile trajectory.

Digital Techniques in Analog Simulation. The use of digital input and output with an analog computer also represents the use of hybrid techniques. Clearly, in many cases conversion equipment would be required. It has been mentioned briefly in Sec. 32-2, that high-speed analog memory (Sec. 11-15) and switching (Sec. 11-8) elements have come into use as components of general-purpose analog computers. The use of such switching, memory, and logic elements is in fact the use of digital techniques with an analog computer. An analog simulation of the flight-control system which includes an airborne digital computer as an element in the loop, i.e., a physical simulation including the actual flight computer, is clearly a hybrid simulation.

Hybrid Systems. In addition to the limited examples of hybridization mentioned in the preceding paragraphs, it is clearly possible to perform complete mathematical simulations involving true hybrid systems, i.e., general-purpose analog and digital computers interconnected by appropriate conversion equipment. Such simulations offer significant advantages as follows:

1. The analog equipment lends itself readily to simulating that portion of the system which includes high frequencies, complex nonlinearities, and/or physical elements of considerable variability in measured characteristics. For example, in the study of the trajectory of an aerospace vehicle the analog computer is ideally suited for the simulation of the airframe with its attendant nonlinearities, atmospheric characteristics which are known only within wide limits of tolerance, and the flight-control system which may include frequencies as high as 10 to 50 cps. An interface with a pilot may also involve analog signals.

2. Elements involving requirements for high-accuracy, drift-free computation and complex decision functions are ideally suited for simulation by a digital computer. For example, in an aerospace mission the computation of the trajectory from acceleration information, a computation of position in space, or the simulation of the airborne digital computer itself are logically suited for a general-purpose digital computer.

Such simulations have been performed to a limited extent for a number of years, and it is expected that true hybrid mission simulations will continue to increase in frequency as hybrid equipment becomes more readily available and more convenient to use. A block diagram of a complete hybrid simulation of a space-vehicle trajectory is shown in Fig. 32-23.

32-6. Summary

In this chapter, various aspects of the simulation method have been presented. The general characteristics of simulation have been outlined, and simulation with analog, digital, and hybrid techniques has been discussed. Each method has been illustrated with a variety of examples. The importance of simulation in system engineering arises from the fact that simulation makes possible the verification of proposed designs before completion of system development, thus resulting in invaluable savings in time and money. The major considerations involved in the use of simulation have been discussed in considerable detail. It has been pointed out that simulation is always partial and requires the selection and isolation of significant variables. Consequently, the results of a particular simulation are generally applicable only to particular aspects of a system design. Since simulation is a tool used in the verification of design hypotheses as well as being a source of new ideas, new designs, and new hypotheses, it forms an invaluable link in the process of system engineering.

REFERENCES

1. Ashby, W. R.: *Design for a Brain*, 2d ed., John Wiley & Sons, Inc., New York, 1960.
2. Rosenblatt, F.: The Perceptron, *Cornell Aeron. Lab. Rept.* VG-1196-G-1, January, 1958.
3. Karplus, W. J.: *Analog Simulation*, McGraw-Hill Book Company, New York, 1958.
4. Kohr, R. H.: Real-time Automobile Ride Simulation, *Proc. Western Joint Computer Conf.*, vol. 17, pp. 285–300, May, 1960.
5. Goode, H. H., and R. E. Machol: *System Engineering*, McGraw-Hill Book Company, New York, 1957 (particularly pp. 273–278 and 403–407 are concerned with simulation).
6. Johnson, C. L.: *Analog Computer Techniques*, 2d ed., McGraw-Hill Book Company, New York, 1963.
7. Sadoff, M. E.: A Study of a Pilot's Ability to Control during Simulated Stability Augmentation System Failures, *NASA Tech. Note* D-1552, November, 1962.
8. Norum, V. D., M. Adelberg, and R. L. Farrenkopf: A Critical Evaluation of a Conductive Surface Analog in the Analog Simulation of Particle Trajectories, *Proc. Third Intern. Analog Computation Meeting*, pp. 79–94, 1961.
9. Proceedings of the Bionics Symposium, *WADD Tech. Rept.* 60-600, September, 1960.
10. Harmon, L. D., J. Levinson, and W. A. Van Bergijk: Analog Models of Neural Mechanisms, *IRE Trans. Inform. Theory*, vol. IT-8, pp. 107–112, February, 1962.
11. Sadoff, M. E., and C. W. Harper: A Critical Review of Piloted Flight Simulator Research, presented at IAS Meeting on The Man-Machine Competition, Seattle, August, 1962.

12. Westbrook, C. B.: Simulation in Modern Aerospace Vehicle Design, presented at AGARD meeting in Brussels, Belgium, April, 1961.
13. Chestnut, H.: Modeling and Simulation, *Elec. Eng.*, August, 1962, pp. 609–614.
14. Bekey, G. A.: The Human Operator as a Sampled-data System, *IRE Trans. Human Factors Electron.*, vol. HFE-3, pp. 43–51, September, 1962.
15. Bekey, G. A., D. Darms, W. A. Manson, and N. S. Assali: Analog Simulation of the Cardiovascular System of the Fetal Lamb, *Proc. Third Ann. Symposium Biomedical Eng.*, San Diego, 1963.
16. Huskey, H., and G. A. Korn: *Computer Handbook*, McGraw-Hill Book Company, New York, 1962.
17. Goode, H. H.: Simulation—Its Place in System Design, *Proc. IRE*, vol. 39, pp. 1501–1506, 1951.
18. Vazsonyi, A.: An On-line Management System Using English Language, *Proc. Western Joint Computer Conf.*, 1961, pp. 17–37.
19. Webb, K. W., and J. Moshman: A Simulation for Optimal Scheduling of Maintenance of Aircraft, paper presented at 16th National Meeting of the Operations Research Society of America, Nov. 11–13, 1959.
20. Monroe, A. J.: *Digital Processes for Sampled Data Systems*, John Wiley & Sons, Inc., New York, 1962.
21. Gerlough, D. L.: Simulation of Freeway Traffic by an Electronic Computer, *Proc. Highway Res. Board*, vol. 35, pp. 543–547, 1956.
22. Moore, C. J., and T. S. Lewis: Digital Simulation of Discrete Flow Systems, *Commun. ACM*, vol. 3, pp. 659–660, 662, 1960.
23. Gerlough, D. L.: A Comparison of Techniques for Simulating the Flow of Discrete Objects, paper presented at the National Simulation Conference, Oct. 23–25, 1958.
24. Taussky, O., and J. Todd: Generation and Testing of Pseudo-random Numbers, in H. A. Meyer (ed.), *Symposium on Monte Carlo Methods*, pp. 15–28, John Wiley & Sons, Inc., New York, 1956.
25. Kahn, H.: Use of Different Monte Carlo Sampling Techniques, in H. A. Meyer (ed.), *Symposium on Monte Carlo Methods*, pp. 146–190, John Wiley & Sons, Inc., New York, 1956.
26. Butler, J. W.: Machine Sampling from Given Probability Distributions, in H. A. Meyer (ed.), *Symposium on Monte Carlo Methods*, pp. 249–264, John Wiley & Sons, Inc., New York, 1956.
27. Zacharias, W. B.: The Purposeful Use of Random Variables, *Proc. Symposium on Digital Simulation Techniques for Predicting the Performance of Large-scale Systems* (unclassified), held at the University of Michigan, May 23–25, 1960, pp. 53–60.
28. Chaitt, M. C., and R. F. Meier: A Modification of the Events Store Control Program, *Proc. Symposium on Digital Simulation Techniques for Predicting the Performance of Large-scale Systems*, held at the University of Michigan, May 23–25, 1960, pp. 15–21.
29. Gerlough, D. L.: Traffic Inputs for Simulation on a Digital Computer, *Proc. Highway Res. Board*, vol. 39, pp. 480–492, 1959.
30. Bekey, G. A., M. J. Merritt, and N. S. Assali: Synthesis and Optimization of a Model of Cardiovascular Dynamics, *Proc. Fourth Intern. Analog Computation Meeting*, 1964.
31. Golomb, S. W.: Sequences with Randomness Properties, (Internal Report) Glenn L. Martin Co., Baltimore, Md., June 14, 1955.
32. Hampton, R. L. T.: A Hybrid Analog-Digital Pseudo-random Noise Generator, *Proc. Spring Joint Computer Conf.*, vol. 25, pp. 287–301, 1964.

Chapter 33

RELIABILITY

HAROLD D. ROSS, JR., *Director of Research Relations, IBM Corporation, Data Systems Division, Poughkeepsie, New York.*

CONTENTS

33-1. Introduction	33-1
33-2. Definitions, Quantitative Expression, and Prediction	33-2
Reliability Definition	33-2
Exponential Failure Law; Mean Time to Failure	33-3
Reliability Prediction	33-4
33-3. The Constituents of Reliability	33-4
Component Parts	33-5
Environments	33-5
Project Organization for Reliability	33-6
33-4. Designing for Reliability	33-6
Component Parts—Selection, Standardization	33-7
Design of Reliable Electronic Circuits	33-7
Use of Redundancy to Increase Reliability	33-9
33-5. Maintainability	33-13
Definition of Maintainability	33-14
Repairability Prediction	33-14
Design for Maintainability	33-15

33-1. Introduction

The reliability of a system is one of the most important attributes contributing to its overall performance and as such must be given adequate consideration at the initial planning stage. At this time, when the first trade-offs between performance, cost, and schedule are weighed, estimates must be made of the reliability required of the system because they usually have a major impact on the realizability of the system as well as on the cost.

Figure 33-1 outlines the process of attaining reliability in a system. It is important to note that considerations affecting reliability extend through the entire life of a system, not just the design phase; they involve manufacture, operation, and maintenance as well. Figure 33-1 further points out that predictions of system reliability are needed during the design phase in order to determine whether operational objectives will be met and to aid in selecting the particular design approaches that will permit achieving them. Supporting this process, there must be adequate data on the reliability aspects of the key ingredients from which the system will be assembled. This includes not only physical parts but the methods of manufacture, maintenance, and operation that will be used, and the people who form a major segment of the system.

Finally, it must be emphasized that the attainment of a high system reliability calls for a sustained effort over long periods of time by a large number of people and organizations. This has led to the establishment of reliability engineering groups in such system projects, which commonly report at a high management level to ensure that reliability receives proper emphasis.

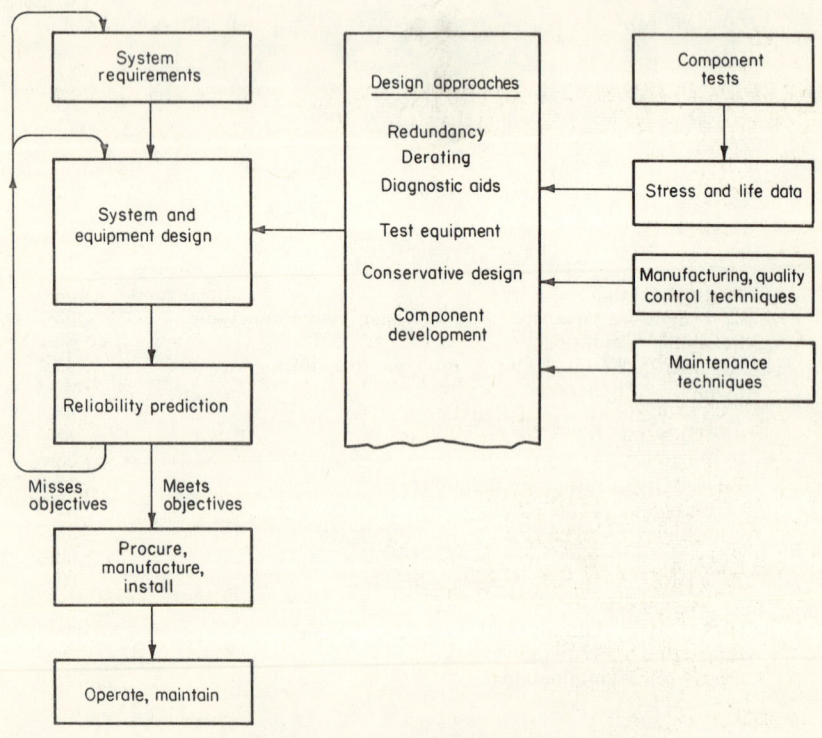

Fig. 33-1. The process of attaining system reliability.

33-2. Definitions, Quantitative Expression, and Prediction

Reliability Definition. The usual definition for reliability is similar to the following: "Reliability is the probability that a specified function will be performed for a specified time under given conditions." It is, in brief, the chance that the item in question will "work." The definition points out the need to know certain basic things:

The required time of operation
The environment of operation
What constitutes successful or adequate operation

The result of applying the above definition is a quantitative value for reliability. For example, a satellite might be required to have a 90 per cent probability of successful operation for a period of 1 year, in orbit.

Several classes of operation may be encountered:

1. Continuous operation for long periods. Examples here would be an air-traffic-control system, a satellite, or communication equipment.

2. Operation for short periods where operation is preceded by a test for correct functioning. An example would be a missile guidance system.

3. One-time-use operation where testing is not practical, such as a solid-propellant rocket.

Exponential Failure Law; Mean Time to Failure. While complex systems or equipments may fail, as a function of time, in a variety of ways, it has been found that electronic equipment tends to have a characteristic failure behavior. This behavior is illustrated in Fig. 33-2 and is characterized by three principal stages.

Stage 1 is the early life of the equipment and exhibits a high but rapidly declining failure rate as the short-lived and marginal components are discovered and eliminated by a "debugging" process. Stage 3 is the period of increasing numbers of failures due to wear-out processes after extended life. Stage 2 is a period of relatively constant rate of failure characteristic of the main operating life of the equipment, in which failures occur randomly and infrequently.

The curve is obtained by plotting smoothed data from equipment life tests in which failures are quickly repaired so as to keep the population constant. The infrequent random failures will smooth to a flat failure-rate curve, finally increasing rapidly in the

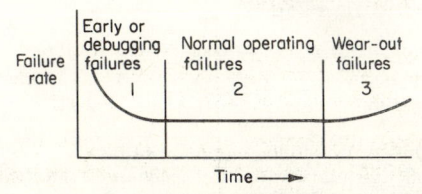

FIG. 33-2. Failure-behavior curve.

wear-out region. The flat portion establishes the normal operating region, and here failures can be considered random and independent, corresponding to a Poisson process. Such a failure behavior is the desired one from the viewpoint of reliability engineering as it permits reasonable predictions of reliability from short-time test data.

In examining the data from life tests, it will be found that for a Poisson process the intervals between failures follow a characteristic pattern: many short intervals, fewer longer ones, and very few extremely long intervals. When the numbers of occurrences of different time intervals are plotted, an exponential curve is formed. (This exponential curve should not be confused with the shape of the ends of the curve of Fig. 33-2—we are speaking only of the flat or stage 2 portion.) This leads to the exponential *failure law*

$$R = e^{-\lambda t} = e^{-t/m}$$

where R = reliability or probability of successful operation for the time t
λ = failure rate
m = inverse of the failure rate or mean time between failures (MTBF)

The system designer is commonly concerned with the probability of an equipment operating without failure for a given period, e.g., for the length of a mission. Where this period is *equal* to the MTBF, the probability (or reliability R) is 0.368. Notice that, to obtain a high value for R, the MTBF must be many times the mission length; for example, for $R = 0.9$, $m \approx 10t$; for $R = 0.99$, $m \approx 100t$.

Reliability engineering is concerned with the distribution of failures in the time domain, not with the distribution of defects in a population as in quality control. It is important to recognize that a major problem is that of determining the actual distribution of failures in the time domain—this usually requires a testing program.

The tests should be carried on long enough to determine the duration of the flat portion of the failure curve; however, there is no need to determine the exact nature of the distribution in stage 3, as prediction is not possible in this region. Many different distributions, calling for different mathematical models, are found in testing electronic parts. Among the models found useful, the Gaussian, Poisson, and Weibull (see Table 38-3) are of particular interest.

Reliability Prediction. As was pointed out in Fig. 33-1, one of the key steps in the attainment of a reliable system is the prediction of its reliability. Such predictions are required for a variety of purposes, e.g., evaluation of alternative designs, to determine whether a proposed design meets the stated requirements of the system, or to determine which portions of a system design are the "weak links" from the standpoint of reliability. Where the objective is a comparison, good results are readily obtainable as long as the analysis is done in the same way in each case and the basic data are comparable. Where the objective is an estimate of the absolute value of the reliability, the problem is more difficult and much care is called for in the preparation of the mathematical model and in the choice of reliability data.

The most generally used prediction technique for predicting the reliability of electronic equipment calls for the following:

1. List the quantities of each different part used.
2. Multiply each quantity by its failure rate and sum the total, to obtain the overall failure rate.
3. Calculate the resulting reliability of the equipment from the exponential law, using the overall failure rate.

The particular expressions used are:

$$R \text{ system} = R_1 \times R_2 \times R_3 \times \cdots \times R_n$$

or

$$R \text{ system} = e^{-\lambda_1 t} \times e^{-\lambda_2 t} \times e^{-\lambda_3 t} \times \cdots \times e^{-\lambda_n t}$$

$$\lambda \text{ system} = \lambda_1 + \lambda_2 + \lambda_3 + \cdots + \lambda_n$$

The technique depends on the following assumptions:

1. All parts have constant failure rates; i.e., they are in the stage-2 period and consequently the exponential law applies.
2. The failure-rate data are available and pertain to the environment found in this equipment.
3. For successful operation of the equipment, every part must operate.
4. Failure of one part is independent of failure of the others.

The validity of these assumptions must be carefully considered when this technique is used, in order to avoid substantial errors. For example, many types of complex equipment, such as electronic computers, do not at all times use every element of the system. This must be taken into account in calculating the system reliability lest the calculated result be unduly pessimistic. Alternate modes of operation, standby equipment, and other redundant elements further complicate the matter. In connection with point 4, many or most component parts of a system will commonly share the same physical environment (e.g., the same power and air-conditioning sources) and in various ways have failure characteristics which are not completely independent. The matter of obtaining reliable and pertinent data on component part and failure rates is one of the chief problems in reliability prediction. Despite these hazards, reliability predictions have been developed to the point of quite good accuracy.

33-3. The Constituents of Reliability

This section is concerned with the "materials" or ingredients with which the system designer has to work in attaining his reliability objectives. It has to do with the nature of the physical parts which are available as well as with the factors which constitute the environment (taken in the broadest sense) in which the physical equipment will be produced and used. To some degree, these elements are under the system designer's control, at least insofar as he can select from or emphasize certain ones, but in general he must take them as given and achieve a suitable design in spite of certain inadequacies which are beyond his control.

Component Parts. The intrinsic reliability of a system is a consequence of the parts which are used in its assembly and the maturity of the design approach which is followed. The first will be considered here; the second is the subject of Sec. 33-4.

Although many kinds of parts (structural, chemical, mechanical, electromechanical, electronic, optical, etc.) may be found in a complex system, the most difficult problems of part reliability have arisen in connection with electronic parts. This seems to be because the processes which determine part life are quite imperfectly known, the large quantities of such parts used necessitates exceedingly low failure rates, and the origin of many electronic parts is in an environment which emphasizes low cost rather than long life. Therefore, this section will deal mainly with electronic parts, although the principal points are broadly applicable; usually there is not time to develop a range of these parts especially for the system under design (unless clearly unusual conditions are expected), and initially, at least, an attempt is made to use readily available components.

The initial problem with component parts, from a reliability standpoint, is to obtain data on the failure rate or probable life of each type of part to be used, *under the conditions of use expected in the system.* The significant conditions will vary with the type of part but will commonly involve such factors as

Voltage	Temperature
Current	Humidity
Power dissipation	Altitude or pressure
Vibration	Cooling
Shock	Radiation

Vast quantities of component-part-failure data are available; finding data which apply to particular conditions of use may be very difficult and require the establishment of special testing programs. Such programs require long periods of time since reliable methods of conducting accelerated tests are not well established. Many sources of data are accessible to the designer, some of which are listed below.

However, in the component-reliability field, there is no substitute for breadth of experience, and reliance is usually put on skilled personnel experienced in this field who are well acquainted with the available sources of data. Typical sources of reliability data are

Electronic Industries Association
Armed Service Technical Information Agency
Aeronautical Radio, Incorporated

Environments. The environment in which the system will be expected to operate is a parameter which can have strong influence on system reliability and must be taken into account by the designer. It is customary to think of several classes of environment such as

Ground-based, fixed	Marine
Ground-based, mobile	Missile
Aircraft	Satellite

As an illustration of the variation of key environmental parameters in the different environments, representative figures are given below:

Vibration: $2g$, satellite; ground-based, fixed. $> 100g$, missile.
Temperature: < -50 to $270°F$, satellite. 0 to $130°F$, ground-based, fixed.
Humidity: 10 to 100 per cent, marine. < 25 per cent, satellite.
Shock: $2g$, ground-based, fixed. $140g$, missile.

These figures are given to show that extreme environments must be given due consideration in system design, inasmuch as some of the conditions listed will greatly shorten the life of certain components. The designer may choose either to adopt more rugged components, at higher cost, or to mitigate the severity of the environment by special temperature regulation, shock mounting, etc. The prediction of mean time to failure for the system must be based on part-failure data which pertain to the particular environment if the prediction is to have any value, as order-of-magnitude differences in failure rate may readily occur among different environments.

An aspect of "environment" affecting the overall availability of a system that may easily be overlooked in the design stage is the caliber of personnel to be used in maintaining the system. Here again, the system designers may have little control over the type of such personnel to be used, although much latitude remains in the breadth of training given and the array of diagnostic aids and test equipment furnished. It will probably become more and more difficult to obtain well-qualified servicing personnel as the number of complex systems increases; it therefore behooves the designer to give thought to the qualifications of the personnel who will perform this vital function.

Project Organization for Reliability. Since reliability is an attribute that results from substantial effort through the design, manufacture, and installation of a system, its attainment is to a high degree dependent on proper project management and organization. It must be appreciated at the start that hard decisions affecting overall costs, performance, and schedules will inevitably have to be made if reliability objectives are to be met; therefore, firm support of reliability goals must be present at top-management levels if they are to be achieved. An effective way of doing this is to set up a reliability manager, reporting at the highest level of the project; this man's chief functions are to set quantitative reliability and maintainability objectives and monitor the progress of design to see that reliability is being given proper attention during design. This man must have strong top-management support if he is to prove effective.

33-4. Designing for Reliability

This section is concerned with specific techniques available for consideration by the system designer in order to attain his reliability objectives. Among the techniques listed are those which aim to prevent malfunctions; those which circumvent or correct malfunctions which do occur; and those which facilitate the detection, diagnosis, or repair of the malfunctions. Inasmuch as most of these schemes contribute some additional expense or complication to the system, it is important to consider the trade-offs which will arise between reliability and various other system characteristics. Examples abound of cases where light weight, low cost, or other factors were emphasized at the expense of reliability and, ultimately, the operational effectiveness of the system. Therefore, attention is needed, at the beginning, to the relationship between reliability and other performance factors in order to obtain optimum overall effectiveness.

Reliability as a function of size or weight is a trade-off that often arises, particularly in airborne or space applications. The importance of size/weight arises in several ways:

At the component-part level, the life of many parts is closely related to their physical size, which varies with the amount of insulation used, temperature reached in operation, etc.

Where severe operating environments are encountered, additional equipment requiring space and weight may be needed to provide an internal environment conducive to long component life.

Where exceptional mean times to failure are a requirement, redundant equipment may be necessary—this also increases size and weight.

Reliability as a function of cost is a very common trade-off problem. Here the question frequently becomes one of whether all relevant costs are being considered; e.g., costs of maintenance, spare parts, and failure of equipment during vital operations. In the case of military equipment, inadequate reliability may often result in loss of the equipment itself without its mission being accomplished. Therefore, it is essential that full operational costs be weighted against the cost of increased reliability in the equipment as originally delivered.

A further step in the process of design for reliability is that of setting goals for the reliability of subsystems or elements of the total system. Today it is common to have a system-reliability goal which is specified in, for example, the contract under which development takes place. In order to have a balanced design, similar and compatible goals must be set for units of the system—a simple process if operation of the system is totally dependent on correct operation of every subsystem and all subsystems are assembled of the same basic components or building blocks with no redundancy. When these conditions are not present, the analysis whereby unit objectives are set may become extremely complicated, especially if several modes of degraded operation are provided for.

Component Parts—Selection, Standardization. Inasmuch as component parts are the most basic determinant of system reliability, great care is required in their selection. This is a highly specialized subject in which actual experience is invaluable; however, certain principles are fundamental to success. Perhaps the foremost of these is that availability of components must govern the design of circuits and equipment rather than establishing a design and then attempting to find suitable components. This principle must be followed because basic component reliability is heavily dependent on continuous, large-volume production and unchanging processes. This implies use of components, wherever possible, that are already in production and whose manufacture is well understood; furthermore these conditions are prerequisite to predicting the reliability of systems yet to be built, from life data on components made in the past. Of course, a pretty good idea of the designs to be used must be in mind in order to select components at all.

A necessary preliminary to component selection is the study of environmental conditions for the system in storage, transit, and operational use. Also, where government or other contracts are involved, there will usually be specifications called out in the contract which have much to say about component-part selection.

Actual selection of a component part, then, may involve the following factors:

Availability of Life-test Data for Reliability Prediction. It must be determined that the data are for an application and environment similar enough to the proposed one to be valid.

Basic-component Design and Manufacture. As requirements grow in severity, it is increasingly necessary to judge whether the basic design and materials in the part are such as to give long life. It may also be necessary to judge whether the manufacturing process is well controlled and such as to give highly uniform output.

Number of Sources of the Part. It is highly desirable to avoid single-source components; if there is no alternative, it is often possible (and under government contracts, often mandatory) to establish another source. Care is needed to ensure that the part specification is adequately detailed to give interchangeable parts from the several vendors.

Vendor Rating. Another well-proved practice is the rating of potential parts vendors by a team of specialists. Such ratings should consider the vendor's past performance, production capacity, quality control, and organization as well as the suitability of the components themselves.

Standardization and approval procedures are also necessary to get maximum benefit from good component selection. Once the best parts are identified, it is of great advantage to minimize the number of different types and values and to use these as widely as possible. A very firm hand is required here, as it is the nature of engineers to continue optimizing their designs by calling for parts that fit their needs a bit better than the standard ones do. An approval procedure must be set up whereby each use of a component part is passed on by an independent person or group as to the appropriateness of that part in that application, whether standard parts were used, etc. Such a group must have firm authority together with acknowledged competence and must not wait for designs to be brought to them but seek out the designers and actively monitor their work. A usual function of this group is to provide a standard-parts book with ratings and application data for the use of designers.

Design of Reliable Electronic Circuits. The type of system with which we are concerned usually contains some amount of electronic circuitry—often this may be the principal constituent of the system, as, for example, in command/control applications where digital computers are widely used. The reliability of electronic circuits, especially digital circuitry, is therefore an exceedingly important factor in overall system reliability. This section deals with some methods for attaining high circuit reliability. Methods which depend on redundancy are dealt with in the following section.

Many of the techniques to be considered will tend to increase circuit complexity; therefore, it should be pointed out here that complexity may, if overdone, lead to a net reduction in reliability. This must be kept in mind as circuit design criteria and methods are considered and checks made to be sure that the total component-part count has not increased more than the reliability-improvement factor. This is an example of what was meant in the flow chart of Fig. 33-1, where an iterative process of design and reliability prediction was shown as a means of converging to the desired system reliability.

Another important factor in achieving the necessary circuit reliability is the need for *circuit standardization*. This is as important as the related point of *component-part standardization*, covered in the preceding section. Here, too, a firm hand is required to see that the total number of different circuit types is kept to a minimum; in spite of the firmest resolve, the number of circuits can easily increase by an order of magnitude during the design stage as allegedly unique situations are encountered for the first time. Such circuit proliferation is most prevalent in nondigital circuitry associated with low-level devices, inasmuch as it has become relatively well accepted to use

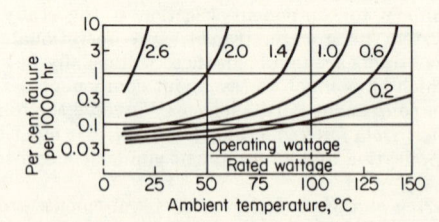

FIG. 33-3. Typical curve of catastrophic failure rate vs. stress level ("derating curve").

a small number of "basic" circuits for logic applications. Control over the addition of new circuit types may rest in a central circuit-design group or in the same organization which controls component parts. Lack of such control inevitably results in available circuit-design talent being spread over so many different circuit types that none gets the exhaustive analysis and test required to meet severe reliability goals; furthermore, large numbers of different circuits greatly complicate manufacture, maintenance, and field-parts stocks and therefore raise the cost of the system.

Derating of component parts, or using them conservatively with large safety factors, is one of the most effective ways of increasing circuit reliability and is universally used. The problem here is to obtain suitable data on the effect of derating on the life of the component part; that is, to determine the relationship between the stress induced in the part by current, voltage, temperature, etc., and the resulting internal strain which shows up in reduced part life. A common way of presenting this sort of data is by means of derating curves which show the proper factor to apply to the rated value to obtain the desired derated value for particular conditions of operation. An example of such a derating curve, for composition resistors, is shown in Fig. 33-3.*

Derating information must be up to date for the particular parts used and must be obtained from parts vendors or testing agencies on as current a basis as possible.

A second technique of reliable circuit design is to make adequate allowance for variation of component *values*. Such variations are the result of manufacturing tolerances, environment, and aging of the component. Variations due to environment are usually of known magnitude and direction, while manufacturing tolerances are statistical and must be handled as such; for example, the output of a component-

* Ref. 3, p. 64.

part manufacturing line may have values distributed over a range of ±5 per cent about the nominal. The combined effect of this manufacturing distribution, effect of temperature, aging, voltage, and so on may result in a +10 per cent, −3 per cent component as far as circuit design is concerned.

Several different approaches to the component tolerance problem in circuit design have been advocated; they vary in complexity and degree of conservatism. An early scheme was to require the circuit to perform within specification with all components at their "worst" value (as far as circuit performance was concerned) of manufacturing variation and one critical component at end-of-life value; or to have *all* components simultaneously at the "worst" end of life value. More recently it has been the feeling that these methods were excessively conservative and that the probability of all components ever being at end-of-life values is exceedingly low. The use of "statistical design" has therefore become popular. One method here is to use a computer to help determine how variations from nominal component values translate into variations in circuit performance, using the statistical distribution of the values of the various components and a Monte Carlo technique (Sec. 32-3). By this method, the designer can simulate the performance of thousands of such circuits without building them and can learn the behavior of the entire group. The method starts by simulating random component values for all the given components and calculating circuit performance for many combinations of random component values (frequently thousands) to obtain a distribution of the desired circuit performance parameter. It was found in one example that the number of logic stages between level setters could be double the number permitted by worst-case design, and with a better knowledge of the frequency of failure. Such computer techniques have also been carried another step to optimize the design by successively varying component values, according to criteria from the designer, to improve circuit performance.

Still another technique that has been frequently used is that of marginal checking. Here, each circuit is provided with a "handle," typically one of the supply voltages, which can be varied over a range in order to degrade the circuit performance intentionally. The assumption here is that a circuit having one or more faulty components will fail at a lesser excursion of the marginal-checking voltage than one with all components within tolerance. Such systems were quite widely used in large vacuum-tube computers, where the tubes tended to fail after a relatively short life and in a gradual manner. Perhaps the most extensive use of marginal checking was in the 50,000-tube AN/FSQ-7 computers in the SAGE air-defense system. In these systems, the computer was divided into several hundred sections with separate marginal-check voltage lines to each section. By computer program control, voltage excursions could be applied successively to different parts of the computer while suitable test programs were run. A substantial quantity of equipment was required for this entire function. The provision of a marginal-checking "handle" on each circuit was also used effectively in optimizing the original design of the circuit.

Use of Redundancy to Increase Reliability. In the design of electronic equipment it may be found that great care in the selection of component parts and design of circuits still does not give sufficiently high reliability in the system. In this event, it will probably be necessary to design some degree of redundancy into the system and thereby circumvent, to a greater or lesser degree, the failures that occur. The techniques of redundancy are of quite recent development, and quantitative means for dealing with them are still rudimentary; however, present trends in apparatus development will lead to a great increase in their importance. This is largely due to the emphasis on increased complexity and greater numbers of circuits and logical elements, which tend to increase the frequency of failure, and a strong push toward reduced size, which makes failure repair a much more difficult process. Indeed, some forms of technology under development, such as cryogenic circuits, may be repairable only by complete replacement.

A number of basic forms of redundancy are available to the system designer. For example, the coding of information, as it is transmitted about the system, may be of a type which includes redundant information in order to detect and/or correct errors as they occur (Sec. 22-15); additional personnel or spare parts may be provided; additional equipment may be provided in an "active" or "standby" fashion. This section will deal with *equipment* redundancy, at both the component-part level and the circuit or functional-element level.

In considering the possible benefits of redundancy, a number of factors are pertinent. The first of these is the *cost* of the additional equipment, for redundancy will almost surely increase this. Others are the range of improvement in reliability that will result; the types of errors which are expected to occur; which reliability factors are affected positively and negatively; and the effect of the additional equipment on size, weight, power required, design time, reduction in performance, etc.

It should also be pointed out that there are some less obvious, negative aspects of equipment redundancy. Among these are the effects of a larger number of electrical connections, a "component" which often receives very little attention in design; and the fact that incipient failure in equipment may be very much harder to detect because the failure is masked by the very presence of redundant equipment.

The first form of redundancy is simply to duplicate the nonredundant equipment totally. Each equipment does the same job, simultaneously, with the output being taken from the one that is operating "correctly;" therefore, in addition to the two basic equipments, there is required some means of telling whether a unit is operating correctly or not, and a switch—either manual or automatic.

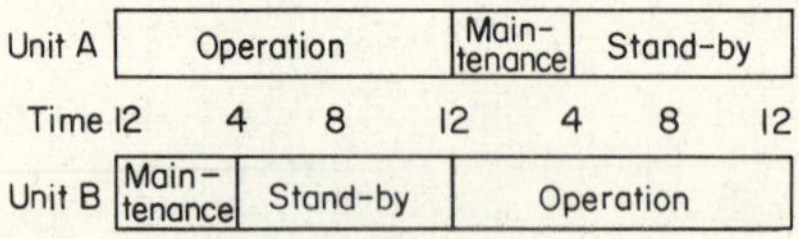

Unit A	Operation	Main-tenance	Stand-by
Time 12	4 8	12	4 8 12
Unit B	Main-tenance	Stand-by	Operation

FIG. 33-4. Duplex-system mode of operation.

The reliability R of a unit was previously said to be expressed by

$$R = e^{-t/m}$$

where m is the mean time to failure for the unit. For the simple actively redundant system described above, the corresponding expression is

$$R = 1 - (1 - e^{-t/m})^2 = e^{-t/m}(2 - e^{-t/m})$$

The improvement in reliability is shown in Fig. 33-5 (second curve from top).

In contrast, the second unit may be arranged to start operating only after the first fails. This configuration, while retaining the same amount of equipment as the previous one, requires the power of only one unit, which may be a considerable advantage. The expression for reliability here is

$$R = e^{-t/m}\left(\frac{1+t}{m}\right)$$

As compared with the previous scheme, this system may require more complex equipment for switchover if the switch must be done quickly; there may also be a severe problem of the total loss of important data which are held in the previously operating unit. Where digital computers are concerned, elaborate means may be needed to ensure the safe retention of these data during switchover. The relative reliability of this system is also shown in Fig. 33-5 (upper curve).*

It should be pointed out that in large, continuously operating systems ($t \gg m$), time must be allowed for periodic maintenance. Where such continuous operation is required, some form of redundancy is obviously necessary; however, the reliability of such a system is reduced to the extent that periodic maintenance cuts down the degree of "standby" protection. Figure 33-4 shows a mode of operation for such a "duplex"

* F. A. Applegate, A Commentary on Redundancy, *Redundancy Techniques for Computing Systems*, p. 375, Spartan Books, Washington, D.C., 1962.

RELIABILITY 33-11

system where 4 hours in each 24 must be devoted to the routine maintenance of each unit. Note that during only 16 hours of each 24 is there a standby facility.

In this type of system, switchovers are regularly made at 12 o'clock to permit scheduled maintenance; unscheduled or emergency switches can be made at any time, except that from 12 to 4 such switches will find no standby system to assume the load.

Yet another form of redundancy may be used in cases where extremely high reliability for a short period is the requirement. This scheme uses three units together with a majority or voting element which automatically selects as the correct output the output which is the same from two or more units. This, of course, requires three times the amount of equipment, which must be in continuous operation, plus the voting element, which may also have to be triplicated if its reliability is excessively low. The expression for reliability here is

$$R = 3(e^{-t/m})^2 - 2(e^{-t/m})^3$$

Figure 33-5 compares the reliability of these four systems as a function of time (note that the systems which employ two units involve some loss of time in switching from one unit to the other plus sufficient additional equipment to determine that incorrect operation is occurring).

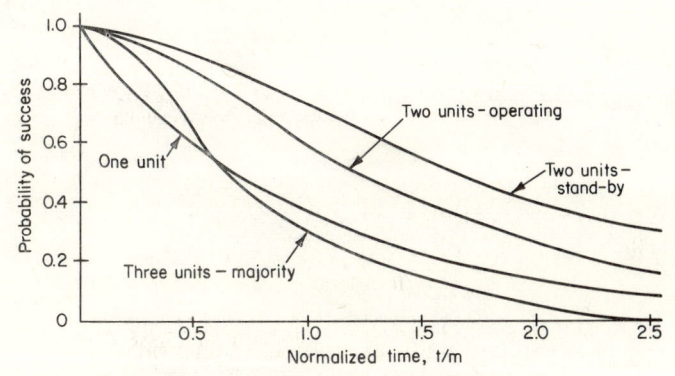

FIG. 33-5. Relative reliability of four redundant systems.

A somewhat different approach to the incorporation of redundancy is to use it at the level of logic circuits, in digital equipment, rather than at the level of complete equipments. In the scheme described below,* logic is constructed in quadruplicate and so connected as to correct any errors that occur, by means of correct signals from the adjacent units in the "quad." Further, the scheme appears able to correct for multiple failures provided they do not occur in successive stages of logic. Figures 33-6 through 33-8 illustrate the method.

Figure 33-6 shows a logical network in original form. Quadruplication first calls for providing four copies of this network in place of the original one. With identical inputs, the four networks will produce identical outputs if no failures have occurred. If a failure does occur, as shown in Fig. 33-7 where the output of the right-hand AND should be one, not zero, the error is corrected at the next OR by a correct signal from the left-hand AND. A similar method permits overcoming wrong OR outputs at an AND circuit farther along. Figure 33-8 shows a complete "quadded" version of Fig. 33-6, with an example of how certain errors may propagate for two stages before being overcome. Figure 33-8 also shows the three different ways of pairing outputs of four circuits; it is necessary, where AND's and OR's alternate, to choose these patterns so that no signal encounters the same pattern twice in successive logic levels. This

* J. G. Tryon, Quadded Logic, *Redundancy Techniques for Computing Systems*, pp. 205–228, Spartan Books, Washington, D.C., 1962.

33-12 SYSTEM TECHNIQUES

prevents an error from propagating unduly far before correction. The technique has been extended to cover the connection of like (AND-AND) units, NOT circuits, flip-flops, registers, and timing chains.

Redundancy can be carried a further step downward, into the circuits themselves, in the form of redundant components. Such redundancy gives the greatest *reliability* payoff but does so at the expense of circuit performance. The approach* is to use a series-parallel ("quad") group of four diodes whenever one diode would normally be

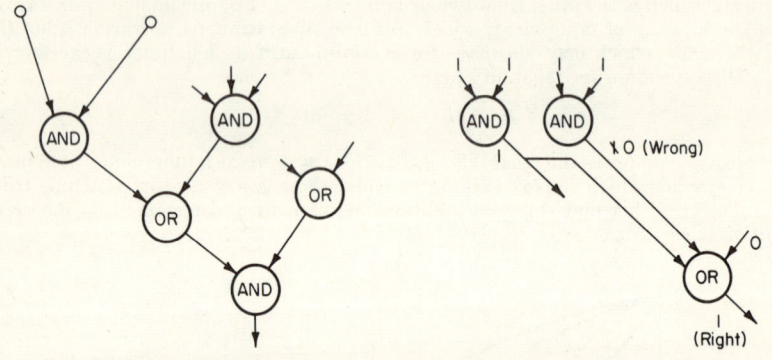

FIG. 33-6. Original nonredundant circuit.

FIG. 33-7. Erroneous 0 from an AND corrected in the next OR.

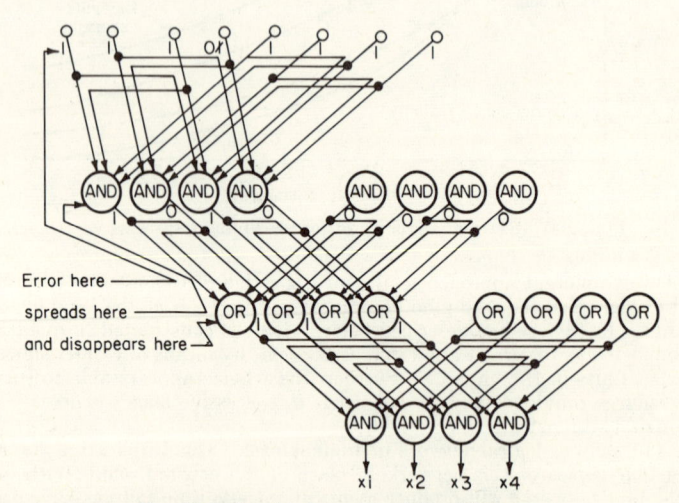

FIG. 33-8. Quadded version of Fig. 33-6.

used, and a similar series-parallel arrangement of four transistors with their associated resistors and capacitors, in place of one, as shown in Fig. 33-9.

Several points should be made about this type of redundancy. First, it is a very effective way of increasing reliability; for example, using reasonable values for component failure rates, the calculated mean time to failure for a computer of 2,000 logic blocks increased by about 50-fold and the reliability of a basic logic block by several

---

* R. M. Fasano and A. B. Lemack, A Quad Configuration—Reliability and Design Aspects, *Proc. Eighth Natl. Symp. Reliability Quality Control, IRE*, 1962, pp. 398–399.

orders of magnitude. Second, the technique makes greater demands on the transistors, reduces the circuits' driving power to one-fourth, reduces speed, and increases power dissipation and overall power required. Finally, it may be difficult to provide ways of verifying that all the redundancy does in fact still exist at the beginning of a mission—the number of test points required may be completely unreasonable.

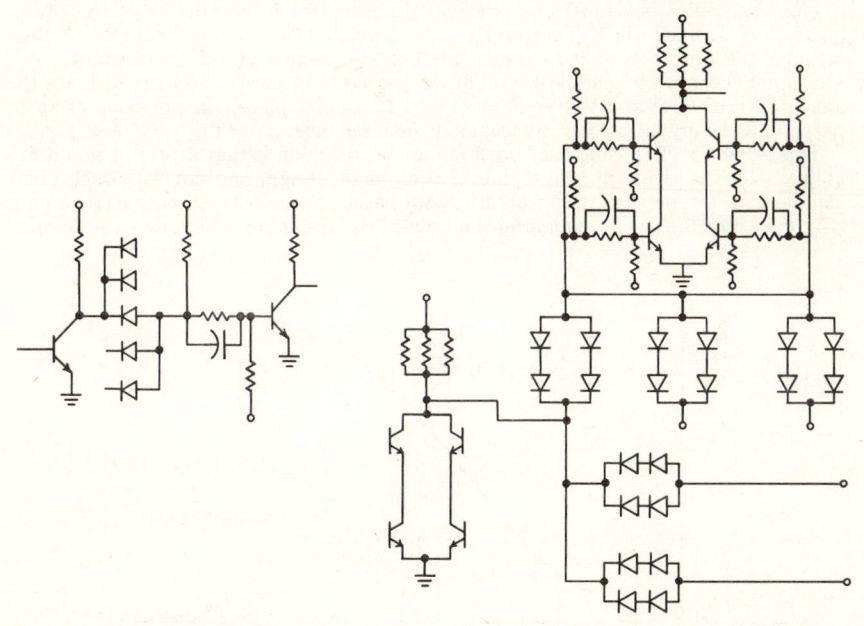

Fig. 33-9. Transistor-diode logic circuit—nonredundant configuration (left), quad redundant configuration (right).

33-5. Maintainability

At the beginning of this chapter it was pointed out that attaining adequate reliability requires consideration of the entire working life of a system, and that keeping the system operating is a major factor in overall system effectiveness. This section deals with the attribute of *maintainability*—its definition and prediction and the things that can be done in equipment design to enhance it. The importance of maintainability to the system designer arises in several ways:

1. It is increasingly recognized that the cost of maintenance after a system is installed may greatly exceed the initial cost; thus maintenance is a major economic factor in considering the feasibility of a proposed system.

2. To the extent that equipment is not operational because of maintenance requirements, it may be necessary to provide costly redundant equipment.

3. The operational effectiveness of a system is greatly influenced by the quality of maintenance it receives, which in turn depends on the availability of skilled personnel. However, the increasing number of systems reduces the likelihood that such personnel will in fact be available.

4. Rapidly growing complexity of systems is greatly increasing the effect of the above factors.

Quantitative methods for dealing with maintainability are at a much earlier stage of development than for reliability but are rapidly evolving along similar lines. As with reliability, we are concerned with a quantitative definition, with prediction, and with trade-offs between this attribute and other parameters of the system. In particular, increased maintainability must be weighed against the cost of special equipment features required to achieve it.

Definition of Maintainability. The definition of maintainability is similar to that for reliability and is as follows: "Maintainability is the probability that, when maintenance action is initiated under stated conditions, a failed system will be restored to operable condition within a specified time."

Maintainability is a function of several primary factors: the design of the equipment, the personnel who will perform maintenance, and the support facilities. Design includes the provision of test equipment, built-in diagnostic aids, accessibility, etc. Personnel encompasses the skill level of the people to be used. Support includes the availability of tools and test equipment at a particular place, the provision of spare parts, and the organization responsible for maintenance.

This section will be concerned with *repairability*, which is that aspect of maintainability which is primarily a function of equipment design, and not with such other aspects as the time required to obtain spare parts. Repairability is concerned only with the *active* diagnosis, repair, and checkout of the system by maintenance personnel.

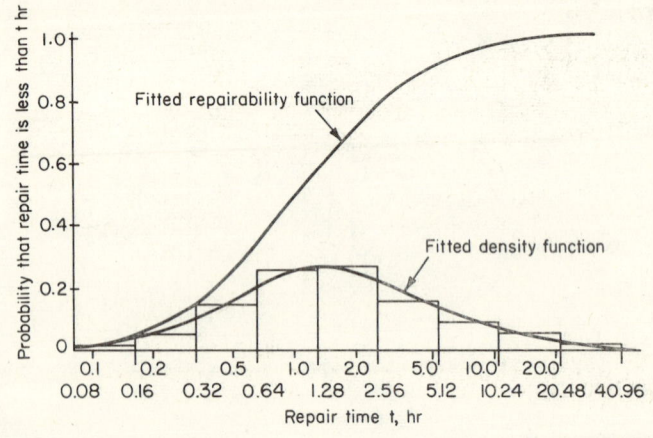

Fig. 33-10. Normal fitted repairability and density functions of transformed log-normal repair times. Histogram shows distribution of observed repair times. (Note that the scale of the abscissa is logarithmic.)

Data are required for the prediction of repairability, just as they were for prediction of reliability. In this case, what is needed is data on the times needed to repair typical units of equipment, where the units in question correspond generally to the ones the maintenance man will be dealing with in the system being designed. Such units may vary from an individual resistor or other electronic part, to a complete radar transmitter-receiver or some module of an electronic computer. It has been found that a log-normal distribution of repair times adequately represents the data because of the wide range of values of repair time commonly found. Such a distribution is shown in Fig. 33-10.

The repair times used are those which affect *system* down time only; this will probably involve only the replacement of a defective unit by a good one. The defective unit may be repaired elsewhere—this process does not affect system operation.

Repairability Prediction. In order to predict* the repairability of an equipment, data must be collected to form distributions of the type shown in Fig. 33-10, i.e., the times required to locate, repair, and verify correct operation. These data should cover as many of the replaceable or repairable elements as possible, under a variety

* R. C. Horne, Measurement and Prediction of Systems Maintainability, *Electronic Maintainability*, vol. 3, pp. 215–229, Engineering Publishers, Elizabeth, N.J., 1960.

of conditions of repair. Such conditions would take into account the adequacy of personnel, availability of test equipment and other aids, accessibility, and other similar factors. (The greater the range of conditions for which data are available, the more generally useful the data will be.) The data will be still more useful if the conditions of operation under which the unit failed are known.

With these data available, prediction becomes a matter of analyzing the quantities of major components or units in the system under design and the conditions of operation and maintenance which are expected to prevail, comparing this with the general repair data obtained above, and computing the overall system repairability in a way similar to the prediction of reliability. In detail, the steps are:

1. Analyze the design, in terms of the units which will be involved in repairing the system, whether these are major equipment units, pluggable circuit boards, individual component parts, or whatever, to get the quantity of each such unit, its conditions of operation, and the conditions under which its replacement will be done.

2. From general failure-rate data, determine the predicted frequency of failure of each unit.

3. List the quantities of each different unit used and

   a. Multiply each quantity by the predicted frequency of failure (failures per 1,000 hr) and by the relative frequency of occurrence of a particular repair-time interval for that type of unit (as obtained from a graph like that of Fig. 33-10).

   b. Sum these products, and plot a repair-time curve for the entire system or equipment under consideration. From this curve, one may read the probability of repairing the system within any prescribed number of hours.

The principal problem with this technique is the lack of data on the relationships between the probability of repair of a particular kind of unit and its basic design, the diagnostic aids available, and the level of maintenance personnel. It is in this respect that the prediction of maintainability has not reached the level that reliability prediction has.

Design for Maintainability. As with the techniques of prediction, so with the techniques of design we find that maintainability lags behind reliability. However, there is one point that clearly applies to both areas—the responsibility for achieving adequate maintainability must be definitely assigned with accountability for results. Frequently this responsibility is given to the reliability manager, since much of the data and techniques are common to both areas. This point is covered in Sec. 33-3.

Listed below are some design features that aim to enhance maintainability and serviceability:

More rapid diagnosis of failure and isolation to a particular unit is assisted by

   Automatic or built-in test equipment
   Marginal checking, to aid in locating circuits which are on the verge of failure
   Simplicity in design
   Diagrams especially adapted to the needs of maintenance personnel
   Numerous and readily available test points

More rapid repair or replacement of a failed unit is facilitated by

   High accessibility of all parts
   Modular, pluggable construction
   Easy identification of units, parts, terminals, etc.
   Absence of hazards such as exposed high voltages, high temperatures, and moving parts

Particular consideration must be given to test equipment. While complex, automatic test equipment may in principle shorten the time for failure diagnosis and location, the maintenance of the test equipment may be a serious problem and offset the supposed advantages of lowered skill requirements and reduced time requirements.

Thus the test equipment, in its design, must be subject to the same quality of design for reliability and maintainability as the main equipment.

Finally, the various diagrams, drawings, schematics, parts lists, etc., that are prepared as aids in manufacturing the equipment are customarily also furnished to field personnel as maintenance aids. Such information falls increasingly short of meeting maintenance requirements, and unique diagrams for maintenance are definitely required. Diagrams for use in maintenance should:

Be readily understood with a minimum of references to other diagrams; i.e., all the information on a particular unit should be on one sheet rather than scattered over several.

Be arranged to set forth the operation of the unit in a simple, logical way. This is especially important in complex digital data-processing equipment.

Show test points, waveforms, locations of units in frames, or component parts on chassis or circuit boards.

REFERENCES

1. Blanton, H. E., and R. M. Jacobs: A Survey of Techniques for Analysis and Prediction of Equipment Reliability, *IRE Trans. Reliability Quality Control*, vol. RQC-11, no. 2, July, 1962, pp. 18–35.
2. Robins, R. S.: On Models for Reliability Prediction, *IRE Trans. Reliability Quality Control*, vol. RQC-11, no. 1, May, 1962, pp. 33–43.
3. Gryna, F. M., N. J. McAfee, C. M. Ryerson, and S. Zwerling (eds.): *Reliability Training Text*, Institute of Radio Engineers, 1960.
4. Hellerman, Leo: A Computer Application to Reliable Design, *IRE Trans. Reliability Quality Control*, vol. RQC-11, no. 1, May, 1962, pp. 9–18.
5. Buelow, F. K.: Improvements in Current Switching, *IRE Trans. Electron. Computers*, vol. EC-9, no. 4, December, 1960, pp. 415–418.
6. Doyle, R. H., R. A. Meyer, and R. P. Pedowitz: Automatic Failure Recovery in a Digital Data Processing System, *IBM J. Res. Develop.*, January, 1959, pp. 2–12.
7. Wilcox, R. H. and W. C. Mann (eds.): *Redundancy Techniques for Computing Systems*, Spartan Books, Washington, D.C., 1962.
8. *Proc. Eighth Natl. Symp. Reliability Quality Control*, IRE, 1962.
9. Ankenbrandt, F. L. (ed.): *Electronic Maintainability*, vol. 3, Engineering Publishers, Elizabeth, N.J., 1960.
10. von Neumann, J.: *Probabilistic Logics and the Synthesis of Reliable Organisms from Unreliable Components*, pp. 43–98, Automata Studies, Annals of Mathematics Study no. 34, Princeton University Press, Princeton, N.J., 1956.
11. Moore, E. F., and C. E. Shannon: Reliable Circuits Using Less Reliable Relays, *J. Franklin Inst.*, vol. 262, part I, pp. 191–208, September, 1956; part II, pp. 281–297, October, 1956.
12. Calabro, S. R.: *Reliability Principles and Practices*, McGraw-Hill Book Company, New York, 1962.

Chapter 34

SYSTEM TESTING

GERARD A. La ROCCA, *Head of Scientific Space Projects, Space and Information Systems Division, North American Aviation.*

CONTENTS

34-1. Test Planning	34-2
34-2. Field test Objectives	34-2
34-3. Test Scheduling	34-3
34-4. Test-group Organization	34-3
34-5. Design of Experiments	34-4
34-6. Reliability Mathematical Model	34-4
34-7. Test-data Flow and Analysis	34-5
34-8. Control of Configuration Changes	34-5
34-9. Personnel Subsystem Testing	34-6
34-10. Environmental Testing	34-6
34-11. Test Facilities	34-11
Environmental Chamber	34-11
Electro Gasdynamic Facility	34-11
Low-density Hypervelocity Tunnel	34-11
Hypersonic True-temperature Tunnel	34-11
Materials Environmental Test Facility	34-12
Ultraclean Rooms	34-12
34-12. Rocket-engine Testing	34-12
34-13. Types of Tests	34-13
34-14. Test Objectives	34-13
Research and Development Tests	34-13
Preliminary Flight-rating Tests	34-13
Research and Development Flight Tests	34-14
Qualification Tests	34-14
Production Quality-assurance Tests	34-14
Operational Flight Tests	34-15
34-15. Flight Testing	34-15
34-16. Total Guidance-system Error	34-15
34-17. Analysis of Inertial-guidance Test Data	34-16
34-18. Coordination and Liaison	34-16

34-2 SYSTEM TECHNIQUES

The system that is discussed in this chapter refers primarily to a large military space system. This example is used since the author's primary experience is with military ballistic missiles and space systems. The problems and techniques brought out in this chapter are comparable with the testing of other large-scale systems.

Field testing of large-scale systems as defined in this chapter is the testing of the complete system in its intended operational environment. In the initial stages of field testing of the system, there is high confidence in the reliability and functional capability of the equipment, subsystems, and components. Technical manuals will have been prepared for operation, maintenance, and repair. Personnel to be used in the test will have been trained for their positions in the operation and support of the system.

34-1. Test Planning

Test planning should begin in the early stages of the design and development phases of the program. There should be a general test plan which will encompass all testing for the project or program. The general test plan includes component, assembly, subsystem, system, and flight tests. The various phases of the test program should be organized to ensure that results of the early test phases are used to the maximum extent in the later phases of testing.

The general field-test plan should contain the following elements:

1. Test objectives
2. Participating agencies and responsibilities
3. Test-force organization and functions
4. Operating procedures, methods, and controls
5. Test schedules

The general field test plan should be followed by detail test plans. A detail test plan will be written for each specific test or flight and contains the following elements:

1. Detailed test objectives
2. Test responsibilities
3. Test support
4. Items in test
5. Test methods
6. Typical measurements
7. Data reduction
8. Criteria for success
9. Test-operations flow chart (locations and functions)
10. Test-operations flow schedule (bar chart)
11. Test-operations PERT network (for complex tests)

34-2. Field-test Objectives

The following is a typical categorization of the general field-test objectives in any large-scale system. The test objectives are listed as primary or secondary. This arrangement is used to provide the test conductor with guides as to the relative importance of malfunctions or deficiencies uncovered during pre-test or pre-flight preparations and checkouts.

1. *Hardware Capabilities and Limitations.* To evaluate the performance, compatibility, and interaction with the facilities, operational equipment, and support equipment, placing special emphasis on determining the inherent capabilities, limitations, and deficiencies of the integrated system.

2. *Reliability.* To provide data for establishing the system in commission rate, and to demonstrate that the system has the reliability goals assigned (see Chapter 33). The number of tests required depends on the "confidence level" that it is desired to attain.

SYSTEM TESTING   34-3

3. *Technical-data Adequacy.* To determine the currency, accuracy, sufficiency, and adequacy of the technical manuals and data required for the operation, inspection, and maintenance of the system.

4. *Evaluation of Operations, Logistics, and Maintenance Plans.*

5. *Safety.* To evaluate all system operations with regard to hazards and safety.

6. *Personnel Subsystem Performance.* To demonstrate that the personnel subsystem performance is adequately supported by equipment design, technical data, job environment, training, and organizational control procedures.

34-3. Test Scheduling

In scheduling the key dates and the time phasing of the test program with respect to the overall systems-development schedule, much depends upon whether urgency dictates a policy of "concurrence," wherein one is committed to production concurrently with development and test, or whether a prototype will be built and tested prior to commitment to production. In any event, the entire formal test program, to include component, subsystem, and system tests, should be integrated into an overall test plan and schedule. Among the principal features of such a test schedule is the need for action very early in the development program to identify long lead time instrumentation, test facilities, and other resources. Significant test milestones should be identified for each phase of testing. A typical systems-development

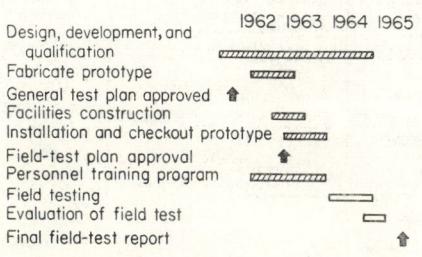

FIG. 34-1. Typical field-test schedule.

schedule showing the time phasing of a test program is shown in Fig. 34-1. The schedules should be used as an active management tool, and updated and formally reviewed by top-level management at least monthly. Probable and actual slippages should be indicated, along with the rescheduled dates.

A new technique called PERT is used for scheduling complex systems testing. PERT enables the scheduling work to be computerized and provides valuable and timely information for test managers and other responsible personnel (see Sec. 36-4).

34-4. Test-group Organization

Specifying the organization which is to carry out the field-test program is a very important function of the test plan. This is particularly true when two or more organizations are involved in the test. Also, it is important to establish a separate organizational entity, temporary though it may be, for the following reasons. First, the managers and systems-test and project engineers should be in a position to be completely objective in their analysis and conclusion. Ideally, these personnel should have no close organizational alliance with the design and manufacturing elements which produced the system. Second, the field test will normally be conducted at a site remote from the main plant, and an independent organization, capable of self-support, is necessary. It should be noted that while the test organization should be organizationally independent and self-supporting, it is functionally interdependent in regard to systems design, operation, performance, and evaluation. Though there are many variations and it is impossible to generalize, a typical test organization is shown in Fig. 34-2.

34-5. Design of Experiments

Because of cost and scheduling limitations, the test-program manager may not have the opportunity to conduct large numbers of tests or to have instrumentation and data recording at every point at all times he wishes. Therefore, he should become familiar with some of the special mathematical techniques for setting up the test program in such a way as to obtain the maximum amount of useful information within the constraints established. In particular, small-sample techniques and analysis of variance are useful tools. The special techniques and mathematics are developed through coordination with the reliability department. There must be confidence

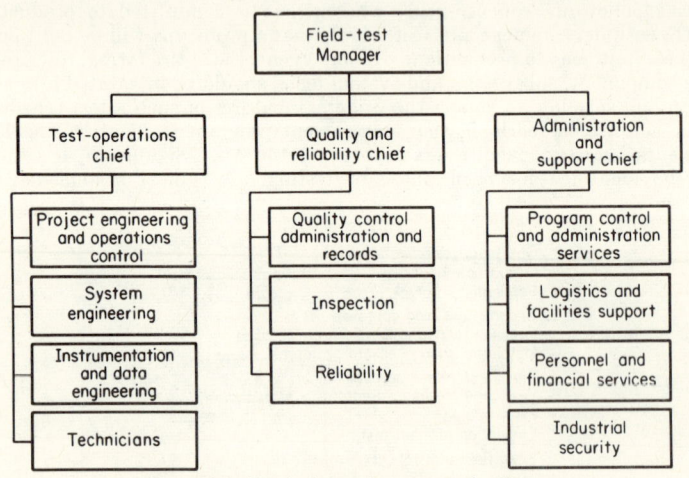

FIG. 34-2. Typical test-force organization.

that the completion of the test program will demonstrate objectively whether or not the test objectives were accomplished, the specifications were met, and the system is ready and suitable for operational use.

34-6. Reliability Mathematical Model

The use of a mathematical model which simulates system operation can be useful as both a diagnostic and predictive tool as an adjunct to the test program. It can also provide an effective means for automated analysis of large volumes of operating and failure data. Proposed changes to the equipment or its mode of operation can be simulated in the model to determine the effect on the reliability of the system and to provide a basis for decision on whether or not to make the change. Low reliability or efficiency can be traced to its cause and the weakest links can be identified and strengthened, or an optimum application of resources for product improvement can be applied.

In the mathematical model, the reliabilities of each individual component or subsystem are entered into a mathematical expression which simulates the system in operation. The output of the model is the theoretical system reliability (Chapter 33). This can be compared with the system reliability inferred from complete system tests, using the ratio of successes to total trials, or a ratio of corresponding operating or down time. Any large difference in the results obtained by these two approaches might indicate that human factors, environment, or maintenance techniques, rather than the hardware alone, are deteriorating the system reliability. The principles described above are portrayed in Fig. 34-3.

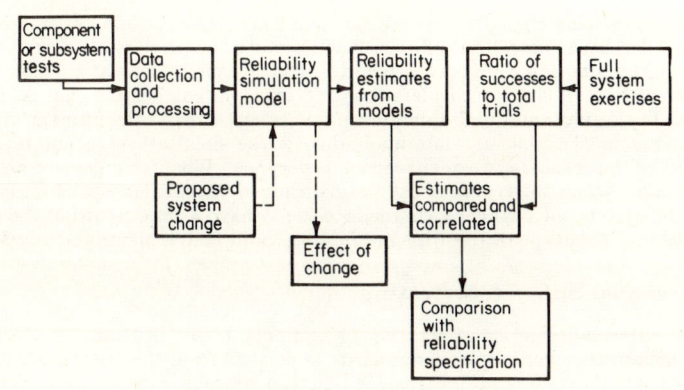

FIG. 34-3. Measurement of system reliability.

34-7. Test-data Flow and Analysis

A field-test program will generate very large quantities of scientific and technical data. The data will vary from oscillograph traces to component-failure reporting forms. Each piece of data should be gathered in response to a specific need in meeting a test objective. Data which do not contribute toward this end should not be recorded. Data should be fully exploited to assist in meeting the originally intended

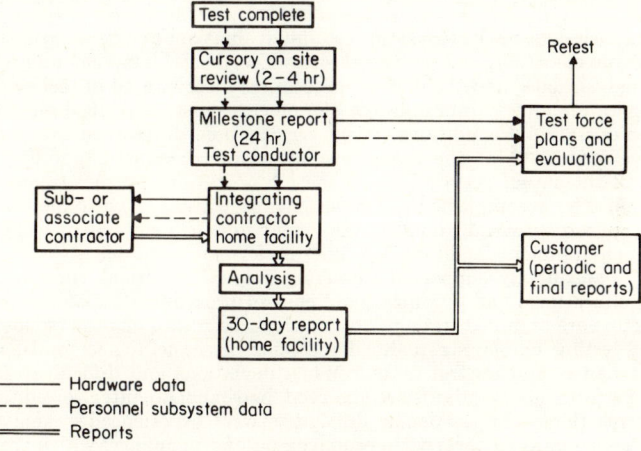

FIG. 34-4. Typical test-data flow and analysis.

objectives. A comprehensive and detailed plan for data processing, flow, analysis, and reporting should be developed. Responsibilities and frequencies and content of reports, as well as routing and levels of review of test reports, must be clearly specified in advance. Decision points for retest and for requests to the home plant for engineering changes must be explicitly stated. A typical example of data flow and analysis is shown in Fig. 34-4.

34-8. Control of Configuration Changes

It is important to maintain positive control over engineering changes to the prototype system during field test. It has been found necessary in large-systems work to require all engineering changes, beyond a certain design or development freeze point, to be controlled by a centralized office, which may use the services of personnel from various departments to form a review board. This is usually done at the home plant rather than at the test site. Considerations or criteria for engineering changes include cost, reliability, safety, and maintainability. It is best to hold the number of equipment changes to the absolute minimum during the field-test program to permit meaningful comparisons between successive test runs. When changes are necessary and are made, accurate records must be maintained, particularly as to when and where in the system the changes were made, so as to have a time record of the system configuration. This type of record is essential to valid test analysis and conclusions.

34-9. Personnel Subsystem Testing

System performance is an integration of equipment and personnel performance, each of which must meet specified standards as derived from the criteria set forth in the system specification and incorporated in the various subsystem performance specifications. To determine that equipment performance attains the required standard as defined in the performance specifications, each component is first tested individually. When this performance is satisfactory the component is tested in the system to determine that it is compatible with and will work in the system. In this manner the complete system is progressively tested to prove that it will function reliably as an integrated hardware system.

After satisfactory component and hardware-system performance has been established, operational and maintenance procedures are applied to the operational equipment under operational conditions. Using these established procedures, the final step in proving the complete system is to test the personnel performance on the operational equipment.

To accomplish personnel subsystem test and evaluation, two types of testing activities are required. First, the personnel subsystem basic data, including functional and task-analysis data, and the derived procedures incorporated in technical publications for operating and maintaining the equipment, must be verified on actual hardware under operational conditions. The verified procedures then are employed as performance standards for the actual evaluation of personnel capability. Second, evaluation of the actual performance of the job operations by the trained operating personnel must be accomplished. The specific objectives of personnel-performance testing are concerned with evaluating the extent to which equipment design and the operational environment facilitate personnel performance and whether the personnel subsystem is adequately supported through personnel selection, manning, training, technical publications, and organizational control procedures. Data for evaluation of these factors are obtained through observation and recording of personnel performance during testing, employing qualified operating personnel to accomplish the operation, maintenance, and control tasks. When deviations and difficulties in meeting prescribed performance standards are observed, personnel are interviewed to ascertain from their reactions why particular difficulties were experienced. Analysis of the data leads to evaluation reports with recommendations for changes which are processed through established change procedures. Approved changes can result in equipment or facility modifications and in revisions to one or more of the personnel subsystem elements. Testing and evaluation of the system, including personnel performance, result in necessary system modifications which are required to establish confidence in the capability of the trained personnel to meet specified job-performance standards.

34-10. Environmental Testing

In a system-development schedule, environmental testing is usually compressed into the time period that remains after the requirements of the production and design groups have been met.

Environmental testing covers a broad spectrum of investigation, from the verification of theoretical and empirical concepts of preliminary design to the flight tests at

SYSTEM TESTING 34-7

such facilities as Edwards Air Force Base, (California), White Sands Missile Range (New Mexico), Western Test Range (California), Eglin Air Force Base (Florida), and the Eastern Test Range (Florida).

Although environmental testing is broad enough to include high-altitude probes into the Van Allen belt, this section has been restricted to the environmental testing that can be done in laboratories on the ground. Some problems can be handled with existing facilities, while in others new capabilities are required.

Below are listed the basic categories of environmental testing.

1. Propulsion
2. Aerospace dynamics
3. Structures
4. Bioastronautics
5. Navigation and guidance

In order to discuss the major problem areas within these categories, that is, to describe the necessary research and development effort, we intend to describe an environmental test program on a hypothetical manned-satellite system (HMSS).

The mission of this hypothetical system is satellite interception and inspection. It is a two-stage rocket-powered satellite. The last stage of our system is recoverable, launched vertically, and will have a glide reentry and landing. The mission will take 3 to 4 days. The crew will be two men. Our orbit will take us out 2,000 miles, with the special requirement to be able to change orbit direction and altitude.

Our first task is to break the mission of HMSS into the various phases of its flight, so that the problems peculiar to a particular speed, altitude, and other environment may be grouped.

The phases of flight will be labeled

1. Launch
2. Boost
3. Orbit
4. Reentry

Some of the major problems encountered during the launch phase are

Propulsion

1. Vibration
2. Combustion instability
3. Freeze-up of components during long holds

Structure

1. Starting impact loads
2. Wind loads
3. Vibration

Bioastronautics

1. Escape

Propulsion problems at launch are different from those of the other phases of flight, because the ambient pressures and temperatures usually do not have to be artificially adjusted. Hence environmental testing covers functional performance, the isolation of failures caused by vibration and noise, combustion instability, and the freeze-up of components located close to cryogenic systems. All these must be investigated to ensure a successful lift-off.

Surface winds have proved to be a definite hazard in launching operations above ground. The Air Force makes extensive use of NASA laboratories for investigations of this problem.

When human beings are aboard at launch, as they would be in the case of our hypothetical system, the safety of the crew in the event of a malfunction of the booster becomes a paramount consideration. Extensive wind-tunnel tests are made to ensure that adequate stability and control will be available during this escape maneuver. Equally important in establishing an escape system is the knowledge of the human tolerance to acceleration under escape conditions and deceleration under reentry conditions. At Holloman Air Force Base, N.M., sleds are used to study both acceleration and deceleration. Human beings and chimpanzees are used as subjects for these tests.

Now let us examine the boost phase. New problems are encountered and some that were present at launch are retained.

Propulsion

1. Base heating

Aerospace dynamics

1. Stability and control at separation
2. Escape system

Structures

1. Buffeting
2. Flutter

Human factors

1. Acceleration

Premature termination of thrust during the boost phase of several missile flights has been caused by hot gases from the main rocket jets or the fuel-rich exhaust of the turbine pumps. In both cases these hot gases have recirculated back into the base of the booster. When this situation occurs, serious damage results to the structure and components located in the base. The severity of base heating depends on a number of factors, including missile configurations and the number of rocket nozzles in the base. Also, the Mach number and altitude have a strong influence on base recirculation.

Investigations of this phenomenon would be conducted in the altitude propulsion facilities for our hypothetical system. Tests of this type have been conducted on models of several existing systems. Because these tests can be conducted at various altitudes, the magnitude and location of base heating can be investigated along the entire trajectory of boost. As a result of such tests, designers can take corrective action before flight testing. These corrective actions are usually a combination of aerodynamic-design changes and the addition of insulation.

During boost the stability and control changes experienced by the vehicle during separation from the booster stages can also prove to be a serious matter. The vehicle and booster can be mounted independent of one another in a wind tunnel. In a test like this the forces and moments are measured for various relative locations of the vehicle and booster. It is important to accomplish these tests under as near to flight conditions as possible, since the shock patterns during separation directly affect the validity of the results.

A significant phenomenon through the transonic range is buffeting. This buffeting results from oscillating shocks. One such test was on the Titan in the 16-ft transonic tunnel at Arnold Engineering Development Center, Tenn. This shock oscillation causes peak loads of approximately three to five times the static loads. At Wright Patterson Air Force Base, Ohio, human beings are used as subjects to study their reactions during extreme buffeting periods.

The glider of our hypothetical system is now in orbit. The principal problems requiring environmental testing are

Propulsion

1. Start and restart
2. Thermal balance
3. Orbit adjust

Structures

1. Meteorite effects
2. Pressure-vessel techniques
3. Thermal equilibrium

Navigation-guidance

1. Altitude and temperature effects
2. Gravity effects on accelerometers
3. Target resolution

Human factors

1. Zero gravity
2. Radiation effects
3. Physiological and psychological

Recognized problems are involved with the operation of propulsion systems for extended and intermittent operation in space. For example, it will be difficult to establish the thermal balance of propellants and other components in an environment of near vacuum pressures, while the temperature is cycling between the direct thermal radiation of the sun on one side and near absolute zero of cold space on the other. Seal materials may outgas and change their properties quite drastically. Pinhole leaks may vent the propellants.

Several failures of satellite systems have occurred that were attributable to a failure of one of the upper stages to ignite.

A test cell to simulate space environments is described below: This cell provides an altitude environment of 350,000 ft in the cell liner. Cold walls of approximately 100°K and radiant-heat panels will provide the temperature variations of space. The engine will be located inside the cell. On firing, the exhaust gases will be carried away through a diffuser. Diffusion and cryopumping will be used to maintain an altitude of over 200,000 ft downstream of the rocket nozzle during firing, while at the same time maintaining 300,000 ft upstream of the rocket nozzle. The facility will permit tests of orbit-adjust rockets down to 1 psi chamber pressure.

Had the mission of our hypothetical system been of much longer duration, say for weeks or months, consideration should have been given to electrical propulsion as a means for effecting changes in orbit. An electrical-propulsion facility would be needed.

Now the structural integrity of the glider itself must be examined. Under bombardment from meteorite particles, severe damage may be sustained. At present, impact work is done under research conditions rather than under system-development conditions; a high-velocity impact range is needed.

As our mission requires the glider to traverse the Van Allen radiation belt, shielding must be incorporated in its structure. Thus the total effects of the space environment on the composite structure must be assessed.

The structural aspects of the glider as a pressure vessel would be examined in any one of several chambers. An aerospace chamber will be able to investigate the structural integrity of the complete vehicle under conditions that very nearly simulate space. This chamber will have a vacuum of 10^{-8} mm Hg or an altitude equivalent of

320 nautical miles. Also, it will be possible to provide a combination of temperature simulation and vibration.

With regard to navigation and guidance, environmental investigations can presently be conducted on the sleds at Holloman for inertial systems. The entire system can be subjected to climatic conditions, including thermal load, pressure, vibration, and acceleration. Currently, we need to be able to calibrate more precisely the accelerometers used to determine one's position in space. This is particularly true for our hypothetical system.

In the bioastronautics area, the need to examine the physiological and psychological effects of space flight on the crew members is already quite apparent. Facilities to perform such examinations on the ground are limited. The effects of weightlessness over prolonged periods are not known, and *time* in such investigations appears to be most important.

Wright Field has resorted to the use of a submersion technique (called hypodynamics) and flight tests with airplanes to acquire preliminary data.

The effect on a crew from the threat of radiation and other multiple stresses will have an influence on their physiological and psychological state. Environmental facilities need to be made available for determining these effects on human beings before flight.

After completing the mission in space, we must return our crew safely to earth. The principal problems that remain to be solved are

Aerospace dynamics

1. Decelerators
2. Parachutes
3. Stability and control

Structures

1. Heating and loading
2. Materials

Bioastronautics

1. Multistress

Propulsion is not included because our space-propulsion tests have verified that we can get into reentry attitude with the initial deceleration provided by retro-rockets.

Our descent takes us from hypervelocity to landing speeds. The complex of wind tunnels in the United States provides us with the aerospace dynamic data we require. In order to ensure proper control of our deceleration, extensive tests are made on lifting surfaces and decelerators. At lower speeds parachutes may be employed.

At present, wind tunnels that operate at hypervelocities do not possess sufficient size, temperature, or running time to determine several major design parameters. Thus it is difficult to assess the validity of the vehicle design for reentry. To overcome our lack of facilities, the required parameters are obtained by superposition of data from a number of different facilities.

Heating-rate data are obtained from wind tunnels. Preliminary tests are made using heat-sensitive paint. From such data as these, heat gauges can be located for quantitative data. Load data are obtained from pressure-distribution tests in wind tunnels. Data are obtained at various combinations of angle of attack and roll angle.

With information obtained in the preceding tests, it is possible to make static structural tests on full-scale components with temperature and aerodynamic load simulation. Loads are applied by means of jacks, and the heat is applied by means of radiant-heat panels. Because the reflectors for the lamps are gold-plated, CO_2 is used for backside cooling. Aeroelastic data are obtained in flutter tests. It is fairly routine in such tests to get to zero damping and then back out; however, occasionally a rather spectacular destruction occurs.

Currently Wright Field can apply combined loads and heat to 2200°F. These capabilities must be extended in the near future. Wind tunnels with true-temperature simulation must be built to obtain data for more efficient spacecraft structures.

The mission of our hypothetical satellite has been completed. The crew made the trip with the knowledge that the technical problems had been fairly well investigated in the environmental facilities of the country.

34-11. Test Facilities

Test facilities presently under study and those which are vitally needed for the United States to regain and maintain leadership in space technology will be described in the following paragraphs.

Modern test facilities are expensive. We must simulate in sealed chambers, and under controlled conditions, the near vacuum of outer space; create relative velocities of at least 36,000 ft/sec; vary temperatures from minus 450 to 28,000°F; simulate the effects of solar pressure; and study the effects of radiation; we must do all these things in order to subject the vehicle to conditions which closely approximate those encountered in actual flight. Development of capabilities, such as guidance, maneuverability and rendezvous in space, life sustenance and human effectiveness in the space environment, selective and variable reentry and soft landings, all require research and development facilities in depth and on a wide front if we are to maintain a place in the successful exploration of space. Some typical aerospace systems facilities are:

Environmental Chamber. A system of this sort provides simulation of most of the environmental conditions existing in space for testing full-scale advanced space vehicles and lunar bases. A vacuum chamber several hundred feet high and some 150 ft in diameter will provide a vacuum of 10^{-8} mm Hg (320 miles altitude). Principal components are the main test chamber, solar simulator, cryogenic pumps, and a heat-sink system along with required access locks, vehicle-preparation facilities, nuclear hot shop, and biomedical and support laboratories.

Electro Gasdynamic Facility (50 Mw). The reentry of vehicles at near orbital velocities into the earth's atmosphere has led to requirements for high-stagnation enthalphy gasdynamic facilities for performing high-temperature hypervelocity research and development work on such problems as aerodynamic heating, structural strength, and flight stability and control. The electric-arc-heated gasdynamic facility, using electric energy to heat the air to 18,000° to produce the high-stagnation enthalphies needed, will be a very useful tool. A facility of this type will perform development work on such problems as aerodynamic heating, structural strength, and flight stability and control information. The facility should have relatively long duration test times necessary for reliable investigations of these hypervelocity problems.

Low-density Hypervelocity Tunnel (LORHO). A low-density hypervelocity tunnel for testing vehicles that operate at speeds up to 28,000 ft/sec and altitudes up to 400,000 ft is another vital tool in environmental testing. This tunnel will permit aerodynamic, thermodynamic, aeroelastic, and structural characteristics of such systems to be determined in their true operational environment. Large-scale configurations or full-scale components can be tested in a hypervelocity airstream to great advantage in a test section approximately 10 ft in diameter. A facility of this type would require a power supply in the neighborhood of 450 Mw. This facility would be able to obtain significant parameters of a sophisticated vehicle capable of both departing and reentering the earth's atmosphere in a safe and routine manner. Specific problems could be investigated or associated with simultaneous effects of high heat-transfer rates, accelerations, real-gas aerodynamic flows, structural deformations, communications, and time-dependent phenomena.

Hypersonic True-temperature Tunnel. For aerodynamic and propulsion tests of recoverable boosters, lifting reentry vehicles, and orbital aircraft, a tunnel to simulate true flight environments to altitudes of 250,000 ft and velocities of 13,000 ft/sec should be of sufficient size (in the neighborhood of 10 ft in diameter). Useful data, such as interaction of aerodynamic heating, aerodynamic forces, and other phenomena encountered in high-speed flight for engines and airframes on these vehicles, would be obtained since they cannot as yet be determined analytically. Some critical problems that can be solved in a tunnel of this nature are

1. Hypersonic engine inlet and exhaust nozzle development
2. Supersonic combustion
3. Engine heat-exchanger development
4. Integrated propulsion-system development
5. Airframe cooling-system development
6. Stability and control problems under combined aerodynamic and thermal loading
7. Aerothermoelastic problems

Materials Environmental Test Facility. A facility for the evaluation of mechanical, physical, and electrical properties of newly developed materials and material systems under complex environmental conditions should simulate as nearly as practicable those to which the materials in advanced weapons systems and space vehicles will be exposed.

Ultraclean Rooms. Ultraclean white rooms are necessary for the testing and assembly of critical, precision-floated gyro devices. They are normally used to provide a minimum of contamination in the assembly of miniature integrating gyros and pendulous accelerometers. These rooms are generally constructed with floors covered with rubber tile, glass and aluminum walls, Plexiglas ceilings, and specially constructed Formica workbenches. Especially clean air is needed. The air may be cleaned through a series of Fiberglas filters and electrostatic filters which are normally designed to remove over 90 per cent of all airborne contamination 1 micron or larger. The absolute filter will remove 99.8 per cent of all particles 0.3 micron or larger. This system is used for the air system in the ultraclean rooms, assistant foreman's office, and ultraclean personnel corridor. This system is improved over the existing clean filtered-air systems by installing a second bank of absolute filters immediately preceding entry into the room. Each ultraclean room has its own temperature control and reheat coil for individual control. Environmental conditions for the ultraclean-room air are as follows:

Temperature, $72 \pm 2°F$
Relative humidity, 25 to 40 per cent
Air pressure, greater than the surrounding areas by $\frac{1}{4}$ in. H_2O pressure
Dust count, expected less than 5,000 particles, 1 micron or larger, per cubic foot

Many other environmental facilities either exist or are under construction throughout industry and in government laboratories. Simulators in the field of bioastronautics, life sciences, six-degree-of-freedom space simulators for training of aerospace pilots, dynamic testing centrifuges for dynamic testing of equipment such as inertial guidance systems, altitude-rocket test stands, electrical-propulsion test chambers, ultrahigh-altitude rocket cells, radiation-weapon test facilities, magnetohydrodynamics laboratories, and the like are becoming more and more necessary as we move into the development of cislunar and planetary space research.

34-12. Rocket-engine Testing

In liquid- and solid-propulsion-system development, designers and engineers invariably are required to go beyond available experimental data and to extrapolate into regimes of performance where there are no test-data points to guide them. The need for test information is taken for granted in missile-propulsion work, but too often it is looked upon as verification or proof of design rather than as a partner in progress. The proper integration of testing data results in rapid progress because experimental data are available during analysis and design development rather than during demonstration testing. If testing is treated as a necessary evil to follow rather than to accompany design and engineering, progress is slow and overall costs are high.

The ballistic-missile programs have demonstrated the thesis that testing is an integral part of development of propulsion systems and has a regenerative, beneficial effect on the progress of the project. These programs have provided information regarding developmental test requirements for both liquid- and solid-propellant propulsion systems. The kinds of tests typically conducted for both liquid and solid systems are listed below, and in the pages that follow a description is given of what the test is and why it is conducted. Some estimate of numbers of tests that should be conducted is provided. However, the numbers will vary dependent upon schedules, costs, availability of test facilities, the nature of problems involved, and the like.

34-13. Types of Tests

The pattern of development of propulsion systems with respect to tests may be listed as follows:

1. Research and development tests
   a. Ground level
   b. Simulated altitude
   c. Special evaluation tests
2. Preliminary flight-rating tests
   a. Ground level
   b. Simulated altitude
3. Research and development flight tests
4. Qualification tests
   a. Ground level
   b. Simulated altitude
5. Production quality-assurance tests
6. Operational flight tests

34-14. Test Objectives

Research and Development Tests. The developmental phase includes design, engineering, fabrication, and many tests. Engineering testing, in contrast to proof or acceptance testing, runs throughout this phase. Designs are tested by work on computers and simulators. Engineering assumptions and compromises are tested by models and mock-ups. Critical components are fabricated and tested to determine performance limitations as well as weight and space requirements. Major components and subassemblies are tested to determine their compatibility in the system.

Along with the above, subscale motors are tested both at ground level and at simulated altitude for the purpose of evaluating materials and component performance and survivability, propellant evaluation, and nozzle performance. As many as 200 tests of this type may be required. Usually 40 or 50 full-scale solid-propellant motors are tested during research and development, of which 10 per cent or more should be tested at simulated altitudes above 100,000 ft. In the case of liquid-propellant systems, many of the subassemblies such as gearboxes, pumps, and valves can initially be tested separately from the combustion chamber and nozzle. Ultimately, however, the entire system must be tested many times as a unit.

A special evaluation test program is required for solid-propellant propulsion systems. For example, the Minuteman program consisted of a series of tests to provide for an evaluation of defective motors, motor-destruct systems, and motor explosive classification and hazards. Very little of this type of testing is generally done for liquid systems, though it should be considered for future systems.

Preliminary Flight-rating Tests. Preliminary flight-rating test (PFRT) programs are conducted for both liquid- and solid-propellant propulsion systems to demonstrate that the propulsion system is ready to be flight-tested. Several important factors are evaluated in determining the state of readiness. For example:

1. Do all the parts fit together and function properly, particularly the GFE (government-furnished equipment)?

2. Is there an acceptable degree of reliability and probability of success?

3. Do the performance characteristics comply with the model specifications and other applicable specifications, and is performance suitable for targeting and trajectory calculations?

4. Is the unit safe to transport and handle?

The foregoing items are generally applicable, but it is necessary that a PFRT requirements document be prepared for each different propulsion system.

For liquid-propellant systems it is generally required to demonstrate the ability of the system to function satisfactorily for six rated durations, i.e., six times the normal powered flight time, and to make at least 10 successful starts.

For solid-propellant systems, using Minuteman as an example, 12 motors for each stage are required to be tested in accordance with a test plan. At least three motors (other than stage 1) for each upper stage should be tested at altitude, including at least one thrust-termination test at altitude, if applicable.

Research and Development Flight Tests. Flight-test operations for both liquid- and solid-propellant systems are a final part of the developmental phase. They are a continuation of the pattern of data gathering to verify or modify engineering assumptions and compromises. Adequate instrumentation, sufficient telemetry equipment and channels so that ground checks can be run in parallel with flight operations, and systems-designed data-acquisition and data-reduction equipment all speed up the development cycle by furnishing test data quickly back into the engineering process.

In general, flight-test instrumentation is directed toward providing data which will define the overall missile performance. It is considered advisable to provide a basic instrumentation list such that the instrumentation installed on propulsion units can accommodate measurements required to accomplish known objectives, provide data regarding certain contingencies or anomalies, and determine catastrophic-failure causes.

Strictly from the point of view of propulsion-system evaluation, 10 or 12 successful flight tests should be adequate, provided the tests are well planned.

Qualification Tests. A qualification test program for both liquid- and solid-propellant systems is designed to verify the acceptability for operational use of

1. The propulsion-system designs to meet specification requirements and expected environments
2. Fabrication, assembly, loading, and handling techniques
3. Quality control and inspection techniques
4. Subcontractor-furnished material and parts

The qualification test program should also be used to determine the suitability for production use of second-source materials and components. An outgrowth of a qualification program is the obtainment of final ballistic parameters for an operational model specification and to furnish data for targeting.

The qualification program is a verification program. Therefore, no tests are made which were not previously conducted in the developmental phase; i.e., this is not a research and development program per se, even though research and development may be incomplete and continuing.

For a liquid-propellant system, it is generally required that the complete system be functionally evaluated for 12 full-duration tests. For solid-propellant systems, using Minuteman as an example, a test program involving 18 motors is required. Some tests for both types of propulsion systems should be conducted at simulated altitude as is done in a PFRT program.

Production Quality-assurance Tests. Solid-propellant motors are drawn from a production line based upon a statistically designed sampling plan such that larger numbers of motors are taken and tested initially, similar in many ways to a moving-line inspection plan. It is poor practice to shorten this phase by skimping on testing and gambling on immediate production, because the product of production methods and tooling must be checked before production for inventory is started.

Production acceptance tests for a liquid-propellant propulsion system are represented by a short series of tests conducted on each unit prior to its incorporation into an operational system.

Operational Flight Tests. This phase of the production program is a final verification and demonstration of the quality of the product of production methods, tooling, and assembly, and that the military requirement has been met. This phase is also used for training of the using service. As many as 10 operational-type missiles may be flown during the first 6 to 12 months of production for inventory.

34-15. Flight Testing

In the testing of ballistic missiles and space vehicles, more and more emphasis must be placed on testing major subsystems and complete systems under environmental conditions in simulated facilities. However, this testing cannot replace the research and development flight-test program which is a necessary and integral part of the development of any successful missile system. In the final analysis, the flight-test program makes available to the developer component-performance data peculiar to the flight environment which cannot be directly obtained in any other manner. Flight testing of complete propulsion systems, reentry systems, and inertial guidance systems is particularly important because these systems contain many components which are sensitive to conditions not readily or completely obtainable in environmental facilities. For example, many components in the guidance system are adversely sensitive to sustained linear accelerations and the complex vibration and acoustic environment of powered flight.

In order to obtain as much development data as possible early in the development cycle, laboratory, sled, and flight-test programs are used in a complementary manner. In a sled test of a system, a vehicle is accelerated on a sled riding on a straight track by large rockets, thus subjecting the test article to high linear accelerations and decelerations. The sled-test program provides for evaluation of accelerometer nonlinearities on a series of sled runs. This sled program provides for deriving quantitative values for the accelerometer coefficients, since the instrumentation system for measuring sled position as a function of time is much more accurate than external data available in flight testing. Furthermore, the separation of accelerometer nonlinear terms is greatly facilitated by the availability of high negative as well as positive accelerations on a sled run. However, even after many sled runs and thorough analysis of the test data, it is not possible to assign reliable values to all the coefficients in the accelerometer-error model. In addition, the sled-test data are of very limited use in determining the coefficients in the gyro and stable-platform error models; therefore, laboratory tilt-table and centrifuge data must be used in conjunction with sled data to piece together a quantitative mathematical description of the inertial components in an acceleration force field.

34-16. Total Guidance-system Error

The total target miss for ballistic vehicles can be broken down into two general classes of errors, those which occur during powered flight prior to burnout, and those which occur after burnout, during the free-flight and reentry periods of flight (Sec. 19-3). Included in the former are the errors arising from

1. Guidance-system maloperation
2. The inability of the control system to follow guidance-system steering commands
3. The inability of the propulsion system to follow guidance-system cut-off commands

It is the purpose of the guidance postflight analysis to identify these errors and to trace them back to their equipment origins in the subsystem. The system analyst has the additional responsibility of determining the source of postburnout errors arising from gravity anomalies during free flight, uncertainties in the position of the target, and reentry errors.

34-17. Analysis of Inertial-guidance Test Data

As with any other intricate electromechanical system, the successful development of an inertial guidance system for a ballistic missile or space vehicle requires a comprehensive system test program which exercises the full dynamic range of system variables in the actual operational environment. A program of this nature serves to isolate any faulty components in the system. After elimination of all potential trouble areas, the test will then provide the statistical parameters, such as accuracy, reliability, environmental tolerances, and other parameters which describe overall performance.

As with all test data, difficulties arise during the data-taking process because of noise and uncertainties associated with the measurement standard. Analysis, coupled with sound technical judgment, must then be exercised, based on imminent knowledge of the system under test and of the test instrumentation. Testing of inertial systems will be used as an example for data analysis which applies to any testing process.

Test data confounded by noise arise from telemetry nonlinearities in the r-f link in flight testing, external instrumentation during flight or ground testing, and unfavorable geometrical factors of the instrumentation. Thus it becomes necessary to develop special system-analysis techniques which optimize the recovery of performance information from noisy and frequently conflicting test data. In analyzing test data for guidance-system performance information, very high quality trajectory data are necessary in order to perform the regression computations necessary to separate overall guidance-system error into individual accelerometer and gyro error sources.

34-18. Coordination and Liaison

Consideration of the tasks, hardware, and personnel involved in the planning and execution of a system test program discloses that there must be a large amount of effort spent in coordination and liaison work. This work is a vital part of system testing and is necessary to accomplish the objectives within the allotted schedules. Some of the major organizations groups with whom detail coordination is required are shown:

Design engineering	Program control
Manufacturing	Contracts
Quality control	Systems integration
Reliability	Logistics
Facilities	Test ranges
Ground-support equipment	Customer

BIBLIOGRAPHY

Test Program Plan for Ballistic Missiles, USAF Ballistic Systems Division, Los Angeles, Apr. 13, 1961.

Brown K., and Peter B. Weiser: *Ground Support Systems for Missiles and Space Vehicles*, McGraw-Hill Book Company, New York, 1961.

Chase, Wilton P.: *Personnel Subsystem Development and Test Program*, Space Technology Laboratories, July 15, 1961.

Cupo, E.: *Proceedings of the Intercommand Working Group on Mathematical Models for Systems Effectiveness*, USAF Space Technology Laboratories, 1961.

Drucker, A. N.: *The Performance Analysis of Ballistic Missile on Space Vehicle Inertial Guidance Systems*, Space Technology Laboratories, May, 1961.

Chapter 35

ECONOMICS

KENNETH E. BOULDING, *University of Michigan.*

CONTENTS

35-1. Relation of Economics to System Engineering	35-1
35-2. Economic Efficiency	35-2
35-3. Rate of Return	35-2
35-4. Marginal Analysis	35-3
35-5. Price Structure	35-4
35-6. Productivity	35-4
35-7. Aggregate-product and -income Concepts	35-5
35-8. Inflation and Deflation	35-6
35-9. Planning of Engineering Projects	35-7
35-10. The Problem of Uncertainty	35-8

35-1. Relation of Economics to System Engineering

The relation of economics to engineering is not unlike that of criticism to literary or musical composition and performance. Engineering proposes but economics disposes. It is not surprising, therefore, to find that the relationship between the engineer and the economist is sometimes a little strained. An economist, indeed, once defined the engineer as a man who spends his life finding out the best way to do something that should not be done at all. By contrast the engineer frequently visualizes the economist as the man who comes along and spoils his beautiful plans for dams and bridges by pointing out the financial obstacles.

The economist then has two functions, both of them rather unpopular, in relation to system engineering. The first is to raise the question of the internal standards by which any particular design or project shall be judged. This is the problem of evaluating the different possible ways of doing a particular thing or designing a particular project. The second function of the economist is to raise the question whether any particular project should be done at all, in the light of alternative uses of the resources and the general overall pattern of the development of the economy. These two functions are closely related. Both involve an ordering of alternatives by a set of ordinal numbers such as best, second best, and third best. In the problem of internal evaluation we are essentially ordering a set of alternative ways of doing a particular project. In the external evaluation we take the best of these ways, according to some criterion, and compare them with the alternative uses of resources in other fields.

35-2. Economic Efficiency

The nature of the criterion of economic efficiency is of course the heart of the problem. The basic concept is that of a process of production or transformation by which certain inputs (resources) are transformed into outputs (product). Engineering concepts of efficiency usually involve the ratio of some kind of output to some kind of input. The economic problem is complicated by the fact that there are usually many outputs and many inputs of different dimensions. Under these circumstances, in order to get a single figure for the efficiency of a process we must reduce these outputs and inputs to a common measure by means of a set of evaluation coefficients. Thus if $x_1, x_2, \ldots, x_n$ are the inputs, and if the resulting outputs are $y_1, y_2, \ldots, y_m$, then in order to get a measure of efficiency we would have to multiply each of these quantities by an evaluation coefficient such as $v_{x_1}, v_{x_2}, \ldots, v_{x_n}; v_{y_1}, v_{y_2}, \ldots, v_{y_m}$. Then a measure of efficiency might be given by the formula

$$E_1 = \frac{\sum_{i=1}^{i=m} y_i v_{y_i}}{\sum_{i=1}^{i=n} x_i v_{x_i}} \tag{35-1}$$

If evaluation coefficients are measured in dollars per unit of input or output, then E_1 measures the dollar's worth of output per dollar's worth of input.

35-3. Rate of Return

Although this is parallel to engineering concepts of efficiency, it is unfortunately not satisfactory to the economist, because inputs and outputs do not occur at the same time but are strung out along a time scale. A process which gives, shall we say, $120 worth of output for each $100 worth of input 1 year after the input is put in is usually more desirable than a process which gives $120 worth of output per $100 of input 20 years after the input has been put in. In spite of, or perhaps because of, the fundamental nature of this problem there is still considerable controversy as to how to handle the time structure in the evaluation of productive processes. One way of handling the problem is to assume that there is a structure of rates of interest in the society and that by a process of compounding or discounting at these rates of interest we can express a value at any given date in terms of an equivalent value at some other date. We can then select a base date, usually at the initiation of a project, and we can reduce the values of all inputs and outputs to the base-date equivalent. Thus the base-date equivalent V_0 of a value V_t at time t years after the base date, assuming annual compounding, is given by the formula

$$V_0 = V_t (1 + i)^{-t} \tag{35-2}$$

When we have the base-date equivalent of the values of all inputs and outputs we can express this either as a ratio of the base-date value of output to the base-date value of input, as in (35-1), or we can express the result in terms of a net value:

$$N_0 = Y_0 - X_0 \tag{35-3}$$

N_0 is the net value at base date, Y_0 is the base-date value of all outputs, and X_0 is the base-date value of all inputs.

An alternative solution is to measure the efficiency of a process by its internal rate of return. The internal rate of return r is that rate of discount or rate of growth at which the net base-date value of all inputs and outputs is zero, the base date being at or before the first input or output (as inputs usually precede outputs, the first item is usually an input). This is given by the following equation:

$$\sum_{i=1}^{i=m} y_i v_{y_i}(1+r)^{-t_i} - \sum_{i=1}^{i=n} x_i v_x (1+r)^{-t_i} = 0 \tag{35-4}$$

The advantage of the internal rate of return is that it does not involve any assumptions about future interest rates, although of course we can never escape the uncertainty involved in postulating future inputs, outputs, and evaluation coefficients. It provides an easy way of comparing projects of different sizes, for the size of a project in itself does not affect the measure. Whatever the theoretical advantages, however, the fact remains that the internal rate of return is very little used in the evaluation of projects, especially in the case of public works.

35-4. Marginal Analysis

After having selected to his own satisfaction a measure of profitability or worthwhileness of a process, the economist usually goes on to assert that the best combination of inputs and outputs in the process is that which maximizes the measure of profitability, which may of course be either the base-date value of net receipts as in (35-3) or the internal rate of return as in (35-4). It is at this point that the economist presents a way of thinking which should be useful to the engineer but which is frequently only frustrating. The frustrations arise perhaps because the economist assumes a production function, that is, he assumes that out of all conceivable combinations of inputs and outputs only a limited set is in fact feasible. For the engineer, however, the production function is not so much given as it is a problem. It is the business of the engineer to improve production functions, that is, to get more output per unit of input. He may not unreasonably be a little irritated, therefore, when the economist blithely assumes a production function as given, and proceeds to argue from there on. The economist, however, is within his rights in insisting that for each possible production function there is still an important problem to solve, which is: which out of the possible combinations of inputs and outputs is the best as judged by the criterion of maximizing profitability or, more generally, worthwhileness?

A formal solution to this problem is given by the marginal analysis, which simply involves maximizing the measure of profitability under the constraint given by an assumed production function and assumed evaluation coefficients or, more generally, assumed functions relating the evaluation coefficients to the quantities of input and output. In terms of the net-present-value criterion of (35-3), it is easy to show that N_0 is maximized when the marginal revenue of each input or output is equal to its marginal cost, both being expressed in discounted (that is, base-date) equivalents. The maximization of the internal rate of return gives essentially the same result, the only difference being that in this case the discounting is performed at the internal rate rather than at external rates of interest. What this means is that, if we take any set of inputs and outputs which is consistent with the production function, and if we suppose that we increase by a small unit the quantity of one of these, we then ask ourselves: is the net addition to the base-date value of the revenues greater than the addition to the base-date value of the costs? If it is, it obviously pays to make the change. Only when it does not pay to make any change—that is, when all marginal costs are equal to the corresponding marginal revenues at the base date—is the profitability at a maximum.

The marginal analysis has a defect of being highly formal and not easy to implement. To some extent it is possible to implement it through the device of linear or other forms of programming (Chapters 25 to 27), though even here there is frequently a difficult problem in information input. It is often quite difficult to find out the marginal rate of transformation of input into output or the marginal rate of substitution of one input or output for another. In general, our information system is much better at giving average figures than marginal figures. Partly this may be because we have not yet learned to think accurately enough in incremental terms. But more fundamentally it is because the information which is required to obtain these marginal quantities can be derived only from a degree of variability in the variables of the system and a stability of its parameters which is not usually obtained in nature.

35-5. Price Structure

What is important here is a mode of thinking about a problem rather than a neat solution. The economist, for instance, points continually to the importance of the evaluation coefficients in determining the optimum solution. The evaluation coefficients are of course "shadow prices," and they usually have some relation to real prices. A change in the price structure can radically change the best solution of engineering problems. Thus if labor is cheap and capital dear, laborers with baskets of earth on their heads may be cheaper than power shovels. The economist frequently accuses the engineer of thinking too much in physical terms and not enough in value terms. The California water problem may be cited as a case in point, where the engineer tends to conceive the problem in terms of taking water from the rainy north of California and transporting it to the dry south, whereas the economist is likely to point out that a more satisfactory solution to the problem might be to raise the price of water to the point where people would economize its use, and where it would pay to introduce water-saving inventions and improvements.

All this points to one of the basic contentions of the economist, which is that the relative price structure from which the system of evaluation coefficients ultimately has to be derived is not arbitrary but fulfills a number of important social functions. Suppose P_i represents a set of prices of all commodities in terms of some common measure of value, or numeraire, such as the dollar. There is a very large number of such price sets, and we suppose that for each one of these there corresponds a set of shortages and surpluses of all the various commodities. A shortage is defined as a situation in which the quantity demanded exceeds the quantity which is offered for sale, and a surplus likewise is a situation in which the quantity offered for sale exceeds the quantity demanded.

We now suppose that there is some "equilibrium" set of shortages and surpluses for which all these values are zero; that is, for each commodity the quantity demanded exactly equals the quantity supplied. The price set which corresponds to this equilibrium shortage-surplus set is the equilibrium price set in the sense of clearing the market. It is better, perhaps, called the normal price set P_n. Under certain circumstances (which are not always realized in practice), this is a true equilibrium set in the sense that, if the price set diverges from P_n, then for those commodities the price of which is "too high" there will be surpluses, and for those whose price is "too low" there will be shortages; the prices that are too high will fall, and those that are too low will rise; and, given a set of limiting circumstances which are not wholly unreasonable, the system will move toward the normal price set.

There are many things in practice, however, which can interfere with this movement, such as monopoly or governmental price fixing. If the actual price set is not equivalent to the normal price set, we can explain this in terms of some kind of resistances or frictions. One weakness of this type of analysis is that it tends to neglect the possibility of a shift in the set functions themselves, that is, in the functions which relate the price sets to the shortage-surplus sets. The propositions of price theory are very simple truths, such as, for instance, if we fix the price of agricultural commodities much above normal we are almost certain to be plagued with agricultural surpluses, and if we fix house rents much below normal we are likely to have a shortage of housing. Such propositions may not be much more elegant than the proposition that if we touch a hot stove we will be burned, but nevertheless simple truths of this order are important for the policy makers.

35-6. Productivity

The impact of the engineer on the normal price structure comes about largely through his impact on the productivity of the productive process in different commodities. There is one price structure P_a which is the same as the relative structure of alternative costs. The alternative cost ratio of two commodities is the amount of one commodity which can be produced with the resources released by the sacrifice of one unit of the other. If by giving up a pound of butter we can release resources which can eventually produce 10 pounds of bread, then the alternative cost ratio is 1 pound of butter to 10 pounds of bread. In the classical model of free markets, flexible prices, and perfect competition, there is a tendency for P_n to approach P_a. Suppose, for instance, that the relative price of bread to butter in the marketplace was 1 pound of butter to 15 pounds of bread, whereas the alternative cost ratio was 1:10. In the marketplace bread would be "cheap" and butter would be "dear." The production of butter would be profitable and the production of bread would be unprofitable relative to some average level of profitability. If there were no obstacles to the flow of resources, this would attract resources into butter production and resources would flow out of bread production. This would diminish the output of bread, which would raise its price, and increase the output of butter, which would lower its price; and this process would go on until the relative prices in the marketplace were approximately equal to the alternative costs. In order to arrest this process we would either have to introduce obstacles to the free flow of resources or we would have to change the set function relating the price structure to demands and supplies, shortages, and surpluses.

When we consider the long-run dynamics of the social system as a whole, another function of the price system emerges which has been badly neglected by economists. This is the function of the direction of technological change. By and large, things which are perceived as "dear" will be economized and those which are perceived as "cheap" will not, not only in their immediate use but also in the attention which is given to the process of technological change. Technology will tend to move in the direction of eliminating the dear commodities and encouraging the use of the cheap ones. We have here a real dilemma. Insofar as economic development involves essentially an increase in the productivity of labor, we want to encourage the development of a laborsaving technology. One way to do this, however, is to make labor dear, that is, to increase real wages. Increasing wages, however, may easily put us in the position of creating unemployment because of the function of the wage system as part of the price system in clearing the market for labor.

35-7. Aggregate-product and -income Concepts

This brings us to that part of economics known as the theory of unemployment or stabilization which is not, perhaps, of direct importance to the problems which the engineer has to solve but which is of enormous importance in determining the general setting in which he has to work. The basic concept here is that of the *gross national product*, or GNP. This is a measure of the total output of the society, eliminating double counting (for instance, we must not count both the wheat and the flour and bread that it goes into). If we subtract the dollar value of the depreciation of physical capital from the gross national product and make some other minor adjustments, we get the *net national product* (NNP). In the United States accounts, the *national income* is defined as the net national product less indirect business taxes, business transfer payments (mostly bad debts), profits of government enterprises less government subsidies, and a statistical discrepancy. Indirect taxes are deducted because the value of the gross national product includes commodities valued at prices which include the indirect taxes. Business transfers consist of items which are deducted from gross receipts as costs in calculating profits, but which do not find their way into the earnings of factors of production. The surpluses and deficits of government agencies are likewise deductions from the total product which do not accrue to any private individual. These deductions then enable us to define the national income in such a way that it can be distributed entirely among the claims of various factors of production. Further deductions on that part of the national income which is not available to persons, such as undistributed corporate profits, reserves for corporate tax liability, and employee contributions to social security and pension funds, gives us *personal income*, and subtracting direct taxes from this gives us *disposable personal income*, the smallest aggregate-income concept recognized by the national accounts. The gross-national-product concept is most relevant to the discussion of the overall level of employment, for we have as a formula

$$\text{GNP} = \begin{pmatrix}\text{hours of labor}\\ \text{performed}\end{pmatrix} \times \begin{pmatrix}\text{average value of}\\ \text{the product of an}\\ \text{hour's labor}\end{pmatrix} \quad (35\text{-}5)$$

If the productivity of labor is constant, as in short periods it is, employment and the gross national product fluctuate proportionally. In considering the determinants of consumption, disposable personal income is the most relevant quantity.

35-8. Inflation and Deflation

For any given society in a given year we can suppose that there is some GNP which is "ideal." It will always be physically possible to produce more than the ideal, but only at an unacceptable sacrifice in leisure. Whatever the GNP is, the total output of the society must be absorbed in some way. There are four principal avenues of absorption: (1) household purchases; (2) government purchases; (3) business accumulations, that is, goods remaining in the hands of businesses; and (4) foreign balance or net exports. If we suppose that all the product originates in the hands of businesses, which means including certain government and household enterprises in this category, then it is easy to see that actual business accumulations—that is, the actual increase in the stocks of goods of all kinds held by businesses, whether in the form of raw materials, inventory of finished product, buildings, or capital equipment—must be equal to the total output—that is, the GNP, less the household and government purchases, less the excess of exports over imports, and less the depreciation of existing capital. Suppose now that at the ideal level of GNP the amount of household and government purchases, etc., produces an actual accumulation of goods in the hands of businesses which is more than businesses want to acquire. These unwanted accumulations can produce a combination of two effects: there can be a downward pressure on money prices, or a cutback in output with resulting unemployment. A fall in prices might diminish the willingness of businesses to increase their accumulations of goods, and unemployment diminishes incomes and diminishes household purchases. Both these results lead to further unwanted accumulations, which may lead to a further decline in prices and more unemployment. This process can go on as in the period from 1929 to 1932 until the gross national product is far below the ideal and there is widespread unemployment and social collapse.

Fortunately, there are fairly easy remedies for this disease, for both the banking and financial system on the one hand and the tax system on the other can be used as stabilizers. We have at present in the United States a powerful built-in stabilizer in the shape of the deductible-at-source progressive income tax and the social security system. If money incomes start to decline, income tax collections decline much more sharply as people get out of the upper brackets into lower brackets, and the Federal government therefore runs a deficit. A deficit of the Federal government, however, increases the stock of money in the hands of the public, and this tends to stop the decline in money incomes. By contrast, if there is inflation and money incomes are rising, more people get in the upper brackets, income tax collections rise more sharply than income, the Federal government begins to run a surplus which removes money from the pockets of the public, and this automatically checks the inflation. This automatic mechanism is not powerful enough to deal with really major disturbances on either side. World War II, for instance, was not financed without inflation, and a combination of circumstances is still possible which would give us a serious depression. The automatic stabilizers, however, can be made more effective and can also be supplemented by conscious policy. This is of great importance in any kind of long-run planning.

35-9. Planning of Engineering Projects

Most of the above considerations apply with particular force to advanced market-oriented economies such as the United States and Western Europe. It must not be thought, however, that economics ceases to be relevant when we move to communist countries or to underdeveloped countries. The fundamental principles of economics apply no matter what the institutional framework of society, but of course the machinery by which these principles are carried out differs a great deal from country to country. In all societies, an engineering project can be realized only if it passes one of two tests: it must either be incorporated into the budget of a larger organization or it must produce products which can sell in the market for enough to cover the cost of production. In socialist economies the budget test is all-important; a socialist economy is a "one-firm state," in which a single economic organization controls all the capital, plans all the investment, and determines the mix of products. Even in a socialist state, of course, market considerations are not irrelevant, but they tend to be reflected in adjustments of the price structure and the structure of taxes and subsidies rather than in changes in the product mix itself. In the socialist state there may be consumer choice but there is no consumer sovereignty. Even in capitalist societies, however, the budget test is an important one, and many engineers work in large organizations, whether private or public, in which the budget test is paramount and conditions are not very much unlike a socialist state.

In the modern world, the difference between the developed countries of both ideological camps and the underdeveloped world is very great, and the gap does not seem to be diminishing. The main contribution of economics to the theory of economic development is to point to the fact that financial rates of return may not be an adequate guide in assessing the priorities for investment. In the poor countries a good deal of social overhead and public investment in the form of education, roads, public health, and even some basic utilities and industries may be necessary to provide a framework for private investment. Economics also points to the importance of "external economies," sometimes called "spillover effects," by which some investments create benefits and opportunities outside the particular industry in which the investment is made. In assessing priorities for investment, preference should be given to those, other conditions being equal, which have the most beneficial external effects.

There are two rough categories of engineering problems for which the relevance of economics may be very different. The first concerns the day-to-day operation of existing installations. Economics may not be very relevant to these problems, though one suspects that sociology is highly relevant, especially the sociology of organization and group relationships. It is, however, in the second kind of engineering, which involves the planning of new installations, that economics is most relevant, simply because there are more complex alternatives, and the problem of choice among alternatives is the heart of economics. One of the major problems of economic engineering of this kind is the sheer cost of preparing alternative plans, especially in large-scale projects extending many years into the future. The economic justification of any plan depends on its comparison with others. Engineers, however, frequently have to make what are fundamentally economic decisions to concentrate on some particular plan at an early stage in the development of a project, usually on very insufficient evidence. Concentration on a particular plan, however, means that other alternative possibilities are likely to be neglected. In the business world and in the market sector of the economy there tends to be a constant search for such alternatives, simply because the discovery of a neglected alternative is likely to be highly profitable to the discoverer. In the world of large-scale organizations, whether governmental or private, neglected alternatives are less likely to be discovered, and they may even be deliberately suppressed if they are upsetting to the existing organization. The problem of how much effort should be devoted to search, either in a particular organization or in society as a whole, is a difficult one which has been given too little attention either by economists or by engineers.

35-10. The Problem of Uncertainty

Perhaps the major problem of this kind of decision making is that of uncertainty. This may be dealt with either by increasing knowledge and thereby diminishing the uncertainty itself, or by protecting ourselves against the uncertainty by being prepared for disappointments and by being flexible. The theory of decision making under uncertainty (Chapter 24), to which game theory (Chapter 23) has made an important contribution, is only in the borderland of economics, but it is nevertheless a subject of great importance to the planning engineer. In the absence of uncertainty, projects are likely to be much more specific and much less adaptable than they need be in the presence of uncertainty. A certain amount of uncertainty may be removed by economic analysis and projections; on the other hand, there is grave danger that we may base decisions on "certainty equivalents" which then become extremely dangerous in the face of actual uncertainty. The strategic planning in national defense is probably the worst example of this fallacy, where we seem to be building into our system a positive probability of irretrievable disaster. Even at less exalted levels, however, engineering projects are often more vulnerable to unexpected, unfavorable changes in the environment than they should be. In the world of finance we take care of uncertainty largely by liquidity. We hold liquid assets partly at least because we do not know the future and we are therefore willing to sacrifice more directly remunerative uses of these assets because of the ability which they give us to react to unexpected changes. Diversification is another method of coping with uncertainty. This is the famous principle of not having all the eggs in one basket. There is a constant tension, therefore, in the design of projects between the desire to design things efficiently for specific uses and the need to build flexibility into the overall enterprise in case specific uses turn out to be useless. There is no point in being highly efficient in the production of something which nobody wants. This perhaps is the last message of the economist to the engineer.

Chapter 36

MANAGEMENT

R. D. O'NEAL, *Bendix Systems Division, Bendix Corporation.*

J. F. CLAYTON, *Bendix Systems Division, Bendix Corporation.*

CONTENTS

36-1. Introduction	36-1
36-2. Theories of Management	36-2
36-3. The Project System	36-3
36-4. PERT	36-5

36-1. Introduction

System engineering may be described as the discipline of utilizing the resources of an organized multitechnology team for the solution of an engineering problem. If it is desired to transform an engineering solution into something useful in the real world, management must be added. Management is that ingredient in our economy which makes productive all our resources both physical and human.

In the years since World War II, both science and technology have expanded at an increasingly rapid pace. The period has been well described by Gaddis as the age of massive engineering.[1]

We are no longer in the Age of Mass Production; rather, we are in the Age of Massive Engineering. Now recently developed knowledge is being massively used to determine specific solutions to new problems in all fields. This is not to say that mass production will no longer be the keystone of our material prosperity. But it does strongly suggest that it is time to concentrate our resources on the management frontier where problems are profound and management knowledge is all too limited.

On the engineering side, the techniques of system engineering have been rushed into the breach to handle the increasing complexity of the nation's large systems undertakings, both military and commercial (see Chapter 1). Despite its overriding importance, less progress has been made on the management side. Management theorists have not been of much help in solving the new and unique problems which massive engineering and especially massive system engineering have posed to American government and industry. Managers of system design, therefore, have had to develop their own approach—the project system. It is this practical project-management system, not yet elevated to the status of a formal management theory, that this chapter will cover.

36-2. Theories of Management

Before discussing the project manager, let us review the theories of management for business in general and discover why something new had to be invented. In the nineteenth century and earlier there was, of course, no need for such nonsense as management theory. The man who owned the business told subordinates what to do and that was that. Very few scientists or engineers were required in the business, and labor relations were simple. In the early part of this century, simultaneously with the emergence of effective labor organizations, it became important for the proprietor to define the function of each employee more precisely. In 1911, Taylor published his famous book[2] which quantified and organized the management function in industry.

Drucker[3] stated it very accurately in 1954.

The emergence of management as an essential, a distinct and a leading institution is a pivotal event in social history. Rarely, if ever, has a new basic institution, a new leading group, emerged as fast as has management since the turn of this century. Rarely in human history has a new institution proven indispensable so quickly; and even less often has a new institution arrived with so little opposition, so little disturbance, so little controversy.

In subsequent years, successful managers added to Taylor's foundation, and the problems of management functions and management-employee relations in the era of mass production seemed to be solved. With few exceptions, all American industry is now based on the 1911 type of scientific management principles from which we have the traditional organization charts, job descriptions, principles of delegation of authority, etc. To industry in the 1940s and 1950s these principles were so familiar that they seemed to be a part of the natural law or at least sanctioned by holy writ. This approach to management is called "traditional" or "classical" or, as Koontz calls it, the management process school.[4]

It perceives management as a process of getting things done by people who operate in organized groups. By analyzing the process, establishing a conceptual framework for it, and identifying the principles underlying the process, this approach builds a theory of management. It regards management as a process that is essentially the same whether in business, government, or any other enterprise, and which involves the same process, whether at the level of president or foreman in a given enterprise. It does, however, recognize that the environment of management differs widely between enterprises and levels. According to this school, management theory is seen as a way of summarizing the organizing experience so that practice can be improved.

With increased research success in psychology and sociology in the past twenty years, new emphasis was given to personal relations in management. At the same time, scientists and especially mathematicians thought that they could contribute a more or less precise mathematical model of the management function. All this effort has resulted in several new management theories which challenge the traditional view. Koontz has aptly called the resulting confusion the management-theory jungle. He defines other existing schools of thought as follows:[5]

The *empirical school*, a branch of the traditionalists, identifies management as a study of experience, sometimes with intent to draw generalizations but often merely as a means of transferring this experience to practitioners and students. It is founded on the premise that if we analyze the experience of successful managers or the mistakes made in management we shall somehow learn the application of the most effective kinds of management techniques.

The second distinct approach to management is concerned exclusively with human relations. Two individual schools of thought have been identified. The *human behavior school* is based on the central thesis that, since managing involves getting things done with and through people, the study of management must be centered on interpersonal relations. This approach is sometimes called the behavioral-sciences approach. This school concentrates on the people part of management. It rests on the principle that people work together in groups to accomplish objectives. Adherents of this school have a heavy orientation toward psychology and social psychology. Some focus attention on the manager as a leader and sometimes equate managership with leadership.

The *social systems school* originally developed by Chester Barnard[6] is closely related to the human behavior school; it looks on management as a system of cultural interrelationships. The theory examines the needs of the individual to overcome through cooperation the biological, physical, and social limitations of himself and his environment. From the total of cooperative systems he defined the "formal organization," which consisted of any cooperative system in which there are persons able to communicate with each other and willing to contribute action toward a conscious common purpose.

A third approach is concerned primarily with the decision-making process, whether it is accomplished intuitively or with mathematical assistance. The *decision theory school* concentrates on the selection of a course of action or of an idea from various possible alternatives (see Chapter 24). In its approach this school may deal with the decision itself, with persons or organizational groups who make the decision, or with an analysis of the decision process.[7,8]

The *mathematical school* are those theorists who see management as a system of mathematical models (Chapter 26) and processes.

There can be no doubt that all these schools have made a substantial contribution to the problem of management for massive engineering. Certainly as the ratio of highly educated engineers and scientists increases relative to the total project work force, interpersonal relations become more important. Also as the content of the product or service becomes more sophisticated, mathematical models of more complexity for the total process of the management function become more meaningful. The practicing manager of system projects, however, has looked in vain among these theories to find the specific solution for his problem.

36-3. The Project System

Program management makes system engineering productive. The program-management concept places together, under single-point direction, all the skills, specialties, talents, and resources required to execute a system mission. It originated with and implements the weapon-system concept of the military services, although there have been notable achievements with program management in commercial enterprises. Program management integrates the functional specialties of engineers, accountants, and administrators to the task of executing a completely integrated product or service. The approach is equally applicable to both government and industry and has been successfully used by both. The program or project (the words are often used interchangeably) is typically organized in a vertical structure by subtasks of the overall system. This contrasts with the functional specialty or horizontal organization in the traditional alignment. It is not unusual for an accountant to be working for an engineer or vice versa if the particular task requires such a combination of skills. Gaddis[9] was one of the first to describe to the business world the new functions and qualifications of project management.

The problem that the program manager and his government or industrial top managers face is that of running a business within a business. What are the criteria for abandoning traditional functional organization? How many programs of this kind can be established within a single corporation or government agency? How should the projects be organized? Should they be miniatures of the traditional organizations of the management process school or should a new blend incorporating concepts from behavioralists and management science be formally included? All these questions are just now becoming the subjects of intensive research. The program-management concept, born in the middle 1950s, is now adolescent. It is still not very sophisticated, but it can read and write and its elders are worried about its future. In its mature years program management promises to have a profound effect on our business economy.

One of the business-research organizations interested in the impact of the program-management approach on the future of American business has brought together practicing program managers and their superiors and asked them to discuss their experiences in the management of such programs as ICBM, aerodynamic missiles, fleet ballistic missiles, and the new space programs. The sum total of the remarks of these successful program executives has been brought together in proceedings[10] which provide an excellent description of the current state of the art in program-management practice. The researchers concluded "It was apparent from the discussion that many of the newer concepts most effective in organizing for goal accomplishment are coming from the advanced-technology industries. They are innovating with respect to managerial processes and organizational relationships and have developed approaches significantly different from traditional concepts."

The forcing function for these innovations is primarily the incredible acceleration of technology in the last 25 years. Discoveries of science are being reduced to practice in shorter and shorter times. The time lag between perception of a need and fulfillment in a sophisticated system is now approximately one-half of what it was a generation ago. In the advanced-technology industries the ratio of engineering to unskilled labor has increased to the point that research and development is now itself a big business.

The management innovations which result from system projects are still changing, are still evolving. No one can say in this age precisely how it must be or where it will lead. It is particularly difficult to express these innovations using the classical arrangement of titles within boxes. The boxes themselves tend to divide and diversify that which should be brought together. Goode and Machol[11] describe the most widely used organizations of system design. Even in the short time since publication of that work further evolution has taken place, and always in the direction of increased centralization of management function. In 1963, an important customer[12] for system projects expressed further concern about the inability of intermediate types of program organization to satisfy requirements. By intermediate we mean the superposition of interdisciplinary system-management objectives on classical functional organizations.

There is, however, a definite direction to this evolution. It is focused on the second of the three top levels of the program hierarchy. That there must be a hierarchical structure is not only acceptable to program managers but also seems self-evident from the inherent nature of systems themselves.

At the first level of program management is the chief executive, who may be called the program manager or director, or some equivalent name. Within the program his function is entirely analogous to that of the chief executive of the business; i.e., he is responsible first for master planning to satisfy needs of his customer and then for the selection and direction of his second-level managers.

At the second level there are two and only two management functions, system engineering and resources allocation. It is the function of system engineering to specify and direct the interdisciplinary design of system components of level three. The resources-allocation manager is responsible for the interrelations among the buyers' dollars, and the sellers' capital, profits, and human skills.

At the third level are the two functions of product design and administrative control. In very sophisticated systems the product at this third level may itself be a system requiring program management. The functions previously described would in that case be reproduced again at this level. It is not unusual in modern system ventures to have three or four tiers of program managers. The chief executive at one level is the product designer of the higher level.

The function of the system-engineering manager at the second level is to translate the complicated and interrelated requirements of the master plan into the familiar language of third-level designers. "Designers" is used here in a very broad sense, for early in the life of the program they may be mathematical theorists, later product engineers, and still later test pilots or astronauts.

In every system program there are four phases (compare Sec. 1-2). These phases are only generally time-sequential, and information feedback flows continuously along them. The four phases are:

Definition of Problem. This phase answers the question: what is the need of the customer which this system proposes to fulfill?

Analytical Solution. Through a process of synthesis, based on either mathematics or experience, description of the system which fulfills these needs and determination of the resources which must be allocated to obtain it.

Mechanization. The physical assembly and integration of system components using specifications from the analytical phase.

Verification. In the actual environment of the customer's world, tests of the system which establish that it does in fact fulfill the original need.

Clearly no individual or single group can perform these varying system-engineering management functions throughout the life of the program. It is one of the most important responsibilities of the program executive to anticipate the transition between his program phases and to adjust his second management level accordingly.

In small programs the system-engineering management is often provided by a single individual. In very large undertakings hundreds may be involved and even separate corporate entities established to fulfill this single requirement.

Of equal importance to system engineering and at the left hand of the program executive is the manager of resources. This function, which cuts across firmly entrenched and traditional departments of finance, accounting, contracts, and personnel, has been slower to develop than its engineering counterpart. The efficiency and economy of business data processing increasingly permit the resources manager to quantify and control the relationships among costs, business performance, and personnel skills. Variously called the business manager or program-control manager, this new program-management function may ultimately have the larger impact on future conduct of business. Johnson et al.[13] discuss this and many other implications of the application of system concepts in their book on the concept of managing by system.

36-4. PERT

The new business and management information systems, now in widespread use in the defense/space industries especially, arose from a need to solve project-management problems. Invented by a prominent management-consultant firm working under Navy sponsorship, the PERT (Program Evaluation and Review Technique) system was first used by the Navy's program organization established to manage the Polaris Fleet Ballistic Missile. PERT was introduced in 1958 when the Polaris program was in the late mechanization–early verification phases. The system was aggressively implemented by Navy management; within a year it was in successful use by all subordinate organizations in the program.

In early 1960, PERT was applied to another major Navy weapons system in the problem-definition phase. Intellectual activities as well as physical operations had to be scheduled and controlled. To the surprise of many managers and engineers the experiment was notably successful. Soon thereafter, the Navy system was adopted by other agencies of the Defense Department, NASA, and most of the principal defense contractors.

Research was initiated to extend the PERT time-only technique to cost and performance dimensions. In 1964, PERT/COST became a standard requirement for all major programs of the Defense Department and NASA.[14]

PERT, and all the variations of critical-path methods, is simply an organized way of scheduling and controlling any project. After the initial shock has worn off of first seeing a fully developed 100-event network, most professional engineers have a very positive reaction to using the system. Although there is considerable paper work, software, and machine time involved, most managers now believe that the use of PERT and PERT/COST results in a net saving to the program.

PERT is useful because it is fundamentally simple. There are three steps: (1) Develop the network of events (Fig. 36-1) from time now to the end objective of the program. (2) Estimate the time to accomplish each activity in the network. (3) Analyze, through simple calculations, the schedule problems in getting from here to there. Obviously PERT does not dictate how to solve problems which may be revealed by this analysis. This is still the manager's job.

In each one of these steps PERT has some rules which must be rather rigorously adhered to if the end result is to be meaningful. In step 1 it is vital that all events in the network meet these two rules. (1) An event is a specific accomplishment required to meet the end objective. (2) The event must be recognizable as occurring at a specific point in time. Thus an event could not be "design box X" but rather event 1 is "begin designing box X" and event 2 "complete design of box X."

An activity is the effort that links two successive events in a PERT network. Clearly an activity cannot be started until its initiating event has been accomplished. Also the terminal event linked by the activity cannot be completed until the activity is finished. Thus the design of box X is an activity.

Intellectual and physical actions, administrative or technological requirements, decision making, research, design, development, production, testing, etc., may be included as events. These are all specific accomplishments required to achieve the end objective.

The second step in the PERT approach is to estimate the time required for the activities. The validity of subsequent analysis is directly associated with the care with which this is done. Time estimates must be made by experts in the activities concerned. It is fundamental to the PERT approach that intermediate supervision,

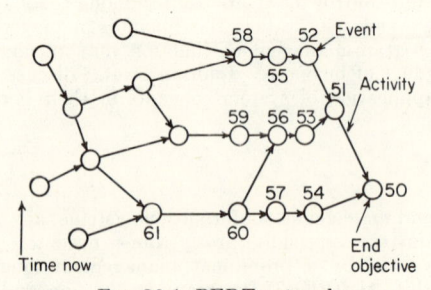

FIG. 36-1. PERT network.

much less top management, not make these estimates. The engineers or administrators who will be asked to do the job must make the estimates. Three time estimates are made for each activity under assumptions of optimistic, pessimistic, and most likely conditions. Notice that PERT recognizes the uncertainties involved in research and development operations. This is the primary reason for its widespread acceptance among professional engineers and scientists.

The conditions under which time is estimated are as follows:

1. *Optimistic* time. The first estimate is an optimistic one, in that it gives the best or shortest time. There is little chance of completing the activity in less than the optimistic time.

2. *Pessimistic* time. The pessimistic time estimate indicates the worst or longest time estimated for the completion of an activity. There is little chance that the activity would require longer than the time allotted by the pessimistic estimate.

3. *Most likely* time. The most likely time estimate represents the activity time, which, in the mind of the estimator, has the greatest probability of occurring. If many knowledgeable people were asked to estimate the most likely time for the completion of a given activity, the value given most often would form the most likely time estimate.

The engineer's three estimates of necessary time are portrayed on the top line of Fig. 36-2. An activity is started; its completion is estimated at some future time depending upon how fortuitously the program develops. Thus the points a, m, and b will correspond to the optimistic, likely, and pessimistic estimates of the engineer.

MANAGEMENT 36-7

The lower portion of Fig. 36-2 shows general characteristics of the probability distribution of these times. The point m is representative of the most probable time. It is assumed that there is relatively little chance that either the optimistic or pessimistic estimates will be realized. Hence small probabilities are associated with the points a and b. No assumption is made about the position of the point m relative to a and b. It is free to take any position between the two extremes—depending entirely upon the estimators' judgment.

Figure 36-3 illustrates the distribution of activity time estimates. Curve A shows a distribution skewed toward the most optimistic time, while curve D shows a distribution skewed toward the most pessimistic time. Curves B and C show a normal distribution of activity time, with curve C showing less "spread" than curve B.

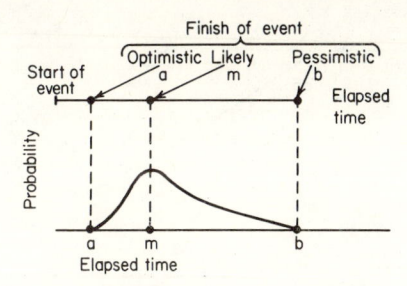

Fig. 36-2. Estimating the time distribution.

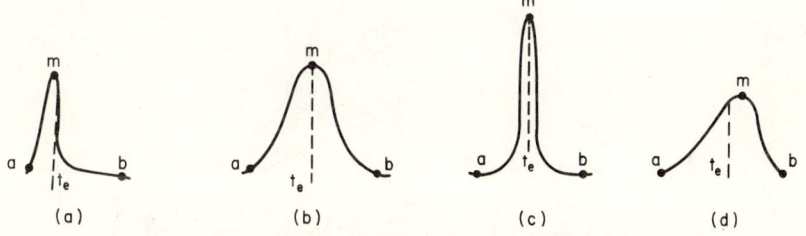

Fig. 36-3. Activity time estimates and mean activity time.

Mathematical and empirical evaluations based on extensive experience with PERT on the Polaris program have verified the validity of the following expressions:

$$t_e = \frac{a + 4m + b}{6} \qquad (36\text{-}1)$$

$$(\sigma_{t_e})^2 = \left(\frac{b - a}{6}\right)^2 \qquad (36\text{-}2)$$

where t_e = expected time to complete an activity

$$(\sigma_{t_e})^2 = \text{variance in } t_e \qquad (36\text{-}3)$$

The expected time is a formal statistical term that translates directly into layman's language. The variance in t_e is a measure of confidence in the expected time. If the variance is small, then the confidence in t_e is high; likewise, if the variance is large, then the confidence in t_e is low. Variance is related to the "spread" shown in curves B and C. In curve B, the confidence in t_e is low, whereas it is high in curve C.

We now have information with which to show the ordered sequence of events in a network from time now to the end objective. We can also show, for each event in

SYSTEM TECHNIQUES

the network, the immediately preceding events and the events which follow, together with the elapsed time estimates in terms of mean time or expected time between events and the variance for that time.

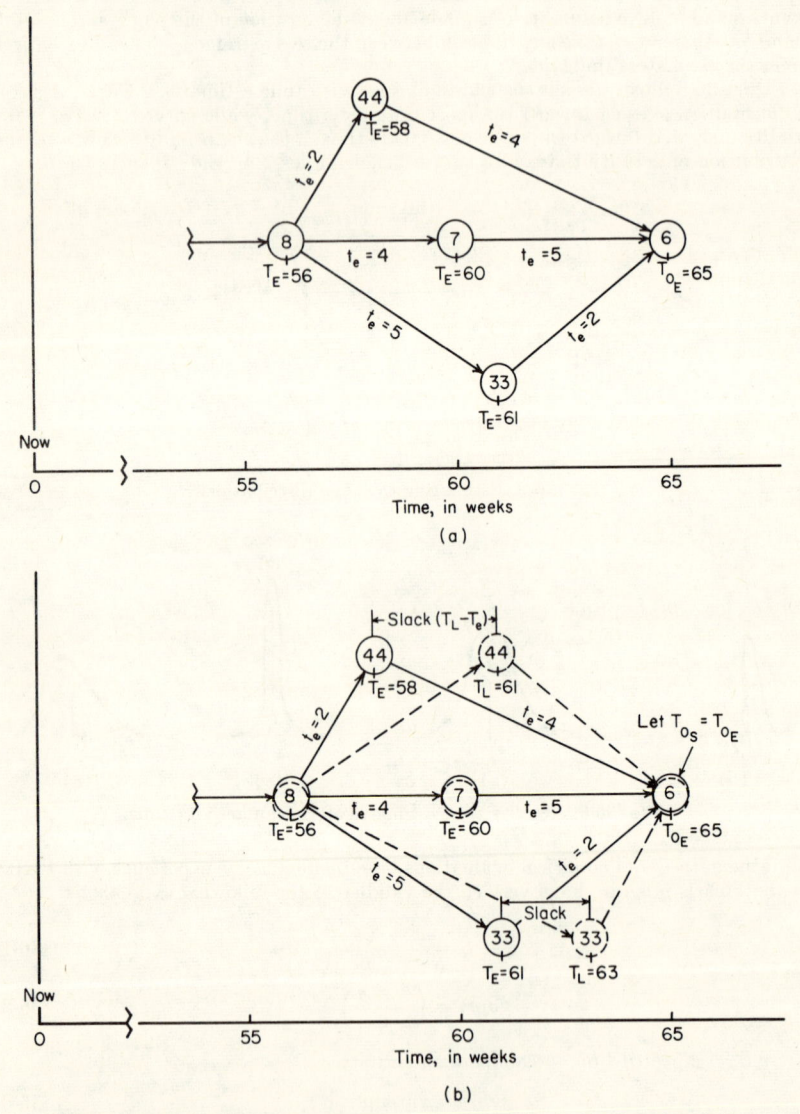

FIG. 36-4. Determination of slack. (a) Calculated t_e. (b) Calculated T_L.

The third step in PERT is to analyze the schedule problem. This is done first by determining the amount of slack associated with the various activities. Identification of slack leads directly to identification of the longest or "critical path" through the network of events between time now and the end objective.

To illustrate this process, let us examine a small portion of a PERT network (Fig. 36-4). Suppose that we are concerned with the completion of event 6, an end objective, starting with event 8, time now. To move from event 8 to event 6 requires three different activity paths, through events 7, 33, and 44. In part a of this figure, the symbols T_E under events 7, 8, 33, and 44 represent the earliest time for the completion of each event in terms of weeks from time now. The symbol T_{O_E} represents the earliest time for completion of the end objective, and t_e, of course, represents the expected time for completing each activity.

To compute the earliest time for each event, we start with its predecessor event and add the connecting t_e. For example, event 44 has as its predecessor event 8, with an earliest time of completion of 56 weeks. The t_e for the 8/44 activity is 2 weeks; therefore, the earliest time for completing event 44 is 56 weeks plus 2 weeks, or 58 weeks. Continuing in this fashion, it can be observed that, of the three paths between events 8 and 6, one will require 6 weeks, one will require 7 weeks, and one will require 9 weeks. Clearly, the 9-week path will determine the value of T_{O_E} of 56 weeks plus 9 weeks, or 65 weeks.

Now, in order to determine slack, we must establish the *latest* time as well as the earliest time for each event. The computation of latest time is suggested in part b of Fig. 36-4.

Recall that, in order to compute T_E (earliest time), we started at the left and worked to the right of the PERT chart, adding the t_e for each activity. The operations associated with computing the latest time T_L for each event are directly analogous, except that we start at the right, at the end objective event; we let the scheduled date for the end of objective be T_{OS}; then in order to find the slack associated with the network, we designate $T_{OS} = T_{O_E}$. Date for completion of the objective event is the scheduled date.

Next, to find T_L for any predecessor event, we subtract from T_{OS} the t_e associated with the particular activity. For example, t_e for activity 44/6 is 4 weeks, so that the latest time for event 44 is 65 weeks, less 4 weeks, or 61 weeks, and the slack time associated with the completion of event 44 is 61 weeks less 58 weeks, or 3 weeks. In a similar fashion the slack time associated with event 33 is 2 weeks.

PERT computer outputs for this small sample network would tell us where the critical or longest path is—the path in which we might have problems—and where there might be some slack that could be better used elsewhere for meeting objective 6 or other objectives of the program. The PERT system does not tell the decision maker what to do, but it does tell him where to look in order to take executive action, where his problem areas are, and where better use of resources might be made.

This example shows how slack exists in a system as a consequence of multiple path junctions which arise when two or more activities contribute to a third. Specifically, it is shown that the activities leading to events 44 and 33 have slack; i.e., events 44 and 33 could be scheduled anywhere within their slack range and still not disturb the expectation of timely accomplishment of the final event at week 65.

For example, suppose we wish to reduce the 9 weeks expected time for moving from event 8 to objective 6, or to increase the probability of completing objective 6 in 65 weeks. The program manager might first take a hard look at the activity leading to the completion of event 44 and raise the question, "Are we spending more resources than necessary to complete event 44 by a scheduled date that is set earlier than necessary?" We might find in this case, for example, that pulling 1,000 man-hours out of event 44 might result in a 2 weeks later completion of event 44 at no loss to the completion of end objective 6, and that better use could be made of these man-hours. The next question might be whether or not this 1,000 man-hours could buy time in the critical path, either to reduce the expected time or to increase the confidence level. If either of these choices were desirable, the resources could be transferred to activities in the critical path, or if not, the same resources could be better expended elsewhere in the program.

Another important output of the PERT system is that it enables comparison of best expectations for meeting objectives with schedule dates for meeting these same objectives. Further, PERT estimates the probability of meeting any objective in a network at any point in time including the schedule time. For example, in Fig. 36-5, event 50 might have been scheduled at time T_{OS}, where T_S = scheduled time for completion of single event and T_{OS} = scheduled time for completion of all events in a network. However, an earliest-time analysis could indicate that the event is expected to occur at time T_{O_E} with the variance T_e^2. The probability distribution of times for accomplishing an event can be closely approximated by the normal probability density. It is therefore possible to calculate the probability that event 50 will have occurred by time T_{OS} as represented in Fig. 36-5 by the shaded area under the curve. What does this example tell the program manager? First, it informs him that the chance of completing objective 50 in 82 weeks, the scheduled time, is only 1 out of 20, or approximately 5 per cent. He then has a choice of several actions; for example, he may slip the 82-week schedule to 92 weeks in order to have a higher confidence in completing the objective by a scheduled date; or he may decide to take action to shorten activity times along the critical path.

Note that, although the probability of meeting this date is unlikely, PERT informs the manager that this is what he may expect if he takes no action. PERT outputs provide realistic probabilities for the completion of scheduled events with which managers can reappraise a given schedule. A word of caution is appropriate here. The PERT "expected time" should not be used actually to schedule the project. It

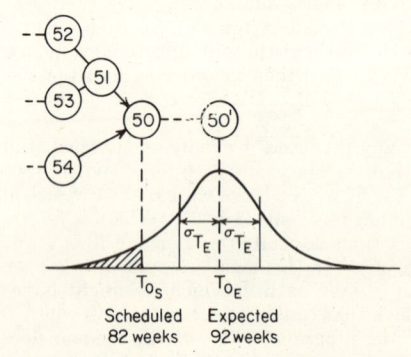

Fig. 36-5. Estimate of probability of meeting scheduled date T_{OS}.

is wrong theoretically; worse still, the project will find itself in the inexorable grip of Parkinson's first law.

Figure 36-6 illustrates the information which might be contained in a PERT computer print-out sheet. This sheet shows the ordered sequence of events in a network from time now to the end objective. It shows the expected times or earliest times by which each event might be completed, and the latest times working backward through the network by which each event must be completed in order to meet the expected time for the end objective. The next column shows the slack. A simple subtraction of the earliest time from the latest time portrays the degree of slack or criticalness. The events which have zero slack are on the "critical path," i.e., one path of effort in the network which is the longest. This chart also shows that two of the events in this network have 8 weeks of slack and two have 27 weeks of slack. It also shows the schedule time translated into weeks from time now. The last column presents probability estimates for meeting the scheduled dates of each event.

Recent advances in high-speed input-output devices for computers, including alphanumeric computer-operated displays, permit program controllers to make rapid assessment of changes. In Fig. 36-7, a change in time estimate is being inserted directly into a PERT computer program, with implications in slack and critical path immediately available on the display.

The PERT/COST system is simply an extension of the basic PERT time procedures described above. Both cost and schedule are planned on a common framework or structure. It is not necessary to develop a cost for each activity of the network, but only that the ultimate assemblage of cost elements bear a defined relationship to the PERT network.

The main stream of research in critical-path methods of management-information systems is toward total resource allocation. More than just the tangibles of time and dollars total resource allocation requires that the quality of technical performance

Event No.	T_E expected	T_L latest	Slack $T_L - T_E$	Schedule T_S	Probability of meeting schedule
50	92	92	0	82	0.05
51	85	85	0	77	0.09
54	74	82	8	73	0.42
52	47	74	27	70	0.99
53	70	70	0	60	0.04
55	35	62	27	55	0.99
56	60	60	0	50	0.02
57	56	64	8	55	0.42
.	.	.	.	.	.
.	.	.	.	.	.
.	.	.	.	.	.
X, now	0	.	.	.	.

Fig. 36-6. Outputs (time is shown in weeks from X or time "now").

Fig. 36-7. PERT network on computer-operated electronic display.

of the activity must be definable and measurable. When total resource allocation can be mechanized in PERT fashion, then the key function of the resource-allocation manager will be possible not only in program-management organizations but throughout American business.

Innovation is both inevitable and desirable in organization of the group and in responsibility of the individual. A manager will know better what he must do and his modern organization will better enable him to do it. More important than all these is the intangible spirit of the organization. Both the practicing manager and the management theorist agree that it is the spirit of an organization that ultimately determines the quality of the job.

REFERENCES

1. Gaddis, P. O.: The Age of Massive Engineering, *Harvard Business Review*, January–February, 1961, p. 138.
2. Taylor, F. W.: *Scientific Management*, Harper & Brothers, New York, 1911.
3. Drucker, Peter F.: *The Practice of Management*, Harper & Row, Publishers, Incorporated, New York, 1954.
4. Koontz, H.: Making Sense of Management Theory, *Harvard Business Review*, July–August, 1962, p. 25.
5. Koontz, H.: Making Sense of Management Theory, *Harvard Business Review*, July–August, 1962, pp. 26–36.
6. Barnard, C.: *The Functions of the Executive*, Harvard University Press, Cambridge, Mass., 1938.
7. Simon, H. A.: *The New Science of Management Decision*, Harper & Brothers, New York, 1960.
8. Luce, R. D., and H. Raiffa: *Games and Decisions*, John Wiley & Sons, Inc., New York, 1957.
9. Gaddis, P. O.: The Project Manager, *Harvard Business Review*, May–June, 1959.
10. Kast, F. E., and J. E. Rosenzweig: *Science, Technology, and Management*, McGraw-Hill Book Company, New York, 1963.
11. Goode, H. H., and Robert E. Machol: *System Engineering*, McGraw-Hill Book Company, New York, 1957.
12. USAF Systems Command: A Summary of Lessons Learned from Air Force Management Surveys, June 1, 1963, AFSCP 375-2, p. 9.
13. Johnson, R. A., F. E. Kast, and J. E. Rosenzweig: *The Theory and Management of Systems*, McGraw-Hill Book Company, New York, 1963.
14. DOD and NASA Guide, PERT COST, Office of the Secretary of Defense and NASA, June, 1962.

Chapter 37

RADIO-TELEMETRY SYSTEMS

CONRAD H. HOEPPNER, *President, Industrial Electronetics Corporation.*

CONTENTS

37-1. Introduction	37-1
37-2. Radio-telemetry Applications	37-3
Real-time Data	37-3
Postflight Data Analysis	37-3
Preflight Missile and Equipment Checkout	37-3
Down-range Data Gathering	37-4
37-3. Problems and Requirements of Telemetry Systems	37-4
37-4. Remote Measurement from Transducer to Analyst	37-6
Restrictions Placed on the Methods of Measurement by the Telemetry System	37-6
Choice of Transducers	37-7
Requirements for Telemetry Measurements	37-8
Telemetry Data Records	37-8
Data Reduction	37-8
Data Analysis	37-8
37-5. Systems of Telemetry	37-9
The Portel Telemetry System	37-9
Pulse-code-modulation Telemetry	37-13
The Standard FM/FM Telemetry System	37-18
The Pulse-duration-modulation/FM Telemetry System	37-21
Pulse-position-modulation/Amplitude-modulation Telemetry System	37-23
37-6. Recent Advances in Telemetry	37-24
Transistorized Circuits	37-24
Large Automatic Tracking Antennas	37-24
Phase-locked Frequency-modulation Discriminators	37-25
Solar Power Sources	37-25
Predetection Recording	37-25
Automatic Data-reduction Equipment and Processes	37-26

37-1. Introduction

The development of radio telemetry has been principally centered around the drone and missile programs of the Armed Forces. It began during World War II and has been growing and expanding to embrace aircraft testing as well as the testing of unmanned vehicles. Early drones were basically piloted aircraft with the pilots replaced by autopilots and remote-control equipment. During test phases a pilot was usually carried for take-offs and landings and to observe the results and to determine the deficiencies of the control equipment. As the drone was used as a weapon it was planned that the pilot would be removed and the equipment would function both automatically and by remote control. It soon became apparent that these missiles could be made smaller or of higher performance or more economically if in their initial design no provision was made for a pilot. One of these early missiles, designated the JB-2, was a copy of the German buzzbomb (V-1). Remote-control equipment from the drone programs was adapted to the JB-2. Telemetry was developed to measure the performance of the control equipment and the missile. With this evolution it is seen that remote-control equipment preceded telemetry by some years. It was basically an "on-off" system rather than permitting proportional control of each channel, whereas telemetry required proportional control. Furthermore, the remote control was an intermittent function, inasmuch as the vehicles were stabilized by internal automatic equipment. Telemetry, on the other hand, was a continuous function.

Many quantities needed measurement in the early missile programs. Measurements were made of the servo-control systems, the remote-control systems, and the

Radio telemeter to transmit vibration from the top wires of the Navy's 2,000,000 watt radio antenna. The telemeter acquires its power from the radio current in the antenna.

guidance systems. As missiles were developed, in-flight tests were made of the structures, aerodynamics, propulsion, inertial platforms and guidance, etc. These measurements were usually linear or shaft positions, motions, rotation, thrust, acceleration, temperature, and pressure, and the development of simple transducers for converting these quantities into electrical analogs followed.

Then, even as piloted aircraft became more and more complex, onboard recording was supplemented by radio telemetry which could be analyzed in the event a plane crashed, and also as an aid to the test pilot in checking out prototype aircraft. This kind of testing for aircraft has not yet been widely accepted; most airframe manufacturers still operate with pilot observations and onboard recordings for their initial flight tests. In 1958, the first radio-frequency allocation for aircraft flight testing was granted.

A forerunner of radio telemetry was the telemetry and supervisory control in the electric- and gas-utility transmission and distribution systems. Very few of these developments were borrowed for telemetry in the missile and aircraft programs. The utility measurements were made slowly, requiring a very narrow band of frequencies for intelligence. With wire connections for wire-borne carriers, there were no problems of fading, and many of the systems were applicable only to 100 per cent continuity links between transmitter and receiver. Only quantity changes, such as increases or decreases, were telemetered, and fades such as occur normally in radio systems would render the data useless. Transducers were large and weighty, made for durability and servicing. They were not considered expendable and were chosen largely with a view toward long life and reliability. Their response was slow and they were generally associated with the supervisory control systems. Instead of borrowing techniques from the utility field, the reverse trend has now appeared in which utility telemetering has borrowed from the techniques developed for radio telemetering.

It is the intent in this chapter to present a survey of radio-telemetry equipment used generally in aircraft, missiles, rockets, and satellites, and not point-to-point telemetry equipment tied together by wire lines. Only telemetry applications which are difficult or impossible to perform by means of wire lines will be considered here. In addition to the telemetry systems themselves, associated data-reduction and data-processing problems and systems are also covered. The systems treated herein are generally characterized by a large number of telemetry measurements transmitted over a single radio carrier. Specific examples from the simplest radio telemeter to a large versatile telemetry system are considered, and characteristics and parameters of the principal telemetry systems in present-day use are presented.

37-2. Radio-telemetry Applications

Real-time Data. In the early missile programs it was desired to obtain a visual indication of such things as missile speed and response to controls. Telemetry was also used extensively in preflight checkout of the missile, inasmuch as the measuring system was already installed and additional connections did not have to be made. Real-time data during flight test have largely disappeared during the last decade because of the rapidity of missile responses and the inability of an operator to analyze the data and take proper corrective measures. It is reappearing, however, with the advent of digital computers, and plans have been laid for using real-time telemetry to program tests of missiles through computers for utilizing a given test vehicle in a number of ways instead of the relatively few ways in a given test which have been customary. In the tactical use of missiles, however, the functioning of the guidance and the determination of hits on a target are important to determine whether another missile should be fired at the same target or another target. Telemetering will also be useful in the threat evaluation of targets when it is used in various reconnaissance functions. A principal real-time use of telemetry has been to determine the velocity of a ballistic missile by integrating the acceleration and thus determining the cut-off point. External instrumentation is also used for this purpose.

Postflight Data Analysis. External instrumentation by means of radar and theodolites can generally determine the position of a test vehicle as a function of time. This can be correlated with the internal telemetry instrumentation by differentiating the positions and integrating the acceleration. This is a basic check on telemetry performance. By far the largest use of telemetry is in postflight data analysis. Data taken by telemetry are used for (1) determining the performance of the vehicle, (2) gathering data for improving the design, (3) studying environmental conditions present in the vehicle, (4) determining causes of failure, (5) determining the time occurrence of events, and (6) various scientific measurements of the upper atmosphere and the space beyond the earth. Postflight data analysis takes many forms, as below. In general, the parameters being measured in the vehicle are recorded as a function of time, and external measurements such as position and velocity are also recorded as a function of time. The parameter time is eliminated and the internal function is plotted against position, velocity, or acceleration. Graphic plots are usually acceptable, but on occasions numeric prints of the data as well as preparation of data for entrance into large-scale digital computers is required. Correlation of functions, or plotting of one function against another, is also important.

Preflight Missile and Equipment Checkout. Telemetry is used extensively for the preflight checkout of a missile and its accompanying equipment. This function could be done by wire line and is done that way in many cases, but telemetry is already connected to many of the parameters which are to be measured in preflight checkout and is therefore a simple and economical expedient. Further, the telemetry may be received at a distance from the missile, and many of these preflight checkout tests may be used as calibrations for the operational telemetry stations. Checkout of missile functions by means of telemetry, of course, also checks out the telemetry equipment. Calibration of the entire measuring system may be accomplished. For example, if a pressure is being measured by the telemetry set, a simulated pressure may be applied to the transducer and read by a calibrated instrument. If this is done at many levels, both linearity and scale of the transducer and telemetry equipment are calibrated. This may be done for many channels of telemetry simultaneously and the calibrations recorded for subsequent use in data analysis. In many cases, telemetry equipment is operated from its own separate power supply and may be used in various missile tests without the complete missile system operating.

Down-range Data Gathering. In the test of a moving vehicle such as a missile or aircraft, a single receiving station is usually not adequate to receive data during the entire flight, particularly in the case of long-range low-altitude missiles. Consequently, additional receiving stations are required along the flight path. These are generally referred to as "down-range" receiving stations; they would not be necessary if all the data could be received near the launching site. These stations must be simple, reliable, and easily maintained, because they are usually located in remote and isolated places where it is difficult to keep qualified technical personnel. The data received at these stations must be electrically reproducible so that they may be transported back to the main data-processing center and used with the same equipment as used by the receiving station located at the launching site. This is usually done with magnetic-tape recorders.

Many problems are associated with down-range receiving stations. One problem is the difficulty of acquiring the telemetering signal. In order to maintain good receiving-station sensitivity, large antennas are used. These automatically track the transmitting signal once they are locked on, but they are rather highly directive and it is difficult to make the initial signal acquisition. Another problem is the transportation of the magnetic tape back to the main processing center, which causes considerable delay when transportation by air or even boat is required. In the near future, it is probable that the data will be transmitted by means of conventional wire or radio facilities. The problem of calibrating the station is also a serious one, since calibration cannot be made in conjunction with the transmitter while the missile is on the ground as is done with the station near the launching site. Special calibrating equipment must be employed, and calibration corrections must be carefully correlated with the data. Another problem is the overlap of reception from two or more stations, since at about the halfway point, the same data are received at two or more stations. A decision must then be made which data to use or whether to run a comparison. In the case of fading data, this may be a difficult problem, and very often the calibrations do not compare well enough for indiscriminate use of either station's data.

37-3. Problems and Requirements of Telemetry Systems

One of the first considerations in the design of a telemetry system is the wide variation of transducers for inputs which is encountered. There are both high-level and low-level voltages; both a-c and d-c from sources of both high and low internal impedance. Some transducers, such as thermocouples, generate their own voltage outputs, while other transducers, such as strain-gauge bridges, need excitation in order to provide a voltage output. It is convenient to measure some quantities by coupling them to a variable resistance, and others may be coupled to a variable reluctance. Encountered more and more frequently are outputs from both digital and analog computers which must be transmitted accurately and at the proper instant. Many scientific devices for exploring the upper atmosphere and the space near the earth provide inputs to the telemeter. These include ionization manometers (devices for measuring cosmic-ray intensity and ion density), mass spectrometer pickups for measuring the impact of micrometeorites, and instruments for measuring the intensity and direction of the earth's magnetic field. Many other quantities such as temperature, pressure, shaft position, vibration, and acceleration are converted to voltage in different ways for transmission via the telemeter. Typical telemetry systems may range from a few dozen of these measurements to several hundreds. All must be properly assembled and "conditioned" for introduction into a telemetry link.

It is a common procedure to group inputs in accordance with their voltage level and impedance, and multiplex similar inputs to a common subchannel. This is done either slowly or rapidly, as the expected data require. In the case of temperatures which vary rather slowly, only a few samples per second are required. In the case of vibration, several thousand samples per second may be transmitted to preserve amplitude, frequency, and phase of the detected signal. Similarly, when many measurements are made in a common location on a missile or test vehicle, it is convenient to group them for transmission to a single subchannel. In multistage rockets, measurements made during the operation of one stage may be transferred to similar measurements on another stage after the first stage is dropped. The levels of voltages, sampling rates or frequencies, accuracies, and numbers of channels are all tabulated and grouped, and then the telemetry system is assembled to provide the desired measurements.

Measurements which are collected from all points of a missile are brought to a single transmission center. This is done by means of connected wiring within the missile. Many other equipments are also operating which provide extraneous signals, constituting noise which may be picked up by these connecting wires. Extreme care must be exercised in assembling a telemetry wiring harness for shielding, proper grounding, etc. Care must also be exercised that the telemetry wiring does not radiate spurious signals to surrounding sensitive equipment. Finally, the resistance and capacitance of the wiring may have a definite effect on the output of the transducer, and must be balanced or otherwise compensated and calibrated.

Environmental conditions within the test vehicle during flight are not only severe but are rapidly changing. Temperatures may increase and decrease several hundred degrees. Accelerations may change by an order of magnitude and vibrations may appear and disappear in many and varied combinations. The effects of these conditions upon the telemetry equipment must be foreseen in the design and layout of the equipment in order to minimize the error which is produced. Often calibration is made during the flight of the missile to determine whether or not changes have occurred as a result of these conditions.

Calibration is perhaps the most valuable tool of the telemetry engineer. It is desirable to make preflight, in-flight, and postflight calibrations of each channel all the way from the input to the transducer to the recorded data, but this is seldom possible, and the engineer must resort to substitutes. Certainly in most cases postflight calibration is impossible because the missile has been destroyed. Preflight calibration is often practicable on only a portion of the telemetry system. For example, instead of subjecting the missile element to the ranges of strains which are expected to be encountered, the strain gauge is replaced by a dummy bridge or several dummy bridges to calibrate the remaining portion of the system. It is the rule rather than the exception that the transducers are calibrated separately and their calibration added to that of the connected system. In-flight calibration of course has the difficulty that the measurement may be interrupted at a critical point during the calibration period. Calibration methods have been devised which do not interrupt measurements, but they are limited in range to prevent the overloading of the channel.

The capacity of a telemeter for handling information usually has a fixed maximum determined by the bandwidth of the radio link employed (see Chapter 22). This maximum may not be entirely utilized in a particular test; on the other hand, it may be insufficient for a particular test, and two or more telemetry links are then employed. In most systems there is provision for flexible interchange between the number of channels on one hand and the number of samples per second per channel or the bandwidth of any particular channel on the other. In more recent telemetry systems such as the pulse-code-modulation system (Sec. 12-3), usually there is a provision for the interchange of the number of channels, the number of samples per second per channel, and the accuracy to which each quantity is measured. The requirements for number of channels, frequency response, and accuracy vary widely from program to program. They also vary as a single program progresses from research through development tests and into final use. In the early stages of an upper-atmospheric research or missile research, a large number of accurate measurements of high-frequency response are usually required. Then as the data become better and better known, these measurements are replaced by monitoring-type measurements to determine failures and the time of occurrence of events. However, when the end product is scientific exploration, the measurements may increase in complexity and accuracy as the program progresses. It is well to provide at the very inception of an experimental program a flexible telemetry system which may expand and contract with the changing need.

After the data have been taken and recorded there are many different uses to which it is put. Surprisingly enough, much the greater portion of the data taken in a guided-missiles program is discarded after an initial look without having further usefulness. Methods and procedures for discarding data are extremely important to the efficient use of a telemetry system. Data are discarded because (1) the measurements are as they were expected, (2) they were taken at a much higher rate than required, (3) the transducer or channel became inoperative, (4) the vehicle did not perform properly, or (5) the data expected were not detected. After the worthless data have been discarded, graphic plots, numeric prints, and computations are made, and the results are usually provided in forms permitting multiple distribution of reports.

Other miscellaneous telemetry problems are associated with the radio link from the moving vehicle to the fixed ground receiving station. The strength of the radio signal of course varies with the distance between transmitter and receiver, and there may be such additional complications as multiple propagation paths. Antennas at both the transmitter and receiver are to some degree directive, and the problem of tracking the missile with the ground receiving antenna and eliminating nulls in the missile transmitting antenna is always present. Polarization of signals may vary with the missile's tumbling and rolling such that a single ground receiving antenna may not provide proper output to the receiver even though the signal strength at the location is adequate. In the past, telemetry data have been characterized by frequent fades of relatively short duration often producing data omissions at times when the data are most required. Much progress has been made in recent years to eliminate this unsatisfactory condition.

It must also be remembered that the weight carried aloft by the missile or satellite is of utmost importance. Therefore, small size and weight of the telemeter are vital; even more important is the power requirement of the telemeter. Until the time of solar cells it was necessary to take aboard the missile the entire primary supply of power for operating the telemeter. Consequently this was minimized to provide just the life required in a particular flight. Constant research has increased the watthours per pound of battery weight considerably, but still the telemeter must be the main factor of battery economy.

37-4. Remote Measurement from Transducer To Analyst

Restrictions Placed on the Methods of Measurement by the Telemetry System. In the preceding paragraphs, a number of the requirements placed upon the telemetry system by the transducers and quantities being measured were described. Unfortunately, the development of telemetry has not been such as to satisfy all the requirements; consequently, a compromise is required between telemetry capabilities and the requirements of measurement. The shortcomings and limitations of the telemetry system place restrictions upon measurements above and beyond those which are encountered in the laboratory when the telemeter is not employed. In the first place, an electrical output from the measuring device is required in order that the intelligence may be placed upon a radio link. Consequently, appropriate transducers are necessary. Also, the telemetry system may not be perfectly stable down to zero frequency, and transducers and methods of measurement must be chosen to minimize the effects of drift.

Overmodulating the subcarrier or the time-division multiplexer (see Sec. 12-3) may also have serious consequences in its effect upon other measurements as well as providing inaccuracies in the offending channel. Provision must be made to prevent overmodulating. If various measuring devices are switched, the switching transients must be minimized or the accuracy of the telemetry system may be impaired. When mechanical commutators or time multiplexers are employed, the measurement of the time of occurrence of an event (such as the reception of a cosmic particle) is made more difficult, and the time ambiguity of the multiplexed system is a serious limitation.

The measurement of a large number of parameters requires extensive and bulky equipment unless similar transducers may be grouped together to minimize the signal conditioning required. This fact generally dictates a relatively standard transducer rather than an optimum for each particular measurement.

The bandwidth of the measurement or the frequency with which the measured quantity changes is also seriously limited by the telemeter. In the FM/FM telemeter, the permissible bandwidth varies from a relatively low value on the lower subcarriers to a reasonably high value (2,300 cycles) on the upper subcarrier. The bandwidth of the measurement must not exceed these limitations lest sidebands be generated in adjacent channels, thereby reducing the accuracy of other measurements. In a time-multiplexed system, the problems of folded data are present when the rate of change is faster in cycles than half the sampling rate. Thus it is not known whether the quantity has reversed itself several times between samples or if there has been no reversal at all. It is generally considered desirable to limit the bandwidth of the data such that this ambiguity is not present. However, with refined techniques of analysis, this is not a rigid requirement.

The form which the end-product data take is also a limitation on measurement. In general, time-history plots of the measured quantity are desirable. The speed at which the recording medium moves is very often a limitation. If sampling is not regular, demultiplexing difficulties are magnified. A calibration is usually measured from the plot and is applied to the data when they are replotted. Calibration corrections selected by other means may be applied before plotting or printing data in many cases. Acquiring the calibration curves in the first place is often a difficult procedure. The transducers are calibrated before they are installed but the remaining system must be calibrated by substitution. This requires accessibility and substitute transducers or signals which may be applied. Consequently, the choice of measuring equipment is often limited by the ability to calibrate it.

Choice of Transducers. Within the limitations outlined above, the transducers must be chosen to match the particular telemeter employed. A variable-reluctance transducer may be quite applicable to an FM/FM telemetry system but may be very difficult to use with a time-multiplex system. Transducers are chosen to match levels and impedances in analog style and must often be interleaved with digital transducers such as shaft encoders and output from digital computers. These choices must be made in a systematic manner to maximize the utility of the telemeter. The accuracy of the transducer may in many cases be limited by the telemeter. When this is the case, it is sometimes possible to use several transducers and spread the range of measurement over several telemetry channels. This is done in a manner similar to the display of a watthour meter, in which the reading of each dial is transmitted over a separate channel. Very often separate measurements of the same quantity are made by separate transducers and separate telemetry channels to detect errors in measurement. This redundant measurement has unfortunately been used too little in radio telemetry and the telemetry system trusted too greatly.

Transducers must also be chosen to measure the desired quantity without measuring other parameters to which they are subjected. In other words, a pressure transducer should measure pressure and not measure the effects of temperature. The two principal offenders are temperature and acceleration; transducers must be carefully chosen to be free of temperature and acceleration effects (unless they are used to measure those quantities). Other parameters which affect transducers to a lesser degree are humidity, aging, and the effect of vibration on hysteresis of the transducer. Tests of all these quantities can be made before the transducer is mounted in the vehicle, and from the results of these tests the proper transducer can be chosen. Also, transducers should be chosen in groups with the same kind of measurement handled in the same manner, thereby simplifying equipment. Each transducer of a group may be subjected to different environmental conditions; therefore, it is not usually possible to have a single "dummy transducer" detect the effects of temperature, acceleration, etc., on the other live transducers being used for measurement.

Requirements for Telemetry Measurements. The principal applications of telemetry are:

1. Scientific exploration of the upper atmosphere and base surrounding the earth
2. Gathering test data on experimental vehicles
3. Gathering operational data from missiles employed in tactical use

Requirements for measurements may be divided into three major classes as follows:

1. The number of channels of measurement required
2. The rate of change (frequency response) of the measured parameter in each channel
3. The accuracy of the data to be measured in each channel

Telemetry Data Records. In a telemetry system, it is nearly always necessary to change the form of the quantity being measured; for example, temperature is reproduced not as a temperature but rather as a number or distance on a graphic plot. The measurement of position is one which is most nearly reproduced in its original form. Principal forms of presentation for telemetry data are:

1. Visually read indicators or dials
2. Oscilloscope presentations for visual inspection
3. Time-history plots which may be pen records, hot-wire-recorder records, oscillographic records, or multistylus-plotter records
4. XY plots in which one function is plotted against another on a permanent recording material
5. Numeric printing in which a number is printed to correspond to the units of the value being measured

Data Reduction. An excessive sampling rate may generate more data than are required for analysis or display. These can be compressed by a reduced plotter or by a sampling technique which chooses only occasional data points among the large number which are recorded.

In the event that many channels are monitored for malfunctions and the data received are exactly as expected, a simple inspection and analysis will reveal this fact; no data need then be displayed or recorded, other than the simple statement that no malfunctions occurred.

The employment of smoothing techniques may also reduce the amount of data for analysis and display. Vibration which may occur on an acceleration channel may be smoothed by a low-pass filter or by subsequent computation. In this process, the statistical averaging of a large number of samples not only reduces the amount of data which need to be analyzed but may also increase the accuracy of the data which have been smoothed. The computer is very often employed, and a least-squares method of curve determination is then usually utilized.

Data Analysis. Once a minimum amount of data has been selected and recorded, it is necessary that it be analyzed to determine the parameters which were originally sought. Other computations, in addition to those mentioned above (such as smoothing), include the determination of a function requiring several measurements such as Mach number or vector acceleration. In the latter, three axes of acceleration are measured and the resultant computed. It is also often desirable to plot one function against another. For this purpose, two or more functions are introduced to the computer, and interpolations and time corrections are made for the recorded data points such that an output function (XY plot) may be generated. The process of data computation by large automatic digital computers is beyond the scope of this chapter.

Once the data have been reduced and/or computed, it is necessary that they be printed and duplicated. Printers are required to operate at very high speed to record all the data that are taken. In fact, one telemetry system operates in excess of 25,000 data points per second. If the data are recorded electrically, speed may be reduced to match conventional plotters. However, high-speed plotters, such as the Radiation, Inc., multistylus recorder, are becoming available for real-time data operation.

The recorded data must be protected against degradation in accuracy due to such factors as dimensional instability of the paper, which may expand or shrink with aging or changes in humidity. This is done by printing the measuring scales on the paper at the same time the function is laid down. It is also desirable that multiple copies of data be made. Copying can be done by appropriate photographic or xerographic processes, usually leading to the production of a master for photo-offset printing by means of which the data may be duplicated manyfold for wide distribution.

Once the data have been duplicated and are available for detailed inspection and analysis, two major agencies are interested in that analysis. One is the contractor who is responsible for the development of the missile or the scientific exploration, and the other is the customer, usually a government agency which desires to monitor the progress of the development it is sponsoring. The contractor is interested in determining the parameters of his vehicle to permit development of a vehicle which fulfills its mission. He is also interested in locating defects so that they may be corrected. And, of course, by means of the measured data he can make improvements as required to fulfill his project. The customer desires to inspect data in order that he may monitor the progress through an independent observation rather than take the contractor's analysis. The customer also supplies the test site and facilities for the test, and he must make the determination as to whether or not these are adequate and useful.

37-5. Systems of Telemetry

In this chapter there will be described in some detail two systems of telemetry, and the parameters of three additional systems which are in general use at missile test ranges. Photographs of complete systems and portions of the systems are also included. Of the two systems to be described in detail, one is a simple, single-channel system, and the other is an advanced, comprehensive, versatile multichannel system.

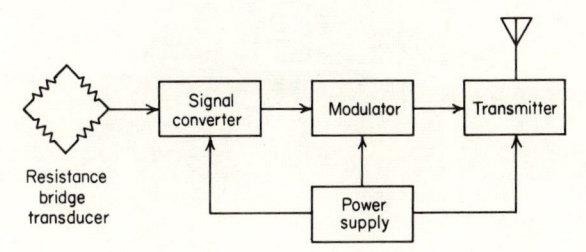

Fig. 37-1. Portel transmitter, block diagram. (*Courtesy of Industrial Electronetics Corporation.*)

These represent the two extremes of telemetry practice. The three other systems described, which are in general use, fall between these extremes.

The Portel Telemetry System. Portel (portable telemetry equipment) was designed as a minimum-volume, economical telemeter for transmission over relatively short distances. The output from this system, normally associated with a resistance-bridge transducer, is usually recorded by a moving-pen recorder. Figures 37-1 and 37-2 are block diagrams of the transmitting and receiving portions, respectively. Figure 37-3 is a photograph of the transmitter and Fig. 37-4 is a photograph of the

37–10　　　　　　　　SYSTEM TECHNIQUES

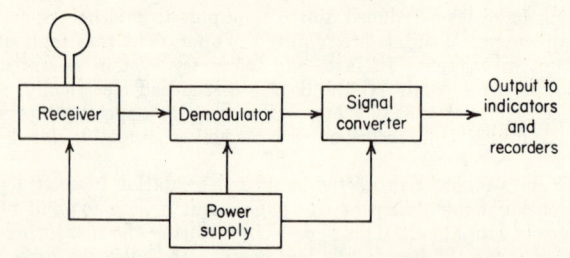

Fig. 37-2. Portel receiver, block diagram.　　(*Courtesy of Industrial Electronetics Corporation.*)

Fig. 37-3. Portel transmitter, photograph.　　(*Courtesy of Industrial Electronetics Corporation.*)

Fig. 37-4. Portel receiving station, photograph.　　(*Courtesy of Industrial Electronetics Corporation.*)

receiving station. Figure 37-5 shows various other forms of transmitters for special applications. This system is of the FM/FM type; i.e., it employs a frequency-modulated subcarrier which, in turn, frequency-modulates a radio carrier. The unbalanced output of the resistance bridge changes the frequency of the subcarrier. A single battery supplies power for the various components. At the receiving station, a radio receiver amplifies the radio-frequency signal and demodulates it. The subcarrier is then fed to a signal converter which provides an electrical output proportional to the unbalance of the resistance bridge. An appropriate power supply here also supplies energy for these functions.

The transmitting unit is designed to be small, light, and efficient, while the receiving station is of conventional rack and panel construction for mounting in a mobile trailer. As may be noted in the illustration, one end cap of the transmitter corona shield is insulated from the remainder of the case in order that it may act as an antenna. The entire transmitter is constructed of solid-state components for a maximum of reliability and a minimum of weight, volume, and especially power. With the battery

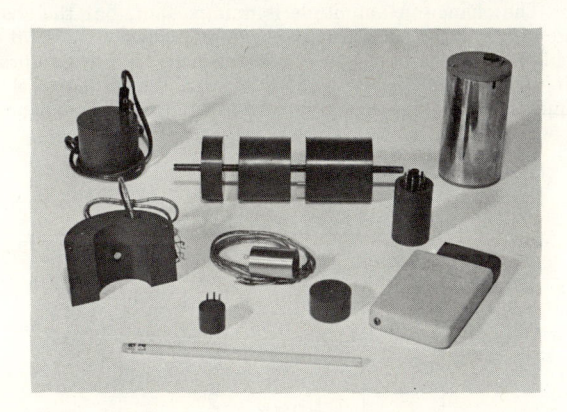

Fig. 37-5. Miniature telemetry transmitters, photograph. (*Courtesy of Industrial Electronetics Corporation.*)

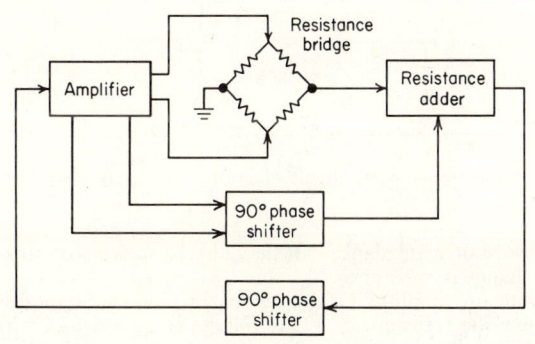

Fig. 37-6. Phase-shift oscillator, block diagram. (*Courtesy of Industrial Electronetics Corporation.*)

supply shown, approximately 200 hr life may be expected. The subcarrier frequency of this particular unit is 2,500 cycles ±15 per cent. The sensitivity is such that a change in resistance of one of the arms of 1 part in 10^4 gives a change of 10 cycles in the subcarrier frequency. This corresponds to the measurement of strain of 25 μin./in. in a 120-ohm gauge with a gauge factor of 2. The zero and full scale are determined at the receiving station upon initial installation, and it is not necessary that the bridge be balanced.

Figure 37-6 is a block diagram of the subcarrier oscillator portion of the transmitter. Excitation is supplied to the bridge from an amplifier, and the output of the bridge is mixed through a resistance-adding circuit with a signal from the amplifier that is shifted 90° from the bridge excitation. Thus the combined signal out of the resistance adder is a sine wave which varies in phase from 0 to 90° from the bridge excitation. This combined voltage is further shifted 90° in a phase shifter such that a total of 180° phase shift is supplied to the input of the amplifier. The amplifier contains an odd number of stages; hence the total shift of the circuit is 360° and the gain is sufficient to cause oscillations. The gain must be carefully controlled and stabilized to prevent oscillations at frequencies other than the design center. This type of oscillator is conventionally referred to as a phase-shift oscillator. Now when the resistance bridge is unbalanced, both the amplitude and phase of the output of the resistance adder change. The change in amplitude is unimportant, but the change in phase causes a change of frequency such that at the new frequency the total phase shift is 180°. It is readily noted that a change of resistance in one direction produces a leading phase shift while an unbalance in the other direction produces a lagging phase shift. Consequently, the frequency is increased or decreased depending upon the

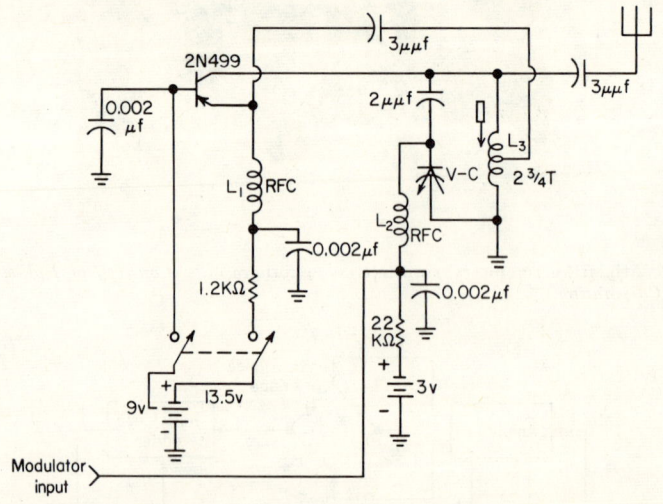

Fig. 37-7. Portel radio transmitter, circuit diagram. (*Courtesy of Industrial Electronetics Corporation.*)

direction of the resistance unbalance. It may also be shown that this shift is a linear function of the change in resistance.

An output from an amplitude-stabilized portion of the amplifier is fed to the modulator of the radio transmitter. This voltage is applied to a varicap which is a portion of the transmitter's tank circuit, changing the capacitance, which in turn changes the frequency at which the oscillator oscillates. Output is coupled from the tank coil through a very small capacitor to the insulated end cap which is the antenna of the transmitter. Linearity of frequency modulation here is also good for small variations of frequencies. The radio oscillator is tunable from 88 to 108 Mc. By simple tank-circuit substitution this frequency may be placed anywhere in the range from 1 to 200 Mc with approximately 20 per cent tuning. Figure 37-7 is a circuit diagram of the transmitter.

The first version of the receiver was designed to be operated in the presence of very high radio-noise fields. Consequently, a shielded-loop antenna was used for input and the entire system was shielded to prevent noise from reaching the intermediate-frequency amplifier. A conventional FM tuner and ratio detector are used to reproduce the subcarrier. Satisfactory operation has been obtained, with the common mode field at the transducer being a factor of 10^8 greater than the signal being measured.

RADIO-TELEMETRY SYSTEMS

This factor can undoubtedly be much greater and is limited only by the shielding and isolation of the radio receiver. The subcarrier output from the receiver is carefully clipped and limited and fed to a phase-locked loop detector. The phase-lock loop was chosen because of its simplicity and its versatility. Also, by adjustment of the feedback filter in the loop (transfer function) the change in output with frequency may be varied to compensate for subsequent nonflat equipment. Thus, in a very simple manner, a rising or falling amplitude vs. frequency characteristic may be

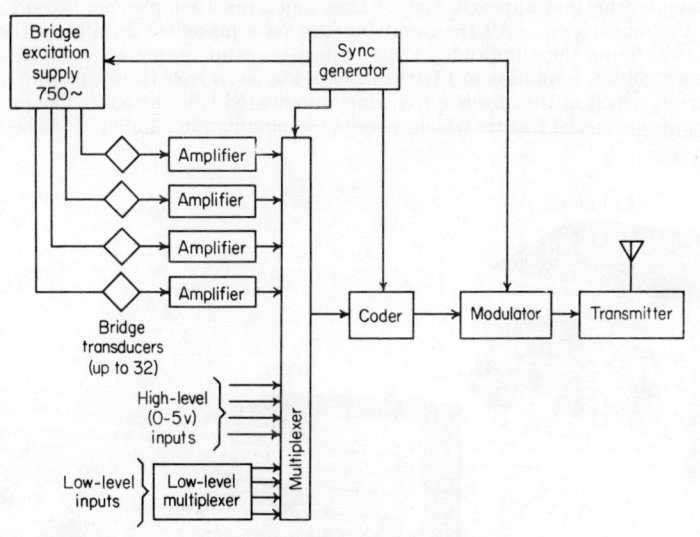

Fig. 37-8. AKT-14, block diagram.

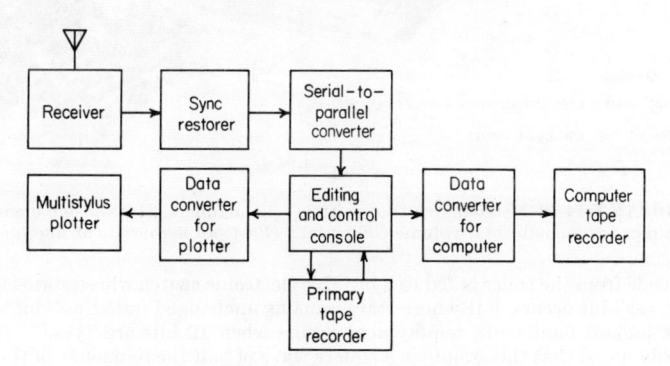

Fig. 37-9. UKR-7, block diagram.

obtained. Furthermore, full-scale voltage may be easily varied and a zero voltage determined quite independently of the bridge balance.

Pulse-code-modulation Telemetry. At the other end of the telemetry-equipment scale lies the AKT-14 UKR-7, pulse-code-modulation telemetry system. This system was developed by Radiation, Inc., under contract from the U.S. Air Force. The salient features of the system are its inherent high accuracy and stability, its large information-handling capacity, its freedom from adjustment and calibration, and its ability to provide data in a form which may be directly assimilated by large general-purpose digital computers. Figures 37-8 and 37-9 are block diagrams of this equipment and Figs. 37-10 and 37-11 are photographs. The AKT-14 equipment is capable of working with a myriad of different transducers. It supplies excitation for resistance bridges and feeds the bridge output into a time multiplexer. In addition, it can handle both high-level and low-level voltage outputs from transducers. Digital transducers may be simply multiplexed into the system following the coder. An a-c supply is used for bridge excitation. This voltage is amplified and sampled by the multiplexer. The samples are extremely short and correspond to the peak of the sine-wave output from the transducer. Each transducer is sampled sequentially, with sample length of approximately 4 μsec, and 24,000 samples per second may be fed to the multiplexer. All the input functions are presented serially to the coder, which transforms the amplitude of the voltage to 8-bit binary code. This process provides a voltage resolution of 1 part in 256. The serial code is then presented to the modulator, which in turn feeds a frequency-modulated transmitter. The modulator incorporates a special feature which permits the modulation bandwidth to be halved.

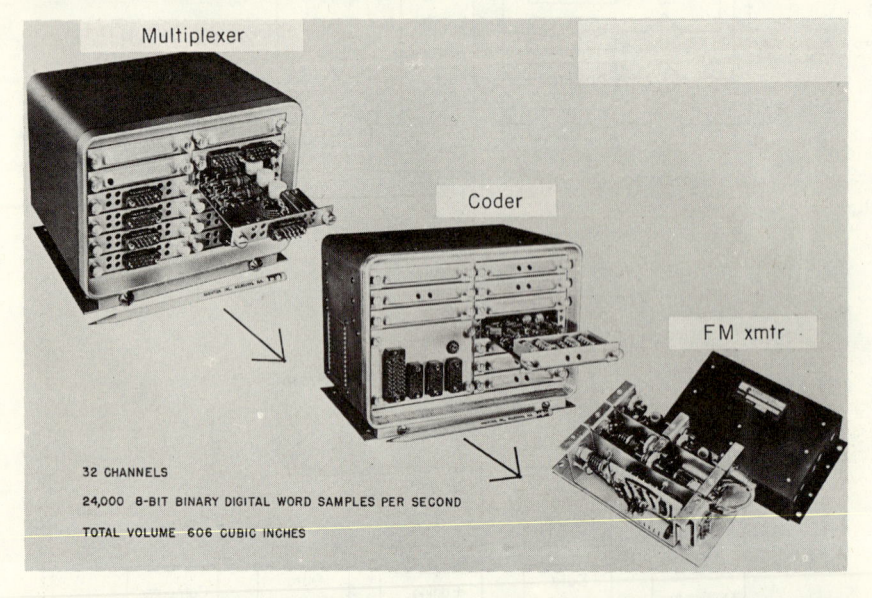

Fig. 37-10. AKT-14 PCM telemetry transmitter. 32 channels, 24,000 8-bit binary digital word samples per second, total volume 606 cu in. (*Photograph courtesy of Radiation, Inc.*)

Output code from the coder is fed to a bistable electronic switch whose state is changed when a "yes" bit occurs, but whose state remains unchanged for a "no" bit. In this case, the highest bandwidth requirement occurs when all bits are "yes." However, it is readily noted that this results in a square wave of half the frequency of the original square wave.

The UKR-7 receiving system receives the transmitted binary-coded signal and feeds it first to a synch-restoring unit. This unit acts as a flywheel during fades in the radio signal to maintain proper functions for the immediate restoration of data at the conclusion of a fade. It also provides the signals necessary to transform the serially transmitted data word to a parallel word. This permits simple recording on multi-track tape recorders and simple conversion operations for plotting and converting data for inputs to computers. If a wideband (video) tape recorder were employed, it could record the output or even the predetector signal from the receiver. However, these recorders are new, expensive, and complex. Consequently, this particular system does not utilize them.

RADIO-TELEMETRY SYSTEMS   37–15

The editing and control console controls the distribution and processing of data. In general practice, all the data are recorded on the primary tape recorder and plotted for quick-look purposes. Particular chosen portions of data are then reproduced from the primary tape recorder for further accurate plotting or conversion to computer inputs. The choice of data to be further processed, once made, is simply transmitted as commands by the editing and control console to provide the data in plotted form or grouped as necessary for further computation in a digital computer. Typical data plots are shown in Figs. 37-12, 37-13, and 37-14.

Fig. 37-11. UKR-7 PCM telemetry receiving station, photograph. (*Courtesy of Radiation, Inc.*)

The parameters of the equipment shown in Figs. 37-10 and 37-11 are as follows:

1. Number of channels, 32
2. Number of samples per second per channel, 750
3. Number of bits per sample, 8
4. Number of bits between samples, 1
5. Total sampling rate, 24,000 samples/sec
6. Total bit rate, 216,000 bits/sec

The subsequent PCM/FM telemetry system has also been developed with the following parameters:

1. Number of channels, 32
2. Number of samples per second per channel, 750
3. Number of bits per sample, 10
4. Number of bits between samples, 1
5. Total sampling rate, 24,000 samples/sec
6. Total bit rate, 264,000/sec

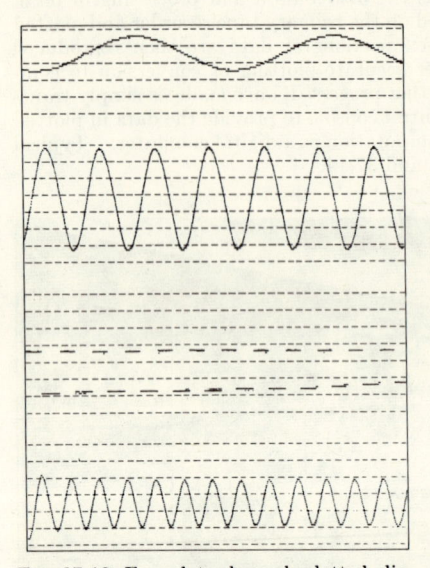

FIG. 37-12. Four data channels plotted, diagram. (*Courtesy of Radiation, Inc.*)

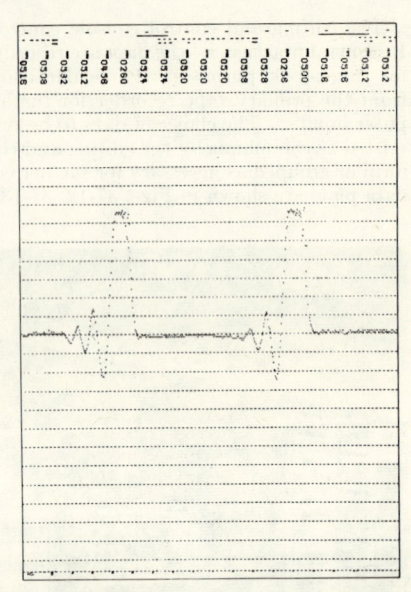

FIG. 37-13. A single data channel plotted and printed. (*Courtesy of Radiation, Inc.*)

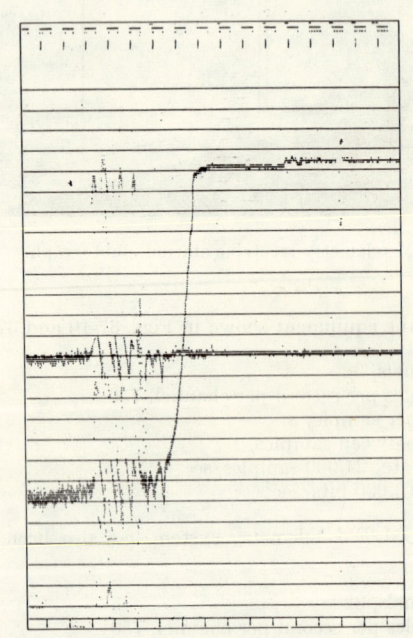

FIG. 37-14. Two data channels plotted. (*Courtesy of Radiation, Inc.*)

RADIO-TELEMETRY SYSTEMS 37-17

Photographs of this equipment are shown in Figs. 37-15 and 37-16. This telemeter was developed for the Holloman Air Force Base and is in use on the sled for the acceleration testing of missile components.

The foregoing descriptions are brief accounts of the lowest to the highest capacity of the telemeters. Also, they are examples of frequency multiplexing and time multiplexing. The use of the subcarrier permits the use of other subcarriers at different

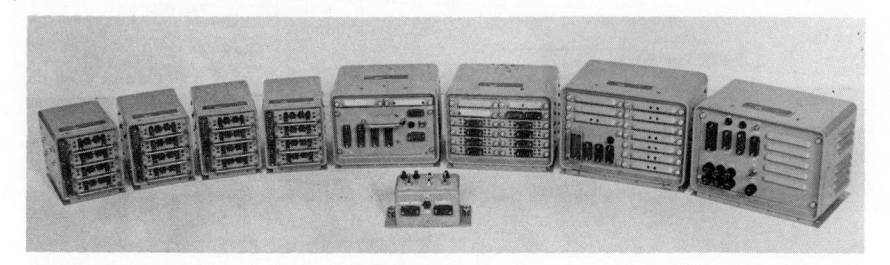

Fig. 37-15. Pulse-code-modulation transmitter, photograph. (*Courtesy of Radiation, Inc.*)

Fig. 37-16. Pulse-code-modulation receiver, photograph. (*Courtesy of Radiation, Inc.*)

frequencies, hence the name frequency multiplexing. Both have their particular best uses, advantages, and disadvantages. Of the systems in use at missile test ranges, one is a frequency-multiplexed system and two are time-multiplexed systems. To the frequency-multiplexed system is usually added, on one or more channels, a time-multiplexed input. This takes the form of a mechanical switch which commutates various inputs to a single channel.

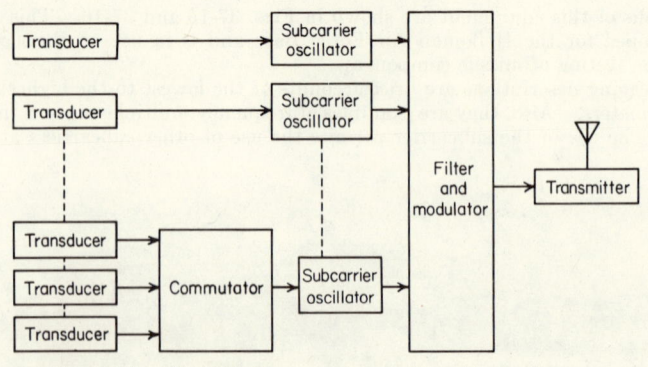

FIG. 37-17. FM/FM telemetry transmitter, block diagram.

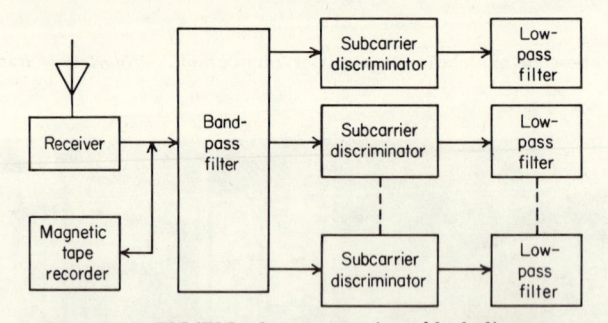

FIG. 37-18. FM/FM telemetry receiver, block diagram.

FIG. 37-19. FM/FM telemetry transmitter for Titan, photograph. (*Courtesy of Electro-mechanical Research.*)

The Standard FM/FM Telemetry System. In Figs. 37-17 and 37-18 are shown block diagrams of the transmitter and receiver sections of the FM/FM telemetry system. Figures 37-19 and 37-20 are photographs of typical telemetry systems. Since there are many varieties of equipment available, no attempt has been made to show a single complete system. As a matter of fact, the use of a single integrated system from a single manufacturer is seldom found in practice; usually a user chooses components from a number of manufacturers and assembles his own system. The system has been in common use for approximately 15 years, and various parameters were "standardized" by members of the Research and Development Board and subsequently by the Inter-Range Instrumentation Group. These standards are presented in Table 37-1.

In the process of recording the subcarriers on magnetic tape at a receiving station, provision may also be made to record a tape-speed-control tone and tape-speed-error-compensation signals, as specified in IRIG standards.

A wide variety of transducers may be utilized with the FM/FM system. Voltage- and current-generating transducers, resistance bridges, and even variable-reluctance transducers may be employed. The latter is generally used as a part of the tank

FIG. 37-20. FM/FM receiving station.

circuit of the subcarrier oscillator. Many forms of subcarrier oscillators are available—voltage controlled, current controlled, resistance bridge, and those using a variable reluctance. Vacuum-tube and transistor models of all types and descriptions are available. They are usually made with output-level adjustments such that a properly emphasized composite signal may be presented to the radio modulator. In choosing a subcarrier oscillator, however, care must be exercised to ascertain that it is available in the particular frequency range desired. Many oscillators are available only for a portion of the standard range.

It will be noted from the block diagram (Fig. 37-17) that the output from the subcarrier oscillators is fed through filters before it reaches the radio modulator. These are not absolutely necessary but are highly desirable in order to perform two functions: (1) to isolate each subcarrier oscillator from another at a different frequency such that its output is transmitted to the modulator instead of dissipated in the other modes; and (2) to prevent overmodulation, or overly rapid modulation, from causing side bands which generate signal frequencies in adjacent channels. These filters are band-pass filters with separate inputs and a common output.

It is necessary that the modulator preserve the linearity of the complex modulating wave. Intermodulation distortion results in spurious signals and data errors; it can be prevented only by maintaining linearity in the complete modulation, radio-frequency, and demodulation system. This is a restriction which is of minor importance in the time-multiplexing systems. It is further necessary for efficient FM operation to reduce the amount of amplitude modulation to an insignificant quantity.

Table 37-1. Standards for FM/FM Telemetry System

Band	Center frequency, cps	Lower limit,* cps	Upper limit,* cps	Max deviation, %	Frequency response,† cps
1	400	370	430	±7.5	6.0
2	560	518	602	±7.5	8.4
3	730	675	785	±7.5	11
4	960	888	1,032	±7.5	14
5	1,300	1,202	1,399	±7.5	20
6	1,700	1,572	1,828	±7.5	25
7	2,300	2,127	2,473	±7.5	35
8	3,000	2,775	3,225	±7.5	45
9	3,900	3,607	4,193	±7.5	59
10	5,400	4,995	5,805	±7.5	81
11	7,350	6,799	7,901	±7.5	110
12	10,500	9,712	11,288	±7.5	160
13	14,500	13,412	15,588	±7.5	220
14	22,000	20,350	23,650	±7.5	330
15	30,000	27,750	32,250	±7.5	450
16	40,000	37,000	43,000	±7.5	600
17	52,500	48,562	56,438	±7.5	790
18	70,000	64,750	75,250	±7.5	1,050
A‡	22,000	18,700	25,300	±15	660
B	30,000	25,500	34,500	±15	900
C	40,000	34,000	46,000	±15	1,200
D	52,500	44,625	60,375	±15	1,600
E	70,000	59,500	80,500	±15	2,100

* Rounded off to nearest cycle.
† The frequency response given is based on maximum deviation and a deviation ratio of 5.
‡ Bands A through E are optional and may be used by omitting adjacent bands as shown in Table 37-2.

Table 37-2. Optional Bands for FM/FM System

Band used	Omit bands
A	13, 15, and B
B	14, 16, A, and C
C	15, 17, B, and D
D	16, 18, C, and E
E	17 and D

Furthermore, many modulators are an inherent part of the frequency-stability system of the radio transmitter and they must be usable with frequency-stabilized transmitters.

Various radio transmitters are in common use. Some have self-excited final-output stages, while others are elaborately crystal-controlled or stabilized through multiplier chains. Output power may vary from a few milliwatts to 100 watts and frequency deviation may vary from 500 Kc down to a few cycles per second. Given in the standard are maximum figures which are provided as services at the various test ranges.

In many cases more than one FM/FM system is used in a single test vehicle and more than one is fed to a single antenna. In this case an r-f multiplexer is used to isolate each transmitter from the other transmitters feeding the same antenna and to match impedances between the transmitters and the antenna. It is desirable to have only a single telemeter antenna on a missile, since cutting through the skin or providing projections is undesirable both structurally and aerodynamically. Consequently, a best location is found for an antenna system and several transmitters are multiplexed to this single system. Noteworthy at this point is the deviation from this requirement by the Russian Sputnik III. Here four antennas, each projecting over a foot from the nose surface, were utilized, greatly reducing the problems of fading, directivity, and radio multiplexing. Subsequent satellites have employed multiple and protruding antennas, greatly relieving the problems of the telemetry engineer.

At the ground station a receiving antenna, which is usually large to give high gain, and therefore directive, tracks the moving missile. Circular polarization is usually employed to receive the signal regardless of the position of the missile. A preamplifier is located near the antenna feed to minimize the noise figure of the system. Enough amplification is included to more than compensate for transmission-line losses. In typical stations a radio-frequency demultiplexer, or channel splitter, is included to permit the simultaneous reception of up to eight telemetry carriers. As many receivers as are required for the transmissions are employed, and these feed separate subcarrier filters. These band-pass filters are required to separate the channels and feed each separate channel intelligence signal to separate subcarrier discriminators.

The discriminators are of many types, including counting discriminators and phase-lock loops. Resonant inductors are seldom used because of the large size that would be required at the low subcarrier frequencies. The intelligence fidelity is maximized by the choice of a proper filter at the output of the subcarrier discriminator. Furthermore, the filter removes the subcarrier frequency before the signal is fed to oscillographs or pen recorders. The signal is also recorded on a magnetic-tape recorder, and in general missile practice, the tape recorder may replace the entire ground station at advance recording bases and missile ranges. When it reproduces the data they are then played through the filters and the discriminators. The tape recorder tends to add an error component in its inherent flutter and wow; this is minimized by the design of the recorder, but it is compensated by the use of a compensating tone, distinct from the subcarrier tone, when the composite subcarriers are recorded. The compensating tone may then be detected and processed in the same manner as the other subcarriers, and the output after demodulation is fed in phase opposition to other subcarriers. This process is not so simple as described, however, because the phase delay through the filters and discriminators of each subcarrier is different. Consequently, the output flutter frequency must be delayed properly to match the particular subcarrier. Furthermore, the delay is variable over the subcarrier frequency range, and compensation cannot be precisely achieved.

The Pulse-duration-modulation/FM Telemetry System. Figures 37-21 and 37-22 depict the operation of this telemetry system, and Fig. 37-23 illustrates a PDM-FM receiving station. This system has been employed at missile ranges since about 1955, and the same type of standards as for the FM/FM system have been prepared. These are included herewith:

Number of samples per frame*	30	45	60	90
Frame rate (frames per second)	30	20	15	10
Commutation rate (samples per second)†	900	900	900	900

* The number of samples per frame available to carry information is two less than the number indicated, because the equivalent of two samples is used in generating the frame-synchronizing pulse.

† Frame rate times number of samples per frame.

37-22　　　　　　　　　SYSTEM TECHNIQUES

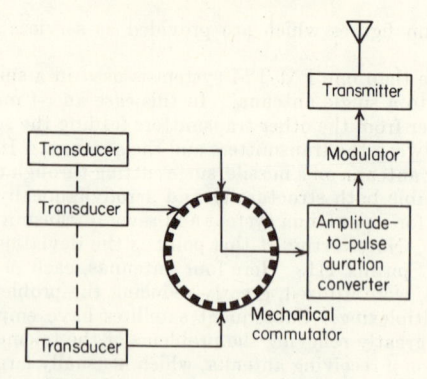

Fig. 37-21. PDM/FM telemetry transmitter, block diagram.

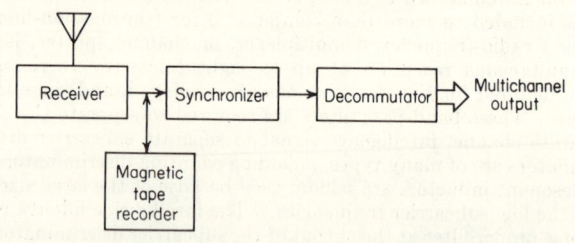

Fig. 37-22. PDM/FM telemetry receiver, block diagram.

Fig. 37-23. PDM telemetry receiving station, photograph. (*Courtesy of Electromechanical Research.*)

The amplitude of the measurands being transmitted in each channel shall determine the duration of the corresponding pulses. The relation between measurands and pulse duration should, in general, be linear.

Minimum pulse duration (zero-level information).......................... 90 ± 30 μsec
Maximum pulse duration (maximum-level information)........................ 700 ± 50 μsec
Pulse rise and decay time (measured between 10 and 90% levels)   10–20 μsec (constant to ±1 μsec for a given transmitting set)

The time interval between the leading edges of successive pulses within a frame shall be uniform from interval to interval within ±25 μsec. This time interval shall have a nominal period equal to 1 divided by the total sampling rate.

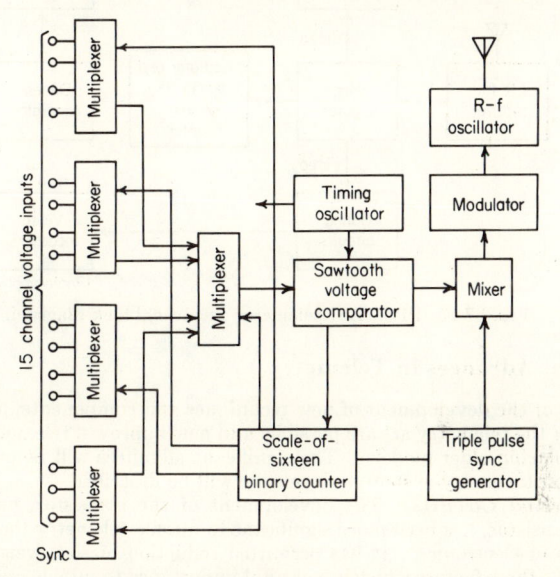

Fig. 37-24. PPM/AM telemetry transmitter, block diagram.

The commutator speed or frame rate shall not vary more than plus 5.0 to minus 15.0 per cent from nominal.

A frame-synchronizing interval equal to two successive pulse time intervals shall exist in the train of pulses transmitted, to be used for synchronization of the commutator and the decommutator.

Many of the parts of the PDM-FM system are common with the FM/FM system, and the description of antennas, radio transmitters, tape recorders, etc., is applicable and will not be repeated here. The system is light, simple, and compact, but not nearly so versatile as the FM/FM system. Transducers are limited to those which produce 0 to 5-volt outputs, or amplifiers must be employed which provide a common level to the commutator. In general, the PDM-FM system may be compared with a single commutated channel of the FM/FM system, and it has often been employed on the 70-Kc subcarrier of the FM/FM system. Variable-reluctance transducers cannot be employed unless their output is in some manner converted to a voltage.

Pulse-position-modulation/Amplitude-modulation Telemetry System. The PPM/AM telemetry system has been employed in the United States principally by the Navy. It has not been chosen as a test-range standard, but receiving equipment

37-24   SYSTEM TECHNIQUES

has been moved into the test ranges as required. There have been further developments of high-capacity PPM/AM telemetry systems which have not been put into use in the United States. The PPM system is very similar to the PDM system, except that instead of using the entire duration of the pulse, only the beginning and the end of the pulse are transmitted as an intelligence marker, and a conversion from voltage amplitude to pulse position is employed. It is also a time-multiplexed system, and equipments with as many as 36 primary channels have been developed, with a total sampling rate of 50,000 samples per second. Typical block diagrams of PPM/AM telemetry transmitters and receivers are shown in Figs. 37-24 and 37-25.

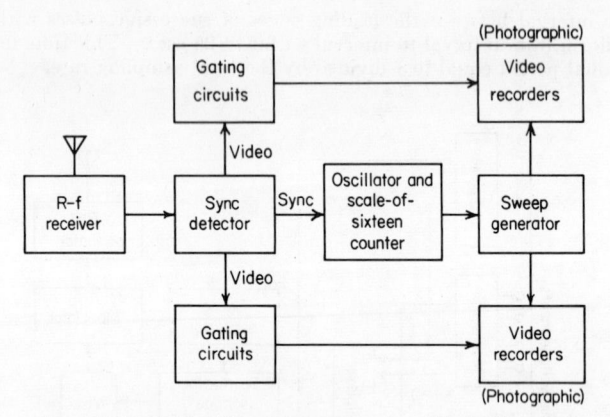

Fig. 37-25. PPM/AM telemetry receiver, block diagram.

37-6. Recent Advances in Telemetry

As a result of the development of new techniques and components, many advances in the state of the telemetry art are possible, and new improved telemetry systems are gradually replacing older models. The significant advances will be enumerated and their effects on telemetry-system performance will be described.

Transistorized Circuits. The development of the transistor, and particularly the silicon transistor, has been more significant to missile telemetry than to any other single branch of electronics. It has permitted reduction of size, weight, and power requirements—three factors which are of vital importance to missile operations. The replacement of the vacuum tube is a gradual process, inasmuch as stable operation over a wide range of temperature operation is more difficult with transistors than with vacuum tubes. High-temperature operation is difficult and high-frequency operation is just being achieved. The use of transistors, however, does make practicable high-capacity systems of pulse-code-modulation telemetry which would otherwise be prohibitively large.

Large Automatic Tracking Antennas. Large, immovable parabolic reflecting antennas were developed for forward-scattering propagation, and the antennas have been adapted to telemetry use. The high gain of the large reflector implies that the beam width is relatively narrow and, therefore, tracking difficulties are presented in missile and satellite operations. It was not until the automatic tracking feature was added to the forward-scatter propagation antenna that it became practical for telemetering from guided missiles. Several automatic tracking telemetry antennas employing 60- and 85-ft reflectors are in service. Their effect upon radio-telemetry reception is phenomenal. Basically, there is a 10-db improvement in reception over previous techniques. This has made data continuous where otherwise there were losses due to fading, or, if the same performance as was previously observed is to be maintained, the missile transmitter power may be reduced by a factor of 10. Figure 37-26 shows a typical large telemetry receiving antenna.

Phase-locked Frequency-modulation Discriminators. Another recent improvement in telemetry receiving techniques has been the phase-locked FM discriminator or detector. The phase-locked principle is one in which the frequency of a local beat-frequency oscillator is varied to correspond to the incoming frequency, thus maintaining the beat frequency between this oscillator and the incoming signal relatively constant so that it may be transmitted through a filter of narrower bandwidth. The local beat-frequency oscillator is of voltage-control design and the control voltage which is generated becomes the demodulated signal. This technique has added another 6 db of improvement to telemetry receiving stations, and it can be continued by further addition to subcarrier oscillators to approximately 15 db and

FIG. 37-26. Large automatic tracking telemetry antenna, photograph. (*Courtesy of Radiation, Inc.*)

even further if the total bandwidth of the telemeter is curtailed. In recent tests it has been shown that, with phase-locked receivers, a 10-mw telemetry transmitter transmitting the lowest six standard subcarriers has been accurately received over several thousand miles.

Solar Power Sources. Also recently developed are "power sources" which transform solar radiant energy directly to electricity (see Sec. 17-5). These devices are particularly useful in satellite telemetry where an indefinite power-source life is required. The cells remain active for long periods in outer space. This power source permits an operating life of telemetry equipment that was impossible a short while ago.

Predetection Recording. In telemetry practice it is usually very desirable to make a primary recording on magnetic tape which is then later reproduced for various processes. In postdetection recording there is an irrevocable loss of signal-to-noise which may be avoided by correlation detectors, etc., if the signal is available for processing before the radio carrier and its modulation have been separated. Furthermore, various systems of telemetry which now require different recorders for proper recording (FM/FM, PDM, and PCM) may be recorded on the same recorder. In fact, the typical video recorder used for television may be adapted to record simultaneously four or more of the typical telemetry signals without regard to the kind of modulation.

Automatic Data-reduction Equipment and Processes. With the advent of the general-purpose digital computer, telemetry data may be processed (multiplied, averaged, smoothed, calibrated, etc.) at high speed and with great accuracy. The principal developments in this field in recent years have been the automatic equipment to bring telemetry data into and out of these digital computers. Also, many smaller special-purpose digital computers have evolved which handle only a specific process such as calibration. These equipments are available in many forms and offer a wide range of adaptability for information handling.

Part VI

USEFUL MATHEMATICS ASSOCIATED WITH SYSTEM ENGINEERING

Part VI

Chapter 38

PROBABILITY

RONALD A. HOWARD, *Massachusetts Institute of Technology.*

CONTENTS

38-1. Introduction	38-1
38-2. The Algebra of Events	38-4
Transmissions through Networks	38-6
38-3. The Foundations of Probability	38-7
Conditional Probability	38-9
Application to Networks	38-9
Bayes's Theorem	38-11
38-4. The Sample Space	38-13
The Probability Tree	38-15
Random Variables	38-16
Expectation	38-17
38-5. Representations of Discrete Random Variables	38-18
Moments	38-18
Cumulative Distributions	38-20
38-6. Continuous Sample Spaces and Random Variables	38-21
The Probability Mass Function Revisited	38-24
Expectations and Moments	38-24
Cumulative Distributions	38-25
38-7. Multidimensional Sample Spaces	38-27
Marginal and Conditional Distributions	38-30
Bayes's Theorem and Independence	38-33
Moments	38-34
Cumulative Probability Distributions	38-35
38-8. Operations on Random Variables	38-35
Changes of Many Variables	38-37
38-9. Operational Transformations	38-41
The Geometric Transform	38-42
38-10. Named Distributions	38-44
38-11. Conclusion	38-48

38-1. Introduction

The analysis of uncertainty seems to be playing a larger and larger role in modern technology. In science, engineering, business, finance, and medicine, we find prob-

38–4   MATHEMATICS ASSOCIATED WITH SYSTEM ENGINEERING

lems that require quantitative reasoning about probabilistic phenomena. The basis for such reasoning is the theory of probability. Probabilistic analysis has risen from its humble beginnings among gambling men in centuries past to become the concern of all educated individuals.

The theory of probability not only enables us to analyze uncertainty, it permits us to make statistical inferences about future behavior. Whole theories of epistemology have been based on probabilistic foundations. Indeed, much of modern statistics is now undergoing a reformulation in terms of an inductive inference system based on probability theory. Today is an exciting time to be a probabilist.

The beginning student of probability must be on guard against a subtle error. This is the feeling that there is something random about the solution of random problems. Students sometimes feel that because they are calculating the result of a random phenomenon, the methods they employ are in some way random. Nothing could be farther from the truth. A problem in probability has an answer that is just as deterministic as a problem in calculus. We obtain the answer, at least theoretically, by a procedure that is as rigorous and logical as the solution of any other problem in mathematics. The purpose of this chapter is to review these logical steps between problem and answer.

38-2. The Algebra of Events

Every science has a language. The language of probability is the algebra of events. We need it to avoid the ambiguity inherent when we express complicated occurrences in the English language. The algebra of events makes every statement crystal clear. This brief summary of the algebra of events is supplemented in Chapter 41.

A convenient set of axioms for the algebra of events is shown in Table 38-1. The axioms in the table are easily interpreted. The quantities A, B, and C represent the occurrences of events A, B, and C. A symbol with a prime is the nonoccurrence of the corresponding event. The expression $A + B$, the logical sum of the events A and B, is the event that either event A or event B or both events have occurred. It is called the "inclusive or" operation, and is sometimes represented by such symbols as $A \cup B$ or $A \vee B$. We shall use the simpler plus sign to indicate this operation because it is in many ways most convenient. We must remember, however, that this operation has different properties from arithmetic addition.

Table 38-1. The Axioms of the Algebra of Events

(1) $A + B = B + A$ .... The commutative axiom
(2) $(A')' = A$
(3) $A + (B + C) = (A + B) + C$ .... The associative axiom
(4) $AB = (A' + B')'$ or by (2) $(AB)' = A' + B'$
(5) $A + (BC) = (A + B)(A + C)$ .... The distributive axiom
(6) $AA' = \emptyset$
(7) $AI = A$
(8) $A + \emptyset = A$

The expression AB is the logical product of the events A and B; it is the event that both A and B have occurred. It, too, has alternate representations like $A \cap B$ or $A \wedge B$, but we shall use standard algebraic product symbolism. The product AB is often called the "and" operation by engineers.

The set of all events that can occur is designated by I; it is the universal event, the event that must occur. Its opposite is the null event $\emptyset$, the event that cannot occur. Notice that $\emptyset$ is written with a stroke to eliminate confusion with the capital "O."

All axioms in Table 38-1 are consistent with the interpretation in terms of events. For example, axiom (4) in the form $(AB)' = A' + B'$ says that the nonoccurrence of the event AB is equivalent to the nonoccurrence of A or the nonoccurrence of B or both nonoccurrences.

The axioms of Table 38-1 have other interpretations than as axioms of an algebra of sets or events. They also represent the relations that must exist between areas in a plane diagram. Such a diagram, called a Venn diagram, is shown in Fig. 38-1. The elements A, B, etc., in the table of axioms now represent the regions marked A, B, etc., in the diagram. The sum of two elements is understood to be the region contained in either or both of them; the product of two elements is the region they share. The entire region of the diagram is I; A' is the region within I, but outside A. The region $\emptyset$ is the region of zero area. With these interpretations all the axioms of Table 38-1 apply to the Venn diagram.

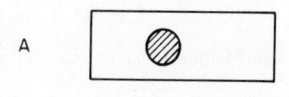

Table 38-2. Some Subsidiary Relations of the Algebra of Events

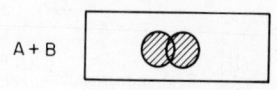

(1') $AB = BA$
(2') ─────
(3') $A(BC) = (AB)C$
(4') $A + B = (A'B')'$ or $(A + B)' = A'B'$
(5') $A(B + C) = AB + AC$
(6') $A + A' = I$
(7') $A + I = I$
(8') $A\emptyset = \emptyset$

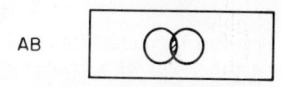

The basic axioms of the algebra may be extended by the usual process of derivation. For example, the following sequence:

Axiom (4) $AB = (A' + B')'$
Axiom (1) $AB = (B' + A')'$
Axiom (4) $AB = BA$

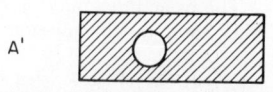

Fig. 38-1. The Venn diagram.

shows that the product operation is commutative. By similar arguments we may write a second table of relations that is analogous to Table 38-1. It is shown in Table 38-2. Each of these relations may be immediately verified by reference to the Venn diagram. Of course, these relations are only some of the useful properties of the algebra of events that we shall want to use. We can derive many more in a very short time.

For example, the most unlikely looking relation,

$$A + A = A \tag{38-1}$$

is readily established.

Statement	Supporting relations
$A + AA' = (A + A)(A + A')$	(5)
$A + \emptyset = (A + A)I$	(6), (6')
$A = A + A$	(8), (7), (1'), (5')

The Venn diagram shows the result immediately. This relation shows, in particular, the necessity of distinguishing in our minds between the present algebra of events and the usual algebra. The relations

$$A + AB = A \tag{38-2}$$
$$A + A'B = A + B \tag{38-3}$$

further emphasize the distinction and the value of the Venn diagram in establishing such relations.

An important concept in the algebra of events is that of "mutually exclusive" events. Two events A and B are mutually exclusive if both cannot occur; that is,

$$AB = \emptyset \tag{38-4}$$

If two such events are represented by regions in a Venn diagram, the regions will not overlap.

38-6 MATHEMATICS ASSOCIATED WITH SYSTEM ENGINEERING

On some occasions it is useful to consider a set of events possessing the property that at least one of them must occur. Such events are said to be "collectively exhaustive"; their sum is the certain event I. Thus a set of events $A_1, A_2, \ldots$ is mutually exclusive if

$$A_i A_j = \emptyset \quad \text{for } i \neq j$$

and collectively exhaustive if

$$A_1 + A_2 + \cdots = I$$

For example, if we let the event A_i be the occurrence of a head for the first time on the ith toss of a coin, then the events $A_1, A_2, \ldots$ would form a mutually exclusive and collectively exhaustive set of events.

Transmissions through Networks. The algebra of events would be of little interest if it did not allow us to solve problems that would be difficult to treat in another way. We find a convincing example in the study of so-called network transmissions. Consider the communication link labeled A in Fig. 38-2. Let A be the event that the link is working, that is, that a closed circuit exists between its ends. Then A' is the event that the link is not working. If we let T be the event of a transmission through a network and T' be the event of no such transmission, then for the simple network of Fig. 38-2, $T = A$, $T' = A'$.

FIG. 38-2. A communication link.
$T = A$
$T' = A'$

The basic links may be combined in many ways. If two are placed in series as in Fig. 38-3, a transmission will exist if and only if both links are working. Thus

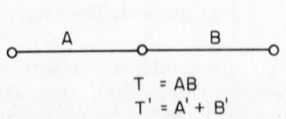

FIG. 38-3. Two links in series.
$T = AB$
$T' = A' + B'$

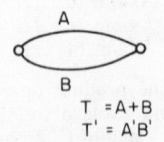

FIG. 38-4. Two links in parallel.
$T = A + B$
$T' = A'B'$

$T = AB$. A transmission will not exist if either or both links are not working; $T' = A' + B'$. We see that the transmission expression of links in series is just the product of the transmission expressions for each link.

Two links can be connected in parallel as in Fig. 38-4. A transmission will exist if either or both links are working; $T = A + B$. A transmission will not exist only if both links are not working; $T' = A'B'$. Thus the transmission expression for links in parallel is the sum of the transmission expressions of the links.

Transmission expressions may be written for very large networks using the basic operations we have just examined. In the context of networks, I is the event of a certain transmission; $\emptyset$ is the event of no transmission. The algebra of events turns out to be identical with the algebra of transmissions through networks.

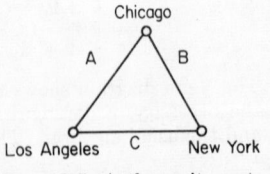

FIG. 38-5. A three-city network.

A simple example may help in illustrating how the algebra is used. Suppose that Los Angeles, Chicago, and New York are connected by the three trunks or links shown in Fig. 38-5. Let us investigate the condition under which each city can talk with the others. Let T_{lc}, T_{cn}, and T_{nl} be the transmissions, direct or indirect, between the pairs, Los Angeles–Chicago, Chicago–New York, and New York–Los Angeles. By application of our basic rules we find

$$\begin{aligned} T_{lc} &= A + BC \\ T_{cn} &= B + AC \\ T_{nl} &= C + AB \end{aligned} \tag{38-5}$$

PROBABILITY

With these relations as our starting point we can derive a number of interesting results. First we can find the conditions under which transmissions will not exist between each pair of cities:

$$T_{lc}' = (A + BC)' = A'(BC)' = A'(B' + C')$$
$$T_{cn}' = (B + AC)' = B'(A' + C')$$
$$T_{nl}' = (C + AB)' = C'(A' + B')$$
(38-6)

Thus Los Angeles and Chicago will not be able to communicate through this network if A is not working and if either B or C or both are not working. Note that complementation of an expression causes each primed event to become unprimed, each sum to become a product, and vice versa. This observation makes complementation a rather simple procedure.

Let us ask under what conditions it is possible for Los Angeles to be able to speak to New York but not to Chicago. This event is represented by $T_{nl}T_{lc}'$. Then we can write

$$T_{nl}T_{lc}' = (C + AB)A'(B' + C')$$
$$= A'C(B' + C')$$
$$= A'B'C$$
(38-7)

We find that this communication situation can arise only when C is the only link working, as is evident from Fig. 38-5.

Of course, this simple example does not require a formal algebra for its solution, but increasing its size by a factor of 2 or 3 would soon make the need for a formal procedure apparent. However, for us, the algebra of events is not an end in itself, but a step on the road to probability.

38-3. The Foundations of Probability

When we discussed the algebra of events we talked about the occurrence or nonoccurrence of a given event. In real life the likelihood that an event will occur can range from virtual impossibility to virtual certainty. The theory of probability provides a framework for assigning real numbers to the likelihood of occurrence of different events so that their likelihoods may be compared and computed. The theory has only three axioms. They rely upon the existence of an algebra of events like that discussed in the previous section so that, in all, probability theory requires about 11 axioms. The probability measure is defined as a real-valued function p defined for all the events that can be constructed by using our algebra. The three axioms the function must satisfy are

(a) For any event A, $p(A) \geq 0$
(b) $p(I) = 1$, a normalization (38-8)
(c) If $AB = \emptyset$ (A and B are mutually exclusive), then $p(A + B) = p(A) + p(B)$

These few axioms seem very little foundation for such a weighty subject, but they are, in fact, enough. The first axiom requires probability to be a nonnegative number. The second normalizes the probability by requiring that the probability of the certain event be 1. The third requires that the probabilities of two mutually exclusive events be additive.

The probability function defined by the three axioms is just the function needed to measure the areas of each region in the Venn diagram of Fig. 38-1. The first axiom says that the area of any region cannot be negative. The second states that the area of the entire diagram shall be considered to be 1. The third is the very reasonable requirement that the area contained in two nonoverlapping regions be the sum of their individual areas.

By using the three axioms we can derive some very important relations. First let us calculate the probability of the event A' if we are given the probability of the event

38-8 MATHEMATICS ASSOCIATED WITH SYSTEM ENGINEERING

A. From relation (6') of Table 38-2 we have

$$A + A' = I \tag{38-9}$$

Since $AA' = \emptyset$ from axiom (6) of Table 38-1, axioms (b) and (c) allow us to write

$$p(A) + p(A') = 1$$

and finally,

$$p(A') = 1 - p(A) \tag{38-10}$$

The probability of the complement of any event is 1 minus the probability of the event.
The probability of the null event $\emptyset$ is easily derived. Since $I' = \emptyset$,

$$p(\emptyset) = 1 - p(I)$$

from (38-10). Then, by axiom (b),

$$p(\emptyset) = 0 \tag{38-11}$$

Axiom (c) shows how to find the probability of the sum of two mutually exclusive events. We can quickly find how to compute the probability of the sum of two events when they are not mutually exclusive. First, we rewrite Eq. (38-3):

$$A + B = A + A'B \tag{38-12}$$

Since $A(A'B) = \emptyset$, it follows from axiom (c) that

$$p(A + B) = p(A) + p(A'B) \tag{38-13}$$

We can write event B in the form

$$B = IB = (A + A')B = AB + A'B \tag{38-14}$$

Then, since $AB(A'B) = \emptyset$, we can again apply axiom (c) to obtain

$$p(B) = p(AB) + p(A'B) \tag{38-15}$$

We can now eliminate $p(A'B)$ between (38-13) and (38-15) to produce

$$p(A + B) = p(A) + p(B) - p(AB) \tag{38-16}$$

The probability of the sum of two events is the sum of their probabilities less the probability of their product. Of course, if the events are mutually exclusive, then the probability of their product will be zero and Eq. (38-16) will reduce to axiom (c).

The Venn diagram may be exploited with success in deriving relations of this type. Recall that $p(A + B)$ is just the area lying in the region of A or the region of B or the region of both A and B. If we computed this area by adding together the areas of regions A and B, we would obtain too large a result because we would be counting twice the area shared by both regions. Thus we have to subtract out this area once to obtain the correct result. This operation is just the one implied by Eq. (38-16).

Simple extensions of these arguments produce many important relations. For example, suppose we wish to calculate the probability of the sum of three events A, B, and C. We write

$$\begin{aligned} p(A + B + C) &= p[A + (B + C)] = p(A) + p(B + C) - p[A(B + C)] \\ &= p(A) + p(B) + p(C) - p(BC) - p(AB + AC) \\ &= p(A) + p(B) + p(C) - p(BC) - p(AB) - p(AC) \\ &\quad + p(ABC) \end{aligned} \tag{38-17}$$

Equation (38-17) requires the interesting Venn diagram shown in Fig. 38-6. The total area within the three regions is equal to the sum of the areas within each of them, less the three lens-shaped areas that were counted twice, plus the small triangular area that was added in three times and then subtracted out three times.

Conditional Probability. One of the more powerful definitions in the theory of probability is that of conditional probability. If A and B are events with $p(A) \neq 0$, then $p(B|A)$, read "the conditional probability of B given A" is defined by

$$p(B|A) = p(AB)/p(A) \tag{38-18}$$

The conditional probability is the ratio of the probability that both events will occur to the probability that the conditioned event will occur. In the Venn diagram it is the ratio of the area shared by both events to the area of the conditioned event.

In a sense, all probabilities are conditional. They are conditioned on the fact that we have defined a set of events I at least one of which is certain to occur. All probabilities we calculate depend upon the fact that we have defined the area of this set of events to be 1. If now we learn that some event A has occurred, then A has become a certain event, and we can renormalize the probabilities of each other event so that they sum to 1 over the area of A. The set of events that we consider to be certain I is thus quite arbitrary. The definition of conditional probability allows us to change the basis for assignment of probabilities at any time.

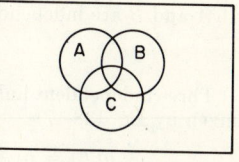

Fig. 38-6. Venn diagram for three events.

If two events A and B are independent, in an intuitive sense, the occurrence of A should have no effect on the occurrence of B. That is,

$$p(B|A) = p(B) \quad \text{or} \quad p(AB)/p(A) = p(B) \tag{38-19}$$

Equivalently,

$$p(AB) = p(A)p(B) \tag{38-20}$$

We shall *define* events to be independent if the probability that both occur is equal to the product of the probabilities that each will occur individually. Note that the definition of independent events rests on the probability function and not on the algebra of events.

To emphasize the definition of independence, let us consider an experiment in which a coin with probability of heads, $0 < p < 1$, is tossed twice. Let A be the event "a head on the first toss." Let B be the event "both tosses are the same." The event A is the set of subevents $\{HH, HT\}$; the event B is the set of subevents $\{HH, TT\}$; the event AB is the event $\{HH\}$. Since p is the probability of a head, $1 - p$ is the probability of a tail. We find

$$p(A) = p^2 + p(1 - p) = p$$
$$p(B) = p^2 + (1 - p)^2$$
$$p(AB) = p^2$$

It follows that if $p \neq \frac{1}{2}$, then $p(AB) \neq p(A)p(B)$ and the events A and B are not independent. However, if $p = \frac{1}{2}$, then $p(AB) = p(A)p(B)$ and A and B are independent. We see that the question of independence can be determined only by knowing the probability function rather than just the definition of the events.

Application to Networks. In Sec. 38-2 we developed an algebra to describe transmissions through networks. The axioms of probability allow us to extend this application to the case of unreliable networks, networks whose links may not function with some probability. We shall let a be the probability of a transmission through the link A shown in Fig. 38-2. We found that two such links in series had a transmission equal to the product of their individual transmissions,

$$T = AB$$

The probability of a transmission through two links in series is then

$$p(T) = p(AB)$$

If the operation of A and B is independent, then $p(AB) = p(A)p(B) = ab$ and we have
$$p(T) = ab \tag{38-21}$$
For two links in parallel we found
$$T = A + B$$
and, therefore, the probability of transmission is
$$p(T) = p(A + B) = p(A) + p(B) - p(AB)$$
If A and B are independent, then
$$p(T) = a + b - ab \tag{38-22}$$

Three independent links A, B, C in parallel would have a probability of transmission given by Eq. (38-17)
$$p(T) = p(A + B + C) = a + b + c - ab - ac - bc + abc \tag{38-23}$$
Of course, the result of (38-23) could have been obtained more quickly by noting
$$\begin{aligned}p(T) &= 1 - p(T') = 1 - p(A'B'C') \\ &= 1 - p(A')p(B')p(C') \\ &= 1 - (1-a)(1-b)(1-c)\end{aligned} \tag{38-24}$$
Some methods for finding the probabilities of transmission will be more direct than others, but all will be correct if we follow the rules.

Suppose that in the three-city example of Fig. 38-5 there are probabilities a, b, c that each of the three trunks A, B, and C will be working after an enemy attack, and that the trunks fail independently. The probability that Los Angeles will be able to talk to Chicago is
$$\begin{aligned}p(T_{lc}) &= p(A + BC) \\ &= p(A) + p(BC) - p(ABC) \\ &= a + bc - abc\end{aligned} \tag{38-25}$$
If $a = b = c = p$, then
$$p(T_{lc}) = p(T_{cn}) = p(T_{nl}) = p + p^2 - p^3 \tag{38-26}$$
If, furthermore, $p = \frac{1}{2}$,
$$p(T_{lc}) = p(T_{cn}) = p(T_{nc}) = \tfrac{5}{8} \tag{38-27}$$
The redundant communication network joining the cities increases the probability that any two will be able to talk after the attack to $\frac{5}{8}$ from the value of $\frac{1}{2}$ that would exist if indirect communication were not possible.

How is the probability that Los Angeles will be able to talk to New York after an attack modified by the knowledge that New York is able to talk to Chicago at that time? Remember that one possible way for Chicago to speak to New York is via Los Angeles. The answer to this question is revealed by the conditional probability $p(T_{lc}|T_{cn})$. By definition (38-18) this probability is
$$p(T_{lc}|T_{cn}) = \frac{p(T_{lc}T_{cn})}{p(T_{cn})} \tag{38-28}$$
The denominator $p(T_{cn})$ is easily found by permutation of letters in Eq. (38-25) to be
$$p(T_{cn}) = b + ac - abc \tag{38-29}$$
The joint probability $p(T_{lc}T_{cn})$ is computed by expressing the event $T_{lc}T_{cn}$ in terms of the events A, B, and C. From Eq. (38-5),
$$\begin{aligned}T_{lc}T_{cn} &= (A + BC)(B + AC) \\ &= AB + AC + BC\end{aligned} \tag{38-30}$$

PROBABILITY   38-11

an expression readily verified by the Venn diagram of Fig. 38-6. Then, by applying (38-17) or by measuring areas in the Venn diagram, we obtain

$$p(T_{lc}T_{cn}) = p(AB) + p(AC) + p(BC) - 2p(ABC) \tag{38-31}$$

because $AB \cdot AC = AB \cdot BC = AC \cdot BC = AB \cdot AC \cdot BC = ABC$

Finally, we use the independence of links to write

$$p(T_{lc}T_{cn}) = p(A)p(B) + p(A)p(C) + p(B)p(C) - 2p(A)p(B)p(C) \tag{38-32}$$

and then substitute the lower-case letters representing the probabilities of transmission in each link:

$$p(T_{lc}T_{cn}) = ab + ac + bc - 2abc \tag{38-33}$$

We can now use the results of (38-29) and (38-33) to write the conditional probability we desire in (38-28),

$$p(T_{lc}|T_{cn}) = \frac{ab + ac + bc - 2abc}{b + ac - abc} \tag{38-34}$$

Equation (38-34) expresses the probability that Los Angeles will be able to talk to Chicago after the attack if it is known that Chicago can talk to New York in terms of the probabilities that each link in the network will be working after the attack. Although the problem considered is very simple, it is difficult to see how this expression could be easily developed by most of us without the guidance of a formal structure.

If $a = b = c = p$, then Eq. (38-34) becomes

$$p(T_{lc}|T_{cn}) = \frac{3p^2 - 2p^3}{p + p^2 - p^3} = \frac{3p - 2p^2}{1 + p - p^2} \tag{38-35}$$

If $p = \frac{1}{2}$, then

$$p(T_{lc}|T_{cn}) = \tfrac{4}{5} \tag{38-36}$$

A comparison of (38-36) with (38-27) shows that, when $p = \frac{1}{2}$, the probability that Los Angeles can talk to Chicago is raised from $\frac{5}{8}$ to $\frac{4}{5}$ by the knowledge that Chicago can talk to New York.

By the symmetry of the problem when $a = b = c = p$, (38-34) gives the probability of communication between any two cities after an attack if it is known that a certain one of them can talk to the third. Note that this is not the same as the probability of communication between any two cities after an attack if it is known that at least one of them, unspecified, can talk to the third. This probability would be represented by an expression like

$$p(T_{lc}|T_{cn} + T_{nl}) \tag{38-37}$$

Bayes's Theorem. Bayes's theorem is the most important single relation in probability theory: it forms the basis for statistical inference. The theorem is easily derived from the basic relations of probability theory. Consider a set of n mutually exclusive and collectively exhaustive events $A_1, A_2, \ldots, A_n$, with known probabilities. By definition

$$A_i A_j = \emptyset \quad 1 \le i \ne j \le n \tag{38-38}$$
$$A_1 + A_2 + \cdots + A_n = I \tag{38-39}$$

Let there be some event B that also lies in I. The situation can be represented by the Venn diagram in Fig. 38-7. The events A_i partition the space I into n regions. The event B occupies some parts of some of these regions. The probabilities or areas of each of the events $A_1, A_2, \ldots, A_n$ have been assumed known. However, the relationships of areas in the diagram cannot be determined unless the area that

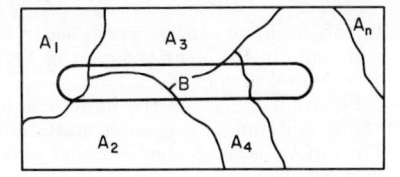

FIG. 38-7. Venn diagram for Bayes's theorem.

B shares with each of the events A_i, $p(A_iB)$, is also known. This information is contained in $p(B|A_i)$ since

$$p(B|A_i) = p(A_iB)/p(A_i) \tag{38-40}$$

and we know $p(A_i)$. Therefore, we shall assume that the conditional probability $p(B|A_i)$ is known for each A_i, $i = 1, 2, \ldots, n$. Now that we know all area relationships in the diagram, the problem remaining is this: What is the probability that A_i has occurred if we know that B has occurred, or, what is $p(A_i|B)$? By definition,

$$p(A_i|B) = p(A_iB)/p(B) \tag{38-41}$$

Equation (38-40) can be used to write (38-41) in the form

$$p(A_i|B) = p(A_i)p(B|A_i)/p(B) \tag{38-42}$$

Equation (38-42) expresses our answer in terms of the known quantities $p(A_i)$ and $p(B|A_i)$ and the quantity $p(B)$ that can be calculated from them as follows: using (38-39) we write

$$B = IB = (A_1 + A_2 + \cdots + A_n)B \tag{38-43}$$

Then

$$B = A_1B + A_2B + \cdots + A_nB \tag{38-44}$$

and

$$p(B) = p(A_1B) + p(A_2B) + \cdots + p(A_nB) \tag{38-45}$$

since

$$A_iBA_jB = \emptyset$$

by (38-38). Finally, (38-40) allows us to write

$$p(B) = p(A_1)p(B|A_1) + p(A_2)p(B|A_2) + \cdots + p(A_n)p(B|A_n)$$
$$= \sum_{i=1}^{n} p(A_i)p(B|A_i) \tag{38-46}$$

When we substitute this result into (38-42) we obtain the most common form of Bayes's theorem

$$p(A_i|B) = \frac{p(A_i)p(B|A_i)}{\sum_{i=1}^{n} p(A_i)p(B|A_i)} \tag{38-47}$$

The conditional probability of A_i given B has been expressed in terms of the known probabilities of the events A_i and the known conditional probabilities of B given A_i. Since $p(A_i)p(B|A_i) = p(A_iB)$, we can write (38-47) in the simpler but less useful form

$$p(A_i|B) = \frac{p(A_iB)}{\sum_{i=1}^{n} p(A_iB)} \tag{38-48}$$

In this form we can see very clearly that the conditional probability of A_i given B represents in the Venn diagram in Fig. 38-7 the area that A_i shares with B divided by the total area of B.

Equation (38-47) is the basic equation of statistical inference. We imagine that nature is in one of n possible states and let A_i be the event that nature is in state i. Then $p(A_i)$ is the probability that nature is in state i based on whatever information is at our disposal.

Now, some event B, statistically related to the state of nature, occurs. The province of statistical inference is this: How does the occurrence of B affect our assignment of probability to the possible states of nature A_i? The question is answered by Bayes's theorem. Equation (38-47) shows that the new probability of A_i, $p(A_i|B)$, also called the posterior probability of A_i because it is after the observance of B, is proportional to $p(A_i)$, the prior probability of A_i, multiplied by the conditional probability of B given A_i. This probability $p(B|A_i)$ is called the likelihood function; it is the probability of observing B if nature were, in fact, in state i. The constant of proportionality in the posterior probabilities is readily established by noting that the denominator of (38-47) is just the sum of the numerator over all values of i. In other words, the posterior probabilities are normalized by the requirement that they sum to 1.

Let us construct an example using the three-city communication problem of Fig. 38-5. We have already found the probabilities of various transmissions under the assumption that links fail independently and that probability of a transmission through any link after an attack is given by the corresponding lower-case letter. Now let us turn the problem around. What is the probability that link A is working if it is known that Los Angeles can talk to Chicago after the attack $p(A|T_{lc})$? Without the knowledge of which cities can communicate, we assigned probability a to $p(A)$. We want to know how this probability has been changed by our knowledge that Los Angeles can talk to Chicago. Bayes's theorem for this problem can be written

$$p(A|T_{lc}) = \frac{p(A)p(T_{lc}|A)}{p(A)p(T_{lc}|A) + p(A')p(T_{lc}|A')} \tag{38-49}$$

The probabilities $p(T_{lc}|A)$ and $p(T_{lc}|A')$ can be found by substituting $a = 1$ and $a = 0$, respectively, in Eq. (38-25)

$$p(T_{lc}|A) = 1 \qquad p(T_{lc}|A') = bc \tag{38-50}$$

Therefore, we can write

$$\begin{aligned} p(A|T_{lc}) &= \frac{a \cdot 1}{a \cdot 1 + (1-a) \cdot bc} \\ &= \frac{a}{a + bc - abc} \end{aligned} \tag{38-51}$$

The probability $p(A|T_{lc})$ could also have been calculated directly from the definition of conditional probability to produce the same result immediately. If $a = b = c = \frac{1}{2}$, then

$$p(A|T_{lc}) = \frac{4}{5} \tag{38-52}$$

The observation that Los Angeles can talk to Chicago has raised the probability that link A is working from $\frac{1}{2}$ to $\frac{4}{5}$.

Thus we have seen how to get a new probability assignment for a set of events as the result of observing some event that is statistically related to that set. All we need is a prior probability distribution over the set and the likelihood function for the process. The mechanics of this operation is becoming known as the province of statistical decision theory (Chapter 24). However, having seen that Bayes's theorem is a direct result of the axioms of probability, we shall leave its inferential implications and continue with the development of the theory of probability.

38-4. The Sample Space

The notion of sample space provides a conceptual framework in which to solve both practical and theoretical probability problems. It is a notion so deceptively simple that its value is often overlooked by the beginning student of the subject.

Suppose that we plan to conduct an experiment with an uncertain outcome. Suppose for the moment that the experiment has only a finite number of possible outcomes. We construct a sample space for the experiment by associating with each possible outcome of the experiment a unique sample point. Thus every outcome is associated with one and only one sample point, and every sample point corresponds to a particular outcome. We assign to every sample point in the sample space the probability to be associated with the corresponding experimental outcome. We have now finished the construction of the sample space for the experiment.

38-14 MATHEMATICS ASSOCIATED WITH SYSTEM ENGINEERING

How is the sample space used? Suppose that we are interested in finding the probability of the event that the experimental outcome will have some particular property. Then we can examine each sample point in the sample space to see whether it has the desired property and therefore causes the event to occur; in this way we determine the set of sample points corresponding to the occurrence of the event. The probability of the event is then computed as the sum of the probabilities associated with each of these sample points.

We see that a sample space is a partition of the outcomes of the experiment into a set of mutually exclusive and collectively exhaustive sample points. Any event defined on the experimental outcomes can be represented as the sum over a set of these sample points. The probability of the event is then the sum of the probabilities of the sample points by the third axiom of probability theory.

The concept is best illustrated by an example. Suppose that the experiment is the roll of a pair of fair dice, one painted red, the other green. The possible outcomes of the experiment are the numbers 1 through 6 on each die, 36 outcomes in all. The sample space for the experiment is shown in Fig. 38-8. The sample point with the check mark represents the outcome 2 on red die, 5 on green die. Since we have said that the dice are fair, we would associate a probability $1/6$ with each number on each die, and since we believe the actions of the two dice to be independent, we would then assign probability $1/36$ to each pair of numbers on the two dice. Therefore, each sample point in Fig. 38-8 is assigned probability $1/36$.

Now we are ready to find the probability of any event defined on the outcome of this experiment. For example, consider the event E_1 that the sum of numbers on the two dice will be 7. We first find the sample points associated with this event; they are the 6 sample points labeled 1 in Fig. 38-8. The probability of E_1 is the sum of the probabilities of these sample points,

$$p(E_1) = 1/36 + 1/36 + 1/36 + 1/36 + 1/36 + 1/36 = 1/6 \qquad (38\text{-}53)$$

the probability of throwing a sum of 7 with the two dice is $1/6$.

Let us consider a more interesting event E_2 the event that the sum of the numbers on the dice is divisible by 3. The 12 sample points corresponding to E_2 are labeled 2 in Fig. 38-8. Since each has probability $1/36$, the probability of E_2 is

$$p(E_2) = 12(1/36) = 1/3 \qquad (38\text{-}54)$$

We could consider other interesting events, such as "the product of the numbers divided by their sum is greater than 2," but it would not add much to our understanding. A more instructive exercise is to change the probabilities to be associated with each sample point in this experiment. Suppose that we are still rolling a red die and a green die together, but that both dice are oddly shaped and biased in favor of larger numbers. The probabilities of throwing each number with each die are, in fact,

$$\begin{array}{lll} p(1) = 1/36 & p(2) = 3/36 & p(3) = 5/36 \\ p(4) = 7/36 & p(5) = 9/36 & p(6) = 11/36 \end{array} \qquad (38\text{-}55)$$

The sample points for the experiment are still the same, but the probability to be associated with them has changed. The probability of the sample point "2 on red die, 5 on green die" is now $p(2)p(5) = 3/36 \times 9/36 = 27/1{,}296$, rather than $1/36 = 36/1{,}296$ as before. The new probabilities of each sample point calculated in this way are shown in Fig. 38-9.

The same sample points are still associated with the events E_1 and E_2, only now the probabilities of those points are different. To find the probability of E_1 we must add the probabilities in Fig. 38-9 for the 6 sample points corresponding to E_1 in Fig. 38-8.

$$p(E_1) = 1/1{,}296(11 + 27 + 35 + 35 + 27 + 11) = 146/1{,}296 \qquad (38\text{-}56)$$

The probability of throwing a sum of 7 with the biased dice is considerably less than the probability of throwing it with the fair dice ($1/6 = 216/1{,}296$).

Finding the probability of E_2, throwing a sum divisible by 3, also requires summing the new probabilities of the 12 old sample points. We find

$$\begin{aligned} p(E_2) &= 1/1{,}296(3+3+9+21+25+21+9+55+63+63+55+121) \\ &= 448/1{,}296 = {}^{28}\!/_{81} \end{aligned}$$

The probability of throwing a sum divisible by 3 is $\frac{1}{81}$ greater than it was with the fair dice.

The Probability Tree. Sometimes it is convenient to draw the sample space in another form, called a probability tree. A probability tree is most helpful when the results of the experiment become known sequentially rather than simultaneously, but it may be used for any experiment. Figure 38-10 illustrates the general form of the tree, although it is drawn for a specific example. We construct such a tree by drawing outward, from a point called a node, a directed line segment or branch for every possible outcome of the first part of the experiment. Each branch is labeled with the outcome and with the probability of that outcome. Then we use the ends of the first branches as the beginning nodes for another set of branches corresponding to each

Number on red die \ Number on green die	1	2	3	4	5	6
1		②			②	①
2	②			②	① ✓	
3			②	①		②
4		②	①		②	
5	②	①		②		
6	①		②			②

FIG. 38-8. Sample space for a roll of two colored fair dice. Each sample point has probability $\frac{1}{36}$.

Number on red die \ Number on green die	1	2	3	4	5	6
1	1	3	5	7	9	11
2	3	9	15	21	27	33
3	5	15	25	35	45	55
4	7	21	35	49	63	77
5	9	27	45	63	81	99
6	11	33	55	77	99	121

FIG. 38-9. Sample space for a roll of two colored biased dice. The probability of each sample point is written in the corresponding box with denominator 1,296 understood.

possible second outcome of the experiment. These branches are labeled with the *conditional* probability of each second outcome given the first outcome. The process is continued until all possible sequences of outcomes of the experiment have been accounted for. The probabilities on the branches diverging from a node are all conditional on the outcomes that preceded that node. The tips of the tree represent the set of sample points for the experiment. The probability of each tip, or sample point, is computed by multiplying together the probabilities on the set of branches that lead from the beginning node of the tree to that tip.

Even when an experiment is not sequential, we can view it as sequential if we find it convenient. For example, in our dice-tossing problem we can say arbitrarily that the number on the red die is revealed before the number on the green die. This convention will not affect any of the probabilities of events based on the original experiment. The probability tree for the toss of the colored dice is shown in Fig. 38-10. Events R_i and G_j denote the events i observed on the red die, j on the green die, respectively. The 36 tips of the tree correspond to the 36 sample points in Fig. 38-8; the sample point indicated by a check in Fig. 38-8 is also indicated by a check in Fig. 38-10. The probabilities of the sample points are computed for the fair dice from

$$p(R_i) = \tfrac{1}{6} \qquad p(G_j|R_i) = p(G_j) = \tfrac{1}{6} \tag{38-57}$$

38-16   MATHEMATICS ASSOCIATED WITH SYSTEM ENGINEERING

Therefore, a probability of $\frac{1}{36}$ will be associated with each tip as in Fig. 38-8. If the dice are the biased variety, then

$$p(R_i) = p(i) \qquad p(G_j|R_i) = p(G_j) = p(j)$$

where $p(1), p(2), \ldots, p(6)$ are specified by Eq. (38-55). The probabilities recorded at the tips of the tree will then be the values given in Fig. 38-9.

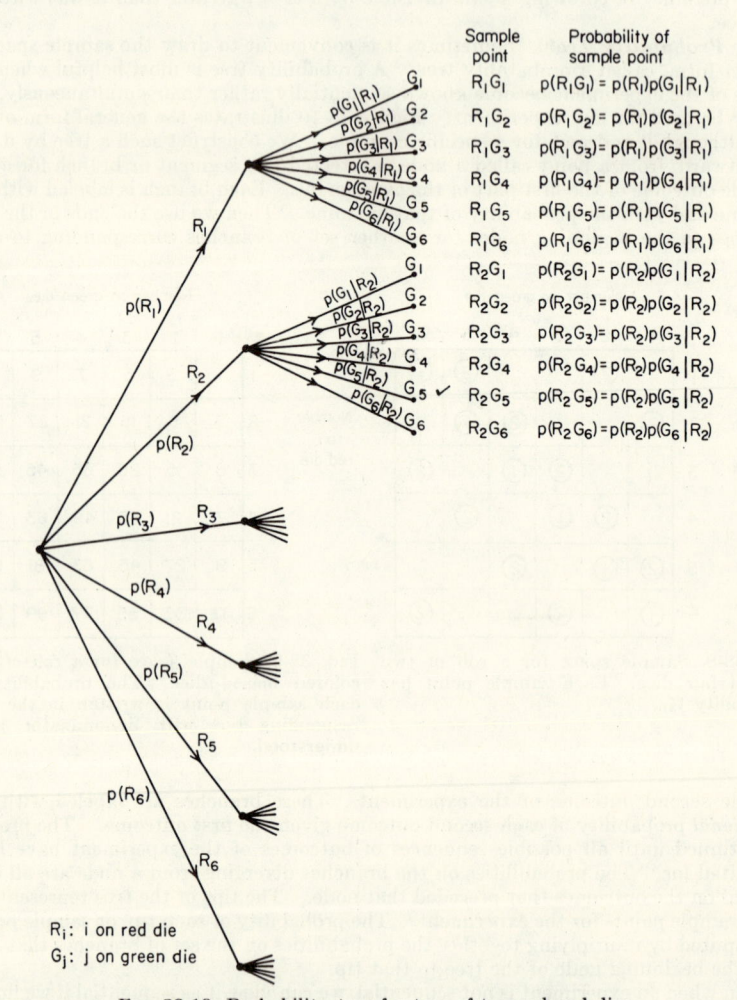

Sample point	Probability of sample point
R_1G_1	$p(R_1G_1) = p(R_1)p(G_1|R_1)$
R_1G_2	$p(R_1G_2) = p(R_1)p(G_2|R_1)$
R_1G_3	$p(R_1G_3) = p(R_1)p(G_3|R_1)$
R_1G_4	$p(R_1G_4) = p(R_1)p(G_4|R_1)$
R_1G_5	$p(R_1G_5) = p(R_1)p(G_5|R_1)$
R_1G_6	$p(R_1G_6) = p(R_1)p(G_6|R_1)$
R_2G_1	$p(R_2G_1) = p(R_2)p(G_1|R_2)$
R_2G_2	$p(R_2G_2) = p(R_2)p(G_2|R_2)$
R_2G_3	$p(R_2G_3) = p(R_2)p(G_3|R_2)$
R_2G_4	$p(R_2G_4) = p(R_2)p(G_4|R_2)$
R_2G_5	$p(R_2G_5) = p(R_2)p(G_5|R_2)$
R_2G_6	$p(R_2G_6) = p(R_2)p(G_6|R_2)$

R_i: i on red die
G_j: j on green die

Fig. 38-10. Probability tree for toss of two colored dice.

The calculations of the probabilities of any events based on the experimental outcome are performed with the tree just as they were in the sample space. The probabilities of the sample points corresponding to the event are summed to obtain the probability of the event. The choice of a sample space in regular or tree form depends upon the structure of the problem. Trees seem to be most useful in sequential and multidimensional problems.

Random Variables. Much of modern probability theory is concerned with operations on random variables. A random variable is a variable whose behavior has been described by a probability function. We can construct a random variable by assigning some real number to each sample point in the sample space. Then each sample point would have associated with it both a probability and a value of the number, called a random variable, associated with the experiment. Every outcome of the experiment will then produce some value of the random variable.

We can illustrate the concept of random variable by using our dice experiment. Let the random variable x associated with the experiment be the product of the numbers on the two dice. The value of this variable for each sample point is shown in Fig. 38-11. We see that the value of the random variable produced by the experiment can be as low as 1 or as high as 36; many but not all of the integers between these numbers could also be produced by the experiment. Much of our later work will be concerned with the description and manipulation of random variables.

Number on red die \ Number on green die	1	2	3	4	5	6
1	1	2	3	4	5	6
2	2	4	6	8	10	12
3	3	6	9	12	15	18
4	4	8	12	16	20	24
5	5	10	15	20	25	30
6	6	12	18	24	30	36

FIG. 38-11. A random variable for the dice experiment. Entry in each box is the value of the random variable; x = product of numbers on dice at the sample point.

Expectation. A very simple but extremely important concept in probability theory is that of the "expectation" of a random variable. The expectation is quickly defined; it is the sum over the entire sample space of the product of the probability and the value of the random variable associated with each sample point. If S_i is the ith sample point, $x(S_i)$ is the value of the random variable at that sample point, and $p(S_i)$ is the probability of the sample point, then the expectation of the random variable x is defined by

$$E(x) = \sum_{\text{all } i} x(S_i)p(S_i) \tag{38-58}$$

The expectation of x is sometimes called the "expected value" of x, the "mean" of x, and the "average value" of x, among other terms. Symbolically it is variously represented by

$$E(x) = \bar{x} = \langle x \rangle \tag{38-59}$$

But no matter what its name or symbol, it is a central concept in probability theory. Although we are getting ahead of our story, it is worth noting that the expectation of the random variable is an especially important number if the experiment is to be repeated many times. If we summed the values of the random variable produced by these experiments and divided by the number of experiments, we would expect the value of this quantity to approach the expectation of the random variable more and more closely as the number of experiments increased.

The expectation of the random variable defined as the product of the numbers on the dice will depend on whether the dice are fair or biased. If the dice are fair then the expectation is calculated by multiplying each value in Fig. 38-11 by the probability of each sample point, $\frac{1}{36}$. We obtain

$$\begin{aligned} E(x) &= \tfrac{1}{36}\ (\text{sum of all entries in array of Fig. 38-11}) \\ &= (\tfrac{1}{36})441 \\ &= 12.25 \end{aligned} \tag{38-60}$$

If the dice are the biased type we considered, then the expectation of x must be calculated by multiplying each value of the random variable in Fig. 38-11 by the corresponding probability of Fig. 38-9 and then summing. We find

$$\begin{aligned} E(x) &= 1/1{,}296 \times 1 + 3/1{,}296 \times 2 + 5/1{,}296 \times 3 + \cdots \\ &= 20.00 \end{aligned} \tag{38-61}$$

38-18 MATHEMATICS ASSOCIATED WITH SYSTEM ENGINEERING

The biased dice produce a considerably higher value of expectation for the product than do the fair dice. The result is reasonable because both dice in the biased pair are biased in favor of yielding higher numbers.

38-5. Representations of Discrete Random Variables

If a sample space has only a denumerable number of sample points, then any random variable associated with that sample space can take on only a denumerable set of values. Such a variable we shall call a discrete random variable. Although we have previously represented this variable by entering its value at each sample point, we have many alternative representations available to us. If all questions about the experiment can be expressed in terms of the values of the random variable, then these alternative representations will be both sufficient and convenient; however, if we ask other kinds of questions, then only the complete sample space will provide the answers.

The alternate representations available are "probability distributions" of the random variable. Note in Fig. 38-11 that, although there are 36 distinct sample points, much fewer than 36 values of the random variable can be generated. Suppose that we marked these values on a line and then drew at each mark a spike whose height was proportional to the probability that that value would be generated by the experiment. A diagram of this type for the tossing of fair dice appears in Fig. 38-12. We shall call it the probability mass function of the random variable because it shows

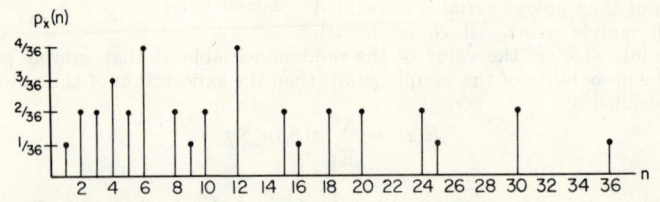

Fig. 38-12. Probability mass function for x, product of numbers on two fair dice.

how the total probability "mass" of one unit is distributed among the possible values of the random variable. The probability mass function is computed by totaling the probability of the sample points associated with each value of the random variable. We shall let $p_x(n)$ be the probability that the random variable x assumes the value n. In many problems, like this one, x can be restricted to be a nonnegative integer. We shall consider cases where this assumption is not satisfied in the next section.

Note that any question about the probability of various products of numbers can be answered by knowing the probability mass function of this random variable. For example, the probability that the product will lie between 14 and 22, inclusive, is just the sum of the heights of the spikes between those two points in the probability mass function of Fig. 38-12,

$$p(14 \leq x \leq 22) = \tfrac{2}{36} + \tfrac{1}{36} + \tfrac{2}{36} + \tfrac{2}{36} = \tfrac{7}{36}$$
$$= \sum_{n=14}^{22} p_x(n) \qquad (38\text{-}62)$$

However, if we should want to ask a question about the sum of the numbers on the dice, then we would have to return to the original sample space or construct a new probability mass function for the sum.

Moments. Let $g(x)$ be any single-valued function of the random variable x. Every point in the sample space could now be labeled with the value of $g(x)$ corresponding to the value of x at that point. Thus $g(x)$ is really just another random variable of the experiment. Consequently, the expectation of $g(x)$ is just the sum over the sample space of its values at each sample point multiplied by the probability

PROBABILITY

of that sample point. However, the total probability that $g(x)$ will take on some value is just the height of the probability mass function at the value of x producing that value of $g(x)$. Therefore, the expectation of $g(x)$ is given by

$$E[g(x)] = \sum_{n=0}^{\infty} g(n)p_x(n) \tag{38-63}$$

Although we work with many different functions g, the most common are those which generate the "moments" of the random variable x. The kth moment of the random variable x is $E(x^k)$ so that

$$E(x^k) = \overline{x^k} = \sum_{n=0}^{\infty} n^k p_x(n) \tag{38-64}$$

The values of $\overline{x^k}$ for small values of k are generally most important. First we note that

$$\overline{x^0} = 1 \tag{38-65}$$

because the sum of the heights of all the spikes in the probability mass function is 1. The mean $\overline{x^1} = \bar{x}$ is the most important moment. It is equal to

$$\bar{x} = \sum_{n=0}^{\infty} n p_x(n) \tag{38-66}$$

in complete agreement with (38-58). If we calculate x by multiplying the height of each spike by the value of x at which it occurs we obtain

$$\bar{x} = 12.25 \tag{38-67}$$

the same result as in (38-60).

The second moment $\overline{x^2}$ is often of interest. It is equal to

$$\overline{x^2} = \sum_{n=0}^{\infty} n^2 p_x(n) \tag{38-68}$$

In fact, we can calculate any moment we wish from (38-64).

All these moments are called moments about the origin ($n = 0$). In some situations we also want to calculate moments about the mean, or expectation, of the random variable x. The kth moment about the mean is defined by

$$E[(x - \bar{x})^k] = \overline{(x - \bar{x})^k} = \sum_{n=0}^{\infty} (n - \bar{x})^k p_x(n) \tag{38-69}$$

We note that the 0th moment about the mean is 1, while the first moment about the mean is zero. The most interesting moment about the mean is the second moment about the mean $\overline{(x - \bar{x})^2}$. It is so interesting that it has been given a special name, the variance. We shall denote the variance of the random variable x by $\check{x}$. We can then write

$$\check{x} = E[(x - \bar{x})^2] = \overline{(x - \bar{x})^2} = \sum_{n=0}^{\infty} (n - \bar{x})^2 p_x(n) \tag{38-70}$$

The square root of the variance is called the standard deviation; we shall denote it by $\overset{s}{x}$.

We can easily derive an important property of the variance

$$\check{x} = \sum_{n=0}^{\infty} (n - \bar{x})^2 p_x(n) = \sum_{n=0}^{\infty} n^2 p_x(n) - 2\bar{x} \sum_{n=0}^{\infty} n p_x(n)$$
$$+ \bar{x}^2 \sum_{n=0}^{\infty} p_x(n) = \overline{x^2} - 2\bar{x}^2 + \bar{x}^2 = \overline{x^2} - \bar{x}^2 \quad (38\text{-}71)$$

The variance of a random variable is the difference between its second moment and its first moment squared. We shall have many occasions to use this relation.

Cumulative Distributions. The probability mass function is not the only useful representation for a discrete random variable. Any random variable can be effectively represented by means of cumulative distributions. Let $P_{x \leq}(y)$ be the probability that a random variable x will take on a value less than or equal to y. The function $P_{x \leq}(\cdot)$ is called the cumulative probability distribution of the random

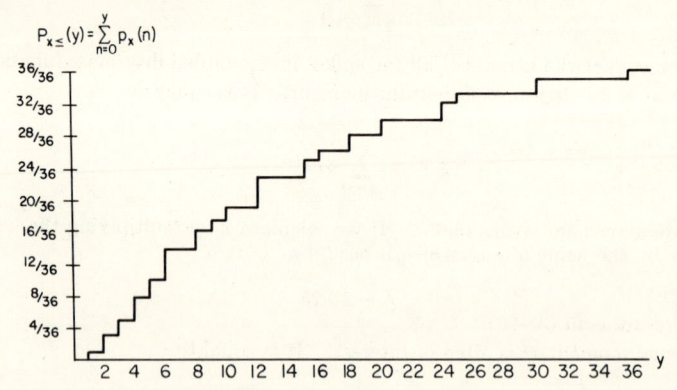

FIG. 38-13. Cumulative probability distribution for x, product of numbers on two fair dice.

variable, or sometimes the distribution function of the random variable. For a discrete random variable, the cumulative probability distribution related to the probability mass function would be

$$P_{x \leq}(y) = \sum_{n=0}^{y} p_x(n) \quad (38\text{-}72)$$

If y lies between two integers, we take the value of the cumulative distribution to be the value at the next lower integer. With this definition $P_{x \leq}(y)$ is just the sum of the heights of all spikes in the probability mass function at or to the left of y; the function is defined for $0 \leq y < \infty$ (and could equally well be defined for $-\infty < y < \infty$). Figure 38-13 shows the cumulative probability distribution for the product of the numbers on two fair dice computed using the probability mass function of Fig. 38-12. The heights of the steps in $P_{x \leq}(\cdot)$ are the heights of the spikes at the corresponding point in $p_x(\cdot)$.

We can calculate the probability that the product will lie between 14 and 22, inclusive, by using the cumulative probability distribution of Fig. 38-13. We write

$$p(14 \leq x \leq 22) = P_{x \leq}(22) - P_{x \leq}(13)$$
$$= {}^{30}\!/_{36} - {}^{23}\!/_{36}$$
$$= {}^{7}\!/_{36} \quad (38\text{-}73)$$

This result agrees with that obtained in Eq. (38-62).

PROBABILITY

In some problems we find it convenient to work with the probability that a random variable will exceed a certain specified value. Let $P_{x>}(y)$ be the probability that the random variable x will be strictly greater than the number y. We shall call the function $P_{x>}(\cdot)$ the complementary cumulative probability distribution of x; its value at any point is 1 minus the value of the cumulative probability distribution at that point,

$$P_{x>}(y) = 1 - P_{x\leq}(y) \tag{38-74}$$

The complementary cumulative probability distribution of x computed using this equation is shown in Fig. 38-14. The complementary cumulative probability distribution has a very important property. Notice the shaded rectangle in Fig. 38-14. Its width is 6 units; its height is the height of the spike at the value 6 in the probability mass function of Fig. 38-12. Its area is then the contribution of this spike in the calculation of the expectation of x. The area under the complementary cumulative

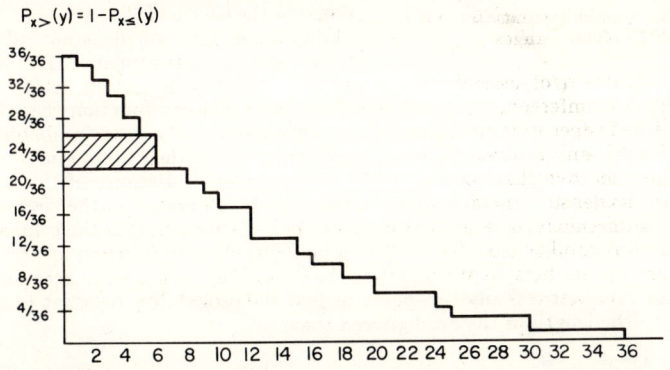

Fig. 38-14. Complementary cumulative probability distribution for x, product of numbers on two fair dice.

probability distribution is just the sum of these contributions for all spikes in the probability mass function. The area under the complementary cumulative probability distribution is therefore just the expectation of the nonnegative random variable x; for example, 12.25 from Eq. (38-60) or (38-67). This result is extremely useful in a variety of practical situations.

38-6. Continuous Sample Spaces and Random Variables

The restriction of the last section that allowed a random variable to take on only nonnegative integer values is burdensome. We could extend it trivially to other discrete values, but we face many probability problems where we can, with advantage, consider the random variable to vary over a continuum. Suppose, for example, that we were required to report the result of spinning a wheel of fortune like that shown in Fig. 38-15. The wheel is perfectly balanced; its circumference is unity. To perform our reporting task we might divide the circumference into the 10 ranges, 0–0.09999 . . . , 0.1–0.19999 . . . , . . . , 0.9–0.99999 . . . . Then we could report in which of these 10 ranges the pointer stopped. Because the wheel is perfectly balanced, we would assign the probability 0.1 to having the pointer stop in each of the regions. We could represent this probability assignment by constructing the prob-

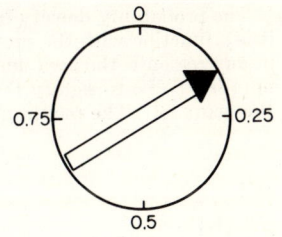

Fig. 38-15. The wheel of fortune.

ability mass function of Fig. 38-16 with spikes of height 0.1 at the midpoint of each interval.

The trouble with this idea is that with this mass function we shall not be able to answer questions like, what is the probability that the pointer will stop in the range from 0.23 to 0.239999. We could answer this question by dividing the circumference into 100 equal arcs and then assign a probability $1/100$ to each of them. But this would not end our difficulties—any question about the position of the pointer that could not be expressed in terms of our chosen intervals could not be answered.

We have the feeling that the answers to questions about the pointer's position should be easy to find if we choose the right method of assigning the amount of probability to each pointer position. We really know the answer we want—we want the probability that the pointer will lie in any arc to be equal to the length of that arc—but the probability mass function does not allow us to make such an assignment. The answer is intuitive. Instead of assigning the probability mass to a certain set of intervals around the circumference, let us define a probability density function that shows the amount of mass per unit circumference at every point. The total probability to be associated with any interval on the circumference would then be the integral of this density function over that portion of the circumference. Because of the balance of the wheel this density function would be the same at all points on the circumference. The only requirements on it are that it cannot be negative and that the total area over it, the total probability mass, be 1. The probability density function for the wheel of fortune that meets these requirements is shown in Fig. 38-17. The area under this density function between any two points is just the probability we want to assign to that range—the length of the arc between them.

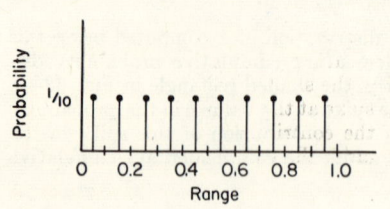

FIG. 38-16. Probability mass function for wheel with 10 equal ranges.

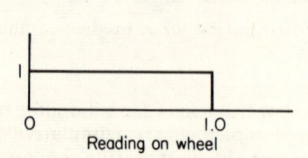

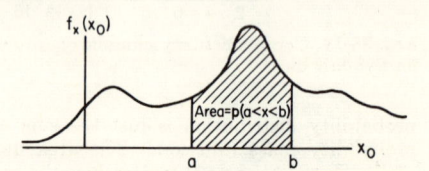

FIG. 38-17. Density function for the wheel of fortune.

FIG. 38-18. A probability density function.

The probability density function of a random variable is thus a very simple idea—it is a function with the property that the probability of finding the random variable in any region is the area under the density function in that region. We shall use the notation $f_x(x_0)$ to signify the density function of the random variable x evaluated at the point x_0. The requirements that a probability density function must satisfy are

$$f_x(x_0) \geq 0 \qquad -\infty < x_0 < \infty \qquad (38\text{-}75)$$

and

$$\int_{-\infty}^{\infty} dx_0 \, f_x(x_0) = 1 \qquad (38\text{-}76)$$

because probabilities cannot be negative and the total probability assigned must be 1.

Of course, not all probability density functions are constants like the one in Fig. 38-17. They can look quite strange, like the one in Fig. 38-18. Nevertheless, even for the density function in Fig. 38-18, the probability that the random variable x will lie between a and b, $b > a$, is the area under the density function between a and b,

$$p(a < x < b) = \int_a^b dx_0 \, f_x(x_0) \qquad (38\text{-}77)$$

This area is shaded in Fig. 38-18.

A random variable described by a probability density function can take on any value between $-\infty$ and $+\infty$. An experiment that produced such a variable could thus have a number of different outcomes in any finite interval. The concept of sample space must be extended to allow for this possibility. The extension is simple. A continuous sample space is a sample space constructed to describe the outcome of an experiment producing continuous variables. It is a region that again has the property that there is a one-to-one correspondence between outcomes of the experiment and points in the region, but now the region is continuous—it has a nondenumerable number of points. Instead of tagging each point with a probability as we did in discrete sample spaces, we construct a probability density function over the region. Any event concerned with the outcome of the experiment can still be represented as a collection of points in the sample space, often an area. The probability of the event is still the sum of the probability of all its associated sample points, but in this case, the probability is the volume of the density function over the area of the event in the sample space. This volume must usually be calculated by integration.

This description of a continuous sample space and the calculations based upon it will have more meaning after we investigate some physical examples, but it is easily interpreted in terms of the wheel of fortune of Fig. 38-15. The experiment of spinning the wheel of fortune will produce as an outcome a number x in the range $0 \leq x < 1$. An appropriate sample space for this experiment is just the line segment between 0 and 1 on the real axis illustrated in Fig. 38-19. The set of sample points associated with the event $0.25 < x \leq 0.75$ is the region of heavy line in this figure. The probability

FIG. 38-19. Sample space for spin of wheel of fortune.

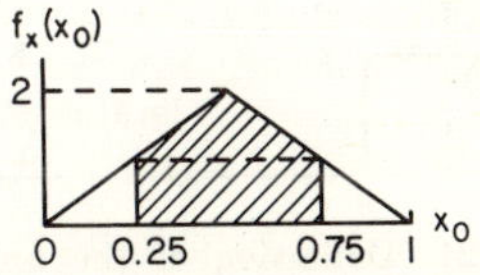

FIG. 38-20. Probability density function for biased wheel of fortune.

of the event is the sum of the probabilities of all sample points in this region. If the wheel is perfectly balanced, the probability density function of Fig. 38-17 will be assigned over the sample space. The probability that x will lie in the desired region is just the area of the density function between 0.25 and 0.75 which is, of course, 0.5.

Suppose, however, that the wheel is not balanced because the pointer has a heavy tip and the wheel is suspended vertically with the orientation of Fig. 38-15. Then we would expect more spins to give values near 0.5; indeed, if there were no friction, all spins would yield 0.5. The probability density function for this wheel might be that shown in Fig. 38-20. Notice that this triangular density function must rise to 2 in order to have an area of 1. The probability that a spin will produce a value between 0.25 and 0.75 is still the amount of probability over that region, but it is now given by the shaded area of the density function of Fig. 38-20. This area is 0.75; it may be calculated by using geometry or calculus.

The idea of a continuous random variable and a continuous sample space is intuitively acceptable, but to make it rigorous requires a branch of mathematics called measure theory. Although a measure-theoretic background is essential for those who want to do research in probability theory itself, there is no necessity that every user of the theory understand it.

The situation is completely analogous to the problem of the structural designer in civil engineering. When the designer uses discrete loads on his structures, he is working with functions just like probability mass functions. When he deals with a distributed load such as a layer of concrete he would think it foolish to attempt to treat it as a very large number of very small discrete loads. Instead, he would adopt the concept of load density in pounds per square foot, and obtain the loading on any area by integrating this density over the area. In performing this operation he is doing the same kind of mathematics we use in working with probability density functions. Structural designers do not seem to suffer from not knowing measure theory; neither shall we in working practical problems. However, to place the issue in proper perspective we must emphasize that measure theory is essential to the theoretical development of probability theory itself.

The Probability Mass Function Revisited. What is the relationship between probability mass functions and probability density functions? They are both useful concepts, but we can make our thinking more general by realizing that every probability mass function can also be considered to be a probability density function. The key to this interpretation is that a finite point of mass produces an infinite density of mass at that point, but an infinite density that integrates to the actual mass at the point over any region containing the point. This is just the concept of an impulse function, a function that is zero everywhere but at one point, is infinite at that point, and yet has a finite area. We shall define the unit impulse function to be the limit of the function shown in Fig. 38-21 as the parameter h approaches zero. We shall denote the unit impulse function by $\delta(x)$; it is drawn as a vertical arrow with ∞ written at the tip and its finite area, 1, written beside it. The probability density function that is itself a unit impulse would have its total probability mass of 1 concentrated at the origin.

We now have the means for representing probability mass functions by probability density functions. All we do is replace the finite mass at a point by an impulse with the same area. When this has been done for all finite mass points, we have a probability density function. If all the finite masses occur at integral values as they did

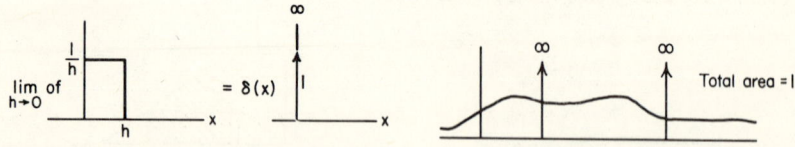

FIG. 38-21. Definition of the unit impulse function.

FIG. 38-22. A mixed density function.

in Sec. 38-5, then the probability mass function is more convenient than the probability density function because we never even need calculus. However, when the finite mass points occur with irregular spacings, then the density-function conversion is useful.

Of course, if a random variable can take on a continuous range of variables, then the density-function concept is indispensable. We sometimes encounter an experiment that can produce either one of a finite set of numbers or a continuous variable covering the same range. The density function for this experiment will be finite at some regions and described by impulses at certain points. Figure 38-22 illustrates such a probability density function, called a mixed density function. The use of the impulse function allows us to treat this type of density function by ordinary calculus rather than the Stieltjes integration used in measure theory.

Expectations and Moments. The expectations or moments of random variables described by density functions are conceptually the same quantities we defined for discrete random variables. The only difference is that the operation required to find them is integration rather than summation. If $g(x)$ is a function of a random variable x, then the expected value of $g(x)$ is defined by

$$E[g(x)] = \overline{g(x)} = \langle g(x) \rangle = \int_{-\infty}^{\infty} dx_0 \, g(x_0) f_x(x_0) \tag{38-78}$$

This expression reduces to that of (38-63) if x is a random variable that can only be a nonnegative integer. The kth moment of x is

$$E(x^k) = \overline{x^k} = \int_{-\infty}^{\infty} dx_0 \, x_0^k f_x(x_0) \tag{38-79}$$

Again we note that the zeroth moment $\overline{x^0}$ is equal to 1, and that the first and second moments $\overline{x^1} = \bar{x}$ and $\overline{x^2}$ are the ones most often computed.

PROBABILITY

The kth moment about the mean is again defined by

$$E(x - \bar{x})^k = \overline{(x - \bar{x})^k} = \int_{-\infty}^{\infty} dx_0 \, (x_0 - \bar{x})^k f_x(x_0) \tag{38-80}$$

The zeroth moment about the mean is equal to 1; the first moment about the mean is equal to zero. The second moment about the mean, the variance $\check{x}$, is again equal to the second moment minus the first moment squared:

$$\check{x} = \overline{(x - \bar{x})^2} = \int_{-\infty}^{\infty} dx_0 \, (x_0 - \bar{x})^2 f_x(x_0) = \overline{x^2} - \bar{x}^2 \tag{38-81}$$

The probability density function for the balanced wheel of fortune, shown in Fig. 38-17, has a mean $\bar{x} = \frac{1}{2}$, a second moment $\overline{x^2} = \frac{1}{3}$, and therefore a variance $\check{x} = \frac{1}{12}$. The probability density function for the biased wheel of fortune, shown in Fig. 38-20, also has a mean $\bar{x} = \frac{1}{2}$, but its second moment $\overline{x^2}$ is $\frac{7}{24}$, and consequently it has a variance $\check{x} = \frac{1}{24}$. In general, distributions with lower variances are more closely concentrated around their means. If we imagine the graph of the density function to be cut out of a metal plate, then the variance of the random variable is proportional to the moment of inertia of the cutout about its center of gravity. This physical model is often helpful in understanding theoretical arguments.

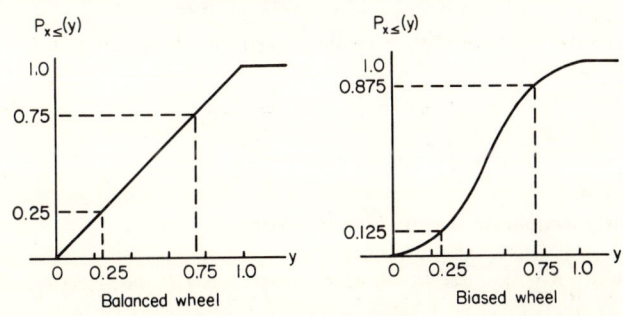

Fig. 38-23. Cumulative probability distributions for wheel of fortune.

Cumulative Distributions. We can easily derive the cumulative distributions associated with a continuous random variable. The probability that a random variable x will assume a value less than or equal to a number y is just the area under the density function between $-\infty$ and y. We again write $P_{x \leq}(y)$ for this cumulative probability; then

$$P_{x \leq}(y) = \int_{-\infty}^{y} dx_0 f_x(x_0) \tag{38-82}$$

The cumulative probability distribution has the following two properties:

$$P_{x \leq}(-\infty) = 0 \qquad P_{x \leq}(\infty) = 1 \tag{38-83}$$

As the argument of the function increases from $-\infty$ to $+\infty$, the cumulative probability distribution rises from 0 to 1. We see from (38-82) that the rise is monotonically nondecreasing because of the nonnegativity of $f_x(\cdot)$.

The cumulative probability distributions of the readings of the balanced and biased wheels of fortune are shown in Fig. 38-23. They were drawn by using the density functions of Figs. 38-17 and 38-20 in Eq. (38-82). Note that since

$$p(0.25 < x \leq 0.75) = P_{x \leq}(0.75) - P_{x \leq}(0.25) \tag{38-84}$$

we obtain 0.50 for this probability for the balanced wheel and 0.75 for the biased wheel in agreement with our earlier results.

We can still define a complementary cumulative probability distribution as 1 minus the cumulative probability distribution. We let $P_{x>}(y)$ be the probability that x will

exceed y; then
$$P_{x>}(y) = 1 - P_{x\leq}(y) \tag{38-85}$$

The complementary cumulative distributions for the wheel of fortune are shown in Fig. 38-24. They still have the important property that the area under them between 0 and ∞ is the mean of a random variable whose range is $(0, \infty)$, as the figure verifies.

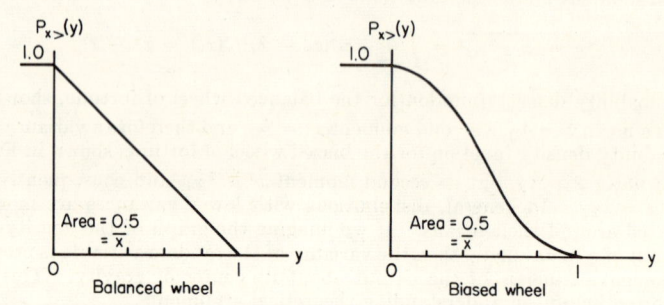

FIG. 38-24. Complementary cumulative probability distributions for wheel of fortune.

However, suppose that we have a random variable whose range is $(-\infty, \infty)$. Its mean is
$$\bar{x} = \int_{-\infty}^{\infty} dx_0\, x_0 f_x(x_0) \tag{38-86}$$
then
$$\bar{x} = \int_0^{\infty} dx_0\, x_0 f_x(x_0) + \int_{-\infty}^0 dx_0\, x_0 f_x(x_0) \tag{38-87}$$

We apparently complicate the equation by writing
$$\bar{x} = \int_0^{\infty} dx_0 \int_0^{x_0} d\alpha\, f_x(x_0) - \int_{-\infty}^0 dx_0 \int_{x_0}^0 d\alpha\, f_x(x_0) \tag{38-88}$$

But after we interchange the order of integration,
$$\bar{x} = \int_0^{\infty} d\alpha \int_{\alpha}^{\infty} dx_0\, f_x(x_0) - \int_{-\infty}^0 d\alpha \int_{-\infty}^{\alpha} dx_0\, f_x(x_0)$$
$$= \int_0^{\infty} dx\, P_{x>}(\alpha) - \int_{-\infty}^0 d\alpha\, P_{x\leq}(\alpha) \tag{38-89}$$

Therefore, the mean of any random variable with range $(-\infty, \infty)$ is equal to the area under its complementary cumulative probability distribution in the range $(0, \infty)$ less the area under its cumulative probability distribution in the range $(-\infty, 0)$.

The definition of the cumulative probability distribution in (38-82) shows that the probability density function is the derivative of the cumulative probability distribution,
$$f_x(x_0) = (d/dx_0) P_{x\leq}(x_0) \tag{38-90}$$

By virtue of (38-85), the density function is the negative of the derivative of the complementary cumulative probability distribution
$$f_x(x_0) = -(d/dx_0) P_{x>}(x_0) \tag{38-91}$$

The density functions and cumulative distributions for the wheel of fortune confirm these findings.

However, suppose we have a mixed density function like the one in Fig. 38-22. The cumulative distributions for such a function will have a jump at every impulse with height equal to the area of the impulse. A cumulative probability distribution that might result is shown in Fig. 38-25. If we desire to find the density function corresponding to this cumulative distribution by differentiation, all we have to do is differentiate the function at all points at which it is differentiable and then supply an impulse at each point of discontinuity with area equal to the height of the discontinuity. With this understanding, we may always use Eqs. (38-90) and (38-91) for finding density functions from cumulative distributions by differentiation.

38-7. Multidimensional Sample Spaces

We often encounter situations where a single random variable is not sufficient to describe the outcome of an experiment. We may need a pair, or three, or even a very large number of random variables to

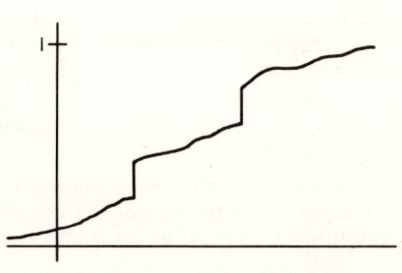

FIG. 38-25. The cumulative probability distribution of a mixed density function.

have a unique description of what took place. We shall illustrate the concepts necessary by using the two-variable case, but they all apply to any number of random variables.

Suppose that the experiment we perform is two spins of the wheel of fortune of Fig. 38-15. The outcome of the experiment is now two random variables, both in the range (0,1). Let x be the outcome of the first spin, y the outcome of the second. Every outcome of the experiment can be represented by a point in that unit square in the x,y plane shown in Fig. 38-26; this square is the sample space for the experiment.

Any event depending on the outcome of the experiment can be represented by an area or a set of areas in this plane. For example, the event that the sum of the squares of the readings is less than $\frac{1}{4}$ is represented by the circular quadrant of radius $\frac{1}{2}$ shown shaded in the figure. Any point in the quadrant meets the requirement; any point outside it does not.

Now we must assign a probability distribution to the sample space. Suppose that the wheel is balanced. Then we know from the density function of Fig. 38-17 that the probability that the first spin will produce a result between a_1 and b_1, $b_1 > a_1$, is $b_1 - a_1$. The probability that the second spin will produce a result between a_2 and b_2, $b_2 > a_2$, is similarly $b_2 - a_2$. Since the spins are assumed to be independent, the probability that the first toss will be between a_1 and b_1 and the second toss will be between a_2 and b_2 is the product of these probabilities $(b_1 - a_1)(b_2 - a_2)$. But from Fig. 38-26, $(b_1 - a_1)(b_2 - a_2)$ is just the area of the event $(a_1 < x < b_1, a_2 < y < b_2)$ in the sample space. Therefore, we want to assign a probability density function of the sample space that will make the probability of any region equal to its area. The probability density function that accomplishes this task is shown in Fig. 38-27 in a

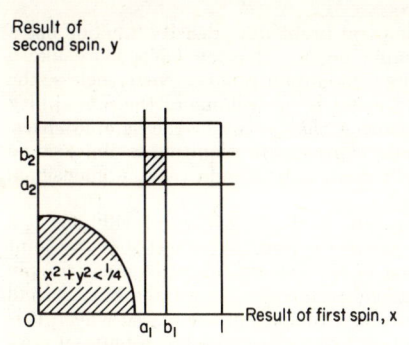

FIG. 38-26. Sample space for two spins of the wheel of fortune.

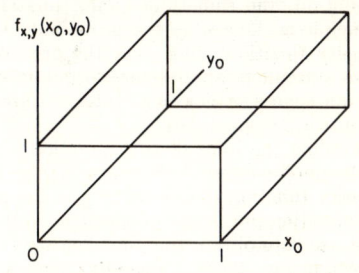

FIG. 38-27. Joint probability density function for two spins of balanced wheel of fortune.

three-dimensional sketch. It is a unit cube, a function whose value is 1 over the unit square. This probability density function assigns probabilities to units of area just as the one-dimensional probability density functions assigned probability to units of length. It is called the joint probability density function for the two random variables x and y. We use $f_{x,y}(x_0,y_0)$ to designate the value of this function at the point x_0, y_0. The joint probability density function for two spins of the balanced wheel of fortune is then

$$f_{x,y}(x_0,y_0) = \begin{cases} 1 & 0 \leq x_0 < 1, 0 \leq y_0 \leq 1 \\ 0 & \text{elsewhere} \end{cases} \quad (38\text{-}92)$$

The probability of any event concerned with the outcome of a two-random-variable experiment is just the volume of the probability density function lying over the area representing that event in the sample space. For example, the volume that lies over

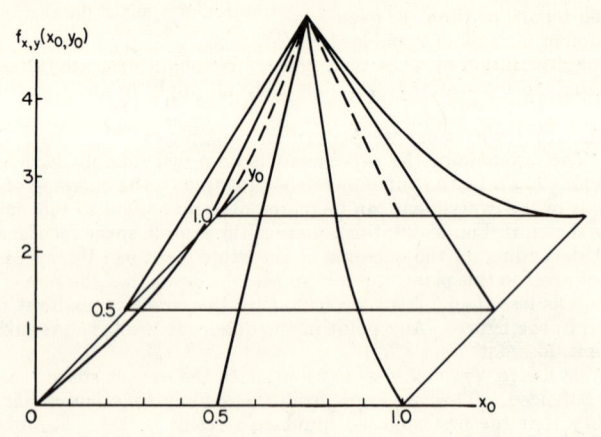

FIG. 38-28. Joint probability density function for two spins of biased wheel of fortune.

the area representing the event $x^2 + y^2 < \frac{1}{4}$ is just the volume of $\frac{1}{4}$ of a right circular cylinder of altitude 1 and radius $\frac{1}{2}$. Therefore,

$$p(x^2 + y^2 < \tfrac{1}{4}) = \tfrac{1}{4}\pi(\tfrac{1}{2})^2 = \pi/16 \quad (38\text{-}93)$$

Of course, it will seldom be the case that the joint probability density function has a form as simple as a constant over the sample space. It often looks more like a mountain, and sometimes, unfortunately, like a mountain range. Nevertheless, the principle is the same—the probability of any event is the volume of the probability density function lying over the area representing that event in the sample space. This statement applies for n-random-variable experiments requiring n-dimensional sample spaces as long as we interpret area and volume to be n and $(n + 1)$-dimensional quantities, respectively.

Finding the joint density function for two spins of the biased wheel will give us a little more exercise. We learned from the case of the balanced wheel that the joint density function of two spins was the product of the density functions for each spin because the spins were independent. We shall generalize this result later, but we shall first use it to obtain the joint density function for two spins of the biased wheel; it is shown in Fig. 38-28. The joint density function looks something like a tent—it rises to a peak of 4 at the center of the unit square and falls to zero at all edges. The analytic expression for this density function is

$$f_{x,y}(x_0,y_0) = \begin{cases} 16x_0y_0 & 0 \leq x_0 \leq \tfrac{1}{2},\ 0 \leq y_0 \leq \tfrac{1}{2} \\ 16x_0(1-y_0) & 0 \leq x_0 \leq \tfrac{1}{2},\ \tfrac{1}{2} < y_0 \leq 1 \\ 16(1-x_0)y_0 & \tfrac{1}{2} < x_0 \leq 1,\ 0 \leq y_0 \leq \tfrac{1}{2} \\ 16(1-x_0)(1-y_0) & \tfrac{1}{2} < x_0 \leq 1,\ 0 < y_0 \leq 1 \\ 0 & \text{elsewhere} \end{cases} \quad (38\text{-}94)$$

The probability of the event $x^2 + y^2 < \tfrac{1}{4}$ is now the volume of the tent that lies over the circular quadrant in the sample space of Fig. 38-26. The volume of this prism-shaped region is not obvious—its calculation would be very difficult for someone who did not know calculus. We shall calculate it, but before we do, there is an important point to be made. The probability part of the problem is over as soon as you have identified the volume of the figure that represents the probability you seek. Many remarks made about "difficult probability problems" really refer to difficult integrations or difficult summations. In a discrete sample space the problem is one of counting; in a continuous sample space it is an integration problem. But no matter which of these it is, *it is not a probability problem*. This remark may be small consolation to the one who has to do the integrating or the summing, but it is, at least, satisfying to have identified the nature of one's difficulty.

To get back to work, the area in the sample space corresponding to the event $(x^2 + y^2 < \tfrac{1}{4})$ lies in the region $(0 \leq x_0 \leq \tfrac{1}{2},\ 0 \leq y_0 \leq \tfrac{1}{2})$ in which

$$f_{x,y}(x_0,y_0) = 16x_0y_0 \quad (38\text{-}95)$$

Therefore

$$p(x^2 + y^2 < \tfrac{1}{4}) = \int dx_0 \int_{x_0^2+y_0^2<\frac{1}{4}} dy_0\, f_{x,y}(x_0,y_0) = \int dx_0 \int_{x_0^2+y_0^2<\frac{1}{4}} dy_0\, 16x_0y_0 \quad (38\text{-}96)$$

and finally,

$$p(x^2 + y^2 < \tfrac{1}{4}) = \int_0^{1/2} dx_0\, 4x_0 \int_0^{\sqrt{1/4 - x_0^2}} dy_0\, 4y_0 = \tfrac{1}{8} \quad (38\text{-}97)$$

The biased wheel causes the value of the probability to fall by a factor of $2/\pi$ from the value $\pi/16$ recorded in Eq. (38-94) for the balanced wheel.

We can solve a famous problem in probability by using the concepts we have already developed. The problem is usually stated in this way: A stick of unit length is broken at random into three pieces. What is the probability that the three pieces can be used to form a triangle?

The problem with this statement illustrates a common source of ambiguity in probability problems—the use of the phrase "at random." The phrase unfortunately has no meaning beyond the fact that a probabilistic process is being used. It is often interpreted as meaning "equally likely" in some sense, but it is often difficult to understand just what is supposed to be equally likely; in any event, a more precise statement is desirable. The safest practice is to avoid use of the phrase entirely and to state, instead, the probability distributions to be associated with the random variables.

Therefore, a more precise statement of the stick-breaking problem is that a stick of unit length has a mark placed on it at a point corresponding to the spin of a balanced wheel of fortune like the one in Fig. 38-15. A second independent spin then determines a second mark on the stick, and finally the stick is cut at the points where the marks are made.

If x is the location of the first mark and y the location of the second, then the triangle problem has the sample space of Fig. 38-26 and the joint probability density function of Fig. 38-27. To solve the problem we must find the points in the sample space that correspond to the possible formation of a triangle. The basic property of the three segments generated that will enable them to form a triangle is that the sum of the lengths of any two must not be greater than the length of the third. An alternate form of this requirement is that no segment can be greater than $\tfrac{1}{2}$ in length.

38-30 MATHEMATICS ASSOCIATED WITH SYSTEM ENGINEERING

The region in the sample space associated with this property of the segments is shown shaded in Fig. 38-29. It consists of two triangular regions symmetrically placed about the line $x_0 = y_0$.

The case where the second cut is to the right of the first, $y > x$, is represented by points above this line. In this case the three segments are x, $y - x$, and $1 - y$ in length. The joint requirement

$$x < \tfrac{1}{2} \qquad y - x < \tfrac{1}{2} \qquad 1 - y < \tfrac{1}{2} \tag{38-98}$$

defines the shaded triangle shown. We can use a similar argument to establish the corresponding region for $y < x$.

The probability of forming a triangle is the part of the volume of the joint probability density function of Fig. 38-27 that lies over this region. Since the area of the region corresponding to the event is equal to the area of a square of side $\tfrac{1}{2}$, and since the height of the density function is one, the volume is $\tfrac{1}{4}$. Therefore, the probability of forming a triangle from a stick broken in this way is $\tfrac{1}{4}$.

There are other ways to solve this problem, some of them tricky and, therefore, fascinating. However, in our professional life we must concentrate on methods and concepts that are *always* appropriate. Probability problems can be interesting mathematical puzzles, but most of the probabilistic problems we face yield only to systematic analysis.

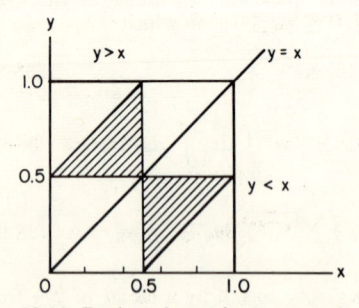

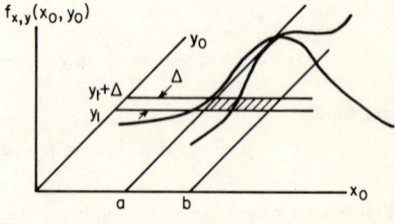

FIG. 38-29. Region of sample space pertinent to forming a triangle.

FIG. 38-30. An arbitrary joint density function.

I often encounter two people arguing about different solutions they have for the "same" probability problem. Since every probability problem has a unique solution, there must be some basic difference in their understanding if we rule out the possibility of numerical mistakes. Possible sources of misunderstanding are the statement of the experiment, the form of the sample space, the assignment of probability to it, the description of events to be investigated, and finally the identification of these events with points in the sample space. The last step of evaluating the probability volume of the events is, of course, a prime suspect for numerical error. With so many places to go wrong it is not surprising that there may be a divergence of opinion on the answer to a problem. The English language is not precise enough for probability calculations; that is why we introduced the algebra of events. But even with the right tools the work requires great care. In perhaps no other field is it so difficult for a man to become conceited about his problem-solving ability.

Marginal and Conditional Distributions. We now have the concepts of multidimensional sample spaces and of joint probability density functions over those spaces. There are several useful ideas that can be derived from these notions. Let us consider the arbitrary joint density function of two variables x and y shown in Fig. 38-30. For convenience we shall assume that x is the age and y is the income for an individual selected from a group. The probability that the individual has an age and income confined to a certain region is just the volume under this density function over that region. Once we know the joint density function we can easily answer any question

PROBABILITY

of this type. Suppose now that an individual is selected from the group and we are asked to find the probability density function of his age, without any regard to his income. A distribution of this type is called a marginal distribution, in this case, the marginal density function of x. A marginal density function is just an ordinary density function, the word marginal referring only to the fact that it was derived from a joint distribution. Let $f_x(x_0)$ be the marginal density function of x evaluated at the point x_0. The probability that x will lie between a and b is then

$$p(a < x < b) = \int_a^b dx_0 \, f_x(x_0) = \int_a^b dx_0 \int_{-\infty}^{\infty} dy_0 \, f_{x,y}(x_0, y_0) \qquad (38\text{-}99)$$

because this probability is also the volume of the shaded region in Fig. 38-30. We see immediately that $f_x(x_0)$ is related to $f_{x,y}(x_0, y_0)$ by

$$f_x(x_0) = \int_{-\infty}^{\infty} dy_0 \, f_{x,y}(x_0, y_0) \qquad (38\text{-}100)$$

The marginal distribution of x is just the joint density function integrated over all other variables but x. Similarly, we have for y the marginal distribution

$$f_y(y_0) = \int_{-\infty}^{\infty} dx_0 \, f_{x,y}(x_0, y_0) \qquad (38\text{-}101)$$

This is the distribution of age regardless of income. Although the names age and income cause us to think of positive variables, all the formulations we develop will be completely general.

Suppose now that when an individual is selected his income is revealed to us. How does this affect the probability distribution of his age? What we want is a conditional probability density function, the density function of age conditional on having a given income. We shall use $f_{x|y}(x_0|y_0)$ to indicate this conditional density function on x when y assumes the known value y_0. The probability that x will lie between a and b when y lies between y_1 and $y_1 + \Delta$ is, by definition of conditional probability,

$$p(a < x < b | y_1 < y < y_1 + \Delta) = \frac{p(a < x < b, y_1 < y < y_1 + \Delta)}{p(y_1 < y < y_1 + \Delta)} \qquad (38\text{-}102)$$

This probability is, in turn, computable from Fig. 38-30 in the form

$$p(a < x < b | y_1 < y < y_1 + \Delta) = \frac{\int_a^b dx_0 \int_{y_1}^{y_1+\Delta} dy_0 \, f_{x,y}(x_0, y_0)}{\int_{y_1}^{y_1+\Delta} dy_0 \, f_y(y_0)} \qquad (38\text{-}103)$$

Now let Δ become arbitrarily small. Equation (38-103) can then be written

$$p(a < x < b | y = y_1) = \int_a^b dx_0 \, f_{x|y}(x_0|y_1) = \frac{\int_a^b dx_0 \, f_{x,y}(x_0, y_1) \Delta}{f_y(y_1) \Delta} \qquad (38\text{-}104)$$

Therefore, we have

$$f_{x|y}(x_0|y_1) = \frac{f_{x,y}(x_0, y_1)}{f_y(y_1)} \qquad (38\text{-}105)$$

The conditional density function of x given y is the joint density function of x and y divided by the marginal density function of y.

For mnemonic reasons, we shall usually write (38-105) in the form

$$f_{x|y}(x_0|y_0) = \frac{f_{x,y}(x_0, y_0)}{f_y(y_0)} \qquad (38\text{-}106)$$

If the value of age x is specified to be x_0, then the conditional probability distribution on income $f_{y|x}(y_0|x_0)$ is

$$f_{y|x}(y_0|x_0) = \frac{f_{x,y}(x_0,y_0)}{f_x(x_0)} \qquad (38\text{-}107)$$

Although (38-106) and (38-107) express one density function as the ratio of two others, they have the same form as the definition for conditional probability and so are easy to remember.

Graphically, both marginal and conditional distributions have simple interpretations. The marginal distribution of a variable is just the projection of the joint density function against the axis representing that variable. The conditional density function, say $f_{x|y}(x_0|y_0)$, is just the slice of the joint density function produced by the plane $y = y_0$, with the result then normalized to have area 1. Visualizing the marginal and conditional distributions implied by a given density function is a good exercise in spatial relations.

An example will illustrate how to derive marginal and conditional distributions from a joint distribution. Suppose that two random variables x and y have the joint density function

$$f_{x,y}(x_0,y_0) = \begin{cases} 6/5 & 0 \le x_0 < y_0 \le 1 \\ 6/5(1 - x_0 + y_0) & 0 \le y_0 < x_0 \le 1 \\ 0 & \text{elsewhere} \end{cases} \qquad (38\text{-}108)$$

This joint density function is shown graphically in Fig. 38-31; it, of course, has total volume 1. We can calculate the probability of any event concerning x and y in the usual way. For example, the probability that y is greater than x is the volume of the flat region above the line $y_0 = x_0$, which equals 0.6.

The marginal density function of x, defined by (38-100), is given by

$$f_x(x_0) = \int_{-\infty}^{\infty} dy_0\, f_{x,y}(x_0,y_0) = \int_0^{x_0} dy_0\, 6/5(1 - x_0 + y_0) + \int_{x_0}^1 dy_0\, 6/5$$
$$= 6/5(1 - x_0^2 + x_0^2/2)$$

or
$$f_x(x_0) = \begin{cases} 6/5(1 - x_0^2/2) & 0 \le x_0 \le 1 \\ 0 & \text{elsewhere} \end{cases} \qquad (38\text{-}109)$$

This marginal density function is also shown in Fig. 38-31. We should always check that our probability density functions integrate to 1, although we shall not always mention that we have done so.

The marginal density function of y is similarly calculated and found to be

$$f_y(y_0) = 6/5(1/2 + y_0 - 1/2 y_0^2) \qquad 0 \le y_0 \le 1 \qquad (38\text{-}110)$$

This function is shown in Fig. 38-31. Note that the marginal density of y is just the marginal density of x rotated about the point $1/2$.

$$f_x(x_0) = f_y(1 - x_0) \qquad 0 \le x_0 \le 1 \qquad (38\text{-}111)$$

The conditional densities are also readily calculated. The conditional density function of x given y, $f_{x|y}(x_0|y_0)$, was defined by

$$f_{x|y}(x_0|y_0) = \frac{f_{x,y}(x_0,y_0)}{f_y(y_0)} \qquad (38\text{-}106)$$

We then use (38-108) and (38-110) to obtain

$$f_{x|y}(x_0|y_0) = \begin{cases} \dfrac{1}{1/2 + y_0 - 1/2 y_0^2} & 0 \le x_0 < y_0 \le 1 \\[6pt] \dfrac{1 - x_0 + y_0}{1/2 + y_0 - 1/2 y_0^2} & 0 \le y_0 < x_0 \le 1 \\[4pt] 0 & \text{elsewhere} \end{cases} \qquad (38\text{-}112)$$

PROBABILITY 38-33

This conditional density function is shown in Fig. 38-31. The form of the curve depends on the value of y_0 selected—it ranges from a rectangle to a triangle. The same calculation using (38-107) yields

$$f_{y|x}(y_0|x_0) = \begin{cases} \dfrac{1}{1 - \frac{1}{2}x_0^2} & 0 \leq x_0 < y_0 \leq 1 \\ \dfrac{1 - x_0 + y_0}{1 - \frac{1}{2}x_0^2} & 0 \leq y_0 < x_0 \leq 1 \\ 0 & \text{elsewhere} \end{cases} \quad (38\text{-}113)$$

which is also plotted in Fig. 38-31. Both conditional density functions must have an area of 1 for any value of the variable on which they are conditioned.

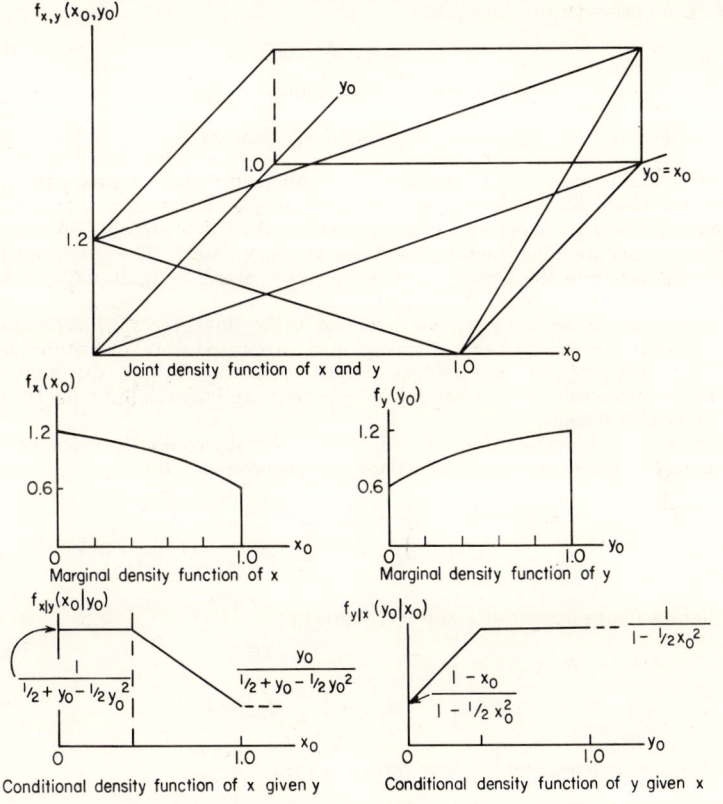

FIG. 38-31. A joint density function and its associates.

Bayes's Theorem and Independence. If we solve for the joint density function $f_{x,y}(x_0,y_0)$ in (38-106) and (38-107), we obtain

$$f_{x,y}(x_0,y_0) = f_{x|y}(x_0|y_0)f_y(y_0) = f_{y|x}(y_0|x_0)f_x(x_0) \quad (38\text{-}114)$$

or

$$f_{x|y}(x_0|y_0) = \frac{f_{y|x}(y_0|x_0)f_x(x_0)}{f_y(y_0)} \quad (38\text{-}115)$$

Finally, we can use (38-101), which defines the marginal density function, to write

$$f_{x|y}(x_0|y_0) = \frac{f_{y|x}(y_0|x_0)f_x(x_0)}{\int_{-\infty}^{\infty} dx_0\, f_{x,y}(x_0,y_0)} = \frac{f_{y|x}(y_0|x_0)f_x(x_0)}{\int_{-\infty}^{\infty} dx_0\, f_{y|x}(y_0|x_0)f_x(x_0)} \quad (38\text{-}116)$$

If we interpret the revelation of y_0 to be the outcome of an experiment providing indirect information about x (such as revealing the income of a person, but not his age), then Eq. (38-116) is the form of Bayes's theorem for continuous random variables. It shows how a prior density function on a random variable $[f_x(x_0)]$ is modified by a likelihood function $[f_{y|x}(y_0|x_0)]$ to produce a posterior density function on the random variable $[f_{x|y}(x_0|y_0)]$. The denominator of the expression is just the integral of the numerator, a normalizing factor to make the posterior density function integrate to 1.

If x and y were not related, then the posterior density function of x would be the same as the prior density function

$$f_{x|y}(x_0|y_0) = f_x(x_0) \quad (38\text{-}117)$$

Equation (38-114) shows that this condition implies

$$f_{x,y}(x_0,y_0) = f_x(x_0)f_y(y_0) \quad \text{all } x_0, y_0 \quad (38\text{-}118)$$

namely, that the joint density function is the product of the marginal density functions. Equation (38-118) is the independence criterion for random variables. Two random variables are independent if and only if their joint density function is the product of their marginal density functions at every point. The definition is completely analogous to the definition of independent events. Again, independence is determined from the probability measure.

We see that the random variables described in Fig. 38-31 are not independent, but that the joint density function for two spins of a balanced wheel of fortune shown in Fig. 38-28 does represent independent random variables. In fact, this joint density function was constructed by assuming independence and then multiplying the density functions for two spins.

Moments. The expectation of a function $g(x,y)$ of two random variables with a joint density function $f_{x,y}(x_0,y_0)$ is defined by extension as

$$E(g(x,y)) = \overline{g(x,y)} = \langle g(x,y) \rangle$$
$$= \int_{-\infty}^{\infty} dx_0 \int_{-\infty}^{\infty} dy_0\, g(x_0,y_0)f_{x,y}(x_0,y_0) \quad (38\text{-}119)$$

The k_1–k_2th joint moment of x and y is defined by

$$E(x^{k_1}y^{k_2}) = \overline{x^{k_1}y^{k_2}}$$
$$= \int_{-\infty}^{\infty} dx_0\, x_0^{k_1} \int_{-\infty}^{\infty} dy_0\, y_0^{k_2} f_{x,y}(x_0,y_0) \quad (38\text{-}120)$$

If x and y are independent random variables, then

$$\overline{x^{k_1}y^{k_2}} = \int_{-\infty}^{\infty} dx_0\, x_0^{k_1}f_x(x_0) \int_{-\infty}^{\infty} dy_0\, y_0^{k_2}f_y(y_0)$$
$$= \overline{x^{k_1}}\,\overline{y^{k_2}} \quad (38\text{-}121)$$

We see that in this case the k_1–k_2th joint moment is the product of the k_1th and k_2th moments of the corresponding marginal distributions.

In particular, we obtain from (38-121) a very important result when $k_1 = k_2 = 1$. The expected value of the product of two *independent* random variables is the product of their individual expected values.

PROBABILITY

When $g(x,y) = ax + by$, we have

$$E(ax + by) = \overline{(ax + by)} = \int_{-\infty}^{\infty} dx_0 \int_{-\infty}^{\infty} dy_0 \, (ax_0 + by_0) f_{x,y}(x_0, y_0)$$

$$= a \int_{-\infty}^{\infty} dx_0 \, x_0 \int_{-\infty}^{\infty} dy_0 \, f_{x,y}(x_0, y_0) + b \int_{-\infty}^{\infty} dy_0 \, y_0 \int_{-\infty}^{\infty} dx_0 \, f_{x,y}(x_0, y_0)$$

$$= a \int_{-\infty}^{\infty} dx_0 \, x_0 f_x(x_0) + b \int_{-\infty}^{\infty} dy_0 \, y_0 f_y(y_0)$$

$$= a\bar{x} + b\bar{y} \tag{38-122}$$

We find that, regardless of the dependence of the random variables, the expected value of a linear combination of random variables is the same linear combination of their expected values. In particular, the expected value of the sum of two random variables is *always* the sum of their expected values. That relation is of great theoretical and practical importance.

Now that we have established the linearity of the expectation operation, we can develop another important result. Let x and y be two random variables and let z be their sum. We already know

$$\bar{z} = \bar{x} + \bar{y} \tag{38-123}$$

but what is the variance of z? We proceed by exploiting the linearity of expectation several times,

$$\overset{\circ}{z} = \overline{z^2} - \bar{z}^2$$
$$= \overline{(x + y)^2} - [\overline{(x + y)}]^2$$
$$= \overline{(x^2 + 2xy + y^2)} - (\bar{x} + \bar{y})^2$$
$$= \overline{x^2} + \overline{2xy} + \overline{y^2} - (\bar{x}^2 + 2\bar{x}\bar{y} + \bar{y}^2)$$
$$= \overline{x^2} - \bar{x}^2 + \overline{y^2} - \bar{y}^2 + 2(\overline{xy} - \bar{x}\bar{y})$$
$$= \overset{\circ}{x} + \overset{\circ}{y} + 2(\overline{xy} - \bar{x}\bar{y}) \tag{38-124}$$

The quantity $\overline{xy} - \bar{x}\bar{y}$ is called the covariance of x and y. If x and y are independent random variables, then $\overline{xy} = \bar{x}\bar{y}$, the covariance is zero, and the variance of the sum is the sum of the individual variances of x and y.

Cumulative Probability Distributions. We can easily define cumulative probability distributions for jointly distributed variables. Let $P_{x \leq, y \leq}(x_c, y_c)$ be the probability that the random variable x will be less than or equal to the number x_c and the random variable y will be less than or equal to y_c.

$$p(x \leq x_c, y \leq y_c) = P_{x \leq, y \leq}(x_c, y_c) = \int_{-\infty}^{x_c} dx_0 \int_{-\infty}^{y_c} dy_0 \, f_{x,y}(x_0, y_0) \tag{38-125}$$

Conversely, the joint density function is derived from the joint cumulative probability distribution by differentiation. If we differentiate (38-125), we obtain

$$f_{x,y}(x_0, y_0) = \frac{\partial^2}{\partial x_0 \, \partial y_0} P_{x \leq, y \leq}(x_0, y_0) \tag{38-126}$$

The joint cumulative probability distribution is related to the marginal cumulative probability distribution by

$$P_{x \leq}(x_c) = P_{x \leq, y \leq}(x_c, \infty) \quad P_{y \leq}(y_c) = P_{x \leq, y \leq}(\infty, y_c) \tag{38-127}$$

38-8. Operations on Random Variables

Much of the everyday use of probability theory is concerned with operations on random variables. We often have to find probability distributions over a new set of

38-36 MATHEMATICS ASSOCIATED WITH SYSTEM ENGINEERING

random variables when the distributions over the old set and the functional relations between the two sets are known. All these operations can be performed with efficiency by exploiting the sample space.

For example, suppose that a random variable x has the density function of the balanced wheel of fortune,

$$f_x(x_0) = \begin{cases} 1 & 0 < x_0 < 1 \\ 0 & \text{elsewhere} \end{cases} \qquad (38\text{-}128)$$

This density function is commonly called a "uniform" distribution between 0 and 1;

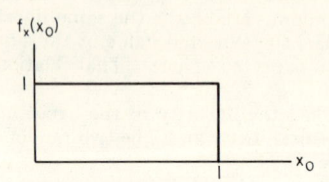

Density function of x, a uniform distribution

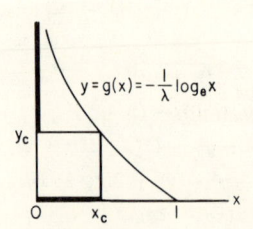

Relationship between y and x

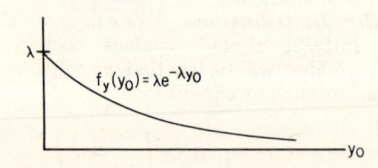

Density function of y, an exponential distribution

FIG. 38-32. A change-of-variable problem.

it is shown in Fig. 38-32. A new random variable y is computed from x according to the function

$$y = g(x) = -\frac{1}{\lambda} \ln x \qquad 0 < x < 1 \qquad (38\text{-}129)$$

The function $g(x)$ that relates y to x is also shown in Fig. 38-32. Our problem is to find $f_y(y_c)$, the density function of y.

We see by inspection of the function relating y and x that the experimental points producing the event $(0 < x < x_c)$ in the x sample space correspond to the points $(y_c < y)$ in the y sample space, where $y_c = g(x_c)$. These corresponding points are

PROBABILITY   38-37

indicated by thickened axes in the plot of $g(x)$. Therefore,

$$p(0 < x < x_c) = p(y_c < y) \tag{38-130}$$

or
$$\int_0^{x_c} dx_0 f_x(x_0) = \int_{y_c}^{\infty} dy_0 f_y(y_0) \tag{38-131}$$

We now differentiate this equation with respect to y_c and obtain

$$f_x(x_c)(dx_c/dy_c) = -f_y(y_c) \tag{38-132}$$
where
$$y_c = g(x_c) \tag{38-133}$$

If we solve (38-133) for x_c in terms of y_c, we obtain

$$x_c = g^{-1}(y_c) \tag{38-134}$$

and can therefore write (38-132) in the form

$$f_y(y_c) = -f_x[g^{-1}(y_c)](d/dy_c)[g^{-1}(y_c)] \quad 0 < y_c < \infty \tag{38-135}$$

Equation (38-135) describes the operations necessary to find the density function of y. Although it is possible to obtain various formulas to solve change-of-variable problems, mistakes are less likely to occur if the problem is worked from first principles. We can now finish our problem very easily. We find from (38-129) that

$$x_c = g^{-1}(y_c) = e^{-\lambda y_c} \quad 0 < y_c < \infty \tag{38-136}$$

and, therefore,

$$dg^{-1}(y_c)/dy_c = -\lambda e^{-\lambda y_c} \tag{38-137}$$

Since the range of $g^{-1}(y_c)$ is (0,1),

$$f_x[g^{-1}(y_c)] = 1 \tag{38-138}$$

and, therefore, from (38-135),

$$f_y(y_c) = \lambda e^{-\lambda y_c} \quad 0 < y_c < \infty \tag{38-139}$$

This density function for y is also plotted in Fig. 38-32; it is the answer to our problem. The density function we obtained for y is very important in probability theory—it is called the exponential density function.

Changes of Many Variables. We may want to change not from one random variable to another, but from one set of many variables to another set. For example, we may want to change from density functions over three Cartesian coordinates to density functions in three-dimensional polar coordinates. The principle involved is always the same—making sure that all events have the same probability in either set of variables.

A particularly common type of change is to find the density function of the answer of an arithmetic operation when the density functions of the operands are known. For example, suppose that x and y are two jointly related positive random variables. How can we find the density function of their product $z = xy$? First, we draw the sample space for the experimental outcome (x,y); it is just the $(x \geq 0, y \geq 0)$ quadrant of the x,y plane shown in Fig. 38-33. The event that the product z will be less than or equal to z_c is represented by the points (x,y) in the shaded region of the figure. The probability of this event is therefore the volume of the joint density function of x and y that lies over this region. We can now write

$$p(z \leq z_c) = \int_0^{z_c} dz_c f_z(z_c) = \int dx_0 \int_{x_0 y_0 \leq z_c} dy_0 f_{x,y}(x_0,y_0) \tag{38-140}$$

or
$$\int_0^{z_c} dz_0 f_z(z_0) = \int_0^{\infty} dx_0 \int_0^{z_0/x_0} dy_0 f_{x,y}(x_0,y_0) \tag{38-141}$$

38-38 MATHEMATICS ASSOCIATED WITH SYSTEM ENGINEERING

Differentiation of this equation with respect to z_c produces

$$f_z(z_c) = \int_0^\infty dx_0 \, (1/x_0) f_{x,y}(x_0, z_0/x_0) \qquad 0 \le z_0 \qquad (38\text{-}142)$$

Equation (38-142) is the answer to the problem of finding the density function of the product in terms of the joint density function of the two numbers multiplied together.

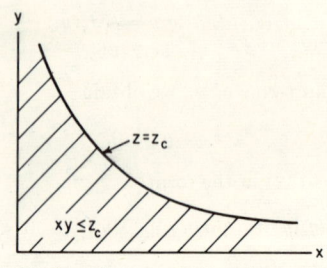

Fig. 38-33. Sample space for product calculation.

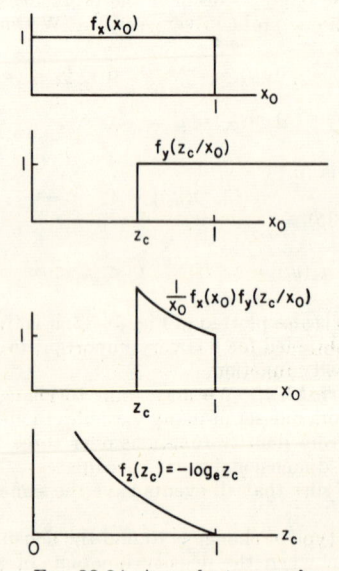

Fig. 38-34. A product example.

If these two numbers x and y are independent, then the joint density function can be factored and we can write our result in the form

$$f_z(z_c) = \int_0^\infty dx_0 \, (1/x_0) f_x(x_0) f_y(z_c/x_0) \qquad 0 \le z_c \qquad (38\text{-}143)$$

Let us suppose that x and y are independent and that both are uniformly distributed between 0 and 1; that is,

$$f_x(x_0) = 1 \qquad 0 \le x_0 \le 1 \qquad f_y(y_0) = 1 \qquad 0 \le y_0 \le 1 \qquad (38\text{-}144)$$

Figure 38-34 shows the functions $f_x(x_0)$, $f_y(z_c/x_0)$, and $1/x_0[f_x(x_0)]f_y(z_c/x_0)$. Since the density function of the product is the integral of the last of these functions, we see that this density function is zero if $z_c > 1$. If $0 \le z_c \le 1$, then by (38-143)

$$f_z(z_c) = \int_{z_c}^{1} dx_0 \, (1/x_0) = -\ln z_c \qquad 0 \le z_c \le 1 \qquad (38\text{-}145)$$

Therefore, the product has a logarithmic distribution as illustrated in Fig. 38-34.

Perhaps the most common arithmetic operation we want to analyze is that of addition. Let z be the sum of two jointly related random variables x and y; $z = x + y$. The sample space for the pair (x,y) is the entire x,y plane. The event that the sum will be less than or equal to a number z_c corresponds to the points lying below the line $x + y = z_c$ in this plane; this region is shown in Fig. 38-35. The probability of this

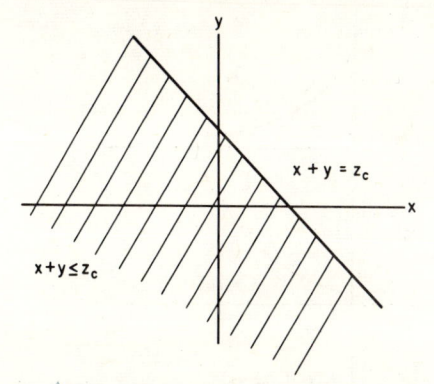

Fig. 38-35. Sample space for sum calculation.

event is again the volume of the joint density function on x and y that lies over this region. Therefore, we can write

$$p(z = x + y \le z_c) = \int_{-\infty}^{z_c} dz_0 f_z(z_0) = \int dx_0 \int_{x_0+y_0 \le z_c} dy_0 \, f_{x,y}(x_0, y_0)$$

$$\int_{-\infty}^{z_c} dz_c f_z(z_0) = \int_{-\infty}^{\infty} dx_0 \int_{-\infty}^{z_c - x_0} dy_0 \, f_{x,y}(x_0, y_0) \qquad (38\text{-}146)$$

Now we differentiate this expression with respect to z_c and obtain as our final result

$$f_z(z_c) = \int_{-\infty}^{\infty} dx_0 \, f_{x,y}(x_0, z_c - x_0) \qquad -\infty < z_c < \infty \qquad (38\text{-}147)$$

The problem of addition is most often encountered in the form where x and y are independent random variables. In this case the joint density function is factorable and we can write (38-147) as

$$f_z(z_c) = \int_{-\infty}^{\infty} dx_0 \, f_x(x_0) f_y(z_c - x_0) \qquad (38\text{-}148)$$

The right-hand side of (38-148) is known as the convolution operation. It is the relation necessary to describe the output of a linear system in terms of its input and impulse response. Because problems in the addition of independent random variables are parts of many problems, Eq. (38-148) is a very important relation.

If both x and y are independent and are exponentially distributed with parameters

λ and μ, respectively, we can easily use the convolution operation to find the density function of their sum. We have

$$f_x(x_0) = \lambda e^{-\lambda x_0} \qquad 0 \leq x_0 < \infty \qquad (38\text{-}149)$$
and
$$f_y(y_0) = \mu e^{-\mu y_0} \qquad 0 \leq y_0 < \infty \qquad (38\text{-}150)$$

The various steps in the calculation are shown in Fig. 38-36. The density function of

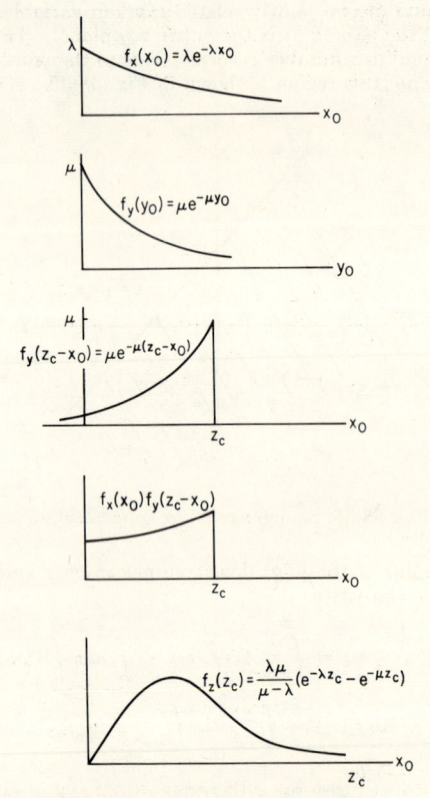

Fig. 38-36. A convolution example.

z at the point z_c is the area under $f_x(x_0)f_y(z_c - x_0)$ between 0 and z_c. We find

$$\begin{aligned}
f_z(z_c) &= \int_0^{z_c} dx_0 \, \lambda e^{-\lambda x_0} \mu e^{-\mu(z_c - x_0)} \\
&= \lambda \mu e^{-\mu z_c} \int_0^{z_c} dx_0 \, e^{(\mu - \lambda) x_0} \\
&= \lambda \mu e^{-\mu z_c} [1/(\mu - \lambda)](e^{(\mu - \lambda) z_c} - 1) \\
&= [\lambda \mu / (\mu - \lambda)](e^{-\lambda z_c} - e^{-\mu z_c}) \qquad 0 \leq z_c \qquad (38\text{-}151)
\end{aligned}$$

This density function for the sum is also shown in Fig. 38-36. We see that it is the sum of two exponentials.

The secret of error-free convolution, or indeed any other operation on random variables, is careful observance of the limits on all functions involved. The habit of writing the domain of every functional expression is worth developing.

38-9. Operational Transformations

Operational transformations of random variables are an important convenience in performing probabilistic calculations. The most important transformation is the expectation of e^{-sx}, where x is a random variable and s is in general a complex variable,

$$E(e^{-sx}) = \overline{e^{-sx}} = \int_{-\infty}^{\infty} dx_0 \, e^{-sx_0} f_x(x_0) \tag{38-152}$$

We call $E(e^{-sx})$ the exponential transform of x and give it the symbol $f_x{}^T(s)$,

$$f_x{}^T(s) = \overline{e^{-sx}} = \int_{-\infty}^{\infty} dx_0 \, e^{-sx_0} f_x(x_0) \tag{38-153}$$

Mathematically, $f_x{}^T(s)$ is known as a bilateral Laplace transform (Chapter 39), but the term exponential transform emphasizes its functional form. Some writers use a similar transform called the "characteristic function."

The exponential transform of a variable that is exponentially distributed is easily calculated. Let

$$f_x(x_0) = \lambda e^{-\lambda x_0} \qquad x_0 \geq 0 \tag{38-154}$$

then
$$f_x{}^T(s) = \int_0^{\infty} dx_0 \, e^{-sx_0} \lambda e^{-\lambda x_0} = \lambda \int_0^{\infty} dx_0 \, e^{-(s+\lambda)x_0}$$
$$= \lambda/(s + \lambda) \tag{38-155}$$

The exponential transform is very useful in calculating the moments of a random variable. If we differentiate (38-153) repeatedly with respect to s, we obtain

$$(d/ds)f_x{}^T(s) = (-1)\int_{-\infty}^{\infty} dx_0 \, x_0 e^{-sx_0} f_x(x_0)$$
$$(d^2/ds^2)f_x{}^T(s) = (-1)^2 \int_{-\infty}^{\infty} dx_0 \, x_0{}^2 e^{sx_0} f_x(x_0) \tag{38-156}$$

$$(d^k/ds^k)f_x{}^T(s) = (-1)^k \int_{-\infty}^{\infty} dx_0 \, x_0{}^k e^{-sx_0} f_x(x_0)$$

Now, by evaluating (38-156) at the point $s = 0$, we can write

$$\int_{-\infty}^{\infty} dx_0 \, x_0{}^k f_x(x_0) = (-1)^k (d^k/ds^k)f_x{}^T(s) \Big|_{s=0} \tag{38-157}$$

or
$$\overline{x^k} = (-1)^k (d^k/ds^k)f_x{}^T(s) \Big|_{s=0} \tag{38-158}$$

Equation (38-158) shows that the kth moment of a random variable can be calculated by evaluating the kth derivative of its exponential transform at $s = 0$ and multiplying by $(-1)^k$. We see immediately that the zeroth moment, which must be 1, is given by $f_x{}^T(0)$. We have, therefore, a check on any exponential transform—it must be equal to 1 when $s = 0$.

We can use (38-158) to find the first two moments of the exponentially distributed random variable whose density function appears in (38-154) and whose exponential transform is given by (38-155). We find

$$(d/ds)f_x{}^T(s) = (d/ds)[\lambda/(s+\lambda)] = -\lambda/(s+\lambda)^2$$
$$(d^2/ds^2)f_x{}^T(s) = (d^2/ds^2)[\lambda/(s+\lambda)^2] = 2\lambda/(s+\lambda)^3 \tag{38-159}$$

Therefore,
$$\bar{x} = (-1)(d/ds)f_x{}^T(s)\Big|_{s=0} = 1/\lambda$$
$$\overline{x^2} = (-1)^2 (d^2/ds^2)f_x{}^T(s) = 2/\lambda^2 \tag{38-160}$$

The variance of x is then

$$\check{x} = \overline{x^2} - \bar{x}^2 = 2/\lambda^2 - 1/\lambda^2 = 1/\lambda^2 \qquad (38\text{-}161)$$

The variance of an exponentially distributed random variable is equal to its mean squared. We can calculate any other moment of the distribution we desire in the same way.

The second important property of the exponential transform is that it greatly simplifies the convolution calculation. Let $z = x + y$ and let x and y be independent. Then,

$$\overline{e^{-sz}} = \overline{e^{-s(x+y)}} = \overline{e^{-sx}e^{-sy}} = \overline{e^{-sx}}\,\overline{e^{-sy}} \qquad (38\text{-}162)$$

where the last step is made possible by the independence assumption. Equation (38-162) shows that

$$f_z^T(s) = f_x^T(s) f_y^T(s) \qquad (38\text{-}163)$$

The exponential transform of the sum of two independent random variables is the product of their exponential transforms.

We can illustrate this property by calculating the density function of the sum of the two independent variables x and y with the density functions shown in Eqs. (38-149) and (38-150). The exponential transforms of each are then

$$f_x^T(s) = \lambda/(s + \lambda) \qquad f_y^T(s) = \mu/(s + \mu) \qquad (38\text{-}164)$$

According to (38-163), the exponential transform of z, their sum, is

$$f_z^T(s) = f_x^T(s) f_y^T(s) = \lambda\mu/(s + \lambda)(s + \mu) \qquad (38\text{-}165)$$

Partial fraction expansion produces

$$\begin{aligned} f_z^T(s) &= \frac{\lambda\mu/(\mu - \lambda)}{s + \lambda} + \frac{-\lambda\mu/(\mu - \lambda)}{s + \mu} \\ &= \frac{\mu}{\mu - \lambda}\frac{\lambda}{s + \lambda} - \frac{\lambda}{\mu - \lambda}\frac{\mu}{s + \mu} \end{aligned} \qquad (38\text{-}166)$$

By using the linearity of the transform relation we can then write

$$f_z(z_c) = \lambda\mu/(\mu - \lambda)(e^{-\lambda z_c} - e^{-\mu z_c}) \qquad 0 \le z_c \qquad (38\text{-}167)$$

in agreement with (38-151).

The Geometric Transform. If a random variable could take on values only at the nonnegative integers, then we found that its description by a probability mass function had certain advantages. We also find it useful to use a special form of the exponential transform when working with such variables. Let $p_n(n_0)$ be the probability that the discrete random variable n takes on the value n_0. The density function for n, $f_n(n_c)$, would then be

$$f_n(n_c) = \sum_{n_0=0}^{\infty} p_n(n_0)\delta(n_c - n_0) \qquad 0 \le n_0 \le \infty \qquad (38\text{-}168)$$

The exponential transform of n is defined by

$$\begin{aligned} f_n^T(s) &= \int_{-\infty}^{\infty} dn_c\, e^{-sn_c} f_n(n_c) = \int_{-\infty}^{\infty} dn_c\, e^{-sn_c} \sum_{n_0=0}^{\infty} p_n(n_0)\delta(n_c - n_0) \\ &= \sum_{n_0=0}^{\infty} p_n(n_0) \int_{-\infty}^{\infty} dn_c\, e^{-sn_c}\delta(n_c - n_0) \\ &= \sum_{n_0=0}^{\infty} p_n(n_0) e^{-sn_0} \end{aligned} \qquad (38\text{-}169)$$

Thus the exponential transform is the weighted sum of exponentials in s. If we now write $z = e^{-s}$ in this expression, then

$$f_n^T(-\ln z) = \sum_{n_0=0}^{\infty} p_n(n_0) z^{n_0} \qquad (38\text{-}170)$$

Finally, we define

$$p_n^T(z) = f_n^T(-\ln z) \qquad (38\text{-}171)$$

and write

$$p_n^T(z) = \sum_{n_0=0}^{\infty} p_n(n_0) z^{n_0} = E(z^n) = \overline{z^n} \qquad (38\text{-}172)$$

Equation (38-172) is the defining equation for the geometric transform of a discrete random variable. Although it is a special case of the exponential transform, it is a special case worth distinguishing. We call it the geometric transform because it is the sum of the heights of the spikes in the mass function weighted by a geometric sequence in z.

We can use the geometric transform to find the moments of a discrete random variable. If we differentiate (38-172) twice with respect to z, we obtain

$$(d/dz) p_n^T(z) = \sum_{n_0=0}^{\infty} p_n(n_0) n_0 z^{n_0-1} \qquad (38\text{-}173)$$

and

$$(d^2/dz^2) p_n^T(z) = \sum_{n_0=0}^{\infty} p_n(n_0) (n_0^2 - n_0) z^{n_0-2} \qquad (38\text{-}174)$$

Now we evaluate these equations at $z = 1$,

$$(d/dz) p_n^T(z) \Big|_{z=1} = \sum_{n_0=0}^{\infty} n_0 p_n(n_0) = \bar{n} \qquad (38\text{-}175)$$

$$(d^2/dz^2) p_n^T(z) \Big|_{z=1} = \sum_{n_0=0}^{\infty} n_0^2 p_n(n_0) - \sum_{n_0=0}^{\infty} n_0 p_n(n_0)$$

$$= \overline{n^2} - \bar{n} \qquad (38\text{-}176)$$

Therefore,

$$\bar{n} = (d/dz) p_n^T(z) \Big|_{z=1} \qquad (38\text{-}177)$$

$$\overline{n^2} = (d^2/dz^2) p_n^T(z) \Big|_{z=1} + (d/dz) p_n^T(z) \Big|_{z=1} \qquad (38\text{-}178)$$

The kth moment of n is therefore a combination of the first k derivatives of $p_n^T(z)$ evaluated at the point $z = 1$. These results could also be derived from Eqs. (38-158) and (38-171) by using the chain rule for differentiation.

The geometric transform also makes the convolution operation simple. Let n and m be independent discrete random variables, and let r be their sum. Then

$$\overline{z^r} = \overline{z^{n+m}} = \overline{z^n z^m} = \overline{z^n}\, \overline{z^m} \qquad (38\text{-}179)$$

and the geometric transform of their sum is the product of their geometric transforms. Since the geometric transform is just a special case of the exponential transform, this result comes as no surprise.

Figure 38-37 shows an example that illustrates the basic properties of geometric transforms. Here we consider two independent random variables with the probability mass functions shown. The geometric transforms of each variable are calculated and used to find the moments and the mass function of their sum. Note that

38-44 MATHEMATICS ASSOCIATED WITH SYSTEM ENGINEERING

the mean and variance of the sum are the sums of those quantities for the two original variables.

38-10. Named Distributions

Up to this point we have concentrated on the general theory of probability without much reference to the specific probability distributions that are in common use. In this section we shall present a number of probability distributions that have proved to be of both theoretical and practical importance. By analogy one might say that we have been discussing automotive engineering—it is now time to mention some particular makes of automobiles.

We shall call the distributions we shall discuss "named" distributions because their recurring use has caused us to assign them names for easy reference. We can illustrate our notation for named distributions by postulating a fictitious named distribution called the "alpha" distribution. The alpha distribution will generally have a

Table 38-3A. Named Con

Name	Density function	Mean
Beta	$f_\beta(x\|r, n) = \dfrac{\Gamma(n)}{\Gamma(r)\Gamma(n-r)} x^{r-1}(1-x)^{n-r-1} \quad 0 \leq x \leq 1$	$\dfrac{r}{n}$
Uniform	$f_u(x\|a, b) = \dfrac{1}{b-a} \quad a \leq x \leq b$	$\dfrac{a+b}{2}$
Gamma	$f_\gamma(x\|\alpha, \beta) = \dfrac{\beta^\alpha}{\Gamma(\alpha)} x^{\alpha-1} e^{-\beta x} \quad \begin{array}{l} \alpha > 0 \\ \beta > 0 \\ x \geq 0 \end{array}$	$\dfrac{\alpha}{\beta}$
Exponential*	$f_e(x\|\mu) = \mu e^{-\mu x} \quad \begin{array}{l} 0 \leq \mu \\ 0 \leq x \end{array}$	$\dfrac{1}{\mu}$
Normal	$f_n(x\|\mu, \sigma) = \dfrac{1}{\sqrt{2\pi}\,\sigma} e^{-\frac{1}{2}\left(\frac{x-\mu}{\sigma}\right)^2} \quad \begin{array}{l} -\infty < \mu < \infty \\ 0 < \sigma \\ -\infty < x < \infty \end{array}$	μ
Lognormal	$f_{ln}(x\|\mu, \sigma) = \dfrac{1}{\sqrt{2\pi}\,\sigma x} e^{-\frac{1}{2}\left(\frac{\ln x - \mu}{\sigma}\right)^2} \quad \begin{array}{l} -\infty < \mu < \infty \\ 0 < \sigma \\ 0 \leq x \end{array}$	$e^{\mu + \frac{\sigma^2}{2}}$
Student's	$f_s(x\|k) = \dfrac{\left(\frac{k-1}{2}\right)!}{\sqrt{k\pi}\,\left(\frac{k-2}{2}\right)!} \dfrac{1}{\left(1 + \frac{x^2}{k}\right)^{\frac{k+1}{2}}} \quad -\infty < x < \infty$	0
Chi-square	$f_{x^2}(x\|\nu) = \dfrac{1}{2^{\nu/2}\,\Gamma(\nu/2 - 1)} x^{\left(\frac{\nu}{2}-1\right)} e^{-\frac{x}{2}} \quad 0 \leq x$	ν
F	$f_F(x\|M, n) = \dfrac{\left(\frac{M+n-2}{2}\right)!}{\left(\frac{M-2}{2}\right)!\left(\frac{n-2}{2}\right)!} \left(\frac{M}{n}\right)^{\frac{M}{2}} \dfrac{x^{\frac{M-2}{2}}}{\left(1 + \frac{M}{n} x\right)^{\frac{M+n}{2}}} \quad \begin{array}{l} 0 \leq x \\ M = 1, 2, \ldots \\ n = 1, 2, \ldots \end{array}$	$\dfrac{n}{n-2}(n > 2)$
Weibull	$f_W(x\|M, \mu) = \mu M x^{M-1} e^{-\mu x^M} \quad \begin{array}{l} 0 \leq \mu \\ 0 \leq x \end{array}$	$\left(\dfrac{1}{\mu}\right)^{\frac{1}{M}} \Gamma\left(\dfrac{1}{M} + 1\right)$
Erlang*	$f_E(x\|k, \mu) = \dfrac{(\mu k)^k}{(k-1)!} x^{k-1} e^{-k\mu x} \quad \begin{array}{l} 0 \leq \mu \\ k = 1, 2, \ldots \\ 0 \leq x \end{array}$	$\dfrac{1}{\mu}$

* See Sec. 28-8.

PROBABILITY

set of parameters $\mu_1, \mu_2, \ldots$ that specify the shape of the distribution; we shall call this set of parameters μ. If the alpha distribution describes the behavior of a continuous random variable then the alpha density function is denoted by $f_\alpha(\cdot|\mu)$. If the alpha distribution describes a discrete random variable, then the alpha probability mass function is denoted by $p_\alpha(\cdot|\mu)$. If "alpha" were a man's name, then the name of the distribution would of course be capitalized and the subscript on the functions would be a capital alpha.

The alpha density function will have moments and moments about the mean. The kth moment of the α distribution will be denoted by ${}^k m_\alpha(\mu)$; thus

$${}^k m_\alpha(\mu) = \int_{-\infty}^{\infty} dx\, x^k f_\alpha(x|\mu) \tag{38-180}$$

As we know, ${}^0 m_\alpha(\mu) = 1$, and ${}^1 m_\alpha(\mu)$ is the mean of the alpha distribution with parameters μ. We shall write the mean simply as $m_\alpha(\mu)$, the 1 being understood.

tinuous Distributions

Variance	Transform	Regenerative?	Comments
$\dfrac{r}{n}\left(1 - \dfrac{r}{n}\right)\left(\dfrac{1}{n+1}\right)$	$\dfrac{\Gamma(n)}{\Gamma(r)} \sum_{j=0}^{\infty} \dfrac{(-s)^j}{j!} \dfrac{\Gamma(r+j)}{\Gamma(n+j)}$	No	
$\dfrac{(b-a)^2}{12}$	$\dfrac{1}{b-a}\left(\dfrac{e^{-sa} - e^{-sb}}{s}\right)$	No	$f_u(\cdot|0, 1) = f_\beta(\cdot|1, 2)$
$\dfrac{\alpha}{\beta^2}$	$\dfrac{\beta^\alpha}{(s+\beta)^\alpha}$	Only if β fixed	
$\dfrac{1}{\mu^2}$	$\dfrac{\mu}{s+\mu}$	No	$f_e(\cdot|\mu) = f_\gamma(\cdot|1, \mu)$
σ^2	$e^{\frac{s^2\sigma^2}{2} - \mu s}$	Yes	Also called Gaussian
$e^{2\mu + \sigma^2}(e^{\sigma^2} - 1)$	....	No	
$\dfrac{k}{k-2}$	....	No	
2ν	$\left(\dfrac{\frac{1}{2}}{s+\frac{1}{2}}\right)^{\frac{\nu}{2}}$	Yes	$f_{x^2}(\cdot|\nu) = f_\gamma\left(\cdot\mid\dfrac{\nu}{2}, \dfrac{1}{2}\right)$
$\dfrac{2n^2(n + M - 2)}{M(n-2)^2(n-4)} \quad (n > 4)$	....	No	
$\left(\dfrac{1}{\mu}\right)^{\frac{2}{M}}\left[\Gamma\left(\dfrac{2}{M}+1\right) - \left(\Gamma\left(\dfrac{1}{M}+1\right)\right)^2\right]$	....	No	$f_W(\cdot|1, \mu) = f_e(\cdot|\mu)$
$\dfrac{1}{k\mu^2}$	$\left(\dfrac{\mu k}{s + \mu k}\right)^k$	No	$f_E \cdot |k, \mu) \leq f_\gamma(\cdot|k, k\mu)$

Table 38-3B. Named Discrete Distributions

Name	Mass function	Mean	Variance	Transform	Regenerative?	Comments
Bernoulli	$p_B(k\|p) = \begin{cases} 1-p & k=0 \\ p & k=1 \\ 0 & k=2,3,\ldots \end{cases} \quad 0 \le p \le 1$	p	$(1-p)p$	$(1-p) + pz$	No	$p_b(k\|1,p) = p_\beta(k\|p)$
Binomial	$p_b(k\|n,p) = \binom{n}{k} p^k (1-p)^{n-k} \quad \begin{array}{l} 0 \le p \le 1 \\ k = 0, 1, 2, \ldots, n \end{array}$	np	$np(1-p)$	$[(1-p) + pz]^n$	Only if p fixed	
Poisson*	$p_P(k\|\mu) = \dfrac{\mu^k e^{-\mu}}{k!} \quad k = 0, 1, 2, \ldots$	μ	μ	$e^{\mu(z-1)}$	Yes	
Geometric	$p_\vartheta(k\|p) = p(1-p)^{k-1} \quad \begin{array}{l} 0 \le p \le 1 \\ k = 1, 2, \ldots \end{array}$	$\dfrac{1}{p}$	$\dfrac{1-p}{p^2}$	$\dfrac{pz}{1-(1-p)z}$	No	
Pascal	$p_{Pa}(n\|k,p) = \binom{n-1}{k-1} p^k (1-p)^{n-k} \quad \begin{array}{l} 0 \le p \le 1 \\ n = k, k+1, \ldots \end{array}$	$\dfrac{k}{p}$	$\dfrac{k(1-p)}{p^2}$	$\left(\dfrac{pz}{1-(1-p)z}\right)^k$	Only if p fixed	$p_{Pa}(\cdot\|1,p) = p_\vartheta(\cdot\|p)$
Hypergeometric	$p_{h\vartheta}(k\|n,r_1,r) = \dfrac{\binom{r_1}{k}\binom{r-r_1}{n-k}}{\binom{r}{n}} \quad k = 0, 1, 2, \ldots, \min(n, r_1)$	$n\dfrac{r_1}{r}$	$\dfrac{r_1 n(r-n)(r-r_1)}{r^2(r-1)}$	$\cdots\cdots\cdots\cdots$	No	

* See Sec. 28-8.

38–46

PROBABILITY

The kth central moment or moment about the mean is signified by ${}^k v_\alpha(\mu)$,

$${}^k v_\alpha(\mu) = \int_{-\infty}^{\infty} dx [x - m_\alpha(\mu)]^k f_\alpha(x|\mu) \tag{38-181}$$

From our earlier work, ${}^0 v_\alpha(\mu) = 1$, ${}^1 v_\alpha(\mu) = 0$, and ${}^2 v_\alpha(\mu)$ is the variance of the alpha distribution. We shall write the variance as $v_\alpha(\mu)$ and understand that the 2 is to be supplied in any ambiguous context. We can then write the almost symmetrical relation

$$v_\alpha(\mu) = {}^2 m_\alpha(\mu) - m_\alpha^2(\mu) \tag{38-182}$$

The variance of the alpha distribution with parameters μ is the second moment of the distribution less the first moment squared.

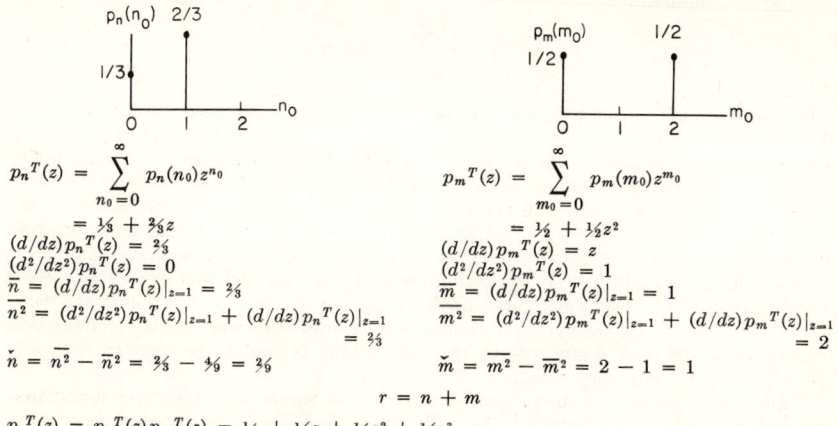

$p_n^T(z) = \sum_{n_0=0}^{\infty} p_n(n_0) z^{n_0}$
$= \tfrac{1}{3} + \tfrac{2}{3} z$
$(d/dz) p_n^T(z) = \tfrac{2}{3}$
$(d^2/dz^2) p_n^T(z) = 0$
$\bar{n} = (d/dz) p_n^T(z)|_{z=1} = \tfrac{2}{3}$
$\overline{n^2} = (d^2/dz^2) p_n^T(z)|_{z=1} + (d/dz) p_n^T(z)|_{z=1}$
$\quad = \tfrac{2}{3}$
$\check{n} = \overline{n^2} - \bar{n}^2 = \tfrac{2}{3} - \tfrac{4}{9} = \tfrac{2}{9}$

$p_m^T(z) = \sum_{m_0=0}^{\infty} p_m(m_0) z^{m_0}$
$= \tfrac{1}{2} + \tfrac{1}{2} z^2$
$(d/dz) p_m^T(z) = z$
$(d^2/dz^2) p_m^T(z) = 1$
$\bar{m} = (d/dz) p_m^T(z)|_{z=1} = 1$
$\overline{m^2} = (d^2/dz^2) p_m^T(z)|_{z=1} + (d/dz) p_m^T(z)|_{z=1}$
$\quad = 2$
$\check{m} = \overline{m^2} - \bar{m}^2 = 2 - 1 = 1$

$r = n + m$
$p_r^T(z) = p_n^T(z) p_m^T(z) = \tfrac{1}{6} + \tfrac{1}{3} z + \tfrac{1}{6} z^2 + \tfrac{1}{3} z^3$

$(d/dz) p_r^T(z) = \tfrac{1}{3} + \tfrac{1}{3} z + z^2$
$(d^2/dz^2) p_r^T(z) = \tfrac{1}{3} + 2z$

$r = (d/dz) p_r^T(z)|_{z=1} = \tfrac{5}{3}$ $\quad \check{r} = \overline{r^2} - \bar{r}^2 = 1\tfrac{1}{9}$
$\overline{r^2} = (d^2/dz^2) p_r^T(z)|_{z=1} + (d/dz) p_r^T(z)|_{z=1} = 4$

FIG. 38-37. An example illustrating the uses of the geometric transform.

If alpha is a probability mass function, then we keep the same notation for the moments, but change the defining equations to

$${}^k m_\alpha(\mu) = \sum_{x=0}^{\infty} x^k p_\alpha(x|\mu) \tag{38-183}$$

and

$${}^k v_\alpha(\mu) = \sum_{x=0}^{\infty} [x - m_\alpha(\mu)]^k p_o(x|\mu) \tag{38-184}$$

The cumulative probability distribution corresponding to the α distribution is denoted by $P_\alpha(\cdot|\mu)$. If α describes a continuous variable, the cumulative probability distribution is defined by

$$P_\alpha(x|\mu) = \int_{-\infty}^{x} dx_0 f_\alpha(x_0|\mu) \quad -\infty < x < \infty \tag{38-185}$$

38-48   MATHEMATICS ASSOCIATED WITH SYSTEM ENGINEERING

but if α describes a discrete random variable, then it is defined by

$$P_\alpha(x|\mu) = \sum_{x_0=0}^{x} p_\alpha(x_0|\mu) \qquad x = 0, 1, 2, \ldots \tag{38-186}$$

The notation for named distributions includes a symbolism for transforms. If the alpha distribution, with parameters μ, describes a continuous random variable, then the exponential transform of the distribution is indicated by $f_\alpha{}^T(s|\mu)$ and defined by

$$f_\alpha{}^T(s|\mu) = \int_{-\infty}^{\infty} dx_0 \, e^{-sx_0} f_\alpha(x_0|\mu) \tag{38-187}$$

If alpha is a probability mass function, then its geometric transform is denoted by $p_\alpha{}^T(z|\mu)$ and defined by

$$p_\alpha{}^T(z|\mu) = \sum_{n_0=0}^{\infty} z^{n_0} p_\alpha(n_0|\mu) \tag{38-188}$$

We often find it useful to know whether a family of distributions has the property that the convolutions of two such distributions with different parameters produce a distribution that is also in the family. The distribution of the sum of independent random variables described by such a distribution would then belong to the same family. We shall call this property the regenerative property of distributions. We can readily determine whether a distribution has this property by seeing if the product of the transforms of the distributions for two sets of parameters is of the same form as the original transform. If the parameters of the two distributions can be completely different and still meet this requirement, then the family is regenerative without qualification. If, however, the requirement is met only when certain parameters are the same for both distributions, then the family is said to be regenerative only when these parameters are fixed.

Table 38-3 summarizes the properties of the most commonly encountered continuous and discrete named distributions. We would find it difficult to work most practical probability problems without referring to at least one of them. And yet the set of these distributions should be expected to change in the course of time as new applications of the theory are investigated. However, it is likely that most of the distributions in Table 38-3 will be important indefinitely.

38-11. Conclusion

We have only begun the study of probability. Many important topics remain to be discussed. For example, there are the many limit theorems of probability that show the form that the probability function will take when an experiment is repeated a very large number of times. These theorems are of both practical and theoretical importance. We also have not been able to begin analyzing stochastic processes, processes where the random variable is a function of time. Stochastic-process models have been applied with success to problems that range from the field of marketing to the field of missile reliability. Stochastic-process theory, itself, is one of the most rapidly growing areas of research in probability theory.

The brief introduction to probability theory presented in this chapter should permit reading the many standard references for specific information, and the many chapters in this handbook which depend on probability theory. In particular, the reader who wishes to delve a little further into the subject should turn to Chapter 28. But probability is not a handbook subject. Most problems require an individual approach based on the fundamental relations we have developed. A cautious approach to probability problems is always the most prudent.

Chapter 39

PROPERTIES OF THE LAPLACE AND RELATED TRANSFORMS

CHARLES L. DOLPH, *Professor of Mathematics, University of Michigan.*

CONTENTS

39-1. Introduction	39-1
39-2. Fundamental Properties of the Laplace Transform	39-3
39-3. The Inverse Laplace Transformation	39-4
39-4. Fourier Transforms	39-7
39-5. Other Integral Transforms	39-11
The Mellin Transform	39-11
The Hankel Transform	39-11
The Stieltjes Transform	39-11
The Stieltjes Inversion Formula	39-12
The Finite Fourier Transform	39-12
The Legendre Transform	39-12
The Finite Hankel Transform	39-12
39-6. Illustrative Examples of the Use of Laplace Transforms	39-13
The Computations of Transforms	39-13
The Solution of Linear Differential Equations with Constant Coefficients	39-13
The Solution of Linear Systems of Ordinary Differential Equations	39-14
Integral-differential Equations of Convolution Type	39-14
Difference-differential Equations with Zero Initial Conditions	39-15
Partial Differential Equations	39-15
Initial-value Problems in Partial Differential Equations	39-17
Appendix 1. An Introduction to Generalized Functions	39-19
Appendix 2. Table of Operations for the Laplace Transformation	39-25
Appendix 3. Table of Laplace Transforms	39-26

39-1. Introduction

The operational method based on the Laplace transformation and its modern counterpart, the theory of semigroups, represents one of the most convenient and powerful mathematical tools for the treatment of a wide variety of mathematical problems. While Laplace (1779) and Cauchy (1823) laid the foundations for the operational method, it was Oliver Heaviside (1850–1923) who systematically applied it and demonstrated its extreme utility for a large class of engineering and mathematical problems. One way the operational methods expounded by Heaviside can be systematically and rigorously developed is by means of the properties of the Laplace transformation, which is formally defined by the relation

$$\mathcal{L}(F) = f(s) = \int_0^\infty e^{-st} F(t)\, dt \qquad (39\text{-}1)$$

To have a mathematical theory it is, of course, necessary to specify the class of admissible functions and the kind of integral that appears in (39-1). Depending upon these choices, one obtains theories at different levels of mathematical sophistication and abstraction. The most sophisticated theory interprets (39-1) as the relation between a semigroup $F(t)$ over a suitable abstract space and its resolvent $f(s)$. This theory is developed at length in Hille and Phillips,[18] and an excellent introductory treatment can be found in Phillips's article.[31] Theories at an intermediate level of abstraction can be found, for example, in the books by Doetsch[7] and Widder.[43] While not so abstract as the theory of semigroups, these treatises still demand a considerable degree of mathematical knowledge and sophistication.

In contrast, most introductory expositions of the theory are based on the Riemann integral, and the class of admissible functions is chosen to consist of piecewise continuous functions of exponential type. This choice will be made here. Specifically we have

Definition 1. A piecewise continuous function $F(t)$, defined for all t, $-\infty < t < \infty$, with $F(t) = 0$ for $t < 0$, will be said to be of the class (E)—of exponential type—if there exists a real σ_F such that

$$\int_0^\infty |F(t)| e^{-\sigma_F t}\, dt$$

exists.

The notation is that used by R. V. Churchill,[5] who has kindly permitted the reprinting of his tables in Appendixes 2 and 3.

While these choices permit a systematic theory, it cannot be too strongly emphasized that the resulting theory contains few necessary and sufficient conditions. In particular, it is sufficient, but not necessary, that $F(t)$ be in the class (E) for the Laplace transform to exist, and there are many functions not in (E) for which $f(s)$ as defined by (39-1) with a Riemann integral is a perfectly well-behaved function. A well-known example is furnished by the function which is the fundamental solution of the one-dimensional heat equation; namely,

$$F(t) = (1/\sqrt{\pi t}) \exp(-a^2/4t) \qquad (39\text{-}2)$$

It has the transform

$$f(s) = (1/\sqrt{s}) e^{-a\sqrt{s}} \qquad a \geq 0,\, s > 0 \qquad (39\text{-}3)$$

as is demonstrated, for example, in Churchill,[5] page 66. This has the consequence that it may happen that the formal use of the method of Laplace transform will produce formulas which have a much wider range of applicability than any step-by-step justification of their derivation would suggest. As such the method can be thought of as a tool for discovery. This frequently occurs in mathematics, and the problem of justifying the end result must often be treated separately from the method which suggested it. The best-known example of this is, perhaps, the formal derivation of the Fourier integral by passage to the limit in a Fourier series, a process extremely difficult to justify in detail, but one which does produce a correct result that is easily justified by other considerations.

In order to present a reasonable theory in the space available, it will be necessary to assume that the reader is familiar with the mathematical terms and techniques through the advanced calculus level as it is to be found, for example, in the book by W. Kaplan.[20] Moreover, since there are now several excellent books (cf. Refs. 5 and

21) which present detailed expositions of the theory involving the above class of functions, no attempt will be made here to do more than occasionally suggest how a given result can be established.

39-2. Fundamental Properties of the Laplace Transform

In order to have a reasonably complete theory, it is necessary to allow the parameters in (39-1) to be complex. We shall consistently write $s = \sigma + i\omega$ and shall assume, for $F(t)$ in the class (E), that $\sigma > \sigma_F$. It is then easy to prove

Theorem 1. 1. The Laplace transform exists for all functions of exponential type as a uniformly convergent integral for $-\infty < \omega < \infty$.

2. The class of exponential functions for which the Laplace transform exists forms a linear space; that is, if $F_1(t)$ and $F_2(t)$ are in the class (E) and c_1 and c_2 are any two complex numbers, then

$$\mathcal{L}(c_1 F_1 + c_2 F_2) = c_1 \mathcal{L}(F_1) + c_2 \mathcal{L}(F_2)$$

3. The transform $f(s)$ is an analytic function for $\sigma > \sigma_F$; i.e., all derivatives with respect to s exist for the complex-valued function $f(s)$ for $\sigma > \sigma_F$.

The basic operational property of the Laplace transform is a direct consequence of the usual integration-by-parts procedure. When this is applied to (39-1) with $F(t)$ replaced by $F'(t)$ the resulting formulas explicitly exhibit the dependence on the initial values in contrast to the formulas used by Heaviside. For simplicity this property will be stated for piecewise smooth and continuous functions, but it is easy to generalize to the case of sectional continuous functions. This case may be found in Churchill,[5] page 8.

Theorem 2. 1. Let $F(t)$, $F'(t)$ be of exponential type and $F(t)$ piecewise smooth and continuous for $t \geq 0$. Further let $\sigma_0 = \max(\sigma_F, \sigma_{F'})$, then

$$\mathcal{L}(F') = s\mathcal{L}(F) - F(0) \qquad \text{for } \sigma > \sigma_0$$

2. Let $F(t), F'(t), \ldots, F^{(n-1)}(t)$ be of exponential type and continuous for $t \geq 0$. Let

$$\sigma_0 = \max(\sigma_F, \sigma_{F'}, \ldots, \sigma_{F^{(n-1)}})$$

then

$$\mathcal{L}(F^{(k)}) = s^k \mathcal{L}(F) - s^{k-1}F(0) - \cdots - F^{(k-1)}(0)$$
$$k = 1, 2, \ldots n \qquad \text{for } \sigma > \sigma_0$$

Further, useful operational properties which are easily derived are contained in

Theorem 3. Let F be in the class (E) for $\sigma > \sigma_F$ and define D_0^{-1} by the relation

$$G(t) = D_0^{-1}F(t) = \int_0^t F(u)\,du$$

then

(1) $\mathcal{L}(D_0^{-1}F) = \mathcal{L}(G) = (1/s)\mathcal{L}(F)$ \qquad for $\sigma > \sigma_F$
(2) $\mathcal{L}[D_0^{-k}(F)] = (1/s^k)\mathcal{L}(F)$ \qquad $k = 1, 2, \ldots$
(3) $\mathcal{L}[F(t-a)] = e^{-as}\mathcal{L}(F)$ \qquad for $a > 0$
(4) $\mathcal{L}(e^{bt}F) = f(s-b)$ \qquad for $\sigma > \sigma_F + \operatorname{Re} b$
(5) $\mathcal{L}(t^n F) = (-1)^n (d^n/ds^n)\mathcal{L}(F)$

One further and most important operational property depends on the notation of the convolution integral.

Definition 2. If $F(t)$, $G(t)$ are piecewise continuous for $t \geq 0$ and identically zero for $t < 0$, the convolution (or Faltung) is defined by the relation

$$F * G = \int_0^t F(u) G(t-u)\, du \qquad (39\text{-}4)$$

The importance of this notation follows directly from (1) of the following theorem, which also contains a statement of the main properties of the convolution.

Theorem 4. Let $F(t)$, $G(t)$, $H(t)$ be in the class (E) and let

$$\sigma_0 > \max\,(\sigma_F, \sigma_G) \qquad \sigma_1 > \max\,(\sigma_F, \sigma_G, \sigma_H)$$

then

(1) $\mathcal{L}(F * G) = \mathcal{L}(F)\mathcal{L}(G)$ for $\sigma > \sigma_0$
(2) $F * G = G * F$
(3) $F * (aG) = (aF) * G = a(F * G)$
(4) $F * (G + H) = F * G + F * H$ for $\sigma > \sigma_1$
(5) $e^{at}(F * G) = e^{at}F * e^{at}G$
(6) $F * (G * H) = (F * G) * H$

The application of the operational properties contained in Theorems 1 through 4 transforms an initial-value problem in ordinary differential equations in t into an algebraic problem in the s plane. Similarly, it transforms an initial-value problem for a partial differential equation in two independent variables, say x and t, into an ordinary differential-equation problem in x depending upon the parameter s. While in some disciplines, such as circuit theory, further analysis can frequently be limited to considerations in the s plane, in general, the problem in the s plane has to be solved and a return made to the t plane by means of the inverse Laplace transformation, which is symbolically written as

$$F(t) = \mathcal{L}^{-1}[f(s)]$$

The first important question which arises about the inverse transform is that of uniqueness. It may be shown that if $F(t)$ is piecewise smooth and continuous then $F(t)$ is uniquely determined by $f(s)$. That is, $f(s)$ determines $F(t)$ uniquely within the class of functions of exponential type except at jump points, or, said in still a different way, $f(s)$ determines $F(t)$ uniquely up to a null function, which is a function, say $N(t)$, with the property that

$$\int_0^T N(u)\,du = 0 \qquad \text{for every } T$$

For almost all applications the addition or subtraction of a null function is of no importance; so that the inverse transform may be taken to determine $F(t)$ uniquely, although strictly speaking it merely determines $F(t)$ to within an equivalent class of functions which differ only by null functions. Since it follows directly from (2) of Theorem 2 that if

$$F_1(t) = \mathcal{L}^{-1}[f_1(s)]$$

and if $F_2(t) = \mathcal{L}^{-1}[f_2(s)]$ then

$$\mathcal{L}^{-1}(c_1 f_1 + c_2 f_2) = c_1 \mathcal{L}^{-1}(f_1) + c_2 \mathcal{L}^{-1}(f_2) = c_1 F_1(t) + c_2 F_2(t) \qquad \text{for } \sigma > \sigma_{F_1},\, \sigma_{F_2}$$

the inverse transforms of many functions may be constructed from tables which contain the transforms of elementary functions. Such a table is appended to this article; while many more extensive tables are available (cf. Doetsch[7]) the entries given here are adequate for many problems. Other methods for obtaining the inverse transform are also available and are discussed in the next section.

39-3. The Inverse Laplace Transformation

One of the most widely used methods for obtaining the inverse Laplace transform is by means of the inversion integral. This integral may be derived from considerations of the Fourier integral or from Cauchy's integral formula and is explicitly given by the formula

$$F(t) = (1/2\pi i) \int_{\gamma - i\infty}^{\gamma + i\infty} e^{st} f(s)\,ds \qquad (39\text{-}5)$$

in which γ is any real number $\gamma > \sigma_F$. The actual evaluation of this integral is usually by means of the theory of analytic functions since, as has been stated, $f(s)$ is

PROPERTIES OF THE LAPLACE AND RELATED TRANSFORMS

analytic for $\sigma > \sigma_F$. In particular then the half plane $\sigma < \gamma$ must contain all singularities of $f(s)$, and if these are isolated, the contour of integration in (39-5) above may often be deformed so as to surround the singularities and the theory of residues used for evaluation of (39-5). Recall that, if $f(s)$ has an isolated singularity at s_0, it will admit a Laurent expansion of the form

$$f(s) = + \cdots + \frac{a_{-k}}{(s-s_0)^k} + \cdots + \frac{a_{-1}}{s-s_0} + a_0 + a_1(s-s_0) + \cdots + a_k(s-s_0)^k + \cdots$$

valid in some annular domain about s_0. The residue of $f(s)$ at s_0 is then given by

$$\operatorname{res}[f(s), s_0] = a_{-1}$$

The calculation of the residues is considerably simplified if the Laurent series contains only a finite number, say n, of negative powers of $(s-s_0)$. In this case, the isolated singularity is said to be a pole of order n. For this case there are simple rules for the calculation of the residue. Thus

(a) If s_0 is an isolated first-order pole (simple pole) then

$$\operatorname{res}[f(s), s_0] = \lim_{s \to s_0}[(s-s_0)f(s)]$$

(b) If s_0 is an isolated simple pole and if $f(s) = p(s)/q(s)$ with $p(s_0) = 0$, $q(s_0) = 0$, $q'(s_0) \neq 0$, then

$$\operatorname{res}[f(s), s_0] = p(s_0)/q'(s_0)$$

(c) If s_0 is an isolated simple pole and $f(s) = p(s)/q(s)$ where p, q have the properties in (b) and if, in addition, $q(s) = (s-s_0)\prod_{k=1}^{n}(s-s_k)$ where the s_k are all the zeros of $q(s)$, then

$$\operatorname{res}[f(s), s_0] = p(s_0) \Big/ \prod_{1}^{n}(s_0 - s_k)$$

(d) If s_0 is an isolated pole of order n, then

$$\operatorname{res}[f(s), s_0] = \lim_{s \to s_0} g^{(n-1)}(s)/(n-1)!$$

where $g(s) = (s-s_0)^n f(s)$.

If one recalls that $f(s)$ is said to have a zero of order k as $|s| \to \infty$ if the function $g(s) = f(1/s)$ has a zero of order k at $s = 0$, then Cauchy's residue theorem implies:

Theorem 5. Let $F(t)$ be in the class (E) for $\sigma > \sigma_F$ and let its Laplace transform $f(s)$ be analytic except at $s_1 \ldots s_n$ and let $f(s)$ have a zero at infinity. Then for $t > 0$

$$f(t) = \sum_{k=1}^{n} \operatorname{res}[f(s)e^{st}, s_k]$$

while $f(0) = \lim_{t \to 0} f(t) = \sum_{k=1}^{n} \operatorname{res}[f(s), s_k] + \tfrac{1}{2}\operatorname{res}[f(s), \infty]$.

When the Laplace-transform method is used to treat partial differential equations, the evaluation of the inversion integral often requires consideration of a denumerably infinite number of isolated singularities, and the function $f(s)$ usually contains parameters such as x, etc. These extra parameters cause no difficulties provided that F is at least continuous in them. In many situations the denumerable singularities can be grouped between successive curves of a given family such as circles, rectangles, and parabolas, and it is possible to adapt Jordan's lemma (see Whittaker and Watson,[42]

p. 115) so as to establish the convergence of the residue series. As a simple example, the following is true:

Theorem 6. Let $f(s)$ be a function for which the inversion integral represents the inverse transform of $F(t)$ and assume that $f(s)$ is analytic in the half plane Re $[s] < \gamma$ except for isolated singular points s_k, $k = 1, 2, \ldots, n, \ldots$. Further, assume that $|f(s)| < M/|s|^k$ whenever s is on the circle C_n defined by $s = r_n$ for some constant M and some $k > 0$. Then the series summed over all $s_k \subset C_n$

$$\Sigma \text{res}\,[f(s)e^{st}, s]$$

converges as $r_n \to \infty$ to $F(t)$ for each positive t.

If the singularities do not stay separated as $n \to \infty$, then much more delicate analysis is required. For an example of the type of treatment necessary in this situation, the interested reader is referred to the paper by Goodrich and Kazarinoff.[14] It is often useful to have available the following:

Theorem 7. Let $f(s)$ be any function of the complex variable s that is analytic with the property that

$$|f(s)| \leq M/|s|^{k+m}$$

in some half plane Re $[s] > \sigma_0$ for $k > 1$ and m any positive integer. Then the inversion integral (39-5) exists along any line $\sigma > \sigma_0$ and converges to the inverse transform $F(t)$ of $f(s)$, and the derivatives satisfy $F^{(n)}(t) = \mathcal{L}^{-1}[s^n f(s)]$, $n = 1, 2, \ldots, m$. Furthermore $F^{(n)}(0) = 0$, $n = 0, 1, 2, \ldots, m$.

Other methods are also available for inverting the Laplace transform. Thus Widder[43,44] proves the following:

Theorem 8. Let $F(t)$ be continuous and have the Laplace transform $f(s)$. Then

$$F(t) = \lim_{k \to \infty}\,[(-1)^k/k!][f^{(k)}(k/t)](k/t)^{k+1}$$

In general, the evaluation of the limit on the right is difficult, but many examples can be found in Ref. 43.

Many problems in engineering and physics require the use of impulses, that is, forces that act for a short time only in the neighborhood of a point. When idealized to instantaneous forces, one would like to introduce, with Dirac, a "function" with the properties that

$$\int_{-\infty}^{\infty} \delta(t)\,dt = 1$$

$$\int_{-\infty}^{\infty} \delta(u)\phi(u)\,du = \phi(0)$$

$$\int_{-\infty}^{\infty} \delta^{(k)}(u)\phi(u)\,du = (-1)^k \phi^{(k)}(0)$$

Unfortunately, as shown by Von Neumann,[41] such a quantity cannot exist as an ordinary function. Recently, beginning with the work of L. Schwartz,[32] quantities possessing such properties as those of the "delta" function have been rigorously and systematically introduced in mathematics under the title of generalized functions. A brief introduction to this subject is contained in Appendix 1. Here results from this appendix will be used in order to write down correct expressions for the Laplace transform of the Heaviside unit function and its derivatives. In particular

$$\mathcal{L}[H(t)] = 1/s$$

$$\mathcal{L}[\delta(t)] = \int_{-\infty}^{\infty} \delta(t)e^{-st}\,dt = 1$$

$$\mathcal{L}[\delta^{(k)}(t)] = \int_{-\infty}^{\infty} \delta^{(k)}(t)e^{-st}$$

$$= (-1)^k (d^k/dt^k) e^{-st}\Big|_{t=0} = s^k$$

$$\mathcal{L}[\delta^{(k)}(t-a)] = (-1)^k (d^k/dt^k) e^{-st}\Big|_{t=a} = s^k e^{-as} \qquad a \geq 0$$

Thus, if $F(t)$ is in the class (E), the Laplace transform of an expression of the form

$$F(t) + \sum_{k=1}^{n} a_k \delta^{(k)}(t - b_k)$$

will exist by linearity and be analytic in a half plane $\sigma_0 > a$. In particular, then, if $F(t)$ is piecewise continuous for $t \geq 0$, equal to zero for $t < 0$, and has only a finite number of jump discontinuities, then its derivative $F'(t)$ can be written in the above form and it follows directly that its Laplace transform is given by

$$\mathcal{L}(F') = s\mathcal{L}(F)$$

Moreover, it can be shown that
$$D_0^{-1}\delta(t) = H(t)$$
$$D_0^{-1}\delta^{(k)}(t - a) = \delta^{(k-1)}(t - a)$$

Moreover, convolutions $F * G$ for generalized functions can be defined. In particular, it can be shown on the basis of the appendix that

$$\delta^{(k)}(t - a) * \delta^{(e)}(t - b) = \delta^{(k+e)}(t - a - b) \qquad a \geq 0, b \geq 0$$
$$F(t) * \delta^{(k)}(t - a) = \delta^{(k)}(t - a) * F(t)$$
$$= F^{(k)}(t - a)$$
$$\mathcal{L}[F * \delta^{(k)}(t)] = \mathcal{L}[F^{(k)}(t)]$$
$$= \mathcal{L}[F]\mathcal{L}[\delta^{(k)}(t)] = s^k \mathcal{L}(F)$$

39-4. Fourier Transforms

Transforms defined in terms of the Fourier integral are closely related to the Laplace transform and in fact are somewhat more fundamental. The Fourier transforms are defined by the relations

$$f(\omega) = \mathcal{F}(F) = 1/\sqrt{2\pi} \int_{-\infty}^{\infty} F(t)e^{-i\omega t} \, dt \qquad (39\text{-}6)$$

$$F(t) = \mathcal{F}^{-1}(f) = 1/\sqrt{2\pi} \int_{-\infty}^{\infty} f(\omega)e^{i\omega t} \, d\omega \qquad (39\text{-}7)$$

Again it is possible to develop a rigorous theory on different levels ranging from what we shall discuss here to transforms of this type over compact groups (cf. Loomis[25]). Perhaps the most widely used satisfactory theory involves the concept of the Lebesgue integral and considers functions whose absolute values are square integrable in this sense from $-\infty$ to ∞. While this theory is beyond the scope of this article, a few brief indications of some of the important results of it will be given at the end of this section.

Definition 3. Let (A) denote the class of piecewise continuous functions $F(t)$ over the interval $-\infty$ to ∞ which are absolutely integrable; i.e., the Riemann integral

$$\int_{-\infty}^{\infty} |F(t)| \, dt$$

is finite.

For this class of functions a theory analogous to that already given for the Laplace transform can be derived. Thus it is not difficult to prove the following:

Theorem 9. Let $F(t)$ be in the class (A); then
1. The Fourier transform $f(\omega)$ of $F(t)$ as defined by (39-6) exists for all ω, $-\infty < \omega < \infty$, and is continuous in ω.
2. The inverse Fourier transform $\mathcal{F}^{-1}(f)$ as defined by (39-7) exists in the sense that

$$\lim_{\alpha \to \infty} (1/\sqrt{2\pi}) \int_{-\alpha}^{\alpha} f(\omega)e^{i\omega t} \, d\omega = (1/\sqrt{2\pi}) \unicode{x2A0F}_{-\infty}^{\infty} f(\omega)e^{i\omega t} \, d\omega *$$

---

* This equation defines the symbol $\unicode{x2A0F}$, which is called a Cauchy principal-part integral.

Moreover, its value is $F(t)$ whenever $F(t)$ is continuous and it is equal to the average value of the left and right limits at any jump point of $F(t)$.

Theorem 10. Let $F(t)$ and $F'(t)$ be in the class (A); then

$$\mathfrak{F}(F') = i\omega\mathfrak{F}(F)$$

Theorem 11. Let $F(t)$ be in the class (A) and let

$$G(t) = \int_0^t F(s)\, ds + a$$

where a is a constant, have an absolutely convergent integral. Then

$$\mathfrak{F}(G) = \mathfrak{F}(F)/i\omega \qquad \omega \neq 0$$

and

$$\mathfrak{F}(G) = \int_{-\infty}^{\infty} G(t)\, dt \qquad \omega = 0$$

Theorem 12. Let $F(t)$ be in the class (A); then

1. $\mathfrak{F}[F(t - a)] = e^{-i\omega a}\mathfrak{F}(F)$
2. $\mathfrak{F}(e^{iat}F) = f(\omega - a)$

The convolution in this theory is defined by the expression

$$F * G = \int_{-\infty}^{\infty} F(u)G(t - u)\, du \tag{39-8}$$

and it can be shown that the following is true:

Theorem 13. Let $F(t), G(t)$ be in the class (A) and let one of them be bounded for all t. Then
1. The convolution as defined by (39-8) is defined and continuous for all t.
2. $F * G = G * F$.
3. $F * (G + H) = F * G + F * H$ if H satisfies the same conditions as G.
4. If F, G, and H are bounded functions in the class (A) then

$$\text{(i)} \quad F * (CG) = (CF) * G = C(F * G)$$

for C any complex number. (ii) $(F * G) * H = F * (G * H)$.

$$\text{(iii)} \quad e^{at}(F * G) = (e^{at}F) * (e^{at}G)$$

5. If $f(\omega)$ and $g(\omega)$ denote the Fourier transforms of $F(t), G(t)$ then

$$(1/\sqrt{2\pi})\mathfrak{F}(F * G) = f(\omega)g(\omega)$$

Theorem 14. Let $F(t), G(t)$ be in the class (A) with respective transform $f(\omega)$ and $g(\omega)$. If $f(\omega)$ and $g(\omega)$ are absolutely integrable on $-\infty < \omega < \infty$ then the convolution

$$F * G = \int_{-\infty}^{\infty} F(\omega)G(\omega - u)\, d\omega$$

exists and is continuous for $-\infty < \omega < \infty$, and moreover

$$\mathfrak{F}(FG) = [f(\omega) * g(\omega)]/\sqrt{2\pi}$$

It is, of course, also possible to define the Fourier transforms of generalized functions. Using the definitions of Appendix 1, it is possible to prove the following:

PROPERTIES OF THE LAPLACE AND RELATED TRANSFORMS 39-9

Theorem 15. If the generalized function $F(t)$ is defined by the sequence $F_n(t)$, the Fourier transform $f(\omega)$ will be defined by the sequence $f_n(\omega)$, where $f_n(\omega)$ is the Fourier transform of $F_n(t)$. Moreover, if $G(t)$ is any good function (see Appendix 1) whose Fourier transform is $g(\omega)$, the following equality will be true:

$$\int_{-\infty}^{\infty} F(t)G(-t)\,dt = \int_{-\infty}^{\infty} f(\omega)g(\omega)\,d\omega$$

Finally, the inverse Fourier transform of the generalized transform $f(\omega)$ will be the generalized function $F(t)$.

As in the theory of Laplace transforms, the theory of generalized functions (cf. Appendix 1) leads to the results that

$$\mathcal{F}[\delta(t)] = 1$$
$$\mathcal{F}[\delta'(t)] = i\omega$$
$$\mathcal{F}[\delta^{(n)}(t-a)] = (i\omega)^n e^{-i\omega a}$$

Although the subject cannot be entered into deeply here, assuming a knowledge of Lebesgue integration convergence of a function $f(\omega,A)$ to a function $f(\omega)$ in the mean is defined by the limit relation

$$\lim_{A \to \infty} \int_{-\infty}^{\infty} |f(\omega,A) - f(\omega)|^2 \, d\omega = 0$$

and a function which has the property that the Lebesgue integral

$$\int_{-\infty}^{\infty} |F(t)|^2 \, dt$$

exists is said to be in the class L_2. Using these concepts it can be shown that the following holds:

Theorem 16. Let $F(t)$ be in the class L_2 and define $f(\omega,A)$ by

$$f(\omega,A) = 1/\sqrt{2\pi} \int_{-A}^{A} F(t)e^{+i\omega t} \, dt$$

then as $A \to \infty$, $f(\omega,A)$ will converge in the mean to a function $f(\omega)$. Also if $F(t,A)$ is defined by the relation

$$F(t,A) = 1/\sqrt{2\pi} \int_{-A}^{A} f(\omega)e^{-i\omega t} \, d\omega$$

then $F(t,A)$ will converge in the mean to $F(t)$. Moreover, Parseval's relation will hold; namely,

$$\int_{-\infty}^{\infty} f(\omega)\overline{g(\omega)} \, d\omega = \int_{-\infty}^{\infty} F(t)\overline{G(t)} \, dt$$

If the functions $f_+(\omega)$ and $f_-(\omega)$ are defined by the relations

$$f_+(\omega) = 1/\sqrt{2\pi} \int_{0}^{\infty} F(t)e^{-i\omega t} \, dt \qquad (39\text{-}9)$$

$$f_-(\omega) = 1/\sqrt{2\pi} \int_{-\infty}^{0} F(t)e^{-i\omega t} \, dt \qquad (39\text{-}10)$$

it can be shown that $f_+(\omega)$ admits a unique analytic extension to the upper half plane which is free from singularities. This extension is made by replacing $-\omega$ in (39-9) above by $-\omega + i\sigma$; $\sigma > \sigma_0 > 0$. Similarly $f_-(\omega)$ admits a unique extension to the lower half plane when $-\omega$ is replaced by $-\omega + i\sigma$ with $\sigma < \sigma_1 < 0$ in (39-10). The resulting formulas are then of the form

If
$$z = \omega + i\sigma$$
$$f_+(z) = 1/\sqrt{2\pi} \int_0^\infty F(t)e^{-i\sigma t}e^{-i\omega t}\,d\omega \qquad \sigma > \sigma_0 > 0$$
$$f_-(z) = 1/\sqrt{2\pi} \int_{-\infty}^0 F(t)e^{-i\sigma t}e^{-i\omega t}\,d\omega \qquad \sigma < \sigma_1 < 0$$

and an inversion theorem for $f(t)$ can be obtained in the form

$$F(t) = 1/\sqrt{2\pi}\left[\int_{-\infty+i\sigma_0}^{\infty+i\sigma_0} f_+(z)e^{+ikz}\,dz + \int_{-\infty+i\sigma_1}^{\infty+i\sigma_1} f_-(z)e^{+ikz}\,dz\right]$$

Moreover, if one forms the difference functions

$$\mathfrak{D}(\omega) = f_+(\omega) - f_-(\omega)$$

the so-called Plemelj formulas can be derived. In one form these are given by

$$f(\omega) = f_+(\omega) + f_-(\omega)$$
$$f(\omega) = 1/\pi i \,P\!\!\!\int_{-\infty}^\infty \mathfrak{D}(u)\,du/(u - \omega)$$

where again the P implies that the integral is to be taken as a principal-part integral as in Theorems 9 and 10. These formulas form the basis for the theory of singular integral equations of the Cauchy type (cf. Muskhelishvili[28]) and also play an important role in various theories involving dispersion relations.

It is now a trivial matter to state the relationship between the Laplace transform and the Fourier transform. Explicitly if $f(s)$ denotes the Laplace transform of F and $f_+(\omega)$ denotes the partial Fourier transform as defined above, this relation is given by

$$f(s) = \mathfrak{L}(f) = \sqrt{2\pi}\,f_+(is)$$

To conclude this section, we draw attention to two important theorems which characterize causality in the transform plane. The first is usually ascribed to Titchmarsh[38] but is also in Paley and Wiener[30] and is the following:

Theorem 17. The class L_2 of functions $F(t)$ that vanish for negative values of their arguments is identical to the class of functions $F(t)$ whose complex Fourier transforms are analytic in the upper half plane $\sigma > 0$ and have a transform such that the Lebesgue integral

$$\int_{-\infty}^\infty |f(\omega + i\sigma)|^2\,d\omega$$

exists. Moreover, the limit

$$f(\omega) = \lim_{\sigma \to 0} f_+(\omega + i\sigma)$$

exists almost everywhere and the real and imaginary parts of the Fourier transform are related by the Hilbert-transform formulas; that is

$$\operatorname{Re} f(\omega) = 1/\pi \int_{-\infty}^\infty [\operatorname{Im} f(x)/(x - \omega)]\,dx$$
$$\operatorname{Im} f(\omega) = -1/\pi \int_{-\infty}^\infty [\operatorname{Re} f(x)/(x - \omega)]\,dx$$

The Hilbert-transform formulas have many important applications to many fields. For example, Tricomi[40] discusses their use in airfoil theory, while Bode[3] makes extensive use of them in various equivalent forms to discuss various relations between the gain and phase of linear networks. In modified form they also form the basis for the dispersion relations in quantum-field theory. In essence, they are direct conse-

PROPERTIES OF THE LAPLACE AND RELATED TRANSFORMS

quences of Poisson's integral formula, and elementary derivations of them can be found in Churchill,[5] Guillemin,[17] or Kaplan.[21]

The second theorem is due to Paley and Wiener[30] and is as follows:

Theorem 18. A necessary condition that a function $\phi(\omega)$ be the modulus of the Fourier transform of a nonidentically zero function $F(t)$ that vanishes on some half axis and is in the class L_2 is that

$$\phi(-\omega) = \phi(\omega)$$
$$\int_{-\infty}^{\infty} |\phi(\omega)|^2 \, d\omega < \infty$$
$$\int_{-\infty}^{\infty} [|\log \phi(\omega)|/(1 + \omega^2)] \, d\omega < \infty$$

This theorem is important in applications since the transforms of physically realizable functions must correspond to functions which are zero up until a signal is applied. Thus failure of the above condition to hold immediately implies that the transfer function cannot be realized physically even though other considerations might have suggested that it had highly desirable properties.

39-5. Other Integral Transforms

Many other transforms are often used in problems of engineering and physics. Here we must be content to give a few examples and some references where detailed treatments can be found.

The Mellin Transform. The Mellin transform is defined by the formula

$$F_m(s) = \int_0^{\infty} f(t) t^{s-1} \, dt$$

and it can be readily shown that if

$$f(\omega) = 1/\sqrt{2\pi} \int_{-\infty}^{\infty} F(t) e^{i\omega t} \, dt$$

is the Fourier transform of $F(t)$ then

$$F_m(s) = \sqrt{2\pi} \, f(-is)$$

Thus it is possible to translate the properties of the Fourier transform into corresponding properties of the Mellin transform. For details see Sneddon[33] or Morse and Feshbach.[27]

The Hankel Transform. The Hankel transform is defined under suitable restriction by the pair of formulas

$$\hat{\phi}(k) = \int_0^{\infty} J_m(kr) \phi(r) k \, dr$$
$$\phi(r) = \int_0^{\infty} J_m(kr) \hat{\phi}(k) k \, dk$$

In these expressions $J_m(v)$ is the mth Bessel function of the first kind. For a detailed derivation and their relation to Fourier transforms, as well as for some of their applications, reference can be made to Sneddon.[33] For more complicated transforms of the same type involving the use of Bessel functions of the second and third kind, reference can be made to Goldstein.[15]

The Stieltjes Transform. The Stieltjes transform is defined for functions $f(t) = 0$ for $t < 0$ by the formula

$$S(F) = \int_0^{\infty} [F(t)/(t + s)] \, dt$$

It is also easy to see that if

$$\mathcal{L}(F) = \int_0^{\infty} e^{-st} F(t) \, dt$$

denotes the Laplace transform of $F(t)$ then

$$S(F) = \mathcal{L}[\mathcal{L}(F)]$$

For a more detailed discussion see Widder.[43]

The Stieltjes Inversion Formula. A closely related result which is of fundamental importance to operator theory and probability theory, the moment problem, the theory of continued fractions, and the theory of Jacobi matrices is the inversion formula of Stieltjes. This result has many slight variants and will be found in Stone[34] and in Neumark[29] in a slightly different form from that to be given here, a form which we have found the most useful. To state the theorem, suppose that we are given a function of the form

$$f(z) = \int_{-\infty}^{\infty} dm(t)/(t-z) \qquad \text{Im } z > 0 \qquad (39\text{-}11)$$

where $m(t)$ is a complex function of bounded variation on the entire real axis which has been normalized by the conditions $m(-\infty) = 0$ and $m(t-0) = m(t)$ for all t, $-\infty \leq t < \infty$. Parenthetically, it should be remarked that a necessary and sufficient condition that a complex function admit the representation (39-11) is that $f(z)$ be analytic in the half plane Im $z > 0$ with the properties that Im $f(z) > 0$ and $\sup |iyf(iy)| < \infty$ there. A proof can be found in Achieser and Glasmann.[1] The Stieltjes inversion formula then says that if t_1, t_2 are any two points of continuity of $m(t)$, then if $z = x + iy$,

$$m(t_2) - m(t_1) = \lim_{y \to 0+} 1/\pi \int_{t_1}^{t_2} \text{Im } f(x + iy) \, dx$$

The conditions on $f(z)$ can be relaxed. For a discussion, see Greenstein.[16]

The Finite Fourier Transform. The finite Fourier transform is defined as the sequence

$$f(n) = 1/\tau \int_0^\tau F(t) e^{-in\omega t} \, dt \qquad \omega = 2\pi/\tau$$

for $n = 0, \pm 1, \pm 2, \ldots$ when $F(t + \tau) = F(t)$ and possesses many properties similar to the Fourier integral transform. When it exists, the inverse transform is given by the series

$$f(t) = \sum_{n=-\infty}^{\infty} f(n) e^{in\omega t}$$

A detailed treatment of this transform can be found, for example, in Kaplan.[20]

The Legendre Transform. The Legendre transform is defined as the sequence

$$l(n) = \int_{-1}^{1} f(x) P_n(x) \, dx$$

where n is a positive integer and $P_n(x)$ is the Legendre polynomial of n degree. As in the case of the finite Fourier integral, inversion is by means of an infinite series; i.e.,

$$f(x) = \sum_{0}^{\infty} [(2n+1)/2] l(n) P_n(x)$$

For more details and some applications, reference may be made to Tranter.[39]

The Finite Hankel Transform. The finite Hankel transform is defined in detail as a function of the application for which it is intended. In the simplest case, it is defined by the formula

$$h(p) = \int_0^1 f(r) J_n(pr) r \, dr$$

PROPERTIES OF THE LAPLACE AND RELATED TRANSFORMS   39-13

where p is a positive root of the equation

$$J_n(p) = 0$$

Under these circumstances the inversion formula is given by

$$f(r) = 2 \sum_0^\infty h(p)[J_n(pr)/J_{n+1}^2(p)]$$

For more details, see Tranter.[39]

Several remarks should now be made. First of all, it should be apparent that the finite Fourier, Legendre, and Hankel transforms represent nothing more than a different way of viewing the known expansion theorem for functions in terms of Fourier series, Legendre polynomials, and Bessel functions. Thus whenever one has an expansion theorem in terms of a set of orthogonal functions one can introduce an analogous finite transform. More generally, however, one can develop operational calculus based on the spectral theory of a given class of operators in a given space. While this cannot be entered into in detail here, an introduction to this subject can be found in Taylor,[36] a detailed treatment for Hilbert spaces in Stone,[34] and a similar treatment for more general spaces in Dunford and Schwartz.[9]

As a final remark, the usefulness of a given transform is considerably enhanced whenever there is a convolution property like that possessed by the Laplace and Fourier transforms. It follows from the discussion above that the Mellin and Stieltjes transform will possess a convolution property; and a convolution property for the Hankel transforms has been derived by Klein[22] and Delsarte,[8] while a convolution property for the Legendre transform has been given by Churchill and Dolph.[6] To the best of the author's knowledge, no convolution property has been given explicitly, on the other hand, for the finite Hankel transform. The most general treatment of the convolution seems to be due to Levitan,[23] to whom we refer the interested reader for further details.

39-6. Illustrative Examples of the Use of Laplace Transforms

The uses to which the above theory can be put are too numerous to permit a systematic treatment here. Some idea of the wide range of its applicability can, however, be gleaned from the following elementary examples.

The Computations of Transforms. As an example, to compute $\mathcal{L}(\cos kt)$ note that $Y = \cos kt$ satisfies $Y'' + k^2 Y = 0$ and the initial conditions $Y(0) = 1$, $Y'(0) = 0$. The operational Theorem 2 gives at once that

$$y(s) = s/(s^2 + k^2)$$

The Solution of Linear Differential Equations with Constant Coefficients. Consider the equation $Y'' + 3Y' + 2Y = F(t)$ for $F(t)$ in the class (E). If, for simplicity, it is assumed that $Y(0) = Y'(0) = 0$ then the same operator rule yields at once the fact that

$$y(s) = \frac{f(s)}{(s+2)(s+1)} = \frac{f(s)}{s+1} - \frac{f(s)}{s+2}$$

Noting that $\mathcal{L}^{-1}(1/s + a) = e^{-at}H(t)$ either from Theorem 3 or from Appendixes 2 and 3 and applying the convolution property of Theorem 4 gives

$$Y(t) = F * e^{-t}H(t) - F * e^{-2t}H(t)$$
$$= e^{-t} \int_0^t F(u) e^u \, du - e^{-2t} \int_0^t F(u) e^{2u} \, du$$

It should be noticed that the roots s_i of

$$s^2 + 3s + 2 = (s+1)(s+2) = 0$$

have the property that Re $s_i < 0$ so that the solutions to the homogeneous equation $Y'' + 3Y' + 2Y = 0$ will tend to zero independently of the values $Y(0)$, $Y'(0)$. Such a system is called stable, and one of the most important questions to be settled in linear systems containing parameters is often that of determining the domains of stability with respect to the parameters involved. Powerful mathematical methods have been developed to aid in this determination, the most often used being the Nyquist criterion or the Hurwitz-Routh criterion. An elementary treatment of these methods can be found in either Kaplan[20] or Guillemin.[17]

The Solution of Linear Systems of Ordinary Differential Equations. Consider, for example, the homogeneous system with arbitrary initial conditions

$$dX/dt = 2X + 2Y - Z$$
$$dY/dt = -2X + 4Y + Z$$
$$dZ/dt = -3X + 8Y + 2Z$$

After Laplace transformation, it becomes

$$(s - 2)x - 2y + z = X(0)$$
$$2x + (s - 4)y - z = Y(0)$$
$$3x - 8y + (s - 2)z = Z(0)$$

Solving this, for example, for $x(s)$ by Cramer's rule yields, if $Y(0) = Z(0) = 0$,

$$x(s) = \frac{\begin{vmatrix} X(0) & -2 & 1 \\ 0 & s-4 & -1 \\ 0 & -8 & s-2 \end{vmatrix}}{\begin{vmatrix} s-2 & -2 & 1 \\ 2 & s-4 & -1 \\ 3 & -8 & s-2 \end{vmatrix}}$$

$$= \frac{s(s-6)X(0)}{(s-6)(s-1)^2}$$

$$X(s) = \left[\frac{1}{s-1} + \frac{1}{(s-1)^2}\right] X(0)$$

Using Appendixes 2 and 3, one finds directly then that

$$X(t) = X(0)(e^t + te^t)$$

This system is not stable, and moreover its solution contains so-called secular terms— the terms of the form te^t in this case—corresponding to the appearance of nonsimple elementary divisors in the determinant of the denominator of the expression for $x(s)$. Here the appearance of nonsimple elementary divisors is handled in an automatic fashion and there is no need for the deep results of this theory. Other systems with constant coefficients can be treated in a similar fashion independent of the appearance of derivatives of different order in any given equation.

Integral-differential Equations of Convolution Type. Consider, for example, the equation

$$aY'(t) + bY(t) = F(t) + \int_0^t G(t-u)Y(u)\,du$$

subject to $Y(0) = 0$. After transformation and use of the convolution, one obtains

$$(as + b)y(s) = f(s) + g(s)y(s)$$

or

$$y(s) = \frac{f(s)}{(as+b) - g(s)}$$

Depending upon the nature of $g(s)$, this may then in turn be inversely transformed by elementary or more complicated means such as the inversion integral.

PROPERTIES OF THE LAPLACE AND RELATED TRANSFORMS **39**–15

Difference-differential Equations with Zero Initial Conditions. Consider, for example, the difference-differential equation

$$Y'(t) - Y(t-1) = aH(t)$$

subject to
$$Y(t) = 0 \quad \text{for } t \leq 0$$

Upon transformation this yields

$$sy(s) - e^{-s}y(s) = a/s$$

from which one obtains

$$y = \frac{a}{s(s - e^{-s})} = \frac{a}{s^2(1 - e/s^{-s})} = \frac{a}{s^2}(1 + e^{-s}/s + e^{-2s}/s^2 + \cdots)$$

Taking the inverse transformation results in

$$Y(t) = a \sum_{n=0}^{\infty} \frac{(t-n)^{n+1}}{(n+1)!} H(t-n)$$

This apparently infinite series is, in reality, a finite sum, since for any given value of t there will always exist integers n, $n+1$ such that the given value of t is bracketed $n < t < n+1$. Thus, for each fixed t all terms $H(t-k)$ for $k \geq n+1$ will be zero since $H(t) = 0$ if $t < 0$.

Partial Differential Equations. The use of Laplace transformation in the solution of problems in partial differential equations often involves more subtle considerations, some of which can be illustrated by the following relatively simple example. Consider the problem described by the partial differential equation

$$U_{xx}(x,t) = U_t(x,t) \tag{39-12}$$

with the initial condition
$$U(x,0) = 0 \tag{39-13}$$

and the boundary conditions
$$U(0,t) = 0 \tag{39-14}$$
$$U(1,t) = F(t) \tag{39-15}$$

Unlike the method of separation of variables, the Laplace-transformation method is not bothered by the occurrence of a time-dependent boundary condition but can be applied directly. The method of separation of variables, to be directly applicable before problem transformation (which is usually possible, as it is here), requires a homogeneous partial differential equation, homogeneous boundary conditions, and inhomogeneous initial conditions. (Cf. Ref. 35, where such transformations are discussed on pages 58–64.) If the Laplace transformation of the partial differential equation and the boundary conditions in the above problem is taken with respect to time, the resulting system becomes

$$su(x,s) = u_{xx}(x,s)$$
$$u(0,s) = 0 \tag{39-16}$$
$$u(1,s) = f(s)$$

The differential equation (39-16) is readily solved when s is treated as a parameter and has a solution

$$u(x,s) = A \sinh \sqrt{s}\, x + B \cosh \sqrt{s}\, x$$

Application of the boundary conditions yields

$$B = 0$$
$$A = f(s)/\sinh \sqrt{s}$$

39-16 MATHEMATICS ASSOCIATED WITH SYSTEM ENGINEERING

so that
$$u(x,s) = \frac{f(s)\sinh\sqrt{s}\,x}{\sinh\sqrt{s}} \qquad (39\text{-}17)$$

Now if $\sqrt{s}$ represents the same branch of the two-valued analytic function $s^{1/2}$ in this expression, the function $\sinh\sqrt{s}\,x/\sinh\sqrt{s}$ is readily seen to be an analytic function of s except at the zeros of the denominator. These are given by $\sinh\sqrt{s} = 0$ or by $s = -n^2\pi^2$, $n = 1, 2, \ldots$. If the inverse transformation of the above expression is taken, the resulting series representation for $\mathcal{L}^{-1}(\sinh\sqrt{s}\,x/\sinh\sqrt{s})$ will fail to converge for $t = 0$. To overcome this difficulty one may write in place of $u(x,s)$ an equivalent expression
$$u(x,s) = sf(s)(\sinh\sqrt{s}\,x/s\sinh\sqrt{s}) \qquad (39\text{-}18)$$

If it is now assumed that $F(t)$ is in the class (E) and has a piecewise continuous derivative, it will follow that
$$sf(s) = L(F') - F(0+)$$
so that use of the convolution property yields
$$U(x,t) = [F'(t) - F(0+)] * \mathcal{L}^{-1}(\sinh\sqrt{s}\,x/s\sinh\sqrt{r}) \qquad (39\text{-}19)$$

The inversion integral may be used to complete the inversion process. Thus
$$\mathcal{L}^{-1}(\sinh\sqrt{s}\,x/s\sinh\sqrt{s}) = (1/2\pi i)\int_{\gamma-i\infty}^{\gamma+i\infty} e^{st}(\sinh\sqrt{s}\,x/s\sinh\sqrt{s})\,ds$$
$$= \sum_{n=0}^{\infty} \operatorname{res}\left(e^{st}\frac{\sinh\sqrt{s}\,x}{s\sinh\sqrt{s}}, -n^2\pi^2\right)$$

The actual evaluation of the residue may be accomplished by rule (b) of Sec. 39-3. Thus if the branch cut of $\sqrt{s}$ is chosen as the positive real axis
$$R_n = \operatorname{res}\left(\frac{\sinh\sqrt{s}\,xe^{st}}{s\sinh\sqrt{s}}, -n^2\pi^2\right) = \operatorname{res}\left[\frac{p(x,s)e^{st}}{q(x,s)}, -n^2\pi^2\right] \quad n = 1, 2, \ldots$$
where $\quad p(x,s) = \sinh\sqrt{s}\,x/s$
and $\quad q(x,s) = \sinh\sqrt{s}$

One obtains the result
$$R_n = \frac{P(x,s)e^{st}}{q'(x,s)}\bigg|_{s=-n^2\pi^2} = (-1)^n \frac{2}{n\pi}\sinh n\pi x e^{-n^2\pi^2 t}$$

On the other hand, it follows directly that
$$\lim_{s\to 0} s(\sinh\sqrt{s}\,x/s\sinh\sqrt{s}) = x$$
so that one has
$$V(x,t) = \mathcal{L}^{-1}(\sinh\sqrt{s}\,x/\sinh\sqrt{s})$$
$$= x + (2/\pi)\sum_{1}^{\infty}[(-1)^n/n]e^{-n^2\pi^2 t}\sin n\pi x$$

Insertion of this result into (39-17) yields the solution in the form
$$U(x,t) = F(+0)V(x,t) + \int_0^t F'(t-)V(x,\sigma)d$$

PROPERTIES OF THE LAPLACE AND RELATED TRANSFORMS   **39**–17

which is one form of Duhamel's formula since $V(x,t)$ may be interpreted as the solution to the given problem when $F(t)$ is taken to be unity. An order argument applied to $u(x,s)$ in (39-18) will yield the fact that

$$|u(x,s)| < M/|s|^k$$

in the right-hand half plane for any fixed k so that Theorem 15 is applicable and yields the result that

$$U(x,0) = 0$$

so that all conditions of the given problem are formally satisfied. The formula for the solution given here can be rigorously established by the use of a modified form of Theorem 14 in which the circles are replaced by a family of parabolas which thread the singularities $s = -n^2\pi^2$. Full details can be found in Churchill,[5] pages 199–213.

Initial-value Problems in Partial Differential Equations. If instead of the initial condition used above, one requires

$$U(x,0) = G(x)$$

and if the boundary conditions are assumed homogeneous

$$U(0,t) = U(1,t) = 0$$

the differential equation after Laplace transformation takes the form

$$u_{xx}(x,s) - su(x,s) = -g(s)$$

and this must be solved subject to the two-point boundary condition

$$u(0,s) = u(1,s) = 0$$

This problem is a typical Sturm-Liouville problem whose resolution involves the introduction of the concept of a Green's function. To motivate this concept, consider the simpler problem of solving

$$d^2Y/dx^2 = -\delta(x - x')$$

[see Appendix 1 for the meaning to be attached to $\delta(x)$] subject to $y(0) = y(1) = 0$.

Using the properties of a generalized function discussed in Sec. 39-3, one has by successive integrations

$$dY/dx = H(x - x') + A$$
$$Y = R(x - x') + Ax + B$$

where

$$R(t) = \int_0^t H(u)\, du$$

is the ramp function

$$R(t) = 0 \quad \text{for } t \leq 0$$
$$R(t) = t \quad \phantom{\text{for }} t \geq 0$$

Imposition of the boundary conditions yields $B = 0$ and $A = -R(1 - x')$. The resulting expression is the Green's function

$$G(x,x') = R(x - x') - R(1 - x')x$$

This may be put into more familiar form by considering the following two cases:
If $x < x'$ then $R(x - x') = 0$ and $R(1 - x') = 1 - x'$ so that

$$G(x,x') = (x' - 1)x$$

If $x > x'$ then $R(x - x') = x - x'$ so that

$$G(x,x') = (x - x') - (1 - x')x = (x - 1)x'$$

The Green's function is readily seen to be continuous at $x = x'$ and to possess a unit jump in its derivative at $x = x'$. Moreover, it may be directly verified that it satisfies the homogeneous differential equation $d^2Y/dx^2 = 0$ at all points $x = x'$. Finally,

the solution to the problem
$$d^2Y/dx^2 = -f(x)$$
is given
$$Y(x) = \int_0^1 G(x,x')f(x')\,dx' \tag{39-20}$$
since
$$d^2Y/dx^2 = \int_0^1 G_{xx}(x',x)f(x')\,dx' = -\int_0^1 \delta(x-x')f(x')\,dx'$$
$$= -f(x)$$

In more complicated cases, the Green's function cannot be so straightforwardly constructed but use must be made of the continuity property, the jump property, and the fact that it is a solution of the homogeneous equation everywhere except at $x = x'$. Alternatively, it may be defined as a solution to the given problem given in the form (39-20), and its other properties may be deduced from this expression. The first approach is systematically treated in Ince[19] and Neumark,[29] while the second method can be found in Tamarkin and Feller.[35] Friedman[12] also gives many examples.

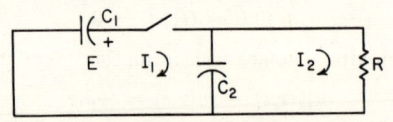

FIG. 39-1. *RC* network.

For the problem stated at the beginning of this discussion, the corresponding expression can be shown, by either method, to be

$$R(x,x',s) = -\frac{\sinh[(1-x)\sqrt{s}]}{s\sinh\sqrt{s}}\sinh x'\sqrt{s} \qquad 0 \le x' \le x$$

$$= -\frac{\sinh[(1-x')\sqrt{s}]\sinh x\sqrt{s}}{s\sinh\sqrt{s}} \qquad x \le x' \le 1$$

so that the solution to the given problem requires the inversion of

$$u(x,s) = \int_0^1 R(x,x',s)g(s)\,dx'$$

and this may be carried out in a manner similar to that used above under Partial Differential Equations. For full details, see Churchill,[5] pp. 219–222.

Generalized Functions in Circuit Theory. As a final example we shall briefly discuss the occurrence of a delta function or, in this context, the occurrence of an impulsive current in a simple network. Consider the network shown in Fig. 39-1. Setting $D_1 = 1/C_1$ and $D_2 = 1/C_2$, Kirchhoff's laws yield the equations

$$D_1\int_0^t I_1(u)\,du - E + D_2\int_0^t [I_1(u) - I_2(u)]\,du = 0$$

$$D_2\int_0^t [I_2(u) - I_1(u)]\,du + RI_2(t) = 0$$

After taking the Laplace transform, these become

$$(D_1 + D_2)[i_1(s)/s] - D_2[i_2(s)/s] = E/s$$
$$-D_2[i_1(s)/s] + (D_2/s + R)i_2(s) = 0$$

Solving by Cramer's rule for $i_1(s)$, one obtains

$$i_1(s) = \frac{\begin{vmatrix} E & -D_2 \\ 0 & D_2/s + R \end{vmatrix}}{\begin{vmatrix} (D_1 + D_2) & -D_2 \\ -D_2 & D_2/s + R \end{vmatrix}}$$

$$= E \frac{Rs + D_2}{[R(D_1 + D_2) + D_2{}^2]s + D_2(D_1 + D_2)}$$

Letting

$$\alpha = R(D_1 + D_2) + D_2{}^2$$
$$\beta = [D_2(D_1 + D_2)]/\alpha \quad \text{and} \quad \gamma = D_2/R - \beta$$

this may be written as

$$i_1(s) = E \frac{R}{\alpha} \frac{[s + D_2/R]}{s + \beta} = E \frac{R}{\alpha} \left[1 + \frac{D_2/R - \beta}{s + \beta} \right]$$
$$= E(R/\alpha)(1 + \gamma/s + \beta)$$

Taking the inverse Laplace transform, one therefore finds

$$I_1(t) = E(R/\alpha)[\delta(t) + \gamma e^{-\beta t}]$$

The delta function occurs since the voltage across a capacitance cannot change instantaneously as it must do once the switch is closed since then the two capacitances are effectively parallel. This simple example, while special, is typical of how impulsive currents can appear in networks involving loops having capacitances and voltages as their only elements. For further details any one of the numerous books such as that by Cheng[4] or Arsac (Ref. 2, Chapter 7) may be consulted.

Appendix 1. An Introduction to Generalized Functions

The precise method used for introducing generalized functions is dependent upon the applications for which the theory is intended. In its most general form to date, it has been systematically developed by Schwartz[32] by the use of the concept of "distributions." This form of the theory makes extensive use of various topological spaces and is much too involved for presentation here. A more readable development using essentially the same approach has been given by Gel'fand and Schilow,[13] and Arsac[2] has given a simplified presentation suitable for many physical applications. An independent and somewhat simpler approach is due to Mikusinski.[26] The relationship between the theories of Schwartz and Mikusinski has been discussed by Temple[37] and Erdelyi,[10] who also has written an elementary presentation of the theory of Mikusinski.[11] Since our interest here will be confined to the Heaviside unit function and its derivatives and the delta function and its derivatives, it is possible to simplify the subject still further and proceed in the manner of Lighthill.[24] Since the presentation here will include only a small part of the material of Ref. 24, the terminology and definitions of this book will be followed precisely in spite of their quaintness. This should enable the interested reader to begin more easily a more detailed study of this subject.

The method used by Lighthill to introduce generalized functions is by means of equivalent classes of Cauchy sequences of functions, and as Temple[37] observed, this is the basic method underlying all the theories of generalized functions which differ only in the topology chosen for the Cauchy sequences. As such the method is similar to Cantor's procedure for the introduction of the real numbers and to the, by now, well-known procedure for completing a metric space containing a dense set of points.

To motivate the discussions, recall that an "equivalence relation" written with the symbol $\sim$ is said to hold for a set S of elements $a, b, c, \ldots$. If given any two elements a, b in S it is possible to decide whether $a \sim b$ or $b \sim a$ and if in addition the "equivalence relation" has the properties of being reflexive, symmetric, and transitive. In symbols,

these relations are, respectively,
1. $a \sim a$.
2. If $a \sim b$ then $b \sim a$.
3. If $a \sim b$ and $b \sim c$ then $a \sim c$.

Recall also Cauchy's criterion for the convergence of a sequence of numbers $\{a_n\}$ namely, if for any given positive ϵ there exists a definite integer m such that for all $n > m$

$$|a_n - a_{n+p}| < \epsilon \qquad p = 1, 2, 3, \ldots$$

the sequence $\{a_n\}$ is said to converge.

Using this notion restricted to rational numbers, Cantor introduced the notion of real numbers by considering equivalent classes of convergent Cauchy sequences of rational numbers. The notion of equivalence in this case is defined as follows:

The convergent Cauchy sequences of rational numbers $\{a_n\}$ and $\{b_n\}$ are said to be equivalent if given any arbitrary positive $\epsilon > 0$ there exists a definite integer m such that for all $n > m$ and all $p > 0$

$$|a_{n+p} - b_{n+p}| < \epsilon$$

It is easily verified that this is an equivalence relation in the sense defined above, and it is also important to notice that there is no requirement that $a_n = b_n$.

Now if the convergent Cauchy sequence of rational numbers $\{a_n\}$ converges to a rational number A in the sense that given any positive ϵ there exists a definite number m such that for $n > m$

$$|A - a_n| < \epsilon$$

one writes $\lim_{n \to \infty} a_n = A$. If no rational number satisfying this last definition exists, one introduces an "ideal element"—in this case the real number—associated with the above convergent Cauchy sequence and all other convergent Cauchy sequences equivalent to it in the above sense. With a suitable definition of "distance" this process is just that used to complete an arbitrary metric space when one is given a dense set of points in it.

To introduce "generalized functions" one of course needs a different definition of the "equivalence relation." To motivate the definition chosen by Lighthill, let us first recall two familiar classical results in which the analogs of a Cauchy sequence of rational numbers occur. If $f(x)$ is a continuously differentiable periodic function of period 2π, then it is well known that it can be represented by its Fourier series, the limit of the partial sums, as $n \to \infty$

$$S_n(t) = a_0/2 + \sum_1^n a_k \cos hx + \sum_1^n b_k \sin kx$$

By the use of complex numbers and geometric series, this sum can be transformed into the Dirichlet kernel, with the result that the statement that a Fourier series represents $f(x)$ is equivalent to the statement that

$$f(x) = \lim_{n \to \infty} 1/\pi \int_{-\pi}^{\pi} f(x') \frac{\sin(n + \tfrac{1}{2})(x' - x)}{2 \sin \tfrac{1}{2}(x' - x)} dx'$$

Now if it were possible to take the limit in the expression under the integral sign, then the resulting expression exhibits a delta-function type of behavior. This may be seen by noting that the expression may be rewritten as

$$f(x) = \lim_{n \to \infty} 1/\pi \int_{-\pi}^{\pi} g(x') \frac{\sin(n + \tfrac{1}{2})(x' - x)\, dx'}{x' - x}$$

where

$$g(x) = f(x') \Big/ \frac{\sin \tfrac{1}{2}(x' - x)}{(x' - x)/2}$$

and using the fact that

$$\lim_{u \to 0} \frac{\sin ku}{u} = k$$

Similarly the solution to the simple heat problem for an infinite rod defined by the equations

$$\partial U/\partial t = k(\partial^2 U/\partial x^2) \qquad U(x,t)_{t=0^+} = f(x)$$

is well known to be given by the integral

$$U(x,t) = \int_{-\infty}^{\infty} \frac{e^{\frac{-(x-x')}{4kt}}}{2\sqrt{\pi kt}} f(x)' \, dx'$$

In order to recover the initial values it is therefore necessary that

$$f(x) = \lim_{t \to 0^+} u(x,t) = \lim_{t \to 0^+} \int_{-\infty}^{\infty} \frac{e^{\frac{-(x-x')}{4kt}}}{2\sqrt{\pi kt}} f(x') \, dx'$$

This again can be justified mathematically for a suitable class of functions, and again if it were possible to take the limit under the integral sign the resulting expression would exhibit a delta-function type of behavior. These two examples suggest the following definitions of Lighthill:[24]

Definition 4. A good function $g(t)$ is one which is everywhere differentiable any number of times and such that it and all its derivatives have the property that

$$|g^{(k)}(t)|/|t|^N \text{ is bounded} \qquad R = 0, 1, 2, \ldots$$

as $|t| \to \infty$ for all N, k.

Definition 5. A sequence $\{f_n(t)\}$ of good functions is called regular if, for any good function $g(x)$ whatever, the limit

$$\lim_{n \to \infty} \int_{-\infty}^{\infty} f_n(t) g(t) \, dt \qquad (39\text{-}21)$$

exists.

Definition 6. Two regular sequences of good functions are called equivalent if, for any good functions $g(x)$ whatever, the limit (39-21) is the same for each sequence.

Definition 7. A generalized function $f(x)$ is defined as a regular sequence $\{f_n(x)\}$ of good functions, but two generalized functions are said to be equal if the corresponding regular sequences are equivalent. Thus each generalized function is really the class of all regular sequences equivalent to a given regular sequence. The integral

$$\int_{-\infty}^{\infty} f(t) g(t) \, dt$$

of the product of a generalized function $f(t)$ and a good function $g(x)$ is defined as

$$\lim_{n \to \infty} \int_{-\infty}^{\infty} f_n(t) g(t) \, dt$$

This is permissible because the limit is the same for all equivalent sequences $f_n(x)$.

The above definitions of Lighthill can be made somewhat less abstract by taking advantage of a method developed by Widlund,[46] who used so-called Hermite functions to characterize both the classes of "good" functions and generalized functions just defined. An earlier and more extensive development by this same method is due to Korevaar,[45] who, however, did not relate his results to those of Lighthill. A Hermite function $\phi_n(t)$ is defined by the relation

$$\phi_n(t) = 1/(2^n n!)^{1/2} \pi^{1/4} e^{-t^2/2} H_n(t) \qquad n = 0, 1, 2, \ldots$$

where $H_n(t)$ is the usual Hermite polynomial given by

$$H_n(t) = (-1)^n e^{t^2} (d^n/dx^n)(e^{-t^2}) \qquad n = 0, 1, 2, \ldots$$

Hermite functions can readily be shown to satisfy the differential equation

$$\phi_n'' + (2n + 1 - t^2) \phi_n = 0$$

and this equation can be used to establish the following characterization of the class of "good" functions:

A function $G(t)$ is a "good" function if and only if it admits the convergent orthogonal expansion

$$G(t) = \sum_{n=0}^{\infty} a_n \phi_n(t)$$

in which the coefficients a_n are such that

$$|a_n| < M(m)/(n+1)^m$$

for arbitrary m and $M(m)$ independent of n.

If one defines (f,G) by the relations

$$(f,G) = \int_{-\infty}^{\infty} f(t)G(t)\,dt$$

then the coefficients in the above expansion are given simply by

$$a_k = (G, \phi_k)$$

Moreover, in this notation Lighthill's class of generalized functions are just the equivalent class of functions $f_n(t)$, $\{f_n(t)\}$ in the class of good functions for which $\lim_{n \to \infty} (f_n, G)$ exists for all good functions $G(t)$ and the idealized element which represents the generalized function $f(t)$ is defined by the relation

$$(f,G) = \lim_{n \to \infty} (f_n, G)$$

The generalized functions $f(x)$ and $g(x)$ are said to be equivalent then if $(f,G) = (g,G)$ for every good function $G(t)$.

Finally, any generalized function in this sense is equivalent to the set

$$\sum_{k=0}^{n} (f, \phi_k) \phi_k(t)$$

in that the relation

$$(f,G) = \sum_{k=0}^{\infty} (f, \phi_k)(\phi_k, G)$$

holds true for all good functions $G(t)$. A necessary and sufficient condition that a sum of the form

$$\sum_{k=0}^{\infty} b_k \phi_k(t)$$

represents a generalized function in this sense is that positive constants K and k independent of the index n exist such that the coefficients satisfy the inequality

$$|b_n| < K(n+1)^k$$

That is, the coefficients cannot grow faster than a fixed power, depending on the generalized function $f(x)$ of the index of the coefficient.

Lighthill also introduces another class of functions by the following:

Definition 8. A fairly good function is one which is everywhere differentiable any number of times and such that it and all its derivatives have the property that

$$|g^{(k)}(k)|/|t|^N \text{ is bounded} \quad k = 0, 1, 2, \ldots$$

as $|t| \to \infty$ for some N.

Definition 9. If two generalized functions $f(t)$ and $h(t)$ are defined by the sequences $\{f_n(t)\}$ and $\{h_n(t)\}$, then their sum $f(t) + h(t)$ is defined by the sequence $\{f_n(t) + h_n(t)\}$. Also, the derivative $f'(t)$ is defined by the sequence $\{f_n'(t)\}$. Also $f(t' + b)$ is defined by the sequence $\{f_n(at + b)\}$. Also $\phi(t)f(t)$, where $\phi(t)$ is a fairly good function, is defined by the sequence $\{\phi(t)f_n(t)\}$.

Definition 10. If $f(t)$ is a function of t in the ordinary sense, such that $(1 + t^2)^{-N} f(t)$ is absolutely integrable from $-\infty$ to ∞ for some N, then the generalized function $f(t)$ is

PROPERTIES OF THE LAPLACE AND RELATED TRANSFORMS

defined by a sequence $\{f_n(t)\}$ such that for any good function $g(t)$

$$\lim_{n\to\infty} \int_{-\infty}^{\infty} f_n(t)g(t)\, dt = \int_{-\infty}^{\infty} f(t)g(t)\, dt$$

Definition 11. If $h(t)$ is an ordinary function and $f(t)$ a generalized function and

$$\int_{-\infty}^{\infty} f(t)g(t)\, dt = \int_a^b h(t)g(t)\, dt$$

for *every* good function $g(t)$ which is zero outside $a < x < b$ where a and b may be infinite provided that the right-hand side exists as an ordinary integral for all good functions $g(t)$, then one has $f(t) = h(t)$, $a < t < b$. [Note that this imposes no restriction on $h(x)$ outside of this interval.]

Definition 12. If $f(x,t)$ is a generalized function of x for each value of the parameter t, and $f(x)$ is another generalized function such that, for any good function $g(x)$

$$\lim_{t\to a} \int_{-\infty}^{\infty} f(x,t)g(x)\, dx = \int_{-\infty}^{\infty} f(x)g(x)\, dx$$

then one says

$$\lim_{t\to a} f(x,t) = f(x)$$

Definition 13. If $f(x,t)$ is a generalized function of x for each value of the parameter t, then one defines

$$\frac{\partial}{\partial t} f(x,t) = \lim_{t_2 \to t_1} \frac{f(x,t_2) - f(x,t_1)}{t_2 - t_1}$$

and

$$\sum_{t=0}^{\infty} f(x,t) = \lim_{n\to\infty} \sum_{t=0}^{n} f(x,t)$$

provided that the limit function exists in each case.

Definition 14. The generalized function $f(x)$ is said to be even or odd if

$$\int_{-\infty}^{\infty} f(t)g(t)\, dt = 0$$

for all odd or even good functions $g(t)$, respectively.

Using these definitions it is not difficult to establish the following:

Theorem 19. If Definition 9 applies then for any good function $g(t)$ it follows that

(1) $$\int_{-\infty}^{\infty} f'(t)g(t)\, dt = -\int_{-\infty}^{\infty} f(t)g'(t)\, dt$$

(2) $$\int_{-\infty}^{\infty} f(at+b)g(t)\, dt = \frac{1}{|a|} \int_{-\infty}^{\infty} f(t)g\left(\frac{t-b}{a}\right) dt$$

(3) $$\int_{-\infty}^{\infty} \{\phi(t)f(t)\}g(t)\, dt = \int_{-\infty}^{\infty} f(t)\{\phi(t)g(t)\}\, dx$$

Since we are primarily concerned with the Heaviside unit function and its derivatives, we note first that the sequence $\{e^{t^2/n^2}\}$ for $t \geq 0$ and all equivalent sequences defines the generalized Heaviside unit function $H(x)$ such that

$$\int_0^{\infty} H(t)g(t)\, dt = \int_0^{\infty} g(t)\, dt$$

Without the restriction $t \geq 0$, the corresponding generalized function $I(t)$ would be the generalization of the unit function. Similarly the sequences equivalent to $\{e^{-nt^2}(n/\pi)^{1/2}\}$

define a generalized function $\delta(t)$ such that

$$\int_{-\infty}^{\infty} \delta(t)g(t)\, dt = g(0)$$

It now follows from Theorem 19 that $H'(t) = \delta(t)$ so that

(1) $$\int_{-\infty}^{\infty} H'(t)g(t)\, dt = -g(0)$$

(2) $$\int_{-\infty}^{\infty} \delta^n(t)g(t)\, dt = (-1)^n g^{(n)}(0)$$

Furthermore it is also easy to prove the following:

Theorem 20. 1. If $f(t)$ is a generalized function and $f'(t) = 0$ then $f(t) = aI(t)$ if t can take on positive and negative values and $f(t) = bH(t)$ if t is restricted to $t \geq 0$.

2. If $f(t)$ is a generalized function and $tf(t) = 0$ then $f(t) = c\delta(t)$.

3. If $f(t)$ is a generalized function and $t^n f(t) = 0$ then $f(t) = \sum_{i=1}^{n} a_i \delta^i(t)$.

Theorem 21. If $f(t)$ is an ordinary differentiable function such that both $f(t)$ and $f'(t)$ satisfy Definition 11, then the derivative of the generalized function formed from $f(t)$ is the generalized function formed from $f'(x)$.

In terms of the characterization of this class of generalized functions by Hermite functions introduced above, it is not difficult to see that if the generalized function $f(t)$ has the representation

$$f(t) \sim \sum_{n=0}^{\infty} a_n \phi_n(t)$$

then its derivative $f'(x)$ has the representation

$$f'(t) \sim \sum_{n=0}^{\infty} [(n/2 + \tfrac{1}{2})^{\frac{1}{2}} a_{n+1} - (n/2)^{\frac{1}{2}} a_{n-1}] \phi_n(t)$$

while $tf(t)$ has the representation

$$tf(t) \sim \sum_{n=0}^{\infty} [(n/2 + \tfrac{1}{2})^{\frac{1}{2}} a_{n+1} + (n/2)^{\frac{1}{2}} a_{n-1}] \phi_n(t)$$

In this representation, the generalized unit function $I(t)$ admits the representation

$$I(t) \sim \sum_{n=0}^{\infty} [(2n)!^{\frac{1}{2}} \pi^{\frac{1}{4}} / 2^{n-1} n!] \phi_{2n}(t)$$

while the generalized delta function has the expansion

$$\delta(t) \sim \sum_{n=0}^{\infty} [(-1)^n (2n)!^{\frac{1}{2}} / 2^n n! \pi^{\frac{1}{4}}] \phi_{2n}(t)$$

To conclude this brief appendix on generalized functions, the representation in terms of Hermite functions makes the Fourier-transform properties stated for them almost trivial once it is noted that the Fourier transform of $\phi_n(x)$ is just $(-i)^n \phi_n(\omega)$. Thus a generalized function in the sense of Lighthill has a Fourier transform which is a generalized function in the same sense, since the presence of $(-i)^n$ clearly cannot affect any convergent properties. Finally, the use of expansions of the form

$$\Sigma\, a_{mnp} \ldots \phi_m(t)\, \phi_n(n)\, \phi_p(z) \ldots$$

makes it possible easily to extend the theory presented here to more than one variable without essentially altering its character.

PROPERTIES OF THE LAPLACE AND RELATED TRANSFORMS 39-25

Appendix 2. Table of Operations for the Laplace Transformation*

	$F(t)$	$f(s)$
1	$F(t)$	$\int_0^\infty e^{-st} F(t)\, dt$
2	$AF(t) + BG(t)$	$Af(s) + Bg(s)$
3	$F'(t)$	$sf(s) - F(+0)$
4	$F^{(n)}(t)$	$s^n f(s) - s^{n-1} F(+0) - s^{n-2} F'(+0) - \cdots - F^{(n-1)}(+0)$
5	$\int_0^t F(r)\, dr$	$\dfrac{1}{s} f(s)$
6	$\int_0^t \int_0^r F(\lambda)\, d\lambda\, dr$	$\dfrac{1}{s^2} f(s)$
7	$\int_0^t F_1(t-\tau) F_2(\tau)\, d\tau = F_1 * F_2$	$f_1(s) f_2(s)$
8	$tF(t)$	$-f'(s)$
9	$t^n F(t)$	$(-1)^n f^{(n)}(s)$
10	$\dfrac{1}{t} F(t)$	$\int_s^\infty f(x)\, dx$
11	$e^{at} F(t)$	$f(s - a)$
12	$F(t - b)$, where $F(t) = 0$ when $t < 0$	$e^{-bs} f(s)$
13	$\dfrac{1}{c} F\left(\dfrac{t}{c}\right)\ (c > 0)$	$f(cs)$
14	$\dfrac{1}{c} e^{\frac{bt}{c}} F\left(\dfrac{t}{c}\right)\ (c > 0)$	$f(cs - b)$
15	$F(t)$, when $F(t + a) = F(t)$	$\dfrac{\int_0^a e^{-st} F(t)\, dt}{1 - e^{-as}}$
16	$F(t)$, when $F(t + a) = -F(t)$	$\dfrac{\int_0^a e^{-st} F(t)\, dt}{1 + e^{-as}}$
17	$F_1(t)$, the half-wave rectification of $F(t)$ in No. 16	$\dfrac{f(s)}{1 - e^{-as}}$
18	$F_2(t)$, the full-wave rectification of $F(t)$ in No. 16	$f(s) \coth \dfrac{as}{2}$
19	$\displaystyle\sum_1^m \dfrac{p(a_n)}{q'(a_n)} e^{a_n t}$	$\dfrac{p(s)}{q(s)}$, $q(s) = (s - a_1)(s - a_2) \cdots (s - a_m)$

* From R. V. Churchill, *Operational Mathematics*, 2d ed., McGraw-Hill Book Company, New York, 1958.

Appendix 3. Table of Laplace Transforms†

	$f(s)$	$F(t)$
1	$\dfrac{1}{s}$	1
2	$\dfrac{1}{s^2}$	t
3	$\dfrac{1}{s^n}$ $(n = 1, 2, \ldots)$	$\dfrac{t^{n-1}}{(n-1)!}$
4	$\dfrac{1}{\sqrt{s}}$	$\dfrac{1}{\sqrt{\pi t}}$
5	$s^{-\frac{3}{2}}$	$2\sqrt{\dfrac{t}{\pi}}$
6	$s^{-(n+\frac{1}{2})}$ $(n = 1, 2, \ldots)$	$\dfrac{2^n t^{n-\frac{1}{2}}}{1 \times 3 \times 5 \cdots (2n-1)\sqrt{\pi}}$
7	$\dfrac{\Gamma(k)}{s^k}$ $(k > 0)$	t^{k-1}
8	$\dfrac{1}{s-a}$	e^{at}
9	$\dfrac{1}{(s-a)^2}$	te^{at}
10	$\dfrac{1}{(s-a)^n}$ $(n = 1, 2, \ldots)$	$\dfrac{1}{(n-1)!} t^{n-1} e^{at}$
11	$\dfrac{\Gamma(k)}{(s-a)^k}$ $(k > 0)$	$t^{k-1} e^{at}$
12*	$\dfrac{1}{(s-a)(s-b)}$	$\dfrac{1}{a-b}(e^{at} - e^{bt})$
13*	$\dfrac{s}{(s-a)(s-b)}$	$\dfrac{1}{a-b}(ae^{at} - be^{bt})$
14*	$\dfrac{1}{(s-a)(s-b)(s-c)}$	$-\dfrac{(b-c)e^{at} + (c-a)e^{bt} + (a-b)e^{ct}}{(a-b)(b-c)(c-a)}$
15	$\dfrac{1}{s^2 + a^2}$	$\dfrac{1}{a}\sin at$
16	$\dfrac{s}{s^2 + a^2}$	$\cos at$

* Here a, b, and (in 14) c represent distinct constants.

† From R. V. Churchill, *Operational Mathematics*, 2d ed., McGraw-Hill Book Company, New York, 1958.

Appendix 3. Table of Laplace Transforms (Continued)

	$f(s)$	$F(t)$
17	$\dfrac{1}{s^2 - a^2}$	$\dfrac{1}{a} \sinh at$
18	$\dfrac{s}{s^2 - a^2}$	$\cosh at$
19	$\dfrac{1}{s(s^2 + a^2)}$	$\dfrac{1}{a^2}(1 - \cos at)$
20	$\dfrac{1}{s^2(s^2 + a^2)}$	$\dfrac{1}{a^3}(at - \sin at)$
21	$\dfrac{1}{(s^2 + a^2)^2}$	$\dfrac{1}{2a^3}(\sin at - at \cos at)$
22	$\dfrac{s}{(s^2 + a^2)^2}$	$\dfrac{t}{2a} \sin at$
23	$\dfrac{s^2}{(s^2 + a^2)^2}$	$\dfrac{1}{2a}(\sin at + at \cos at)$
24	$\dfrac{s^2 - a^2}{(s^2 + a^2)^2}$	$t \cos at$
25	$\dfrac{s}{(s^2 + a^2)(s^2 + b^2)} \; (a^2 \neq b^2)$	$\dfrac{\cos at - \cos bt}{b^2 - a^2}$
26	$\dfrac{1}{(s - a)^2 + b^2}$	$\dfrac{1}{b} e^{at} \sin bt$
27	$\dfrac{s - a}{(s - a)^2 + b^2}$	$e^{at} \cos bt$
28	$\dfrac{3a^2}{s^3 + a^3}$	$e^{-at} - e^{at/2}\left(\cos \dfrac{at\sqrt{3}}{2} - \sqrt{3} \sin \dfrac{at\sqrt{3}}{2}\right)$
29	$\dfrac{4a^3}{s^4 + 4a^4}$	$\sin at \cosh at - \cos at \sinh at$
30	$\dfrac{s}{s^4 + 4a^4}$	$\dfrac{1}{2a^2} \sin at \sinh at$
31	$\dfrac{1}{s^4 - a^4}$	$\dfrac{1}{2a^3}(\sinh at - \sin at)$
32	$\dfrac{s}{s^4 - a^4}$	$\dfrac{1}{2a^2}(\cosh at - \cos at)$
33	$\dfrac{8a^3 s^2}{(s^2 + a^2)^3}$	$(1 + a^2 t^2) \sin at - at \cos at$
34*	$\dfrac{1}{s}\left(\dfrac{s - 1}{s}\right)^n$	$L_n(t) = \dfrac{e^t}{n!} \dfrac{d^n}{dt^n}(t^n e^{-t})$
35	$\dfrac{s}{(s - a)^{3/2}}$	$\dfrac{1}{\sqrt{\pi t}} e^{at}(1 + 2at)$
36	$\sqrt{s - a} - \sqrt{s - b}$	$\dfrac{1}{2\sqrt{\pi t^3}}(e^{bt} - e^{at})$

* $L_n(t)$ is the Laguerre polynomial of degree n.

Appendix 3. Table of Laplace Transforms (Continued)

	$f(s)$	$F(t)$
37	$\dfrac{1}{\sqrt{s}+a}$	$\dfrac{1}{\sqrt{\pi t}} - ae^{a^2 t}\,\text{erfc}\,(a\sqrt{t})$
38	$\dfrac{\sqrt{s}}{s-a^2}$	$\dfrac{1}{\sqrt{\pi t}} + ae^{a^2 t}\,\text{erf}\,(a\sqrt{t})$
39	$\dfrac{\sqrt{s}}{s+a^2}$	$\dfrac{1}{\sqrt{\pi t}} - \dfrac{2a}{\sqrt{\pi}}\,e^{-a^2 t}\int_0^{a\sqrt{t}} e^{\lambda^2}\,d\lambda$
40	$\dfrac{1}{\sqrt{s}\,(s-a^2)}$	$\dfrac{1}{a}\,e^{a^2 t}\,\text{erf}\,(a\sqrt{t})$
41	$\dfrac{1}{\sqrt{s}\,(s+a^2)}$	$\dfrac{2}{a\sqrt{\pi}}\,e^{-a^2 t}\int_0^{a\sqrt{t}} e^{\lambda^2}\,d\lambda$
42	$\dfrac{b^2-a^2}{(s-a^2)(b+\sqrt{s})}$	$e^{a^2 t}[b - a\,\text{erf}\,(a\sqrt{t})] - be^{b^2 t}\,\text{erfc}\,(b\sqrt{t})$
43	$\dfrac{1}{\sqrt{s}\,(\sqrt{s}+a)}$	$e^{a^2 t}\,\text{erfc}\,(a\sqrt{t})$
44	$\dfrac{1}{(s+a)\sqrt{s+b}}$	$\dfrac{1}{\sqrt{b-a}}\,e^{-at}\,\text{erf}\,(\sqrt{b-a}\,\sqrt{t})$
45	$\dfrac{b^2-a^2}{\sqrt{s}\,(s-a^2)(\sqrt{s}+b)}$	$e^{a^2 t}\left[\dfrac{b}{a}\,\text{erf}\,(a\sqrt{t}) - 1\right] + e^{b^2 t}\,\text{erfc}\,(b\sqrt{t})$
46*	$\dfrac{(1-s)^n}{s^{n+\frac{1}{2}}}$	$\dfrac{n!}{(2n)!\,\sqrt{\pi t}}\,H_{2n}(\sqrt{t})$
47	$\dfrac{(1-s)^n}{s^{n+\frac{3}{2}}}$	$-\dfrac{n!}{\sqrt{\pi}\,(2n+1)!}\,H_{2n+1}(\sqrt{t})$
48†	$\dfrac{\sqrt{s+2a}}{\sqrt{s}} - 1$	$ae^{-at}[I_1(at) + I_0(at)]$
49	$\dfrac{1}{\sqrt{s+a}\,\sqrt{s+b}}$	$e^{-\frac{1}{2}(a+b)t}\,I_0\!\left(\dfrac{a-b}{2}t\right)$
50	$\dfrac{\Gamma(k)}{(s+a)^k(s+b)^k}\;(k>0)$	$\sqrt{\pi}\left(\dfrac{t}{a-b}\right)^{k-\frac{1}{2}}e^{-\frac{1}{2}(a+b)t}\times I_{k-\frac{1}{2}}\!\left(\dfrac{a-b}{2}t\right)$
51	$\dfrac{1}{(s+a)^{\frac{1}{2}}(s+b)^{\frac{3}{2}}}$	$te^{-\frac{1}{2}(a+b)t}\left[I_0\!\left(\dfrac{a-b}{2}t\right) + I_1\!\left(\dfrac{a-b}{2}t\right)\right]$
52	$\dfrac{\sqrt{s+2a}-\sqrt{s}}{\sqrt{s+2a}+\sqrt{s}}$	$\dfrac{1}{t}\,e^{-at}I_1(at)$

* $H_n(x)$ is the Hermite polynomial, $H_n(x) = e^{x^2}\dfrac{d^n}{dx^n}(e^{-x^2})$.

† $I_n(x) = i^{-n}J_n(ix)$, where J_n is Bessel's function of the first kind.

Appendix 3. Table of Laplace Transforms (Continued)

	$f(s)$	$F(t)$
53	$\dfrac{(a-b)^k}{(\sqrt{s+a}+\sqrt{s+b})^{2k}}\ (k>0)$	$\dfrac{k}{t}e^{-\frac{1}{2}(a+b)t}I_k\left(\dfrac{a-b}{2}t\right)$
54	$\dfrac{(\sqrt{s+a}+\sqrt{s})^{-2\nu}}{\sqrt{s}\sqrt{s+a}}\ (\nu>-1)$	$\dfrac{1}{a^\nu}e^{-\frac{1}{2}at}I_\nu\left(\dfrac{1}{2}at\right)$
55	$\dfrac{1}{\sqrt{s^2+a^2}}$	$J_0(at)$
56	$\dfrac{(\sqrt{s^2+a^2}-s)^\nu}{\sqrt{s^2+a^2}}\ (\nu>-1)$	$a^\nu J_\nu(at)$
57	$\dfrac{1}{(s^2+a^2)^k}\ (k>0)$	$\dfrac{\sqrt{\pi}}{\Gamma(k)}\left(\dfrac{t}{2a}\right)^{k-\frac{1}{2}}J_{k-\frac{1}{2}}(at)$
58	$(\sqrt{s^2+a^2}-s)^k\ (k>0)$	$\dfrac{ka^k}{t}J_k(at)$
59	$\dfrac{(s-\sqrt{s^2-a^2})^\nu}{\sqrt{s^2-a^2}}\ (\nu>-1)$	$a^\nu I_\nu(at)$
60	$\dfrac{1}{(s^2-a^2)^k}\ (k>0)$	$\dfrac{\sqrt{\pi}}{\Gamma(k)}\left(\dfrac{t}{2a}\right)^{k-\frac{1}{2}}I_{k-\frac{1}{2}}(at)$
61	$\dfrac{e^{-ks}}{s}$	$S_k(t)=\begin{cases}0\text{ when }0<t<k\\1\text{ when }t>k\end{cases}$
62	$\dfrac{e^{-ks}}{s^2}$	$\begin{cases}0 & \text{when }0<t<k\\t-k & \text{when }t>k\end{cases}$
63	$\dfrac{e^{-ks}}{s^\mu}\ (\mu>0)$	$\begin{cases}0 & \text{when }0<t<k\\\dfrac{(t-k)^{\mu-1}}{\Gamma(\mu)} & \text{when }t>k\end{cases}$
64	$\dfrac{1-e^{-ks}}{s}$	$\begin{cases}1\text{ when }0<t<k\\0\text{ when }t>k\end{cases}$
65	$\dfrac{1}{s(1-e^{-ks})}=\dfrac{1+\coth\frac{1}{2}ks}{2s}$	$1+[t/k]=n$ when $(n-1)k<t<nk$ $(n=1,2,\ldots)$
66	$\dfrac{1}{s(e^{ks}-a)}$	$\begin{cases}0\text{ when }0<t<k\\1+a+a^2+\cdots+a^{n-1}\\ \quad\text{when }nk<t<(n+1)k\\ \quad(n=1,2,\ldots)\end{cases}$
67	$\dfrac{1}{s}\tanh ks$	$M(2k,t)=(-1)^{n-1}$ when $2k(n-1)<t<2kn$ $(n=1,2,\ldots)$
68	$\dfrac{1}{s(1+e^{-ks})}$	$\dfrac{1}{2}M(k,t)+\dfrac{1}{2}=\dfrac{1-(-1)^n}{2}$ when $(n-1)k<t<nk$
69	$\dfrac{1}{s^2}\tanh ks$	$H(2k,t)$

39-30 MATHEMATICS ASSOCIATED WITH SYSTEM ENGINEERING

Appendix 3. Table of Laplace Transforms (Continued)

	$f(s)$	$F(t)$
70	$\dfrac{1}{s \sinh ks}$	$F(t) = 2(n-1)$ when $(2n-3)k < t < (2n-1)k$ $(t > 0)$
71	$\dfrac{1}{s \cosh ks}$	$M(2k, t+3k) + 1 = 1 + (-1)^n$ when $(2n-3)k < t < (2n-1)k$ $(t > 0)$
72	$\dfrac{1}{s} \coth ks$	$F(t) = 2n - 1$ when $2k(n-1) < t < 2kn$
73	$\dfrac{k}{s^2 + k^2} \coth \dfrac{\pi s}{2k}$	$\lvert \sin kt \rvert$
74	$\dfrac{1}{(s^2+1)(1-e^{-\pi s})}$	$\begin{cases} \sin t \text{ when} \\ \quad (2n-2)\pi < t < (2n-1)\pi \\ 0 \text{ when} \\ \quad (2n-1)\pi < t < 2n\pi \end{cases}$
75	$\dfrac{1}{s} e^{-(k/s)}$	$J_0(2\sqrt{kt})$
76	$\dfrac{1}{\sqrt{s}} e^{-(k/s)}$	$\dfrac{1}{\sqrt{\pi t}} \cos 2\sqrt{kt}$
77	$\dfrac{1}{\sqrt{s}} e^{k/s}$	$\dfrac{1}{\sqrt{\pi t}} \cosh 2\sqrt{kt}$
78	$\dfrac{1}{s^{\frac{3}{2}}} e^{-(k/s)}$	$\dfrac{1}{\sqrt{\pi k}} \sin 2\sqrt{kt}$
79	$\dfrac{1}{s^{\frac{3}{2}}} e^{k/s}$	$\dfrac{1}{\sqrt{\pi k}} \sinh 2\sqrt{kt}$
80	$\dfrac{1}{s^\mu} e^{-(k/s)} \;(\mu > 0)$	$\left(\dfrac{t}{k}\right)^{(\mu-1)/2} J_{\mu-1}(2\sqrt{kt})$
81	$\dfrac{1}{s^\mu} e^{k/s} \;(\mu > 0)$	$\left(\dfrac{t}{k}\right)^{(\mu-1)/2} I_{\mu-1}(2\sqrt{kt})$
82	$e^{-k\sqrt{s}} \;(k > 0)$	$\dfrac{k}{2\sqrt{\pi t^3}} \exp\left(-\dfrac{k^2}{4t}\right)$
83	$\dfrac{1}{s} e^{-k\sqrt{s}} \;(k \geqq 0)$	$\operatorname{erfc}\left(\dfrac{k}{2\sqrt{t}}\right)$
84	$\dfrac{1}{\sqrt{s}} e^{-k\sqrt{s}} \;(k \geqq 0)$	$\dfrac{1}{\sqrt{\pi t}} \exp\left(-\dfrac{k^2}{4t}\right)$
85	$s^{-\frac{3}{2}} e^{-k\sqrt{s}} \;(k \geqq 0)$	$2\sqrt{\dfrac{t}{\pi}} \exp\left(-\dfrac{k^2}{4t}\right)$ $\quad - k \operatorname{erfc}\left(\dfrac{k}{2\sqrt{t}}\right)$
86	$\dfrac{ae^{-k\sqrt{s}}}{s(a+\sqrt{s})} \;(k \geqq 0)$	$-e^{ak}e^{a^2 t} \operatorname{erfc}\left(a\sqrt{t} + \dfrac{k}{2\sqrt{t}}\right)$ $\quad + \operatorname{erfc}\left(\dfrac{k}{2\sqrt{t}}\right)$

Appendix 3. Table of Laplace Transforms (Continued)

	$f(s)$	$F(t)$
87	$\dfrac{e^{-k\sqrt{s}}}{\sqrt{s}\,(a+\sqrt{s})}\ (k \geqq 0)$	$e^{ak}e^{a^2 t}\operatorname{erfc}\left(a\sqrt{t}+\dfrac{k}{2\sqrt{t}}\right)$
88	$\dfrac{e^{-k\sqrt{s(s+a)}}}{\sqrt{s(s+a)}}$	$\begin{cases} 0 & \text{when } 0 < t < k \\ e^{-\frac{1}{2}at}I_0(\tfrac{1}{2}a\sqrt{t^2-k^2}) \end{cases}$ when $t > k$
89	$\dfrac{e^{-k\sqrt{s^2+a^2}}}{\sqrt{s^2+a^2}}$	$\begin{cases} 0 & \text{when } 0 < t < k \\ J_0(a\sqrt{t^2-k^2}) & \text{when } t > k \end{cases}$
90	$\dfrac{e^{-k\sqrt{s^2-a^2}}}{\sqrt{s^2-a^2}}$	$\begin{cases} 0 & \text{when } 0 < t < k \\ I_0(a\sqrt{t^2-k^2}) & \text{when } t > k \end{cases}$
91	$\dfrac{e^{-k(\sqrt{s^2+a^2}-s)}}{\sqrt{s^2+a^2}}\ (k \geqq 0)$	$J_0(a\sqrt{t^2+2kt})$
92	$e^{-ks} - e^{-k\sqrt{s^2+a^2}}$	$\begin{cases} 0 & \text{when } 0 < t < k \\ \dfrac{ak}{\sqrt{t^2-k^2}}J_1(a\sqrt{t^2-k^2}) & \text{when } t > k \end{cases}$
93	$e^{-k\sqrt{s^2-a^2}} - e^{-ks}$	$\begin{cases} 0 & \text{when } 0 < t < k \\ \dfrac{ak}{\sqrt{t^2-k^2}}I_1(a\sqrt{t^2-k^2}) & \text{when } t > k \end{cases}$
94	$\dfrac{a^\nu e^{-k\sqrt{s^2+a^2}}}{\sqrt{s^2+a^2}\,(\sqrt{s^2+a^2}+s)^\nu}$ $(\nu > -1)$	$\begin{cases} 0 & \text{when } 0 < t < k \\ \left(\dfrac{t-k}{t+k}\right)^{\frac{1}{2}\nu} J_\nu(a\sqrt{t^2-k^2}) & \text{when } t > k \end{cases}$
95	$\dfrac{1}{s}\log s$	$\Gamma'(1) - \log t\ \ [\Gamma'(1) = -0.5772]$
96	$\dfrac{1}{s^k}\log s\ (k > 0)$	$t^{k-1}\left\{\dfrac{\Gamma'(k)}{[\Gamma(k)]^2} - \dfrac{\log t}{\Gamma(k)}\right\}$
97*	$\dfrac{\log s}{s-a}\ (a > 0)$	$e^{at}[\log a - \operatorname{Ei}(-at)]$
98	$\dfrac{\log s}{s^2+1}$	$\cos t\operatorname{Si} t - \sin t\operatorname{Ci} t$
99	$\dfrac{s\log s}{s^2+1}$	$-\sin t\operatorname{Si} t - \cos t\operatorname{Ci} t$
100	$\dfrac{1}{s}\log(1+ks)\ (k > 0)$	$-\operatorname{Ei}\left(-\dfrac{t}{k}\right)$

* For definition of this function and other integral functions, see, for instance, Jahnke and Emde, "Tables of Functions."

Appendix 3. Table of Laplace Transforms (Continued)

	$f(s)$	$F(t)$
101	$\log \dfrac{s-a}{s-b}$	$\dfrac{1}{t}(e^{bt} - e^{at})$
102	$\dfrac{1}{s}\log(1+k^2s^2)$	$-2\,\mathrm{Ci}\left(\dfrac{t}{k}\right)$
103	$\dfrac{1}{s}\log(s^2+a^2)\ (a>0)$	$2\log a - 2\,\mathrm{Ci}\,(at)$
104	$\dfrac{1}{s^2}\log(s^2+a^2)\ (a>0)$	$\dfrac{2}{a}[at\log a + \sin at - at\,\mathrm{Ci}\,(at)]$
105	$\log \dfrac{s^2+a^2}{s^2}$	$\dfrac{2}{t}(1-\cos at)$
106	$\log \dfrac{s^2-a^2}{s^2}$	$\dfrac{2}{t}(1-\cosh at)$
107	$\arctan \dfrac{k}{s}$	$\dfrac{1}{t}\sin kt$
108	$\dfrac{1}{s}\arctan \dfrac{k}{s}$	$\mathrm{Si}\,(kt)$
109	$e^{k^2s^2}\mathrm{erfc}\,(ks)\ (k>0)$	$\dfrac{1}{k\sqrt{\pi}}\exp\left(-\dfrac{t^2}{4k^2}\right)$
110	$\dfrac{1}{s}e^{k^2s^2}\mathrm{erfc}\,(ks)\ (k>0)$	$\mathrm{erf}\left(\dfrac{t}{2k}\right)$
111	$e^{ks}\mathrm{erfc}\,\sqrt{ks}\ (k>0)$	$\dfrac{\sqrt{k}}{\pi\sqrt{t}\,(t+k)}$
112	$\dfrac{1}{\sqrt{s}}\mathrm{erfc}\,(\sqrt{ks})$	$\begin{cases}0 & \text{when } 0<t<k \\ (\pi t)^{-\frac{1}{2}} & \text{when } t>k\end{cases}$
113	$\dfrac{1}{\sqrt{s}}e^{ks}\mathrm{erfc}\,(\sqrt{ks})\ (k>0)$	$\dfrac{1}{\sqrt{\pi(t+k)}}$
114	$\mathrm{erf}\left(\dfrac{k}{\sqrt{s}}\right)$	$\dfrac{1}{\pi t}\sin(2k\sqrt{t})$
115	$\dfrac{1}{\sqrt{s}}e^{k^2/s}\mathrm{erfc}\left(\dfrac{k}{\sqrt{s}}\right)$	$\dfrac{1}{\sqrt{\pi t}}e^{-2k\sqrt{t}}$
116*	$K_0(ks)$	$\begin{cases}0 & \text{when } 0<t<k \\ (t^2-k^2)^{-\frac{1}{2}} & \text{when } t>k\end{cases}$
117	$K_0(k\sqrt{s})$	$\dfrac{1}{2t}\exp\left(-\dfrac{k^2}{4t}\right)$
118	$\dfrac{1}{s}e^{ks}K_1(ks)$	$\dfrac{1}{k}\sqrt{t(t+2k)}$
119	$\dfrac{1}{\sqrt{s}}K_1(k\sqrt{s})$	$\dfrac{1}{k}\exp\left(-\dfrac{k^2}{4t}\right)$
120	$\dfrac{1}{\sqrt{s}}e^{k/s}K_0\left(\dfrac{k}{s}\right)$	$\dfrac{2}{\sqrt{\pi t}}K_0(2\sqrt{2kt})$

* $K_n(x)$ is Bessel's function of the second kind for the imaginary argument.

Appendix 3. Table of Laplace Transforms (Continued)

	$f(s)$	$F(t)$
121	$\pi e^{-ks} I_0(ks)$	$\begin{cases} [t(2k-t)]^{-\frac{1}{2}} & \text{when } 0 < t < 2k \\ 0 & \text{when } t > 2k \end{cases}$
122	$e^{-ks} I_1(ks)$	$\begin{cases} \dfrac{k-t}{\pi k \sqrt{t(2k-t)}} & \text{when } 0 < t < 2k \\ 0 & \text{when } t > 2k \end{cases}$
123	$-e^{as} \operatorname{Ei}(-as)$	$\dfrac{1}{t+a}\ (a > 0)$
124	$\dfrac{1}{a} + s e^{as} \operatorname{Ei}(-as)$	$\dfrac{1}{(t+a)^2}\ (a > 0)$
125	$\left(\dfrac{\pi}{2} - \operatorname{Si} s\right) \cos s + \operatorname{Ci} s \sin s$	$\dfrac{1}{t^2 + 1}$

REFERENCES

1. Achieser, N. I., and I. M. Glasmann: *Theorie der linearen Operatoren in Hilbert Raum*, Mathematische Lehrbücher, vol. 4, Akademie-Verlag GmbH, Berlin, 1954.
2. Arsac, J.: *Transformation de Fourier et théorie des distributions*, Dunod, Paris, 1961.
3. Bode, H. W.: *Network Analysis and Feedback Amplifier Design*, D. Van Nostrand Company, Inc., Princeton, N.J., 1945.
4. Cheng, D. K.: *Analysis of Linear Systems*, Addison-Wesley Publishing Company, Inc., Reading, Mass., 1959.
5. Churchill, R. V.: *Operational Mathematics*, 2d ed., McGraw-Hill Book Company, New York, 1958.
6. Churchill, R. V., and C. L. Dolph: Inverse Transforms of Products of Legendre Transforms, *Proc. Am. Math. Soc.*, vol. 5, pp. 93–100, 1954.
7. Doetsch, G.: *Theorie und Anwendung der Laplace-transformation*, Dover Publications, Inc., New York, 1943.
8. Delsarte, J.: Sur une extension de la formule de Taylor, *J. Math. Pures et Appl.* (9), vol. 17, pp. 213–231, 1936.
9. Dunford, N., and J. T. Schwartz: *Linear Operators*, Part I, John Wiley & Sons, Inc., New York, 1958.
10. Erdelyi, A.: From Delta Functions to Distributions, in E. F. Beckenbach (ed.), *Modern Mathematics for the Engineer*, Second Series, pp. 5–50, McGraw-Hill Book Company, New York, 1961.
11. Erdelyi, A.: *Operational Calculus and Generalized Functions*, Holt, Rinehart and Winston, Inc., New York, 1962.
12. Friedman, B.: *Principles and Techniques of Applied Mathematics*, John Wiley & Sons, Inc., New York, 1956.
13. Gel'fand, I. M., and G. E. Schilow: *Verallgemeinerte Funktionen (Distributionen)*, Deutscher Verlag der Wissenschaften, Berlin, vol. I, 1960; vol. II, 1962.
14. Goodrich, R. F., and N. D. Kazarinoff: Scalar Diffraction by Elliptic Cylinders and Prolate Spheroids Whose Eccentricity Is Almost One, *Proc. Cambridge Phil. Soc.*, vol. 59, pp. 167–182, 1963.
15. Goldstein, S.: Some Two-dimensional Diffusion Problems with Circular Symmetry, *Proc. London Math. Soc.*, vol. 34, p. 51, 1932.
16. Greenstein, D.: On the Analytic Continuation of Functions Which Map the Upper Half Plane into Itself, *J. Math. Anal. Appl.*, vol. 1, p. 355, 1960.
17. Guillemin, E. A.: *The Mathematics of Circuit Analysis*, John Wiley & Sons, Inc., New York, 1949.
18. Hille, E., and R. S. Phillips: Functional Analysis and Semi-groups, *Am. Math. Soc. Colloq. Publ.*, vol. 31, Providence, 1957.
19. Ince, E. L.: *Ordinary Differential Equations*, Dover Publications, Inc., New York, 1944.
20. Kaplan, W.: *Advanced Calculus*, Addison-Wesley Publishing Company, Inc., Reading, Mass., 1952.
21. Kaplan, W.: *Operational Methods for Linear Systems*, Addison-Wesley Publishing Company, Inc., Reading, Mass., 1962.
22. Klein, J. S.: Some Results in the Theory of Hankel Transforms, Thesis, University of Michigan, 1958.
23. Levitan, B. M.: The Application of Generalized Displacement Operators to Linear Differential Equations of the Second Order, *Usp. Math., Nauk*, NS4, no. 1, vol. 29, pp. 3–112, 1949.
24. Lighthill, M. J.: *An Introduction to Fourier Analysis and Generalized Functions*, Cambridge University Press, New York.
25. Loomis, L. H.: *An Introduction to Abstract Harmonic Analysis*, D. Van Nostrand Company, Inc., Princeton, N.J., 1953.
26. Mikusinski, J.: *Operational Calculus*, 5th ed., Pergamon Press, New York, 1959.
27. Morse, P. M., and H. Feshbach, *Methods of Theoretical Physics*, vol. I, McGraw-Hill Book Company, New York, 1953.
28. Muskelishvili, N. I.: *Singular Integral Equations*, Erven P. Noordhoff, NV, Groningen, Netherlands, 1953.
29. Neumark, M. A.: *Lineare Differentialoperatoren*, Akademie-Verlag GmbH, Berlin, 1960.
30. Paley, R. E. A., and N. Wiener: Fourier Transforms in the Complex Domain, *Am. Math. Soc. Colloq. Publ.*, vol. 19, Providence, 1934.
31. Phillips, R. S.: Semi-group Methods in the Theory of Partial Differential Equations, in E. F. Beckenbach (ed.), *Modern Mathematics for the Engineer*, Second Series, pp. 100–132, McGraw-Hill Book Company, New York, 1961.

32. Schwartz, L.: *Théorie des distributions*, Hermann & Cie, Paris, vol. I, 1950, vol. II, 1951.
33. Sneddon, I. N.: *Fourier Transforms*, McGraw-Hill Book Company, New York, 1951.
34. Stone, M. H.: Linear Transformations in Hilbert Space and Their Applications to Analysis, *Am. Math. Soc. Colloq.*, vol. 15, reprinted 1951.
35. Tamarkin, J. D., and W. Feller: *Partial Differential Equations*, Brown University, Providence, 1941.
36. Taylor, A. E.: *Introduction to Functional Analysis*, John Wiley & Sons, Inc., New York, 1958.
37. Temple, G.: Generalized Functions, *Proc. Roy. Soc. (London)*, Ser. A, vol. 228, pp. 175–190, 1955.
38. Titchmarsh, E. C.: *Introduction to the Theory of Fourier Integrals*, Oxford University Press, Fair Lawn, N.J., 1937.
39. Tranter, C. J.: *Integral Transforms in Mathematical Physics*, John Wiley & Sons, Inc., New York, 1951.
40. Tricomi, F. G.: *Integral Equations*, Interscience Publishers, Inc., New York, 1957.
41. Von Neumann, J.: *Mathematical Foundations of Quantum Mechanics*, Princeton University Press, Princeton, N.J., 1955.
42. Whittaker, E. T., and G. N. Watson: *A Course of Modern Analysis*, The Macmillan Company, New York, reprinted 1945.
43. Widder, D. V.: *The Laplace Transform*, Princeton University Press, Princeton, N.J., 1941.
44. Widder, D. V.: *Advanced Calculus*, 2d ed., Prentice-Hall, Inc., Englewood Cliffs, N.J., 1961.
45. Korevaar, J.: Pansions and the Theory of Fourier Transforms, *Trans. Am. Math. Soc.*, vol. 91, p. 53, 1959.
46. Widlund, O.: On the Expansion of Generalized Functions in Series of Hermite Functions, *Kgl. Tek. Hogskol. Handl. (Trans. Roy. Inst. Technol., Stockholm)*, no. 173, 1961.

Chapter 40

NUMERICAL ANALYSIS

CLAY L. PERRY, *University of California, San Diego, La Jolla, California.*

MANUEL ROTENBERG, *University of California, San Diego, La Jolla, California.*

CONTENTS

40-1. Introduction	40-2
40-2. Solution of Simultaneous Algebraic Equations	40-2
Matrix Formulation	40-3
Matrix Inversion	40-3
Iterative Methods	40-4
40-3. Solutions of Nonlinear Equations	40-5
One Equation in One Unknown	40-6
Polynomial Equation in One Unknown	40-7
Simultaneous Nonlinear Equations	40-7
40-4. Interpolation	40-8
Notation of Finite Differences	40-8
Interpolation Formulas of Polynomial Type	40-9
Interpolation with Fourier Series	40-14
Numerical Integration	40-14
Numerical Differentiation	40-16
40-5. Numerical Solution of Ordinary Differential Equations	40-16
Notation and Definitions	40-16
Numerical Solution	40-17
Starting Methods	40-18
Methods of Higher Accuracy	40-19
Predictor-Corrector Methods	40-19
Runge-Kutta and Other One-step Methods	40-21
Approximation Errors	40-22
40-6. Least-squares Approximation	40-24
The Principle of Least Squares	40-24
Statement of the Problem	40-24
Errors in the Coefficients	40-25
Error at a Given Point	40-26
Average Error over an Interval	40-26
Orthogonal Polynomials	40-26
40-7. Minima and Maxima of Functions	40-26
40-8. Monte Carlo Methods	40-28
Distributions of Random Variables	40-28
Numerical Integration	40-28
Discussion	40-30

40-1. Introduction

Numerical analysis is concerned with the theory behind the methods of finding numerical solutions to mathematically formulated problems. The advent of general-purpose computers has had a profound effect on the practicality of numerical methods and has removed much of the tedium formerly associated with computation. New methods of finding numerical solutions are being invented at a rapid rate and old methods are becoming better understood.

The tools of the engineer using numerical methods include descriptions of numerical procedures and their properties, tables of values of functions, a desk calculator for short computations, and a general-purpose digital computer with a good subroutine library. Experience is one of the most important factors in numerical analysis. While reading and analysis provide some insight, practical experimentation with the numerical methods to be applied is also very desirable. Experimental application of numerical methods of interest to representative problems with known solutions frequently provides useful information on computation errors and on the applicability of the method. It is dangerous to assume, however, that numerical experimentation with a number of special cases will provide adequate information about the general applicability of a numerical method. It is generally advisable to discuss the problem with a numerical analyst before large computations are undertaken.

References. Textbooks such as Buckingham's *Numerical Methods*,[3] Hamming's *Numerical Methods for Scientists and Engineers*,[21] Hildebrand's *Introduction to Numerical Analysis*,[23] Householder's *Principles of Numerical Analysis*,[25] Kopal's *Numerical Analysis*,[34] and Lanczos's *Applied Analysis*[37] are very valuable general references. Hamming's book[21] provides an excellent introduction to the basic concepts of Fourier approximations. *Survey of Numerical Analysis*[58] provides a summary of work in several areas of numerical analysis and contains a large number of references. *Modern Computing Methods*[44] by the staff of the British National Physics Laboratory is an excellent source of numerical methods for problems in linear algebraic equations and some other topics and has an annotated general bibliography summarized by topic. Gelfond's *Calculus of Finite Differences*,[64] Henrici's *Elements of Numerical Analysis*,[65] Householder's *Theory of Matrices in Numerical Analysis*,[66] Stiefel's *An Introduction to Numerical Mathematics*,[68] and Wilkinson's *Rounding Errors in Algebraic Processes*[69] also contain information on numerical methods useful to engineers.

Computer programs for several of the numerical methods described in this chapter can be found in the *Communications of the Association for Computing Machinery*,[7] in *Numerische Mathematik*,[47] and in the book edited by Ralston and Wilf.[49]

The mathematical literature contains descriptions of many numerical methods that are not included in textbooks. *Computing Reviews*,[8] *Mathematics of Computation*,[41] and *Mathematical Reviews*[40] provide reviews of these methods and a guide to their location. Householder's bibliographies[24,25] of books and articles in numerical analysis contain numerous references to descriptions of numerical methods.

Most of the tables of function values published before 1962 are summarized in *An Index of Mathematical Tables*.[11] Statistical tables are summarized in *Guide to Tables in Mathematical Statistics*.[18] The three review journals[8,40,41] mentioned above contain reviews of recent tables. Page 94 of *Survey of Numerical Analysis*[58] has a list of mathematical tables that will be of use to most groups applying numerical methods. Mathematics and statistics dictionaries[28,29,33] are helpful in understanding new terms. The *Handbook of Mathematical Functions* is an outstanding reference to the most commonly used formulas and tables.

40-2. Solution of Simultaneous Algebraic Equations

The numerical problems involved in solving simultaneous linear algebraic equations are embodied in a vast literature. Excellent compilations of methods and bibliographies can be found in Faddeeva,[10] Householder,[24,25] and *Modern Computing Methods*.[44] The computer programming aspects of these problems can be found in Ralston and

Wilf.[49] Methods appropriate for use with a desk calculator are described in Buckingham,[3] Fox,[15] and Scarborough.[53] This section describes direct and indirect methods for solving systems of linear algebraic equations.

Matrix Formulation. The problem is to find values of the n variables x_1, $x_2, \ldots, x_n$ which satisfy the n simultaneous equations

$$\begin{aligned} a_{11}x_1 + a_{12}x_2 + \cdots + a_{1n}x_n &= b_1 \\ a_{21}x_1 + a_{22}x_2 + \cdots + a_{2n}x_n &= b_2 \\ &\vdots \\ a_{n1}x_1 + a_{n2}x_2 + \cdots + a_{nn}x_n &= b_n \end{aligned} \qquad (40\text{-}1)$$

The coefficients a_{ij} and b_i are assumed to be given. If the determinant $D(\mathbf{A})$ of the matrix $\mathbf{A}$ of coefficients a_{ij} is not zero, then (40-1) has a unique solution. If all the b_i are zero the system of equations is *homogeneous* and will not have a solution other than the zero solution (all $x_i = 0$), unless $D(\mathbf{A}) = 0$. When the determinant $D(\mathbf{A})$ is small compared with the size of the elements, the equations are *ill conditioned* and the solution may be difficult to compute accurately. Other formulations for the problem may lead to better-conditioned systems.

The set of equations (40-1) is conveniently represented as a single matrix equation

$$\mathbf{A}\mathbf{x} = \mathbf{b} \qquad (40\text{-}2)$$

where $\mathbf{A}$ is the matrix of coefficients a_{ij}, and $\mathbf{x}$ and $\mathbf{b}$ are column vectors whose components are x_i and b_i, respectively. The problem of solving (40-1) can be resolved into finding a matrix $\mathbf{A}^{-1}$, called the inverse of $\mathbf{A}$, such that $\mathbf{A}^{-1}\mathbf{A} = \mathbf{I}$. The identity matrix $\mathbf{I}$ is the matrix which has ones on the diagonal and zeros elsewhere. The solution is found by multiplying both sides of (40-2) by $\mathbf{A}^{-1}$, resulting in

$$\mathbf{x} = \mathbf{A}^{-1}\mathbf{b} \qquad (40\text{-}3)$$

Matrix Inversion. The basic method for finding A^{-1} is the Gauss elimination method. An $n \times 2n$ (n rows and $2n$ columns) augmented matrix is written as follows:

$$\begin{pmatrix} 1 & 0 & 0 & 0 & \cdots & 0 & a_{11} & a_{12} & \cdots & a_{1n} \\ 0 & 1 & 0 & 0 & \cdots & 0 & a_{21} & a_{22} & \cdots & a_{2n} \\ \vdots & & & & & & \vdots & & & \vdots \\ 0 & \cdots & & 0 & 0 & 1 & a_{n1} & a_{n2} & \cdots & a_{nn} \end{pmatrix} \qquad (40\text{-}4)$$

In determining the inverse, any row of (40-4) may be multiplied by any nonzero number, and any row may be subtracted or added to another. By repeatedly performing such operations, (40-4) may be transformed into the form (40-5). For example, if $a_{11} \neq 0$, the first row can be multiplied by $1/a_{11}$. Then subtraction of the modified first row multiplied by a_{i1} from the ith row, for $i = 2, 3, \ldots, n$, produces the desired $(n + 1)$st column of (40-5).

$$\begin{pmatrix} c_{11} & c_{12} & \cdots & c_{1n} & 1 & 0 & \cdots & 0 \\ c_{21} & c_{22} & \cdots & c_{2n} & 0 & 1 & \cdots & 0 \\ \vdots & & & \vdots & & & & \\ c_{n1} & c_{n2} & \cdots & c_{nn} & 0 & 0 & \cdots & 1 \end{pmatrix} \qquad (40\text{-}5)$$

The left half of matrix (40-5) is $\mathbf{A}^{-1}$; the right half is the identity matrix $\mathbf{I}$. The kth column, $1 \leq k \leq n$, of (40-5) is the solution of the system of linear equations (40-1) with the column of constants $b_i = 1$ for $i = k$, and $b_i = 0$ for $i \neq k$.

The same sequence of row operations that transformed the matrix (40-4) into (40-5)

40-4 MATHEMATICS ASSOCIATED WITH SYSTEM ENGINEERING

may be used to transform matrix (40-6)

$$\begin{pmatrix} s_1 & b_1{}^{(1)} & \cdots & b_1{}^{(m)} & a_{11} & a_{12} & \cdots & a_{1n} \\ s_2 & b_2{}^{(1)} & \cdots & b_2{}^{(m)} & a_{21} & a_{22} & \cdots & a_{2n} \\ \cdot & \cdot & & \cdot & \cdot & \cdot & & \cdot \\ \cdot & \cdot & & \cdot & \cdot & \cdot & & \cdot \\ \cdot & \cdot & & \cdot & \cdot & \cdot & & \cdot \\ s_n & b_n{}^{(1)} & \cdots & b_n{}^{(m)} & a_{n1} & a_{n2} & \cdots & a_{nn} \end{pmatrix} \quad (40\text{-}6)$$

into the matrix (40-7)

$$\begin{pmatrix} s_1' & x_1{}^{(1)} & \cdots & x_1{}^{(m)} & 1 & 0 & \cdots & 0 \\ s_2' & x_2{}^{(1)} & \cdots & x_2{}^{(m)} & 0 & 1 & \cdots & 0 \\ \cdot & \cdot & & \cdot & & & & \\ \cdot & \cdot & & \cdot & & & & \\ \cdot & \cdot & & \cdot & & & & \\ s_n' & x_n{}^{(1)} & \cdots & x_n{}^{(m)} & 0 & 0 & \cdots & 1 \end{pmatrix} \quad (40\text{-}7)$$

The last n columns of (40-7) form the identity matrix $\mathbf{I}$ and are the same as the last n columns of (40-5). The $(k+1)$st ($k = 1, \ldots, m$) column of (40-7) is the solution of the system of equations (40-1) with the column of constants $b_i = b_i{}^{(k)}$; $i = 1, 2, \ldots, n$. Thus this transformation of (40-6) into (40-7) may be used to find simultaneously the solution to m systems of linear algebraic equations (40-1) without determining the inverse matrix $\mathbf{A}^{-1}$. If s_i in (40-6) is the sum of the other $m+n$ elements $b_i{}^{(k)}$ and a_{ij} of the ith row of (40-6), then in the transformed matrix (40-7), s_i' is the sum

$$s_i' = 1 + \sum_k x_i{}^{(k)}$$

of the other elements in the ith row of (40-7). This first column of (40-6) and of (40-7) is usually included for checking purposes when desk calculators are used.

The accumulation of rounding errors, in the above-described row operations, depends on the order in which the nonzero elements are eliminated. Dwyer,[9] Householder,[25] and *Modern Computing Methods*[44] describe methods for minimizing these errors.

Small errors in the matrix inverse may be reduced by the Hotelling method. Let the approximate inverse of $\mathbf{A}$ be denoted by $\bar{\mathbf{A}}^{-1}$. Then matrix multiplication yields

$$\mathbf{A}\bar{\mathbf{A}}^{-1} = \mathbf{I} - \boldsymbol{\Delta} \quad (40\text{-}8)$$

where $\boldsymbol{\Delta}$ is a matrix whose elements are small compared with 1. The true inverse can be found by either of the following expansions:

$$\begin{aligned} \mathbf{A}^{-1} &= \bar{\mathbf{A}}^{-1}(\mathbf{I} + \boldsymbol{\Delta})(\mathbf{I} + \boldsymbol{\Delta}^2)(\mathbf{I} + \boldsymbol{\Delta}^4) \cdots (\mathbf{I} + \boldsymbol{\Delta}^{2k}) \cdots \\ &= \bar{\mathbf{A}}^{-1}(\mathbf{I} + \boldsymbol{\Delta} + \boldsymbol{\Delta}^2 + \cdots) \end{aligned} \quad (40\text{-}9)$$

These series will converge if the elements of the $n \times n$ matrix $\boldsymbol{\Delta}$ are less than $1/n$. Accuracy is improved if double-precision accumulation is used in the evaluation of the terms in (40-9); cf. *Modern Computing Methods*.[44]

Iterative Methods. Iterative methods of solution of (40-1) have the advantage that only one equation of (40-1) need be operated on at a time. Hence very large systems of equations can be solved with a relatively small machine with good intermediate storage facilities. Iteration methods are very useful in improving the accuracy of approximate solutions obtained by direct methods (see *Modern Computing Methods*[44] and *Varga*[59]).

The "ordinary" iterative method consists of first guessing the solution to (40-1). If the equations are arranged so the largest coefficient is on the main diagonal then an easy first guess is $x_i{}^{(1)} = b_i/a_{ii}$. The second iterate $x_1{}^{(2)}$ is found by substituting $x_i{}^{(1)}$, $i = 2, 3, \ldots, n$, into the first equation of (40-1); $x_2{}^{(2)}$ is found by substituting

$x_i^{(1)}$, $i = 1, 3, 4, \ldots, n$, into the second line of (40-1), and so on. When the last line is finished, $x_i^{(2)}$, $i = 1, 2, 3, \ldots, n$, will have been found and the process is repeated to find the third iterates $x_i^{(3)}$. Provided certain conditions on the matrix **A** are met,[10,25] the process will converge to the solution of (40-1).

The Gauss-Seidel iterative method is a variant of the ordinary iterative method and converges for all matrices **A**, which are *positive definite*, although this is not a necessary condition for convergence.[25,58] In the ordinary iterative method, a new approximate solution $x_k^{(n+1)}$ was set aside until all the $x_i^{(n+1)}$ were found. In the Gauss-Seidel scheme, the new $x_k^{(n+1)}$ replaces the old approximation $x_k^{(n)}$ immediately and is used in finding the new $x_j^{(n+1)}$, $j > k$.

To summarize, for the ordinary iterative method one obtains the successive approximations from

$$x_k^{(n+1)} = (-1/a_{kk})[a_{k1}x_1^{(n)} + \cdots + a_{k,k-1}x_{k-1}^{(n)} + a_{k,k+1}x_{k+1}^{(n)} + \cdots + a_{k,n}x_n^{(n)} - b_k] \quad (k = 1, 2, \ldots, n)$$

while for Gauss-Seidel,

$$x_k^{(n+1)} = (-1/a_{kk})[a_{k1}x_1^{(n+1)} + \cdots + a_{k,k-1}x_{k-1}^{(n+1)} + a_{k,k+1}x_{k+1}^{(n)} + \cdots + a_{k,n}x_n^{(n)} - b_k] \quad (k = 1, 2, \ldots, n)$$

The inverse of a large matrix can be built up from knowledge of the inverse of smaller matrices by the *bordering method*.[10] An $n \times n$ matrix $\mathbf{A}_n$ is considered to be an $(n-1) \times (n-1)$ matrix $\mathbf{A}_{n-1}$ with an added row and column:

$$\mathbf{A}_n = \begin{pmatrix} a_{11} & a_{12} & \cdots & a_{1,n-1} & a_{1n} \\ \vdots & & & \vdots & \vdots \\ a_{n-1,1} & a_{n-1,2} & \cdots & a_{n-1,n-1} & a_{n-1,n} \\ a_{n1} & a_{n2} & \cdots & a_{n,n-1} & a_{nn} \end{pmatrix}$$

or, in an obvious notation,

$$\mathbf{A}_n = \begin{pmatrix} \mathbf{A}_{n-1} & \mathbf{u}_{n-1} \\ \mathbf{v}_{n-1} & a_{nn} \end{pmatrix}$$

Here, $\mathbf{u}_{n-1}$ and $\mathbf{v}_{n-1}$ are respectively column and row vectors, each with $(n-1)$ elements, and a_{nn} is a number. The inverse of $\mathbf{A}_n$, in an analogous notation, is

$$\mathbf{A}_n^{-1} = \begin{pmatrix} \mathbf{P}_{n-1} & \mathbf{r}_{n-1} \\ \mathbf{q}_{n-1} & 1/\alpha_n \end{pmatrix}$$

where

$$\mathbf{P}_{n-1} = \mathbf{A}_{n-1}^{-1} + \frac{\mathbf{A}_{n-1}^{-1}\mathbf{u}_{n-1}\mathbf{v}_{n-1}\mathbf{A}_{n-1}^{-1}}{\alpha_n}$$

$$\mathbf{q}_{n-1} = -\frac{\mathbf{v}_{n-1}\mathbf{A}_{n-1}^{-1}}{\alpha_n}$$

$$\mathbf{r}_{n-1} = -\frac{\mathbf{A}_{n-1}^{-1}\mathbf{u}_{n-1}}{\alpha_n}$$

$$\alpha_n = a_{nn} - \mathbf{v}_{n-1}\mathbf{A}_{n-1}^{-1}\mathbf{u}_{n-1}$$

Notice that one needs to find the inverse of only $\mathbf{A}_{n-1}$, which is smaller than the original matrix $\mathbf{A}_n$. If $\mathbf{A}_{n-1}$ cannot be inverted easily one tries to invert the smaller matrix $\mathbf{A}_{n-2}$. Indeed, one may start by inverting the 1×1 matrix $\mathbf{A}_1 = a_{11}$.

40-3. Solutions of Nonlinear Equations

This section describes methods for solving general equations or systems of equations. The equations may involve algebraic expressions and transcendental expressions but not derivatives or integrals of the variables. Explicit formulas for the solution of nonlinear equations are known for only a few cases, such as polynomial equations, in one variable, of degree less than 5. In the general case, solutions are obtained iteratively or possibly by graphical or nomogram procedures if only one or two variables are involved. Rough approximations to the solutions may be found by one method, such as a graph or a numerical survey, and then a second method, such as Newton's method, can be used to refine the solution. Care must be taken to ensure that solutions are not overlooked.

One Equation in One Unknown. Graphical methods are useful in determining the properties of a function and in determining the number and approximate location of solutions of equations in one real variable. Solutions of the equation $f(x) = 0$ are determined by the intersections of the graphs $y = f(x)$ and the axis $y = 0$. Some computation may be saved if $f(x)$ is of a special form. For example, if

$$f(x) = f_1(x) - f_2(x)$$

then the intersections of the graphs $y = f_1(x)$ and $y = f_2(x)$ are the solutions of $f(x) = 0$.

The equation $f(x) = 0$ is said to have a simple solution at $x = x_0$ if $f(x_0) = 0$ and the derivative $f'(x_0) \neq 0$. In this case $f(x)$ changes sign at x_0. If $f(a)$ and $f(b)$ are of opposite sign and $f(x)$ is a continuous function, then there is at least one solution of $f(x) = 0$ in the interval between a and b. In the *bisection method*, $f(x)$ is evaluated at the center of the interval. Either the value of the function at this center point is of opposite sign to one of the values at the end points, or the function is zero and a solution is found. By successively bisecting the subinterval in which the function changes sign, location of a solution becomes known with increasing accuracy. The number of iterations to achieve a given accuracy is known in advance and is independent of the function $f(x)$, provided only that the function can be evaluated accurately enough to determine the sign correctly. After n bisections, the interval which contains the root is reduced by a factor of 2^n.

Approximation of the function $f(x)$ by an interpolation polynomial and solution of the interpolation polynomial may be used to obtain successive approximation procedures. In practice polynomial approximations of degree 1 or 2 are used.

In the method of *false position*, linear interpolation is used. If $f(a)$ and $f(b)$ are of opposite sign, then the linear-interpolation approximation for a solution of $f(x) = 0$ is

$$c = [af(b) - bf(a)]/[f(b) - f(a)] \qquad (40\text{-}10)$$

and c is between a and b. If $f(c) \neq 0$ then the interval (a,c) or (c,b) on which $f(x)$ changes sign is used to obtain the next approximation to the solution by linear interpolation, and the process is continued until the interval that contains a solution is small.

The *Newton-Raphson* method, or more commonly Newton's method, uses the first two terms of the Taylor-series expansion to obtain an iterative process for obtaining solutions to $f(x) = 0$. The $(n + 1)$st approximation $x^{(n+1)}$ to the solution is given by

$$x^{(n+1)} = x^{(n)} - f(x^{(n)})/f'(x^{(n)}) \qquad (40\text{-}11)$$

Convergence of the iterates is discussed in Hildebrand[23] and Ostrowski.[48] The root should be simple; otherwise $f'(x)$ approaches zero as the root is approached and computational difficulty may arise. The Newton method (40-11) is applicable to location of complex roots as well as real roots.

Other successive-substitution methods may be obtained by deriving from $f(x) = 0$ an equation $x = g(x)$, with corresponding solutions, and using the iteration

$$x^{(n+1)} = g(x^{(n)}) \qquad (40\text{-}12)$$

The iterations will converge if $|g'(x)| \leq k < 1$ within an interval $|x - \alpha| \leq |x^{(1)} - \alpha|$ containing the solution α and the initial approximation $x^{(1)}$. The rate of convergence depends on the size of $g'(x)$. When $g'(x) \neq 0$ and does not vary appreciably near the solution, the Aitken δ^2 extrapolation[23,25] for the solution may be used to accelerate convergence of the iterates. The Aitken extrapolation $\bar{x}$ for the solution α is

$$\bar{x} = \frac{x^{(n)}x^{(n+2)} - x^{(n+1)}x^{(n+1)}}{x^{(n+2)} - 2x^{(n+1)} + x^{(n)}} \qquad (40\text{-}13)$$

When $g(x) \equiv x - f(x)/f'(x)$, the iterations (40-12) are Newton's method. In this case $g'(\alpha) = 0$ and the successive iterates converge like $|x^{(n+1)} - \alpha| \leq c|x^{(n)} - \alpha|^2$, for some constant c; i.e., the convergence is quadratic. Householder[25] describes other choices for $g(x)$ for which the final convergence may be faster than Newton's method.

Polynomial Equation in One Unknown. Many methods are available to solve polynomial equations,

$$f(x) = a_n x^n + a_{n-1} x^{n-1} + \cdots + a_1 x + a_0 = 0 \qquad a_n \neq 0 \qquad (40\text{-}14)$$

Besides Newton's method discussed above other methods such as Graeffe's root-squaring technique and Bairstow's and Muller's methods (cf. Hildebrand[23]) are available. The latter methods do not fail when the roots are not simple. When roots are very close together, Graeffe's method is of particular importance. Bairstow's method and Muller's method obtain the roots in pairs. Once the approximate value of a root is known, Bairstow's method will converge more quickly than Newton's. All the methods mentioned above can be used for finding complex roots. If the coefficients a_i of (40-14) are real then the initial approximation for determining a complex root by Newton's method must be complex.

All n of the real and complex roots x of the polynomial equation (40-14) satisfy the following inequalities. These and other bounds may be found in Wilf.[61]

$$|x| \leq \max_{n > i > 0} \{|a_0/a_n|, 1 + |a_i/a_n|\} \qquad (40\text{-}15)$$

$$|x| \leq 2 \max_{n > i > 0} \{|a_{n-i}/a_n|^{1/i}\} \qquad (40\text{-}16)$$

$$|x| \leq \max_{n > i > 0} \{|a_0/a_1|, 2|a_i/a_{i+1}|\} \qquad (40\text{-}17)$$

Simultaneous Nonlinear Equations. A system of m nonlinear equations in n unknowns may have several solutions, a unique solution, or none at all. When no solution exists, the system is said to be an inconsistent system. Inconsistent systems may arise, for example, when parameters in the system of equations have measurement errors. In such cases, least-squares or other statistical methods are used to obtain approximate solutions.

The system of equations may be represented by

$$f_i(x_1, x_2, \ldots, x_n) = 0 \qquad i = 1, 2, \ldots, m \qquad (40\text{-}18)$$

The function

$$g(x_1, x_2, \ldots, x_n) = f_1^2 + f_2^2 + \cdots + f_m^2 \qquad (40\text{-}19)$$

is nonnegative and $g = 0$ when and only when all the functions $f_i = 0$. Thus the absolute minima of the function g are solutions to the system of equations (40-18). The methods of Sec. 40-7 may be used to determine the minima of (40-19). If the minimum of g is greater than zero, then the points where the minimum occurs are called the *least-squares solutions* of (40-18).

The solutions to the system of equations (40-18) can be found by minimizing other functions which are positive except when all $f_i = 0$. For example, the zero minima of the function

$$h(x_1, x_2, \ldots, x_n) = |f_1| + |f_2| + \cdots + |f_m| \qquad (40\text{-}20)$$

are solutions to the system of Eqs. (40-18). This function h has sharper minima than the function g. Since h usually has discontinuous derivatives at the minimum, difficulties occur in the direct application of Newton's method.

Successive linearization is another general method of determining solutions to the system of nonlinear equations (40-18) when $n = m$. The equations are approximated, at an initial estimate $x_1^{(1)}, x_2^{(1)}, \ldots, x_n^{(1)}$ of the solution, by the linear terms of the Taylor-series expansion. The resulting system of linear algebraic equations is

$$0 = f_i(x_1^{(1)}, x_2^{(1)}, \ldots, x_n^{(1)}) + \sum_{k=1}^{n} (x_k - x_k^{(1)}) \frac{\partial f_i(x_1^{(1)}, x_2^{(1)}, \ldots, x_n^{(1)})}{\partial x_k}$$

$$i = 1, 2, \ldots, n \qquad (40\text{-}21)$$

The solution of this system of linear algebraic equations is then used as the next approximation $x_1^{(2)}, x_2^{(2)}, \ldots, x_n^{(2)}$ to the solution of the system (40-18). This process may be repeated using the new approximation as the point at which the Taylor-series linear approximation is obtained. The resulting procedure is Newton's method for systems of equations.

40-4. Interpolation

Interpolation is the process of finding approximate values of a function at points other than those at which the function is known. The information used in the interpolation formula may include values of the function and perhaps derivatives and differences of the function at a discrete number of points. Interpolation formulas are usually based on approximating the desired function by a linear combination of known functions. Most of the frequently used approximations for a desired function $f(x)$ may be represented as

$$f(x) \doteq c_1 p_1(x) + c_2 p_2(x) + \cdots + c_n p_n(x)$$

The functions $p_i(x)$ are usually easily computed ones such as polynomials or exponentials. The constants c_i are determined from information concerning the function $f(x)$ and the properties of the chosen functions $p_i(x)$.

The standard reference work in the field of interpolation is Steffensen.[55]

Notation of Finite Differences. A point at which a function is tabulated is denoted by x_i. The range of the subscript is $0, 1, \ldots, N$, or sometimes $-N, -N+1, \ldots, 0, 1, \ldots, N$, as noted. The value of the function at x_i is denoted by y_i.

The forward difference of y_i is defined as

$$\Delta y_i \equiv y_{i+1} - y_i \qquad (40\text{-}22)$$

Differencing can be done repeatedly:

$$\Delta^s y_i \equiv \Delta^{s-1}(y_{i+1} - y_i) \equiv \sum_{r=0}^{s} (-1)^{s-r} \binom{s}{r} y_{i+r}$$

where $\binom{s}{r}$ is the binomial coefficient $\dfrac{s!}{r!(s-r)!}$.

Similarly, the backward difference is defined as

$$\nabla y_i \equiv y_i - y_{i-1} \qquad (40\text{-}23)$$

and the higher-order backward differences are defined as

$$\nabla^s y_i \equiv \sum_{r=0}^{s} (-1)^r \binom{s}{r} y_{i-r}$$

Many formulas are conveniently expressed in terms of central differences,

$$\delta y_i \equiv y_{i+\frac{1}{2}} - y_{i-\frac{1}{2}} \qquad (40\text{-}24)$$

In general, *even*-order central differences of a function evaluated at an integral subscript involve tabulated ordinates only. For example,

$$\delta^2 y_i \equiv y_{i+1} - 2y_i + y_{i-1}$$

Moreover, *odd* central differences of a function evaluated at a half-integral subscript are also a function of tabulated values only. For example,

$$\delta y_{i+\frac{1}{2}} \equiv y_{i+1} - y_i$$

We shall also use the mean central difference

$$\partial^n y_k \equiv \tfrac{1}{2}[\delta^n y_{k+\frac{1}{2}} + \delta^n y_{k-\frac{1}{2}}] \tag{40-25}$$

In cases where a function is not tabulated at equally spaced intervals the notion of *divided differences* is convenient:

$$y[x_i] \equiv y_i$$
$$y[x_0,x_1] \equiv \frac{y[x_0] - y[x_1]}{x_0 - x_1} \tag{40-26}$$
$$y[x_0,x_1,x_2] \equiv \frac{y[x_0,x_1] - y[x_1,x_2]}{x_0 - x_2} \quad \text{etc.}$$

If the function y is tabulated at equally spaced intervals, then the divided differences are related to the other kinds of differences. Let the constant interval $x_{i+1} - x_i$ be h; then

$$\Delta^s y_k \equiv s!h^s y[x_k, \ldots, x_{k+s}]$$
$$\nabla^s y_k \equiv s!h^s y[x_k, \ldots, x_{k-s}]$$
$$\delta^{2m+1} y_{k+\frac{1}{2}} \equiv (2m+1)!h^{2m+1} y[x_{k-m}, \ldots, x_k, \ldots, x_{k+m}, x_{k+m+1}]$$
$$\delta^{2m} y_k \equiv (2m)!h^{2m} y[x_{k-m}, \ldots, x_k, \ldots, x_{k+m}]$$

The nth difference of a finite sum of functions is equal to the sum of the nth difference of each of the functions. The nth difference of a polynomial of degree n is a constant.

Interpolation Formulas of Polynomial Type. Many forms of interpolation formulas result from the desire to maximize the accuracy of the interpolation (or extrapolation) with as few terms as possible. The most commonly used formulas are tabulated in Table 40-1 accompanied by comments on their usage. Trigonometric polynomials are frequently used to approximate functions which are not analytic.

In the case of formulas which apply to equally spaced intervals h, the notation

$$\theta \equiv (x - x_0)/h$$

is convenient. Notice then that $\theta - 1 \equiv (x - x_1)/h$ and, in general,

$$\theta - n \equiv (x - x_n)/h$$
$$\theta + n \equiv (x - x_{-n})/h$$

The approximation errors are given in terms of a polynomial $P_n(x)$:

$$P_n(x) \equiv (x - x_0)(x - x_1) \cdots (x - x_n) \equiv h^{n+1}\theta(\theta - 1) \cdots (\theta - n)$$

The quantity $y^{(m)}(\xi)$ also appears in the error formulas. This stands for the mth derivative of the function y taken at some point ξ which lies in the interval in which all the relevant points $x_0, x_1, \ldots, x_n, x$, lie. For estimates of the maximum error, ξ may be taken as the point at which $|y^{(m)}(x)|$ is maximum.

If the derivative as well as the value of a function $y(x)$ is known at m points, then an interpolation formula can be derived which has an error corresponding to that of a $(2m - 1)$ polynomial fit of the data:

$$y(x) = \sum_{k=1}^{m} y_k H_k(x) + \sum_{k=1}^{m} y_k' H_k^*(x) + \text{error} \tag{40-27}$$

where
$$H_k(x) = [1 - 2L_i'(x_i)(x - x_i)][L_i(x)]^2$$
$$H_k^*(x) = (x - x_i)[L_i(x)]^2$$
and
$$\text{Error} = \{[P_n(x)]^2/(2n)!\} y^{(2n)}(\xi) \tag{40-28}$$

The function $L_i(x)$ is defined in Table 40-1. Equation (40-27) is known as Hermite's interpolation formula. In general, the formulas involving derivatives as well as values of functions are known as *osculatory interpolation formulas*. For methods with higher derivatives, see Fort.[14] For numerical aids in computing the coefficients H_k and H_k^*, see Salzer.[52]

Table 40-1. Interpolation Formulas

Lagrange's formula:
$$y(x) = y_0 L_0^{(n)}(x) + y_1 L_1^{(n)}(x) + \cdots + y_n L_n^{(n)}(x) + \text{error}$$

where $L_k^{(n)}(x) \equiv \prod_{\substack{j=0 \\ j \neq k}}^{n} \dfrac{x - x_j}{x_k - x_j}$

$\text{Error} = P_n(x) \dfrac{y^{(n+1)}(\xi)}{(n+1)!}$

Comments: Tables of $L_k^{(n)}$ exist (see Refs. 56 and 63). Unfortunately Lagrangian coefficients all have to be changed if n is changed

Newton-Lagrange's formula in terms of divided differences:
$$y(x) = y_0 + (x - x_0)y[x_0, x_1] + (x - x_0)(x - x_1)y[x_0, x_1, x_2] + \cdots + (x - x_0)(x - x_1) \cdots (x - x_{n-1})y[x_0, x_1, \ldots, x_n] + \text{error}$$

$\text{Error} = P_n(x) \dfrac{y^{m+1}(\xi)}{(n+1)!} = P_n(x)y[x_0, x_1, \ldots, x_n, x]$

Comments: This is simply a regrouping of the terms in the preceding formula

Lagrange's formula for functions of period D (Hermite's formula):
$$y(x) = y_0 \mathcal{L}_0^{(n)}(x) + y_1 \mathcal{L}_1^{(n)}(x) + y_2 \mathcal{L}_2^{(n)}(x) + \cdots + y_n \mathcal{L}_n^{(n)}(x)$$

where $\mathcal{L}_k^{(n)} = \prod_{\substack{j=0 \\ j \neq k}}^{n} \dfrac{\sin\,(2\pi/D)(x - x_j)}{\sin\,(2\pi/D)(x_k - x_j)}$

Comments: See Ref. 51 for tables of $\mathcal{L}_k^{(n)}(x)$

Newton's forward-interpolation formula:
$$y(x_0 + \theta h) = y_0 + \theta \Delta y_0 + \dfrac{\theta(\theta - 1)}{2!} \Delta^2 y_0 + \cdots + \dfrac{\theta(\theta - 1) \cdots (\theta - n + 1)}{n!} \Delta^n y_0 + \text{error}$$

$\text{Error} = \dfrac{P_n(x)}{(n+1)!} y^{(n+1)}(\xi)$

Comments: Newton's formula has the advantage that it is simple to add new terms to the interpolation formula. The forward-difference formula is useful when interpolation is to be done near the beginning of a table

40–10

Newton's backward-interpolation formula:

$$y(x_n - \theta h) = y_n + \theta \nabla y_n + \frac{\theta(\theta+1)}{2!}\nabla^2 y_n + \cdots + \frac{\theta(\theta+1)\cdots(\theta+n-1)}{n!}\nabla^n y_n + \text{error}$$

$$\text{Error} = \frac{h^{n+1}}{(n+1)!}\theta(\theta+1)\cdots(\theta+n)y^{(n+1)}(\xi)$$

This formula is useful when interpolation is to be done near the end of a table

---

Newton's forward-interpolation formula for functions of two variables:

$$f(x_0 + \theta h, y_0 + \mu k) \doteq f(x_0, y_0) + \theta \Delta_h f_0 + \mu \Delta_k f_0 + \frac{\theta(\theta-1)}{2!}\Delta_{hh}f_0 + \theta\mu \Delta_{hk}f_0 + \frac{\mu(\mu-1)}{2!}\Delta_{kk}f_0$$
$$+ \frac{\theta(\theta-1)(\theta-2)}{3!}\Delta_{hhh}f_0 + \frac{\theta(\theta-1)\mu}{2!}\Delta_{hhk}f_0 + \frac{\theta\mu(\mu-1)}{2!}\Delta_{hkk}f_0 + \frac{\mu(\mu-1)(\mu-2)}{3!}\Delta_{kkk}f_0 + \cdots \quad \text{etc.}$$

where $f_0 \equiv f(x_0, y_0)$
and $\Delta_h f_0 = f(x_0 + h, y_0) - f(x_0, y_0)$
$\Delta_k f_0 = f(x_0, y_0 + k) - f(x_0, y_0)$
$\Delta_{hh}f_0 = f(x_0 + 2h, y_0) - 2f(x_0 + h, y_0) + f(x_0, y_0)$
$\Delta_{hk}f_0 = f(x_0 + h, y_0 + k) - f(x_0 + h, y_0) - f(x_0, y_0 + k) + f(x_0, y_0)$

This is the two-dimensional analog of Newton's forward-interpolation formula

---

Gauss's forward-interpolation formula:

$$y(x_0 + \theta h) = y_0 + \theta \delta y_{\frac{1}{2}} + \frac{\theta(\theta-1)}{2!}\delta^2 y_0 + \frac{\theta(\theta-1)}{3!}\delta^3 y_{\frac{1}{2}} + \frac{\theta(\theta^2 - 1^2)(\theta-2)}{4!}\delta^4 y_0$$
$$+ \cdots + \underbrace{\frac{\theta(\theta^2-1^2)(\theta^2-2^2)\cdots(\theta^2-(m-1)^2)(\theta-m)}{(2m)!}\delta^{2m}y_0}_{n = 2m \text{ if } n \text{ is even or } n = 2m+1 \text{ if } n \text{ is odd}}$$
$$+ \frac{\theta(\theta^2-1^2)\cdots(\theta^2 - m^2)}{(2m+1)!}\delta^{2m+1}y_{\frac{1}{2}} \qquad (n \text{ even})$$
$$\frac{\theta(\theta^2-1^2)\cdots(\theta^2-m^2)}{(2m+1)!}\delta^{2m+1}y_{\frac{1}{2}} + \text{error} \qquad (n \text{ odd})$$

$$\text{Error} = \begin{cases} \dfrac{h^{2m+1}\theta(\theta^2-1^2)\cdots(\theta^2-m^2)}{(2m+1)!}y^{(2m+1)}(\xi) & (n \text{ even}) \\[1em] \dfrac{h^{2m+2}\theta(\theta^2-1^2)\cdots(\theta^2-m^2)(\theta-m-1)}{(2m+2)!}y^{(2m+2)}(\xi) & (n \text{ odd}) \end{cases}$$

Gauss's formula is useful when interpolations are likely to range on both sides of x_0. When it is terminated with an even difference $2m$, it agrees with the function at x_0, $x_{\pm 1}, x_{\pm 2}, \ldots, x_{\pm m}$. Otherwise, when it is terminated with an odd difference $2m+1$, it agrees with the function at x_0, $x_{\pm 1}, \ldots, x_{\pm m}, x_{m+1}$

Table 40-1. Interpolation Formulas (Continued)

	Comments
Gauss's backward-interpolation formula: $y(x_0 - \theta h) = y[x_0 + (-\theta)h]$ Completely similar to Gauss's forward formula with θ replaced by $-\theta$ and $y_{1/2}$ by $y_{-1/2}$	The backward formula is identical to Gauss's forward formula when terminated with an even difference $2m$: both formulas agree with the points at $x_0, x_{\pm 1}, \ldots, x_{\pm m}$. If the backward formula is terminated with an odd difference $2m + 1$, it agrees with the point x_0, $x_{\pm 1}, \ldots, x_{\pm m}, x_{-m-1}$
Stirling's formula: $y(x_0 + \theta h) = y_0 + \theta \partial y_0 + \dfrac{\theta^2}{2}\delta^2 y_0 + \dfrac{\theta(\theta^2 - 1^2)}{3!}\partial^3 y_0 + \dfrac{\theta^2(\theta^2 - 1^2)}{4!}\delta^4 y_0 + \cdots$ $\qquad + \dfrac{\theta^2(\theta^2 - 1^2)\cdots(\theta^2 - (m-1)^2)}{(2m)!}\delta^{2m} y_0 \qquad (n\text{ even})$ $\qquad + \dfrac{\theta(\theta^2 - 1^2)\cdots(\theta^2 - m^2)}{(2m+1)!}\partial^{2m+1} y_0 \qquad (n\text{ odd})$ $\qquad + \text{error}$ where $n = 2m$ if n is even or $n = 2m + 1$ if n is odd $\text{Error} = \begin{cases} \dfrac{\theta^2(\theta^2 - 1^2)(\theta^2 - 2^2)\cdots(\theta^2 - m^2)}{h^{2m+1}(2m+1)!} y^{(2m+1)}(\xi) & (n\text{ even}) \\ \text{or} \\ \dfrac{\theta(\theta^2 - 1^2)(\theta^2 - 2^2)\cdots(\theta^2 - m^2)}{h^{2m+2}(2m+2)!} [(\theta - m - 1)y^{(2m+2)}(\xi_1) + (\theta + m - 1)y^{(2m+1)}(\xi_2)] & (n\text{ odd}) \end{cases}$ where ξ_1 and ξ_2 are two independent points in the interpolation range	Stirling's formula is most accurate when the point to be interpolated lies in the range $x_0 \pm \tfrac{1}{2}h$. See (40-25) for the definition of $\partial^n y_0$ and (40-24) for $\delta^n y_0$

Bessel's formula:

$$y(x_0 + \theta h) = \partial y_{1/2} + (\theta - 1/2)\partial y_{1/2} + \frac{\theta(\theta-1)}{2!}\partial^2 y_{1/2} + \frac{\theta(\theta-1)(\theta-1/2)}{3!}\partial^3 y_{1/2}$$

$$+ \cdots + \frac{\theta(\theta^2 - 1^2)\cdots[\theta^2 - (m-1)^2](\theta - m)}{2m!}\partial^{2m} y_{1/2} \quad (n \text{ even})$$

$$+ \frac{\theta(\theta^2 - 1^2)\cdots[\theta^2 - (m-1)^2](\theta - m)(\theta - 1/2)}{(2m+1)!}\partial^{2m+1} y_{1/2} \quad (n \text{ odd})$$

where $n = 2m$ if n is even, or $n = 2m + 1$ if n is odd

$$\text{Error} = \begin{cases} h^{2m+1}\dfrac{\theta(\theta^2-1^2)\cdots[\theta^2-(m-1)^2](\theta-m)}{2(2m+1)!}[(\theta+m)y^{(2m+1)}(\xi_1) + (\theta-m-1)y^{(2m+1)}(\xi_2)] & (n \text{ even}) \\ \text{Same as Gauss above} & (n \text{ odd}) \end{cases}$$

Bessel's formula is designed to give accuracy symmetrically about the point midway between x_0 and x_1

Everett's formula:

$$y(x_0 + \theta h) = (1-\theta)y_0 - \frac{\theta(\theta-1)(\theta-2)}{3!}\delta^2 y_0 - \frac{(\theta+1)\theta(\theta-1)(\theta-2)(\theta-3)}{5!}\delta^4 y_0 - \cdots$$

$$- \frac{(\theta+m-1)(\theta+m-2)\cdots(\theta-m-1)(\theta-m-2)}{(2m+1)!}\delta^{2m} y_0 + \theta y_1 + \frac{(\theta+1)\theta(\theta-1)}{3!}\delta^2 y_1$$

$$+ \frac{(\theta+2)(\theta+1)\theta(\theta-1)(\theta-2)}{5!}\delta^4 y_1 + \cdots + \frac{(\theta+m)(\theta+m-1)\cdots(\theta-m)}{(2m+1)!}\delta^{2m} y_1 + \text{error}$$

$$\text{Error} = h^{2m+2}\frac{\theta(\theta^2-1^2)\cdots(\theta^2-m^2)(\theta-m-1)}{(2m+2)!}y^{(2m+2)}(\xi)$$

This formula is useful because in many tables the first few central differences of even order are given

Steffensen's formula:

$$y(x_0 + \theta h) = \frac{(\theta+1)\theta}{2!}\partial y_{1/2} + \frac{(\theta+2)(\theta+1)\theta(\theta-1)}{4!}\partial^3 y_{1/2} + \cdots + \frac{(\theta+m+1)(\theta+m)\cdots(\theta-m)}{(2m+2)!}\partial^{2m+1} y_{1/2}$$

$$- \frac{\theta(\theta-1)}{2!}\delta y_{-1/2} - \frac{(\theta+1)\theta(\theta-1)(\theta-2)}{4!}\delta^3 y_{-1/2} - \cdots - \frac{(\theta+m)(\theta+m-1)\cdots(\theta-m-1)}{(2m+1)!}\partial^{2m+1} y_{-1/2} + \text{error}$$

$$\text{Error} = h^{2m+3}\frac{\theta(\theta^2-1^2)\cdots(\theta^2-(m+1)^2)}{(2m+2)!}y^{(2m+2)}(\xi)$$

This is the analog of the preceding formula for odd differences

Interpolation with Fourier Series.

Finite Fourier series are frequently used in the approximation of periodic functions. If $f(t)$ has a period $2L$ then a finite Fourier-series approximation that is equal to $f(t)$ at the $2N$ points in the period, that is, at $t = 0, L/N, 2L/N, \ldots, [(2N - 1)/N]L$ is

$$f(t) \doteq \alpha_0/2 + \sum_{j=1}^{N-1} [\alpha_j \cos(\pi jt/L) + b_j \sin(\pi jt/L)] + \alpha_N \cos(\pi Nt/L) \qquad (40\text{-}29)$$

where N is an integer, and

$$\alpha_0 = 1/N \sum_{k=0}^{2N-1} f(kL/N)$$

$$\alpha_j = 1/N \sum_{k=0}^{2N-1} f(kL/N) \cos(\pi jk/N)$$

$$\alpha_N = 1/N \sum_{k=0}^{2N-1} (-1)^k f(kL/N)$$

$$b_j = 1/N \sum_{k=0}^{2N-1} f(kL/N) \sin(\pi jk/N)$$

Methods for simplifying computation of these coefficients are described in Chapter 6 of Hamming[21] and in Lanczos.[37] In Whittaker and Robinson[60] there are descriptions of worksheets for convenient hand computation of the coefficients.

Care must be exercised when differentiating Fourier series, since a poorly convergent or divergent series may result. If a function is approximated by a finite Fourier series of N terms, then a convergent approximation for the derivative is obtained by multiplying the kth term, $k = 1, 2, \ldots, N$, of the finite series

$$f(x) = a_0 + \sum_{k=1}^{N} (a_k \sin kx + b_k \cos kx)$$

by the Lanczos σ_k factor

$$\sigma_k = \frac{\sin(k\pi/N)}{k\pi/N}$$

and then taking the derivative of the series term by term. The effect of this modification is to smooth the derivative at points where the overshoot (the Gibbs phenomenon) is large.[21,37]

Numerical Integration.

Numerical integration can be approached by fitting a polynomial through a selected number of points and integrating the polynomial. Thus, if there are two points y_0 and y_1, the linear-interpolation formula is

$$y(x) \doteq y_0 + x(y_1 - y_0) \qquad (40\text{-}30)$$

and the integral of this linear approximation, applied to n equal intervals $(x_i - x_{i-1} \equiv h)$ yields the trapezoidal rule

$$\int_{x_0}^{x_n} y(x)\, dx \doteq (h/2)y_0 + h \sum_{k=1}^{n-1} y_k + (h/2)y_n \qquad (40\text{-}31)$$

Integration formulas are summarized in Table 40-2.

Table 40-2. Numerical-integration Formulas

Formula	Comments
Trapezoidal rule: $$\int_{x_0}^{x_n} y(x)\, dx \doteq (h/2)(y_0 + y_n) + h \sum_{k=1}^{n-1} y_k$$ Error $= (n/12)h^3 y''(\xi)$ where $x_0 \leq \xi \leq x_n$	To estimate the maximum error, use the value of ξ where $\lvert y''(\xi) \rvert$ is largest
Simpson's rule: $$\int_{x_0}^{x_{2m}} y(x)\, dx \doteq (h/3)(y_0 + 4y_1 + 2y_2 + 4y_3 + \cdots + 4y_{2m-1} + y_{2m})$$ Error $= (2m/180) h^5 y^{(4)}(\xi)$ $x_0 \leq \xi \leq x_{2m}$	Simpson's rule is always used with an odd number of points. It is probably the most frequently used of all formulas

In general,

$$\int_{x_0}^{x_n} y(x)\, dx \doteq \alpha h \sum_{k=0}^{n} \beta_k y_k + E_n \qquad (40\text{-}32)$$

where E_n is the error committed. The formulas for odd n (except $n = 1$) yield error estimates which are larger than the error estimates for the formula for $n - 1$ and are therefore omitted. Notice that if the integral extends over (x_0, x_m), where $x_m = x_0 + n(m/n)h$, the formulas must be applied m/n times and the error must also be multiplied by m/n. Table 40-3 lists the coefficients α, β, and E_n in (40-32) for various values of n. Notice that the missing β_i can be supplied by symmetry.

Table 40-3. Numerical-integration Coefficients

n	α	β_0	β_1	β_2	β_3	β_4	β_5	E_n
1	1/2	1	1					$(1/12)h^3 y''(\xi)$
2	1/3	1	4	1				$(1/90)h^5 y^{(4)}(\xi)$
4	2/45	7	32	12	32	7		$(8/945)h^7 y^{(6)}(\xi)$
6	1/140	41	216	27	272	...	...	$(9/1,400)h^9 y^{(8)}(\xi)$
8	4/14,175	989	5,888	-928	10,496	$-4,540$	...	
10	5/299,376	16,067	106,300	$-48,525$	272,400	$-260,550$	427,368	...

The coefficients are symmetrical about $\beta_{n/2}$. Dots in the above table indicate missing coefficients.

It is sometimes the case that an integral is to be evaluated and the integrand is given in analytical form. Since the points at which the integrand is to be evaluated can be chosen arbitrarily, one can optimize the accuracy obtained with a given number of points by selecting both the location of the points and the weights given at each point. An n-point formula can be made to give the same order of error as a $2n - 1$ point formula with points spaced uniformly. This method of numerical integration is known as Gaussian quadrature.

In the following, the interval to be integrated over is assumed to be $(0,1)$. If the interval of the original variable x' is (a,b) then a change of independent variable must be made: $x = a/(a - b) - x'/(a - b)$. The integration formula is then

$$\int_0^1 y(x)\,dx = w_1 y(x_1) + w_2 y(x_2) + \cdots + w_n y(x_n) \tag{40-33}$$

The weights w_k and the points of evaluation x_k are given below for $n = 3, 4, 5$, and 6 points in the interval $(0,1)$. Further values are tabulated in Kopal,[34] Krylov,[35] Scarborough,[53] and Moors.[45] The location of the points is frequently given for the interval $(-1,1)$, in which case the points x_k are the zeros of the Legendre polynomials.

$n = 3$:
$\quad x_1 = 0.1127016654 \qquad\qquad w_1 = w_3 = 0.2777777777$
$\quad x_2 = 0.5$
$\quad x_3 = 0.8872983346 \qquad\qquad w_2 = 0.4444444444$

$n = 4$:
$\quad x_1 = 0.0694318442 \qquad\qquad w_1 = w_4 = 0.1739274226$
$\quad x_2 = 0.3300094782 \qquad\qquad w_2 = w_3 = 0.3260725774$
$\quad x_3 = 0.6699905218$
$\quad x_4 = 0.9305681558$

$n = 5$:
$\quad x_1 = 0.0469100770 \qquad\qquad w_1 = w_5 = 0.1184634425$
$\quad x_2 = 0.2307653449 \qquad\qquad w_2 = w_4 = 0.2393143352$
$\quad x_3 = 0.5$
$\quad x_4 = 0.7692346551 \qquad\qquad w_3 = 0.2844444444$
$\quad x_5 = 0.9530899230$

$n = 6$:
$\quad x_1 = 0.0337652429 \qquad\qquad w_1 = w_6 = 0.0856622462$
$\quad x_2 = 0.1693953068$
$\quad x_3 = 0.3806904070 \qquad\qquad w_2 = w_5 = 0.1803807865$
$\quad x_4 = 0.6193095930$
$\quad x_5 = 0.8306046932 \qquad\qquad w_3 = w_4 = 0.2339569673$
$\quad x_6 = 0.9662347571$

Numerical Differentiation. Differentiation formulas in terms of tabular differences are easily derived directly from Newton's formula. Notice that the formulas are not functions of the variable x, since it is assumed the derivatives are to be computed at tabular points. If the point of interest is not tabulated, then the derivative can be found by differentiating one of the interpolation formulas in Sec. 40-3 under Interpolation Formulas of Polynomial Type. In the following, the error is given in square brackets. The error in the approximation for the derivative may be large, particularly when a high-order polynomial approximation is used. Further formulas can be found in Hildebrand.[23]

Three-point formulas:

$$y_{-1}' = (1/2h)(-3y_{-1} + 4y_0 - y_1) + [(h^2/3)y'''(\xi)]$$
$$y_0' = (1/2h)(-y_{-1} + y_1) - [(h^2/6)y'''(\xi)]$$
$$y_1' = (1/2h)(y_{-1} - 4y_0 + 3y_1) + [(h^2/3)y'''(\xi)]$$

Five-point formulas:

$$y_{-2}' = (1/12h)(-25y_{-2} + 48y_{-1} - 36y_0 + 16y_1 - 3y_2) + [(h^5/5)y^{(5)}(\xi)]$$
$$y_{-1}' = (1/12h)(-3y_{-2} - 10y_{-1} + 18y_0 - 6y_1 + y_2) - [(h^4/20)y^{(5)}(\xi)]$$
$$y_0' = (1/12h)(y_{-2} - 8y_{-1} + 8y_1 - y_2) + [(h^4/30)y^{(5)}(\xi)]$$
$$y_1' = (1/12h)(-y_{-2} + 6y_{-1} - 18y_0 + 10y_1 + 3y_2) - [(h^4/20)y^{(5)}(\xi)]$$
$$y_2' = (1/12h)(3y_{-2} - 16y_{-1} + 36y_0 - 48y_1 + 25y_2) + [(h^5/5)y^{(5)}(\xi)]$$

40-5. Numerical Solution of Ordinary Differential Equations

Notation and Definitions. An ordinary differential equation can be expressed as

$$F(x,y,y',y'',\ldots,y^{(n)}) = 0$$

where F is the function describing the relation between the variables, x is the independent variable, y is the dependent variable, y' is the first derivative of y with respect to x, and $y^{(n)}$ is the nth derivative. The highest derivative determines the order of the equation. A solution of the differential equation, in an interval $a < x < b$, is a function $y = f(x)$, for which the nth derivative exists, such that the differential equation becomes an identity for all x in the interval (a,b) when y and its derivatives are replaced by the function $f(x)$ and its derivatives. A unique solution is obtained generally if values of $y, y', \ldots, y^{(n-1)}$ are assigned at some initial point x_0.

Ordinary differential equations have only one independent variable x, but a system of ordinary differential equations may determine the values of many dependent variables $y_1(x), y_2(x), \ldots, y_m(x)$. A system of m first-order equations can be represented as

$$F_1(x, y_1, \ldots, y_m, y_1', \ldots, y_m') = 0$$
$$F_2(x, y_1, \ldots, y_m, y_1', \ldots, y_m') = 0$$
$$\vdots$$
$$F_m(x, y_1, \ldots, y_m, y_1', \ldots, y_m') = 0$$

(40-34)

The system of equations (40-34) can be expressed in the concise notation of vectors as

$$\mathbf{F}(x, \mathbf{y}, \mathbf{y}') = \mathbf{0} \qquad (40\text{-}35)$$

The vector function $\mathbf{F}$ and the vectors $\mathbf{y}$, $\mathbf{y}'$, and $\mathbf{0}$ are represented by boldface symbols to distinguish them as vectors. Equations (40-34) are then the equations for the components of (40-35).

Notice that once an arbitrary number of dependent variables is admitted in a system of differential equations such as (40-34), one need not consider equations of higher order than first. For example, the second-order linear inhomogeneous differential equation

$$d^2y/dx^2 + g(x)(dy/dx) + f(x)y = 0$$

can be written as a system of two first-order equations

$$dy_1/dx - y_2 = 0$$
$$dy_2/dx + g(x)y_2 + f(x)y_1 = 0$$

The methods discussed in the following sections apply when $\mathbf{y}$ is considered to be a vector-valued function with one or more components. In general, when $\mathbf{y}$ has more than one component, the error formulas as described do not apply.

Numerical Solution. When a differential equation is solved numerically, the numerical approximation to the solution is found only at discrete abscissas x_n ($n = 0, 1, 2, \ldots, N$). The approximate solution at these points is denoted by $\mathbf{y}_n$ and the exact solution by $\mathbf{y}(x_n)$. The *truncation* or *discretization error* is $\mathbf{e}_n \equiv \mathbf{y}_n - \mathbf{y}(x_n)$. In practice, the solution of the discrete equations that approximate the differential equations is not evaluated exactly. The approximate solution as evaluated, $\tilde{\mathbf{y}}_n$, contains *rounding errors* which are defined as $\mathbf{r}_n = \mathbf{y}_n - \tilde{\mathbf{y}}_n$. In cases where it is necessary to denote a particular component of the solution, two subscripts are used, with the first subscript denoting the component and the second subscript denoting the value for the independent variable. The step size $x_{n+1} - x_n$, between successive abscissas, is usually assumed to be a constant h independent of n.

The simplest case of (40-35) is

$$\mathbf{y}'(x) = \mathbf{f}(x, \mathbf{y}) \qquad (40\text{-}36)$$

It is assumed that the initial condition $\mathbf{y}(x_0) = \mathbf{y}_0$ is given. Simple Taylor-series expansion may be used to obtain an approximate expression for $\mathbf{y}(x_{n+1})$

$$\mathbf{y}(x_{n+1}) = \mathbf{y}(x_n + h) = \mathbf{y}(x_n) + h\mathbf{y}'(x_n) + (h^2/2!)\mathbf{y}''(x_n) + \cdots$$

so a step-by-step solution of (40-36) can be written as

$$y_{n+1} = y_n + h\mathbf{f}[x_n, y_n] = y(x_{n+1}) + e_{n+1} \qquad (40\text{-}37)$$

where the term e_{n+1}, which is neglected in each step of the computation, is the accumulated truncation error, and for the first step is of the order $(h^2/2)y''(x)$, in this case. Equation (40-37) is known as the Euler method. A detailed description of the growth of errors may be found in Henrici.[22] By using the above Taylor-series expansion to find $y(x_{n-1})$ in terms of $y(x_n)$, and then eliminating the terms $(h^2/2!)y''(x_n)$, another formula, more accurate than (40-37), may be obtained:

$$y_{n+1} = y_{n-1} + 2h\mathbf{f}[x_n, y_n] = y(x_{n+1}) + e_{n+1} \qquad (40\text{-}38)$$

where the truncation error term e_{n+1}, neglected in the numerical calculation, is of the order $(h^3/3)y'''(x_n)$ for the first step. One difficulty in using (40-38) is in getting "started," as discussed below, since one needs two initial values, say y_0 and y_1. A second difficulty in using (40-38) is the growth of errors; see Hamming.[21]

The extension of (40-37) and (40-38) to higher-order equations, or, equivalently, to a set of simultaneous equations, is straightforward. The equations for the m components of the vector form of (40-38) become

$$y_i(x_{n+1}) \doteq y_{i,n+1} = y_{i,n-1} + 2hf_i(x_n, y_{1,n}, \ldots, y_{m,n}) \qquad i = 1, 2, \ldots, m \qquad (40\text{-}39)$$

Example. To solve

$$d^2y/dx^2 = x \ln(y) \quad \text{or} \quad dy_1/dx = y_2 \quad dy_2/dx = x \ln(y_1)$$

These formulas (40-39) for numerical solution of an equivalent system of two first-order equations are

and
$$\begin{aligned} y_1(x_{n+1}) &\doteq y_{1,n+1} = y_{1,n-1} + 2hy_{2,n} \\ y_2(x_{n+1}) &\doteq y_{2,n+1} = y_{2,n-1} + 2hx \ln y_{1,n} \end{aligned} \qquad (40\text{-}40)$$

Suppose the initial values $y_1(x_0)$ and $y_2(x_0)$ are given, and the additional starting values $y_1(x_1)$ and $y_2(x_1)$ have been found; then simple successive substitutions in (40-40) allow one to extend the numerical solution recursively until the desired final value x_N for the independent variable is reached.

Starting Methods. Extensions to higher than rudimentary accuracy, of multistep methods such as (40-39), have the disadvantage that they need more initial values than are usually given with the original differential equation. One may use a formula such as (40-37) to start off, but this may introduce undesirable initial errors due to low-order truncation error. A Taylor series retaining terms to at least the order of the truncation error of the numerical integration scheme or use of a single-step method as described below under Runge-Kutta and Other One-step Methods is often the correct approach to finding starting values.

Suppose we are given the initial value $\mathbf{y}(x_0) = \mathbf{y}_0$ and wish to integrate

$$\mathbf{y}' = \mathbf{f}(x, \mathbf{y}) \qquad (40\text{-}41)$$

If f is differentiable then repeated differentiation gives

$$\mathbf{y}'' = \mathbf{f}'(x, \mathbf{y}) = \boldsymbol{\phi}(x, \mathbf{y}) = \partial \mathbf{f}/\partial x + \sum_{i=1}^{m} (\partial \mathbf{f}/\partial y_i)(dy_i/dx)$$

$$\mathbf{y}''' = \boldsymbol{\phi}'(x, \mathbf{y}) = \partial \boldsymbol{\phi}/\partial x + \sum_{i=1}^{m} (\partial \boldsymbol{\phi}/\partial y_i)(dy_i/dx) \qquad (40\text{-}42)$$

All these derivatives may be evaluated [by substituting from (40-41) into (40-42)] since at the point x_0, the value of $\mathbf{y}$ is given. The Taylor expansion for $\mathbf{y}(x)$ in terms of $\mathbf{y}(x_0)$ is

$$\mathbf{y}(x) = \mathbf{y}(x_0) + (x - x_0)\mathbf{y}'(x_0) + [(x - x_0)^2/2!]\mathbf{y}''(x_0) + \cdots \\ + [(x - x_0)^n/n!]\mathbf{y}^{(n)}(x_0) + \cdots$$

In particular, the values of $\mathbf{y}(x_1)$, $\mathbf{y}(x_2)$, ... can be calculated. Indeed, if enough terms in this expansion are known, this series solution to (40-41) can be used to evaluate the solution out to $x = x_N$. There are four difficulties, however. The series may converge slowly or not at all for large x_N. The derivatives may be complicated or impossible to express. The high-order derivatives may not exist. Errors may grow rapidly.

The single-step methods described below can also be used to obtain starting values. Most of these single-step methods do not require derivatives other than $\mathbf{y}'$.

Methods of Higher Accuracy. The methods just described can be generalized to methods that include more terms and usually produce more accurate results when y is a smooth function. The most popular of the more accurate numerical methods are the multistep methods, combinations of multistep methods including the predictor-corrector methods, and the single-step methods of Runge-Kutta type.

The multistep methods for numerical solution of $\mathbf{y}' = \mathbf{f}(x,\mathbf{y})$ with initial condition $\mathbf{y}(x_0) = \mathbf{y}_0$ are based on an interpolation or extrapolation formula of the following type:

$$\mathbf{y}_{n+1} = a_0 \mathbf{y}_n + a_1 \mathbf{y}_{n-1} + \cdots + a_k \mathbf{y}_{n-k} + h[b_{-1}\mathbf{y}_{n+1}' + b_0 \mathbf{y}_n' + \cdots + b_k \mathbf{y}_{n-k}'] \\ + h^2[c_{-1}\mathbf{y}_{n+1}'' + c_0 \mathbf{y}_n'' + \cdots + c_k \mathbf{y}_{n-k}''] \quad (40\text{-}43)$$

where a_i, b_i, and c_i are constants and the solution is to be approximated at $x_n = x_0 + nh$. The choice of the coefficients determines the truncation and rounding errors and the rate of propagation of these errors. The multistep method (40-43) requires starting values $(x_0,\mathbf{y}_0)$, $(x_1,\mathbf{y}_1)$, ..., $(x_k,\mathbf{y}_k)$ which are found by the methods described under Starting Methods above and Runge-Kutta and Other One-step Methods below. Succeeding values are found by successively solving (40-43) for the next value $\mathbf{y}_{n+1}$. If both b_{-1} and c_{-1} are zero, then (40-43) is an extrapolation formula and the solution for $\mathbf{y}_{n+1}$ is trivial. Otherwise (40-43) is an interpolation formula and the solution for $\mathbf{y}_{n+1}$ is a nonlinear problem when $\mathbf{f}(x,\mathbf{y})$ is a nonlinear function of y. Such problems are solved iteratively by methods described in the following paragraphs.

The multistep methods extrapolate from points of the solution which have been found previously, or interpolate between known points and estimated values of the solution. Their principal disadvantage is the difficulty mentioned earlier: getting started. Also, some multistep methods introduce undesirable growth of error. Detailed descriptions of multistep methods may be found in Hamming,[21] Henrici,[22] Kopal,[34] and Collatz.[6] Table 40-4 contains pairs of multistep methods. These equations may be used individually or in pairs as described in the next section.

Predictor-Corrector Methods. For differential equations $\mathbf{y}' = \mathbf{f}(x,\mathbf{y})$ with complicated functions f, the efficiency of the numerical integration scheme depends primarily on the number of evaluations of $\mathbf{f}(x,y)$ required for each step. Predictor-corrector methods use a pair of multistep formulas, with truncation errors of similar order, to minimize the required number of evaluations of functions. One such pair is

Predictor:
$$\mathbf{y}_{n+1} = \mathbf{y}_{n-1} + 2h\mathbf{y}_n' = \mathbf{y}(x_{n+1}) - \{(h^3/3)y'''(\xi_1)\} \quad (40\text{-}44)$$
Corrector:
$$\mathbf{y}_{n+1} = \mathbf{y}_n + (h/2)(\mathbf{y}_{n+1}' + \mathbf{y}_n') = \mathbf{y}(x_{n+1}) - \{-(h^3/12)y'''(\xi_2)\} \quad (40\text{-}45)$$

where the quantity in brackets is the local truncation error. Equations (40-44) and (40-45) are used together as follows: Suppose the initial condition is $\mathbf{y}(x_0) = \mathbf{y}_0$, and that $\mathbf{y}_1 \doteq \mathbf{y}(x_0 + h)$ has been found by a starting procedure. Then the value of $\mathbf{y}(x_2)$ is estimated by evaluation of the right-hand side of (40-44). Next, a new estimate for $\mathbf{y}_2$ is obtained by evaluating the right-hand side of the corrector equation (40-45) using x_1, $\mathbf{y}_1$, x_2, and the predicted value for $\mathbf{y}_2$ just obtained from the predictor equa-

Table 40-4. Predictor-iterated Corrector Methods

$y' = f(x,y)$, $x_{n+1} - x_n \equiv h > 0$ $A \equiv \partial f/\partial y$ $q \equiv hA$

Formula	Local truncation error	Comment*
Predictor: $y_{n+1} = y_{n-1} + 2hy_n'$	$(h^3/3)y'''(\xi_1)$ $x_{n-1} < \xi_1 < x_{n+1}$	The predictor used alone is convergent and has weak stability (Ref. 22, pp. 240–242) as $h \to 0$. Propagation errors increase if $A > 0$ or $A < 0$. The relative error increases if $A < 0$ (see Hamming[21] and Henrici[22])
Corrector: $y_{n+1} = y_n + (h/2)(y_{n+1}' + y_n')$	$-(h^3/12)y'''(\xi_2)$ $x_n < \xi_2 < x_{n+1}$	Iteration of the corrector converges if $|q| < 2$. The corrector is always relatively stable. Propagated errors decrease if $A < 0$. The predicted and converged corrected values differ by $\tfrac{5}{12} y'''(\xi_3)$. This quantity is used to control the truncation error and as a check for gross computations. If more than two corrections are required the step size should be reduced
Adams-Bashford Predictor: $y_{n+1} = y_n + (h/24)(55y_n' - 59y_{n-1}' + 37y_{n-2}' - 9y_{n-3}')$ Corrector: $y_{n+1} = y_n + (h/24)(9y_{n+1}' + 19y_n' - 5y_{n-1}' + y_{n-2}')$	$\tfrac{251}{720} h^5 y^{(5)}(\xi_1)$ $x_{n-3} < \xi_1 < x_{n+1}$ $-\tfrac{19}{720} h^5 y^{(5)}(\xi_2)$ $x_{n-2} < \xi_2 < x_{n+1}$	Predictor used alone is stable. The propagated errors do not increase if $-\tfrac{3}{10} \leq q \leq 0$. Iteration of the corrector converges if $|q| < \tfrac{3}{5}$. The corrector is stable and propagated errors do not increase if $-3 \leq q \leq 0$. The predicted and corrected values differ by $\tfrac{3}{8} h^5 y^{(5)}(\xi_3)$
Milne Predictor: $y_{n+1} = y_{n-3} + (4h/3)(2y_n' - y_{n-1}' + 2y_{n-2}')$ Corrector: $y_{n+1} = y_{n-1} + (h/3)(y_{n+1}' + 4y_n' + y_{n-1}')$	$\tfrac{28}{90} h^5 y^{(5)}(\xi_1)$ $x_{n-3} < \xi_1 < x_{n+1}$ $-\tfrac{1}{90} h^5 y^{(5)}(\xi_2)$ $x_{n-1} < \xi_2 < x_{n+1}$	This is the Milne method; see Milne.[43] The predictor used alone is weakly stable as $h \to 0$. Propagated errors *increase* if $q < 0$. Iteration of the corrector converges if $|q| < 3$. The corrector is weakly stable. Propagated errors increase if $A < 0$. The difference between the predicted and corrected values is $\tfrac{29}{90} h^5 y^{(5)}(\xi_3)$. This value is used to control the truncation error and perhaps to interpolate for higher-order truncation error. This method is not used for numerical integration over large regions when A is negative

* The comments concerning increase of errors are for the case q = constant and y a single real variable.

tion (40-44). If the new estimate for y_2 differs appreciably from the previous estimate for y_2 then the right-hand side of (40-45) is reevaluated using the latest estimate for y_2, until the iterates converge. The rate of convergence is determined by the size of $h[\partial \mathbf{f}(x,y)/\partial y_i]$. If more than two evaluations of (40-45) are required, the size of h should be halved. This requires the use of new starting values to obtain $\mathbf{y}$ at the midpoints.

In the special case that y''' is a *slowly changing function* an estimate of the size of the local truncation error is obtained by subtracting the predicted $(\mathbf{y}_{n+1})_p$ and corrected $(\mathbf{y}_{n+1})_c$ estimates for the solution, obtained from (40-44) and (40-45), respectively,

$$y(x_{n+1}) \doteq (\mathbf{y}_{n+1})_p + (h^3/3)\mathbf{y}'''(\xi_1)$$
$$y(x_{n+1}) \doteq (\mathbf{y}_{n+1})_c - (h^3/12)\mathbf{y}'''(\xi_2)$$
$$(\mathbf{y}_{n+1})_c - (\mathbf{y}_{n+1})_p = -5/12 h^3 \mathbf{y}'''(\xi)$$
$$x_{n-1} \leq \xi \leq x_{n+1}$$

This error estimate may be used to control the size of truncation error by indicating when the step size h should be halved or doubled. An abrupt change in the error estimate indicates either a computational blunder or changes in the function requiring special attention. The error estimate can also be used to interpolate for an estimate for y_{n+1} with higher-order truncation error:

$$\mathbf{y}_{n+1} = (\mathbf{y}_{n+1})_c - \tfrac{1}{12}h^3 \mathbf{y}'''(\xi) \doteq \tfrac{4}{5}(\mathbf{y}_{n+1})_c + \tfrac{1}{5}(\mathbf{y}_{n+1})_p$$

Methods using a modification of this technique, as described by Hamming,[22] are summarized in Table 40-5.

Table 40-5. Predictor Single-correction Methods

Formulas	Local truncation error
Adams-Bashford method: $p_{n+1} = y_n + (h/24)(55 y_n' - 59 y_{n-1}' + 37 y_{n-2}' - 9 y_{n-3'})$ $m_{n+1} = p_{n+1} - \tfrac{251}{270}(p_n - c_n)$ $c_{n+1} = y_{n-1} + (h/24)[9f(x_{n+1},m_{n+1}) + 19 y_n' - 5 y_{n-1}' + y_{n-2}']$ $y_{n+1} = c_{n+1} + \tfrac{19}{270}(p_{n+1} - c_{n+1})$	$251/720\, h^5 y^{(5)}$ $-19/720\, h^5 y^{(5)}$
Milne method: $p_{n+1} = y_{n-3} + (4h/3)(2 y_n' - y_{n-1}' + 2 y_{n-2}')$ $m_{n+1} = p_{n+1} - \tfrac{28}{29}(p_n - c_n)$ $c_{n+1} = y_{n-1} + (h/3)[f(x_{n+1},m_{n+1}) + 4 y_n' + y_{n-1}']$ $y_{n+1} = c_{n+1} + \tfrac{1}{29}(p_{n+1} - c_{n+1})$ $p_n - c_n$ is used to control truncation error. The interval is modified when $p_n - c_n$ changes significantly	$28/90\, h^5 y^{(5)}$ $-1/90\, h^5 y^{(5)}$ See Hamming[21] for discussion of error growth

The principal disadvantages of predictor-corrector methods are the difficulty involved in obtaining starting values for the initial computation and reduction of step size. The advantages of the methods are the small number of evaluations of the functions and the convenient estimate of truncation errors.

Some predictor-corrector combinations are unstable. That is, the propagation of truncation and rounding errors grows rapidly even when the step size is small. These methods should not be used for numerical solutions over large intervals unless the solution grows more rapidly than the error propagates. When the solution grows more rapidly than the errors the method is said to be *relatively stable*.

The Adams-Bashford predictor-corrector pair is the most popular method. This method is stable if h is small enough. The rounding error and truncation error are larger than those for the Milne predictor-corrector pair, but in the Milne method the propagated errors oscillate and grow when $\partial f/\partial y < 0$.

Runge-Kutta and Other One-step Methods. In the numerical solution of $\mathbf{y}' = \mathbf{f}(x,y)$ by one-step methods, the approximate solution $\mathbf{y}_{n+1} \doteq \mathbf{y}(x_{n+1})$ at

$x_{n+1} = x_n + h$ is calculated from x_n and y_n only. Starting procedures are not required and the interval size may be changed easily. One-step methods are also frequently used to obtain starting values for the predictor-corrector methods. One-step methods are usually based on Taylor series evaluated at points in the interval (x_n, x_{n+1}).

One of the simplest one-step methods is

$$\mathbf{y}_{n+1} = \mathbf{y}_n + (h/2)[\mathbf{f}(x_n, \mathbf{y}_n) + \mathbf{f}(x_n + h, \mathbf{y}_n + h\mathbf{f}_n)] \tag{40-46}$$

for which the local truncation error is $h^3(f''/12 - f_y f'/4)$ in the case where f and y are not vectors. This method is obtained from Taylor-series expansion at x_n and x_{n+1}. The numerical approximation for the solution of the differential equation is obtained by evaluating (40-46) once for each integration step. This involves two evaluations of $\mathbf{f}(x,y)$ for each step. Thus if $\mathbf{f}(x,y)$ is complicated, the amount of computation work involved is about the same as that for the predictor-corrector method with two evaluations of the corrector (40-45).

The most popular single-step method is the Runge-Kutta fourth-order method in Table 40-6. This method involves four evaluations of $\mathbf{f}(x,y)$ and is thus comparable in computation time with the Adams-Bashford fourth-order method with an average of four evaluations of the corrector.

A disadvantage of the one-step methods is the amount of computation required, which becomes important when $\mathbf{f}(x,y)$ involves a substantial number of arithmetic operations. This disadvantage is balanced by the convenience in changing interval size, in the simplicity of starting the solution, and in the behavior of error growth.

Collatz[6] and Henrici[22] describe several one-step methods for numerical integration of differential equations including methods for equations of special form such as $y'' = f(x,y)$, $y''' = f(x,y)$.

Approximation Errors. The error in the numerical solution of differential equations is composed of the *local truncation errors* due to replacement of derivatives by differences, the *local rounding errors* due to computations using a finite number of digits, and the *propagated errors*, that is, the errors in subsequent steps due to the effect of local errors in each integration step. In simple cases the accumulated error at $x = b$ can be expressed as

$$E(b) \doteq (\epsilon/h)u + h^p v \tag{40-47}$$

where u and v are slowly varying functions of h and b. The first term represents contribution of the local rounding errors ϵ and increases as the step size h decreases. The second term represents the contribution of the truncation errors and decreases as h decreases. Equation (40-47) indicates that an optimal interval size exists. For smaller step sizes the increase in rounding-error contribution offsets the decrease in truncation-error contribution. As described by Henrici[22] and F. John,[30] knowledge of the form such as (40-47) of the total error can be used to estimate the total error. Assume that the computations are performed with a large enough number of digits that the contribution of the rounding errors is negligible and assume further that v does not change appreciably as a function of h. Then the difference of two approximations for $y(b)$, using two different interval sizes can be used to determine the approximate size of contribution of the truncation errors. This is called Richardson's extrapolation method; see Henrici.[22]

It is known that some numerical-integration schemes have small local truncation errors but propagate errors with an undesirable rate of increase. For example, the multistep method

$$y_{n+1} = 4y_n - 3y_{n-1} - 2hf(x_{n-1}, y_{n-1}) \tag{40-48}$$

has a small local truncation error, but an error increases by a factor of about 3 in each integration step. In this case, as h approaches zero, the solution to (40-48) does not converge to the solution of $y' = f(x,y)$. Method (40-48) is very unstable. In using numerical methods for integration of differential equations it is desirable to use well-tested methods and to estimate the errors.

Table 40-6. One-step Methods

$$y' = f(x,y) \qquad x_{n+1} = x_n + h \qquad f_n \equiv f(x_n, y_n) \qquad f_y \equiv \partial f / \partial y$$

Formulas	Local truncation error	Comments
Taylor expansion: $y_{n+1} = y_n + hf_n + (h^2/2)f_n' + \cdots + (h^p/p!)f_n^{(p-1)}$	$[h^{p+1}/(p+1)!]f^{(p)}$	The high-order derivatives may become complicated expressions to evaluate
Improved Euler method: $y_{n+1} = y_n + (h/2)[f_n + f(x_n + h, y_n + hf_n)]$	$h_3(\tfrac{1}{12}f'' - \tfrac{1}{4}f_y f')$	One method of controlling error is to obtain two approximations to the solution by numerical integration with two increment sizes. Another method is to reduce the interval size when the second difference $y_{n+2} - 2y_{n+1} + y_n$ increases
Modified Euler method: $y_{n+1} = y_n + hf[x_n + h/2, y_n + (h/2)f_n]$	$h^3(-\tfrac{1}{24}f'' - \tfrac{1}{8}f_y f')$	See improved Euler method
Runge-Kutta fourth-order method: $\mathbf{k}_1 = hf_n$ $\mathbf{k}_2 = hf(x_n + h/2, y_n + \tfrac{1}{2}\mathbf{k}_1)$ $\mathbf{k}_3 = hf(x_n + h/2, y_n + \tfrac{1}{2}\mathbf{k}_2)$ $\mathbf{k}_4 = hf(x_n + h, y_n + \mathbf{k}_3)$ $y_{n+1} = y_n + \tfrac{1}{6}(\mathbf{k}_1 + 2\mathbf{k}_2 + 2\mathbf{k}_3 + \mathbf{k}_4)$	$h^5\{\tfrac{1}{2.800}f^{(iv)} - \tfrac{1}{5\,76}f_y f''' \\ + \tfrac{1}{2\,88}(f_y{}^2 - f_{xy} - f_{yy}f)f'' \\ + \tfrac{1}{1\,92}(2f_{xy}f_y + 3f_{yy}f_y f \\ - 2f_y{}^3 + f_{yy}f_x)f'\}$	One method of controlling error is to obtain two approximations to the solution by numerical integration with two interval sizes. Another method is to reduce the step size when the fourth difference $y_{n+1} - 4y_n + 6y_{n-1} - 4y_{n-2} + y_{n-3}$ increases. See Collatz[6] and Henrici[22] for estimates of the total error

Liniger[38] indicates that propagation errors due to truncation do not increase when the Runge-Kutta methods described above are used to integrate the equation $y' = A(x)y$ if $q = hA$ satisfies the following conditions:
1. For the improved Euler and modified Euler methods: $-2 \leq q \leq 0$.
2. For the Runge-Kutta method: $-2.785 \leq q \leq 0$.

40-6. Least-Squares Approximation

The Principle of Least Squares. This section describes the use of the principle of least squares for obtaining estimates for a function $Y(x)$ for which approximate values are known. If each member of a set of N measured values $y_i = y(x_i)$ is obtained from a normal distribution, with standard deviation σ_i centered about some "true" values Y_i, then the probability that the "true value" at X_i lies between Y_i and $Y_i + dY_i$ is

$$P(Y_i)\, dY_i = (\sigma_i \sqrt{2\pi})^{-1} \exp\left[-\frac{(Y_i - y_i)^2}{2\sigma_i^2}\right] dY_i \qquad i = 1, 2, \ldots, N$$

The most probable guess of the function Y is then determined when

$$\chi^2 \equiv \sum_{i=1}^{N} \frac{(Y_i - y_i)^2}{\sigma_i^2} = \text{minimum} \qquad (40\text{-}49)$$

Equation (40-49) is Legendre's principle of least squares.

A more complete discussion of least-squares approximation is given in Whittaker and Robinson[60] and Hildebrand.[23]

Statement of the Problem. It is assumed that the "true" values Y_i lie on a curve of the form

$$Y(x) = \sum_{k=0}^{n} a_k \varphi_k(x) \qquad n < N - 1 \qquad (40\text{-}50)$$

The functions φ_k are completely arbitrary and are chosen for the requirements of the problem. For example, they may be the kth powers of x, or the kth polynomial of a set of polynomials. The a_k are assumed to be determined by imposing condition (40-49) in the form

$$\partial \chi^2 / \partial a_k = 2 \sum_{i=1}^{N} w_i \left[\sum_{j=0}^{n} a_j \varphi_j(x_i) - y_i \right] \varphi_k(x_i) = 0 \qquad k = 0, 1, 2, \ldots, n \qquad (40\text{-}51)$$

Since χ^2 is a quadratic polynomial in the a_i, its derivatives are zero at the minimum. Equations (40-51), slightly rearranged, are known as the *normal equations* for the a_k:

$$\sum_{j=0}^{n} a_j \left[\sum_{i=1}^{N} w_i \varphi_j(x_i) \varphi_k(x_i) \right] = \sum_{i=1}^{N} w_i y_i \varphi_k(x_i) \qquad k = 0, 1, \ldots, n \qquad (40\text{-}52)$$

The *weights* w_i are estimates of the reciprocals of the squares of the standard deviations

$$w_i \doteq (1/\sigma_i^2) \qquad i = 1, 2, \ldots, N$$

The normal equations are often written in matrix form

$$\mathbf{Ha} = \mathbf{b} \qquad (40\text{-}53)$$

where the elements of $\mathbf{H}$ and $\mathbf{b}$ are

$$H_{jk} = H_{kj} = \sum_{i=1}^{N} w_i \varphi_k(x_i) \varphi_j(x_i) \qquad j, k = 0, 1, \ldots, n$$

$$b_k = \sum_{i=1}^{N} w_i y_i \varphi_k(x_i) \qquad k = 0, 1, \ldots, n \qquad (40\text{-}54)$$

and **a** is the column vector whose elements are a_k. The solution to (40-53) is

$$\mathbf{a} = \mathbf{H}^{-1}\mathbf{b}$$

Thus the problem resolves into one of inverting the matrix H. If the φ_k are chosen to be orthogonal functions with respect to the weight w, then H is diagonal and, therefore, its inversion is trivial. In this case, the coefficients a_k of (40-50) are, for the discrete case,

$$a_k = (1/\gamma_k) \sum_{i=1}^{N} y(x_i)\varphi_k(x_i)w(x_i)$$

where

$$\gamma_k = \sum_{i=1}^{N} \varphi_k^2(x_i)w(x_i) \qquad k = 0, 1, \ldots, n$$

or, for the continuous case,

$$a_k = (1/\gamma_k) \int_a^b y(x)\varphi_k(x)w(x)\,dx$$

where

$$\gamma_k = \int_a^b \varphi_k^2(x)w(x)\,dx$$

depending on whether the φ_k are defined over the discrete set $1(1)N$ or over the continuous interval (a,b). Examples of orthogonal polynomials and the associated values of γ_k are listed in Table 40-7.

Table 40-7. Orthogonal Polynomials

Polynomial	(a,b)	$w(x)$	Recursion relation	γ_r
Legendre	$(-1,1)$	1	$P_{r+1}(x) = \dfrac{2r+1}{r+1}xP_r(x) - \dfrac{r}{r+1}P_{r-1}(x)$ $P_0(x) = 1,\ P_1(x) = x$	$\dfrac{2}{2r+1}$
Laguerre	$(0,\infty)$	$e^{-\alpha x}$	$L_{r+1}(\alpha x) = (1 + 2r - \alpha x)L_r(\alpha x) - r^2 L_{r-1}(\alpha x)$ $L_0(\alpha x) = 1,\ L_1(\alpha x) = 1 - \alpha x$	$(r!)^2/\alpha$
Hermite	$(-\infty,\infty)$	$e^{-\alpha^2 x^2}$	$H_{r+1}(\alpha x) = 2\alpha x H_r(\alpha x) - 2r H_{r-1}(\alpha x)$ $H_0(\alpha x) = 1,\ H_1(\alpha x) = 2\alpha x$	$(2^r r!/\alpha)\sqrt{\pi}$
Chebyshev	$(-1,1)$	$(1-x^2)^{-\frac{1}{2}}$	$T_{r+1}(x) = 2xT_r(x) - T_{r-1}(x)$ $T_0(x) = 1,\ T_1(x) = x$	$\pi\ (r=0)$ $(\pi/2)(r \neq 0)$
Gram	$(1,N)$	1	$\phi_r(n) = \sum_{k=0}^{r}(-1)^{k+r}\dfrac{(k+r)^{(2k)}(n-1)^{(k)}}{(k!)^2(N-1)^{(k)}}$ $n = 1, 2, \ldots, N;\ r = 0, 1, \ldots, N-1$ where $s^{(t)} \equiv s(s-1)\cdots(s-t+1)$ $s^{(0)} \equiv 1$ for $s \geq 0$ $0^{(t)} \equiv 0$ for $t > 0$	$\dfrac{1}{2r+1}\dfrac{(N+r)^{(r+1)}}{(N-1)^{(r)}}$

Errors in the Coefficients. It was assumed that the data could be fitted with a curve of the form (40-50), and the coefficients a_k were determined from the normal equations (40-52) or (40-53). The question thus arises as to how accurately the a_k are determined.[23] Assume perfect measurements could be made without error and the resulting data could be represented by

$$F(x) = \sum_{k=0}^{n} A_k \varphi_k(x) \qquad (40\text{-}55)$$

40-26 MATHEMATICS ASSOCIATED WITH SYSTEM ENGINEERING

In this case, a definition of the uncertainty in the a_k can be made:

$$(a_k - A_k)^2 \equiv (\Delta a_k)^2 \equiv \frac{(H^{-1})_{kk}\chi^2}{N - n - 1} \tag{40-56}$$

where $(H^{-1})_{kk}$ is the kth diagonal component of the inverse of H [see (40-54)], χ^2 is defined in (40-49), N is the number of experimental points, and $n + 1$ is the number of parameters available in (40-50). The approximations used in obtaining (40-56) tend to make $(\Delta a_k)^2$ a conservative estimate of the actual error.

Error at a Given Point. The square of the error at a point x is defined as

$$[\Delta Y(x)]^2 = [F(x) - Y(x)]^2$$

where $F(x)$ is defined in (40-55) and $Y(x)$ in (40-50). It is possible to make a conservative estimate of this error (after averaging over the errors introduced by the a_k):

$$[\Delta Y]_{av}^2 \equiv \frac{\chi^2}{N - n - 1} \sum_{k=0}^{n} \sum_{l=0}^{n} (H^{-1})_{lk} \varphi_k(x) \varphi_l(x)$$

Average Error over an Interval. Over the continuous interval (a,b) the average error is

$$[\Delta Y]^2 = \frac{1}{b - a} \int_a^b [\Delta Y(x)]^2 \, dx$$

$$= \frac{1}{(b - a)(N - n - 1)} \sum_{i=1}^{N} w_i[y_i - Y(x_i)]^2 \sum_{\substack{k=0 \\ l=0}}^{n} (H^{-1})_{lk} \int_a^b \varphi_k(x) \varphi_l(x) \, dx$$

which when the $\varphi_k(x)$ are polynomials becomes

$$[\Delta^2(x)]_{av} = \frac{1}{(b - a)(N - n - 1)} \sum_{i=1}^{N} w_i[y_i - Y(x_i)]^2 \sum_{\substack{k=0 \\ l=0}}^{n} \frac{(H^{-1})_{lk}}{l + k + 1} (x_b^{l+k+1} - x_a^{l+k+1})$$

Orthogonal Polynomials. Orthogonal polynomials have the property

$$\int_a^b \varphi_k(x) \varphi_j(x) w(x) \, dx = \delta_{kj} \gamma_k$$

or, in the discrete case

$$\sum_{i=1}^{N} \varphi_k(x_i) \varphi_j(x_i) w(x_i) = \delta_{kj} \gamma_k \qquad k, j = 0, 1, \ldots, n$$

The Kronecker delta δ_{kj} is unity if $k = j$ and zero otherwise. Some common orthogonal polynomials are shown in Table 40-7.

40-7. Minima and Maxima of Functions

Minimization and maximization procedures are basic to system engineering and are discussed in numerous parts of this handbook, notably in Chapters 23 through 30.

This section describes methods for locating extreme values of a real-valued function of one or more real variables. Changing the sign of a function interchanges the roles of the extreme values. For example, the minima of $f(x_1, x_2, \ldots, x_n)$ are the maxima of $-f(x_1, x_2, \ldots, x_n)$. Thus it is sufficient to describe methods for finding the minima of functions.

A function which is continuous assumes its minima in, or on the boundary of, the region in which it is defined. If the function is linear, the minimum value is assumed on the boundary of the region. For linear functions the most efficient procedures for determining the minima are those of linear programming; see Chapters 25 and 26.

If the function $f(x_1, x_2, \ldots, x_n)$ has a continuous first derivative with respect to each variable, these partial derivatives are zero at the minima which occur within the region. A point at which the derivative in every direction is zero may be an absolute maximum point, a local maximum point, an absolute minimum point, a local minimum point, or a saddle point. If the function has continuous second derivatives and the first derivative with respect to each variable is zero at a point and the matrix of elements $a_{ij} = \partial^2 f / \partial x_i \, \partial x_j$ is positive definite at the point, then the point is a minimum point. The first derivatives of f may not be zero at minima points that are on the boundary of the region.

In the location of minima of a nonlinear function of general type, approximate locations of the minima are usually determined first. These approximate locations may be determined from a knowledge of the mathematical properties of the function, from a rough search of the region in which the function is defined, or by arbitrary choice. Accurate locations for the minima are then determined by successive approximation procedures such as the *gradient methods* (including the method of *steepest descent*), *relaxation methods*, the use of interpolation functions, direct-search methods, and the method using steps in random directions.

Let $(x_1{}^{(k)}, x_2{}^{(k)}, \ldots, x_n{}^{(k)})$ be the coordinates of the kth approximation to the location of a minimum point. The coordinates of the $(k + 1)$st approximation to the location of the minimum point may then be represented as

$$x_i{}^{(k+1)} = x_i{}^{(k)} + z_i{}^{(k)} \qquad i = 1, 2, \ldots, n$$

In *relaxation methods*, each step is in a direction parallel to a coordinate axis x_i. Thus in determining the location of the next approximation, only one coordinate changes. In the simplest of these relaxation procedures, the successive directions are the directions of the coordinate axes taken in cyclic order. The new value of the only coordinate that changes, say x_j, is the minimum of f in that direction. In the Southwell relaxation, the derivatives $\partial f / \partial x_i$ are evaluated at the approximation. The coordinate that is changed to obtain the next approximation is the coordinate x_j for which $\partial f / \partial x_j$ is largest. The value for $z_j{}^{(k)} = (-\partial f / \partial x_j)/(\partial^2 f / \partial x_j{}^2)$ and all other $z_i{}^{(k)} = 0$. Variants of these methods are described in the expository review by Spang.[54]

In the method of *steepest descent*, the step to the next approximation for the minimum is in the direction of the maximum rate of change of the function, i.e., in the direction of the gradient. The step size is determined by $\bar{\lambda}$, the smallest positive λ for which the function of a single variable

$$g(\lambda) = f(a_1 - \lambda b_1, a_2 - \lambda b_2, \ldots, a_n - \lambda b_n) \qquad (40\text{-}57)$$

assumes its first minimum. In (40-57), the a_i are the coordinates $x_i{}^{(k)}$ of the most recent approximate location for the minimum of the function and the b_i are the directional derivatives $\partial f / \partial x_i$ evaluated at that point. The step sizes are then

$$z_i{}^{(k)} = -\bar{\lambda} b_i \qquad (40\text{-}58)$$

The derivatives b_i may be difficult to compute. Finite differences or other interpolation schemes may then be used to estimate the derivatives; however, some analysts have indicated that direct-search procedures are preferable in this case; see Spang.[54]

Other gradient methods may be obtained by multiplying the components $\partial f / \partial x$ of the gradient vector by the elements of a positive-definite matrix. In the Newton method the step sizes are

$$z_i{}^{(k)} = -\sum_{j=1}^{n} d_{ij}(\partial f / \partial x_j)$$

where d_{ij} are the elements of the inverse of the matrix whose elements are $\partial^2 f/\partial x_i \, \partial x_j$. This method converges very rapidly near the minimum when the matrix d_{ij} is positive-definite.

An interesting summary of the use of sequential decision processes for locating the extreme values of complicated functions and functions with random measurement errors is contained in an article by M. Flood.[39]

40-8. Monte Carlo Methods (Calculations Involving Random Variables)

Problems of system design may involve quantities whose values are determined by chance. Examples of such quantities include random loads, electronic noise, and measurement errors. These variables may be a number, a set of numbers, vector, or function. One of the methods for numerically treating problems involving random variables is called the *Monte Carlo method* by numerical analysts or *statistical sampling method* by statisticians, and is discussed in Sec. 32-3.

Monte Carlo methods involve statistical sampling procedures and are experimental in nature. In addition to computer simulation of systems, Monte Carlo methods have been applied to completely deterministic problems, such as the inversion of matrices, solution of partial differential equations, location of extrema, and numerical integration. In these cases the deterministic problem is replaced by a probability model such that the expected value of the answer to the probability model is the solution of the deterministic problem.

In a Monte Carlo computation, statistical results are obtained by repetitive sampling procedures. The probability that the results of the Monte Carlo computation are within a given interval of the theoretical results is a function of the size of the sample.

Distributions of Random Variables. The heart of a Monte Carlo computation lies in the random selection of numbers from a given probability distribution. In practical computations these numbers are obtained from printed tables or from computer operations that produce pseudorandom numbers with the properties expected from numbers obtained by random selection. Tables of uniformly distributed numbers may be obtained in books like the Rand table[50] or others.[19,57] Several computer procedures have been developed for producing very long sequences of suitable pseudorandom numbers.

One of the best-known and most efficient computer methods of generating a sequence of uniformly distributed pseudorandom numbers r_i with a desk calculator or decimal computer involves only simple multiplication. Start with

$$r_1 = 0.00403\ 53607$$
$$r_{i+1} = \text{F.P.} (403\ 53607 \cdot r_i)$$

where F.P. (x) denotes the fractional part of x. The numbers r_i repeat after a cycle of exactly 50 million, that is, $r_{50,000,001} = r_1$. Subsequences of the r_i appear to be uniformly distributed on the interval $0 < r < 1$. See Eq. (32-12) and Table 32-3.

Sequences of numbers with other than uniform distributions can be obtained by transforming the numbers obtained from a uniform distribution. Table 40-8 summarizes methods for obtaining such distributions.

Numerical Integration. Estimation of the value of a definite integral provides an example of a Monte Carlo computation and the interpretation of the results. In practice Monte Carlo computations have frequently been used to estimate the value of integrals in six or more dimensions. In fewer dimensions it is usually substantially more economical to use numerical-integration formulas. The example considered here is the Monte Carlo estimation of the definite integral

$$I = \int_0^1 \int_0^1 \cdots \int_0^1 F(x_1, x_2, \ldots, x_n) \, dx_1 \, dx_2, \ldots, dx_n$$

It is assumed that the integrals of F and F^2 are finite. The integrand is evaluated for

Table 40-8. Methods of Obtaining Nonuniform Distributions of Random Numbers from Uniform Distributions

x is a number from the desired frequency distribution
r is a number from a uniform frequency distribution on the interval $0 \leq r \leq 1$

Desired frequency distribution	Transformation of uniform distribution
General distribution $f(x) \qquad a \leq x \leq b$	$x = F^{-1}(r)$ where $F(y) = \int_a^y f(x)\,dx$ and F^{-1} is the inverse function to F (inverse of cumulative-distribution method)
General distribution $f(x) \qquad 0 < x < 1$	Choose two new random numbers r_1 and r_2 from a uniform distribution. If $r_1 < [f(r_2)/m]$, where $m \geq \max f(x)$ on $0 \leq x \leq 1$, then $x = r_2$. If not, then repeat the above process of selection and test (rejection method)
Uniform distribution; $a \leq x \leq b$	$x = (b - a)r + a$
Exponential distribution $e^{-\alpha x} \qquad 0 \leq x$	$x = -\tfrac{1}{2} \log_e r$ or $x = -\tfrac{1}{2} \log_e (1 - r)$
Normal distribution $(\sigma \sqrt{2\pi})^{-1} e^{-(x-m)^2/(2\sigma^2)} \qquad -\infty \leq x \leq \infty$	Two numbers from a normal distribution are generated from two numbers for a uniform distribution by $x_1 = \sigma(-2 \log_e r_1)^{1/2} \cos (2\pi r_2) + m$ $x_2 = \sigma(-2 \log_e r_2)^{1/2} \sin (2\pi r_1) + m$
Approximation for normal distribution $f(x) \doteq (\sigma \sqrt{2\pi})^{-1} e^{-x^2/(2\sigma^2)}$ $-\sigma \sqrt{3N} \leq x \leq \sigma \sqrt{3N}$	$x = \sigma \sqrt{12N} \left[(1/N) \sum_{i=1}^{N} (r_i - \tfrac{1}{2}) \right]$ N is usually chosen as 10 or 12
$1 - \|x\| \qquad -1 < x < 1$	$x = r_1 + r_2 - 1$ or $x = r_1 - r_2$
$\tfrac{1}{2}(\tfrac{3}{2} - \|x\|)^2;\ \tfrac{1}{2} \leq \|x\| \leq \tfrac{3}{2}$ and $\tfrac{1}{2}(\tfrac{3}{2} - 2x^2);\ -\tfrac{1}{2} \leq x \leq \tfrac{1}{2}$	$x = r_1 + r_2 + r_3 - \tfrac{3}{2}$ or $x = r_1 + r_2 - r_3 - \tfrac{1}{2}$
Distribution for $x = \cos \pi r$ and $y = \sin \pi r$ $(2/\pi)(1 - x^2)^{-1/2} \qquad -1 \leq x \leq 1$ $(4/\pi)(1 - y^2)^{-1/2} \qquad 0 \leq y \leq 1$	Method 1: Choose two new random numbers r_1 and r_2. Define $S = r_1^2 + r_2^2$. If $S \geq 1$ or $S = 0$ start over. Otherwise $x = (r_1^2 - r_2^2)/S$, $y = 2r_1 r_2/S$ Method 2: $\qquad x = \sin \pi r,\ y = \cos \pi r$ This method requires evaluation of trigonometric functions
Chi-square distribution with $2n$ degrees of freedom $[1/2^n \Gamma(n)] e^{-\tfrac{1}{2}x} x^{n-1} \qquad 0 \leq x \leq \infty$	$x = -2 \sum_{i=1}^{n} \log_e r_i$

a random sample of N points. The estimate for the value of the integral is

$$I \doteq \bar{F} = (1/N) \sum_{i=1}^{N} F_i$$

where F_i denotes the evaluation of the function at the ith point selected at random, with uniform probability, in the region of integration, e.g., each coordinate chosen independently from a uniform distribution in the interval (0,1). An estimate of the variance of the sample values F is

$$\sigma^2(F) = [1/(N-1)] \left(-N\bar{F}^2 + \sum_{i=1}^{N} F_i^2 \right)$$

An estimate for the variance of the estimate $\bar{F}$ for the integral is

$$\sigma^2(\bar{F}) = (1/N)\sigma^2(F)$$

For large samples the probability is about 0.7 that Monte Carlo estimate $\bar{F}$ will be within $\pm \sqrt{(1/N)\sigma^2(F)}$ of the theoretical mean I, and about 0.95 that the estimate $\bar{F}$ will be within $\pm 2\sqrt{(1/N)\sigma^2(F)}$ of I. Reduction of the size of the confidence interval for the result by a factor a requires an increase in the number of sample points by a factor of a^2.

Discussion. The efficiency of statistical estimation procedures for deterministic problems and for simulations may be substantially increased by techniques which exploit qualitative information of the problem. H. Kahn summarizes six of these techniques in the book edited by Meyer.[42] Other references containing general information on Monte Carlo methods include the book on Monte Carlo methods for problems in nuclear physics by Cashwell and Everett;[4] and the collections of papers in the books edited by Alder, Fernbach, and Rotenberg;[1] by Beckenbach;[2] by Householder, Forsythe, and Germond,[26] and by Meyer.[42] Hull and Dobell[27] have developed a list of 149 references for random-number generation.

REFERENCES

1. Alder, B., S. Fernbach, and M. Rotenberg (eds.): *Methods in Computational Physics*, vol. 1, Academic Press Inc., New York, 1963.
2. Beckenbach, E. F. (ed.): *Modern Mathematics for the Engineer, Second Series*, McGraw-Hill Book Company, New York, 1961.
3. Buckingham, R. A.: *Numerical Methods*, Pitman Publishing Corporation, New York, 1957.
4. Cashwell, E. D., and C. J. Everett: *A Practical Manual on the Monte Carlo Method for Random Walk Problems*, Pergamon Press, New York, 1959.
5. Chase, P. E.: Stability of Predictor-Corrector Methods for Ordinary Differential Equations, *J. Assoc. Comput. Mach.*, vol. 9, 1962.
6. Collatz, L.: *The Numerical Treatment of Differential Equations*, 3d ed., Springer-Verlag OHG, Berlin, 1960.
7. *Communications of the Association for Computing Machinery*, vol. 1– , 1958–
8. *Computing Reviews*, Association for Computing Machinery, vol. 1– , 1960–
9. Dwyer, P. S.: *Linear Computations*, John Wiley & Sons, Inc., New York, 1951.
10. Faddeeva, D. K., and V. N. Faddeeva: *Computational Methods of Linear Algebra*, W. H. Freeman and Company, San Francisco, 1963.
11. Fletcher, A., J. Miller, L. Rosenhead, and L. Comrie: *An Index of Mathematical Tables*, Addison-Wesley Publishing Company, Inc., Reading, Mass., 1962.
12. Forsythe, G. E., and P. C. Rosenbloom: *Numerical Analysis and Partial Differential Equations*, John Wiley & Sons, Inc., New York, 1958.
13. Forsythe, G. E., and W. R. Wasow: *Finite-difference Methods for Partial Differential Equations*, John Wiley & Sons, Inc., New York, 1960.
14. Fort, T.: *Finite Differences and Difference Equations in the Real Domain*, Oxford University Press, Fair Lawn, N.J., 1948.

15. Fox, L.: *Practical Solution of Linear Equations and Inversion of Matrices*, Applied Mathematics Series, National Bureau of Standards, no. 39, Washington, D.C., 1954.
16. Fox, L.: *The Numerical Solution of Two-point Boundary Problems in Ordinary Differential Equations*, Oxford University Press, Fair Lawn, N.J., 1957.
17. Fox, L.: *Numerical Solution of Ordinary and Partial Differential Equations*, Pergamon Press, New York, 1962.
18. Greenwood, J. A., and H. O. Hartley: *Guide to Tables in Mathematical Statistics*, Princeton University Press, Princeton, N.J., 1962.
19. Hald, A.: *Statistical Tables and Formulas*, John Wiley & Sons, Inc., New York, 1952.
20. Hammersley, J. M.: Monte Carlo Methods for Solving Multivariable Problems, *Ann. N.Y. Acad. Sci.*, vol. 86, pp. 844–874, 1960.
21. Hamming, R. W.: *Numerical Methods for Scientists and Engineers*, McGraw-Hill Book Company, New York, 1962.
22. Henrici, P.: *Discrete Variable Methods in Ordinary Differential Equations*, John Wiley & Sons, Inc., New York, 1962.
23. Hildebrand, F. B.: *Introduction to Numerical Analysis*, McGraw-Hill Book Company, New York, 1956.
24. Householder, A. S.: Bibliography on Numerical Analysis, *J. Assoc. Comput. Mach.*, vol. 3, no. 2, pp. 85–100, 1956.
25. Householder, A. S.: *Principles of Numerical Analysis*, McGraw-Hill Book Company, New York, 1953.
26. Householder, A. S., G. E. Forsythe, and H. H. Germond (eds.): *Monte Carlo Method*, Applied Mathematics Series, National Bureau of Standards, vol. 12, 1951.
27. Hull, T. E., and A. R. Dobell: Random Number Generators, *SIAM Rev.*, vol. 4, p. 23, 1962.
28. *International Dictionary of Applied Mathematics*, D. Van Nostrand Company, Inc., Princeton, N.J., 1960.
29. James, G., and R. C. James (eds.): *Mathematics Dictionary*, D. Van Nostrand Company, Inc., Princeton, N.J., 1959.
30. John, Fritz: *Advanced Numerical Analysis*, New York University, Institute of Mathematical Sciences, 1956.
31. Jordan, Charles: *Calculus of Finite Differences*, 2d ed., Chelsea Publishing Company, New York, 1950.
32. Kamke, E.: *Differentialgleichungen Lösungsmethoden und Lösungen*, Chelsea Publishing Company, New York, 1959.
33. Kendall, M. G., and W. R. Buckland: *A Dictionary of Statistical Terms*, 2d ed., Hafner Publishing Company, Inc., New York, 1960.
34. Kopal, Z.: *Numerical Analysis*, 2d ed., John Wiley & Sons, Inc., 1961.
35. Krylov, V. I.: *Approximate Calculation of Integrals*, The Macmillan Company, New York, 1962.
36. Kunz, K. S.: *Numerical Analysis*, McGraw-Hill Book Company, New York, 1957.
37. Lanczos, C.: *Applied Analysis*, Prentice-Hall, Inc., Englewood Cliffs, N.J., 1956.
38. Liniger, W.: *Zur Stabilität der numerischen Integrationsmethoden für Differentialgleichungen*, L. Speich, Zürich, 1957.
39. Machol, R. E. (ed.): *Information and Decision Processes*, McGraw-Hill Book Company, New York, 1960.
40. *Mathematical Reviews*, American Mathematical Society, Providence, R.I., vol. 1– , 1940–
41. *Mathematics of Computation* (formerly *Mathematical Tables and Other Aids to Computation*), The National Research Council, vol. 1, 1943.
42. Meyer, H. A. (ed.): *Symposium on Monte Carlo Methods*, John Wiley & Sons, Inc., New York, 1956.
43. Milne, W. E.: *Numerical Solution of Differential Equations*, John Wiley & Sons, Inc., New York, 1953.
44. *Modern Computing Methods*, National Physical Laboratory, Her Majesty's Stationery Office, Teddington, England, 1961.
45. Moors, B. P.: *Valeur approximative d'une intégral définie*, Gauthier-Villars, Paris, 1905.
46. Murphy, G. M.: *Ordinary Differential Equations and Their Solutions*, D. Van Nostrand Company, Inc., Princeton, N.J., 1960.
47. *Numerische Mathematik*, Springer-Verlag OHG, Berlin, vol. 1– , 1959–
48. Ostrowski, A. M.: *Solution of Equations and Systems of Equations*, Academic Press, Inc., New York, 1960.
49. Ralston, A., and H. S. Wilf (eds.): *Mathematical Methods for Digital Computers*, John Wiley & Sons, Inc., New York, 1960.

50. Rand Corporation: *A Million Random Digits with 100,000 Normal Deviates*, The Free Press of Glencoe, New York, 1955.
51. Salzer, H. E.: Coefficients for Facilitating Trigonometric Interpolation, *J. Math. Phys.*, vol. 27, p. 274, 1948.
52. Salzer, H. E.: New Formulas for Facilitating Osculatory Interpolation, *J. Res. Natl. Bur. Std.*, vol. 52, pp. 211–216, 1954.
53. Scarborough, J. B.: *Numerical Mathematical Analysis*, 5th ed., The Johns Hopkins Press, Baltimore, 1962.
54. Spang, H. A.: A Review of Minimization Techniques for Nonlinear Functions, *SIAM Rev.*, vol. 4, no. 4, 1962.
55. Steffensen, J. F.: *Interpolation*, 2d ed., Chelsea Publishing Company, New York, 1950.
56. *Tables of Lagrangian Interpolation Coefficients*, National Bureau of Standards, Columbia Press Series, vol. 4, Columbia University Press, New York, 1944.
57. Tippett, L. H. C.: Random Sampling Numbers, *Tracts for Computers*, no. 15, Cambridge University Press, New York, 1927.
58. Todd, J. (ed.): *Survey of Numerical Analysis*, McGraw-Hill Book Company, New York, 1962.
59. Varga, R. S.: *Matrix Iterative Analysis*, Prentice-Hall, Inc., Englewood Cliffs, N.J., 1962.
60. Whittaker, E. T., and W. Robinson: *The Calculus of Observations*, Blackie & Son, Ltd., London, 1948.
61. Wilf, H. S.: *Mathematics for the Physical Sciences*, John Wiley & Sons, Inc., New York, 1962.
62. Wilkinson, J. H.: *The Algebraic Eigenvalue Problem*, Oxford University Press, Fair Lawn, N.J., 1963.
63. Abramowitz, M., and I. A. Stegun: *Handbook of Mathematical Functions with Formulas, Graphs and Mathematical Tables*, U.S. Government Printing Office, Washington, D.C., 1964.
64. Gelfond, A. O.: *Calculus of Finite Differences*, Dunod, Paris, 1963.
65. Henrici, P.: *Elements of Numerical Analysis*, John Wiley & Sons, Inc., New York, 1964.
66. Householder, A. S.: *Theory of Matrices in Numerical Analysis*, Blaisdell Publishing Company, New York, 1964.
67. *Journal of the Society for Industrial and Applied Mathematics*, Series B, 1964–
68. Stiefel, E. L.: *An Introduction to Numerical Mathematics*, Academic Press, New York, 1963.
69. Wilkinson, J. H.: *Rounding Errors in Algebraic Processes*, Prentice-Hall, Inc., Englewood Cliffs, N.J., 1963.

Chapter 41

THE PROPOSITIONAL CALCULUS*

ARTHUR W. BURKS, *Professor of Philosophy, University of Michigan.*
RICHARD A. LAING, *Logic of Computers Group, University of Michigan.*

CONTENTS

41-1. The Propositional Calculus................................ 41-1
41-2. Truth Tables... 41-2
41-3. Disjunctive Normal Form................................ 41-3
41-4. Boolean Algebra.. 41-4
41-5. Testing for Validity..................................... 41-4
41-6. Duality.. 41-5
41-7. Using Logic to Design Circuits........................... 41-6
41-8. Minimality for a Single Function......................... 41-8
41-9. Higher Logics.. 41-9

41-1. The Propositional Calculus

The propositional calculus is the two-valued logic of the connectives "either . . . or," "and," "not," etc. The calculus contains the following symbols: *propositional variables* p, q, r, s, etc.; the *truth values* "true" (or 1 or t) and "false" (or 0 or f); and the *logical connectives*, of which the most common are shown in Table 41-1. A more elementary presentation of the related "algebra of events" is given in Sec. 38-2.

These logical connectives are to some extent interconvertible; for instance, employing just negation and disjunction, or just negation and conjunction, or the nan function alone, all the remaining connectives can be expressed.

Well-formed formulas of the propositional calculus may be constructed by compounding the symbols of the calculus in an intuitively meaningful way, e.g., "$p \vee q$" (p or q). Complex combinations of the basic symbols may require the use of parentheses as punctuation, but extra parentheses are usually omitted when no confusion will result. Also the convention that negation is weaker (has a smaller scope of influence) than a conjunction, and conjunction weaker than disjunction allows the dropping of additional parentheses; e.g., "$p \vee \sim qr$" means $[p \vee ((\sim q)r)]$ rather than

---

* This chapter is a revision of materials on logic and computers in H. Huskey and G. A. Korn (eds.), *Computer Handbook*, pp. 14-13 to 14-23, McGraw-Hill Book Company, New York, 1961.

$[p \lor \sim(q \cdot r)]$ or $[(p \lor \sim q)r]$. This is analogous to the familiar algebraic convention that $a + b \times c$ means $a + (b \times c)$ and not $(a + b) \times c$.

The propositional calculus has applications in the analysis of complex assertions in both natural and formal languages. It has proved useful in the description and analysis of complicated systems (although often in its Boolean counterpart: see Sec. 41-4), and has been found particularly valuable in digital-computer-system design and description, because the fundamental logical and switching circuits used in digital computers are usually devices (1) whose input and output wires have two significant states (e.g., a high and a low voltage) which can be associated with the values "true" and "false" of the calculus, and (2) whose output is a logical function of the inputs (e.g., an output of a component might be "high" just in case *either* the one input *or* the other input is high). The synthesis of complex digital-computer circuits is greatly facilitated by this association of computer element wires with propositional variables, electrical states of wires with truth and falsehood, and the logical functions realized by computer elements with the logical connectives of the propositional calculus. This application of the calculus may be illustrated with the idealized computer conjunction element of Fig. 41-1. Each input wire is labeled with a variable which is interpreted to mean that the labeled wire is activated (high); e.g., p is true if the wire labeled p

Table 41-1. Logical Connectives

Negation.............................	Not p	$\bar{p}, \sim p, p'$
Conjunction...........................	p and q	$p \cdot q, pq, p \,\&\, q, p \land q$
Inclusive disjunction..................	p or q (or both)	$p \lor q, p + q$
Conditional implication................	If p then q	$p \supset q, p \to q$
Biconditional equivalence..............	p just in case q	$p \equiv q, p = q$
Exclusive disjunction, inequivalence...	p or q but not both	$p \not\equiv q$
Nan...................................	Not both p and q	$p \,\underline{\&}\, q$
Nor...................................	Neither p nor q	$p \,\underline{\lor}\, q$

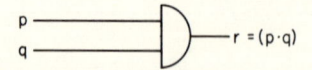

Fig. 41-1. Computer element for $r = (p \cdot q)$.

is high; otherwise p is false. The output wire is labeled with the corresponding logical function $(p \cdot q)$ or with another variable (r) which is set equal to this function. This is the *normal interpretation;* the *dual interpretation,* in which "true" is associated with a low voltage, may also be used, but confusion can result from mixing the two interpretations.

41-2. Truth Tables

A formula of the propositional calculus which is true for all possible values of the variables is called a *valid formula* or *tautology*.

A formula may be tested for validity by evaluating it for all possible values of its variables; $p \lor \bar{p}$ is thus valid (a tautology) since for all possible consistent substitutions of truth values for its variables, the resulting expression is true (i.e., $0 \lor \bar{0}$ and $1 \lor \bar{1}$ are both true). Thus the formula at the head of the rightmost column of Table 41-2 is a tautology; under every possible consistent substitution of truth values for the two variables of the formula, the formula is true. The four possible combinations of truth values that the two variables may take are listed beneath p and q at the left of the table, and the values these combinations produce in the whole formula are shown (at the far right) always to be 1. The intervening columns record the systematic analysis of the pertinent constituents of the whole formula.

THE PROPOSITIONAL CALCULUS   41–3

Note that the two columns headed by the variables p and q contain all four different combinations of 2 bits; hence in general the truth table for an n-variable formula will contain 2^n rows. Each row is headed by a specific n-bit binary number giving the particular assignment of truth values for the variables for that row. This assignment of all possible combinations of truth values to the variables can also be represented by a conjunction of all the variables where the variables are negated or not negated according to the value they are to take for the row to be tested; see the leftmost column of the truth table above. Such conjunctions of all the variables of a formula are called *basic conjunctions*.

Table 41-2. Truth Table for $(\overline{pq}) \equiv (\bar{p} \vee \bar{q})$

	p	q	$p \cdot q$	$(\overline{pq})$	$\bar{p}$	$\bar{q}$	$\bar{p} \vee \bar{q}$	$(\overline{pq}) \equiv (\bar{p} \vee \bar{q})$
$\bar{p}\bar{q}$	0	0	0	1	1	1	1	1
$\bar{p}q$	0	1	0	1	1	0	1	1
$p\bar{q}$	1	0	0	1	0	1	1	1
pq	1	1	1	0	0	0	0	1

41-3. Disjunctive Normal Form

A formula which is false for all possible values of the variables is called a contradiction; e.g., $p \cdot \bar{p}$. It is clear from an inspection of a truth table that every logical function which is not a contradiction is equivalent to a disjunction of those basic conjunctions of its variables for which it is true; thus the formula evaluated in Table 41-2 is equivalent to $\bar{p}\bar{q} \vee \bar{p}q \vee p\bar{q} \vee pq$ (or, as we have said, it is true under all combinations of values for the variables). For a further example, $\overline{pq}$ is equivalent to $\bar{p}\bar{q} \vee \bar{p}q \vee p\bar{q}$; in this case the formula is "contingent" since it is neither tautologous nor contradictory, and its value will vary with the assignment of values to its variables [it will in fact have the value 0 only for the case where both p and q are 1, which is of course precisely the conjunct which is missing from the disjunction (of correct conjuncts)]. Such a disjunction of all the basic conjuncts of the variables for which the formula is true is called the "*expanded*" *disjunctive normal form* of the formula.

The disjunctive-normal-form expression of a tautology of n variables will contain all 2^n basic conjunctions of those n variables, and so one method of testing for tautology-hood is to expand a formula into disjunctive normal form to see if the expansion contains all the basic conjunctions.

For a specific and systematically applied ordering of the basic conjunctions, a disjunctive normal form may be expressed more briefly by writing the corresponding column of the truth table; e.g., $\overline{pq}$ may be expressed by the binary number 1110 (in Table 41-2, see the column under $\overline{pq}$). For a particular ordering of the variables, the disjunctive normal form may be expressed arithmetically by using the binary numbers which head the rows of the truth table opposite the corresponding basic conjunctions which are true; e.g., $\overline{pq}$ is expressed by 00 $\vee$ 01 $\vee$ 10.

The following is an important consequence of the fact that every logical function may be expressed in disjunctive normal form. Suppose it is desired to realize an arbitrary logical expression A and that there are available wires representing not only the variables of A but also the negations of these variables; the latter would be true, for example, if the desired circuit were driven by symmetrical flip-flops, or if the conjunction elements have inhibitory as well as activating inputs. Under these circumstances any logical function A may in principle be exemplified in a circuit consisting of conjunction elements (one for each basic conjunction in the disjunctive normal form expansion of A) driving a disjunctive element (with as many inputs as there are basic conjunctions in the disjunctive normal form expression of A). It should be noted that such a circuit has a *depth* of two elements; i.e., there are two logical elements between each input and the output. Circuits constructed by the use of addi-

41-4 MATHEMATICS ASSOCIATED WITH SYSTEM ENGINEERING

tional logical elements and with a depth greater than two may, of course, be more economical.

41-4. Boolean Algebra

The propositional calculus is one interpretation of Boolean algebra, and certain useful associations can be made. The propositions of the calculus can be viewed as the elements of the algebra, and the product, sum, and complement can be interpreted as symbolizing conjunction, disjunction, and negation. If the "equals" of the algebra is interpreted as symbolizing "equivalence" in the logic, then all the axioms and theorems of Boolean algebra become logically true propositions of the calculus.

As seen in Sec. 41-3, all the expressions of propositional calculus may be put in the disjunctive normal form, which employs only "and," "or," and "not." For each expression in normal form, equivalent Boolean expressions can be obtained by interpreting conjunctions, disjunctions, and negations as product, sum, and complement, or by actually replacing the logical connectives with more familiar symbols for the Boolean operations (such as $\cap$, $\cup$, $^-$, or $\cdot$, $+$, $-$). For those whose background makes Boolean algebra the more familiar or natural system, the particular associations between the propositional calculus and Boolean algebra are given in Table 41-3.

Table 41-3. Associations between Boolean Algebra and Propositional Calculus[*]

Boolean	Logical
$p \cap q$	$p \cdot q$
$p \cup q$	$p \vee q$
$\bar{p}$	$\bar{p}$
$\bar{p} \cup q$	$p \supset q$
$(\bar{p} \cup q) \cap (p \cup \bar{q})$	$p \equiv q$
$(p \cap \bar{q}) \cup (\bar{p} \cap q)$	$p \not\equiv q$
$\bar{p} \cup \bar{q}$	$p \mathbin{\&} q$
$\bar{p} \cap \bar{q}$	$p \triangledown q$

[*] For a more detailed and complete exposition of the relationships between Boolean algebra and propositional calculus see The Algebra of Classes, Appendix A, especially pp. 328–329 of Ref. 1 or Chapter XI, Algebra of Classes, especially Sec. 4, Application to Logic, in Ref. 2. For relations between the axiom schemes of the two systems see Ref. 3.

41-5. Testing for Validity

It has already been mentioned that one can test a formula for validity by constructing a truth table for that formula. This procedure is purely routine and hence can readily be mechanized. This can be done by connecting up a circuit which realizes the formula under investigation, and testing the output of this circuit for all possible inputs; or by programming the problem on a general-purpose digital computer; or by means of a logic machine which stores the formula and determines its truth value for every possible assignment of truth values to the variables.

The truth-table method of testing validity is tedious and impractical for expressions with a large number of variables; e.g., for six variables, 2^6 or 64 rows are needed. A procedure which relies on ingenuity is generally quicker for the human being (as contrasted to a machine). Simple formulas may be tested intuitively, and a list of simple tautologies such as those in Table 41-4 is easily remembered after a little practice. With a working stock of simple tautologies one can establish other tautologies by using the following principles:

Interchange of Equivalents. If $A \equiv B$ is a tautology, and D results from C by replacing A by B in C, then $C \equiv D$ is a tautology; e.g., by the use of DeMorgan's theorem we can establish that $[p \cdot \overline{(qr)}] = [p \cdot (\bar{q} \vee \bar{r})]$.

Principle of Substitution. The result of substituting any well-formed formula for a variable of a tautology is a tautology; e.g., the result of substituting $r \cdot s$ for p in the first form of the law of excluded middle is the tautology $r \cdot s \vee \overline{(rs)}$.

THE PROPOSITIONAL CALCULUS

Modus Ponens. If $A \supset B$ is a tautology, and A is a tautology, then B is a tautology; e.g., $p \supset (q \supset p)$ is a tautology and so is $r \vee \bar{r}$, and by substituting the latter for p and using modus ponens we derive the tautology $q \supset (r \vee \bar{r})$.

For a fuller treatment of these methods see any standard textbook in symbolic logic, e.g., Ref. 1. For a more elementary treatment, see Ref. 4.

Table 41-4. Tautologies

Double negation.............. $p \equiv \bar{\bar{p}}$
Excluded middle.............. $p \vee \bar{p} \qquad (p \vee \bar{p}) \equiv 1$
Contradiction................. $(p\bar{p}) \equiv 0$
Commutation.................. $(p \cdot q) \equiv (q \cdot p)$
$\qquad (p \vee q) \equiv (q \vee p)$
Association.................. $[p \cdot (qr)] \equiv [(pq) \cdot r]$
$\qquad [p \vee (q \vee r)] \equiv [(p \vee q) \vee r]$
Distribution................. $[p \cdot (q \vee r)] \equiv [pq \vee pr]$
$\qquad [p \vee qr] \equiv [(p \vee q) \cdot (p \vee r)]$
DeMorgan's theorem........... $\overline{(pq)} \equiv (\bar{p} \vee \bar{q})$
$\qquad \overline{(p \vee q)} \equiv (\bar{p}\bar{q})$
Absorption................... $(p \vee 1) \equiv 1 \qquad (p \cdot 1) \equiv p$
$\qquad (p \vee 0) \equiv p \qquad (p \cdot 0) \equiv 0$
$\qquad (p \vee p) \equiv p \qquad (p \cdot p) \equiv p$
$\qquad p \equiv (p \vee pq) \qquad p \equiv p(p \vee q)$
$\qquad (pq \vee \bar{p}r) \equiv (pq \vee \bar{p}r \vee qr)$
Simplification............... $(p \vee q) \equiv (p \vee \bar{p}q)$
$\qquad (pq) \equiv [p \cdot (\bar{p} \vee q)]$
Multiplication............... $[(p \vee q) \cdot (r \vee s)] \equiv (pr \vee ps \vee qr \vee qs)$

41-6. Duality

Another principle useful for constructing tautologies is a generalization of DeMorgan's theorem called "duality." The dual A^d of a formula A is formed by interchanging the following symbols in pairs:

$$\begin{array}{cc} 0 & 1 \\ p & \sim p \\ \vee & \cdot \\ \equiv & \not\equiv \\ \triangledown & \& \end{array}$$

The principle of duality states that the following are tautologies:

$$\bar{A} \equiv A^d \qquad A \equiv \bar{A}^d$$

Hence if $A \equiv B$ is a tautology so is $A^d \equiv B^d$, and if A is a tautology so is $\bar{A}^d$.

It is worth noting that a shift from the *normal* to the *dual* interpretation of the propositional calculus constitutes an interchange of each variable with its dual, and hence results in an interchange of dual connectives. Thus applying duality to

$$r \text{ is high} \equiv (s \text{ is high } or \ t \text{ is high})$$

we get

$$r \text{ is low} \equiv (s \text{ is low } and \ t \text{ is low})$$

since $\sim(p \not\equiv q) \equiv (p \equiv q)$ is a tautology. This sometimes is expressed by saying that a "positive-or" is a "negative-and."

The disjunctive normal form expression $\bar{A}$ of a formula A is a disjunction of the basic conjunctions for which A is false. The dual $\bar{A}^d$ is a conjunction of disjunctions and states the conditions under which A is true; it is called the (*expanded*) *conjunctive normal form* of A. Thus if A is $\bar{p}q \vee p\bar{q}$, $\bar{A}$ is $\bar{p}\bar{q} \vee pq$, $\bar{A}^d$ (and hence the conjunctive normal form of A) is $(p \vee q) \cdot (\bar{p} \vee \bar{q})$. The conjunctive normal form has the same utility as the disjunctive normal form.

41-7. Using Logic to Design Circuits

A common design problem is that of constructing a circuit to realize a given formula. For example, in a binary-system adding circuit, the sum digit s_i and carry-out digit c_{i-1} are certain functions of the addend digit a_i, augend digit b_i, and carry-in digit c_i:

$$s_i = (a_i \not\equiv b_i \not\equiv c_i) \tag{41-1}$$
$$c_{i-1} = (a_i b_i \lor a_i c_i \lor b_i c_i) \tag{41-2}$$

and the problem may be to construct a net to perform these operations. The expression $a \not\equiv b \not\equiv c$ is defined to mean either $(a \not\equiv b) \not\equiv c$ or $a \not\equiv (b \not\equiv c)$, since these are always equivalent. If there are circuit building blocks available corresponding to the logical connectives used in the formulas, then the solution consists of a straightforward interconnection of these elements according to the pattern of the formulas; see Fig. 41-2 for a net realizing c_{i-1}, where the rightmost element is a disjunction (buffer) circuit.

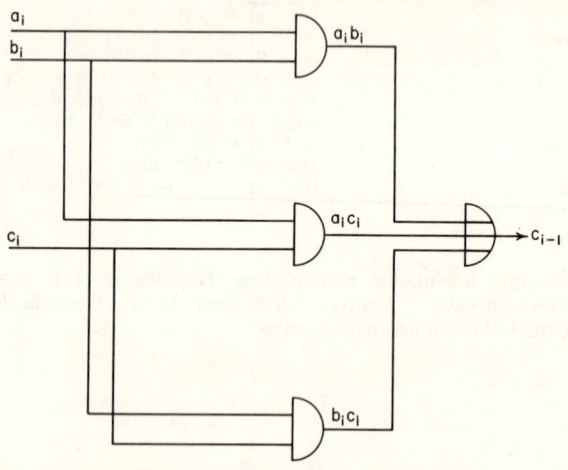

FIG. 41-2. Net realizing carry-out digit in addition.

There may not, however, be circuit components which directly realize the connectives used in the formula. Thus we may wish to construct a circuit for s_i out of conjunction and disjunction elements, where the conjunction elements may have inhibiting inputs (represented by small circles; see Fig. 41-3) as well as activating inputs. In this case the given formula is transformed by any of the aforementioned techniques into an equivalent one containing only the allowed connectives. Thus it may be shown that the disjunctive normal form of s_i is

$$a_i \bar{b}_i \bar{c}_i \lor \bar{a}_i b_i \bar{c}_i \lor \bar{a}_i \bar{b}_i c_i \lor a_i b_i c_i \tag{41-3}$$

either by the use of the tautology

$$(p \not\equiv q) \equiv (p\bar{q} \lor \bar{p}q) \tag{41-4}$$

together with the tautologies previously cited, or by proof of the fact that $p_1 \not\equiv p_2 \not\equiv p_3 \ldots \not\equiv p_n$ is true just in case an odd number of the variables are true.

Whenever there is a one-to-one correspondence between logical connectives and available circuit elements it is easy to pass from formula to circuit and vice versa. Indeed, formula (41-3) and Fig. 41-3 say essentially the same thing, one in logical

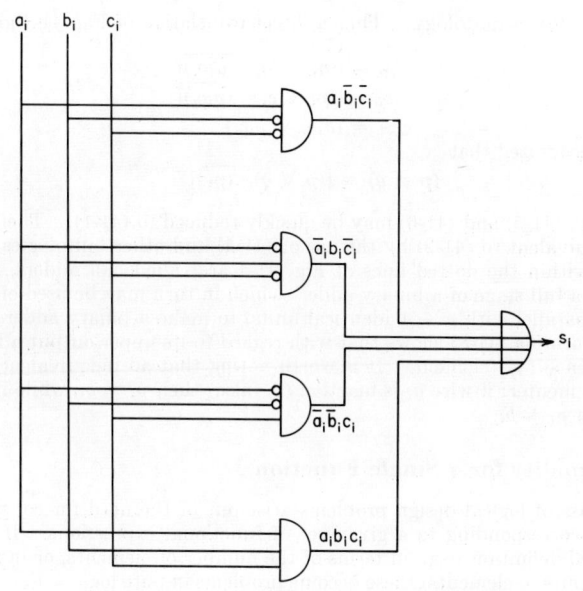

FIG. 41-3. Net realizing sum digit in addition.

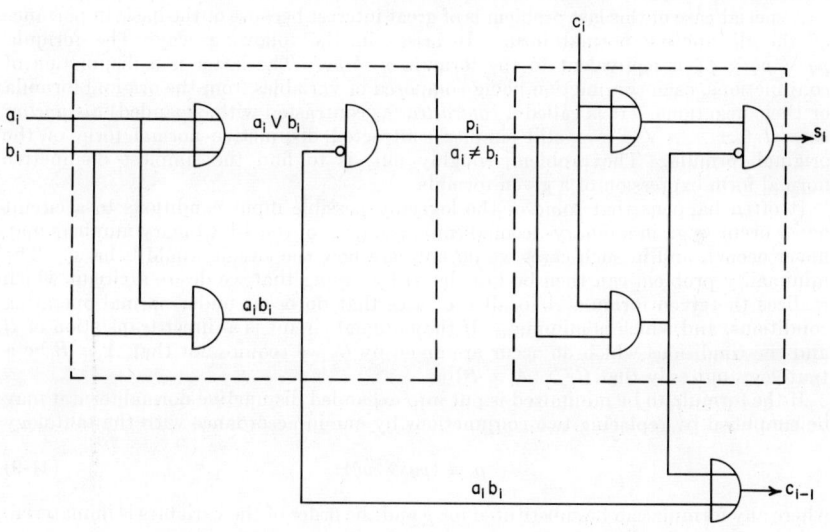

FIG. 41-4. Net realizing sum and carry digits in addition.

symbols, the other diagrammatically. We shall speak of such transformations from logical formulas to diagrams as *direct translations*.

Another problem, this time of analysis, is the following: Given a net, does it realize a specified formula; e.g., does Fig. 41-4 realize s_i and c_{i-1}? A related problem is that of determining whether two nets realize the same formula. Because of the ease with which direct translation between formulas and net diagrams may be made, both these problems reduce to the question of whether two formulas A and B are equivalent, i.e.,

41-8 MATHEMATICS ASSOCIATED WITH SYSTEM ENGINEERING

whether $A \equiv B$ is a tautology. Thus a direct translation of Fig. 41-4 gives

$$p_i = [(a_i \vee b_i) \cdot \overline{(a_i b_i)}] \tag{41-5}$$
$$s_i = [(p_i \vee c_i) \cdot \overline{(p_i c_i)}] \tag{41-6}$$
$$c_{i-1} = (a_i b_i \vee p_i c_i) \tag{41-7}$$

When it is recognized that

$$(p \not\equiv q) \equiv [(p \vee q) \cdot \overline{(pq)}] \tag{41-8}$$

is a tautology, (41-5) and (41-6) may be quickly reduced to (41-1). Then (41-7) may be shown equivalent to (41-2) by the use of (41-4) and other tautologies.

The nets within the dotted lines of Fig. 41-4 are called half adders, since two in tandem give a full stage of a binary adder (which in turn may be used either sequentially, or in parallel with $n - 1$ identical units, to make a binary adder for an n-bit number). Formula (41-8) shows that with regard to its upper output a half adder is an inequivalence ($\not\equiv$) element. It is worth noting that an inequivalence element is also a complementer: if wire a_i is inactive (a_i false) then $p_i \equiv b_i$, while if a_i is active (a_i true) then $p_i \equiv \bar{b}_i$.

41-8. Minimality for a Single Function

A large class of logical design problems arise out of the need for constructing the simplest net corresponding to a given set of functional expressions. If "simple" is given a logical definition (e.g., in terms of the number of elements, or in terms of the number of inputs to elements) these become problems in pure logic. For example, the problem of constructing a minimal net realizing the formula A is the logical problem of determining which of the formulas equivalent to A has this minimal property.

A special case of this last problem is of great interest because of the basic importance of the disjunctive normal form. It arises in the following way: The formula $pq \vee p\bar{q} \vee \bar{p}\bar{q}$ is equivalent to the formula $p \vee \bar{p}\bar{q}$. The latter is a disjunction of conjunctions, each conjunction being composed of variables from the original formula or their negations. It is called a *contracted* (as contrasted with expanded) *disjunctive normal form*. $p \vee \bar{q}$ is a still simpler contracted disjunctive normal form of the original formula. The problem, roughly put, is to find the simplest disjunctive normal form expression of a given formula.

It often happens that some of the logically possible input conditions to a circuit never occur (e.g., in a binary-decimal conversion, six of the 4-bit binary numbers may never occur), and in such cases we do not care how the circuit would behave. The minimality problem can then be broadened by saying that we desire a circuit which realizes the given formula A in all the cases that do occur under normal operating conditions, and which is minimal. If the minimal circuit is a direct translation of B and the conditions which do occur are given by C, we require not that $A \equiv B$ be a tautology but only that $C \supset (A \equiv B)$ is.

If the formula to be minimized is put into expanded disjunctive normal form it may be simplified by replacing two conjunctions by one in accordance with the tautology

$$p \equiv (pq \vee p\bar{q}) \tag{41-9}$$

where any formula can be substituted for p and the order of the variables is immaterial, e.g.,

$$abd \equiv (abcd \vee ab\bar{c}d)$$

Similarly it may be possible to replace four, eight, sixteen, . . . conjunctions by use of the tautologies

$$p \equiv (pqr \vee pq\bar{r} \vee p\bar{q}r \vee p\bar{q}\bar{r})$$
$$p \equiv (pqrs \vee pqr\bar{s} \vee pq\bar{r}s \vee p\bar{q}rs \vee pq\bar{r}\bar{s}$$
$$\vee p\bar{q}\bar{r}s \vee p\bar{q}r\bar{s} \vee p\bar{q}\bar{r}\bar{s}) \quad \text{etc.}$$

respectively, but clearly these reductions can also be accomplished by repeated applications of (41-9). A common method of solving this particular minimality problem is systematically to test for all possible simplifications of this kind, and then to make a minimal selection from the conjunctions that result. A nonoccurring case (or conjunction) may be used to find a simpler conjunction, but it need not be represented in the final disjunction. There are many ways of systematizing this simplification process, some of which use truth tables.[5,6,7] The method illustrated next combines features of several of these.

The expression to be minimized is the disjunction of
(1) $\bar{a}bc$
(2) $ab\bar{c}$
(3) $\bar{a}\bar{b}c$

with the following cases never occurring:
(4) abc
(5) $\bar{a}\bar{b}\bar{c}$

Starting with (1) we compare it successively with the terms following it until we can use the tautology $p \equiv (pq \lor p\bar{q})$. Combining (1) with (3) and (4), respectively, we get
(6) $\bar{a}c$ From (1) and (3)
(7) bc From (1) and (4)

A check is placed opposite (1) and (3) to show that they will not be in the minimal formula. (The nonoccurring cases will not be in the minimal formula; so they are checked at the outset.) Warning: Comparisons of (1) with the other formulas cannot be stopped after (6) is found because (1) may be useful again, as it was in finding (7). Applying the above procedure to (2) we get
(8) ab From (2) and (4)

and so we check (2). (3) and (5) give
(9) $\bar{a}\bar{b}$ From (3) and (5)

(4) and (5) give no further simplification, nor do (6), (7), (8), or (9).

The desired minimal formula is a simplest disjunction of some or all of the unchecked formulas [these are (6), (7), (8), and (9) in our example] which is equivalent to the original formula [in our example this is the disjunction of (1), (2), and (3)]. If a basic conjunction of the original formula is unchecked [this case does not occur in our example] it will appear in the final formula, while if it is checked some unchecked formula derived from it must appear. Since (6) was derived from (1) and (3) while (8) was derived from (2), $a\bar{c} \lor ab$ is the minimal formula in our example.

If the formula to be minimized is given in contracted disjunctive normal form it must be transformed into its expanded disjunctive normal form before the foregoing technique is applicable. Quine has devised a method in which this expansion is unnecessary.[8] See also Refs. 9 and 10. It is a useful method but not always superior to the one just illustrated because in some cases the easiest way to find which of the unchecked formulas is to be used in the minimal expression involves expanding the given expression into expanded disjunctive normal form. As formulated, Quine's method is not applicable to problems with nonoccurring cases, but it may be extended to this type of problem by combining it with the method illustrated above (with some modifications of each).

41-9. Higher Logics

Frequently in system description, synthesis, or analysis, timed action of components must explicitly be taken into account. For the description of such systems, propositional calculus (or its underlying mathematical structure, Boolean algebra) is inadequate. In the applications so far discussed, it was assumed that information flow in the system as well as all the concomitant switching took place during a single interval of time. To enable us to deal explicitly with systems with memory and time delay, we must augment the propositional calculus. We can do this by adding to our calculus a symbol for moments of time [e.g., (t)] which will be affixed to our variables [e.g., $p(t)$] and which will take as values the nonnegative integers. We shall say that, for example, $p(t)$ is true just in case the wire labeled p is active at time t. In order to represent the presence of timed action diagrammatically we shall require an additional element, the delay element, shown in Fig. 41-5.

The use of time variables and the delay element in expressing a system which models an abstracted behavioral property of the human nervous system is shown in Fig. 41-6. This circuit behaves somewhat like a certain small portion of the human nervous system. It expresses the familiar phenomenon which takes place when a cold object (a small piece of ice, for instance) is pressed against the skin. If the cold object is held against the skin for only an instant, the object feels as if it were hot. If, however, the cold object remains in contact with the skin for more than an instant, the nervous system registers correctly that the object is cold. (In Fig. 41-6, s indicates the stimulus line from the object, and h and c indicate, respectively, the hot and cold signals registered in the central nervous system.)

FIG. 41-5. The delay element, symbolizing "whenever $p(t)$ is true, then $q(t + 1)$ is true." (q can be considered false at time zero.)

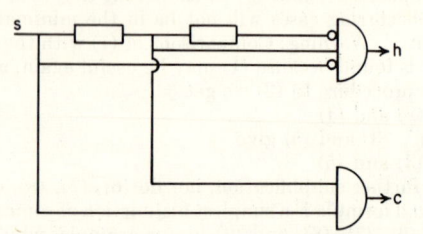

FIG. 41-6. Net utilizing delay element. Behavior conditions:

$$h(t + 2) = \overline{s(t)} \cdot s(t + 1) \cdot \overline{s(t + 2)}$$
$$c(t + 1) = s(t) \cdot s(t + 1)$$

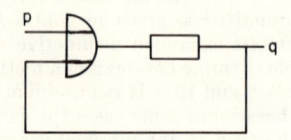

FIG. 41-7. Feedback element.

The use of symbols for moments of time provides a convenient means of expressing not only delayed timed action but feedback (or memory) as well, as exemplified in Fig. 41-7.

Our logical calculus may be further augmented by *quantifiers* ($\forall$, $\exists$). $\forall t$ and $\exists t$ may be read, respectively, as "for every time t" and "there is a time t." The symbol $<$ ("less than") may also be added. The behavior of the logical network of Fig. 41-7 can then be expressed by the formula

$$q(t) = (\exists x)[p(x) \cdot (x < t)]$$

which means q is energized at time t, just in case there was a time x, earlier than t, at which p was activated.

Higher logics possessing not only the above properties but also containing various additional relations and arithmetic operations have been employed in the description and design of complex systems. For many of these logics there is no systematic method (such as the truth table) for determining valid expressions. Indeed, for some of these logics it has been shown that there can exist no general decision procedure for telling whether an arbitrary expression in the logical language is true or false. The systematic application of such logics to the design of systems poses difficult problems, the overcoming of which is the object of much contemporary research.

For the use of logics beyond the propositional calculus in (especially digital-circuit) system design, see Refs. 11 and 12.

REFERENCES

1. Copi, Irving M.: *Symbolic Logic*, The Macmillan Company, New York, 1959.
2. Birkhoff, Garrett, and Saunders MacLane: *A Survey of Modern Algebra*, rev. ed., The Macmillan Company, New York, 1953.
3. Leblanc, Hugues: Boolean Algebra and the Propositional Calculus, *Mind*, vol. 71, no. 283, pp. 383–386, July, 1962.
4. Quine, W. V.: *Methods of Logic*, rev. ed., Holt, Rinehart and Winston, Inc., New York, 1959.
5. The Staff of the Harvard University Computation Laboratory: *Synthesis of Electronic Computing and Control Circuits*, Harvard University Press, Cambridge, Mass., 1951.
6. Veitch, E. W.: A Chart Method for Simplifying Truth Functions, *Proc. Assoc. Comput. Mach.*, 1952.
7. Quine, W. V.: The Problem of Simplifying Truth Functions, *Am. Math. Monthly*, vol. 59, pp. 521–531, 1952.
8. Quine, W. V.: A Way to Simplify Truth Functions, *Am. Math. Monthly*, vol. 62, pp. 627–631, 1955.
9. McCluskey, E. J.: Minimization of Boolean Functions, *Bell System Tech. J.*, vol. 35, no. 6, pp. 1417–1444, November, 1956.
10. Ghazala, M. J.: Irredundant Disjunctive and Conjunctive Forms of a Boolean Function, *IBM J. Res. Develop.*, vol. 1, no. 2, pp. 171–176, April, 1957.
11. Burks, A. W., and Hao Wang: The Logic of Automata, *J. Assoc. Comput. Mach.*, vol. 4, pp. 193–218, 279–297, 1957.
12. Church, Alonzo: *Application of Recursive Arithmetic to the Problem of Circuit Synthesis*, summaries, 1957, 2d ed., 1960, Summer Institute for Symbolic Logic, Institute for Defense Analysis, Princeton, N.J.

INDEX

A scope, 14-59
Absorptance, 15-3
Absorption (of infrared), 15-6
Absorptivity, spectral, 17-18
Accelerometer, 19-24
Access, asynchronous, 8-23
   balanced, 8-13
   complex, 8-12
   cyclic, 8-12
   random, 8-11, 8-16
   serial, 8-12
   uniform, 8-12
   (*See also* Storage)
Access times, 8-11, 8-12, 8-16, 8-23
Accuracy (*see* Error)
Ackeret formula, 18-13, 18-14
Acoustics, 2-5, 2-8
Active region, transistor, 9-5
Active systems, infrared, 15-2
Activity, in PERT, 36-6
Adaline, 30-16, 30-17
Adams-Bashford method, 40-20, 40-21
Adaptive control, 30-1 to 30-20
   human, 31-15 to 31-17
Adaptive system, 24-29, 24-30
Adder, 9-17, 9-18
   binary, 41-8
Addition, 10-8, 10-16
Adler tube, 14-46
Advection, 5-8
Aerodynamics, 6-12, 18-1 to 18-18
Aeronomy, 6-2
Air bearing table, 32-24
Air masses, 5-7
Aircraft, propulsion of, 20-4
Airfoil, subsonic, 18-6
   supersonic, 18-13
   transonic, 18-10
Airglow, 6-3
Aitken extrapolation, 40-6
Albedo, 5-3, 5-4
Algorithmic completion, 26-2, 26-5, 26-15, 26-26
Allocation processes, 27-5
Alphabet, 22-22
Ambiguity, 14-13, 37-7
Ambiguity function, 14-10
Ammonia, 20-28
Amplitron, 14-22 to 14-24
Analog computer (*see* Computer)
Analog memory, 11-20
Analog-to-digital conversion, 11-10, 11-11, 12-8, 32-29
AND, in the algebra of events, 38-4
   implementation of, 9-8, 9-10, 9-13, 9-14, 9-16, 9-17, 9-21, 9-24, 9-27
   in the propositional calculus, 41-2
Anomaly (orbital), 17-5
Antenna, 14-30 to 14-40
   automatic tracking, 37-24
   telemetry, 37-21

Anticyclone, 5-7
Aphelion, 7-23
Apogee, 7-23, 17-4
Apparent solar time, 7-24
Area rule, 18-11
Argon, 21-33, 21-36
Arithmetic, 8-18, 8-23, 8-25, 8-50
   (*See also* Multiplication; Addition; etc.)
Arithmetic unit, 8-6
Array antenna, 14-37
Arrival distribution, 28-14
Arsenic trisulfide, 15-14
Artificial variable, 25-7, 26-6
Ascending node, 7-23, 17-4
Aspect ratio, 18-6
   effective, 18-14
Asserting amplifier, 9-20
Associative law, 41-5
Associative memory, 8-13, 8-20
Asteroid, 7-32
ASTRAC, 11-23
Astronomical unit, 7-23
Astronomy, 7-1 to 7-36
Asymptotic equipartition property, 22-18
Atmosphere, 5-1 to 6-17
   upper, 6-1 to 6-17, 37-8
Atmospheric drag, 17-7
Atmospheric transmission, of infrared, 15-6
   of radio waves, 14-55 to 14-58
ATR, 14-29
Attack, angle of, 18-5
Attenuation, of a communications signal, 13-7, 13-8, 13-13, 13-14
   of infrared, 15-6
   of radar, 14-47, 14-56
   transmission-line, 14-63
Autobarotropic, 5-7
Automata, learning, 30-14
Automotive vehicles, propulsion of, 20-2
Autopilot, rate, 19-6
Auxiliary storage, 8-10
Avalanche, in diodes, 9-4
Average, 38-17
Average sample number, 24-13, 24-25
Axioms of algebra, 38-4
   of Boolean algebra and propositional calculus, 41-3
   of probability, 38-7

B scope, 14-60
BCD (binary-coded decimal), 8-11
Backscatter (*see* Cross section, radar)
Backward difference, 40-8
Backward-wave amplifier, 14-44
Backward-wave device, 14-22
Ballistic coefficient, 17-8
Ballistic guidance systems, 19-2, 19-14 to 19-25
Band limitation, 13-14

1

INDEX

Bandwidth, in communications systems, **12-1** to **12-10**, **13-4** to **13-8**
    critical, **31-9**
    in radar, **14-6**, **14-9**, **14-10**, **14-21**, **14-22**
    in satellite TV systems, **17-27**
    in telemetry, **37-5**
Barium fluoride, **15-14**
Baroclinic, **5-7**
Barrier potential, **21-24**, **21-25**
Base, of a transistor, **9-4**
Base-date equivalent, **35**-2
Basic variable, **25**-2
Batch arrivals, **28-24**
Bathyscaph, **2-6**
Bathythermograph, **2-6**
Battery, **17-23**, **20-32**, **20-33**
Bayes decision rule, **24-12**, **24-15**
Bayes equivocation, **24-12**
Bayes estimator, **24-27**, **24-28**
Bayes risk, **24-11**, **24-16**
Bayes rule, **30-12**
Bayes system, **24-11**
Bayes theorem, **38-11** to **38-13**, **38-33**
Beam, antenna, **14-30**
Beam-rider navigation, **19-13**
Beamwidth, **14-32**
Bernoulli distribution, **38-46**
Bessel's interpolation formula, **40-13**
Beta distribution, **38-44** to **38-45**
Biconjugate circuit, **13-16**
Binary detection, **24-14**
Binomial distribution, **38-46**
Bisection method, **40-6**
Bistatic system, **14-3**, **14-5**
Bit-configuration representation of traffic, **32-16**
Bjerknes circulation theorem, **2-11**
Blackbody, **16-3**
    in infrared, **15-3**, **15-4**
    at microwave frequencies, **14-42**
    sun considered as, **5-4**
    in thermionic emission, **21-29**
Blackout, **6-16**
Block diagram, **29-7**
Blood circulation, simulation of, **32-12**
Bolometer, **15-16**, **15-18**, **15-25**, **15-26**, **16-8**, **17-27**
Boltzmann's constant, **21-9**
Boolean algebra, **38-4ff**, **41-4**
Boolean operations, **8-19**
Bordering method, **40-5**
Bottleneck, **28-8**
Boundary layer, aerodynamic, **18-8**
Brake horsepower, **20-9**
Brake mean effective pressure, **20-10**
Branch signal, **29-8**
Brayton cycle, **20-16**, **20-17**, **20-34**
Bréguet's formula, **20-5**
Bremsstrahlung radiation, **21-39**
Brightness, **16-4**, **16-9**
Brownian motion, **28-3**
B-scan, **14-59**
Buffered input-output, **8-7**
Buffeting, **34-8**
Busy period, **28-22**
Button plasma gun, **20-35**
Bytes, **8-19**

Cadmium telluride, **15-16**
Calcium fluoride, **15-14**
Calibration, **37-4** to **37-7**
Camber line, **18-6**, **18-14**
Canard, **18-3**
Candle, **16-5**, **16-6**
Canonical form, linear program, **25-2**
Capacity, of storage, **8-11**

Carbon dioxide, **6-4**
Cards, magnetic, **8-12** to **8-16**
Car-following theories, **4-11**
Carnot cycle, **21-4**, **21-5**
Carry digit, **41-6**, **41-7**
Cascade graph, **29-13**
Cassegrain system, **14-34**
Catadioptric, **15-13**
Cauchy principal-part integral, **39-7**
Cavity, resonant, **14-22**
C-band, **14**-4
CCS, **13-20**
Ceiling (of an aircraft), **20-5**
Celestial equator, **7-23**
Celestial pole, **7-23**
Celestial sphere, **7-23**, **17-3**
Central difference, **40-8**
Central processor, **8-6**, **8-17**
Centralization, **1-8**
Centrifugal forces, **2-10**
Centrifuge, human, **32-23**
Cesium, in magnetohydrodynamics, **21-33**, **21-36**
Cesium bromide, **15-16**
Cesium iodide, **15-16**
Cesium vapor, in thermionic inversion, **21-25**
Chamber pressure, **20-26**
Channel, binary erasure, **22-13**
    binary symmetric, **22-13**
    capacity, **22-6**
    communications, **12-2**
    discrete, capacity, **22-12**
    service (see Queue theory)
Chapman-Kolmogorov equations, **28-3**
Character (data unit), **8-11**, **8-18**
Character recognition, **8-28**
    (See also Pattern recognition)
Characteristic function, **38-41**
Characteristic velocity, **20-7**, **20-25**, **20-26**
Chebyshev polynomial, **40-25**
Checkout, preflight, **37-3**
Chi-function, **14-10**
Chirp, **14-11**, **14-67**
Chi-square distribution, **32-11**, **38-44** to **38-45**, **40-29**
Chlorine trifluoride, **20-29**
Choke joint, **14-30**
Chopper stabilization, **10-5**
Chromaticity diagram, **31-6**
Chromosphere, **7-15**, **7-29**, **7-31**
Chunk of information, **31-11**
Circuits, logical, **9-1** to **9-27**
Circulation (atmosphere), **5-5**
City, **4-1** to **4-20**
Climb, rate of, **20-4**
Closed-loop system, **29-3**, **30-1**, **30-2**
Clouds, **5-7**, **5-12**
Cloud-seeding, **5-9** to **5-11**
Clutter, **14-9** to **14-18**, **14-53**, **14-54**, **15-3**
Coalition, **23-15**
Coaxitron, **14-19**, **14-20**
Code, Shannon-Fano, **22-24**
Code converter, **8-26**
Codes, uniquely decipherable, **22-24**
Coding, in communications systems, **12-2** to **12-4**, **12-8** to **12-10**, **13-6** to **13-8**
    auditory, **31-7** to **31-10**
    in human memory, **31-10**
    in information theory, **22-7**, **22-22** to **22-27**
    visual, **31-5** to **31-7**
Coefficient of drag, **6-12**
Coherence, in decision theory, **24-7**
    in radar, **14-7**, **14-70**
    (See also Detection, coherent)
Coherent estimation, **24-28**
Coherent integration, **14-7**
Cold front, **5-7**

INDEX

Collectively exhaustive events, 38-6
Collector (of a transistor), 9-4
Collision-course navigation (*see* Proportional navigation)
Collision cross section, 21-35, 21-39
Collision frequency, 6-7, 6-16
Colloid propulsion, 20-41
Coma, 7-32
Comb filter, 14-15
Comentropy, 22-9
Comet, 7-13, 7-21, 7-32, 7-33
Common base, 9-5
Common emitter, 9-5
Communication, theory of, 22-4
Communications, 2-5, 12-1 to 13-24
Commutative law, 41-5
Comparator, 11-10 to 11-12, 11-16 to 11-20, 11-23, 29-2, 29-7
Comparison, 8-18
Competition, in economics, 35-4
  theory of, 23-1 to 23-20
Competitive design, 1-6
Competitive systems, 1-4
Complementation, 38-7
Complementing amplifier, 9-20
Component, 1-7
Compounding, 35-2
Compressibility, 18-10
Computation, 10-1
Computer, 1-4
  analog, 10-1 to 11-26, 37-4
    repetitive, 10-3
    in simulation, 32-27
  digital, 8-1 to 9-27, 37-3
    compared to analog, 10-2
    in radar systems, 14-59
    in simulation, 32-27
    standby, 33-10
    system design, 41-2
    in telemetry, 37-4
  (*See also* Numerical analysis)
  (*See also* Simulation)
Conditional distribution, 38-30 to 38-33
Conditional probability, 38-9
Conduction band, 21-6, 21-10
Conflict, 23-1
Congestion, 4-2
  theory of (*see* Queue theory)
Conical scan, 14-35
Conjunctive normal form, 41-5
Connected graph, 23-2
Constant-bearing navigation, 19-4
Constant false alarm rate, 14-41
Constellation, 7-23
Constrained regressions, 26-15
Contact potential, 21-24, 21-25
Continuum flow, 6-13
Control, aerodynamic, 18-3, 18-9
  remote, 37-2
Control system, analog computer in, 11-6
  final-value, 19-1
  manual, 31-15
Control theory, 29-1 to 29-51
  dynamic programming in, 27-4 to 27-6
Control unit, 8-6
Controller, digital, simulation of, 32-10
  digital computer, 8-7, 8-17, 8-20 to 8-27, 8-30
Control-signal synthesis, 30-9
Convergence, Cauchy, 39-20
Convergence in the mean, 39-9
Convolution, 38-39, 38-42, 38-43, 39-3 to 39-4, 39-13, 39-14
Convoy, 1-9
Core, 8-13
Core storage, 8-12

Coriolis force, in meteorology, 5-3, 5-6, 5-11
  in the ocean, 2-10, 2-11
Corona, 6-4, 7-14 to 7-16, 7-31
Correlation coefficient, 22-5
Cosmic rays, 7-19
Cost/effectiveness, 1-7, 1-8
Cost form, 25-2
Cost function, 24-10
  constant, 24-27
Cost, PERT, 36-10
Costs, alternative, 35-5
Coulomb friction, 10-12
Counter, 9-24
Covariance, 38-35
Cowell's method, 17-6
Cramér-Rao inequality, 24-26
Criterion (*see* Measure)
Criterion function, 27-3
Critical frequency, 6-14
Critical path, 36-8 to 36-10
Crossed field, 14-24
Crossed-field motor, plasma, 20-36
Crossed-field tube, 14-21, 14-22, 14-24
Cross section, radar, 14-47 to 14-55, 14-61
Crosstalk, 12-5, 12-7, 12-9, 13-8
Crowbar circuit, 14-28
Crust (of the earth), 3-4
Cryostat, 15-21
Cumulative probability distribution, 38-20, 38-25, 38-35
  complementary, 38-21
Cumulonimbus clouds, 5-8
Cut-off frequency, 9-6
Cut-off region, 9-5
Cybernetics, 1-10
Cycle time, 8-12
Cyclone, 5-7
Cyclotron radiation, 21-39

D-c amplifier, 10-4 to 10-7
D region, 6-14
d'Alembert's paradox, 18-6
Damping, precession, 17-11 to 17-12
Dark adaptation, 31-6
Data manipulation, 8-5ff
Data processing, 8-5ff
  real-time, 37-3
  telemetry, 37-26
  test, 34-5
  (*See also* Information processing)
Data reduction, 37-8
Data retrieval, 8-28
Data transmission, 12-1
Debugging, 33-3
Decipherability, 22-24
Decision functions, simulation of, 32-15
Decision making, human, 31-11 to 31-14
Decision making under uncertainty, 35-8
Decision mechanisms, 14-6
Decision processes, multistage, 27-1
  sequential, 40-28
Decision rule, 24-8
Decision theory, 14-5, 24-1 to 24-31, 38-13
Declination, astronomical, 7-26, 17-3
  magnetic, 3-5
Decoder (*see* Coding)
Definition (of system), 1-12
Deflation, 35-6
Degenerate solution, 25-5
Delay, in logic, 41-10
Delay line, 14-15 to 14-18
Delay time, 9-6
Delta function (*see* Impulse function)
Delta wing, 18-14

DeMorgan's theorem, 41-5
Density function, conditional, 38-31
  joint, 38-28
  marginal, 38-31
  mixed, 38-24, 38-27
  in probability, 38-22, 38-24
Density-impulse parameter, 20-26
Dependent node, 29-9
Depletion layer, 9-3, 9-4, 21-11 to 21-14, 21-25
Derating, 33-8
Detection, in adaptive systems, 24-30
  binary, 24-6
  coherent, 24-18, 24-22, 24-23
  at a distance, 14-3
  incoherent, 24-19, 24-22
  multiple alternative, 24-6, 24-23
  of radar signals, 14-5 to 14-9
  signal, 24-6, 24-14 to 24-25
  simulation of, 11-12, 11-13
  (*See also* Range equation)
Detection-effectiveness parameter, 31-5
Detectivity, 15-20
Detectors, infrared, 15-16
  radiation, 16-7
  thermal, 16-7
Deuterium, 21-38
Diesel-cycle engine, 20-11
Differences, 40-8ff
Differential analyzer, 10-3, 11-2, 11-20
Differential equations (*see* Equations, differential)
Differentiation, 10-8
Diffraction, 16-13
Diffusion capacitance, 9-4
  region, 21-13
Digital computer (*see* Computer)
Digital-to-analog conversion, 11-10, 12-8, 32-29
Diode, in analog computers, 10-10, 10-11, 10-14, 11-7, 11-14, 11-19
  in modulator, 14-27
  properties of, 9-3 to 9-4
  redundancy of, 33-12
  in satellite power supply, 17-24, 17-25
  (*See also* Tunnel diode)
Diode amplifier, 14-44, 14-46
Diode logic, 9-8
Diode matrix, 9-8
Dip, magnetic, 3-5
Direct analog, 10-3
Discounting, 35-2
Discriminator, phase-locked, 37-25
  telemetry, 37-21
Disjunction, 41-2
Disjunctive normal form, 41-3, 41-8
Disk and wheel integrator, 10-16
Disks, magnetic, 8-12 to 8-16
Display, alphanumeric computer-operated, 36-10 to 36-11
  in optical systems, 16-12
  in radar systems, 14-59 to 14-60
  in simulator, 32-24
  visual, 31-7, 31-8
Display design, 31-7
Display quickening, 31-16
Distortion, nonlinear, 13-8, 13-13
Distribution, probability, 38-18
Distribution function, 38-20
Distributive law, 41-5
Divided difference, 40-9
Division, 10-11
Doping, 9-3 to 9-7
Drag, induced, 18-6
Drag coefficient, 5-13, 18-5ff
Drift, 10-5
Drift, spin-axis, 17-12
Drum, magnetic, 8-11 to 8-14, 8-30

Dual-slicer circuit, 11-11, 11-13
Duality, in game theory, 23-11
  in linear programming, 26-10 to 26-15
  in propositional calculus, 41-5
Dump circuit, 14-28
Duplex computer (*see* Multicomputer)
Duplex-system operation, 33-10
Duplexer, 14-29
Dynamic programming, 27-1 to 27-8, 30-9
Dynamic range, 14-41

E layer, 6-13
E region, 6-15
Earth, 7-29, 7-30
ERF (*see* Error function)
Earthquake, 2-13
Echo sounders, 2-6
Echoing area, radar, (*see* Cross section, radar)
Ecliptic, 17-3
Ecliptic plane, 7-23
Economics, 35-1 to 35-8
Efficiency, of coding, 22-23
  conversion, 21-16
  economic, 35-2
  thermionic, 21-29
  thermodynamic, 21-3 to 21-5
Electrical propulsion, 20-31
Electrochemical potential (*see* Fermi level)
Electromagnetic radiation, 7-15
  (*See also* Radar; Infrared; Optical sensors)
Electron gas, 21-6
Electrothermodynamic thrust device, 20-34
Elements, abundance of, 7-19
Elsasser model, 5-5
Emissivity, 15-3, 16-3, 16-4
  spectral, 17-18
Emissivity coefficients, thermal, 21-28 to 21-29
Emittance, 15-3, 15-4, 16-4, 16-7
Emitter (of a transistor), 9-4
Enciphering (*see* Coding)
Encke's method, 17-6
Encoding (*see* Coding)
End instruments, 13-3
Energy conversion, 21-1 to 21-42
Energy gap, 21-7
Entropy, conditional, 22-10
  of information, 22-6ff
  maximization of, 22-16
  thermodynamic, 21-4
Environment, computer, 8-5
  system, 2-1 to 7-36
Environmental chamber, 34-11
Environmental effect on reliability, 33-5
Environmental simulation, 32-33
Environmental testing, 34-6 to 34-11
Ephemeris, 7-23
Ephemeris time, 7-24
Equalization, of a signal, 13-14, 13-15
Equation of state, air, 6-5
  sea water, 2-8
Equations, differential, 40-16 to 40-23
    solution on analog computer, 11-4 to 11-7
  nonlinear, 40-5
  simultaneous algebraic, 40-2 to 40-5
Equatorial coordinate system, 7-26
Equilibrium, aerodynamic, 18-9
  temperature, 17-18
  (*See also* Stability)
Equinox, 17-3
Equivalence, biconditional, 41-2
Equivalence relation, 39-19
Equivocation, 22-11, 24-10, 24-11
Ergodic chain, 22-19
Erlang, 13-20
Erlang distribution, 13-21, 13-22, 38-44 to 38-45

INDEX

Erlang function, 28-15, 28-16, 28-20, 28-28
Error, in analog computers, 11-3
  in differentiation, 40-16
  discretization, 40-17
  in integration, 40-15, 40-19, 40-20
  in least squares, 40-25 to 40-26
  in manual systems, 31-16
  propagated, 40-22
  in radar measurement, 14-9
  rounding, 40-17
    in matrix inversion, 40-4
  truncation, 40-17 to 40-22
Error correction, 12-3, 12-10
Error function, 10-13
Error signal, in feedback systems, 29-2ff
  in radar tracking 14-35
Estimation, interval, 24-7
  parameter, 24-28
  point, 24-7
  waveform, 24-29
Estimation theory, 24-26
Ethylene oxide, 20-30
Euler method, 40-18, 40-23
Evaluation, 1-6
  economic, 35-1
Evaluation coefficient, 35-2ff
Evaluation functions, 24-9
Evapograph, 15-16
Event, in PERT, 36-6
Events of low probability, 1-7
Everett's interpolation formula, 40-13
Excluded middle, 41-5
Exosphere, 6-3, 7-2, 7-6
Expectation, 38-17 to 38-19, 38-24, 38-44
  of a linear combination, 38-35
  of multivariable function, 38-34
Expected value (*see* Expectation)
Explorer I, 17-2
Exponential distribution, 38-44 to 38-45
  convolution of, 38-39
  derivation of, 38-36 to 38-37
  entropy of, 22-16
  generation of random deviates from 32-21, 40-29
  in queue theory, 28-15ff
  (*See also* Poisson distribution)
Exponential failure law, 33-3
Exponential type, function of, 39-2
Extensive form, game, 23-1
External storage (*see* Storage, external)
Extremization, 26-2
  (*See also* Maxima)
Eye, 16-8, 16-12

F distribution, 38-44 to 38-45
F region, 6-15
Facilities, computer, 8-5
Failure, mean time to, 33-3
Fairly good function, 39-22
Fall time, 9-6
False alarm, 14-8, 14-9, 14-41, 14-62, 15-27, 24-15, 24-16, 24-22, 31-5
  (*See also* Constant false alarm rate)
False position method, 40-6
Fanjet engine, 20-24
Far field, 14-36
Faraday effect, 14-58
Feasible program, 25-1
Feedback, 29-1 to 29-51, 30-1
Feed-forward system, 29-46
Fermi-Dirac distribution, 21-6
Fermi function, 21-8 to 21-10
Fermi level, 21-7
Ferrites, 9-7

Fetch, 2-11
Fictitious play, method of, 23-12
Fidelity of simulation, 32-24
Field test, 34-1 to 34-16
Flight table, three-axis, 32-6 to 32-7
Flight trainer, 32-24
File processing, 8-46
Filter, infrared, 15-13, 17-27
  linear, 24-20, 24-24
  matched, 14-5 to 14-18, 14-40 to 14-41, 14-59
  optimum averaging, 30-4
Filtering, spatial, 15-16
Finite differences, 40-8
First-come-first-served, 28-25
Fixed-lead navigation, 19-4
Flame holder, 20-22
Flicker frequency, critical, 31-6
Flip-flop, 9-22, 9-24, 9-27, 11-23
Floating instrument platform (FLIP), 2-6
Floating point, 8-18, 8-19
Flow, aerodynamic, types of, 6-13
Flow diagram, 29-6
Flow graph, 29-7, 29-11
Fluorine, 20-28
FM/FM telemetry, 37-7, 37-10, 37-18 to 37-21
  standards for, 37-20
Focusing, of antenna, 14-34
  of electron beam, 14-21
  infrared, 15-13
Fog, 5-8
Foot-candle, 16-6
Foot-lambert, 16-6
Forbush decrease, 7-20
Forecasting, weather, 5-14, 5-15
Formulation of the problem, 1-6
Forward difference, 40-8
Fourier series, 40-14
Fourier transform, 39-7 to 39-11, 39-24
  finite, 39-12
Four-wire transmission system, 13-16
Fraunhofer zone, 14-36
Free molecule flow, 6-13
Frequency multiplexing, 37-17
Fresnel law, 15-12
Fresnel zone, 14-35
Friction velocity, 5-12
Front, in meteorology, 5-7
Frost prevention, 5-10
Fuel cell, 20-32, 20-33, 21-5
Function generator, 10-11

Gain, antenna, 14-31 to 14-33
Gain margin, 29-33
Gal, 3-6
Galactic coordinate system, 7-25, 7-26
Galactic space, 7-22
Galaxy, 7-22, 7-23, 7-34, 7-35
Game, finite, 23-2
  regression, 26-19
Game theory, 23-1 to 23-20
Games against nature, 23-17
Gamma distribution, 38-44 to 38-45
Gamma (of a sensor), 16-9
Gas turbine, 20-15 to 20-20, 20-32
Gasdynamic facilities, 34-11
Gates, 9-13
Gather-write, 8-20, 8-28
Gating (*see* Switching)
Gauss elimination method, 40-3
Gauss's interpolation formula, 40-11, 40-12
Gauss-Seidel method, 40-5
Gaussian distribution, 38-44 to 38-45
  entropy of, 22-16
  generation of random deviates from, 38-44 to 38-45

INDEX

Gaussian noise, 12-8, 22-16, 24-20, 24-28
   generation of, 32-10
Gaussian process, 24-16, 24-20
   entropy of, 22-20 to 22-22
Gaussian quadrature, 40-15
General-purpose digital computer, 8-5
Generalized function, 39-6 to 39-7, 39-19 to 39-24
Geocorona, 6-5
Geoid, 3-1ff
Geomagnetic coordinate system, 7-4
Geomagnetism (*see* Magnetic field)
Geometric distribution, 28-24, 38-46
Geophysical space, 7-2
Geopotential altitude, 6-6
Geostrophic approximation, 5-6
Germanium, 9-3, 15-16
Glass, optical, 15-14
Glauert-Prandtl rule, 18-10, 18-11
Globar, 15-4
GNP (*see* Gross national product)
Goal-programming model, 26-17
Good function, 39-21
Gradient method, 40-27
Gradient wind, 5-6
Gram polynomial, 40-25
Graph, 23-2
Graph determinant, 29-15, 29-16
Gravimeter, 3-6
Gravitation, 3-6
Gravity, differential, 17-13
Gravity, zero, 32-24, 34-9
Gravity model, 4-6
Green's function, 39-17
Grid-controlled tube, 14-19
Gross national product, 35-5, 35-6
Groundwater, 3-5
Guard bands, 12-4
Guidance, 19-1 to 19-25
Guidance-system test, 34-15 to 34-16
Guided-missile simulation, 32-29
Gutenberg-Wiechert discontinuity, 3-4

H-R diagram, 7-23, 7-32, 7-34
Hail, 5-10
Half adder, 41-8
Hall effect, 21-38
Ham-Lang system, 29-46, 29-47
Hamming code, 22-25
Hankel transform, 39-11
   finite, 39-12
Hardware, 1-4
Harmonic analysis, 2-12
Hearing, 31-7 to 31-10
Heating, aerodynamic, 18-16
Heaviside layer, 6-13
Heavy water, 21-38
Helicopter, 20-5
Hermite function, 39-21
Hermite polynomial, 39-21, 40-25
Hermite's interpolation formula, 40-9, 40-10
Hertzsprung-Russell plot (*see* H-R diagram)
Heterosphere, 6-5
Hick's law, 31-12
High-energy protons, 7-10
High-traffic design, 1-6
Hilbert transform, 39-10
Hit equation, 19-16
Holding time, telephone call, 13-20
   (*See also* Queue theory)
Hole-electron pair, 9-3
Hole gas, 21-6
Homeostat, 32-4
Homing guidance system, 19-2 to 19-14
Homopause, 6-4

Homosphere, 6-4
Horizon coordinate system, 7-25, 7-26
Hotelling method, 40-4
Hour angle, 7-26
Hubble's constant, 7-24, 7-35
Human engineering, 31-3 to 31-19
Human operator, simulation of, 32-27
Hybrid analog-digital simulation, 32-28 to 32-30
Hybrid circuit, 13-16
Hybrid computer, 11-1 to 11-26
Hybrid engine, 20-27
Hydrazine, 20-28, 20-30
Hydrodynamics, 18-3
   (*See also* Ocean)
Hydrogen, 20-28
Hydrogen peroxide, 20-29, 20-30
Hydrostatic equation, 6-5
Hydrostatic law, 5-7, 7-8
Hyperexponential distribution, 28-15, 28-16, 28-24
Hypergeometric distribution, 38-46
Hypersonic airframes, 18-16
Hypersonic tunnel, 34-11
Hypervelocity tunnel, 34-11
Hypodynamics, 34-10
Hypothesis, class, 24-7
   null, 24-7
   simple, 24-7
Hypothesis testing, 24-6

ICAO, 6-9
ICBM (*see* Ballistic guidance systems; Missile)
Ideal observer, 14-6, 24-16, 31-5
Ideal receiver, 14-6
Identification, in adaptive control, 24-30, 30-2 to 30-5
Idle period, 28-22
Idle time, 28-12
Illuminance, 16-6, 16-7
Image force, 21-22
Image sensing, 17-25
Impulse, specific, 20-6, 20-8, 20-21, 20-23 to 20-28, 20-34 to 20-37, 20-40
   total, 20-26
Impulse function, 38-24, 39-6, 39-18
Imputation, 22-14 to 22-16
Inclination, magnetic, 3-5
Incoherent (*see* Coherence; Detection, incoherent; Integration, incoherent)
Inconsistent system of equations, 26-3, 40-7
Independence, 38-9, 38-33, 38-34
Index cycle, 5-6
Index of difficulty of movements, 31-14
   of performance, 30-5
   of refraction, 6-14, 15-12
Index residue, 29-15
Indicated horsepower, 20-10
Indicated mean effective pressure, 20-10
Indirect analog, 10-3
Indium antimonide, 15-16
Inertial guidance, 19-24 to 19-25
Infeasibility form, 25-8
Inflation, 35-6
Information, average, 22-9
   mutual, 22-5
   unit of, 22-8
Information compression, human, 31-13
Information filtering, human, 31-13
Information loss, 24-10, 24-12
Information measure, 22-5 to 22-6, 22-9
Information processing (*see* Data processing)
   human 31-3 to 31-19
Information set, 23-2
Information theory, 22-1 to 22-28
Infrared, 15-1 to 15-28

INDEX

Infrared sensors, in space technology, 17-27
Inhibit, 9-20, 9-21, 11-17
  (*See also* NOT)
Input-output model, 4-14
Instructions, 8-19
Integration, analog, 10-8
  incoherent, 14-7, 14-17
  numerical, 40-14 to 40-16
Integrator, analog, 11-12
Integrity, 1-4
Intelligibility, speech, 31-9
Interest, rate of, 35-2
Interference, aerodynamic, 18-8
Interference filter, 16-7
Interlock, 8-30
Internal-combusion engine, 20-9
Internal storage (*see* Storage, internal)
International geophysical year, 3-1
Interplanetary gas, 7-13
Interplanetary missions, **20**-7
Interpolation, 40-8 to 40-16
Interpretability, 31-7
Interruption, in digital computer operation, 8-20 to 8-30
Interstellar space, 7-22
Inventory control, 8-47
Inversion, meteorological, 5-12
Inverter, 9-11 to 9-15, 9-18, 9-20, 9-27
  (*See also* NOT)
Ion propulsion system, 20-37
Ion rocket, **20**-8
Ion source, 20-39
Ionosphere, 6-2, 6-13 to 6-16, 7-6 to 7-7
  propagation in, 14-57
Irradiance, 16-3, 16-4
Irreducible code, 22-24
Irtran-1, 15-14
Irtran-2, 15-13
Isocline, 3-5
Isodynamic lines, 3-6
Isogonic lines, 3-5

JP fuel, 20-17
Jet engine, **20**-17
Julian date time, 7-24
Jupiter, 5-2, 7-29 to 7-31, **20**-7

K-band, 14-4
Karman ogive, 18-12
Kel-F, **15**-16
Kell factor, 17-27
Kennelly layer, 6-13
Kernel of a game, 23-10
Kernel strategy, 23-10
Kinematic viscosity, 6-9
Kirchhoff's law, 15-3
Klystron, 14-21, 14-24, 14-64
Knudsen number, 6-12
KRS-5, 15-13, 15-16
KRS-6, 15-16

L-band, 14-4
Lagrange's interpolation formula, 40-10
Lagrange's planetary equations, 17-7
Laguerre polynomial, 40-25
Lambert, 16-6
Lambert's law, **15**-4, 16-5
Land-use model, 4-12
Laplace transform, 39-1 to 39-7
  operations for, 39-25
  table of, 39-26 to 39-33
  use of, 39-13 to 39-19
Lapse rate, 5-2, 5-12
Laser, 12-4

Latency, 8-12, 8-17
Launching (satellite), 17-14
Leakage transmission, 29-40
Learning, psychology of, 30-11
Learning system, 24-30 to 24-31, **30**-1 to **30**-20
Least-squares approximation, 40-24 to 40-26
Least-squares principle, **26**-16
Least-squares solution, 40-7
Legendre polynomial, **40**-25
Legendre transform, 39-12
Legibility, 31-7
Lens, infrared, 15-12
  microwave, 14-37, 14-39
Library, 8-16
Lifetime (of a satellite), 17-8
Lift coefficient, 18-5ff
Lift-to-drag ratio, 20-5
Light, visible, 16-1ff
Likelihood, maximum (*see* Maximum likelihood)
Likelihood function, **38**-13, **38**-34
Likelihood ratio, 24-15 to 24-17, 24-24, 24-25
Likelihood ratio test, 24-13
Limen, difference, 31-5
Limiting, 10-10 to 10-12, 10-16
Linear-beam tube, 14-20, 14-24
Linear programming, 25-1 to 26-30
  on analog computer, 11-14
  in economics, **35**-3
  solution of game by, 23-11 to 23-12
Linear space, 39-3
Line of nodes, 7-24
List processing, 8-20
Lithium fluoride, 15-14
Loading error, 10-7
Locating industrial sites, 4-8
Lock-out, 8-28
Logic, 9-1 to 9-27
  time-dependent, 41-9
  (*See also* Boolean algebra)
Logical connective, **41**-1, **41**-2
Logical operations, 8-19
Log-normal distribution, 33-14, **38**-44 to **38**-45
Look-ahead, 8-22, 8-23
Loop transmittance, 29-9
LORHO, 34-11
Loss function, 24-9, 24-10
Low-energy protons, 7-10
Lucite, 15-16
Lumen, 16-5 to 16-7
Luminance, 16-5 to 16-9
Luminosity, 7-24, 7-33
Luminous flux, 16-4 to 16-7
Lunar-landing simulator, 32-34
Luneberg lens, 14-39

Mach cone, 18-14
Mach number, 6-12, **18**-2ff
  critical, 18-6
  effect on efficiency, 20-24
  in MHD generators, 21-33 to 21-36
Mach wave, 18-13
Machine variables, 11-2
Macroinstructions, 8-25
Magnetic circuit elements, 9-7
Magnetic disk, 8-12 to 8-16
Magnetic field, in crossed-field tubes, 14-21
  of earth, 3-5 to 3-6, 7-2 to 7-6, 17-13
  effect on cosmic rays, 7-17 to 7-20
  in protonosphere, 7-8
  in interstellar space, 7-22
  of the sun, 7-13, 7-14, 7-29
  of Venus, 7-31
Magnetic logic, 9-18 to 9-22
Magnetic-pulse amplifier, 9-19
Magnetic storm, 7-17

INDEX

Magnetic tape, 8-7 to 8-31, 32-10, 37-4, 37-19, 37-21, 37-25
Magnetic torquing, 17-13
Magnetohydrodynamic conversion, 21-32 to 21-38
Magnetohydrodynamic propulsion, 20-35
Magnetohydrodynamics, 21-5
Magnetometer, 3-6
Magnetron, 14-22
Magnitude, of a star, 7-23
Main sequence, 7-24, 7-33, 7-34
Maintainability, 33-13 to 33-16
   definition of, 33-14
Majority carrier, 9-3, 9-4, 21-11
Majority logic, 9-9, 33-11
Management, 36-1 to 36-12
Management by exception, 1-8
Management-information system, 36-11
Man-in-the-loop simulation, 32-23
Man-machine systems, 31-3
Mantle, 3-4
Mapping radar, 14-66
Mapping systems, 14-5
Marginal analysis, 35-3
Marginal checking, 33-9, 33-15
Marginal distribution, 38-31
Maria, 7-31
Marine propulsion system, 20-4
Mark sensing, 8-28
Markov process, 28-2, 30-14, 30-16
   continuous-time, 28-12
Markov sources, entropy of, 22-18
Mars, 5-2, 7-29 to 7-31, 20-7, 20-9
Maser, 14-44, 14-46, 14-47
Mason's reduction formula, 29-15
Mason's reduction theorem, 29-24
Mass function, probability, 38-18, 38-24
Master program, 8-17
Matched filter (*see* Filter, matched)
Mathematical model (*see* Model)
Matrix, diode, 9-8
   ill conditioned, 40-3
Matrix game, 23-5, 23-9
Matrix inversion, 40-3
Matrix-switch structure, 8-8
Maxima and minima, 40-26 to 40-28
   (*See also* Extremization; Optimization; Minimization)
Maximin strategy, 23-6
   (*See also* Minimax)
Maximization of profitability, 35-3
Maximum likelihood, 24-26 to 24-29, 30-13
Maximum-remembering circuit, 11-11, 11-13
Maxterms, 9-9, 9-10
Maxwell-Boltzmann distribution, 21-6
Mean (*see* Expectation)
Mean free path, 6-7
Measure of effectiveness, 1-10
Measure of profitability, 35-3
Measure of value, 21-2 to 21-3
Megalopolis, 4-4
Mellin transform, 39-11
Memorandum representation of traffic, 32-18
Memory, in analog computers, 10-2, 11-20, 11-23
   in analog systems, 32-29
   in digital systems, (*see* Storage)
   human, 31-10 to 31-11
   in logical systems, 41-10
Mercury, 5-2, 7-29, 7-30, 7-31
Mesopause, 6-3
Mesosphere, 6-3, 6-4
Message, 12-1ff, 13-3, 13-4, 13-20, 22-7ff
Meteor, 7-21, 7-32
Meteorite, 7-21, 34-9, 37-4
Meteorology, 5-1 to 5-18
Methodology of system engineering, 1-3 to 1-13

Micrologic, 9-27
Micrometeorite (*see* Meteorite)
Microprogramming, 8-25
Microwave tubes, 14-19
Mictopause, 6-4
Milky Way, 7-35
Milne method, 40-20, 40-21
Minimax decision rule, **24**-12
Minimax problem, 30-6
Minimax regret, 23-19
Minimax strategy, 23-6
Minimax theorem, 23-8, 23-10, 23-11, 26-12
Minimization (*see* Extremization; Optimization)
   of logical form, 41-8
Minimum latency coding, 8-12
Minority carrier, 9-3, 9-4, 21-11
Minterm, 9-9, 9-10
Mirror machine, 21-40
Miss, in detection, 24-15
Miss distance, (*see* Guidance)
Missile, 37-3, 37-8
   ballistic, 19-14
   guided, 19-1
   (*See also* Guidance)
Mission velocity, 20-7
Mixed strategy, 23-3, 23-6
MMH, 20-29
Mock-up, 1-6
   in simulators, 32-24 to 32-27
   in system test, 34-13
Model, circuit, 29-6
   continuous, 32-4
   development of, 32-2
   logical, 32-16
   mathematical, 1-6, 1-10 to 1-11, 29-5 to 29-6
   as an analog, 32-7
   in analog computing, 11-1
   of human operator, 32-27
   in management, 36-3
   for reliability, 33-3, 34-4
   in system test, 34-4
   scale, 32-4
   stochastic, 32-2
Model equivalences, 26-2
Modeling, 26-1 to 26-3
   (*See also* Simulation)
Model-making theory, 26-28
Model-reference system, 30-10
Modular construction, 33-15
Modularity, 8-8
Modulation, 12-5, 14-11
   (*See also* Pulse modulation)
Modulator, 14-25 to 14-28
   telemetry, 37-20
Module, 8-11, 9-27, 11-23, 11-24
Modus ponens, 41-5
Moho, 3-4
Mohorovicic discontinuity, 3-4
Moments of a random variable, 38-18 to 38-20, 38-24, 38-41
Monomethyl hydrazine, 20-29
Monopulse, 14-34, 14-35
Monopulse comparator, 14-36
Monostatic system, 14-3, 14-5, 14-29
Monte Carlo, 11-23, 32-19 to 32-22, 33-9, 40-28
Moon, 7-29, 7-30
   spectral radiance of, 15-5
Moonlight, 16-6
Motive barrier, 21-21
Movement, speed and accuracy of, 31-14 to 31-15
Moving target indication, (*see* MTI)
Moving-base simulator, 32-26
MTBF, 33-3
MTI, 14-9 to 14-18, 14-53, 14-64
Multiaperture device, 9-22
Multicomputer, 8-8, 8-10

INDEX

Multiloop system, 29-42 to 29-50
Multipath, 12-10
Multiplexing, 8-28, 8-30, 12-4, 13-4, 13-13
  frequency-division, 12-4
  time-division, 12-5
Multiplication, 10-7, 10-9, 10-18
Multiplicative congruential method, 32-20
Multiprocessing, 8-24
Multiprocessor, 8-10
Multirunning, 8-23
Multisequencing, 8-24
Multistage process, 27-2
Multistep method, 40-19
Multistylus recorder, 37-9
Multivibrator, 11-11 to 11-15, 11-24
Munsell color chips, 31-6
Mutually exclusive events, 38-5

Nadir, 7-24
Nan, 41-2
NAND, 9-13 to 9-16
National income, 35-5
Navigation, 19-1 to 19-25
Near field, 14-35
Negative resistance, 9-6, 14-46
Neptune, 5-2, 7-29, 7-30, 7-32
Nernst glower, 15-4
Net national product, 35-5
Network, communication, 13-17, 13-23
  PERT, 36-6
  unreliable, 38-9
Network analyzer, 32-5
Network transmission, 38-6
Neural-net model, 30-12
Neuron, 11-14, 11-17, 30-16
Neutron albedo, 7-10
New York City, 4-2
Newton-Raphson method, 40-6
Newton's interpolation formula, 40-10, 40-11
Newton's linearization method, 40-6, 40-8
Neyman-Pearson test, 24-16
Niobium, 21-40
Nitrogen tetroxide, 20-28
Nitromethane, 20-30
Node, 38-15
Node absorption, 29-14
Node signal, 29-8
Node splitting, 29-14
Noise, in adaptive systems, 30-3 to 30-4
  in automatic control systems, 29-32, 29-43
  channel, 22-16
  in communication systems, 12-8, 12-9, 12-10, 13-8
  cosmic, 14-42
  in decision theory, 24-7ff
  encoding to combat, 12-6 to 12-9, 22-24
  entropy of, 22-11
  $1/f$, 15-17
  Gaussian, 12-8, 22-16, 24-20, 24-28
  generation-recombination, 15-18
  in guidance system, 19-12 to 19-13
  in infrared detectors, 15-17 to 15-27
  loudness of, 31-9
  in manual control systems, 31-15
  in measurements, 32-2
  photon, 15-18, 15-21
  in radar detection, 14-5 to 14-9, 14-41
  in radar receivers, 14-41 to 14-47
  in semiconductors, 15-17
  in sensory perception, 31-4
  in telemetry, 37-5
  in testing, 34-16
  white, 32-10
  (*See also* Error; Signal-to-noise ratio)

Noise current in semiconductors, 15-17
Noise equivalent power, 15-19
Noise factor (*see* Noise temperature)
Noise figure (*see* Noise temperature)
Noise generator, 11-12, 11-23, 32-10
Noise temperature, 14-42, 14-44, 14-45
Noiseless channel, 22-14
Non-Archimedean weight, 26-3
Noncoherent (*see* Coherence; Detection, incoherent; Integration)
Non-zero-sum game, 23-16
NOR, 9-13 to 9-16, 9-27, 41-2
Normal distribution (*see* Gaussian distribution)
Normal equations in least squares, 40-24
Normal form, game, 23-3
NOT, 9-11, 9-20, 41-2
Nova, 7-34
n-person game, 23-13 to 23-15
NPN transistor, 9-4
N-on-P silicon cell, 17-23
N-type semiconductor, 9-3, 21-6, 21-10
Nuclear propulsion (*see* Rocket engine, nuclear; Ramjet, nuclear; Turbine, nuclear; Turboelectric power plant, nuclear)
Nuclear radiation, 9-7
Nuclei, condensation, 5-9
Numerical analysis, 40-1 to 40-32

Objective form, 25-1
Oblateness (of the earth), 17-8
Occlusion, 5-7
Ocean, 2-3 to 2-15
Ocean-atmosphere system, 2-7
Ocean currents, 2-10
Ocean hemisphere, 3-3
Octal, 8-11
Oddments, 23-9
Off-line operation, 8-7
One-step methods, 40-21 to 40-23
Ooze, 2-12, 2-14
Open-loop system, 29-3
Operands, 8-18
Operating characteristic function, 24-25
Operation of a system, 28-1
Operational amplifier, 10-4 to 10-7, 11-2
Operations in a digital computer, 8-18
Operations research, 1-8 to 1-10
Optical filters, 16-7
Optical materials, 15-12
Optical properties (of sea water), 2-9
Optical sensors, 16-1 to 16-15
Optical systems, infrared, 15-9
Optics, geometrical, 14-48
  physical, 14-48
Optimal feasible solution, 25-2
Optimal policy, 27-3
Optimal strategy, 23-10
Optimality, criteria of, 24-5
  principle of, 27-4
Optimization, adaptive, 30-5
  algorithmic, 26-15
  under fluctuations, 4-10
  under inequality constraints, 26-2
  system, 1-9, 21-2 to 21-3
    by decision theory, 24-9 to 24-13
  of urban subsystems, 4-7
  (*See also* Extremization; Maximization; Minimization; Suboptimization)
Optimum, local, 30-8
Optimum decision process, 24-13
OR, 9-8 to 9-27, 38-4, 41-2
Orbital mechanics, 17-3
Organ-pipe scanner, 14-36
Orographic lifting, 5-8

Orthochromatic film, 16-9
Orthogonal polynomials, 40-25, 40-26
Oscillator, 14-18
  phase-shift, 37-11 to 37-12
  subcarrier, 37-19
Osculatory interpolation formula, 40-9
Otto-cycle engine, 20-11
Overflow, 8-21, 8-24
Overhead, in digital computer operation, 8-29, 8-31, 8-45
Oxygen, 5-2, 20-29
Oxygen difluoride, 20-28
Ozone, 5-4, 6-3, 6-4

Padé approximation, 10-14 to 10-16
Pair generation, electron-hole, 21-11
Panchromatic film, 16-9
Pandemonium, 30-12, 30-15, 30-16
Paper tape, 8-21
Parallax, 7-24
Parallel operation, 8-22, 10-2, 11-3
Parameter estimation, 24-6
Parameter margin, 29-33, 29-42
  infinite return difference, 29-39
  unity return difference, 29-34
Parametric amplifier, 14-44 to 14-46
Parity check, 8-27, 22-26
Part (of a system), 1-5, 1-7
Pascal distribution, 38-46
Patchbay, 11-2
Patchboard, 11-23, 11-24
Path factor, 29-15
Pattern recognition, 24-31, 30-11 to 30-17
  (*See also* Character recognition)
Patterning, temporal, 31-10
Pauli exclusion principle, 21-8
Payroll computation, 8-47
Perceptron, 30-14 to 30-16, 32-4
Perchloryl fluoride, 20-28
Perfect information, in game, 23-8
Periclase, 15-14
Perigee, 7-24, 17-4
Perihelion, 7-24
Period (of an earth satellite), 17-8
PERT, 36-5 to 36-11
Perturbation theory, 17-3 to 17-10
Phased array, 14-37
Phases (of system design), 1-5, 36-4
Photochemical sensor, 16-1, 16-9
Photoconductive cell, 16-8, 16-10, 16-12
Photodetectors, 15-16
Photoduodiode, 15-25
Photoelectric sensor, 16-1, 16-8
Photoemissive tube, 16-8, 16-10
Photographic emulsion, 16-9
Photometry, 16-4
Photomultiplier, 15-25
Photon, properties of, 16-2
Photopic vision, 31-5
Photosphere, 7-15, 7-29
Photothermal sensors, 16-1
Photovoltaic cell, 16-8, 16-11, 16-12, 20-33
Pinch instability, 21-40
Ping-pong, 9-23
Pitching moment, 18-4
Plan position indicator, 14-59, 14-60
Planck's constant, 21-9
Planck's law, 15-3, 16-2
Planetary atmospheres, 5-2
Planets, 5-2, 7-27
Plankton, 2-12
Plasma, in electrical propulsion systems, 20-34 to 20-40
  extraterrestrial, 7-14 to 7-17

Plasma, as ideal gas, 21-6
  MDH, 21-32
  in a thermionic converter, 21-25
  thermonuclear, 21-38
Plasma frequency, 6-14
Plasma gun, 20-35
Plasma jet, 20-34
Plasma oscillations, 21-25
Plasma rocket, 20-8
Play of a game, 23-2
Plemelj formulas, 39-10
Pluto, 7-29, 7-30, 7-32
PN junction, 9-3
P-on-*N* silicon cell, 17-23
Pointer, moving, 31-8
Poisson distribution, 28-14, 38-46
  applied to telephone traffic, 13-20 to 13-22
  generation of random deviates from, 32-21
  in queue theory, 28-8, 28-14 to 28-27
  in reliability theory, 33-3
  stuttering, 28-24
  (*See also* Exponential distribution)
Polarization, 14-33, 14-48, 14-53, 14-58
Polish notation, 8-25
Pollaczek-Khinchin formula, 28-24
Polymorphic structure, 8-10
Population potential, 4-6
Portel, 37-9
Postdetection integration, 14-17
Posterior probability, 38-12
Potential temperature, 5-11
Potentiometer, 10-7
Power supply, satellite, 17-22
Poynting-Robertson effect, 7-22
PPI, 14-59, 14-60
Prandtl-Meyer fan, 18-13
Precession, 17-10 to 17-12
Precession of the equinoxes, 7-24
Precipitation, 3-5, 5-7
Predetection processing, 14-17
Predetection recording, 37-25
Prediction, 14-59
Predictor display, 31-16
Predictor-corrector method, 40-19, 40-20
Preemptive preference, 26-3
Preemptive priority, 28-26
Prefix code, 22-24
Preliminary design, 1-5
Price set, normal, 35-4
Price structure, 35-4
Principles of system design, 1-7
Printer, digital, 8-28
Prior density function, 38-34
Prior probability, 38-12
Priority in queues, 28-26
Proactive interference, 31-11
Probability, 38-3 to 38-48
Probability distributions, named, 38-44 to 38-48
Product, logical, 38-5
Program evaluation and review technique, 36-5
Program management, 36-3 to 36-5
Project system, 36-3 to 36-5
Propellant-feed system, 20-27
Propellants, internal-combustion engine, 20-15
  rocket, 20-27 to 20-30
Proper motion, 7-24
Proportional navigation, 19-5 to 19-13
Propositional calculus, 41-1 to 41-11
Propulsion, 20-1 to 20-42
Propulsion test, 34-7 to 34-9, 34-12 to 34-13
Propulsive efficiency, 20-21, 20-22, 20-27
n-Propyl nitrate, 20-30
Protection, in digital computers, 8-24 to 8-25, 8-28
Protonosphere, 6-5, 7-6

INDEX

Protosphere, 6-5
Pseudorandom numbers, 32-20, 40-28
Psuedorandom sequences, 32-21
Psychophysics, 31-4
P-type semiconductor, 9-3, 21-6
Pulse amplitude modulation, 12-5
Pulse code modulation, 12-5, 12-7
Pulse compression, 14-11, 14-67
Pulse Doppler, 14-14
Pulse frequency modulation, 12-5
Pulse time modulation, 12-5, 12-6
Pulse width modulation, 12-5
Pulse-code-modulation telemetry, 37-13
Pulsed signal, 14-3
Pulse-duration-modulation/FM telemetry, 37-21
Pulsejet engine, 20-23, 20-24
Pulse-position-modulation/amplitude-modulation telemetry, 37-23
Punched cards, 8-7, 8-14 to 8-16, 8-26, 8-29
Punched tape, 8-14 to 8-16
Pure strategy, 23-3
Pursuit navigation, 19-2 to 19-6

Quadding, 33-11 to 33-13
Quality control, 33-3
Quantifiers, 41-10
Quantum detectors, 16-7
Quarter-square principle, 10-10
Quartz crystal, 14-18
Queue, 8-22, 8-24, 8-31
Queue discipline, 28-16, 28-21, 28-25 to 28-27
Queue theory, 28-1 to 28-30
Queuing of vehicles, 4-12
Quiet, in detection, 24-15
Quiet sun, 7-15

Radar, 14-1 to 14-70
Radar range equation, 14-60 to 14-63
Radiance, 15-4
Radiant flux, 16-3 to 16-5
Radiation belts, 7-8
Radio, 12-1
Radio astronomy, 7-22
Radio properties (of sea water), 2-9
Radioisotope power supply, 20-32, 20-33
Radiometry, 16-3
Radiosonde, 5-14
Rail vehicles, propulsion of, 20-2
Rail-type motor, plasma, 20-36
Rain, 5-8
  effect on radiation, 14-52, 14-53, 14-57
Raindrops, 14-50
Rainmaking, 5-9
Ramjet engine, 20-22 to 20-24
  nuclear, 20-23
  supersonic, 20-24
Ramsauer effect, 21-25
Random access (*see* Access, random)
Random numbers (*see* Pseudorandom numbers)
Random processes, simulation of, 32-10
  (*See also* Monte Carlo)
Random signals (*see* Noise)
Random variable, 38-16
  continuous, 38-21
  discrete, 38-18
  operations on, 38-35
Random variables, in Monte Carlo, 40-28
  sum of, 38-39
Random walk, 28-3 to 28-5
Range equation, infrared, 15-27
  radar, 14-60 to 14-63

Rankine cycle, 20-16, 20-17, 20-34
Rate of climb, 20-5
Rationality, 23-14
Rayleigh limit (diffraction), 16-13
Rayleigh region, 14-49
Rayleigh scattering, 5-4
Reaction time, 31-12
Readers, magnetic, 8-28
  optical, 8-28
Read-out, 11-3
Real time, 11-5
Real time simulation, 11-19, 32-23, 32-28
Receiver, radar, 14-40 to 14-47
  telemetry, 37-12, 37-14, 37-18 to 37-19, 37-24
Reciprocating engine, 20-9 to 20-15, 20-23
Reciprocity theorem, 14-30
Recognition, 24-31
Recombination, electron-hole, 21-11
Rectifier, 9-3
Red fuming nitric acid, 20-28
Red giant, 7-34
Redundancy, minimum, 22-25
  for reliability, 33-9 to 33-13
  in telemetry, 37-7
Reentry, 34-11
Reflecting systems, optical, 15-13
Refracting systems, 15-13
Regenerative gas turbine, 20-16
Regenerative property of distributions, 38-48
Regulation, of a signal, 13-14, 13-15
Regulator, voltage, 17-25
Reinforcement learning, 30-13
Relativistic solar events, 7-18
Relaxation method, 40-27
Relay, 17-1
Reliability, 33-1 to 33-16
  of a computer system, 8-31
  definition of, 33-2
  of electronic circuits, 33-8
  prediction of, 33-4
  in system test, 34-2, 34-4
Reliability engineering group, 33-2
Reliability manager, 33-6, 33-15
Remanence, 9-7
Repairability, 33-14
Resistive sheet, 32-5
Resolution, and ambiguity, 14-10
  Doppler, 14-10
  high, with synthetic antenna, 14-67 to 14-70
  in infrared systems, 15-2, 15-3, 15-27
  in mapping radars, 14-55
  in optical systems, 16-13 to 16-15
  of radar targets, 14-9 to 14-18
  in satellite TV systems, 17-26, 17-27
  and sensitivity, 14-62
Resolver, analog, 10-14
Response time, 31-16
Restricted-basis-entry procedures, 26-24, 26-27, 26-28
Retentivity, 9-7
Reticle, tracking, 15-13
Retroactive interference, 31-11
Return, rate of, 35-2
Return difference, 29-10
  null, 29-28
Reusable storage, 8-6
Reynolds number, 6-12, 18-2, 18-3
Richardson's extrapolation method, 40-22
Right ascension, 7-26, 17-3
Ring counter, 9-25
RIR-2, 15-14
Rise time, 9-6
Risk, conditional, 24-10
Risk minimization, 24-11
Rocket, electromagnetic, 20-35

INDEX

Rocket engine, 20-24 to 20-41
  electrical, 20-31
  ion, 20-37 to 20-40
  liquid-propellant, 20-27
  nuclear, 20-30
  solid-propellant, 20-30
  test of, 34-12
Rocket propulsion, 20-6
Rocket sled, 32-23
Rolling moment, 18-4
Rolling resistance, 20-2
Rooted tree, 23-2
Rossby waves, 5-6
Rotating joint, 14-30
Roughness, 5-13
Rounding error, 40-22
Routines, 8-6
Runge-Kutta method, 40-21 to 40-23
Rutile, 15-14

S-band, 14-4
Saddle point, 23-6
Saddle-point theory, 26-12
Salinity, 2-7
Sample space, 38-13
  continuous, 38-23
  multidimensional, 38-27
Sample-and-hold, analog, 32-11, 32-12
Sampled-data system, simulation of, 32-10
Sampling, 12-5
  continuous, 24-7, 24-19
  discrete, 24-7, 24-19
Sampling theorem, 12-5
Sapphire, 15-14
Satellite, telemetry from, 37-3
Satellite computer, 8-10
Satellite system, 17-1 to 17-30
  testing of, 34-7
Satellites (*see* Moon; Planets)
Saturation region, 9-5
Saturn, 5-2, 7-29, 7-30, 7-32
Scale, moving, 31-8
Scale factors, in analog computers, 10-7, 11-5
Scale height, 6-6
Scaling, in analog computation, 11-5
Scanning, antenna, 14-37, 14-60 to 14-63
  event (in simulation), 32-22
  periodic (in simulation), 32-22
Scattering, of infrared, 15-9
  (*See also* Cross section, radar)
Scattering matrix, 14-48
Scatter-read, 8-20, 8-28
Scheduling, concurrent, 34-3
Scientific management, 36-2
Scotopic vision, 31-6
Scylla machine, 21-40
Sea states, 2-11
Search, in adaptive decisions, 30-8
Search systems, 14-5, 14-58
Sears-Haack body, 18-12
Seebeck coefficient, 21-14 to 21-21
Seiche, 2-12
Seismic waves, 3-4
Selector, 13-5
Selector circuit, 11-10, 11-12
Self-analysis, 8-21
Self-calibration, 29-19
Semiactive guidance, 14-5
Semiconductor, voltage profiles, 21-5 to 21-7
Semiconductors, 9-2, 9-3
Sensitivity, 29-10 to 29-11, 29-19 to 29-42
  differential, 31-5
  topological interpretation of, 29-24
  zero, 29-44

Sensitivity graph, canonical, 29-26
Sequence counter, 8-17
Sequential decision process, 24-13
Sequential detection, 24-25
Sequential test, truncated, 24-13
Serial operation, 10-2
Service facility, 28-8
Service period, 28-23
Service system, stochastic, 28-8
Servomechanism, 11-9
  (*See also* Control theory)
Shapley value, 23-15
Shapley-Snow theorem, 23-10
Sheath, plasma, 21-25 to 21-28
Shift register, 9-24 to 9-27, 11-23
Shock stall, 18-11
Shock wave, 4-11, 18-11, 18-13, 18-16
Sialic rocks, 3-4
Side lobes, 14-32, 14-38, 15-3
Sidereal period, 7-25
Sidereal time, 7-25
Sideslip, angle of, 18-5
Signal, definition, 13-3
  minimum detectable, 24-21
  properties of, 13-4 to 13-7
  stochastic, 24-20, 24-23
Signal detection (*see* Detection)
Signal extraction, 24-6, 24-25
Signal-flow graph (*see* Flow graph)
Signal-to-noise ratio, 24-16, 24-17
  in adaptive systems, 30-3
  in communication systems, 12-3, 12-6 to 12-9, 13-8
  in infrared systems, 15-27
  in optimum detectors, 24-21, 24-22
  in radar detection, 14-5 to 14-8, 14-40, 14-60 to 14-63
  in satellite TV system, 17-27
  sensitivity to mismatch, 14-18
  in telemetry, 37-25
Silica, fused, 15-14
Silicon, 9-3, 15-14
Sima layer, 3-4
Simplex method, 25-1 to 26-30
  in solving games, 23-11
Simpson's rule, 32-12, 40-15
Simulation, 10-1 to 10-2, 32-1 to 32-31
  analog, 11-3 to 11-25, 32-4 to 32-14
  digital, 32-15 to 32-22
  direct, 32-7
  importance in system engineering, 1-6, 1-11
  of manned systems, 32-22 to 32-28
  mathematical, 32-3, 32-7 to 32-8, 32-10
  uses of, 32-3
  (*See also* Model)
Simulator, direct analog as, 10-3
Simultaneous operations, 8-29
  (*See also* Multiprocessing; Multirunning)
Single-thread design, 1-6
Sink, heat, 21-4
Sink node, 29-9
Sink temperature, high, 21-21
Skin friction, 18-8
Sky, spectral radiance of, 15-5
Slack time, 36-8 to 36-10
Slack variable, 25-7, 26-6
Slip flow, 6-13
Snow, 5-9
Snowbursts, 5-10
Sodium chloride, 15-16
Solar beam, 7-16
Solar constant, 5-4
Solar cell, 17-22, 17-23, 21-5
Solar flare, 7-16
Solar power supplies, 20-33

INDEX

Solar radiation, 5-4
Solar system, 5-2, 7-27, 7-29
Solar wind, 7-14
Sole, 14-21, 14-22
Solid-state devices, 9-2
Solstice, 17-3
Sone, 31-5, 31-6
Sonic barrier, 18-10
Source, heat, 21-4
Source node, 29-9, 29-12
Southwell relaxation, 40-27
Space, 7-1 to 7-36
Space charge, 14-23
Space vehicles, propulsion of, 20-6
Space-charge wave, 21-25
Spacecraft (*see* Satellite)
Specific impulse (*see* Impulse)
Specific power, 20-8
Spectral filter, 15-13
Spectral lines, 15-4
Spectrophotometric filter, 16-7
Speed of sound, 18-2
Spinning-body dynamics, 17-10
Spool, 20-15
Spread F echo, 6-15
Square loop, 9-7
Square-loop ferrite, 9-22
Square wave, 10-15
Stability, in adaptive systems, 30-10
  aerodynamic, 18-9
  of the atmosphere, 5-2, 5-7, 5-8, 5-13
  economic, 35-6
  of feedback systems, 29-33 to 29-50
  of integration procedure, 40-20 to 40-22
  of linear systems, 39-14
  in manually controlled systems, 31-16
  of plasma, 21-40
  of radar transmitter, 14-18
  of satellites, 17-2
  traffic-stream, 4-11
Standard atmospheres, 6-9
Standard deviation, 38-19
Standardization, 33-7
Standby operation, 33-10
Star, 7-32
Star catalog, 7-32
State representation in simulation, 32-16
State of a system, 28-2
Statistical decision theory (*see* Decision theory)
Statistical inference, 38-12
Steepest descent, 11-17, 40-27
Steerable array, 14-40
Stefan-Boltzmann law, 15-3, 16-3
Steffensen's interpolation formula, 40-13
Stellerator, 21-40
Steps (of system design), 1-5 to 1-6
Stieltjes transform, 39-11
Stipulations vector, 26-4
Stirling's interpolation formula, 40-12
Stochastic inputs, 1-4
Stochastic matrix, 28-5
Stochastic model, 32-2
Stochastic process (*see* Random processes; Monte Carlo)
Storage, asynchronous, 8-12
  erasable, 8-12
  external, 8-6, 8-14, 8-16, 8-21, 8-25 to 8-28
  fixed, 8-12
  internal, 8-6, 8-10
  (*See also* Access; Memory)
Strategy, 23-3ff
Stratocumulus, 5-9
Stratopause, 6-3
Stratosphere, 6-3, 6-4
Strontium titanate, 15-14

Structural variable, 26-6
Structure (digital computer), 8-6
Student's distribution, 38-44 to 38-45
Sturm-Liouville problem, 39-17
Subjective probability, 31-13
Suboptimization, 1-8 to 1-10, 28-12
Subsystem, 1-5, 1-7
Sudden ionospheric disturbance (S. I. D.), 6-16
Sum, logical, 38-5
Sum digit, 41-6
Sun, 5-4, 7-15, 7-27 to 7-30, 7-33
  spectral radiance of, 15-5
Sunlight, 16-6
Sunspot, 6-12, 7-29
Superadditivity, 23-14
Superconductor, 21-40
Supercooled water, 5-8
Superheterodyne receiver, 14-40
Supernova, 7-34
Supervisor routine, 8-22
Surface tension, 2-9
Surface water, 3-5
Surprisal, 24-10
Surveillance radar, 14-64
Sutherland formula, 18-2
Sweepback, 18-11
Swept wing, 18-11
Switching, in analog integrators, 11-3
  automatic, in telephone systems, 13-21
  circuit, 13-3
  in communication systems, 13-3, 13-4, 13-17 to 13-19, 13-23, 13-24
  four-wire, 13-16 to 13-17
  in hybrid computers, 11-8 to 11-11, 11-20, 11-23 to 11-24
  in logical circuits, 9-5 to 9-7, 9-16 to 9-23
  in magnetic elements, 9-7, 9-19 to 9-22
  message, 13-3
  in radar modulator, 14-25 to 14-28
  in repetitive analog computers, 11-4, 11-20
  in transistors, 9-5 to 9-6, 9-16
  in tunnel diodes, 9-6 to 9-7, 9-18, 9-23
Switching time, 9-6, 9-7
Synapse, 11-14, 11-17
Synchronization, 8-20, 12-6
Synnoetics, 1-10
Synodic period, 7-25
Synoptic meteorology, 5-14
System, military space, 34-2
System analysis, 1-4, 1-11
System comparison, 24-9
System design, infrared, 15-27
System engineering, in project organization, 36-4
  relation of economics to, 35-1
System evaluation, 24-9
System optimization (*see* Optimization)
System structure, optimum, 24-4
System test, partial, 32-5
System testing, 34-1 to 34-16
System transfer function, 29-9

Table look-up, 8-19
Table searching, 8-20
Tableau, simplex, 25-8
Tail (aerodynamics), 18-3
Tape, magnetic (*see* Magnetic tape)
Tautology, 41-6
Telegraphy, 12-1, 12-2
Telemetry, 37-1 to 37-26
Telephone calls, frequency of, 4-7
Telephone systems, 4-2, 13-1 to 13-24
Telephony, 12-1, 12-2
Television, 12-1, 12-2
  satellite surveillance, 17-25
  in simulator, 32-24

INDEX

Tellurium, 15-14
Temperature, kinetic, 21-9
  scale, 6-7, 6-10
Temperature control, 17-18 to 17-22
Template matching, 30-11
Terminals, communication, 13-15
Test, qualification, 34-14
  quality-assurance, 34-14
Test equipment, 33-15
Test facilities, 34-11
Test objectives, 34-2 to 34-3, 34-13 to 34-15
Test organization, 34-3
Test planning, 34-2
Testing, 34-1 to 34-16
  accelerated, 33-5
  flight, 37-2
  of personal performance, 34-6
Tetrafluorohydrazine, 20-29
Texas tower, 2-7
Thermal control, 17-18 to 17-22
Thermal detectors, 15-16
Thermionic conversion, 21-5, 21-21 to 21-32
Thermionic emission, 21-22
Thermistor, 15-18
Thermodynamic efficiency, 21-5
Thermodynamics, atmospheric, 5-2, 5-3
  first law, 21-3
  of the ocean, 2-9
Thermoelectric conversion, 21-5 to 21-21
Thermonuclear energy conversion, 21-38
Thermonuclear explosion, 7-11
Thermonuclear reaction, 7-15
Thermosphere, 6-3
Threshold, in Bayes decision rule, 24-15
  differential, 31-5
  in radar detection, 14-6
  in sensory perception, 31-5
Threshold circuit, 9-9
Threshold decision device, 14-15
Threshold detection, 24-16
Threshold logic, 30-14
Thrust coefficient, 20-25, 20-26
Thyratron, 14-27
Time delay, 10-14, 10-15, 14-11
  simulation of, 32-10
Time multiplexing, 37-17
Time scale, 11-3, 11-6
Time sharing, 8-30
Time-division multipliers, 10-11
Timer circuit, 11-12
Tiros, 17-1
Tolerance, component, 33-9
Tools (of system design), 1-5
TR (*see* Duplexer)
Tracking, of displays, 31-8
  simulation of, 32-27 to 32-28
Tracking feed system, 14-35
Tracking radar, 14-65 to 14-66
Tracking reticle, 15-13
Tracking system, data processing in, 14-60
  definition of, 14-5
Track-while-scan, 14-59
Trade-off analysis, 1-7
Trade-off involving reliability, 33-1, 33-6
Traffic, 4-10
  simulation of, 32-16
Traffic engineering, 13-19
Trajectories, ballistic, 19-14
  optimal, 27-5
Transducer, in telemetry, 37-2, 37-4, 37-7, 37-14
Transduction (*see* Energy conversion)
Transfer function, evaluation of, 29-11
  human operator, 31-15 to 31-16
  optical, 16-14
  synthesis of analog, 10-8 to 10-9

Transfer ratio, 9-5
Transform, exponential, 38-41
  geometric, 38-42
  Laplace, bilateral, 38-41
Transforms, 39-1 to 39-35
Transinformation, 22-6, 22-12ff
Transistor, 9-2ff, 33-12, 37-24
Transistor amplifier, 29-25
Transistor characteristic curves, 9-5
Transition, rate of, 28-12
Transition probability, 28-2
Transmission, data, 13-6
Transmittance, 29-9
Transmitter, radar, 14-18
  telemetry, 37-10 to 37-12, 37-14 to 37-18,
    37-20 to 37-23
Transonic aerodynamics, 18-10
Transportation model, 4-16
Transportation route planning, 4-9
Trapped radiation, 7-10
Trapezoidal rule, 40-14, 40-15
Traveling-wave tube, 14-21, 14-24, 14-44, 14-45
Tree, in list processing, 8-20
  probability, 38-15
  rooted, 23-2
Triangular wave, 10-15
Tristimulus specification, 31-6
Tritium, 21-38
Trochoidal trajectories, 21-39
Tropical year, 7-25
Tropopause, 6-3
Troposphere, 6-3
Truth table, 41-2
T-strap motor, plasma, 20-36
Tungsten, 21-31
Tunnel, 4-11
Tunnel diode, 9-6, 9-16, 9-23, 14-44, 14-46
Tunneling, 9-7
Turbine, nuclear, 20-32
Turboelectric power plant, nuclear, 20-33
Turbojet engine, 20-23, 20-24
Turboprop engine, 20-23
Turbulent transfer, 5-13
21-cm line, 7-23
TWT (*see* Traveling-wave tube)

UDMH, 20-29
Ultraclean room, 34-12
Uncertainty, in economics, 35-8
  (*See also* Probability; Decision theory)
Uniform distribution, 38-44 to 38-45
  entropy of, 22-16
  generation of random deviates from, 40-29
U. S. standard atmosphere, 6-8 to 6-11
Universal time, 7-25
Unsymmetrical dimethylhydrazine, 20-29
Uranus, 5-2, 7-29, 7-30, 7-32
Urban areas, 4-1 to 4-20
Utility, 31-14
  nontransferable, 23-17
Utility function, 23-4
Utilization factor, 28-21, 28-23

Valence-bond band, 21-6, 21-10
Valid formula, 41-2
Validity, testing for, 41-4
Value (*see* Measure of value)
Value of a game, 23-9, 23-10
Van Allen belts, 7-32
Vapor pressure of water, 5-8
Varactor, 14-46
Variable-stability airplane, 32-26
Variance, 38-19, 38-42, 38-45
Venn diagram, 38-5

INDEX

Venus, 5-2, 7-29, 7-30, 7-31, 20-7
Vernal equinox, 7-24
Vibration, 32-24
Vibration testing, 17-16
Vidicon camera, 17-25, 17-26
Viscosity, of air, 6-8, 18-2
   of sea water, 2-8
Visibility, 31-7
Vision, 31-5 to 31-7
Visual acuity, 16-12
   (*See also* Eye)
Volumetric efficiency, 20-9, 20-10
von Karman number, 5-12

Waiting line (*see* Queue theory)
Waiting time, 8-12
Warm front, 5-7
Water, vapor pressure of, 5-8
   sea, density of, 2-8
   (*See also* Clouds; Rain, etc.)
Water cycle, 2-10
Water vapor, 6-4
Wave, ocean, 2-11
Wave drag, 18-11, 18-14
Wave number, 16-2
Weapon-system concept, 36-3
Wear-out failure, 33-3
Weather modification, 5-10
Weber's law, 31-5

Weibull distribution, 38-44 to 38-45
Well-formed formula, 41-1
Whitcomb rule, 18-11
White dwarf, 7-33, 7-34
White room, 34-12
Wien's displacement law, 15-3, 16-3
Windows, infrared, 15-6
Wing (aerodynamics), 18-3, 18-10
   (*See also* Airfoil)
Word (data unit), 8-11, 8-18
Work function, 21-20, 21-23
Working storage, 8-10
Worst-case design, 33-9

X band, 14-4
Xerograph, 8-28

Yawing moment, 18-4
Yo-yo speed reduction, 17-13

Zener effect, 9-4
Zenith, 7-24
Zero-sum game, 23-3 to 23-16
Zipf's law, 4-5
Zodiacal light, 7-21, 7-24
Zonal flow, 5-6